MODERN INDUSTRIAL HYGIENE

Volume 1, 2nd Edition

Recognition and Evaluation of Chemical Agents

Jimmy L. Perkins
University of Texas Health Science Center – Houston
School of Public Health
San Antonio Regional Campus

ACGIH®
Cincinnati, Ohio, U.S.A.

ISBN: 978-1-882417-75-9

Printed in the United States of America

ACGIH®
1330 Kemper Meadow Drive
Cincinnati, Ohio 45240-4148
U.S.A.
Telephone: (513) 742-2020
Fax: (513) 742-3355
E-mail: publishing@acgih.org
http://www.acgih.org

CONTENTS

CONTRIBUTORS

Mark Nicas, PhD, MPH, CIH
School of Public Health
University of California
Berkeley, CA
Chapter 15

Elaine Symanski, PhD
University of Texas
Health Science Center
School of Public Health
Houston, TX
Chapter 17

FOREWORD

Each time I teach my course *Risk Recognition, Evaluation and Control in the Workplace for Physical, Chemical, Biological and Radiological Hazards (Comprehensive Industrial Hygiene Review)*, the students ask: "Is there a textbook that goes with this course?" At present, and even more so in the future, the best answer may well be *Modern Industrial Hygiene* by Professor Jimmy Perkins. This book, Volume I, is the first of four books that, when completed, will address the *complete* subjects that make up the science and the art of industrial hygiene.

This first volume covers the subjects of, for example:

- What is industrial hygiene?
- The history of industrial hygiene;
- Toxic chemicals in the workplace;
- Occupational exposure limits;
- Exposure routes;
- The physical nature of gases, vapors and aerosols; and
- Methods and strategies of qualitative exposure assessment.

Some of the most current and complex information can be found in three subject areas. These are [1] processes and their chemical hazards, [2] sampling strategies and statistics, and [3] field and laboratory analytical chemistry, as follows:

- Industrial processes and their toxic chemicals, which chapter includes: abrasive blasting; boilers and furnaces; confined spaces; diesel engines; electroplating; grinding, polishing and buffing; laboratories; machining and cutting oils; metal surface treatments; painting; plastics; and welding and allied processes;
- Statistical approaches to the treatment of field and laboratory data, which chapters focus on mathematical modeling, including: emission rate modeling; dilution and mixing; analysis of distributions; variability and confidence intervals; analysis of

variance and covariance; and determinants of exposure, all of which, taken together can guide the industrial hygienist in preparing strategies for quantitative surveying of the workplace

- Methods of sampling, analysis and monitoring of gases and vapors, which chapters focus on: laboratory analysis; sampling of gases, vapors and aerosols; real time monitoring of air contaminants; and quality control and quality assurance in analytical laboratories.

Volume 2 in the *Modern Industrial Hygiene* series was published by ACGIH in 2003. It deals with "Biological Aspects" which includes the subject of ergonomics. Volume 3 will deal with "Control of Chemical Agents." Volume 4 will deal with "Physical Agents" including noise, vibration, pressure, heat and cold, and ionizing and non-ionizing radiation. It is important to note that the four volumes in this series are, and will be, a tour-de-force for Professor Perkins. Although he has had authors and assistance with many of the chapters, all of the material was either edited or has been written by him. Professor Perkins' mastery of the body of knowledge that *is industrial hygiene* will be a benchmark for us all.

PREFACE

Most Industrial or Occupational Hygienists, being scientists, probably remember working the math or chemistry problems that were part of their textbooks. The questions at the end of chapters in biology texts could be particularly troublesome as they were often open-ended and sometimes subjective. No doubt we agonized over these; nonetheless, many were thought provoking. The questions and problems were an important part of the learning process. They emphasized the most important material, an increasingly important process as the sea of information continues to expand. They also forced us to use the tools that we had just read about; many times the uses that the questions forced were not the uses that we envisioned upon first reading the material. Thus, the learning process was further advanced one very important step.

But the learning process continues even after a formal degree program. Many hygienists would undoubtedly say that on-the-job-training is ultimately far more extensive than that gained while in university. So what tools are available to the hygienist during this advanced stage of the learning process? Often the tools that were learned in college become more refined in one's mind as the professional career continues. In other words, in the classroom we learn the concept in a general sort of way. We think that we know how to apply it, and we usually think that we have a firm understanding of it. After all we passed the exam! But along comes reality and experience, both of which cause us to question the perceptions that we formed in the classroom. Because of contrary field observations, we may even question the basic tenets of the concept or the underpinnings of some long-held notion or axiom. It is at these moments, common for some, rare for many, but nevertheless experienced by everyone, that we need a thorough review of the axiom and the events that led to its becoming a part of the lore of hygiene.

It is the above process, occurring continuously throughout one's lifetime that constitutes learning, or more aptly understanding. The formal student must observe information from various angles in order to gain, not an understanding, for that usually comes with maturity and repeated application, but rather an appreciation for the multifaceted nature of information. The student gains this appreciation by viewing the information in different contexts. A good teacher provides one context, a good textbook the second, and a good set of questions forces the third. Heretofore, the hygiene student has not had access to a general textbook written from this perspective and containing questions and problems and their answers.

The practicing hygienist continues to learn by refining his or her understanding through experience and review. Reviewing old concepts with the advantage of experience often sheds a different meaning on the concepts, causing them to be understood as never before or to be questioned as unrealistic in light of observation. Furthermore, as information content increases, we seek to summarize greater amounts of information into small packages so that it can be assimilated. This often requires that assumptions be made, and with experience the assumptions are often called into question. In order for the experienced hygienist to take full advantage of this stage of the learning process he or she must possess an adequate reference book, one that explains the assumptions and provides the references upon which the concepts and axioms are based. The hygienist has not had such a reference tool.

Thus, the two guiding lights in the preparation of this book have been 1) to prepare a true textbook that is well illustrated, contains examples, and has questions and answers and 2) to construct a resource for the practicing hygienist that will allow many readings of the same material, each giving a new meaning based on the most recent field experiences. And when the hygienist asks what are the roots of this assumption, or this observation, or this statement, the references to find answers to these questions are there.

* * * * * * * * * *

This volume represents only a portion of the practice of hygiene: the tools needed for recognition and evaluation of chemical agents. This leaves enough material for at least three more volumes. Volume 2 includes a wide range of biological issues including the recognition, evaluation, and control of biological agents; risk assessment, epidemiology, and the biological basis for standards; the biological basis for the science of ergonomics and control techniques for the ergonomic related diseases; biological monitoring; and dermal exposure assessment. A third volume, expected in 2008, focuses on controls for chemical agents. Unlike previous texts on this subject, all control techniques from ventilation to behavior modification will receive treatment in one text. Volume 4[1] will address all of the physical agents, including noise and vibration, heat and cold, and ionizing and non-ionizing radiation. All aspects of the physical agents from recognition to control will be addressed.

This volume, in addition to containing the information on recognition and evaluation of chemical agents, also serves as an introduction to the field. The first section has three chapters that describe the discipline. The second section describes the disease agents of the workplace and thus the physical and biological agents are introduced; the various exposure routes are explained. The third section relates to the basic tools used for assessing the workplace. Here the statistical basis of measurements is explored, the exposure limit is discussed, and the impetus for much of what goes on in the field, federal legislation and regulation, is discussed. The section also contains two chapters on the physical properties of the chemical agents. This information, extracted from physical chemistry and aerosol physics, is generally not addressed in other hygiene texts. Section IV is again general to the field in that it presents the hygiene

---

[1] The completion of this volume is on hold or perhaps in doubt due to the recent passing of Dr. John Leonowich. John was one of the world's leading authorities on ionizing and non-ionizing radiation.

process; background, walkthrough, and quantitative surveys. The statistical aspects of data analysis are reviewed here in detail and hopefully in an explanatory nature as found in no other text. Chapters on mathematical and statistical modeling of exposures is new for this edition.

The last two sections are devoted to the sampling and analysis of the chemical agents. Relative to analysis there are no doubt other texts that are more detailed on the various analytical techniques such as gas chromatography, but there should be more than enough here to educate the hygienist as to the principles applied by the analytical chemist. Thus, the informed hygienist can ask the right questions. Relative to air sampling, there are again texts with more thorough coverage of, for example, filtration or adsorption. But the material here is far more detailed than previously presented in any text intended for the general hygiene audience.

A word about provincialism. Rather than use the term industrial hygiene or the term occupational hygiene, I have usually opted for just hygiene, readily admitting that it may be even less meaningful that the first two. I have tried (I suspect at times in vain) to be sensitive to the needs of my foreign colleagues. As such, while the units used are not always *Systeme Internacionale* (SI), metric units are always given in addition to the English units. Hopefully, one day soon the latter will disappear from the earth! Relative to legislation, regulation, the historical perspective, and other social and cultural issues, I have tried to include some information from countries such as the United Kingdom, Canada, and Australia. On occasion a few other perspectives are given. However, I must mostly admit failure here as the U.S. situation changes too rapidly to be tracked, let alone the issues in other countries.

There are extensive changes to this edition. First, more than 11 years of references have been added and the information updated. Numerous errors have been corrected. My sincere apologies to those who were not able to recognize these in the first edition. I suspect there are more to be found so read everything with a jaundiced eye and do not hesitate to tell me of errors. Finally, two new chapters are added in the exposure assessment section. Dr. Mark Nicas kindly added a chapter on modeling of exposures, an area of application whose importance is increasing in this field as well as throughout the broader field of environmental health. Dr. Elaine Symanski has done an admirable job of simplifying the numerous issues around the factors that affect exposure analysis and the statistical means for relating these.

In summary, the book is an attempt to provide reference material to the practicing hygienist and a textbook to the student in a format not previously seen in the field. The need for a delicate balance is recognized and hopefully achieved.

Dedication

to the memory of Dr. H. Kenneth Dillon

ACKNOWLEDGMENTS

A work this size can not be accomplished without the efforts of many, regardless of what the title page may imply. Thus I wish to acknowledge the help of all those who contributed, even if your name is not specifically listed here because of the author's memory loss. The UAB program in industrial hygiene began in 1978 with the solitary efforts of Dr. Vernon E. Rose. His dedication to the quality of that program was certainly a part of the inspiration for this work. Later Dr. Ken Dillon came to the faculty. Many of the mathematical and physical chemistry derivations contained herein would not exist were it not for Ken's willingness to explain things to this biologist. Unfortunately, Ken died in a tragic accident while attending the American Industrial Hygiene Conference and Exposition in Atlanta in 2004. He is missed; this volume is dedicated to his memory.

There were numerous students along the way who made significant contributions, but the most significant contribution from the group was their desire to understand; I hope that this helps in their quest. Several students deserve special recognition for there individual contributions; Jennifer Martin, Keith Vescial, Judith McBride, Melanie Isabel, Jennifer Ehlert, and Sherry Hall did literature searches, reviewed chapters, prepared some figures, and extracted data for the first edition and some of that is repeated in this edition. David Duncan's masters project is largely presented in Appendix B, and Nancy Hopson Maher diligently answered all of the questions at the end of the chapters – any errors, however, are mine. Meredith Perkins helped in extracting the data for Appendix A. Brenda Wade, Joan Ohrn, and Sara Chalkley were instrumental in preparing the first edition and that work was revised for this edition. Rita Williams and the staff of ACGIH® are to be commended for their work in typesetting and editing this edition.

Each chapter was reviewed by one or two of the following individuals: Chuck Adkins, Susan Arnold, Paul Baron, Lisa Brosseau, Brian Davies, Gene Feigley, Larry Ferguson, Maureen Hamilton, Martin Harper, Bill Hinds, Mike Jayjock, Steve Levine, Bob Lieckfield, Mort Lippmann, Claudio Lungu, Zack Mansdorf, John Mulhausen, Gary Nelson, Mark Nicas, Robert Phalen, Will Popendorf, Vern Rose, Mark Stenzel, David Sylvain, Jim Vincent, Ted Zellers, and John Zey. They are a wonderful group of professionals and academicians, but any errors that remain are mine. A special thanks to anyone forgotten.

I

INDUSTRIAL HYGIENE – THE DISCIPLINE

1

WHAT IS AN INDUSTRIAL HYGIENIST?

Most people have an understanding of the roles, duties, salaries, backgrounds, and many other aspects of the familiar professions. Even elementary-school-aged children can give a reasonable answer when asked, "What does a lawyer do?" The same can be said for the roles of physicians, biologists, engineers, nurses, accountants, architects, and so on. These professions are known to most people. However, the public has very little knowledge of the profession of industrial hygiene, even though that knowledge is increasing. Although there are over 10,000 industrial hygienists in the United States today, and thousands more throughout the world, there were as few as 4,000 in the United States in 1980, and in 1970 fewer than 1,500. The industrial hygiene profession grew rapidly in the 1980s, but has leveled off since 1995. Possible reasons for this are explored in subsequent chapters.

Lack of public appreciation for what an industrial hygienist does may be tied to the complexity of the industrial hygienist's job. The hygienist's job is complex because he or she has to be an expert or at least have competence in many technical disciplines including chemistry, physics, biology, engineering (mechanical and chemical and specific topics such as ventilation), mathematics (especially statistics), mathematical modeling, biomechanical hazards (ergonomics), toxicology, physical hazards, risk assessment, etc. In addition, the hygienist must have competence in the social sciences including: persuasion; problem solving; negotiation; education; and a facilitation. He must be able to understand the complex and technical jargon and convert it to something that is meaningful and understandable to the worker and to technical and management non-health professionals. He or she is the interface between the health, engineering, and management aspects of what goes on, not only in industry, but in many aspects of society.

Consequently, it would be more useful to give examples of situations in which an industrial hygienist's expertise would be valuable, rather than to try to define what an industrial hygienist does in broad and general terms. Several examples have been prepared to illustrate the activities a hygienist performs; the types of problems that he or she addresses; the solutions he or she identifies and facilitates; the value added to the company's (and society's) assets in assuring workers' well-being and the contribution to the business such as increased productivity, reduced costs and minimal liability. When reading the examples below, consider the background or mental tools that the person addressing the problems must possess, the variety of the problems, and the necessity for the industrial hygienist to cooperate with other health, engineering, and physical sciences professionals in solving these problems. The problems below are not given in any particular order.

- A large office building housing 151 persons is leased by an accounting firm that performs all types of accounting services for general industry as well as individuals. Several of the employees have complained about the quality of the air and symptoms that they think are due to air quality. The symptoms include stuffiness, scratchy throat, watering eyes, runny nose, sneezing, and fatigue. The complaints appear to be coming from the secretarial staff although no one has kept an accurate record of them nor systematically assessed all employees. Management has ignored the complaints, as the building is only one year old and was designed according to what the architect called *the latest building design code*. Business is going quite well for the firm, and it is difficult for any of the managerial staff to deal with this problem. Finally, a vice-president is assigned to find a consultant who can investigate the problem and put the complaints to rest. Who should be called?

- In the past month, a company manufacturing chemical specialty products for home use has received a number of complaints involving a new product that recently has been marketed for providing a stain-proof finish to upholstery and carpets. The company is rather small, not a Fortune 500 company, but it did spend time and money to see that the product was *properly* tested prior to being marketed. The testing was done at a local university with which management has a relationship. Although the consumer complaints are not homogeneous, there are some common threads among them. An engineer on the production staff has been assigned to determine if the complaints are more than a coincidence. She has contacted the university personnel who conducted the original research, and they have informed her that if the product is being used according to the label's instructions, it should not be causing these symptoms. However, even the university personnel admit that the similarity of the symptoms suggests that the problems are more than coincidence. In order to investigate the situation further, who should the engineer call?

- A foundry that manufactures heavy machinery parts has recently begun using a new protective coating. The coating is placed at strategic places on various parts so that during the *carbonizing process* these portions of the parts are not carbon-treated. The new protective coating does not leave any kind of residue on the parts, and it protects better than the old coating. However, during the first two weeks of its use, three department employees have complained of dizziness and fatigue. The superintendent has contacted the coating distributor, who has assured him that there is nothing in the coating that should cause those symptoms, provided that the coating is being used as instructed. However, during the third week of use, an employee faints while using the protective coating. When he is revived, the employee is taken to a local physician who conducts pre-employment physicals for the company, and the physician recommends that he not return to work for several days. Who should the superintendent contact to determine if the protective coating is leading to exposures that might be causing health problems, and, if so, the best way to control these problems?

- The daughter of a 78-year-old woman is concerned about her mother's health, and a nagging question keeps bothering her about whether the air in her mother's home is the culprit. The mother recently fell and broke her hip after complaining for several days of dizziness and light-headedness. Prior to this accident, she had been in good health, according to her physician. In addition, in a recent checkup, the physician could find nothing wrong with her that might cause the dizziness, and during her stay in the hospital for treatment of the hip fracture, she did not experience dizziness or light-headedness. Because the accident took place early in the heating season, and the furnace had recently been activated for the first time, the daughter was concerned about potential problems associated with the furnace. She contacted the company that services the furnace for her mother each year and was assured that the furnace had been inspected at the end of the heating season last spring and that it had been in perfect working condition. The company agreed to send out a representative, but it was the beginning of the season, so it would be two weeks before this could happen. The daughter then called someone in the local public health department, who agreed that she should be concerned and suggested that she call someone who could examine the air quality in the house. Who did the health department recommend that she call?

- A physician in a community of 50,000 people has, in the last month, seen three cases of an unusual disorder called peripheral neuropathy – a disorder related to an inability to control the extremities of the body. Although the first case was not surprising, the second concerned the physician, and the third was cause for alarm. He determined that all three were employed at the same manufacturing plant. This company manufactures chemical products. Remembering an environmental epidemiology lecture that the physician had in a public health course 12 years earlier, he decided to call the local public health department. The director of the county health department, a physician, was also concerned but was uncertain about the proper course of action. She suggested calling the state epidemiologist at the capitol. Once notified, the state epidemiologist told the physician that an investigation should be undertaken as soon as possible. However, in this particular case, the state had no statutory authority to undertake the investigation, and would first have to obtain the company's permission to enter the premises. The state epidemiologist remembered that the National Institute for Occupational Safety and Health should be notified, and would send a team consisting of an epidemiologist, a physician, and other health professionals who could enter the site under the statutory authority of the Occupational Safety and Health Act. What other health professionals do you think would be on this team?

- An engineer for a company that manufactures fire extinguishers has been given the responsibility to investigate new and alternative methods for applying protective coatings to the outside of fire extinguishers. These coatings need to be more effective, less costly, and safer than the spray system currently in use. She has communicated with a foreign company that offers a *powdered paint* system, which utilizes less coating material than the old system, gives a harder protective finish, and is promoted as being safer than liquid spray systems. The foreign company has assured the engineer that the system, once installed, meets all of the regulations regarding ventilation and exposure levels of the foreign country. However, the foreign company is not certain about the system's ability to meet U.S. regulations. Who should the engineer contact in order to determine the ability of the foreign system to meet U.S. safety and health regulations?

- Maintenance personnel at a university are required to maintain fans, located on top of buildings, that draw air from laboratory hoods and exhaust it to the outside. The fans must be periodically inspected, and, as necessary, belts must be replaced and bearings lubricated. Of course, any fan that is being serviced by the maintenance personnel must be de-energized and locked in the *off* position. However, these personnel are concerned that other fans that are not in the *off* position but are exhausting to the roof of the building may be emitting poisonous gases or vapors into the air near where they are working. The supervisor has informed the personnel that the maintenance procedures have been conducted safely for over 20 years, and that there is nothing to be concerned about. However, on one occasion, a worker detected an odor of *rotten eggs* emanating from an exhaust stack while he was maintaining a nearby fan. Nevertheless, the maintenance supervisor claims that this has occurred before, and that exposure to the *rotten egg smell* is not hazardous. The maintenance workers feel their health is at stake and there must be someone with whom they can discuss this potentially hazardous situation. Who should they call?

- An insurance company that has been insuring businesses for fire, casualty, and theft for over 50 years has determined that insuring employers to provide worker's compensation benefits to injured or ill employees could be a lucrative business. Under the worker's compensation system, the employer provides medical benefits and time off or monetary benefits to workers who become injured or ill on the job. It is a no-fault system. Obviously the object of the insurance company is to charge rates such that, on average, it will collect more money than it pays out in benefits. If the insurance company can selectively insure companies that have good injury and illness records and, more important, appear to have workplaces' that are not apt to cause injury or illness to workers, then the insurance company is more likely to be profitable. The company currently employs fire and other safety inspectors (for example, a chemist who specializes in explosions) but realizes that it needs a different type of person to evaluate individual plants to determine their propensity to cause illness to workers. What type of person do they need to hire?

- An attorney was approached by two potential plaintiffs who claimed that while they were performing a job for a construction company, it became very difficult for them to breathe, and approximately two days later, pneumonia-like symptoms developed, which had persisted for two months. Their doctor was not sure if the symptoms were related to the operation at the construction company; however, the workers were certain that the symptoms were related to the event. The doctor also was not certain whether the symptoms would persist, or whether they would gradually improve. The lawyer obtained information that indicated that the workers were cleaning some copper and other metal surfaces with an acidic cleaning solution. She called an occupational physician who was an expert witness in occupational exposure claims. The occupational physician was not certain what might have caused the symptoms; however, he said that it was a good possibility that the symptoms were related to the event. He also indicated that, in order to make a case, the attorney would have to determine what caused the symptoms. The physician suggested that the mixing of the acid with impurities in the metal could have caused the symptoms. He recommended that the attorney contact other health professionals, who could give her information about the evolution of gases and vapors during this type of cleaning process. Who should the attorney call?

- A small restaurant is located next to a dry-cleaning establishment. Occasionally the smell of dry cleaning fluids becomes so strong in the restaurant that customers complain and even leave. In addition, the employees at the restaurant have complained to management about the dry-cleaning smells. Management has contacted the local health department, which inspects the restaurant for food sanitation purposes. The sanitation inspector determined that there was nothing that could be done about the dry cleaning odors, but said that they should not be a hazard to the employees or to customers. She was not certain if the restaurant owner could do anything legally about the problem, but suggested that the owner contact a lawyer. The restaurant owner contacted several attorneys, but none of them was interested in taking on the case, nor could they recommend any particular legal remedy for the restaurant owner. However, one attorney remarked that there was a type of *environmental engineer* who could evaluate the inside of the restaurant to determine what levels of dry cleaning solvents might be in the air. However, the attorney could not remember the exact name of the profession, and was not even certain if it was an engineering profession. She suggested that the restaurant owner look through the telephone directory to determine the appropriate person. Who should the owner call?

- A group of 12 physicians have formed a partnership and will provide services to a large number of patients. They plan to establish their own clinic containing X-ray equipment, equipment for diagnostic procedures, and other typical clinic instruments. In addition, some of the staff will administer other types of radiographic medicines. Several of the physicians have had experience with clinical situations such as this and realize that they need to comply with state and federal laws for radiation. One physician, in particular, is concerned about other potential hazards in the clinic, such as the chemicals handled during the diagnostic procedures. This physician is concerned that there may be federal or state laws regarding proper use of these chemicals. The physicians want to establish the clinic properly and make certain that their employees are protected from any hazards. Who should they call?

- An employer who owns a machining shop has recently had three worker's compensation insurance claims filed against him by employees who have claimed noise-induced hearing loss from working at the machine shop. The employer realizes that there may be excessive noise because approximately two years earlier an inspector from his worker's compensation insurance carrier took some measurements with an instrument for about 15 minutes and told the employer that the employees might need to wear hearing protection in certain areas. However, the employer did not pursue the matter any further, and though he did provide the employees with hearing protectors, the employees did not wear the devices. The employer now believes that hearing protectors may not be the best answer, and wonders if there is something that can be done to the machinery to make it quieter and reduce the need for hearing protection. Who should he call?

- During a 24-hour period, three hotel occupants became very sick and were hospitalized. Within two days, one died, and the others remained in critical condition. The attending physicians were uncertain of the cause of the symptoms. A type of pneumonia had formed that the physicians had not seen before. The hotel management was extremely worried about the risk to other guests and employees of contracting the disease. In addition, the local public health department, which had been notified, was quite concerned. However, no one had any idea of how to go about determining the cause of the disease or estimating whether and when it might occur again. Who should be called?

- A large research hospital has successfully demonstrated a new technique for preparing a drug that is critical in the care of patients suffering from a disease that is normally fatal after about two years. The new technique for preparing the drug allows it to enter the bloodstream more rapidly and, consequently, to be more effective in treating the patient. The university is working with a drug company through the proper Food and Drug Administration (FDA) channels to gain approval for the drug. The FDA is concerned about side-effects caused by the new preparative technique. Consequently, the drug must be tested in volunteer patients for six months under carefully controlled conditions. Although the physicians are extremely enthusiastic about the new drug, several of the nursing staff who must prepare the drug are not so enthusiastic. During the new preparative technique, the nurses suspect that they are being exposed to the drug. They have taken their concerns to the chief hospital administrator, who also is somewhat concerned about the nurses' exposure. The nurses have told the administrator that the drug is known to cause birth defects in laboratory animals, and that if any nurse of child-bearing age should have an abnormal child, the hospital would be open to a very large lawsuit. Upon notifying the responsible physician, the administrator encounters a great deal of resistance in changing any of the procedures. The physicians are in a hurry to get the drug approved so that it may do the most amount of good for patients. Of course, they are also interested in being the originators of a new preparatory technique that could save thousands of lives. The administrator is not certain what he should do. Who should he call?

- The superintendent of schools of a large system is concerned about the physical integrity of the building envelope of the schools in the system. Ten of the 11 buildings were constructed in the 1970s, and at a recent conference he learned that other school buildings constructed during that time often leaked and the wet surfaces led to the growth of mold. However, he is not aware of a profession that ascertains whether mold exists on building surfaces or in indoor environments. A mold scientist (mycologist) at a local university told him that there are people who could do these assessments, but the mycologist could not remember the name. Who should the superintendent call?

In every case, the advice of a hygienist should be sought. Not all hygienists can solve each of the above problems immediately. However, all should be able to find a solution given some time, or at least be able to put the inquirer in touch with an expert in a related area who can solve the problem. Even though the range of problems presented above is very wide and should indicate the wide range of skills that the industrial hygienist must possess, many industrial hygienists, particularly those employed by smaller plants, may also need to be responsible for other areas. For example, consider the following problems that a hygienist may be called upon to solve.

- A small company has decided to begin a new process that appears to be quite lucrative. Other processes that the company is currently using to manufacture several products do not generate any hazardous waste. However, from this new process, fairly large quantities of a waste product will be generated. Depending upon how the waste product must be handled, the profits of the new process could be severely diminished; consequently, the company may decide not to enter into the process. Normally, an environmental engineer would be called upon to determine how the waste product should be handled, and if it could be treated on-site at less expense than, for example, by sending it off-site. However, the company is small and does not employ an environmental engineer. Consequently, management has called upon the company industrial hygienist to determine solutions.

- A company that manufactures refrigerators at a site in Mexico has been notified by air pollution authorities that it should begin to control its emissions of Freon 113. Previous inspections by air pollution authorities indicated no emissions that needed to be controlled; so the company has not been concerned about air pollution problems. Two industrial hygienists are employed at the U.S. corporate offices. Company management is aware that chloro fluoro carbons (CFCs) are under tight control because of worldwide concern about stratospheric ozone reduction. However, plant management thought that only CFCs used in the internal parts of refrigeration equipment were being controlled in Mexico, not those expended in the manufacturing process. The plant manager has asked one of the industrial hygienists to look into the situation and determine what should be done, what the emission rates of the CFCs are, and whether a substitute chemical is available.

- After an inspection by a team of state engineers, a small manufacturer is ordered to begin controlling wastewater discharges from the plant site. The plant manager is extremely confused about the order because she knows of no water discharges from the site. She has called the plant industrial hygienist in to discuss the order with him. The industrial hygienist informs her that even though no effluents are directly placed in the two stream beds on plant property, regulation requires the control of storm water runoff from plants where hazardous materials may be on the ground or other surfaces around the plant. In this particular case, there is the potential for lead to be in the soil as a result of the manufacturing process. The plant manager has directed the industrial hygienist to determine under what circumstances the runoff needs to be controlled, what the various alternatives are for controlling the runoff, or if it might be more feasible to clean the surrounding soil areas and thus to do away with the need to control storm water runoff.

- A small manufacturing plant containing many ventilation systems has recently become concerned about energy utilization. The cost of electricity has doubled in the past few months as a result of a rate increase imposed upon users to pay for new power plant construction. The plant manager realizes that fans contained in the ventilation system require a great deal of energy during their operation. He also realizes that there are other sources of energy use within the plant, but has called upon the industrial hygienist to investigate ways that the plant may reduce its energy requirements. The industrial hygienist informs him that it may be possible to recirculate exhaust air that has been previously heated in the winter or cooled in the summer. The exhaust air would, of course, have to be cleaned before being recirculated. The plant manager had not thought about the energy cost caused by the exhaust of conditioned air. He has instructed the industrial hygienist to look into ways to recirculate the exhaust air and to look into other ways throughout the plant in which energy may be conserved. This includes the possibility of utilizing a computer-controlled climate control system.

- A small plant employs an industrial hygienist and an industrial hygiene technician, both of whom also serve as safety coordinators for the facility. Worker's compensation claims have been handled in the past by the personnel director and his staff. The industrial hygienist also works closely with the personnel director on worker's compensation matters. The plant manager is concerned about the increasing cost of worker's compensation in spite of steady or decreasing injury and illness rates. The personnel director and his staff are currently heavily involved in negotiations regarding a new computerized personnel system. Consequently, the plant manager has asked the industrial hygienist to look into the question of rising worker's compensation costs. Specifically, he has asked the industrial hygienist to inquire about those costs with the current insurance carrier, to contact other plants and determine what costs are, and to ask other insurance firms about costs. The plant manager has also asked the industrial hygienist to look into the possibility of the company's becoming self-insured.

The industrial hygienist has many responsibilities and is expected to be able to solve, or at least to demonstrate a knowledge of solutions for, many different kinds of problems because he or she serves as the interface between the engineers who design a product, the workers who manufacture and use products, and the members of management who seek a profit on products. In addition, the industrial hygienist serves as a liaison between physicians or other clinical health care providers and the workplace environment.

While an industrial hygienist employed by a company must always keep profit in mind, his or her perspective on profit may be considerably different from that of management. It is in the hygienist's, workers', management and stockholders' best interest to have well and productive workers, and a financially successful company, along with minimal liability. However, not all managers and stockholders have a clear understating that the company benefits form a safe and healthy (and respected) workforce. More than any other professional, the hygienist realizes that the workforce, including both management and those being managed, are resources in which considerable capital has been invested. In many cases, he or she also has to consider the health and well-being of consumers and the community at large. The health of each of these populations can be considered as a resource with economic value or as an individual right that the hygienist has a moral obligation to protect. Either way, the task is large and demanding and requires knowledge from many fields.

QUESTIONS

1.1 Based on the tasks that an industrial hygienist might perform, in what fields of study should he or she have knowledge? Try to list at least seven. What two fields do you think might dominate this knowledge?

1.2 Based on your current knowledge of what an industrial hygienist does, list names other than *an industrial hygienist* that would describe this person. Do you see any potential problems with the current name?

1.3 Based on the broad depth of knowledge that industrial hygiene requires, would it be better to obtain an undergraduate or a graduate degree in the profession? Why?

2

INDUSTRIAL HYGIENE – HISTORICAL PERSPECTIVE

At the beginning of the twentieth century, most persons who had concerns about health matters arising in and out of industry were a handful of physicians and engineers. The physicians led the way; several physicians in the United States made studies in the early part of the century that pointed out the appalling conditions that existed in some workplaces.

Prior to the twentieth century, physicians had made observations about the conditions of health in working persons in Europe. The physicians did an admirable job of identifying the hazardous conditions that existed in many workplaces and pointing out the effects on health caused by exposure to them. The profession that came to be known as industrial hygiene[1] remained in this discovery mode well into the twentieth century. It was not until perhaps the 1930s or 1940s that engineers began to take an active role in finding ways to control the hazards found in industry. Hence, the hygiene profession really came of age only when it began to do something about the exposures that physicians had been discovering. This is not to say that nothing happened prior to 1940. Indeed Carter (2004) points out that by 1920 the basic tenets of the profession had been established. For example, the advantages of ventilation, respirators and work practices had been reported. Dose response and sampling and analytical methods were in use. The essential elements of recognition, evaluation and control were discovered. However, these concepts were not widely used until much later.

Interestingly, today, very few hygienists come from the medical profession. Most physicians who are interested in occupational health become occupational physicians, and there are relatively few of them. Similarly, relatively few hygienists now come from the field of engineering. The majority of hygienists in the United States come from chemistry and biology. Instead of becoming specialized as others have done in many other fields, most hygienists are generalists. Once there were physicians who specialized in the recognition of diseases and

---

[1] In the United States, the profession is known as industrial hygiene. However, in the rest of the world, in particular Great Britain and such parts of the British Commonwealth as Australia and Kenya, it is known as occupational hygiene. From here forward, occupational or industrial hygiene will be referred to as hygiene and the term hygienist will be used for the practitioner.

engineers who specialized in the control of hazards leading to disease. Now most hygienists have a comprehensive ability to recognize, evaluate, and control exposures that lead to occupational disease.

OCCUPATIONAL HEALTH

Occupational health, a field of endeavor in which many different professionals practice, is part of a larger field known as environmental health, and of an even larger field known as public health.

Hygiene was used in preference to health in the common vernacular in the later 19th and early 20th century (Bergey, 1901; Broadhurst, 1908). Thus public hygiene was a government effort. Hygiene means the conditions or practices that cause or effect health. Occupational health is somewhat broader than occupational hygiene, but one should not construe that fact to mean that hygienists are concerned only with creating the conditions for health, and not concerned with such things as finding the determinants of disease. In its narrowest sense, hygiene means controlling the factors that cause disease, whereas hygienists also are taught to recognize hazards that cause disease and to evaluate their potential to cause disease.

There is a need to recognize the differences between occupational health (OH) and industrial or occupational hygiene. First, OH is a field, a field of endeavor or an area of practice. Hygiene is a discipline or a profession like law or medicine. Disciplines are taught. A discipline is an area in which one practices (Merriam-Webster, 1991). The first physicians and engineers who became interested in worker health practiced their professions in the context of occupational health. At some point, a group of engineers and physicians (with perhaps a few chemists and biologists) asked the question, are we practicing our profession (engineering or medicine) in the field of occupational health, or have we defined a new discipline or profession? The answer to the latter was *yes*, and the hygiene discipline was born, although it still struggles somewhat to be recognized as such.

Today, members of many professions, including hygienists, engineers, physicians, and nurses, practice in the field of occupational health. Specialized names such as ergonomist, occupational physician, and occupational health nurse have been developed. People from other professions also practice in the field of occupational health, such as chemists, toxicologists, epidemiologists, and lawyers. So long as each is called chemist, toxicologist, or the like, then each maintains an allegiance to his or her first profession. A chemist who does research on methods to detect workplace chemicals usually remains a chemist first.

However, many persons who have training (in order of frequency) in biology, chemistry, engineering, physics, and medicine have taken subsequent specialized training in hygiene and now consider themselves hygienists. A chemist who becomes a hygienist must look beyond the chemical aspects of occupational health and consider aspects of many other disciplines in the context that each gives to worker health. To repeat, the hygienist is a generalist, who applies knowledge from many disciplines.

Members of many professions practice in the context of occupational health and are part of a team, including hygienists, occupational physicians, occupational health nurses, occupational epidemiologists, occupational toxicologists, and others. Hence, the hygienist is part of the occupational health team. When safety professionals and ergonomists are included in the team, it is usually termed the occupational safety and health team.

The difference between occupational health and occupational safety is defined by convention. Occupational safety also is a field in which members of several professions practice. By convention, those agents in the workplace that tend to cause illness are studied in the field of occupational health, whereas those agents or human acts that may cause injury are studied in the field of occupational safety. The two important words to differentiate are injury and illness. Illness is associated with occupational health and injury is associated with occupational safety. There is considerable overlap in these two definitions. Chapter 4 discusses agents that cause injury and illness in the workplace, and gives a further differentiation of occupational health and safety.

OCCUPATIONAL HEALTH PRIOR TO 1900

From the early 1900s back through time, very few people actually devoted their careers to working in occupational health. However, several notable individuals made observations about occupational health. Most of these individuals were physicians who made observations concerning unhealthy conditions in the workplace and described the disease that they thought resulted from those unhealthy conditions. It appears that the control of those conditions, to the extent that control took place, was left to enlightened individuals, probably the early forerunners of supervisors, foremen, and engineers, who stumbled upon or in some cases systematically discovered means to make work easier or presumably more healthy.

The very earliest records of occupational disease (as reported by Patty without reference, 1958) would have been those of Hippocrates, who made observations about lead poisoning in the fourth century, B.C. Hence, written observations concerning the unhealthiness of work have been made for more than 2,000 years. It is important to note three other points about Hippocrates' work. First, his observations were concerned with lead, which down through the ages has been the most studied occupational hazard. We continue to make discoveries about its effects and continue to reduce exposure levels that were previously thought to be safe. Second, Hippocrates made far reaching observations on the subject of medicine and essentially began the process of turning medicine from an art into a science. Third, Hunter (1978) points out, very succinctly, that the medical principles and the eating, drinking, and exercise habits that Hippocrates recommended, were meant for the upper class. People who worked in the *mechanical arts* were at first slaves, and once freed were still looked upon as fourth-class citizens. This social dichotomy between the haves and have-nots was, and in some cases still is, the biggest obstacle to improving worker health. In Hippocrates' time if a worker complained about conditions, there were many others to take his place; a situation that still exists worldwide. If management does not respect the managed as human beings with equal rights and privileges, then it will not be sincere in providing healthy working conditions, even if it complies with the *letter of the law*.

Galen (130-200), Celsus (first century), and Pliny the Elder (first century) also made references to occupational health. Galen and Celsus were physicians; but Pliny the Elder was not one. However, the latter wrote an encyclopedia of natural sciences (Patty, 1958) in which he referred to the use of animal bladders being worn over the faces of miners to avoid the effects of mercury poisoning. It is not hard to imagine the appalling conditions that existed in many unventilated mines. Mining of metals and metallurgy were the subjects of the first thorough observations on the diseases of occupations. Hunter (1978) points out that by the year 965, mines were established in regions historically associated with Germany, and mining and mining industries remained an important part of commerce for some time. This industry and its tremendously cruel effects on the health of miners, their families, and, in some cases, owners as well, became the subject of observations on the relationship of disease and occupation.

In 1473 Ellenbog (Ramazzini, 1964; Rosen, 1964) described the hazards and mechanisms for the control of mercury and lead. Almost 100 years later (1556) *De Re Metallica* by Georgius Agricola (1494-1555) was published posthumously. It was an incredible 12-volume treatise on the mining industry that only Agricola could have written. His training was broad and included medicine. In 1526 he was appointed physician to the famous mining town of Joachimsthal in Bohemia. *De Re Metallica* (translated by Hoover and Hoover, 1950) reports on every aspect of mining. Volume 6 leads with a very brief (five pages from a 600-page treatise) description of diseases and injuries of miners and methods for their prevention. What appears to be arsenical poisoning was well described. In a region of the Carpathian mountains, it was stated that women resided "who have married seven husbands, all of whom this terrible consumption has carried to a premature death." Although Agricola recognized some of the maladies of mining, his observations were simplistic, even if advanced for his day. He blamed some cases of death on demons, regarding them to be preventable by prayer and fasting, but it is doubtful that fasting aided the miners' already deteriorating health. Indeed, fasting causes metals to be released from body reservoirs such as bone and kidney tissue. On the other hand, Agricola showed incredible insight into the roots of poor workplace hygiene: " . . . for we should always devote more care to maintain our health, that we may freely perform our bodily functions, than to making profits." This quote applies equally well today.

Another man of medicine plied his trade as a contemporary of Agricola. Paracelsus (1493-1541) wrote an entire treatise on diseases of mine and smelter workers (Hunter, 1978). Unlike Agricola, he had no formal training, but he had experience in chemistry and metallurgy and became physician of the mining town of Villach. He is best known for abandoning the humoral theory of medicine and adopting specific remedies for diseases. Some were quite amazing, including opium, mercury, lead, and arsenic. Paracelsus said that "it is the dose only that makes the thing not a poison" (Ottoboni, 1984). He taught in German, not Latin, circulated freely among the masses, and in general did everything he could to buck the scholastic etiquette of the day. For this he was exorcized from academic medicine. He is often cited as the father of toxicology.

More than a century after Paracelsus, the next scientist to coalesce many years of knowledge on occupational health was Ramazzini. A physician, he went beyond Paracelsus and Agricola in his treatise of 1700, *De Morbis Artificum Diatriba* (Discourse on the Disease of Workers). Certainly other descriptions following Paracelsus' description of the diseases of scholars, seafarers, soldiers, salt workers, lawyers, and other occupations had been prepared (Rosen, 1964). However, Ramazzini brought all of them together in one treatment. As a result he has been called the father of occupational medicine (Hunter, 1978).

Ramazzini's second edition (1713) covered 54 occupations. His descriptions of conditions, hazards, diseases, and prevention are, in some cases, amazingly accurate to this day. Hippocrates had formulated a series of questions that physicians should ask of their patients. Ramazzini added an indispensable one:

When a doctor visits a working class home, he should be content to sit on a three legged stool, if there isn't a gilded chair, and he should take time for his examination; and to the questions recommended by Hippocrates he should add one more – What is your occupation?

From this quote, modern physicians can gain insight on more than one point. Although his book seems timeless in its ability to describe diseases and make observations about their control, it is somewhat sad to observe that after almost 300 years, it is still difficult to find general physicians who will ask questions of their patients regarding their occupations. In other words, occupational medicine and the propensity for work to cause disease and ill effects on health still is not adequately recognized by the medical profession.

INDUSTRIAL REVOLUTION

Most Americans probably would be surprised to learn (or relearn) that shortly before the American Revolution for independence, the industrial revolution began in Great Britain. Historians commonly cite 1760 to 1830 as the years for that revolution in Great Britain, but in 1830 the United States was still very much an agricultural economy with little mining or manufacturing. Its cotton was mostly milled in England. Its chemicals, what little it used, were made in Europe, as were its metal products. Not until 1860 would the United States enter the industrial revolution on a large scale, a revolution that ended about 1900 (Kazis and Grossmann, 1982).

Industrialism led to several important changes in the way work was done. First, it was divided. No longer would one craftsman spend two months making a rifle. Instead, many workers would make only small parts of rifles, over and over again. Second, it was mechanized. Third, it called for mass labor – hence, children, women, and anyone else who could perform menial, repetitive, mundane tasks were acceptable and often dispensable. Hunter quoted Pierre Hamp, of this era, as saying:

We live on the sufferings of others. Everyone makes life a torment for some of his fel-

> low men. How many people earn their living pleasantly? Many do so in unpleasant and often intolerable conditions. To love one's occupation is to be happy, but where are the occupations which one can love?

To this, the editors (Raffle et al., 1987) of *Hunter's Diseases of Occupations* have added, "Yet it might be said, in an age of unemployment, that one kind of work, *however unpleasant*, is better than none" (emphasis added). Certainly, workplace conditions have changed for the better,[2] but should work necessitate suffering in unpleasant surroundings? Are all workers entitled to safe and healthful working conditions?

The unsafe and unhealthful conditions brought on by the industrial revolution did not become common in the US until the last half of the 1800s; and because these conditions first existed in Europe, Europeans led the way in attempts at improving them and in research. Thus physicians were employed at textile mills as early as 1789; Sir Robert Peel became the British father of industrial legislation when he introduced the Health and Morals of Apprentices Act of 1802 as a result of working conditions in his own textile mills. This act was followed by the 1819 and 1825 Factories Acts, which, like their predecessors, were largely child labor laws and were not enforced (Luxon, 1984). The 1833 Factories Act added the teeth of enforcement or inspection – an important step that set up an adversarial relationship between government and industry, which has developed full strength in the United States.

Child labor laws were not enacted by most states in the U.S. until 1913 (Kazis and Grossmann, 1982). This meant that until then, six-year-old children in southern mills worked 13-hour days with a 35-minute lunch break. A strong national law was not enacted until 1938.

Great Britain also produced Charles Turner Thackrah (1795-1833), who went one step further than Ramazzini. Whereas the latter realized the importance of occupation to health, Thackrah devoted his career to the prevention of diseases caused by occupation and wrote extensively on the subject, including problems of child labor, tuberculosis in tailors, lung disease in miners and metal grinders, and lead poisoning in glaze dippers. His untimely death from tuberculosis at age thirty-seven retarded the prevention movement.

By 1895 Great Britain had passed several other Factories Acts, each strengthening the former, and in that year for the first time certain occupational diseases, became reportable. Every physician had to report to the government patients with diseases believed to be caused by lead, phosphorus, arsenic, or anthrax. Others were added later. In 1898 Thomas Legge (1863–1932) became the first medical inspector of factories, and other notable British figures

---

[2] For a historical perspective on working conditions around 1900, students are encouraged to read Alice Hamilton's *Exploring the Dangerous Trades* (1943), Upton Sinclair's *The Jungle* (1905), and the first four chapters of *Hunter's Diseases of Occupations* (1978). For a more recent perspective, read *The Hawks Nest Incident* by Martin Cherniack (1986) and Paul Brodeur's *Expendable Americans* (1971). The historical bibliography at the back of this chapter lists other articles of interest. All are enlightening, but, in particular, those that are not from peer-reviewed journals should be read with a jaundiced eye.

soon followed, each associated with seminal research on a disease and work: Edgar Collis (silica), John Sydney Henry (coal tar pitch-skin cancer), E.R.A. Merewether (asbestos), and Ethel Browning (solvents). Interestingly, New York appointed a medical inspector in 1907, but the idea did not catch on.

Unlike in the UK where occupational health was highly dependent on government, in the US there was a mixed dependence on industry and local and state government to accomplish workplace hygiene. While success of this model was quite variable and ultimately deemed unacceptable, it allowed the hygiene discipline to develop at a faster pace, from both a practice and an academic perspective, than elsewhere in the world. On the other hand, by 1900 the UK had a system of injury and illness statistics that was only approached by a few states in the U.S. and non-existent at the federal level (Luxon, 1984). By the early 1900s, U.S. laws, mostly at the state level, were reasonably in line with those in Europe in only a few states. By 1970 the U.S. had accelerated past the rest of the world – at least in legal protection of workers' health.

As a final note on the nineteenth century, little has been written in the English language about the research and legislative efforts of other European countries. However, they certainly made their contribution. For example, the German K.D. Lehmann did extensive studies on the toxicology of industrial gases in the late nineteenth century (Patty, 1958; Hamilton, 1948), and the French chemist and pharmacist Jean Baptiste Alphonse Chevalier (1793–1879) made a significant contribution. No doubt there were many other contributions, such as those of F.F. Eresiman (1842–1915) of Russia (Schilling, 1981).

1910–1940

It is not certain who coined the phrase *industrial or occupational hygiene*; however, prior to 1900 it does not appear to have been used. (There was an Office of Industrial Hygiene and Sanitation in the U.S. government by 1914 (Felton, 1976).) Nevertheless, there were a number of individuals, particularly physicians and engineers, who practiced their profession in the context of occupational health during this time.

Most industrial hygienists who are familiar with the work of Alice Hamilton would call her the first hygienist. Alice Hamilton (1869–1970) was a pioneer in hygiene and occupational medicine in the United States. She stated in 1910, when she entered the field, that there was essentially no American literature to be reviewed. She spent two years reviewing German and English literature prior to beginning her studies (Hamilton, 1948). Alice Hamilton vigorously promoted hygiene in industry. Early in her career, she was a physician first and industrial hygienist second. Today, most hygienists use their backgrounds (e.g., chemistry and biology) as stepping-stones toward their ultimate profession. However, the process by which hygiene became a discipline and a profession took place over approximately 40 years. Its roots, at least in the United States, are in the work of Alice Hamilton and others and began approximately in 1910.

Prior to 1910, no federal legislation had been passed in the United States for the specific purposes of protecting workers' health. Thus, while the Reverend Charles Brigham of the Michigan State Board of Health prepared an essay entitled "The Influence of Occupations Upon Health" as early as 1875, it was not until the pioneering work of Alice Hamilton in Illinois in 1910 to 1915 and the work by Dr. Emery Hayhurst on Ohio industries in 1913 and 1914[3] (Trasko, 1955), that American society began to be awakened to the grim realities of health in the workplace.

Even though Pliny the Elder had pointed out the effects of lead mining upon health as early as the fourth century B.C., the work of Dr. Hamilton again pointed to the fact that lead was the number one cause of extremely disabling disease, or even death, in industry. In fact, the agents of disease that were studied by Dr. Hamilton continue to be agents of disease of the workplace as we enter the 21st century. In addition to lead, Dr. Hamilton pointed out diseases associated with mercury, carbon monoxide, phosphorus, aluminum, beryllium, cadmium, chromium, benzene, methanol, carbon disulfide, and synthetic rubber. It is interesting to point out that the disease associated with aluminum described by Dr. Hamilton was found in miners who were exposed to aluminum dust, not through the mining of aluminum ore (bauxite), but rather through the use of alumina dust to cover the surface of mines for the prevention of silicosis (caused by silica). This process has now stopped, and aluminum is relatively well controlled in other industries. Control of exposures and a relatively low hazard level for aluminum makes it the only agent studied by Dr. Hamilton that is of little hazard in today's industry[4].

Dr. Hamilton's studies and those of others were remarkably revealing of the extent to which disease existed in industry. Nevertheless, very little was done in terms of legislation to control workplace exposures. One of the most oft-cited pieces of legislation for the control of a specific occupational disease is an act of 1912 that placed a heavy tax on the use of yellow phosphorus, also called white phosphorus. The phosphorus was used in the manufacture of matches and caused an extremely disfiguring disease of the jawbone among match workers, known as phossy-jaw. Essentially, the bone deteriorated to the point where, in some cases, workers lost their entire jawbone. However, what is interesting about this particular episode was that even though the disease itself was so disfiguring and a substitute (sulfur) was available for the manufacture of matches, the U.S. government did not choose to place an outright ban on the use of phosphorus. With the exception of food additives that cause cancer in animals (and even the relevant Delaney clause of the Food, Drug, and Cosmetic Act of 1958 has been neutered), this trend of restricting, as opposed to banning, substances prevails in the United States. Even pesticides that have had effects on humans and wildlife (e.g., the chlorinated hydrocarbons) are regulated closely but are not banned from commerce (EPA, 1990).

---

[3] A two-page report in 1903 issued on Factory Sanitation (Doehring, 1903) is the only other exception.

[4] While there is an issue with occupational asthma in the aluminum smelting industry, the causative agent is debated.

WORKERS' COMPENSATION

Instead of trying to control occupational agents of disease through legislation, laws were enacted to compensate workers suffering from occupational injuries. This effort was spearheaded by industry, in particular US Steel, and began at the federal level, when limited benefits were provided in 1908 for U.S. civil service employees. This workers' compensation system was based on a German model and was attractive to workers in that it was a no fault system that seemed to guarantee benefits. On the other hand, employers could budget a fairly constant cost of 1% of operating revenues. Thus it seemed to provide little incentive for reducing injuries. Nevertheless, in 1911, the first state workers' compensation system was formed, and by 1920 all but six states had adopted it. In 1948 Mississippi became the last state to enact a worker's compensation law (Felton, 1976; Berman, 1977).

Many of these laws were originally intended to compensate workers for *injuries* suffered on the job, but gradually various states began compensating workers for *illnesses* contracted on the job. Some states limited compensation to specific diseases such as silicosis and lead disease, whereas other states allowed compensation for any illness that could be associated with the workplace.

The workers' compensation acts had both advantages and disadvantages. This form of no-fault insurance allowed a worker to be compensated for medical benefits and time off from the job and, in some cases, given a monetary grant where disabling disease or injury occurred, but the legislation took away most other legal remedies that the worker had. Prior to workers' compensation acts, a worker could sue in court for damages due to injury or illness arising from the workplace. The employer had three defenses, which were often liberally used. The first was the fellow servant defense: a worker could not be compensated for an injury if another worker's (read: supervisor's) negligence contributed to the injury. The second defense was the assumption of risk. This defense was most reflective of the attitudes of society at the time. It stated that the employee assumed the risk of the hazards inherent in a job. In other words, the employee knew, or should have known that the job was risky. Of course, at that time many workplaces were full of hazards that were largely acknowledged because they were so obvious; hence a worker would have to know about them. The third defense was that of contributory negligence. In this case, if the employee was negligent in any way, he or she would collect nothing, regardless of the employer's negligence (Corn, 1978). Eventually reforms in some states weakened these defenses, the public became more aware of tragic conditions, and worker successes and awards in courts increased. This created another reason for employers to favor the relatively constant cost of the compensation system (Berman, 1977).

INDUSTRIAL HYGIENE UNITS

Another phenomenon that occurred between 1910 and 1940 was also largely a function of the states rather than the federal government. During this time, many states established industrial hygiene units within departments of public health or departments of labor, the majority of them located in departments of public health (Trasko, 1955; Patty, 1958). In addition to their roles as inspectors, individuals in these units also had responsibilities for conducting studies of workplace diseases and, in some cases, conducting physical examinations of workers in certain industries. By 1909 at least 21 of the states had enacted laws that attempted to regulate various working conditions. Some of the laws dealt with ventilation, whereas others pertained to dust control and general sanitation. Of course, enforcement lagged, either because 1) there were no inspectors and industry was simply expected to comply with the laws, 2) inspectors were political appointees who had no training in what they were to inspect, or 3) violations would not stand in court. (To this end it is interesting that factory inspectors in the UK are able to try their own cases before the bar (Luxon, 1984)).

These industrial hygiene units ebbed and flowed until a point midway through the Great Depression. In 1935 the Social Security Act was passed, and one of its provisions made funds available for the expansion of public health programs, which included industrial hygiene programs. This Act alone probably did more to promote industrial hygiene in the United States than any other Act up until the 1970 Occupational Safety and Health Act. By today's standards the amounts of money provided for industrial hygiene programs were not great. For example, from 1947 to 1950, one million dollars in federal funds was allocated. Nevertheless, there were 30 state and local units established by 1939. Hence, industrial hygiene flourished because of an Act that allowed the states to hire personnel to investigate, prevent, and control various diseases that had long been established as some of the most debilitating human diseases.

As further testimony to the expanding development of the industrial hygiene profession at this time, several organizations were established in the late 1930s. In 1937, the Federal Division of Industrial Hygiene held a seminar that led to the establishment in 1939 of the American Conference of Governmental Industrial Hygienists.[5] Until late in the 20th century, members of this organization had to have a government affiliation. In that same year the American Industrial Hygiene Association was established, and it was open to all persons having an interest in industrial hygiene. It is probably impossible to pick an exact point in time when engineers with an interest in hygiene in industry became industrial hygienists. This was a gradual process by which the principles for recognizing, evaluating, and controlling the agents of disease in industry gradually were developed. However, it seems likely that the establishment of the two organizations in 1939 and the impetus added to the development of the field by the Social Security Act can be cited as the beginnings of the industrial hygiene profession.

It should be noted that the British Occupational Hygiene Society was formed in 1953. Another reason for the time lag may be that European countries were far more concerned with the two great wars than the United States, which was not as deeply involved. In fact, the development of the British Industrial Health Research Board in 1918 was for the primary purpose

[5] Under the Social Security Act of 1935, many industrial hygienists were loaned from the federal government to state and local authorities for purposes of setting up state and local hygiene programs.

of increasing British wartime output by finding ways to make workers most efficient through healthful conditions (Warner, 1974). Up to 1939, the Health Board published 80 reports. Nevertheless, in 1942 Merewether conceded the American leadership in *industrial hygiene engineering* (Patty, 1958) and a similar conclusion was drawn by Luxon in 1984.

1940 TO THE ENVIRONMENTAL MOVEMENT

World War II led to a tremendous expansion of industrial hygiene in the United States, in Great Britain, and probably throughout the world. One reason for this, as noted by the British Industrial Health Research Board, was that efficiency of production was increased if workers remained healthy and did their job with the least amount of nuisance and discomfort. Some of the beginnings of the hygiene movement in Great Britain can be traced to similar needs during World War I. Much of the early effort in Great Britain for improving efficiency in World War I was centered around fatigue. It was found that for women engaged in turning shell cases, their productivity could be increased by 11% by decreasing their hours of work by 12%, from 68 to 60 hours per week. Of course, fatigue is a form of discomfort and a lack of well-being that is thus detrimental to health. The study of fatigue in World War I led naturally to the study of other forms of detriment to health. It should be noted that this emphasis on behavior of the worker as a contributor to health and safety placed the burden for health and safety on the worker and detracted from the need to use engineering controls to remove hazards from the workplace.

In the United States, the state efforts that had begun with the Social Security Act were tremendously expanded during the World War II years. The number of state and local units increased to 47 in 38 states. As mentioned earlier, $1,000,000 was provided to the states from 1947 to 1950. During the same time, 70 federal hygienists were loaned to state and local industrial hygiene agencies. As a result, all but two states (Delaware and Nevada) were engaged in hygiene work on at least a limited basis by 1950, and the number of professional personnel in state and local agencies reached 425 (Trasko, 1955).

The industrial development of the United States was a direct result of World War II (Aftalion, 1991). Prior to World War I, the United States had no organic chemicals industry. Because of the depression, the organic chemicals industry was slow to develop, however, after World War II, the U.S. organic chemicals industry expanded significantly. Even though industry was expanding substantially throughout the 1950s, the number of industrial hygiene professionals at state and local agencies was actually declining during the early 1950s due to a reduction in funding under the Social Security Act. Interestingly, however, the role that industrial hygiene came to play in the fields of community air pollution and radiological health – fields which were receiving increased funding – caused an overall increase in the number of hygiene professionals during the 1950s.

Many modern students of hygiene wonder why there is at least a minor emphasis in industrial hygiene on community air pollution and radiological health when these fields are not necessarily encountered in the professional careers of industrial hygienists. The explanation for this can be traced to the 1950s when these two fields were first developing. The first Air Pollution Control Act was passed in 1952, and in addition to providing funds to the federal government for conducting research, it gave federal money to the states for setting up air pollution control programs (Stern et al., 1973). In general, the states used the existing industrial hygiene programs to provide these services. By the same token, with the development of the atomic and hydrogen bombs during and after World War II, increased emphasis was placed upon radiological health in the United States. Once again, it was left to state and local public health departments, and specifically their industrial hygiene functions, to study and control radiological health exposures. For example, during the 1950s the fluoroscope, a shoe-size measuring device utilizing X-rays, was banned as a result of studies done by industrial hygiene personnel (Heimann and Trasko, 1964).

One of the most significant developments of this period was the publication in 1948, with a second edition in 1958, of *Industrial Hygiene and Toxicology* by Frank A. Patty. This two-volume treatise was a milestone in the discipline of industrial hygiene. Although several other books had been published on occupational diseases and on industrial toxicology, no large works had appeared specifically on industrial hygiene since the work of Chenoweth et al. (1938). Patty added the mortar to the bricks. Industrial hygiene had not only become a profession, but also a science with a set of principles and practices, which were fully described in Patty's work. If one uses the scientific method – beginning with precise observations, the collection of further observations based upon preliminary observations, the classification or organization of the facts, stating hypotheses based upon these relations, and testing the hypotheses by the previously mentioned method – then one is conducting science. Hygienists were conducting science by Patty's time and had carved themselves a niche within the natural sciences. No longer were they engineers, chemists, and physicians; they were hygienists.

As the 1950s came to a close, the United States began an era dedicated to the establishment of laws and characterized by a social conscience regarding the adequate provision of a safe, healthful, and pleasing environment to its citizens – the environmental movement. Many laws relating to the environmental movement were passed during the 1960s, and, indeed, the overall awareness of society with respect to the status of the environment was increased manyfold. The culmination of this era was the various pieces of legislation of the early 1970s concerning air pollution, environmental policy, and occupational safety and health.

The Occupational Safety and Health Act of 1970 was preceded by the Mine Health and Safety Act of 1969. Prior to that, the first serious effort at a national occupational safety and health code occurred in 1968, when it appeared as if Congress would pass occupational safety and health legislation under a Democrat president and Congress. Although that legislation failed, the success of the Federal Mine Safety and Health Act, which was spurred on by a series of mining accidents claiming many lives, led to final passage of the Occupational Safety and Health Act on the last day of 1970 under a Republican administration (Mintz, 1988).

The act was largely supported by industry, and although to some this may seem odd, it has become an important lesson of American history, realizing that it is much easier to answer to one boss than to many. Consequently, industry has supported legislation relative to the environment on several occasions in order to avoid variable regulation by the individual states. More confederate systems exist in Canada (Verma, 1996) and Australia (Quinlan and Bohle, 1991), but each has far fewer than 50 states and 300 million people.

By 1968, 42 states had some form of industrial health regulation, but in many states it was quite poor. Overall there was only one state-employed health professional for every 21,000 workers. Only 14 states regulated noise, which was one of the most pervasive occupational health hazards (Page and O'Brien, 1973). There was also a great deal of clamor from labor unions for passage of the legislation.

Passage of the Occupational Safety and Health Act has been cited as the most important event in the history of occupational health in the United States. The Health and Safety Act of Great Britain of 1974 and various acts of Australian states between 1972 and 1989 represent movements similar to the Occupational Safety and Health Act in size and scope, but much different in philosophical approach, as will be seen in Chapter 11 (Quinlan and Bohle, 1991).

However, other developments that took place mainly in the 1960s had an impact on industrial hygiene that cannot be overestimated. These developments, along with the passage of the Occupational Safety and Health Act, led to the exponential growth and importance of the industrial hygiene discipline in the 1970s and 1980s.

Three of the most important developments are related to the way the hygienist evaluates the workplace. By the 1960s, approximately 40 years had passed during which the development of techniques for evaluating airborne chemical exposures had ebbed and flowed. There were impingers, electrostatic precipitators, and filtration techniques for dusts that were reasonably adequate; but the evolution of these techniques had been slow. Techniques to measure exposures to organic chemicals were particularly cumbersome and inaccurate in the 1960s and archaic by today's standards. The development in the 1960s and the common usage by the 1970s of the charcoal tube for collection and the gas chromatograph for evaluation of organic chemicals appear to be inevitable results of a rapidly growing organic chemical industry. Synchronously, the development of a battery-operated portable pump for moving air through a sampling medium (i.e., charcoal) could be worn by workers and was also a great step. Although the development of each of these three tools took place in the 1960s, each was fully developed, adequately documented in the literature, and available to hygienists by 1970. Thus, these three developments, along with the Occupational Safety and Health Act, burst open the doors for U.S. hygienists and led to exponential growth in the population of the profession. This is best shown by the growth of membership in the American Industrial Hygiene Association (AIHA) (Figure 2-1).

1970 AND BEYOND

The number of hygienists in the United States, as evidenced by membership in AIHA, increased exponentially from 1970 to the late 1990s. Likewise, the number of educational pro-

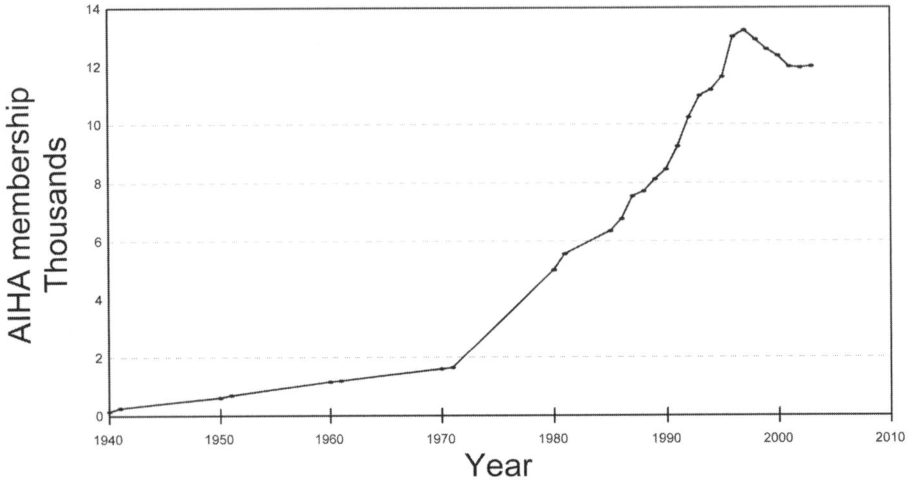

Figure 2-1. Growth in AIHA membership 1940–2003.

grams throughout the country graduating hygiene professionals has increased. Patty pointed out in 1958 that industrial hygiene academia had grown from the very first program in the world, developed at Harvard in 1918 (which, by the way, employed Alice Hamilton), to a list prepared by the Committee on Professional Education of the American Public Health Association (1947) that included nine universities in North America: Columbia, Harvard, Johns Hopkins, Toronto, Yale, and state universities of California, Michigan, Minnesota, and North Carolina offering degrees at the graduate or the undergraduate level for industrial hygienists and occupational safety and health professionals. The London School of Hygiene and Tropical Medicine began offering degrees in 1959 (Rappaport and Flynn, 2003). But there could easily be over 100 institutions now offering baccalaureate, graduate, or associate degrees in some aspect of occupational safety and health. There are apparently about 30 programs that offer graduate degrees in industrial hygiene (for a partial listing of all programs see Halten, 1990).

An important development of the 1970s, which led to high-quality education of graduate-level hygienists as well as other safety and health professionals, was the creation of Educational Resource Centers by the National Institute for Occupational Safety and Health (NIOSH, 2004). This effort has grown from the 8 resource centers originally established in 1976, to 16 current ERCs. In addition, there are 15 industrial hygiene programs that are not part of an ERC but are funded by this process. These training funds are used to purchase equipment and other supplies necessary for the education of industrial hygienists, to fund teaching-faculty salaries, and to provide stipends and pay tuition and fees for hygiene students. Similar funding is also available to occupational medicine, occupational nursing, safety, and ergonomics programs. In order to be classified as an Educational Resource Center, an institution or a consortium of institutions must have training established in three of the four core areas of occupational safety and health (industrial hygiene, nursing, medicine, and safety).

With the tremendous growth of the profession in the 1970s and 1980s, the passage of the Occupational Safety and Health Act in the United States, and increasingly complex technological advancements made in the evaluation and control of workplace environments have come changes in the field of industrial hygiene. The changes that have occurred may not be apparent to many of the professionals in the field. Early physicians were devoted to the recognition of diseases in industry and the causes of those diseases. With time, engineers became interested in the control of agents causing those diseases. As the discipline developed, engineers and chemists became predominant in the field. Recognition of disease remained a major aspect of the profession, even though engineers and chemists often lacked adequate training in the medical and biological sciences. They usually received that *training* through experience.

However, since 1970 practitioners in the field have gradually drifted away from an emphasis on the recognition of disease, and for a very good reason. Imagine going into a workplace in 1940 armed with meager evaluation technologies and limited knowledge of what effects the substances in that workplace might be having on the workers. In addition, written information or research in these areas was quite limited. If the hygienist determined that carbon tetrachloride was present in the workplace, he would have limited resources for determining what levels were there and what levels were safe or acceptable. If the clock is turned back to 1920, the situation is more difficult. Indeed until 1940, hygienists had very little information regarding what was an appropriate level of exposure in the workplace. Hence, hygienists had to *study* the workplace.

A result of the establishment of the American Conference of Governmental Industrial Hygienists (ACGIH®) in 1939 was the publication of exposure guidelines for various substances. These guidelines evolved over 30 years until they were adopted by the Occupational Safety and Health Administration (OSHA) as one of its first administrative acts in the early 1970s[6]. The work of ACGIH® in this area and the establishment of health standards by OSHA continued throughout the 1970s and 1980s until now there are recommended guidelines of presumably safe exposure levels for approximately 600 compounds or elements and their related compounds (e.g., lead and its compounds). Now there are thousands of compounds for which there are airborne exposure levels established as *safe*.

As a consequence, the old approach toward research into the causes of health effects and safe exposure levels by everyday practitioners is emphasized less. Today's practicing professionals apply the established and recommended exposure limits. Research aspects have been left largely to large research organizations such as the National Institute for Occupational Safety and Health, a few academicians, and large companies. Of the more than 10,000 industrial hygienists in the United States, probably fewer than 5% are actively engaged in the activities that caused industrial hygiene to be established. Even so, this 5% (500) is equal to the total

[6] Actually OSHA was allowed to adopt consensus standards (which the TLVs® were not) and government standards under the OSHAct. The 1968 TLVs® had been adopted under the Walsh-Healy (federal contracts) Act and thus were government standards.

number of hygienists in 1950 in the United States (Heimann and Trasko, 1964). This is perhaps the biggest change that occurred in the 1970s and 1980s – the evolution from research to application.

THE FUTURE

Evolution towards application has been prompted by regulation. Hygienists, particularly those in industry, found themselves challenged by new regulations in the 1970s for which they had the responsibility of compliance. Although this certainly caused the numbers of professionals in the field to expand, it has also caused hygienists to take a more reactive, rather than proactive, approach to their practice.

Hygienists trained now and in the future must take a hard look at what role they will play and which services they will provide in their professional careers. To simply learn the tools of the trade and apply those tools without questioning them or trying to improve them means that these industrial hygienists will be performing at a technician level and will be doing very little to advance their profession. The industrial hygienist of the future must continue to develop new tools and refine old ones. He or she must be adept at controlling exposures, but must be able to recognize hazards that previously were not recognized. Finally, to get management's cooperation, the hygienist must be able to communicate well, orally and in writing. This latter step will be the most important one in terms of achieving our goals because good communicators, regardless of their backgrounds, are seen as peers by management. Ultimately, some are promoted to management levels outside of hygiene, where their viewpoints have a wider audience and greater influence.

Since the very essence of what hygiene is about is health, hygienists of the future must pay particular attention to the study of disease, its causes, and ways to prevent it. Professor Theodore Hatch, an eminent academician and professional of industrial hygiene, said as early as 1960,

> Without attempting to become a physician or physiologist, the industrial hygienist must nevertheless acquire an independent capacity to understand and evaluate the human side of the man/environment equation in order to most wisely employ his primary skills in physical science and engineering, to recognize, evaluate, and control the work environment. (Hatch, 1960)

In other words, simply applying the rules and using established tools can be done by anyone; however, the industrial hygienist who contributes the most to the worker and to the profession must understand the basis of health so that his or her application of industrial hygiene principles will be most effective. This not only makes the study of a biology book mandatory, but requires going to the field, observing workers and what they are doing, asking questions about their health, evaluating their exposure levels, and correlating health with exposure levels.

Finally, this effort requires work within the occupational health team. Professor Hatch also said:

> No physical measurements of dustiness, of radiation, heat, noise, or light, of chemical fumes or vapors, no matter how precisely these are made, have direct biological meaning but are useful only in proportion to the goodness of their respective calibrations, so to speak, against appropriate biological scales of human health. Thus, standing alone on its foundation of physical science and engineering, industrial hygiene would have little meaning. Through teamwork, however, between the medical scientists and physicians on the one hand, and the physical scientists and engineers, on the other, it has developed on a base, which provides the link between environment and man. It may be noted here that progress on the medical side is equally limited without such a common base in both the physical and biological sciences. The medical diagnostic techniques, like the procedures of the industrial hygienist, when used alone, will not tell us that an observed state of ill health was necessarily caused by a particular industrial exposure. Such interpretation is permitted only when the diagnostic findings can be evaluated in relation to quantitative description of the conditions of exposure and both examined in the light of previously established correlations between similar environmental stresses and human response.

Industrial hygienists are the cement between the physical, biological, and medical sciences. We must do more than measure exposure levels and simply compare them to presumed safe exposure standards. We must go a step further and interact with the medical professionals in order to determine if effects are occurring at the levels that we have measured, no matter how safe those levels are currently presumed to be. If hygienists take this latter step, then they are doing more than simply being technicians; they are being inquisitive scientists who are doing their best to promote health. This will improve society's awareness of the important role played by hygienists.

INTERNATIONAL EFFORTS

There have been several key developments at the international level which are important not only from a historical perspective but as useful resources for practicing hygienists. In discussing the history of the hygiene movement, the International Labor Organization (ILO) is a key player on the world stage. The ILO was formed in 1919 and was the only survivor of the failed League of Nations. In 1946 it became the first specialized agency of the United Nations (UN). It has produced many standards and research publications that are used throughout the world (ILO, 2004). Similarly, the World Health Organization (WHO) of the UN has had an important place in occupational safety and health around the world and produced publications on the subject.

Two other organizations are worth mentioning. The International Commission on Occupational Health (ICOH) and the International Occupational Hygiene Association (IOHA) are world wide organizations with different missions but similar visions. The IOHA is interested in, among other issues, promoting certified or credentialed hygienists throughout the global economy. The ICOH promotes the broader field of occupational health and has roots that trace to 1902. In 2003, IOHA and ICOH signed a joint declaration urging the ILO and WHO to further promote hygiene in the workplace (IOHA/ICOH, 2003; Zalk, 2003). It should be mentioned that ICOH, in the views of some, is made up of industry representatives that present biased information to leaders and workers in developing countries on the hazards of their products (LaDou, 2007).

Although the following is by no means complete it contains contributions from individuals of several countries on the current and historical efforts in occupational health and hygiene in developed countries.

Occupational Hygiene in Australia[7]. A Scientific Officers in Occupational Hygiene meeting was held in May 1960 following a request to the Occupational Health Committee of the National Health & Medical Research Council in 1959. This group became the focus of occupational hygiene in Australia. In December 1979 a proposal to form a professional organization was made and thus, in February 1980, the Australian Institute of Occupational Hygienist (AIOH) was born, with 29 members. The AIOH has grown to a vibrant organization boasting nearly 500 members in 2006.

From its outset the AIOH established strong links with AIHA, ACGIH® and British Occupational Hygiene Society (BOHS). In the early 1980s, the AIOH came to an agreement with the American Board of Industrial Hygiene (ABIH) to award certification maintenance points for various AIOH functions and started to run the ABIH's core and comprehensive examinations in Australia. In 1982, the first ABIH Exam was held in Melbourne, Australia and 100% of the first batch of candidates passed the examinations, a feat which has never since been repeated. By the mid 1980s the AIOH had representatives on numerous national and state government committees on education and training, chemicals and standards development; and on committees established by other organizations such as National Association of Testing Authorities, Australian Institute of Petroleum, Australian Mining Industry Council, Australian Mesothelioma Register, ACGIH®, and later the Australian government's National Occupational Health and Safety Commission.

In 1987 the AIOH was one of the eight initial signatories to the agreement that formed IOHA. The AIOH President of the day, Pam De Silva, went on to become the second president of IOHA. In 1993 the AIOH had reached a size where it was able to establish a permanently staffed office to carry out the administrative functions of running the Institute. It became an incorporated association and had matured to a point where it was run as a business as well as a professional learned society. By this time the Institute was producing its own publications

---

[7] Contributed by Linda Apthorpe, Pickford & Rhyder Consulting and Honorary Secretary of AIOH.

and running professional development programs.

In 1999 the AIOH Council commenced work on establishing a certification scheme for the maintenance and recognition of professional competence for hygienists in Australia. This scheme came to fruition in 2004 and was recognized by IOHA in 2006.

Occupational Hygiene in Canada.[8] Canada, like the U.S. and many other developed countries, has a complex arrangement of legislation regarding protection of employees from biological, chemical and physical agents as well as occupational safety, public health and environmental protection. Canada is the second largest country on earth, though much smaller by population (36th at 33.1 million). Canada is composed of 10 provinces and 3 territories. Jurisdiction to regulate and enforce Occupational Health and Safety (OHS) legislation is held by either a single ministry at either the federal or provincial level, usually the Ministry of Labour. However, there are circumstances where either the Ministry of Health or Environment may become involved in matters concerning biological, chemical and physical agents.

Federal OHS legislation covers federal employers such as the federal government or postal workers and selected categories of non-government national employers, such as the aviation industry. All other employers fall under the OHS legislation for the provinces or territories where they operate.

On a pragmatic level, here are examples of how the legislation applies:

- Employers who fall under the national OHS framework of the *Canadian Labour Code* have one set of legislation to apply to their employees across Canada, regardless of location.
- Private or government employers within a particular province, for example Ontario, would have to follow the *Occupational Health & Safety Act of Ontario.*
- An employer, such as a private company, that has offices in Vancouver, Calgary, Winnipeg, Toronto, Montreal and Halifax would have to manage their operations to meet all 6 distinct provincial OHS legal environments.

Fortunately, there is a high degree of similarity between provinces and territories; however, there are always sufficient minor and major legal nuances to make cross country OHS management a challenge for any employer. In terms of legislative structure, the structure of OHS laws across Canada is quite consistent:

- The federal and provincial or territorial governments each have an overarching *Occupational Health and Safety Act* or equivalent that lays out powers of the government, to whom the law applies and then the general rights and duties of a series of Regulations are enabled by the Act. An example would be the *Control of Exposure to Biological and Chemical Agents*, regulation 833 under the *Ontario Occupational Health and Safety Act*.

[8] Submitted by Kirby James, BSc, MHSc, CIH, currently (2007) on sabbatical from EHS professions.

- Schedules are published in an Act or Regulation that itemize scope of the legal coverage.
- *Guidelines*, referenced in an Act or Regulation, are quite often standards of care from a third party, non-governmental organization such as ACGIH® or Canadian Standards Association (CSA). The work of these organizations is referenced in regulations as either a required performance reference or in some cases, as a minimum standard of care. In this latter example, an employer may choose another course of action as long as it meets or exceeds the standard of care referenced in the Guidelines.

In terms of punitive clauses within OHS legislation there have been some notable enhancements within the last 5 years, including the addition of flagrant disregard for OHS legislation as an offense under the Federal Criminal Code. The federal, provincial and territorial OHS legislation has authority to fine and/or imprison offenders, either employer or employee. A very strong message has been delivered to the executive level of organizations about personal culpability. Enforcement of OHS laws in Canada uses both the judicial system as well as *ticketing* to dissuade employers. It is an important note that OHS law in Canada follows the premise of *reverse onus* meaning that the accused must prove their innocence and the government need only prove an offence occurred. This is one of very few areas within Canadian law where the fundamental principle of innocence has been reversed for the greater protection of employees.

It would be fair to say that on balance the infrastructure of Canada is generally adequate though rarely *world class*; a balance has been reached by governments to protect employees from exposure and harm for the most part, while not creating notable obstacles to employers. There is always more that can and should be done and there are always those more sensitive or susceptible to harm. For these individuals the laws of Canada have caveats to protect them, though the people affected rightly feel they may not be adequate. Still, based on a detailed working knowledge of the legal requirements in the U.S., European Union (EU), United Kingdom, China, Switzerland and Australia, Canada's OHS infrastructure is quite comparable in terms of effectiveness and legal rigor.

Canada, like most countries, has not yet recognized the professional practice of occupational hygiene in a manner that assigns exclusivity to practice or professional title. That said, the profession of occupational hygiene has a strong history in Canada and remains a notable presence in the protection of society.

The Registered Occupational Hygienist (ROH) and Registered Occupational Hygiene Technologist (ROHT) are the only certifications of their kind offered to Occupational or Industrial Hygiene specialists in Canada. They are administered by the Canadian Registration Board of Occupational Hygienists (CRBOH), the national certifying body for the occupational hygiene profession. The CRBOH was founded in 1986 as the Canadian equivalent to the American Board of Industrial Hygiene.

In 1986, the CIH was widely recognized as the international benchmark for the profession. In offering the ROH it was Canada's intent to address several perceived shortcomings to the CIH examination process. Prerequisites to sit for the ROH or ROHT exams were comparable to the CIH or technologist designations. The main difference between the Canadian and American offerings is the examination process. Unlike the CIH, applicants to the ROH exam complete an essay and an oral board exam in addition to the multiple choice exam. Like the ABIH, the CRBOH also created a code of ethics, which came into effect in 1992, that shared many of the aspects that such codes have across many professions. While the ROH has not displaced the CIH in Canada as the predominant occupational hygiene certification, it remains as a very respectable credential.

Until 2001, occupational hygiene membership associations in Canada were exclusive to a province or region without the benefit of a national organization. However, following the American Industrial Hygiene Conference and Exposition (AIHce) in New Orleans in 2001, a national umbrella occupational hygiene association was created, the Canadian Council of Occupational Hygiene (CCOH), known in French as Counseil Canadien d'Hygiene du Travail (CCHT). Unlike provincial or regional associations, the CCOH has the distinct role to represent the profession federally and internationally including at the International Occupational Hygiene Association (IOHA). It also serves to bring collaboration between the provincial and regional professional associations.

Occupational hygienists and hygiene technologists in Canada have numerous local organizations to which they can have membership to represent their interests and provide ongoing educational opportunities. In Ontario, Canada's most populated province, the Occupational Hygiene Association of Ontario (OHAO) provides the traditional local association services, such as representation at governmental discussions on the profession, coordinating input to new legislation and ongoing professional development of members. Outside of Ontario all other associations that serve occupational hygienists (Alberta, Atlantic Provinces (serving Nova Scotia, New Brunswick, Prince Edward Island and Newfoundland and Labrador), British Columbia /Yukon, Manitoba, and Quebec) are chapters of AIHA.

In conclusion, Canada has a very skilled cadre of certified and non-certified professionals dedicated to occupational hygiene. Many hold American and Canadian certifications and association memberships, in typical fashion for a country that embraces diversity.

Occupational Hygiene in Germany[9]. The German Occupational Hygiene Society (Deutsche Gesellschaft für Arbeitshygiene, DGAH) was founded in Cologne in 1990 by only 12 hygienists. At that time occupational hygiene was not implemented in West German legal requirements and only few companies – mostly international companies – were hiring occupational hygienists. In East Germany, Occupational Hygiene was a well known profession and education and training for occupational hygienists was promoted. This structure was lost dur-

[9] Contributed by Kurt Leichnitz and Hans Pfeil, Deutsche Gesellschaft für Arbeitshygiene, www.dgah.de.

ing Reunification. Now the DGAH has about 90 members and is working to promote the profession and reinstall and develop education and training for hygienists within the reunited Germany.

The activities of the DGAH are, for example, yearly conferences about all parts of occupational hygiene and publishing a *DGAH Brief* (newsletter on occupational hygiene) as well as cooperating with Occupational Safety and Health Associations on various levels. DGAH is a member of IOHA. In this partnership, the DGAH participates in global activity on occupational health and safety. This includes the cooperation with ILO and WHO.

Members of DGAH are representing occupational health, safety, and hygiene in various committees within the German regulations setting bodies. Together with official government institutes, occupational hygiene education and training developments are occurring. New technologies continue to present new risks, thus challenging the occupational hygienist to develop effective solutions. Therefore, an essential part of DGAH's activity is the adaptation of occupational health measures to the changing technological world and the variations of socioeconomic conditions.

Occupational Hygiene in South Africa[10]. Professional occupational hygiene structures in South Africa have evolved along similar lines to those in other countries except that the numbers are much smaller. One significant difference may be that there has been, and still is to some extent, a strict separation of the profession into mining and non-mining practitioners.

The Occupational Hygiene Association of South Africa (OHASA) was established in 1983 to bring together anyone interested in occupational hygiene practice. This was also the year that new health and safety legislation, the Machinery and Occupational Safety Act, was introduced in South Africa. Elements of occupational hygiene practice were specifically referred to for the first time.

A Technical Committee was established in 1987 by the Department of Labor to investigate the training and regulation of occupational health and safety practitioners. Their findings, published in 1992, included the recommendation that lead bodies for occupational health and safety be established to determine standards for these professions.

Around that time a working group of concerned occupational hygienists came together to identify how best to achieve the training and regulation needs for ensuring the professionalism of occupational hygienists in South Africa. This included standards and mechanisms for the education and training, professional development, registration, and disciplinary procedures.

The proposed registration of occupational hygienists created a great deal of debate. Until then there were only a few occupational hygienists with formal training. However, all were generally knowledgeable about workplace health risks. The practitioners without formal training were naturally concerned that any registration system would leave them out. In February 1992, a separate body for the professional registration of occupational hygienists was elected

[10] Contributed by Mr. Rob Ferrie.

to oversee the formation of this registration body.

In February 1994, the Institute of Occupational Hygienists of Southern Africa (IOHSA) was launched with a set of by-laws to ensure it operated in an independent manner. The Institute formed a Board to set standards for practice, process and evaluate applications, and to hold oral and written examinations. A *grandfather clause* exempted those hygienists who had been practicing for more than 5 years from the need to hold post-graduate qualifications. They were, however, required to pass a written examination and an oral competency interview.

The Occupational Health and Safety Act was implemented in 1993 and replaced the Machinery and Occupational Safety Act. For the first time in South African health and safety legislation, occupational medicine and occupational hygiene received equal attention with safety. Under this Act, the Regulations for Hazardous Chemical Substances were brought into place in 1995. These are similar to the UK Control of Substances Hazardous to Health (COSHH) Regulations. In these regulations the Department of Labor required that compliance air monitoring had to be performed by an *approved authority* and that in order to obtain approval the person had to be registered with IOHSA as an occupational hygienist. This legal recognition encouraged hygienists to register with the Institute and membership grew quickly.

Finally in 1999, OHASA and IOHSA were amalgamated as the Southern African Institute for Occupational Hygiene (SAIOH) that had ordinary membership for those interested in occupational hygiene and registered membership for those seeking professional recognition.

The three main objectives of the Institute are a) to promote and develop the profession of occupational hygiene to meet the needs of industry and the community at large, b) to promote and encourage the study of and teaching and training in occupational hygiene, and c) to regulate the profession of occupational hygiene with particular reference to professional registration, use of title and ethical conduct of members of the Institute. As of the end of March 2006 SAIOH had a total of 192 registered members. Unfortunately, this very small number of occupational hygiene specialists is totally inadequate for the 283,782 employers registered with the South African Compensation Commissioner in March 2005 (this excludes mining operations).

Occupational Hygiene in Sweden[11]. The National Institute of Public Health in Stockholm, founded in 1937, included a small Department of Occupational Hygiene, which was the body for further development in Sweden. In 1966 this department was included in the new National Institute of Occupational Medicine. Of growing importance were the Departments of Occupational Medicine, which gradually were founded at the regional hospitals covering all Sweden. Occupational hygiene in industry mainly developed after 1945. The work in Sweden on occupational health problems was initially dedicated to studies on the pathology and diagnostics of occupational diseases. Technical investigations began later. One early step was the initiative in 1947 to study the formation of welding fumes. Chemicals in focus during this period were carbon monoxide, lead, hard metals, nickel, benzene, carbon

[11] Contributed by Associate Professor Hans Welinder, Department of Occupational and Environmental Hygiene, University Hospital, Lund, Sweden.

tetrachloride, trichloroethylene, mercury, cadmium, and synthetic rubber-chloroprene. Up until 1970, studies were mainly focused on health problems in big industries and on very broad exposure investigations over the whole country relative to traditional industrial hazards like noise, crystalline silica, asbestos, and lead. The Workers Protection Fund in 1972 and an agreement in 1976 between the employer's organizations and the trade unions on a protocol for the development of occupational health services for all workplaces advanced the field considerably. However, today there is no academic course specifically directed to the training of specialists within occupational hygiene.

In 1957 the Swedish safety engineers formed the organization Swedish Association for Occupational Safety and Health, with 700 members today. Occupational hygienists, with an academic background, formed the Swedish Society for Occupational and Environmental Hygienists in 1980, with 150 members today. The organization is a member of IOHA and has introduced certification for occupational hygienists.

From a rather slow start with modest resources, occupational hygiene in Sweden has developed impressively and can now provide the society a broad and deep competence to map problems in the working environment. The academic status has improved considerably as many hygienists have acquired doctorate degrees and associate professorships. The progress of occupational hygiene in Sweden has also been demonstrated by an increasing involvement in international scientific cooperation. Swedish occupational hygienists have served as presidents of IOHA (Mrs. Linnea Lillienberg) and of the Section of occupational hygiene within ICOH (Mr. Håkan Westberg).

Occupational Hygiene in Thailand[12]. In 1964 manganese poisoning cases were found in a dry cell battery manufacturing plant. This was recognized as the first occupational disease outbreak in Thailand. The National Occupational Health Committee was established in order to get cooperation among the related ministries, i.e., the Ministry of Public Health, Ministry of Interior, Ministry of Industry and the University Department. As a result, the Ministry of Industry issued the Industry Act, in order to control and supervise occupational health and safety in factories.

In 1972, the labor law was amended and a workmen's compensation fund established by the Department of Labor, Ministry of Interior. The first occupational health and safety regulations were issued in 1976, including the most significant announcement, "Safety at Work for Employee," specifying that employers with 100 or more employees shall have at least one safety officer (person with a B.Sc. in occupational health and safety or who passed a 180-hour training course). However, the safety law was not strongly enforced until 1992 when the Ministry of Labor was established and a new Labor Protection Act 2541, B.E. was promulgated in 1998.

In 1969, an Occupational Health Program (B.Sc.) was launched in the Faculty of Public

[12] Contributed by Dr. Wantanee Phanprasit, Head, Department of Occupational Safety and Health, Faculty of Public Health, Mahidol University.

Health, Mahidol University. Most of the first graduates worked in government and state enterprises such the Electricity Generating Authority of Thailand. As of 2005 there were 883 graduates. In 1982, the department began offering the M.Sc. in Industrial Hygiene; to date (2006), 129 individuals have graduated from this program. Hygiene and safety officers are in demand due to better law enforcement and rapid growth of the industrial sector in the past two decades. Thus, at least 12 academic institutions run occupational health or similar programs at all levels – bachelor's, master's and doctoral – to attempt to meet this demand.

Thailand has two main OH&S professional associations, the Safety and Health at Work Promotion Association (SHAWPAT) and the Occupational Health and Safety at Work Association (OHSWA). SHAWPAT was established in 1987 by approval of the cabinet council, when the Department of Labor, Ministry of Interior organized "National Safety Week." The objectives of the association are to promote occupational health and safety; to distribute knowledge and experience related to OH&S; to cooperate with private and governmental sectors and other organizations to organize or arrange OH&S activities; to support and coordinate with private and governmental sectors and other organizations in order to promote OH&S.

OHSWA was created by Mahidol's OH&S alumni in 1982. The objective of the assembly was to discuss and exchange OH&S knowledge and experience among members; to provide technical service and support to industries, state enterprises, government and labor unions; to disseminate information and create awareness in OH&S including public safety; to conduct research and develop best practice and standards; to promote professional skills; and to collaborate among members and concerned organizations.

Occupational Hygiene in the United Kingdom[13]. Although interest in the effects of work and workplace conditions on health had been the subject of study in the UK for at least two centuries[14], the seeds of a distinct discipline only emerged after the Second World War when scientists, engineers and physicians from backgrounds as diverse as railway engineering, electrical engineering, the coal, steel, chemical and nuclear industries, and academia recognized that they had common interests in the recognition, assessment and control of agents in the workplace. Inspiration for the formation of the British Occupational Hygiene Society (BOHS) came from visits made to the U.S. by practitioners in the UK, who experienced the energy and enthusiasm of AIHA, and witnessed first-hand the fruitful cooperation that AIHA fostered among the many professions involved in occupational hygiene, and the value of an annual conference as a forum for the exchange of ideas and the dissemination of information. The inaugural meeting of BOHS took place in April 1953. The Society has grown steadily since its formation, and today has approximately 1500 members from the UK and countries around the world.

---

[13] Contributed by Hugh Dalrymple, Head of Health, Safety & Employee Services at Fujitsu Services Ltd., BOHS Representative on the IOHA Board of Directors, and a Past President of BOHS.

[14] See other parts of this chapter for history of UK OHS efforts.

In 1958, BOHS published the Annals of Occupational Hygiene, an international journal on occupational hygiene, which continues to be published today and has built a substantial reputation for the quality of its papers on fundamental research and on the practical application of occupational hygiene techniques. During the 1960s, BOHS's Standards and Technology Committees also published seminal papers on hygiene standards for exposure to chrysotile and amosite asbestos dust, cotton and flax dusts, and to wide band noise, and a series of technical guides on workplace contaminant control, sampling and other techniques.

Since its formation, BOHS has organized scientific meetings and conferences, and currently organizes several each year nationally or through its regional groups throughout the UK. BOHS' Annual Conference is one of the key events of the occupational hygiene year, attracting several hundred participants from many countries. The International Symposium *Inhaled Particles*, held approximately every 5 years is a major event for researchers and practitioners alike with an interest in the effects of particles on health.

During the mid-1960s, a group of occupational hygiene practitioners came together to form GOPOH, the Group of Practicising Occupational Hygienists. GOPOH had no formal links to BOHS, though all its members were members of BOHS. Until the mid-1970s, there was much debate about whether BOHS should remain a learned society, whether it should move towards becoming a body to represent the professional interests of occupational hygienists, or whether it should attempt both. As a result, an Institute of Occupational Hygienists (IOH) was formed independently of BOHS in 1975.

In 1967, BOHS set up the British Examining Board in Occupational Hygiene (BEBOH) to define levels of competence in occupational hygiene practice and to certify attainment by individuals of these standards. These exist today as the Diploma of Professional Competence in Occupational Hygiene (Dip Occup Hyg) and the Certificate of Operational Competence in Occupational Hygiene (Cert Occup Hyg). In addition, a system of subject-specific Preliminary Certificates was developed, and the successors of these continue in modular form.

A single body promoting occupational hygiene and representing the profession was formed in 2003, when IOH (by then known as BIOH) merged with BOHS, with the affairs of the occupational hygiene professionals being looked after by the Faculty of Occupational Hygiene, within the enlarged Society. Details of the current organization, its structure and activities, including membership information and details of conferences, meetings, qualifications, Continuing Professional Development and contacts in the Regions of the UK, can be found at www.bohs.org.

Outside the societies and institutes formed for and run by those with an interest in occupational hygiene, occupational hygiene is alive and well and being practiced to a far greater extent than ever before in every branch of occupational health and safety. Legislation on asbestos, chemical control, noise, vibration and many other areas in which the occupational hygienist has traditionally practiced, has lead to these areas coming more into the mainstream of health and safety practice in the UK.

In common with experiences in most other developed countries, the expansion of interest in occupational hygiene, health and well-being has lead to the diminution of the boundaries between the different health and safety related professions, and ironically, although more people than ever before have elements of occupational hygiene in their work, in many quarters, the practice and profession of occupational hygiene is not being recognized readily as a discrete profession or science.

In the UK today, the major sectors of employment for occupational hygienists are consultancy, government and the oil, chemical and pharmaceutical industries. The transformation of the UK economy from a manufacturing to a service economy has lead to a reduction in the number of occupational hygienists employed in many other areas of industry. However, many practitioners now have international roles, and have moved into related areas in health, safety and environmental management, product stewardship and regulatory activities, where their knowledge and deep understanding of occupational hygiene principles is finding good use in the protection of health and work and in communities.

A note on hygiene in developing countries. Not much is published on hygiene activities at the professional or government level in developing countries. LaDou (2007) points out that there are 35,000 multi national companies with 147,000 foreign affiliates. Many of these affiliates are in developing countries where labor costs and government regulation are less than elsewhere. Fortunately 70% of the commerce is attributed to the top 500 companies, and many of these either set internal standards or follow best practices of European or North American governments. Nonetheless, the notion that occupational health is not important in developing countries is formed against the background of other overwhelming health issues, such as vector borne diseases. Spee (2006) reports that in Africa, work related factors account for twice as many lost years of life expectancy per thousand people as in Europe and North America; and this is in a society where life expectancies are already shortened, relative to the developed world, by the aforementioned infectious and vector borne diseases and other factors such as nutrition.

The *informal* and contract industries in many of these countries are significant. These are small operations with less than 20 employees, often family members. Contracts go from giant corporations through middlemen down to the lowest bidder. Child employment is common, as are long working hours, 7 days per week. Many view this employment (adults or children) as fortunate, as many others are unemployed. As hygienists, we have an obligation to improve working conditions for all workers. In some developing countries a network is developing through which hygienists in developed countries can work. For example these organizations are members of IOHA.

- Brazilian Association of Occupational Hygienists
- Hong Kong Institute of Occupational and Environmental Hygiene
- Malaysian Industrial Hygiene Association
- Mexican Industrial Hygiene Association (Asociación Mexicana de Higiene

Industrial – AMHI)

- Polish Association of Industrial Hygienists (Polskie Towarzystwo Higienistów Przemysowych – PTHP)
- Southern African Institute for Occupational Hygiene (SAIOH)
- Taiwan Occupational Hygiene Association (TOHA).

However, there are hygienists working in many other countries such as Venezuela, Columbia, and South Korea. The future of occupational health lies in these countries.

QUESTIONS

2.1 What are the four primary disciplines of the occupational health team?

2.2 Name three major historical events (prior to 1970) in the progression of industrial hygiene as a discipline in the United States.

2.3 Discuss the need for an engineering, chemistry, and biology basis for industrial hygiene based on your current knowledge.

2.4 Describe four periods in the history of occupational health, and summarize what happened in each.

2.5 Discuss your concept of a worker's right to a safe and healthful workplace. Should a workplace be spotless? Hazard-free? Justify your answer.

2.6 What are the advantages and disadvantages of worker's compensation to the employee and the employer?

2.7 What were the three employer defenses in worker injury lawsuits prior to worker's compensation?

2.8 What is AIHA? ACGIH®? BOHS? AIOH? OSHA? HSE? (Hint: Some are not United States organizations.)

REFERENCES

Aftalion, F.: A History of the International Chemical Industry, pp. 102-206. University of Pennsylvania Press, Philadelphia (1991).

Agricola, G.: De Re Metallica. Foreword and translation by H.C. Hoover and L.H. Hoover, pp. vi-xv and 214-218. Dover, New York (1950).

Berman, D.M.: Why work kills, International Journal of Health Services, 7:63-87 (1977).

Bergey, D.H.: Principles of Hygiene. Saunders, Philadelphia (1901).

Broadhurst, J.: Home and Community Hygiene. Lippincott, Philadelphia (1908).
Chenoweth, L.B.; Machle, W.; Schnieder, H.: Industrial Hygiene. Crofts, New York (1938).
Corn, Y.K.: Historical aspects of industrial hygiene – I: Changing attitudes toward occupational health. Am. Ind. Assoc. J. 39:695-699 (1978).
Doehring, C.F.L.: Factory Sanitation and Labor Protection. Bulletin No. 44 , pp. 2-3. U.S. Department of Labor, Washington, DC, USA (1903).
EPA: Suspended; Canceled; and Restricted Pesticides. http://www.epa.gov/opprd001/rup/rupjun03.htm U.S. Environmental Protection Agency, Washington, DC, USA (1990).
Felton, J.S.: 200 Years of Occupational Medicine in the U.S. J. Occup. Med. 18:809-817 (1976).
Halten, J.B.: Graduate Level Environmental (Public) Health Curricula Outside Schools of Public Health. Bureau of Health Professions, Health Research Services Administration, Public Health Services, Washington, DC, USA (1990).
Hamilton, A.: Forty years in the poisonous trades. Am. Ind. Hyg. Assoc. Q. 9:5-17 (1948).
Hatch, T.F.: Training in industrial hygiene in the U.S. J. Occup. Med. 2:321-324 (1960).
Heimann, H.; Trasko, V.M.: Evolution of occupational health programs in state and local governments. Pub. Hlth. Rep. 79:942-946 (1964).
Hunter, D.: The Diseases of Occupations, 6th ed., pp. 12-7. Hodder and Stoughton, London (1978).
ILO, History of the International Labor Organization, http://www.ilo.org/public/english/about/history.htm (2004).
IOHA/ICOH: IOHA/ICOH Declaration on occupational hygiene. Ann. Occup. Hyg. 47:521-2 (2003).
Kazis, R., Grossmann, R.L.: Fear at Work: Job Blackmail, Labor and the Environment, pp. 156-189. Pilgrim Press, New York (1982).
LaDou, J.: International occupational and environmental health, In Environmental and occupational medicine, W.N. Rom, S.B. Markowitz (eds.), 4th ed. Lippincott, Williams and Wilkins, Philadelphia, PA, USA (2007).
Luxon, S.G.: A history of industrial hygiene. Am. Ind. Hyg Assoc. J. 45:731-739 (1984).
Merriam-Webster: Webster's Ninth New Collegiate Dictionary. G.C. Merriam, Springfield, MA (1991).
Mintz, B.W.: Occupational safety and health: The federal regulatory program – a history, in Fundamentals of Industrial Hygiene, 3rd ed., B.A. Plog (ed.). National Safety Council, Chicago (1988).
NIOSH, Training Programs and Educational Resource Center website. U.S. National Institute for Occupational Safety and Health, http://www.cdc.gov/niosh/oep/training.html (2004).
Ottoboni, M.A.: The Dose Makes the Poison. Vincente, Berkeley, CA (1984).
Page, J.A.; O'Brien, M.W.: Bitter Wages, pp. 167-189. Grossman, New York (1973).
Patty, F.A.: Industrial Hygiene and Toxicology, 2nd rev. ed., Vol. I – General Principles, pp. 1-17. Wiley Publishing, New York (1958).
Quinlan, M.; Bohle, P.: Managing Occupational Health and Safety in Australia, a Multidisciplinary Approach, 194-241. McMillan, Melbourne (1991).
Raffle, P.A.B.; Lec, W.R.; McCallum, R.I.; Murray, R.: Hunter's Diseases of Occupations, p. 38. Edward Ame London (1987).
Ramazzini, B.: Diseases of Workers, translated by W.C. Wright. Hafner, New York (1964).
Rappaport, S.M.; Flynn, M.: Commentary: Two seminal contributions of S.A. Roach to the evaluation and control of hazardous substances in air. Ann. Occup. Hyg. 47:343-348 (2003).
Rosen, G.: Introduction to Diseases of Workers, Ramazzini, pp. v-ix. Hafner, New York (1964).
Schilling, R.S.F.: Occupational Health Practice, pp. 1-26. Butterworths, London (1981).
Spee, T.: Occupational hygiene in Africa. Ann. Occ. Hyg. 50:431-435 (2006).
Stern, A.C.; Wohlers, H.C.; Boubel, R.W.; Lowry, W.P.: Fundamentals of Air Pollution. Academic Press, New York (1973).
Trasko, V.M.: Industrial hygiene milestones in government agencies. Am. J. Pub. Hlth. 45:39-46 (1955).
Verma, D.K.: Occupational health and safety initiatives and trends in Canada, in particular in Ontario. Ann. Occup. Hyg. 40:477-485 (1996).
Warner, C.G.: Fifty years of British occupational hygiene. Am. Ind. Hyg. Assoc. J. 34:381-390 (1974).
Zalk, D.M., Letter to editor: IOHA/ICOH declaration on Occupational Health, Ann. Occ. Hyg. 47:519 (2003).

HISTORICAL BIBLIOGRAPHY

Anonymous, Vinyl chloride: Can the worker be protected? NEJ Med. 294:653-657 (1976).
Anonymous: The Cleveland disaster. JAMA 92:1676 (1929).
Brodeur, P.: Expendable Americans. Viking Press, New York (1971).
Carter, T.: British Occupational Hygiene Practice 1720-1920. Ann. Occ. Hyg. 48:299 (2004).
Cherniak, M.: The Hawk's Nest Incident: America's Worst Industrial Disaster. Yale Univ. Press, New Haven, (1986).
Felton, J.S.: Phosphorus – A classical occupational disease. Am. J. Ind. Med. 3:77-120, (1982).
Hamilton, A.: Exploring the Dangerous Trades. Northeastern University Press, Boston (1943).
Hamilton, A.: Forty years in the poisonous trades. Am. Ind. Hyg. Assoc. Q. 9:5-17 (1948).
Jaeger, R.J.: Kepone chronology. Science 193:94, (1976).
Martland, H.S.: Occupational poisoning in manufacture of luminous watch dials. JAMA 92:466 (1929).
McCord, C.P.: What happened at Gauley Bridge 1930-1935? Ind. Med. & Surg. 30:145-149, (1961).
Mehta, P. S.; Mehta, A.S.; Mehta, S.J.; Makhijani, A.B.: Bhopal Tragedy's Health Effects: A Review of Methyl Isocyanate Toxicity. JAMA. 264: 2781-2787 (1990).
Melicow, M.: Percivall Pott (1714 –1788) – 200th anniversary of first report of occupation-induced cancer of scrotum in chimney sweepers (1775). Urology 6:745-749 (1975).
Roberts, W.: Phosvel: A tale of missed cues. Columbia Journalism Review, pp. 23-27 (July/August, 1977).
Sinclair, U.: The Jungle. NAL Penguin, New York (1905).
Weinstein, A.; Wilson, R.J.: Freedom and Crisis. Vol. 2, Random House, New York, pp. 544-561 (1974).
Whorton, D.; Milby, T.H.; Krauss, R.M.; Stubbs, H.A.: Testicular function in DBCP exposed pesticide workers. J. Occup. Med. 21:161-166 (1979).

3

PRINCIPLES AND FOCUS OF INDUSTRIAL/ OCCUPATIONAL HYGIENE

The modern hygienist has many responsibilities and duties that require the hygienist to have a varied background with adequate training in each of the natural sciences and mathematics. In addition, the well-trained hygienist should have adequate preparation in how to apply the natural sciences and mathematics to the recognition, evaluation, and control of disease agents in the workplace. The American Industrial Hygiene Association (AIHA) has defined industrial hygiene as:

> That science and art devoted to the anticipation, recognition, evaluation, and control of those environmental factors or stresses arising in or from the workplace, which may cause sickness, impaired health and well-being, or significant discomfort among workers or among the citizens of the community.

THE ESSENCE OF HYGIENE

Consider the above definition. First, hygiene is defined as *that science and art*. A more appropriate term for the field may be *that applied science*. Hygiene, like engineering, is an applied science. Mechanical engineering is the application of physics, chemical engineering the application of chemistry and physics, and biophysical engineering the application of biology and physics. Hygiene is the application of all the natural sciences and mathematics. Application of science is *art*, and, as with any applied science, there are times when the practitioner comes across a problem for which adequate theory has not been developed in the natural sciences. In these cases, the art is particularly *freelance*. The practitioner is in a situation that requires a decision based upon his or her best available knowledge. This occurs frequently in hygiene. The most dangerous aspect of this is that the hygienist may not later question that decision. Hence, the decision is set, and several years later, disease may develop in the workplace because the decision was the wrong one. Decision making based on limited available knowledge is inherent in any applied science. The process will work well if the scientist constantly seeks new knowledge and re-evaluates the decision based on the new knowledge.

It might also be worthwhile in the definition to state exactly what the science is applying.

Hygiene applies the concepts of the natural sciences and mathematics. This may be why most early hygienists were engineers. Each of the natural sciences as it relates to hygiene is considered below.

The Sciences

BIOLOGY

As hygiene is a health profession, biology is the underpinning of hygiene. Even though the other natural sciences are necessary, biology is the foundation upon which decisions made by the hygienists must ultimately be based. Because the hygienist is concerned with the health of workers and the community at large, he or she must have at least a basic understanding of the anatomy and physiology of the human body, molecular and cellular biology, genetics, embryology, and biochemistry. This is not to imply that the hygiene student should have a biochemistry course at the undergraduate level. However, an undergraduate course containing the essentials of biochemistry, such as those contained in a molecular biology course, is important. If the student entering the field of hygiene does not possess these concepts, then remedial review of the necessary material is appropriate.

In addition to the above subjects dealing with the biology of the human body, other aspects of biology are important to hygienists. Some of the agents that cause disease in the workplace are biological agents (bacteria and other microorganisms; see Chapter 4); so the hygienist must have a working knowledge of microbiology. In addition, the most knowledgeable hygienist will have a universal perspective on humans and the planet Earth, or a sound understanding of the principles of ecology.

CHEMISTRY

A fundamental knowledge of the principles of inorganic and organic chemistry is necessary in order for one to adequately function as a hygienist. The most important aspect of these two fields is nomenclature. The hygienist simply cannot function without being able to imagine the structure of a named chemical of concern. Furthermore, the basic principles of chemistry, such as the way that reactions occur, reaction rates, equilibrium, and so on, are necessary for an understanding of the basic chemical processes that occur in industry.

The hygienist must also understand many other basic principles of physical chemistry, such as the ideal gas law, which are often taught in high school. Many of these concepts may be considered to be physical chemistry, even though a physical chemistry course is not necessary for hygiene work. Nevertheless, some of the principles learned in physical chemistry are applied in hygiene (see Chapters 8 and 9). These include the application of properties such as vapor pressure and boiling point and the concepts of solubility, adsorption, absorption, catalytic reaction, and others.

Finally, the field of analytical chemistry, which is an applied chemistry field, is extremely important in hygiene. Even though many hygienists will never be directly involved in the analysis of the air samples they collect, they nevertheless must have an understanding of what occurs in the analytical chemistry laboratory in order to properly interpret the results. Only a basic understanding of analytical chemistry, as would be acquired in a practical undergraduate chemistry lab, is necessary upon entering a hygiene graduate training program.

PHYSICS

Some of the most important disease agents that occur in the workplace are termed physical agents (see Chapter 4), including such agents as noise and radiation. A basic understanding of these agents must be gained from undergraduate physics. In addition, a very important technique for controlling hazardous agents in the workplace is ventilation. This requires an understanding of the physical principles of fluid dynamics, which will be gained in most graduate degree programs in hygiene.

MATHEMATICS

Hygienists must be proficient in combining variables into physically meaningful equations and performing calculations. For the most part, the calculations are not difficult, requiring no more than algebra and occasionally trigonometry. Hygienists must have a working concept of logarithms and should gain a good understanding of the application of statistics in graduate course work. The latter subject is often misused or underutilized in hygiene. Although knowledge of calculus is not necessary for one to function as a practicing hygienist, individuals who anticipate entering doctoral and/or research programs should have mastered calculus through course work in differential equations.

Anticipation, Recognition, Evaluation, and Control

The AIHA definition goes on to state that these areas of science are used to anticipate, recognize, evaluate, and control disease agents in the workplace. The word *anticipate* was added to the original definition. It may or may not be necessary, as it is implied in recognition; one recognizes that a hazard may exist in the future or in the present or past.

What, then, does it mean "to recognize or anticipate a disease agent in the workplace"? In everyday life we constantly recognize agents that may cause injury and occasionally agents that may cause disease. For example, a roller skate left unattended on a floor is a hazard that may cause a fall. This is a mechanical agent of injury. A wary parent may recognize that playground equipment a child is using is not safe – perhaps a chain is about to break. This is recognition of a mechanical agent. We also recognize health hazards akin to those that occur in industry. For example, we may recognize that if petroleum-based paints are used in a closed room, the vapors from the paint may somehow affect our health. So, the job of recognizing health hazards in industry is easy – or is it? Would you be able to recognize that a person welding in the presence of chlorinated solvents might be exposed to a poisonous gas called phosgene? Or that a person restoring old artwork might be exposed to a gas called arsine?

There are many levels of recognition of disease agents in the workplace. In some cases, it is fairly easy to see a situation and recognize that it is unsafe. In other cases, the hygienist must use the scientific method – the consideration of multiple facts and observations leading to a hypothesis, testing the hypothesis, reformulation and retesting if necessary – in order to most clearly recognize a health hazard. In other cases, the hygienist can only estimate that a hazard may exist. In many cases the hygienist will be able to spot a hazard only after sufficient experience. There are many hazards in the workplace, and most of them have not been documented.

What is evaluation? We often evaluate hazards. Stepping outside on a winter day to determine how to dress provides a qualitative evaluation of temperature. If we had a thermometer, we could go one step further and make a quantitative evaluation.

Once the hygienist has recognized that a health hazard might exist, it may be obvious that the situation is hazardous enough to warrant immediate control. The hygienist immediately should do something to control the situation. However, for the most part, the leap from recognition to control (with a quick qualitative evaluation step) is not possible.

How could a hygienist, while walking through a plant, determine that worker exposures to trichloroethylene were unsafe or hazardous? With experience, there are some techniques that will allow the hygienist to make qualitative judgments about hazards. Usually these judgments are that exposures are very low or not – nothing more definitive. However, this dichotomy can be useful. For further distinction, the evaluation step must be quantitative. This means that the hygienist usually will need to quantitatively analyze the levels of contaminants in workplace air to determine the degree of hazard. In the case of physical agents such as noise and radiation, the types of analyses have the same principles but are based on physics rather than chemistry. In the case of biological agents, the types of analyses are based on microbiology.

The word *hazard* has been used repeatedly. 2,3,7,8-Dibenzo-dioxin is extremely toxic in mice. Strychnine and the toxin produced by the botulinum bacterium are fatal to humans in small doses. Are they hazardous? Yes. Toxicity or potential to cause harm is hazardous. However, for that harm to occur, there must also be an exposure at a significant level. So, if a chemical is toxic (hazardous), and it is handled in such a way that an exposure may occur that could lead to a significant outcome, then that situation is risky, i.e., there is a significant chance or risk of an unacceptable health outcome. The word *significant*, is used several times. Certainly death is a significant outcome; but, in some cases, a reversible skin rash may be the outcome of concern. A significant exposure would be one that leads to a significant risk of disease. Clearly an exposure which causes illness in everyone so exposed would be significant; but what if only 1 in 10 or 1 in 100, or 1 in 10,000 were affected? Which of these is significant? The answer is not clear and deals with the subject of risk management, which is further discussed in Chapter 10. Risk and hazard are relative terms.

What is control? As mentioned earlier, we often control hazards in everyday life. The application of seatbelts to control injuries arising from automobile accidents is a great example of the recognition, evaluation, and control process. Undoubtedly, soon after the first automobile was driven, someone recognized this type of transportation to be a more hazardous form of conveyance than the horse and buggy. However, it was not until accident and injury statistics from automobile accidents were analyzed that a public outcry occurred, and automobile conveyance became commonly thought of as a significant hazard. Although the quantitative automobile accident evaluation step probably took place sometime in the 1950s or earlier, it was not until the early 1960s that the control process began to take place.

This scenario is not unusual since the process of hazard evaluation is often subjective. Inherent in the hazard evaluation step is the concept of risk. No matter what we do to the process of automobile conveyance, it will probably never be risk-free from injury or even death. Therefore, the question arises as to what is an acceptable risk. It was not until the 1960s that society put enough pressure on automobile manufacturers, through government intervention, to cause seatbelts to be installed in automobiles. However, even in the presence of seatbelts, individuals still choose to take risks rather than control exposure to the hazard associated with being thrown about in an automobile like an unguided missile.

There are many techniques available to the hygienist for controlling disease agents in the workplace. As noted earlier, an important one of these is ventilation. Essentially, the principle of local ventilation is to control the disease agent at its source or its origin, and thus to remove it from the workplace. Other techniques control the hazard at the worker. These include the use of respirators or facemasks with absorbents or filters. Also included would be chemical protective clothing for controlling the exposure of skin to chemical agents. Finally, there are various types of *administrative* controls, which are ways of reducing the amount of time that the employee is exposed to a hazard, or changing the way that the employee performs a job to reduce the exposure. A very simple example of this would be rescheduling an operation that must take place outdoors, from the afternoon to the morning, in order to avoid exposure to heat which may lead to heat stress.

The Agents of Disease

According to the AIHA definition the hygienist *recognizes, evaluates, and controls those environmental factors or stresses. . . .* What are the environmental factors or stresses? Typically, hygienists recognize, evaluate, and control biological, chemical, and physical agents in the workplace. These are the hazards that workers face. Some hygienists also include mechanical and ergonomic agents among the types of agents that they evaluate. Mechanical agents are the specific concern of a safety professional, and ergonomic agents are the specific concern of an ergonomist, but in many plants these agents must be addressed by the hygienist.

Safety professionals are concerned about those hazards in the workplace that can lead to injury. These include *slip, trip, and fall agents*, for example, slippery surfaces. They also include sharp surfaces or points that can cause abrasions or cuts to the skin, and pinch-points and other hazards that can cause crushing of human appendages. Very often equipment used in industry must exert a great deal of force in order to get the job done. For example, presses that stamp out pieces of metal in certain shapes will attempt to press a human appendage into the appropriate shape, should it become lodged inside the press. The safety professional must guard against this. Safety professionals are also concerned about fire and explosion hazards.

Ergonomists are concerned about the human/machine/environment interface and may be thought of, somewhat loosely, as specialized safety professionals. Often, in industry, the engineer will design a machine to do a job and put it into an environment suited to the product. Sometimes no consideration is made for the human being who must run the machine in the chosen environment. Is a gymnast required to operate the machine? Or does the machine produce so much heat that it is impossible for the human being to withstand the heat stress? Essentially, the principle upon which ergonomics is founded is that the machine must be designed so that the human being (i.e., the human factors) will be accommodated.

These principles apply not only to machines but to any other processes that involve humans. Consider the assembly line worker in a chicken processing plant who must constantly perform the same motion with wrist and elbow in order to cut a particular part off the chicken. Thousands of such movements may be made repetitively in a single day. Although it may not be possible to decrease the number of movements the worker needs to make, the ergonomist would be concerned about the way in which the movements are made. The role of the ergonomist is to put as little stress as possible on the human in the act of performing work.

Therefore, because specific professions deal with mechanical and ergonomic agents in the workplace, they are typically not included in the definition of hygiene. Nevertheless, many hygienists practice as ergonomics and/or safety professionals in addition to being hygienists. It is not uncommon for the hygienist to perform these roles, particularly in a small plant.

Of the agents that the hygienist is concerned with on a daily basis, the chemical agents occur frequently and receive much attention. Therefore, this entire volume is devoted to chemical agents. However, physical hazards are encountered quite frequently; noise is prevalent throughout most workplaces and consequently is a hazard that the hygienist must constantly recognize. Biological agents such as bacteria and viruses are not as common as the other two types of agents; however, concern about these types of agents has increased markedly over the past 20 years.

The Ecological Perspective

The last part of the AIHA definition implies that the hygienist is not only concerned about agents that cause disease or discomfort among workers, but also about the citizens of the community. This passage implies that the hygienist must appreciate the principles of ecology. Irish

(1973) eloquently pointed this out over 30 years ago at the height of the environmental movement. The need for an ecological perspective is another reason for the biological basis of hygiene. Ecology comes from the Greek for *study of house*. On the surface, this definition does not appear to have much relationship to either ecology or hygiene, but consider a house. Contained within it is the food that a family eats. In addition, one could look at the interrelationships between family members, the design of the house, its landscaping, the way the family interacts with its neighbors, and the way the house interacts with the rest of the neighborhood (e.g., how it fits into the architectural and environmental setting). One could also look at the way in which the house and its surroundings have changed or blended into nature. Hence, the *study of house* actually connotes a great deal. Ecology is the study of the way things fit together, of the essence of nature, of the interrelationship between all things, living and nonliving.

The hygienist cannot perform his or her duties adequately if his or her concern stops at the boundary of the plant or at that portion of the hallway where the hygiene offices stop and the air pollution control engineer's or the physician's offices begin. Often, the hygienist controls chemical agents in the workplace by the use of ventilation. If the air containing the contaminants is removed from the workplace and no provision is made to remove the contaminants from the air, the contaminants simply escape to the outside. Might they have an effect on communities downwind?

Often, workers leave the workplace in the same clothes in which they have performed their jobs. It has been documented more than once that workers can carry home with them contamination that may affect family members (Knishkowy and Baker, 1986). However, the link between a worker and his or her family with respect to safety and health goes beyond the work clothing issue. Workers may suffer genetic disease, or, in the case of females, unborn fetuses may suffer ill effects caused by the workplace. Hence, the health of the worker is both directly and indirectly tied to the health of the family.

Typically, Americans work a 40-hour week. Simple arithmetic will show that there are 168 hours in a week. Hence, workers are on the job for less than 25% of the week. What do they do in their spare time? Do they have noisy hobbies, such as shooting rifles or doing woodworking in the garage? Do they live in neighborhoods that may be close to plants with processes similar to the one where they work? Consequently, is it possible for a worker to be exposed to contaminants for 168 hours per week instead of only 40?

For these and many other reasons, the hygienist must have an ecological perspective. He or she must have a cognizance of the interrelationships of the plant and workers, but the interrelationships of the plant and community must be added to the matrix. The principles of ecology state that every living thing is linked to every other living thing, and that these are in turn linked to nonliving portions of the ecosystem such as air, water, and land. The hygienist must realize that the plant in which he or she works, or the governmental agency for which he or she works, is intricately involved with all other aspects of the nonliving environment. In addition, the work is intricately linked to all other living things in the ecosystem. In controlling workplace health hazards to protect the health of a worker, the hygienist may inadvertently expose nonworkers. In controlling a workplace health hazard, a hygienist may inadvertently create a water or air pollution problem. The hygienist must understand, appreciate, and absorb the principles of ecology and practice them on a daily basis.

THE SAFETY AND HEALTH TEAM

In one way it is unfortunate that many hygienists are employed in small industries where the only other health professional present may be an occupational health nurse, if that. In some situations, the hygienist may have the company of a safety professional. Of course, it is most unfortunate that some large companies do not employ any members of a safety and health team.

In any case, each hygienist must be aware of and benefit from the knowledge that he or she is part of an occupational safety and health team, even though it may not be housed at his or her workplace. This team is minimally composed of physicians, nurses, hygienists, and safety professionals. Fortunately, in many industries all four of these professions work together as part of one organizational group. However, even if this is not the case, the hygienist should be aware of community resources that are available and utilize them to the fullest extent.

Even the hygienist at a small plant must depend upon the services that can be provided by a physician or a nurse and the services that may be provided by a safety professional. For example, consider the hygienist who works alone without any of the other members of the team. In fact, if disease or injury arises at the plant, the first organization the hygienist will want to contact will be a health care organization employing nurses and physicians. Hence, it is to the hygienist's advantage to establish contacts and make agreements with these organizations before it is necessary to do so in an emergency. Likewise, even though there may be no safety professionals at a plant, the hygienist will find it necessary to establish contacts with the local fire department. Fire departments employ specialized safety professionals. It is much too late to first contact the fire department after a fire has started.

Without belaboring the point, it should be obvious that the hygienist must function as part of a health and safety team, and that formal contacts must be made with the other members of the team whether or not those members exist at the plant. Moreover, in addition to the typical members of the occupational safety and health team, there are members of other allied professions with whom the hygienist may need to make contact. Each of these allied professions represents a discipline in which the hygienist should have certain minimal training. Without that minimal training, the hygienist has little idea of how the allied professional may be able to help him or her in protecting worker health.

For example, the hygienist should have training in epidemiology. This training should allow him or her to do preliminary epidemiology work, to interpret epidemiological studies, and to apply them to the workplace setting. Beyond this, however, the hygienist may require the services of an epidemiologist. In fact, the hygienist may require the services of the following allied professionals:

- Epidemiologist
- Toxicologist
- Ergonomist
- Air quality engineer
- Water quality engineer
- Sanitary engineer
- Hazardous waste engineer
- Water quality biologist (ecologist or applied limnologist)
- Acoustical engineer
- Mechanical engineer (ventilation)
- Health physicist

The hygienist often must have some training or knowledge of the work that is performed by each of these professionals. Actually, the properly trained hygienist will have considerable knowledge in many of these fields. In addition, many hygienists have acquired knowledge of occupational health and safety law and, by virtue of having worked in industry or through formal training, will have acquired some knowledge of management. The hygienist is truly a jack-of-all-trades, and hopefully a master of some. Hence, the hygienist serves as an interface between all of these professionals.

Consider the health and safety team with its four member professions. The physician employed or retained by the company should have some knowledge of hygiene, but unfortunately very few do. Similarly, many nurses who work in industrial plants have not had the benefit of any formal occupational health training. Most safety professionals have been brought up through the ranks of industry management or a union, and their safety training has been on the job without emphasis on hygiene. The hygienist, properly trained, will have had a safety course, and an adequate biological background containing information about occupational diseases and their occurrence, origins, symptoms, and in some cases, treatment. Furthermore, the hygienist will have had an epidemiology course, toxicology course work, some information about ergonomics, a general environmental health course containing principles of air and water pollution control, and other course work related to the above listed areas. Therefore, the hygienist can best serve as an interface to all of these related professions.

It should be noted that the health physicist and the general environmental health scientist share this interface with the hygienist. Properly trained health physicists (with training emphasizing radiation hazards) and general environmental health scientists (with training emphasizing community health as compared to workplace health), like hygienists, are health professionals who have backgrounds in the natural sciences and engineering.

COURSE OF STUDY

One might ask how best to prepare academically for a career in hygiene. First, graduate training (at least in the United States) is becoming a prerequisite. In 1971, 56% of respondents certified by the American Board of Industrial Hygiene (ABIH) recommended that future certification minimally require a master's degree (Committee, 1971). Although there is a role for persons with bachelor's degrees in hygiene, several points should be recognized. First, bachelor's degree programs should contain an adequate foundation in the natural sciences and mathematics. Hygiene courses should not be substituted for *science* courses. Second, degree status is reflected in salary and potential for advancement. Based on 2,192 hygienists surveyed by AIHA (2004), the average salary for bachelor's degree holders was $75,000, and for master's degree holders $84,000. Finally, it should be noted that the Accreditation Board for Engineering and Technology (ABET), in cooperation with the Academy of Industrial Hygiene (AIH, a component of AIHA composed solely of certified industrial hygienists), accredits master's level industrial hygiene graduate programs (Garber, 1988) and more recently bachelor's programs. ABIH gives one year of credit toward certification for master's degrees and 6 months credit for bachelor's degrees from ABET-accredited programs.

What should a master's curriculum contain? The answers are as varied as the number of programs. It should minimally conform to earlier recommendations (Committee, 1971) shown in Table 3-1. However, 35 years later a few subjects, or at least new perspectives, should be added. For example, new students are competent in the use of computers, however, hygiene applications should be addressed in the curriculum. Biological agents and non-industrial ventilation are more important in light of indoor air quality concerns. Hazardous waste management has become more important than ambient air pollution control in most hygienists' practice; however, indoor (domicile and office building) air pollution is a major focus. Epidemiology and ergonomics are no longer optional. Consequently, a thorough degree plan should perhaps look like that of Table 3-2.

One sometimes controversial area is the thesis, project, or field study. Whatever it is called, it should have the following objectives:

1. Application of the scientific method.
2. Application of hygiene principles.
3. Improvement of report-writing skills.
4. Improvement of organizational skills.

Most students need work in each of these areas. Although students may think they under-

Table 3-1
Courses recommended by AAIH, 1971

Required
Air Pollution
Air Sampling and Analysis
Biostatistics
Non-Engineering Controls
Industrial Ventilation
Fundamentals of Industrial Hygiene/Occupational Health
Seminar
Health Physics
Noise and Vibration
Toxicology
Thesis or Project
Electives
Environmental Health
Ergonomics
Safety
Epidemiology
Biological Agents

stand the scientific method and hygiene principles, putting them into practice is another matter. All students need practice in writing reports and communication of findings to relevant groups, and a chance to independently organize a project from start to finish when only a letter grade is at stake instead of lives or health.

Since the 1990s the roles of hygienists around the world have broadened (Guillemin, 2006). In addition to the broadened tasks relative to the non-workplace environmental media such as hazardous waste handling and air pollution permitting, non worker populations are more frequently the responsibility of the hygienist. In the U.S., asbestos issues are waning but mold issues are important; both involve populations such as building residents and school children. Other hygienists have become responsible for consumer product safety. In 2007 more and more hygienists find themselves responsible for wellness, health promotion, and *presenteeism*[1] programs (Guillemin, 2006).

[1] Presence at work as opposed to absence, but without expected productivity, likely due to poor working conditions, only one of which might be exposures to disease agents.

Table 3-2
A thorough outline of study for the Master's degree in industrial hygiene

Fundamentals of Industrial Hygiene
Air Sampling and Analysis
Control of Chemical Agents
Industrial Hygiene Case Studies
Physical Agents Effects and Control
Biological Agents Effects and Control
Fundamentals of Occupational Safety/Ergonomics
Biostatistics and Applications to Hygiene
Principles of Occupational Epidemiology
Essentials in Toxicology and Occupational Diseases
Principles of Risk Assessment
Air Pollution
Hazardous Waste Management
Field Interdisciplinary Studies
Preceptorship/Internship/Cooperative work-study
Research Thesis

This broadening of the hygienist's role requires a new educational model (Vincent, 2005). Given just the traditional responsibilities of hygienists, many hygiene degree programs are loaded with courses with little opportunity for electives. How then can we train students to be prepared to undertake these new responsibilities? Vincent makes the case that the problem may be with the word *train*. Perhaps in the past, programs have concentrated on producing a graduate who has the skills to do all the traditional tasks, i.e., trained. It might be better that the graduate is given basic tools and that efforts are placed on helping each person to develop conceptual frameworks, i.e., 'to be educated'.

For example the goal of a controls course might be to create a conceptual framework of how variables, definitions, and tools fit together for solving occupational health problems with ventilation. Each student creates a framework in his mind, but hopefully all such frameworks tend to yield good solutions to problems. Another course across campus may help students create a framework for solving ambient air pollution problems. We need to educate students on connecting these frameworks, i.e., not necessarily giving them all the tools but educating them to use their faculties to solve future, unanticipated problems by using analogies and relationships that derive from the principles of natural sciences and engineering. One course that deals with control principles applied to all kinds of environmental health problems should be adequate to guide a student through a career when educated by way of examples that require critical thinking and creative solutions. Other examples for connecting conceptual frameworks can be made for all of the recognition, evaluation, and control steps and sub-steps.

Vincent (2005) makes the point that hygienists are becoming risk managers. Of course we were always occupational health risk managers, but now the term is being applied to broader health risks. Hygienists will continue to go down this path if the job market requires it, and they will self train where their academic training fails them. Thus, academic programs and professional societies need to evolve with the profession.

THE RELATIONSHIP OF HEALTH, SAFETY AND PRODUCTIVITY

On the surface it would seem that the duties of the hygienist are at odds with the profit motives of corporations in a capitalist economy. Adequately controlling health hazards in industry requires money. Corporations are in the business of making money and wish to reduce capital and recurring expenses in order to increase profits. So hygienists are only employed because of legal requirements placed upon industry...or are they?

The primary responsibility of the hygienist is to protect the health of the worker and the community. One must do so regardless of the cost to industry, or at least must make recommendations to management based on sound scientific judgement and not economics. Of course, this must be tempered with reason and a good understanding of economics. In order to carry out the job and maintain allegiance to the primary responsibility, the hygienist must be convinced that the principles of hygiene do not conflict with management goals or the principles of economics. If the hygienist has done the job adequately, he or she will also be able to convince management that this is the case. If conducted properly, the various aspects of an adequate hygiene program will accomplish at least the objectives in Table 3-3.

While in some cases it is difficult to quantitatively show the benefit of hygiene in these terms, it is certainly not difficult to imagine that workers who work (1) in an environment with acceptable levels of exposure and a management commitment to continuously decrease exposures, and (2) among health professionals who have an outgoing and readily apparent concern for the employees' health, will increase the probability of achieving each of these goals.

On the other hand, a little creativity can show the benefits of safety and health to profits. Procter & Gamble Company spent $10 million on worker's compensation in 1989. The average for the industry was $60 million. Had they spent the additional $50 million, the company's sales would have had to increase by $1 billion to maintain profit levels (Fulwiler, 1993). Figure 3-1 shows how Procter & Gamble's key elements program pays off. Each plant rates its current status in areas such as standards, training, and safety performance relative to expectations (scale of 0-10; 10 is expected). As can be seen, the higher the rating, the lower the incidence rate of accidents and injuries. The point is that highly ethical companies can make more money than less ethical companies, regardless of the conventional wisdom.

An approach similar to Procter & Gamble's has been presented by Brennecke (1997),

Table 3-3
Objectives that are often accomplished but not documented by hygiene programs

Reduce health care costs
Reduce workers' compensation costs
Increase productivity
Decrease lost time or time away from the job
Increase efficiency of output
Increase morale

again for injuries. Table 3-4 shows the amount of sales needed for a company to make up lost resources caused by injuries or illness. For example, if injuries cost $25,000 per year and the company's profit margin is 12% then the company needs to increase sales by $208,000 to offset the injury-related losses. What constitutes a loss? The direct losses such as medical costs are fairly easy to quantitate. However, Table 3-5 shows that there are many others that are more difficult to quantitate. Of course even quantifying the true number of illnesses is a challenge. For example in a workplace where asthma is related to occupational exposure, one would need to estimate the proportion of all asthma caused by the workplace and then have a surveillance program in place to determine when people were missing work due to asthma. Other survey work might estimate how productivity is affected by workers who experience effects while on

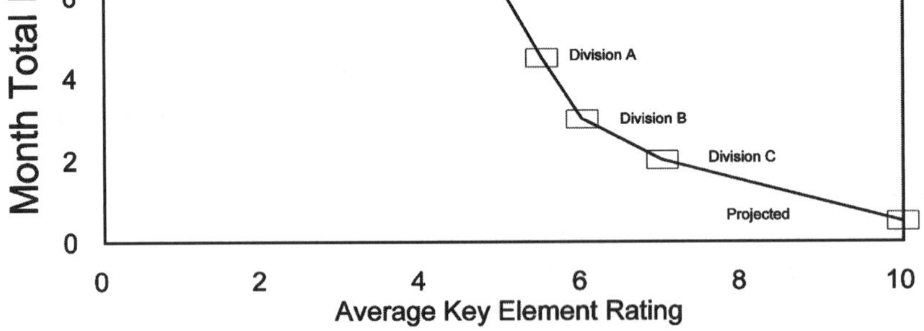

Figure 3-1. Key element ratings vs. total incident rate at Procter & Gamble Company (from Fulwiler, 1991, with permission).

Table 3-4
Examples of injury and illness costs[1]

Direct Cost to the Employer
Medical costs
Temporary disability payments
Permanent disability payments
Rehabilitation
Indirect Costs to the Employer
Repair or replacement of damaged tools, equipment or spoiled work
Interim equipment rentals
Expenses for emergency supplies and equipment (first aid, bloodborne pathogens, spill clean up)
Project delays
Time spent by employees and supervisors responding to the incident
Time spent by employees, supervisors, management and administrative personnel investigating the incident and writing reports
Time lost through sympathy, curiosity, discussions, telling similar stories, swapping opinions and Monday morning quarterbacking
Lost efficiency and/or overtime to make up for project delays or to maintain a project schedule until the injured employee returns
The cost of hiring, training and orienting replacement employee
Decreased productivity of the injured employee on return to work
Lost business and goodwill
Legal costs
OSHA fines
Other Costs
Economic loss to the disabled employee's family
Increases in future insurance premiums

[1] Adapted from Brennecke (1997).

Table 3-5
Additional sales needed to offset incident costs based on profit margin[1]

Yearly cost of incidents	Profit margin 1%	3%	4%	12%
$5,000	$500,000	$167,000	$125,000	$41,700
$10,000	$1,000,000	$333,000	$250,000	$83,300
$25,000	$2,500,000	$833,000	$625,000	$208,000
$50,000	$5,000,000	$1,667,000	$1,250,000	$417,000

[1] Adapted from Brennecke (1997).

the job. These are doable tasks that require measurements and assumptions, but still doable.

It could be argued that such resources would be better spent in controlling the problem rather than assessing its costs. This is true. However, management may balk at the control costs, particularly if no specific standards are being violated. Even a ballpark estimate of the costs is appropriate as these data will help demonstrate the value of the hygiene effort to the company.

Fulwiler (2002) has also argued that hygienists need to do a better job of showing management that they are technology enablers. In other words, we not only prevent illnesses but, in some cases, if it were not for the solutions of hygienists, a process or an industry would not exist. He cites a specific example of how the use of enzymes in the detergent industry caused an occupational disease with which the unions were no longer willing to deal. His group found a solution while another company did not. Consequently his company's brand went from number two to number one. To some degree these enabling events are being accomplished by hygienists daily; however, they are not being documented.

Another approach, particularly in large companies, is to create programs that achieve safety and health goals through internal competition (Krueger and Vigezzi, 2002). In a company with 100 plant sites, find the 10-20 that are performing best or are following some model program. Then convince management to set company goals, even management bonuses, based on the model program. The selling of the program to management will likely need to be based on quantitative estimates of savings, unless management is so enlightened to see the inherent value in a safe and healthy workforce.

CODE OF ETHICS

If the hygienist can do an adequate job of sales and education to convince management that the objectives in Table 3-3 can be accomplished, then management also will be keenly interested in employee health, and the hygienist's job will become easier. If not, then the hygienist may become tempted to alter his or her primary responsibility and even begin to overlook health problems among the workforce. If this happens, the hygienist has only to look at the code of ethics to modify his or her behavior.

The ABIH, AIHA, ACGIH®, and AIH have adopted the same code of ethics explaining how a hygienist should behave. This code essentially mirrors what has been written above. The code is reproduced in an appendix to this chapter. A revised code is expected to appear in 2007.

A MODIFIED DEFINITION AND NAME FOR INDUSTRIAL/OCCUPATIONAL HYGIENE

Considering all of the above, a slightly modified definition of hygiene may be instructive for the beginning hygiene student. Consider the following:

Industrial/occupational hygiene is an applied science utilizing the concepts of the natural sciences and mathematics and the principles of engineering in recognizing, evaluating, and controlling biological, chemical, and physical agents of disease or discomfort that may arise in or from the workplace and affect the health or standard of living of workers or the community at large.

Now that we have defined industrial or occupational hygiene, we might ask does the name fit the definition? Apparently not. In 1975, Cudworth and Watt (1975) reported that 60% of AIHA members (1,121 respondents) indicated that *industrial hygiene* did not effectively describe their functions. Only 36% said it was adequate. On the other hand, 50% of respondents said that AIHA should not change its name.

The subject came up again in 1991. This time the association voted on whether to keep the current name or adopt the name American Industrial and Environmental Health Association. The vote was close, but a two-thirds majority (not achieved) was needed for change (Sheriff, 1991). (However, the joint journal of the AIHA and ACGIH® is now called *Journal of Occupational and Environmental Health.*)

The previous year members of a group of *outside stakeholders*, various industries, academic institutions, medical and legal professions, media, allied associations, and members of Congress and government, were interviewed. Many thought that industrial hygiene was an outdated term, that hygiene implied cleanliness and industrial implied manufacturing, that the term hygienists should be abandoned, and that the industrial arena should be broadened (Henshaw, 1990a). All these seem like good ideas, but no one has invented an acceptable name.

It is much the same the world over. North Americans, Swiss, Norwegians, Spanish, and Italians use the name industrial hygiene, while British, Australians, Swedes, Germans, Dutch, Finns, and Flemish use occupational hygiene (Henshaw, 1990b). *Occupational* seems more explanatory, but many hygienists think that the field has broadened to include the health of communities, consumers, and groups other than occupations. In the United States, hygienists are sometimes confused with dental hygienists or persons in professions related to janitorial services – important professions, but not what we are about.

The name needs to be changed. It should be simple (preferably one word), meaningful (have roots in what we do), unique (not confused with other professions), and catchy (something that sticks in your mind when you hear it on television). Perhaps a discussion with classicists would evolve a Greek or Latin word. That would seem most logical, producing something like *biology* or *ergonomics*.

QUESTIONS

3.1. Explain why biology is the underpinning of hygiene. Be more specific than to say it is health related.

3.2. Think of five potential hazards that we (the average person) recognize, evaluate, and control at least once per year and state how they are controlled.

3.3. The text states that the hygienist is the interface between many different professions. Why can't the engineer or physician serve as this interface?

3.4. Can you explain why the failure to have an ecological perspective has resulted in partial solutions to many environmental problems that has led to more problems, often involving different media (land, water, air)?

3.5. What is the primary responsibility of the hygienist as stated in the code of ethics.

REFERENCES

AIHA: 2004 Salary Calculator (available to members only) American Industrial Hygiene Association, http://www.aiha.org/Content/MemberLogin/MembershipCenter.htm (2004).

Brennecke, R.A.: Contribute to the corporate bottom line. The Synergist AIHA Newsletter, August (1997).

Committee on Recruitment and Training, American Academy of Industrial Hygiene: Report of training requirements. Am. Ind. Hyg. Assoc. J. 32:271-273 (1971).

Cudworth, A.L.; Watt, R.W.: A summary of results of the public relations and communications survey of the American Industrial Hygiene Association. Am. Ind. Hyg. Assoc. J. 31:247-258 (1975).

Fulwiler, R.D.: Health and Safety's Stewardship of Key Business Values: Employees, Public Trust, and Responsibility to Shareholders, TMEI Occasional Paper Series #5.91. Minerva Educational Institute, Xavier University, Cincinnati (1993).

Fulwiler, R.D.: Industrial Hygiene: A Technology Enabler. AIH Diplomate Newsletter, June/July (2002).

Garber, B.T.: An update on the accreditation of academic programs in industrial hygiene. Am. lnd. Hyg. Assoc. J. 49:A688-A690 (1988).

Guillemin, M.P.: Occupational hygiene educational status: Global trends and a global future. Ann. Occ. Hyg. 50:645-649 (2006).

Henshaw, J.L.: President's page. Am. Ind. Hyg. Assoc. J. 51:A723 (1990a).

Henshaw, J.L.: President's page. Am. Ind. Hyg. Assoc. J. 51:A794 (1990b).

Irish, D.D.: The significance of the occupational environment as a part of the total ecological system, In The

Industrial Environment: Its Evaluation and Control. National Institute for Occupational Safety and Health, Cincinnati (1973).
Knishkowy, B.; Baker, E.: Transmission of occupational disease to family contacts. Am. J. Ind. Med. 9:543-550 (1986).
Krueger, K.E.; Vigezzi, M.J.: The Global Star Program. AIH Diplomate Newsletter, Winter (2002).
Sheriff, R.E.: President's page. Am. Ind. Hyg. Assoc. J. 52:A378 (1991).
Vincent, J.H.: Graduate education in occupational hygiene: a rational framework. Ann. Occ. Hyg.49:649-659 (2005).

APPENDIX 3-1. CODE OF ETHICS FOR THE PRACTICE OF INDUSTRIAL HYGIENE[2]

Introduction

The American Board of Industrial Hygiene (ABIH) is a voluntary, non-profit, professional credentialing organization. ABIH certifies qualified industrial hygienists engaged in the practice of industrial hygiene, and who have met the professional knowledge standards established by the Board of Directors. Regardless of any other professional affiliation, the ABIH Code of Ethics (Code) applies to: each individual certified by the ABIH as a Certified Industrial Hygienist (CIH) or a Certified Associate Industrial Hygienist (CAIH) (certificants); and, each individual seeking ABIH certification (candidates). The Code serves as the minimal ethical standards for the professional behavior of ABIH certificants and candidates.

The Code is designed to provide both appropriate ethical practice guidelines and enforceable standards of conduct for all certificants and candidates. The Code also serves as a professional resource for industrial hygienists, as well as for those served by ABIH certificants and candidates.

Preamble/General Guidelines

The ABIH is dedicated to the implementation of appropriate professional standards designed to serve the public, employees, employers, clients and the industrial hygiene profession. First and foremost, ABIH certificants and candidates give priority to health and safety interests related to the protection of people, and act in a manner that promotes integrity and reflects positively on the profession, consistent with accepted moral, ethical and legal standards.

As professionals in the field of industrial hygiene, ABIH certificants and candidates have the obligation to: maintain high standards of integrity and professional conduct; accept responsibility for their actions; continually seek to enhance their professional capabilities; practice with fairness and honesty; and, encourage others to act in a professional manner consistent with the certification standards and responsibilities set forth below.

---

[2] Note, this code is current as of late 2007 (http://www.abih.org/members/documents.html#ethicscode).

I. **Responsibilities to ABIH, the profession and the public.**

A. Certificant and candidate compliance with all organizational rules, policies and legal requirements.

1. Comply with laws, regulations, policies and ethical standards governing professional practice of industrial hygiene and related activities.

2. Provide accurate and truthful representations concerning all certification and recertification information.

3. Maintain the security of ABIH examination information and materials, including the prevention of unauthorized disclosures of test information.

4. Cooperate with ABIH concerning ethics matters and the collection of information related to an ethics matter.

5. Report apparent violations of the ethics code by certificants and candidates upon a reasonable and clear factual basis.

6. Refrain from public behavior that is clearly in violation of professional, ethical or legal standards.

II. **Responsibilities to clients, employers, employees and the public.**

A. Education, experience, competency and performance of professional services.

1. Deliver competent services with objective and independent professional judgment in decision-making.

2. Recognize the limitations of one's professional ability and provide services only when qualified. The certificant/candidate is responsible for determining the limits of his/her own professional abilities based on education, knowledge, skills, practice experience and other relevant considerations.

3. Make a reasonable effort to provide appropriate professional referrals when unable to provide competent professional assistance.

4. Maintain and respect the confidentiality of sensitive information obtained in the course of professional activities unless: the information is reasonably understood to pertain to unlawful activity; a court or governmental agency lawfully directs the release of the information; the client or the employer expressly authorizes the release of specific information; or, the failure to release such information would likely result in death or serious physical harm to employees and/or the public.

5. Properly use professional credentials, and provide truthful and accurate repre-

sentations concerning education, experience, competency and the performance of services.

6. Provide truthful and accurate representations to the public in advertising, public statements or representations, and in the preparation of estimates concerning costs, services and expected results.

7. Recognize and respect the intellectual property rights of others and act in an accurate, truthful and complete manner, including activities related to professional work and research.

8. Affix or authorize the use of one's ABIH seal, stamp or signature only when the document is prepared by the certificant/candidate or someone under his/her direction and control.

B. Conflict of interest and appearance of impropriety.

1. Disclose to clients or employers significant circumstances that could be construed as a conflict of interest or an appearance of impropriety.

2. Avoid conduct that could cause a conflict of interest with a client, employer, employee or the public.

3. Assure that a conflict of interest does not compromise legitimate interests of a client, employer, employee or the public and does not influence or interfere with professional judgments.

4. Refrain from offering or accepting significant payments, gifts or other forms of compensation or benefits in order to secure work or that are intended to influence professional judgment.

C. Public health and safety.

1. Follow appropriate health and safety procedures, in the course of performing professional duties, to protect clients, employers, employees and the public from conditions where injury and damage are reasonably foreseeable.

Effective Date: May 25, 2007

II

DISEASE AGENTS OF THE WORKPLACE

4

DISEASE AGENTS OF THE WORKPLACE

The focal point of industrial hygiene is the disease agents that the hygienist studies. The hygienist must have a broad knowledge of and appreciation for the complexity and the wide range of disease agents that can occur in both the workplace and the general environment. The agents are conveniently and logically divided into three groups: biological, chemical, and physical. The basis for understanding the nature of these three groups is in the three natural sciences: biology, chemistry, and physics. Herein lies the single most important reason why the industrial hygienist must have a good grounding in the natural sciences. In this chapter biological and physical agents are introduced, and detailed discussion is given for chemical agents. For a more detailed overview of the biological agents see *Modern Industrial Hygiene, Volume 2, Biological Aspects*.

BIOLOGICAL AGENTS

For the vast majority of the disease agents of biological origin, the fundamentals of microbiology are necessary in order for one to understand how these agents affect the worker. Fundamental microbiology concepts give the hygienist an appreciation for mechanisms of transmission of the agents, agent life histories, signs and symptoms, human immune system interactions with the agents, and the ecological relationships of the agents to the environment. This last aspect is particularly important for understanding how and when the biological agents arise in the workplace and the best way to control them.

As with any classification system, agent classification based upon biological, chemical, and physical properties does have some overlap. However, the three areas can be defined such that the lines of demarcation are fairly well characterized. For example, consider cotton dust, an important disease agent that affects the lungs of textile and other workers. One popular theory of the cause of this disease is that a class of chemicals, arising from bacteria present on the plant material (the bract), which is harvested with the cotton, is responsible for the asthma-type reaction that many workers have upon inhaling the cotton dust (Jacobs, 1989; Douwes et al., 2003). The chemicals originating in the bacteria are endotoxins. Hence, the question may be asked, are endotoxins biological agents or chemical agents?

Consider linemen who string telephone line in the eastern United States. An important disease agent for these workers is poison ivy. Poison ivy is not a microbiological agent – the specific cause of the skin reaction caused by poison ivy is a chemical; but poison ivy is biological in nature. Once again, is the disease caused by poison ivy the result of a biological or a chemical agent?

Another skin disease associated with bakers is manifested among a small portion of bakers who handle the spice cinnamon. The cinnamon causes dermatitis, which can be quite severe (Birmingham, 1977). The cinnamon is of biological origin; however, a specific chemical from the cinnamon causes the reaction.

Pharmaceutical workers often are exposed in their work environment to chemicals derived from naturally occurring organisms. One example is penicillin, the antibiotic, which is derived from fungi. In the case of penicillin, the chemical is specifically concentrated from the biological source; the pharmaceutical company extracts and concentrates the penicillin from a fungus. In this case we might consider that a biological agent is producing a chemical agent.

There are many chemicals that investigators have extracted, purified, and/or concentrated from biological sources, and they may be considered chemical agents. In some cases, researchers have even synthesized in the laboratory a synthetic product whose structure originates with a natural compound. Perhaps it is belaboring the point, but in essence, all biological agents are chemical agents. One or a few chemicals, such as the endotoxins, ultimately cause a disease.

Although plants such as poison ivy or cinnamon are important biological agents, the number of workers who come in contact with them and the severity of the disease that they produce make them of less importance than the microbiological agents. By virtue of their size, microbiological agents readily penetrate the lungs, and once there, either reproduce and cause infectious disease, or with little or no reproduction, cause allergic or hypersensitive reactions by virtue of their irritation of the lung tissues and the corresponding immunological response of the human being. Hence, the microbiological agents are the most important of the biological agents. These two disease pathways (infection and allergic reaction) must be considered in the case of each microbiological agent.

Classical Biological Agents

For many years, the biological agents were something of a novelty in the field of industrial hygiene. It was widely acknowledged, in certain industries, that agents of biological origin were of particular importance; in general industry, however, the biological agents were rarely considered. Farmers, slaughterhouse workers, poultry workers, meat packers, swine confinement workers, hunters and trappers, sanitation workers, hospital workers, bird breeders, and others were commonly known to be exposed to microbiological agents in their workplaces (see Table 4-1). It was realized, then, that through contact with the biological agents, the diseases that they caused could manifest themselves among the workers. Table 4-1 lists a

Table 4-1
Classical biological agent-induced infectious occupational diseases[1]

Group	Disease	Source	Occupations at risk
Viruses	Rabies	cats, dogs, wild animals	vets, spelunkers, farmers, ranchers
	Orf	sheep, goats	shepherds, shearers, vets, stockyard workers
	Milker's Nodule	cattle	dairy operators, veterinarians, ranchers
	Newcastle Disease	birds	poultry handlers, vets
	Hepatitis	man	health-care workers
	AIDS	man	health-care workers
Rickettsia	Rocky Mountain Spotted Fever	ticks	outdoor workers in wooded areas
	Q Fever	ticks	placental tissue of domestic animal handlers
	Psittacosis	birds	pet shop owners, poultry handlers, vets, taxidermists, zookeepers
Bacteria	Tetanus	various	those with risk of skin-breaking injury by contaminated objects
	Anthrax	livestock	goat hair, wool and hide handlers
	Brucellosis	mainly livestock	packing house employees, livestock handlers
	Leptospirosis	rodents, livestock, dogs	farm workers, livestock handlers, troops, sewer workers
	Plague	rodents, lagomorphs	Western U.S. – shepherds, ranchers, farmers, field scientists
	Tuberculosis	man	health-care workers
	Mycobacteriosis	soil, milk, water	Gulf Coast fishermen, pet store owners, health-care workers
	Erysipeloid	surfaces	animal, meat, poultry, fish handlers
	Tularemia	ticks, wild and some domestic animals	deer, rabbit, animal handlers, forestry workers
	Lyme Disease	ticks, wild animals	workers in certain localized wooded areas
Fungi	Candidiasis	man	dishwashers, cooks, bartenders, bakers, meat handlers
	Aspergillosis	composting vegetation	farmers, grain handlers, hemp workers
	Coccidiomycosis	soil	farmers, excavators, troops, textile workers
	Histoplasmosis	bird-dropping enriched soil	farmers, spelunkers, pigeon handlers
	Mycetoma	soil, composting vegetation	farmers
	Sporotrichosos	sphagnum moss	gardeners, farmers
	Chromomycosis	soil, composting vegetation	farmers – Southern U.S.
	Dermatophytoses	man, sometimes animals	pet hide and animal handlers, farmers, athletes, life-guards

[1] Adapted from Barber and Husting (1977) and Chin (2000).

few diseases that are of fairly recent recognition, such as acquired immune deficiency syndrome (AIDS) and Lyme disease. The classical biological agents are all infectious. (For a good review of these agents and the exposed groups, see Barber and Husting, 1977 and for a more thorough listing see Douwes et al., 2003).

Until recent years, very few biological agents were recognized as allergens. In a number of occupations it was known that biological agents could be the cause of irritating lung and sometimes skin diseases. Those affected occupations included farmers, textile workers, hemp workers, and others. However, it was not until relatively recently that the diseases caused among these professions were tied together by common disease agents. Possibly some of the diseases are caused by endotoxins derived from bacteria or by chemicals derived from fungi found to occur naturally on once-living organic matter such as cotton, hay, and hemp (Jacobs, 1989).

Recent Importance of Biological Agents

Recently, biological agents have taken on new importance. In the mid-1970s, a number of energy-related events took place that caused the tightening of buildings. The tightening process included not only office buildings, but also homes. The basic idea was to remove any cracks or openings from the building and tighten the seals around windows and doors so that leakage of conditioned air would be minimized. Unfortunately, this process also had the unwanted effect of trapping pollutants inside the building that previously were of no concern because they were kept at low concentrations by the natural ventilation occurring in loosely constructed buildings. The loose construction also included operable windows, which in most multi-story buildings are now a thing of the past.

The tightening of buildings has caused an increase in concentration of chemical agents as well as biological agents, leading to a set of conditions known as tight building syndrome. (In the last 20 years volumes have been published on this subject; see, for example, ASHRAE-IAQ symposia such as 1987). When the concentrations of many biological agents reach a critical level, they become important because they are capable of causing a hypersensitive immune system response (i.e., an allergic response) in some individuals. For the most part, the agents causing these symptoms are thought to be fungi and bacteria (Weissman and Schuyler, 1991).

However, one other source of allergy-type reactions is also of major importance and belongs to neither of these two groups. This is the house dust mite (Weissman and Schuyler, 1991). Human beings are constantly sloughing off skin cells. It is estimated that a human being loses millions of skin cells per day. These skin cells settle on various surfaces, including the insulated interior of ventilation ducting. There, skin cells become the favorite food of the house dust mite (*Dermatophagoides* spp.). The mite thrives on the conditions present in the ventilation ducting and also on carpeting and upholstery. As long as the mite stays within the ductwork it is of no problem. However, for various reasons, the mite can become airborne. It is the exoskeleton of the mite that causes the allergic reaction. Hence, dead mites are as capable of producing disease as live mites.

For the bacteria and the fungal agents that can cause indoor air pollution, there usually must be a source of food present to cause the growth of these agents (Spengler, 1991). This may mean an organic material that has been wetted and may remain damp, and can include ceiling tiles, carpeting, and upholstery. A common growth medium is the condensation pan located at the bottom of most air conditioning systems. These pans collect the condensate from the evaporator coils and normally drain freely. However, if they do not drain freely, a growth of microorganisms develops. Then once the pans begin to dry, the microbiological growth can sporulate, sending thousands of spores into the ventilation system. Humidifiers are also common sources for growth.

In summary, the biological agents are now considered to be a much more widespread and important group of disease agents than they seemed in the past. Although the specific agents of infectious disease that occur in certain workplaces are still of concern and must be controlled, it is now more commonplace to find concern for biological agents that cause nonspecific immunological disorders in a variety of unrelated workplaces. These workplaces can be as different as office buildings and foundries.

PHYSICAL AGENTS

Although certain of the infectious biological agents are associated with specific industries or worker groups, that is not the case for most of the physical agents. Because of the widespread use of electromagnetic energy and the propensity of most types of industrial equipment to produce noise, vibration, and heat, the physical agents are pervasive throughout industry. Of the physical agents, perhaps only excess constant atmospheric pressure is often associated with specific industries. Most frequently, hyper-barism, or high atmospheric pressure, is associated with two specific work groups, divers and workers in caissons, or tunnel workers.

The physical agents may be classified as in Table 4-2. Of these agents, noise is by far the most important. It is probably the most common single disease agent. OSHA (BNA, 1983) estimated that 5.1 million U.S. workers were being exposed to noise in excess of 85 decibels, a level considered to cause excess risk for hearing loss. Later work by NIOSH (as cited in NIOSH, 1998) indicated that in a wide range of surveyed industries, approximately 17% of workers experienced exposure to noise at levels less than 85 dBA at least once per week for 90% of the year. NIOSH estimates that more than 30 million U.S. workers are exposed to hazardous noise. It is estimated that the risk for hearing loss for lifetime exposure above 85 dBA is at least 15% (NIOSH, 1998).

Heat stress is probably equally pervasive to noise in summer, especially in the Southern states and subtropical countries. Tolerance for heat stress can be acquired through acclimatization, which constitutes biochemical and physiological changes that allow the body to maintain an acceptable core temperature under a heat load. Radiative (infrared electromagnetic energy),

Table 4-2
The Physical Agents

Ionizing Radiation	*Molecular-Kinetic Energy*
x-rays	convective heat
gamma rays	conductive heat
alpha particles[a]	*Pressure, intermittent waves*
beta particles[a]	noise
neutrons[a]	vibration
Non-Ionizing Radiation	*Pressure, constant or near constant*
ultraviolet light	hyper- and hypobarism
visible light	
infrared light & heat	
radio frequency and microwaves	
magnetic fields	

[a] Although not true *radiation*, these particles are classified as radiation.

convective (heat transfer from a liquid or gaseous surround, e.g., air), and conductive (heat transfer from a solid medium) heat contribute to a heat load. The latter two are caused by kinetic energy in the molecules of the medium being transferred to the molecules of the body.

The non-ionizing radiations are differing wavelengths of electromagnetic (EM) energy that can cause a variety of health effects, ranging from the sunburn and skin cancer effects of ultraviolet light to the increase in kinetic energy imparted to molecules by microwaves. Microwaves affect the body much as they do a piece of meat that is being cooked in a microwave oven.

With the exception of ultraviolet light, most non-ionizing radiation is not present in densities great enough to cause harm unless they are concentrated by humans. Lasers, microwave relay towers, radar, and radio and television transmission towers are examples of concentrated sources of EM energy. Lasers direct a beam of EM energy that remains concentrated along its course. The other sources radiate outward in all (or many) directions from a central point; so the EM field becomes weaker as one moves away from the source.

Magnetic fields are receiving increased interest as a result of studies indicating that they may cause an increased risk for cancer, particularly among children (NAS Committee, 1997). Magnetic fields are created with electrical current. So, power lines and even the wiring in the home are a source of magnetic fields. There are many types of magnetic fields – constant versus pulsed, for example – and this issue has caused the issue of magnetic fields and cancer to be quite complex.

Ionizing radiation has the ability to create ions by stripping electrons from molecules. Of course, the ions are quite reactive, and their reaction causes tissue damage. Ionization of DNA can lead to mutation and cancer.

The various forms of electromagnetic energy are related by their physics. However, they can cause considerably different disease manifestations. In the case of X-rays or gamma rays, cancer and/or other seriously disabling or fatal diseases can occur at doses that may be encountered in the workplace. However, in the case of infrared radiation, doses commonly encountered in the workplace are very unlikely to be the cause of fatalities.

THE ESSENCE OF ALL DISEASE AGENTS

A knowledge of physics is essential to an understanding of the physical agents, and perhaps a good way to think of all agents is according to the force that is instrumental in their causing disease. This approach also is instructive in an understanding of the nature of each agent.

The four categories of forces are gravitational, electromagnetic, and strong and weak nuclear forces (Hawking, 1988). Interestingly, physicists (and most of us) understand that these forces exist, but do not firmly grasp the nature of each of them. Take gravity, for instance. This force holds our feet to the ground, but how does it exert itself, and how is it related to the other three forces? The strong nuclear force holds protons together against an electromagnetic force that repels particles of like charge. The weak nuclear force causes decay particles (such as protons, electrons, and neutrons) to be ejected from an atom and become ionizing particles.

The electromagnetic force is the most important to hygienists. It is responsible for gamma rays, ultraviolet light, infrared light, and all other energy forms of the electromagnetic spectrum (see Figure 4-1). The electromagnetic force also causes the attraction or the repulsion of molecules and imparts kinetic energy to them. Hence, it is responsible for conductive and convective heat that conveys energy from molecule to molecule.

The electromagnetic force also is partially responsible for such hazards as noise and vibration. Both are the result of pressure waves through a medium. In the case of noise, the medium is air. The amount and speed of noise or sound produced is dependent on the elasticity and mass of the medium (Peterson and Gross, 1978). The elasticity of the medium is partly a function of the attraction of molecules (EM force) in the medium. The inertia of the medium is also a function of gravity. The stronger the force of gravity is, the denser the medium and the greater the impact of the pressure waves on the body. Of course, on earth this force is relatively constant.

Alpha and beta particles are caused by the weak nuclear force. Like gamma rays, they are called ionizing radiation for their ability to strip electrons from molecules, creating an ion. However, we might think only of gamma rays as *radiation*, as that term is often associated with the EM spectrum. Certainly alpha and beta particles are not part of the EM spectrum, but they each have the ability to cause EM energy to be formed when they interact with matter. That EM energy is a photon, and if it has high enough energy, it too can ionize a molecule.

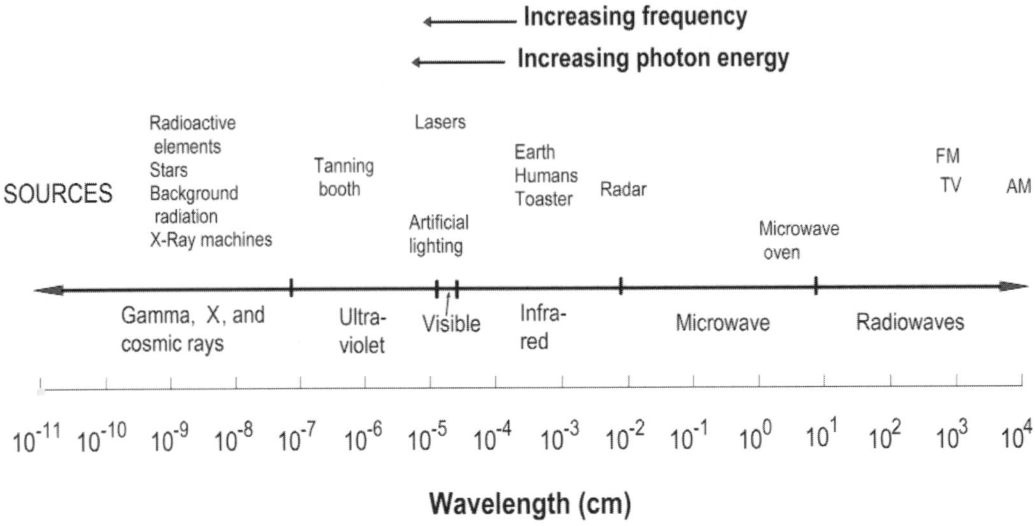

Figure 4-1. The electromagnetic spectrum.

The *action* exerted by the chemical agents is actually derived from the EM force; and because biological agents are *organized* chemical agents, they too are derived from the EM force.

CHEMICAL AGENTS

Although there seem to be no readily available statistics to support this statement, it must be recognized that the majority of hygienists spend most of their time dealing with chemical agents. Some hygienists specialize in biological agents or specific physical agents, such as ionizing radiation or noise, but many hygienists are generalists who work for specific industries, or private or government consultants that generalize in many industries.

This preoccupation with chemical agents is not without just cause. First, chemical agents are prevalent throughout most workplaces. Even in a setting such as a hospital where it might be expected that the hygienist would be preoccupied with biological agents, there are many chemicals that require the majority of the hygienist's time. Second, there are a large number of chemical agents, certainly many more than the physical agents and considerably more than the biological agents. In fact, there are more than 75,000 chemicals in U.S. workplaces (EPA, 2007).

Additionally, it is estimated that about 700 new chemicals are manufactured and placed in commerce each year. This is not surprising, as the Chemical Abstract Service (CAS, 2007) listed 29 million chemicals that had been synthesized by the end of 2006. This number appears to be increasing at a rate of about 1 million per year since 1991. Research chemists in industry have a vast stockpile of chemicals upon which to draw for the chemicals of the future.

Of course, with more than 75,000 chemicals in commerce, the hygienist must have some mechanism for placing the many chemicals into categories based on similar chemical and/or toxicological properties. Organic chemistry provides one type of hierarchical classification. It is also instructive to sort the chemical agents according to how they are used. Each of these approaches has utility, so that the hygiene student must be familiar with organic chemistry nomenclature and with common industrial processes and operations. Knowledge of industrial processes will allow the hygienist to draw conclusions about the potential hazard of a chemical through knowledge of how it is used. Similarly, if the hygienist can classify the chemical into an organic chemical grouping, then a preliminary assessment can be made of its chemical and toxicological properties.

Although many books have been written on the various processes that occur in industry, and the next chapter is devoted to this subject, it is very difficult for the hygienist to read about a particular process and understand exactly how it works without having seen it. The practicing hygienist gains from experience a knowledge of industrial processes that cannot be appreciated through reading. On the other hand, the ability of a hygienist to rapidly place an organic chemical into a particular nomenclature scheme, and from that to derive a preliminary estimation of the chemical's properties, is something that can be learned through study. Hence, it is of the utmost importance that the hygiene student be able to rapidly construct, or at least estimate, the structure of a substance, place it in a chemical grouping, and then, based on knowledge of group properties, estimate physical, chemical, and toxicological properties for the chemical.

Classification by Chemistry

Classification of a chemical agent by its chemistry can be performed in a dichotomous fashion. The first step in the process is to determine whether the chemical is *organic* or *inorganic*. Inorganic and organic chemicals have important chemical differences that determine properties such as the ability of the chemical to be absorbed into the body, dissolve and/or disassociate in water, and the physical form in which it is likely to become airborne. Basic to this understanding is that covalent bonds are typically associated with organic chemicals, and ionic bonds are more typically associated with inorganic chemicals. Although this does not always hold true, it is consistent enough that it can account for many of the differences in chemical and physical properties of the two groups.

THE INORGANIC CHEMICALS

Examination of the periodic table of the elements (see Figure 4-2) for a tally of the elements that might be considered to be metals will show the percentage of metals is quite high. In fact, it will show that the vast majority of the elements are classified as metals. These metals can form many compounds and are probably responsible for a large number of the 75,000+ chemicals in commerce. For example, Table 4-3 shows the number of compounds of several metals including mercury (Hg) and lead (Pb).

Commonly, a compound such as lead sulfate owes its toxicity to the cation or the metallic element. Although some of the toxicity is owed to the sulfate ion (for example the ability of the lead to disassociate from the anion and be adsorbed), so much of the toxicity is owed to the lead that it is very common for inorganic lead compounds to be grouped together and given one *toxicity rating.* The same is true for many of the other metal compounds. Hence, in the case of the inorganic metal compounds, it is the cation, or the metal element, that is usually of most importance.

Figure 4-2, showing the periodic table, has 11 shaded elements on the right-hand side of the table. To the left of these, all of the elements, with the exception of hydrogen, are considered to be metals. A metal is a substance that forms positive ions when in solution and hydroxides rather than acids in solution with water. Most have high electrical and heat conductivity (Lewis, 1997). While these properties have meaning, particularly from a toxicological perspective, the definition begs the question, what else is in the periodic table? Since 75% of all the elements are metals it may be instructive to classify the remaining elements of the periodic table.

At the far right-hand side of the periodic table is the column of noble gases. In general, these elements have little importance in hygiene. They are also termed inert gases, and while this is not strictly true (xenon, radon, and krypton are slightly reactive, Lewis, 1997), from a toxicological perspective, these gases can be considered of extremely low toxicity (the exception being the radioactivity of radon). Essentially the only mode of toxicity is to displace oxygen from the lungs. They are not metals and none of the metals is gaseous.

The second column from the right has very important consequences in hygiene. This is the group of halogens. These elements are extremely reactive, and include chlorine, an important element in hygiene and industrial processes. Halogens are the most electronegative elements in the periodic table and have a high attraction for metals, particularly those on the far left-hand side of the table. The halogens form acids when they react with hydrogen or water. Hydrogen chloride, hydrogen bromide, and hydrogen fluoride are all acidic, gaseous compounds. These compounds are termed acidic gases, or acid gases, and because they are in the gaseous state (and can be in solution with water), have the ability to easily enter the respiratory tract and cause severe damage. In addition to forming compounds with the metals, the halogens are also important in the organic chemical industry. In particular, fluorine, chlorine, and bromine are used to form compounds of great industrial importance that will be discussed later (see Table 4-3).

Of the remaining non-metals in the periodic table, the others can be placed into two groups. First are the semiconductors. Once called the metalloids, they are the shaded elements to the right. They include boron, sulfur, silicon, arsenic, tellurium, phosphorous, carbon, selenium, germanium, antimony, and polonium. The semiconductors are an interesting group of

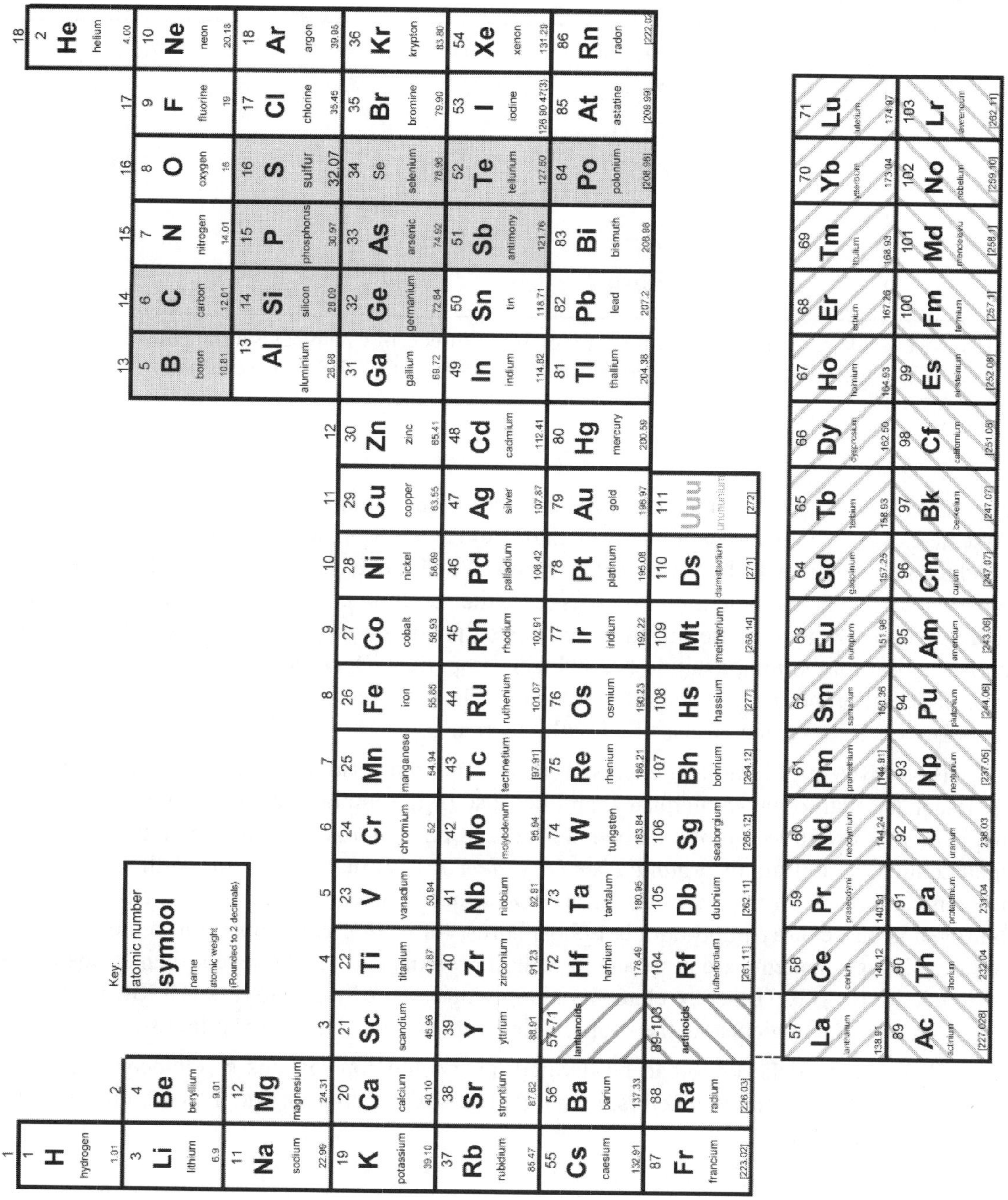

Figure 4-2. The periodic table of the elements.

Table 4-3
Number of compounds in RTECS containing certain elements

C	144,704	F	7,940
O	125,711	Na	3,371
N	123,711	K	463
C1	46,419	Hg	384
S	31,727	Cr	124
Br	7,573	Pb	115
P	4,959	Mn	73

Based on 158,902 compounds in the Registry of Toxic Effects of Chemical Substances (Inquiries Service, Canadian Centre for Occupational Health and Safety, CCOHS, www.ccohs.ca).

elements in that they sometimes take on the properties of metals and at other times they do not. Arsenic can exist both as a cation and an anion, and is perhaps the most important of the semiconductors for hygienists because of its toxicity. When it exists as an anion in the presence of a particularly toxic cationic metal such as lead, the resulting compound, lead arsenate, is particularly toxic. Note at the top of these elements that aluminum is not a semiconductor, but a metal; and at the bottom, astatine has halogen properties. Because the semiconductors often take on metal properties, particularly toxicological properties, some of them, such as arsenic, are sometimes conveniently classified with the metals.

This leaves three nonmetal elements (H, O, N) in the periodic table, all of them commonly associated with organic chemistry; and because of the importance of organic chemistry to hygiene, it is useful to remove carbon, sulfur, and phosphorus from the semiconductors and include them in a group of *organic chemistry elements*. Virtually all of the millions of organic compounds are formed with C, H, and O primarily, and to a lesser extent S, P, and N. Nitrogen is particularly important, but phosphorus, for example, is important in the pesticides industry. These elements are represented well in proteins and biochemicals such as adenosine triphosphate or ATP. These six elements along with three of the halogens (Cl, Br, F) may be considered by the hygienist as the primary elements of the petrochemical industry.

Thus, the periodic table can be divided neatly into four groups of elements: the large group of metals and semiconductors, the noble gases, the halogens, and the organic chemistry elements (C, O, H, P, S, N).

Metals

There are many metal elements; they form thousands of compounds; they have been industrially important since the Bronze Age; they are now ubiquitous in industry and nature; many are essential to life; and most, including some of those essential to life, are quite toxic in doses caused by exposures that are common in industry. For these reasons, the metals are extremely important.

Table 4-4 shows the most common metals in the earth's crust. Not surprisingly, the most common have been incorporated, via evolution, into biochemical processes. Table 4-4 also shows those elements that are essential to human life. For example, chromium when taken in insufficient doses will cause impaired glucose metabolism (Nestle, 1983); however, too much ingested chromium may cause kidney failure (Goyer and Clarkson, 2001). It may be important to point out this ambiguity to workers when training them on the effects of metals, but certainly one should not to try to convince them that "a little exposure is good for them."

Many of the processes in which metals are commonly found are familiar ones such as mining. Others include refining, smelting, and roasting of ores; welding and milling; and forging, casting, and alloying. However, metals also have many specialty uses such as in pigments, catalysts, and organic chemicals such as pesticides and flame retardants.

It is not sufficient to say that an element such as chromium is toxic to humans. Several important factors influence toxicity, and the hygienist must be aware of the major considerations. These are as follows:

- Solubility in water and lipid
- Route of exposure
- Route of metabolism
- Valence

Solubility. A compound poorly soluble in water and lipid, particularly under gastrointestinal (GI) conditions, will not be absorbed, and will present little or no toxicity via the oral route. For example, $BaSO_4$ (barium sulfate) is used in large quantities to serve as an energy absorber in X-ray images of the GI tract. Being insoluble (or at least of very low solubility), it passes through the GI tract with no systemic toxicity. If the radiologist chose BaCl (barium chloride) by mistake, death would no doubt occur. The latter is water soluble.

Organic metal compounds are soluble in the lung and intestine (and often across the skin); so organic lead and mercury compounds (e.g., methyl mercury) tend to be more toxic than, or as toxic as, inorganic compounds (e.g., mercuric chloride). These organic metal compounds are said to be lipid-soluble.

Exposure Route. The previous discussion alluded to the importance of exposure route in metal toxicity. Insoluble oxides of nickel, zinc, cadmium, and manganese may cause lesions such as cancer in the case of Ni and Cd (Goyer and Clarkson, 2001) in the respiratory tract. Although they are *insoluble*, oxides of Cd, Zn, Mn, and other metals are better known for their systemic toxicity than local effects. Thus, even though insoluble in the GI tract, some metallic

Table 4-4
Most common metals in Earth's upper crust as characterized by high calcium-granitic rock[1]

Metals	Content[2]	Essentiality[3]
Silicon	31.4	
Aluminum	8.2	
Iron	3.0	YES
Sodium	2.8	YES
Calcium	2.5	YES
Potassium	2.5	YES
Magnesium	0.9	YES
Titanium	0.3	
Manganese	540.0	YES
Strontium	440.0	
Barium	420.0	
Zirconium	140.0	
Rubidium	110.0	
Vanadium	88.0	
Zinc	60.0	YES
Yttrium	44.0	
Copper	30.0	YES
Lithium	24.0	
Chromium	22.0	YES
Gallium	17.0	
Nickel	15.0	
Lead	15.0	
Scandium	14.0	
Boron	9.0	
Thorium	9.0	
Cobalt	7.0	YES
Uranium	2.0	
Cesium	2.0	
Beryllium	2.0	
Molybdenum	< 1	YES
Selenium	< 1	YES
42 remaining metals and semi-conductors < 1%		

1 Values are percent or parts per 100 above double line and parts per million below double line.

2 Content data derived from Ragland (1978) and essentiality data from Nestle (1983).

3 Essential to human life.

oxides are adsorbed across the lung (Goyer and Clarkson, 2001).

Route of Metabolism. Some metals are essential, and nonessential metals may mimic their role in metabolism and disrupt the process. Lead may take the place of calcium in its role in the central nervous system or of zinc in heme metabolism. The toxic metal usually has the same valence as the essential metal. It should be noted that many of the *transition elements* (columns 3-12 in Figure 4-2) have multiple valence states and so are adept at this role. These metals are often termed the heavy metals, a name often implying toxicity. However, by definition, heavy metals include all but sodium, lithium, and beryllium (Lewis, 1997), and the latter is one of the most toxic metals.

Valence. The importance of valence in mimicking metabolism of essential elements has been mentioned. However, for at least one other metal, valence is important from another perspective. Chromium has three valences (2, 3, 6). As alluded to earlier, oxides of chromium cause lung cancer. However, only Cr^{+6} compounds appear to do this. Thus CrO_3 is carcinogenic, but Cr_20_3 is not.

Some Important Metals

Metals can be important for several reasons. Metals such as iron and aluminum are a part of virtually everything we touch, and so are ubiquitous in industry (see Table 4-5); but relative to other metals they are of low toxicity. From a hygiene point of view, those metals with relatively high use and high toxicity are of constant concern. Table 4-5 shows the most highly used metals and, of these, seven may be classed as extremely toxic. The remainder are of lesser toxicity, about 10 to 1,000 times less toxic than the seven. Several of the seven are human carcinogens. For an excellent discussion of the toxicity of these and other metals see Goyer and Clarkson (2001). The extremely toxic metals are discussed below in order of consumption.

Lead. Other chapters allude to ancient uses and health effects of lead. Even though it is probably the most widely and longest-studied metal, levels of presumed safe exposure continue to be lowered, as evidenced by EPA's standard for lead in drinking water (EPA, 1988). It is commonly found in industry in paints, batteries, solder, steel, and often alloys. It is an excellent preservative in paints, but spray painting causes unacceptable exposures to it. Although many users of paints require no-lead substitutes, leaded paints have the habit of mysteriously appearing. The removal of old paint using abrasive blasting or heat (blowtorch) is particularly hazardous. At unacceptable levels of exposure, it causes anemia, renal impairment, and severe central nervous system effects. However, at lower levels, more subtle behavior effects can occur, such as learning disabilities, particularly in children.

Chromium. The +3 valence is essential in animals, but the +6 valence is a lung carcinogen in humans. It is widely used in chrome plating, stainless steel, alloys such as nickel-chromium, printing, tanning, dyes and pigments, photography, and paints. It is also known to cause dermatitis, nasal inflammation, and nasal septa perforation.

Nickel. Like Cr+6, nickel is a lung carcinogen. Workers in sintering or ore roasting oper-

Table 4-5
U.S. metals production and consumption[1] (metric tons)

Metal	Year	Primary production	Secondary production	Apparent[2] consumption	World production
Beryllium	2000		180	300	226
Platinum group[3]	2000	13		238	365
Selenium	2002			742	1,480
Mercury	2002				1,800
Gold	2002	298	38	264	2,550
Cadmium	2000	1,890	425	2,010	19,700
Silver	2002	2,580		5,340	20,000
Arsenic	2000			24,400	27,900
Cobalt	2002	0	2,800	17,100	40,300
Vanadium	2000	0	0	3,520	43,000
Tungsten	2002	0	4,380	11,900	59,100
Antimony	2000	20,900	7,920	49,400	118,000
Molybdenum	2000	41,100	33,800	29,100	129,000
Tin	2002	0	3,590	57,500	249,000
Nickel	2002	0	99,800	221,000	1,340,000
Lead	2002	262,000	1,070,000	1,510,000	2,910,000
Chromium	2001			325,000	3,970,000
Titanium O_2	2002	1,410,000	1,530,000	1,120,000	
Zinc	2002	182,000	113,000	1,170,000	8,360,000
Magnesium	2002	409,000	26,400	976,000	11,600,000
Copper	2001	1,630,000		2,500,000	13,700,000
Aluminum	2002	3,668,000	3,450,000	7,530,000	24,000,000
Iron/Steel	2002	40,200,000		42,000,000	604,000,000

[1] From Kelly et al. (2003). Blank cells indicate data were not available or were withheld because they are proprietary.

[2] AC = primary production + secondary recovery (old scrap) + imports – exports ± adjustments for inventory (stock) changes.

[3] Pt, Pd, Ir, Os, Rh, Ru

ations are particularly susceptible to the carcinogen. It is an essential element for humans. Most nickel is used in steel and other alloys. Other uses include plating, coins, catalysts, vessels, nickel-cadmium batteries, circuits, and glass. Like chromium, it also causes dermatitis (nickel itch).

Arsenic. This ancient poison is not a true metal, but a semiconductor. It is largely derived as a by-product of ores containing metals such as copper and lead, and is used in lead alloys, antifreeze, tanning, semiconductors and, most importantly, pesticides. Any process that liberates H+ in the presence of arsenic forms arsine, AsH_3, a particularly hazardous, gaseous form of arsenic. Like Cr and Ni, As is a carcinogen, particularly in combination with lead. Arsenic causes a variety of skin lesions and accumulates in skin, hair, nails, and it is a skin carcinogen. It has numerous systemic effects, including reproductive effects.

Cadmium. Like the previously discussed metals, it is likely a carcinogen and has severe effects on the kidneys. Cd is a common contaminant in zinc and lead ores. It is widely used in electroplating. Fumes formed from the welding or heating of galvanized iron (that contains cadmium) can be fatal. In conjunction with kidney disease it may cause hypertension on prolonged exposure. Cd affects calcium metabolism, leading to bone pain and/or osteoporosis. A severe environmental contamination episode in Japan lead to *itai-itai* or ouch-ouch disease, describing the symptoms of this bone toxicity. It is also used in alloys, solders (beware of *silver* solder containing Ag/Cd/Zn alloy), pigments, and nickel-cadmium batteries.

Beryllium. Like cadmium, Be is a modern metal; its extensive use is recent; and its ability to form lightweight alloys has made it a space age metal. Of the metals discussed, it has the lowest acceptable level of exposure (0.002 mg/m^3, OSHA, 2007), although OSHA has a Cd standard of 0.005 mg/m^3 (OSHA, 2004). Its greatest effect is on the lungs, where it forms granular lesions leading to berylliosis. It also causes contact dermatitis, is a suspected carcinogen, and may cause numerous systemic effects, particularly on the liver. It is used in the nuclear and space industries, ceramics and rocket fuel, and for non-spark tools. Next to lithium it is the lightest metal.

Mercury. A wide range of forms including organic forms translate to a wide array of toxic effects. Unlike the other six metals discussed here, Hg gives no evidence of carcinogenicity. Numerous central nervous system effects, including the *mad hatter's syndrome*, are attributed to mercury. It is also toxic to the kidneys. Only mercury, cesium, francium, and gallium are liquids among the metals. Consequently, mercury vapor is particularly hazardous, as unacceptable concentrations are easily evolved from small amounts of the liquid, such as from broken barometers, thermometers, and hospital and electrical equipment. It is used in electrical apparati, in chlor-alkali cells to produce Cl_2 from seawater, in NaOH plants, as a catalyst for polyurethane foams, and in dental preparations. Its use in chlor-alkali cells in Japan led to contamination of water and fish with biotransformed methyl mercury, which led to *Minimata disease*, named for the contaminated bay. Its use in drugs, pesticides, and paints has decreased, but the hygienist should be wary.

The Inorganic Gases

A number of inorganic gases have widespread use in industry. Many of these gases are building blocks in the construction of more complex chemicals. In addition, some of them are by-products of industrial processes that, while not of great importance in the manufacture of a product, can be of great importance in terms of the worker's health. The more important ones are described in Table 4-6.

ORGANIC CHEMICALS

Defining organic chemistry is a bit like defining the class of metals. Although the compounds of organic chemistry have certain similarities, there are gray areas or compounds that seem to fit neither organic nor inorganic chemistry. Lewis (1997) defines organic chemistry as "[embracing] all compounds of carbon except such binary compounds as the carbon oxides [CO, CO_2], the carbides, carbon disulfide, etc.; such ternary compounds as metallic cyanides, metallic carbonyls, phosgene ($COCl_2$), carbonyl sulfide (COS), etc.; and the metallic carbonates." Such a definition results from the gradual evolution of organic chemistry from its roots as the study of chemicals derived from living organisms.

Carbon tetrachloride is not derived from living organisms, but can be produced by heating methane in the presence of chlorine. However, it can also be produced by heating CS_2 and Cl_2 in the presence of Fe. Phosgene is prepared industrially from CO_2 and Cl_2, but can be produced by heating organochlorine compounds such as trichloroethylene ($Cl_2 - C = C - ClH$) in the presence of O_2. Trichloroethylene and CCl_4 are considered organic. CS_2, CO_2, $COCl_2$ are not. The definition, widely used, must be memorized.

Of the millions of compounds that have been synthesized, the majority are organic compounds. Methane, CH_4, is the simplest of the organic compounds, and while the many millions include some incredibly complex compounds that must be modeled on a computer in order to examine their structures, the majority, which have important uses in industry, are relatively simple. Therefore, it is necessary for the hygienist to memorize only a few simple rules for organic chemistry nomenclature.

The organic chemical industry throughout the world is based on petroleum and, to some extent, coal. Petroleum is composed of hydrocarbons. In addition, some salts, sulfur compounds, and other *impurities* are found in petroleum, and are partially removed during refining. What is left is a complex mixture of hundreds of hydrocarbon compounds containing only trace amounts of CH compounds that also contain N or O (Lewis, 1997).

These hydrocarbons can be broken down into two groups: the aliphatic and the aromatic or ring compounds. The aliphatic compounds have no aromatic ring structure, that is, no six-carbon ring structure with *shared* electrons. The basic building block of all aliphatic compounds is methane. The basic building block of all aromatic compounds is benzene. It is important not to overemphasize the dichotomy. Although it is useful, the differentiation

Table 4-6
Important inorganic gases

Name and structure	Major uses
Respiratory irritants	
Ammonia (NH_3)	Fertilizers, synthesis, refrigerant
Chlorine (Cl_2)	Organochlorine synthesis, H_2O treatment, food processing
Fluorine (F_2)	Fluoride synthesis, fluorocarbon synthesis, water treatment
Hydrogen chloride[1] (HCL)	Synthesis, cleaning, ore reduction; food processing
Hydrogen fluoride[1] (HF)	Synthesis, Al production, pickling stainless steel, etching glass
Nitric oxide (NO)	Nitric acid synthesis, bleaching rayon, by-product of some processes such as heating $NaNO_3$
Nitrogen dioxide[1] (NO_2)	Nitric acid, fertilizers, explosives, synthesis, metallurgy, reprocessing spent nuclear fuel
Ozone (O_3)	Water/waste treatment, bleaching, bacteriacide, synthesis
Phosgene ($COCl_2$)	Synthesis, by-product of heating or UV interaction with some chlorinated hydrocarbons
Phosphine (PH_3)	Synthesis, semi-conductors
Sulfur dioxide[1] (SO_2)	Synthesis, paper, metals, bleaching, food additive, fertilizers, catalyst, etc.
Blood and respiratory-related toxicants	
Arsine (AsH_3)	Synthesis, electronics
Carbon monoxide (CO)	Synthesis, fuels, metallurgy, by-product of combustion
Hydrogen cyanide[2] (HCN)	Organic synthesis
Hydrogen sulfide (H_2S)	Purification of acids, precipitation, by-product of anaerobic decomposition
Carcinogens	
Nickel carbonyl[3] ($Ni(CO)_4$)	Nickel processing
Radon (Rn)	Medicine, radiography, naturally occurring radioactive gas
Asphyxiants and anesthetics	
Carbon dioxide (CO_2)	Refrigeration, beverages, propellant, synthesis, fire extinguishing, many others, by-products of combustion and respiration
Nitrous oxide (N_2O)	Anesthetic, propellant, leak detection

[1] Acid gases evolve from their sister acids, such as hydrochloric acid.

[2] Boiling point is 26.4°C.

[3] Boiling point is 43°C.

between, for example, ketones and acids may be more important than whether they are aliphatic or aromatic.

From these simple compounds, all other hydrocarbons can be made. The structures of the aromatic compounds in petroleum can contain up to four or five joined rings. However, the more complex organic chemicals produced in the chemical industry are generally not made from complex petroleum hydrocarbons. Instead, in the process of producing organic chemicals from petroleum, generally the complex compounds within the petroleum are first broken down into more simple compounds, and then the process of building the more complex compounds begins. The aromatic ring compounds are typically *cracked* into smaller compounds, the most common of which is ethylene. The basic building block ethylene may be used as an example of how compounds are made in the organic chemical industry (Table 4-7). Ethylene is the most widely manufactured or produced organic chemical in the U.S. and the world (CEN, 2004).

From this brief presentation it can be seen that the many organic chemicals are produced in a systematic fashion starting with simple building blocks and adding chemical groups according to the rules of organic chemistry. In a similar manner the naming of organic compounds takes place. By understanding IUPAC (International Union of Pure and Applied Chemists) rules for naming compounds, one can break down the individual parts of a very long chemical name and ascertain the structure, or conversely a compound may be named by dissecting the structure into basic building blocks.

Organic Chemistry Nomenclature

Figure 4-3 shows the basic organic classes that can be made by the addition of simple groups or radicals. It is important to note that these chemical classes may span both aliphatic and aromatic compounds. For example, a two-carbon structure containing a hydroxyl group is called ethanol, and by virtue of the hydroxyl group is an alcohol. It follows then that all aliphatic compounds having a single substitution of an OH – group are alcohols and have properties related to the alcohols. However, a simple alcohol created by the addition of an OH – group to the benzene aromatic ring has properties that are considerably different from those of the aliphatic alcohols. This aromatic alcohol is known as phenol. This example shows that in some situations the functional group must be considered along with the remaining structure of the compound to ascertain properties. There are exceptions to every rule.

Unfortunately, many common names for organic chemicals were derived prior to the IUPAC system. These are still widely used in the vernacular. Hence, xylene is dimethylbenzene and, more specifically, there are three isomers of xylene: 1,2-, 1,3-, and 1,4-dimethylbenzene. Rarely does one use the term 1,4-dimethylbenzene; so the industrial hygienist must know the common names. Compounds that are usually referred to by common rather than IUPAC names and are commonplace in industry are shown in Figure 4-4. Their structures are worth knowing.

There are many other compounds that are referred to by common names. A good exam-

Table 4-7
Major ethylene derivatives and uses[1] – Processes in italics (with % of ethylene production used) followed by petrochemical intermediates and products in order of synthesis with (uses)

I. Polymerization (43%)
- A. polyethylene, low density (film and sheet; injection molding; paper, wire, and cable coatings)
- B. polyethylene, high density (blow molding; injection molding; pipe, conduit, tubing; film and sheet)
- C. polyethylene copolymers (film; electrical inshot-melt coatings and adhesives)
- D. ethylene-propylene co- and terpolymers (automotive parts; tires; hose and tubing; thermoplastic modifiers; wire and cable)

II. Oxidation (23%)
- A. ethylene oxide (surfactants)
  - 1. ethylene glycol (antifreeze)
    - a. polyesters (filament; fiber; film)
  - 2. di- and triethylene glycol (dehydration agent; solvents)
    - a. polyols
      - (1) polyurethane and polyesters (filament; fiber; film; flexible and rigid foams; coatings; elastomers; adhesives; sealants)
    - b. morpholine (rubber chemicals; corrosion inhibitors; optical brighteners)
      - (1) glycol ethers (solvents)
  - 3. ethanolamines (liquid detergents; acid gas absorbent; textile chemicals)
- B. acetaldehyde and vinyl acetate
  - 1. poly vinyl acetate (emulsion paints; adhesives; textile chemicals; paper and surface coatings)
  - 2. poly vinyl alcohol (adhesives; textile chemicals; coatings; emulsifiers; safety glass)
  - 3. polyvinyl chloride-acetate copolymers (phonograph records; floor tiles; solution coatings; PVC pipe)
  - 4. acetic anhydride
    - a. cellulose acetate (fibers; cigarette filters; plastics)
    - b. acetyl salicylic acid (aspirin)
  - 5. acetic acid
    - a. acetate esters (solvents)
  - 6. chloroacetic acid (herbicides)
    - a. cellulose ethers (detergent promoters; textile chemicals)
  - 7. pentaerythritol (alkyd resins; synthetic lubricants)
  - 8. peracetic acid (bleaching agent; catalyst)

Table 4-7 (Cont.)
Major ethylene derivatives and uses[1] – Processes in italics (with % of ethylene production used) followed by petrochemical intermediates and products in order of synthesis with (uses)

III. Halogenation–hydro halogenation (16%)

A. ethylene dichloride
   1. trichloroethylene (metal degreasing solvent)
   2. perchloroethylene (dry cleaning and metal cleaning solvent)
   3. methyl chloroform (metal degreasing and cleaning solvent)
   4. vinyl chloride monomer
      a. poly (vinyl chloride) (pipe and tubing; sheet; construction materials; textile and paper coatings; film; insulation and packaging materials)
   5. vinylidene chloride
      a. poly (vinylidene chloride) and copolymers (film; coatings for food packaging)
   6. ethylenediamine (carbamate fungicides; chelating agent)

B. ethyl chloride (refrigerant; solvent; ethylation agent)
   1. tetraethyl lead (gasoline additive)

C. ethylene dibromide (lead scavenging agent)

D. vinyl bromide (flame retardant for fibers and plastics)

E. ethyl bromide (ethylation agent; solvent; refrigerant)

IV. Alkylation (9%)

A. ethylbenzene, styrene
   1. polystyrene-GP high, expanded (packaging; housewares; appliances; TV cabinets; furniture; lighting; construction; toys)
   2. acrylonitrile-butadiene-styrene copolymers (pipe and fittings; appliances; automotive use; business machines; telephones; electrical and electronic application)
   3. styrene – butadiene latexes (textile and paper coatings; emulsion paint; adhesive foam products)
   4. styrene – acrylonitrile copolymers (appliances; automotive use; housewares; furniture; packaging)
   5. styrene – butadiene elastomers (tires; mechanical rubber goods; footwear; adhesives; chewing gum)
   6. unsaturated polyesters (reinforced boats, bathtubs, tanks, panels, and trays; non-reinforced decorative and electrical applications)

B. ethyl toluene, vinyl toluene
   1. unsaturated polyesters (coating resins; electrical insulation)

C. ethyl anilines (dyes; pharmaceuticals; pesticides)

Table 4-7 (Cont.)
Major ethylene derivatives and uses[1] – Processes in italics (with % of ethylene production used) followed by petrochemical intermediates and products in order of synthesis with (uses)

D. 1,4-hexadiene
   1. ethylene-propylene terpolymers (automotive parts; tires: hose and tubing; wire and cable)

E. aluminum alkyls (reaction promoters, initiators, catalysts)

V. Oligomerization (4%)

A. α-olefins (plasticizers; surfactants)

B. linear primary alcohols (plasticizers; surfactants)

VI. Hydration (4%)

A. ethanol (toiletries; cosmetics; industrial solvents and thinners; detergents; flavors; disinfectants)
   1. single-cell protein – yeast (animal feed)
   2. ethyl acetate (solvent)

B. ethyl amines (corrosion inhibitor; rubber chemicals; insecticides)

C. acetaldehyde

D. ethyl ether (solvent; medicinals; pharmaceuticals)

VII. Oxo-reaction (1%)

A. propionaldehyde
   1. propionic acid (grain preservatives; cellulose plastics; herbicides)
   2. n-propyl alcohol (herbicides; solvent)

B. Other (ripening agents for fruit and vegetables; herbicide)

---

[1] Adapted from Kniel et al. (1988).

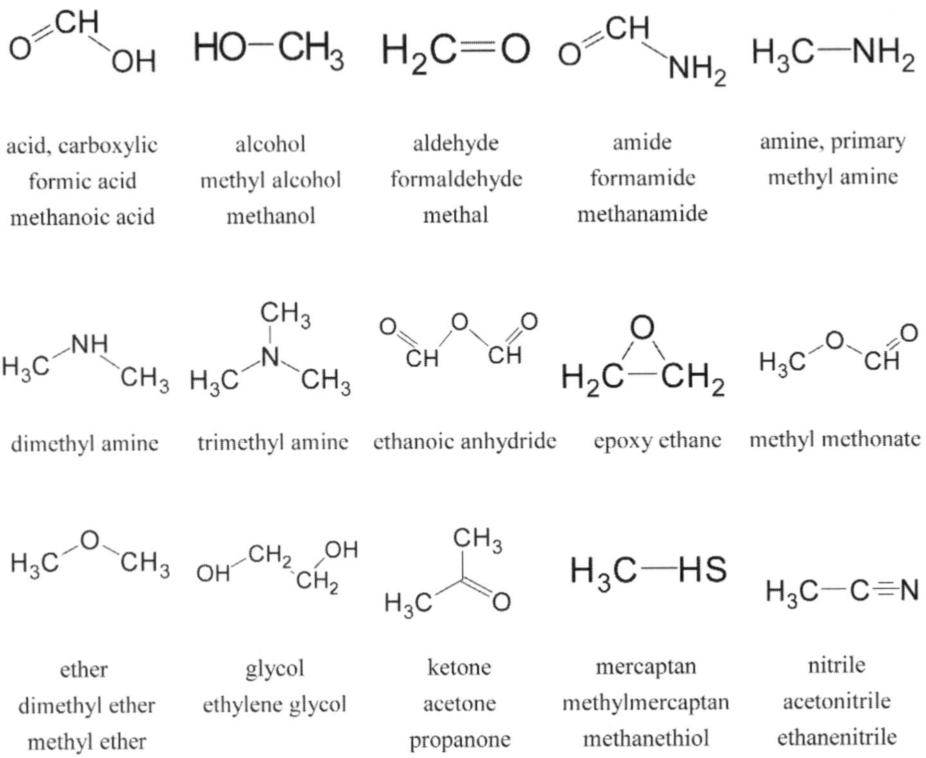

Figure 4-3. Common chemical classes, simplest structures in those classes, and nomenclature.

ple is the large number of acids (mostly aliphatic) that are important in biochemistry (e.g., lactic acid). Most of these are important only in certain segments of industry such as pharmaceuticals. The dicarboxylic acids are also referred to by common names. Two of these are phthalic acid (Figure 4-4) and maleic acid (Figure 4-6). Figure 4-5 shows some cyclic ethers, a group with wide ranging properties. Many are solvents, while the more complex, TCDD and TCDF are very toxic by-products of combustion of chlorinated compounds. Figure 4-7 shows some common radical names that are used to form compound names of compounds that are frequently encountered.

Classification by Usage

While it is important that the hygienist be able to construct a compound from a name, it is also important that the hygienist be able to narrow the possibilities for identifying the chemical composition of a product when the only information given is how it is used. For example, a worker describes to the hygienist symptoms that he had after using a solvent. The hygienist should think of the various types of solvents that may be candidates and then ask for informa-

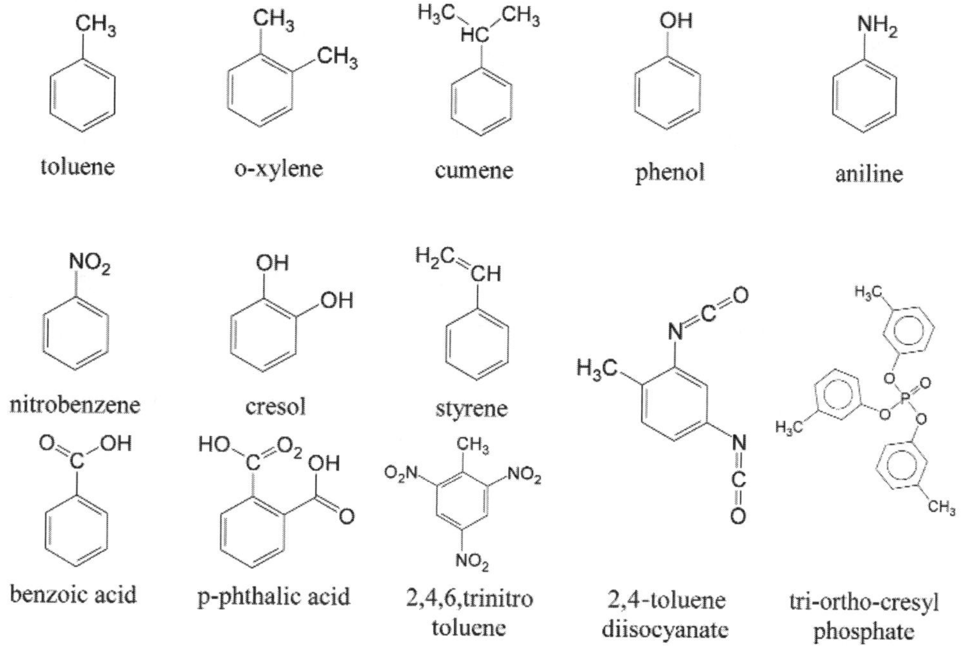

Figure 4-4. Structures and common names of some aromatic compounds.

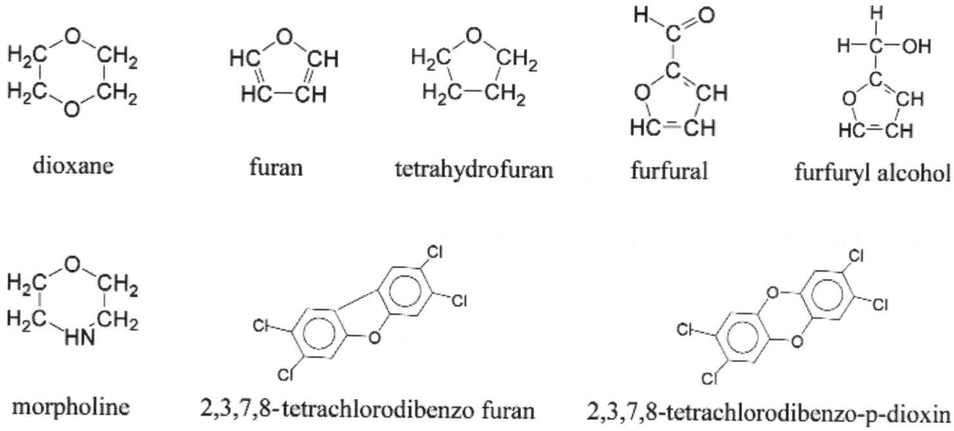

Figure 4-5. Structures and common names of some cyclic ethers.

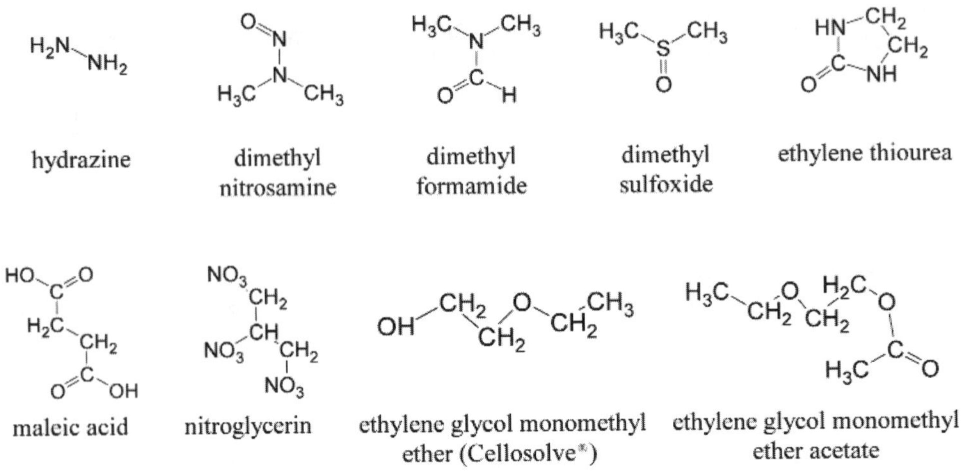

Figure 4-6. Structures and common names of some relatively common industrial chemicals.

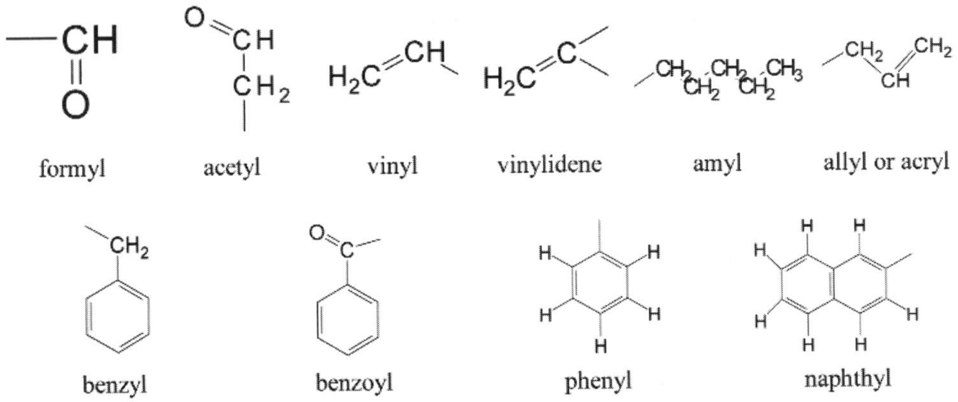

Figure 4-7. Structures and common names of some radicals.

tion about how the solvent was used, its odor, appearance, brand name, etc. This will give clues as to what the solvent may have been. The hygienist may find out that the solvent was being used to strip paint from surfaces. Previous experience and knowledge tell the hygienist that the solvent probably contained methylene chloride or one of the glycol ethers and may have contained an alcohol such as methanol, a ketone such as acetone, and an aromatic hydrocarbon such as toluene.

The hygienist may be confronted with a problem concerning pesticides in which she questions the worker about the use of the pesticide. From this she determines that the pesticide was most likely an insecticide. Based on her knowledge of insecticides, she determines that the compound was most likely an organophosphorus compound, or it may have been a carbamate, but not likely an organochlorine compound. The hygiene student must be familiar with the various use categories of chemicals and the chemicals that comprise those use categories.

SOLVENTS

The solvents are the most important classification of chemicals used in industry and unfortunately also vary the most. It is important to note that the word *solvent* often conveys different meanings to the layperson and the hygienist. For example, a solvent does not need to be a liquid at room temperature. Cyclohexanol, a solid at room temperature, is used as a solvent in several industrial processes. Often industrial processes are run at elevated temperatures. Solvents also may be gaseous in nature. Of course, solvents may be inorganic as well as organic. Indeed, water is the most common solvent, not only for everyday uses but also in industry. A solvent, then, is simply a compound that has intermolecular forces that are similar to those of another compound. Thus, the solvent and the solute (that which is being dissolved) have a common affinity for one another and readily mix.

The idea of intermolecular forces is very important to the hygienist and is discussed in Chapter 8. When these forces are similar for two compounds, the compounds tend to go into solution. The old adage *like dissolves like* applies. So a polar compound is a good solvent for another polar compound. Alcohols tend to be readily dissolved in other alcohols or in other compounds that have high hydrogen bonding or polar forces.

A special group of solvents, widely used, is the petroleum distillates. They are distilled directly from petroleum and are, therefore, non-polar aliphatic and aromatic hydrocarbons. What makes them important to our discussion is that they are not pure, but are found in mixtures, and the names are often confusing. As can be seen in Table 4-8, the various distillates are separated by their boiling ranges. Note the following: (1) the carbon number for each (boiling point increases with number of carbons); and (2) the class of components in each (aromatics are more likely in the higher boiling point fractions). Generally speaking, organic compounds that are liquid at room temperature have probably been used as solvents. This means compounds with molecular weights from 65 (methanol) up to about 193 (1,1,2,2-tetrachloroethane). However, some groups are more common than others, while certain specialty

Table 4-8
Physicochemical characteristics of petroleum naphthas and thinners[1]

Common Name	Alternate Name	B.P. °C(°F)	Mol. Wt.	Carbon Number	Class of Components[2]
Crude oil	Earth oil, petroleum	<0 to >1000 (<32 to >2000)	N.A.	C_1–>C_{50}	About 300 organic substances identified, some heavy metals
Gasoline	Benzin, petrol, motor or aviation gasoline	32–210 (90–410)	~100	C_4–C_{12}	n- and isoparaffins, olefins, aromatics
Petroleum ether	Ligroin, petroleum benzin	30–60 (86–140)	~75	C_5–C_6	Paraffins (pentanes, hexanes, isohexanes)
Rubber solvent		45–125 (113–257)	84 – 97	C_5–C_8	Paraffins, monocycloparaffins, olefins (trace), benzene, alkylbenzenes
VM & P naphtha	Range of 80 thinner	95–160 (203–320)	87 – 114	C_7–C_{11}	Paraffins, mono- and dicycloparaffins, benzene (trace), alkylbenzenes
Mineral spirits (petroleum spirits)	Refined petroleum solvent, White	150–200 (302–392)	144 – 169	C_9–C_{12}	Paraffins, naphthenes, olefins, aromatics
Stoddard solvent	White spirits	160–210 (320–410)	~140	C_9–C_{12}	Paraffins, mono- and dicycloparaffins, benzene (trace), alkylbenzenes
High flash naphtha		150–204 (302–400)	~140	C_7–C_{12}	Paraffins, naphthenes, aromatics
140° flash naphtha	Aliphatic solvent naphtha	185–207 (364–403)	154	C_5–C_{13}	Paraffins, mono- and dicycloparaffins, benzene (trace), alkylbenzene
Aromatic petroleum naphtha	Coal tar naphtha	93–315 (200–600)	~140	C_8–C_{13}	Paraffins, alkylbenzenes
High aromatic naphtha	High aromatic solvent	184–206 (364–403)	~140	C_8–C_{13}	Paraffins, mono- and di- and tricyclic naphthenes, alkylbenzenes and naphthalenes, olefins

Table 4-8 (Cont.)
Physicochemical characteristics of petroleum naphthas and thinners[1]

Common Name	Alternate Name	B.P. °C(°F)	Mol. Wt.	Carbon Number	Class of Components[2]
Thinner 40		187–231 (368–447)	~148	C_8–C_{13}	Paraffins, mono- and dicycloparaffins, mono- and diolefins
Thinner 50		98–105 (208–221)	97	C_6–C_8	Paraffins, olefins, naphthenes, aromatics
Thinner 60		128–159 (263–319)	120	C_5–C_{10}	Paraffins, monocycloparaffins, alkylbenzenes
Thinner 70		157–211 (315–411)	132	C_5–C_{12}	Paraffins, monocycloparaffins, alkylbenzenes
Thinner 80		97–142 (206–288)	106	C_6–C_9	Paraffins, dicycloparaffins, alkylbenzenes
Kerosene	Stove oil	163–288 (325–550)	~180	C_{10}–C_{16}	Aliphatics, mono- and dicycloparaffins, alkylbenzenes
Deodorized kerosene		208–272 (406–522)	179	C_6–C_{14}	Paraffins, mono- and dicycloparaffins, aromatics

1 Adapted from EPA (1977) and NIOSH (1977).
2 Paraffins = alkanes, olefins = alkenes

solvents are gaining in popularity. Chlorinated hydrocarbons, shown in Figure 4-8, are extremely common but becoming less so because of their toxicity and environmental consequences. For example, from 1981 to 1991 sales of perchloroethylene went from 691 to 240 million gallons and for methylene chloride from 891 to 389 million gallons. Sales of acetone, butanol and ethylene glycol have increased (CEN, 2004).

In addition to the use of alcohols, ketones, hydrocarbons, and chlorinated hydrocarbons as solvents, the following specialty solvents are becoming more popular: dimethyl formamide, tetrahydrofuran, dimethylsulfoxide, the esters, and the glycol ethers (Cellosolves®) and their esters. Some of these are shown in Figure 4-6.

MONOMERS

One of the most important industries in the world today involves taking small, simple

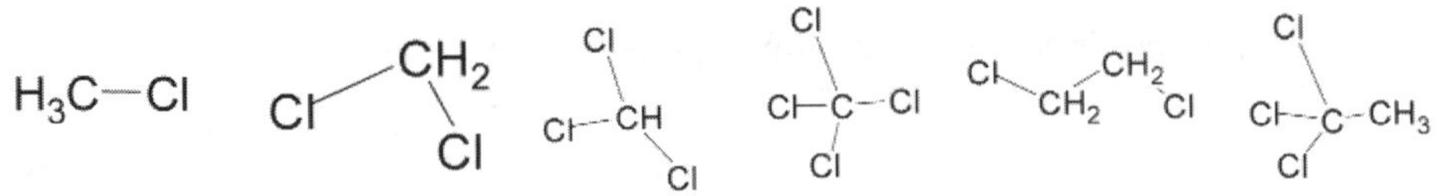

methyl chloride
methylene chloride, dichloromethane, methylene dichloride
chloroform, trichloromethane
carbon tetrachloride
ethylene chloride, 1,2-dichloroethane
methyl chloroform, 1,1,1-trichloroethane

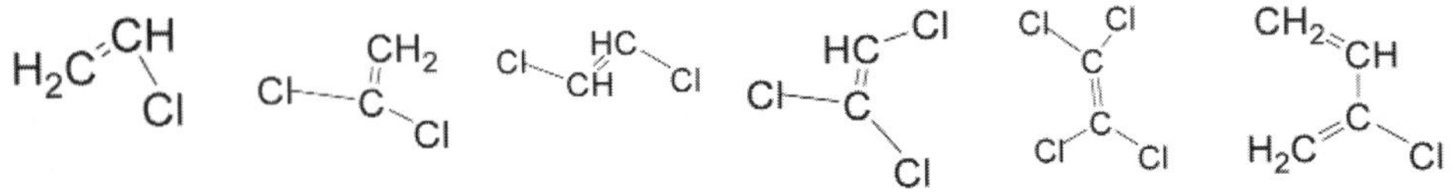

vinyl chloride, chloroethylene
vinylidene chloride, 1,1-dichloroethylene
1,2-dichloroethylene
trichloroethylene
perchloroethylene
chloroprene, 2-chloro-1,3-butadiene

Figure 4-8. Eleven of the more common halogenated hydrocarbons or alkyl halides.

molecules and building from them much larger compounds termed polymers. Some of the monomers used to manufacture the polymeric substances are of great concern to the hygienist. The monomers most often encountered in the polymer industry and their associated polymers are shown in Figure 4-9. It should be noted that Figure 4-9 shows only some of the thousands of manufactured polymers made from many fewer monomers. In addition, many names actually refer to families of polymers. For example, a polyester is the product of any dialcohol and diacid.

Polymers may be thermosetting, in which case the chains are cross-linked and held in place; or they may be thermoplastic, in which case the chains are only attracted to each other by intermolecular forces, and thus the polymers may melt and be reset or reshaped. A thermosetting polymer that is not crosslinked is called a resin and is often very viscous. Thermoplastic polymers are often produced as pellets or sheets and are then molded or formed with heat and pressure.

It may be thought that the monomers are only of concern during the actual manufacture of the polymer compound or resin, however, when the polymers are reprocessed or crosslinked at sites different from the manufacturing site, unreacted monomer may offgas. It is possible that a significant amount of the monomer does not completely react during the polymerization process. For thermoplastic polymers, unreacted monomer may then be released during subsequent heating. Also, the monomer may be released whenever degradation of the polymer occurs. Landfills are known to release vinyl chloride gas, occasionally in considerable quanti-

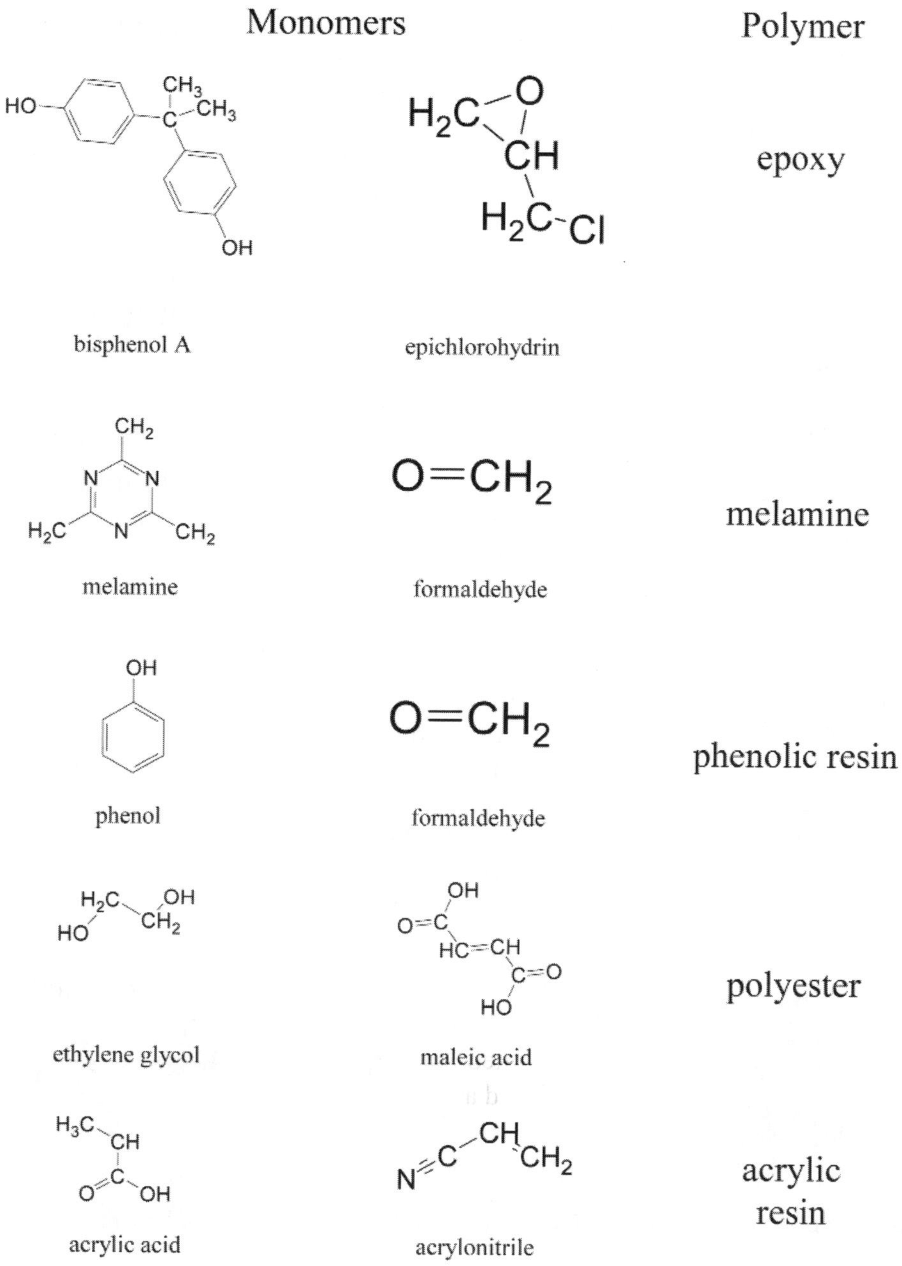

Figure 4-9. Monomers used to produce some of the more common polymers.

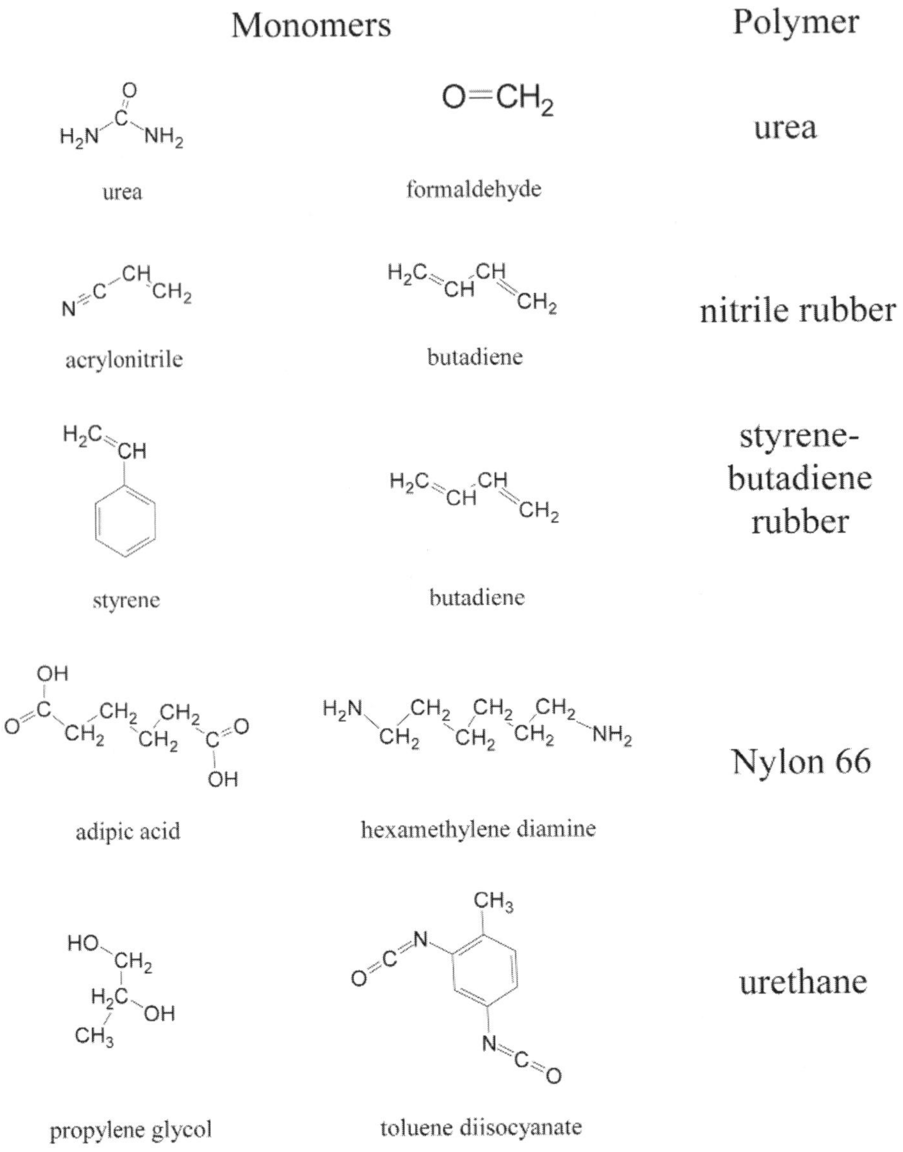

Figure 4-9 (Cont.). Monomers used to produce some of the more common polymers.

ties, which comes from the breakdown of polyvinyl chloride (PVC) in the landfill as well as newspaper print, which contains carbon and chlorine used to bleach the paper (Stearns and Beizer, 1985). Several of the monomers shown in Figure 4-9 have been associated with important toxicological outcomes, particularly cancer.

PESTICIDES

Although this is one of the most specialized use categories, every hygienist should have some knowledge of pesticides, as they are among the most toxic compounds that have been purified or synthesized. This should not be surprising because pesticides are designed to kill a pest. What is important for the hygienist to realize is that the term pesticide alone has very little useful meaning. Specifically, the hygienist needs to know what type of pesticide is being used. Figure 4-10 shows some of the various kinds of pesticides. The class names are quite simple in that they convey the type of pest that is being controlled. Those compounds that control pests with biologies most similar to humans are those that generally are most acutely toxic to humans. However, even those that control organisms with far less biological similarity to humans (e.g., herbicides) also can have significant effects on humans.

Of these, the most important are the insecticides, again based on volume of use and toxicity. The insecticides are largely composed of three groups of compounds, examples of which are shown in Figure 4-10. At one time the organochlorine compounds were the most important, as it was widely thought that because of their low acute toxicities they were safest. However, it has since been determined that they have grave consequences on wildlife due to their longevity in the ecosystem, and can have chronic toxicities in humans. The organophosphorus compounds have now become the most important insecticides. These compounds are based upon organophosphorus war agents first developed in Germany prior to World War II. Some of the early casualties associated with these compounds were laboratory chemists who were doing research on the compounds (Carson, 1962).

Also of rising importance is a fourth group of insecticides based on pyrethrins, called pyrethroids. The former are naturally occurring compounds, derived from plants related to *Chrysanthemum*, that break down relatively rapidly. The pyrethroids are synthetic and while based on pyrethrin, their chemistry is quite complex and variable. They are not without toxicity to humans (Ray, 1991).

In the U.S., approximately 900 active ingredients registered by the Environmental Protection Agency for use as pesticides are used to make about 20,000 pesticide products (EPA, 2001). Workers may be exposed to these compounds during manufacture or the formulation process, which are the two most obvious processes. However, workers may also be exposed during mixing or spraying of the compounds, or after application has occurred when the worker must re-enter a farm field where a pesticide has been applied.

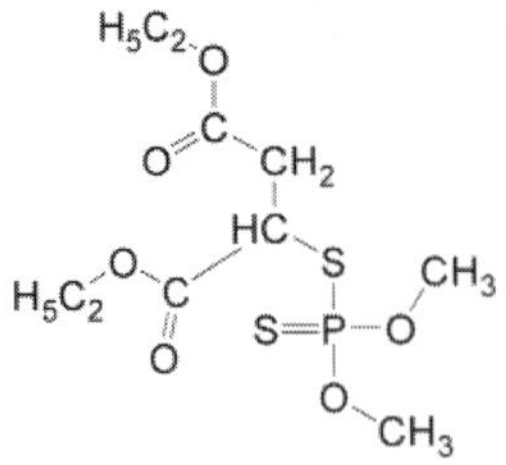

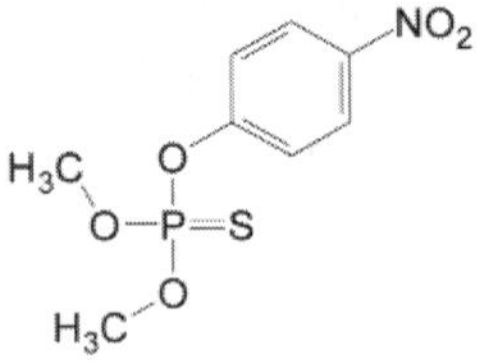

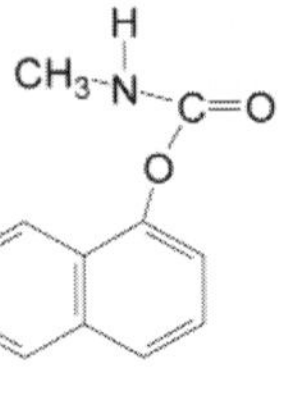

Malathion (organophosphorus) Parathion (organophosphorus) Carbaryl (carbamate)

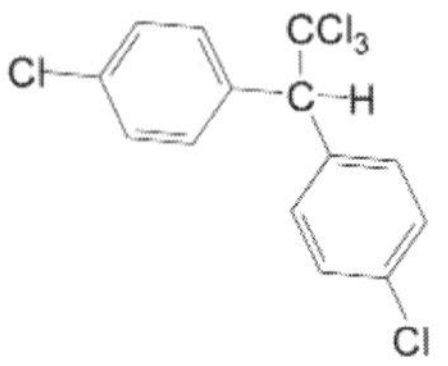

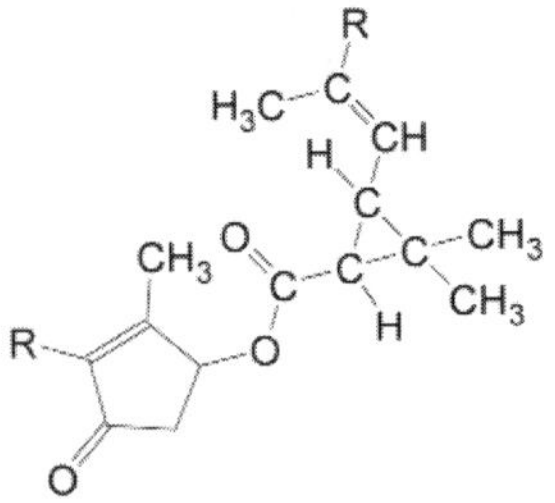

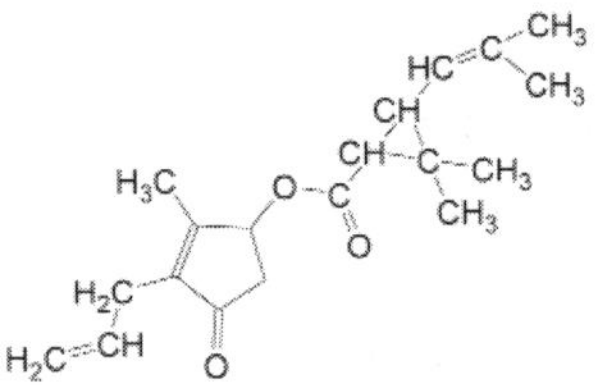

DDT (organochlorine) pyrethrum (pyrethrins) Allethrin (pyrethroids)

Herbicides

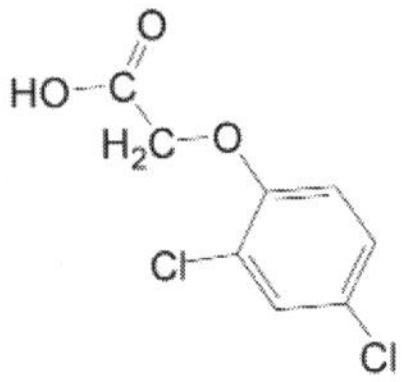

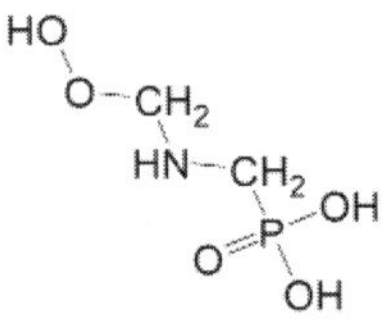

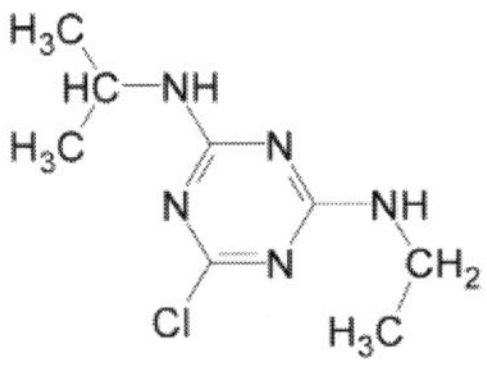

2,4-D (dichlorophenoxy acetic acid) Glyphosate (Roundup®) Atrazine

Fungicide moth balls insect repellent

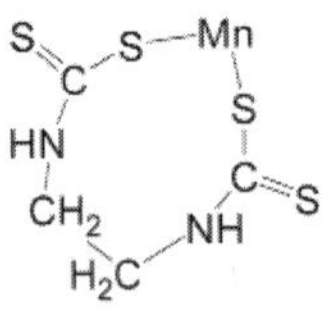

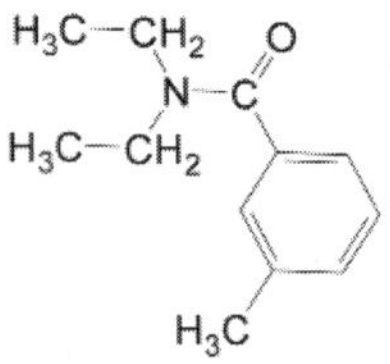

Maneb p-chlorodibenzene DEET (n,n-diethyl toluamide)

Figure 4-10. Some characteristic pesticide structures.

OTHER USE CATEGORIES

There are many other categories of organic compounds according to use. Some of these are dyes, pigments, and adhesives. One could also list surfactants, detergents, accelerants, preservatives, flocculants, and many others. What is important for the hygienist to learn is that these use categories are not composed of mysterious chemicals about which the hygienist should know nothing. Rather, the use categories commonly contain chemicals that have common chemical properties. After all, the purpose of the use category is to accomplish some end-goal that will require certain chemical properties. Hence, the hygienist, upon encountering a particular compound that has a particular use, should investigate its chemistry and the chemistry of other similarly used compounds to determine how they are manufactured and what the common properties are. This information may be of use in evaluating and controlling the substance, particularly for substituting a less hazardous chemical for a hazardous one.

Physical Forms of Chemical Agents

It is not uncommon to hear people speak of diesel fumes or gasoline fumes or hood fumes or fumes from a fire. Very seldom does the layperson use the term vapor, and a fog is something that rolls in from the sea. However, each of these terms has a different and specific meaning in hygiene. The physical form in which a chemical agent can exist or does exist in a particular airborne circumstance is extremely important, as it denotes how the worker may be exposed. For this reason, the hygienist must be quite accurate in the way that he or she describes the atmospheric physical nature of a chemical agent.

The various types of physical states for airborne chemical agents are given in Table 4-9. There are two main categories. First, there are the solid and liquid states that exist in the atmosphere, broadly termed aerosols or particulates. Note that whether the particle is liquid or solid does not matter; both are termed aerosol or particulate. The second main category is the gases, of which vapors are a part.

The solid and liquid forms are further divided according to their origin. For solid forms, there are dusts, fumes, and smokes. Fibers are a special class of dusts. The most important differentiation among the three is particle size, as this determines the depth to which the particle penetrates the lungs. Figure 4-11 shows sizes of various particulates common to industry and to everyday life. In general terms, sizes range from the smallest fumes to smokes to the largest solid particles, the dusts. However, there can be considerable overlap among these sizes, as is shown. Of course the bacteria, viruses, and other microbiological agents are also particulate matter. Very often they are attached to solid or liquid particulates. Their sizes are also shown. The airborne liquid particulates are mists, fogs, and smokes. Note that a smoke can be either liquid or solid.

There are two forms of gases to differentiate. This differentiation is strictly arbitrary and based upon the fact that room temperature, also classified as normal temperature, is 70°F (21°C) in the United States, particularly among engineers, and 25°C (77°F) in the rest of the

Table 4-9
Physical forms and sizes of air contaminants[1]

States – Sources	Relative Sizes (µm)
Solids or liquids suspended in air (particulates or aerosols)	
Solids	
Dusts – mechanical action	> 1
Smokes – incomplete combustion of organics	0.01 – 1
Fumes – condensation of vapor	0.001 – 1
Fibers – mechanical action, aspect ration > 3 to 20[2]	> 0.003
Liquids	
Mists – mechanical action on or condensation of liquids	0.01 – 200
Fogs – a visible mist	0.01 – 200
Smokes – incomplete combustion of organics	0.01 – 1
Airborne molecules (gases and vapors)	
Gases – reactions, naturally occurring	0.0002 – 0.0005
Vapors – evaporation, sublimation from liquids and solids	> 0.0005

[1] Adapted from Hinds (1999) and McDermott (1985).
[2] Fibers are at least three times longer than their diameters, by definition.

world and among U.S. chemists.[1] Hence, a substance that, at equilibrium, is in a gaseous state at room temperature and standard atmospheric pressure is considered a gas. Vapors are a special form of gas; these substances evaporate from liquids (and sublime from solids) at room temperature. In other words, the substance comprising a vapor, at equilibrium, can exist as a liquid or a gas at room temperature and standard atmospheric pressure. The vapor evaporates from the surface of the liquid. Organic compounds that are solids at room temperature will sublimate some of their molecules as vapors. The examples below should give a better understanding of the application of these definitions.

Gasoline. When liquid gasoline or petrol evaporates, it releases vapors. It is concise to designate vapors as more than one type of compound evaporating from a liquid or solid surface and to reserve the term vapor for a single compound evaporating from a liquid or solid surface. When gasoline burns, gases such as CO_2, CO, SO_2, and NO_X are produced, depending on combustion temperature and availability of O_2. If the burning process is fairly complete, very few of the hydrocarbons will escape unburned, and few vapors and smokes will be emit-

[1] Hygienists tend to use 25° for chemistry problems and 21°C or 70°F for engineering problems such as ventilation.

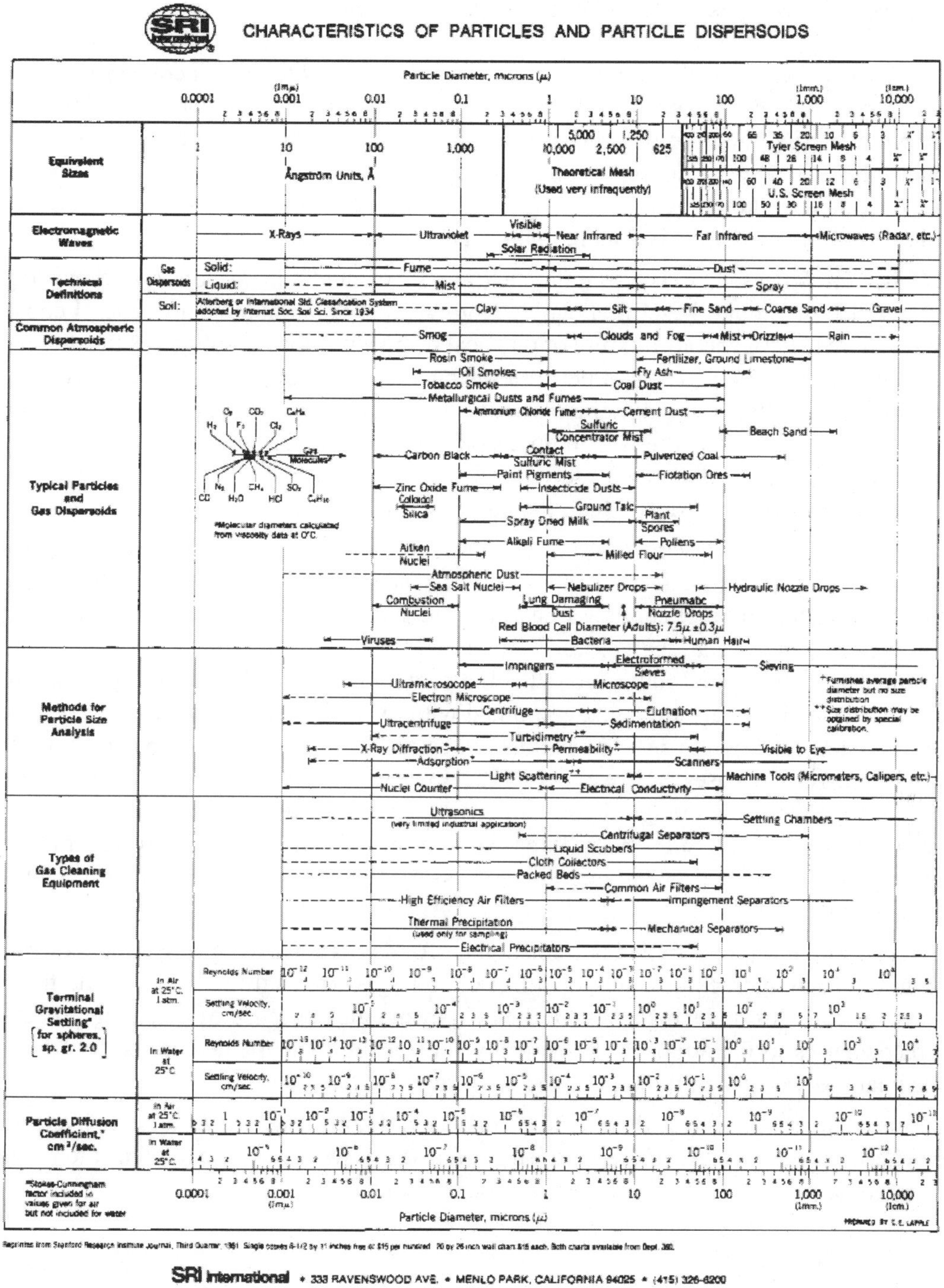

Figure 4-11. The relative sizes and other characteristics of aerosols and gases (from Department 300, Stanford Research Institute, Menlo Park, CA, with permission).

ted, as gasoline is composed of relatively light hydrocarbons; however, in the internal combustion engine, very fine particulate smokes are produced, as are unburned hydrocarbons. Some of the latter may be vapors that remain as vapors as they cool; others will condense on the particulate matter, largely carbon. Gasoline contains some inorganic compounds of metals and sulfur, which during the burning process are converted to their oxide form and upon entering the atmosphere usually condense onto other particulate matter; thus, while some fume formation occurs, on the whole, the term gasoline fumes is misleading. Since the products are quite complex, *gasoline emissions* is a more appropriate term.

Diesel Fuel. Diesel fuel is a complex mixture of heavier hydrocarbons than those of gasoline. Nevertheless, liquid diesel fuel evolves into vapors. Upon combustion, diesel emissions contain a more heterogeneous mixture of physical forms than does gasoline. As the diesel compounds are larger and heavier, and because of the operating temperature and pressure of the diesel combustion engine, the diesel emissions contain more smokes composed of carbon particles and adsorbed hydrocarbons than do gasoline emissions. Diesel particulates also contain the inorganic matter contained in diesel fuel such as metals, which form oxides upon combustion. For both gasoline and diesel fuel, the combustion products are best termed emissions.

Welding. There are several types of welding processes, some of which can be quite complex; however, simply stated, the welding process consists of the melting of two pieces of metal. When metals are melted, they first vaporize, and upon entering the atmosphere, which is relatively rich in oxygen, they oxidize. The molecules then coalesce to form small particulate matter, which is termed fume. Note that the evolution of fume is essentially a combustion process because the metals are being combined with oxygen. Hence, the term fumes is generally reserved for the combustion of metals, whereas the incomplete combustion of organic materials generally releases smoke and vapors. However, the evaporation or incomplete combustion of high molecular weight hydrocarbons is sometimes referred to as fumes (e.g., as in polymer fume fever); so smokes or fumes may be produced from burning organic matter, depending on the particle size emitted, the process employed, and the molecular weight of the parent compounds.

Acids. The term acid fumes is a misnomer. Acids may evolve gases, such as hydrogen chloride gas from hydrochloric acid or sulfur dioxide gas from sulfuric acid. In addition, they may evolve either mists or fogs. Where the acid evaporates because of a heating process and then recondenses in the atmosphere, typically fogs are formed. If the acid is disturbed by a mechanical process, generally larger droplets form – mists.

This chapter emphasizes that the hygienist should know the structure, and physical form of the chemical agent that is of concern. This information will allow the hygienist to recognize a potential hazard, and to evaluate and control it efficiently.

QUESTIONS

4.1 Name an important biological agent or group of agents affecting indoor air quality.

4.2 What is the most common single disease agent in industry?

4.3 Which of the four forces is least important to understanding the basic nature of occupational disease agents? Explain why.

4.4 Classify the elements into four groups.

4.5 Generally describe the raw and intermediate materials in the production of a polymer.

4.6 Explain how a metal fume is formed. What is its approximate particle size?

REFERENCES

ASHRAE: Practical Control of Indoor Air Problems: Proceedings of the ASHRAE Conference, IAQ 87. American Society of Heating, Refrigerating, and Air-Conditioning Engineers, Arlington, VA (1987).

Barber, T.E.; Husting, E.L.: Biological hazards, In Occupational Diseases. M.M. Key, A.F. Henschel, J.N. Ligo, and I.R. Tabershaw (eds.). National Institute for Occupational Safety and Health, U.S. Government Printing Office, Washington DC (1977).

Birmingham, D.J.: Dermatoses, In Occupational Diseases, M.M. Key, A.F. Henschel, J. Butler, R.N. Ligo, I.R. Tabershaw (eds.). National Institute for Occupational Safety and Health, Cincinnati, OH, USA (1977).

BNA: OSHA issues hearing conservation rules, sees savings through performance approach. Occup. Safety Hlth. 12: 827, Bureau of National Affairs, Washington, DC (1983).

Carson, R.: Silent Spring. Fawcett Crest, New York (1962).

CAS: The Latest CAS Registry Number and Substance Count. Chemical Abstract Service, Columbus, OH http://www.cas.org/cgi-bin/regreport.pl (2007).

CEN: Facts & figures for the chemical industry. Production: Some Improvement, Chemical & Engineering News, 82 (27):23-63 (2004).

Chin, J. (ed.): Control of Communicable Diseases Manual, 17th ed. The American Public Health Association, Washington, DC (2000).

Douwes, J.; Thorne, P.S.; Pearce, N.; Heederik, D.: Biological Agents: Recognition, In Modern Industrial Hygiene, Volume 2, Biological Aspects, J.L. Perkins, (ed.). ACGIH® (2003).

EPA: Petroleum refining industry, In Industrial Process Profiles for Environmental Use, EPA-600/2-77-02 3C. Environmental Protection Agency, Research Triangle Park, NC (1977).

EPA: National primary drinking water regulations: Maximum contaminant level goal and national primary regulations for lead and copper. United States Environmental Protection Agency. Fed. Reg. 53:31516 (1988).

EPA: Office of Pesticide Programs FY 2001 Annual Report. U.S. Environmental Protection Agency, Washington, DC (2001).

EPA: Toxic Substances Control Act Chemical Substance Inventory. Note that paper version was last updated in 1990, see http://www.epa.gov/oppt/newchems/invntory.htm United States Environmental Protection Agency, Washington, DC, USA, (2007).

Goyer, R.A.; Clarkson, T.W.: Toxic effects of metals, In Casarett and Doull's Toxicology. 6th ed., C.D. Klaasen, M. (ed.). McGraw-Hill, New York (2001).

Hawking, S.W.: A Brief History of Time. Bantam, New York (1988).

Hinds, W.C.: Aerosol Technology: Properties, Behavior, and Measurement of Airborne Particles, pp. 6–8. Wiley Publishing, New York (1999).

Jacobs, R.R.: Airborne endotoxins: An association with occupational lung disease. Appl. Ind. Hyg. 4: 50-56, 1989.

Kelly, T.; Buckingham, D.; DiFrancesco, C.; Porter, K.; Goonan, T.; Sznopek, J.; Berry, C.; Crane, M.: Historical Statistics for Mineral and Material Commodities in the United States, U.S. Geological Survey Open-File Report 01-006, Version 8.3, Online Only, USGS http://minerals.usgs.gov/minerals/pubs/of01-006/#data (2003).

Kniel, L.; Winter, O.; Chung-Hu, T.: Ethylene, In The Kirk-Othmer Encyclopedia of Chemical Technology, 3rd ed., volume 9. M. Grayson, ed., pp. 424-426. John Wiley & Sons, New York (2001).

Lewis, R.J. Sr.: Hawley's Condensed Chemical Dictionary, 13th ed. Wiley, New York (1997).

McDermott, H.J.: Handbook of Ventilation for Contaminant Control, 2nd ed., p. 142. Butterworth Publishers, Boston (1985).

NAS Committee: Possible Health Effects of Exposure to Residential Electric and Magnetic Fields, Committee on the Possible Effects of Electromagnetic Fields on Biologic Systems. National Research Council, National Academy Press, Washington, DC, USA, (1997).

Nestle, M.: Nutrition, In Harper's Review of Biochemistry, D.W. Martin, P.A. Mayes, V.W. Rodwell (eds.). Lange Medical Publications, Los Altos, CA (1983).

NIOSH: Occupational Exposure to Refined Petroleum Solvents. National Institute for Occupational Safety and Health, Cincinnati, OH (1977).

NIOSH: Criteria for a Recommended Standard: Occupational Noise Exposure, Publication 98-126. National Institute for Occupational Safety and Health, Cincinnati, OH (1998).

OSHA: Code of Federal Regulations 29, part 1910.1000, table Z-2. U.S. Government Printing Office, Washington, DC http://www.osha.gov/index.html (2007).

Peterson, A.P.G.; Gross, E.E. Jr.: Handbook of Noise Measurement, 8th ed. Gen Rad, Concord, MA (1978).

Ragland, P.E.: Geochemical distribution of elements, In The Encyclopedia of the Geological Sciences, p. 201, Daniel N. Lupedes (ed.). McGraw-Hill, New York (1978).

Ray, D.E.: Pesticides derived from plants and other organisms, In Handbook of Pesticide Toxicology, J.W. Hayes and E.R. Laws (eds.). Academic Press, San Diego (1991).

Spengler, J.D.: Sources and concentrations of indoor air pollution, In Indoor Air Pollution, J.M. Samet and J.D. Spengler (eds.). The Johns Hopkins University Press, Baltimore (1991).

Stearns, R.P.; Beizer, M.B.: Landfill gas is a growing concern, World Wastes, pp. 34-36 (February 1985).

Weissman, D.N.; Schuyler, M.R.: Biological agents and allergic diseases, In Indoor Air Pollution, J.M. Samet and J.D. Spengler (eds.). The Johns Hopkins University Press, Baltimore (1991).

5

COMMON INDUSTRIAL PROCESSES AND THEIR ASSOCIATED AGENTS

While total non farm employment in the U.S. increased from 91 million in 1980 to 130 million in 2004, production workers employed in manufacturing declined from 13.6 million to 10 million (BLS, 2004). Even though manufacturing is declining in the United States, this industry group continues to employ significant numbers, and no doubt the numbers of manufacturing employees is increasing worldwide, simply due to worldwide economic and population growth.

The processes, or in some cases equipment or systems, described in this chapter are common throughout the manufacturing sector. No matter how small the plant, it can be expected that at least one of the described processes will be present, and frequently many will be. Therefore, it is important that the hygienist have at least a baseline working knowledge of these operations before entering any plant.

Much of the information upon which this chapter is based is taken from the seven general references listed at the end of this chapter. Other less common processes may be found in these texts. In addition, where appropriate, references to specific processes are cited.

ABRASIVE BLASTING

Description

In abrasive blasting an abrasive material is propelled toward a surface and upon impact loosens and removes a surface coating. In the U.S., the most common form of this is sandblasting. Paint, other polymeric or synthetic coatings, rust, scale, and fused sand (from foundry molds) are commonly removed in this way. The process is very common in bridge repair, shipbuilding and repair, building construction, founding, and metal finishing. However, it is also common in any industry where steel surfaces must be maintained. For example, in the petrochemical industry countless pipelines and vessels are in a constant state of blasting and recoating.

The materials used for blasting are quite varied. Sand, or crystalline silica,[1] is plentiful and cheap and frequently used. However, it is quite toxic, and more expensive substitutes include aluminum oxide grit and shot, metallurgic slag, steel shot and grit, and pumice. Synthetic materials include silicon carbide and fused aluminum oxide. Softer, and in many cases less hazardous, materials include nut shells, glass beads, ground corncobs, frozen carbon dioxide pellets, and plastic pellets. The above are the most common ones, but there are other materials used as abrasives (for example, in sandpaper), and these may appear in specialized blasting operations. They include corundum, emery, feldspar, and diamond.

Abrasive blasting operations may take place in the open, as on the side of a bridge, or in an enclosure. Outdoor operations may take place in partial enclosures, particularly if the coating contains lead. Indoor operations should always be enclosed, particularly if sand is used. Of course, this requires that the operator, if inside the enclosure, wear respiratory protection. In the United States abrasive blasting hoods used specifically for this task are the norm.

Hazards

Burgess (1991) aptly points out that chemical hazards associated with blasting may arise from the base metal, the coating, or the abrasive. In addition, noise is a hazard.

BASE METAL

Blasting of mild steel, iron, or aluminum should not pose a problem. All other materials, such as stainless, galvanized, or specialty steel, have the potential to release substantial amounts of such metals as lead, nickel, manganese, chromium, and cadmium. In addition, blasting on such base metals as beryllium likely will pose an unacceptable risk.

COATING

The most significant hazard posed by coatings is in the form of lead-based paints. Documented exposures to lead during the blasting of such materials have revealed levels ranging from 500 to 10,000 mg/m^3. Exposures to sandblaster helpers of 90 to 560 μg/m^3 have also been documented (NIOSH, 1991). An airborne level of 50 μg/m^3, 8-hr time weighted average, is the current U.S. standard for exposure to lead.

Other paint pigments such as chromium are also hazardous. However, the base polymer itself should not be. Fungicidal and other additives to paints containing mercury or tin may also be hazardous.

Other unwanted coatings such as scale (a calcareous deposit) and rust (oxides of iron)

[1] Other mineral names such as agate, quartz, tripoli, sandstone, and novaculite as well as many trade names also apply to silica.

generally are not significantly hazardous. However, oxides of the more toxic metals such as lead and even zinc are hazardous in blasting operations.

BLASTING AGENT

Silica is an extremely serious lung fibrogenic dust, causing a disease called silicosis, and a suspected lung carcinogen. Although there are standards for silica exposure, and blasters are required to wear blasting helmets, the standards do not appear to set an adequate risk level, and/or they are not met. The U.S. standard for silica is essentially unchanged from 1970 and, no doubt, inadequately controls risk. In1974, the one million U.S. sandblasters experienced 59,000 cases of silicosis; an approximate lifetime risk of 6 in 100. Data such as these have caused a number of European countries to ban the use of silica in this operation (Landrigan, 1987). OSHA considered banning silica in sandblasting, but subsequently withdrew its intent (Burgess, 1995).

In addition, older style blasting hoods do not appear to be effective (Glindmeyer and Hammad, 1988). In one study of lead exposures during blasting operations, levels outside the blasting hood were 640 to 1400 $\mu g/m^3$, whereas inside the hood they were only reduced to 230-860 $\mu g/m^3$ (Adkison, 1989). Although a similar reduction may not necessarily have occurred for the workers' silica exposures, nonetheless the above reduction (a factor of 2 to 3) points out the inadequacies of these older abrasive blasting hoods for protection against any but the most innocuous contaminants.

Other blasting agents also pose serious hazards. Metal slags, also known as dross or cinder, contain silicates, a non-crystalline and lower-hazard material than silica. In addition, slags, which float to the surface and are skimmed off of molten metals during smelting, may contain the primary or other hazardous metal. Their effects, when breathed, may include lung cancer and chronic bronchitis (Stettler et al., 1988). Most of the other agents do not appear to have caused problems, but a constant vigil should be kept because, regardless of the agent, the process naturally leads to very high exposures.

BOILERS AND FURNACES

Description

There is a *boiler* in virtually every U.S. home – the hot water heater. Moreover, there is likely a furnace for heating air or an oven for baking. Their operating principles are similar to those of devices used in industry; however, unlike the devices used in industry, home ovens and hot water heaters may operate for 20 years without maintenance, usually because of lower temperatures. (Hot water heaters with tanks do not actually boil the water.) On the other hand, industrial boilers, furnaces, ovens, and kilns require considerable maintenance, and it is in this operation that hazards exist. Nevertheless, the maintenance operations occur infrequently enough that unless an inquiry is made by the hygienist, hazards may be missed.

Boilers are devices for heating water. Steam or hot water may be produced. Basically, a boiler is a system of tubes in an open space. Either the tubes are filled with water and the space is filled with hot combustion gases, or vice versa. It is a rather simple operation with most problems either arising from buildup of ash or slag from the combustion of the fuel or buildup of scale from impurities (largely calcareous or siliceous minerals) in the water.

Furnaces are also known as ovens or kilns and usually heat something other than water although their purpose may be to dry or remove water from a product. Although the operating temperature may be very low (50°C), it can also be hot enough to melt some metals (2000°C). While furnaces are common in the various metals industries, they are also used to cure plastics in a variety of products ranging from toys to electrical wiring. They are used to dry a wide variety of products; to manufacture carbon black and coke (both are forms of carbon) from natural gas, oil, or coal; for baking bread; and for curing painted surfaces.

Hazards

Boilers and furnaces typically require refractory or insulation in order to operate efficiently and preserve their metal exteriors. Asbestos, fiber glass, or ceramics may be used. The latter may contain silica. During maintenance these materials are often disturbed if not entirely replaced. Where coal or oil is burned, hazards result from the fly ash that is collected from the stack and the bottom ash or unburned material that does not go up the stack. The metal content in these materials can be hazardous in maintenance operations. In particular, vanadium exposures are associated with oil-fired units.

Of particular concern are the chemicals used to treat boiler water or to remove scale from boilers. These include various acids, potentially including mixtures yielding hydrofluoric acid; conditioning chemicals including sodium hydroxide (caustic), hydrazine, sodium sulfite, amines, and chelating agents such as ethylenediaminetetraacetic acid (EDT A). Wetting agents used in cleaning may cause allergic reactions. Boilers may be blasted with steel shot, which is not a significant health hazard; however, other blasting agents may be used and should be suspected.

As a result of the wide variety of furnace uses, the hazards are variable. These hazards may occur during operation as well as maintenance. Solvents and other organics driven off during drying and curing operations must be controlled for explosion or flammability. Even so, inhalation exposures may still be significant and overlooked. Virtually any type of furnace operation involving the heating of metals so that they are malleable requires engineering control of emissions.

CONFINED SPACES

Description

Confined spaces, by their very nature, allow for the accumulation of air contaminants, which may be toxic or may act to reduce the concentration of oxygen to undesirable levels. Confined spaces also restrict rescue efforts and may result in death of poorly prepared or ill-trained rescuers.

A confined space is generally:

- Enclosed;
- Large enough for a person to enter and perform work;
- Restricted or limited in means of entry or exit;
- Occupied rarely.

In addition, it will have one or more of the following:

- Flammable or explosive conditions;
- Dusty conditions limiting visibility to less than five feet;
- Oxygen concentration of less than 19.5%;
- A substance in an unacceptable concentration relative to health;
- A condition that is immediately dangerous to life; that could result in irreversible bodily harm, eye damage, irritation, or other conditions that could impair escape;
- A bulk material (sand, sandblasting material, etc.) that could engulf a person;
- An internal configuration that could keep an entrant from escaping, or a floor that slopes sharply to a cone shape;
- Any other recognized serious safety or health hazard such as the potential for electrocution or dismemberment (OSHA, 1993).

Such spaces exist in the majority of workplaces. As pointed out, they are rarely occupied. However, when they are occupied, generally for maintenance or to correct a malfunction, they present important health and safety hazards.

Hazards

In a general sense, the hazards are described in the above definition. Unfortunately, it is difficult to be more specific than this. Because the spaces are rarely entered, most health hazards are usually of an acute nature. Where combustion processes are involved, one should always suspect carbon monoxide. Sewers or other areas where anaerobic respiration in the presence of sulfur (by bacteria) takes place, will contain hydrogen sulfide. These may contain high levels of hydrogen. In combustion areas or sewers, carbon dioxide may be in high enough concentrations (> 20,000 ppm or 2 %) to significantly displace oxygen. Oxygen displacing CO_2, in combination with O_2 diminution caused by combustion or respiration, may cause O_2 levels to fall below 19.5%.

In any confined space where there is a source of combustible gases such as hydrocarbons, accumulation to significant explosive levels is possible. In many confined spaces there is generally little air movement so that over time gases and vapors may sort vertically according to density. This may exacerbate explosion hazards as well as inhalation hazards caused by solvents that may be used to remove organic substances or in applying coatings.

No doubt confined spaces are hazardous. The U.S. government has estimated that there are about 240,000 workplaces with 4.8 million confined spaces that employ 12.2 million workers. It has further estimated that its confined space standard will prevent 63 deaths and almost 13,000 injuries annually (OSHA, 1993).

DIESEL ENGINES

Description

It is commonly known that diesel engines are widely used as a means of conveyance; in particular this involves workers in the trucking, rail, and bus industries. From an industrial point of view, diesels are often associated with mobile heavy equipment in the mining and construction industries. However, diesels are also commonly used in forklifts and farm equipment, as well as in many stationary industrial applications such as electrical power and steam generation.

The main advantages of diesel over gasoline engines are that diesel engines burn cheaper fuel (at least this is often the case) and consume less fuel per horsepower per hour (Encyclopedia Britannica, 1987). Because diesel fuel is less volatile and more dense than gasoline, higher pressures or compression is required to combust it relative to gasoline. This means that ambient nitrogen and oxygen more readily combine to form oxides of nitrogen in the combustion process. For engines up to about 5,000 horsepower, diesel is the most common industrial power source.

Hazards

It is well known that diesel emissions are irritating and unpleasant to the nose. In addition, the engines may produce a noise hazard. Until the 1970s, when toxic effects such as carcinogenicity began to be studied, these were probably the biggest known drawbacks to diesel engines.

The emissions contain many pollutants, including carbon dioxide and monoxide, oxides of nitrogen and sulfur, hydrocarbons (such as formaldehyde, benzene, phenol, and acrolein) and poly nuclear aromatic hydrocarbons (PNAHs). Particulates can contain unburnt liquid fuel, other liquid hydrocarbon combustion products, and solid carbonaceous particles. The particles are composed of solid carbon cores, forming chains or clusters, to which are adsorbed as many as 18,000 different compounds. More than 95% of the particles are < 1 μm in diameter, and 15 to 65% of their mass is adsorbed compounds, largely PNAHs (NIOSH, 1988a). Even though the list of constituents is impressive, they are generally found in very low concentrations unless ventilation is particularity lacking. In some situations, the aldehydes can be found at irritating levels (Scheepers and Bos, 1992a).

A number of the PNAHs are carcinogens, and consequently diesel emissions have long been suspected of being lung carcinogens. Evidence indicates that diesel emissions are carcinogenic in the lungs of laboratory animals, and recent epidemiological data confirm this in a cohort of railroad workers exposed to diesel emissions (Garshick et al., 1988) and among truck drivers (Steenland et al., 1990). An excellent review of diesel emission formation and toxicity can be found in Scheepers and Bos (1992a,b). Two more recent meta-analysis of diesel emissions epidemiological studies have concluded that the emissions are human lung carcinogens with a relative risk of about 1.4 (Lipsett and Campleman, 1999; EPA, 2002). Finally, several case reports have indicated the potential for diesel emissions to cause asthma (Wade and Newman, 1993).

ELECTROPLATING

Description

Prevention of corrosion and rust has historically been the primary reason for applying protective coatings to metal parts. However, there is no doubt that a piece of chrome-plated steel, a silver-plated serving dish, or a gold-plated ring is more aesthetically pleasing than the base metal would be. Hence, decoration is also an important reason for coating metals. More recently the technique has been applied to plastic and rubber parts that have been economically substituted for metal parts, particularly in automobiles and appliances. Other reasons for plating or coating include improving hardness and wear resistance, improving electrical properties such as resistance, and providing a substrate for soldering.

The plating process is carried out in an electrolytic bath, as shown in Figure 5-1. In an electrolytic process, current passes from an anode to a cathode. The anode is usually composed of the same metal that is being plated to the workpiece, although it may be steel in the gold process or lead in the chrome process (Gabe, 1978). The cathode is the part(s) that is (are) being plated (the workpiece). These are suspended from the work bar (cathode) directly or in a basket. The bath is one of four types, as shown in Table 5-1. The most universal baths are cyanide (complex alkaline) and simple acid. However, chromium plating is widely conducted and relies on a complex acid bath. The bath is composed chiefly of water and a salt of the metal

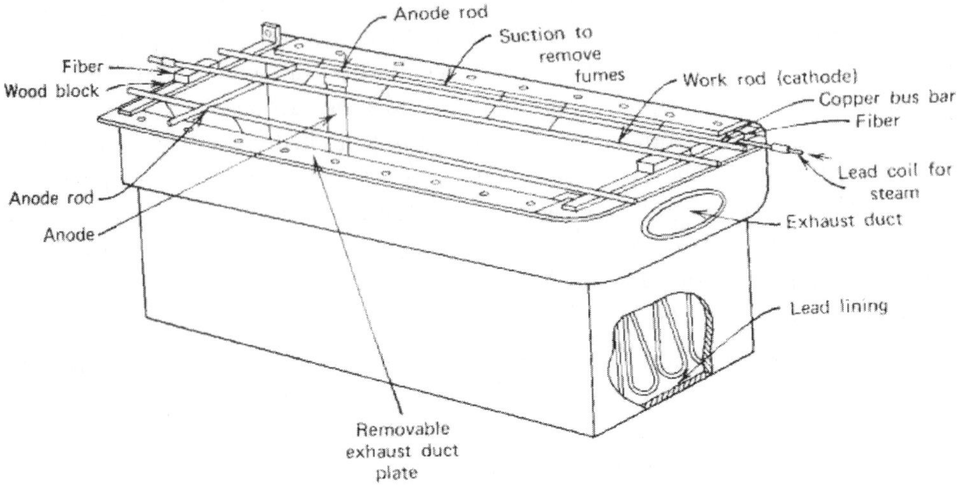

Figure 5-1. A typical electrolytic bath (from Burgess, 1995, with permission).

to be plated. If the salt is copper cyanide, for example, then cyanide ion will form when the salt dissociates. Table 5-1 should be used as a general guide, and more specific information should be obtained for any particular bath. In addition, one should be aware of various bath additives for adjusting pH, and other aspects of water chemistry.

An important event in electroplating is the dissociation of water, releasing oxygen at the anode and hydrogen at the cathode. In some processes, cathode current efficiency is great, and little dissociation takes place (see Table 5-2). In others, for example, the chrome process, dissociation is great, resulting in a fine mist or fog above the tank. Other process factors that affect this are temperature (some processes operate at temperatures as high as 90°C) and mechanical agitation or rotation of the parts to improve coverage (Gabe, 1978; NIOSH, 1984a).

In anodizing, the process is reversed from electroplating; the workpiece is the anode. The purpose of anodizing is to impart a hard and corrosion-resistant, oxidized surface to the metal. Typically a sulfuric or chromic acid bath is used, and misting is high because the purpose is to generate oxygen at the workpiece.

Hazards

Unfortunately, the most commonly used plating metals – chromium, nickel, zinc, and cadmium – have moderate to high toxicities. However, even the less toxic and less used metals – copper, silver, and gold – are plated in *baths* that contain hazardous components. These include cyanide, acids, and alkalis.

Table 5-1
Basic electroplating baths (data from Gabe, 1978)

Type	Description	T°C	Metals
Simple Acid	Metal cation in solution with sulfate, fluoborate, sulfamates, etc.	20–50	Cu, Sn, Zn, Ni, Co, Pb, Fe
Complexed Acid	Metal is anion and is deposited through intermediate stage such as dichromate ($H_2Cr_2O_7$) and chromate formation in chromic acid baths.	40–60	Ni, Cr, Sn
Simple Alkaline	Best example is stannate bath in which potassium stannate and water form a hydroxide.	80–90	Sn, Ni
Complex Alkaline	Generally cyanide baths requiring alkaline electrolytes to avoid hydrocyanic acid formation.	20–70	Ag, Au, Cd, Cu, Zn, Brass, Bronze

Table 5-2
Bath efficiency factors (from NIOSH, 1984a)

Deposited Material	Bath Composition	Cathode Current Density (amp/ft^2)	Cathode Current Efficiency (%)
Brass	Copper–Zinc	5 – 35	50 – 90
Bronze	Cyanide	5 – 35	70 – 90
Cadmium	Cyanide	–	90
Chromium hardplate	Chromic acid	100 – 430	12 – 15
Copper (Rochelle)	Cyanide	20 – 40	50
Nickel Plate	Watts	10 – 100	95 – 98
Silver	Cyanide	5 – 150	99
Tin and Tin Alloy	Stannate	30 – 400	60 – 90
	Acid (sulfate)	10 – 400	100
	Acid (fluoborate)	75 – 130	100
Zinc	Cyanide	20 – 50	75 – 95

Table 5-3
Airborne contaminants released by electroplating, electropolishing, and electroless plating operations (adapted from Table 13.70.07, Industrial Ventilation Manual, ACGIH®, 2007)

Process	Type	Notes	Component of Bath That May be Released to Atmosphere (18)	Physical and Chemical Nature of Major Atmospheric Contaminant	Usual Temp.
Electropolishing	Aluminum	1	Sulfuric, Hydrofluoric Acids	Acid Mist, Hydrogen Flouride Gas, Steam	140–200
	Brass, Bronze	1	Phosphoric Acid	Acid Mist	68
	Copper	1	Phosphoric Acid	Acid Mist	68
	Iron	1	Sulfuric, Hydrochloric, Perchloric Acids	Acid Mist, Hydrogen Chloride Gas, Steam	68–175
	Monel	1	Sulfuric Acid	Acid Mist, Steam	86–160
	Nickel	1	Sulfuric Acid	Acid Mist, Steam	86–160
	Stainless Steel	1	Sulfuric, Hydrofluoric, Chromic Acids	Acid Mist, Hydrogen Fluoride Gas, Steam	70–300
	Steel	1	Sulfuric, Hydrochloric, Perchloric Acids	Acid Mist, Hydrogen Gas, Steam	68–175
Strike Solutions	Copper		Cyanide Salts		
	Silver		Cyanide Salts		
	Wood's Nickel		Nickel Chloride, Hydrochloric Acid	Hydrogen Chloride Gas, Chloride Mist	70–90
Electroless Plating	Copper		Formaldehyde	Formaldehyde Gas	75
	Nickel	2	Ammonium Hydroxide	Ammonia Gas	190
Electroplating Alkaline	Platinum		Ammonium Phosphate, Ammonia Gas	Ammonia Gas	158–203
	Tin		Sodium Stannate	Tin Salt Mist, Steam	140–170
	Zinc	3	None	None	170–180

Table 5-3 (Cont.)
Airborne contaminants released by electroplating, electropolishing, and electroless plating operations (adapted from Table 13.70.07, Industrial Ventilation Manual, ACGIH®, 2007)

Process	Type	Notes	Component of Bath That May be Released to Atmosphere (18)	Physical and Chemical Nature of Major Atmospheric Contaminant	Usual Temp.
Electroplating Fluoborate	Cadmium		Fluoborate Salts	Fluoborate Mist, Steam	70–170
	Copper		Copper Fluoborate	Fluoborate Mist, Steam	70–170
	Indium		Fluoborate Salts	Fluoborate Mist, Steam	70–170
	Lead		Lead Fluoborate–Fluoboric Acid	Fluoborate Mist, Hydrogen Fluoride Gas	70–90
	Lead–Tin Alloy		Lead Fluoborate–Fluoboric Acid	Fluoborate Mist	70–100
	Nickel		Nickel Fluoborate	Fluoborate Mist	100–170
	Tin		Stannous Fluoborate, Fluoboric Acid	Fluoborate Mist	70–100
	Zinc		Fluoborate Salts	Fluoborate Mist, Steam	70–170
Electroplating Cyanide	Brass, Bronze	4,5	Cyanide Salts, Ammonium Hydroxide	Cyanide Mist, Ammonia Gas	60–100
	Bright Zinc	5	Cyanide Salts, Sodium Hydroxide	Cyanide, Akaline Mists	70–120
	Cadmium	5	None	None	70–100
	Copper	5,6	None	None	70–160
	Copper	5,7	Cyanide Salts, Sodium Hydroxide	Cyanide, Alkaline Mists, Steam	110–160
	Indium	5	Cyanide Salts, Sodium Hydroxide	Cyanide, Alkaline Mists	70–120
	Silver	5	None	None	72–120

Table 5-3 (Cont.)
Airborne contaminants released by electroplating, electropolishing, and electroless plating operations (adapted from Table 13.70.07, Industrial Ventilation Manual, ACGIH®, 2007)

Process	Type	Notes	Component of Bath That May be Released to Atmosphere (18)	Physical and Chemical Nature of Major Atmospheric Contaminant	Usual Temp.
	Tin – Zinc Alloy	5	Cyanide Salts, Potassium Hydroxide	Cyanide, Alkaline Mists, Steam	120–140
	White Alloy	5,8	Cyanide Salts, Sodium Stannate	Cyanide, Alkaline Mists	120–150
	Zinc	5,9	Cyanide Salts, Sodium Hydroxide	Cyanide, Alkaline Mists	70–120
Electroplating Acid	Chromium		Chromic Acid	Chromic Acid Mists	90–140
	Copper	10	Copper Sulfate, Sulfuric Acid	Sulfuric Acid Mist	75–120
	Indium	12	None	None	70–120
	Indium	13,14	Sulfamic Acid, Sulfamate Salts	Sulfamate Mist	70–90
	Iron		Chloride Salts, Hydrochloric Acid	Hydrochloric Acid Mist, Steam	190–210
	Iron	12	None	None	70–120
	Nickel	3	Ammonium Fluoride, Hydrofluoric Acid	Hydrofluoric Acid Mist	102
	Nickel and Black Nickel	12,15	None	None	70–150
	Nickel	9,12	Nickel Sulfate	Nickel Sulfate Mist	70–90
	Nickel	13,14	Nickel Sulfamate	Sulfamate Mist	75–160
	Palladium	15	None	None	70–120
	Rhodium	12,17	None	None	70–120
	Tin		Tin Halide	Halide Mist	70–90
	Tin	12	None	None	70–120

Table 5-3 (Cont.)
Airborne contaminants released by electroplating, electropolishing, and electroless plating operations (adapted from Table 13.70.07, Industrial Ventilation Manual, ACGIH®, 2007)

Process	Type	Notes	Component of Bath That May be Released to Atmosphere (18)	Physical and Chemical Nature of Major Atmospheric Contaminant	Usual Temp.
	Zinc		Zinc Chloride	Zinc Chloride Mist	75–120
	Zinc	12	None	None	70–120

1. Arsine may be produced due to the presence of arsenic in the metal or polishing bath.
2. Alkaline bath
3. On magnesium
4. Also Copper-Cadmium Bronze
5. HCN gas may be evolved due to the acidic action of CO_2 in the air at the surface of the bath.
6. Conventional cyanide bath
7. Except conventional cyanide bath
8. Albaloy, Spekwhite, Bonwhite (alloys of copper, tin, zinc)
9. Using insoluble anodes
10. Over 90°F
11. Mild organic acid bath
12. Sulfate bath
13. Sulfamate bath
14. Air agitated
15. Chloride bath
16. Nitrite bath
17. Phosphate bath
18. Hydrogen gas also released by many of these operations.

Table 5-3 gives the common hazards. As previously mentioned, review of specific processes should be made to note any process changes or additives that could lead to other hazards. Finally, electroplating is an end process. Other important steps in the treatment of metals (and equally hazardous from a hygiene point of view) are the surface preparation steps. These are discussed below in the sections on *Metal Surface Treatment*.

GRINDING, POLISHING, AND BUFFING

Description

Very often in the manufacture (founding, milling, extrusion, molding, etc.) of a metal or plastic product, appendages or other imperfections must be removed before the final product can be finished. For example, metal castings produced in foundries will have appendages known as gates or risers caused by the hole necessary to pour the mold. Other surface imperfections may also exist, depending on the process and the desired quality of the finish. Grinding is used for rough finishes. A finished fire hydrant will have been ground but not buffed or polished. Polishing removes much smaller imperfections that may be thought of as equivalent to scratches. Buffing is used to provide a lustrous surface, as on a piece of jewelry or a stainless steel ball bearing. Grinding removes a significant amount of the workpiece, polishing less, whereas in buffing only insignificant amounts are removed.

The processes are accomplished with abrasives; some are identical with those used in abrasive blasting; although it would be uncommon to find silica used. The abrasive is incorporated in a wheel, disk, sphere, or other object, depending on the nature of the work. A bonding material serves to hold the abrasive in place, so that as the work proceeds, both the abrasive and the bonding agent are worn off, revealing fresh abrasive.

The common abrasives and bonding agents are shown in Table 5-4. The codes may be used to identify the makeup of a particular grinding wheel. The most common abrasives used in grinding are the synthetics, aluminum oxide, silicon carbide and diamond. The most common bonding agents are vitrified glass (in a process whereby silicate and feldspar are heated and pressured to form an amorphous structure) and phenol-formaldehyde resin, a plastic.

For polishing, other, finer abrasives may be used, including chalk, pumice, and tripoli (a form of silica). In buffing operations, rouge or tin putty is generally used although fine aluminum oxide may be used on harder metals. Red rouge is composed of fine particles of iron oxide, and green rouge contains particles of chrome oxide (Cr_2O_3). The rouge abrasive is incorporated in clay, grease, or wax in the shape of a bar.

Hazards

In grinding and polishing the hazard is largely associated with the workpiece and not the wheel (unless the wheel explodes!). The base metal is, of course, important, and, for example, unacceptable airborne levels of lead can easily be achieved in grinding on brass (zinc-copper alloy often containing $< 10\%$ amounts of lead). But, the silica hazard is equally important when grinding on castings that have been formed in silica molds. Without adequate ventilation, this operation often exceeds acceptable levels for silica.

Polishing operations are generally less hazardous than grinding because relatively small amounts ($\ll 0.1$ mm) of base material are removed in polishing. Nevertheless, particularly toxic base metals should be of concern.

In buffing, the base metal is of little concern. Large amounts of the rouge may be used. Green rouge, while containing chrome, consists of the trivalent type and thus is considerably less hazardous than the hexavalent species.

Table 5-4
Grinding wheel nomenclature for the products of one company based on ANSI (1990)

5-4a A typical wheel marking includes the following information or codes

Abrasive	Grit	Grade	Structure	Bond Type
WA	46	K	8	V

5-4b Abrasive and bond characteristics (other abrasives include D, diamond; other bonds include E, shellac and S, silicate)

Abrasive		Bond Type	
Symbols	**Characteristics**	**Symbols**	**Characteristics**
A	Regular aluminum oxide, hard and tough, suitable for mild steel. Cast-iron, etc. in rough grinding.	V (Vitrified)	Most common bond for precision grinding porosity and strength. Wheels made with this bond give high stock removal and their rigidity helps to attain high precision. Not affected by water, acid, oils or ordinary temperature variations.
10A	The prime regular aluminum oxide, hard and tough. It is suited for cylindrical grinding of unhardened steels.		
WA, 38A	White aluminum oxide, hard and brittle. Suitable for all kinds of steel grinding.		
32A	Strong sharp mono crystalline aluminum oxide. Suitable for ail kinds of tool precision grounding.	B (Resinoid)	Used for wheels in fabrication shops, foundries, billet shops, sharpening and gumming saws, and in many precision applications. These wheels excel in rapid stock removal as well as in applications requiring better finishes. This bond is also restricted to reinforced resinoid products, cut-off wheels, or cut-off wheels, and for snagging with portable grinders.
FA, 57A	Semi-friable, with a grinding action. Intermediate A and WA. Suitable for cylindrical grinding of steel parts.		
PA	Pink aluminum oxide is suitable for tool grinding and surface grinding.		
RA	Ruby aluminum oxide, hard and less brittle than PA. Suitable for high alloy steel, etc., in precision grinding.		
PSA	Special pink abrasive, similar to 32A. Suitable for all kinds of hard steels in precision grinding.	R (Rubber)	Used for most centerless feed wheels, precision ball race grinding wheels, and for snagging with portable grinders Used where finish is important and for cut-off wheels where burr and burn must be held to a minimum.
AZ	A modified high zirconia aluminum oxide, used in foundries snagging cast iron and steels for heavy duty work.		

5-4b (Cont.) Abrasive and bond characteristics (other abrasives include D, diamond; other bonds include E, shellac and S, silicate)

Abrasive		Bond Type	
Symbols	**Characteristics**	**Symbols**	**Characteristics**
C	Black silicon carbide. Used for grinding gray iron, non-ferrous metals and nonmetallic materials such as glass, rubber, stones, etc.	Mg (Magnesia Oxychloride)	A notable bond used for cutlery or particularly thin workpieces, marble surface grinding and super finishing.
GC	Green silicon carbide is well suited for special alloy grinding such as tungsten carbide, cold-rolled steels, sapphire, glass, gem and granite.		

5-4c Grade indicates the relative strength of the bond that holds the abrasive grains in place

Very Soft	Soft	Medium	Hard	Very Hard
DEFG	HIJK	LMNO	PQRS	TUVWXYZ

5-4d Grain size (a symbol after grain size indicates mixed grain from 1, fine to 7, coarse)

Coarse	Medium	Fine	Very Fine	Ultra Fine
10 12 14	30 36 46	70 80 90 100	240 280 320 400	1000 1200 1500 2000
16 20 24	54 60	120 150 180 220	500 600 700 800	2500 3000 4000

5-4e Structure (dense to open)

Symbol	0	1	2	3	4	5	6	7	8	9	10	11	12	13	14
Abrasive Ratio %	62	60	58	56	54	52	50	48	46	44	42	40	38	36	34

LABORATORIES

Description

Perhaps laboratories need little description because virtually any hygienist or hygiene student will be familiar, to some extent, with laboratory operations via his or her science background. However, few realize that laboratory operations are found in virtually every industry. Research laboratory operations are associated with such workplaces as pharmaceutical companies, universities, hospitals, and clinical and chemical assay laboratories. However, product quality control, product analysis, and product development laboratories exist in very many industries.

Product quality control and analysis laboratories are probably the most important. Tests may include chemical analyses, microbiological assays, and tests of physical properties or engineering tests such as tensile strength.

Product development laboratory operations may fit within a typical lab hood or be of a larger pilot scale. In many cases the products, if not the raw materials, will be unfamiliar and therefore particularly dangerous in these operations.

The respect for safety and health in a laboratory setting may be high, corresponding to a high level of concern on the part of management. However, laboratories can also be among the most dangerous areas of a plant for a number of reasons. Laboratory personnel may develop a false sense of security based on their professional training and consequently may disregard safety and health warnings. Few academic training programs for chemists and biologists incorporate health and safety issues. Personnel often must handle materials on a temporary basis, and taking the time to learn safe handling procedures may be viewed as nonproductive or of low priority. In short, scientists and technicians may be too occupied with their research to consider safe practices. Personnel are often exposed to many hazardous substances or other hazards over the course of a career, and although the dose of any one substance may be small, taken together the effect of the hazards may be significant.

For the above reasons, usual health and safety standards should not be applied to laboratories. Instead an *as low as reasonably achievable* or *ALARA* approach should be taken. For example, all work involving chemicals should be performed in a laboratory hood regardless of the hazard or exposure.

Hazards

The hazards are impossible to enumerate. In addition to chemical hazards, the hygienist should look for ionizing radiation sources, lasers, and biological agents. Among the chemical agents, be aware that laboratory techniques occasionally prescribe the use of chemicals that have been abandoned in manufacturing because of their toxicity. A reagent that has not been in industrial use for 20 to 30 years may continue to be used in the laboratory even though its toxicity is such that it has been widely substituted in industry. Carcinogens such as β-naphthylamine and benzidine are good examples. Compressed gases are common, and workers must take adequate precautions against exposures when changing empty tanks.

MACHINING AND CUTTING OILS

Description

Generally, the purpose of machining is to reshape or reform a workpiece. The workpiece may be metal, plastic, wood, or other material. Metal workpieces are most common and are thought of in conjunction with machining; however, plastic workpieces are becoming more common. Woodworking is generally considered separately because the hazards are different; in particular, no cutting fluids are used. Metal pressing (forging) and rolling (rolling mill) may also be considered as metal machining, particularly because cutting or machining oils are used; however, they are generally considered separately. Similar processes to pressing and rolling that are used for plastics are termed extrusion, and are discussed later in this chapter.

Machining involves lathes, drills, millers, shapers, planers, and surface grinding. The latter does not generally involve cutting oils and was discussed above in this chapter. Conventional tool machining is most popular, but electrochemical and electric discharge machining are increasing in importance.

In conventional tool machining, generally a special tool or the workpiece turns, or both turn in contact with each other. Because the tool is harder than the workpiece, it shears off a portion of the workpiece. Friction is great, and the interface comes under extremes of temperature and perhaps pressure. For this reason, a coolant and a friction reducer or cutting oil are needed.

The hygienist is concerned with the chemical nature of the cutting fluid, the workpiece, and the cutting tool. The workpiece may be almost any metal or plastic although mild steel is most common. The cutting tools may be of high carbon alloyed steels containing vanadium, chromium, and manganese; high speed steels of manganese and tungsten; cobalt steels; cast alloys of tungsten, chromium and cobalt; and tungsten carbide.

The cutting oils[2] are more complicated, as can be seen in Table 5-5. Each is composed of a base oil (or water in the case of synthetics) and a number of additives. Indeed, the exact formulation is often more complex than Table 5-5 indicates as, for example, two or more germicides, corrosion inhibitors, polar additives, or base oils may be used. The formulation is often proprietary, but because the additives become unbalanced during use, they must be augmented over time. Thus, some of the raw materials are often on hand at the site for this purpose.

Cutting oils generally are not discarded, only replenished. Because the oils and some of the additives are biodegradable, germicides or biocides are needed to control odor as well as to help preserve the nature of the formulation. Protection of the workers from inhalation of biological aerosols is a secondary but important benefit. Too much biocide may lead to other health problems.

Hazards

The cutting tool is not of concern, as only small amounts are lost over time. On the other hand, the workplace is of particular concern if the common metals in it are of high toxicity. Beryllium is a good example. Plastics and the fumes that are formed when they are machined

[2] Sometimes these are confused with mineral oils. The latter are organic petroleum derivatives (see IARC, 1984) that are sometimes used in cutting oils (see Table 5-5).

Table 5-5
Composition of cutting fluids (NIOSH 1978)

I. Mineral oil
  1. Base 60–100%, paraffinic or naphthenic
  2. Polar additives
     a. Animal and vegetable oils, fats, and waxes to wet and penetrate the chip/tool interface
     b. Synthetic boundary lubricants: esters, fatty oils and acids, poly or complex alcohols
  3. Extreme pressure (EP) lubricants
     a. Sulfur-free, or combined as sulfurized mineral oil or sulfurized fat
     b. Chlorine, as long-chain chlorinated wax or chlorinated ester
     c. Combination: sulfo-chlorinated mineral oil or sulfo-chlorinated fatty oil
     d. Phosphorus, as organic phosphate or metallic phosphate
  4. Germicides

II. Emulsified oil (soluble oil) – opaque, milky appearance
  1. Base: mineral oil, comprising 50–90% of the concentrate; in use the concentrate is diluted with water in ratios of 1:5 to 1:50
  2. Emulsifiers: petroleum sulfonates, amine soaps, rosin soaps, naphthenic acids
  3. Polar additives: sperm oil, lard oil, and esters
  4. Extreme pressure (EP) lubricants
  5. Corrosion inhibitors: polar organics, e.g., hydroxylamines
  6. Germicides
  7. Dyes

III. Synthetics (transparent)
  1. Base: water, comprising 50–80% of the concentrate; in use the concentrate is diluted with water in ratios of 1:10 to 1:200. True synthetics contain no oil. Semisynthetics are available that contain mineral oil present in amounts of 5–25% of the concentrate
  2. Corrosion inhibitors
     a. Inorganics: borates, nitrites, nitrates, phosphates
     b. Organics: amines, nitrites (amines and nitrites are typical and cheap)
  3. Surfactants
  4. Lubricants: esters
  5. Dyes
  6. Germicides

may result in polymer fume fever (DuPont, 1970). Coolants may exacerbate the polymer fume fever problem (NIOSH, 1987). Noise is also a constant hazard.

The most important hazards are from the cutting oils, which have caused a wide variety of skin diseases including oil acne, contact dermatitis, skin and scrotal cancer, and oil melanosis. The last is a loss of pigmentation, largely seen in Mediterranean people. The cancers have been reported mostly in Europe (Key et al., 1983).

Some of the additives, particularly biocides (including glutaraldehyde and formaldehyde) and corrosion inhibitors, are pulmonary irritants. Under normal conditions they are present in low concentrations and should cause only minimal problems. However, where the old adage "a little is good, so a lot is better" is followed, when workers replenish these oils, pulmonary irritation can develop. In addition, heavy workloads or sensitive individuals may lead to complaints of pulmonary problems.

A large concern in the past was the formation of carcinogenic nitrosamines as a result of the addition of sodium nitrite and triethanolamine to cutting fluids. A 1983 study (NIOSH, 1983) found nitrosamines in 18 of 32 cutting oils examined. Although most manufacturers of cutting oils should be aware of and are currently avoiding this problem, the potential for nitrosamine formation should not be overlooked.

METAL SURFACE TREATMENT – INORGANIC

Description

This section and the following section deal with the treatment of metal surfaces, often for the purpose of applying a subsequent coating such as electroplating or painting, or so that welding, brazing, or soldering can take place. Consequently, wherever painting or electroplating is done, surface treatment will usually take place. The two most common surface treatment processes are degreasing (an organic chemical process) and pickling (generally an inorganic chemical process).

The *pickling* process usually involves an acid bath of sulfuric, hydrochloric, or phosphoric acid. Nitric acid, a strong oxidizer, is used more sparingly than these acids, and hydrofluoric acid, in small amounts, is sometimes used in conjunction with the others. Some authors also include alkaline and hot salt baths in the pickling process description, but the term appears to have been derived from the early use of vinegar (acetic acid, Burgess, 1991). The term *bright dip* is also used for specific pickling operations.

The primary purpose of acid pickling is to remove rust (iron oxide) and scale. The latter term seems to apply to any unwanted metal oxide, as well as to calcareous deposits from water systems (Lewis, 1993). There are four classes of acids and bases used. Inhibited, non-oxidizing acids, such as HCl, H_3PO_4, and some H_2SO_4, may be used on all iron and steel (except 300 series stainless), all copper alloys, and template. Similarly, organic acids and alkaline solutions may be used on the above metals. Nitric acid (HNO_3) may be used only on 300 series stainless containing chromium. Only certain formulations of HNO_3, organic acids, and alkalies are used on aluminum. Finally, none of the above is used on galvanized steel (for cleaning purposes); but for all of the metals above, organic solvents (degreasing) may be used for removing grease and grime (Schneberger, 1985).

Alkaline techniques employ sodium hydroxide, potassium hydroxide, sodium carbonate, sodium meta- or ortho-silicate, trisodium phosphate, borax, and tetrasodium pyrophosphate. This process generally does a better job than acid pickling on greases, paints, and other organics, but is slower in removing scale. For this reason electrolytic, heated baths are used, followed by a water quenching operation.

Molten *salt* baths are also used in metal treatment. They are usually hot (430–540°C) baths of sodium hydroxide. They have the advantage of being able to remove oxides of such metals as chromium, nickel, iron, and molybdenum without attacking the base metal. In addition thcy will remove oils, greases, carbon, and even sand from castings. Subsequent to the bath, the workpiece must be quenched; the generation of steam and thermal shock helps remove the scale. This is followed by acid pickling and a water rinse.

Phosphoric acid salts are used on iron and steel and their alloys (particularly mild and galvanized steel) in a process called phosphating. The salts are generally of iron, manganese, ammonium, zinc, or sodium and the bath is usually heated. The process affects the appearance, hardness, electrical properties and, to some degree, the corrosion resistance. Some metal surfaces are prepared this way prior to painting, particularly in the automotive industry.

Hazards

The most important hazards are the acids or the bases that make up the baths. In any of the pickling operations, the quality of the final product as well as the hazard to the worker is determined by acid or alkaline concentration, bath temperature, processing time, contamination of the bath, additives (particularly inhibitors) present, and agitation of the solution. Additives can be particularly important, in that some create a surface foam that tends to decrease the evolution of acid mist from the bath. The faster the pickling rate is, the faster the evolution of hydrogen gas, and therefore, the greater the production of acid mists. In electrolytic alkaline cleaning, the amount of current applied and agitation are also important factors. For electrolytic alkaline baths, the amount of current is also of importance in exposure. Generally, the faster the process, the higher the concentration and the temperature must be, and therefore, the greater the likelihood of a higher airborne concentration of mists and gases.

METAL SURFACE TREATMENT – ORGANIC

Description

Organic chemicals, historically hydrocarbons and chlorinated hydrocarbons, are used to remove greases, oils, and other organic-based grime from metals and plastics. This process, called degreasing, is probably the most wide-ranging process in industry. At one time every automotive garage and fueling station used an open bucket of gasoline or kerosene to clean or degrease engine parts, tools, and hardware. Now the systems are slightly more sophisticated and often employ recycled solvents, but the process is still degreasing. On the other end of the spectrum, a variety of organic chemical mixtures for cleaning circuit boards, chips, and other electronic components has been spawned in the electronics industry.

Prior to 1987, the degreasing process was fairly simple to describe. It could be broken down into two somewhat distinct processes. Cold degreasing has remained relatively unchanged. In this simple process, a solvent is used to spray, soak, or in any way coat a metal part to clean it. Typically, kerosene or mineral spirits such as Stoddard solvent (a C7-C11 petroleum fraction) are used, although other more specialized solvents such as ketones, aromatic hydrocarbons, and even heterogeneous mixtures may be found in use. The old gasoline station buckets have been replaced for the most part by safety containers (commonly referred to as parts washers) that allow the part to be sprayed or soaked. When not in use, or during soaking, these containers can be closed to control vapors for both safety and health reasons. Although some cold degreasing operations employ chlorinated solvents, this is unusual and not recommended, for reasons to be discussed.

Hot or vapor degreasing is generally the more hazardous of the two processes, and on a per-operation basis is carried out on a larger scale. In the vapor degreasing process, a large tank is used, as shown in Figure 5-2. The solvent is placed in the bottom of the tank and boiled. The solvent vapors rise into the work zone, or the region where the parts are introduced. The parts are typically lowered into the tank in a basket with a hoist. Large parts may be lowered with chains or slings. The relatively cool parts cause the vapor to condense, and the solvent and grease fall to the bottom. At this point the parts appear dry and are finished. However, parts with surface irregularities will trap solvent, which may be removed with the parts from the tank.

Two important design features conserve solvent and therefore lower the workers' breathing concentration. First, a set of condensation coils containing cold water or another coolant rings the inside of the tank. These coils cause a cool layer of air to form, which condenses the solvent so that it falls. Second, a freeboard typically extends about three feet (one meter) above the condensation coils. This reduces air crosscurrents. Nevertheless, some solvent is lost, particularly in lowering and raising parts and when air cross currents are present.

Before 1987[3], the properties of the solvent (e.g., boiling point and vapor pressure) determined the design of the degreaser. For this reason, solvent substitution cannot be performed without redesign. However, the Montreal Protocol of 1987 required the phase-out of production of many chlorofluorocarbons (CFCs) and other ozone depleting substances by the year 1996 (Seidel, 1991). Certain aliphatic compounds containing chlorine, fluorine, and bromine

[3] The Montreal Protocol, phasing out the production and use of certain ozone depleting chemicals, was signed in 1987. However, it was modified with an earlier schedule in 1992 and regulations in the U.S. and other countries did not go into affect until mid 1990s.

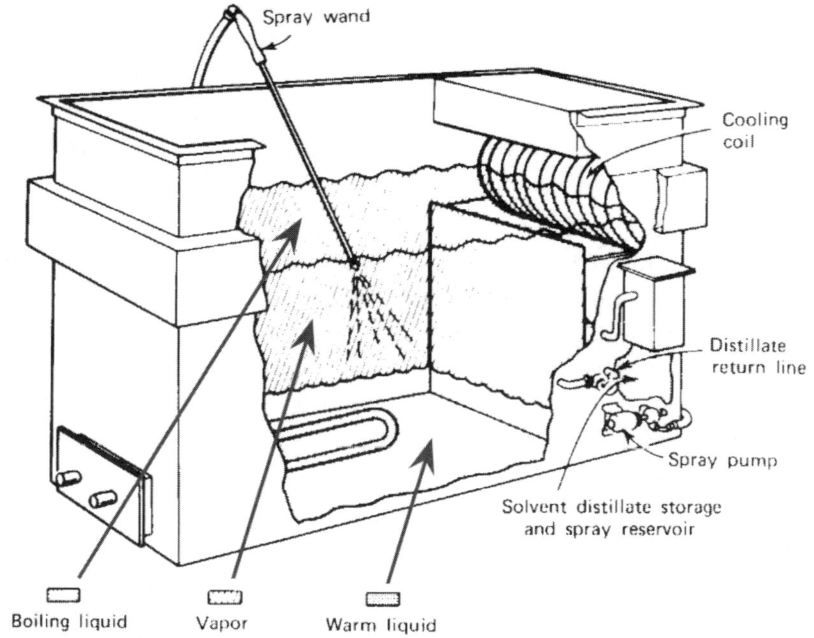

Figure 5-2. A typical vapor degreaser (from Surface Preparation and Finishes for Metals, J.A. Murphy, McGraw-Hill, 1971, with permission).

can remain stable in the lower atmosphere long enough so as to increase the likelihood that they will enter the stratosphere, where they react with ultraviolet light, ionize, and destroy ozone. In the stratosphere, ozone absorption of harmful ultraviolet light reduces the risk of skin cancer and other diseases in humans and is also beneficial to non-human species.

Two important vapor degreasing compounds belong to this group of long-lived chemicals: Freon 113™ (1,1,2-chloro-l,2,2-fluoroethane) and l,l,l-trichlorethane. Other vapor degreasing solvents – perchloroethylene, trichloroethylene, and methylene chloride – are not long-lived but are considerably more hazardous to the worker than the former solvents. Thus, when the health hazards of these latter three were discovered in the 1970s and early 1980s, regulations became stricter. For example, trichloroethylene is a class 2 carcinogen in the European Union and all three are Hazardous Air Pollutants in the U.S. under the National Emission Standards for Hazardous Air Pollutants of the 1990 Clean Air Act. In the 1970s and 1980s there was a change to the two, above-mentioned ozone depleting substances. As a result of the Montreal Protocol in the mid 1990s, there is now a movement back to perchloroethylene, trichloroethylene, and methylene chloride. For example, at the time of the production ban on 1,1,1-TCE there were around 3,000 vapor degreasers in the south coast basin of California. After 1996 there were fewer than 250 and most of those used perchloroethylene. Since 2002 the Southern California district controlling air pollution has banned the use of perchloroethylene and other chlorinated solvents in degreasers.

Other substitutes are available. These include hydrofluoroethers (e.g., methoxy – nonafluorobutane or HFE 7100), blends of hydrofluoroethers, N-methyl pyrrolidone, p-chlorobenzotrifluoride, N-propyl bromide, 1,2-butylene oxide, and hydrofluorocarbons such as HFC 4310. Hydrochlorofluorocarbons such as HCFC 141b (1,1-dichloro-1-fluoroethane) were banned as of 2003. Restrictions are changing constantly and so one should refer to EPA's SNAP program for restrictions on ozone depleting chemicals (EPA, 1996).

Perhaps the most widely used substitutes are water-based solutions containing various additives or alkaline solutions (based on caustic soda and detergent with phosphate and sodium meta-silicate additives) and d-limonene, a derivative of citrus peel. These solutions do not utilize old style vapor technology. No one method and not even a few approaches are achieving dominance; so the hygienist must always be on the lookout for new hazards. These and other formulations are being tried in the metals industry. None seems to have the universal acceptance or application of 1,1,1-trichloroethane.

In the electronics industry the major application is in the removal of rosin flux (essentially the gooey organic substance left when the terpenes are distilled from pine tar) from circuit boards. The flux is necessary to obtain a good lead solder junction. Historically, a number of chlorinated solvents were used in these degreasing tanks. By 1987 mixtures of Freon 113 with either methylene chloride, acetone, methanol, or ethanol were most popular. Later the use of glycol ethers became common, but concern over their reproductive effects has essentially stopped that use.

Currently there appear to be three approaches: alterative solvents for removing rosin fluxes (e.g., 1,1-dichloro-1-fluoroethane[4] and n-propyl bromide), use of aqueous fluxes removed with aqueous solutions, and use of *no-clean* fluxes. Certainly the last two appear to be most attractive, but both leave residues on the equipment that in some applications are unacceptable. Aqueous solvents include mixtures based on an amine, such as triethanolamine, glycol ethers, and water. Also long-chain esters and alcohols are being used. Emulsion cleaners utilize an emulsifier that allows polar solvents such as water and dibasic acid esters (for example, see Axarel™ products of DuPont) to be mixed with non-polar mineral spirits.

Hazards

Regardless of the degreasing equipment or process, solvent vapors are released. The use of 1,1,1-trichlorethane and Freon 113 was prompted to some extent because their toxicities are lower than those of perchloroethylene, trichloroethylene, and methylene chloride. So, all else being equal, higher vapor levels outside the degreaser could be allowed without the need for ventilation control.

---

[4] This compound is technically an HCFC and not a CFC. The more bonds to hydrogen on these simple hydrocarbons, the faster they degrade in the troposphere and therefore are less likely to reach the stratosphere. All are scheduled for phase out in the next few decades under the Montreal Protocol.

In the newer approaches, perhaps the glycol ethers have been of the most concern. Certain of these compounds do seem to be associated with reproductive effects (Rudolph and Swan, 1986; Welch et al., 1988; Pastides et al., 1989). The propyl-based glycol ethers (until recently the most popular were methyl- and ethyl-based) may be the least toxic of the group. The alkaline-based immersion processes are heated and therefore, depending on the workload, may require evaluation of sodium hydroxide mists. Some emulsion solutions may be of low hazard when used in a soak tank. However, others containing compounds such as phenols or cresylic acid should be evaluated. The verdict on d-limonene is still out. Although it is *natural*, recent animal evidence indicates carcinogenic potential (NTP, 1990). Virtually all degreasing systems pose a hazard via dermal exposure, either as a cause of dermatitis or through the ability of many organic compounds to be absorbed through the skin.

In short, cold degreasing processes employing hydrocarbons should pose no threat so long as fire safety systems are used, lids are tightly in place whenever possible, and the use is occasional. On the other hand, all vapor degreasers generally require evaluation regardless of the solvent. For some newer solvents such as d-limonene, evaluation with respect to toxicity is incomplete. Consequently, they should be controlled closely until toxicity data are developed.

PAINTING

Description

A better name for this process would be coating because it covers a number of coatings that are not typically considered to be paints. There are a wide variety of techniques used to apply coatings, as well as a wide variety of coating types. Some pertinent terms are defined in Table 5-6. Although many construction, particularly single-family dwelling, applications of coatings use latex formulations, industrial applications tend to be oil-base. Latex paints are water suspensions; although they contain water, they also contain organic solvents. Oil-base coatings contain petroleum-derived constituents, including organic solvents, but no water.

Paints consist of a vehicle, made up of binder and usually solvent; pigments; and a broad group of compounds, usually in small amounts, termed additives. The binder is an organic polymer and today it is usually synthetic. Natural binders, such as rosin or shellac, usually are derived from trees and now account for less than 10% of all binders. The binder cures by oxidation, solvent drying, heat, or some combination of these, and the hardened substrate binds the pigments, or in the case of varnish simply provides a protective coating. Figure 5-3 shows the variety of binders or *film-formers* that are used as part of the vehicle.

Figure 5-3 also shows the variety of the pigments used, and it is in this area that the most important hygiene concerns are usually found. Although solvent exposures are now considered to be quite important, and exposures to components of some binders, particularly epoxies and urethanes, can be significant, historically chrome, cadmium, and especially lead pigments caused very significant effects among painters. Efforts are now being made to provide substi-

Table 5-6
Types of coatings

Paint	A solution, emulsion, or suspension of inorganic or organic pigments. A polymeric binder to hold pigments in place and provide a protective surface. Solvents are normally present, but are not in powdered paints. Various additives include biocides, drying agents, corrosion inhibitors, fire retardants, fillers, and others.
Varnish	A non-pigmented solution of solvents, and a binder of natural oils, resins, and/or synthetic polymers that *dries* by solvent evaporation and air oxidation of the binder.
Lacquer	A solution of solvents and, usually, natural resins such as shellac in which the solvent evaporates and no further curing or chemical reaction takes place.
Enamel	A pigmented varnish.
Stain	Essentially a paint with low solids content.
Primer	Essentially a paint having a high concentration of corrosion inhibitors (for metals) and/or additives to improve adhesion or sandability.

tutes for these pigments, particularly lead, but they continue to pose hazards in the spray application of modern paint systems. In addition, they are particularly hazardous during removal procedures such as blasting, sanding, grinding, and heating.

The wide range of solvents and additives is also impressive. It should be noted that latex paints contain coalescing solvents, in addition to the water base, but they are generally present in concentrations of less than 15% (Hansen et al., 1987), whereas oil-base coatings may contain more than 75% solvent. Some of the most important additives, from a hygiene point of view, are the biocides, which are usually fungicides. These include tributyl tin, copper and zinc naphthenate, copper oxide (Burgess, 1991), and phenyl mercury compounds (Peterson, 1984). In addition, proprietary biocides used in latex paints may release significant quantities of formaldehyde (Hansen et al., 1987). Drying agents added to paints are generally metal catalysts used to promote oxidation or curing. They include compounds of cobalt, lead, barium, calcium, cerium, iron, zinc, and zirconium. They are generally of low concentration in the coating.

Two of the binders deserve special attention – epoxies and polyurethanes. Both are generally two-part systems. Epoxies are usually composed of a polymer of epichlorohydrin and bisphenol A, as shown in Figure 5-4. This component is the epoxy. Excess epichlorohydrin is used to cause unreacted epoxide groups to be at the end of the chain. Thus, unreacted epichlorohydrin exists in the liquid epoxy resin. This is then added to a hardener or polyamine, which reacts with the epoxide groups. Typical polyamines include diethylene triamine, triethylene tetramine, and tetraethylene pentamine. Both the polyamines and the bisphenol A can be substituted, based on cost, but not the epichlorohydrin (Bilmeyer, 1984). Baked-on epoxy sys-

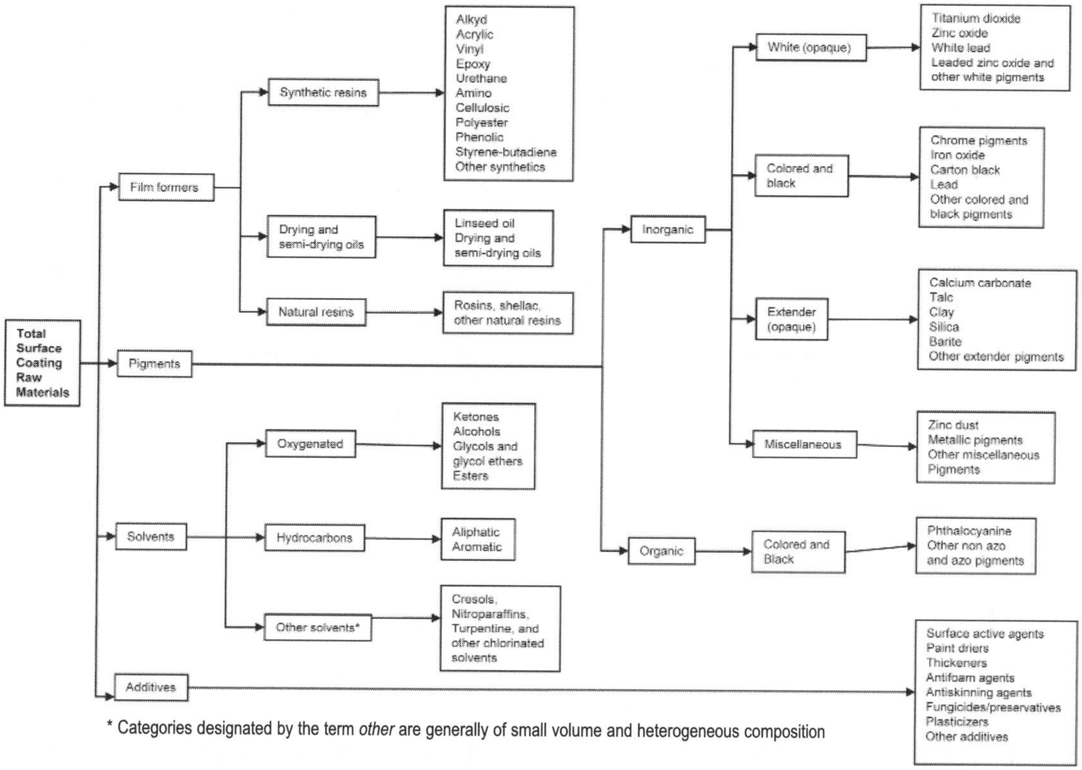

Figure 5-3. Raw materials used in coatings (from NIOSH, 1984).

tems use less toxic polyamide curing agents, but they require lower molecular weight epoxies, so that there are more unreacted chain-end epoxides on a volume basis. In addition, one-part systems exist in which the epoxy is reacted with a drying oil that is a non-saturated fatty acid. Usually with the aid of a catalyst, the drying oil cures or hardens at the double bonds by air-induced oxidation. Because the drying oil is pre-reacted with the epoxy, one-part systems are generally safer to apply than two-part systems.

Polyurethanes are generally composed of a diisocyanate, such as toluene diisocyanate, and a polyol, such as a polyester made from adipic acid and ethylene glycol, as shown in Figure 5-4. These components together form a prepolymer. As with the epoxy system, excess isocyanate groups exist in this prepolymer or first component of the two-component system. The prepolymer then is reacted at the isocyanate groups with a second component. The second component may be a diamine or a glycol or even water vapor from the air, which reacts upon drying. As with epoxies, a drying oil may be used to form a one-part system.

Coatings may be applied by brushing, rolling, dipping, or wiping. However, some type of spray process is the most common and potentially the most hazardous method. There are sev-

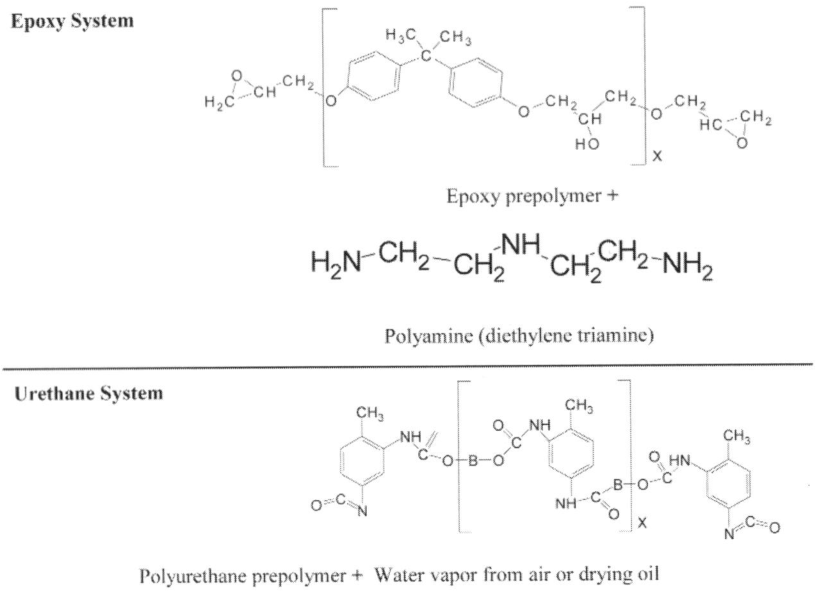

Figure 5-4. Examples of one type of urethane and epoxy systems. There are several variations of each.

eral types of spraying techniques. The most common utilizes compressed air to force the coating through an orifice in a spray gun, resulting in a droplet having a size dependent on orifice size, air pressure, paint viscosity, and other factors. The compressed air causes overspray, the bouncing of the paint from the surface or the blow-by of the paint caused by the buoyancy of the air. This wastes product and exposes the worker to the coating, so other techniques have been invented; but certain types of jobs require compressed air spraying.

Airless spraying results from mechanical pressure being applied to the paint. Control of droplet size is generally less adequate with this method than with compressed air. If a conductive surface is being sprayed, then better application is achieved if the droplets and the surface have opposite charge. Airless or compressed air techniques may be used in this electrostatic fashion, in which case the gun is used to impart a charge to the droplets.

Powder paint operations are now widely used; the paint contains no solvent. Typically the solid or viscous particles are sprayed on electrostatically and robotically. The parts then enter an oven where a heat and/or radiation curing process smooths the paint and completes any necessary chemical reactions. The painting must be performed in well-controlled ventilation booths.

Hazards

Totally reacted long-chain polymers do not appear to be hazardous although many are untested. New polymers are currently exempt from the U.S. EPA's regulations under the Toxic

Substance Control Act. Binders and monomers that are not totally reacted, such as the epoxies and polyurethanes, can present a significant risk. Unreacted monomers, such as vinyl compounds in vinyl paints and acrylic acid esters in acrylic paints, may not pose a significant risk based on concentration. On the other hand, free epichlorohydrin or diisocyanate may pose a significant airborne hazard, particularly some of the isocyanates, which are sensitizers at very low concentrations. In two-part systems, the unreacted epoxide or isocyanate ends of the prepolymers are immune sensitizers. In addition, in the epoxide system the polyamines are sensitizers (Peterson, 1984), The inhalation hazard is obviously greatest in spraying operations; but even in all other applications the skin must be protected, as these compounds cause allergic contact dermatitis, and the individual can be sensitized by this route as well.

Because of problems caused by these systems in the past, formulation changes have been made. However, the hygienist should always be aware that changes are rarely universal (even if mandated by government, compliance is rarcly 100%), and all improvements are rarely equal. So just as you think the epoxy system is under control, the purchasing agent slips in a cheaper brand that does not incorporate the chemical changes that decreased toxicity. Relatively recent studies have shown that modern epoxy (Rempel et al., 1991) and polyurethane (Patterson et al., 1990; Reh and Lushniak, 1990; Decker, 1991) paint systems exist that are capable of causing high exposure levels and health effects, even when brushing processes are used (Huang et al., 1991).

Chrome, cadmium, and lead pigments must be monitored. The use of lead as a pigment in spray painting operations will almost always lead to hazardous exposures. The same can probably be said for cadmium with recent changes in U.S. federally acceptable levels. Chromium is a lung carcinogen in the hexavalent oxide form, and the hexavalent water-insoluble pigments cannot be ruled out from causing a similar risk. Surprisingly, 12,700 California workers were exposed to lead-based paints in 1991, an indication that lead is still prevalent in paints (Sharp et al., 1991). This date, 1991, is almost 20 years after lead was removed from household paint and 13 years after U.S. OSHA posed significant regulations on occupational exposure to lead.

The number of solvents is too large for listing. The hydrocarbons are the most common, and petroleum distillate mixtures are often used. There is some confusion about the nomenclature of these products, which is defined in Chapter 4, Table 4-8. The various hydrocarbons have relatively low toxicity as solvents go, with the exception of hexane and benzene. These compounds should not be present in significant quantities in coatings from reputable companies. If in doubt, consult the product literature or contact the company.

It has long been known that many solvents have anesthetic effects at high concentrations. However, it is now recognized that painters with long-term exposures to paint solvents may experience mood and behavioral effects such as increased accident rates (Hunting et al., 1991; Triebig et al., 2000), effects on autonomic nerve conduction velocities (Murata et al., 1991), effects on verbal learning, memory, and cognitive function (Hanninen et al., 1991) and neuroticism (Chen et al., 2001). Other studies have failed to find such effects among paint manufacturing employees (Bleecker et al., 1991). Other studies have found only minor effects in painters (De Rosa et al., 1985). Nonetheless, animal studies show behavioral effects when all other factors are controlled (Goldberg et al., 1964; Silverman and Williams, 1975; Savolainen et al., 1977, 1979).

With respect to the additives, tributyl tin may be a hazard during spraying of marine and other coatings that must withstand a harsh environment. Tributyl tin compounds are important skin and eye irritants (Sheldon, 1975). Mercury compounds in latex paints may be a potential problem for homeowners; levels in one study were about 2% of the federal occupational standard for airborne levels (Beusterien et al., 1991; Hefflin et al.,1993). Formaldehyde may also be a concern in latex paints (Hansen et al., 1987; van Faassen and Borm, 1991).

PLASTICS PROCESSING

Description

As late as 1970, a short description of industrial processes such as this chapter probably would not have contained a section on plastics. Metals operations would have been far more important. However, that scenario has changed considerably. The number of products that were once metallic and are now plastic is so large that plastic operations now seem to be more numerous than metal operations such as forging and founding. The products include such diverse items as automobile interiors, food containers, piping, appliance cabinets, and a wide range of products once made from nonmetallic materials (such as packaging of all sorts that was once made of paper and items once made of wood such as boats and furniture).

The processes used to manufacture plastic items have some similarities to the metals manufacturing processes. In some cases, molds are used and liquid or molten plastics are added to the molds. In other cases, sheets are manufactured in a manner similar to that of the rolling mills used to make steel or aluminum sheet. Other products are made in a fashion similar to that of the stamping, pressing, or forging metal operations.

However, the diversity of processes for manufacturing plastics products exceeds that of metals, largely because plastics are far more versatile than metals. Because the diversity of plastic processes is very great, these processes are widely distributed in industry. A brief description of most plastics processing operations is given below. For a more thorough description, see Mutchler and Proskie (1984); the majority of this discussion is taken from that reference.

As a preface, it is important to note that, by and large, the polymerization process or the process leading to the production of a polymer (see Table 5-7 for a definition of many of the terms used in this section) occurs in the petrochemical industry. Thus, the specialty industries that manufacture finished product often work with liquid polymeric resins or solids; the solids come in sheets, pellets, granules, flakes, etc., and therefore must be reshaped. These are usually thermoplastic polymers (see Table 5-7). In a few cases, the specialty industry works with monomers or partially polymerized resins. These processes usually lead to thermosetting plas-

Table 5-7
A glossary of polymer terms

Crosslinking	A chemical reaction between polymer chains by which a thermoplastic resin is converted to thermoset. The crosslinking can be promoted by pressure and heat or the addition of a crosslinking agent, such as styrene in the polyester-styrene system. The process is also known as vulcanizing.
Curing	The evaporation of solvent from or the crosslinking of a polymer.
Elastomer	A polymer that, at room temperature, rapidly returns to its original dimensions after being deformed (e.g., stretched), also rubber.
Filler	A usually cheap, *inert* material added to lower costs or modify strength or other properties.
Monomer	A relatively small molecule capable of reacting to form a polymer.
Plastic	A polymeric material that at some stage in its manufacture was shaped by flow.
Plasticizer	A compound incorporated in a polymeric material to increase its plasticity (flexibility) or workability. Dioctylphthalate is a widely used plasticizer.
Polymer	A molecule generally consisting of one or more repeated monomers.
Resin	A solid or viscous polymeric material that is further processed into a plastic or rubber.
Stabilizer	An ingredient added to retard degradation due to heat, oxidation (antioxidant), hydrolysis, photolysis, etc.
Thermo-plastic	A plastic that can be repeatedly melted or dissolved and reshaped. It is a solid because of intermolecular forces of attraction between polymer chains.
Thermo-setting	An infusible plastic. It cannot be melted or dissolved without degradation because of crosslinks or chemical bonds between polymer chains.

tics. This differentiation is important because fully polymerized materials are typically of quite low toxicity. The following are the most commonly used processes.

EXTRUSION

A powdered or granular resin is melted and forced through an opening. The opening determines the shape of the product, which is typically tubes or rods such as piping, fibers, and wire insulation. Sheets are also made; but most commonly the extrusion process is used to make a solid form such as a pellet, which will then be made into a finished product by another process. The purpose of this intermediate step is to mix a powdered polymer with additives such as plasticizers, antioxidants, stabilizers, lubricants, fillers, and colorants in a process called *compounding*.

CALENDARING

Closely related to extrusion, this process is one in which the resin is forced between successively smaller spaces in between rollers, much as in a steel rolling mill. The plastic may be coated onto fabric. PVC is the most common resin used in this sheet forming process.

MOLDING

There are at least five principal types of plastics processes that utilize a mold. In some cases, the mold is removed to reveal the product. In others, the mold is part of the product as either the exterior (e.g., foam insulation blown inside a refrigerator door) or the interior (e.g., a wooden portable building coated with a fiberglass reinforced resin exterior). In the molding processes, monomers or partially polymerized polymers are often used.

In *injection* molding, molten plastic is forced between the male and female parts of a closed mold to make a cup, box, cap, appliance housing, telephone set, automotive steering wheel, toy, or one of many other products. In *blow* molding only the female or the exterior portion of the mold is used. A softened plastic tube is inserted into the mold, and the inserted end of the tube is closed; air is forced into the other end to form a bottle or similar container in the shape of the mold. Polyethylene, polyvinyl chloride, high density polyethylene, polystyrene, and polycarbonate are typically used.

In *compression* molding, a partially polymerized resin such as phenol-formaldehyde and additives are charged into a mold, and then heat and pressure are applied. The heating causes cross-linking to complete the polymerization process. *Transfer* molding is similar except that the resin is preheated and then forced into the mold. This allows more even heating and therefore more even curing.

Perhaps the most ubiquitous and important processes in terms of health concerns are the *casting* and *foam processing* molding operations. In the casting process, liquid resin or monomer is charged to the mold, and the mold is heated to achieve cure. Unlike compression or transfer molding, no or little pressure is involved. Both finished and partially finished products of many types are made in this way. Perhaps the best known products, Lucite™ and Plexiglas™ are made by this process. The same monomer, methyl methacrylate, is also used in the dentist's office to cast the missing portions of broken teeth. Polyurethane (see description of polyurethane chemistry under *Painting*) and polyester-styrene are also used in this process to cast such items as simulated-marble bathroom countertops in numerous small shops.

In foam processing and several of the molding operations, several different polymers may be used, but the polymer is given a cellular or internal bubble structure through the use of a blowing or foaming agent. In some cases, the foaming agent is produced as a by-product of the polymerization reaction, for example, carbon dioxide. In other cases, nitrogen is injected or produced in the reaction as a result of the addition of one of many nitrogen-containing blowing agents (see Mutchler and Proskie, 1984, for a list). Volatile blowing agents such as trichlorofluoromethane were used in the past but were phased out in response to the Montreal Protocol and replaced by HCFCs, HFCs, and hydrocarbons such as cyclopentane (see above, "Metal Finishing – Organic").

The final molding process is called *laminating*. In this process, a mold or a form is used and the resin is applied to the form. The form may be removed once the product cures, or left as part of the product as in an acrylic laminated tabletop, but several layers are applied, which may include the mold. Polyester-styrene fiberglass-reinforced boat hulls, storage buildings, and many other products are made this way, often in small shops. Usually the product is built up in several coats, with curing in between each. For this reason and because some of the polymers used are the same, this process has some commonalities with painting. Phenol-formaldehyde resins also are widely used in this process in items such as plywood.

There are at least five polymers, mentioned above, that are of extreme importance in terms of widespread use and hazard level. They are often handled in a partially polymerized state, and are referred to here as the unfinished resins. The reader is referred to the painting section for a description of the polyurethane and epoxy polymerization processes. Similar raw materials are used in laminating, casting, and foam operations. In addition, polyester-styrene is another commonly used polymer system, as is polymethyl methacrylate. Examples of the chemical reactions are given in Figure 5-5. The fifth system is phenol-formaldehyde resins, and it is not depicted in Figure 5-5 because its chemistry can be quite variable, depending on the desired properties. Basically, the processor receives a partially polymerized polyol resin. Further formaldehyde addition is needed to complete the polymerization. This is accomplished by adding hexamethylene tetramine, which, with heat, decomposes to formaldehyde and ammonia. The resin is widely used in making cardboard, plywood, and other wood products, as well as many other laminates and some injection-molded products.

WELDING

There are several processes that are used to join two pieces of plastic. One approach is to use a solvent to dissolve two chemically similar polymers along the junction. When the solvent dries, the two pieces are bonded via intermolecular forces. As an example, the acrylics such as Plexiglas can be welded with short-chain chlorinated hydrocarbons such as ethylene dichloride. Microwave electromagnetic radiation is also used in combination with mechanical pressure. Frequencies are typically in the 10 to 100 mHz range, and are set at 13.56, 27.12, and 40.68 mHz in the United States (Largent and Olishifski, 1988). Ultrasonic pressure energies and mechanical pressure are also used. Thermoplastic resins are usually involved, as each of the techniques works by breaking intermolecular forces and then allowing them to reform in a rearranged state.

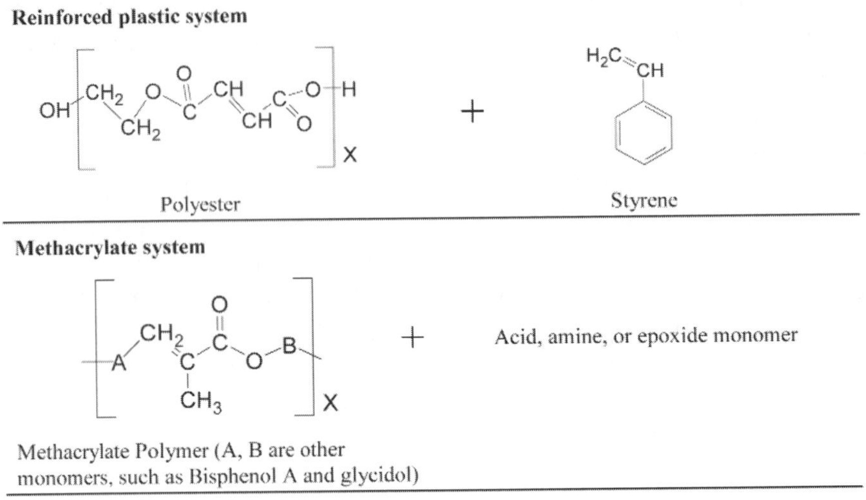

Figure 5-5. Examples of one type of reinforced plastic (styrene crosslinked polymer) and a methacrylate system.

Hazards

The hazards are as varied as the polymers and processes. Perhaps a good place to start is with the most common generic hazard – dust. Many of the processes generate plastic dust, particularly if fine-grained resins are fed to the extrusion and other machinery. These dusts should be composed of high molecular weight polymer molecules, and for this reason they are often labeled inert or nuisance dusts. Unfortunately, few have been evaluated for toxicity. Of a more certain concern are those dusts generated in operations where partially uncured resin may be sanded, grinded, or may in some other way become airborne. The most important processes in which this may happen involve the five unfinished resins mentioned previously. However, it is possible to have free monomer in otherwise solid and supposedly finished resin. Two polymer groups that have been suspected of this in the past are polyvinyl chloride and those polymers involving acrylonitrile. Although it is likely that significant amounts of free monomer have been controlled by the manufacturer, the hygienist still should be aware of the possibility of bad batches or poor quality control.

Where unfinished resins are used, the potential for exposure to monomer or reactive resins via the dermal or respiratory routes exists. Such compounds as styrene, polyamines, diisocyanates or epichlorohydrin (either free or attached to prepolymers), formaldehyde, and methyl methacrylate are quite toxic, and all but formaldehyde are skin allergens. Other additives are also possible in these unfinished resin systems. One of the most important groups is the peroxide catalysts such as benzoyl peroxide and methylethylketone peroxide.

In many cases, polymers are heated to the point of decomposition. In some cases, this occurs purposely, as in cleaning out plugged equipment or using a heated wire to cut a plastic film. In other cases, such as fires or improperly controlled welding units, unintentional decomposition may occur. The nature of the decomposition products is dependent on chemistry of the polymer and additives. Hydrocarbon polymers such as polyethylene decompose into considerably less hazardous products than do polymers containing nitrogen, chlorine, and fluorine. Polymers such as Teflon™, polyvinylchloride, and polyurethane can decompose into extremely irritating products (Johnston et al., 1988; Orzel et al., 1989). In the case of polyfluorocarbon decomposition products, the term polymer fume fever has been used to describe the symptoms associated with exposure. For a good review of decomposition products of polymers, such as polypropylene, acrylics, polyurethane, and phenol-formaldehyde, see the *Journal of the American College of Toxicology 7(2)*. Generally, additives are present in small concentrations with regard to hazardous combustion products. However, inorganic additives such as lead stabilizers in PVC and pigments such as cadmium may be important. A wide range of solvents is used in welding operations, but their use in cleanup and maintenance operations is more important. Often maintenance operations are uncontrolled, and thus solvent exposures can be high. Finally, there is often a noise hazard associated with processing equipment and a burn hazard associated with microwave welding equipment.

WELDING AND ALLIED PROCESSES

Description

Welding is a materials joining process that produces coalescence of materials by heating them to suitable temperatures, with or without the application of pressure or by the application of pressure alone, and with or without the use of a filler material (AWS, 1983a). As such, welding includes two processes that are often described separately: soldering and brazing. In addition, an allied process, cutting of metals, should be included because welding and cutting are often coupled and the hazards are similar. Although there are several differences between soldering, brazing, and other welding processes, the main difference is temperature of operation, and this in turn affects hazard.

SOLDERING

In the soldering process, a filler material (solder) is used to join two pieces of metal. The solder is melted, and the melting temperature may be as high as 450°C (840°F). However, as Table 5-8 shows, the most popular solder formulations rarely require such high temperatures. Although the metal workpieces and the solder will exceed these temperatures, the metal workpieces do not melt as they do in most welding processes.

In order for the solder to adhere to the metal, first the oxide film or tarnish must be removed from the metal. A flux is used to do this. The flux contains an acid or base that dissolves the tarnish. When the solder is applied, it displaces this solution. Fluxes may be inor-

Table 5-8
Common solder alloys sold by AIM, Inc, Cranston, Rhode Island, USA

Metal		Components (%)					Melting
		Ag	Cu	Pb	Sb	Sn	Point (°C)
Tin	Sn99					99.9	232
	Sn96.5	3.5				96.5	221
	Sn95	5				95	221–240
	Sn90			10		90	183–213
	Sn70			30		70	183–193
	Sn63			37		63	183
	Sn62	2		36		62	179
	Sn60			40		60	183–188
	Sn50			50		50	183–212
	Sn40			60		40	183–247
	Sn35			65		35	183–250
	Sn30			70		30	183–257
	Sn20			80		20	183–280
	Sn10			90		10	275–302
	Sn5			95		5	308–312
	Sn3			97		3	314–320
Silver	Ag03A	2.5	0.8		0.5	96.2	217
	SAC305	3.0	0.8			96.5	217–218
	Ag2.5	2.5		97.5			304
	Ag5.5	5.5		94.5			305–364
Antimony	Sb5				5	95	232–240
	Sb2			63	2	35	185–243
	Sb1			79	1	20	184–270
Lead	Pb88	2		88		10	268–290
	Pb68			68	2	30	185–243
	Pb80	2		80		18	252–260
	Pb94	1.5		93.5		5	305–306
	Pb93	2.5		92.5		5	299–304
	Pb95	2		95		3	299–305
	Pb96			96	2	2	252–295

ganic acids or salts such as hydrochloric or hydrofluoric acid or zinc chloride, but the most common flux is rosin, an organic derivative of pine trees. Rosin contains abietic acid, which dissolves the tarnish. It is thinned with alcohols, glycols, or hydrocarbons, and may include organic chloride activators to aid in the solution process. There are a wide range of other organic fluxes, as shown in Table 5-9. These fluxes are often in water-soluble formulations. Therefore, they may see greater future use in the electronics industry, where fluxes are often removed after soldering, and several organic solvents are being phased out (see section on "Metal Cleaning – Organic").

BRAZING

Brazing occurs at temperatures greater than 450°C (840°F), and like soldering requires a filler material and does not melt the joined metals. However, a flux may not bc uscd if a vacuum or reducing atmosphere is applied. Filler materials may contain cadmium, silver, zinc and, copper, for the most part, but may also include phosphorus, nickel, chromium, beryllium, magnesium, and lithium. Fluxes are composed of borates, fluorides, chlorides, acids such as boric, and bases such as potassium or sodium hydroxide (NIOSH, 1988b). Often the term silver soldering is used for brazing because of the high content of silver in many brazing fillers (AWS, 1980). However, in general, silver is not the most important hazard, but cadmium is. Brazing alloys without cadmium are available, and while cost and temperature of application need to be considered, the cost effectiveness compared to controlling cadmium exposures to current occupational exposure limits is worth considering, given the high toxicity of cadmium.

Table 5-9
Types of organic, nonrosin fluxes

Type	Typical Fluxes	Vehicle
Acids	Lactic, oleic, stearic, glutamic, phthalic	Water, organic solvents, petrolatum paste, polyethylene glycol
Halogens	Aniline hydrochloride, glutamic acid hydrochloride, bromide derivatives of palmitic acid, hydrazine hydrochloride or hydrobromide	Water, organic solvents, polyethylene glycol
Amines and Amides	Urea, ethylenediamine, mono- and triethanolamine	Water, organic solvents, petrolatum paste, polyethylene glycol

Adapted from Burgess (1991) and Manko (1979).

WELDING

Unfortunately, whereas soldering and brazing are two of the principal types of welding processes, all other types of welding also are classified as welding; so welding is both a major category and a subcategory. Therefore, all welding processes not including soldering and brazing are distinguished and referred to in this section as HT (high temperature) welding. This is not the only thing confusing about welding, as there are more than 50 types of HT welding in five categories (see Figure 5-6), and most of the categories have a common name in addition to their American Welding Society approved name. This is further demonstrated in Figure 5-7, which also includes some cutting and brazing processes.

The most common type of HT welding, arc welding, relies on an electrical arc or current; generally the workpiece is one electrode, and a filler rod serves as a second electrode. The current causes the workpiece and the filler to coalesce or melt together. Figure 5-7 shows that the three most common types of welding are in the arc welding (AW) category. The three types are differentiated according to how the weld is shielded and what type of electrode rod is used.

In the most common type of welding, shielded metal AW, the filler rod (commonly referred to as the welding rod) serves as the electrode and is coated with a solid flux that vaporizes at the work surface to form a gaseous shield, protecting the weld from oxidation. In gas metal AW, the filler rod still serves as the electrode but is not coated. Instead a gas such as argon or carbon dioxide is used to shield the weld. In gas tungsten AW, a gas is also used to shield the weld, but a non-consumable tungsten electrode is used, and a separate filler rod is fed to the weld. Welding rods are made from a great variety of materials, and each rod is given an AWS classification. (For a complete list of rod materials, classifications, and fumes recommended for air analysis, the reader is encouraged to see AWS, 1983b and 2003.) Copper, chrome (IV), manganese, zinc, and nickel are the more common constituents of welding rods.

There are several other important types of HT welding. Oxyfuel (OF) welding is very similar to OF cutting. A fuel (usually acetylene) and oxygen blend is used to coalesce the metals with or without a filler. In submerged arc, or SAW, the process is submerged in a granular flux. The flux not only shields the weld but reduces the amount of fume, UV light, and other hazards. Finally, in resistance welding, the resistance to current flow through a workpiece and added pressure produce a weld.

CUTTING

There are several cutting techniques that closely parallel welding processes. The difference is that the heat is used to separate the pieces rather than join them. These include the common oxy-acetylene torch (simple use of heat from a flame), air carbon arc, and plasma arc. In the latter, a gas stream is heated to ionization, thus forming a plasma. The plasma jet cuts the metal and blows away the molten metal. In air carbon arc cutting, a metal is cut with the heat from a carbon electrode induced arc and blown away with a jet of air.

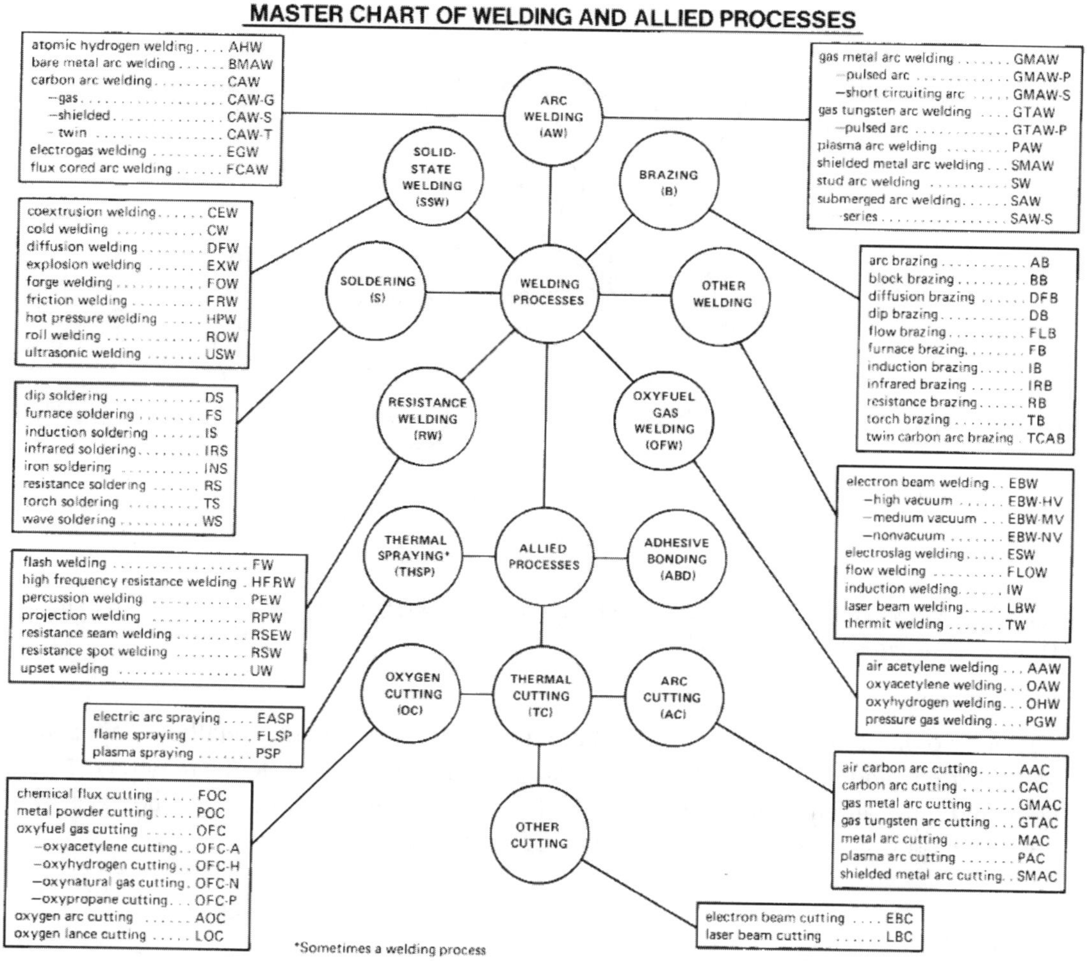

Figure 5-6. Master chart of welding and allied processes (from AWS, 1980, with permission).

Hazards

In all of the above processes, the hazards can be placed in one of three groups: base metal, fillers, and fluxes. The hazards are listed in Table 5-10 and gaseous emissions from the various types of HT welding are shown in Table 5-11.

FLUXES

Most of the soldering fluxes, including the rosin flux, have the potential to be irritating to skin and lungs. Most of the brazing fluxes also can be irritating to skin and lungs; the amine

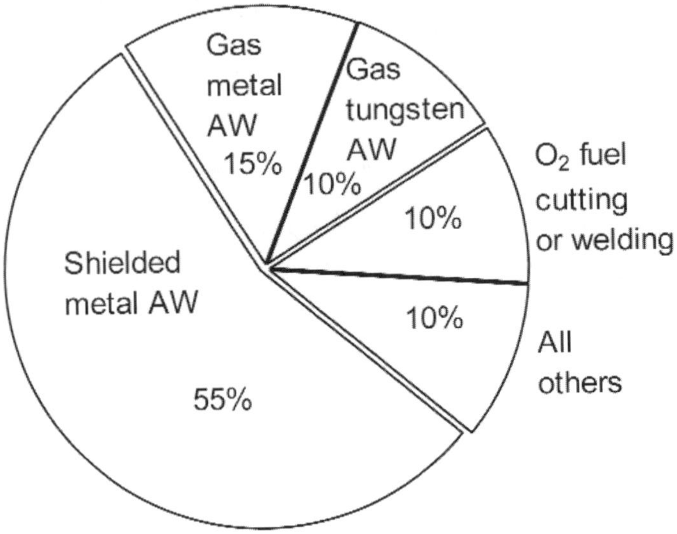

Figure 5-7. The most common welding and cutting processes and their estimated prevalence in industry (from AIHA, 1984, with permission).

and chloride organic fluxes are quite so. Many welding fluxes contain metallic components as outlined earlier, and common SMAW fluxes contain fluoride-producing minerals. NIOSH (1988b) and other authorities (AIHA, 1984) make no mention of fluorine gas; however, AWS lists it as a hazard in three types of welding (see Table 5-11).

BASE METAL

The base metal does not pose a hazard in soldering or brazing, as temperatures are considerably above the melting point. In HT welding of carbon steel, if iron concentrations are acceptable, then other metals in the steel should not be significant. However, in poorly ventilated situations even minor contaminants in the iron may reach hazardous levels. In alloy welding and welding of coated steels, such as galvanized steel, additives such as chromium, cadmium, nickel, lead, zinc, and copper, among others, may reach hazardous levels.

FILLERS

In soldering, lead may pose a hazard in poorly controlled conditions. In brazing, cadmium is the most common hazard. In HT welding, the filler rods are often largely composed of iron, but many specialized rods exist, as outlined in AWS (1983b). The hazardous agents in these fillers include chrome (IV), manganese, and nickel.

Table 5-10
Specific welding processes and associated hazardous agents[1]

Process	Hazardous Agent
Brazing/cadmium filler	Cadmium
Oxy-fuel cutting, welding	Carbon monoxide Nitric oxide (NO) Nitrogen dioxide (NO_2)
Gas metal arc welding – Al or Al-Mg	Ultraviolet (UV) radiation Ozone
GMAW – stainless steel	Hexavalent chromium (VI) Nickel Ozone
GMAW, all types using carbon dioxide	Carbon monoxide
Gas tungsten arc welding – Al or Al-Mg	UV radiation
Shielded metal arc welding, low-hydrogen electrodes	Fluorides UV radiation
SMAW – iron or steel	Iron oxide UV radiation
SMAW – stainless steel	Chromium (VI) Nickel UV radiation
Plasma cutting – aluminum	Noise Ozone

[1] From NIOSH (1988b)

Table 5-11
Gases associated with specific arc welding processes[1]

Welding process	Common Gases			
	CO	F	NO_x	O_3
Shielded metal arc welding		X	X	
Flux cored arc welding	X	X	X	
Gas metal arc welding				
Argon helium shielding			X	
CO_2 shielding	X		X	
Gas tungsten arc welding			X	X
Submerged arc welding		X		

[1] From AWS (1983) with permission

QUESTIONS

5.1 A manufacturer takes metal stock, shapes it, and creates a final finished (painted) product. What processes are likely to be involved?

5.2 Describe the makeup of paints.

5.3 Explain the dilemma in using halogenated hydrocarbons, in terms of toxicity and environmental consequences.

5.4 Describe the major types of HT welding in a comparative fashion.

REFERENCES

ACGIH®: Industrial Ventilation: A Manual of Recommended Practice for Design, 26th Ed. American Conference of Governmental Industrial Hygienists, Cincinnati, Ohio (2007).

Adkison, P.D.: Complying with regulations on lead paint removal from utility structures. J. Protective Coatings and Linings 6(10):33-37 (1989).

AIHA: Welding Health and Safety Resource Manual. American Industrial Hygiene Association, Akron, OH (1984).

ANSI: Standard Markings for Identifying Grinding Wheels and Other Bonded Abrasives, B74.13.1990. American National Standards Institute, New York (1990).

AWS: Welding Terms and Definitions, A3.0-80. American Welding Society, Miami, FL (1980).

AWS: Effects of Welding on Health III. American Welding Society, Miami, FL (1983a).

AWS: Evaluating Contaminants in the Welding Environment: A Sampling Strategy Guide, AWS Fl.3-83. American Welding Society, Miami, FL (1983b).

AWS: Safety and Health Fact Sheets, 3rd Edition. American Welding Society, Miami, FL (2003).

Beusterien, K.M.; Etzel, R.A.; Agoes, M.M.; Egeland, G.M.; Socie, E.M.; Rouse, M.A.; Mortensen, B.: Indoor air mercury concentrations following application of interior latex paint. Arch. Env. Contam. Tox. 21:62-64 (1991).

Bilmeyer, F.W.: Textbook of Polymer Science, 3rd ed. Wiley, New York (1984).

Bleecker, M.L.; Bolla, K.I.; Agnew, J.A.; Schwartz, B.S.; Ford, D.P.: Dose-related subclinical neurobehavioral effects of chronic exposure to low levels of organic solvents. Am. J. Ind. Med. 19:715-728 (1991).

BLS: Bureau of Labor Statistics Data, U.S. Department of Labor, Bureau of Labor Statistics, Washington DC, http://www.bls.gov/ces/cesbtabs.htm (2004).

Chen, R.; Dick, F.; Semple, S.; Seaton, A.; Walker, L.G.: Exposure to organic solvents and personality. Occupational and Environmental Medicine 58:14 (2001).

Decker, J.A.: Health hazard evaluation report no. HETA-90-368-2137. National Institute for Occupational Safety and Health, Cincinnati, OH (1991).

De Rosa, E.; et al.: The industrial use of solvents and risk of neurotoxicity. Ann. Occup. Hyg. 29:391-397 (1985).

DuPont: Teflon fluorocarbon resins – safety in handling and use. Fluorocarbons Division, E.I. DuPont de Nemours and Co., Wilmington, DE (1970).

Encyclopedia Britannica: Diesel engine, In The New Encyclopedia Britannica, 15th ed., 4:85-86, 18475-477. Encyclopedia Britannica, Chicago, IL (1987).

EPA: Significant New Alternatives Policy (SNAP), Protection of Stratospheric Ozone: Listing of Substitutes for Ozone-Depleting Substances, Federal Register 61:25604 http://www.epa.gov/ozone/snap/ (1996).

EPA: Health Assessment Document for Diesel Engine Exhaust, EPA/600/8-90/057F. United States Environmental Protection Agency, Washington, DC (2002).

Gabe, D.R.: Principles of Metal Surface Treatment and Protection, 2nd ed. Pergamon Press, New York (1978).

Garshick, E.; Schenker, M.B.; Munoz, A.; Segal, M.; Smith, T.J.; Woskie, S.R.; Hammond, S.K.; Speizer, F.E.: A retrospective cohort study of lung cancer and diesel exhaust exposure in railroad workers. Am. Rev. Respir. Dis. 137:820-825 (1988).

Glindmeyer, H.W.; Hammad, Y.Y.: Contributing factors to sandblasters silicosis: Inadequate respiratory protection equipment and standards. JOM 10:917-921 (1988).

Goldberg, M.E.; Johnson, H.E.; Pozzani, U.C.; Smyth, H.F., Jr.: Behavioral response of rats during inhalation of trichloroethylene and carbon disulphide vapours. Acta Pharrmacologica et Toxicologica 21:36-44 (1964).

Hanninen, H.; Antti-Poika, M.; Juntunen, J.; Koskenvuo, M.: Exposure to organic solvents and neuropsychological dysfunction: A study on monozygotic twins. Brit. J. Ind. Med. 48:18-25 (1991).

Hansen, M.K.; Larsen, M.; Cohr, K.H.: Waterborne paints: A review of their chemistry and toxicology and the results of determinations made during their use. Scand. J. Work Environ. Hlth. 13:473-485 (1987).

Hefflin, B.J.; Etzel, R.A.; Agocs, M.M.; Stratton, J.W.; Ikawa, G.K.; Barsan, M.E.; Schutte, G.R.; Paschal, D.C.; Kozman, A.R.: Mercury Exposure from Exterior Latex Paint, Appl. Occup. Env. Hyg. 8: 866 (1993).

HSE: Cadmium in silver soldering or brazing, HSE Engineering Sheet No 31. United Kingdom Health and Safety Executive http://www.hse.gov.uk/pubns/eis31.pdf (1999).

Huang, J.; Wang, X.P.; Chen, B.M.; Veda, A.; Aoyama, K.; Matsushita, T.: Immunological effects of toluene diisocyanate exposure on painters. Arch. Env. Contam. Tox. 21:607-611 (1991).

Hunting, K.L.; Matanoski, G.M.; Larson, M.; Wolford, R.: Solvent exposure and the risk of slips, trips, and falls among painters. Am. J. Ind. Med. 20:353-370 (1991).

IARC: Mineral oils, In IARC Monographs on the Evaluation of the Carcinogenic Risk of Chemicals to Humans, Vol. 33. International Agency for Research on Cancer, Lyon, France (1984).

Johnston, P.K.; Doyle, E.; Orzel, R.A.: Acrylics: A literature review of thermal decomposition products and toxicity. J. Am. Coll. Tox. 7:139-200 (1988).

Key, M.M.; Taylor, J.S.; Yang, C.: Grinding and cutting fluids, In Encyclopedia of Occupational Health and Safety, Vol, I, 3rd ed., L. Parmeggiani (ed.). International Labour Office, Geneva, Switzerland (1983).

Landrigan, P.J.: Silicosis. Occupational Medicine: State of the Art Reviews 2:319-326 (1987).

Largent, E.J.; Olishifski, J.: Nonionizing radiation, In Fundamentals of Industrial Hygiene, 3rd ed., B.A. Plog (ed.). National Safety Council, Chicago, IL (1988).

Lewis, R.J., Sr.: Hawley's Condensed Chemical Dictionary, 12th ed. Wiley, New York (1993).

Lipsett, M.; Campelman, S.: Occupational exposure to diesel exhaust and lung cancer: a meta-analysis. Am. J. Pub.

Hlth. 89:1009-1017 (1999).
Manko, H.H.: Solders and Soldering. McGraw-Hill, New York (1979).
Morris, M.; Wolf, K.: Alternatives to Perchloroethylene Vapor Degreasing for Plating Operations: Case Study Conversions, Institute for Research and Technical Assistance, Produced for the California Environmental Protection Agency, Department of Toxic Substance Control, Sacramento, CA (February 2003).
Murata, K.; Ariki, S.; Yokoyama, K.; Maeda, K.: Autonomic and peripheral nervous system dysfunction in workers exposed to mixed organic solvents. Int. Arch. Occup. Environ. Hlth. 63:335-340 (1991).
Murphy, J.A.: Surface preparation and finishes for metals: A manual covering the planning, operations, and economics associated with modern metals. McGraw-Hill, New York, NY (1971).
Mutchler, J.E.; Proskie, K.G.: Plastics processing operations, In Industrial Hygiene Aspects of Plant Operations: Vol. 2, Unit Operations and Product Fabrication, L, J, Cralley and L, V. Cralley (eds.). Macmillan, New York (1984).
NIOSH: Guidelines for the Control of Exposure to Metalworking Fluids, no. 78-165. National Institute for Occupational Safety and Health, Cincinnati, OH (1978).
NIOSH: N-Nitroso Compounds in the Factory Environment. National Institute for Occupational Safety and Health, Cincinnati, OH (1983).
NIOSH: Control Technology Assessment: Metal Plating and Cleaning Operations. National Institute for Occupational Safety and Health, Cincinnati, OH (1984a).
NIOSH: Recommendations for Control of Occupational Safety and Health Hazards, Manufacture of Paint and Allied Coating Products. National Institute for Occupational Safety Health, Cincinnati, OH (1984b).
NIOSH: Health hazard evaluation report no. HETA-84-496-l766. National Institute for Occupational Safety and Health, Cincinnati, OH (1987).
NIOSH: Current Intelligence Bulletin No. 50: Carcinogenic Effects of Exposure to Diesel Exhaust. National Institute for Occupational Safety and Health, Cincinnati, OH (1988a).
NIOSH: Criteria for a Recommended Standard: Welding, Brazing, and Thermal Cutting. National Institute for Occupational Safety and Health, Cincinnati, OH (1988b).
NIOSH: Request for assistance in preventing lead poisoning in construction workers, 91-116. National Institute for Occupational Safety and Health, Cincinnati, OH (1991).
NTP: Toxicology and Carcinogenesis Studies of D-limonene (CAS No. 5989-27-5) in F344/n Rats and $B_6C_3F_1$ Mice (gavage studies). National Toxicology Program, Research Triangle Park, NC, USA (1990).
Orzel, R.A.; Womble, S.E.; Ahmed, F.; Brasted, H.S.: Flexible polyurethane foam: A literature review of thermal decomposition products and toxicity. J. Am. Coll. Tox:. 8: 1139-1175 (1989).
OSHA: Preamble to OSHA final rule for permit-required confined spaces. Fed. Reg. 58:4462 (Jan. 14, 1993).
Pastides, H.; Calabrese, E.J.; Hosmer, D.W., Jr.; Harris, D.R., Jr.: Spontaneous abortion and general illness symptoms among semiconductor manufacturers, In Hazard Assessment and Control Technology in Semiconductor Manufacturing, pp. 53-73. Lewis Publishers, Chelsea, MI (1989).
Patterson, R.; Nugent, K.M.; Harris, K.E.; Eberle, M.E.: Immunologic hemorrhagic pneumonia caused by isocyanates. Am Rev. Resp. Dis. 141:226-230 (1990).
Peterson, J.E.: Painting and coating, In Industrial Hygiene Aspects of Plant Operations: Vol. 2. Unit Operations and Product Fabrication, L. J. Cralley and L. V. Cralley (eds.). Macmillan, New York (1984).
Reh, C.M.; Lushniak, B.D.: Health hazard evaluation report no. HETA-87-350-2084. National Institute for Occupational Safety and Health, Cincinnati, OH (1990).
Rempel, D.; Jones, J.; Atterbury, M.; Balmes, J.: Respiratory effects of exposure of shipyard workers to epoxy paints. Brit. J. Ind. Med. 48:783-787 (1991).
Rudolph, L.; Swan, S.H.: Reproductive hazards in the microelectronics industry. State of the Art Reviews: Occupational Medicine 1, no. 1:135-143 (1986).
Savolainen, H.; Pfaffli, P.; Tengen, M.; Vainio, H.: Biochemical and behavioral effects of inhalation exposure to tetrachloroethylene and dichloromethane. J. Neuropath. Exp. Neur. 36:941-949 (1977).
Savolainen, H.; Pfaffli, P.; Helojoki, M.; Tengen, M.: Neurochemical and behavioral effects of long-term intermittent inhalation of xylene vapour and simultaneous ethanol intake. Acta Pharmacologica et Toxicologica 44:200-207 (1979).

Scheepers, P.T.J.; Bos, R.P.: Combustion of diesel fuel from a toxicological perspective II: Toxicity. Int. Arch. Occup. Environ. Hlth. 64:163-177 (1992a).

Scheepers, P.T.J.; Bos, R.P.: Combustion of diesel fuel from a toxicological perspective I: Origin of incomplete combustion products. Int. Arch. Occup. Environ. Hlth. 64:149-161 (1992b).

Schneberger, G.L.: Metal surface treatments, In Kirk-Othmer Concise Encyclopedia of Chemical Technology. Wiley Publishing, New York (1985).

Seidel, S.R.: Stratospheric ozone. EPA J. 17:34-38 (1991).

Sharp, D.S.; Perkins, C.; Samuels, S.J.; Rosenberg, J.; Rudolph, L.: Lead use in California industry: Its prevalence and health implications. Am. Ind. Hyg. Assoc. J. 52:409-416 (1991).

Sheldon, A.W.: Effects of organotin anti-fouling coatings on man and his environment. J. Paint Tech. 47:54-58 (1975).

Silverman, A.P.; Williams, H.: Behavior of rats exposed to trichloroethylene vapour. Brit. J. Ind. Med. 32:308-315 (1975).

Steenland, N.K.; Silverman, D.T.; Hornung, R.W.: Case-control study of lung cancer and truck driving in the Teamsters Union. AJPH 80(6):670-674 (1990).

Stettler, L.E. et al.: Fibrogenicity and carcinogenic potential of smelter slags used as abrasive blasting substitutes. J. Tox. Env. Hlth. 25, no. 1:35-36 (1988).

Triebig, G.; Nasterlack, M.; Hacke, W.; Frank, K.H.; Schmittner, H.: Neuropsychiatric symptoms in active construction painters with chronic solvent exposure. Neuro Toxicology 21: 791 (2000).

van Faassen, A.; Borm, P.J.A.: Composition and health hazards of water-based construction paints: Results from a survey in the Netherlands. Env. Hlth. Persp. 92:147-154 (1991).

Wade. J.F., III; Newman, L.S.: Diesel asthma: Reactive airways disease following overexposure to locomotive exhaust. JOM 35(2):149-154 (1993).

Welch, L.S.; Schrader, S.M.; Turner, T.W.; Cullen, M.R.: Effects of exposure to ethylene glycol ethers on shipyard painters: II. Male reproduction. Am. J. Ind. Med. 14, no. 5:509-526 (1988).

General References

Burgess, W.A.: Potential exposures in the manufacturing industry – their recognition and control, In Patty's Industrial Hygiene and Toxicology, 4th ed., Vol. I, part A, G.D. Clayton, F.E. Clayton (eds.). Wiley Publishing, New York (1991).

Burgess, W.A.: Recognition of Health Hazards in Industry: A Review of Materials and Processes, 2nd ed. Wiley Publishing, New York (1995).

Cralley, L.V.; Cralley, L.J.: Industrial Hygiene Aspects of Plant Operations: Vol. 1, Process Flows. Macmillan, New York (1982).

Cralley, L.V.; Cralley, L.J.: Industrial Hygiene Aspects of Plant Operations: Vol. 2. Unit Operations and Product Fabrication. Macmillan, New York (1984).

Cralley, L.V.; Cralley, L.J.: Industrial Hygiene Aspects of Plant Operations: Vol. 3. Engineering Considerations in Equipment Selection, Layout, and Building Design. Macmillan, New York (1985).

Kirk-Othmer: Concise Encyclopedia of Chemical Technology, 4th ed. Wiley Publishing, New York (2003).

Stellman, J.M. (ed.): Encyclopedia of Occupational Health and Safety, 4th ed., Vols. 1-4. International Labour Office, Geneva, Switzerland (1998).

6

EXPOSURE ROUTES

INTRODUCTION

We have seen that there are about 80,000 chemicals in American industry to which workers may be exposed. Even though many are solids and liquids, all have the potential to become airborne and take on certain physical properties that allow their entry into the lungs. The lungs are an important route of entry for chemicals – in the workplace, the most important. But how do chemicals get from the air passages of the lungs to the various body organs or the eventual site of toxic action? Do the various physical states in which chemicals can exist affect their ability to enter the body via the lungs? The answers to these questions are not so obvious. It may be thought that the discussion should end with these questions because the inhalation route is of obvious importance. But we might also ask, by what other routes may chemicals enter the body?

The student with little experience with settings that have significant airborne chemical exposures may consider the above question only in terms of his or her own perspective. Inhalation is probably most often considered as a typical route of exposure for the majority of xenobiotics. However, most people receive significant portions of their total dose of xenobiotics (chemicals not associated with normal life functions) via ingestion, by eating foods or drinking water containing these chemicals. In some cases, the chemicals are placed on the foods purposely (pesticides); others are the result of legal contamination, for example, of water (permitted discharges); whereas others are formed in the environment with building blocks that exist naturally or are released by humans (for example, the combination of chlorine with methane to form chlorinated hydrocarbons in water).

Although the ingestion route of chemical exposure may be most significant for an office worker, it can also be important for a foundry worker. Certainly a foundry worker is exposed to fumes and dusts via the inhalation route, but some of these particulates may enter through the mouth, be impacted in the saliva, and then swallowed. Prior to 1920 this was thought to be the sole mechanism of lead toxicity. In the medical realm, it was not even considered that lead could be absorbed via the lungs (Hamilton, 1948).

There are other ways to ingest toxicants on the job. Workers with contaminated hands may touch food that is then swallowed. Hands that contaminate tobacco are another source leading to ingestion. Well into the 1930s industrial management thought these routes were the major routes for lead absorption. In other words, the lead poisoning problem could be solved if workers would just take a bath! Government entities believed the same to be true as late as 1940 for other chemicals such as benzene (see Hamilton, 1948, for the historical references).

Interestingly, particles that first enter the lungs can be coughed back up into the mouth or pharynx (on phlegm) and swallowed. The ingestion route is sometimes overlooked as an important exposure route for workplace contaminants.

There is a third route that often is more important than ingestion, and in certain cases, by no means infrequent, it may be more important than inhalation. Most people do not think of the skin as a route of entry because intact skin is thought to be a good barrier that protects the internal body organs from the outside world. However, skin is an important route of entry for many chemicals, since some chemicals can be readily absorbed through the skin, gaining entry to the blood, and can then be distributed throughout the system. The skin does not need to be damaged in order for this process to occur, although damaged skin is more susceptible to absorption of chemicals than intact skin.

Other routes of entry would include any body portal. Not so obvious, but included in this group, would be the eyes and ears. The eyes essentially can be considered part of the skin in terms of their ability to absorb contaminants, and because they constitute such a small area, they are not of general importance. Chemical injury to the eye is, of course, of great importance; however, this is considered a direct effect upon the eye and rarely leads to exposure of internal body organs. It has been thought by some that the ears (via the thin eardrum or tympanic membrane) can play an important role in absorption of compounds from the outside air. This is particularly mentioned when workers must wear respirators to enter areas that have high levels of contamination (Birkner, 1980). Normally, the eardrum provides an effective means of blocking the path of airborne contaminants. Actually, the eardrum, like the skin, absorbs contaminants; but even a perforated eardrum allows relatively small amounts of even highly contaminated air to enter the lungs or the digestive tract via the eustachian tubes (Ronk and White, 1985). Hence, the three major routes for contaminants to gain entry to the body in order of general importance are inhalation, dermal absorption, and ingestion (Josephson, 1983).

This chapter provides only a summary examination of the three exposure routes. Volume 2 of Modern Industrial Hygiene, Biological Aspects, addresses the inhalation route in more detail in Chapters 2 and 3 and addresses the dermal route in more detail in Chapter 4. Readers are directed to those chapters for additional figures and explanations.

INHALATION EXPOSURE ROUTE

It is tempting to think of the inhalation route as composed chiefly of the lung surfaces.

However, this route includes the entire respiratory system, which constitutes a larger portion of the body than the lungs. In order to understand how contaminants first enter the mouth or the nose and eventually gain entrance to the bloodstream to be distributed throughout the body, one must understand the anatomy and physiology of the respiratory tract. Fortunately, many airborne particles cannot penetrate deeply into the lung to be absorbed, because of their size. On the other hand, gases and vapors have a relatively easy path through the lungs to the circulatory system, from which they can be distributed throughout the body.

Ventilation Rate

An important concept for exposure to airborne contaminants is ventilation rate. This is the volume of air entering the lung per unit time, typically L/min or m^3/hr. This value will effect exposure since the greater the ventilation rate, the more particles and/or gases and vapors enter the lung in a given time. Ventilation rates (VR) vary not only with work rate, but also sex, age, weight, and a host of other factors. While the literature is replete with data on energy or oxygen consumed at varying work rates, there are fewer examples of ventilation rates for varying work rates. Silverman et al. (1951) determined that average ventilation rates for students ranged from 10.3 ± 1.7 L/min at rest to over 100 L/min at stationary bicycle resistance-distance rates of 1660 kg-m/min for 10 minutes. The sedentary rate (10.3) is slightly higher than that typically cited (7.5 L/min, 0.45 m^3/hr); however, the ventilation measuring apparatus in Silverman's study caused some resistance to breathing.

EPA (1997) reviewed the literature and could find only one study of VR in workers. These results are presented in Table 6-1. EPA also reported VR for adults in qualitative activity categories that give rough ideas of associated work loads (Table 6-2). This table also shows VR for short term (e.g., a task for a few minutes to an hour) versus long term or 24-hour periods that would include sleeping and other sedentary, non-work activities.

It should be noted that O_2 consumed is often reported in the literature and that this is not the same as ventilation rate. O_2 consumed or V_{O2} is the O_2 used to do work and produce heat. Roughly, the body consumes about 20% of the O_2 that enters the lung and O_2 is about 20% of the air that enters the lung. Thus, V_{O_2} is about 1/25th the VR. This latter value is an approximation and also is not constant across people or work effort. Table 6-3 can be used to determine VR based on work rates. For example, for moderate work, Table 6-3 reports about 300 watts consumed for an average individual (mid point of range). Using constants of conversion, we can calculate the VR as follows:

$$300\ W \times \frac{1\ Kcal/\text{min}}{70\ W} \times \frac{1\ L_{O_2}}{5\ Kcal} \times 25 = 21\ l/\text{min}\ ,\ 3.5\ m^3/hr$$

Note that while 70 W = 1 Kcal/min, 1 L O_2 used per Kcal burned depends on the molecule being metabolized, but is a widely used approximation.

Table 6-1
Distributions of individual and group ventilation rates (VR) (m^3/hr)[a] for outdoor workers (EPA, 1997 and Linn, 1993)

Population[b]	Mean ± SD	1% tile	50% tile	99% tile
All Subjects (n[c]= 19)	1.68 ± 0.72	0.66	1.62	3.90
GCW[d]/Laborers (n = 5)	1.44 ± 0.66	0.48	1.32	3.66
Iron Workers (n = 3)	1.62 ± 0.66	0.60	1.56	3.24
Carpenters (n = 11)	1.86 ± 0.78	0.78	1.74	4.14

[a] 16.67 L/min = 1 m^3/hr
[b] Each group or subgroup mean was calculated from individual means, not from pooled means
[c] n = number of individuals by job
[d] GCW = general construction worker

Table 6-2
Summary of recommended values for inhalation for adults, 19-65+ yrs (from EPA, 1997)

Population	Mean rate	
Long-term Exposures	*m^3/day*	*L/min*
Females	0.48	1.32
Males	0.60	1.56
Short-term Exposures	*m^3/day*	*L/min*
Rest	0.4	6.7
Sedentary Activities	0.5	8.3
Light Activities	1	16.7
Moderate Activities	1.6	26.7
Heavy Activities	3.2	53.3

Table 6-3
Levels of metabolic rate (M) (adapted from ISO, 1989)

Class	Range M (W)	Examples
Resting	M ≤ 117	Resting
Low metabolic rate	117 < M ≤ 234	*Sitting at ease*: light manual work (writing, typing, drawing, sewing, book-keeping), hand and arm work (small bench rook, inspection, assembly or sorting of light material), arm and leg work (driving vehicle in normal conditions, operating foot switch or pedals). *Standing*: drill (small parts), milling machine (small parts), coil winding, small armature winding, machining with low power tools, casual walking (speed up to 2.2 mph or 3.5 km/h).
Moderate metabolic rate	234 < M ≤ 360	*Sustained hand and arm work*: hammering in nails, filling. *Sustained arm and leg work*: offload operation of lorries, tractors or construction equipment. *Sustained arm and trunk work*: work with pneumatic hammer, tractor assembly, plastering, intermittent handling of moderately heavy material, weeding, hoeing, picking fruit or vegetables. Pushing or pulling light weight carts or wheelbarrows, forging, walking at a speed of 2.2 mph or 3.5 km/h.
High metabolic rate	360 < M ≤ 468	*Intense arm and trunk work*: carrying heavy material, shoveling, sledge hammer work, sawing, planing or chiseling hard wood, hand mowing, digging, pushing or pulling heavily loaded handcarts or wheelbarrows, chipping castings, concrete block laying, walking at a speed of 3.4 to 4.4 mph or 5.5 km/h to 7 km/h.
Very high metabolic rate	M > 468	*Very intensive activity at fast to maximum pace*: working with an axe, intense shoveling or digging, climbing stairs, ramp or ladder, walking quickly with small steps, running, walking at speed greater than 4.4 mph or 7 km/h.

Respiratory Tract Anatomy

The respiratory tract is a highly branching system of tubes.[1] The respiratory tract here is divided into three portions simply referred to as upper, middle (tracheobronchial) and lower (pulmonary, alveolar or gas exchange) regions.

UPPER

Air comes into the system via either the nose or the mouth. It appears that the process of evolution created the nose as the major route of entry for inhaled air, as it contains two very important physiological functions. First, the section of the respiratory tract located directly behind the nose, the nasopharyngeal cavity, contains many folds and bifurcations of skin, called turbinates. These essentially act as a heat exchanger, using the heat of the body to warm the air that is being breathed. They are extremely efficient and able to warm extremely cold external air to body temperature prior to its reaching the lungs. Interestingly, when one is exercising in cold air and must breathe through the mouth, this process will not take place, and, in some individuals, cold-air-induced asthma may occur.

The second important physiological property of the nose is that all of the folds of skin are lined with very small hairs, called cilia. Beneath these cilia are epithelial (skin) cells, which are capable of secreting mucus. Any particles caught by the mucus are moved with the mucus by the beating of the cilia; the particles are moved along through the nose and either blown out or swallowed. Neither of these processes may take place if air is breathed through the mouth.

Air that enters either the nose or the mouth next enters the pharynx, the tube located at the convergence of the nasal and oral cavities. The pharynx is essentially an extension of the esophagus or the food tube, which leads to the stomach. However, air does not normally enter the stomach. Instead, air taken in upon inhalation must enter into a second tube, called the trachea or the windpipe. During breathing, both the trachea and the esophagus are in an open position. However, the flow of air is through the trachea, as a negative pressure is being developed in the lungs on inhalation, and the flow is out of the trachea, as a positive pressure is developed upon exhalation. The problem, then, comes when it is time to swallow saliva, food, or liquid to the stomach. When this occurs, two things should happen. First, the epiglottis, a small triangular-shaped flap, covers the trachea. Second, the entire throat is thrust upward toward the base of the tongue to further cover the tracheal opening. Hence, with only a few exceptions, the passage of solids or liquids to the stomach takes place with little interruption of the breathing process. If breathing and swallowing are attempted at the same time, then whatever is being swallowed will enter the trachea, and a violent reaction will occur as the trachea attempts to expel its contents.

Just below the epiglottis and in the uppermost region of the trachea is the larynx or the

[1] For further information on this subject, see Task Group (1966), Bouhuys (1974), Brain (1977), Menzel and Amdur, (1986), and Olishifski and Benjamin (1988).

voice box. This organ is composed of two folds of skin or membranes, called the vocal cords. As air rushes past them, and they are stretched or tightened, sounds are produced that are further modified by the tongue, the sinuses, the oral cavity in general, and the lips. This relatively simple organ is the basis for speech. Although it seems inconsequential in its location, with respect to environmental contaminants, one should keep in mind that it is located in the respiratory tract and, therefore, is exposed to the contaminants that enter the body by this route. Loss of speech through a disease such as cancer of the larynx is particularly debilitating.

MIDDLE

Below the larynx, the trachea or windpipe continues for approximately 11 to 12 centimeters (see Table 6-4), at which point it splits into right and left main trunks. These two main trunks are called the bronchi, one going to the right lung and the other to the left lung. The two bronchi are not symmetrical, as indeed the two lungs are not symmetrical. The right lung is considerably larger than the left lung, mainly because of the position of the heart, which is located on the left side of the body. The right bronchus is shorter and because it must supply a larger lung, it is also wider than the left bronchus. More important, the right bronchus continues in approximately the same direction as the trachea and, consequently, most aspirated material that enters the respiratory tract ends up in the right lung. Once the bronchi enter the lungs, then many thousands of branches, or bifurcations, take place. The smallest branches of this bronchial tree are termed bronchioles.

LOWER

At the terminal ending of each bronchiole are located very small sacs that further divide into polyhedral-like structures with one open face, called the alveoli. These sacs are the functional part of the respiratory system in that the exchange of gases across their membranes takes place and allows the process of biochemical respiration to take place. The walls of the alveoli are only one cell thick. Beyond that wall are capillaries whose interstitial fluid comes in contact with the alveolar wall. Table 6-4 shows that there are approximately 300 million of these tiny bulbs, and 8 million sacs exist, having a total cross-sectional area of approximately 1.2 square meters. On the other hand, the trachea is approximately 2.5 square centimeters in cross-sectional area. Therefore, since the product of velocity and area (ventilation rate or volumetric flow rate) should be a constant at any point in the system, air entering into the trachea at a given speed has its speed reduced by a factor of approximately 5,000 by the time it reaches the alveolar sacs. Table 6-4 shows that for an inspiratory flow rate[2] of 60 L/min (very intense work based on earlier discussions), the velocity at the alveolar sac is essentially nil (0.09 cm/s). This fact will be important in a later discussion of how contaminants are deposited in the lungs.

---

[2] A sedentary ventilation rate is about 10 L/min (CDC, 1996). A 60 L/min VR is equivalent to walking at 2–4 miles per hour (3.5 – 6 km/h).

Table 6-4
The Anatomy of the Human Lung (adapted from Lippmann, 1977)[1]

Airway and Generation		Number/ Generation	Diameter (mm)	Length (mm)	Total Cross Section (cm^2)	Velocity (cm/s)	Residence Time[2] (ms)
Trachea	0	1	18	120	2.5	390	30
Main bronchus	1	2	12	48	2.3	430	11
Lobar bronchus	2	4	8.3	19	2.1	460	4.1
Segmental bronchus	3	16	4.5	13	2.5	390	3.2
Bronchi with cartilage in wall	8	260	1.9	6.4	6.9	140	4.4
Terminal bronchus	11	2×10^3	1.1	3.9	20	52	7.4
Bronchioles with muscle in wall	14	16×10^3	0.74	2.3	69	14	16
Terminal bronchiole	16	66×10^3	0.60	1.6	180	5.4	31
Respiratory bronchi	18	260×10^3	0.50	1.2	53	1.9	60
Alveolar duct	21	2 × 106	0.43	0.7	3.2×10^3	0.32	210
Alveolar sac	23	8 × 106	0.41	0.5	12×10^3	0.09	550
Alveoli		300 × 106	0.28	0.23			

[1] Volume 4800 cm^3 at about three-fourths maximal inflation.
[2] At a flow rate of 60 L/min

Interactions with Gases and Vapors

Gases and vapors are composed of molecules that can be considered well mixed with the oxygen, nitrogen, and other components of air.[3] Contaminant gas and vapor molecules behave similarly to air molecules. Consequently, unless the contaminant molecules diffuse to and dissolve in or react with the tissues of the respiratory tract, it can be assumed that the concentration that reaches the lower region of the lungs is essentially equal to the concentration of the contaminant gas or vapor in the environmental air. However, even if the contaminant gas or vapor is not reactive, there will be some absorption of the contaminant gas or vapor in the tissues of the respiratory tract, and that will be dependent upon the chemical properties of the contaminant gas or vapor (see Chapter 9 for a discussion of molecular interactions). The great-

[3] There are certainly exceptions to this in cases where workers may be very near a source or in very stable (non turbulent) environments where mixing is poor. However, even in a concentration of 10,000 ppm there are still 99% *air* molecules and for our purposes all these molecules are well mixed.

est absorption will occur for those compounds that have similar chemical properties to the fluids that line the respiratory tract. Consequently, their concentrations at the alveoli will be less than the concentrations at the nose dependent on the degree of that interaction. This reduction would be most prevalent for water-soluble gases and vapors.

Reactive gases and vapors, particularly acids and bases and reactive organic compounds such as aldehydes, will tend to react upon contact with respiratory tract surfaces and, consequently, be reduced in concentration at the alveoli. Although this makes the reactive gases and vapors less available to the bloodstream, there still is the important potential for toxicity at the site of their reactivity. Furthermore, although all gases and vapors will be absorbed to some extent in the areas of the lung above the gas exchange region, absorption into the mucosal lining will not necessarily reduce the overall body burden, as the material may be readily absorbed from the mucosa to the underlying cells and be distributed throughout the body. Even dissolved compounds that escape upper respiratory absorption may be swallowed with the mucus and absorbed in the gastrointestinal tract.

Respiratory Tract Aerosol Penetration and Deposition[4]

PENETRATION

As one inhales through the mouth or nose, particles in the air may penetrate or enter into the respiratory system. Whether a particle enters is dependent on its inertia or its tendency to continue along a certain path or at a certain speed. That inertia is in turn a function of its momentum or $m \times v$. Thus, if all the particles are moving at a certain velocity, for example the speed of the air moving into the nose as one inhales, then the particles will differ in momentum as a result of differences in mass. Heavier particles have greater inertia, that is they resist a change in velocity (i.e., speed or direction) more so than lighter particles. The smaller the particle, the greater is the likelihood of changing its angular momentum so as to make a turn and follow the air streamlines as they move into the nose. Larger particles have greater inertia and thus resist this change in angular momentum; they tend to impact on the face or miss the face and the nose altogether. In Chapter 10 the size-dependent penetration of particles into the respiratory system is discussed and in Chapter 24 size-selective air sampling methods that mimic the penetration to the respiratory tract are discussed. Here we discuss the deposition of particles in the respiratory tract – the process that can lead to disease. Like penetration, it is a function of particle size or inertial forces.

MECHANISMS OF DEPOSITION

When polluted air is inspired by a worker, the contaminants do not reach every portion of

[4] For a full discussion of this subject the reader should see Chapter 2 of Volume 2. In addition, penetration is further discussed in Chapters 10 and 24.

the respiratory tract. It is important to consider gases and vapors differently from aerosols in this respect. The availability of aerosols to different regions of the respiratory tract is dependent on particle size. On the other hand, the ability of gases and vapors to penetrate deeply into the lung is a function of the reactivity, solubility, and diffusivity of the gas or the vapor.

While penetration is the fraction of a particle of a certain size that makes its way to a particular region of the lung, deposition is the removal of particles from the air to the respiratory tract surface. A particle cannot be deposited if it does not penetrate. A particle that does penetrate may travel on to a lower level of the respiratory tract or be exhaled. Deposition is the critical variable in disease causation as it is the action that leads to disease.

For aerosols, the mass concentration breathed at the nose is considerably reduced compared to the concentration that reaches the alveoli. This is so because there are several physical phenomena (inertia is one) that cause the various particles to interact with the walls of the respiratory system and be lost from the airstream. These five main physical forces are impaction, sedimentation or gravitational settling, Brownian motion or diffusion, interception, and electrostatic attraction. The first three are the most important (see Figure 6-1).

A particle in a moving airstream has inertia. In other words, the moving particle tends to "want to move" in the same direction and speed in which it is currently moving. As the airstream bends or moves along the contours of the respiratory system, the particle will tend to continue in a tangential direction to the curve or bend in flow. This tendency is dependent on particle velocity and size.[5] Smaller particles are more likely than larger ones to change

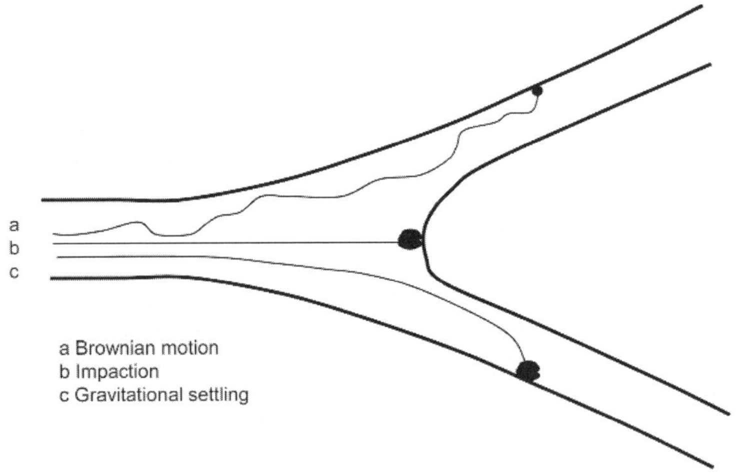

Figure 6-1. Three mechanisms of particle deposition in the lung.

[5] Note: For constant density particles, mass is proportional to size or volume of the particle; see Chapter 8. Although we discuss *size* here, it is actually particle mass that determines momentum and thus inertia.

direction with the airstream and move around impediments. As both size (mass) and velocity increase, momentum increases.

Impaction is the direct contact of a particle upon the sides of the respiratory tract, as the particle fails to stay in its airstream at a bend or a turn. Interception is similar to impaction except that the side of a particle tangentially contacts the wall of the respiratory tract as the particle does not completely make the turn around a bend or an obstacle. These two processes are more important for larger particles than small ones, causing them to be deposited in the upper portions of the respiratory tract.

Gravitational settling is the loss of particles to the walls of the respiratory tract because the velocity of the airstream has slowed to such an extent that the particle simply falls out due to gravity. Due to the exponential increase in the cross-sectional area of the airways, air velocity at the trachea is reduced about 5,000-fold at the alveoli, and even at the terminal bronchiole it has been reduced 70-fold. The velocities in Table 6-4 may be more impressively slow if expressed in more familiar terms. For example, 5.4 cm/s is equivalent to 0.054 m/s, 0.003 km/hr, or 0.002 mi/hr; 0.09 cm/s is equivalent to 0.0009 m/s, 0.00005 km/hr, or 0.0003 mi/hr. Remember also that these are maximum velocities for the above-defined work rate. Thus, sedimentation is effective in the smaller passages; here only smaller particles have penetrated; however, the biggest of these is affected by gravitational settling.

Brownian diffusion is the erratic movement of very small particles caused by the impaction of gas molecules upon their surfaces. The rate of Brownian diffusion for small particles (0.1 μm) has been estimated as 0.03 cm/s in the alveoli. This is an approximate average (Hinds, 1982). As Table 6-4 shows, velocities due to bulk airflow in the alveoli are around 0.09 cm/s. Therefore, for a 0.1 μm particle, diffusion is the predominant method of deposition in this region, with sedimentation also playing a role (Hinds, 1982).

Electrostatic forces are those forces due to negative or positive charges imparted to particles and the resultant repulsion from a surface of like charge or attraction to a surface of differing charge. They may play a minor role in deposition throughout the airways system.

PATTERNS OF DEPOSITION

With the above in mind, imagine 100 large particles (> 10 μm) penetrating into the nose in a single inhalation. A great majority of these will be impacted and deposited upon the surface of the nasal membranes as the air twists and turns throughout the turbinates. The particles that remain suspended will be impacted or intercepted on the various surfaces in the upper to middle region (trachea, larynx, and pharynx and the bronchi). As progress is made into the middle region, sedimentation begins to play an important role, accounting for about 40% of deposition of 10 μm particles (Lippmann, 1977). It is important that not all 100 of the particles will be deposited in any one area; instead a majority are deposited in the upper region, and fewer are deposited to the lower region of the respiratory tract. All of these large particles are deposited prior to reaching the lowest or alveolar region; however, some will be exhaled.

Experimental data and predictive models for humans show how these processes work for particles of various sizes and for the three major regions of the respiratory system. Those regions are shown in Figure 6-2. One model of deposition throughout the respiratory system is shown in Figure 6-3; other models have differences. However, the take home messages are these (Raabe, 1984; Raabe and Stuart, 1999; Stahlhofen et al., 1989; Heyder et al., 1986).

1. Almost all particles > 5 µm that penetrate are deposited.
2. As size decreases, deposition decreases to a minimum (~10%) at around 0.4 µm, i.e., these particles are very likely to be exhaled.
3. At a breathing rate of 45 L/min (medium-heavy work), almost all particles > 10 µm are deposited in the upper region. Smaller particles tend to pass on; however, at higher breathing rates deposition increases for smaller particles in this region.
4. Penetration to the middle region increases as size decreases below 10 µm. For particles between 5 and 10 µm, and low to medium breathing rates, virtually all are deposited in the middle region if they reach that far (i.e., down to the terminal bronchioles).

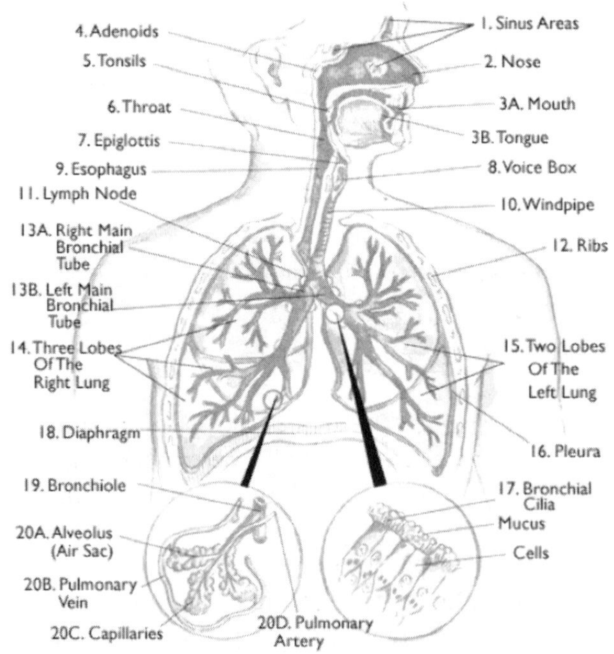

Figure 6-2. The parts of the respiratory system. The respirable penetration region contains the smallest bronchioles and the alveoli; the thoracic region is above the respiratory region up to the throat; the head region is that above the throat (reprinted with permission of American Lung Association, © 2006).

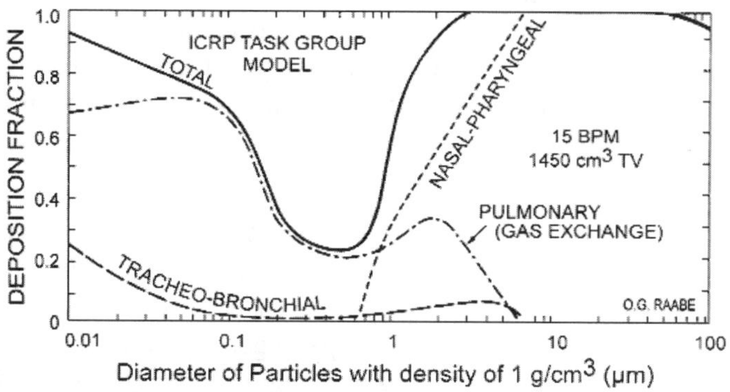

Figure 6-3. Total and regional deposition fractions for various sizes of inhaled airborne spherical particles with physical density of 1.0 g/cm³ in the human respiratory tract as calculated by the International Commission of Radiological Protection (ICRP) Task Group on Lung Dynamics (Morrow et al., 1966) for nasal breathing at a rate of 15 breaths per minute (BPM) and tidal volume (TV) of 1450 cm³ (Raabe, 1982, taken from Raabe and Stuart 1999 and ACGIH®, with permission).[6]

5. Of particles that do penetrate to the gas exchange region, most are not deposited.
6. If one breathes through the mouth, instead of the nose, all of the above changes.

WHY BOTHER?

Given the variability in penetration and deposition with variables such as particle size, sex, age, and mouth versus nose breathing, it seems almost hopeless to create an approach that takes into consideration the most important size fraction for a given disease and contaminant. However, it is important to try to understand where the various particle sizes are deposited. For example, if a compound causes bronchial carcinoma, then it must be of a size that can reach and be deposited in the bronchial region in order to cause damage. Typically these cancers occur in regions of naturally high deposition caused by bifurcations or junctions in the bronchi. If an industrial process produces very fine particles, say, in the range of 0.01 to 1 µm, or very course particles, say, in the range of 10 to 50 µm, either would be less hazardous relative to bronchial carcinoma (on an equal mass exposure basis) than a process producing particles in the 1 to 10 µm range. On the other hand, particles in the 0.01 to 1 µm will have their greatest

[6] The model shown in Figure 6-3 is based on variables such as nose breathing. For mouth breathing and other parameter changes, the curves shift. Furthermore, while the model fits experimental data well, there are some discrepancies. (Further details on the modeling of deposition are in Chapter 2, Volume 2. Also see Raabe, 1984; Raabe and Stuart, 1999; Stahlhofen et al., 1989; Heyder et al., 1986).

effect in the alveolar region. Examples include fibrosis inducing dusts such as silica and asbestos and cigarette smoke. Exposure to nanoparticles is a concern for the future. These particles have sizes (< 0.1 μm or < 100 nm) that allow easy penetration to the alveolar region, and more importantly they have the ability to be absorbed onto the blood and lymph systems. Their use is increasing rapidly and there are many chemistries and applications in use or planned (HSE, 2004).

Particles that are deposited in the lower or gas exchange region are quite important because of the physiology of that region. From the small bronchioles that have 1 mm diameters, upward through the trachea, and throughout the nasal passages, a membrane of ciliated epithelial cells exists that is covered by mucus. As passages become smaller than 1 mm, these cells gradually disappear; none occur in the gas exchange region (Menzel and Amdur, 1986). The cilia are in a constant fluid-like beating motion, moving the mucus material toward the epiglottis at a speed of 1 to 3.5 cm/min. There it is coughed, spat, or usually swallowed. This mucociliary blanket is extremely efficient at removing particles, but it does not occur in the lower or alveolar region. Instead, the alveolar region is dependent upon macrophage cells derived from the immune system to attack foreign particles, surround them, and absorb them into their cellular structure. These cells may then migrate to the bronchioles and be moved on the mucociliary blanket, or they may pass through the alveolar membrane into the lymph system; they may also enter the blood. If the contaminant is toxic to the macrophage cell, the cell may be destroyed and its contents spilled into the alveoli. These contents may be toxic to the alveoli and subsequently ingested by another macrophage.

Particles that are deposited directly onto the mucociliary blanket are removed from the lungs in a matter of minutes to hours if they are not dissolved. On the other hand, particles that are ingested by macrophage may require days to be removed from the lungs; and, in other cases where they are not successfully ingested by macrophage, the particles may remain in the lungs for years. For these reasons, the ability of contaminants to reach the gas exchange region of the lung is extremely important. Particles greater than 5 to 10 μm in diameter are very unlikely to reach the gas exchange region of the lungs because they are deposited in higher regions. However, of particles that do reach the alveoli, those with diameters of 3 to 4 μm are most readily deposited in the gas exchange region (Chan and Lippmann, 1980).

DERMAL EXPOSURE ROUTE[7]

Disease of the skin is the most often reported chemically caused occupational disease in the United States. Of course, occupational skin disease, while not easy to diagnose, is more easily observed or diagnosed than a disease of the liver, for example. Furthermore it is easier to implicate the workplace as a cause of skin disease, whereas the cause of a disease of, for example, the central nervous system is more difficult to pinpoint and must occasionally be

[7] The reader is directed to Chapter 3 of Volume 2 for more detail on dermatotoxicology.

determined (or estimated) in court.

Both the skin and the lungs are essentially *external* to the body and therefore exposed to environmental contaminants. The total surface area of the alveoli has been estimated as 35 m^2 during exhalation and about twice that during inhalation. The alveolar region may be as thin as 0.36 μm (Menzel and Amdur, 1986). On the other hand, the total skin surface is about 1.8 m^2, and the stratum corneum (skin layer that serves as a barrier) thickness is about 15 μm on average[8] (Scheuplein and Blank, 1971; ICRP, 2002).

The skin may seem much less *fragile* than the lungs; but in many cases this is a misconception. Also, the skin can receive direct exposures to quantities of liquids and solids that are very large relative to airborne exposures in the lungs. Thermodynamically, a liquid is considered to behave like a saturated vapor relative to skin absorption. Thus a fairly representative chemical like octane (molecular weight of 114 and a vapor pressure 10 mmHg at 25°C, see Appendix A) would have a saturated vapor concentration of about 13,000 ppm (see Chapter 8 for details on this calculation). This airborne concentration would have the same potential as a liquid exposure. Such a concentration would be highly unusual except perhaps in the case of a leak on a really hot day or from a hot process. Airborne exposure levels that are considered safe for most aliphatic hydrocarbons, and which are not likely to be exceeded, at least not by much, are < 500 ppm (ACGIH®, 2006).

The skin was once thought to be an adequate protective barrier to the absorption of chemical and biological agents unless it was somehow damaged or diseased. Furthermore, the preoccupation with the respiratory tract as the most important route of entry for most workplace contaminants often led to an assumption that the skin was not a particularly important route of entry. After all, unless one was knowledgeable about the microstructure and biochemistry of the skin, it was difficult to understand how a chemical could be transferred across such a substantial membrane and enter the bloodstream in any quantity that might be important.

It is now realized that many contaminants, depending upon their physical and chemical characteristics, can readily pass through the skin and enter the bloodstream. In fact, simply submersing one's hand in a chemical for a short period of time may allow an amount of chemical to enter the bloodstream equivalent to that absorbed in breathing the chemical from the air for 8 hours at a concentration considered to be a threshold for toxicity (Dutkiewicz and Tyras, 1967, 1968; Engstrom et al., 1977).

The two routes of exposure must, therefore, be considered together. If the industrial hygienist assessed only one route of exposure (usually inhalation) and deemed the situation to be safe, the addition of the second route of exposure might cause the situation to be unsafe by a considerable margin.

---

[8] Thickness varies from about 9 μm on the scrotum to over 600 μm on the palms of some hands (ICRP, 2002).

Anatomy and Physiology of the Skin

Figure 6-4 shows that the skin is essentially composed of multiple layers of cells interspersed with sweat glands, sebaceous glands, and hair follicles. These latter structures are commonly termed appendageal structures. The various layers of the skin cells can be divided into two main layers, the dermis and the epidermis.

The epidermis, or uppermost layer of the skin, contains the most important layer for providing a defense to entry of chemicals. The stratum corneum, which is the very uppermost layer of skin cells at the top of the epidermis, is the body's main barrier to the outside world. This layer is composed of dead skin cells that have undergone a process whereby cells from the layers below are dehydrated, compacted, and keratinized (converted to a tough, fibrous,

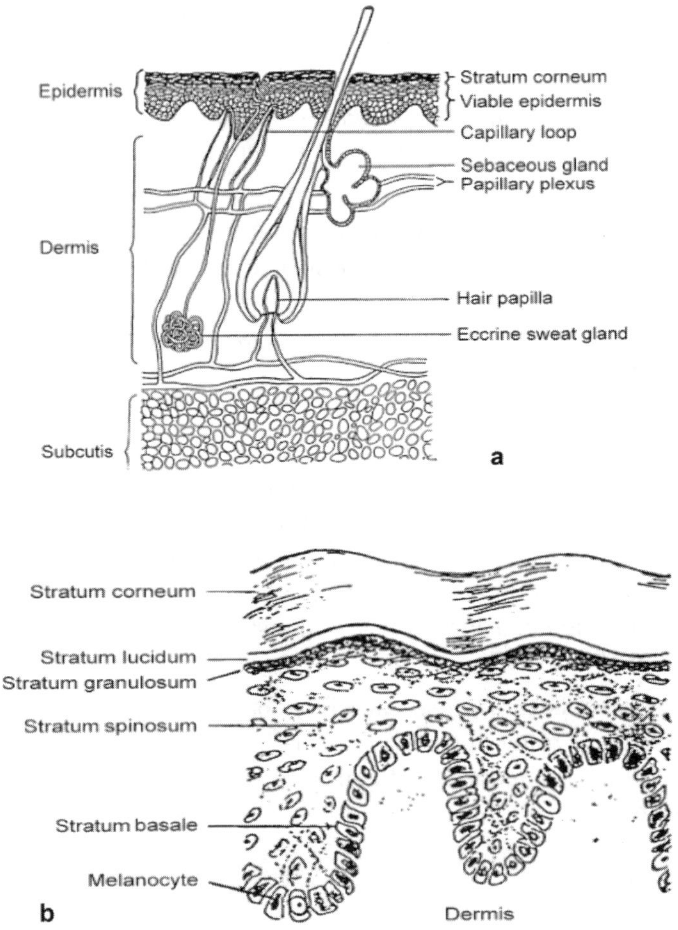

Figure 6-4. The anatomy of the skin showing a) epidermis and dermis and b) detail of the epidermis (from ICRP, 2002, with permission).

sulfonated protein). The keratin (also the substance of hair and fingernails) layers are interspersed with thin lipid layers.

Most of the remaining layers of the epidermis are in various stages of a slow death. The very bottom layer, the basal layer, is alive and metabolically active. Over a period of approximately two to four weeks, cells from the basal layer slowly move toward the surface of the skin, and during that process they undergo a series of transformations. By the time the cells are midway through the process, they are essentially dead. However, the transformation process to the stratum corneum cell type, which is considered to be the body's barrier, is not complete until the cell layer is keratinized.

Beneath the epidermis lies the dermis, a complex organ consisting of many structures including blood vessels. If a chemical is able to get through the stratum corneum, its passage through the remainder of the epidermis can be relatively easy, as the barrier properties of the stratum corneum are not formed in those layers. Once past the epidermis, the chemical can enter the circulatory system.

The major appendages on the surface of the skin include hair follicles, sebaceous or oil-secreting glands, and sweat glands. In general, a hair follicle is associated with a sebaceous gland. For every square centimeter of skin there are approximately 15 sebaceous glands and 100 sweat glands. On average their openings are quite small; they make up only 0.1 to 1% of the skin's surface.

It is important, then, to consider the nature of the stratum corneum. It consists of overlapping layers of corneocytes – the cells that are the end products of the process of keratinization. The spaces between the corneocytes are filled with a lipophilic matrix. Each corneocyte contains keratin filaments, which are in a matrix of lipid and non-fibrous protein. Keratin is relatively polar, whereas the lipophilic material is less polar. The stratum corneum consists of structures with two polarities that can allow either water soluble or fat soluble chemicals to partition into it.

Dermal Absorption

It is now thought, although still open to debate, that the migration of most chemical contaminants across the skin proceeds mainly through the intercellular route, i.e., through the lipid layers (Elias et al., 1987). Although lipid is commonly termed non-polar, this is only a relative term. Lipids or fats are oxygenated, usually through an ester linkage or phosphoryl groups. Some contain nitrogen, and thus all have some degree of polarity. They are thought to arrange themselves in an organized fashion with their polar ends together and their non-polar ends together (Martin, 1983; Mayes, 1983). During stratum corneum formation, lipids are *squeezed* from the cells and made less polar by enzymes (Elias et al., 1987). Even so, some polarity remains.

The two most important factors affecting permeation are solubility and molecular size. Molecules that have both polar and nonpolar properties may permeate most rapidly (i.e., those molecules having a bipolar nature, similar to lipid, are most soluble in lipid).

The permeation of dilute solutions of chemicals has been studied widely in the application of drugs to the skin. In these studies, Fick's first law of diffusion is used to quantify the rate of permeation across the skin. Fick's first law states that the flux rate across the skin is proportional to the thickness of the skin and the concentration gradient across it (see Chapter 8 and Volume 2, Chapter 4). The concentration gradient is the difference between the concentration on the side being exposed and the concentration on the opposite side. The concentration on the exposed side is not the concentration in the liquid or the gas phase but rather the concentration in the topmost exposed skin surface. This concentration is determined by the solubility of the chemical in the stratum corneum.

When the skin is exposed to concentrated industrial chemicals such as industrial solvents, the lipophilic nature of the skin is changed, as the solvent may dissolve the lipid. Therefore, the chemical nature of the skin and its thickness may change with time, and thus a Fick's law application to skin in the laboratory may have little to do with the real world situation.

The human skin permeation properties of relatively few industrial chemicals have been studied. Although animal models have been used to mimic human skin, there are significant differences between most types of animal and human skin, which make animal results difficult to interpret. Some animal models, such as the miniature swine appear to be reasonably compatible with human skin (Reifenrath and Hawkins, 1986). Even though only about 150 chemicals have been tested for permeation through human skin, correlation equations do a fairly good job of predicting the permeation of other chemicals. Given the 1000-fold variation in human skin permeability (from one site or one individual to another), all of the data, including that from animal and mathematical models, are relatively rough estimates (see Volume 2, Chapter 4 for more details).

Absorption of Metals

Hopefully the above discussion has allowed the reader to envision how an organic solvent coming into contact with the skin can pass through the stratum corneum via solution and diffusion and enter the blood. However, it may be difficult to imagine how an ionic chemical such as a metal cation can cross the skin and enter the bloodstream. It is widely known that metal ions such as chromium and nickel cause allergic reactions in some persons at the skin surface. In order for the allergic reactions to take place, the metal ion must pass the stratum corneum to combine with a specific protein and form an antigen (Jacobs, 1990). The immune system then reacts to the antigen to cause the allergic response. Nickel and chromium have been relatively well studied regarding their ability to cross the stratum corneum (Samitz et al., 1967; Samitz and Katz, 1976). However, less is known about the ability of other metal cations such as lead and beryllium to cross the skin and enter the bloodstream. Lead and beryllium do not cause allergic reactions at the skin; consequently, skin permeation of these cations would not be detected in this manner. At least one report has shown the ability of elemental mercury to be absorbed through the skin and cause systemic effects (Maranzana and Finulli, 1965) and a recent report shows that lead oxide can also penetrate the skin (Larese-Filon et al., 2006).

The study of nickel and chromium has elucidated some of the important factors involved in the transmigration of metal cations. First of all, the metallic compound must disassociate, so that the metal ion is released. Hence, penetration of chromium as a sulfate is negligible, that of the nitrate is slight, and that of the chloride is considerably higher than the others. Each of these three species of chromium is of the trivalent form. Hexavalent chromium is more soluble in water than the trivalent forms and therefore permeates better.

One might expect that acidic conditions would increase ionization and therefore penetration. Most detergents are alkaline and would not be expected to dissolve metal compounds. Figure 6-5 shows that to be the case. Thus, while the detergent is not likely to increase disassociation of metallic compounds, it is also not likely to dissolve them into water and remove them. A NIOSH designed acidic detergent (see Figure 6-5) did a better job than Ivory® liquid (Larese-Filon et al., 2006) and in another study did a better job than seven other commercial detergents (Esswein and Boeniger, 2005).

Although absorption of contaminants through the skin is now widely accepted as an important route of exposure, more remains to be learned about this process and about the relative permeation rates for the many industrial organic chemicals. Regardless of the lack of firm information, the dermal route and its contribution to the total body burden should be considered in all exposure assessments.

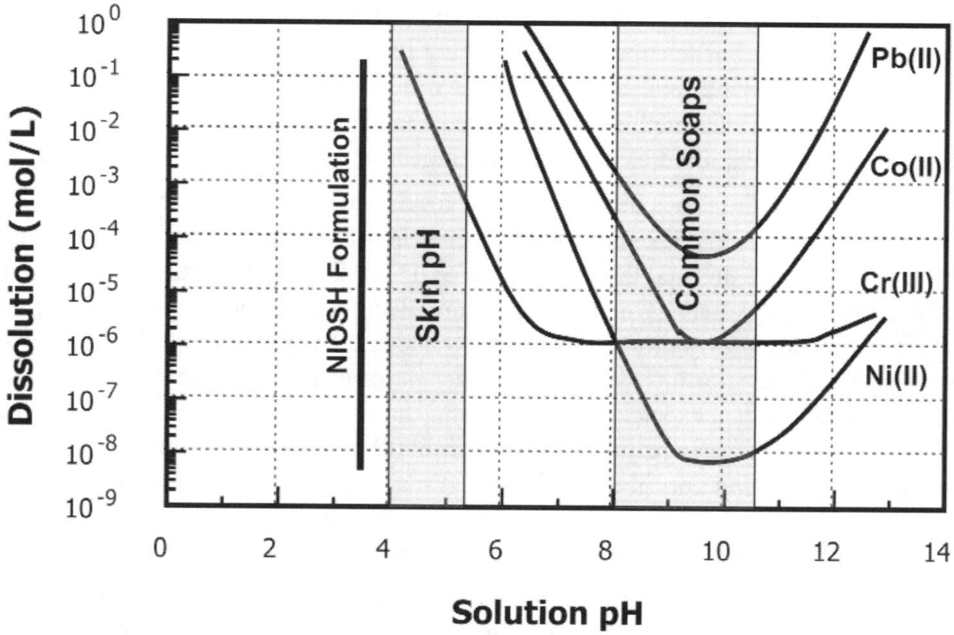

Figure 6-5. Dissolution of elements (either as hydroxide or oxide) in water versus pH (courtesy, Dr. Mark Boeniger, based on mathematical models from Dyer et al., 1998).

INGESTION EXPOSURE ROUTE

As will be seen in later chapters, the industrial hygienist can monitor airborne concentrations of contaminants in order to determine the potential importance of the inhalation route of exposure; and, similarly, the industrial hygienist can monitor concentrations of contaminants on surfaces and perform other studies in order to obtain a less certain estimate of the importance of the dermal exposure route. A quantitative assessment of the importance of the dermal exposure route is much harder than an assessment of the inhalation route. For the ingestion route, however, virtually no tools are available to the industrial hygienist for assessing its importance. The hygienist may postulate that the worker is ingesting a contaminant. (Of course, it should be realized that virtually any solid particulate will be ingested as it is moved upward from the respiratory tract. However, the material ingested in this way would still be considered part of the body burden originally gained by inhalation.) The portion of the body burden that the worker may gain as a result of touching surfaces and then touching food or tobacco is difficult to assess. Nevertheless, for contaminants such as lead and other compounds (Esswein and Boeniger, 2005), it is widely known that the ingestion route can be quite important. In the case of lead, acceptable body burdens are quite low; so even a small contribution via ingestion can be important.

The tools available to the hygienist allow direct implication of the ingestion route for only a few chemicals. For these chemicals, methods are available that allow determination of body burdens of chemicals by examination of blood, urine, exhaled air, or other body fluids or tissue. (See for example the Biological Exposure Indices, BEI®, ACGIH®, 2006.) Comparison of the body burden with airborne exposure levels may lead to the conclusion that the inhalation route cannot account for the total body burden. As a consequence, either the dermal route or ingestion is implicated. This method may be used for lead; and because lead (in the form of lead oxide) is not well absorbed by the skin, the ingestion route usually is implicated.

In some exposure situations the ingestion of food, water, etc., as a contributor to the body burden of a contaminant may be significant even when there is occupational exposure. The U.S. EPA has recognized this and the importance of understanding ingestion intake for the purposes of exposure assessment. The EPA Exposure Factors Handbook (EPA, 1997) contains compilations of intake information for water; fish and shellfish; meat and dairy; grains; fruits and vegetables; and breast milk. In addition, there is information on daily intake of soil and pica (non-nutritive substances such as paint, string, hair, animal droppings, insects, and soil) which is often important to ingestion exposure in children, but may be useful for workplace determinations of ingestion contribution, as well. Finally, it should be mentioned that the EPA Handbook also contains physiological information such as breathing rates; however, the ICRP 89 publication (ICRP, 2002) is a more thorough compilation.

It is important to realize that ingestion and dermal absorption can be important routes of exposure for compounds that are solids at room temperature but are heated, sometimes to liquefaction, in the industrial process. Even in so-called closed systems, these hot compounds will escape to the atmosphere, where they are rapidly cooled and condense on surfaces. There they can easily contaminate employees' hands and clothing. However, gaseous contaminants are not likely to be ingested or dermally absorbed because of their volatility. Volatile liquid contaminants are also likely to evaporate from surfaces and, as a consequence, not to build up over time as would nonvolatile or semi-volatile (vapor pressure < 1 mmHg) compounds.

QUESTIONS

6.1. Make a comparison of the relative magnitude of exposures to xenobiotics, their routes of entry, the types of xenobiotics, and their sources for an urban-living office worker in a large city and a rural-living employee of a foundry. State whatever assumptions you make about sources of drinking water, nearness to work, etc.

6.2. For compound X, an airborne exposure level that is said to cause an acceptable risk level (1 in 10,000) of disease after repeated exposure is 10 mg/m^3. At this exposure level a worker respires 10 m^3 in a day (8 hours).

   a. What will his or her total inhalation dose be for one day?

   b. Of this *dose*, will all of it reach the site of action, say, the nephrons of the kidney? If not, what happens to the portion that does not reach the site of action?

   c. Briefly, what are the difficulties in applying the *safe* dose calculated above to the dermal route of entry? If the dose were applied to the skin, would the same amount reach the site of action as when it entered the lungs? Explain your answer.

REFERENCES

ACGIH®: 2005-2006 Threshold Limit Values for Chemical Substances and Physical Agents and Biological Exposure Indices. American Conference of Governmental Industrial Hygienists, Cincinnati, OH (2006).

American Lung Association: Human Respiratory System, New York http://www.lungusa.org/site/pp.asp?c=dvLUK9O0E&b=22576 (2006).

Birkner, L.R.: Respiratory Protection: A Manual and Guideline, p. 15. American Industrial Hygiene Association, Fairfax, VA (1980).

Brain, J.D.: The respiratory tract and the environment. Env. Hlth. Persp. 20: 113-126 (1977).

Bouhuys, A.: Breathing: Physiology, Environment and Lung Disease, p. 30. Grone & Stratton, New York (1974).

CDC: Physical Activity and Health: A Report of the Surgeon General. Chap. 3. U.S. Department of Health and Human Services, Centers for Disease Control and Prevention, National Center for Chronic Disease Prevention and Health Promotion, Atlanta, GA (1996).

Chan, T.L., Lippmann, M.: Experimental measurements and empirical modeling of the regional deposition of inhaled particles in humans. Am. Ind. Hyg. Assoc. J. 41:399-408 (1980).

Dutkiewicz, T., Tyras, H.: A study of the skin absorption of ethylbenzene in man. Brit. J. Ind. Med. 24:330-332 (1967).

Dutkiewicz, T., Tyras, H.: Skin absorption of toluene, styrene, and xylene by man. Brit. J. Ind. Med. 25:243 (1968).

Dyer, J.A.; et al.: A Practical Guide for Determining the Solubility of Metal Hydroxides and Oxides in Water, Environ. Progress 17:1-8 (1998).

Elias, P.M.; et al.: The stratum corneum two-compartment model and its functional implications. Pharmacol. Skin 1: 1-9, Karger, Basel (1987).

Engstrom, K., Husman, K., Riihimaki, V.: Percutaneous absorption of m-xylene in man. Int. Arch. Occup. Environ. Hlth. 39:181-189 (1977).

EPA: Exposure Factors Handbook, Volume I – General Factors, August 1997. U.S. Environmental Protection Agency, Washington, DC http://www.epa.gov/ncea/pdfs/efh/front.pdf (1997).

Esswein, E., Boeniger, M.F.: Preventing the toxic handoff. Occ. Haz. September: 53-61 www.occupationalhazards.com (2005).

Hamilton, A.: Forty years in the poisonous trades. Am. Ind. Hyg. Assoc. Q. 9(1):5-17 (1948).

Heyder, J.; Gebhart, J.; Rudolf, G.; Schiller, C. F.; Stahlhofen, W.: Deposition of particles in the human respiratory tract in the size range 0.0005–15 μm. J. Aerosol Sci. 17(5):811-825 (1986).

Hinds, W.C.: Aerosol Technology: Properties, Behavior, and Measurement of Airborne Particles, p. 141. Wiley Publishing, New York (1982).

HSE: Nanoparticles: An occupational hygiene review. R.J. Aitken, K.S. Creely, C.L. Tran. UK Health and Safety Executive Report 274 http://www.hse.gov.uk/research/rrpdf/rr274.pdf (2004).

ICRP: Basic Anatomical and Physiological Data for Use in Radiological Protection: Reference Values, Pub. 89, J. Valentin (ed.). Annals of the ICRP, The International Commission on Radiological Protection 32 (3-4):1-277 (2002).

ISO: ISO 7243, Hot Environments – Estimation of the Heat Stress on Working Man Based on the WBGT Index (Wet Bulb Globe Temperature). Geneva, Switzerland (1989).

Jacobs, R.R.: Overview of skin structure, function, and toxicology, in Chemical Protective Clothing, Vol. 1, J.S. Johnson, K.J. Andersen (eds.). American Industrial Hygiene Association, Fairfax, VA (1990).

Josephson, J.: Exposure pathways of workplace contaminants. Environ. Sci. Technol. 17:168a-172a (1983).

Larese-Filon, F.; Boeniger, M.; Maina, G.; Adami, G.; Spinelli, P.; Damian, A.: Skin Absorption of Inorganic Lead (PbO) and the Effect of Skin Cleansers. J. Occup. Environ. Med. 48:692–699 (2006).

Linn, W.S.; Spier, C.E.; Hackney, J.D.: Activity patterns in ozone-exposed construction workers. J. Occ. Med. Tox. 2(1):1-14 (1993).

Lippmann, M.: Regional deposition of particles in the human respiratory tract, In Handbook of Physiology, Section 9: Reactions to Environmental Agents, D.H.K. Lee (ed.). American Physiological Society, Bethesda, MD (1977).

Maranzana, P.; Finulli, M.: Intossicazione mercuriale conseguente a penetrazione accidentale di mercurio metallico nel sottocutaneo. Med. Lavoro 56:357-366, English abstract (1965).

Martin, D.W.: Membranes, In Harper's Review of Biochemistry, D.W. Martin, P.A. Mayes, V.W. Rodwell (eds.). Lange Medical Publications, Los Altos, CA (1983).

Mayes, P.A.: Lipids, In Harper's Review of Biochemistry, D.W. Martin, P.A. Mayes, V.W. Rodwell (eds.). Lange Medical Publications, Los Altos, CA (1983).

Menzel, D.B.; Amdur, M.O.: Toxic responses of the respiratory system, In Casarett and Doull's Toxicology, 3rd ed., C.D. Klaassen, M.O. Amdur, J. Doull (eds.). Macmillan, New York (1986).

Morrow, P.P.; Bates, D.V.; Fish, B.R.; et al.: Deposition and retention models for dosimetry of the human respiratory tract, Report of the International Commission on Radiological Protection, ICRP, Task Group on Lung Dynamics. Health Phys. 12:1273-207 (1966).

Olishifski, J.B.; Benjamin, G.S.: The lungs, In Fundamentals of Industrial Hygiene, 3rd ed., B.A. Plog (ed.). National Safety Council, Chicago, IL (1988).

Raabe, O.G.: Deposition and clearance of inhaled aerosols, In Mechanisms of respiratory toxicology, H. Witschi, P. Nettesheim, eds. CRC Press, Boca Raton, FL, USA (1982).

Raabe, O.G.: Size-selective sampling criteria for thoracic and respirable mass fractions. Ann. Am. Conf. Ind. Hyg. 11: 53-65 (1984).

Raabe, O.G., Stuart, B.O.: Sampling criteria for the thoracic and respirable fractions, In Particle Size-Selective Sampling for Particulate Air Contaminants, J.H. Vincent (ed.). ACGIH® Cincinnati, OH (1999).

Reifenrath, W.G.; Hawkins, G.S.: The weanling Yorkshire pig as an animal model for measuring percutaneous penetration studies, in Swine in Biomedical Research, Vol. I, M.E. Tumbleson (ed.). Plenum Publ., New York (1986).

Ronk, R.; White, M.K.: Hydrogen sulfide and the probabilities of *inhalation* through a tympanic membrane defect. JOM 27:337-340 (1985).

Samitz, M.H.; Katz, S.A.: Nickel-epidermal interactions: Diffusion and binding. Env. Res. 11:34-39 (1976).

Samitz, M.H.; Katz, S.; Shrager, J.D.: Studies of the diffusion of chromium compounds through skin. J. Invest. Dermatol. 48:514-520 (1967).

Scheuplein, R.J.; Blank, I.H.: Permeability of the skin. Physiol. Rev. 51(4):702-747 (1971).

Silverman, L.; Lee, G.; Plotkin, T.; Sawyers, L.A.; Yancey, A.R.: Air flow measurements on human subjects and without respiratory resistance at several work rates. AMA Arch. Ind. Hyg. Occup. Med. 3:461-478 (1951).

Stahlhofen, W.; Rudolf, G.; James, A.C.: Intercomparison of experimental regional aerosol deposition data. J. Aerosol Med. 2(3):285-308 (1989).

Task Group on Lung Dynamics: Deposition and retention models for internal dosimetry of the human respiratory tract. Hlth. Phys. 12:173-207 (1966).

III

TOOLS FOR ASSESSING THE WORKPLACE

7

APPROXIMATING THE TRUTH

Industrial hygiene is an applied science. One of the purposes of any science is to seek the truth, or to approximate the truth. Applied scientists are in the business of measuring things. In order for those measurements to be acceptable, they must lie within a reasonable margin of truth. What is an acceptable approximation? What factors affect the approximation? How should the final answer be written?

Think, as a scientist, about the problem of measuring something. A surveyor may be confronted with the job of measuring the height of a mountain peak. This is considerably different from measuring the diameter of an atom; however, both are measurement problems, and similar kinds of issues must be addressed. For example, in what units should the measurements be made? how accurate should the measurements be? how should this accuracy be stated? how many times should the measurement be made? and how similar should the repeated measurements be? The answers to many of these questions are subjective. In some cases, a government body or perhaps another quasi-regulatory agency has set guidelines or standards to be followed with regard to the questions; however, in making these determinations, the governing body has made a subjective decision about how the question should be answered. In other words, the acceptability of the technique that one uses, or the adequacy of the answer that one obtains when measuring something is in the eyes of the beholder.

ACCURACY

Many people use the words *accuracy* and *accurate*; few, however, have a concrete notion of the implications of these words. "That is an accurate description of the process." To many people, the implication of this statement is that the description of the process is without error. Anyone reading the description should arrive at a concept of the process similar to concepts arrived at by other persons reading the description. Notice that the word *similar* was used in the previous sentence as opposed to *identical*. The layperson may be tempted to use words such as *identical*. However, the scientist tends more toward the use of such words as *approximately*, *equivalently*, *similarly* or *no significant difference*.

What is accurate? If something is accurate, is it exact? Is it without error? "These automobiles have accurate speedometers." Does this mean that the speedometer gives the exact speed, or that it gives a speed reading that is acceptable relative to the truth? Are the speedometers always close to the truth? How close? If they are not exact, then who determines what is close enough? Hopefully these questions have made you realize that your concept of accuracy may need some adjustments.

Imagine that we need to measure a person's height with a meter stick, but the meter stick has no divisions on it. If we measure the person's height to the nearest meter, then the maximum error we can make is approximately 0.5 m. For example, a person of 1.45 m height might be labeled 2 m rather than 1 m. For the sake of demonstration, say that a female we are measuring is actually 1.8 m, but she is measured as 2 m.

Now consider the surveyor who measures the elevation of a mountain peak. Assume that the mountain peak is 5,192.5 m. The surveyor employs the same criteria used for measuring the girl's height (measurement to the nearest meter). He measures the peak numerous times and reports an average measurement of 5,193 meters. He has made an absolute error of 0.5 m. The error is 2.5 times greater than the error made in measuring the 1.8 meter female (this error was only 0.2 m). On the other hand, one would intuitively say that the surveyor's measurement was considerably more accurate than the measurement of the female's height. As can be seen, the use of individually held concepts about accuracy can cause a great deal of misunderstanding. Accuracy should be defined in terms of the context of what is being measured. These examples of the measurement of heights of two separate objects do not provide fair comparisons, as the errors (0.5 m and 0.2 m) are compared to the heights of two very different objects. So we should compare measurement accuracies for similar objects or in some way normalize (i.e., equalize) the accuracy indices. Statistical concepts that can be used for this are discussed later.

Precision and Bias

An example widely used to facilitate understanding of the concept of accuracy is the use of the bull's-eye. Figure 7-1 shows four bull's-eyes. For each of the four, six shots have been fired at the center of the bull's-eye. For these purposes, the truth is the center of the bull's-eye.

If we consider that each of the four bull's-eyes represents the marksmanship of four different marksmen, which would be most accurate? Intuitively, most people would say that Figure 7-1a represents the most accurate marksman. This is true, but why? In order to answer the question, consider marksmen B through D. Which is the most accurate? To attempt an answer to this question, we should consider two components of accuracy. These two components, like accuracy, are relative, and must be defined in the context of what is being measured. The two components are precision and bias, or equivalently random and systematic error.

Precision, or random error, is the tightness or the repeatability of a series of measurements of the same attribute. In terms of marksmen B through D, Marksman B has the most precise shooting ability, in that all six shots are clustered most closely together. He repeatedly hits approximately the same area of the target. However, Marksman B does not hit the center of the

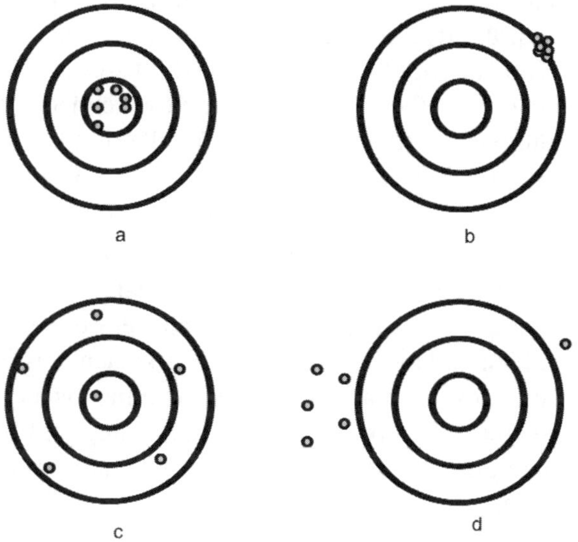

Figure 7-1. Four bulls eyes demonstrating a) precise, unbiased, and accurate, b) precise and biased, c) imprecise and unbiased, and d) biased and imprecise with a possible outlier.

bull's-eye, whereas in one out of six tries, Marksman C does hit the center of the bull's-eye. From this illustration, it might seem that Marksman C is the better shot. This brings us to the concept of bias.

Bias, or systematic error, is not related to the repeatability of measurements, but rather is related to the tendency of measurements relative to the truth. This is an important differentiation. Precision has nothing to do with the truth. A marksman can be extremely precise and yet never hit the target. Similarly, a scientist can consistently get similar answers but be a long way from the truth. Such a scientist is quite precise but is highly biased.

In the definition of bias the word *tendency* was used – the tendency of values relative to the truth. In order to determine bias, we need some estimator of tendency. For example, Marksman B tends to hit in the upper right quadrant of the bull's-eye, whereas the shots of Marksman C are scattered equally around the bull's-eye. For the marksmen, we may define their tendency as the average Cartesian center of the six shots. This is like an average shot. If something is measured repeatedly (say, the volume of water in a beaker), and the error associated with the repeated measurements is symmetrical – in other words, the likelihood of measuring too high is equal to the likelihood of measuring too low – then we can use a measurement of *central* tendency in order to characterize the bias.

Several measures of central tendency are used in statistics. The most common of these is the mean, which has similarity to a Cartesian center. However, the use of the mean may not always give us the best estimate of central tendency. For example, consider Marksman D. If we find the Cartesian center (equivalent to the mean) of his six shots, it will be to the left of the bull's-eye, to the right of the five clustered shots. However, his tendency is clearly well to the left of the bull's-eye. The one shot to the right of the bull's-eye may be aberrant (caused by something unusual). We can use other measures of central tendency to discover this effect. Two other measures are the median (the value in the middle of all values), and the mode (the value that occurs most frequently). The median and the mode are hard to envision for a bull's-eye example, but should be about the same in a series of symmetrically distributed numerical measurements.

In summary, the precision, or random error, is an indication of the similarity of measurements. The bias, or systematic error, is an indication of how close the measurements tend to be to the truth. So Marksman B is the most precise, and Marksman C is unbiased. However, our original goal was to find the most accurate marksman. Accuracy is a combination of both precision and bias. So we need a means for quantitating precision and bias, and for combining precision and bias in meaningful ways to find accuracy so that reasonable or acceptable values for each parameter can be defined.

Application of Statistics

The application of statistical methods and concepts must be made in order to quantify precision, bias, and accuracy. Although no attempt will be made at this point to give any statistical foundation, it is important that the reader understand that statistical methods and concepts are being utilized. In the previous discussion the statistical concepts mean, median, mode, and measure of central tendency were used. To demonstrate these concepts, consider the following 10 repeated measurements given by a radar speed gun: 64, 65, 60, 62, 68, 67, 67, 64, 64, 66. The true speed of the automobile is always 60.

The mean, or average, of these measurements is the sum of the measurements divided by the total number of measurements or

$$\bar{X} = \sum \frac{x_i}{n} \qquad \text{[Equation 7-1]}$$

where x_i values are the individual measurements, n is the number of measurements, and $\bar{x}$ is the mean. In this case, the mean is equal to 64.7. The true speed of the car and the mean are used to determine the bias of the measurement method. Bias is defined as

$$\left[\frac{(MEAN - TRUTH)}{TRUTH}\right] \times 100\% \qquad \text{[Equation 7-2]}$$

So, the bias of the radar gun as determined by the 10 measurements is 7.8%. In other words, at least for the 10 measurements that were made, the radar gun tends to be 7.8% higher than the truth. It is important when using the bias equation to construct it as shown, so that the sign of the bias is known. In the example above, the bias is positive, indicating that the radar gun measurement tends to be too high. If the terms in the numerator are transposed, then the sign of the bias will be incorrect.

Establishing a formal relationship for precision is a bit more difficult than for bias. In the case of precision we need an index of the repeatability of measurements. In other words, we would like to know the average deviation or discrepancy. Remember, we are interested in the average deviation, not from the truth, but from the average of the measurements. The precision is the average deviation from the average measurement. We could simply subtract each measurement from the average and then determine the average deviation. However, if the deviations are symmetrical and the sign of the deviation is conserved, then the average deviation will be zero. In other words, for 10 measurements there would be five negative deviations and five positive deviations of the same magnitude, and the resulting value of zero for the average deviation would be misleading. In order to avoid this problem, a new term is derived, called the standard deviation (S) rather than average deviation:

$$S = \left[\frac{\sum (\bar{X} - x_i)^2}{n - 1} \right]^{0.5} \qquad \text{[Equation 7-3]}$$

As can be seen, the deviations are squared, the average of the squares is found, and then the square root of the average is taken. The average is obtained by dividing by *n-1* rather than *n* for a statistical reason. We are estimating the true standard deviation (σ) using an estimate of the mean ($\bar{x}$). Thus we are using an estimate to obtain an estimate. The use of *n-1* penalizes us one *degree of freedom*. However, as *n* gets larger, the difference between *n* and *n-1* becomes smaller. So when $n = 30$, $\sqrt{n-1}$ and $\sqrt{n}$ are essentially equivalent. The units of the standard deviation (kilometers/hour, for example) are the same as the original units of the measurement.

Stating the standard deviation alone as a measure of precision often is not adequate. For example, consider the measurement of the height of a person and the elevation of a mountain peak. If told that the standard deviation of the measurements of these two objects is 0.1 m, would you say that the precision of the two measurement systems is equal? Intuitively, one should think that the precision of the method for measuring the mountain peak is considerably greater than that of the method for measuring the height of a person, as the standard deviation for the mountain peak method is only 0.1 m over, say, 5,000 m, whereas the standard deviation for the height of the person is also 0.1 m, but over a height of only 2 m. A way for adjusting or normalizing the standard deviation – that is, making its value relative to the measurement values – is needed. The relative standard deviation (*RSD* or S_r), also called the coefficient of variation (CV), accomplishes this.

$$CV = \frac{S}{\overline{X}} \times 100\% \qquad \text{[Equation 7-4]}$$

If values of precision (CV) are calculated for the people heights and mountain heights, using the previous values, we have 0.1/2 × 100% = 5% and 0.1/5,000 × 100% = 0.002%, respectively. Intuitively, these values seem more representative. Similarly, we can calculate the standard deviation of the radar gun as 2.45, and the CV or precision as 2.45/64.7 × 100% = 3.8%. We now must derive a concept and an equation for accuracy that combine the information gained from precision and bias, but to do this the concept of normal distribution must be introduced.

Normal Distribution

Consider a measurement method that has no bias; however, as with most measurement methods, repetitive measurements do not give the same results, and consequently the CV is greater than zero. Let us say that the precision of this measurement method is 10% (as defined by a coefficient of variation). Because there is no bias, the CV indicates that the average deviation from the truth is 10%. Can this value relate how large a single deviation may be or how frequently a deviation of 10%, 15%, or even 20% may occur? Of course, the smallest deviation that may occur is 0%, but it would be useful to know, for example, that some fraction of the deviations were less than some value. Assume that 95% of all deviations were within ± 20%. If we knew this, then we could make a statement about the precision of any given measurement. For example, if we measured an object one time, we could state that there was a 95% probability for the single measurement to lie within "± 20% of the theoretical mean of measurements." As there is no bias (the mean and the truth are equal), we can substitute "± 20% of the true value." A way to derive this useful information about the magnitude and the frequency of deviations is based upon statistics and involves the normal distribution.

Statistics is the science of describing or making inferences about a group of observations. The universe of possible observations from which the observations are drawn is first defined by the scientist as the population. Typically, the observations are made for only a subset of the population, known as a sample. Inferences for the entire population are then made by constructing a model, based on the sample, that represents the overall population. Models are used in virtually all branches of science and are simplifications of something more complex. The representations can be physical scale models, or they can be abstract. In the case of statistics, the models are mathematical abstractions and typically take the form of a mathematical equation relating two or more variables. The most widely used model in statistics incorporates two variables: $f(x)$ is the relative likelihood with which a certain value of the variable x occurs. An example would be the height of people, in which case x represents the values for the various heights that may occur, and $f(x)$ represents the relative likelihood of occurrence of those heights.

Figure 7-2 shows a frequency plot of repeated measurements of the constant concentration of a gas in a closed, unventilated space. Because the gas concentration is constant, the variability in the measurements is caused by random error in the measurement method. The shape of the plot is typical of randomly varying repeated measurements, in that it is peaked, and the peak is located in the center (i.e., the shape is symmetrical). Although several mathematical models may be used to approximately describe the shape of Figure 7-2, the model that has been developed most fully in statistics is the normal distribution. It is a mathematical equation relating $f(x)$ to x and having two parameters, as follows:

$$f(x) = \frac{1}{\sigma\sqrt{2\pi}}\exp\left[\frac{1}{-2\sigma}(x-\mu)^2\right] \qquad \text{[Equation 7-5]}$$

As Figure 7-2 shows, the typical shape of the normal distribution is bell-shaped or Gaussian. All normal curves have a similar shape to that shown in Figure 7-2. There are only two characteristics of the curves that vary: the width of the curve and the location of the curve along the x-axis. These characteristics are controlled by the two parameters μ (mu) and σ (sigma) in Equation 7-5, where μ is the location parameter or the mean of the distribution and can be estimated by $\bar{x}$. As μ changes, the curve moves left or right. As σ becomes smaller, the shape of the curve becomes more narrow and more peaked – σ being the true standard deviation, which can be estimated by s, as previously defined.

The Greek symbols σ and μ are used to indicate the population parameters, the model parameters or the truth, whereas the symbols $\bar{x}$ and s are used to indicate estimates of μ and σ. These estimates are typically derived from a sample of the population and can be used to con-

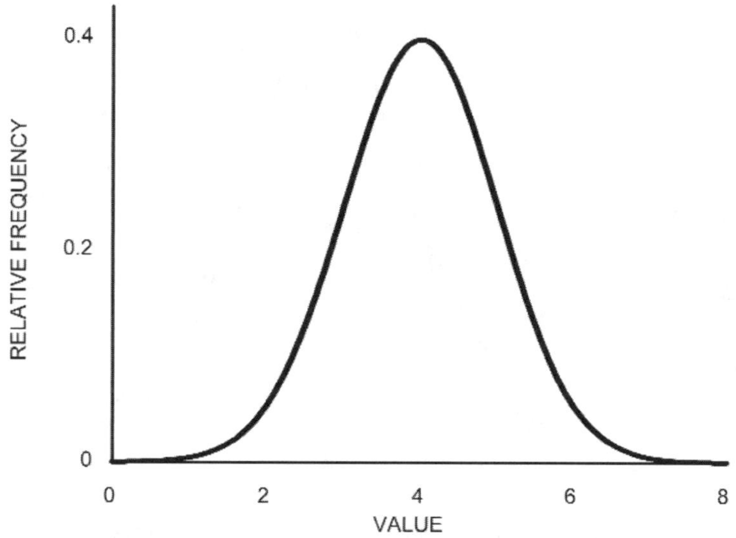

Figure 7-2. A normally distributed distribution of gas concentrations from an enclosed mixed space.

struct a normal distribution model.

Examine the two curves in Figure 7-3, which show a frequency distribution for a sample of a population and the normal distribution model based on the estimated parameters ($\bar{x}$, s) derived from that sample. The two curves have obvious differences. Why, then, is the normal distribution used as a model for something that it does not exactly replicate? First, the sample is not likely to be exactly like the population from which it is drawn. The smaller the sample size, the more likely it is that there will be differences. Second, a model allows the use of statistical techniques, shortcuts, so if the true population frequency distribution is close to normal, we are willing to accept the associated error caused by small departures from normality.

When measurements of a particular object are taken repeatedly, it is often assumed that the variability about the mean of those measurements is symmetrical or bell-shaped, that is, modeled by a normal distribution. Indeed, in measuring an object repeatedly, it seems reasonable that an individual measurement is as likely to be higher than the average as it is to be lower; hence, a symmetrical shape is expected.

There are many advantages in using the normal distribution as a model, which can be seen in their application to determining the precision, bias, and accuracy of a measurement method. These are the important features:

1. The mean and the standard deviation can be determined quite easily, and are the two parameters needed to characterize the model.

2. Regardless of the values of the mean and the standard deviation, all normal distributions have the following characteristics:

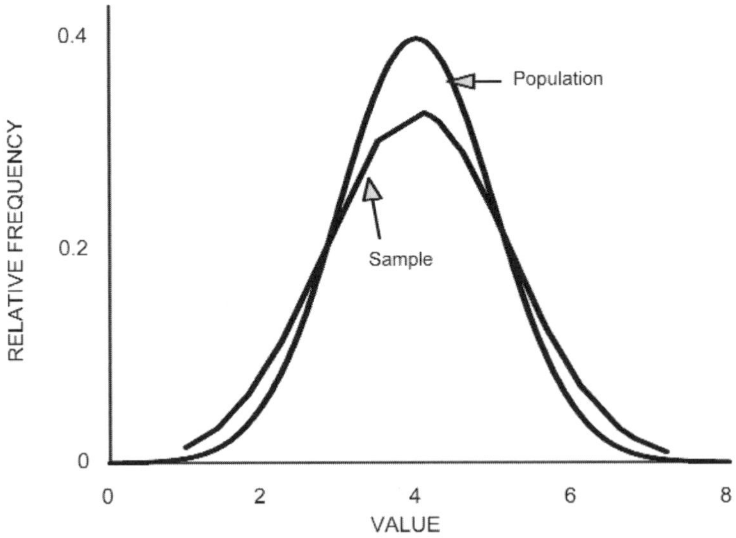

Figure 7-3. A frequency distribution for a sample and the model distribution for the population based on the mean and standard deviation estimated from the sample.

a. Plus and minus one standard deviation from the mean includes 68% of the population values; i.e., these values represent the 16th and 84th percentiles of the distribution.

b. Plus and minus two (strictly, 1.96) standard deviations from the mean includes 95% of the population values; i.e., these values represent the 2.5th and 97.5th percentiles of the distribution.

3. The mean is equal to the median is equal to the mode. All three measures of central tendency have the same value.

Many other attributes of the normal distribution are useful, particularly for testing hypotheses about a population. However, for our purposes, the above properties will suffice.

A Formal Definition of Accuracy

As stated earlier, it would be convenient if we had some way of making a statement about the precision of a method, such as "95% of all measurements lie within 20% of the mean of measurements." By using the normal distribution, we are able to construct such a statement. The precision statement is equivalent to finding the percentiles of a normal distribution. As shown in Figure 7-4, the 2.5th percentile of a normal distribution lies 1.96 or approximately two standard deviations below the mean, and the 97.5th percentile lies two standard deviations above the mean. This indicates that 5% of the population lies outside of these two percentiles, and 95% of the population lies within the two percentiles. Consequently, $\overline{x} \pm 2s$ gives the 2.5th

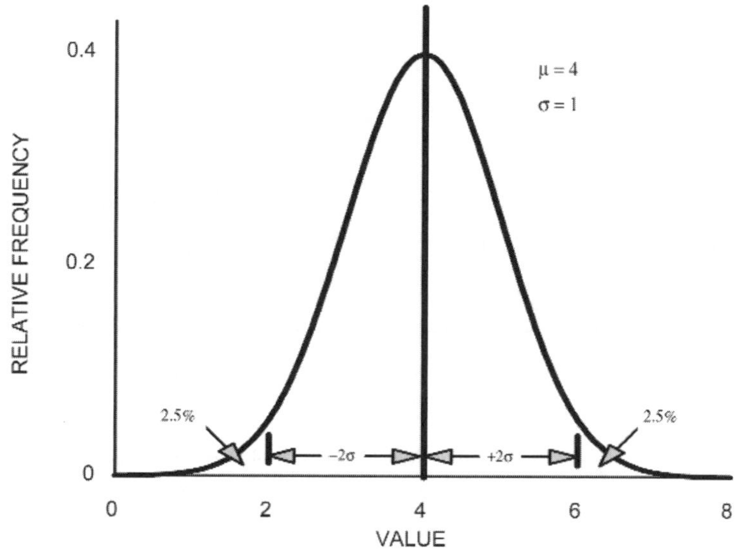

Figure 7-4. A normal distribution with $\mu = 4$ and $\sigma = 1$ showing the 2.5th and 97.5th percentiles.

and 97.5th percentiles of the distribution. It is also sometimes convenient to write the 2.5th and 97.5th percentiles as $\bar{x} \pm 2(CV)(\bar{x})$ if we know the CV[1]. This allows a precision statement to be made as follows:

> 95% of repeated measurements of the same quantity lie within ± 2 *CV*% of the mean of measurements.

In the example of the radar speed gun, the *CV* is 3.8%; so 95% of all measurements could be expected to lie within 7.6% of the mean of measurements, 64.7. If the measurement method has no bias, then the precision statement is also an accuracy statement, and the precision statement can be interpreted to mean that a single measured speed, using the radar gun, would have a 95% likelihood of being within ± 7.6% of the true speed.

However, this radar gun does have a bias (in this case positive); so the precision statement must be adjusted by a factor equivalent to the bias, as the bias will shift the mean of the measurements away from the true value. Figure 7-5 shows that for a measurement method having a bias in the positive direction, the 97.5th percentile on the distribution will be an additional increment (equal to the bias) farther from the true value than it would be if the method had no bias. For this reason, overall accuracy of a measurement method must consider both bias and precision.

Using the radar gun example, an approximate accuracy statement can be made with the calculated precision and bias using the following equation (Taylor et al., 1977):

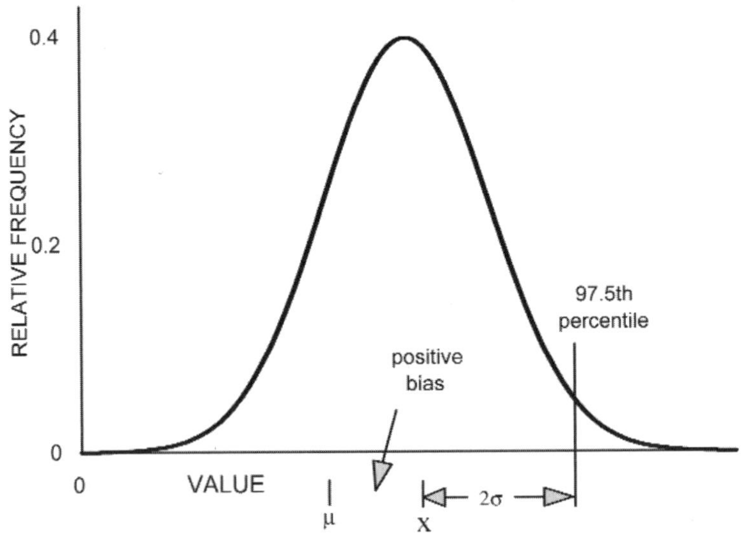

Figure 7-5. A sample distribution with mean = X relative to the true mean (μ). μ-X is the bias of the estimate.

---

[1] $(CV)(\bar{x}) = s$

$$A = 2\ CV \pm |BIAS| \qquad \text{[Equation 7-6]}$$

where A = accuracy. For the radar gun data, the overall accuracy of the gun is 2 (3.8) + 7.8 = 15.4%. So an individual measurement of speed using the gun has approximately a 95% probability of being within 15.4% of the true speed. Note that in reality the bias and the precision are estimates and therefore have uncertainty, i.e., the estimate of accuracy (15.4%) is an estimate based on two other estimates. Thus, a more formal calculation of accuracy is more complicated than this (see discussion in Chapter 25, "Accuracy" for more details).

It should also be apparent now that precision and accuracy statements must be made in terms of distribution percentiles. Because the normal distribution's tails are asymptotic, one can never make a statement that entails 100% of the probability. To increase the probability beyond 95%, a heavy penalty is paid for each small increment. For example, to increase the probability to include 99.7% of the distribution, $\pm 3\sigma$ are needed. Beyond this, only fractions of a percentile are gained for each standard deviation.

The radar gun has a precision of 3.8%, a bias of 7.8%, and a 95% accuracy of 15.4%. Are these values good? Does your answer depend on whether you are speeding? Would your answer be different if you were a radar gun manufacturer? Would it help if you were familiar with similar data for other guns? What is precise, and what is accurate? What is an acceptable level of bias? These are relative terms that must be defined. An accuracy of ± 25% may be adequate for the measurement of some objects; however, in other cases an accuracy of ± 1% may be necessary. In the case of many measurement methods used by hygienists, an accuracy of ± 25% is commonly accepted as accurate (Taylor et al., 1977).

Notice, from Equation 7-6 and Figure 7-6, that if the bias exceeds two times the CV, then most of the population of repeated measurements will not include the true value. In general, biases in measurement methods, particularly this large, are not acceptable. Biases are systematic errors associated with some part of the measurement system. The system includes not only the instrument but also the person operating the instrument, certain environmental factors, and so on. Many measurement methods have biases with known causes; these should be eliminated. For example, the marksman who is biased may need to adjust her scope or aim so as to compensate for the bias. Where cause is not known, it is common to correct for the bias by adjusting measured values. If the bias is large and the causes are not explainable, the method may not be acceptable.

MEASUREMENT ERROR

By now you should be convinced that the act of measuring an object is by its very nature associated with measurement error. These errors are of two types, systematic error, caused by a consistent and reproducible error in the measurement system, and random error, which is an inherent error in any measurement method and is caused by numerous factors that are unpredictable for a given measurement. The control of random error is a cost-benefit problem; it often can be reduced, but the additional precision may not be worth the cost. Systematic and

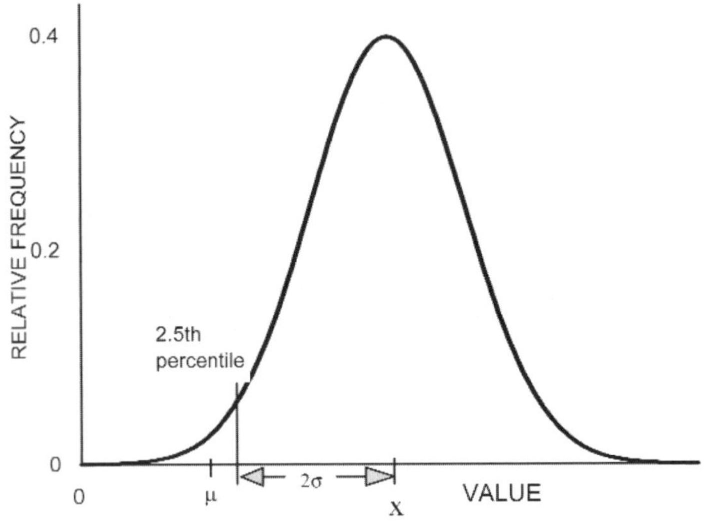

Figure 7-6. A sample distribution with mean = X, relative to the true mean μ, showing that in this case the bias exceeds 2σ.

random error together constitute the overall error of the measurement method or the measurement error.

So a single measurement is not perfect; error will be associated with it. Of course, measuring only one member of a population for purposes of estimating a parameter of the population, such as the mean, is not a good technique. The more measurements taken from the population, the more accurate is the estimate of any parameter, such as the mean.

Hygienists commonly make population parameter estimates. For example, the average exposure of a group of workers to a chemical on a given day, the average exposure to a chemical for a single worker over a number of days, or the average exposure for a group of workers to a chemical over a number of days, may be evaluated. Each combination of day (or other appropriate period of measurement) and worker constitutes a member of the population of exposures. The overall population may contain hundreds of such person-day exposures. Rather than measure each exposure, the hygienist uses statistical sampling tools. These allow estimates to be made from only a sample of the population. Each member of the sample is measured and subject to the usually random error of the measurement method.

STATISTICAL SAMPLING ERROR

Consider a jar filled with black and red beans. There are 95 black beans and 5 red beans. Not knowing the exact number of each type of bean and not wishing to count all of the beans, we could select a sample of the beans from the jar and then estimate the proportion of the two types of beans. Obviously if we pick a sample of 10 beans from the jar, it is possible that all 10 may be black. This would lead us to the conclusion that 100% of the beans are black. The difference between what we conclude from a sample and reality or truth is called sampling error.

The hygienist is confronted with sampling error when only a sample of the possible exposures in a population is measured. The goal of any scientific measurement is to control all sources of error in a cost-effective manner. These include measurement error as well as statistical sampling error. However, sampling error constitutes the larger source of error associated with most industrial hygiene measurements.

As with measurement error, we can use statistics to help us estimate the magnitude of the sampling error. In the case of measurement error, the normal distribution was used. However, for statistical sampling error associated with measuring workplace chemical exposures, the normal distribution generally is not appropriate – the distribution of exposures over time or across workers is commonly not symmetrical (Esmen and Hammad, 1977; NIOSH, 1977; Rappaport, 1991). The hygienist may wish to measure things other than exposures to chemicals. For example, exposures to physical agents such as noise may be normally distributed (Behar and Plener, 1984).

Ways of using each of the various statistical distributions are discussed in detail in Chapters 9, 14, 16, and 17. However, it is important now to realize that in accounting for statistical sampling error associated with industrial hygiene exposure assessment, a distribution other than the normal is needed. The distribution that is commonly used is the lognormal distribution. Like the normal distribution, the lognormal distribution has many applications in industrial hygiene as well as other fields.

LOGNORMAL DISTRIBUTION

The good news is that the lognormal distribution is related to the normal distribution; a lognormally distributed population of observations can be made normally distributed by plotting the logarithms of the values. The bad news is that the similarities end here. The various tools that have been derived from the normal distribution for testing hypotheses, making inferences, and describing characteristics of populations are not so well developed for the lognormal distribution, and those that have been derived are rather complex. Nevertheless, failure to use the lognormal, when it is appropriate, can lead to important errors both in the testing of hypotheses and in describing the parameters (such as mean and variance) of a population.

As with the normal distribution, the lognormal is a mathematical model that is represented by an equation:

$$f(x) = \frac{1}{x\sigma_L\sqrt{2\pi}} \exp\left[-\frac{1}{2\sigma_L^2}(ln x - \mu_L)^2\right] \qquad \text{[Equation 7-7]}$$

If Equations 7-5 and 7-7 are compared, one can see several similarities. In the case of the normal probability density function (Equation 7-5), the two parameters are σ and μ. In the case of the lognormal, the two parameters are σ_L and μ_L. These two parameters are not the mean and the standard deviation of the members of the population, but rather the mean and the standard deviation of the logarithms of the members of the population.

Whereas the normal distribution is characteristically bell shaped, the shape of the lognormal distribution cannot be characterized so easily. Figure 7-7 shows that the shape of lognormal distributions can vary considerably. In fact, some lognormal distributions can be so slightly skewed (asymmetrical) that the normal model is adequate for characterizing them. As shown in Figure 7-7, as the geometric standard deviation and geometric mean change, the shape of the distribution changes even if the value for the mean remains constant. As the geometric mean increases, the geometric standard deviation *decreases* for a constant mean.

In order to develop equations for the lognormal, some terms must be defined. It is important that the reader memorize the definitions of these terms in order not to be confused.

1. x_i is defined as any observation from lognormal population.

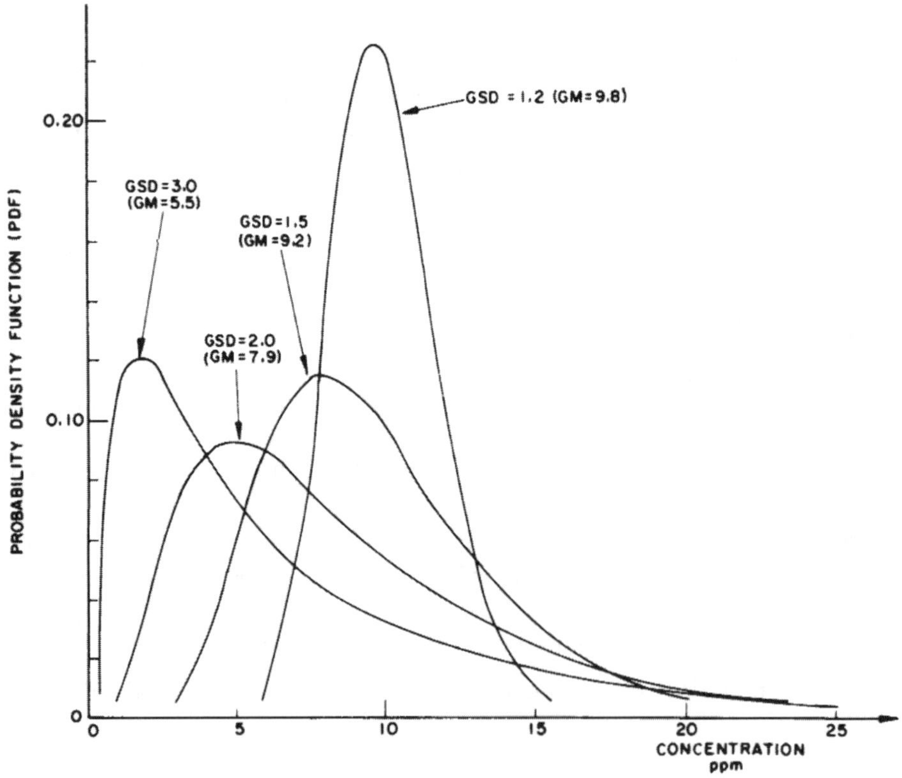

Figure 7-7. Four lognormal distributions with varying GSD and GM, but a constant mean value of 10 ppm (from NIOSH, 1977).

2. x_{L_i} is defined as the logarithmic transformation of x_i.

3. The true distribution mean and the standard deviation of the x_i are defined as μ and σ.

4. The true distribution mean and the standard deviation of the x_{L_i} are μ_L and σ_L.

5. The true geometric mean of the x_i is defined as

$$\mu_g = \exp(\mu_L) \quad \text{[Equation 7-8]}$$

6. The true geometric standard deviation of the x_i (GSD) is defined as

$$\sigma_g = \exp(\sigma_L) \quad \text{[Equation 7-9]}$$

7. $\mu \neq \mu_L \neq \mu_g$ and $\sigma \neq \sigma_L \neq \sigma_g$.

There are six parameters that are important to an understanding of the lognormal distribution, whereas for the normal only two parameters are defined. However, any two of the six lognormal parameters may be used to define a lognormal distribution, as the interconvertibility of the equations in Table 7-1 shows.

Figure 7-8 shows the mean, geometric mean, and mode of a lognormal distribution. They are not equal as they are for the normal distribution. The mode is the value that occurs most frequently in a distribution. The median of a lognormal is equal to μ_g, the geometric mean. This is derived as follows. We define ln $x_i = x_{L_i}$. Because the x_{L_i} are normally distributed, the median x_{L_i} = mean x_{L_i} or μ_L. Now $\exp(\mu_L) = \mu_g$, but this exponential transformation does not change the relative placement of the values. In other words, μ_L is the median or middle value of the x_{L_i}, and so $\exp(\mu_L)$ is still a median or middle value. The geometric mean, μ_g, is the median of a lognormal, not its mean. The geometric mean is always larger than the mode, and the true mean is always larger than the geometric mean, as Figure 7-8 shows.

Among the terms for variation or spread of a lognormally distributed population, the geometric standard deviation is more widely used than σ or σ_L. For a lognormal distribution, σ or σ_L gives us information about the spread of the distribution; however, as shown previously, without normalizing the standard deviation (e.g., calculating the coefficient of variation), we do not have any information concerning the relationship of the spread (variance) of the distribution to the mean. In other words, a standard deviation of 1 will have different importance if the mean is 1,000 than if the mean is 10. Standard deviations are positive values with no theoretical bound.

On the other hand, the geometric standard deviation has considerably different characteristics from the standard deviation. It is a measure of relative variability similar to the CV. (It can be related to the CV as in Table 7-1). For lognormal distributions of importance to hygienists, the geometric standard deviation ranges from approximately 1.2 up to values of 10, and most are in the range of 1.5 to 3.5 (NIOSH, 1977; Rappaport, 1991; Buringh and Lanting, 1991). *This is true regardless of the mean or its units.* To further examine this characteristic of the lognormal distribution and the geometric standard deviation, see Table 7-2. For the first

Table 7-1
Equations for parameters of the lognormal distribution

Given	To Obtain	Use
μL	$\mu_g =$	$\exp(\mu_L)$
μ_c, σ_c	$\mu_g =$	$\mu_c^2 / (\mu_c^2 + \sigma_c^2)^{0.5}$
σ_L	$\sigma_g =$	$\exp(\sigma_L)$
μ_c, σ_c	$\sigma_g =$	$\exp[(\ln[1 + \sigma_c^2/\mu_c^2])^{0.5}]$
μ_L, σ_L	$\mu_c =$	$\exp[\mu_L + 0.5\sigma_L^2]$
μ_g, σ_L	$\mu_c =$	$\mu_g \exp[0.5\sigma_L^2]$
μ_L, σ_L	$\sigma_c =$	$([\exp(2\mu_L+\sigma_L^2)][\exp(\sigma_L^2)- 1])^{0.5}$
μ_g, σ_L	$\sigma_c =$	$(\mu_g^2[\exp(\sigma_L^2)][\exp(\sigma_L^2) - 1])^{0.5}$
μ_g	$\mu_L =$	$\ln \mu_g$
μ_c, μ_L	$\mu_L =$	$\ln \mu_c - 0.5(\sigma_L^2)$
σ_g	$\sigma_L =$	$\ln \sigma_g$
μ_c, σ_c	$\sigma_L =$	$(\ln[1 + \sigma_c^2/\mu_c^2])^{0.5}$

μ_c, σ_c: mean and standard deviation of concentration
μ_g, σ_g: geometric mean and geometric std. dev. of concentration
μ_L, σ_L: mean and standard deviation of the natural logs of concentration

two columns of concentrations shown, one contains observations that are 1,000 times the size of the other. Consequently, the mean and the standard deviation of the second sample are 1,000 times larger than the first. In the third and fourth columns, the logarithms are given for the two sets of concentrations, and column three reflects the natural logs of column one. Note that the standard deviations of the logarithms for the two samples are equal, and consequently the geometric standard deviations (the exponentials of the standard deviations of the logs) are also equal. Hence, even though the magnitudes of the two data sets are considerably different, the geometric standard deviations are the same; so the geometric standard deviation is independent of the order of magnitude of the values in a population.

THE RELATIONSHIP OF SAMPLING AND MEASUREMENT ERRORS

Although the concept will be revisited later, it is important for the purposes of completing the ideas presented in this chapter to consider the magnitude of statistical sampling error associated with a geometric standard deviation in the range of 1.5 to 3.5 compared to typical measurement error values. Disregarding for the moment the units of what we are measuring, consider a measurement method that has a precision or CV of ± 10% (bias corrected or more often near 0). This measurement error is typical for industrial hygiene airborne concentration

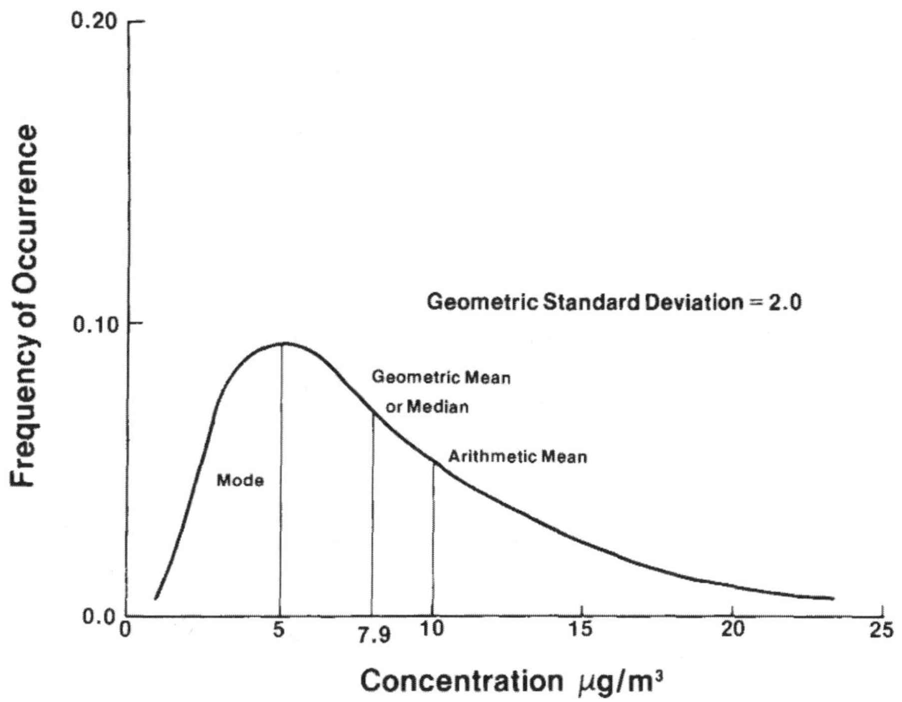

Figure 7-8. Lognormal distribution showing relative positions of the three measures of central tendency: mean = 10 μg/m³ and geometric standard deviation of 2.0 μg/m³. Note that mode ≠ median ≠ mean.

measurement methods. Now suppose we use this measurement method to measure the values in a statistical sample from a population of concentrations that has a geometric standard deviation of 2.7. Our purpose is to estimate the mean of the population. There are two sources of error in the estimate to be concerned with: that caused by the measurement technique and that caused by sampling the population.

The relationship between these two sources of error was shown (Nicas et al., 1991). When the measurement error is CV = 10% and the GSD of the population of exposure concentrations is 2.7, the variability associated with the measurement method is only 1.6% of the total variability. Hence, sampling variability is 98.4%. (This is further discussed in Chapter 16.)

The overall error in estimating a parameter of a distribution such as the mean can be reduced by using a more precise measurement method or taking more measurements (i.e., increasing sample size or n). However, the emphasis on increasing n is more important, as it has the greater effect on reducing overall estimation error. The conclusion is that although it is reasonable to use measurement methods that are precise and unbiased, it is more important to take statistical samples that are of an adequate size to overcome the substantial effects of statistical sampling error.

Table 7-2
Demonstration of the independent nature of GSDs

Concentration	Conc*1000	ln conc	ln (conc*1000)	
1	1000	0.000	6.908	
1.1	1100	0.095	7.003	
1.2	1200	0.182	7.090	
1.3	1300	0.262	7.170	
1.4	1400	0.336	7.244	
1.5	1500	0.405	7.313	
1.6	1600	0.470	7.378	
1.7	1700	0.531	7.438	
1.8	1800	0.588	7.496	
1.9	1900	0.642	7.550	
0.303	**302.8**	**0.215**	**0.215**	**standard deviation**
		1.240	**1.240**	**Geometric SD**

The question of adequate sample size is a constant consideration for all scientists, and its importance does not escape the hygienist. In order to arrive at adequate sample sizes for hygiene work, one must understand the implications of the lognormal distribution and the relationship of sampling error to measurement error.

The truth is never found. We can only approximate or estimate it. Along these lines it will be seen in later chapters that, as with all other aspects of science, theories cannot be proved. Scientists form hypotheses, and if those hypotheses survive a number of tests, then they become theories. When theories are tested often enough and they fail to be rejected, then they are often accepted but not proven.

Hygienists, being applied scientists, often must use theories and even hypotheses in their everyday work as though they were truth. However, no scientist, applied or theoretical, should fall into the truth trap. To accept theories as truth and continuously use them without question is not the role of a scientist. A technician, who simply follows a recipe for accomplishing some end, may be so confident that the recipe (method) works, that he or she never questions its results, but accepts as truth the theories upon which the recipe is based. The role of the scientist is to question, reformulate, and retest hypotheses. The role of the applied scientist is to utilize accepted theories and even hypotheses but constantly to be on guard for the slightest flaw in any results that may reveal a flaw in theory.

QUESTIONS

7.1 What are reasonably acceptable values for precision, bias, and accuracy? Explain your answers.

7.2 Ten measurements are made of the concentration of an airborne contaminant in a sealed chamber. The true concentration in the chamber is 10 mg/m^3. The measurements are 9.2, 9.8, 10.7, 10, 9.9, 9.9, 10.4, 10.1, 10.3, and 9.7. Does the measurement technique meet the NIOSH accuracy criterion of ± 25% for 95% of measurements?

7.3. A measurement technique is developed by your company for measuring the concentration of a proprietary compound. Under laboratory conditions, repeated measurements of a concentration of 100 mg/m^3 give these results: 120, 121, 128, 119, 119, 131, 117, 118, 119, 120. What are your recommendations regarding further development of this method? Show your calculations.

7.4. Two measurement methods are compared. One gives a precision of 3% and a bias of 10%. The second gives a precision of 8% and a bias of 0%. Which is better? Justify your answer and think about the controlability of bias and precision.

7.5. Ten values are derived as a sample from a lognormal distribution: 10, 11, 28, 12,12, 20, 21, 12, 23, 19. What are the GSD, mean, and geometric mean of this sample? Calculate the mean by using Equation 7-1.

7.6 What is the relationship(s) between systematic error, random error, precision, bias, and accuracy?

7.7 Contrast measurement error and sampling error.

REFERENCES

Behar, A.; Plener, R.: Noise exposure-sampling strategy and risk assessment. Am. Ind. Hyg. Assoc. J. 45:105-109 (1984).

Buringh, E.; Lanting, R.: Exposure variability in the workplace: Its implications for the assessment of compliance. Am. Ind. Hyg. Assoc. J. 52:6-13 (1991).

Esmen, N.A.; Hammad, Y.Y.: Log-normality of environmental sampling data. J. Env. Sci. Hlth. A12(1&2):29-41 (1977).

NIOSH: Occupational Exposure Sampling Strategy Manual. National Institute for Occupational Safety and Health, Cincinnati, OH (1977).

Nicas, M.; Simmons, B.P.; Spear, R.C.: Environmental versus analytical variability in exposure measurements. Am. Ind. Hyg. Assoc. J. 52:553-557 (1991).

Rappaport, S.M.: Assessment of long-term exposures to toxic substances in air. Am. Occup. Hyg., 35: 1-121 (1991).

Taylor, D.G.; Kupel, R.E.; Bryant, J.M.: Documentation of the NIOSH Validation Tests. National Institute for Occupational Safety and Health, Cincinnati, OH (1977).

8

THE PHYSICAL NATURE OF GASES AND VAPORS

Chapter 6 indicated that the hygienist is most commonly concerned with the inhalation route of exposure. Therefore, it is necessary to understand the behavior of air mixed with gases and/or vapor as well as mixtures of particles and air, or aerosols, which are discussed in the next chapter. Many of the laws, relationships, and definitions learned in physical science or physical chemistry apply to the work of the hygienist; so the most pertinent are reviewed here. (Remember that vapors are a special class of gases; see Chapter 4.)

Air is a compressible fluid that behaves for most practical purposes as an ideal gas, i.e., one that behaves according to a simple kinetic theory in which the gas molecules are assumed to have virtually no volume and to have no forces of attraction acting between them. The gas molecules are assumed to be in perpetual random motion, constantly colliding with one another and with solid objects such as the walls of a container in which they are confined. It is these collisions which exert gas 'pressure'. The average velocity and kinetic energy of the molecules depends on the temperature of the gas. As the gas is heated, the molecules move faster and collide with greater force; as the temperature is lowered, the molecules move more slowly eventually coming to a halt at a temperature of –273.15°C (–459.4°F), which is known as absolute zero.

Dry air is composed mostly of nitrogen and oxygen with only about 1% of other gases, mostly argon (Chapter 4, Table 4-5). Moist air can contain significant amounts of water vapor depending on the air temperature, for example at 40°C, (104°F) saturated air at normal atmospheric pressure contains about 7.3% water vapor by volume.

DENSITY AND SPECIFIC GRAVITY

Density is the mass of a substance per unit volume. At 4°C water has a density of 1 g/ml. Many hydrocarbon liquids used in industry have densities just less than that of water, say 0.7 to 1 g/ml, whereas halogenated hydrocarbon liquids have densities generally greater than one, to about 1.5 g/ml. Specific gravity is defined as the density of a substance divided by the density of water. Water has a specific gravity of one and is the liquid against which all other liquid and *solid* specific gravities are compared.

Gas densities are also expressed in units of mass/volume. At 70°F and one atmosphere (ATM) pressure, air has a density of 0.0749 lb/ft^3; at 25°C, it has a density of 1.187 kg/m^3. The specific gravity of gases and vapors is relative to air, not water. Specific gravities of important liquids, gases and vapors are found in Appendix A.

CONCENTRATION

Terms such as molal, molar, and normal are used to express concentrations in liquid solution. However, in hygiene work emphasis is on airborne concentrations; terms such as parts per million (ppm) and parts per billion (ppb) are used to express the concentration of gas mixtures. Although these terms have been popularized by the news media, they are often confused by the public.

A ppm is a part of something in one million total parts. A part is often left undefined, but it is commonly taken to be a unit of volume; thus one liter (1 L) of vinyl chloride mixed with 999,999 L of air has a concentration of 1 ppm. Note that the units (L) cancel. However, ppm can be expressed in terms other than volume. A ppm by mass is not equal to a ppm by volume. A ppm by mole is equal to a ppm by volume if the gases can be assumed to be ideal. The terms parts per trillion (ppt), ppb, and numerous others have readily meaningful definitions. Note that a percent is a part per hundred (pph).

The argument has been made before, usually to downplay the relevance of some environmental standard, that a ppm is an extremely small quantity. The elusive dry martini is often given as an analogy. Indeed, a one ppm martini would represent the driest martini in history, about one ounce of vermouth in a 10,000 gallon tank of gin (or about 30 ml in 38,000 L). Although this may seem like a dry martini, the analogy is totally irrelevant to what 1 ppm represents to the body's defense mechanisms. It can be shown that during an 8-hour shift a person would likely breathe 10 m^3 of air doing light-medium work. If the person were breathing 1 ppm of some contaminant, this would represent a volume of

$$10\ m^3\ mixed\ air \times \frac{1\ ml\ contaminant}{1\ m^3\ mixed\ air}\ (ppm) = 10\ ml\ contaminant$$

This seems like a small volume, but it will be shown later that a mole of any ideal gas occupies about 24 L. By coupling this with a knowledge of Avogadro's number we can calculate that the person breathes

$$10\ ml \times \frac{1\ L}{1000\ mL} \times \frac{1\ mole}{24\ L} \times \frac{6.022 \times 10^{23}\ molecules}{1\ mole} \cong 10^{18}\ molecules$$

This seems as excessive as the martini seemed dry. However, both calculations are misleading, as different contaminants are eliminated from the body in different ways and at different rates. The relevance of the dose is only important compared to the number of defense molecules (enzymes, for example) that the body can generate for purposes of eliminating that substance.

While the ppm concept may be appropriate to express the concentration of one gas mixed with another, it does not apply when solid or liquid particles are suspended in air. It seems inappropriate to give the volume of a solid relative to the volume of a gas. A more useful expression is the mass of particulate in a volume of air. This is a type of density term. Generally, milligrams or micrograms of particulate per cubic meter of air is used. Although ppm is not used to express aerosol concentrations, mg/m^3 is also used to express gas concentrations, in addition to ppm. As will be seen later, the two terms are interconvertible for gases and vapors; and one is dependent on temperature and pressure, but the other is not.

ABSOLUTE TEMPERATURE

By the ideal gas law, when a 273 ml volume of gas at 0°C is heated to 25°C, the volume of the gas increases by 25 ml. Similarly, if the gas is cooled to -50°C and does not liquefy, then its volume decreases to 223 ml. Since this linear relationship holds, then the gas should have no volume at -273°C (precisely -273.15 °C). In order to accommodate this linear relationship, the Kelvin temperature scale was created:

$$°K = °C + 273.15$$

The absolute temperature is used in almost all pressure, volume, or temperature relationships involving gases. There is an equivalent relationship in the English measurement system that allows one to avoid the two-step conversion of °C to °K. Instead:

$$°R = °F + 460$$

where °R is the absolute zero Rankin scale and can be used equivalently to the Kelvin scale.

PRESSURE

Numerous units are used to express gas pressure. The measured pressure is often compared to the pressure exerted by the gases of the atmosphere; so atmospheric pressure must be defined. Because it varies with altitude, latitude, and temperature, it must be defined relative to these terms. An atmosphere of pressure is defined as the *average* pressure exerted by the atmosphere at sea level and at 45° latitude. Pressure is commonly thought of as force applied over a unit of area; therefore, one atmosphere is 14.7 lb/in^2 or 1.013×10^5 Pascals. One atmosphere of pressure will support a column of mercury 760 mm in height or a column of water 407.7 inches (1035.6 cm) in height as used in a manometer. This implies that a column of air reaching from sea level to the vacuum of space applies the same force on the earth's surface as the column of mercury or the column of water, provided that all three columns have the same footprint or surface area. Several other units of pressure are used:

1 atm = 760 mmHg = 407.7 in H_20 = 14.7 1b/in^2 = 760 torr = 1 bar = 101.3kP = 101300 Pascals

STANDARD AND NORMAL TEMPERATURE AND PRESSURE (STP AND NTP)

Realizing that many physical properties are dependent on temperature and pressure, chemists have defined a standard state or set of conditions as STP or 0°C (273°K) and 1 atm. However, hygienists are usually interested in another set of conditions, namely 25°C or about room temperature and 1 atm. In order to differentiate this set of conditions from STP, it is commonly referred to as normal temperature and pressure (NTP). However, the differentiation is not always applied; sometimes NTP is referred to as STP. To make matters worse, American engineers have defined standard as air at 70°F and one atmosphere. This would be better if it were called *normal air* and defined at 77°F because 25° C = 77°F. The *standard air* definition is mostly important in problems dealing with ventilation control of the workplace atmosphere (see Volume 3). In addition, there is some confusion over standard pressure. One atmosphere is 101.3 kPa, but this is sometimes rounded to 100 kPa. Sometimes one bar is used, which is also 100 kPa. Consequently, we will define NTP and refer to it in this book as 25°C and 1 atm (101.3 kPa).

GAS LAWS[1]

Boyle's Law — The Relationship between Gas Pressure and Volume

When pressure is applied to a gas, the volume of the gas is reduced (the gas becomes squeezed). If the pressure of a gas is doubled, its volume is halved (providing the temperature remains constant). This relationship is expressed mathematically by Boyle's law,

$$P_1V_1 = P_2V_2 \qquad \text{[Equation 8-1]}$$

where P is the pressure and V the volume for conditions 1 and 2.

Example 8-1. If you have 1 L of air trapped in a piston at a pressure of 1.5 atm (P = 1.5 atm, V_1 = 1 L) and you increase the pressure to 5 atm, then the volume of the gas will become, from Equation 8-1:

$$1.5\ atm \times 1\ L = 5\ atm \times V_2$$

[1] Professor Chris Gray of Deakin University Australia is gratefully credited for much of this section.

$$V_2 = 0.3\ L$$

The calculations are similar in UK/US units. For example if we start with 1 ft^3 of air at 1.5 atm pressure, then at 5 atm the volume would become:

$$1.5\ atm \times 1\ ft^3 = 5\ atm \times V_2$$
$$V_2 = 0.3\ ft^3$$

Charles's Law — The Relationship between Gas Volume and Temperature

Like most other substances, air expands when heated. For practical purposes all gases expand equally when their temperature is raised; the volume of any gas is doubled if the temperature is raised from 0°C to 273°C (if the pressure remains constant). Starting at 0°C, the volume of a gas reduces by 1/273 for each 1°C reduction in temperature, which implies that at −273°C the volume of an ideal gas should be zero. In practice, real gases liquefy before the temperature reaches this low value, but in the simple kinetic theory, this temperature of −273°C (absolute zero) is that at which the gas molecules come to rest (i.e., have no kinetic energy) and therefore occupy virtually no space. This leads to the 'absolute' or 'Kelvin' scale of temperature on which absolute zero (or zero Kelvin, written 0°K) is −273°C. Ice melts at 273°K and water boils at 373°K. On the Fahrenheit scale absolute zero is at minus 460°F. This is zero degrees Rankin, 0°R. Using the Kelvin scale of temperature, Charles's law states that the volume of any gas is proportional to the (absolute) temperature, or

$$\frac{V_1}{T_1} = \frac{V_2}{T_2} \qquad \text{[Equation 8-2]}$$

where V is the gas volume and T is its absolute temperature.

Example 8-2. If we start with 1 L of air at 10°C and heat it to a new temperature of 500°C in a piston with no applied force (so that its pressure remains at one atm), then its volume will increase as follows, according to Equation 8-2:

$$T_1 = 10 + 273 = 283°K$$
$$T_2 = 500 + 273 = 773°K$$
$$\frac{1\ L}{283°K} = \frac{V_2}{773°K}$$
$$V_2 = \frac{1 \ x\ 773}{283} = 2.73\ L$$

The calculations are similar in UK/US units. For example if we start with 1 cubic foot of air at 50°F (50 + 460°R), then at 932°F (932 + 460°R) the volume would become 2.73 ft^3.

Remember, always carry out gas calculations in *absolute* temperature units.

Combined Gas Equations

In real world problems involving gases it is common to find that temperature and pressure vary concurrently. To deal with such problems it is useful to combine Boyle's and Charles's laws as,

$$\frac{P_1V_1}{T_1} = \frac{P_2V_2}{T_2} \qquad \text{[Equation 8-3]}$$

This equation allows one changed property of a gas to be predicted, provided that the other two properties are known at both conditions.

This equation strictly applies to ideal gases but air can be considered to approximate to an ideal gas for our purposes. The term ideal is often overlooked. The relationship assumes that the gas molecules are not attracted to one another and that they occupy no volume. For most true gases, the assumptions do not cause an important error, but for vapors, the size of the error increases with the size of the strength of the attraction of the molecules for each other. The latter is caused by dipole moments or polarity and will be discussed later. Hygienists usually assume ideality in their calculations without questioning the assumption. This is further discussed below under "Molar Volume."

Example 8-3. If we had 28 L (1 ft^3) of gas at a pressure of 101.3 kPa (1 atm) in a steel cylinder at 25°C (77°F) and this was heated to 900°C (1652°F), the volume would not be able to rise ($V_1 = V_2$), so the pressure would rise instead as

$$\frac{101.3}{298} = \frac{P_1}{1173}, \ or\ P_1 = 398 \text{ kPa } (SI)$$

$$\frac{1}{537} = \frac{P_1}{2112}, \ or\ P_1 = 3.96\ atm\ (UK/US)$$

Effect of Pressure and Temperature on Concentration

Equation 8-3 can be useful in one's thinking about the effect of temperature and pressure on concentration terms. For ppm, the volume terms in the numerator and denominator are equally affected by temperature and pressure. For example, a balloon inflated to 1 m^3 at sea level contains 1 ml of SO_2. If the balloon is taken to 1,000 m elevation and the same temperature, it will expand because of decreased atmospheric pressure. However, the SO_2 will also proportionally expand, and 1 ppm SO_2 will continue to exist in the balloon. Suppose that we express the volume of SO_2 as a mass (0.38 mg) and then take the balloon to 1,000 m. Now the balloon expands, and we have something less than 0.38 mg/m^3, as the SO_2 mass has become *diluted* in a bigger air volume.

The effect of pressure and temperature on mg/m^3 concentrations is generally not great for typical ambient conditions, i.e. within ±10% of NTP on the absolute temperature scale.

However, effects on pressure and temperature compared to NTP caused by elevation (e.g., 4500 m or 15,000 ft in Chile) or by seasonal extremes of temperature (e.g., 40°C ~100°F in Australia in February) can be significant. The previous paragraph might lead one to believe that the safest thing to do is express gas and vapor concentrations in ppm and forget mg/m³, but the opposite is true. It is the mass of a contaminant in a given breathed volume that causes the biological effect. Hence, assuming that breathing rates are constant at different temperature and pressure conditions, concentration expressed as mg/m³ represents the mass breathed regardless of atmospheric conditions; but ppm represents a changing mass as conditions change. In other words, two persons breathing 1 ppm at the same rates, but at different elevations, are not getting the same dose; but at 1 mg/m³ they are (see also Stephenson and Lillquist, 2001).

However, there is a complicating factor. If the worker at altitude is acclimatized, he breathes faster than the one at sea level so as to increase O_2 intake (see, e.g., Lahiri, 1977 or Frisancho, 1979). Thus, there is a physiological consideration in addition to the physics. In these situations, it has been observed that health effects (fibrosis of the lung, in particular) occur more rapidly at higher elevations all else being equal (Dümmer, 1998). There are really three questions related to T and P: when to adjust the volume collected by a pump, when to adjust a safe exposure level (an occupational exposure limit) that is predicated on exposure in NTP conditions, and when to adjust a measured concentration. This issue is discussed further with interpretation of TLVs® (Chapter 10) and constant flow apparati for pumps (Chapter 20).

Ideal Gas Law

The ideal gas law is derived from the combination of Boyle's law, Charles's law, and Avogadro's law. The latter states that volume is proportional to the number of molecules or moles. The three laws can be combined as

$$PV = nRT \qquad \text{[Equation 8-4]}$$

where R is the ideal gas constant and has units of $0.08206 \ \frac{atm\text{-}L}{°K\text{-}mole}$.

Density and the Ideal Gas Law

The relationship of the density of a gas to other variables can also be derived from the Ideal Gas Law. First, because density (ρ) equals mass/volume, ρ equals the product of the number of moles of a gas and its molecular weight (mass per mole) divided by the gas volume, or $\rho = nMW/V$. (Prove this by canceling the units.) If Equation 8-4 is multiplied by MW and rearranged, then:

$$\frac{MW \times P}{RT} = \frac{nMW}{V} = \rho \qquad \text{[Equation 8-5]}$$

Not surprisingly then, ρ is dependent upon *MW, P,* and *T*, and at constant *P* and *T* the ideal gas density varies only with *MW*. Therefore:

$$\frac{MW_1}{MW_2} = \frac{\rho_1}{\rho_2} \qquad \text{[Equation 8-6]}$$

and the density of one *ideal* gas may be estimated from another if the molecular weights are known.

Air has a density of about 1.2 kg m^{-3} (~ 0.0749 lb/ft^3) at sea level at 21°C (70°F). This is about 830 times less than the density of water which is 1000 kg m^{-3} (~62.42 lb/ft^3) but it is still significant. The air in a typical factory may have a mass of several tons and a fan which moves air at 1 m^3 sec^{-1} (~2119 cfm) is moving over 4 metric tons of air (about 10,000 lb) every hour.

The density (ρ) of air depends on its temperature, pressure, and humidity. At normal conditions 25°C (77°F) and 100,000 Pa (401.6 in.w)[2], dry air has a density of 1.169 kg m^{-3} (0.07297 lb/ft^3). Its density at other conditions can be calculated from Equation 8-5. For example

$$\rho \propto \frac{1}{V}$$

Thus if the pressure is doubled and, therefore, the volume is halved, the density must be doubled. Based on Equation 8-5, the combined gas equation can be rewritten:

$$\frac{\rho_1 T_1}{P_1} = \frac{\rho_2 T_2}{P_2} \qquad \text{[Equation 8-7]}$$

THE EFFECT OF WATER VAPOR ON AIR DENSITY

If water vapor is added to air, the air density is *reduced*. This may not be obvious at first until it is remembered that water molecules in moist air are replacing air molecules and not adding to them, otherwise the pressure would be increased. Air molecules are mostly nitrogen with a molecular weight of 28 while water molecules have a molecular weight of 18, hence water vapor-laden air is less dense than dry air. (The average molecular weight of dry air is actually 28.96).

As an extreme example, dry air at 99°C (210.2°F) has a density of about 0.93 kg m^{-3}, (~0.058 lb/ft^3) whereas saturated air at 99°C is mostly water vapor and has a density of about 0.58 kg/m^3 (~0.036 lb/ft^3). Under normal atmospheric conditions, the water vapor content of air does not decrease air density by more than a few percent, but saturated air from a hot tank

---

[2] 100,000 Pa is not quite the old standard atmosphere (101325 Pa). Thus, one commonly sees that a standard atmosphere is 407 in water at this old standard, as opposed to 401.

or drying oven could have a much reduced density and this could significantly affect the performance of the ventilation system.

If you know the water content (w, volume %) of a particular volume of air then you can calculate the density of the moist air as follows:

$$\rho_{moist} = \rho_{dry}\left(100 - \frac{0.379\ w}{100}\right) \qquad \text{[Equation 8-8]}$$

This expression is based on the molecular weight of water (18) and the average molecular weight of air (28.96).

Similarly if you know the total air pressure (P_T) and the partial pressure of water vapor (P_{H20}) then you can use the expression:

$$\rho_{moist} = \rho_{dry}\left(P_T - \frac{0.379\ P_{H_2O}}{P_T}\right) \qquad \text{[Equation 8-9]}$$

Example 8-4. What would be the density of moist air at 200°C (392°F), 1 *atm* if the partial pressure of water vapor was 0.4 *atm*? First we calculate the density of dry air at 200°C, 1 *atm* from Equation 8-5. Note that $P_1 = P_2 = 1\ atm$. The density of dry air at 0°C and 1 *atm* is 1.2929 kg m^{-3}, hence:

$$\rho_2 = \frac{\rho_1 T_1}{T_2} = \frac{1.2929\ \text{kg m}^{-3} \times 273\ \text{K}}{(200 + 273)\text{K}} = 0.746\ \text{kg m}^{-3}$$

Next we adjust this for the moisture content using Equation 8-9:

$$0.746\ \text{kg m}^{-3}\left(\frac{1\,atm - 0.379 \times 0.4\,atm}{1\,atm}\right) = 0.633\ kg\ \text{m}^{-3}$$

Molar Volume

As R is constant for an ideal gas, if P and T are held constant in Equation 8-4, then the volume that one mole of any ideal gas occupies, or the molar volume, may be found. The standard (273°K, 1 *atm*) molar volume MV_{STP} may be determined as:

$$MV_{STP} = 0.08206\ \frac{L\text{-}atm}{°K\text{-}mole} \times \frac{273\ °K}{1\ atm} = 22.4\ L/mole$$

This is less used by hygienists than the normal molar volume, calculated as above but at 298°K to give 24.45 L/mole. In working at conditions considerably different from normal, Equation 8-3 may be used to find a new molar volume, or the ideal gas constant, R, may be used as above. The ideal molar volume at NTP can be used by hygienists to convert ppm to mg/m³, or vice versa, if the molecular weight of the gas is known. For example, for benzene (B):

$$1\ ppm = \frac{1\ \mu L_B}{L_{air}} \times \frac{10^3 L_{air}}{1 m^3_{air}} \times \frac{1\ L_B}{10^6 \mu L_B} \times \frac{mole}{24.45\ L_B} \times \frac{78\ g}{mole} \times \frac{10^3 mg}{g} = 3.2\ mg/m^3$$

The student should understand all of the above steps and the unit cancellations. However, because all the exponents cancel, there is a shortcut to this calculation:

$$1\ ppm = \frac{MW}{24.45}\ mg/m^3 \qquad \text{[Equation 8-10]}$$

where MW is the molecular weight of the gas or the vapor. Of course, the equation can also be reversed to find ppm. Unfortunately, the calculation is often confused, and the MW/24.45 term is inverted. Therefore, it may be helpful to remember an index value. For example, 1 ppm benzene ≅ 3 mg/m³. By remembering this, one can always set up the fraction involving MW and 24.45 in the correct manner, if one also remembers that the MW for benzene is 78.

Table 8-1 shows that even though we assume that all gases and vapors behave ideally (i.e., that 24.45 is their molar volume), this is not the case. The error is small in *most* cases «5%, but as the molecule increases in size, the error can become significant. For example, for diphenyl the error is about 8.5%. This departure from ideality is *diluted* when one is converting ppm to mg/m³ or vice versa, provided that the concentration is not large, say, greater than 10,000 ppm (or 1%). This is because at a concentration of 10 ppm benzene, a benzene molecule *sees* another benzene molecule in only about 1 in 100,000 collisions, and the attraction is small. The error made by assuming ideality can be important when almost pure vapors are being used, for example, as laboratory calibration gases. The error in assumption of ideality, regardless of size, does err on the safe side relative to exposures; the ppm concentration calculated by using 24.45 L/mole is overestimated.

Dalton's Law of Partial Pressures

In a gas mixture such as air, the total pressure of the gas can be thought of as being the summation of the 'partial pressures' of the components of the gas. For example, in dry air the total pressure is equal to the sum of the partial pressures of nitrogen, oxygen, argon, carbon dioxide and other trace gases. The partial pressure of each of the gases depends on the number of molecules of each in the mixture, and this is proportional to the volume of each gas in the mixture. For example, in a gas mixture of 50% (by volume) of nitrogen and oxygen at 1 standard atmosphere, the partial pressure of each gas will be 0.5 atmospheres.

If we add water vapor or any other gas or vapor to air, the proportion of oxygen and nitrogen in the air must go down. For example, if we add argon to air in an unpressurized container until argon represents 50% of the total by volume then the oxygen concentration would be reduced to one half, i.e., to about 10.5%. As the total pressure in an open container would

Table 8-1
Molar volumes[1] (L/mole) at NTP for some vapors and gases

Acetone	23.98
Benzene	23.82
μ-Butane	23.98
Cresol	23.28
Decane	22.63
Diphenyl	22.37
Ethanol	24.04
Ethyl mercaptan	24.07
Ethylene dichloride	23.86
Heptane	23.39
Propionic acid	23.73
Propyl amine	23.94
Xylene	23.32

[1] Determined using van der Waal's equation and constants from Lange (1979).

remain constant at 1 atmosphere as we add other gases, the partial pressure of the original gases must be reduced. This occurs because the air is displaced from the container by the added gases.

In a closed container such as a gas cylinder or pipeline, if we add gases to the mixture the total pressure must rise as none of the gases can escape, i.e., we will have more gas occupying the same space. In this case, the partial pressures of the original gases will remain the same and the total pressure rise will be equal to the additional partial pressure of the added gases. For example, if we start with air in a closed container at 1 atmosphere and an oxygen partial pressure of 0.21 atmosphere, and then introduce argon until the total pressure is doubled to 2 atmospheres, the oxygen partial pressure will remain the same, i.e., 0.21 atmospheres and the argon partial pressure will be 1 atmosphere.

If a mixture of gases is ideal, then the pressure (and volume) of each is proportional to n, the number of moles of each. Or, rearranging Equation 8-4:

$$n_a = \frac{V_a P_a}{RT_a}$$

where *a* refers to a component of a mixture. Dalton's law states that:

$$P_a + P_b + P_c = P_T \qquad \text{[Equation 8-11]}$$

Vapor Pressure

At constant total volume (V_T) and temperature (T) of a system:

$$P_A \propto n_A$$

where A represents component A, or:

$$P_A V_T = n_A RT$$

At constant total pressure (P_T) and temperature of a system:

$$P_A \propto n_A, \text{ or}$$
$$P_T V_A = n_A R T$$

Combining the two equations yields:

$$P_A V_T = P_T V_A$$

Rearranging yields:

$$\frac{P_A}{P_T} = \frac{V_A}{V_T}$$ [Equation 8-12]

where we may recall that V_A/V_T equals the concentration of component *A* (e.g., ppm). From Equation 8-12, P_A/P_T is also equal to the concentration of *A*.

If the volume above a liquid were evacuated and then through evaporation the pressure above the liquid were allowed to equilibrate, the pressure on the side of the container would be equal to the pressure at the surface of the liquid. This is called saturated vapor pressure, and is the maximum pressure that the vapor evaporating from a liquid can exert at a given temperature.

Because P_A/P_T equals the concentration of *A*, (Equation 8-4), the maximum (saturated) concentration that evaporation from a liquid may achieve is:

$$\frac{P^\circ}{P_T} = C_s$$ [Equation 8-13]

where P° is the vapor pressure at some temperature, C_s is the saturated concentration, and P_T is usually atmospheric pressure, e.g., 1 atm or 760 mmHg at standard or normal conditions. Thus, at a given temperature C_s cannot be exceeded for that vapor.

It is important to note that vapor pressures are dependent on temperature. The vapor pressure of all liquid substances increases with temperature up to the critical point. The temperature at which the vapor pressure of a liquid equals 1 atm is the normal boiling point of the liquid.

VAPOR PRESSURE PREDICTION

Knowledge of the vapor pressure of a chemical can be extremely valuable, as it relates to the propensity for the chemical to volatilize and pose an inhalation hazard. However, vapor pressures vary with temperature. The relationship of vapor pressure to temperature is not linear, but generally takes on a classical Arrhenius form:

$$lnP° = A – B/T \qquad \text{[Equation 8-14]}$$

where *A* and *B* are constants. This form is often applicable over small ranges of *T* and is known as the Clasius-Clapeyron equation. Other forms also have been applied, such as the Antoine equation:

$$lnP° = A - \frac{B}{T + C} \qquad \text{[Equation 8-15]}$$

where C is also a constant. Sometimes Equations 8-14 and 8-15 are both referred to as Antoine equations. The constants are available in Appendix B for many compounds.

EVAPORATION OF MIXTURES: IDEAL SOLUTIONS

In an ideal gas, the intermolecular forces are assumed to be weak or negligible. However, since liquids only exist because of intermolecular forces, the use of the term ideal solution may appear incorrect. If the solution is comprised of two liquids which are similar in molecular size and intermolecular interactions then the term ideal solution may be used.

In a pure liquid the molecules are attracted to each other by intermolecular forces — the stronger the forces, the lower the vapor pressure. If the molecules of A are put into another liquid B to from a dilute solution, then the vapor pressure of the solution depends upon the forces acting between molecules of A and the surrounding B molecules. If A and B are very similar (e.g., hydrocarbon analogues such as heptane and octane) the A-A forces and the A-B forces are also similar and the vapor pressure will be determined by **Raoult's law** which states that in ideal solutions of two liquids, the total vapor pressure *P* is equal to the sum of the two partial pressures, i.e., $P_A + P_B$. If X_A and X_B are the mole fractions of the components, then the mole fraction of one component in a mixture is equal to the number of moles of that component divided by the total number of moles of all components in the mixture. Raoult's law stated for component A is

$$P_A = X_A P°_A \qquad \text{[Equation 8-16]}$$

where $P°_A$ is the vapor pressure of pure A. Exactly the same result is obtained for the component B.

The partial and total vapor pressures as a function of composition at a given temperature for an ideal system (to which the Raoult's equation is applicable) is shown in Figure 8-1 for toluene and benzene. The plot of the partial pressure of each constituent against its respective mole fraction in the liquid is a straight line passing through the origin. The total vapor pressure is equal to the sum of the partial vapor pressures and is also a straight line. Ideal or near-ideal solutions are mixtures of similar compounds such as benzene/toluene[3] and hexane/heptane.

Example 8-5. A solvent contains 0.1 mole fraction benzene in toluene. What is the partial pressure of benzene over the mixture? $P°_{benzene} = 100$ mmHg, and $P°_{toluene} = 8$ mmHg.

Partial pressure of benzene: $P_B = X_B\,P°_B = 0.1 \times 100$ mmHg $= 10$ mmHg

EVAPORATION OF MIXTURES: NON-IDEAL SOLUTIONS

If the two components of a liquid mixture are different in character (particularly polarity), the molecular forces between the different molecules will not be equal. The result will be that the presence of B will affect the escaping tendency of the A molecules, and vice versa;

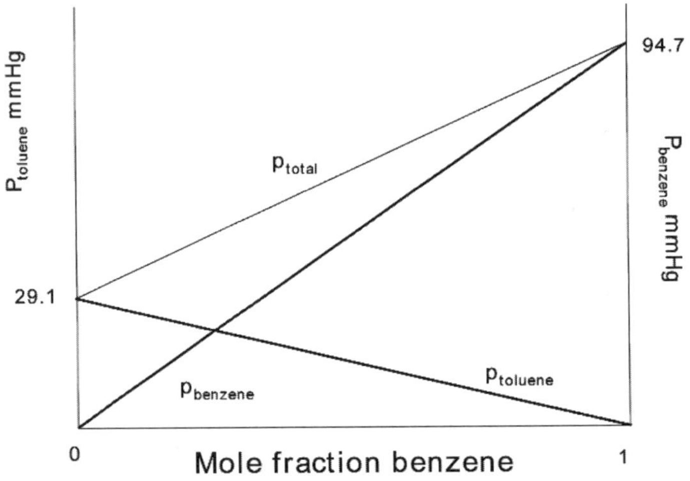

Figure 8-1. Experimental vapor pressures of mixtures of benzene and toluene which form a near ideal mixture that follows Raoult's law. The x-axis is the fraction of the solution that is composed of benzene (adapted from Castellan, 1983).

[3] This system is nearly ideal. However, it is not exactly ideal. The activity coefficient (see later discussion) for this mixture is 2 for benzene at any concentration of less than 1% benzene in toluene (Hirata, M., S. Ohe, and K. Nagahama, Computer Aided Data Book of Vapor-Liquid Equilibria. Kodansha Limited, Elsevier Scientific, Tokyo, 1975). In a broader sense, because most mixtures of organics (or of an organic in water) are more volatile than predicted by Raoult's law, errors using Raoult's law will be common and potentially dangerous. In this case, by a factor of two at most, and by a factor of 1.1 for the more dilute solution in Example 8.5 that follows.

Raoult's law will then not be applicable. A mixture of two dissimilar liquids would, therefore, behave in a non-ideal manner.

If A and B are dissimilar (e.g., toluene and water) then the A-B forces are weaker than the A-A forces so that the escaping tendency of A will be increased. The partial vapor pressure of A will consequently be greater than is to be expected from Raoult's law; such behavior is known as a positive deviation from the ideal law. When one component of a two component mixture exhibits positive deviations, the other component must do the same; the non-ideal system as a whole is then said to show positive deviations from Raoult's law. Examples of such mixtures include alcohols with hydrocarbons.

It is also possible for negative deviations from ideal behavior to occur when the A-B attractive forces are greater than the A-A or the B-B forces. This is not as common as positive deviation but occurs with mixtures of pyridine and acetic acid, chloroform and ethyl ether, hydrochloric acid and water. The type of vapor pressure curves obtained at a given temperature for mixtures of this kind is shown in Figure 8-2 for chloroform and acetone.

Deviations from ideal behavior are expressed by the activity coefficient[4] (γ) in the modified form of Raoult's law equation:

$$P_A = X_A \gamma_A P^\circ_A \qquad \text{[Equation 8-17]}$$

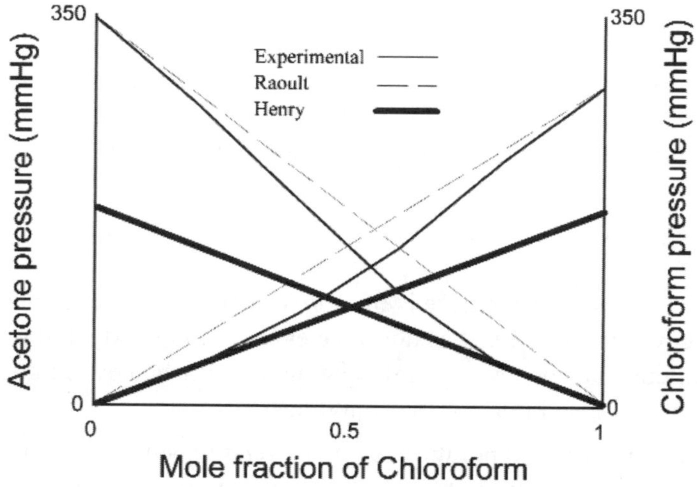

Figure 8-2. Experimental vapor pressures of an acetone-chloroform mixture versus those predicted from Raoult's and Henry's laws. The results depart negatively from ideality or Raoult's law. The x-axis is the fraction of the solution that is composed of chloroform (adapted from Castellan, 1983).

[4] For more on the use of activity coefficients and prediction of activity coefficients see Popendorf (2006, pp. 150-158).

Example 8-6. The vapor pressure of acetone is about 6645 Pa at 20°C. Acetone in a hydrocarbon mixture at a mole fraction of 0.1 might be expected to have a partial pressure of 0.1 × 6645 = 664.5 Pa, but if acetone deviates positively from Raoult's law, and $\gamma_{acetone} = 6$, then

$$P_{acetone} = X_{acetone}\,\gamma_{acetone}\,P°_{acetone} = 0.1 \times 6 \times 6645 = 3987 \text{ Pa}$$

Example 8-7. An adhesive based on toluene used in a continuous flow process is replaced with a water-based adhesive containing only 0.001 mole fraction toluene. By how much is the potential exposure to toluene reduced? $P°_{toluene} = 2924$ Pa at 20°C, $\gamma_{toluene}$ in water at 0.001 mole fraction = 1700.

$$P_{tol} = X_{tol}\,\gamma_{tol}\,P°_{tol} = 0.001 \times 1700 \times 2924 = 4971\ Pa$$

The partial pressure of toluene over the original adhesive was 2924 Pa. Therefore, the potential exposure is actually increased by changing to the water-based adhesive because of the large positive deviation from Raoult's law.

HENRY'S LAW

Although the activity coefficient of a solute *A* (γ_A) varies with composition there is a region at low concentration where it may be assumed to be approximately constant. This is the Henry's law region in which the following relationship approximately holds:

$$P_A = X_A\,k_A\,P°_A \qquad \text{[Equation 8-18]}$$

where k_A is the Henry's law constant for the solute *A* in a particular solvent. Figure 8-2 illustrates the Henry's law approximation for dilute chloroform-acetone mixtures and also the slight deviation from Raoult's law behavior for non-dilute mixtures.

EVAPORATION RATE

The evaporation rate of a substance is of more direct importance in determining airborne concentrations than vapor pressure. If a substance evaporates quickly compared to the ventilation rate in a room, then the concentration will be high. Published evaporation rates for organic solvents are often expressed relative to n-butyl acetate.

In the absence of relative evaporation rate data, relative vapor pressure can provide a fair approximation. The theoretical evaporation rate in still air *(R)* is given by

$$R = k\,P°\,D \qquad \text{[Equation 8-19]}$$

where k = an empirical constant, $P°$ = vapor pressure, and D = diffusivity of the vapor in air. D can be estimated (see for example Graham's law in a later section). This equation indicates that large molecules diffuse more slowly and, therefore, evaporate more slowly in a homologous series.

In still air, molecules evaporate from the liquid surface and produce a saturated vapor in the 'boundary layer' from where they diffuse relatively slowly into the bulk air. In practical situations in industry, evaporation rates are usually much higher than would be expected from simple diffusion calculations because air movements transfer vapor-laden air from the boundary region into the bulk air. Two important ways in which this occurs are free and forced convection. Free convection is the result of buoyancy in the vapor which can be caused by temperature gradients; if the liquid is warmer than its surroundings, convection currents are established which carry vapor away from the liquid surface. Forced convection is caused by air movements which disturb the boundary layer, for example drafts across the liquid surface caused by natural air movement or ventilation. Free and forced convection are usually the dominant mechanisms of mass transfer in the evaporation of liquids in the workplace. Empirical equations are available to predict the evaporation rate of liquids from the vapor pressure, surface area, temperature and air speed, and these have been embodied in computer programs.

HUMIDITY

Relative humidity *(RH)* is the amount of water in the air (concentration) relative to the maximum amount it can hold at that temperature. So it is the ratio of two concentrations, two masses, or two pressures, the denominator being the saturated value or concentration (C_s), which is temperature-dependent. The saturated vapor pressure for water at a given temperature can be used to find C_s, and if the *RH* has been determined, then the ambient concentration of water can be calculated:

$$RH = \frac{C_a}{C_s} \times 100\% \qquad \text{[Equation 8-20]}$$

C_a is usually in the g/m³ range and thus far more concentrated than most industrial air contaminants (mg/m³ or less). Absolute humidity is simply C_a and should be differentiated from *RH*. Whereas C_a is an absolute, *RH* is temperature-dependent. The vapor pressure of water is often determined from psychrometric charts and other functions such as that in Figure 8-3.

VISCOSITY OF AIR

Fluids such as water and air exhibit viscosity, which is a resistance to shearing motion. By this we mean that adjacent layers of fluid do not slide freely past one another. When liquids are flowing through pipes there is friction between the moving liquid and the stationary walls of the pipe that slows the liquid layer closest to the walls. Adjacent liquid layers are also slowed because of friction between them and the slower layers nearer to the walls. This results in a velocity profile in which the liquid at the center of the pipe moves much quicker than that

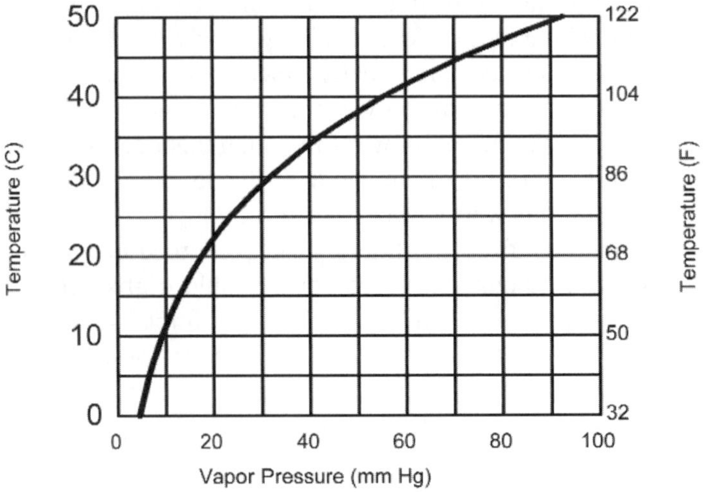

Figure 8-3. Saturated vapor pressure of water versus temperature. To find pressures at relative humidities other than 100% multiply the pressure at 100% RH by the %RH.

close to the walls.

You are familiar with viscous fluids such as syrup and less viscous (mobile) liquids such as alcohol and water. You may also be aware that in general liquids become less viscous at higher temperature; the oil in your car engine, for example, 'thins' when the engine gets hot. Gases such as air also exhibit viscosity, although this is a different phenomenon from that in liquids. In gases there is no actual friction between adjacent layers. However, molecules can diffuse from one layer to the next and this results in transfer of kinetic energy. Close to the walls of a pipe in which a gas is flowing, gas molecules can diffuse to the walls and lose forward velocity by collision with the irregular surface. This slows the gas layers closest to the walls. Adjacent layers are slowed when molecules with less forward velocity diffuse into them and vice versa. The rate of molecular diffusion in gases increases with temperature and, therefore, gases *increase* in viscosity with temperature.

Viscosity (η) is defined mathematically as shearing stress (force divided by area) divided by transverse velocity gradient (velocity divided by distance), or

$$\eta = \frac{F \div d^2}{v \div d} \qquad \text{[Equation 8-21]}$$

The viscosity of a fluid is illustrated in Figure 8-4. The top plate is sliding to the right at velocity v while the bottom plate is fixed. The fluid between the plates has a thickness of d and an area in contact with the plates of A. The force exerted on the plate to obtain the velocity of v is F.

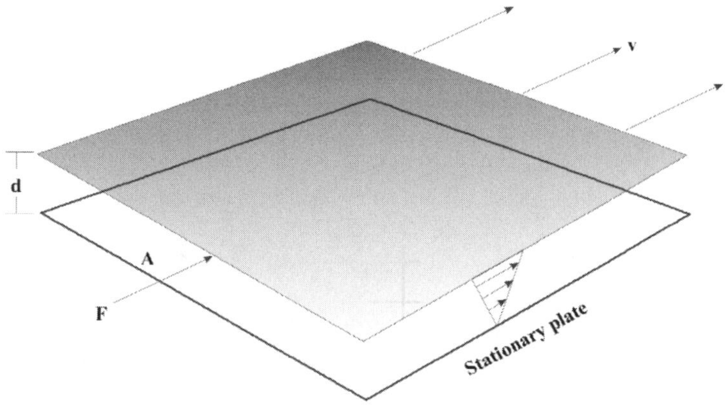

Figure 8-4. The variables (v, A, d, and F) that effect viscosity or the fluid resistance between two plates. The triangle with arrows shows that the velocity is greatest furthest from the stationary plate. See Equation 8-21.

The SI unit of viscosity is newton-seconds per square meter (N-s m^{-2}). This may also be expressed as pascal seconds (Pa-s). The viscosity of air at 25°C is about 1.84×10^{-5} Pa-s rising to about 3.7×10^{-3} Pa-s at 600°C (see Figure 8-5a, b).

POLARITY AND SOLUBILITY

The deviation of a gas from ideality is dependent on two assumptions: that the gas occupies no volume, and that the individual molecules have no interactions or attractions for one another. Attempts to correct for these assumptions are typified by the van der Waal's equation:

$$\left(P + \frac{n^2 a}{V^2}\right)(V - nb) = nRT \qquad \text{[Equation 8-22]}$$

where a and b are constants depending on molecular size and attraction. Note its similarity to Equation 8-4.

Molecular interactions are caused by electrostatic attractions between molecules. They are sometimes referred to as bonds but are not nearly so strong as bonds, nor should they be confused with covalent or ionic bonds (see Table 8-2). Instead these interactions are caused by dipoles, the locations of differing polarity on molecules. A strongly polar molecule has both strong negative and strong positive charges, as the two are usually in an overall balance across the molecule. The positive end of one molecule is attracted to the negative end of another. Therefore, molecules of the same compound tend to form the strongest attractions, as their opposite charges are most likely to be matched. Similarly, molecules that are from different compounds but have similar charges are also attracted. The more that molecules are attracted,

a)

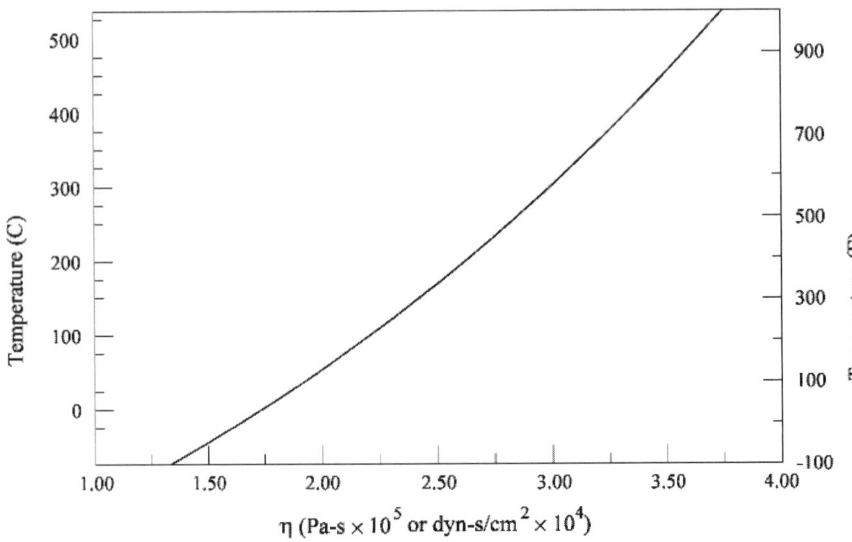

b)

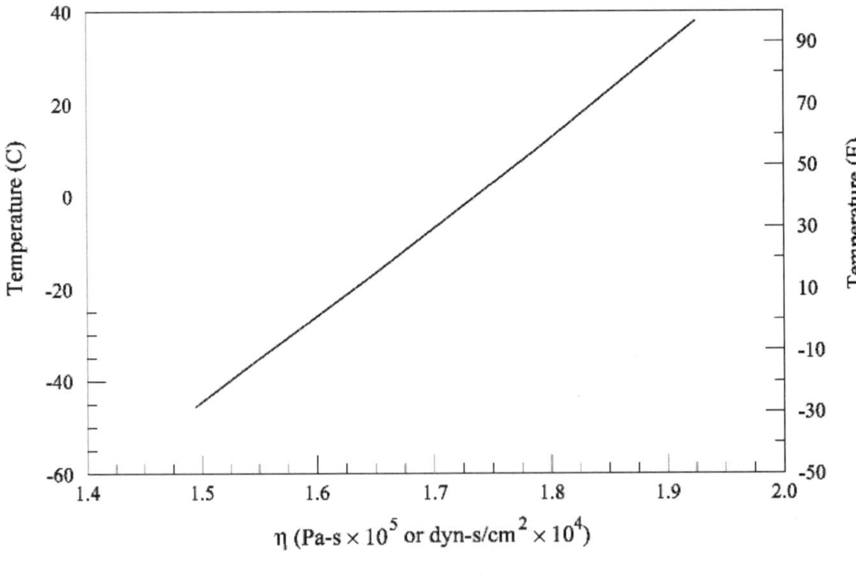

Figure 8-5 a and b. Viscosity of dry air versus temperature for 2 scales. Data from Sutherland's formula (Crane, 1988): $\eta = \eta_0*(a/b)*(T/T_0)^{3/2}$, where $a = 0.555T_0 + C$, $b = 0.555T + C$, η = viscosity in centipoise, η_0 = reference viscosity at reference temperature (0.01827 centipoise), T = input temperature (°R), T_0 = reference temperature (524.07°R), C = Sutherland's constant, 120 for air.

the smaller the volume that they occupy – one cause of a departure from ideality.

There are three main types of forces that cause polarity in organic molecules. The first of these causes momentary dipoles to form. It has been referred to as van der Waals, London, or dispersion forces. Imagine the electron cloud around a molecule such as hexane. At any instant in time it is doubtful that the electrons are evenly distributed. Where too few electrons exist, there is a weak positive charge; and where too many exist, the charge is negative. The locations of the poles are ephemeral; so they are called momentary dipoles. These weak attractive forces cause hydrocarbons such as pentane to exist as a liquid at room temperature.

Pentane (five carbons) is the smallest aliphatic hydrocarbon that exists as a liquid; but the smallest ketone, acetone (three carbons), also exists as a liquid. Acetone's molecular weight (58) is the same as that of butane (four carbons), which exists as a gas at NTP. The difference in the two states can be explained by the stronger attractive forces existing in acetone.

Oxygen, fluorine, and nitrogen atoms are strongly electronegative. As a result they tend to pull electrons toward them. Chlorine and sulfur are also thought of as quite electronegative, but they are larger than the other three, and the attraction is spread over a larger surface area. Hence O, F, and N cause the formation of permanent dipoles and polarity. This is a polar force, and it is present in such compounds as aldehydes, ketones, amines, and some fluorocarbons.

This force is also present in alcohols, but the specific arrangement of atoms in alcohols leads to a force known as hydrogen bonding, which is several times stronger than other permanent dipoles (see Table 8-2). Hydrogen bonding is strongest in water and helps to give that compound its peculiar properties, such as great surface tension. A hydroxyl group (OH^-) contains an oxygen, which tends to control hydrogen's electron, leaving a strongly polar proton. The hydroxyl group accounts for hydrogen bonding; so this force is strongest in alcohols and glycols.

Hydrocarbons such as pentane have dispersion forces only, whereas many molecules possess all three forces to varying degrees. The three forces combined make up the cohesive energy density *(CED)*. The square root of the *CED* has been termed the solubility parameter. The square root of each force has also been termed a parameter (e.g., the dispersion parameter); so

$$\delta_d^2 + \delta_p^2 + \delta_h^2 = CED^2 \qquad \text{[Equation 8-23]}$$

where δ is a solubility parameter, and *d, p, h* are dispersion, polarity, and hydrogen bonding, respectively. The three parameters have been termed the three-dimensional solubility parameter, or 3DSP (Hansen, 1968). They can be used to predict solution by using the old adage "like dissolves like." In other words, the more similar that two compounds are, the more likely it is that there will be attraction and therefore solubility. The likelihood of solubility increases as the *solubility distance*, or *A*, decreases between the two molecules. This can be calculated,

Table 8-2
Types of bonds and interactions and their disassociation energies

Type	Disassociation energy (kcal/mole)
Covalent bond	50–200
Ionic bond	10–20
Hydrogen	3–7
Polar	1.5–3
Dispersion	0.5–2

From Bilmeyer (1984).

using Cartesian geometry, as:

$$[(\delta_{d1} - \delta_{d2})^2 + (\delta_{p1} - \delta_{p2})^2 + (\delta_{h1} - \delta_{h2})^2]^{1/2} = A \qquad \text{[Equation 8-24]}$$

A knowledge of solubility or polarity has many applications in hygiene work, including use in selecting protective clothing that will not be attacked by chemical(s), predicting the service life of respirator charcoal cartridges, anticipating the efficacy of charcoal as an air sampling medium, and selection of appropriate gas chromatography columns, among others. The 3DSP values for many industrially important compounds are given in Appendix C.

GRAHAM'S LAW

Diffusion is a simple process in which molecules tend to even their distribution over space. It is due to the random kinetic movement of the molecules and their bumping into one another. Thermodynamically, it is a relatively ordered state transforming to a more chaotic one.

Graham's law states that the diffusion rate of an ideal gas is inversely proportional to the square root of its density. Because density is proportional to *MW*, the square root of *MW* is inversely proportional to the diffusion rate (R_d), and:

$$\frac{R_{d_1}}{R_{d_2}} = \left(\frac{MW_2}{MW_1}\right)^{0.5} \qquad \text{[Equation 8-25]}$$

So, for example, methane (*MW* = 16) diffuses approximately 1.6 times faster than propane (MW = 42). The relationship assumes ideality and approximates best for members of a homologous series, e.g., the alkanes. Diffusion coefficients with application to Fick's law have been estimated from empirical relationships (see Chapter 22).

FICK'S LAWS OF DIFFUSION

Graham's law (Equation 8-25) gives the diffusion rate of one type of molecule if the rate for another molecule is known. In other words, the rates are relative. We can say, for example, that molecule A diffuses at a rate twice as fast as that of molecule B. However, in many cases we need to know, quantitatively, how diffusion proceeds for a particular molecule in a particular environment. In hygiene work, usually a one-dimensional treatment of diffusion is sufficient. For example, we are often interested in the rate of diffusion or flux across a membrane or a space, from point *A* to point *B*, as shown in Figure 8-6. These applications are most common for passive dosimeters, a type of air sampling device (see Chapter 22) and for diffusion of chemicals through protective clothing (see, e.g., Perkins, 1990).

Fick's first law of diffusion can be used to predict the flux or rate of diffusion from point *A* to *B*. It states that the flux rate from *A* to *B* is proportional to the concentration gradient:

$$F = -D\frac{dC}{dx} \qquad \text{[Equation 8-26]}$$

or the change in concentration across the distance *x*. The quantity *D* is a proportionality constant and is known as the diffusion coefficient. When the flux rate is in units of mg/cm-sec, *D* has units of cm^2/sec, C is in mg/cm^3, and *x* is in cm. Therefore, *F* is the mass of compound passing across some area per unit time. The distance, *x* is important only in terms of establishing the concentration gradient. If the concentrations at *A* and *B* are similar, then the gradient will be low, and F will be small.

There are unique values of *D* for benzene diffusing through water, or air, or polypropy-

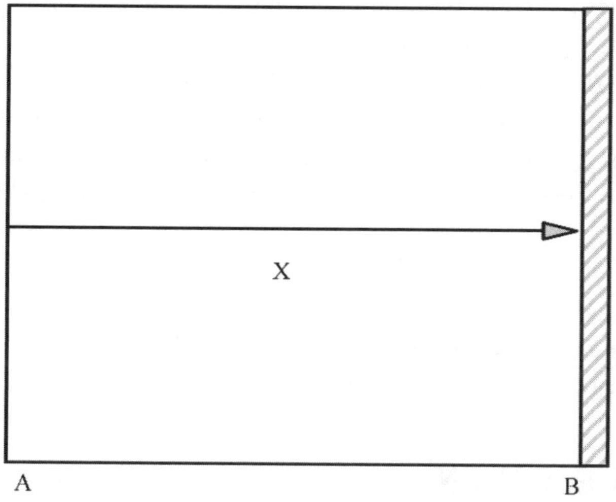

Figure 8-6. Depiction of a diffusion process across path length X, from point A to point B. Point B might represent a sorptive surface.

lene; each is a proportionality constant. The values for D must be derived empirically or from experimentation, empirical equations for diffusion coefficients may be used, such as the Hirschfelder equation. These applications are discussed in Chapter 23.

PERMEATION

Permeation across a membrane is treated as a special case of diffusion. It is relatively easy to imagine and quite easy to measure the concentration gradient across an air space, A to B. But how would one measure the concentration gradient across the thickness of a membrane, say, a rubber glove, if one side were exposed to a liquid chemical? The concentration on the exposed side is the concentration in the uppermost layer of the polymer, right at the surface, not in the air or liquid that is outside the polymer. We can measure this concentration if Henry's law (Equation 8-18) applies.

For a gas permeant, Henry's law may be used to substitute for dC in Equation 8-26:

$$F = \frac{Dk_H}{LdP} \qquad \text{[Equation 8-27]}$$

If we are dealing with a liquid (and using Equation 8-18), P exerted by the liquid is equal to the saturation vapor pressure P_v of the liquid. (Thermodynamically, it can be said that a liquid and its saturated vapor have the same chemical potential. Also, k_H can be defined as the saturated vapor pressure of the liquid divided by the solubility of the liquid in the polymer, or $P°/S$ (Lyman et al., 1982). So substituting $P = P°$ and $k_H = P°/S$, we have $C = P°(P°/S)$, or $C = S$, if we assume that the downstream concentration is 0. In other words, the concentration just at the surface of the polymer is equal to the solubility of the liquid in the polymer, and on the opposite side of the polymer the concentration is 0. The concentration on the exposed side can be determined by immersing the polymer in the liquid for a period of time and measuring the weight gain in the polymer, or the amount of liquid absorbed.

For applications involving polymer membranes, such as protective gloves and permeation monitors (Chapter 23), the concentration at the downstream point *(b)* is often set to zero, and the distance x is defined as the polymer thickness, L. Therefore Equation 8-27 can be redefined as:

$$F = -\frac{DS}{L} \qquad \text{[Equation 8-28]}$$

Finally, a new constant, P, or permeability, can be defined as DS. P is a constant for any polymer-solvent system regardless of the thickness of the polymer.

THE CONTINUITY EQUATION

Airflow is important in hygiene work. Air is moved in order to dilute or remove air pollutants. Air is moved with small pumps through filters and other devices in order to sample air contaminants. One equation is of particular importance in defining airflow through a pipe or a tube. The continuity equation is based on the conservation of mass. It states that the mass flowing at one place in a contiguous system is equal to the mass flowing at any other place in the system. The mass flow through the system is ρVA, where ρ is the density of the flowing fluid, usually air, V is the velocity, and A is the cross-sectional area at a point in the system. Generally, we assume that the density of air is constant for any point in the system.[3] Therefore:

$$A_1 V_1 = A_2 V_2 \qquad \text{[Equation 8-29]}$$

In other words, the flow at point 1 equals that at point 2. AV has units of volume per time or flow rate (Q); so:

$$Q = VA \qquad \text{[Equation 8-30]}$$

Example 8-8. Air flowing at 200 cm^3/min through a tube with a cross-sectional inside area of 0.1 cm^2 has what velocity?

$$Q = VA\ ,\ V = \frac{Q}{A} = \frac{200\ \frac{cm^3}{\min}}{0.1\ cm^2} = 2{,}000\ \frac{cm}{\min}$$

QUESTIONS

8.1. Calculate the height in a water manometer showing standard pressure. Use knowledge that standard pressure is 760 mmHg, and the specific gravity of mercury is 13.58. Show unit cancellation.

8.2. An analytical method requires that a 25 L air sample be collected over an 8-hour period. What flow rate is needed?

8.3. One ounce (28.3 g) of Hg is spilled in an unventilated room measuring 3 × 3 × 10 m. What is the maximum concentration of Hg that can occur in the room? The $P°$ of Hg is 0.0013 mmHg.

8.4. The vapor pressure of a compound at 25°C is 41 mmHg. What is the saturated vapor concentration of the compound?

8.5. What is the density of air at 300°F and 1.2 *atm*?

---

[3] There are some high velocity situations where this is not a good assumption. See Chapter 20 on pumps and the section on critical orifices.

8.6. The lower flammable limit (LFL) is the minimum concentration of a combustible gas or vapor that will ignite in the presence of NTP air and a source of ignition. It is usually quite high, in the percent range. A monitoring instrument reads 5% LFL, and for the compound in question the LFL is 5%. What concentration of the compound (ppm) is being measured?

8.7. A liquid mixture of two compounds (*50:50* by volume) has vapor pressures of X = 5 mmHg and Y = 50 mmHg. Assuming they behave ideally, what would you expect the relative airborne concentrations (*X:Y*) to be. What law is your estimation based on?

8.8. In a simulated experiment, H_2S is released from a point source and is first detected 500 ft from the source 5 minutes later. Under similar conditions, how long would you expect it to take to detect SF_6 at the same concentration? Assume ideality and the detectors have equal sensitivities.

8.9. Given the following solubility data, which two of the three compounds would you expect to be most soluble in one another? Prove this using Equation 8-25.

Chemical	δ_d	δ_p	δ_h
Hexane	4.9	0.0	0.0
Ethyl benzene	7.8	0.6	1.4
Aniline	19.4	5.1	10.2

Note that the solubility parameters (δ) are in units of $(J/cm^3)^{0.5}$.

8.10. The flux rate across a space at point *a* is 1 $mg/cm^2/sec$ for a compound when the concentration is 100 ppm at point *a* and 0 ppm at point *b*, and there is a linear gradient from *a* to *b*. What is the flux rate when the concentrations are $C_a = 100$, $C_b = 60$?

8.11. What are the relative diffusion rates for ethane and normal octane.

8.12. Given a 10 mg/m^3 concentration of methyl ethyl ketone, calculate the concentration in ppm.

8.13. If one collects 1 mg of benzene in a 20 L air sample at 40°C and 790 mm of mercury, what is the ppm concentration?

8.14. One pint (473 ml) of toluene spills in a 10 × 10 × 10 m unventilated room. What is the approximate concentration in both mg/m^3 and ppm if all of the toluene evaporates? (Hint: Convert liquid volume to vapor volume.)

8.15. For propionic acid, *a* is 28 and *b* is 8,000 in the Clasius-Clapeyron equation. Calculate the vapor pressure at 30°C, and then calculate the concentration of a saturated vapor at that temperature.

8.16. What is the absolute humidity in mg/m^3 under conditions of 75% relative humidity at 70°F?

8.17. Show that 1 ppm by volume is equivalent to 1 ppm by mole. Assume ideality.

REFERENCES

Bilmeyer, F.W.: Textbook of Polymer Science, 3rd ed. Wiley Publishing, New York (1984).

Castellan, G.W.: Physical Chemistry, 3rd edition. Addison-Wesley, Reading, MA, USA (1983).

Crane: Flow of fluids through valves, fittings, and pipe, Technical Paper No. 410 (TP 410). Crane Company, New York http://www.cranevalve.com/tech.htm (1988).

Dümmer, W.: Prevention of Occupational Hazards at High Altitudes, In Encyclopaedia of Occupational Health and Safety, J.M. Stellman (ed). International Labour Office, Geneva, Switzerland (1998).

Frisancho, R.A.: Human Adaptation: A Functional Interpretation, pp. 101-116, 117-138. C.V. Mosby Co., St. Louis (1979).

Hansen, C.M.: The universality of the solubility parameter, In IEC Product/Research and Development, 8:2-11(1968).

Lahiri, S.: Physiological responses and adaptations to high altitude, In International Review of Environmental Physiology II, (15), pp. 217-251, Robershaw, D. (ed.). University Park Press, Baltimore (1977).

Lange's Handbook of Chemistry, J.A. Dean (ed.). McGraw-Hill, New York (1979).

Lyman, W.J.; Reehl, W.F.; Rosenblatt, D.H.: Handbook of Chemical Property Estimation Methods: Environmental Behavior of Organic Compounds, pp. 11-15. McGraw-Hill, New York (1982).

Perkins, J.L.: Solvent-polymer interactions, In Chemical Protective Clothing, J.S. Johnson, K.J. Anderson (eds.). American Industrial Hygiene Association, Fairfax, VA (1990).

Popendorf, W.: Industrial hygiene control of airborne chemical hazards. CRC Taylor and Francis, Boca Raton, FL, USA (2006).

Stephenson, D.J.; Lillquist, D.R.: The effects of temperature and pressure on airborne exposure concentrations when performing compliance evaluations using ACGIH® TLVs® and OSHA PELs. Appl. Occ. Env. Hyg. 16:482-486 (2001).

9

BASIC AEROSOL PHYSICS

Aerosols have a tremendous range of sizes, as shown in Chapter 4, Figure 4-11. The very smallest aerosols approach the size of gas and vapor molecules. These particles tend to behave much like gases. They essentially do not fall out of the atmosphere; they move with speeds similar to those of gas molecules; and their direction of movement is controlled by collisions with gas molecules more than anything else. If all aerosol particles of interest were in this size range, their behavior would be fairly easy to describe. However, aerosols come in many sizes. Until recently, particles smaller than about 0.1 μm were not considered important since they contributed little to the total mass. However, with the increasing use in industry of nanoparticles, which are as small as 0.01 μm (10 nM), the small end of the relevant size range has expanded. In addition, it is now recognized that the number of particles (or at least their total surface area) may be a better index of toxicity than the mass that is inhaled. On the high end, aerosols larger than 100 μm in diameter do not easily penetrate the head (get into the nose or mouth), unless breathing rates are high or outdoor wind velocities are toward the face (indoor velocities are generally too small; see Chapter 24 for more detail). Therefore, for hygiene purposes, the range of interest is roughly 0.01 to 100 μm; however, larger particles can be important.[1]

Although aerosol physics is a vast field,[2] we will concentrate on several key aspects that have application to basic hygiene problems. The questions fall into three main areas: How do particles move under differing forces? Since particles usually come in various sizes, how are the sizes distributed? What happens to visible light that is incident on a particle?[3] Regarding the movement of particles, we are interested in what happens to particles without velocity that

[1] For application to outdoor air pollution, other sizes are also important. Both smaller and larger sizes cause reduction of visibility and, therefore, require control. Recent information about ultrafine particles (< 0.1 and often < 0.01 μm) indicate that they are formed in the atmosphere and may be of physiological importance.

[2] Most of this discussion is taken from Hinds, (1999) which is recommended reading for a thorough discussion of the subject. Other important references have been published including Reist (1984) and Baron and Willeke (2001).

[3] This question has a role in detecting particles.

are released into still air, as well as particles released with great initial velocity, such as from a grinding wheel. We can use information about the movement of particles to design instruments for collecting them as well as learning how they are collected by the respiratory tract. We can make inferences from particle size distributions about the physiological importance of a collection of particles. Finally, an understanding of the manner in which light interacts with particles can be used to size them and measure their concentration.

PARTICLE MOTION

Reynolds Number

The motion of particles is largely dependent on the medium in which they exist; for our purposes, air. Air molecules bombard particles and cause friction or resistance on their movement, also known as drag. A particle ejected at high speed into still air imparts some of its momentum to the air until it comes to rest. The more viscous the fluid, the faster this transfer occurs. The viscosity of a fluid such as air can be thought of in connection with the ability of one moving plate to continue moving relative to a second stationary plate when the plates are separated by a fluid such as air. The air resists the movement, and so a constant force, *F*, has to be applied to continue the movement. The force is dependent on the area of the plates, *A*, the distance between the plates, *y*, and the velocity of the moving plate, *V*, or:

$$F = \frac{\eta A V}{y} \qquad \text{[Equation 9-1]}$$

where η is a coefficient of viscosity dependent on the nature of the fluid. To help relate this to particles think of the moving plate as a particle.

Road asphalt or bitumen is thought of as viscous, and as it is heated, it becomes less so. This is so because the attractions between the molecules are decreasing with temperature. On the other hand, air viscosity increases with temperature. In the case of air, molecular attractions are unimportant at normal temperatures. Instead, as air is heated, there are more molecular collisions with a surface, creating greater friction. The viscosity of air (η) is 0.000181 dyne-s/cm^2 at 20°C.

If the same particle that we have been discussing is ejected into air at a relatively low speed, then the air tends to streamline around it in a uniform manner. The movement is said to be laminar. If the speed is increased, then the particle leaves a wake of turbulence behind it. Similarly, if the particle is still and the air moving, then the faster the air moves, the more developed the turbulence. It is hard to imagine a particle standing still under such conditions, but we could use another object, such as the inside of a pipe through which air is flowing. As the air speed increases, the turbulence caused by its interaction with the pipe increases.

The propensity for turbulence to occur can be found by comparing the forces that resist inertia (of the particle or the air, depending on your frame of reference) with the forces that resist friction. (In either case, it is caused by the air molecules themselves.) This propensity can be related as the Reynolds number (Re):

$$Re = \frac{\rho V d}{\eta} \qquad \text{[Equation 9-2]}$$

where ρ is the density of the gas, *V* is the relative velocity between gas and object, *d* is the diameter of the object (a pipe or a particle), and η is the viscosity of the gas. Notice that inertial forces (mass × velocity) are in the numerator and resistance forces are in the denominator. *Re* is unitless.

Re provides a convenient way to determine if turbulent flow exists. These empirically determined ranges for pipe flow or particle motion are shown in Table 9-1. If the density (0.0012 g/cm^3) and viscosity (0.000181 dyne-sec/cm^2) of air at 20°C are used in Equation 9-2, then:

$$Re = 6.6\ Vd \qquad \text{[Equation 9-3]}$$

In order to have laminar flow for a 1 μm particle, *V* cannot exceed 1,500 cm/s or about 3,000 ft/min, where *d* is in cm. For a 10 μm particle these values would be reduced by one-tenth.

Stokes's Law

Newton was the first to characterize the factors that affect the uniform motion of a particle through a fluid. He showed that the force of drag was related to the velocity and the diameter of the particle and the density of the gas by a coefficient, the drag coefficient (C_D). He also found that C_D was approximately a constant (0.44). However, Newton was largely interested in cannonballs (when he was not interested in planets). For smaller particles, or large particles at very high velocities, C_D is not a constant, as shown in Figure 9-1. Only at very high *Re* is C_D approximately constant. For many applications hygienists are interested in $Re < 1$, the so-called Stokes region, as shown in Figure 9-1. Stokes showed that, in this region, the C_D was a simple function of *Re*, $C_D = 24/Re$. Stokes derived the relationship by taking a slightly differ-

Table 9-1
Reynolds numbers for various flow regimes in pipes or for moving particles

Flow Regime	Pipes	Particles
laminar	< 1	< 2000
transitional	1–1000	2000–4000
turbulent	> 1000	> 4000

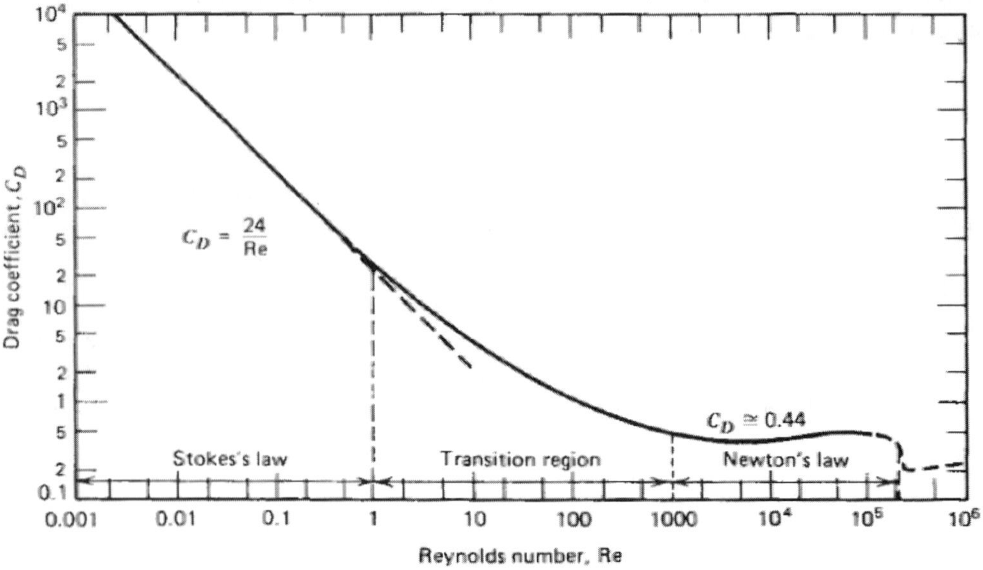

Figure 9-1. Drag coefficient versus Reynolds number for spheres (from Hinds, 1982, with permission).

ent approach to Newton; he showed that the drag force (F_D) followed this function.

$$F_D = 3\pi\eta Vd \qquad \text{[Equation 9-4]}$$

In Stokes's law, frictional factors are emphasized more than the inertial ones in Newton's approach.

Settling Velocity

If a particle is released into still air with essentially no momentum, as might occur in an industrial process, it may be of interest to know how quickly the particle settles. When the particle is first released, it accelerates because of the force of gravity, $F = mg$, or mass × acceleration due to gravity. Eventually it reaches a constant velocity when the $F_D = mg$, that is, the force reacting downward equals the drag force resisting downward motion.

Thus, by equating Equation 9-4 with mg, we can solve for V, in which case V is the terminal settling velocity, V_{TS}, or

$$V_{TS} = \frac{\rho_p d^2 g}{18 \eta} \qquad \text{[Equation 9-5]}$$

where g is the acceleration due to gravity. Substituting for g (981 cm/s^2) and η at 20°C, and assuming that the particle has a density equal to that of water or 1 g/cm³, we have:

$$V_{TS} = 0.003d^2 \qquad \text{[Equation 9-6]}$$

where d is in μm and V_{TS} is in cm/s. The equation is only applicable in the Stokes region. For a 100 μm particle, V_{TS} would be 30 cm/s. Therefore, Re would be 2 (see Equation 9-3) or just outside of the Stokes region.

The equation is applicable for particles as small as 1 μm. Below this value the particles actually settle faster than the equation would predict,[4] because of something called slip, which occurs because of violation of the assumption that the relative velocity at the particle surface is zero. In other words, the air molecules are being *dragged* with the particle. As particles become smaller, this is more likely. If the Cunningham correction factor (C_c) is applied to correct this error ($0.003\ d^2/C_c$), Equation 9-6 can be used accurately down to $d = 0.1$ μm (see Hinds, 1999 for details). A more complicated correction factor allows calculations for all particle sizes. Finally, Equation 9-6 assumes spherical particles, so discrepancies usually must be accounted for, as described next.

Aerodynamic Diameter

For the sake of illustration, imagine an irregularly shaped particle that has density of 1 and a V_{TS} of 10 cm/s. Falling with the particle are an infinite number of water droplets of different diameters. Water droplets are essentially spherical and they have unit density. Therefore, Equation 9-6 can be applied to them without violating any assumptions. Since there are an infinite number of water droplets, at least one of them will have the same V_{TS} as the particle. The diameter of that water droplet is the aerodynamic diameter of the particle. The aerodynamic diameter (d_a) of a particle is the diameter of a unit density sphere (water is unit density) with the same V_{TS} as that particle. Therefore, we can apply Equation 9-6 to non-unit-density particles of irregular shape if we write:

$$V_{TS} = 0.003\ SG\ d^2_a \qquad \text{[Equation 9-7]}$$

where SG is the specific gravity of the particle relative to water. Aerodynamic diameter is an important aerosol concept and is used in many applications.

Tranquil Versus Stirred Settling

Equation 9-7 applies to a particle settling in still air. Air often is not still but rather is mixed by turbulence, a process referred to as eddy diffusion. Occasionally there are direct applications of Equation 9-7 such as dust settling in a confined space, for example, a grain stor-

---

[4] Remember that this is still air. Small particles are easily tossed upward if the air is moving with turbulence.

age bin. But in most cases, Equation 9-7 is an approximation, as some turbulence will exist. Equation 9-7 is an ideal situation, a boundary. A second boundary also can be envisioned. Imagine a room containing an aerosol, the particles all about the same size, that is vigorously mixed such that the concentration remains uniform throughout the room. Figure 9-2 shows that the concentration in the room does change with time but remains uniform over space. At the other boundary (that modeled by Equation 9-7), Figure 9-2 shows that the concentration is either the beginning value, C_0, or 0. If we examine the change in concentration at a point *P* in either room, the change over time appears as in the figure.

Although both boundaries have assumptions that are not often met, the stirred situation is closer to reality in most turbulent environments. Therefore, it would be useful to derive an equation for the stirred case. In this case, the velocity caused by settling is being enhanced or reduced by the vertical component of the turbulent diffusion.[5] The rate of removal of the particles is not as in the tranquil situation, but instead is dependent on the number of particles left at any time. Therefore, the rate decays with time, and in this ideal situation the concentration never reaches 0. An exponential decay function describes the change in concentration with time:

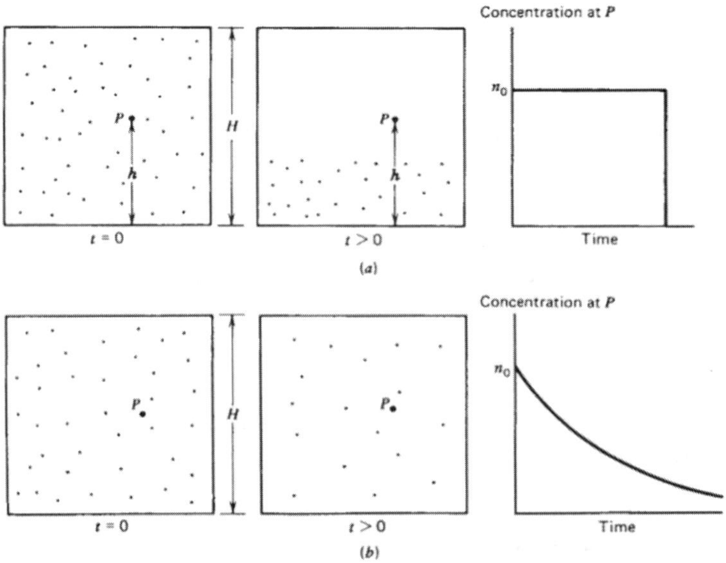

Figure 9-2. Two ideal boundaries for particle settling in a room. a) Tranquil. b) Stirred (from Hinds, 1982, with permission).

[5] Turbulent diffusion should not be confused with molecular diffusion. The former is dependent on convective currents; it is bulk air movement. The latter is a net effect of the summation of random movements of molecules owed to their individual kinetic energies.

$$\frac{C_t}{C_0} = e^{\frac{-V_{TS}\,t}{H}} \qquad \text{[Equation 9-8]}$$

where t is the elapsed time in seconds, and H is the height (cm) where the concentration is measured. The key assumption for this equation is that mixing is uniform throughout the room, leading to a uniform concentration. Where mixing is less complete than this, concentrations will decline faster; for when the concentration has reached 0 throughout the room in the tranquil case, the concentration is still $1/e$ of the original concentration in the stirred case.

Horizontal Stopping Distance

Several applications in hygiene work depend on knowledge of the distance a particle will travel when it is released with some momentum ($V_0 \times m$). For example, will a grinding wheel throw a particle outside of the capture zone of an exhaust hood? Can a particle of sawdust be ejected from a sawblade into an employee's eye? To what distance can a dentist's drill eject a particle from a patient's mouth?

If a particle is ejected from a sawblade, then its initial velocity (V_0) is the velocity of the tip of the sawblade. A 10 in (25 cm) diameter saw blade moving at 10,000 revolutions per minute (RPM) has a surface speed of

$$V = RPM\,\pi\,d_w \qquad \text{[Equation 9-9]}$$

or 26,000 ft/min (13,000 cm/sec). The particle will immediately begin to decelerate. If it is ejected in the horizontal plane, then gravity will also accelerate it in the vertical direction. However, we can consider its movement only along the horizontal axis, and ask at what point will its velocity in the horizontal direction (V_h) be reduced to 0. The distance that it has traveled when $V_h = 0$ is called the horizontal stopping distance (S). In the Stokes region, S can be defined as

$$S = \frac{V_0\,\rho_p\,d^2\,C_c}{18\,\eta} \qquad \text{[Equation 9-10]}$$

However, substitution into Equation 9-10 will show that for particles in the Stokes region ($Re < 1$), S is typically < 1 cm. Generally, much larger particles outside the Stokes region are of interest. An empirical expression has been used to determine S for particles having $Re <$ 1500:

$$S = \frac{\rho_p\,d}{\rho_g}\left[Re_0^{1/3} - \sqrt{6}\,\arctan\frac{Re_0^{1/3}}{\sqrt{6}}\right] \qquad \text{[Equation 9-11]}$$

where ρ_g = density of the gas, and Re_0 is the initial *Re*. Figure 9-3 shows values of S calculated for 1 μm diameter particles using Equation 9-11.

Inertial Impaction

Particle size selective sampling of air can be accomplished if one understands the principles of inertial impaction. If a particle is moving in the downward y-direction and the airstream is forced to change directions at 90°, then the particle will have a tendency to continue traveling in the y-direction. If the tendency is great enough, then the particle will impact the surface that causes the airstream to change directions. The surface can be used to capture the particle. Figure 9-4 shows how the process works. Essentially, if the stopping distance of the particle as it exits the nozzle exceeds the distance between the nozzle and the surface, then the particle will be impacted.

The tendency of the particle to impact the surface is a function of its forward momentum or inertia. Particle stopping distance can be thought of as inertial potential, and thus an impaction parameter can be written as the ratio of the stopping distance to the radius of the jet (see Figure 9-4) through which the air is forced, or:

$$Stk = \frac{\rho_p \, d_p^2 \, V \, C_c}{9 \, \eta \, d_j} \qquad \text{[Equation 9-12]}$$

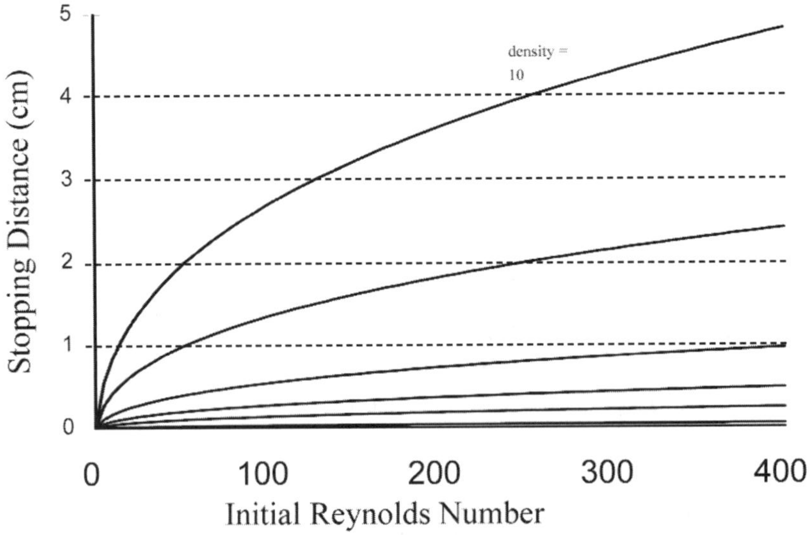

Figure 9-3. Horizontal stopping distance (S) versus Reynolds number for a 1 μm sphere as a function of 6 particle densities (0.1, 0.5, 1, 2, 5, 10). The curves can be used for other particle diameters by multiplying the y-axis by diameter.

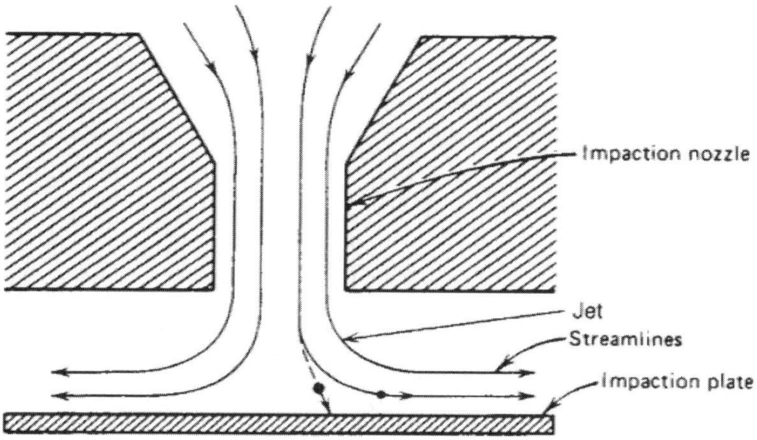

Figure 9-4. Cross-sectional view of an impactor (from Hinds, 1982, with permission).

where the subscripts refer to particle and jet, and *Stk* is the impaction parameter or Stokes number. Collection efficiency curves are often plotted as in Figure 9-5. As the *Stk* value approaches 0, particles are more inclined to laminar tracking with the streamlines.

PARTICLE DISTRIBUTIONS

A device that employs inertial separation principles can be used to separate particles into size fractions. Then, depending on the analytical technique used, the particles in each size range may be counted (usually electronically), or their total mass may be assayed. Depending on which technique is used, a frequency distribution can be constructed, as shown in Figure 9-6. The distribution of particle size by *mass* shows the fraction of the total mass collected in each size range. The distribution by *count* shows the fraction of the total number or the count of particles in each size range. The diameter in the middle of that distribution of counts within each size range is the median or the count median diameter (CMD). Note that the range sizes are not necessarily equal. However, to standardize the size ranges, in either case, the fraction is divided by the width (μm) of the size range. Otherwise, the height of the bars would depend on the width of the bars.

Either distribution is skewed like the lognormal distributions presented in Chapter 7. Therefore, the lognormal distribution often is used to characterize particle size distributions. Remember that there are several ways to measure central tendency of a distribution, such as the mean, mode, and median (the geometric mean for a lognormal). Couple this with two ways of characterizing a particle size distribution (by mass and by count),[6] and there is a confusing

[6] Actually there are at least two others: surface area and length.

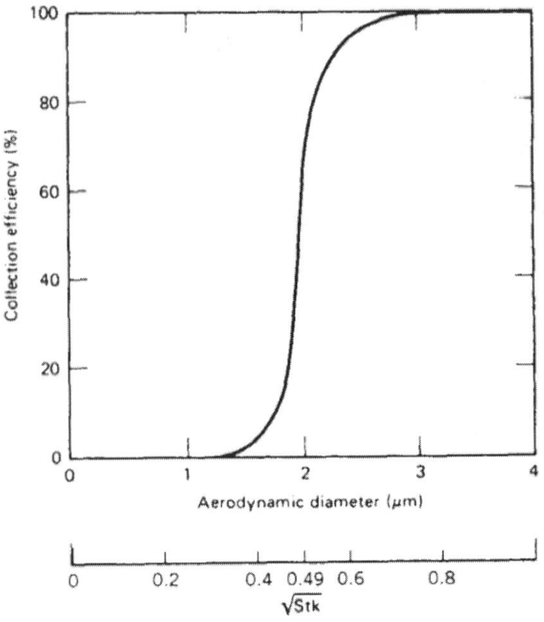

Figure 9-5. Typical impactor efficiency curve (from Hinds, 1982, with permission).

array of calculation techniques and terms for characterizing the central tendency of the distribution. These methods are not discussed at length here, but can be found in Hinds (1999). In addition, if one of the many averages (any of the measures of central tendency) and the geometric standard deviation (GSD) are known, the other averages can be calculated. The coefficients needed to convert CMD to the other most common forms are given in Table 9-2. The general form of the equation for making these conversions is

$$\frac{d_A}{CMD} = e^{b \ln^2 \sigma_g} \qquad \text{[Equation 9-13]}$$

where d_A is the central diameter to be calculated, σ_g is the GSD of the CMD distribution, and b is a value given in Table 9-2.

Formalities aside, what are the most common ways of calculating average diameter and what do they mean? First, monodisperse aerosols have one diameter; they are commonly produced in the laboratory for experimentation and calibration. In the plant atmosphere, disperse aerosols exist, and these aerosols have a size range, depending on the physical process that produces them.

If the size data are characterized by count, a count median diameter can be determined as the geometric mean of the count distribution. This value can be calculated as

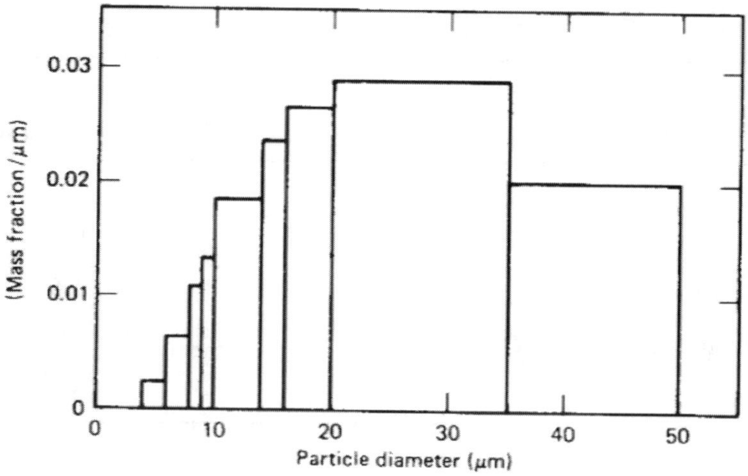

a

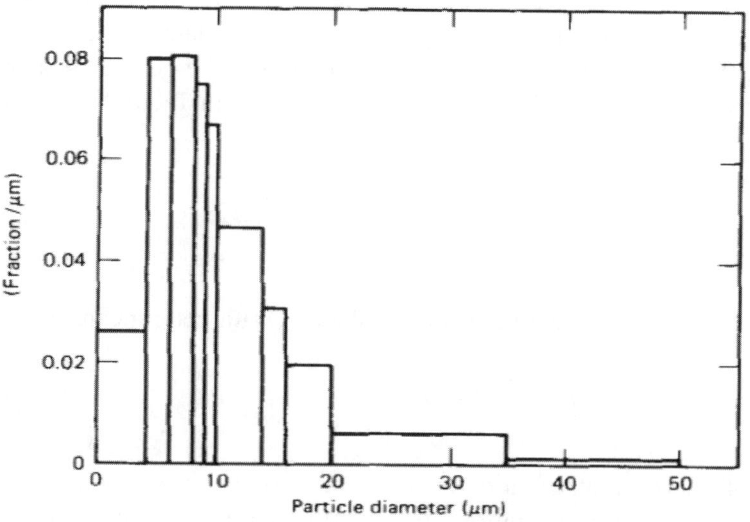

b

Figure 9-6. Frequency distributions of particle size by a) mass and b) count (from Hinds, 1982, with permission).

Table 9-2
Coefficients for Equation 9-13 for the most common average diameters

To Convert from *CMD* to	Value of b in Equation 9-13
Mode, $\hat{d}$	-1
Count mean diameter, $\bar{d}$	0.5
Diameter of average mass, $d_{\bar{m}}$	1.5
Mass median diameter, ***MMD***	3
Mass mean diameter, d_{mm}	3.5

$$CMD = \exp \frac{\sum n_i \ln d_i}{N} \qquad \text{[Equation 9-14]}$$

where the subscript i refers the i size ranges, n the counts, and d the midpoint diameters for each size range. CMD also often is found by plotting the size data on log-probability paper, as shown in Figure 9-7. The median is at the 50th percentile. The geometric standard deviation of the count distribution is needed to make the conversions in Table 9-2. It can be calculated from the count data as

$$GSD_c = \exp \left[\frac{\sum n_i(\ln d_i - \ln CMD)^2}{N - 1} \right]^{1/2} \qquad \text{[Equation 9-15]}$$

The count mean diameter can be determined as an arithmetic mean:

$$\bar{d} = \frac{\sum n_i d_i}{N} \qquad \text{[Equation 9-16]}$$

The mode of the count distribution is approximately the midpoint of that size range with the highest count, or the peak in Figure 9-6b. Finally, there is a diameter of average mass or volume. It is determined from the count distribution even though it has the term mass. It is calculated as

$$d_{\bar{m}} = \left[\frac{\sum n_i d_i^3}{N} \right]^{1/3} \qquad \text{[Equation 9-17]}$$

The diameter of average mass is also referred to as the third moment average; it is calcu-

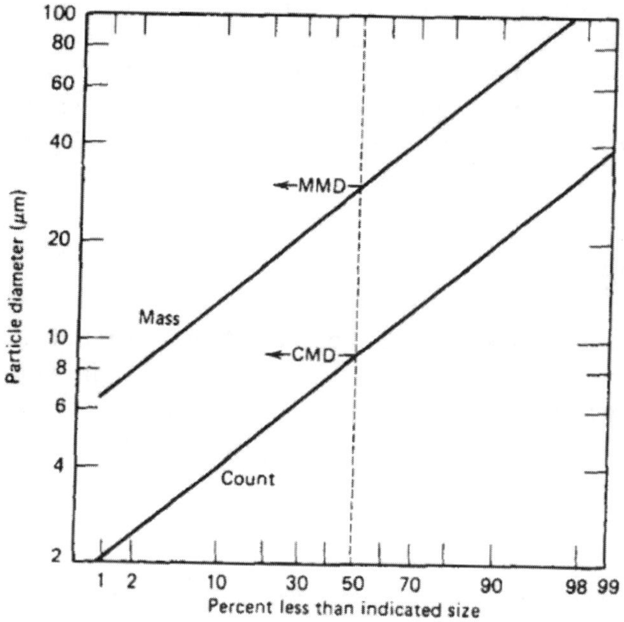

Figure 9-7. Log-probability plots showing the CMD and MMD (from Hinds, 1982, with permission).

lated by determining the volume occupied by the particles, a function of mass. It is called the diameter of average mass because mass (m) equals volume times density, which can be rearranged to

$$d^3 = \frac{6\ m}{\rho_p\ \pi} \qquad \text{[Equation 9-18]}$$

In other words, mass is a function of d³.

Two more averages can be calculated based on the mass distribution: the mass median diameter and the mass mean diameter. The mass median diameter (MMD), like the CMD, is a geometric mean. It is the particle size for which half the particles are smaller and half are larger, but unlike the CMD it is based on mass being assayed in each size distribution, not count:

$$MMD = \exp\left[\frac{\sum m_i \ln d_i}{M}\right] \qquad \text{[Equation 9-19]}$$

The mass mean diameter (d_{mm}) can be calculated as

$$d_{mm} = \frac{\sum m_i\ d_i}{M} \qquad \text{[Equation 9-20]}$$

In summary, methods for determining five averages have been given, and by using Table 9-2, count averages can be converted to mass averages. The two most common averages for characterizing a distribution of particle sizes are the CMD and the MMD. The relationships of the various averages are shown in Figure 9-8. This figure applies to a specific set of data with GSD = 1.89. The specific order of the means will remain the same regardless of GSD; however, as the GSD becomes smaller, the differences between the means become smaller.

PARTICLE OPTICS

Several instruments used for field and/or laboratory research that count particles, or in some cases measure particle mass, rely on the interaction of light with particles.[7] Typically, these instruments have a light source of known strength. The light from any source, for example, a light bulb, is non-polarized. This means that the light waves oscillate in all planes perpendicular to the beam. Polarized light oscillates in only one plane. Light sources in most

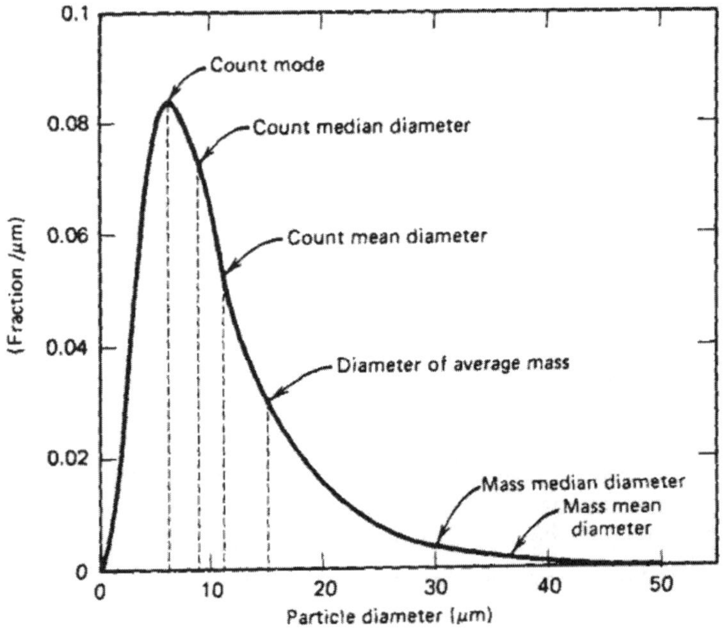

Figure 9-8. Particle size distribution (CMD = 9.0 μm, GSD = 1.89) based on count showing the relative position of various average diameters (from Hinds, 1982, with permission).

[7] The use of optical properties to measure particles is explored more fully in Chapter 27.

aerosol measuring instruments are filtered to make them polarized or a laser light source is used, which allows their interaction with particles to be more easily studied.

When light interacts with a particle, it can be absorbed, reflected, refracted, diffracted, or re-emitted at a different wavelength (thermal, Raman, fluorescence, or phosphorescence). These interactions are shown in Figure 9-9. The first four are used most widely in particle measuring instruments. Refraction occurs when light passes through a particle and changes speed. Note that the wavelength is not altered, but that the direction of the light is. The ability of any material to refract light is characterized by its refractive index, or the ratio of speed of the light in a vacuum to the speed in the medium. The value is always greater than one for non-absorbing materials, and for some of these materials, such as air, it is very close to one. For absorbing materials, the refractive index involves imaginary numbers. Diffraction also causes light to change direction and occurs when light just interacts with the edge of a particle.

Extinction

If a source of parallel light beams is shown through an aerosol, some of the light will be absorbed, and other rays will be scattered (diffracted, reflected, and refracted). Thus, a detector on the opposite side of the source will not detect all of the light that is emitted. This extinction of the light is dependent on the number, shape, size, and refractive index of the particles.

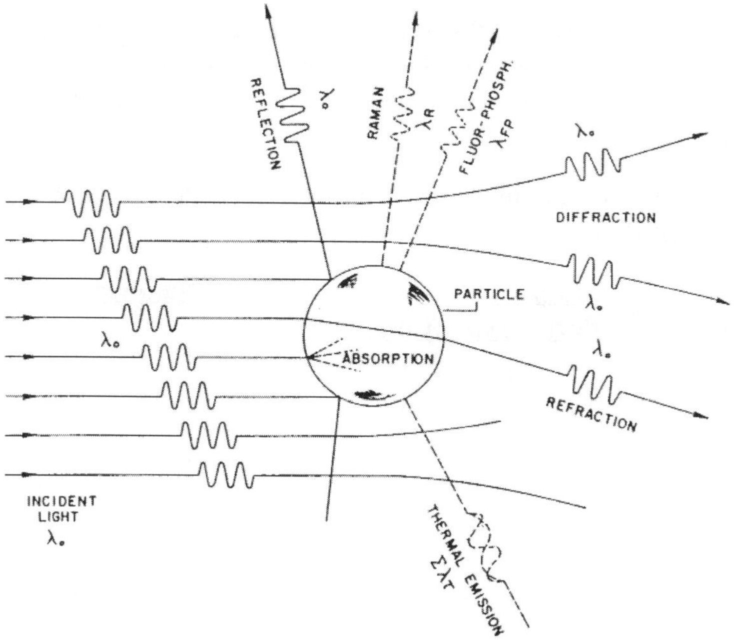

Figure 9-9. The resulting interactions of particles and light.

Scattering

Instead of measuring the effect of the particles on the light at a position downstream or, 0° from the source, the effects of scattering can be measured at other angles. Forward-scattered light deviates only slightly from its incident direction, whereas back-scattered light is reflected back toward the source (> 90°). A single particle as small as 0.1 µm causes scattered light that can be detected. Indeed, many instruments count particles one at a time in this fashion.

QUESTIONS

9.1 For 50 µm particles, what is the highest velocity in air that they can have and remain in the Stokes region?

9.2 What is the terminal settling velocity for 10 µm aerodynamic diameter sawdust particles having a specific gravity of 0.16?

9.3 During a 1-hour interval the measured concentration of the sawdust particles (see question 9.2) at a height of 6 ft or 180 cm does not appear to change significantly from a value of 5mg/m³. Assume that stirred settling adequately characterizes the setting. The room size is 1,000 m³. What rate of sawdust generation (g/hr) is needed to account for the difference in concentration calculated by assuming stirred settling and the measured result?

9.4 A sawblade rotates at 10,000 rpm, has a diameter of 5 in or 12.5 cm, and produces dust with the characteristics described in question 9.2. Will the dust be ejected to an employee's eye if the employee's eye remains a minimum of 3 ft or 90 cm from the blade?

9.5. For the following particle size distribution calculate the CMD, count mean, MMD, and the GSD_c.

Midpoint Particle size (µm)	Number
1	4
3	7
5	4
8	2
10	1

REFERENCES

Baron, P.A.; Willeke, K.: Aerosol Measurement: Principles, Techniques, and Applications, 2nd Edition, Wiley Publishing, New York (2001).

Hinds, W.C.: Aerosol Technology: Properties, Behavior, and Measurement of Airborne Particles, Wiley Publishing, New York (1982).

Hinds, W.C.: Aerosol Technology: Properties, Behavior, and Measurement of Airborne Particles, Second edition, Wiley Publishing, New York (1999).

Reist, P.C.: Introduction to Aerosol Science, Macmillan, New York (1984).

APPENDIX 9-1. Density of Some Common Aerosol Materials, g/cm³ (data from Hinds, 1982)

Solids	
Aluminum	2.5–2.8
Aluminum oxide	4.0
Ammonium	1.8
Asbestos	2.4–3.8
Asbestos, chrysotile	2.4–2.6
Carnuba wax	1.0
Coal	1.2–1.8
Fly ash	0.7–2.6
Fly ash cenospheres	0.7–1.0
Glass, common	2.4–2.8
Granite	2.4–2.7
Iron	7.0–7.9
Iron oxide	5.2
Lead	11.3
Limestone	2.1–2.9
Marble	2.4–2.8
Methylene blue	1.26
Natural fibers	1–1.6
Paraffin	0.9
Plastics	1–1.6
Pollens	0.45–1.05
Polystyrene	1.05
Polyvinyl toluene	1.03
Portland cement	3.2
Quartz	2.6
Sodium chloride	2.2
Sulfur	1.9–2.1
Starch	1.5
Talc	2.6–2.8
Titanium dioxide	4.3

Uranine dye	1.53
Wood (dry)	0.4–1.0
Zinc	6.9
Zinc oxide	5.6
Liquids	
Alcohol	0.79
Dibutyl phthalate	1.045
Dioctyl phthlate (DOP; di-2(ethylhexyl) phthalate)	0.983
Dioctyl sebecate	0.915
Hydrochloric acid	1.19
Mercury	13.6
Oils	0.88–0.94
Polyethylene glycol	1.13
Sulfuric acid	1.84
Water	1.00

10

OCCUPATIONAL EXPOSURE LIMITS AND THRESHOLD LIMIT VALUES

HISTORICAL PERSPECTIVE

In the early 1900s, when Dr. Alice Hamilton first began to explore the dangerous trades in the United States, she had considerable difficulty in convincing plant management that the contaminated air being breathed by the workers had a greater effect on their overall health status than the amount of contamination being ingested by them as a result of contaminated hands and contaminants entering the mouth. This was particularly true for lead. The concept that vapors or very small particles could cross the lung and enter the bloodstream was foreign to many people. Dr. Hamilton and others before her realized that if airborne concentrations of contaminants could be limited, then it was likely that workers could labor in contaminated atmospheres with little risk of harm (Hamilton, 1948).

The understanding of what constitutes a safe exposure to an airborne contaminant has evolved, with the evolution largely influenced by the development of occupational exposure limits (OELs). The very first OELs were developed for gases and vapors known to potentially cause death at concentrations that could readily occur in the workplace. These included compounds such as carbon monoxide, chlorine, ammonia, hydrogen cyanide, hydrogen chloride, bromine, benzene, carbon tetrachloride, and carbon disulfide. The very first OEL was developed for carbon monoxide by Max Grouper of Germany (Cook, 1987), based upon his study of carbon monoxide exposure of hens and rabbits. The statement made by Grouper in summarizing the effects of carbon monoxide gives us a clear indication of his understanding of OELs – an understanding still used by most of the public and a great many hygienists today. Grouper stated that "the boundaries of injurious action of carbon monoxide lie at a concentration of all probability of 500 parts per million but certainly not less than 200 parts per million." The key words are "certainly not less than 200 parts per million." This indicates that workers can be exposed to concentrations of less than 200 parts per million and not suffer injurious effect. It was an absolute statement, namely, that *all* workers could be exposed continuously at that level without suffering injurious effect. It was the concept of a no-effect level.

Interestingly, as early as 1912 Kobert published a list of OELs that appeared to recognize the difficulty in assigning a single exposure value as a no-effect level (Cook, 1987). In Kobert's tables, values for 20 substances are listed under four headings: (1) rapidly fatal to man and animal, (2) dangerous in 0.5 to 1 hour, (3) 0.5 to 1 hour without serious disturbances, and (4) only minimal symptoms observed. This latter category, only minimal symptoms observed, is essentially a long-term OEL. Note that the author stated that at the concentration value published in this fourth column, some symptoms would be observed; however, he termed them minimal. Kobert may have been the first to recognize that OELs cannot be absolute, but rather that for any exposure value there may be some symptoms shown by a small proportion of the workforce.

Various lists of OELs have been published in the United States and other countries, particularly Germany. Although most of the early lists contained acutely hazardous gases and vapors, some aerosols began to make their way onto the lists because of their great toxicity (e.g., silica and lead).

Prior to 1930, none of the published lists were enforced by any regulatory agency. A list had been published in 1921 by the U.S. Bureau of Mines, and other lists had been published by employees of the U.S. Public Health Service. However, it appears that these lists carried no regulatory enforcement. On the other hand, in 1930 a Soviet ministry of labor decree contained OELs for 12 industrial substances. According to the International Labor Office, these became enforceable in the Soviet Union in 1933 (Cook, 1987). There is some question about the degree to which they were enforced. In the Soviet Union, OELs tended to be set at a level at which exposure would result in no deviation of the normal state of the organism. This level includes harmful deviations and deviations thought to be inconsequential in the United States and the *western world*. Consequently, OELs in the Soviet Union (and now Russia) have been set at levels that are commonly one to three orders magnitude lower than in other countries.

In the United States, the Commonwealth of Massachusetts first published a list of enforceable OELs in 1937; and Germany did so in 1938. From this beginning, the adoption of OELs became common in many countries throughout the world. For example, in 1977 the International Labor Office published OELs either prescribed or recommended by 16 European countries, Japan, the United States, and the Council of Europe. A 1980 addendum to the publication had OELs for 1,178 substances (ILO, 1980). In 1987, Cook published a compendium of OELs that included OELs for 45 countries and 66 states, provinces, and other political jurisdictions of the United States and Canada.

Thus, for various reasons, OELs became valuable tools in the industrial hygiene profession. To gain an appreciation for the value of OELs, consider the hygienist working in the United States prior to 1921. Although Kobert's list of 20 substances had been published in Germany in 1912, it is unlikely that most U.S. hygienists would have had access to it. Furthermore, although the U.S. Bureau of Mines published a list in 1921, it contained only 33 substances, and there were certain other substances of interest to the industrial hygienist.[1] Thus, the number of published OELs was limited.

Carcinogens and Thresholds

As OELs continue to be developed throughout the world by occupational health professionals from all backgrounds, two central questions remain unanswered: what exactly is an OEL, and what should it do? There has been a difference of opinion on the international level regarding these questions. For example, Magnuson (1965) stated that the Soviet OEL was set at a level where exposure would result in no deviation in the normal state of the organism. On the other hand, in the United States some OELs were idealized as levels below which no harmful effects would occur over a working lifetime, though certain minor physiological changes might occur without health hazard. Many today would disagree with the absoluteness of the words *no harmful effects*, particularly when applied to a biologically diverse population of workers.

There has been constant conflict over what is considered biologically acceptable as an OEL and what can be achieved in industry at reasonable economic cost. This is best exemplified in a quote from Suma'mur of Indonesia (as quoted by Cook, 1987): "Levels should be selected to secure the continuity of existing production processing for the employment objective but minimize or if possible eradicate hazards of occupation." Note that production rather than health seems to be the first priority.

The controversy over whether a single OEL value could protect all workers came to a head as the mechanisms of cancer-causing chemicals were elucidated. The *one-hit theory* stated that, theoretically, a single molecule could cause genetic changes leading to cancer. Therefore, applying a threshold or no-effect OEL to these compounds is in violation of that theory. In 1979 the German commission stated that there was not sufficient knowledge to justify assigning OELs to carcinogens (Cook, 1987). The same no-threshold doctrine was applied by the National Institute for Occupational Safety and Health in recommending OELs for carcinogens (NIOSH, 1978). That agency set OELs at the lowest value that could be detected with analytical methods. It was, therefore, assumed that if any carcinogen could be detected, it existed, but that if it could not be detected, its existence could not be shown.

Unlike the compounds of concern of the early1900s, carcinogens had an effect after many

---

[1] Not every hygienist appreciated the need for OELs. In 1927, Cook suggested at the National Safety Conference that control measures should include air analysis for comparison with OELs to determine whether those measures were adequate – a common practice in modern-day industrial hygiene. However, a question arose from some members in the audience about how it could be done. In other words, a whole new concept had been introduced – the hygienist was now confronted with a limited number of OELs. Consequently, a major role of hygiene would be to develop the data used to set OELs. Some of these data were developed by governmental agencies, whereas others were established as internal limits of private corporations.

years of exposure to relatively low doses. Nevertheless, the concept of an OEL as an absolute threshold continued to be held by many hygienists. It was the recognition of the mechanisms of carcinogens, the development of the concepts of risk assessment, and a greater appreciation for the biological diversity of the human species that caused changes in the OEL concept. The changes have been gradual and subtle, and so the purpose of OELs continues to be somewhat elusive; clearly they are not absolute thresholds of safe dose for everyone, but the risk levels associated with each are often undefined.

In summary, the absolute threshold concept was first applied to acutely hazardous substances. Immediate and often irreversible effects could be seen if the threshold was violated or later found to be inappropriate. As the chronic nature of the toxicity of many compounds (culminating with carcinogens) has been recognized, the application of the threshold concept has become more difficult.

HUMAN BIOLOGICAL DIVERSITY

One of the most striking things about life is its diversity, not only between species but within them. The human species offers many examples of the vast differences that can occur between two individuals. These differences in genetics, biochemistry, physiology, anatomy, and other aspects of the human body should adequately demonstrate to anyone who has observed them that no two human beings are alike (even *identical* twins experience different environments and mutation rates); and when an adequate sized population of human beings – say, for the sake of argument, more than 100 persons – is taken into consideration, tremendous diversity can exist.

One of the most important areas of biological diversity for the hygienist and all occupational health professionals is the response of the body's defense systems to the attack of an outside disease agent, be it biological, chemical, or physical. In Chapter 8, the tremendous number of molecules that one may be exposed to by breathing small concentrations of gases or vapors for an 8-hour workday was derived. However, there are numerous bodily defense mechanisms that destroy or excrete molecules or repair damage caused by molecules; nevertheless, there is great variability in the adequacy of the defense mechanisms. Therefore, it should be relatively easy to understand that every individual has a different threshold concentration of response to a given chemical for a given effect, and that individual threshold can vary from day to day. However, it might be difficult to understand the integration of that concept. For example, over a large population (100,000), for a given chemical, there may not be a single concentration that is without any consequential effect for the entire population. More specifically, while we may theorize that even the most sensitive person in this group must express a threshold, finding that threshold is of no practical purpose and would be next to impossible to determine, at least with today's technology. This implies that in order for the entire human population to be protected (i.e., to have a zero percent risk of effect), the exposures must be zero. In American society this would be impossible unless we completely dismantled the petroleum, coal, and wood distillate industries and all industries dependent upon them. Even after we did so, there would be many other naturally occurring chemicals in the environment that would be capable of causing disease. We must, therefore, recognize that all enterprise has associated risk and it is the acceptable risk that drives the acceptable exposure level. The problems of defining *acceptable* risk were recognized early in the OEL-setting process (Key, 1973) and continue to be discussed (Gough, 1991; Fairhurst, 1995; and Jayjock et al., 2001).

OELs of any kind cannot be absolute safeguards. Instead, they must be regarded as acceptable risk levels. For example, it is estimated that workers who are exposed to one part per million of benzene for 8 hours each day over a working lifetime (45 years) have about a 1 in 100 chance of contracting leukemia, a blood cancer (OSHA, 1987; ACGIH®, 2007). There are two ways to look at this: On the one hand, 99 of the 100 workers will not, on average, contract leukemia. From this point of view it may seem that a great deal is being done to protect the workers. On the other hand, one of the 100 workers will, on average, contract leukemia. If you are this unlucky person, then all of the possible understanding of biological diversity and acceptable risk that you can muster will do little to relieve your suffering.

In a democratic society, the hygienist, realizing it or not, often works under the premise that OELs, at least those adopted by the government, reflect a suitable tradeoff that has society's approval. In other words, the benefits provided by the industrial process offset the associated costs in human life, injury, or illness. Although the hygienist may use this as a preliminary assumption, he or she should nevertheless endeavor to reduce workplace exposures beyond OELs for these reasons: (1) societal values change, (2) the actions of government do not always adequately reflect the views of society, (3) the cost of the exposure may be underestimated, and (4) the calculated risk almost always becomes greater as more information is discovered about the chemical.

A Biological Basis for OELs[2]

OELs should have a biological basis. For example, OELs set for aerosols should be related to the sizes of aerosols that can penetrate the respiratory tract to the area in which the response is elicited. If the response is a systemic one, then it may be reasonable to believe that absorption can occur throughout the respiratory tract, and consequently, the total mass of the aerosol inhaled is important. On the other hand, if the agent elicits a response only at the deepest portions of the lung, then only that fraction of the aerosol that can penetrate into those portions should be considered.[3]

Most importantly, the OEL should be based on the manner in which the human body

---

[2] The reader is instructed to consult Volume 2, Chapters 11 and 12 for more information on this subject and on Risk Assessment.

[3] A case can be made that the portion that actually deposits in the alveolar region will be an index of the dose that causes the response. More on this later.

absorbs, metabolizes, and excretes the toxic material. This problem has two aspects: toxicodynamics and toxicokinetics. The dynamics of a toxicant relate to its effect on the body, such as the occurrence of cancer, high blood pressure, or heart muscle sensitization. The dynamics include what systems are affected and at what doses. The doses are determined from a dose-response relationship; the latter is used to establish a *safe* exposure level – the OEL.

The kinetics are also very important to the setting of OELs. They tell us how quickly chemicals are absorbed and excreted, and how often exposures should be monitored. The kinetics establish the rate at which the toxicant reaches a target site. This is dependent on the absorption, distribution, metabolism, and excretion of the toxicant. The kinetics determine the averaging time for exposure monitoring. It will be seen later that when workers work unusual workshifts, i.e., 12 hours, their exposure levels must be adjusted, and this too is based on kinetics. A model will be used to develop a concept of the kinetics of a toxicant.

KINETICS – A PHYSICAL MODEL

Consider a water bucket that has a hole at the bottom. As the bucket fills, water is leaking out of the bottom. Think of the leak in the bottom as the excretion mechanism of the human body and of the top opening as the various routes of exposure by which a chemical agent may gain entrance to the body. A line drawn around the perimeter of the bucket at a particular height, say 10 cm, above the leak represents the level at which the *individual's* biological response becomes unacceptable – a threshold.

An individual water molecule has a greater residence time in the bucket if the leak is very small as opposed to being very large. This residence time is essentially independent of how much water is in the bucket because as the water level increases, the pressure also increases, and the leak rate increases. If the bucket is filled to the line and then the time that it takes for the bucket to half-empty is measured, this can be called the residence half-time. The half-time is related to the size of the leak.

Now imagine that we measure the rate at which the water is going into the bucket, and the rate is not constant but varies from minute to minute. If we know the average rate at which the water enters the bucket, then we subtract from that the average rate at which water leaves the bucket. This gives us the rate at which the water height rises or falls – that is, the average rate at which the water accumulates.

The rate at which the water enters the bucket is effectively the dose or the exposure rate. If this rate varies from minute to minute, then we must consider how the variation in this *exposure* rate affects the probability that the water level will go beyond the line. It turns out that the way in which we measure the exposure rate affects our ability to detect a transgression of the line.

To make the problem more realistic and relevant to the human body, let's assume we cannot see the line. We will also assume that all individuals (buckets) have the same *excretion* rate. (This common excretion rate is essentially assumed. Since we cannot measure each individual's excretion rate, we use an average or a *worst case* value such as the first or fifth percentile, at least in animal studies.) If we determine the *exposure* rate, we know the rate at which the water level accumulates. If it rises for some length of time, *t*, at a rate, *r*, then r × *t* tells us how full the bucket is at time *t*. In other words, we can calculate whether the line is exceeded, even though we cannot *see* the line.

Now, we just need to know the *rate* of fill, and unfortunately we cannot determine this directly. Instead, we have a meter that measures the *amount* (mass or volume) of water that has entered during some time period. Then by dividing by the *time* over which the meter measures, we have the rate of water inflow. Of course, the rate is averaged over that time period. Finally, for reasons shown later, we can only read the meter every 15 minutes or every 8 hours.

In the case of 15 minute readings, we can find the average inflow rate, and then the assumed outflow rate is subtracted to find the rate of rise or fall. Since we have set the rate of outflow as a constant, the rate of inflow alone is indicative of the rate of rise or fall. Initially, however, we are only interested in the rate of rise, as this phenomenon causes the line to be exceeded.

At the end of the 15 minute period we can make a meter reading, do a calculation, and decide if the water is above the line. However, the calculation only tells us about the water level at the end of the 15 minute period. Is it possible that the line was exceeded during the 15 min period, but we could not detect it because we were not *looking* often enough? Clearly the answer is *yes*.

We need to consider one more analogy regarding the shape of the bucket. Some buckets are tall and narrow, others are wide, and there are all combinations in between. The shape of the bucket represents the size of the body's capacity for the *chemical*. Some chemicals have few storage sites but react rapidly at the site of action. The other extreme is chemicals that are stored in bone or fat, which are relatively large reservoirs. Finally, people come in various sizes and have varying capacities based on physical size. However, for simplicity, we will assume that everyone weighs 70 kg in order to take size out of the equation. If the bucket is tall and narrow relative to the leaking hole, then it empties very fast, but it can also fill very fast. Therefore, if we do not look often enough we will miss an exceedance of the threshold. If the bucket is wide, then it has greater capacity than a narrow bucket, and an increase in exposure rate for even a relatively long period (say, a day) may not cause the level to exceed the line; and then overnight excretion reduces at least some of the loading of the day-long exposure.

The problem is one of how often we need to check the rate of fill or the exposure rate, and this depends on the size of the bucket and the leak rate. Both of these variables are reflected in the residence half-time concept. Chemicals with short half-times have narrow reservoirs relative to their leak rates. Chemicals with long half-times have wide reservoirs relative to their leak rates.[4] The degree of fluctuation in the water level is a function of the exposure rate and the length of the half-time.

---

[4] Most chemicals are in between. The bucket concept gets across the point about half-times of chemicals in the body and how exposure rates affect body burdens, but in reality, it is a gross oversimplification of how body burden leads to toxicological effects.

A MATHEMATICAL MODEL

Roach (1966, 1977) worked out this problem quantitatively. He showed that if the goal is not to allow an unnoticed fluctuation in exposure that causes an exceedance of the threshold, then we need to *look* with a minimum periodicity. In other words, the period of time over which we average the exposure rate should not exceed $3T_{0.5}$ or three times the half-time. Also, he showed that if we average for less than $0.33T_{0.5}$, we gain nothing in terms of detecting exceedances. So if a chemical has a half-time of 1 hour, we should average the exposure over no more than 3 hours so as not to miss any *peaks* that might cause an exceedance of the threshold.

Roach's model is based on relatively simple assumptions, but they are reasonable. Unfortunately, the guiding principle – measurement time (t) $< 3T_{0.5}$ – is difficult to put into practice for several reasons. First, toxicant half-times in humans are not widely known. Second, many chemicals have several body reservoirs with differing $T_{0.5}$ associated with each target organ; these are often incompletely known. Finally, it would be difficult to monitor the exposures of workers exposed to mixtures. Often the components of a mixture can be sampled from the air by using the same device, say, a charcoal tube. However, if we set $t = 3T_{0.5}$, we would need a separate charcoal tube for each component of the mixture, as each would have a different value t, if their $T_{0.5}$ values were different.

Although it is doubtful that Roach's arguments were considered in setting compromise values of t, values have been set that have become universally used. For compounds with medium to long half-times, t = 8 hours or one normal shift; for compounds with short half-times, t = 15 minutes; and for compounds with very short half-times, t is instantaneous. In the latter case, the exposure level is not allowed to exceed, at least in theory, some value, a true MAC or maximum allowable concentration.

There are no steadfast rules for assigning compounds to the three *t's* or averaging times. In fact, as stated earlier, $T_{0.5}$ is not always available. Instead, the averaging time is probably based on analogy to other compounds and experience in the workplace. However, consider the potential ramifications of having only three averaging times relative to Roach's arguments. Compounds with $T_{0.5} > 5$ minutes would be adequately monitored with t =15 minutes. Ideally, compounds having $T_{0.5} < 5$ minutes should be monitored instantaneously. Compounds with $T_{0.5} > 2.33$ hours (140 minutes) would be adequately monitored at t = 8 hours. If Roach's arguments are adequate, then the current approach to setting OELs as either 15 minutes or 8 hours time weighted averages (TWA) could allow biological thresholds to be exceeded if either $T_{0.5} < 5$ minutes and t = 15 minutes, or $T_{0.5} < 140$ minutes and t = 8 hours. For example, consider a compound with $T_{0.5}$= 100 minutes. If t is set at 15 minutes, then we err on the safe side by averaging too often. However, if t = 8 hours, then even if we find that the exposure rate averaged over 8 hours is acceptable, exposures could be occurring that would cause the biological threshold to be exceeded, and they would not be detected. Ideally, t = 15 minutes should be applied to all compounds having 5 minutes $< T_{0.5} <$ 140 minutes.

DYNAMICS

The dynamics of the toxicant must be studied in order to determine the mechanism by which it causes effect. How does mercury affect the nervous system? If chemical A depresses enzyme B what will that lead to? Once the effect(s) is (are) determined, a dose-response curve for each effect, like the one shown in Figure 10-1 can be constructed. This curve is usually derived from animal studies, but may be derived from epidemiological studies of humans. From it an acceptable dose or exposure can be determined. In deriving an OEL from the dose response curve, one of the most important aspects is the determination of whether there is a *threshold* of response for the effect of interest.

Threshold or Not?

There are two basic approaches to interpreting dose-response data: threshold and no threshold. Historically, toxicologists assumed that thresholds existed for all effects at the population level. In other words, below some dose anyone could be exposed without the effect occurring.

With the understanding of carcinogen mechanisms in the 1970s, it was recognized that a single molecule could cause a mutation, and that this single mutation could conceivably escape defense mechanisms and ultimately lead to a tumor. It was further recognized that carcinogens might have no threshold. In practice, if the data in Figure 10-2 are for a carcinogen, we assume some curve (e.g., a straight line) back to the origin from the lowest dose, and then use that line (actually its slope) to find the dose that corresponds to an acceptable risk. This is commonly done for carcinogens. Conversely, if one first picks a dose (OEL) then the risk at that dose or exposure level can be determined. For example, one may determine that a risk of 1 per 1000

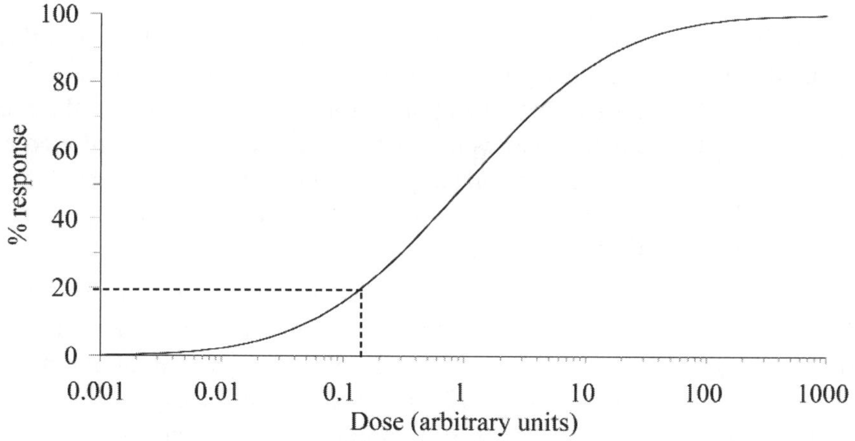

Figure 10-1. Dose response curve.

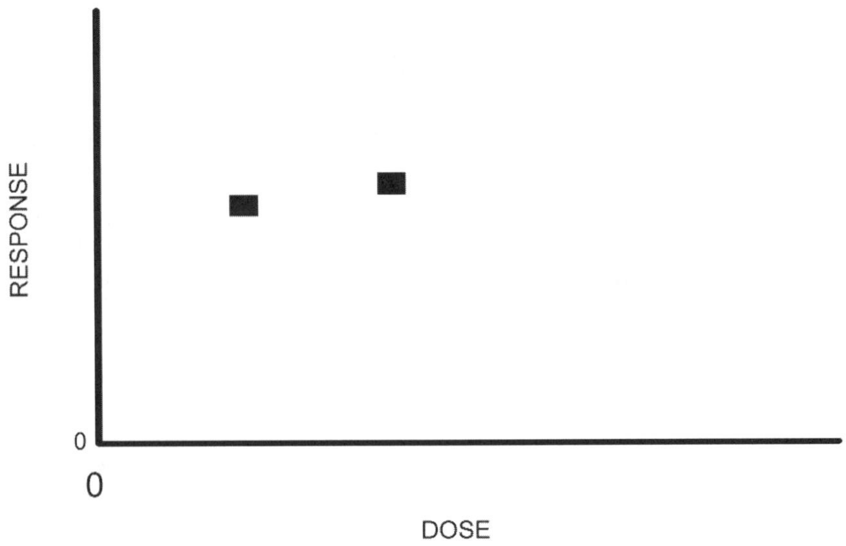

Figure 10-2. Experimental results at two doses. Note that a straight line relationship does not go through the origin nor indicate a threshold.

or 0.1% is acceptable, and for that risk we determine from the dose response that a lifetime of exposure to 1 ppm would be correspondingly acceptable, or that an OEL of 1 ppm is feasible (OSHA has to consider technological and economic feasibility) and that does gives a risk of 1 per 1000 such exposures. In the no threshold approach we admit that there is residual risk if there is exposure. The concept that there is residual risk associated with an OEL is a difficult one for many people to swallow in a world where many people assume that certain activities should be risk free.

For non carcinogens the approach continues to be the determination of a threshold and then application of uncertainty or safety factors. Uncertainty factors (UF) generally range from 1–10 and there can be as many as four UFs applied for reasons that reflect the quality of the data or its applicability to the human exposure scenario. The data may be plotted as in Figure 10-3. In this animal study, there is a dose that appears to generate no response – the no observable effect level or NOEL. Rather than setting the OEL at the NOEL, we apply safety factors to improve the odds that the OEL will be *safe*.

Remember that just because we do not detect a response at a particular dose does not mean that the effect cannot occur at that or even a lower dose. Among other things NOELs are dependent on sample sizes. If I look for an effect from a dose given to 50 animals, the smallest average risk that I can detect is 1 in 50 (2%). If the risk of the effect is actually 1 in 100, then I am not likely to determine that there is an effect; there will appear to be a threshold. In the limit we cannot establish a population threshold unless we test every member (or at least a

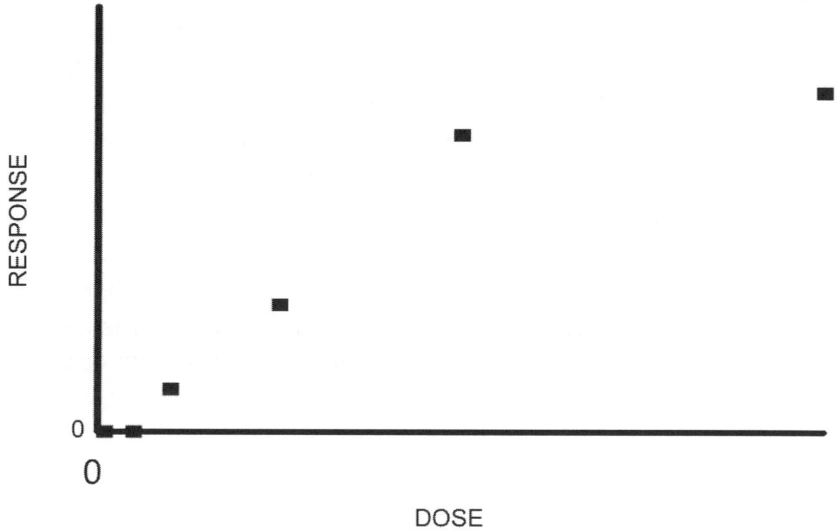

Figure 10-3. Typical dose response for animal data showing a non zero dose at which the response is 0 – a threshold.

very large sample) of the population (Jayjock et al., 2001). In that regard, recent air pollution studies (e.g., Schwartz et al., 2002) have failed to reveal a threshold for such endpoints as daily deaths caused by exposure to small particles in ambient air, implying that even very small increases in small particle concentrations will effect at least a few people. The key to these studies has been large sample sizes. The Schwartz et al. study included the populations of the counties and surrounding counties of six U.S. cities including Boston and St. Louis over decades, i.e., 1000s of persons and deaths. Such sample sizes are never achieved in animal studies and even large epidemiological studies of workers rarely include such large populations. Hence, in most of the studies that are used to set OELs, limited sample sizes may be the main reason that we see thresholds.

The population threshold is the threshold for the most sensitive member. This would seem to argue against thresholds, at least in a practical sense. On the other hand, a strong argument can be made for thresholds based mechanisms of repair (Daston, 1993; Rogers and Kavlock, 2001). If a threshold exists for the mechanism, then that would imply a threshold for the effect. For example a chemical inhibits an enzyme and this causes an effect, say a malformation in fetal development. If we lower the dose and see no malformation but continue to see a reduced but detectable level of enzyme depression, does this not infer a threshold for that effect? The problem with this argument is that we can get into a never ending circle in that the lowered enzyme depression may be causing some other, more subtle, even delayed, effect for which we are not looking or that the original effect cannot be detected due to sample size. Obviously this argument – pro and con thresholds – can and will go on for sometime.

In contrast to the approach for carcinogens, when we use the NOEL and UF approach we do not quantitate risk, i.e., we do not know the odds that a worker will be harmed when exposed at the OEL. Furthermore, the UF is often not applied in any way that reflects sample size. Intuitively, it seems that the sample size (animal or human data) used to estimate the NOEL would drive at least one UF. Jayjock et al. (2001) have derived a way to estimate a risk level when UFs are used. This will be discussed more fully later.

Fairhurst has reviewed uncertainty factors that have been applied to OELs developed by the UK. Where human data have been available, uncertainty factors have been 1 or 2. For OELS based on animal data the UF has ranged from a low of 1, where there were supporting human data, to 60; however, the vast majority were < 10. He finds a consistency in application (i.e., precision) but there is a lack of data for determining appropriate application of UFs (i.e., do they protect and are they biased?).

THE IDEAL EXPOSURE LIMIT

Exposure limits are not perfect. Some hygienists and occupational health professionals regard current exposure limits as absolutely safe exposure levels. The preface of the ACGIH® *TLVs® and BEIs®* Book states that the TLVs® are levels which *will protect nearly all workers*. However, this qualitative phrase is sometimes ignored, in light of long-held notions about the absoluteness of standards. The risk associated with many OELs is not stated (other than in a qualitative way, as in *nearly all workers*). The associated risk of disease when workers are exposed at the OEL can be considerable. It has been shown that in the past, the most widely used OELs, the TLVs®, in some cases may actually protect fewer than half of exposed workers from the effect of concern (Roach and Rappaport, 1990). What then are the attributes of an adequate OEL?

Documentation

The rationale for the OEL should be clearly documented. In the *Documentation*, all information considered relevant to the development of the OEL should be given. This information is generally in two forms: (1) studies of humans who have been exposed to the compound and (2) studies of animal exposures. The various studies of exposures, whether they are of animals or humans, should lead to the development of a dose-response curve (Figures 10-1 through 10-3). The response may be death, it may be irreversible lung damage, it may be cancer, or it may be as simple as skin irritation. In the case of a Russian dose-response curve, it could be a small and difficult to assess behavioral change.

Which Effect?

The effects that the OEL protects against should be stated. Consider the situation where the hygienist is seeing effect *A* in the workforce even though exposures in the workplace are below the OEL. Properly prepared documentation reveals to you that effects *B*, *C*, and *D* were studied, and the OEL is designed to protect against those effects. No mention is made of effect *A*; so you realize that it may be an unrecognized effect, or it may be caused by a confounding factor. On the other hand, in a poorly prepared documentation, effect *A* may be mentioned, but in the final analysis its relationship to the OEL is not stated. This can leave the reader confused.

For overt or gross effects such as death or tumor formation, the effect can be easily seen and quantitated. However, other effects are not so easily seen or quantitated, and the importance of these effects, when they are noticed, is often difficult to interpret. For example, in an animal study, an effect may be changes in the levels of a particular hormone. If the hormone levels are not examined, the effect goes unnoticed. If they are examined, what constitutes a significant effect? Hormone levels are constantly changing naturally for reasons that are not fully understood. Even so, the chemically induced changes may be important. The history of toxicology is that we are constantly finding new effects that we did not previously see.

Dose Response

To the extent possible dose response curves should be constructed. Ideally, from a dose-response curve, the risk of effect (i.e., the percent affected) can be determined for a given dose. We need to realize that an effect may not be seen at a given dose because of statistical chance. If 100 animals are studied and the incidence of a particular effect is 1%, then on average the effect will be seen in one animal out of 100. However, by chance, the effect may not show up in the experiment, and what if the incidence of the effect is 0.1% or 0.01% or there are only 20 animals in each group? Is an effect rate of 0.1% important?

Method of Extrapolation and Assumptions about Thresholds

Information is never available that allows one to construct a complete dose-response curve. This is related to the issues raised above regrading sample size. The highest dose that does not cause a response is often thought of as a threshold, but of course it may be a real threshold or an artifact of sampling size. Disregarding threshold detection for the moment, there are other reasons why construction of complete dose response curves can be difficult. It is not uncommon to have incomplete or less than optimal information, but the available data and any assumptions that have been made in deriving a dose response relationship should be clearly stated.

For example, Figure 10-2 shows information for only two responses at two doses. Information may be needed about doses lying in between the two, and consequently an interpolation can be made that includes an assumption about the shape of the curve, that is, the relationship of dose and response. More often, particularly for carcinogens, extrapolation to doses below the lowest dose that has been studied, and associated curve-shape assumptions, must be made. These should be stated.

When UFs are used these should be stated along with identifying why that value was used and why a UF was needed. This approach clearly reveals all the decisions that have gone into the establishment of the OEL so that anyone who wishes to question the OEL or perhaps justify a lower or even higher number is provided with the information necessary for doing so.

Defining Acceptable Risk

If the OEL sets an acceptable standard of exposure then the definition of *acceptable* should be stated. Dose-response curves are used to arrive at an acceptable dose or exposure level. Arriving at the acceptable dose requires that one first settle upon an acceptable risk. For carcinogens where a probability based OEL is derived, this can be done more easily than in the case of the application of UFs to a NOEL. In the former case if, an acceptable risk is 20 in 100 (see dotted line in Figure 10-1), then the acceptable dose or exposure standard would be about 175. Several regulatory agencies of the U.S. government have adopted acceptable risk criteria. The Occupational Safety and Health Administration (OSHA) utilizes an acceptable risk level of 1 in 1,000, or 0.1%. However, as will be seen, OSHA has not achieved this ideal value in setting its OELs. Still, even in the United States, the OELs derived by OSHA are not the most widely used limits; rather the TLVs® of ACGIH® are.

HISTORY OF THE TLV®

In the United States and indeed in many parts of the world, a set of OELs that are published annually by ACGIH®[5] has become widely used. These OELs are published by committees[6] of the ACGIH®, which works on a voluntary basis.

Two important committees were established in 1941. In addition to the previously mentioned committee of ACGIH®, a committee of the American National Standards Institute (ANSI) – then known as the American Standard Association – was also established. This ANSI committee is known as Z-37. Both the ACGIH® and ANSI committees develop OELs for workplace contaminants. Unlike ACGIH®, ANSI is a consensus organization composed of individuals from industry, government, labor, and other concerned parties. On the other hand, the ACGIH® committee is composed only of industrial hygienists, occupational epidemiologists, occupational physicians, and toxicologists. There has been effort in the recent past to maintain an equal mix (25%) of each of these disciplines. Historically, the members were

---

[5] The terms ACGIH® and TLV® are registered trademarks of ACGIH®.

[6] There are three occupational exposure guideline committees — TLVs® for Chemical Substances, TLVs® for Physical Agents and Biological Exposure Indices (BEI®). These committees are separate from each other and run as free-standing committees.

employed by government, with the exception of some industrial consultants. About 1998, ACGIH® changed its bylaws to allow industry or Associate Members to be full, voting members of any of the ACGIH® committees. At one time, the limit for these types of members on a committee was 25%; recently this was changed to 50%. However, the TLV® Committee Chairs have tried to keep the Associate Member numbers to 25%, recognizing the importance of limiting conflicts of interest.[7]

From 1940–1970, both of these organizations were instrumental in establishing OELs. However, because the ANSI standards were consensus in nature, and therefore took longer than those of ACGIH® to formulate, ACGIH® quickly outpaced ANSI in terms of the number of OELs established. Before 1970, neither organization's OELs had a legal basis in all U.S. states; they were only guidelines. (Note the TLVs® as published by ACGIH® have always been guidelines only).

In 1942, the committee of ACGIH® published a table of 63 substances with maximum allowable concentrations. Interestingly, this report contained this statement: "The table is not to be construed as recommended safe concentrations. The material is presented without comment." This may be interpreted as meaning that the values given in the report were not absolute, but rather should be interpreted as *ball park* guidelines. These values were starting points for the hygienist to control the workplace.

By 1945 a member of the ACGIH® committee had published a list of 132 industrial compounds and recommended OELs (Cook, 1945, 1946). For some of these, OELs had been taken from other sources, whereas for others the OELs were based upon the author's interpretation of toxicity data. As had been the case since the early 1900s, most of the OELs were based upon toxicity work conducted with animals. Very few data from human experience in the workplace were available.

Cook's list was modified and added to the ANSI list of 13 mineral dusts and the 1942 ACGIH® list, and the result was published by the ACGIH® subcommittee on threshold limits in 1946. The final list contained 131 gases, vapors, dusts, fumes, and mists. Interestingly, even though the committee called itself the subcommittee on threshold limits, the OELs were referred to as MACs or maximum allowable concentrations. This latter term was developed in the German literature and continues to be used by the Germans.

The OEL list published by ACGIH® continued to grow, and the limits began to be called threshold limit values or TLVs®. It was not until 1962 that the TLV® committee published *Documentation of the Threshold Limit Values for Substances in Workroom Air*, a publication

[7] Around 1998, ACGIH® adopted a conflict of interest policy, to manage the various levels of conflict encountered by each committee member working on substances. Members with high levels of conflict are excluded from preparing *Documentation* and commenting and voting on the proposed TLVs®. Low and medium levels of conflict may result in similar, although lesser types of exclusions. Oral declarations of conflict are made to the full committee on an annual basis. Conflict discussions are usually held at every subcommittee meeting.

now titled *Documentation of the Threshold Limit Values and Biological Exposure Indices* and in its 7th edition. It provides a short synopsis of the rationale for each TLV®. The TLV® list now contains about 700 entries and is updated annually by ACGIH® (ACGIH®, 2007). The actual number of chemicals addressed is greater than 700, as there are a number of entries such as *mercury and its inorganic compounds*. In addition, there are TLVs® published for various physical agents such as radiation. Finally, a subcommittee of the TLV® committee has published a number of Biological Exposure Indices (BEIs®). These are values that relate *acceptable* levels of certain contaminants in biological tissues such as blood and urine. Thus, the TLV® committee has provided a great deal of information to the hygiene community in the United States and throughout the world.

Many countries now use the TLVs® as official OELs for enforcement purposes. The federal government of the United States and many individual states have adopted the TLVs® as official government standards (see Chapter 11).

CHARACTERISTICS OF TLVs®

Averaging Times

As with most OELs, there are several TLV® types:

- the TLV®-time weighted average (TLV®-TWA),
- the TLV®-short-term exposure limit (TLV®-STEL), and
- the threshold limit value-ceiling (TLV®-C).

Unfortunately, the TLV®-TWA and the TLV®-STEL terminology is a bit confusing; both are time weighted averages. Nevertheless, many people do not think of the STEL as a time weighted average. The TLV®-TWA is an 8-hour time weighted average, whereas the STEL is a 15-minute average. The TWA aspect of a TLV® denotes that exposure values are averaged over the time period, and the average of exposures is compared to the TLV®. A TLV®-TWA is to be compared to an exposure measurement averaged over an entire work shift of about 8 hours. Since the TLV®-STEL is averaged over a much shorter period (15 minutes), measurements made for these shorter periods reveal larger fluctuations in exposure. These larger fluctuations are most important, toxicologically, for acute or fast-acting toxins. The ceiling is an instantaneous value, which can be difficult to measure. Nevertheless, it has biological plausibility, particularly for substances that may elicit a near immediate, irreversible response when a threshold concentration is exceeded, i.e., compounds with $T_{0.5} < 5$ minutes.

It is important to note that some substances may have both the TLV®-TWA and the TLV®-STEL. This implies that these substances have two important reservoirs in the body, each with its own $T_{0.5}$. The TLV®-TWA is used to control the more chronic aspect of the chemical's toxicity. However, because the TLV®-TWA also allows exposure to become quite high during the course of the 8 hours averaging time (ACGIH® does place an upper boundary on *excursions* above the TLV®-TWA, to be discussed later), it is necessary to impose a STEL for chemicals that can also cause an acute response at levels considerably higher than the TLV®-TWA. This concept is demonstrated in Figure 10-4. Even though the 8-hr TWA concentration for the exposure is less than the TLV®-TWA, a high peak occurs that yields a 15-min TWA that exceeds the TLV®-STEL.

ACGIH® states that the STEL should never be exceeded. However, as an example, consider the following. If 10 workers perform the same job, then they are exposed for 10 × 32 discrete 15 minute periods in an 8-hour day. During a year (250 working days), this would be 80,000 discrete 15 minute periods. It seems unlikely (and later chapters will show that it is unlikely) that all of the 80,000 short-term concentrations are below the STEL, no matter how well the process is controlled.

ACGIH® requires that exposures occurring at levels between the TLV® and the STEL "should be no longer than 15 minutes [perhaps meaning each *excursion* above the TLV®], should occur no more than 4 times per day, and at least one hour should elapse between each excursion." This requirement is difficult to monitor and may be over burdensome in terms of achieving only a small incremental protection of the worker. As one increases control of the workplace to meet this STEL requirement, the probability of exceeding the TLV®-TWA is reduced to near zero (Tuggle, 2000). Presumably, the purpose of the requirement is to allow biological clearance between these higher exposures, but there is no biological basis stated for it.

Even for those TLV®-TWA substances without a TLV®-STEL, ACGIH® states that exposures during the 8-hour averaging period should not be unrestrained; that is, excursions above

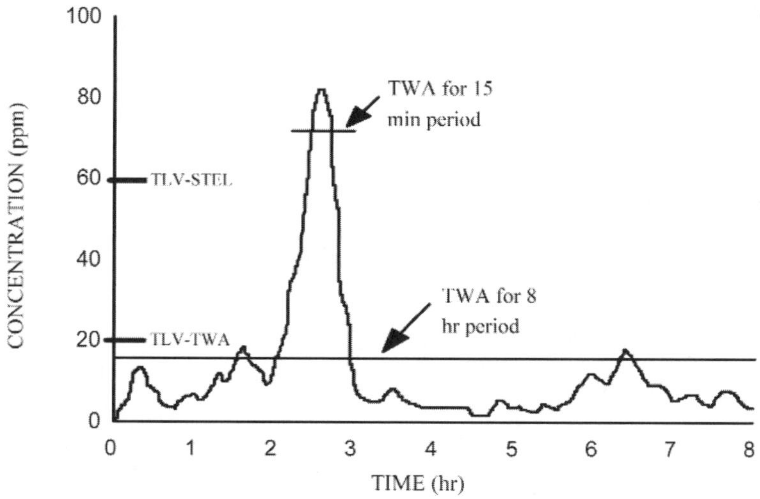

Figure 10-4. Concentration profiles showing the 8-hr TWA and a 15-min TWA for a substance having a TLV®-TWA of 20 ppm and a 15-min TLV®-STEL of 60.

the TLV®-TWA should be limited. For example, consider Figure 10-5, where for a very short period of time the exposure has reached a level more than three times the TLV®-TWA ; but because exposures are low for the remainder of the 8-hour period, the TLV®-TWA is not exceeded on an 8-hr TWA basis. ACGIH® states that "excursions in worker exposure levels may exceed three times the TLV®-TWA for no more than a total of 30 minutes during a work day, and under no circumstances should they exceed 5 times the TLV®-TWA provided that the TLV®-TWA is not exceeded."

The argument for the above statement is not biologically based; it is statistically based on the idea that well controlled workplaces should have geometric standard deviations (GSDs) that do not exceed 2. In order to understand this argument, one must go back to Chapter 7 and consider the discussion of the lognormal distribution. ACGIH® assumes that workplace exposures are lognormally distributed; that is, a lognormal model of workplace exposure adequately describes the frequency of occurrence of exposure values. ACGIH® states that the typical workplace exposure GSD is between 1.5 and 2. However, GSDs are often greater than 2 and can reach as high as 10, and a more reasonable average GSD for *8-hour* exposures might be 2.7 (Buringh and Lanting, 1991). Therefore, average GSDs for short-term (< 8 hours) exposures would be > 2.7 (Spear et al., 1986; Francis et al., 1989). Nevertheless for the sake of discussion, ACGIH® states correctly that for any lognormal with a GSD of 2.0, 5% of all values will exceed 3.13 times the geometric mean (remember this is the median, not the true mean). Five percent of a 480-minute workday is 24 minutes. ACGIH® is allowing 30 minutes of exposure to exceed three times the TLV®. By statistical chance alone, this will occur if average exposures are controlled at the TLV®-TWA and their GSD is 2.0. ACGIH® recognizes that some workplaces will have GSDs that exceed 2.0, and if worker effects are not seen when the

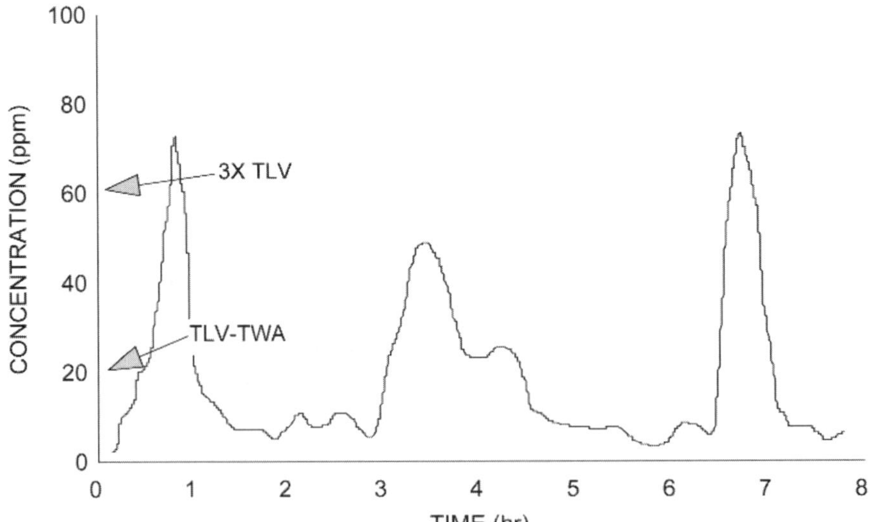

Figure 10-5. Concentration profile showing two excursions beyond 3X TLV® over an 8-hour period.

TLV®-TWA is met, then excursion practices can be modified from these recommendations.

In addition, ACGIH® recommends that exposures not ever exceed five times the TLV®. If it is reasonable to expect that the TLV® may be exceeded by a factor of three for 30 minutes in a workday, then would it be reasonable to expect that the TLV® would also be exceeded by a factor of five for at least some portion of the day, even if it is as short as 1 minute? The answer is *yes*. However, few hygienists concern themselves with, or have the tools to measure, exceedances of the 5 × TLV® excursion limit. It can be shown that, if the true GSD of short-term exposures is three, then about 7% of all values will, in fact, exceed 5 times the geometric mean.

Skin Notation

Some of the TLVs® have a *skin notation*. There is often a great deal of confusion about the meaning of the skin notation. Some think it means that the chemical causes skin irritation or dermatitis. Others think that the measurement of exposure should include measurement of exposure via the skin. However, this is very difficult to do. The meaning of the skin notation is simply that *the chemical may be absorbed through the skin in significant quantities*. Airborne exposures, controlled at the TLV®, may be too high, if the worker's skin is also being exposed. The body burden will be increased by the skin exposure, and it will not be reflected in the measurement of airborne concentrations. Hence, the worker may be overexposed even though the TLV® guideline is being met.

The skin notation is qualitative. In reality, virtually any organic compound can pass through the skin and enter the bloodstream, given enough residence time on the skin. Because the criteria used by the TLV® committee to arrive at the skin notation are not given, and generally no, or only slight, justification is given in the *Documentation* for each chemical, it is difficult to know whether the presence or lack of a skin notation is justified (Scansetti et al., 1988). In essence, skin notations should not be equal, since different chemicals have differing potentials to penetrate the skin, but there are few data for constructing a hierarchy of skin notations (see Volume 2, Chapter 3 for greater detail on skin exposure and skin notations).

In essence, a more formalized approach to setting standards for skin exposure (Fenske and Van Hemmen, 1994) is needed. The Dutch have proposed a notation when the amount absorbed by the hands and forearms in 1 hour can be expected to equal or exceed 10% of the amount that would be inhaled and absorbed at the TLV® (de Cock et al., 1996). While data to assess this criterion are sparse, methods for skin exposure assessment are available (see, e.g., Ness, 1994 and Chapter 3 of Volume 2), and the literature is continuing to reveal that the skin route is the most important route for some chemicals (see, e.g., Christiani et al., 1986).

Aerosol penetration and deposition in the respiratory tract

PRINCIPLES

Aerosols will penetrate and deposit in the lungs as a function of size. In Chapter 6, the deposition of aerosols in the human respiratory tract was discussed. Here penetration is discussed because OELs for aerosols are based on penetration to the respiratory tract. If the aerosol exerts its toxicity regardless of its deposition site (i.e., it is absorbed and becomes a systemic toxin like a pesticide, or it causes local effects like inflammation caused by respiratory irritants such as some pollens) then all the deposited aerosol becomes the dose. On the other hand, if the effect is fibrosis in the gas exchange region or bronchial carcinoma, then the dose is related only to that portion of the aerosol that actually penetrates to and deposits in these regions. Thus, aerosol TLVs® are differentiated according to where their site of action is.

Ideally, it is the amount that is *deposited in the lung* that makes the dose and therefore this is the amount that we should assay as an index of exposure (Esmen et al., 2002). However, scientists in the U.S. and other nations adopted *conventions* for size selective sampling that are based on penetration, not deposition. In other words, OELs that are based on these conventions require the use of air samplers that sample the air the way that the aerosol would penetrate to different regions of the respiratory tract. Various reasons have been given for this (Solderholm and McCawley, 1990).

- Instruments for measuring penetration curves already exist.
- The deposited fraction is a more difficult fraction to measure than that which penetrates to a particular region.
- More than one filter would likely need to be analyzed to quantitate the deposited fraction.
- The penetrating fraction is always larger than the deposited fraction, thus erring on the safe side.
- The deposition curves are dependent on more variables (such as age, breathing rate, mouth vs. nose, hygroscopic nature of the aerosol, etc.) than are the penetration curves and thus would be harder to agree upon.

However, there are instances when the differences between penetration and deposition are quite important. For example, one can argue that since penetration overestimates deposition, epidemiological studies, when the exposure results are based on penetration, are flawed and may be unable to find important effects (Hewitt, 1991). Esmen et al. (2002) used a computer to create pairs of size distributions that would have the same sampled mass based on penetration, but have different mass median aerodynamic diameters and GSDs. They determined, through Monte Carlo simulations, that even though the penetration sampling results would indicate equivalent exposures, the actual mass of aerosol deposited could be as much as ten fold different between the two aerosol distributions. Johnson and Esmen (2004) showed that, as expected, penetration was always larger than deposition for a wide range of aerosol distributions, however, exposure levels based on penetration would often be misclassified. In epidemiological studies, workers are often classified as low, medium, high or some other convention in terms of exposure. When an exposure is misclassified, there is less accuracy in the determination of the dose-response and a greater likelihood of not finding significant results when they may in fact exist. Today, nanoparticle technology makes these issues quite important, since deposition is always less than 100% for particles < 2μm. Nanoparticles are < 0.1 μm (HSE, 2004) and deposition would be on the order of 10-40%, while penetration would be 100%. Exposure to these particles is just beginning and therefore epidemiological work will one day be based on exposure data collected today.

Each size-selective penetration convention dictates the construction and/or operation of a sampling device that is used to sample only that fraction of the atmospheric particles that meet the size convention. Manufacturers are driven by demand. Thus, if the U.S. OSHA or the UK HSE has standards that require an older device A and the TLVs® require a newer device B, the hygienist must make a decision. Of course, that decision will drive what the manufacturers do.[8] As a result, in the last 15 years, there has been a wide range of devices available, and these are discussed more fully in Chapter 24.

While the dose-response, size-dependent concept seems straightforward enough, it has a 30 year history that is complicated by past practices and slow government action (Kenny, 2000; and Kenny and Ogden, 2000). This is further complicated by the fact that the U.S. EPA has adopted a different convention for ambient particulate sampling than the occupational conventions that are used essentially worldwide, and many other governments are adopting that convention for ambient air monitoring. It should also be noted that, while it was relatively difficult to determine what the size-driven aerosol penetration and deposition curves look like, the task of determining which fraction is important for each aerosol, e.g., lead oxide versus calcium sulfate, is also difficult. For example, asbestos and other fibers largely effect the lungs and thus it would be practical to use a thoracic-based, size-selective sampler that would only collect fibers of a size that would penetrate to the lungs (about 3 μm diameter) (Baron, 1996). However, this has not been done. Therefore, although the TLV® committee adapted the size selective sampling convention over 20 years ago, it has not applied it to all aerosols. Thus, the TLVs® are a mix of old and new conventions.

THE NEW AND OLD CONVENTIONS

Historically, the TLV® committee did recognize the importance of size selection. Until 1993, all TLVs® for aerosols were based on either the *total* fraction or *respirable* fraction. Simplistically, it was often stated that respirable particles had aerodynamic diameters smaller than 10 μm. Similarly, the total fraction was in fact not total, as the particles enter a device which selects a portion of the particles, and that portion is dependent on how the device is operated.

However, the TLV® committee proposed a refinement of particle size-selective sampling criteria in 1992, which went into effect in 1994. This set of conventions has also been adopt-

[8] All particulate TLVs® are being converted to the new convention as they are revised.

ed in Europe (CEN, 1993) and internationally (ISO, 1995). There are three particulate mass (PM) fractions based upon three regions of the respiratory tract considered to be of physiological importance. In turn, the OEL for an aerosol should be set based on the region of physiological importance for that substance. The required efficiency of a collector that is used to sample any one of three fractions is shown in Tables 10-1 through 10-3 and Figure 10-6. The proposed efficiencies closely mimic the average penetration of each particle size to that lung region (Phalen et al., 1986; Soderholm, 1989) based on a long history of work largely carried out in the UK (Kenny and Ogden, 2000). The particle masses are:

1. *Inhalable*: This mass should be reflective of the particle size that typically gains entrance to the nose and mouth. The IPM-TLVs® are applied to aerosols that are hazardous throughout the respiratory tract. The old total fraction has greatest similarity to this fraction, but the two are considerably different. It appears from the TLV® list that many of the old total designations have been converted to inhalable. However, aerosols which have no designation remain, and by default these fall in the *total* designation and require that sampling methodology until ACGIH® can address them.

2. *Thoracic*: This mass reflects the particulate that typically travels past the nasopharynx with little deposition and is, therefore, available to the lung from the thoracic region down to the alveoli. This fraction is probably closest to the old respirable fraction. For both, particles with aerodynamic diameters of 10 μm have a 50% chance of being included (see Table 10-2 and Figure 10-6). This is called the

Table 10-1
Inhalable (IPM) size selective criteria (ACGIH®, 2007)

Particle Aerodynamic Diameter (μm)	Inhalable Particulate Mass (%)
0	100
1	97
2	94
5	87
10	77
20	65
30	58
40	54.5
50	52.5
100	50

Table 10-2
Thoracic (TPM) size selective criteria (ACGIH®, 2007)

Particle Aerodynamic Diameter (μm)	Inhalable Particulate Mass (%)
0	100
2	94
4	89
6	80.5
8	67
10	50
12	35
14	23
16	15
18	9.5
20	6
25	2

Table 10-3
Respirable (RPM) size selective criteria (ACGIH®, 2007)

Particle Aerodynamic Diameter (μm)	Inhalable Particulate Mass (%)
0	100
1	97
2	91
3	74
4	50
5	30
6	17
7	9
8	5
10	1

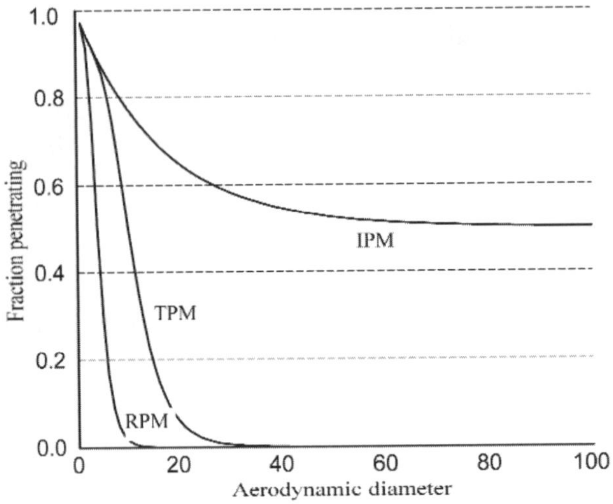

Figure 10-6. The inhalable (IPM), thoracic (TPM), and respirable (RPM) mass fraction sampling conventions based on aerosol penetration to those regions of the respiratory tract.

median cut-point. The TPM-TLVs® are applied to aerosols that primarily effect the tracheobronchial region. It appears from the TLV® list that one aerosol as of 2005 (sulfuric acid) carries the thoracic designation. Baron has recommended the application of the thoracic criterion to sampling of fibers such as asbestos (Baron, 1996). In the U.S., the EPA has used a thoracic fraction called PM_{10} since 1986 and, in other countries, this is also being adapted to outdoor air standards for the general population (Kenny and Ogden, 2000).

3. *Respirable*: This mass reflects aerosols that generally pass the nasopharynx and the tracheobronchial tree and are available to the alveoli or gas-exchange region. The RPM-TLVs® are applied to aerosols whose site of action is the gas exchange region of the lung. It should be noted that this fraction, although termed respirable, is not the same as the old respirable fraction. The old fraction had a cut-point of 10 µm while the new fraction has a cut-point of 4 µm. It appears from the TLV® list that the old respirable fractions have been converted to the new designation.

Importantly, the new criteria for size selectivity of ACGIH® parallel criteria of the International Standards Organization (ISO, 1995) and European Standardization Committee (CEN, 1993). However, the European Union Chemical Agents Directive (which writes OELs) has left adoption of the criteria vague (Kenny and Ogden, 2000).

A WORD OF CAUTION ABOUT THE CONVENTIONS

The conventions indicate how much of each particle size should enter the various regions of the respiratory system (see Figure 10-6). The reader should be thinking that these must be average values. Since the penetration of a particle to the respiratory tract depends on the particle's inertia and that depends on momentum ($m \times v$), then the particle's velocity must depend on things like breathing rate and, for inhalability, wind speed and direction. For example, if the wind were blowing into the face of a worker, larger particles would be more likely to enter the mouth than if the worker were perpendicular to the wind direction, requiring the particle to turn 90° to enter the mouth. On the other hand, once the particle enters the respiratory system, wind speed and direction have no bearing on the penetration to the lungs, but penetration to different regions of the lung is dependent on work and breathing rates.

Thus, the inhalable convention was determined by sampling aspirated particles from behind the mouth of a mannikin (nose versus mouth breathing gave about the same results). The wind speeds were varied from 1-4 m/s (200-800 f/min). The inhalable values in Figure 10-6 are the averaged results for these wind speeds. In addition, the mannikin was rotated through 360° thus averaging out orientation, based on the idea that workers move about. Finally, the breathing rate was sinusoidal and at a rate of 20 l/min (light to moderate work, see Chapter 6, Table 6-2). For more details see Vincent (1999). As can be seen in Figure 10-7, the inhalable convention fits for these wind speeds but underestimates penetration of large particles when higher, outdoor-like wind speeds occur. In addition, the convention makes no promises about particles larger than 100 μm, that is simply where the data stop.

Since the adoption of the convention, work has been performed to determine typical wind speeds in workplaces. They rarely exceed 0.3 and are typically < 0.1 m/s or about 20 f/min; this is near the lower range of most hand-held anemometers (Baldwin and Maynard, 1998). This is classified as low air movement by aerosol scientists. Creating a relatively uniform particle distribution in such an atmosphere and determining its concentration is difficult. Nonetheless, elegant systems have been designed to determine how much aerosol this same mannikin might aspirate under conditions of low air movement. Figure 10-8 shows a proposed inhalable convention (Aitken et al., 1999) for low air movement compared to the current convention. Given that people working outside can experience wind speeds (and therefore inhale very large particles from the ensuing blowing of materials) that exceed the original tested wind speeds, it has been proposed that the inhalable convention actually be a series of curves for varying conditions (Kenny, 2000). For example, a 36 km/hr wind (about 20 mph) would be 10 m/s.

Particulate Not Otherwise Specified and Nuisance Dust

The majority of dusts produced in the workplace can be expected to have a direct effect upon the respiratory tract and/or be absorbed across the respiratory tract and cause systemic effect at another organ site. Historically, the *TLVs® and BEIs®* Book contained a published list

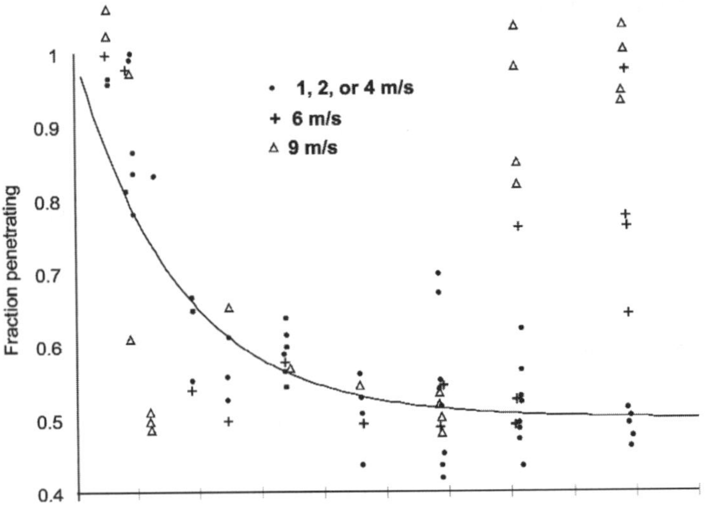

Figure 10-7. Calculated conventional IPM curve (line) and experimental data for 5 wind speeds (1-9 m/s, from Vincent, 1999, with permission).

of *nuisance dusts*. These dusts were thought to be a nuisance to the worker, that is, to get into the eyes, ears, and nose, cake up on the skin, and reduce visibility (ACGIH®, 1987). The nuisance dust list was last published in 1987. Subsequently, ACGIH® has incorporated most of the old *nuisance* dust listing into the body of the TLV® listing and also developed a new category, particulate not otherwise specified (PNOS). The current *TLVs® and BEIs®* Book (2007) states that PNOSs: (1) do not have a TLV®, (2) are insoluble or poorly soluble in water (or lung fluid), and (3) have low toxicity. ACGIH® also states that it believes that even inert and insoluble particles may have toxicity. Therefore even particulates that meet the above three criteria should be controlled to *guideline*[9] levels.

These factors give an excellent example of the need to have adequate training in using TLVs®. Unfortunately, there is a tendency for some users of TLVs®, including hygienists, to assume that non-listed particulates are PNOS, when, in fact, the 10 mg/m³ *guideline* for inhalable dust (or 3 mg/m³ *guideline* for respirable dust) may not be sufficient. The fact that a particulate does not have a TLV® does not mean that it is automatically a PNOS. Instead, it must meet the above three conditions. Interpreting these conditions is not easy. Unevaluated PNOSs will have various particle size distributions and, depending on those distributions and the site of action, either the 3 mg/m³ guideline for respirable dust or 10 mg/m³ guideline for inhalable dust will have to be applied. In short, hygienists should be extremely reluctant to apply the ACGIH® guideline for PNOSs. But the alterative is to find data to support an OEL or derive

[9] In the preface to the *TLVs® and BEIs®* Book, it is stated repeatedly that TLVs® are guidelines, not standards. Yet for the PNOS it is stated that the value of 10 mg/m³ for inhalable dust is a guideline, not a TLV®. This seems confusing.

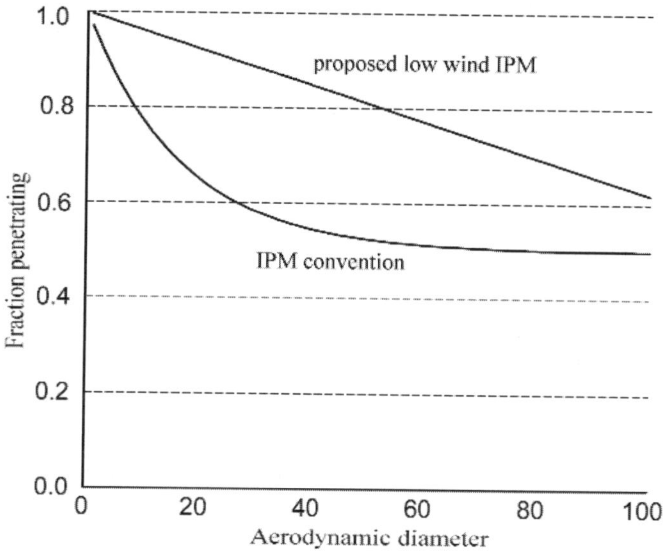

Figure 10-8. The inhalable (IPM) mass fraction sampling convention based on aerosol penetration to the respiratory tract at wind speeds of 0.5-4 m/s and a proposed convention for more typical workplace wind speeds of ≤0.1 m/s.

the data, something that not all companies are willing to support.

An airborne particulate concentration of 10 mg/m³ is generally an extremely *dusty* situation, which most hygienists would recognize as an uncontrolled situation. Regardless of the toxicity of the particulate, lack of control that leads to 10 mg/m³ or greater concentrations is simply not good hygiene. Often, simple control mechanisms can be implemented to reduce particulate concentrations that are this high. Where this is not possible, it is recommended that workers wear respiratory protection until the toxicity of the particulate can be fully evaluated.

Mixtures

As often as not, a worker will be exposed to more than one substance at a time in the workplace. How should the exposure guidelines for the various substances be applied? The most important consideration is whether the substances have similar toxicological pathways and end results. For example, two compounds that have the same metabolic pathway (i.e., use the same biochemical means for conversion to less harmful and/or excretable substances) may have eventual effects on differing organ systems. Nevertheless, because the metabolic pathway is the same, the doses of the two substances essentially act additively. Once a threshold is reached, the metabolic pathway may be overloaded, causing the substance to accumulate in the tissues. There are many other possible ways in which two or more substances may act in an additive manner in causing an eventual toxic effect.

Where it is assumed that the two or more substances may cause an additive effect, the following equation may be used to determine compliance with a *mixture TLV®*:

$$\frac{C_1}{TLV_1} + \frac{C_2}{TLV_2} + \ldots \frac{C_n}{TLV_n} \leq 1 \qquad \text{[Equation 10-1]}$$

where C is the airborne concentration of each component (1, 2 . . . *n*) of the mixture, and compliance is assumed when the sum is unity or less.

The *TLVs® and BEIs®* Book has a column – "TLV® Basis" – that can be used as a preliminary guide for grouping chemicals based on similar effects. However, it does not list all effects, only the most important ones, but the list is up-to-date. Another approach for categorizing effects was taken by U.S. OSHA (1979) (also cited in Paustenbach, 1985 and Craig et al., 1999). Paustenbach also lists the 500+ chemicals and their primary effects as categorized by OSHA; however, this approach is now quite out of date.[10]

In some cases, it is relatively easy to determine that two or more compounds will have additive effects. For example, two chlorinated solvents known to affect the liver can be considered additively. Two metallic aerosols that are both known to cause metal fume fever should be considered additively. However, other situations are not so easy to interpret. For example, consider exposure to trichloroethylene (TCE) and carbon monoxide (CO). CO effectively reduces the amount of available oxygen in the blood by combining with hemoglobin to form carboxyhemoglobin; thus, less hemoglobin is available for oxygen transport. TCE at relatively high concentrations is a known heart sensitizer. It causes the cardiac tissue to be more sensitive to the arrythmogenic effects of endogenous epinephrine and epinephrine-like drugs (Balazs et al., 1986). Hence, one substance affects the heart muscle, the other the cardiovascular system. Could a reduction of oxygen to the heart and other organs be further affected by irregular heartbeats or vice versa? Are these substances additive in their toxic effect? In situations such as this, the hygienist should err on the safe side in assuming that the answer is *yes*.

Equation 10-1 is based on the potential for additive effects, but the exact relationship for combining the effects of the two toxins can be quite variable. For example, the two toxins may interact in a multiplicative manner. Assume that individual doses of 5 units of substance A or 50 units of substance B cause a response of one unit. If the two doses are administered together, the additive response would be two units. However, in a multiplicative system, the effect caused by the two doses would be greater than two. This is called synergism. Ethanol and carbon tetrachloride both affect the liver, but together they act synergistically. The effect of

---

[10] The tables were originally used by OSHA to adjust for long work schedules. It no longer makes adjustments in all cases, except where the lead standard specifically allows (OSHA, 1997). See also the Unusual Work Schedule section later in this chapter.

administering doses of two chemicals may also be antagonistic; in other words, in the above example, the effective dose may be less than two units.

In addition to the potential for additive effects caused by the exposure to two or more substances, there is also the possibility of additive effects caused when a worker is exposed to a single substance and a physical agent, the most common and important such agents being heat stress and noise. However, radiation, abnormal pressure such as that at high altitude, and even psychological stress may reduce the body's normal defense mechanisms to chemical substances. Furthermore, individual workers may be suffering from certain physiological or physical stresses, such as lack of sleep, which may make them more susceptible to the effects of a toxic substance. It should be obvious that the number of factors that can make an individual more susceptible to a toxic chemical is only limited by the imagination. This is yet another reason for not making direct application of TLVs®, but rather applying a safety factor wherever possible.

Roach (1994) elegantly argued that the approach to mixtures based upon additive similar effects ignores the likelihood that at exposures below the TLV® there is risk. In other words, the TLVs® are not absolute thresholds. Therefore, since a risk is a risk, regardless of the organ system, the risk should be added as in Equation 10-1 for all hazardous agents a worker is exposed to, regardless of similar effect, so as to determine *detriment to health*. However, he further argues that based on the lognormal distribution of tolerance, an exponent could be applied:

$$\left(\frac{C_1}{TLV_1}\right)^s + \left(\frac{C_2}{TLV_2}\right)^s + \ldots \left(\frac{C_n}{TLV_n}\right)^s \leq 1 \qquad \text{[Equation 10-2]}$$

where s is conservatively set to 2 but could be higher in theory.

Carcinogens

Carcinogens are chemicals that are capable of causing cancer by themselves or in conjunction with other chemicals or environmental factors such as diet, viruses, or ultraviolet light. There are many types of cancer that can be caused by environmental chemicals; so this is a very diverse group of chemicals.

Cancer results from a complex process. Unlike many other occupational diseases, cancer involves a period between exposure and disease that can be quite long. In other words, before the disease actually manifests itself and becomes clinically evident, or before changes in cell morphology occur that may signal a later change to cancerous cells, the biochemical process that leads to these changes often will have begun many years earlier, when exposure was first occurring. The interim period between exposure and gross changes in cell morphology, which may be 30 or more years, is called the latency period. This latency period is a major difference between cancer and other chronic diseases such as fibrosis of the lungs, degeneration of the nephrons of the kidneys, or cirrhosis of the liver. For the latter three disease classes, even though clinical signs may not appear for many years after first exposure (sometimes 20 years in the case of asbestosis and silicosis), changes in cell morphology and/or tissue damage usually begin relatively soon after exposure.

Theoretically, a single asbestos fiber could cause a microscopically small region of fibrosis, or a single cadmium ion could destroy a cell of the kidney. However, these areas would be extremely small, and the organs would have considerable capacity remaining. So, even the most sensitive individual could receive some dose of cadmium before a critical level of kidney damage occurred. (Of course, *critical* can be difficult to define.) On the other hand, it is conceptually possible that one molecule of vinyl chloride could affect the DNA of a single liver cell, and that damage could go unrepaired. If the cell multiplies (i.e., does not die), other events occur at a later time (e.g., decay of homeostatic factors, later mutations to the cells, or exposure to other chemicals that promote cancer), this can lead to one of the daughter cells multiplying uncontrollably. This scenario, unlike the one above, implies no threshold of safe exposure. It should be added that there are numerous arguments against this one-hit or no-threshold scenario (e.g., Pitot and Dragan, 1996).

In the mid 1970s, the no-threshold concept for carcinogens became important, and this situation affected the way in which OELs were set. NIOSH (1978), for example, opted for a policy of lowest detectable limit. The smallest concentration that could be analytically detected over the averaging time, using a reliable method, became the recommended OEL. If the chemical could not be detected, then the concentration could not be differentiated from zero. United States government agencies including the Environmental Protection Agency now set OELs based on acceptable risk levels, even though a zero threshold is still one of their assumptions for defining the dose-response curve.

Other countries have taken different approaches that eventually have led to the same endpoint and are reflective of a country's evolution of thought. For example, the German Commission for Investigation of Health Hazards of Chemical Compounds in the Work Area of the German Scientific Society believes that there is not sufficient knowledge to justify assigning OELs for carcinogens. However, *technical guiding concentrations* have been published for suspected carcinogens (Henschler, 1979). Adherence to these values will reduce risk but not completely eliminate it.

ACGIH® recognizes five classes of carcinogens. This approach is not atypical, compared to regulatory or other agencies (see e.g., IARC, 1982). The classes are shown in Figure 10-9. For compounds in the final three categories (A3-A5), cancer is not the basis for the value of the TLV®. Table 10-4 shows that for these latter 3 classes, cancer is never listed as the basis. But the Al and A2 categories generally do indicate carcinogen potential, which may affect the TLV®. There is no consistency in indicating the carcinogen potential. For example, ACGIH® is more likely to have recommended an OEL than not for the A1 and A2 categories (see Table 10-4). Therefore, for A1/A2 compounds for which cancer is listed first as the basis for the TLV® (and a TLV® exists), one might expect a clear link between the TLV® and human and/or animal cancer studies. In other words, the result (TLV®) should reflect the bases for the TLV®

ACGIH® Carcinogen Classifications

TLV-A1:	Confirmed Human Carcinogen: The agent is carcinogenic to humans based on the weight of evidence from epidemiologic studies.
TLV-A2:	Suspected Human Carcinogen: Human data are accepted as adequate in quality but are conflicting or insufficient to classify the agent as a confirmed human carcinogen; OR, the agent is carcinogenic in experimental animals at dose(s), by route(s) of exposure, at site(s), of histologic type(s), or by mechanism(s) considered relevant to worker exposure. The A2 is used primarily when there is limited evidence of carcinogenicity in humans and sufficient evidence of carcinogenicity in experimental animals with relevance to humans.
TLV-A3:	Confirmed Animal Carcinogen with Unknown Relevance to Humans: The agent is carcinogenic in experimental animals at a relatively high dose, by route(s) of administration, at site(s), of histologic type(s), or by mechanism(s) that may not be relevant to worker exposure. Available epidemiologic studies do not confirm an increased risk of cancer in exposed humans. Available evidence does not suggest that the agent is likely to cause cancer in humans except under uncommon or unlikely routes or levels of exposure.
TLV-A4:	Not Classifiable as a Human Carcinogen: Agents which cause concern that they could be carcinogenic for humans but which cannot be assessed conclusively because of a lack of data. *In vitro* or animal studies do not provide indications of carcinogenicity which are sufficient to classify the agent into one of the other categories.
TLV-A5:	Not Suspected as a Human Carcinogen: The agent is not suspected to be a human carcinogen on the basis of properly conducted epidemiologic studies in humans. These studies have sufficiently long follow-up, reliable exposure histories, sufficiently high dose, and adequate statistical power to conclude that exposure to the agent does not convey a significant risk of cancer to humans; OR, the evidence suggesting a lack of carcinogenicity in experimental animals is supported by mechanistic data.

Figure 10-9. The ACGIH® categories for carcinogenicity for chemical substances.

Table 10-4
Number ACGIH® TLV® entries with a TLV® and classified as carcinogen (number with no TLV®) based on 2004 TLVs®

	ACGIH® Carcinogen Classification				
Cancer listed as basis?	**A1**	**A2**	**A3**	**A4**	**A5**
No	1	2	93 (6)	164 (2)	3
Listed, but not first	1	9			
Listed first	11 (3)	7 (12)			
Total	16	30	99	166	3

(the research data). However, this is difficult to determine, at least for the following chemicals: beryllium, bis-chloromethyl ether, chromium, coal tar pitch, nickel, wood dust, butadiene, ethylene oxide, and chromates of zinc, calcium, strontium, and lead.

It is recognized that it takes both an adequate exposure and hazard assessment to make a statement about risk or dose response. However to state, for example, that the exposure data are not adequate for doing so, and then to provide a TLV® anyway, that has no clear derivation, is not fair to the user. On the other hand, the TLV® committee has made important strides in an effort to get more dose response and risk information to the user as *Documentation* for asbestos, benzene, dichlorobutene, vinyl chloride, ethylene oxide, and chromium compounds indicate (ACGIH®, 2007). Given that the committee is composed of volunteers and our thoughts on cancer continue to evolve, it may be some time before this process is completed.

Unusual Work Schedules

For reasons of convenience, sample averaging periods for associated OELs of 8 hours and 15 minutes have been adopted for most compounds. Ideally, however, the length of time over which exposures are monitored, for purposes of comparison to an OEL, should be dependent upon the half-life of the chemical agent in the body and the actual length of the worker's working day. If the worker works for 8 hours, it is then assumed that the worker has at least 16 hours for the body to excrete and metabolize the absorbed compounds. During the course of the workday, the worker is absorbing compound and excreting it. So the body burden will build up over the work day. However, the goal of the OEL should be not to allow the body burden to exceed that body burden which is associated with an unacceptable risk of adverse affect. If the worker works a shift longer than 8 hours, or works for 8 hours and is off for only 8 hours and then works another 8 hours, it is obvious that the worker's body may not have sufficient time for the compound to be metabolized and excreted.

Any kind of work regimen other than 8 hours on and 16 hours off, five days per week, is considered to be an unusual workshift. Unusual workshifts increase the risk that the worker's body burden of the compound will become unacceptable. Several researchers have derived equations for adjusting OELs for unusual work schedules that usually require knowledge of the biological half-life and other variables (Paustenbach, 2003 and Verma, 2000). These tend to be difficult to use, mainly because they require input data (e.g., $T_{0.5}$) that are sometimes unknown. The *TLVs® and BEIs®* Book discusses the use of two procedures. A simplistic yet adequate approach is given by Brief and Scala (1975). The technique of Brief and Scala, in addition to being the easiest to use, is also the most conservative of all models. The Brief and Scala reduction factor (RF) for the OEL is

$$RF = \frac{8}{h_d} \times \frac{24-h}{16} \qquad \text{[Equation 10-3]}$$

for a 5-day work week where h_d is the length of the shift in hours, and

$$RF = \frac{40}{h_w} \times \frac{168-h_w}{128} \qquad \text{[Equation 10-4]}$$

for a seven day work week, where h_w is the number of hours exposed per week. These equations are reliable and conservative. They are applicable for compounds with half-lives of $4 < T_{0.5} < 400$ hours. Outside this range no adjustments are necessary.

The second model discussed by ACGIH® is the Quebec model (Brodeur, 2001). It tends to give values nearer the pharmacokinetic based models (i.e., based on $T_{0.5}$). The authors have placed compounds into categories based on half-life (see Table 10-5) and compiled a table of adjustment factors (F_A) that depend on category and shift length (see Appendix 10-1). The F_A are used to adjust the TLV® for compounds with short term effects by the formula

$$F_A = \frac{8}{H_d} \qquad \text{[Equation 10-5]}$$

and for compounds with long term effects by the formula

$$F_A = \frac{40}{H_w} \qquad \text{[Equation 10-6]}$$

where H_w and H_d are the number of hours worked in a week and a shift or day, respectively. To use the formulas, first find the adjustment category (AC) for the compound in Appendix 10-1. Then the AC is used to determine which of the above two equations are used to calculate the F_A to be used to adjust the TLV®. Note: adjusted TLVs® should never be lower than their original counterparts; when shifts are complicated as in 8 hours one day and 10 the next, always use the most conservative adjustment factor; and the adjusted TLV® is compared to the TWA

Table 10-5
Adjustment Categories (AC) for unusual work schedules and equations 10-5 and 10-6

AC	Definitions	Adjustment
I-a	Substances with a ceiling value	
I-b	Irritating or foul-smelling substances	
I-c	Simple asphyxiants; substances that are a safety hazard; very weak health hazard and whose half-life is less than 4 hours. Substances for which the TLV® is based on technological limitations	none
II	Substances exerting ill-effects after short-term exposure	day-based adjustment (Equation 10-5)
III	Substances exerting ill-effects after long-term exposure	week-based adjustment (Equation 10-6)
IV	Substances exerting ill-effects after either short- or long-term exposure	Use Equations 10-5 and 10-6 and apply the more severe of the two

Adapted from Brodeur et al., 2001. For a calculation tool go to http://www.irsst.qc.ca/en/ut_vema.htm.

for the entire shift regardless of length.

Notice of Intended Changes

When the TLV® committee proposes to make additions, deletions, or changes to the *TLVs® and BEIs®* Book, these are placed in a section entitled "Notice of Intended Changes," after a fairly lengthy process that is detailed in Figure 10-10. The changes remain in that section for a period of two years, during which the TLV® committee is open to comments from the public regarding whether the changes are appropriate. During this time, the *Documentation* upon which the changes are proposed is available from the TLV® committee. Any hygienist who has concerns with compounds for which changes are recommended should obtain the *Documentation* in order to determine the foundation for the changes. On reviewing the *Documentation*, the hygienist should make his or her own judgment as to whether the changes are appropriate; in particular, the hygienist should consider whether reduction in the TLV® level has been made to the extent necessary in light of newly available information. It is advisable that the hygienist make arrangements to follow the new TLV® guidelines prior to the end

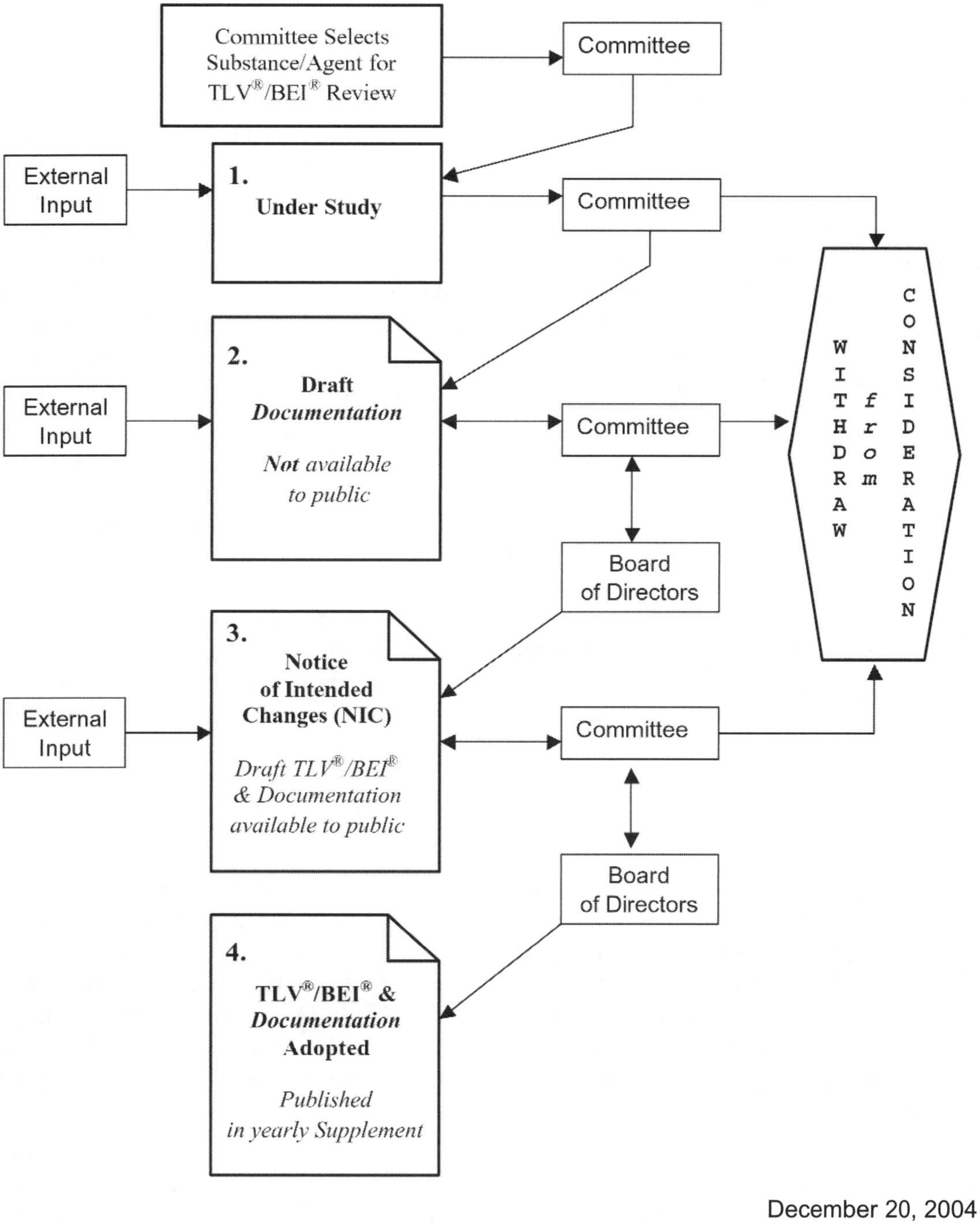

Figure 10-10. The TLV®/BEI® Development Process Flow Chart.

of the two year period.

Unusual Ambient Conditions

The effect of temperature and pressure or air density on concentration was discussed in Chapter 8. The TLVs® for gases and vapors can be expressed in ppm by volume at NTP or as mass per volume (mg/m³). Remember that as density changes, ppm concentration does not, but mass per volume does. The *TLVs® and BEIs®* Book instructs the user to compare all measured concentrations (aerosols and gases) expressed in mass/volume directly with the airborne concentration; and the volume in the denominator of the latter should not be adjusted when sampling conditions are not NTP. In other words, make all comparisons to the TLV® expressed as mass/volume. Of course, most hygienists do the opposite – compare measured concentrations to TLVs® or other OELs expressed as ppm without making corrections to volumes collected at non NTP conditions (see also Stephenson and Lillquist, 2001).

The rationale for the ACGIH® guidance is as follows. Workers are acclimatized. That is, they breathe about the same volume each day whether under a high pressure or a low pressure weather system, or whether the temperature is 10°C or 30°C. This may or may not be true. A construction worker, having worked all winter in Minnesota, may breathe more rapidly when he encounters the odd heat wave in May that drives temperatures to 100°F (39 °C). However, the difference in breathing rates would be quite small. Similarly, the hygienist that chooses to make comparisons of uncorrected ppm concentrations to TLVs® expressed in ppm does not make a large error. For example, consider that the 100°F day also is coupled with low pressure. Both cause the air to be less dense – less O_2 per liter. How much less? Let's assume that the pressure is 29 in Hg, relatively low. Then,

$$\frac{29}{29.92} \times \frac{70+460}{100+460} = 0.92$$

and the air is 92% of the density of air at NTP. We might assume that the unacclimatized worker breathes 1/0.92 harder on this day. However, TLVs® are not so accurate that we should take this small difference into consideration, particularly for a day or two. As we will see later, sampling and analytical methods have accuracies of $> \pm$ 10%. Making a correction of the volume for a ppm comparison or adjusting a TLV® for the worker in Minnesota is a bit like carrying out a calculation to 3 significant figures when the values in the calculation have only two significant figures. Therefore, for all but the most extreme temperature and pressure conditions, no adjustment to the volume is necessary.

For a more extreme condition, consider the miner in Chile or the western U.S. who works in a fly in and fly out operation at 11,000 ft (3,300 m). After two weeks on the job, he flies home to sea level. This worker is not acclimatized and will breathe faster at elevation. His dose will be larger compared to the same worker, doing the same job, at sea level. Therefore, the TLV® should be adjusted downward for these situations (Dümmer, 1998). Finally, pumps used to collect air samples that are calibrated in the lab will respond differently at non NTP (outside the laboratory) conditions, depending on the type of pump. This correction factor is discussed in Chapter 20.

Unlisted Substances

The temptation to place unlisted particulates in the PNOS dust category is analogous to the situation for other compounds. There are approximately 700 TLVs®, and if one counted all of the exposure guidelines that have been produced by various worldwide governmental agencies and volunteer agencies, perhaps 1,200 exposure guidelines would be found (Cook, 1987). However, with 80,000+ compounds in U.S. commerce, there is obviously a great number of compounds for which no TLV® or other exposure guideline exists. When no exposure guideline can be found for a substance, the hygienist should not make the assumption that the substance is not toxicologically important. The only solution is to revert to pre-TLV® practices, by doing a thorough literature search and attempting to derive an exposure guideline based on animal or human toxicological information and/or analogy. In some cases, original research may need to be performed, either on laboratory animals or epidemiological studies of the workplace, or both. Once a tentative limit is set, it should be constantly tested with environmental and medical monitoring.

Substances of Variable Composition

Previously the TLV® committee recognized two processes that release by-products to the workplace air and are of variable chemical composition; polytetrafluoroethylene decomposition products and welding fume. Polytetrafluoroethylene or PTFE is a type of fluorocarbon that, when heated or oxidized, can decompose into a number of different products. It is recognized that the decomposition products can elicit a toxic effect in the lungs that can be quite debilitating (NIOSH, 1977). However, there appeared to be insufficient information to propose any TLV®. For welding fumes-total particulate (NOC), which was previously given a TLV® of 5 mg/m^3, it is best to monitor the components of the fume and apply a mixture TLV® where appropriate.

INTERPRETING THE TLVs®

A More Rational Basis

For the TLVs®, acceptable risk levels have been stated qualitatively (for an exception, see the *Documentation* for benzene in *Documentation of the TLVs® and BEIs®* (ACGIH®, 1997). Although the *nearly all workers* phrase makes the enlightened hygienist aware that some risk is implied with each TLV®, the lack of a definition for the word *nearly* is less than optimal. Some may interpret this to mean 1 in 1,000 workers will suffer; others will interpret it to mean

5 in 100.

TLVs® and many other standards are based on the concept of thresholds. Remember this involves an extrapolation downward from a NOEL using uncertainty factors. Previously, the risks associated with exposures at these OELs was assumed by many to be zero. In the case of the TLVs®, it was assumed to be non-zero but it was not quantitated. However, a NOEL is an observation that has error. This error is dependent on sample size (statistical sampling error as in Chapter 7) and the shape of the dose response curve. Thus NOELs, assumed zero response rates, actually could represent true response rates that are as high as 3–21% depending on these factors[11] (Leisenring and Ryan, 1992). When we divide by uncertainty factors we reduce these possible response rates. However, these uncertainty factors do not inform the user of the likely residual risk that may remain even after uncertainty factors are applied (Vincent, 1998).

Recent work (Jayjock et al., 2001) has produced an elegant way to quantitate the risk associated with a TLV®, or for that matter any OEL or other standard that is based on the threshold concept (e.g., the U.S. EPA acceptable tolerance or food intake for a pesticide). Jayjock et al. have suggested several approaches to estimating the residual risk at a NOEL-based OEL (see Figure 10-11). First the data from an experiment are modeled. Software is available for this (see e.g., EPA, 2005). These models are not new; they have been used by toxicologists for this kind of data for decades. As with any mathematical model, its 95% (or some other value) upper confidence limit (UCL) can be calculated. (Again, the software often does this.) These approaches give the standard setting body a better idea of the appropriateness of the OEL and give the user a better level of confidence.

Use and Misuse

TLVs® have been and continue to be widely used by hygienists. If applied wisely, knowledgeably, and with adequate safety factors, they are quite useful. No doubt as the *Documentation* process continues to improve, the utility of TLVs® will also increase. This usefulness is very much a function of the manner in which the hygienist approaches the use of TLVs®. To do so haphazardly invites problems. The hygienist should carefully read the preface to the *TLVs® and BEIs®* Book before using it, and should refer to that preface often. The preface states that the TLVs® are guidelines or recommendations and are not fine lines between safe and dangerous concentrations. The hygienist must be convinced that the TLVs® or any OELs are not magical numbers, absolute values of safety, which can be applied without thought. If this were the case, then university training for hygienists would not be appropriate; anyone with minimal training can compare a number to a number. Most importantly, the user

[11] Think of running an exposure experiment with 20 animals at a given dose and the true risk is 1 in 20. The first time you may observe no effects (0 in 20). Then run the experiment over and over. If it is run enough, the average risk will be 1 in 20, but the outcome in any one experiment could range from 0 to perhaps 4 or 5 or more, depending on the number of experiments and other factors.

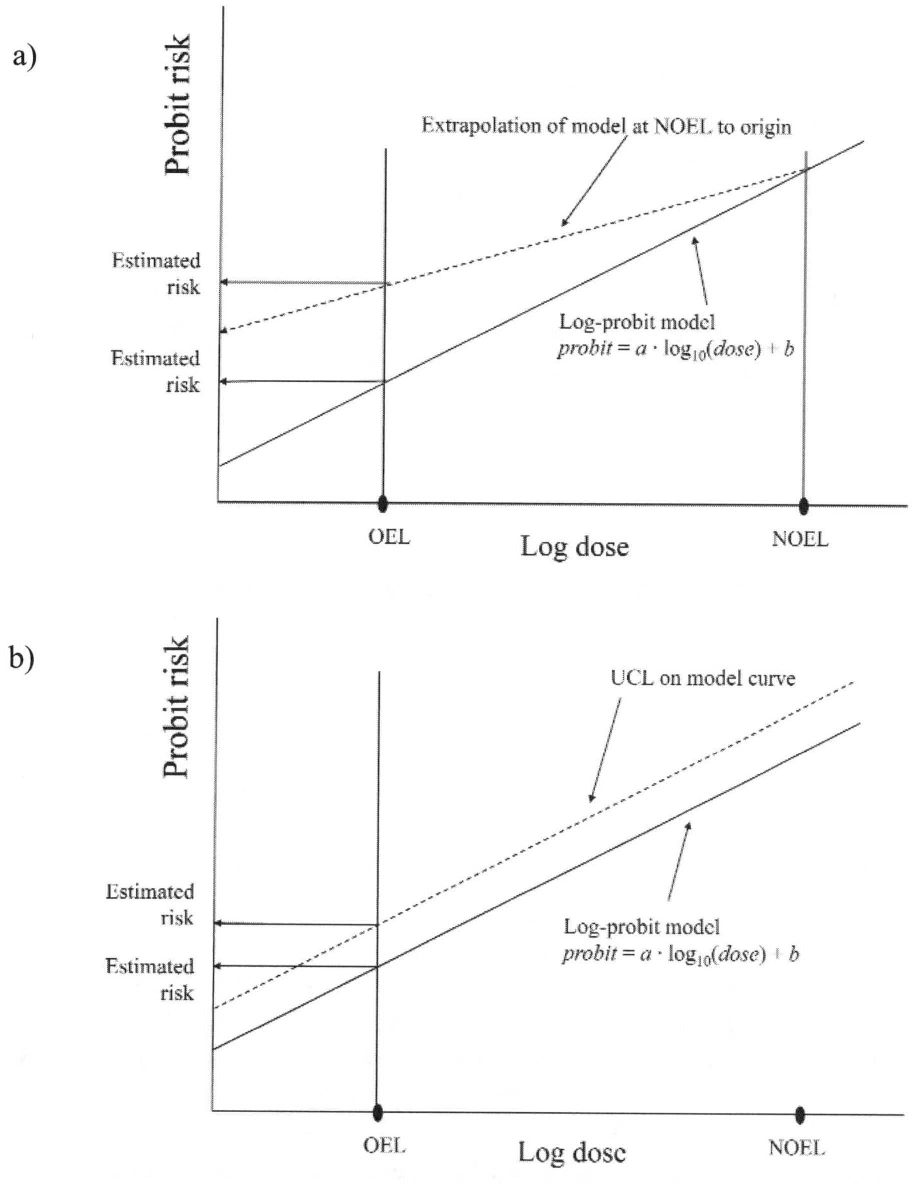

Figure 10-11. a) Log-probit dose response model fitted to animal data showing the NOEL and a proposed OEL based on dividing the NOEL by uncertainty factors. The estimated risk at the OEL can be extrapolated directly from the model or from a line that is extrapolated from the NOEL dose on the model curve back to the origin (0,0). The latter gives the more conservative estimate of risk. b) Same as above except an upper confidence limit line for the modeled relationship (it might be the 90th, 95th, 99th, etc.) is calculated and the estimated risk is extrapolated from this line giving a more conservative estimate of risk.

should read the *Documentation* and not rely solely on the number, i.e., the conclusions of the committee.

The policy statement in the *TLVs® and BEIs®* Book states that TLVs® are not to be used as legal standards, without full compliance with applicable regulatory procedures. The ACGIH® does not advocate their use as such; but as scientific opinions. However, as will be seen in the next chapter, they have been adopted by the U.S. government and governments throughout the world, and in many cases *full compliance with applicable regulatory procedures* does not amount to more than a government edict saying the TLVs® are now law. Unfortunately, this usage has given them a standing beyond guidelines. A legal standard cannot be a guideline; it is a thin blue line; an employer who crosses it is in violation and thus subject to penalty. This unfortunate level of acceptance further promotes the myth that TLVs® are *safe levels of exposure*. After all, if a government body adopts a particular number as a standard, doesn't it imply to most people that levels below the standard are always safe, and those above are always unsafe?

TLVs® should not to be used as guidelines for controlling atmospheres other than the normal workplace environment. These include communities in which air pollution exists, and unusual workplace exposure conditions such as unusually long shifts, or even continuous exposure such as might be experienced in a two weeks on, two weeks off work regimen. However, they are now widely used to set community air pollution standards for communities near hazardous waste sites, for indoor air quality in homes, and other circumstances. States often use them as such by dividing by 4.2 (the ratio of a 40-hour work week to a full 7 days or 168 hours) and then applying safety factors of 10 × 10 (Paustenbach, 1997). The TLVs® are divided by 420 regardless of the biological basis for the TLV®, based on the realities that full time exposure allows no time for clearance and the considerably greater diversity of sensitivities in the general population compared to the relatively healthy working population.

The *TLVs® and BEIs®* Book states that TLVs® should not be used by anyone untrained in the discipline of industrial hygiene. This implies that hygienists and similarly trained occupational health professionals should have an understanding of the limitations of TLVs®. This should include the knowledge that TLVs® must be adjusted downward based on exposure variability in the workplace, adequacy of the *Documentation*, and the nature of the workforce, among other factors. As we will see in subsequent chapters, there is a rational basis for controlling workplace exposure levels to approximately one-fourth of the TLV® or less.

Finally, although not common, a TLV® may be set at a level which is not measurable, nor technologically or economially feasible. They are one expression of scientific opinion. They are guidelines.

Legal Challenges[12]

In recent years, several lawsuits have been filed seeking to enjoin ACGIH® from proposing, adopting, or publishing TLVs® for various chemical agents including 1-bromopropane, copper, silica, sodium sesquicarbonate (Trona), and synthetic vitreous fibers (specifically refactory ceramic fibers). The claims in these suits have included allegations that ACGIH® is either a government agency or functioning as a federal advisory committee; that the TLVs® are not based on good science; and that ACGIH® has an obligation to follow the requirements of the Federal Administrative Procedures Act.

During the course of these lawsuits, the courts have recognized that ACGIH® is not a federal agency; that ACGIH® is not a federal advisory committee; that ACGIH® does not have to follow the Federal Administrative Procedures Act; and that ACGIH® has a First Amendment right to publish its scientific opinions. During the time when some of these lawsuits were pending, the U.S. House of Representatives Subcommittee on Workplace Protections held hearings questioning OSHA's use of ACGIH® TLVs® as part of the regulatory process. Questions were asked regarding whether ACGIH® TLVs® are standards and whether ACGIH® follows the consensus standard making process.

Although it is ACGIH®'s policy not to comment on legislation pending before the Congress or regulations proposed or pending before federal or state regulatory agencies, ACGIH® did testify before the Subcommittee on Workplace Protections explaining that the ACGIH® TLVs® are designed not as standards, but as guidelines for use by industrial hygienists. ACGIH® also prepared a position paper differentiating the TLV® guidelines from regulatory standards. A copy of the position paper was sent to OSHA.

In its testimony before the Congress and in its position paper regarding the purpose of the TLVs® and BEIs®, ACGIH® has reached out to provide a comprehensive explanation of the proper use of TLVs® and BEIs® and the procedures that are followed in developing TLVs® and BEIs®. Hopefully, such action will result in a better understanding of how TLVs® and BEIs® can be used to promote safety in the workplace. To the extent that ACGIH® has been able to better explain what it does and how it does it, the organization is stronger. At the same time, defending any association in complex litigation is both time consuming and costly. When a non-profit professional society such as ACGIH® has to defend itself against charges made by entities with much greater resources, the non-profit organization is vulnerable. However, the organization has an obligation to defend its position. ACGIH® will continue to defend its right to publish TLVs® and BEIs® based on the best science available.

[12] This synopsis is taken from many articles that have appeared in issues of the *ACGIH® Today!* newsletter (www.acgih.org).

OTHER SOURCES FOR OELs

Although the ACGIH® TLVs® are used widely throughout the world, many governments have set unique OELs for some compounds and used the TLVs® to augment them. For example, in the United States, OSHA has set OELs for 12 compounds (29 CFR 1910.1001–1050) and adapted TLVs® for many others.

NIOSH has published more than 80 criteria documents since its inception. These have a chapter entitled "Basis for the Standard" that fully describes the standard-setting rationale. Unfortunately, most were published prior to 1979; so they are outdated and do not contain risk assessments.

Finally, other governments, such as those of Great Britain (Clark and Lewis, 1996), Germany, Australia, and Canada (Vincent, 1998) and the European Union through its Scientific Committee for Occupational Exposure Limits (SCOEL) publish OELs, which are reviewed in the next chapter. See Cook (1987) for a complete listing of OELs. A more up to date listing for 31 countries can be found in *Documentation of the TLVs® and BEIs® with Other Worldwide Occupational Exposure Values CD-ROM – 2007*, ACGIH®, Publication #0107CD.

QUESTIONS

10.1. Using the most recent *Documentation* for the TLVs® that you can find,[13] choose two compounds and review the *Documentation*. In a few sentences for each compound, describe the health effects upon which the TLV® is based and the associated level of risk at the TLV® or the apparent no observable effects level (NOEL) and uncertainty factors that were used to derive the TLV®. In other words, can you determine how many workers are expected to suffer the effect when exposed at the TLV® or can you determine that the TLV® is set at a NOEL in humans or animals or if uncertainty factors are applied? Describe any difficulties that you might have.

10.2. Presume that the 8-hr TWA-TLV® for a compound is 10 ppm, and this sets a risk level of 1 in 100 for persons working 8 hours a day, 5 days a week, for 45 years. Assume also that the relationship of years exposed at the TLV® (i.e., dose) is linear with effect. In other words, if a group of 200 workers is exposed for 22.5 years, one worker would be expected to contract the disease, on average. Based on the following distribution of exposure years for a cohort of workers, how many cases of the disease would you expect to find?

[13] The complete set of *Documentation* is about $1000. However, members of ACGIH® are entitled to five free *Documentation* downloads per year.

Years exposed	Number of workers
1	28
2	13
3	14
4	16
5–9	162
10–14	111
14–19	268
20–29	138
30–39	74
40–44	11
> 45	3

10.3. Often the rate of excretion (the leak rate in the bucket) can be modeled by an exponential decay equation such as

$$m_t = m_0 e^{-T_{0.5}k}$$

where m_t is the mass or dose remaining in the body after time t, m_o is the initial mass, $T_{0.5}$ is the half-time, and k is an empirical elimination constant. If the half-time is the time at which $m_t/m_o = 0.5$, derive a general equation in its simplest form for the $T_{0.5}$ that can be used for any compound.

10.4. Some compounds have both a TLV®-TWA and a TLV®-STEL. Assuming that the compound would need to have at least two $T_{0.5}$'s for two different effects, then according to Roach's recommendations, what are the ranges in which those two $T_{0.5}$'s should lie?

10.5. For a liquid mixture of three components in equal concentration the following data are given:

Component	Vapor pressure (mmHg) @ 25°C	TLV® (mg/m^3)
A	100	100
B	10	50
C	1	0.1

Assume that Raoult's law applies to the concentration above the mixture. Calculate a TLV® for the mixture, and assume that you will use a measuring instrument that will not differentiate between the three, but will give the same signal (output) for each compound on a molar basis.

10.6. A monodispersed aerosol having an aerodynamic diameter of 5 μm is present in a workplace. The aerosol causes bronchial carcinoma. In order to mimic the human respiratory system, about what percentage of this aerosol should be collected by a sampler?

10.7. A compound (such as a metal) has a half-life of 3 days. Persons exposed to this compound work 10-hour shifts, 5 days/week. According to Brief and Scala, the TLV® should be reduced by what factor? According to Roach, the averaging time should be 9 days ($3T_{0.5}$). Based on this information, what is wrong with the Brief and Scala approach?

10.8. Three chemicals having additive effects, hexane, octane, and heptane, have TLVs® of 50, 100, and 400 ppm, respectively. In the workplace, ppm concentrations of 25, 75, and 10, respectively, are found. Has the additive TLV® for the mixture been exceeded? Should you use the ppm or mg/m^3 concentration units?

REFERENCES

ACGIH®: Documentation of the Threshold Limit Values and Biological Exposure Indices with other Worldwide Occupational Exposure Values, American Conference of Governmental Industrial Hygienists, Cincinnati, OH (2007).

ACGIH®: Threshold Limit Values and Biological Exposure Indices for 1987-88, American Conference of Governmental Industrial Hygienists, Cincinnati, OH (1987).

ACGIH®: 2007 Threshold Limit Values for Chemical Substances and Physical Agents and Biological Exposure Indices, ACGIH®, Cincinnati, OH (2007).

Aitken, R.J.; Baldwin, P.E.J.; Beaumont, G.C.; Kenny, L.C.; Maynard, A.D.: Aerosol inhalability in low air movement environments. J. Aerosol Sci. 30(5):613-626 (1999).

Balazs, T., Hanig, J.P., Herman, E.H.: Toxic responses of the cardiovascular system, in Casarett and Doull's Toxicology, 3rd ed., C.D. Klaassen, M.O. Amdur, and J. Doull, (eds.). Macmillan, New York (1986).

Baldwin, P.E.J.; Maynard, D.: A survey of wind speeds in indoor workplaces. Ann. Occup. Hyg. 42(5):303-313 (1998).

Baron, P.A.: Application of the thoracic sampling definition to fiber measurement. Am. Ind. Hyg. Assoc. J. 57:820-824 (1996).

Brief, R.S., Scala, R.A.: Occupational exposure limits for novel work schedules. Am. Ind. Hyg. Assoc. J. 76:467-476 (1975).

Brodeur, J., Vyskocil, A., Tardif, R., Perrault, G., Drolet, D., Truchon, G., Lemay, F.: Adjustment of permissible exposure values to unusual work schedules. Am. Ind. Hyg. Assoc. J. 62:584-594 (2001).

Buringh, E., Lanting, R.: Exposure variability in the workplace: Its implication for the assessment of compliance. Am. Ind. Hyg. Assoc. J. 52:6-13 (1991).

CEN: Workplace atmospheres: Size Fraction Definitions for Measurement of Airborne Particles in the Workplace. EN 481. European Standardization Committee, Brussels, Belgium (1993).

Christiani, D.C.D. Kriebel, Fox, N.J., Baker, E.L.: Persistently elevated polychlorinated biphenyl levels from residual contamination of workplace surfaces. Am. J. Ind. Med. 10:143-151 (1986).

Clark, K., Lewis, G.: An introduction to U.K. health and safety legislation. The Synergist, October (1996).

Cook, W.A.: Occupational Exposure Limits-Worldwide. American Industrial Hygiene Association, Fairfax, VA (1987).

Cook, W.A.: Maximum allowable concentrations of industrial atmospheric contaminants. Ind. Med. Ind. Hyg. Suppl. 6:936-946 (1945). Also as Appendix C, pp. 235-258 in C.D. Bridges: Job Placement of the Physically Handicapped. McGraw-Hill, New York (1946).

Craig, D.K., Baskett, R.L., Davis, J.S., Dukes, L., Hansen, D.J., Petrocchi, A.J., Powell, T.J., Sutherland, P.J., Tuccinardi, Jr., T.E.: Recommended default methodology for analysis of airborne exposures to mixtures of chemicals in emergencies. Appl. Occup. Env. Hyg. 14(9):609- 617 (1999).

Daston, G.P.: Do thresholds exist for developmental toxicants? A review of the theoretical and experimental evidence, in Issues and Review in Teratology, H. Kalter (ed), Plenum Press, New York (1993).

de Cock, J., Heederik, D., Kromhout, H., Boleij, J.S.M.: Short communication, strategy for assigning a "skin notation": a comment. Ann. Occup. Hyg. 40(5):611- 614 (1996).

Dümmer, W.: Prevention of Occupational Hazards at High Altitudes, in Encyclopaedia of Occupational Health and Safety, Stellman, J.M. (ed.), International Labour Office, Geneva, Switzerland (1998).

EPA: Benchmark Dose Software, National Center for Environmental Assessment, U.S. Environmental Protection Agency, Washington, DC (2005). http://cfpub.epa.gov/ncea/cfm/recordisplay.cfm?deid=20167

Esmen, N.A., Johnson, D.L., Agron, G.M.: The variability of delivered dose of aerosols with the same respirable concentration but different size distributions. Ann. Occup. Hyg. 46(4):401-407 (2002).

Fairhurst, S.: The uncertainty factor in the setting of occupational exposure standards. Ann. Occup. Hyg. 39(3):375-385 (1995).

Fenske, R.A., Van Hemmen, J.J.: Occupational skin exposure to chemical substances: Setting limits. Ann. Occup. Hyg. 38:333-336 (1994).

Francis, M.S. Selvin, Spear, R., Rappaport, S.: The effect of autocorrelation on the estimation of workers' daily exposures. Am. Ind. Hyg. Assoc. J. 50:37-43 (1989).

Gough, M.: Zero risk or acceptable risk. Am. Ind. Hyg. Assoc. J. 52:556-560 (1991).

Hamilton, A.: Forty years in the poisonous trades. Am. Ind. Hyg. Assoc. J. 9:5-17 (1948).

Henschler, D.: Maximum allowable concentration values in the Federal Republic of Germany, pp. 83-86, in Trans. XLI Annual Meeting, American Conference of Governmental Industrial Hygienists, Chicago, IL (1979).

Hewitt, P.; McCawley, M.A.: A microcomputer spreadsheet technique for analyzing multimodal particle size distributions. Appl. Occup. Env. Hyg. 6:865-873 (1991).

HSE: Nanoparticles: An occupational hygiene review, R.J. Aitken, K.S. Creely, C.L. Tran, UK Health and Safety Executive Report 274 (2004). http://www.hse.gov.uk/research/rrpdf/rr274.pdf

IARC Monographs: Evaluation of the Carcinogenic Risk of Chemicals to Humans, Suppl. 4. World Health Organization, Lyon, France (1982).

ILO: Occupational Exposure Limits for Airborne Toxic Substances, 2nd (revised) ed., Occupational Safety and Health Series No. 376. International Labor Office, Geneva, Switzerland (1980).

ISO: Air Quality Particle Size Fraction Definitions for Health-Related Sampling. Standard 7708. International Organization for Standardization. Geneva, Switzerland (1995).

Jayjock, M.A., Lewis, P.G., Lynch, J.R.: Quantitative level of protection offered to workers by ACGIH® threshold limit values occupational exposure limits. Am. Ind. Hyg. Assoc. J. 62 (2001).

Johnson, D.L.; Esmen, N.A.: Method-induced Misclassification for a Respirable Dust Sampled Using ISO/ACGIH/CEN Criteria. Ann. Occup. Hyg. 48:13-20 (2004).

Kenny, L.C.: The international conventions for health-related sampling of aerosals – A Review of Current Status and Future Evolution. Appl. Occup. Env. Hyg. 15(1): 68- 71 (2000).

Kenny, L.C., Ogden, T.L.: Twenty-five years of inhalable dust. Am. Ind. Hyg. Assoc. J. 44(8):561- 563 (2000).

Key, M.M.: Health standards and standard selling in the United States. Ann. N. Y. Acad. Sci.. pp. 707-711 (1973).

Leisenring W., Ryan, L.: Statistical properties of the NOAEL, Regul. Toxicol. Pharmacol.15(2 Pt 1):161-71(1992).

Magnuson. H.J.: Soviet and American Standards for Industrial Health. Arch. Environ. Health 10:543-545 (1965).

Ness, S.A.: Surface and Dermal Monitoring for Toxic Exposures. Van Nostrand Reinhold, New York (1994).

NIOSH: A Recommended Standard for Occupational Exposure Acrylonitrile. National Institute for Occupational Safety and Health, 78-116, Cincinnati, OH (1978).

NIOSH: Criteria for a Recommended Standard...Occupational Exposure to Decomposition Products of Fluorocarbon Polymers. National Institute for Occupational Safety and Health, U.S. Government Printing Office, Cincinnati, OH (1977).

OSHA: Field Operations Manual, Occupational Safety and Health Administration, Washington, DC, (1979).

OSHA: Occupational exposure to benzene, final rule, 29 CFR Part 1910. Fed. Reg. 52:34460-34578 (1987).

OSHA: Standard Interpretations, 01/23/1997 – Calculation methods used under the air contaminants standard for extended work shifts. http://www.osha.gov/pls/oshaweb/owadisp.show_document?p_table=INTERPRETATIONS&p_id=22333 (1997).

Paustenbach, D.J.: The history and biological basis for occupational exposure limits, in J.L. Perkins, (ed.) Modern Industrial Hygiene, Volume 2, ACGIH®, Cincinnati, OH, USA (2003).

Paustenbach, D.J.: Occupational exposure limits, pharmacokinetics, and unusual work schedules, in Patty's Industrial Hygiene and Toxicology. Vol. 3a, 2nd ed. J. Cralley and V. Cralley (eds.). Wiley, New York (1985).

Phalen, R.F., Hinds, W.C., John, W., Lioy, P.J., Lippmann, M., McCaley, M.A., Rabbe, O.G., Soderholm, S.C., Stuart, B.O.: Rationale and recommendations for particle size-selective sampling in the workplace. Appl. Ind. Hyg. 1:3-14 (1986).

Pitot, H.C., Dragan, Y.P.: Chemical Carcinogenesis, in Casarett and Doull's Toxicology, 5th edition. C.D. Klassen (ed.), McGraw Hill, New York (1996).

Roach, S.A.: A more rational basis for air sampling programs. Am. Ind. Hyg. Assoc. J. 1-12 (1966).

Roach, S.A.: A most rational basis for air sampling programmes. Ann. Occup. Hyg. 20:65-84 (1977).

Roach, S.A.: On assessment of hazards to health at work. Am. Ind. Hyg. Assoc. J. 55(12):1125- 1130 (1994).

Roach, S.A., Rappaport, S.M.: But they are not thresholds: A critical analysis of the documentation of threshold limit values. Am. J. Ind. Med. 17:727-753 (1990).

Rogers, J.M., Kavlock, R.J.: Developmental Toxicology, in Casarett and Doull's Toxicology, 6th edition. C.D. Klaassen (ed), McGraw-Hill, New York (2001).

Scansetti, G., Piolatto, G., Rubino, G.: Skin notation in the context of workplace exposure standards. Am. J. Ind. Med. 14:725-732 (1988).

Schwartz, J., Laden, F., Zanobetti, A.: The concentration response relationship between $PM_{2.5}$ and daily deaths. Env. Hlth. Pers. 110:1025-1029 (2002).

Soderholm, S.C.: Proposed international conventions for particle size-selective sampling. J. Occup. Hyg. 33:301-320 (1989).

Soderholm, S.C.; McCawley, M.A.: Should dust samplers mimic human lung deposition? Appl. Occup. Environ. Hyg. 5(12):829-833 (1990).

Spear, R.C., Selvin, S., Francis, M.: The influence of averaging time on the distribution of exposures. Am. Ind. Hyg. Assoc. J. 47:365-368 (1986).

Stephenson, D.J., Lillquist, D.R.: The effects of temperature and pressure on airborne exposure concentrations when performing compliance evaluations using ACGIH® TLVs® and OSHA PELs, Appl. Occ. Env. Hyg. 16:482-486 (2001).

Tuggle, R.M.: The relationship between TLV®-TWA compliance and TLV®-STEL compliance. Appl. Occ. Env. Hyg. 15:380-386 (2000).

Verma, D.K.: Adjustment of occupational exposure limits for unusual work schedules. Am. Ind. Hyg. Assoc. J. 61:367- 374 (2000).

Vincent, J.H.: International occupational exposure standards: A review and commentary. Am Ind. Hyg. Assoc. J. 59:729- 742 (1998).

Vincent, J.H.: Sampling criteria for the inhalable fraction, in Particle size-selective sampling for particulate air contaminants, J.H. Vincent (ed.), ACGIH®, Cincinnati, OH, USA (1999).

11

FEDERAL LEGISLATION AND REGULATION RELATIVE TO INDUSTRIAL HYGIENE

THE UNITED STATES OCCUPATIONAL SAFETY AND HEALTH ACT AND ITS CONSEQUENCES

Introduction

To U.S. hygienists, the most important piece of federal legislation is the Occupational Safety and Health Act of 1970. Prior to 1970 only a few states regulated safety and health in the workplace. This situation was cited, along with others, as a major reason for the need for federal regulation of the workplace. There was an understanding among many occupational health professionals, and indeed politicians, that injuries in the workplace, in particular, were on the rise, even though the factual information used to support that understanding was sometimes questioned. Illness data were more unreliable, if not nonexistent. The accident rate was rising; however, at that time, accident and injury statistics were less reliable. Finally, the worker's compensation system was doing little to promote safety and health in the workplace because the awards made to individual workers were relatively small, and because employers could budget for worker's compensation insurance each year, it was an easily justified cost of business; nonetheless, worker's compensation rates were on the increase.

However, the accident, injury, and death statistics that were available were appalling. For example, it was estimated that approximately 20,000 workers died, 3 million were disabled and about 100 million injuries occurred each year in U.S. workplaces (Mintz, 1988). A number of factors must be considered when these types of statistics are reported. Questions that should be asked about these statistics include the following:

1. Do the data include deaths, illnesses, or injuries caused by effects on health arising from the workplace or caused only by unsafe acts? Occupational disease data have always been difficult to formulate and collect. They are typically considerably underestimated (Rose, 1977; OTA, 1985).

2. How is accident or injury defined? For example, there are *lost-time* accidents and *first-aid* accidents; the latter is a subset of the former (Brauer, 1990).

3. Do the statistics include data related to injuries or accidents associated with the operation of motor vehicles while on the job, heart attacks, homicide, or employers of fewer than 10 persons?[1] Motor vehicle accidents account for about 30% of workplace deaths (BLS, 2004), heart attacks account for 11% (OTA, 1985) and small employers report 1 to 20% (OTA, 1985).

4. Are the data collected by random sample or by voluntary provision of information? This too has varied considerably.

These data and other perceptions about health and safety in the workplace led the U.S. Congress in 1968 to consider a bill on occupational safety and health. Originally, there was strong opposition from the business community. The idea that industry can police itself has often been cited as a reason for non regulation throughout the history of the industrial revolution. Another tactic used successfully to prevent regulation is the concept that regulation would be most burdensome upon small businesses, driving many of them out of business and reducing competition. These and other arguments were used to defeat the 1968 bill (Mintz, 1988).

However, in 1969 a series of coal mine accidents occurred that changed the general attitude of the country toward safety and health legislation. This led to the passage in 1969 of two important statutes, the Coal Mine Safety Act and the Construction Safety Act. By 1970, industry, Congress, and a new Republican administration were willing to agree on safety and health legislation, and the Act was clearly supported by industry. One incentive for industry was the belief was that one law would be better than many different State laws. As usual, this legislation was a compromise; however, it has never been opened for general review by an act of Congress, even though there have been attempts to do so.[2] Nevertheless, Congress has changed the thrust of the legislation by limiting the way in which the Occupational Safety and Health Administration may use the funds that are appropriated to it.

The Occupational Safety and Health Act and OSHA

When it went into effect on April 28, 1971, the Occupational Safety and Health Act created a new administrative unit within the Department of Labor, the Occupational Safety and

---

1 Small employers are often exempted from record-keeping duties.

2 It has been amended 18 times. Most amendments have been to reflect changes in other statutes. Public Law 101-508 (Deficit Reduction Act) amended the penalty structure. Penalties for serious violations were modified from up to $1,000 to up to $10,000. The penalty for willful violations was changed from $10,000 to $70,000 (personal communication, C. Adkins, OSHA).

Health Administration (OSHA). To lead this new administration, an Assistant Secretary for Labor, OSHA, was to be appointed. OSHA's main duty would be to serve as workplace police over the 6 million workplace establishments in the United States. There was a precedent for establishing the enforcement agency within the Department of Labor, as the Department of Labor had housed a Bureau of Labor Standards, which had responsibility for administering the Walsh-Healey Act, the Long-Shoreman and Harbor Workers Act, and the Construction Safety Act. It is important to point out that the Walsh-Healey Act, or the Public Contracts Act of 1936, with amendments, provided for federal occupational safety and health standards for federal government contractors and their employees.

Of course, there continued to be state agencies, particularly in those states having safety and health legislation, that were doing the job that OSHA was now designed to do. Acknowledging this, Congress set forth provisions in the Act to allow OSHA to enter into agreements with individual states that would allow them to administer their own safety and health matters. The negotiations with the states, in determining what would be required of them and how they would go about enforcing their standards, have been tumultuous at times. However, it appears that the number of states that have approved state plans has stabilized at roughly half of them. These approved states receive up to 50% of their funding from the federal government, and, although their standards need to be as stringent as those of OSHA, this also affords them the possibility of having standards more stringent than those of OSHA. Overall, however, the Act has led to relatively uniform standards across the country, similar to those of individual European countries but unlike those of less centralized countries such as Australia and Canada.

Another important aspect of state responsibility has been the evolution of the free, on-site consultation function (OSHA, 2005a). Throughout the 1970s, several members of Congress and business organizations argued for the provision of free on-site consultation for selected businesses. This was particularly important because it was clear that, upon finding a violation, OSHA would be forced to issue a citation. Hence, many businesses wanted consultation services provided that would allow them to discover their problems before an OSHA visit. Now all 50 states have such programs, funded mostly by the federal government. They have been called 21 (d) (formerly, 7(c)1) agencies with respect to the clause of the OSHA Act that allows OSHA to enter into contracts with state agencies.

The services provided by these state agencies vary from state to state; however, in general, a small employer may request that the state agency come to his or her establishment and either do a walkthrough inspection of the entire establishment or inspect one specific process. Upon finding any violations, the consultant instructs the employer as to what needs to be done to correct the deficiency and is given a timetable for doing so. If the deficiency is not corrected, and as appropriate, if all negotiations have failed, then it is reported to OSHA. The definition of the size of employer eligible for this service varies from state to state.

Standards, Regulations, and Rules

As with all enabling legislation, the Occupational Safety and Health Act simply provided for the establishment of OSHA and gave it certain conditions to follow in conducting its business. It was then up to OSHA to promulgate the standards, rules, and regulations by which industry would have to operate in order to comply successfully with the Occupational Safety and Health Act.

When OSHA was first established, it had no standards to enforce; however, a provision of the Act made it relatively easy for OSHA to adopt *start-up* standards. These standards could be adopted from any national consensus standards or pre-OSHA standards which were issued under the Walsh-Healey Act, the Long-Shoreman and Harbor Workers Compensation Act, and the Construction Safety Act.

There are several national consensus standards organizations, including the American National Standards Institute (ANSI), the National Fire Protection Association (NFPA), and the American Society for Testing and Materials (ASTM). Generally, these organizations are composed of committees whose members have interests in a specific subject such as the regulation of chemicals in workplaces or protective clothing for firefighters. The members of the committee are typically representatives of labor, government, industry, and academia. Hence, once a standard is passed by such an organization, it is considered consensus by virtue of the make-up of the committee. Virtually all of the safety-related standards that were originally adopted by OSHA were consensus standards of ANSI, ASTM, NFPA, the American Welding Society, the Compressed Gas Association, and Underwriter's Laboratories. These standards were adopted by OSHA only one month after it was created.

Many of these standards were never intended to be enforced. In other words, some of the provisions were meant by the standards organization to be voluntary or best practice but not necessarily required for safe workplaces. Therefore, many of the standards as adopted and enforced by OSHA were controversial in that they required what seemed to be mundane provisions. As a result much lobbying of Congress occurred. OSHA developed a poor reputation, and depictions of the requirements of the Act, although somewhat exaggerated (see Figure 11-1), were produced. Approximately 600 standards that contained the word *should* rather than *shall* were removed or deleted in November 1978 (Fed. Reg., 1978).

PERMISSIBLE EXPOSURE LIMITS

The OSHA standards of particular importance to hygienists are shown in Table 11-1. From a health point of view, the more important standards were first adopted under the Walsh-Healey Act. The standards adopted from Walsh-Healey were designated Tables Z-l and Z-3 of 29 CFR 1910.1000 (this means volume 29, part 1910.1000 of the *U.S. Code of Federal Regulations*). These are standards for acceptable levels of exposure in the workplace to various substances, and are termed permissible exposure limits (PELs). Substances in Tables Z-1 and Z-3 were first published as the 1968 Threshold Limit Values of the American Conference

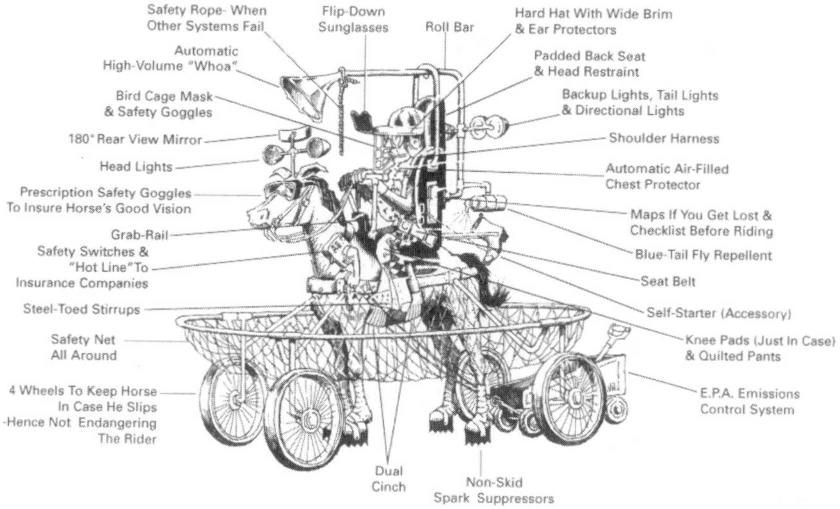

Figure 11-1. The celebrated cowboy following an OSHA inspection (original artist unknown).

of Governmental Industrial Hygienists.[3]

The Table Z-2 standards of 29 CFR 1910.1000 were originally ANSI standards. The Z-2 table is shown as Table 11-2. Table Z-2 is confusing because the second column is not a true ceiling. It can be seen that there are actually three values for each substance. The first, an 8-hr TWA, is similar to the ACGIH® TLV-TWA. However, the acceptable ceiling concentration is not analogous to the ACGIH® ceiling. Unlike the ACGIH® ceiling, the Z-2 ceiling can be exceeded for the period of time shown in the last column. The third column is in fact a true ceiling, as it is never to be exceeded (a near impossibility; see Chapter 16).

COMPREHENSIVE STANDARDS

There are 26 other standards that deal with specific substances and are shown in Table 11-3. These comprehensive standards have similarities that should be discussed. First, each of the comprehensive standards, with the exception of the *13 carcinogens* (1910.1003–1910.1016)

[3] In 1989 the Z-1 and Z-3 tables were updated and combined into Table Z-1-A (Fed. Reg., 1989). The changes largely reflected the 1987–1988 ACGIH® TLVs®, except in a few cases where they reflected National Institute for Occupational Safety and Health (NIOSH) Recommended Exposure Limits (REL). In addition, a few of the old PEL values remained unchanged. However, after OSHA published these standards, they were challenged in court and vacated based on a lack of evidence that the new values would substantially reduce risk (BNA, 1992a). Some states have a process for keeping OELs up to date, for example, California (Cohen et al., 2007).

Table 11-1
OSHA standards of general interest to hygienists

STANDARD	SUBJECT
1903	Inspections, Citations, and Proposed Penalties
1904	Recording and Reporting Occupational Injuries and Illnesses
1910.20	Access to Employee Exposure and Medical Records
1910.94	Ventilation
1910.95	Occupational Noise Exposure
1910.96	Ionizing Radiation
1910.97	Non-ionizing Radiation
1910.101–110	Hazardous Materials
1910.120	Hazardous Waste Operations and Emergency Response
1910.134	Respiratory Protection
1910.1000	Air Contaminants
1910.1001–1910.1101	Standards for Specific Air Contaminants
1910.1200	Hazard Communication
1910.1450	Occupational Exposure to Hazardous Chemicals in Laboratories

and Coke Oven Emissions (1910.1029), contains a PEL. In addition to the PEL, some standards may contain a short-term exposure limit (STEL), which is equivalent to the ACGIH® 15-min STEL. However, in the case of some OSHA standards (see, e.g., benzene) the STEL may be for periods longer than 15 minutes. No comprehensive OSHA standards contain ceilings.

The comprehensive standards contain an environmental exposure limit termed the action level. The action level is usually set at one-half the PEL (lead is one exception – 0.6 PEL). Although NIOSH originally developed the concept of an action level for other purposes (see Chapter 14), OSHA has used it exclusively as a means of causing employers to perform specific provisions of each standard even though employee exposures may be less than the PEL. Specifically, if employee exposures exceed the action level, then the employer must perform periodic measurement of exposure levels, provide medical surveillance for employees, and perform employee training. Note that, conversely if an employer performs exposure monitoring and finds that levels are less than the action level, and the employer deems that no process changes will take place that may change exposure levels, then the employer does not need to perform further exposure monitoring, nor must the employer provide medical surveillance or training for that particular substance. Hence, the action level (AL) causes three levels of compliance: exposures < AL, no action necessary; AL < exposures < PEL, action necessary with

Table 11-2
OSHA 29 CFR 1910.1000 Table Z-2

Substance	8-hr TWA	Acceptable Ceiling Conc.	Acceptable max. peak above the ceiling conc. for an 8-hr shift	
			Conc.	Max. Duration (min)
Benzene[a] (Z37.40–1969)	10 ppm	25 ppm	50 ppm	10
Beryllium and compounds (Z37.29–1970)	2 ug/m^3	5 ug/m^3	25 ug/m^3	30
Cadmium fume[b] (Z37.5–1970)	0.1 mg/m^3	0.3 mg/m^3		
Cadmium dust[b] (Z37.5–1970)	0.2 mg/m^3	0.6 mg/m^3		
Carbon disulfide (Z37.3–1968)	20 ppm	30 ppm	100 ppm	30
Carbon tetrachloride (Z37.17–1967)	10 ppm	25 ppm	200 ppm	5 in any 4 hr
Chromic acid/chromates (Z37–7–1971)	1 mg/ m^3			
Ethylene dibromide (Z37.31–1970)	20 ppm	30 ppm	50 ppm	5
Ethylene dichloride (Z37.21–1969)	50 ppm	100 ppm	200 ppm	5 in any 3 hr
Fluoride as dust (Z37.28–1969)	2.5 mg/m^3			
Formaldehyde: see 1910.1048				
Hydrogen fluoride (Z37.28–1969)	3 ppm			
Hydrogen sulfide (Z37.2–1966)		20 ppm	50 ppm	10, once only if no other meas. exp. occurs
Mercury (Z37.8–1971)	1 mg/m^3			
Methyl chloride (Z37.18–1969)	100 ppm	200 ppm	300 ppm	5 in any 3 hr
Methylene chloride (Z37.23–1969)	100 ppm	1,000 ppm	2,000 ppm	5 in any 2 hr
Organo (alkyl) mercury (Z37.30–1969)	0.01 mg/m^3	0.04 mg/m^3		
Styrene (Z37.15–1969)	100 ppm	200 ppm	600 ppm	5 in any 3 hr
Tetrachloroethylene (Z37.22–1967)	100 ppm	200 ppm	300 ppm	5 in any 3 hr
Toluene (Z37.12–1967)	200 ppm	300 ppm	500 ppm	10
Trichloroethylene (Z37.19–1967)	100 ppm	200 ppm	300 ppm	5 in any 2 hr

[a] This standard applies to the industry segments exempt from the 1 ppm 8-hr TWA and 5 ppm STEL of the benzene standard at 1910.1028.

[b] This standard applies to any operations or sectors for which the Cadmium standard, 1910.1027, is stayed or otherwise not in effect.

[57 FR 42388, Sept. 14, 1992; 58 FR 21781, April 23, 1993; 58 FR 35340, June 30, 1993]

Table 11-3
Specific substances with comprehensive standards

STANDARD	SUBSTANCE
1910.1001	Asbestos, tremolite, anthophylite, and actinolite (eff. 7/21/86)
1910.1003	4-Nitrobiphenyl
1901.1004	alpha-Naphthylamine
1910.1005	[Reserved]
1910.1006	Methyl chloromethyl ether
1910.1007	3,3'-Dichlorobenzidine (and its salts)
1910.1008	bis-Chloromethyl ether
1910.1009	beta-Naphthylamine
1910.1010	Benzidine
1910.1011	4-Aminodiphenyl
1910.1012	Ethylenimine
1910.1013	beta-Propiolactone
1910.1014	2-Acetylaminofluorene
1910.1015	4-Dimethylaminoazobenzene
1910.1016	N-Nitrosodimethylamine
1910.1017	Vinyl chloride
1910.1018	Inorganic arsenic
1910.1025	Lead
1910.1026	Chromium (VI)
1910.1028	Benzene
1910.1029	Coke oven emissions
1910.1043	Cotton dust
1910.1044	1,2-dibromo-3-chloropropane
1910.1045	Acrylonitrile
1910.1047	Ethylene oxide
1910.1048	Formaldehyde
1910.1050	Methylenedianiline
1910.1051	1,3-Butadiene
1910.1052	Methylene Chloride

regard to medical monitoring, exposure monitoring, and training; and exposures > PEL, all of the above and, the employer must take action to abate the overexposure.

Other Provisions

For each of the standards in Table 11-3, in addition to the PELs, other provisions generally include:

1. Provisions for the frequency with which exposures must be monitored and the sampling and analytical methods that will be used to perform the monitoring.

2. Provisions for medical monitoring of employees and the minimum requirements for the contents of the medical, historical, and physical examinations.

3. Provisions for the frequency with which employees will be trained and the content of the training.

4. Any requirements for biological monitoring of employees using specimens such as blood or urine.

5. Methods by which the employer is allowed to comply with the standard, such as the use of engineering controls, personal protective equipment including respirator and chemical protective clothing, and administrative controls.

6. Provisions for the labeling of containers and the posting of signs for regulated areas.

7. A schedule by which employers must comply with the standard.

For the substances listed in Table 11-3, very complete and detailed standards exist that provide a great deal of information to the employer. However, for the more than 500 entries in Tables Z-1, Z-2, and Z-3, no supplementary requirements or information exists. The OELs contained in these tables are somewhat *performance-based* standards.[4] Essentially, the employer is given a standard with which he or she should comply; the means of compliance are left up to the employer. On the other hand, the standards listed in Table 11-3 are more *design or specification* standards; to a greater extent, the employer is told how to go about complying with the standard.

Another important dichotomy for differentiating standards is the vertical versus the hori-

[4] Well, not exactly if one reads the fine print. Paragraph 1910.100 (e) of the OSHA standards requires employers to use engineering or administrative controls to control overexposures where feasible, but where these are not successful, respiratory protection can be used but must be done in accordance with 1919.134. To the extent that one can choose between the various types of administrative and engineering controls, the standards are performance driven.

zontal standard. The standards for the individual compounds in Table 11-3 are vertical in that they apply generally to all industries. (Note: In a few of the standards there are exceptions, such as construction.) On the other hand, Table 11-3 also lists a standard for coke oven emissions. Obviously, this standard applies only to coke ovens (an industrial process) and is a horizontal standard.

HAZARD COMMUNICATION

Of the health standards promulgated by OSHA, the Hazard Communication Standard (1910.1200) has had the most far-reaching effects on the greatest number of employees, employers, and hygienists. This standard, sometimes referred to as *Employee Right to Know*, has three major aspects: the training of employees, the provision of material safety data sheets (MSDS) by manufacturers to the users of hazardous materials, and the labeling of pipes and containers containing hazardous materials.

As an example of the effect of hazard communication, consider that prior to 1987 an employer using several solvents listed in the Z tables (e.g., perchloroethylene, trichloroethylene and methylene chloride) had a reasonable chance of effectively avoiding OSHA health regulations. Although the Z tables require control of the substances, they require no monitoring. Consequently, the employer may presume compliance. For reasons shown in Chapter 14, even if there is an OSHA inspection, and exposure levels are sometimes over the PELs, OSHA may not discover this.

Although Hazard Communication did not require monitoring, it was no longer possible for the employer to ignore health effects in the workplace. The employer was now responsible for obtaining a MSDS for many workplace chemicals and training employees on how to interpret the MSDS. As Table 11-4 shows, in 2004, Hazard Communication was the second most often violated standard, and the most frequently violated general industry and health standard. It is much easier for OSHA to detect a violation of Hazard Communication than that of a PEL.

The history of this regulation is complex and several provisions were contested in court (Mintz, 1988). The original document was promulgated in 1983 (Fed. Reg., 1983) and revised in 1986 (Fed. Reg., 1986), though all employees were not covered until 1989. Several provisions were controversial. First, the trade secret provision, which guarded trade secrets, was too restrictive. It was changed to reflect U.S. patent law. Any chemical component was not a trade secret if it could be discovered by *reverse engineering*. Given modern chemical technology, this definition covers about anything. However, unenlightened manufacturers continue to claim *trade secret* without regard to the ease with which their product recipes can be discovered. This means that a hygienist or any other person including an employee must request disclosure of the chemical ingredients in writing. Usually the best way to get the manufacturer's attention is to threaten discontinuance of purchase and follow-through if necessary. The standard originally applied only to manufacturing employers. However, it now applies to non manufacturing employers as well. This was actually an important change. By 1983, right-to-know

Table 11-4
OSHA Standards (29 CFR) most frequently cited in year ending September, 2006 (OSHA, 2006) (Note that 1910 is general industry and 1926 is construction)

Standard Section	Number times cited	Penalties ($US)	Subject
1926.0451	8736	7623566	General requirements, scaffolding
1926.0501	6278	7476759	Duty to have fall protection
1910.1200	6198	1232440	Hazard communication
1910.0134	3794	1003293	Respiratory protection
1910.0147	3456	3107174	The control of hazardous energy (lockout/tagout)
1910.0178	2842	1877819	Powered industrial trucks
1910.0305	2634	1202072	Wiring methods, components, and equipment for general use
1910.0212	2552	3270731	General requirements for all machines
1926.1053	2294	1149220	Ladders
1910.0303	1986	1008597	General requirements, electrical
Health related standards not in top 10			
1910.0132	1661	1091156	Personal protective equipment, general requirements
1910.1030	1214	510575	Bloodborne pathogens
5A0001	1163	2245018	General duty clause/OSH Act
1910.0146	1024	896995	Permit-required confined spaces
1910.0095	869	588393	Occupational noise exposure
1926.1101	756	547470	Asbestos

laws had sprung up in several states and municipalities. Each was different; so industry was faced with compliance with several laws. The courts ruled that the federal rule preempted state and local laws where there was overlap. So long as the federal rule exempted non manufacturing employers, state laws could cover such employers. Hence, the federal rule was changed to cover all employers.

The rule applies only to *hazardous* chemicals. The manufacturers determine hazard relative to chemical and/or physical properties (e.g., explosion hazard) based on rather complex criteria. However, as a baseline, all chemicals on the following lists are automatically deemed hazardous:

1. OSHA Z-tables and all CFR part 1910 chemicals with comprehensive standards.

2. ACGIH® TLVs®.

3. National Toxicology Program Annual Report on Carcinogens.

4. International Agency for Research on Cancer Monographs.

The latter two lists relate to cancer hazard and are reviewed in Chapter 12. If either organization lists the substance as a human or a potential human carcinogen, and if the chemical is a product component at a concentration of greater than 0.1%, then the manufacturer must state this. Each non carcinogen must be identified if it is present at concentrations greater than 1%. Although manufacturers may not reveal the identity of hazardous chemicals under trade secret provisions, they still must state the hazards of each hazardous component.

If the chemical is not on the above lists, then manufacturers must do a hazard evaluation or assume hazard by default. For liability purposes (and because risk is hazard × exposure and exposure is out of the manufacturer's direct control), many manufacturers are issuing MSDS for all products.

The MSDS was originally required in the Maritime or Longshoreman Standards (29 CFR 1918) and was often considered a fairly useless and incomplete document. However, many MSDS are now vastly improved, as manufacturers realize that, for liability reasons, complete and forthright information (i.e., an adequate warning) is not only good business but ethically correct. This can sometimes make MSDSs technical and difficult for even hygienists to interpret (Kolp et al., 1993).

Once the employer obtains the MSDS from the manufacturer, each employee with potential exposure to the product must be trained regarding:

1. techniques to detect the presence or release of a chemical

2. chemical, physical, and health hazards

3. control methods

4. details of the employer's written hazard communication program including:

   a. labeling system

   b. MSDS interpretation

   c. how employees access hazard information.

Enforcement

The major duty of OSHA is to enforce the standards that it promulgates. To perform this duty, it set up 10 regional and 84 area offices as of 2005. The area offices are typically located in large cities, and the number varies from time to time, depending on levels of appropriation and political considerations. Operating from the area office is an area director, the official who administers the enforcement activities for a specific geographical area. Under each area director, there are one or more Team Leaders that lead the Strategic Team and the Response Team. Teams consist of both Industrial Hygienists and Safety Specialists. Safety Specialists typically conduct safety inspections but are trained to recognize potential Health Hazards. Identified potential Health Hazards are referred to an Industrial Hygienist for a complete evaluation. All health inspections are conducted by hygiene compliance officers.

Every plant hygienist has at least some concern about OSHA inspections and should be aware of the procedures followed by the OSHA officer. First, the hygienist should become familiar with the OSHA *Field Inspection Reference Manual* (OSHA, 2005b), formerly the *Field Operations* Manual and the OSHA *Technical Manual* (OSHA, 2005c). Remember that OSHA had to interpret the enabling legislation, the Occupational Safety and Health Act, in order to determine what standards it should set. Both employers and inspection officers must interpret those standards. Rather than have each inspection officer make a different interpretation, OSHA has set forth guidelines for inspection policy in its *Field Inspection Reference Manual*. The manual contains general instructions and policies on field inspection operations, with emphasis on both safety and health hazards. On the other hand, the Technical Manual addresses technical issues and practices to be followed by OSHA compliance officers. For example, if the plant hygienist questions the way an OSHA compliance officer takes an air sample, then the Technical Manual should be consulted in order to determine if the compliance officer was operating under OSHA procedures.

According to the OSHA *Field Inspection Reference Manual*, the OSHA hygienist first does a background survey (see Chapter 12) to become familiar with a particular industry and its processes in preparation for the site visit. OSHA inspections are unannounced; however, it has been determined in the courts that employers may require a search warrant prior to allowing an OSHA inspector onto the plant site. In effect, this only delays the inspection, as OSHA officers generally will obtain the search warrants. In some states this may be done in a short period of time. The OSHA officer carries credentials that should be examined by plant management.

Upon arriving at the plant, the compliance officer first conducts an opening conference. If the employees are represented by a union, OSHA will inform management that the union should be notified and invited to attend the inspection. Where no union is represented, OSHA informs management that employees should be given the opportunity to observe the inspection. It is in the best interest of a plant hygienist to include employee representatives in OSHA inspections and all hygiene operations wherever possible. This improves the rapport between the hygienist and the employees as well as the efficacy of hygiene programs, as it gains the trust of the employees.

The OSHA officer will ask for information from plant management, including all records associated with hygiene and safety and information concerning what kinds of activities the industry performs, or products that it makes, process flow diagrams, plant layout, and so on. At a minimum, the OSHA officer will inspect OSHA Form 300, Log of Work Related Injuries and Illnesses. This is a summary of events that are reported individually on OSHA Form 301 (OSHA, 2005d).

Records examination is an important part of an OSHA inspection. These records include not only Forms 300 and 301 but also records associated with specific standards, such as medical and air monitoring records required for comprehensive health standards, records of training required for hazard communication and also in comprehensive standards, collections of material safety data sheets for the hazardous substances handled within the plant and required under the hazard communication standard, records of respirator fit testing and use, and others, required by individual standards, where applicable. When companies fail to keep records at multiple plant locations (or, for that matter, have other types of similar violations at multiple sites), these violations can be termed egregious by the OSHA regional headquarters office. As a result fines may be multiplied many times, ultimately reaching millions of dollars. There is no doubt that record-keeping is important, and because of the monetary fine that may be assessed to a company, the hygienist should make every effort to see that records are kept in an adequate and organized fashion. Computerized records are recommended.

Once the OSHA officer is familiar with the plant layout, the processes and chemicals used, and the available records, a walkthrough inspection is begun. The OSHA hygienist performs the walkthrough survey essentially as outlined in Chapter 13. This is a reconnaissance survey. The OSHA officer gathers information that will help in the preparation for possible air sampling or exposure assessment. This information includes employee interviews that are conducted in private. Often management balks at the prospect of allowing employees to talk directly to an OSHA officer, out of management's presence. However, this should not be contested, as OSHA has established the right to do so.

After the walkthrough survey, the OSHA officer is prepared to take any air samples of the workplace air environment that may be necessary. In addition, the OSHA officer may use information from the walkthrough survey to plan surveys of airflows in ventilation systems, radiation, lighting, heat stress, and noise surveys. It is this survey portion of the inspection that often takes the greatest amount of time during a health inspection. Depending on the size of the plant, the number of individual operations within it, the number of employees within each operation, and the number of hazardous substances within each, the survey portion of the OSHA inspection may last from a few days to several weeks, and, in some cases, months.

When the OSHA officer has compiled the results of the survey, a closing conference is held. Very often during a health inspection, the OSHA officer will first hold a preliminary closing conference, which takes place at the end of the survey portion of the inspection. Upon obtaining and analyzing the results of the survey (a process that may take several months), the OSHA officer conducts the final closing conference, which should also include employees. If the results are fairly simple to present, a telephone call or a letter may be all that is necessary in the final closing conference. During this conference, any alleged violations will be presented to plant management, along with a time schedule for abating the hazards. Throughout the inspection process, OSHA will supply general procedures or methods by which the hazards can be controlled.

PENALTIES, CITATIONS, AND VIOLATIONS

OSHA may issue citations based on six types of violations, shown in Table 11-5. Obviously, the descriptions are subject to considerable interpretation; guidance detail is given in the *Field Inspection Reference Manual*. In 2005, the average proposed penalty for a serious violation was $842 and for a willful violation, $41,875 (personal communication, Mr. Charles Adkins, OSHA). These values were up substantially from values of the 1970s and 1980s (see Figure 11-2) as a result of OSHA's increased penalty schedule and the issuance of penalties based on egregious violations. The latter are multiple violations of the same rule when in each case the employer makes a conscious decision to violate, and the specific standard prohibits individual acts or a single course of action (BNA, 1993). For example, if the employer fails to record a certain kind of injury, then each case of failure to record is a potential willful violation. On the other hand, if the employer does not have a respirator policy, then this is only one potential violation, as the employer has not made the intentional decision not to write a policy for each respirator-wearing employee. Violations of record-keeping for injuries and illnesses are most frequently being termed egregious.

Apparently, in general, the state-plan states (those states with approved plans to run their own OSH programs) have not been as aggressive in assessing penalties as the federal OSHA. That has changed recently and state plan penalties are approaching those of federal OSHA. States had average penalties for willful violations of $27,778 and for serious violations of $797 violations in 2005 (personal communication, Mr. Charles Adkins, OSHA).

In some cases, serious, willful, or repeat violations may be considered as causing imminent danger. The OSHA officer would bring an imminent danger situation to the attention of the area director who might then issue an imminent danger notice. This determination would be made only if the OSHA officer detected a situation that could lead immediately to serious harm or possible death for the employee. Consequently, the hazard would need to be removed immediately, as opposed to waiting for abatement to be scheduled. This citation is not issued frequently, and when it is used, it may be necessary for the area director to go to U.S. District Court to cause the employer to perform the necessary immediate abatement action.

The *de minimis* violation requires a citation even though no penalty is proposed. A ladder having 24 in. (61 cm) spaces between rungs would require a penalty, whereas one with 13 in. (33 cm) spaces would require a citation but no penalty, as the standard requires 12 in. (30 cm) spaces.

Many years ago it was common for OSHA to return to the plant for a follow-up visit at the end of the scheduled abatement period. However, it has become more common now for OSHA officers to depend upon plant management to send a written response to the abatement requirements. If the written response is satisfactory, very often follow-up inspections are not conducted. One reason for the discontinuance of many follow-up inspections is that OSHA officers commonly have found compliance with abatement requirements.

For the plant hygienist, the process of OSHA inspection may not end with the closing conference. If alleged violations are cited by the OSHA officer, plant management may choose

Table 11-5
Violations, penalties, and descriptions under the U.S. OSH Act

Violation	Penalty	Description
De Minimis	$0	Violation with no direct or indirect relationship to safety or health.
Other than Serious	$0 – 7,000	Probable resulting accident or illness would not cause death or serious physical harm.
Serious	$1,500 – 7,000	Substantial probability of death or serious physical harm.
Willful	$5,000 – 70,000	Intentional violation of act or plain indifference to requirements.
Willful/Criminal	Assigned by courts	Willful violation leading to death.
Repeat	$200 – 70,000	Repeat violation of a substantially similar condition which was previously abated.
Failure to Abate	< $7,000 / day	Does not abate violation by scheduled date.

to contest the citation. Any aspect of any citation including the time period for abatement, may be contested by the employer. In addition, employees or their representatives may contest the time period for abatement. There are specific procedures for filing the contest, which must be done within 15 working days after the citation is received.

If the employer wishes to contest a citation, the first step is to have a meeting with the OSHA area director. This informal meeting allows management and OSHA to clear up any misunderstandings that may exist about the citation, particularly any misinterpretations of OSHA regulations. Should the informal meeting not prove fruitful, the next step for plant management is to contest the citation through the appeals procedure provided by the Occupational Safety and Health Act.

Via this procedure, the citation is first reviewed by the Occupational Safety and Health Review Commission, which was created by the Occupational Safety and Health Act. It has no administrative connection to the Department of Labor; this was one of the compromises made in the original passage of the Act. The Commission is composed of three members, who are political appointees. However, the first-time review of a citation is performed by an administrative law judge who works for the Commission. During the hearing before the administrative law judge, both OSHA and representatives of management may testify. If management does not testify, it is assumed by the law judge that anything stated by OSHA is in fact true. Consequently, it is in the best interest of management to testify, and the company hygienist may be expected to testify in cases of alleged health violations.

Once the administrative law judge makes a decision, the three-board commission has the prerogative of reviewing the decision. This action may take place at the request of only one

a)

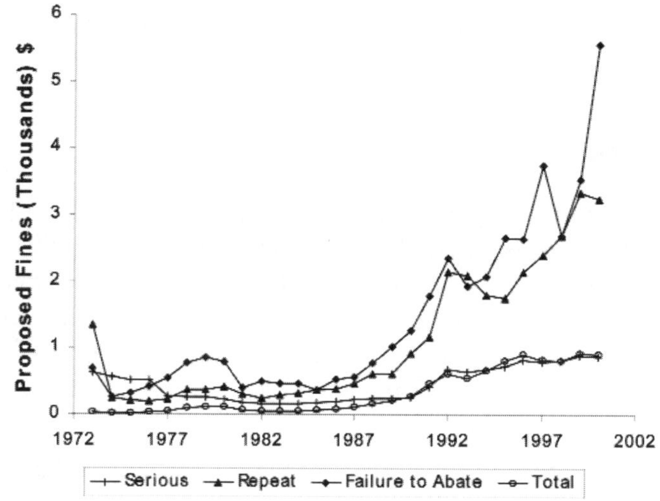

b)

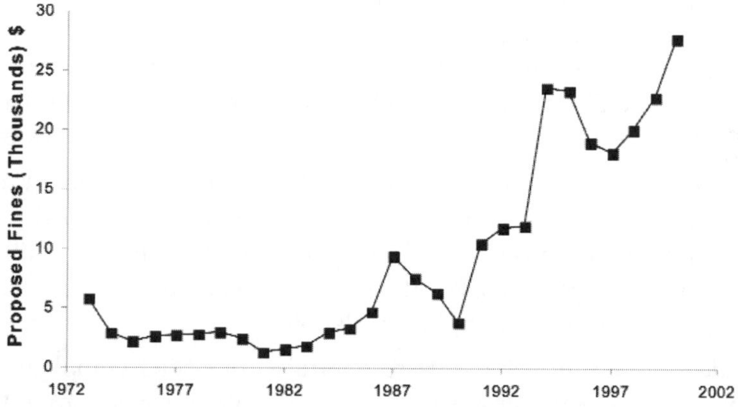

Figure 11-2. a) Average penalties proposed by federal OSHA for four different violations. b) Average penalties proposed by federal OSHA for willful violations.

member of the Commission. Should the employer or OSHA disagree with the decision of the law judge or the Commission (Note: The Commission may decide not to review.), the next course of action would be to appeal to the U.S. Court of Appeals and ultimately the Supreme Court.[5] The Supreme Court and the U.S. District Courts have made some important decisions regarding OSHA enforcement.

ENFORCEMENT PRIORITIES

It is interesting to quantify the power that OSHA and the states actually wield.[6] Federal OSHA had 971 inspection officers in 2000 and 395 were hygienists (OSHA, 2005e). These numbers have been generally declining since 1980. These 395 hygienists were responsible for conducting 8,724 health inspections in fiscal 2000, which required an average of 36 hours each. The number of inspections may sound impressive, but federal OSHA is responsible for about 3.6 million workplaces and 55 million workers in 27 states.[7] Obviously, priorities must be set about who will be inspected.

There are four general types of inspections that OSHA has delineated in order to help set priorities: imminent danger, fatality/catastrophe, complaints/referrals, and programmed. Priority is given by OSHA to each inspection in the order given here. In other words, time is used for programmed inspections only if some of the other three types are not backlogged.

The imminent danger inspection is scheduled when the compliance officer finds an imminent danger and the area officer agrees and issues a notice. It is scheduled immediately and is relatively rare. Safety or health inspections prompted by fatal accidents or catastrophes accounted for fewer than 5% of total 2000 inspections or just over 1,000 federal inspections. In the state-plan states, accident inspections were more than three times as frequent.

The third category, complaints/referrals, actually involves a variety of triggering actions. Employee complaints can be formal (written) and receive highest priority. Non-formal (oral) employee complaints receive lower priority. In 2000, just less than 25% of all inspections were prompted by employee complaints. Even if part were prompted by disgruntled employees, the figure speaks poorly for employee/employer relations.

Also included in the third category are referrals from other compliance officers (e.g., a safety officer may refer a health problem noticed on a safety inspection), referrals from other

---

[5] Information on finalized citations are available on the OSHA website and copies of the complete final citations can be obtained from OSHA through the Freedom of Information Act (FOIA) by submitting a request to OSHA.

[6] Most of the data used in this section were obtained from DOL (2002).

[7] Of the 50 states, 23 have approved plans. These plans account for 40% of the workforce and employers. However, the state-plan states conducted 60% of all 2001 health inspections (almost 21,000 total). The state plans found the majority of all violations were willful violations.

agencies such as the state consulting agencies, and from news media or employer reports of serious accidents. It should be noted that it is OSHA policy not to schedule an inspection based solely on an employer's inquiry.

Programmed inspections receive the most resources, and it is here that OSHA sets true priorities. Policy states that the top priority for safety inspections is triggered by worse than average injury/illness statistics. For health inspections, priority is given to those industries with a previous history of serious OSHA health citations. Consequently, OSHA will periodically *target* certain industries for health or safety inspections. Construction, maritime, logging, and oil and gas exploration are always considered high hazard activities.

The target or programmed industry priorities are generally set at the area level and reviewed at regional and national levels. A list of scheduled inspections is prepared each fiscal year and is divided into two cycles or parts. Work on the second cycle does not begin until the first cycle is completed.

DUTIES OF EMPLOYERS AND EMPLOYEES UNDER THE ACT

The Act details two duties of employers and one duty of employees. The first duty of employers, even though stated explicitly, is in fact implicit: that the employer shall comply with all Occupational Safety and Health Standards promulgated under the OSH Act. The only duty prescribed for employees is that they, too, shall comply with the safety and health standards that are applicable to an employee's conduct.

One of the most important aspects of the Act, particularly with respect to enforcement, is the second required duty on the part of the employer: "Furnish to each employee a place of employment which is free from *recognized hazards* that are causing, or are likely to cause, death or serious physical harm to his employees." (emphasis added) This is section 5(a)1 of the Act and is often cited as the general duty clause. It is important; it gives OSHA the ability to cite an employer for violation of any recognized code of conduct. For example, if the employer is violating a consensus standard that has not been accepted under OSHA as a rule or a regulation, the employer may be cited. Such citations are open to contest; however, in many cases OSHA has established that it has the authority to cite employers for violations of accepted conduct. The key phrase is *recognized hazard*. OSHA has been particularly prone to use the clause when a hazard, not regulated under OSHA, has nevertheless become the subject of media attention. A good example is the failure of employers to prevent repetitive trauma disorders (injuries), often to the wrist, caused by the repetition of bodily movements. This clause had been used in a widely varying manner by OSHA under different presidential administrations.

This clause should not be of undue concern to the hygienist. Indeed, it is both reasonable and ethical for the hygienist to do everything possible to make certain that the workplace is in compliance with any applicable and recognized standard of conduct. Most conscientious hygienists probably will choose to go beyond recognized standards, whether consensus or not.

LIMITS ON THE JURISDICTION OF OSHA ENFORCEMENT

Although it would seem on the surface that every worker in the United States should have protection under the Occupational Safety and Health Act, for various reasons this is not the case. Up until 1980, the degree of protection from workplace hazards for federal employees was largely up to the individual administrators of the federal agencies. Then, in February 1980, President Carter signed an executive order requiring all federal agencies to comply with OSHA standards, requiring OSHA to conduct on-site inspections of federal facilities, and requiring each federal agency to establish a safety and health committee. As a result of this executive order, a great increase in federal activity has taken place in the area of safety and health (Mintz, 1988).

However, the same cannot be said for state employees (Foster, 1993). In those states where OSHA state plans are in effect, state employees do receive protection under the Act. However, in all states except New York and Connecticut where state plans are not in effect, state employees have no legal protection under the Occupational Safety and Health Act. In some states (e.g., Alabama), an employee of the State Department of Transportation or even a local police department does not receive protection directly from the Act, but is protected only to the extent that management of that agency is willing to comply voluntarily with federal standards. There are roughly 10 million employees or about 9% of the U.S. workforce who fall into this category. Fortunately for many state employees, particularly those in hazardous operations (e.g., university laboratories), at least partial application of federal OSHA regulations has been made. However, in most states without state plans, compliance with OSHA standards by state agencies is uncoordinated at best.

In addition to the above areas, OSHA jurisdiction is also limited by the Act in those areas where federal agencies in existence prior to 1970 had statutory control of worker safety and health for a segment of the workforce. Specifically, the Act states that it shall not apply to *working conditions* in those cases where another federal agency has statutory authority to "prescribe or enforce standards or regulations affecting occupational safety or health." For example, the Department of Transportation has responsibility through the Coast Guard and the Federal Aviation Administration for the safety and health of some public transportation workers. Transportation, mining, and nuclear workers are most affected by this clause. This aspect of the law has at times been controversial (Mintz, 1988).

OSHA usually has tried to avoid any duplication of effort or failure of employees to be covered by safety and health regulations by entering into agreements with other federal agencies. For example, in 1974 OSHA and EPA agreed that EPA would regulate safety and health for agricultural workers who must re-enter cropped fields after pesticide spraying. This came after OSHA tried to issue an Emergency Temporary Standard for field worker re-entry (Ashford, 1976; Mintz, 1988). The jurisdictions have further evolved as follows: Workers applying or mixing any pesticide, including applicators of pesticides to homes, are the responsibility of EPA. The manufacture and the formulation of pesticides are the responsibility of OSHA.

Other employers have been exempted by appropriations riders (i.e., statements concerning what OSHA can or cannot spend money on). OSHA cannot perform programmed inspections of small employers (less than ten employees), but can perform other inspections, such as imminent danger (Foster, 1993).

National Institute for Occupational Safety and Health (NIOSH)

In addition to creating OSHA, the Act also created another important agency, the National Institute for Occupational Safety and Health (NIOSH). Whereas OSHA provides enforcement and promulgation of standards, NIOSH's role is mainly to conduct research. In April 1971, NIOSH was created from previous Public Health Service units that had been in existence since the early part of the 20th century (see Chapter 2).

Under the OSH Act, NIOSH is given several specific duties:

- *To prepare criteria for recommended standards for submission to OSHA*. These documents, known as criteria documents, contain information similar to that contained in a comprehensive OSHA standard for a substance; they also include the criteria upon which the recommendations are made. Approximately 90 of these documents have been issued since the inception of NIOSH, the vast majority of which are now of historical importance only because most were issued prior to 1979.

- *To conduct health hazard evaluations (HHE) at the request of employees or employers*. Since its inception, NOSH has conducted more than 1,900 HHEs, mostly at the request of employees or their representatives. When NIOSH receives a request to conduct such an evaluation, a team composed of any or all of the following is assembled: epidemiologist, physician, hygienist, safety professional, and nurse. This team enters the workplace (it is given the right to do so by the Act) and performs research to investigate the complaints filed. Each of these documents is published by NIOSH and can be obtained from the NIOSH website (http://www.cdc.gov/niosh/hhe/default.html). The studies provide a great deal of information about individual industrial operations and the kinds of exposure levels that may be expected in those operations. They are a valuable resource to the hygienist.

- *To prepare a registry of toxic substances*. This document, now known as the *Registry of Toxic Effects of Chemical Substances* (RTECS), has grown to over seven volumes. It is available from several vendors. An up-to-date list of vendors is kept at the NIOSH website (http://www.cdc.gov/niosh/rtecs/RTECSaccess.html). The huge document is essentially a compilation of abstracted information on studies of toxicity and summary opinions concerning toxicity, such as those that may be given by federal administrative authorities in the form of standards.

- *To conduct industry wide studies*. NIOSH conducts horizontal studies of industries for the purposes of ascertaining the hazards in those industries, the state of the art of control techniques utilized, and other aspects. Typically, these studies are used for control technology assessment. In other cases, epidemiological studies have been made of an entire industry in order to determine health effects that may be associated with particular aspects of the industry. These studies are published by NIOSH.

- *To conduct training*. NIOSH is responsible for the training of occupational health personnel throughout the country. This means the training of hygienists, physicians, nurses, and safety processionals and members of other professional groups who may have an interest in OSH. Although NIOSH conducts individual short courses (largely in Cincinnati, OH) for the above-mentioned professions, its biggest effort in the area of training is that of the Educational Resource Centers described in Chapter 2.

In addition to the duties outlined in the Occupational Safety and Health Act. NIOSH also has two other important roles to play in the field of occupational health and safety. First, as originally required under the Coal Mine Safety and Health Act and under the OSH Act, NIOSH performs testing and certification of respirators. Secondly, under the Coal Mine Safety and Health Act and the Mining Safety and Health Act of 1977, NIOSH has specific research responsibilities in the area of mining. These responsibilities are largely carried out at the Morgantown, West Virginia laboratory. In addition to the research responsibilities, NIOSH is responsible for issuing criteria documents to the Mining Safety and Health Administration (MSHA). The MSHA is required to act upon these documents within 60 days under the MSH Act, whereas OSHA is not required to act upon NIOSH criteria for recommended standards.

KEY OCCUPATIONAL HEALTH STANDARDS ACTIVITIES IN OTHER COUNTRIES

Given that major international corporations are instrumental in shaping the world's economy, the European Union is functioning in terms of setting standards for a large portion of the industrialized world, and both of these factors are influencing the setting of occupational health standards throughout the world, several key activities outside the U.S. must be presented. While the European Union is currently undertaking several activities that may eventually force the U.S. to increase standard activities or at least modify current standards, the roots of EU activities are of course in the member nations and the United Kingdom Health and Safety Executive (HSE). The German MAK Commission has made important contributions in this regard as well.

The SCOEL and OELs

The Scientific Committee on Occupational Exposure Limits (SCOEL) was created by the European Commission in 1995 (EU, 1995). In particular SCOEL was established to give advice on the setting of OEL values based on scientific data and where appropriate proposes 8-hr TWAs, STELs, and *biological limit values*. SCOEL recommends *limit values, or OELVs* in an agreed-to short summary document. The Commission makes this document public for health based scientific comments and further data. After about six months, SCOEL reviews the comments, makes appropriate changes and adopts the final version.

SCOEL makes recommendations to the Commission on *health-based* OELs, meaning that the literature leads the committee to the identification of a clear threshold dose, below which exposure to the chemical in question is not expected to lead to adverse effects. (Apparently this is the legal wording.)

The European Commission may use the scientific advice from SCOEL to make proposals for OELVs. In this sense SCOEL plays a similar role to NIOSH in the U.S., which recommends exposure limits to OSHA. OELs based solely on scientific considerations are considered as *adaptations to technical progress* and are *indicative, or IOELVs*, and feasibility factors (including socio economic and technical) are not taken into account. For any IOELV, Member States of the EU must establish a "national exposure limit value, taking into account the Community IOELV, determining its nature in accordance with national legislation and practice." In other words, each country uses the IOELV as a guide in the context of that country's legislative practices. To date there are 30 substances and 3 classes of substances for which IOELVs are adopted (EU, 2006).

Limits passed by the Commission that also take into account socio-economic and technical feasibility factors become directive proposals for the European Council and Parliament to consider and when passed are binding for all member counties. Interestingly, OSHA has never adopted a NIOSH recommended limit, and the European Council and Parliament have not adopted any binding OELVs.

REACH

The European Commission proposed a new EU regulatory framework for the Registration, Evaluation, and Authorization of Chemicals (REACH) in October 2003, and this was approved by the European Council of Environmental Ministers and the European Parliament to take effect June of 2007. The regulation requires more detailed and earlier identification of the properties of chemical substances, and gives greater responsibility to industry to manage the risks from chemicals and to provide safety information on the substances. Substance manufacturers and importers will be required to gather information on the properties of these chemicals and to register this information in a central database, within a newly established Chemicals Agency. This agency will also coordinate the in depth evaluation of suspiciously hazardous chemicals and run a public database. The Regulation also calls for the progressive substitution of the most dangerous chemicals when suitable alternatives have been identified. Industry will have to show that the use of their chemicals in articles is safe. This may involve labeling and instructions for safe use.

This sounds much like portions of the Hazard Communication Regulation of U.S. OSHA of 1986 and the Toxic Substances Control Act of 1976 administered by EPA. However, OSHA does not have the ability to restrict or authorize highly hazardous materials, and while EPA does have that power, it is used sparingly (see later discussion). This regulation, at least on the surface, appears to have greater powers and potential for restricting chemical uses. The obligations are to be phased in, with the 1500 or so most hazardous substances receiving first consideration for authorization or restriction.

MAK

Outside of the TLV® committee of the ACGIH®, the MAK Commission of Germany[8] has been the most active organization in the world relative to the establishment of OELs. To date it has established OELs for more than 800 substances and classes. The MAK values are available on the Internet (http://www.mrw.interscience.wiley.com/makbat/makbat_contents_fs. html).

OTHER FEDERAL LEGISLATION OF IMPORTANCE TO INDUSTRIAL HYGIENE

Since 1970 a number of laws and subsequent regulations have been passed that have direct or indirect application to the work of hygienists. In the following synopses an attempt is made to give a brief description of each and its impact on the hygienist. Many are far more complicated than the OSH Act; so the synopses are in no way intended to be complete. For further information the reader should refer to such references as the following titles of the Code of Federal Regulations which can be found at the website for each respective agency.

10 Energy (Nuclear Regulatory Commission)

30 Mineral Resources (Mine Safety and Health Administration)

40 Environmental Protection (Environmental Protection Agency)

46 Shipping (Department of Transportation)

49 Transportation (Department of Transportation)

[8] It is now formally called the Commission for the Investigation of Health Hazards of Chemical Compounds in the Work Area and is incorporated into the standard setting process as part of German law.

Most of the following discussion is taken from Sullivan et al. (1997), which is a good resource.

Clean Air Act

The Clean Air Act was first passed in 1955 and has been amended numerous times. The amendments of 1990 have had a far-reaching impact on air quality. Industrial hygiene has a direct effect on air pollution because the frequent method of controlling an airborne hazard is to capture it by using ventilation techniques and to emit it to the outside air. With few exceptions, that can be done legally for most contaminants. However, the tightening of air quality standards will increase the number of compounds for which a permit may be acquired and that may require controlling those so that they are not emitted to the atmosphere. This is currently or will be felt most in the many urban areas that will have to control emissions of volatile organic chemicals, as they are precursors of ozone. The major features of air quality legislation are set out in the following paragraphs.

National Ambient Air Quality Standards (NAAQS). These standards (see Table 11-6) originally applied to over 400 air quality control regions of the country. If a region does not meet one of the standards, then the 1990 amendments give a timetable for compliance. For example, a region failing to meet the carbon monoxide standard would need to control automobile emissions, as they account for over 90% of CO emissions. Techniques for control can include car-pooling and public transportation.

In 1996 EPA revised the ozone and particulate matter (PM) standards. The ozone standard became an 8-hr TWA as opposed to the old 1-hr TWA based on the likelihood of chronic effects. The $PM_{2.5}$ standard was added in recognition of residual effects caused by exposures in the gas exchange region of the lungs (see Chapter 10 on particle size effects). As of 2003 more than 120 areas did not meet the ozone standards; this includes more than half of the US population. For $PM_{2.5}$, about 40 areas did not meet the standard (EPA, 2005a).

The NAAQS criteria pollutants are generally associated with combustion sources. There are a few exceptions. Ozone is dependent on hydrocarbons and NO_2, both of which come from combustion. Hydrocarbons also are fugitive emissions from many different processes as disparate as petrochemical plants, dry cleaners, house painting, and filling the gasoline tank on a car. Lead was a serious airborne hazard when leaded fuels were burned. That is no longer the case. The *air quality index* is a somewhat complexly derived number based on the NAAQS (except lead), which is given on most afternoon weather reports.

New Source Performance Standards (NSPS). The 1970 Act required that new or modified processes comply with what became known as *Best Available Control Technology* (BACT). Under those provisions, a standard would apply to a source category such as fossil-fuel-fired steam generators or sulfuric acid plants. Generally, the technologies addressed the control of one of the NAAQS pollutants. However, other pollutants were addressed, such as fluorides from aluminum reduction. Later, as ozone was recognized as a primary problem in

Table 11-6
National ambient air quality standards and sources of pollutants

Pollutant	Averaging Time	Concen-tration	Design standard[1]	Major Sources
Carbon monoxide	8 hours 1 hour	9 ppm 35 ppm	Second highest value over 2 yrs.	Transportation
Lead	3 months	1.5 µg/m³	Quarterly daily average	Transportation, smelting
Nitrogen dioxide	Annual	100 µg/m³	Annual mean of quarterly daily average	Combustion
Ozone	1 hour	0.12 ppm	3 year average of annual fourth highest value	Atmospheric photochemical reactions (UV light, NO_2, VOCs)
	8 hours	0.084 ppm		
Particulates (PM_{10})	Annual	50 µg/m³	3 year mean of average of quarterly means in one year	Coal, diesel combustion, mechanical processes
	24 hours	150 µg/m³	99th %tile of 24-hr mean conc.	
($PM_{2.5}$)	Annual	15 µg/m³	3 year mean of average of quarterly means in one year	Atmospheric reactions of SO_2, NO_2, NH_3, etc., combustion
	24 hours	65 µg/m³	98th %tile of 24-hr mean conc. averaged over 3 yrs.	
Sulfur dioxide	Annual	80 µg/m³	Annual daily mean	Coal combustion, copper smelting

[1] Exceedance of the design standard means non-attainment of the standard for a county or region.

many areas, hydrocarbon control increased in importance.

State Implementation Plan (SIP). Relative to NAAQS and NSPS, one might ask, how did the NAAQS apply to existing sources, and how could a region achieve NAAQS? This was basically left to the states. States had to develop plans including mechanisms for negotiating compliance schedules with major existing sources.

Prevention of Significant Deterioration (PSD). This controversial aspect of the regulations is derived from the 1970 requirement that the government should *protect and enhance* air quality. The courts interpreted this to mean that existing air quality was to be protected. However, opponents cried that the alternative was no growth. In 1977, Congress formalized the PSD concept by classifying regions into Class I, or little deterioration allowed, such as national parks; Class II, or moderate deterioration; and Class III, deterioration allowed up to the NAAQS.

Motor Vehicle Emission Control. It has long been recognized that individuals, as a result of driving automobiles, account for most hydrocarbon emissions (as on road vehicles have been controlled, off road engines have become equally important) and CO emissions. Consequently, since 1961 California and since 1963 the United States have required some form of vehicle emission control.

Permits. Prior to the 1990 amendments, a permitting process existed for all water pollution sources but not for air pollution sources. Under these amendments, states issue permits to thousands of businesses that never before considered air pollution. Each business must inventory its sources to determine if it emits more than target levels for several classes of air pollutants defined by the amendments. Hygienists must play an important role in implementing this amendment.

National Emission Standards for Hazardous Air Pollutants (NESHAPS). The 1970 amendments required standards for substances having such high public health risk that they must be controlled to the lowest possible level. Only five pollutants (asbestos, mercury, beryllium, vinyl chloride, and benzene) were designated NESHAPS pollutants. The standards are emissions-control-technology-based (with the exception of asbestos, which is work-practice-based) but apply only to certain sources. Not all sources emitting benzene, for example, are covered. The NESHAPS apply to existing as well as new or modified sources.

The 1990 amendments drastically changed NESHAPS. The Hazardous Organic NESHAPS (HON) amendment called for the promulgation of standards for 189 chemicals in what are now identified as 370 sources (BNA, 1992b). The standard defines the Maximum Available Control Technology, or MACT, and requires the technology, or its equivalent, that is now controlling emissions in the best 12% of establishments. If the MACT standards do not control exposures to the compounds to acceptable disease risk levels (one in one million), new standards must be proposed within eight years.

Another important feature of the 1990 amendments for hygienists is the requirement of standards for the 189 compounds when emitted from *area sources*. These are small but widely distributed sources such as dry cleaners, woodstoves, and gasoline stations. There are other provisions in the 1990 amendments that, in most cases, have only a minor impact on hygienists.

Even though the emission of many pollutants to the outside air was generally not legally impeded prior to 1990, it was ethically impeded in many cases from the standpoint of risk to the community. The 1990 amendments make virtually every hygienist accountable to community air pollution regulations.

Resource Conservation and Recovery Act (RCRA), Comprehensive Environmental Research, Compensation, and Liability Act (CERCLA), and Emergency Planning and Community Right-to-Know Act (SARA-III)

These three acts are related to the practice of industrial hygiene. RCRA was passed in 1976 and amended in 1986. It set standards for the operation of hazardous waste disposal facilities that must be permitted by EPA. These include landfills, treatment lagoons, incinerators, recyclers, *landfarming* operations, storage facilities, transporters, and generators. Prior to about 1980 when the first regulations were passed, incidents such as Love Canal (New York), Times Beach (Missouri), and Valley of the Drums (Kentucky) were caused when little attention was paid to the proper disposal, treatment, or recycling of hazardous wastes.

Therefore, in 1980 CERCLA was passed specifically to clean up *abandoned* hazardous waste sites. The term *abandoned* includes those sites that existed before RCRA but that could not be or were not brought up to RCRA regulations. Thousands of these sites exist. Many have been identified by EPA; in other cases, the states or individual companies are coordinating cleanup.

OSHA standard 29 CFR 1910.120, known as HAZWOPER, specifically addresses hazardous waste operations and emergency response. It covers all cleanup operations, permitted operations at treatment, storage, or disposal facilities, and emergency responses for releases or threats of release of hazardous substances. The standard covers safety and health written programs, training, medical surveillance, engineering control, site characterization and analysis or evaluation, respirators, chemical protective clothing, exposure monitoring, and decontamination.

When CERCLA was amended in 1986, a third title was added. The amendments were termed the Superfund Amendments and Reauthorization Act (SARA); and Title III of SARA is the Emergency Planning and Community Right-to-Know Act, often referred to as SARA-III. This Act has some commonalities with OSHA's worker right-to-know or Hazard Communication standard. As the OSHA standard provides a means for informing the worker about chemical hazards, SARA-III does so for the community via state and local emergency planning committees and local fire departments. The MSDS required under OSHA at each plant must be provided to the above agencies; or if a list of chemicals is provided, the MSDS must be available on demand.

Other requirements under the law involve special reporting requirements for approximately 400 *extremely hazardous substances*. Unexpected releases of these compounds that exceed a triggering quantity must also be reported. Finally, intentional releases of these substances via waste streams, stacks, or fugitive emissions must be reported. The latter quantities are entered into a computerized database that is available to the public (EPA, 2005b). This database has made public some extremely large emissions of organic compounds, some of them carcinogens, and has caused considerable debate, especially in affected communities.

The main impetus for SARA was the disaster at Bhopal, India, which killed over 2,000 people, and a considerably less major, but nevertheless important, incident in West Virginia, which involved the same chemical (methyl isocyanate) and company (Union Carbide).

Toxic Substances Control Act (ToSCA)

Various laws were passed in the 1970s addressing air, water, and land pollution. However, there was concern that regulation was necessary at the manufacturing level, that is, before the chemical became a pollutant in one of the three media. In this particular case, Congress was concerned with new chemicals for which little developed risk information would exist, and with previously manufactured chemicals for which the risk of deleterious health effects seemed high, but complete information was lacking.

Consequently, ToSCA required that EPA complete an inventory of all currently used chemicals. In 2005, that inventory contained about 75,000 chemicals that were used in commerce, with about 500 to 1,000 new chemicals added each year (EPA, 2005c).

Anyone wishing to manufacture or import a chemical not on the list must issue a premanufacture notice (PMN). In the 25 years ending in 2004, EPA had received over 32,000 PMNs. The PMN requires submission of health and environmental data including occupational health data. Based on this submission, EPA may ban or control manufacture, processing, and use of the chemical. Use includes controls such as ventilation, respirators, or chemical protective clothing. If EPA cannot make a decision based on the PMN, it will request more information via testing of the chemical. The required testing may impinge on the duties of the company hygienist.

ToSCA is potentially far-reaching. Its goal is to identify and regulate unreasonable risks to health. This implies worker health as well as consumer or community health. Implicitly, EPA, under ToSCA, has the power to promulgate OSH rules and regulations. In order to coordinate this overlap, a memorandum of understanding was entered into between EPA and OSHA (EPA, 1986).

As of 2004, of the 32,559 PMNs submitted, 3,812 received regulatory action, and in more than 300 cases, further testing was volunteered. The remainder apparently were approved (EPA, 2005d).

Federal Insecticide, Fungicide, and Rodenticide Act (FIFRA)

Under this Act, EPA both registers and restricts the use of pesticides. It is of obvious interest to companies manufacturing or formulating pesticides, but of equal importance to companies employing sprayers or mixers of pesticides. Commercial sprayers of certain chemical pesticides in residential or private buildings must have training and certification under EPA regulations. Pesticide users, who only spray pesticides at buildings that are owned by their employer and are not occupied by leasers or tenants, are not required to obtain the certification (Miller, 1991). However, it would be wise to obtain training similar to that required for commercial sprayers.

Federal Mine Safety and Health Act

This 1969 Act was amended in 1977; however, it originally set the stage for the passage of the 1970 OSH Act. Like the OSH Act, it created an agency in the Department of Labor – the Mine Safety and Health Administration (MSHA), which replaced the old Mining Enforcement and Safety Administration of the Interior Department.

A mine is broadly defined as all underground and surface mines and their attached facilities, including roads, ore processing facilities, tailing ponds, buildings, and so on. Many industrial processes such as smelting and refining are often associated with mines and are covered by MSHA, not OSHA. The Act covers construction of mines as well as mining and radiation hazards that one might otherwise assume are covered by the Nuclear Regulatory Commission.

Energy Reauthorization and Atomic Energy Acts

The Nuclear Regulatory Commission (NRC), under these Acts, is the primary federal agency dealing with all aspects of ionizing radiation, including workplace exposure due to likely sources of radiation such as a nuclear power plant; workplace exposures to small sources of radiation such as tank gauges or diagnostic X-ray machines are also included. Essentially, the NRC (or approved states) regulates radiation sources by licensing them. OSHA assumes that the licensees are operating properly unless they find something suspicious. OSHA does enforce standards for human exposure to the sources, but these are adopted from NRC or original Atomic Energy Commission standards.

QUESTIONS

11.1. Discuss the current status or OSHA health standards. Are they adequate? Why?

11.2. State an acceptable risk level for an occupational disease, such as 1 in 1,000, and defend it. Should risk levels be different for different populations, that is, workers versus the general public?

11.3. Should OSHA issue standards by industry (horizontal) or by chemical (vertical), or generic standards covering a wide array of chemicals and industries, such as hazard communication or respirator use? Defend your answer.

REFERENCES

Ashford, N.: Crisis in the Workplace: Occupational Disease and Injury. MIT Press, Cambridge, MA (1976).

BLS: National census of fatal occupational injuries in 2004. News USDL 05-1598, August 25, United States Department of Labor, Bureau of Labor Statistics Washington, DC http://www.bls.gov/iif/oshcfoi1.htm (2004).

BNA: Eleventh circuit court vacates OSHA's 1989 Standard, remands to agency. Occupational Safety and Health Reporter 22:219-220 (1992a).

BNA: Proposal on hazardous organic emissions, final early reduction rules issued by EPA. Occupational Safety and Health Reporter 23:1707-1708 (1992b).

BNA: OSHA's egregious penalty policy for record keeping violations upheld. Occupational Safety and Health Reporter 22:1601-1602 (1993).

Brauer, R.L.: Safety and Health for Engineers. Van Nostrand Reinhold, New York (1990).

Cohen, R.; Steinmaus, C.; Quinlaw, P.; Roberts, T.; Cooper, M.: Keeping up with Science: Developing PELs in California. The Synergist, p. 32-34 (March 2007).

EPA: Memorandum of understanding between the Environmental Protection Agency and the Department of Labor, EPA Agreement No. PW 16931704-01-1. Environmental Protection Agency, Washington, DC (1986).

EPA: Green Book Nonattainment Areas for Criteria Pollutants. U.S. Environmental Protection Agency, Washington, DC http://www.epa.gov/oar/oaqps/greenbk/ (2005a).

EPA: Toxic release inventory. Environmental Protection Agency, Washington, DC http://www.epa.gov/tri/ (2005b).

EPA: Toxic Substances Control Act Chemical Substance Inventory. Environmental Protection Agency, Cincinnati, OH http://www.epa.gov/opptintr/newchems/invntory.htm (2005c).

EPA: Summary of accomplishments: New Chemicals Program Activities Through September 30, 2004. Environmental Protection Agency, Washington, DC http://www.epa.gov/opptintr/newchems/accomplishments.htm (2005d).

EU: Commission Decision of 12 July setting up a Scientific Committee on Occupational Exposure Limits to Chemical Agents. OJ L 188, 9.8.1995, p. 14-15 http://ec.europa.eu/employment_social/health_safety/scoel_en.htm (1995).

EU: Commission Directive 2006/15/EC. Official Journal of the European Union: 9.2.2006, http://eurlex.europa.eu/LexUriServ/site/en/oj/2006/l_038/l_03820060209en00360039.pdf (February 7, 2006).

Federal Register 43: 49726. U.S. Government Printing Office, Washington, DC (1978).

Federal Register 48: 53280. U.S. Government Printing Office, Washington, DC (1983).

Federal Register 51: 34590. U.S. Government Printing Office, Washington, DC (1986).

Federal Register 54: 2332. U.S. Government Printing Office, Washington, DC (1989).

Foster, J.F.: Who's covered? Job Safety Health Q. 4:15-19 (1993).

Kolp, P.; Sattler, B.; Blayney, M.; Sherwood, T.: Comprehensibility of Material Safety Data Sheets. Am. J. Ind. Med. 23:135-141 (1993).

Miller, M.L.: Federal regulation of pesticides. In Environmental Law Handbook, 11th ed. J.G. Arbuckle, et al. (eds.). Government Institutes, Rockville, MD (1991).

Mintz, B.W.: Occupational safety and health: The federal regulatory program – a history. In Fundamentals of Industrial Hygiene, 3rd ed., B. A. Plog (ed.). National Safety Council, Chicago, IL (1988).

OSHA: Consultation Program. Occupational Safety and Health Administration, Washington, DC http://www.osha.gov/dcsp/smallbusiness/consult.html# (2005a).

OSHA: Field Inspection Reference Manual, Occupational Safety and Health Administration, Washington, DC http://www.osha.gov/Firm_osha_toc/Firm_toc_by_sect.html (2005b).

OSHA: Technical Manual. Occupational Safety and Health Administration, Washington, DC http://www.osha.gov/dts/osta/otm/otm_toc.html (2005c).

OSHA: Forms for Recording Work-Related Injuries and Illnesses. Occupational Safety and Health Administration, Washington, DC http://www.osha.gov/recordkeeping/new-osha300form1-1-04.pdf (2005d).

OSHA: Occupational Safety and Health Administration, Washington, DC (2005e).

OSHA: Most frequently cited standards. Occupational Safety and Health Administration, Washington, DC http://www.osha.gov/pls/imis/citedstandard.html (2006).

OTA: Preventing Illness and Injury in the Workplace, OTA-H-250. Office of Technology Assessment, U.S. Government Printing Office, Washington, DC (1985).

Rose, V.E.: Reliability and Utilization of Occupational Disease Data. National Institute for Occupational Safety and Health, Cincinnati, OH (1977).
Sullivan, F.P.; et al.: Environmental Law Handbook, 14th ed. Government Institutes, Rockville, MD, USA (1997).

IV

THE FRAMEWORK FOR ASSESSING THE WORKPLACE

12

GATHERING INFORMATION ABOUT THE WORKPLACE

SURVEYS

The manner in which the hygienist goes about trying to discover the nuances and subtleties of an individual plant or a plant operation is part of the overall hygiene process. Historically, this process, which is closely related to recognition, evaluation, and control steps, has been divided into three or four types of surveys. For example, in 1958, Patty described four types of surveys:

- The standardized inspection was described as the gathering of information on a standardized form about the demographics of a plant.
- The preliminary or observational survey, also called the walkthrough survey, entailed walking through the plant, interviewing workers and management, and gathering information about the processes and agents.
- The investigational or appraisement survey, now termed the in-depth survey or the exposure assessment phase, had as its goal the gathering of quantitative information about the workplace, including exposure values for various agents, that often must be obtained with instrumentation.
- The combined hygiene and medical survey was the integration of an in-depth survey with medical examinations given by physicians.

This last survey is not performed much anymore except by government agencies or for epidemiological studies. Not given in Patty's list of surveys, but somewhat analogous to the standardized inspection, is a survey commonly referred to as a background survey. This survey is essentially a literature search but often includes a search for information via contact with other hygienists or members of management, in person or by telephone.

The three main types of surveys commonly employed today are the background survey, the walkthrough survey, and the in-depth survey. They comprise a logical progression from the gathering of historical information to the qualitative examination of a plant using all of the human senses, followed by an in-depth quantitative survey for the agents thought to be most important as a result of the first two surveys. This progression takes us through the recognition and evaluation steps. However, as one might imagine, in not all cases will each of the elements of each of the surveys or even all of the surveys need to be conducted in order to accomplish the hygienist's ultimate goal of controlling agents in the workplace. Although the extent of the surveys may vary, the hygienist must possess a minimum amount of information before specific recommendations for controls can be made.

Most frequently, information gathering is the collection of information associated with instrumentation. Generally, this information is in the form of airborne concentrations of workplace agents, including biological, chemical, and physical agents. It also includes noise levels, calculation of heat stress indices, and electromagnetic radiation levels, as well as other information such as air temperature, pressure, and wind speed readings, measurements of air movement associated with ventilation systems, relative humidity, and surface contamination levels. This information cannot be reliably measured with any of the human senses but must be quantitated with the use of analytical survey tools, which can be quite simple or extremely complex.

Even though analytical survey information is the basis upon which hygienists often make judgments about the need for controls, information gathered without the use of analytical survey tools is important. This information includes data on the plant and its processes, the potential disease agents present, the number of employees in the plant and other demographics, and historical information about exposure levels. All of this information is gathered with the senses, and it is usually the information upon which the hygienist plans the in-depth survey needed to gather current exposure information.

INFORMATION-GATHERING ROLES OF HYGIENISTS

The hygienist is a detective, seeking information about workplace disease agents that may be improperly controlled and therefore causing disease among workers. Hygienists have many different roles; some, such as government inspectors, find themselves in a different type of industry almost daily. Others, working for only one corporation, may be responsible for the health and safety of employees working in a complex variety of industrial processes due to the diverse nature of many corporations. Still others may work only at the level of a single plant, but that plant may contain many operations, each with its own agents and propensity to cause disease.

To give an idea of the diversity of workplaces, consider the following. There are 6 million workplaces in the United States. The *Standard Industrial Classification (SIC) Manual*[1]

[1] This code has now been replaced by the North American Industry Classification System (NAICS).

(OMB, 1987; 1997) has codes for 467 different types of workplaces in manufacturing alone. The *SIC Manual* also includes numerous other workplace settings, in agriculture and forestry, mining, construction, transportation and utilities, wholesale and retail trade, insurance and real estate, general services, and public administration. Plants in a single type of industry will have similarities; however, no two plants are exactly alike.

Because of the diversity of workplace settings and of the roles of the hygienist, the types of sensory information gathered vary in importance. The government inspector, whether a compliance officer or a consultant, has responsibilities common to the insurance inspector or the insurance risk manager and the private consultant. Each is expected to go into a vast array of industries and conduct surveys for any type of agent that exists. Consequently, they are expected to have some knowledge about these industries. However, it is obvious that they cannot have the first-hand knowledge necessary to conduct a good hygiene survey for every possible industry that they may enter; thus, for hygienists in these roles, literature searches and telephone interviews are often extremely important prior to their entering the plant. The hygienist must be well prepared upon entering the plant, or will discover, upon trying to plan the in-depth survey, that critical information was not gathered. Hygienists in these roles often will follow each of three survey steps: the background survey, the walkthrough survey, and the in-depth survey. However, the in-depth survey often will be finite or time-limited. Therefore, these hygienists can only gain a snapshot view of the exposures that exist on a limited number of days; and, based on this information, conclusions must be drawn about the need for controls.

On the other hand, the hygienist assigned to a specific plant has a considerably different role. Although responsible for only one plant, the hygienist must know that plant extremely well. The in-depth survey is often a continuing survey that requires considerable resources. Sampling of the workplace environment is conducted periodically throughout the year in order to detect normally occurring variability in concentrations brought on by deliberate or inadvertent process changes, which may include substitution of new raw materials, changes in materials suppliers, production rates, process methods, equipment, engineering controls, maintenance schedules, worker training, etc. Consequently, the plant hygienist is in the mode of conducting background, walkthrough, and in-depth surveys simultaneously in different locations of the plant. One of the hygienist's greatest sources of background information is plant documents outlining processes and the materials used in those processes. The hygienist relies heavily on information such as purchasing documents and process flow-charts. The trust developed with equipment operators can provide most valuable information about the adequacy of a particular control, toxic effects that may be caused by a particular agent, or the supervisory staff's real commitment to safety and health.

Hence, for the plant hygienist, the industrial hygiene process of recognition, evaluation, and control and the process of background, walkthrough, and in-depth surveys may not appear to be followed precisely because of his or her familiarity with the plant. However, if the entire sequence of knowledge gathering about a particular process or agent is observed, it will reveal that a good plant hygienist is in fact following the industrial hygiene process even though knowledge appears to be gained sporadically over long periods of time. Indeed, the plant hygienist may go through the survey process several times before getting approval for needed controls for a given process.

The corporate-level hygienist may be even more specialized than the plant hygienist. This is particularly the case with regard to a team of corporate hygienists. Very often a corporate hygienist is given responsibility for a particular aspect of industrial hygiene such as the analysis of exposure records, the preparation and monitoring of written programs, training, hazard communication, and so on. Obviously, for these specialized hygienists, the hygiene process may be somewhat irrelevant. However, in order to do their job properly, they must have knowledge of the processes involved at the individual plants within the corporation, the manner in which the workers accomplish their jobs, and other types of information that normally would be gathered in the survey process.

In some cases, the corporate hygienists are not specialized because there is only one corporate hygienist or because each corporate hygienist is assigned responsibility for individual plants or regions rather than individual program aspects, even though the hygienist's office may be at headquarters. These hygienists tend to function rather like consultants and often have the same perspective on the survey process that a government or private consultant would have.

There are other examples of hygiene roles that demonstrate that the practice of hygiene cannot always exactly follow the three-step process outlined earlier. Nevertheless, all hygienists regardless of their roles, must gather adequate information in order to meet their primary goal of controlling exposures to prevent disease.

TYPES OF INFORMATION AVAILABLE TO THE HYGIENIST

The information that the hygienist gathers during the recognition and evaluation steps can be simply categorized as sensory and nonsensory. The latter category constitutes information that can be gathered only with the help of analytical survey instruments, and is commonly gathered during the walkthrough and in-depth surveys. The sensory information is gathered via seeing, smelling, and hearing, but touching can be important, and tasting is an integral part of odor detection. The sensory information can be divided into two parts: information gathered before entering the plant and that gathered during the walkthrough and in-depth surveys. The remainder of this chapter will consider the types of information that can be gathered before entering the plant. This information is largely obtained through literature searches and telephone interviews.

Sensory Pre-Visit Information – The Background Survey

A hygienist needs certain types of information in order to ask intelligent questions upon arriving at the plant, including the seven basic categories of information described below. Not every type of information will be necessary for the hygienist's purposes, but it is a good idea for the hygienist to have a checklist such as that in Figure 12-1 that can be used to assure that important information is not overlooked. The purpose of Figure 12-1 is to prompt questions, information requests, literature searches, and so forth. It could be used for the background survey, as well as the walkthrough.

Processes. The hygienist must have information about the common or specific kinds of processes that take place within the plant. Common processes take place in a wide variety of plants and, consequently, are not specific to a particular type of industry. Many are described in Chapter 5. For example, welding, cutting, soldering, painting, degreasing, abrasive blasting, grinding, buffing, polishing, and diesel combustion can be found in a wide variety of industries. The gathering of information about these processes can be somewhat different from the technique needed to gather information about processes associated only with a specific type of industry. For example, distillation takes place in the petrochemical industry, electric arc furnaces can be found in secondary smelters, and pulping is generally used in the paper industry. Each of these processes is specific to the industry in which it exists.

Materials. Raw materials, products, and by-products are associated with each process. Raw materials go into the process, and they are often converted to form desired products and by-products that often are not wanted, or at least are not the primary reason for the process. The hygienist must be aware of the materials in each of the three categories. In addition to raw materials that are modified in the process, there are other materials that are used in the process that do not become part of the end product. Very often these are materials used to maintain equipment or to finish the product, such as greases, oils, emulsifiers, solvents, catalysts, etc.

Material Properties. In addition to a simple listing of the raw materials, products, and by-products, the hygienist must know the properties of the materials. Types of information often needed include vapor pressure, the regulatory or voluntary occupational exposure limits (OELs), the physical state of the materials as they exist in and before entering the process, the toxic effects of the materials, fire and explosion propensity of the materials and, where possible, information regarding the magnitude of exposure levels that can be expected for the materials in the process.

Employee Demographics. It is often important to know information about the employees who work in a particular industry or with a particular plant process. Information such as the number of workers, gender of the workers, race, educational level, age, socio-economic and other factors can be important in helping the hygienist determine how to conduct the in-depth survey of the plant. In addition, this information can be useful in interpreting the results of the in-depth survey. For example, one might use the information to determine why unexpected health effects are occurring in a particular population of workers.

The Nature of the Work. The hygienist must have a feel for the type of job that the worker must perform. For example, is shift work involved that may require workers to change their sleeping habits every week or may require workers to work a 16-hour shift once a week? What is the magnitude of productivity? Are the workers considered to be highly productive? Is the

Reminder Checklist

	Applicable?	Not applicable
Processes		
Continuous/batch/task-oriented	☐	☐
Production variables	☐	☐
Heat stress	☐	☐
Noise	☐	☐
Materials		
Major and minor raw materials	☐	☐
Special materials – catalysts, solvents	☐	☐
What happens to by-products, solid, liquid wastes	☐	☐
Are products potentially hazardous	☐	☐
Material Properties		
Make list of PELs	☐	☐
Are any materials solids or gases?	☐	☐
Are aerosols produced?	☐	☐
Dermal absorption potential	☐	☐
Vapor pressure data	☐	☐
Employee Demographics		
Number of employees in each production area	☐	☐
Number of workers for each job description per shift	☐	☐
Sex, age, race, socio-economic, and education data	☐	☐
Nature of Work		
Number of shifts, times, hours	☐	☐
Specialty jobs or infrequent tasks	☐	☐
Union organized? How?	☐	☐
Safety, Health Committees?	☐	☐
Control Measures		
Standard operating procedures	☐	☐
Training	☐	☐
Ventilation – records of evaluation	☐	☐
Chemical protective clothing	☐	☐
Medical monitoring	☐	☐
Are engineering techniques such as substitution possible?	☐	☐
Records		
OSHA 300/301	☐	☐
Medical summaries	☐	☐
Past hygiene reports	☐	☐
Training content	☐	☐

Figure 12-1. Checklist of reminders for background walkthrough survey.

work dirty? Do the workers bargain through a union? Does the work take place inside or outdoors? Is the work repetitive or varied?

Control Measures. Before entering the plant, the hygienist should have some knowledge of appropriate and acceptable control measures for a given process. For example, some processes may be considered difficult to control through engineering because of their sporadic nature and, consequently, respirators may be appropriate. Other processes may have a history of being well controlled with ventilation. By knowing information about the expected types of controls, the hygienist can ask appropriate questions about the presence and adequacy of those controls while visiting the plant.

Safety and Health Records. The hygienist can learn a great deal from plant records. Often the hygienist will have to examine these records upon entering the plant. However, in some cases, summaries of the records can be reviewed before entry into the plant. The advantage of this is that it allows the hygienist to review the records in an unhurried manner, interpret them, and ask appropriate questions about them upon entering the plant. Examples of such records include accident records, records of previous hygiene surveys, OSHA inspections, epidemiological studies, and medical reports or summaries of medical surveys.

SOURCES OF INFORMATION

The job of information gathering can be greatly facilitated if the hygienist is aware of the most reliable sources from which to obtain the information. Much of it can be gathered from standard reference sources, which the hygienist should keep in a personal library. Other information is specialized and will require that the hygienist conduct either a library or a computer search. Finally, some of the information can be obtained only by contacting individuals who are experts in the particular area, agency or organization being surveyed. Trade associations are a good source of information and may be the sole source for certain types of information. Sources can be either general or plant-specific. General sources can be used for a wide variety of industries and processes, whereas plant-specific information usually has to be gained from the individual plant.

General Information Sources

The following reference works are particularly helpful. (See references at end of chapter).

ACGIH® TLVs®. These were described fully in Chapter 10 and are of obvious importance.

Burgess (1995). This very useful text contains information on general processes such as welding, degreasing, sand-blasting, etc. The processes are described in terms of machinery, raw materials, and products. The disease agents (usually chemical agents) and the physical agents most prominently associated with those processes are also described.

Computerized Information. There are a wide range of companies that offer various occupational safety and health data via Internet. Information includes access to *RTECS* (NIOSH, 2007a), material safety data sheets, the OSHA IMIS (Integrated Management Information System) that includes all OSHA inspection reports, and others. Because the list of companies offering such services changes constantly, no attempt is made to list them here. An Internet search is fruitful and many companies display at trade shows.

Cralley and Cralley (1984). This three-volume set gives information about specific industries such as refining, aluminum reduction, secondary smelting, etc. There is a chapter associated with each industry, and very often process descriptions are diagramed with schematics and details. Unfortunately, the series is now out of print.

Document Series. There are several document series that can be very useful to hygienists:

- *NIOSH Criteria for a Recommended Standard (NIOSH, 2007b)*. . . . Many of these are now out of date, having been published in the 1970s. However, they are useful starting places for background information on toxicity, uses, controls, and historical concentrations for almost 100 substances or groups of substances and other topics.

- *IARC Monographs on the Evaluation of Carcinogenic Risk of Chemicals to Humans.* More than 80 volumes have been published (2005), providing evaluations of the carcinogenicity of hundreds of chemicals. A scheme for classifying the chemicals according to potency is given.

- *Annual Report on Carcinogens.* This report, in its 11th edition (even though they are called annual they are published every 2-4 years), provides a synopsis of the results of carcinogenicity tests sponsored by the National Toxicology Program (NTP, 2005). A classification scheme is also used. This document and the IARC documents are cited in OSHA standards, e.g., see Hazard Communication, 29 CFR 1910.1200

- *IPCS Environmental Health Criteria (various dates).* Health and environmental data are provided for 231chemicals or classes of chemicals. Ambient air, water, and exposure routes other than those found in the occupational setting are also considered.

Hawley's Condensed Chemical Dictionary. This volume (see Lewis, 2002) has thousands of entries on organic and inorganic chemicals, their hazards, and their uses. It is very useful for determining structures of organic chemicals. It also contains material properties information such as boiling point and density, but little vapor pressure information. In addition to chemical names, many chemical terms are defined, such as polymer, Houdry cracking process, humectant, and Hooke's law. Famous people in chemistry are also described.

Journals and Periodicals. There are numerous journals related to the occupational safety and health field that can be used by the hygienist to gain specific information about industries and processes. Journals are important because they provide up-to-date information, which is often not incorporated into standard reference texts until it is three or more years old. In addition, very often reference texts only give small amounts of summary information from individual journal articles. The journal articles themselves can supply details that can be helpful to the hygienist. In addition, journal articles are an excellent source for retrieving information about previous exposure levels in similar industries or processes. Important journals, with their publishers in parentheses, are as follows.

- *American Industrial Hygiene Association Journal* (AIHA, Fairfax, VA)
- *Applied Occupational and Environmental Hygiene* (ACGIH®, Cincinnati, OH)

  Note: in January 2004 the above two journals were combined into the *Journal of Occupational and Environmental Health* and can be obtained from either organization.

- *Annals of Occupational Hygiene* (British Occupational Hygiene Society, Derby, UK)
- *Scandinavian Journal of Occupational Health* (published by the occupational health agencies of the four Scandinavian countries, http://www.sjweh.fi/)
- *Journal of Occupational and Environmental Medicine* (American College of Occupational and Environmental Medicine, Arlington Heights, IL http://www.acoem.org/joemjournal.aspx)
- *Occupational and Environmental Medicine* (BMJ Publishing Group, London, http://oem.bmjjournals.com/)
- *American Journal of Industrial Medicine* (Wiley Interscience, New York, http://www3.interscience.wiley.com/cgi-bin/jhome/34471)
- *Archives of Environmental Health* (published by Heldref, http://www.heldref.org/aeh.php)
- *Journal of Exposure Analysis and Environmental Epidemiology* (International Society of Exposure Analysis, http://www.nature.com/jea/index.html)
- *Risk Analysis: An International Journal* (Society for Risk Analysis, http://www.sra.org/journal.php)

In addition to the above journals, some other periodicals can be valuable. The best known of these are published by the Bureau of National Affairs, Washington, DC. For hygienists, the most important periodical is the *Occupational Safety and Health Reporter* of the BNA, a weekly publication that provides up-to-date information on legislative and regulatory actions, judicial proceedings, important studies, and new publications. In addition to the weekly periodical, other valuable reference materials are included with the subscription. OSHA standards are included in their entirety and are updated. All pertinent legislation is included. Lists of government and legislative agencies are given with addresses and telephone numbers.

Kirk-Othmer (2004). This encyclopedia is available in two forms. A condensed version gives summary information about the specific industrial processes, whereas the 26-volume version offers a very detailed description of various processes. The descriptions include general processes as well as those associated with specific industries. This work is written from an engineering point of view, and consequently gives less health information than do the previous two texts. However, considerably more detail about actual processes and their materials is given in Kirk-Othmer. An electronic version is available through many libraries.

Merck (O'Neil et al., 2001); Sax (Lewis, 2004); Proctor and Hughes (Hathaway, 2004). These volumes contain information similar to that in *Hawley's Condensed Chemical Dictionary*; however, their listings are not so extensive as the chemical dictionary and contain only chemical names. On the other hand, the health information contained in these volumes is more extensive than that in the dictionary.

NIOSH, Other Documents. NIOSH publishes a wide range of documents that can be useful to the hygienist in addition to the *RTECS* and criteria documents. Two important NIOSH documents are published because of legislative mandate; the more important of these (NIOSH, 2007c) is the Health Hazard Evaluations (HHEs). Employees or employers can request that NIOSH conduct a health hazard evaluation of their plant. Each HHE is followed by a written report. More than 1900 of these evaluations are available at the NIOSH website and can be searched by keyword. The documents are extremely helpful for providing information about the processes, materials, and exposures encountered in a particular industry. They are also valuable historical documents, useful for determining industry changes.

Although they are fewer in number than the HHEs, NIOSH also conducts industry-wide studies (see NIOSH, 2007d). Reports on industry-wide studies are typically more detailed and can have a control, technology slant or examine the incidence and prevalence of acute and chronic disease and the nature and extent of acute and chronic responses to potentially hazardous agents in the work environment.

NIOSH also publishes analytical methods (NIOSH, 2007e) for determining airborne concentrations of chemicals (and a few for other media and agents), which are discussed in more detail in Chapter 24 and can be found on the website listed in the references section.

NIOSH Pocket Guide (2007f). The NIOSH Pocket Guide to Chemical Hazards (NPG) is intended as a source of general industrial hygiene information on several hundred chemicals/classes for workers, employers, and occupational health professionals. The NPG does not contain an analysis of all pertinent data, rather it presents key information and data in abbreviated or tabular form for chemicals or substance groupings (e.g., cyanides, fluorides, manganese compounds) that are found in the work environment. Data include synonyms, structures, exposure limits, chemical and physical properties, incompatibilities and reactivities, measurement methods, control methods, and health hazards. The information found in the NPG should help users recognize and control occupational chemical hazards. The book is very useful, but information regarding control and measurement method recommendations should be used cautiously. For control questions there are no right answers, only the best answers for the situation. The recommendations may not be best for your situation, particularly since some were generated by using computerized design logic. The physical properties from the Guide are reproduced in Appendix B to this volume.

Odor Thresholds (AIHA, 1989). This document is the most complete compendium of odor thresholds available to the hygienist. The information is given in two formats, one yielding information on threshold data that have met certain criteria and the other containing all odor threshold data from previous studies. The result is an excellent review of odor threshold data on a chemical-by-chemical basis and a derivation of the geometric mean odor threshold indicating the concentration at which approximately 50% of the population can detect an odor. (See Chapter 13 for more detail on odor thresholds.)

OSHA Standards. All U.S. occupational health and safety regulations are contained in two volumes (OSHA, 2007a, b), and are updated in hardcopy about every two years (more frequently via the Internet). Therefore, new standards in the interim period can only be obtained from OSHA, via the Internet, or in the Federal Register. Generally, one copy of a new standard is available free of charge from the U.S. Government Printing Office, Washington, DC, or the area OSHA office.

RTECS (NIOSH, 2007a). The *Registry of Toxic Effects of Chemical Substances* is also called the "Toxic Substances List" (a misnomer). Under the Occupational Safety and Health Act, NIOSH is required to compile the document. It contains unique chemical entries and is one of the best references available for chemical synonyms. The most important property of this publication is that it contains summarized information regarding the toxicity of each substance, such as LD_{50}s (lethal oral doses to 50% of animals), LC_{50}s (lethal concentrations to 50% of animals), carcinogenicity, mutagenicity, and teratogenicity. The references for this information are given. Some physical property information is given, as are chemical formulas and information about health standards including Environmental Protection Agency, Department of Transportation, and OSHA standards. *RTECS* contains over 160,000 entries and is no longer available in hardcopy.

Trade Associations. Most industries have formed trade associations. Individual member companies pay dues to fund these organizations, and the resulting associations perform various services. One such service is the dispensing of information to the public about the industry. Some trade organizations have safety and health committees that produce safety and health materials about the particular industry. There are many trade associations in the United States and some of the more important ones in terms of safety and health information are listed below. A more extensive listing can be found in the BNA Occupational Safety and Health Reference File.

- American Chemistry Council, formerly Chemical Manufacturers Association (Washington, DC)

- American Petroleum Institute (Washington, DC)
- American Welding Society (Miami, FL)
- American Forest & Paper Association (Washington, DC)
- Synthetic Organic Chemicals Manufacturers Association (Washington, DC)

Plant-Specific Information Sources

Very often the hygienist is able to obtain from the individual plant a great deal of information on materials, processes, production rates, demographics, and records, prior to entering the plant. Various types of information may be available, including the following:

Advertising and Promotional Information. Most plants have pamphlets and brochures about the company that contain production and other valuable information. Often, demographic and process information can be gained from this literature. Such promotional information is commonly available to the public.

Process/Training Documents. Plants often possess documentation that is used to train operators and supervisory management in the workings of the plant. These documents are generally not available to the public but may be obtained by the hygienist. They contain valuable information about processes, materials, standard operating procedures, and the work tasks that are to be performed by the operators. Very often the work practices or procedures are specifically outlined and, by knowing them, the hygienist can determine if the procedures are being followed properly. In addition, the information informs the hygienist about what the operators should know with respect to safety and health.

Records. The various records that plants can be expected to have include the following:

- OSHA form 301 describing accidents and form 300 summarizing accidents that have occurred in the past year.
- OSHA inspection reports.
- Reports of previous hygiene surveys conducted by consultants, on-site hygienists, government consultation agencies, or insurance inspectors.
- Medical reports containing summaries of medical information (individual identifiers should be removed).
- Reports on demographics. The personnel or some other appropriate department will often print reports on numbers of employees, age, race, sex, etc. In addition, production data often are available in report form.
- Material safety data sheets. It is now required in the United States, as well as many

other countries that material safety data sheets for each hazardous chemical used or produced at a particular manufacturing plant be held on-site. These documents should contain chemical-specific information and can be used by the hygienist for the background survey. The information should include: identity of chemical(s) in the product; physical and chemical characteristics; physical hazards such as fire and explosion; health hazards; exposure limits; assessment of carcinogenicity; safe handling, control, and emergency procedures; and the name and address of the manufacturer.

- Written programs. It is required by law in the United States that companies utilizing respirators for control of occupational hazards have a written program available that documents various aspects of respirator use. In addition, there are other written programs and written recordkeeping requirements mandated by law. In many plants, written programs exist for the overall occupational safety and health program; these are not mandated by law (as of 2007) but exist as a matter of good policy. The following types of written programs and procedures are often available and should be sought:

    - Respiratory protection*
    - Chemical protective clothing
    - Confined space entry*
    - Hearing conservation*
    - Hazard communication*
    - Industrial hygiene monitoring or exposure assessment
    - Medical monitoring**
    - Overall occupational safety and health

    * required
    ** required for some substances

SUMMARY

The above is not an exhaustive compilation of the information sources used by every hygienist. However, most hygienists utilize information from all of these sources at one time or another, and often on a regular basis. The reference texts or texts similar to them should be a standard part of every hygienist's library. In addition, a hygienist should subscribe to one or more journals, and should at least read the abstracts monthly to determine what may be happening in areas of interest. Weekly periodicals are extremely important for keeping the hygienist up-to-date on government matters. Finally, all hygienists should now be computer literate. For purposes of conducting efficient literature searches and managing the information that hygienists now collect, there is no substitute for computer systems. Although hygienists in various types of roles must modify the survey process to suit their needs, there is, nevertheless, always a need to perform literature searches and track down information via telephone or computer, or by any other means necessary regardless of one's role. A hygienist who has performed an adequate background survey will be able to perform as an astute detective upon entering the plant. This properly *educated* hygienist will then be ready to begin collecting additional sensory and/or analytical instrumentation information.

QUESTIONS

12.1. Is the background survey generally part of the recognition, evaluation, or control step of industrial hygiene?

12.2. An OSHA compliance officer needs to plan a survey of an electronics plant that manufactures printed circuit boards. She has not entered such a plant before. A new corporate hygienist is also planning a survey of the same plant, one of 13 plants in his company. Each of the 13 plants is an electronics components plant, but the end products are quite different. How will the background surveys for the two hygienists differ?

12.3. Regarding levels of safe or acceptable exposures for chemical contaminants, what documents would you review for the background survey? Assume that you are working in the United States, and you know from a previous review which particular compounds are of interest.

12.4. A compound called propylene glycol monoethyl ether is being used at your facility. It has no ACGIH® TLV® or OSHA PEL. Where can you find information about its toxicity?

12.5. As a consultant, you are asked to evaluate a new plastics processing operation. The process uses proprietary chemicals and is patented. Discuss some resources/techniques you can use to find out about the process and its hazards.

REFERENCES

AIHA: Odor Thresholds for Chemicals and Established Occupational Health Hazards. Am. Ind. Hyg. Assoc. Fairfax, VA (1989).

Burgess, W.A.: Recognition of Health Hazards in Industry. Wiley. New York (1995).

Cralley, L.V.; Cralley, L.J. (eds.): Industrial Hygiene Aspects of Plant Operations, Vols. 1-3. Macmillan, New York (1984). Out of print.

Hathaway, G.J.; Proctor, N.H.: Proctor and Hughes' Chemical Hazards of the Workplace, 5th Edition. Wiley-Interscience, NY, NY (2004).

IARC: Monographs on the Evaluation of the Carcinogenic Risk of Chemicals to Humans. International Agency for Research on Cancer, WHO, Lyon, France (2005) http://www-cie.iarc.fr/monoeval/allmonos.html.

IPCS: Environmental Health Criteria. International Programme on Chemical Safety, World Health Organization Publications, 1211 Geneva 27, Switzerland. http://www.who.int/ipcs/publications/ehc/en/ (var. dates).

Kirk-Othmer: Concise or unabridged Encyclopaedia of Chemical Technology, 5th ed., 27 vols. Wiley, New York (2004).

Lewis, R.J., Sr.: Hawley's Condensed Chemical Dictionary, 14th ed. Wiley, New York (2002).

Lewis, R.J., Sr.: Sax's Dangerous Properties of Industrial Materials, 11th ed. Wiley, New York (2004).

O'Neil, M.J.; Smith, A.; Heckelman, P.E.; Budavari, S. (eds.): Merck Index: An Encyclopedia of Chemicals, Drugs, and Biologicals, 13th ed. Wiley, New York (2001).

NIOSH: Registry of Toxic Effects of Chemical Substances, National Institute for Occupational Safety and Health, Cincinnati, OH. http://www.cdc.gov/niosh/rtecs/default.html (2007a).

NIOSH: Criteria For a Recommended Standard ... National Institute for Occupational Safety and Health, Cincinnati, OH. http://www.cdc.gov/niosh/critdoc2.html (2007b).

NIOSH: Health Hazard Evaluation ... National Institute for Occupational Safety and Health, Cincinnati, OH. http://www.cdc.gov/niosh/hhe/default.html (2007c).

NIOSH: Publications Catalog. National Institute for Occupational Safety and Health, Cincinnati, OH. available electronically through NIOSHTIC-2 http://www2.cdc.gov/nioshtic-2/Nioshtic2.htm (2007d).

NIOSH: Manual of Analytical Methods, National Institute for Occupational Safety and Health, Cincinnati, OH. http://www.cdc.gov/niosh/nmam/ (2007e).

NIOSH: Pocket Guide to Chemical Hazards, No. 97-140. National Institute for Occupational Safety and Health, Superintendent of Documents, U.S. Government Printing Office, Washington, DC (2004). http://www.cdc.gov/niosh/npg/npg.html (2007f).

NTP: Annual Report on Carcinogens, 11th ed. National Institute of Environmental Health Sciences, Research Triangle Park, NC http://ntp.niehs.nih.gov/ntpweb/ (2005).

OMB: Standard Industrial Classification Manual. U.S. Office of Management and Budget, for sale by National Technical Information Service. Springfield, VA (1987).

OMB: North American Industry Classification System. U.S. Office of Management and Budget, for sale by National Technical Information Service, see also http://www.bls.gov/bls/naics.htm (1997).

OSHA: Code of Federal Regulations, Vol. 29, Parts 1900 to 1910 (§§ 1901.1–1910.999). U.S. Government Printing Office, Washington, DC. http://www.osha.gov/ (2007a).

OSHA: Code of Federal Regulations, Vol. 29, part 1910 (§1910.1000 to end). U.S. Government Printing Office, Washington, DC. http://www.osha.gov/ (2007b).

Patty, F.A.: Industrial Hygiene and Toxicology, 2nd (revised) ed., Vol. I. Wiley, New York (1958).

13

QUALITATIVE SURVEYING – THE WALKTHROUGH

INTRODUCTION

The hygienist follows a process similar to that of a detective solving a crime. For the hygienist, the crime is an inadequately controlled disease agent. The detective, upon first hearing or suspecting that a crime has been committed, seeks background information about the crime or persons committing it. For example, a file on a suspect that indicates previous crimes committed would be considered background information. The hygienist gathers this type of information in the background survey. Then, equipped with the background information, the detective arrives at the scene of the crime and uses the five senses to try to reconstruct the crime. The hygienist performs the same tasks in the walkthrough survey. The detective then gathers pieces of evidence that will give information not obtainable with the senses. Typically these pieces of evidence will be analyzed at the crime lab. The hygienist performs the same task in the in-depth survey by collecting samples of air and other media.

The manner in which the walkthrough survey is conducted depends upon the role of the hygienist (plant hygienist, consultant, etc.). The hygienist's role will usually dictate whether the walkthrough and in-depth surveys are on a continuing and open-ended basis or are finite in time. Walkthrough surveys can last for a period of a few minutes to many days, and can be conducted continuously or sporadically over a much longer period of time (a year). Often the nature of the walkthrough and the nature of the in-depth surveys are interdependent.

In-depth surveys made by plant hygienists can be defined to include a year or more of exposure time. Considerable walkthrough time may be necessary to plan such a survey. As will be seen in the next chapter, an in-depth survey that takes place over a very short period of time and is not performed on a continuous basis is only a snapshot of the overall worker exposure pattern. From this point of view, the word *in-depth* is misleading, and *quantitative* is more descriptive of the survey that gathers data after the walkthrough survey. This usage implies that the walkthrough survey is qualitative in nature. Although some quantitative data, such as air velocity readings or direct instantaneous readings of airborne concentrations, may be taken on the walkthrough, most of the information is sensory and qualitative. Even some of the instrumentation data is often semiquantitative at best.

Regardless of the continuity or the duration of the walkthrough survey, the results of the walkthrough are only as good as the preparation of the hygienist. The preparation of the hygienist is only as good as the background information that the hygienist has gathered about the processes. There are many reasons for this; however, perhaps one example will indicate how an inadequately prepared hygienist can easily miss important workplace exposures. In the smelting of metals, *removing the dross* is a relatively short-term task that varies in its frequency of repetition. The dross is the buildup of impurities on the surface of the molten metal, and it must be removed before the metal is poured into molds. During a short walkthrough survey of the secondary smelter, it is very likely that a hygienist would not observe the removal of the dross, and consequently, would not see a task that may be the most important task in the operator's overall exposure. If the hygienist has not performed an adequate background survey to determine how the process works, then he or she may not observe the removal of the dross or ask questions about it. In industry, many operations are not continuous, but rather are task-oriented. It is, therefore, very likely that in a short-duration walkthrough in which a snapshot is obtained, many tasks and other processes will be missed. Maintenance tasks are almost always missed. These omissions can be discovered only through knowledgeable questioning based on the background survey.

QUALITATIVE SURVEYING

Comparing the Background Survey with Reality

The hygienist should try to walk through the process from its beginning to its end. Plants are not necessarily laid out in a logical manner; so if the walkthrough is started at the *open door*, it is very likely that one will not be observing the process flow from raw materials through the various stages that lead to the end product. If the hygienist has performed an adequate background survey, then he or she will be able to determine the appropriate starting place for the process (or at least ask intelligent questions about it) and anticipate the various steps along the way.

In walking through the process, the hygienist should constantly compare preconceptions about the process with reality. Is the process developing in the manner that was determined from the written process descriptions? Are the materials that are being used those indicated in written process descriptions? Are the workers' work practices and the written standard operating procedures for manufacturing the product the same as those described in the literature? If the hygienist observes discrepancies, these discrepancies should be questioned. For example, the hygienist may observe that a particular raw material is not being used; another has been substituted for it. Upon inquiring, the hygienist may find that the new material is cheaper than the old one. This leaves open the possibility that the new material is more toxic than the old one, as price appears to be the main selection criterion. On the other hand, the hygienist may be informed that the new material is being used because the supplier deems it to be less toxic. Now the hygienist would like to have information from the supplier that documents why the new material is less toxic. Could it be that the new material has a lesser degree of acute toxicity. However, upon further research the hygienist determines that the new material has a high risk of chronic toxicity.

Whom to Interview

What we read in a textbook gives us a frame of reference that we can modify according to what we actually observe. However, observations can take us only so far, particularly in a short walkthrough survey; so we must depend heavily upon information supplied to us by plant personnel. Obtaining good, thorough, honest and, most important, unbiased information from plant personnel requires communication skills that are not possessed by everyone but that can be developed. One of the most important requirements for good communication is that the parties gain each other's trust. Gaining the trust of various plant personnel, particularly those of a different gender, race, or socioeconomic background, requires a great deal of skill. Some of us develop this skill naturally through our social lives; however, others must work hard to obtain this skill. The development of communication skills for the practicing hygienist is extremely important and cannot be overlooked.

The types of personnel with whom the hygienist must speak will vary based on the hygienist's role. In a government compliance inspection, the hygienist usually has a preliminary meeting with management prior to conducting a survey. Consequently, the highest levels of plant management are usually the necessary starting point for interviews. Some of the questions asked of upper plant management will be similar to those for supervisory management or even operators. However, each group of personnel will have a different perspective, and the hygienist must sort out these various perspectives to obtain a true picture.

Never jump to conclusions, and try to adjust for any built-in biases. It is very difficult for a hygienist, in a snapshot interview, to become completely cognizant of the various types of interrelationships that have developed between management and labor. These interrelationships can affect the answers given to the hygienist. Hence, the hygienist must do his or her best to ascertain what biases may exist among the various plant personnel, and then to interpret the answers according to the perceived biases.

In addition to the upper levels of plant management, other levels of both line and staff management should be interviewed. Line management are those personnel who have a direct responsibility for production or some aspect of production and are part of the pyramidal hierarchy from the plant manager down to the lowest level of the workforce. Staff managers are peripheral managers who often report directly to either the plant manager or an assistant to the plant manager and have only peripheral responsibilities for production; that is, they generally have no direct line to the operator or the production line employee. Usually a plant hygienist is considered part of staff management.

Members of line management who should be questioned by the hygienist include managers responsible for production in those areas of the plant where the hygienist has a concern, and the foreman and supervisors beneath this manager who have the greatest degree of interaction with the workers. These members of management often include engineers responsible for controlling process parameters through the equipment operators.

Members of staff management who should be interviewed include engineers responsible for establishing process parameters. Design engineers can explain how a process is expected to work; supervisory engineers can explain how it has been modified to work effectively. Disagreement between the two can occur. Very often there is a supervisor at the process level. For example, a line person is responsible for production output of the electric arc furnace of a secondary smelter. Typically each shift will have a foreman who manages the employees on that shift. In addition, there may be a staff engineer(s) responsible for research on improvements or refinements, air pollution controls, new additives, etc. for the process. In some companies, the supervisor may tell you that things run in a certain way that may be true on the day shift, as that is the time when the supervisor typically works; but the foremen on the night and graveyard shifts may share different versions of what should be the same process procedures. The staff engineer may have yet another view.

Chemists are often responsible for quality control (QC) when a product can be tested by using chemical methods. Engineers may be responsible for QC when physical properties of the product are important. In either case, QC personnel can inform the hygienist about changes in product quality that may reflect equipment malfunction and help to explain high exposure levels.

Other occupational safety and health professionals may be interviewed as available. A safety manager or representative and/or a nurse is present in many plants. These personnel tend to have a better understanding, appreciation, and perception of risk than persons without OSH training. In addition, they possess knowledge of accidents, injuries, and illnesses that may not be in the written records. In some cases, the hygienist may choose to interview a contracting physician who may not be located on-site but may appear at the plant on certain days of the week or the month. These persons will have differing perspectives from those of management and the workforce.

Often the operators represent the most important source of information to the hygienist. Not surprisingly, a supervisory or design engineer may tell a hygienist that a process should operate in a certain way, and then the hygienist may find an operator performing the process differently. There are a number of reasons for this. It may be that the operator has found it easier to perform the process his or her own way, even though that may affect the quality of the product or may affect exposure levels. On the other hand, it may be that the operator and the immediate foreman have determined that a better product can be made if the process is modified accordingly. In any case, the hygienist needs to determine exactly how the process works and the reasons for any modifications. The hygienist can then use his or her detective abilities to determine the likelihood that modification is affecting exposures.

Most important, the operators are responsible for the process on a day-to-day, even a minute-by-minute, basis. Simply because they are not there to observe it, the line supervisors may not have full knowledge or recollection of exactly how frequently a spill, a malfunction, an off-specification product, or other events occur that can lead to high exposures. Operators usually have an accurate perspective on the frequency of such events.

Of course, such perspectives can be tainted by any operator bias. An operator who does not like the job and thinks it is dirty or may be harming his or her health (regardless of the truth) may be biased to think that spills occur much more often than they do. On the other hand, a supervisor who thinks that work is available to those who wish to get paid, and that people should not complain about their jobs, may believe that spills are very infrequent, and that those that do occur are of no important consequence when, in fact, this is far from the truth. The hygienist must sort out the various perspectives. The hygienist must be a communicator, and communication requires not only getting your point across, but adequately interpreting the points made by others.

Questions to Ask

When the various workplace personnel are being interviewed, the hygienist should anticipate the kinds of questions that must be asked to obtain needed information. It is best to make a list of the questions prior to interviewing the personnel. Some questions will be asked repeatedly of line management, staff management, and workers, as well as safety and health professionals; other questions will be more specific to a particular group of personnel. What follows is a partial listing of the kinds of questions that one should ask. It is not meant to be an exhaustive list, but a review of it should remind anyone conducting a walkthrough survey for a particular situation that there are other relevant questions.

1. What is the status of current production rates? Are they normal? Are changes expected in the near future or have changes occurred recently?

2. Is the process currently operating as it normally does? Have modifications been made recently or are they planned in the future? Are the current raw materials the same as have always been used? When were recent changes made? Are the temperature, pressure, and other critical operation parameters at normal status?

3. Are the number of workers associated with the particular process at normal levels? Is anyone absent? Are there unusual rates of absenteeism for the particular process?

4. Does the process operate similarly on all shifts? Are maintenance or other activities carried out on one particular shift? Are all shifts fully staffed?

5. What is the maintenance schedule for the equipment? How is maintenance performed?

6. How are special procedures carried out? How is sweeping performed? How is spill response conducted?

Using the Senses

Most of us are endowed with the ability to taste, smell, see, hear, and touch. Although it seems like a relatively simple and straightforward idea that one should use these senses when conducting a walkthrough survey, it is amazing how often something will be overlooked in a plant because the senses did not detect it. Efficient use of the senses often comes only with experience. However, even the most inexperienced hygienists will conduct a more thorough walkthrough survey if they concentrate particularly hard on using the various senses at each juncture in the survey. Such a focus is particularly important because it is easy for one of the senses to be so overwhelmed that the others are not used. For example, in an industrial situation where noise is particularly high, the hygienist may be so overwhelmed with the noise that the other senses do not detect a particular odor in the area, or the hygienist does not see a particular work practice that could lead to a significant exposure or an improperly operating ventilation system. Therefore, it is important that the hygienist be cognizant of the ability of strong stimuli to overwhelm a particular sense at the expense of others. Take your time and allow all of the senses to survey the situation.

ODOR THRESHOLDS

Although sight and hearing certainly are two of the more important senses to use in the workplace, the hygienist has historically relied upon the combined senses of taste and smell. These two senses are closely related, as they can both be stimulated by airborne molecules. In 1958, Patty went so far as to recommend that hygienists be able to detect the odors of the major industrial gases and vapors, and even to be able to estimate their concentrations by odor. It is possible to do this; some persons are trained to recognize odors and their concentrations. However, even the senses of trained persons are subject to the day-to-day variability caused by environmental fluctuations (temperature, humidity, etc.), and changes in the person's biology. For example, a simple cold or allergic reaction can completely change one's ability to detect a particular odor.

Historically, the detection of odors was relied on fairly heavily to estimate workplace concentrations of gases and vapors. Fifty years ago, instrumentation for such purposes was either nonexistent, inaccurate, or cumbersome. Though instrumentation has vastly improved, odor detection continues to be used as a means of recognition (identification) and even evaluation (quantitation) of gases and vapors.

Quantitating gas/vapor concentrations by using odor detection relies mainly on the concept of odor threshold. If the odor of a chemical can be detected only at concentrations above 100 ppm, then if the exposure limit is 500 ppm and the odor cannot be detected in the workplace, the concentration that exists must be below the odor threshold and thus acceptable. Unfortunately, this inexact concept is the concept of odor threshold that too many occupational health professionals still use. Because of individual variability, in reality there is no single concentration for a given chemical that can be said to be the odor threshold for a population of individuals. The same is true for an individual. What may be a person's odor threshold today will change tomorrow due to weather, biological variables, and changes in the setting in which the odor exists.

These points can be illustrated by odor threshold data, which were reviewed by a committee of AIHA (AIHA, 1989). The first part of the review consisted of screening those studies that met certain criteria. A portion of those acceptable studies was then used to determine a geometric mean air odor threshold. However, there are actually two odor thresholds that can be estimated, one based on detection of the odor and the other based on recognition of the odor as a specific chemical. Usually the detection threshold is lower than the recognition threshold.

Table 13-1 shows some examples of data from the AIHA publication. No attempt was made to select data that had particularly wide ranges. Nevertheless, it can be seen that the median (geometric mean) odor thresholds are difficult to pinpoint. Furthermore, it is important to point out that the ranges (last column) do not represent absolute maxima or minima for individuals. Rather they are ranges of medians taken from different test panels. Hence, the odor thresholds of some individuals can be expected to be above or below the *range of acceptable values* and even outside the *range of all referenced values*. The *geometric mean air odor threshold* is essentially a median of medians, so about 50% of persons will have odor thresholds greater than or less than this value. Furthermore, if the distribution of odor thresholds is skewed right, as is most likely, the 95th percentile will be farther from the median than the 5th percentile (see Chapter 7 for discussion of lognormal distribution).

The reliance on odor detection as a tool in walkthrough surveys is largely a choice of the individual hygienist. Several uses of odor thresholds are appropriate, and others are not so appropriate in light of Table 13-1. They can be classed as qualitative (i.e., what chemical is present) and quantitative (i.e., what concentration exists).

With respect to qualitative uses, some might argue that if the background survey is conducted properly, one will know about all of the materials in the workplace and not have to rely on odor to detect what is there. However, compounds are frequently found in the workplace that may not come to light in a background survey. For example, the material safety data sheet for a compound may be unavailable or inadequate to properly identify the compound. Finally, one may need to identify by-products or reactants as a result of an unusual episode or a previously undocumented process. Any of the above scenarios might be at least partially solved with odor detection.

Quantitatively, odors and odor thresholds may be used to help plan an in-depth survey, recommend controls, or declare a workplace in compliance with a PEL. The use of odors and odor thresholds to help plan an in-depth survey is generally appropriate, but the latter two uses could be a bit risky. However, any of the above are best accomplished if the hygienist deter-

Table 13-1
Representative critiqued odor threshold (thld) values[a]

Name	TLV®-1992 (ppm)	Geometric Mean Odor Thld[b] (ppm)	Type of Thld[c]	Odor Character	Range of Acceptable Values[d] (ppm)	Range of All Referenced Values (ppm)
Acetone	TWA = 750 STEL = 1000	62 130	d r	sweet fruity	3.6–653 33–699	0.40–800
Aniline	TWA = 2	2.4	d	pungent oily	0.58–10	0.012–10
Benzene	TWA = 0.1	61 97	d r	aromatic sweet	34–119	0.78–160
1,3-Butadiene	TWA = 10	0.45 1.1	d r	aromatic rubber	* *	0.099–76
Isoamyl Acetate	TWA = 100 STEL = 125	0.22	d	banana	0.0034–209	0.00075–209
Dichloropropane	TWA = 75 STEL = 110	0.26 0.52	d r	sweet	* *	0.26–0.52
Trichloroethylene	TWA = 50 STEL = 200	82 110	d r	ether solvent	* *	0.5–167

[a] Taken from AIHA (1989).
[b] Geometric mean based on AIHA critiqued work of Gemert and Nettenbraiter (1977).
[c] Detection threshold indicated by d; recognition threshold indicated by r.
[d] * indicates only a single accepted value, shown in column three.

mines his or her own odor threshold, something rarely done now. Even so, the hygienist should realize that the actual concentration at which he or she first detects an odor will vary from day to day and from place to place.

Odors can help in the planning of the in-depth or quantitative survey. In order to calibrate equipment for this survey, it is often necessary to first estimate the concentration that will be measured. The nature of the use of the chemical, its physical properties such as vapor pressure and odor information from the walkthrough survey, will help in this process.

The other quantitative uses of odors and odor thresholds relate to deciding to use controls – emergency controls or long-term controls – or that there is no need for controls (an in-compliance workplace). By examining the data in Table 13-1, examples for each of these decisions can be found.

1. Workplace in compliance. In the case of dichloropropane, the ACGIH® TLV® is 75 ppm, the range of all referenced odor threshold values is 0.26 to 0.52. (These data came from only one study.) If this study is reliable, then the workplace is very likely to be in compliance with the TLV® if no chemical odor can be detected. However, the data for acetone represent what is more often the norm, that is, the TLV® is within or close to the range of all referenced values. When using this approach, the hygienist should be confident of at least two pieces of information: the adequacy of the TLV® and that the workplace conditions were average or normal on the day of observation (odor detection).

2. Need for emergency respirator. The data in Table 13-1 for benzene indicate an emergency or out-of-control situation if an odor can be detected. In this case, if the process producing the benzene odor is frequent or continuous, and benzene can be smelled, then an average concentration greatly exceeding 0.1 ppm is likely.

3. Need for long-term controls. In this example, if one were certain that the long-term average concentration of benzene was greater than 0.1 ppm, then emergency use of respirators would be appropriate; but, in addition, plans for reducing the concentration by using engineering controls should be prepared. Odor thresholds do not provide the degree of certainty necessary for a large expenditure. Instead, while respirators are being used, an in-depth survey should be completed to ascertain, more certainly, the workplace concentrations. In essence, odor should not be used for recommending long-term controls unless abatement of an unpleasant, but not necessarily hazardous, odor is warranted.

The use of odors and odor thresholds requires a few more words of caution. The ability to detect an odor and to gauge its concentration at levels above the odor threshold, if reliably determined by the hygienist, represents only a snapshot of the concentration that exists at that particular time. For this reason, a decision not to conduct an in-depth survey for a particular compound simply because the hygienist cannot smell an odor is not, by itself, justifiable. The propensity of a compound to enter the air as a result of its physical properties and how it is used in the process should also be considered.

The sense of smell is easily fatigued if it is continuously stimulated or overstimulated. It is well known, for example, that hydrogen sulfide has a rotten-egg odor and that, at low concentrations, it can be smelled relatively easily. However, as concentrations increase, the sense of smell becomes overstimulated, and the brain blocks the stimulus, thereby causing the person to think that there is no longer any hydrogen sulfide. Similarly, workers who work day after day with a particular compound suffer odor fatigue as a result of their continuous exposure. Exposures from the day before can cause a two- to threefold increase in odor thresholds the next day (Ahlstrom et al., 1986). Therefore, the odor thresholds can be considerably higher than those of a population that has not been continuously stimulated by the chemical.

Finally, when the AIHA publication was prepared (1989), odor thresholds could be designated according to the AIHA criteria for only 110 compounds. Although other publications may contain odor threshold lists, at present only this limited AIHA listing can be used reliably. Other lists often report only a single value, and Table 13-1 shows that single values are very misleading representations of odor thresholds.

Looking for the Unexpected

In order to expand the space and time framework associated with a particular walk-through survey, the hygienist must be particularly alert. With respect to expanding the time framework, the hygienist must ask questions about future and past operations in order to determine how they compare to current status. With respect to the space framework, the hygienist will see what he or she is taken to see, or what he or she asks to see. Consequently, what one asks to see is extremely important, since very often workplace personnel will not take the hygienist to particular areas, either because they simply forget, or because they do not think that those areas are particularly important. In some cases, there may even be an attempt to hide certain areas from the hygienist. The hygienist should be particularly keen to ask questions about intermittent and discontinuous operations that may not be occurring during the walk-through. Maintenance is one of the most common and important operations. In some large plants (particularly in the petrochemical and metals industries) a full-scale maintenance operation requiring days to weeks (a *turn-around*) will take place approximately yearly. Exposures (at least to benzene) are often highest at these times (OSHA, 1987). Other maintenance activities are more frequent but also more subtle. A good example is the replacement of empty gas cylinders that may have once contained chlorine or ethylene oxide. Very specific procedures are commonly needed regarding the closing of valves and the removal of hoses if high exposures are to be avoided.

In some processes, the hygienist must be aware of pre-treatment and batch activities. For example, a particular operation that occurs in the beginning of a process may be performed repeatedly over a period of time until a certain amount of raw material has been stockpiled. The stockpile is then used for weeks until it is almost depleted, at which time the repeated batch process takes place again. Consequently, upon observing the process, the hygienist may assume that the raw material is prepared somewhere else. This assumption should not be made.

On the opposite end of the process, the hygienist should be particularly concerned about the disposal of unwanted by-products. Particularly, the hygienist should be aware that all control efforts typically lead to the accumulation of contaminants for disposal. For example, ventilation systems used for control of aerosols will accumulate aerosols within the ductwork, in settling bins, or in other types of air cleaners, such as filtration units. Someone must maintain these units and consequently encounter the accumulated contaminants. Solid and liquid contaminants may be disposed of on-site (although, at least in the U.S. and Europe, this is uncommon today) or may be accumulated on-site and then shipped off to another site. Particular workers are often associated with the waste operations. The hygienist should be aware of regulations regarding hazardous waste and should inquire particularly about workers who may have to handle hazardous waste at the work site.

Laboratory facilities are also often overlooked. Most industrial plants have quality control laboratories, each with its own set of hazards, that the hygienist must examine. In addition, many industrial plants have small-scale versions of processes or proposed processes. These pilot plants are often laboratory-scale. Very often adequate measures for controlling exposures are not taken; pilot plants are often permanent, but operations change frequently, and therefore, expenditures for a control that may be needed only for a few weeks may be limited.

Warehouses for storing both raw and end products can be important sources of exposure, particularly if spill procedures are not adequate, or if the product *cures* or *off gases* during storage. In addition, many warehouses utilize forklifts or other motorized conveyances that can cause contamination of the air with carbon monoxide or other important contaminants. Battery-operated forklifts are not without their problems, as the batteries must be recharged, and the charging rooms can be hazardous if evolving gases (e.g., H_2) are not controlled.

The hygienist should seek out unusual places. In essence, anywhere that human beings go, the hygienist should also go or at least take a peek. Furthermore, even places where workers do not commonly go, but that nevertheless are places that may be associated with the overall *health* of the workplace, should be examined. For example, for many buildings the roof is not a place where workers commonly go. However, the roof is often the point at which ventilation systems exit the building, and where *fresh* air is taken into the building. Therefore, the possibility exists that contaminants being exhausted from the building can be reentrained in the building. The hygienist can only determine this by going onto the roof and examining the ventilation systems. The hygienist should also enter basement and attic areas and, if it is appropriate, look inside tanks or other confined spaces where workers may enter. It may not be necessary that the hygienist do exactly what the workers do, but the hygienist should at least see all areas where workers enter in order to spot any potential hazards that he or she could not otherwise find. To simply ask workers or other personnel what the inside of something looks like or what is on the roof is not good enough, as these personnel are not trained hygienists and consequently may not know what is important.

SETTING PRIORITIES FOR IN-DEPTH SURVEYS

The ultimate decision to be made from the hygiene survey process is whether controls are adequate for each of the hazards found in the workplace. After the walkthrough survey, it may be obvious to the hygienist that controls are not adequate for certain disease agents, and therefore, no further evaluation is necessary before certain control procedures are put into place. Simple control procedures such as retraining of workers and enforcement of or changes in standard operating procedures can be initiated at minimum expense. However, for some contaminants, further evaluation must take place in order to determine whether more controls are necessary. Further evaluation usually requires the use of instrumentation to quantitate the concentrations of airborne contaminants. Even when controls seem to be in order, and exposures appear to be acceptable, documentation of airborne concentrations is often necessary for future reference and legal considerations.

With respect to those contaminants that may need further evaluation, and where resources are limited (often the case), the hygienist must set priorities for the contaminants that are to be considered in the in-depth survey. For the plant hygienist, often the most important contaminants are surveyed first, and those thought to be of less importance are surveyed as time and resources permit. But how can priorities for the in-depth survey be set? What factors should one consider in determining how to select potential hazards for the in-depth survey? The hygienist must perform a qualitative risk evaluation. Those compounds of higher perceived hazard should be surveyed first. The difficult part is to determine how to assess qualitative risk and what factors should be considered in qualitative risk formulation.

Setting Priorities with Qualitative Risk Evaluation

Risk is the product of hazard and exposure. Hazard for airborne contaminants is essentially toxicity. A qualitative risk assessment involves estimating risk based upon qualitative estimates of hazard and exposure. The qualitative risks are then used to set priorities for quantitative assessment. A wide range of qualitative risk evaluation techniques have been formulated, largely for determining acute or safety hazards in workplace settings (see, e.g., Gressel and Gideon, 1991). By acute hazards we mean explosion, fire, etc. These techniques range from simple to complex, with the degree of complexity having a direct correlation with the difficulty of obtaining the data needed for the qualitative risk evaluation. Although these qualitative risk evaluation techniques do not have direct application to qualitative risk evaluation for exposure to chemical contaminants, they can be adapted to it, and a review of the techniques may be appropriate before one enters into a complex priority-setting procedure. However, for more simple priority-setting procedures, as might take place after a walkthrough survey, a procedure that considers the most important aspects of chemical risk is appropriate.

For a consulting or regulatory hygienist, the priority-setting process may often be cognitive only. However, a written description of the rationale for including or eliminating certain contaminants from the in-depth evaluation is warranted. If nothing else, ethical and future legal considerations make this a good practice. Obviously the contaminants for which there is the greatest risk should receive attention; however, what is the compound with the most health risk in a workplace setting? Is it the compound having the greatest toxicity? Is it the compound to which the greatest number of people are exposed? Is it the compound used with the greatest frequency? Or in the greatest tonnage? Is it the compound with the greatest vapor pressure? There are a number of factors that can be considered in the setting of priorities. Frequently, the three most important factors are the toxicity of the compound, its propensity to enter the workplace air, and the number of workers who are exposed to it. The latter two are exposure variables. Time of use is also a consideration, and if such data are gathered during a careful walkthrough, this can be considered in the priority setting.

TOXICITY

The toxicity of a compound can be roughly indexed with available exposure limits. In other words, for priority-setting purposes, a compound with an OEL of 400 ppm may be thought to be twice as toxic as a compound with an exposure limit of 800 ppm. TLVs® and other exposure limits can be used for such purposes if one realizes (1) the limitations of those exposure limits and (2) the limitations of the comparison process. The *TLV® and BEIs®* Book states that TLVs® are not a *relative index of toxicity*. Since further details are not given to explain this statement, we must try to interpret logically what constraints it places on the use of TLVs® in priority setting.

Regarding point (1) above, the risk of disease associated with exposure to two compounds at their respective TLV® concentrations, although not stated, should be presumed different. In theory, this should not be true (the acceptable risk level, such as 1 in 1,000, should be about constant), but in fact residual risk levels are not likely the same for two chemicals at their respective TLVs® (per discussion in Chapter 10). The risk is not stated in the TLV® *Documentation*; therefore, relative risk is probably not reflected in a ratio of two TLVs®.

Relative to point (2) above, the toxicities of acute and chronic toxins should not be compared. Comparisons should be made only for compounds having similar target organs or compounds from the same family (e.g., halogenated aliphatic hydrocarbons). Also, if two compounds do in fact have the same stated relative risks of 1 in 100 for a specific neurological disorder at TLVs® of 1 and 10 ppm, respectively, then we might say that the two compounds have a tenfold difference in neurological toxicity. However, one or both of the compounds may have effects on other organ systems, e.g., the reproductive system. It is likely that the relative risks of reproductive effects at the respective TLVs® would not be 1 in 100, nor would the reproductive risks for the two compounds likely be equal.

The Threshold Limit Values for Chemical Substances (TLV®-CS) Committee defines the TLV® for each compound to be that level of exposure which will protect *nearly all workers*. One would expect that for two compounds having similar effects (e.g., nephritis), their TLVs® would be good approximate indices of toxicity, i.e., the word *nearly* would have the same meaning for each TLV®, but we do not know this. For these reasons, the difficulties of using the TLVs® or PELs as relative indices of toxicity or of residual risk at the OEL are evident. Unfortunately, they are the only indices available.

Therefore, without having a quantitative baseline residual risk definition for TLVs®, their use as relative indices for toxicity in priority setting must be undertaken cautiously. A safety factor should be used. For example, compounds having TLVs® that differ by a factor of 4 or less might be considered to have equivalent toxicities.

TENDENCY TO BECOME AIRBORNE – VOLATILITY

The tendency of a compound to enter the air is directly related to its vapor pressure. For example, a compound with a vapor pressure of 50 mm or mercury (~ 7000 Pa) at standard temperature, potentially would be twice as likely to become airborne as one with a vapor pressure of 25 mm of mercury, all else being equal. Under similar conditions, the first compound would be expected to form concentrations in the air approximately twice as high as those of the second compound. The potential for mixtures to form azeotropes and thereby reduce vapor pressures is a complicating factor. However, by assuming that volatility is proportional to pure compound vapor pressure, the actual volatilities of mixtures are at least overestimated in most cases. There are a few cases where the opposite is true (Glasstone, 1946). For a good listing of binary azeotropes see Gordon and Ford (1972).

TENDENCY TO BECOME AIRBORNE – DUSTINESS

For aerosols, particularly those derived from solids, vapor pressure is not of great importance, and it can be misleading.[1] If the vapor pressure of solids were used in priority setting, no doubt most solids would be of low priority. Dustiness is the property of solids that best relates to potential exposure; but unfortunately it is dependent on several factors, and thus little information is available on it. However, in some situations, particularly where two solid materials are being evaluated for future use in a process, or where differing formulations of a product are being evaluated for consumer use, it may be appropriate to evaluate dustiness. Several techniques are available, as the idea has received renewed interest (Elder et al., 1974; Heitbrink et al., 1989; Cowherd et al., 1989; Plinke et al., 1991; Carlson et al., 1992; Breum, 1999; and Brouwer et al., 2006). Reist and Creed (1994) have proposed a method for determining the respirable mass fraction of a bulk dust sample. Liden (2006) has reviewed pertinent issues and attempts on the part of the UK HSE and the CEN to devise dustiness standard methods.

VHI

The concepts of toxicity and volatility can be combined to give an index of the risk posed by a compound. The highest, or saturated, concentration *(C_s)* of a particular compound that can be reached in workplace air can be calculated from its vapor pressure *(P^o)* at a given temperature and ambient atmospheric pressure *(P_a)*:

$$C_s = \frac{P^\circ}{P_a} \qquad \text{[Equation 13-1]}$$

If C_s is less than the TLV® for a compound, that would indicate that the TLV® concentra-

[1] See Chapter 24 for greater detail. Solids can have vapor pressures that make them semi-volatile and of presumed low vapor concentration. However, they may also be aerosolized in processing and thus exposures to vapor and aerosol occur.

tion could not be reached unless the compound were heated to temperatures above ambient, or somehow physically or mechanically divided such that solid particulate matter or liquid droplets were formed. Similarly, two compounds having a similar TLV® but with one having a vapor pressure five times that of the other would have different potential risk levels.

The concepts have been combined into an index. A vapor hazard[2] index (VHI), as proposed by Popendorf (1984), considers vapor pressure and toxicity. The vapor hazard index is calculated as:

$$VHI = \log_{10}\left(\frac{C_s}{TLV}\right) \qquad \text{[Equation 13-2]}$$

This index can be multiplied by the number of workers exposed and further adjusted by the time of exposure in order to set priorities. Table 13-2 shows vapor hazard indices that have been calculated for fictitious chemicals representing a wide range of industrial compounds. When the VHI is less than zero, the TLV® concentration of the vapor cannot be reached unless the compound is heated above ambient temperature. (Note: Remember the vapor pressure is temperature-dependent, and so Table 13-2 assumes that ambient conditions are 25°C. If other conditions exist, the vapor pressure should be adjusted accordingly.)

An important caution about the use of vapor pressure for assessing airborne propensity should be noted. It is best illustrated with an example. The vapor pressures of polychlorinated biphenyls are quite low («1 mmHg), and so airborne levels are also quite low («10 μg/m^3) in jobs related to electric transformer repair. Yet serum PCB levels can be elevated (Christiani et al., 1986). The exposure is coming via the skin from contaminated surfaces. Thus even though the fugitive emissions are small for such compounds, the emissions appear to condense on surfaces and accumulate. Other compounds in this class would include, for example, pentachlorophenol, some insecticides, and plasticizers. In addition, higher molecular weight compounds that are aerosolized and sprayed (pesticides for example) will drift to both desired and undesired surfaces. Their transferability to the skin of an operator from those surfaces may contribute substantially to exposure (see e.g., Ramwell et al., 2006). The use of surface wipe samples is a technique for assessing the relative contamination of surfaces (ASTM, 2003 and OSHA, 1999).

NUMBER EXPOSED AND FREQUENCY OF EXPOSURE

In order to set priorities for an in-depth survey, the VHI can be modified by using the number of workers exposed and the time of exposure. Consider again the data in Table 13-2. Chemical A has the highest VHI because it also has the highest vapor pressure and lowest

[2] This is better thought of as a risk index since toxicity is hazard, i.e., the potential to cause disease, and volatility is an indicator of exposure potential. Risk is the product of exposure and hazard (see Chapter 10 and Chapters 11 and 12 of Volume 3).

Table 13-2
Example of setting priorities using the Vapor Hazard Index

CHEMICAL	A	B	C
Vapor Pressure @ 25°C (mmHg)	365	113	121
TLV® (8 hr-TWA, ppm)	1	5	350
Fraction of Workforce Exposed	0.1	0.05	1
Averaging Time Fraction	0.06 (30 min)	1	1
VHI	5.7	4.5	2.7
Priority Score[1]	0.014	0.23	2.7

[1] VHI × fraction of workforce exposed × fraction OEL averaging time (8 hours in this case) that exposure takes place.

TLV®. However, only 10% of the workforce is exposed to chemical A, and they are exposed only for a period of 30 minutes per 8-hour shift. These last two factors cause chemical A to have the lowest priority score. The priority score was determined by adjusting the VHI by the fraction of the total workforce exposed and the fraction of the 8-hour day during which exposure takes place. The total workforce can be defined in several ways. It should be defined relative to that area of the plant for which priorities are being set. For a consultant or a compliance officer, this may be the entire plant. For a plant hygienist, a priority scoring system may be developed for each operation, process, area, division, operating unit, etc. In any case, it is important that the workforce be defined so that the denominator used to calculate the fraction of the workforce exposed is constant for each chemical being considered in the prioritization.

It should also be noted that that fraction of the workforce exposed can lead to some unwanted results in extreme cases. For example, in a workforce of 1,000, if only one worker/job is exposed to a very toxic compound, then perhaps that exposure should be evaluated regardless of the priority score.

Finally, note that for each chemical in Table 13-2, an 8-hr TWA-TLV® was used. If any of the chemicals had 15-min STELs or ceilings only, they would have to be considered separately. When both an 8-hr and a 15-min OEL exist for a chemical, the priority score can be evaluated for each. Therefore, if chemical A had a 15-min STEL of 5 ppm, the VHI would decrease, but the fraction of the averaging time would be one.[3] So the priority score would increase.

[3] Although 30/15 = 2, it may not make sense to allow the fraction of the averaging time to exceed one, since for compounds with STELs the value could be as high as 32 if exposure occurred for the entire day. The purpose of the fraction of averaging-time weighting is just to allow an adjustment to be made when exposures are less than 15 minutes or less than 8 hours, and thus TWA exposures for the full period (8 hours or 15 minutes) are affected by the actual exposure time.

The priority scores for A, B, and C, if ranked from high to low, differ by an order of magnitude. Therefore, errors associated with any of the data (particularly TLVs®) would not likely affect the priority score. On the other hand, if priority scores of, for example, 5 and 3 had been obtained, they should be considered equivalent because of inaccuracies associated with using TLVs® as indices of toxicity and vapor pressures to relate volatility for mixtures.

Setting Priorities Based on Modeling Physical and Chemical Data

The conclusion that a chemical does not pose an airborne exposure threat because its saturated vapor concentration is less than the OEL (VHI < 0), is a *back of the envelope calculation*. It is also a simple form of modeling. In Volume 3, we will discover that the median concentration of a contaminant in a space can be predicted with information about the flow rate of outside air to the space and the generation rate of contaminant in the space. Of course the median concentration is just one value of all the possible concentrations in that space, depending on how well it is mixed, but if the median is much lower than the OEL, we feel pretty confident that workers in the space are not overexposed.

There are approaches to estimating exposures other than to actually collect parcels of air and measure the concentrations. These modeling approaches are becoming more sophisticated. However, even the simplest model can give us *ballpark* estimates that are inexpensive. It has even been argued that if exposures vary considerably in a workplace and only a few measurements will likely be made, then mathematical models can be used to estimate the exposure distribution with more accuracy and less cost (Nicas and Jayjock, 2002).

Models for estimating ambient concentrations of pollutants have been developed at least since the 1960s. The U.S. EPA and other environmental agencies around the world have lead the way (EPA, 2005). Generally, these models have been developed to estimate ambient exposures; however, several have application to industrial hygiene problems. In Europe, the European Union has developed an exposure model that is used in regulatory matters to estimate occupational exposure (Northage, 2005).

There is now a wide range of efforts with respect to modeling occupational exposure (e.g., Flynn et al., 1999). Relatively simple models have been used to predict exposures as a function of room size and distance from a source (Cherrie, 1999). Models for concentrations arising from spills or evaporation have been developed (Smith, 2001; Reinke and Brosseau, 1997; and Lennert et al., 1997). Computational fluid dynamics (Bennet et al., 2000 and Feigley et al., 2002) have been used to visually simulate room concentrations and make comparisons with mathematical models that are based on diffusion. Markov chain theory has been used to model exposure probability distributions (Nicas, 2000). The physical and chemical aspects of spray painting have been empirically determined to derive a model for spray painting exposure (Flynn et al., 1999). A chemical mass balance receptor model has been developed for an electroplating operation to predict the contribution from each source made to several receptors or workers (Wadden et al., 1998). Mathematical modeling has been used to optimize job rotation so as to achieve a minimum average noise exposure (Tharmmaphornphilas et al., 2003).[4]

Modeling will become more prominent in hygiene work in the future. Resources are always limited, and it is simply a matter of understanding (and not being intimidated by) models that leads to their application to estimate exposure, determine the most cost effective controls, estimate the effects on human physiology of differing exposure profiles, and find solutions to a host of other hygiene issues. Further details are found in Chapter 15.

An Approach to Setting Priorities Based on Potential Risk and Uncertainty

AIHA (1998, 2006) has proposed an approach to priority setting that involves the gathering of mostly subjective data on the workplace, workforce, and the agent (essentially all the types of information that have been discussed in this and the previous chapter) to make an initial determination of whether exposures are acceptable, uncertain, or unacceptable. Sometimes quantitative exposure data may be available to add to this process, and over time as the assessment program matures and more data are accumulated, the process becomes less subjective and more objective. This is the iterative process that a plant hygienist should use. Exposures that are unacceptable should be controlled; those that are acceptable should be reassessed periodically or anytime conditions change. When time and resources permit acceptable exposures should be quantitatively assessed to verify the initial subjective finding and increase certainty. Finally, uncertain exposures require quantitative exposure assessment in order to make a certain finding of acceptability.

Priorities can be set based on risk and uncertainty (U). Remembering that risk (R) is the product of exposure (E) and hazard (H), then the priority (P) for further assessment is

$$P = R \times U = E \times H \times U \qquad \text{[Equation 13-3]}$$

In setting priorities, an initial exposure rating (E) should be made for each exposure. Tables 13-3 and 13-4 show two approaches recommended by AIHA (1998). Note that all of the values in Tables 13-3 and 13-4 are acceptable – below the OEL. However, there may be uncertainty associated with these ratings, i.e., these determinations can be made based on no actual exposure measurements or modeling results, only subjective reasoning, or mounds of data as we will discuss in the next chapters. However, whatever the inputs are, they should be documented.

The hazard might be assessed based on a logic like that shown in Table 13-5. Note that this determination can also have uncertainty. The total uncertainty in both H and E might be assigned to one of three categories, each with a score as shown in Table 13-6. Thus, from Equation 13-3, priorities with a score of 0 would be certain and require no further assessment. All other priorities would be ranked based on their score. The higher the score the greater the

[4] It should be noted that job rotation *spreads the dose* and so if later we find that the compound causes toxic effects at lower levels than once thought, more cases of disease will occur.

need to gather quantitative data. How we gather this quantitative data and further assess uncertainty and the acceptability of exposure is the subject of the next three chapters.

SUMMARY

Having completed the walkthrough survey, the hygienist has collected all of the sensory information necessary to go about the task of collecting quantitative data with appropriate analytical instrumentation. Note that the hygienist also should have anticipated the need for and collected other information that might be necessary for the future implementation of controls. Once the analytical results have been returned to the hygienist from the laboratory, the hygienist can use the sensory information and the quantitative data to make a decision about necessary controls.

Table 13-3
Qualitative exposure rating based on average exposure* (adapted from AIHA, 1989)

Category	Description
1	> OEL
2	50–100% of OEL
3	10–50% of OEL
4	< 10% of OEL

* Exposure can be by air, dermal, or ingestion routes. Rating should not take into consideration the use of personal protective equipment. OEL is the acceptable average concentration of an environmental agent with cumulative effects.

Table 13-4
Qualitative exposure rating based on estimate of 95 %tile of exposures* (adapted from AIHA, 2006)

Category	Description
1	95 %tile > OEL; > 5% exceedance
2	0.5 OEL < 95 %tile > OEL; > 5% exceedance of 0.5 OEL
3	0.1 OEL < 95 %tile > 0.5OEL; > 5% exceedance of 0.1 OEL
4	95 %tile < 0.1 OEL; > rare exceedance of 0.1 OEL

* Exposure can be by air, dermal, or ingestion routes. Rating should not take into consideration the use of personal protective equipment. Agent may be of chronic or acute toxicity.

Table 13-5
AIHA qualitative health-effect rating (adapted from AIHA, 2006)

Category	Health effect
0	Reversible effects of little concern or suspected adverse health effects
1	Reversible health effects of concern
2	Severe, reversible health effects of concern
3	Irreversible health effects of concern
4	Life threatening or disabling injury or illness

Table 13-6
A method for semi-quantifying uncertainty (adapted from Mulhausen and Damiano, 1989 and AIHA, 2006)

Uncertainty	Criteria	Score
Certain	Highly quantitative data with tight confidence limits; judgment made on a very good understanding of the risk (hazard and exposure) factors	0
Uncertain	Some quantitative data, good walkthrough and background survey data, perhaps surrogate comparison data from other processes or industries; the judgment is made on some data but clearly more is needed	1
Very, highly uncertain	No quantitative data, poor comparison data, uncertain background and walkthrough data; judgments made on little exposure or health data	2

The important step of planning the in-depth survey cannot be overlooked. Though the background and walkthrough survey can take a considerable amount of the hygienist's time, the in-depth survey is usually much more time-consuming. The hygienist who jumps to conclusions during the walkthrough survey, hurriedly gathers the appropriate instruments and then goes back to the field to collect quantitative information probably will be wasting time and resources. Information collected in this way usually cannot be analyzed in such a way that one can draw meaningful conclusions.

Having set priorities for the compounds to be analyzed in the in-depth survey(s), the hygienist must formulate a question or a hypothesis about each of them regarding their acceptability. Those questions and how they are answered are the subject of Chapter 14.

QUESTIONS

13.1. Based on Chapter 5, prepare a list of questions that you would ask of employees at a facility that performs welding. You will ask the questions during the walkthrough, and assume that during the background survey very little information was discovered about the welding performed in this type of manufacturing facility.

13.2. Two plants manufacture the same product. In one plant, the plant manager signs all safety and health directives as she would sign a production directive. In the second plant, safety and health directives are signed by the safety/personnel officer. Will you approach the walkthrough survey in these two plants differently? How?

13.3. You have just been hired as the first hygienist at a 5,000-employee petroleum refinery that has 32 process units. Generally, exposures are to hydrocarbons, but other contaminants, process additives, catalysts, etc., are used, such as methyl ethyl ketone, ammonia, nickel, and phenol. Examine literature resources on petroleum refining and hydrocarbons. Where should you place most monitoring emphasis in your first year? Do you need a formal scheme such as the VHI or AIHA approach for making this decision? Why?

REFERENCES

Ahlstrom, R.; Berglund, B.; Berglund, U.; Lindvall, T.; Wennberg, A.: Impaired odor perception in tank cleaners. Scand J. Work Environ. Hlth. 12:574–581 (1986).

AIHA: Odor Threshold for Chemicals with Established Occupational Health Standards. American Industrial Hygiene Association, Fairfax, VA (1989).

AIHA: A strategy for assessing and managing occupational exposures, 2nd ed., Mulhausen, J.R. and J. Damiano (eds.). American Industrial Hygiene Association, Fairfax, VA (1998).

AIHA: A strategy for assessing and managing occupational exposures, 3rd ed., J.S. Ignacio and B. Bullock (eds.). American Industrial Hygiene Association, Fairfax, VA (2006).

ASTM: Standard Practice for Collection of Settled Dust Samples Using Wipe Sampling Methods for Subsequent Determination of Metals, ASTM D6966-03. ASTM International, West Conshohocken, PA, USA (2003).

Bennett, J.S.; Feigley, C.E.; Khan, J.K.; Hosni, M.H.: Comparison of mathematical models for exposure assessment with computational fluid dynamic simulation. Appl. Occup. and Environ. Hyg. 15:131-144 (2000).

Breum, N.O.: The rotating drum dustiness tester: Variability in dustiness in relation to sample mass, testing time, and surface adhesion. Ann. Occup. Hyg. 8:557- 566 (1999).

Brouwer, D.H.; Links, I.H.M.; De Vreede, S.A.F.; Christopher, Y.: Size Selective Dustiness and Exposure; Simulated Workplace Comparisons. Ann. Occup. Hyg., Vol. 50, No. 5, pp. 445-452 (2006).

Carlson, K.H.; Herman, D.R.; Markcy, T.F.; Wolff, R.K.; Dorato, M.A.: A comparison of two dustiness evaluation methods. Am. Ind. Hyg. Assoc. J. 53:448-454 (1992).

Cherrie J.W.: The effect of room size and general ventilation on the relationship between near and far-field concentrations. Appl. Occup. Environ. Hyg. 14:539-546 (1999).

Christiani, D.C.; Krichel, D.; Fox, N.J.; Baker, E.L.: Persistently elevated polychlorinated biphenyl levels from residual contamination of workplace surfaces. Am. J. Ind. Med. 10: 143-151 (1986).

Cowherd, C., Jr.; Grelinger, M.A.; Englehart, P.; Kent, R.F.; Wong, K.F.: An apparatus and methodology for predicting the dustiness of materials. Am. Ind. Hyg. Assoc. J. 50: 123-130 (1989).

Elder, J.C.; Etlinger, H.J.; Nelson, R.Y.: Chamber studies of visibility-reducing aerosols. Atmospheric Environment 8:1035-1048 (1974).

EPA: Regulatory Environmental Modeling. U.S. Environmental Protection Agency, http://cfpub.epa.gov/crem/ (2007).

Feigley, C.E.; Bennett, J.S.; Khan, J.; Lee, E.: Performance of deterministic workplace exposure assessment models for various contaminant source, air inlet, and exhaust locations. AIHAJ 63:402-412 (2002).

Flynn, M.R.; Gatano, B.L.; McKernan, J.L.; Dunn, K.H.; Blazicko, B.A.; Carlton, G.N.: Modeling breathing-zone concentrations of airborne contaminants generated during compressed air spray painting. Ann. Occup. Hyg. 43:67-76 (1999).

Gemert, L.J.V.; Nettenbreijer, A.H.: Compilation of odor threshold values in air and water. Zeist and Voorburg, The Netherlands: Central Institute for Nutrition and Food Research TNO (CIVO-TNO)/National Institute for Water Supply (RID) (1977).

Glasstone, S.: Textbook of Physical Chemistry, 2nd ed., p. 720. D. Van Nostrand Company, New York (1946).

Gordon, A.J.; Ford, R.A.: The Chemists Companion, p. 24. Wiley, New York (1972).

Gressel, M.G.; Gideon, J.A.: An overview of process hazard evaluation techniques. Am. Ind. Hyg. Assoc. J. 52:158-163 (1991).

Heitbrink, W.A.; Todd, W.F.; Fischbach, T.J.: Correlation of tests for material dustiness with worker exposure from the bagging of powders. Appl. Ind. Hyg. J. 4:12-16 (1989).

Lennert, A.L.; Nielson, F.; Breum, N.O.: Evaluation of evaporation and concentration distribution models – a test chamber study. Ann. Occup. Hyg. 41:625-641 (1997).

Liden, G.: Dustiness testing of materials handled at workplaces. Ann. Occup. Hyg., Vol. 50, No. 5, pp. 437-439 (2006).

Nicas, M.: Markov modeling of contaminant concentrations in indoor air. AIHAJ 61:484-491 (2000).

Nicas, M.; Jayjock, M.: Uncertainty in exposure estimates made by modeling versus monitoring. AIHA Journal (63):275-283 (2002).

Northage, C.: EASEing into the future. Ann. Occup. Hyg. 49:99-101 (2005).

OSHA: Occupational exposure to benzene: final rule. 29 CFR Part 1910. Fed. Reg. 52:34540 (1987).

OSHA: OSHA Technical Manual, Section II, Chapter 2, Sampling for surface contamination. U.S. Occupational Health and Safety Administration, Washington, DC, USA http://www.osha.gov/dts/osta/otm/otm_ii/otm_ii_2.html#2 (1999).

Patty, F.A. (ed.): Industrial Hygiene and Toxicology. Wiley, New York (1958).

Plinke, M.A.E.; Leith, D.; Holstein, D.B.; Boundy, M.G.: Experimental examination of factors that affect dust generation. Am. Ind. Hyg. Assoc. J. 52:521-528 (1991).

Popendorf, W.: Vapor pressure and solvent vapor hazards. Am. Ind. Hyg. Assoc. J. 45:719-726 (1984).

Ramwell, C.T.; Johnson, P.D.; Corns, H.: Transferability of six pesticides from agricultural sprayer surfaces. Ann. Occup. Hyg., Vol. 50, No. 3, pp. 323-329 (2006).

Reist, P.; Creed, D.: Use of a sedimentation method for determining respirable mass fraction in a bulk dust sample. Am. Ind. Hyg. Assoc. J. 55:619-625 (1994).

Reinke, P.H.; Brosseau, L.M.: Development of a model to predict air contaminant concentrations following indoor spills of volatile liquids. Ann. Occup. Hyg. 41:415-435 (1997).

Smith, R.L.: Predicting evaporation rates and times for spills of chemical mixtures. Ann. occup. Hyg. 45:437-445 (2001).

Tharmmaphornphilas, W.; Green, B.; Carnahan, B.J.; Norman, B.A.: Applying mathematical modeling to create job rotation schedules for minimizing occupational noise exposure. AIHAJ 64:401-405 (2003).

Wadden, R.A.; Liao, S.; Scheff, P.A.; Franke, J.E.; Conroy, L.M.: Application of receptor modeling to indoor air emissions from electroplating. Appl. Occup. Environ. Hyg. 13:829-838 (1998).

14

QUANTITATIVE SURVEYING – EXPOSURE ASSESSMENT INTRODUCTION

INTRODUCTION

Airborne concentrations of chemicals may be monitored by using the methods described in Parts V and VI to determine employee exposures to a chemical. The chemical(s) to be monitored may be chosen according to priorities set in Chapter 13. Although there can be various reasons for monitoring employee exposures or taking measurements of airborne contamination levels, compliance monitoring is a common reason and receives the most discussion in the industrial hygiene and occupational health literature. At first glance, it may appear that monitoring for compliance with standards can be easily accomplished. However, a number of issues are open to discussion.

It seems straightforward that when an employee is overexposed, there is noncompliance with a legal or recognized Occupational Exposure Limit (OEL). However, the meaning of the word *overexposure* is widely debated. As an example, the current Occupational Safety and Health Administration (OSHA) PEL for lead is 50 $\mu g/m^3$ averaged over an 8-hour period. If an employee's exposure is monitored every day for a year and it is determined that on only one of those days his exposure is greater than 50 $\mu g/m^3$, is the employee overexposed? Regardless of your answer, if OSHA monitors on that day, a citation may be issued.

It should be obvious that an acceptable risk level must be defined. Is it acceptable for an employee to be exposed at levels greater than 50 $\mu g/m^3$ for 5% of the time? Is it more acceptable for an employee to be exposed at levels greater than 50 $\mu g/m^3$ for 1% of the time? Although neither of these scenarios may seem palatable, it must be understood that no exposure situation is risk-free. Because there are many factors that will cause exposures to vary, there is always some chance, no matter how small, that an employee will be exposed to concentrations above a standard, even when the standard appears to be relatively quite high. Furthermore, when the employee is exposed below the OEL, risks of unwanted effects still remain. We must also consider how the OEL is set. Is it based on average exposures that were received by historic worker groups and their disease outcomes; or is it based on some other index of exposure, e.g., the highest exposures that those groups received; or is it based on animal studies?

The health professional also must consider the adequacy of current recommended, consensus, and regulatory OELs with regard to overexposures. Ideally, an OEL for a chemical would protect all workers who come into contact with that chemical from adverse health effects. However, the reality is that human variation of response to chemical exposures is large, individual responses can vary from day to day, and chemical exposure concentrations vary with time, production, process, and other factors. Consequently, when an OEL is set at an apparently low level, there is always a chance, no matter how small, that some individuals will not be protected from the associated disease.

Adding to this imperfect situation is the fact that most OSHA PELs for chemical exposures are about 40 years out of date. Although the *TLVs® and BEIs®* Book is republished yearly, only a small number of the TLVs® are updated each year. These OELs are based upon the health effects information that was available when the TLV® was last updated. Regardless of the volume of health effects data that are available for a particular chemical, more adequate data are always needed. This need is particularly apparent for human health effects data, since in many cases, OELs must be set by using animal data extrapolated to humans from animals.

Safety factors are sometimes applied to standards for chemical exposures. However, these safety factors should not be looked upon as a means for over-exposing employees without adverse effects, but rather should be seen for what they are: uncertainties about the risk of human health effects because of inadequacies in health effects data.

The previously stated objective for most monitoring efforts is to determine the state of compliance. The exposure assessment strategy must determine if a particular employee should be monitored, or if the particular job should be monitored. In the latter case, representative measurements must be taken from the job category and extrapolated to each employee in the job category. This can further complicate the attainment of the objective.

In addition, one must consider the various factors that affect an employee's exposure variability. Are there day-to-day or short-term variations? Process changes? Season-to-season productivity changes? Do the employee's exposure levels change with experience? Do exposures change with weather patterns? The answer to these questions is often *yes*, which complicates the exposure assessment strategy.

The person in charge of monitoring employee exposures can expect little help in terms of guidance for developing exposure assessment strategies. OSHA has not issued any guidance in this regard. The National Institute for Occupational Safety and Health (NIOSH) has issued an Employee Exposure Determination and Measurement Strategy, which will be discussed in detail later. The authors of this scheme are to be commended for their pioneering efforts; however, this first effort has received many comments of support, modification, and disagreement since its publication. This strategy is currently used in most OSHA comprehensive standards. AIHA has also developed some guidance in the area, which is discussed later. However, none of these efforts has reduced the considerable amount of thought that must go into the design of an exposure assessment strategy.

One goal of the NIOSH scheme was to give OSHA compliance officers and employers a relatively easy method to determine when sampling should be done and how frequently. Taken at face value, it may be interpreted as requiring little professional judgment to administer the scheme. Perhaps therein lies the biggest fallacy in the use of this strategy or any other *cookbook* technique. Regardless of how simple or complicated exposure assessment strategy decisions may be, there will always be exceptions and unusual circumstances. Professional judgment must be used to determine the course of action to take when those exceptions arise. Monitoring of employee exposures and parallel health effects is not an area for the amateur or untrained person.

Exposure assessment strategies are further complicated by current OSHA enforcement strategy. Because employee exposure variability exists in every workplace and is usually lognormally distributed, the odds of finding a single employee exposure that is greater than the standard are increased the more frequently the employees are monitored, even though the arithmetic average of exposures may be below the standard. In addition, OSHA (and state programs), which has responsibility for 6,000,000 American workplaces, must set priorities, but is seldom able to take a large number of measurements in a given workplace. Consequently, two conclusions can be drawn: the fewer employees that are monitored by an employer makes it more likely that the employer will appear to be in compliance; and, the chances of not receiving a citation for being out of compliance are generally in the employer's favor, since OSHA's resources for taking measurements are limited.

SAMPLES AND POPULATIONS

Hygienists often use the words *sampling strategy*, but use of the word *sample* often causes confusion in the occupational health literature (Leidel and Busch, 1985; Rappaport, 1987). In the statistical sense, the word sample means a group of one or more items selected from a parent population; however, hygienists and chemists often use the word in a different sense. For example, a hygienist often uses sample to mean the collection of a parcel of air or the collection of contaminant from a parcel of air. However, here the word sample will be reserved for use in the statistical sense, whereas the hygiene sample will be termed a measurement. All exposure measurements taken by hygienists are exposure estimates. In this context, it is important to point out that an exposure estimate may be made from one or more exposure measurements. Hence, an estimate of employee exposure for a given day may be determined by measuring the employee's exposure on that day. On the other hand, the exposure estimate for a group of workers performing the same job may be determined by monitoring randomly selected workers on randomly selected days. Each of these examples constitutes a statistical sample of measurements.

A population is a group of objects from which a sample is drawn. Population definitions must be very specific for the circumstances; a complete description of the population must be given. For example, one may be interested in the population of exposures for workers who perform a particular job on the third shift during summer. The connotation is that exposures on the third shift during summer differ from exposures on other shifts during other seasons. Another definition for a population may be the population of exposures for a given worker throughout his or her working lifetime. Note that there are many ways to define a population of exposures because there are many variables that affect exposures.

REASONS FOR EXPOSURE MONITORING

Before one can determine how a population of exposures is to be sampled, one first must determine why the population is being sampled. An imperfect reason for sampling a population is that *some information is better than none*. A haphazard and presumably expedient decision to gather data is a potential waste of time and resources (Leidel and Busch, 1985). At the very least, the population of exposures to be sampled should be defined. Once the population's timeframe, location, tasks, and other determinant variables have been decided (these usually involve space or time), then the advice of a statistician or the hygienist's use of statistical knowledge is necessary.

The opposite extreme is to contemplate indefinitely the design of a study. It should be remembered that the best time to seek statistical advice is before the study begins in order to evaluate various scenarios for defining populations and determining sample sizes. The final selection of sample size is often a compromise between the ideal statistical sample size and what can be obtained, given the resources. The point of determining the ideal is to set goals for statistical power and sensitivity or accuracy. If a smaller sample size is selected, the adequacy of what will be attained as compared to the original goals may be analyzed. If little information or decision power will be obtained, then time and resources should not be wasted. For example, one may determine that an adequate sample size for a given project with stated statistical goals is 50. However, time and resources may allow the collection of only 20 measurements. In addition, it may be determined that 20 measurements will allow only a very small probability of rejecting the null hypothesis; for example, the mean of the population of exposures equals or is greater than the OEL. Consequently, in this case, it may be better to redefine the population (if that is reasonable) or to take an even smaller number of measurements for a pilot study. The pilot study could be used to better estimate or to reaffirm the population parameters (mean and standard deviation) and, subsequently, to estimate the necessary sample size. See also discussion in Chapter 16 of Bayesian methods.

There are reasons for monitoring occupational exposures other than to determine compliance with legal, consensus, or corporate OELs. The following is a brief discussion of a few of those situations. Others might include responding to litigation or worker complaints or making workers feel more at ease in an obviously safe (low risk) situation. The remainder of the discussion in the chapter is devoted to exposure monitoring for comparison with a target value or OEL (i.e., compliance).

Epidemiological Studies

There are various kinds of epidemiological studies (see Volume 2), but one important differentiation is between those that look at worker exposures that have occurred (retrospective) and those that look at worker exposures that will occur (prospective). Retrospective studies require the use of past exposure monitoring records, which are often inadequate; or if no records are available, the reconstruction of exposures is necessary, based upon extrapolation from current conditions to past conditions. There have been a wide range of approaches to this problem (e.g., Benke et al., 2001; Ramachandran and Vincent, 1999; Stewart et al., 2003; Tielemans et al., 2002; and Wu et al., 1999). In prospective studies an interesting issue is whether peak exposures (during a workday) are important in causing acute but also chronic effects, and if so, what aspect of peak exposures is the driving force. Thus, in a prospective epidemiological study, the measurement of peaks may be important (Peller et al., 2004).

Evaluation of Control Measures and Determining Control Strategies

Control measures are generally divided into three major categories: engineering, personal protective equipment, and administrative. The evaluation of controls often takes the form of before and after studies of exposure levels. Concentrations may be monitored at fixed points within the workplace, or in the worker's breathing zone. Examples of control strategies that may require evaluation include engineering controls such as ventilation, administrative controls such as reassignment of workers to different job locations, and work practice administrative controls such as the use of wet versus dry sweeping. There are many other examples of controls that may be evaluated, including the before-and-after effects of process modifications.

In many cases, the determination of what to control is not easy. For example, a worker group may have many different tasks, each contributing differently to the daily exposure. Perhaps the workers in the group have mean exposures that are very similar, but the variability of exposure for individual workers varies considerably. A control that effects all workers equally may do little to control the high variability that only a portion of the workers display. A better strategy would be to determine how to control this variability for the select workers only; but first, it must be determined that a pattern exists. There are many variables that effect exposure. Could it be that production, season, shift, or even gender among a long list of possibilities, effects exposure? If so, then an appropriate control strategy might be based on finding these determinants of exposure.

Reentrainment Studies

When contaminants are exhausted from a work area or building, they should not be reentrained in the intake air system or via natural ventilation. Simulation studies may be conducted using suitable tracers such as SF_6 to determine if reentrainment can occur and to what extent. After release of the simulant, exposures are monitored in the work zone. This may be considered as a special case of an *adequacy of controls* study.

Process Evaluations

Different processes that accomplish similar end products may be evaluated in a before-and-after situation. However, in the case where differing processes may exist in different plants or in different parts of a plant, processes may be evaluated in the same timeframe rather than before and after. This type of evaluation is useful because time-related variables can be controlled.

Comparison of Measurement Methods

A common practice of hygienists is to make *field* comparisons of measurement methods. In the 1980s, this was particularly true for comparison of passive dosimeters with pump-driven, active monitoring techniques. Comparison of other measurement methods are also possible, such as comparison of two types of impingers, two methods for counting asbestos fibers that may require collection on different filters, two techniques for monitoring cotton dust, or two techniques for monitoring an organic vapor (e.g., the use of activated charcoal versus a synthetic polymer).

Research

There are many examples of occupational exposure measurement studies performed for purposes of research. One example might be the measurement of occupational exposures in different plants within the same industry or in different industries, to determine and compare variability in the population of exposures for those groups. The end result may help determine the economic and technologic feasibility of an OEL.

Confined Space Monitoring

Often confined spaces or special maintenance areas must be monitored to determine the need for respirator use, ability to perform hot work (welding or cutting), or identification of *hot spots* that may require further ventilation. This work is always done before workers enter such an area. When the hazards are acute (such as explosion), adequate safety factors must be built into OELs (in this case flammability limits), and some form of subsequent continuous monitoring is usually necessary.

Task Monitoring

For various reasons it may be appropriate to measure exposures for the duration of specific tasks (Lowman, 2005). For example, if a worker performs three tasks repetitively each day, then it may be determined that one task is more hazardous than the others. An 8-hr TWA or integrated exposure measurement would not indicate this. Tasks may also be monitored for purposes of estimating worst-case exposures. For example, if one uses a worst-case task exposure estimate combined with a worst-case estimate of the number of tasks performed per day, then a worst-case estimate of daily exposure can be made. Even in the design of a strategy for monitoring exposures for compliance with say 8-hr TWAs, short-term and task measurements may be made and used to calculate long-term averages, sometimes at little additional cost.

A Note about Wipe Sampling

Sometimes a hygienist may take wipe samples by wiping a filter material over a surface. There are standard techniques used for this; some are contained in the *OSHA Field Inspection Reference Manual*.

The purpose of the wipe sample is not to predict airborne concentrations. In fact, studies clearly show that there is no relationship between surface and airborne concentration levels (Caplan, 1993). Instead, wipe samples may indicate the hygienic condition of a plant or an area or may indicate the potential for contaminated surfaces to lead to skin exposure from contact. However, both scenarios are difficult to prove without appropriate experimental controls, standardization, and characterization of acceptable contamination levels. Most surfaces are contaminated, but they may not be polluted (i.e., have an unacceptable risk). For further discussion, the reader should see McArthur (1992) and Lichtenwalner (1992).

Although the above reasons for monitoring are by no means a complete list, the point is that each has a different objective and, therefore, a different hypothesis to be proved or tested. Consequently, the exposure assessment strategy will also vary.

HISTORICAL PERSPECTIVE

A first attempt to conscientiously design an exposure assessment strategy may lead one to conclude that no one has ever attempted the feat before. There is no easy-to-follow and error-free decision-making process for which the input variables are clearly defined and the output is easily interpretable. AIHA (1998) has published a nice, thorough guide, but it is book-length. A discussion of this and other strategies and an overall strategy based on Chapters 14 through 17 of this Volume will be presented in Chapter 18.

Prior to the Occupational Safety and Health Act of 1970, hygienists often investigated obviously high exposures to major disease-producing occupational agents such as lead, mercury, asbestos, and silica; sophisticated exposure assessment strategies were usually not necessary, in that overexposures were easily documented. In addition, there were very few monitoring techniques available. The gas chromatograph, charcoal tube, and personal sampling pump did not come into widespread use until the 1970s. Consequently, in many cases the idea of an exposure assessment strategy was far advanced. The need to devise monitoring techniques was far more urgent than deciding how often to use them. Nevertheless, much useful information was derived in the years prior to 1970.

Perhaps the best way to view the historical advancements in occupational exposure assessment strategy is to address the *five questions* of Hosey (1973): where, how long, when, who, and how many? (*What to sample?* may also be added to the list, but the question is addressed in the previous chapter as priority setting.) Hosey gave general answers to the above questions that reflected the available knowledge at that time. Since then, much has been written about the questions of who and how many. Information about the other three questions has a considerably older and certain history; so many hygienists no longer see the importance of the questions of where, how long, and when.

Where?

Over 50 years ago, Oldham and Roach (1952) argued that the most representative manner for estimating exposures of coal miners to coal dust was for a member of the monitoring team to follow an individual worker with a monitoring instrument. It was argued that measurements taken in general fixed areas around the coal mine face were not necessarily reflective of exposures to individual workers, and data were given to support this claim. Thus, Oldham and Roach modified a concept first presented by Patterson (1939), which later became the *breathing zone* concept; that is, samples taken near the worker's nose or mouth (Roach et al., 1967). This was usually accomplished by the hygienist following the worker while holding a monitoring probe in the worker's breathing zone.

Once portable monitoring devices were available, however, the breathing zone concept was modified by the *personal sample* concept. (Note that it is difficult to be consistent in the use of the word *sample* as previously defined.) This is now the most often used technique. The worker wears the monitoring instrument, with the probe usually attached to the collar.

Although most health personnel taking exposure measurements use the personal monitoring concept without question, several studies have shown that the exact location of the probe within the breathing zone can have an effect on the results. A number of these studies have shown that resuspension of dust from clothing can act to positively bias lapel results relative to results taken at the nose, mouth, or forehead (Chatterjee et al., 1969; Butterworth and Donoghue, 1980; Cohen et al., 1982; McGuire, 1974, as reported by Cohen et al., 1982; Martinelli et al., 1983; Bohen, 1983, as reported by Cohen et al., 1984). The issue of whether the mere wearing of the device causes exposure results to be higher has also been raised (Cherrie et al., 1994 and Cherrie, 2003).

In one study performed for the purpose of comparing right versus left lapel, dominant versus non-dominant side, and side oriented to a dust source versus opposite side, significant effects were found (Hall, 1991). In two separate studies (contained in one paper) reporting right and left lapel data, both significant and nonsignificant results were seen (Butterworth and Donoghue, 1980). Similar mixed results were reported in a study of welders (Goller and Paik,

1985). Simpson has shown that the lapel is an inappropriate location for solderers (Simpson, 2005). Guffey et al. (2001) presented results that hinted that, in some conditions, the chest location oversamples the nose. Eriksson et al., 2005 found no left versus right effects among workers exposed to styrene. It may be that effects, if they occur, are more likely for aerosols than gases and vapors.

No doubt the situation is complex and depends on many factors such as the dominant side of the worker (handedness), how the worker is oriented to the work, the physical state of the contaminant, and the contributions of other, nearby sources. The potential for bias in monitoring the breathing zone cannot be ruled out and no doubt deserves further research. Until then, hygienists may consider alternating left and right lapels in an attempt to average out any bias.

In summary, the current answer to the question of *Where to sample?* is the personal sample when worker exposures are being estimated. Area or fixed point monitoring and breathing zone monitoring are still used, depending on the goals of the exposure assessment strategy and limitations that may be placed on it.

How Long?

Bloomfield (1933) and Dallavalle (1940) were among the first to use the time-weighted average (TWA) concept. They pointed out that if a worker were exposed to five different types of dusty areas, his exposure in each dusty area should be weighted according to the time spent in that area. Thus, the question of how long to sample was often answered by the amount of time the worker spent in a particular area. The development of personal monitoring devices has eliminated the need to consider the amount of time a worker spends in each area, as these devices integrate concentration with respect to time.

There were very few OELs 70 years ago, yet many hygienists have lost sight of the fact that the 8-hr TWA or 15-min TWA concept has not always existed. Today, the question of *how long* is often answered by the nature of the OEL to which the monitoring results are to be compared. Consensus, legal, and corporate OELs are often couched in terms of an 8-hr or a 15-min TWA, or ceiling. The time weighted average concept involves the integration of concentration with the stated time. The ceiling concept, in its simplest form, incorporates the idea of an instantaneous monitoring result; therefore, it requires a continuous monitoring instrument.

As most occupational health exposure standards follow these three time formats, many hygienists consider the question of how long in light of the nature of the OEL. However, Roach (1966) first recommended that this question be considered in light of the time-associated body burden and clearance mechanism for the particular contaminant. He recommended an optimum sampling period of one-tenth of the half-time (the half-time being the time at which one-half of an initial dose to the body would be cleared from the bloodstream or the target organ). The goal was to set exposure standards that protect against exceeding a critical body burden in light of the rate at which the body can clear itself of the contaminant. Essentially a *mass balance* approach was used, whereby the rate of accumulation equals the rate of exposure minus the rate of clearance. Roach stated that a large safety margin was represented in the recommended one-tenth of the half-time *monitoring time*, and that a duration of one-third or one-fifth of the half-time would also be appropriate (with a smaller safety factor). He gave sampling periods ranging from 2 minutes for acetone, ammonia, and other irritants up to 10 workshifts for such cumulative toxins as lead and mineral dust. (For a detailed discussion of body burdens and half-times, see Chapter 10 in this Volume and Chapters 11 and 12 in Volume 2.)

A later paper by Roach (1977) considered the smoothing effects of ventilation and measurement devices in addition to that of biological response. Roach's model showed that the sampling period should not exceed 3*T*, where *T* is the biological half-time. *T* can be defined relative to any organ or site or relative to a reactive metabolite. In the case of carcinogens, it may be best defined relative to the repair rate of damage to DNA or some other similar parameter. Roach's concept is of great importance, but requires more discussion and thought about its application to compounds whose pharmacodynamics and kinetics are not straightforward. Also, his concept would be difficult to implement in legal OELs because each compound would likely have a different averaging time, making monitoring difficult.

Two other considerations in response to the question of how long involve assessment of exposures to compounds with multiple standards and situations where workers are exposed during novel workshifts or schedules. The first problem arises when workers are exposed to compounds that may have both acute and chronic effects at levels that can be encountered in the workplace. For example, over the short-term, higher concentrations of many organic solvents will cause central nervous system effects that may alter workers' performance or, in the worst case, lead to death. On the other hand, long-term exposures to some organic solvents at much lower doses have chronic effects on other organ systems, such as the blood-forming organs and the liver. Consequently, there appears to be a need for two standards, one addressing chronic effects and the other addressing acute effects. A biological half-time exists for each pathway, requiring multiple averaging times.

It should be pointed out that recent short-term exposure limits (STEL) of OSHA may not necessarily be promulgated on this premise (OSHA, 1988). In promulgating the asbestos STEL, OSHA stated that risk of *chronic* health effects will be further reduced. The STEL has no relationship to any acute effects from asbestos, but is simply meant to control short, task-oriented exposures that contribute significantly to chronic exposures.

Regarding novel work schedules (e.g., four 10-hour shifts per week) various strategies have been worked out for adjusting OELs (see, e.g., Roach, 1978 and Paustenbach, 1985). Usually, the numerical value of the OEL is adjusted for acute toxicants, and the averaging time is typically taken as shift length, for chronic toxins. (This is further discussed in Chapter 10 in this Volume and Chapters 11 and 12 in Volume 2.)

When?

Whether performing random sampling or worst-case sampling, the question of *when* is important. Worker exposures vary in both time and space. Workers picked at random will tend to give an unbiased estimate of the average exposure over *space*. Similarly, days on which sampling takes place should be randomly chosen in order to avoid time-related biases. Seasonal changes in production, downtime for maintenance, within-week production cycles, and many other factors can cause day-to-day variability in exposure levels. In order to avoid biases related to these determinant variables, the group of workers should be concisely defined and, just as important, the time period (e.g., one year) should be defined so that random sampling can take place across time and space.

Worst-case sampling does not normally utilize random sampling concepts, in that a single worker or a few workers thought to have the highest exposures are chosen for monitoring. Therefore, the answer to the *when* question is, ideally, when the exposures are expected to be the greatest. Hygienists who are most likely to use worst-case sampling (compliance officers, corporate-headquarters-based hygienists, and insurance company risk managers) usually do not have the luxury of taking multiple samples over many days so as to characterize the full exposure distribution. However, the odds of obtaining a worst-case sample are increased if one can sample on the day of the predicted maximum exposure. This is an important defect in most worst-case strategies and leaves those hygienists with time constraints in a quandary.

Who?

Oldham and Roach (1952) first addressed the question of who to sample by recommending that a random sample be taken of all workers to be studied (i.e., in the population). They recommended that all workers in a study be listed, that the number of measurements that could be taken based upon the amount of resources available to the monitoring team be determined, and that the number of measurements be applied to the worker list on a random basis.

The concept of random sampling has been taken beyond the original idea of Oldham and Roach. Stratified random sampling, a further division of a population of workers into groups that have or are thought to have equivalent exposures, followed by random sampling within each group, has been recommended (Roach et al., 1967). This concept was further developed and given the name *exposure zone concept* by Corn and Esmen (1979). Later, the same concept was referred to as the *homogeneous exposure group* or HEG (AIHA, 1991). In essence, the idea behind a stratified random sampling strategy or the HEG concept is to refine the definition of the exposed population by using determinant variables such as shift, job description, area of the plant, or location within the process. The HEG should contain workers that have equivalent risks of occupational disease.

Spear (1991) has shown that even though workers may be classified subjectively (often based on job description) as being in the same HEG, the individual risks to the workers may be quite different. This is presumably a result of work practices and other variables controlled by individual workers, and not by the job. Rappaport (1991) has further formalized the rules for grouping workers and shown that assigning workers to groups of equivalent risk often requires a very broad definition of equivalent. (There is more about equivalency in Chapter 17.)

For these reasons and others that are discussed in Chapters 15 and 16, the word *homogeneous* is really unattainable and, therefore, the term *similar exposure group* (SEG) is now recommended (AIHA, 1998).

NIOSH (1977) formalized a technique known as the Employee Exposure Determination and Measurement Strategy, which primarily involves the concept of sampling the worst-case exposure. Although the worst-case concept had no doubt been used previously, NIOSH was the first to document its use in a stepwise procedure for determining number of measurements to be taken and who to sample. Basically, the philosophy of the NIOSH strategy (see later discussion) is that by selecting a worst-case exposure situation, one may rule out the need to do further sampling if the worst-case exposure is acceptably low.

How Many?

The concept of exposure distributions, introduced in Chapter 7, has a direct relationship to this question: how many exposures should be measured? Oldham (1953) first recognized that the variability of dust concentrations at a coal mine face was lognormally distributed. Since then, other researchers have affirmed this observation for other exposures (NIOSH, 1975; Esmen and Hammad, 1977; Kromhout et al., 1987; and Waters et al., 1991). These confirmations of the tendency towards lognormality are important, as statistically designed exposure assessment strategies must utilize the appropriate statistical distribution model in order to most accurately or cost effectively answer the question of how many.

This question was first given lengthy consideration by NIOSH (1977). The *Occupational Exposure Sampling Strategy Manual* received criticism because of the apparently large sample sizes that were recommended in order to satisfactorily characterize populations of exposures. However, it should be pointed out that prior to this time little consideration had been given to the adequate statistical characterization of populations of exposures. Reliable, inexpensive, and routinely used monitoring techniques and requirements for controlling workplace exposures to levels that would adequately protect employees' health had only recently come into being. The sudden convergence of these factors had an impact on hygiene that continues today.

The idea that monitoring of one or two worker-days[1] will *always* adequately answer a question regarding overexposure is no longer acceptable. Hygienists and others responsible for monitoring occupational exposures must consider adequate statistical design in their studies, must pose properly worded statistical hypotheses or questions, and must be willing to change their exposure assessment strategies as more information regarding the *five questions* becomes available.

Although concerns remain about current practices regarding the questions of *where, what, and how long,* the remainder of this chapter (and the next) will be largely devoted to the two

[1] One worker monitored, on one day is one worker-day.

remaining questions of who and how many. They are virtually inseparable and cause the greatest concern, confusion, and debate for hygienists.

NIOSH SAMPLING STRATEGY

One of the first formal attempts to answer the questions of who and how many was made in the NIOSH *Recommended Employee Exposure Determination and Measurement Strategy*. An equivalent to this scheme is used by OSHA in all of its comprehensive standards for compounds such as asbestos. This strategy is shown in Figure 14-1. The first decision that must be made concerns the release of the chemical into the workplace air. Unless the chemical has a very low vapor pressure or is contained in a closed system, it probably will be released into the workplace air.[2]

The second decision requires that the health professional make a professional judgment regarding whether an employee may be exposed at a concentration greater than the action level (0.5 OEL). Perhaps the decision whether to sample is the most difficult one to make. Even for experienced hygienists, exposure levels can often be surprisingly low or high. Factors that may be considered in making this decision include:

- Vapor pressure of the chemical
- Duration time of exposure to the chemical (i.e., task duration)
- Process temperature
- Likelihood of work habits to have a great effect on exposure
- Magnitude of the occupational exposure standard

Even when these factors are considered, it is difficult to predict when exposures will exceed one-half the OEL. More importantly, it must be determined whether the employee will be exposed to the chemical at concentrations greater than 0.5 OEL. This implies that, on any day, any exposure greater than the action level is significant. If this requirement is taken at face value, the decision is fairly easy; it can be assumed, in most cases, that an exposure greater than the action level will occur on one day in a given year, certainly in a working lifetime. Hence, this literal interpretation is not valid.

The third decision requires the selection of maximum risk (worst-case) employees for monitoring. The worst-case concept, first used by Roach et al. (1967), has obvious advantages and at least these three disadvantages:

1. In many cases, it will be impossible to determine a worst-case exposure for that

---

[2] The hygienist should beware of probable skin exposure from contaminated surfaces for low vapor pressure compounds having low airborne concentrations.

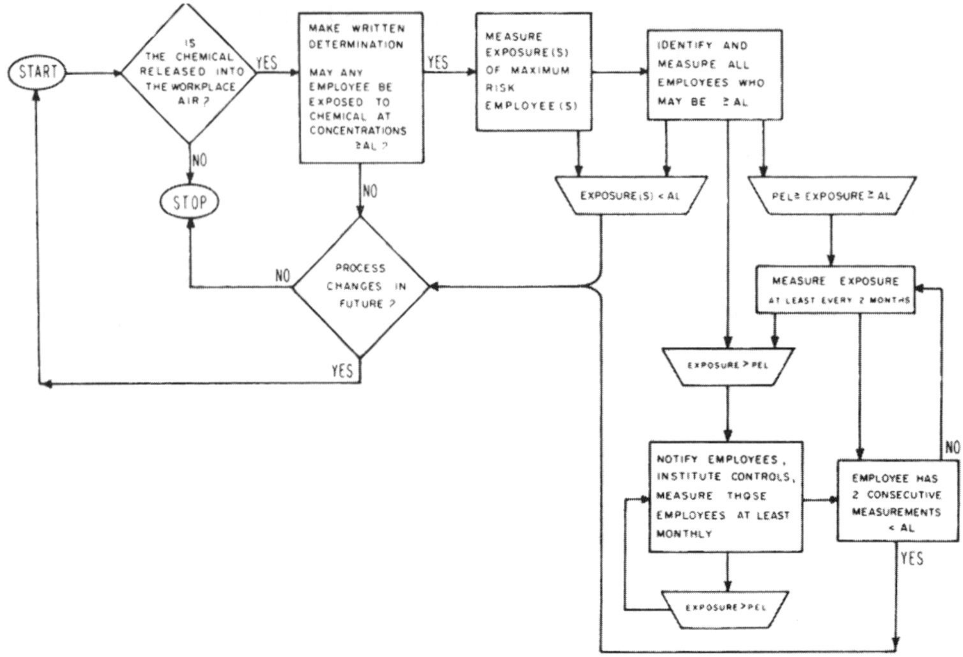

Figure 14-1. The NIOSH recommended employee exposure determination and measurement strategy (from NIOSH, 1977).

day. In other words, there may be many workers doing approximately the same job, so that each will appear to have equal likelihood of being worst-case on a particular day. The alternative is to randomly sample the group, using a sample size that will practically assure that at least one worker is sampled who meets the definition of worst-case. Table 14-1 can be used when the definition of worst-case is *top 10%*. In other words, if a group of workers of size 50 is being considered, then a random sample of size 18 will provide a 90% probability that at least one worker having an exposure in the top 10% of exposures on that day will be sampled.

2. Even if one manages to find the worst-case exposure for that day, it is virtually impossible to extrapolate to other days. What may be worst-case today may be average tomorrow.

3. Subjective selection of workers for a worst-case assessment is not random sampling, and therefore, the results cannot be used to determine other parameters of the exposure distribution. All employees who may be above the action level must be identified and their exposures measured. This implies that random sampling is

Table 14-1
Size of partial sample needed to be 90% confident of obtaining a top 10% measurement[1]

Size of Population N	Number of Required Measurements[2]
8	7
9	8
10	9
11–12	10
13–14	11
15–17	12
18–20	13
21–24	14
25–29	15
30–37	16
38–40	17
50	18

[1] Adapted from NIOSH, 1977.
[2] n = N if N < 7

not performed. Instead, every employee with potential overexposure must be monitored.

In the fourth decision, if exposures of these worst-case employees are between the AL and the PEL, then measurement of their exposures must be continued on a bimonthly basis. If exposures are greater than the PEL, then exposures must be monitored on a monthly basis until they are reduced below the PEL. However, one should realize that, if it is preliminarily determined that exposures of some employees are between the AL and the PEL, then it is likely that exposures of some employees in that group will be between the AL and the PEL in any month. In fact, it is likely that exposures occasionally will be above the PEL because of inherent variability in the exposure distribution. Consequently, one of the biggest concerns about this particular portion of the strategy is that the employer may find himself or herself in a decision loop from which escape is unlikely unless the ramifications of the strategy are understood.

Tuggle (1981, 1982) first recognized this loop potential and proposed the use of the term *acceptable risk*. He stated that a certain percentage of overexposures must be tolerated because of the nature of exposure variability. Tuggle made several recommended changes to the NIOSH strategy, including the use of tolerance limits (see Chapter 7 and Equation 15-4) and termination of monitoring based on at least two results, not one. Tuggle pointed out that termination of monitoring based on one result leaves a high probability that sampling will cease for populations having a large percentage (say, 35%) of exposures greater than the OEL. By requiring that two results be < AL, this probability is decreased. Both these points were important contributions. The tolerance limit approach did lead us away from worst-case to statistically based sampling. Nevertheless, Tuggle's approach required large sample sizes (50–400), was flawed (Selvin et al., 1987), and did not justify the risk level that it proposed, that is, the 95% tolerance limit implies that 5% overexposures is acceptable.

Rock (1982) pointed out that exposures to acute substances, greater than the OEL, should be tolerated in a different manner from overexposures to substances having chronic effects. This concept was first used by Roach (1966). For an acute effect, Rock recommended a modification of Tuggle's tolerance limit concept; only one overexposure out of 250 would be tolerated. The selection of one overexposure out of 250 is arbitrary and was made for purposes of illustration. Its appropriateness for acute toxins has been further discussed (Rappaport et al., 1981a; Selvin et al., 1987).

Rappaport et al. (1981b) argued similarly to Rock (1982) that air sampling in the assessment of continuous exposures to acutely toxic chemicals must allow for some exceedance rate of overexposures. They suggested an exceedance (or tolerance) rate of 5%, provided that exposure limits incorporate appropriate safety factors, which are relatively large for potent acute toxins but would be smaller for irritants. Their logic, like that of Rock (1982) and Tuggle (1981, 1982), was that an absolute standard for an acutely toxic substance is unworkable, as any distribution of exposures would be deemed out of compliance if a sufficiently large number of measurements were collected.

However, Rappaport (1987) later argued that extrapolations from the main body of exposure data out into the unknown regions of the right tail of the distribution (say, the 95th percentile) requires large sample sizes and depends critically upon assumptions concerning the independence of measurements, the stationarity (changing of the distribution over time), and continuity (whether there is in fact more than one distribution) of the exposure series, and any parametric models (e.g., the lognormal) that might have been used. Thus, considerable resources must be expended just to determine what exposures might be excessive under some set of unknown circumstances. Finally, using a tolerance limit or an exceedance rate approach, one must feel comfortable in allowing, for example, 5% overexposures at unknown concentrations. Therefore, Rappaport argued that monitoring resources for acute toxins (in particular those with ceiling values) would be best invested in continuously monitoring sensors, which may warn workers of excessive releases, or should be used to make changes in the process or controls in order to lower exposure levels. This is a relatively simple yet elegant solution. The solution for chronic toxins is considerably more complex.

In summary, worst-case sampling reveals nothing about the exposure distribution over time or space; it only gives limited information about exposures, and the mean or other parameters cannot be estimated. On the other hand, random sampling over time and space will allow estimates of the mean as well as any percentiles of the exposure distribution. So, for example, the 95th percentile could be estimated as worst-case if this information were important in assessing risk. Inspection officers often do not have time to do a statistically based study over time, nor does the current approach to standards enforcement warrant statistical-based studies (more will be said about this later). On the other hand, corporate-based hygienists should use statistically based sampling wherever possible, as it affords greater flexibility than worst-case sampling. In Chapters 15 and 16, statistically based strategies for chronic and mildly acute (8-hr TWA and 15-min STEL) toxicants are discussed.

QUESTIONS

14.1. Give five examples of exposure populations. Be careful to define each in terms of space and time.

14.2. Is an employee who is exposed to lead at concentrations greater than 50 $\mu g/m^3$ (the PEL) on one day out of 30 days overexposed to lead (based on discussion of risk in Chapter 10)?

14.3. Give five possible answers to the question of *how long*? Consider potential legal requirements such as PELs, unusual circumstances, and the reasons discussed for conducting exposure monitoring.

14.4. Explain why the NIOSH Strategy (Chapter 14, Figure 14-1) is an advantageous approach for OSHA compliance officers to follow. Contrast it with random sampling over time.

REFERENCES

AIHA: A strategy for occupational exposure assessment, N.C. Hawkins, S.K. Norwood, J.C. Rock (eds.). American Industrial Hygiene Association, Fairfax, VA (1991).

AIHA: A strategy for assessing and managing occupational exposures, 2nd ed., J.R. Mulhausen, J. Damiano (eds.). American Industrial Hygiene Association, Fairfax, VA (1998).

AIHA: A strategy for assessing and managing occupational exposures, 3rd ed., B. Bullock, J.S. Ignacio (eds.). American Industrial Hygiene Association, Fairfax, VA (2006).

Benke, G.; Sim, M.; Fritschi, L.; Aldred, G.; Forbes, A.; Kauppinen, T.: Comparison of occupational exposure using three different methods: Hygiene panel, job exposure matrix (JEM), and self reports. Appl. Occup. Environ. Hyg. 16:84-91 (2001).

Bloomfield, J.J.: Preliminary surveys of the industrial environment. Publ. Hlth. Rep. 48:1343 (1933).

Butterworth, R.; Donoghue, J.K.: Contribution of activity released from protective clothing to air contamination measured by personal air samplers. Hlth. Phys. 18:319-323 (1980).

Caplan, K.J.: The significance of wipe samples. Am. Ind. Hyg. Assoc. J. 54:70-75 (1993).

Chatterjee, B.B., et al.: The location of personal sampler filter heads. Am. Ind. Hyg. Assoc. J. 30:643-645 (1969).

Cherrie, J.W.: The beginning of the science underpinning occupational hygiene. Ann. Occup. Hyg. 47:179-185 (2003).

Cherrie J.W.; Lynch, G.; Bord, B.S.; Heatherfield, P.; Cowie, H.; Robertson, A.: Does the wearing of sampling pumps affect exposure? Ann. Occup. Hyg. 38:827-838 (1994).

Cohen, B.S., et al.: Exposure estimates from personal lapel monitors. Am. Ind. Hyg. Assoc. J. 43:239-243 (1982).

Cohen, B.S., et al.: Bias in air sampling techniques used to measure inhalation exposure. Am. Ind. Hyg. Assoc. J. 45:187-192 (1984).

Corn, M.; Esmen, N.A.: Workplace exposure zones for classification of employee exposures to physical and chemical agents. Am. Ind. Hyg. Assoc. J. 40:47 (1979).

Dallavalle, J.M.: Averaging and weighting of dust exposures. Publ. Hlth. Bull. No. 250, App. I (1940).

Eriksson, K.; Liljelind, I.; Fahlen, J.; Lampa, E.: Should styrene be sampled on the left or right shoulder? — An important question in employee self-assessment. Ann. Occup. Hyg., Vol. 49(6):529-533 (2005).

Esmen, N.A.; Hammad,Y.Y.: Log-normality of environmental sampling data. J. Environ. Sci. Hlth. A12 (1&2):29-41 (1977).

Goller, J.W.; Paik, N.W.: A comparison of iron oxide fume inside and outside of welding helmets. Am. Ind. Hyg. Assoc. J. 46:89-93 (1985).

Guffy, G.E.; Flanagan, M.E.; van Belle, G.: Air sampling at the chest and ear as representative of the breathing zone. AIHAJ 62:416-427 (2001).

Hall, S.C.: Comparison of right and left lapel sampling results, Master's Research Project. University of Alabama at Birmingham, School of Public Health (1991).

Hosey, A.D.: General principles in evaluating the occupational environment, In The Industrial Environment – Its Evaluation and Control. National Institute for Occupational Safety and Health, Cincinnati, OH (1973).

Kromhout, H.; Oostendorp, Y.; Heederik, D.; Boleij, J.S.M.: Agreement between qualitative exposure estimates and quantitative exposure estimates. Am. J. Ind. Med. 12:551-562 (1987).

Leidel, N.A.; Busch, K.A.: Statistical design and data analysis requirements, In Patty's Industrial Hygiene and Toxicology, L.J Cralley, L.V. Cralley (eds.). Wiley, New York (1985).

Lichtenwalner, C.P.: Evaluation of wipe sampling procedures and elemental surface contamination. Am. Ind. Hyg. Assoc. J. 53:657-659 (1992).

Lowman, K.: Exposure Assessment, What's Next? Task-Based Sampling and Real-Time Air Monitoring. The Synergist 16(11):43 (2005).

Martinelli, C.A., et al: Monitoring real-time aerosol distribution in the breathing zone. Am. Ind. Hyg. Assoc. J. 44:280-287 (1983).

McArthur, B.: Dermal measurement and wipe sampling methods: A review. Appl. Occup. Env. Hyg. 7:599-606 (1992).

NIOSH: Exposure Measurement Action Level and Occupational Environmental Variability, N.A. Leidel, K.A. Busch, W.E. Crouse (eds.). National Institute for Occupational Safety and Health, Cincinnati, OH (1975).

NIOSH: Occupational Exposure Sampling Strategy Manual, N.A. Leidel, K.A. Busch, J.R. Lynch (eds.). National Institute for Occupational Safety and Health, Cincinnati, OH (1977).

Oldham, P.D.: The nature and the variability of dust concentrations at the coal face. Brit. J. Ind. Med. 10:227 (1953).

Oldham, P.D.; Roach, S.A.: A sampling procedure for measuring industrial dust exposure. Brit. J. Ind. Med. 9:112 (1952).

OSHA: Occupational exposure to asbestos, tremolite, anthophyllite, and actinolite. Fed. Reg. 53:35610 Occupational Safety and Health Administration (1988).

Patterson, H.S.: Bull. Insn. Min. Metall., No. 417 (1939).

Paustenbach, D.J.: Occupational exposure limits, pharmacokinetics, and unusual work schedules, In Patty's Industrial Hygiene and Toxicology, Vol. 3A, 2nd ed., L.J. Cralley, L.V. Cralley (eds.). Wiley, New York (1985).

Peller, L.; Burstyn, I.; de Pater, N.; Kromhout, H.: Characteristics of peaks of inhalation exposure to organic solvents. Ann. Occ. Hyg. 48:643-652 (2004).

Ramachandran, G.; Vincent, J.: A Bayesian approach to retrospective exposure assessment. Appl. Occ. Env. Hyg. 14:547-557 (1999).

Rappaport, S.M.: Biological considerations for designing sampling strategies. Symposium on Advances in Air Sampling, Pacific Grove, California. American Conference of Governmental Industrial Hygienists, Cincinnati, OH (1987).

Rappaport, S.M.: Review: Assessment of long-term exposures to toxic substances in air. Ann. Occup. Hyg. 35:61-121 (1991).

Rappaport, S.M.; Selvin, S.; Spear, R.C.; Keil, C.: An evaluation of statistical schemes for air sampling, In Chemical Hazards in the Workplace. American Chemical Society, New York (1981a).

Rappaport, S.M.; Selvin, S.; Spear, R.C.; Keil, C.: Air sampling in the assessment of continuous exposures to acutely toxic chemicals. Part I-Strategy. Am. Ind. Hyg. Assoc. J. 42:831 (1981b).

Roach, S.A.: A more rational basis for air sampling programs. Am. Ind. Hyg. Assoc. J. 27:1 (1966).

Roach, S.A.: A most rational basis for air sampling programs. Ann. Occup. Hyg. 20:65 (1977).

Roach, S.A.: Threshold limit values for extraordinary work schedules. Am. Ind. Hyg. Assoc. J. 39:345 (1978).

Roach, S.A.; Baier, E.J.; Ayer, H.E.; Harris, R.L.: Testing compliance with threshold limit values for respirable dusts. Am. Ind. Hyg. Assoc. J. 28:543 (1967).

Rock, J.C.: A comparison between OSHA-compliance criteria and action-level decision criteria. Am. Ind. Hyg. Assoc. J. 43:297 (1982).

Selvin, S.; Rappaport, S.; Spear, R.; Schulman, J.; Francis, M.: A note on the assessment of exposure using one-sided tolerance limits. Am. Ind. Hyg. Assoc. J. 48:89 (1987).

Simpson A.T.: Comparison of sampling positions when measuring personal exposure to solder fume. Ann. Occup. Hyg. 49:439-442 (2005).

Spear, R.C.: Individual versus group differences in exposure and risk, In Exposure Assessment for Epidemiology and Hazard Control, S.M. Rappaport, T.J. Smith (eds.). Lewis Publishers, Chelsea, MI (1991).

Stewart, P.A.; Lees, P.S.; Correa, A.; Breysse, P.; Gail, M.; Graubard, B.I.: Evaluation of three retrospective exposure assessment methods. Ann. Occup. Hyg. 47:399-411 (2003).

Tielemans, E.; Marquat, H.; de Cock, J.; Groenwold, M.; Van Hemmen, J.: A proposal for evaluation of exposure data. Ann. Occ. Hyg. 46:287-297 (2002).

Tuggle, R.M.: A NIOSH decision scheme. Am. Ind. Hyg. Assoc. J. 42:493 (1981).

Tuggle, R.M.: Assessment of occupational exposure using one-sided tolerance limits. Am. Ind. Hyg. Assoc. J. 43:338 (1982).

Waters, M.A.; Selvin, S.; Rappaport, S.M.: A measure of goodness-of-fit for the lognormal model applied to occupational exposures. Am. Ind. Hyg. Assoc. J. 52:493-502 (1991).

Wu, J.; Milton, D.K.; Hammond, S.K.; Spear, R.C.: Hierarchical cluster analysis applied to workers' exposures in fiberglass insulation manufacturing. Ann. Occ. Hyg. 43:43-55 (1999).

15

QUANTITATIVE SURVEYING – APPLICATION OF MATHEMATICAL MODELING TO ESTIMATE AIR CONTAMINANT EXPOSURE

Mark Nicas, PhD, MPH, CIH

INTRODUCTION

The traditional approach to characterizing exposure intensity to air contaminants is via collecting air samples from a subject's breathing zone or from a fixed site close to the subject's location. However, situations arise in which direct monitoring is not practical or possible. For example, we might desire exposure estimates for a discontinued operation, for a future operation under consideration, or for planning emergency response to an accidental release. We also might seek to prioritize our monitoring activities, or to decide whether monitoring an operation is needed in the first place. In such cases, mathematical modeling can be used to estimate exposure intensity.

There are different types of algorithms that can be termed a *mathematical model*. For example, a regression model is derived by statistically relating a series of measured exposure values to sets of observed exposure factors associated with the measurements. A regression model does not describe the physical processes by which exposure values are generated; rather, it predicts exposure values based on observed statistical associations. In contrast, this chapter focuses on physical-chemical models which try to describe the physical processes that lead to airborne exposure. Such models use simplifying assumptions to keep the mathematics relatively simple. As one tries to better capture physical reality and avoid simplifying assumptions, the price paid is greater mathematical complexity. The need to find the right balance between a realistic model and a tractable model is captured by a quote attributed to Albert Einstein: "Things should be kept as simple as possible, but not simpler."

This chapter describes the basic elements of a physical-chemical model: 1) the instantaneous contaminant mass emission rate function; 2) the pattern of contaminant dispersion in air; and 3) the location of the individual relative to the point(s) of contaminant emission. Also described in detail are select models that the author has found useful and/or that are found in the industrial hygiene literature.

CHEMICAL EMISSION RATE FUNCTIONS

The instantaneous emission rate is denoted *G(t)*, with a mass time^{-1} unit such as mg min^{-1}. The parenthetical t indicates that the emission rate may vary with time. When a constant rate is assumed, we use the notation *G(t)* = *G*. Although an emission rate is seldom constant, it may be reasonable to assume constant emission in a given application, for example, if we want to estimate a full-shift TWA exposure level and are not concerned with short-term exposure intensity.

Constant Emission Based on Mass Balance

Sometimes a material mass balance can be used to estimate a constant emission rate. For example, if we know the mass of solvent that evaporated during a time interval T, the *constant* emission rate is:

$$G,\ mg\ min^{-1} = \frac{\text{Mass evaporated (mg) in the interval}}{T\ (min)} \qquad \text{[Equation 15-1]}$$

Constant Emission from a Tank

A problem that has been studied by different investigators is the vapor emission rate from a pure liquid with a fixed surface area and temperature, as would apply to an open surface tank or a large pool of spilled liquid. An algorithm developed by Hummel (1996) for predicting a constant emission rate is

$$G,\ mg\ min^{-1} = \frac{165.6\ MW^{0.833}\ P_V \left(.0345 + \frac{1}{MW}\right)^{0.25} A}{T^{.05}} \sqrt{\frac{U}{LP_{atm}}} \qquad \text{[Equation 15-2]}$$

where P_v = vapor pressure of evaporating chemical (Pa) [Note: 1 atm = 101,325 Pa]

MW = molecular weight of the chemical (g mol^{-1})

T = temperature of the liquid (Kelvin) [Note: Kelvin = 293.3 + °C]

U = air speed over the liquid pool (m sec^{-1})

P_{atm} = atmospheric pressure (Pa)

A = surface area of liquid pool (m^2)

L = surface length of liquid pool (m)

The surface length of the liquid pool, L, is the longest dimension of the surface. For example, if an open surface tank is 3 m × 1 m, then $L = 3$ m. Air is presumed to move in the same direction as the longest length aspect.

Constant Emission due to Evaporation from a Bottle

The emission of chemical from an open bottle due to molecular diffusion can be treated as a constant rate. The neck of the bottle is considered a diffusion path of length L (m), and is not subject to air motion. The cross-sectional area of the neck is denoted A (m^2). For a pure chemical, the contaminant concentration in the bottle headspace (at the start of the diffusion path) is assumed to equal the chemical's saturation concentration in air, denoted C_{sat} (mg m^{-3}). It is assumed that once molecules reach the top of the bottle's neck (at the end of the diffusion path), they are immediately dispersed such that the chemical concentration at the top of the neck is essentially zero. The presumed constant emission rate from the bottle is

$$G,\ mg\ min^{-1} = \frac{A \times D_M \times C_{sat}}{L} \qquad \text{[Equation 15-3]}$$

where D_M (m^2 min^{-1}) is the chemical's molecular diffusion coefficient in air. If a published D_M value is not available, it can be estimated using the Fuller (Fuller et al., 1969) equation

$$D_M,\ m^2\ min^{-1} = \frac{0.6 \times T^{1.75}}{P_{atm}} \times \frac{\sqrt{.0345 + \frac{1}{MW}}}{\left[2.70 + \left(\sum V_{chem}\right)^{1/3}\right]^2} \qquad \text{[Equation 15-4]}$$

where $\sum V_{chem} = \sum_{i=1}^{k} n_i \times v_i$

k = number of different elements in the molecule

n_i = number of atoms of the ith element in the molecule

v_i = atomic diffusion volume (cm^3 mol^{-1}) of the ith element

T = air temperature (Kelvin units)

MW = molecular weight of the chemical (g mol^{-1})

P_{atm} = atmospheric pressure (Pa) [Note: 1 atm = 101,325 Pa]

The following are some atomic diffusion volumes to use in the Fuller equation: Carbon, 15.9; Hydrogen, 2.31; Oxygen, 6.11; Chlorine, 21.0; Fluorine, 14.7; Bromine, 21.9; Iodine, 29.8; Sulfur, 22.9; Nitrogen, 4.54.

The chemical's saturation concentration in air is:

$$C_{sat} \; in \; mg \; m^{-3} = \frac{1.605 \times 10^4 \times P_V \times MW}{T} \quad \text{[Equation 15-5]}$$

where P_v is the chemical's vapor pressure (mmHg) at temperature T (Kelvin units), and MW is the chemical's molecular weight (g).

Constant Emission due to Filling a Container

The saturation concentration concept can be applied to estimating an emission rate associated with filling a container with liquid. A simplifying assumption is that the contaminant concentration in the container headspace immediately reaches the chemical's saturation level in air. Let F (m^3 min^{-1}) denote the volumetric fill rate of the container. During the filling period (equal to the container volume in m^3 divided by F), the presumed constant emission rate is

$$G, \; mg \; min^{-1} = C_{sat} \times F \quad \text{[Equation 15-6]}$$

If the container is empty prior to filling, it is unrealistic to assume that a saturation concentration in headspace air is instantly achieved. Instead, it might be better to model the headspace concentration as linearly increasing from zero to C_{sat} over the filling period. In that case, the average emission rate over the filling period would be one-half the value computed by Equation 15-6. Also note that the filling method can affect the headspace concentration. For example, in top splash loading, liquid falls from the fill pipe at the top of the container and splashes onto the liquid below. Splashing can create liquid aerosol inside the headspace, which increases the chemical's mass concentration in headspace air.

Exponentially Decreasing Emission Rate

This function applies when there is a finite mass of chemical in the source, such that the source is not replenished as mass is lost. The exponentially decreasing emission rate is

$$G(t), \; mg \; min^{-1} = G_0 \; exp(-\alpha \times t) \quad \text{[Equation 15-7]}$$

where G_0 $mg \; min^{-1}$ is the emission rate at time zero, and α is a first-order rate constant with the unit min^{-1}. The highest emission rate occurs at the outset. If M_0 (mg) denotes the initial chemical mass present in the source, the initial emission rate at time zero is $G_0 = \alpha\ M_0$. According to Equation 15-7, there is always some emission, even after a very long time has passed; however, if the product $\propto \times t$ is greater than 10, the predicted emission rate is at least five orders of magnitude less than the initial rate. The exponentially decreasing emission rate function may be appropriate for organic solvent evaporation from small spills (Keil and Nicas, 2003), for the evaporation of a minor constituent from a liquid mixture in a tank, (Nicas et al., 2006) and for off gassing of sorbed contaminants (Nicas, 2003).

Linearly Decreasing Emission Rate

This function also applies when there is a finite mass of chemical in the source. Modeling the emission rate as linearly decreasing avoids having to assume that emission never ceases as in the exponentially decreasing emission rate function. The linearly decreasing emission rate is

$$G(t),\ mg\ min^{-1} = G_0 - \left(\frac{G_0}{t_{max}}\right) \times t \qquad \text{[Equation 15-8]}$$

where $G_0\ mg\ min^{-1}$ is the emission rate at time zero, and t_{max} is the time (min) at which all emission ceases. The function applies over the specified time interval $0 \le t \le t_{max}$. This function has been applied to solvent evaporation from painted surfaces (Haberlin and Heinsohn, 1993).

Increasing Emission Rate Approaching Constant Emission

This function pertains when evaporation occurs while a liquid is still being applied. For example, spraying a solvent-containing coating takes one hour, and solvent emission from the applied coating is not instantaneous. Two assumptions are that the chemical of interest is applied at a constant rate, I ($mg\ min^{-1}$), and the applied chemical simultaneously evaporates from the surface with first-order rate constant α (min^{-1}). The emission rate function is

$$G(t),\ mg\ min^{-1} = I \times [1 - \exp(-\alpha \times t)] \qquad \text{[Equation 15-9]}$$

As time passes, the term $\exp(-\alpha \times t)$ goes to zero, and the emission rate becomes the constant value $I\ mg\ min^{-1}$.

Decreasing Emission Rate Approaching Constant Emission

The last function considered pertains to when the emission rate decreases due to a *back-pressure* exerted by the chemical in room air (Jayjock, 1994). To explain, consider a chemical in the liquid or solid phase. The rate at which chemical molecules leave the liquid or solid phase and enter room air (the number of molecules per unit time) tends to be constant if the temperature and exposed surface area of the chemical remain constant. As the chemical enters room air, thereby increasing its concentration and partial pressure in air, there is an increase in the rate at which the airborne chemical molecules return to the liquid or solid phase (*condense*). This increasing rate of return decreases the net rate at which the chemical is emitted into room air; the net rate is the rate of emission into room air minus the rate of return to the liquid or solid phase. Because the net emission rate decreases as the chemical's partial pressure in air increases, the airborne chemical is said to exert a *backpressure* on the liquid or solid phase chemical.The emission rate function is

$$G(t) = G_0 \times \left(1 - \frac{C(t)}{C_{sat}}\right) \qquad \text{[Equation 15-10]}$$

where G_0 mg min^{-1} is the emission rate at time zero (assuming zero concentration in air), C_{sat} mg m^{-3} is the chemical's saturation concentration in room air, and C(t) mg m^{-3} is the chemical's concentration in room air at time t (min). G(t) decreases as the airborne concentration C(t) increases and approaches C_{sat}. If the chemical concentration in air reaches a steady-state value, the emission rate also reaches a steady-state value given by

$$G_{SS},\ mg/min \ = \ G_0 \times \left(1 - \frac{G_0}{G_0 + Q \times C_{sat}}\right) \qquad \text{[Equation 15-11]}$$

To promote accuracy, it would be best for all emission rate functions for gas-phase air contaminants to account for a backpressure effect. One advantage of incorporating backpressure is that it prevents predicting an airborne concentration exceeding C_{sat}, which is generally impossible. However, if the expected concentration values are far below C_{sat}, the error made by ignoring the backpressure effect is small. For example, toluene has a saturation vapor pressure of 21 mmHg at 20°C. For a total room pressure of 760 mmHg, $C_{sat} \approx$ 28,000 ppm. It would be unusual (fortunately) for the toluene concentration in an indoor workplace to approach 5% of this value; therefore, if backpressure is ignored, the error made in overestimating the toluene emission rate should be less than 5%. Still, the importance of accounting for backpressure is best made on a case-by-case basis.

There are other variable emission rate functions that may be appropriate depending on the application; for example, a chemical might be released in a cyclical or sinusoidal fashion due to a process cycle. Note that if a variable emission rate function is complex, we might choose to consider the function over a sequence of short intervals, and for each interval assign a constant emission rate corresponding to the total mass emitted during the interval, as shown in Equation 15-1.

DISPERSION PATTERNS AND WORKER LOCATION

Four patterns of contaminant dispersion in room air are considered. Because worker location relative to the emission source is integral to the first two constructs, the issue of location is discussed at the same time.

Well-Mixed Room (WMR)

In a WMR, the chemical instantaneously and uniformly mixes throughout room air, such that the chemical's airborne concentration is the same at all points in the room. Therefore, worker location relative to the source is not a factor, because the exposure level is presumed to be the same at all locations. The room parameters are the volume V (m^3) and the supply/exhaust air rate is Q (m^3 min^{-1}). To be strict, we should consider only the non-occluded air volume, which is the total room volume minus the volume of the solid objects in the room. In general, the occluded volume is only a small percent of the total room volume, so this fine point is usually ignored.

Perfect mixing is physically unrealistic, and can lead to seriously underestimating exposure intensity for a worker located close to an emission source. However, the WMR construct is reasonable for estimating exposure in three situations: 1) for individuals in the room at some distance from a point source, say, > 2 meters; 2) for all individuals in the room about two minutes after emission from a point source ceases; and 3) for all individuals in the room if the emissions are from a large surface area (for example, painted walls) or from numerous emission sources spread throughout the room.

It has become common to use a mixing factor m in the WMR construct to account for imperfect air mixing. The supply/exhaust air rate Q is multiplied by m to give an *effective* supply/exhaust rate, or $Q_{EFF} = m \times Q$, where $0 < m \leq 1$. The value $m = 1$ is perfect mixing, while $m < 1$ is imperfect mixing. Some texts present m in terms of a safety factor K, where $K = 1/m$. In a ventilation application, if one requires a certain Q_{EFF} value, the design airflow rate is recommended to be increased K-fold, or $Q_{DESIGN} = K \times Q_{EFF}$.

In the author's judgment, the mixing factor should not be used with the WMR dispersion construct for three reasons. First, it violates the fundamental principle of conservation of mass for the chemical in room air and for the airflows in the room (Heinsohn, 1991; Nicas, 1996). Second, assigning a value for m basically involves guessing (Heinsohn, 1991). Third, estimating a value for m based on measurements of concentration decay over time can still lead to gross errors (Nicas, 1996). Where it is important to account for imperfect air mixing, such as for estimating exposure near an emission source, the near field/far field construct or a turbulent eddy diffusion construct should be used.

Near Field/Far Field

The principles underlying this construct were first described by W.C. Hemeon (1963) half a century ago, although the time-dependent concentration equations were developed more recently (Nicas, 1996). The room is treated as containing two contiguous zones – the near field (NF) zone surrounds the emission source and contains the breathing zone of the individual whose exposure is to be estimated, and the far field (FF) zone comprises the rest of the room. The respective NF and FF volumes, V_{NF} (m^3) and V_{FF} (m^3), sum to the room volume. Air within each zone is assumed to be perfectly mixed, but there is a limited airflow between the two zones. Room supply/exhaust airflows into and out of the FF zone at rate Q ($m^3\ min^{-1}$). The interzonal airflow rate β ($m^3\ min^{-1}$) depends on the conceptual NF zone geometry (which involves the individual's distance from the source) and a random air speed s ($m\ min^{-1}$) near the source. One study reported that indoor air speeds are lognormally distributed with a geometric mean of 3.7 $m\ min^{-1}$ (12 fpm) and a geometric standard deviation of 2 (Baldwin and Maynard, 1998).

Consider a near field which is a hemisphere centered on the emission point. The hemisphere's radius r (m) is set to include the individual's breathing zone. The free surface area *(FSA)* (m^2) of the near field is that conceptual surface through which air can flow; for a hemisphere with its base on a solid surface, $FSA = 2\ \pi\ r^2$. The interzonal airflow rate is

$$\beta,\ m^3\ min^{-1} = \tfrac{1}{2} \times FSA \times s \qquad \text{[Equation 15-12]}$$

The factor ½ is used because in a small section of free surface area, air cannot simultaneously flow into and out of the NF zone. Assuming equality of airflow into and out of the NF zone also maintains a mass or volume balance of air in this zone. This balance is analogous to the equality in the airflow rates Q into and out of the larger room.

The NF/FF construct leads to predicting a higher chemical concentration in the NF zone than in the FF zone, which is intuitively reasonable. A drawback is that the NF/FF construct predicts an unrealistic discontinuity in the concentration across the NF zone boundary. A turbulent eddy diffusion dispersion pattern, which is discussed next, produces a continuous concentration gradient with distance away from the source.

Turbulent Eddy Diffusion without Advection

Turbulent eddy diffusion involves the random motion of parcels (eddies) of air which carry chemical molecules (or particles) away from the source. Turbulent diffusion is distinct from molecular diffusion which involves the random motion of individual molecules due to their kinetic energy. Turbulent diffusion is characterized by a diffusion coefficient D_T ($m^2\ min^{-1}$). Although D_T has the same length^2 time^{-1} unit as the molecular diffusion coefficient D_M, the magnitude of D_T is much greater than the magnitude of D_M, in general. D_M values are on the order of 0.001 $m^2\ min^{-1}$, while D_T values are on the order of 0.1 to 10 $m^2\ min^{-1}$ (Keil, 2000).

This dispersion construct requires specifying the individual's position relative to the source. Without advective airflow (the bulk movement of air in a given direction, on average), an equal concentration surface is predicted at a fixed radial distance from the emission point. Due to this symmetry, the worker's location can be expressed either as a radial distance r (m) from the source, or in the alternative, the location can be expressed as a Cartesian coordinate position (x, y, z) relative to the emission point which is taken to be at the origin (0, 0, 0).

Turbulent Eddy Diffusion with Advection

The dispersion pattern is turbulent diffusion with an advective airflow along one of the room axes, such that contaminant movement via directional airflow is superimposed on diffusion. Turbulent diffusion is again characterized by the coefficient D_T, but a mean air speed U (m min^{-1}) along the room axis is also specified. The individual's location must be expressed as a Cartesian coordinate position (x, y, z) relative to the emission source taken to be at the origin (0, 0, 0).

SPECIFIC MODELS

Because any emission rate function can be matched with any dispersion construct, there are numerous models available. In general, models that assume a constant emission rate and/or the WMR dispersion construct are relatively simple mathematically, and usually have a closed-form solution for the time-dependent contaminant exposure concentration, $C(t)$ mg m^{-3}. A closed-form solution means there is single equation such that if we insert model input values, we obtain an output number. Models without closed-form solutions are solved numerically via a computer code. In this section, some select models with closed-form solutions are presented and illustrated. For models without closed-form solutions, a computer code is presented that implements a numerical method.

WMR with a Constant Emission Rate

Assume that a room with volume V (m^3) and supply/exhaust air rate Q (m^3 min^{-1}) contains a source emitting a chemical at constant rate G (mg min^{-1}). Let C_{IN} (mg m^{-3}) denote the chemical concentration in room supply air, where $C_{IN} \geq 0$. The first-order rate of contaminant removal from room air via exhaust ventilation is $Q \div V$ (min^{-1}), which is the number of air changes per minute. Let k_L (min^{-1}) denote a possible first-order rate of chemical elimination by some other mechanism, where $k_L \geq 0$. For example, a chemical vapor may adsorb onto room surfaces or be degraded to a different chemical species, and a particulate contaminant can settle due to gravity as well as deposit onto room surfaces. Let C_0 (mg m^{-3}) denote the chemical concentration in room air at time zero, where $C_0 \geq 0$. The equation which predicts $C(t)$, for $t \geq 0$, is

$$C(t),\ \text{mg m}^{-3} = \frac{G + C_{IN}Q}{Q + k_L V}\left[1 - \exp\left(-\frac{Q + k_L V}{V} \times t\right)\right] + C_0 \exp\left(-\frac{Q + k_L V}{V} \times t\right)$$

[Equation 15-13]

Note that incorporating a non-ventilatory loss mechanism (summarized by the k_L term) may not be straightforward, because the loss pathway may be reversible. For example, molecules that adsorb onto a surface at the present time may desorb back into room air at a future time. To properly account for the reverse process, one needs to consider the rate constant which would apply for the reverse process (which may not be the same as k_L), and the model should include a separate compartment that represents the state (sometimes termed a *sink*) to which the contaminant is lost and from which it can be remitted. The details of such modeling are beyond the scope of this chapter.

If the emission process continues for a long time such that $t \to \infty$, the chemical concentration reaches a steady-state value, or

$$C_{SS},\ \text{mg m}^{-3} = \frac{G + C_{IN}Q}{Q + k_L V}$$

[Equation 15-14]

Equation 15-14 signifies that the steady-state concentration does not depend on the initial concentration value C_0. To illustrate the model, consider a process that emits toluene vapor at constant rate G = 1000 mg min^{-1} into the air of a WMR room with V = 100 m^3 and Q = 10 m^3 min^{-1}; this supply/exhaust air rate corresponds to 0.1 air change per minute. The supply air does not contain toluene (C_{IN} = 0), and the only loss mechanism of toluene vapor from room air is the exhaust airflow (k_L = 0). At time zero, the initial toluene concentration in room air is zero. By inserting these input values into Equation 15-13, we obtain

$$C(t),\ \text{mg m}^{-3} = 100 \times [1 - \exp(-0.1 \times t)]$$

By inserting the input values into Equation 15-14, we obtain C_{SS} = 100 mg m^{-3} (equal to 27 ppm at NTP conditions). The solid line curve in Figure 15-1 labeled $G(t) = G$ shows the predicted concentration profile from t = 0 to t = 120 min.

Now, assume that at t = 120 min, toluene emission ceases. The solid line curve in Figure 15-1 shows the exponential purging of toluene vapor from room air in the period t = 120 min to t = 180 min, referenced to time zero. The general equation for purging of contaminant from room air by dilution ventilation (given no new contaminant emission) is

$$C(t),\ \text{mg m}^{-3} = C_0 \exp\left(-\frac{Q}{V} \times t\right)$$

[Equation 15-15]

where C_0 (mg m^{-3}) is the contaminant concentration in room air when the purging begins. In the Figure 15-1 scenario, the purging equation for 120 min < $t \le$ 180 min is

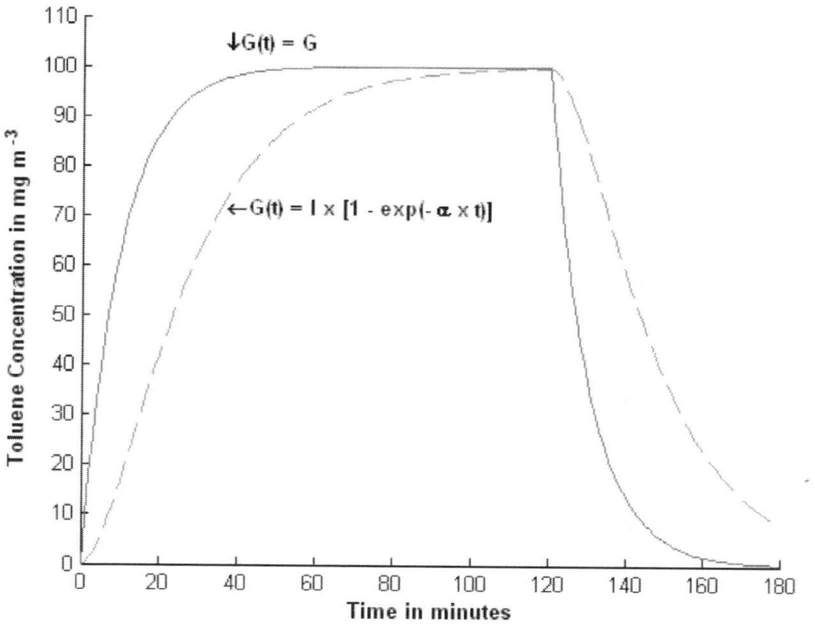

Figure 15-1. The predicted toluene concentration in a well mixed room with $V = 100$ m³ and $Q = 10$ m³ min⁻¹, for two alternative emission rate functions. One rate is constant at 1000 mg min⁻¹, and corresponds to the solid line curve labeled $G(t) = G$. The other rate increases from an initial value of zero to a final constant rate of 1000 mg min⁻¹, and corresponds to the dashed line curve labeled $G(t) = I \times [1 - exp(-\alpha \times t)]$. At 120 minutes, de novo application of toluene ceases. For the solid line curve, it is assumed that new toluene emission also ceases at 120 minutes. For the dashed line curve, it is assumed that residual liquid toluene on a surface continues to evaporate after 120 minutes.

$$C(t), \text{ mg m}^{-3} = 100 \times \exp(-.01 \times [t - 120])$$

The reason that 120 is subtracted from t in the power of the exponent term is to maintain the original reference time scale.

WMR with an Exponentially Decreasing Emission Rate

Assume that a room with volume V (m³) and supply/exhaust air rate Q (m³ min⁻¹) contains a source emitting a chemical at an exponentially decreasing rate $G(t) = \alpha\, M_0 \exp(-\alpha \times t)$ in mg min⁻¹. For simplicity, we do not consider the previous terms C_0, C_{IN}, and k_L, although they could be included. The equation which predicts $C(t)$, for $t \geq 0$, is

$$C(t), \text{mg m}^{-3} = \frac{\alpha M_0}{Q - \alpha V}\left[\exp(-\alpha \times t) - \exp\left(-\frac{Q}{V} \times t\right)\right] \quad \text{[Equation 15-16]}$$

Because the emission rate is always decreasing, the chemical concentration does not reach a steady-state value (other than zero). However, a peak concentration C_{peak} (mg m^{-3}) develops, and the time at which it occurs, denoted t_{peak} (min), is

$$t_{peak}, \text{min} = \frac{V}{\alpha V - Q} \times \ln\left(\frac{\alpha V}{Q}\right) \quad \text{[Equation 15-17]}$$

To compute C_{peak}, we insert t_{peak} into the $C(t)$ equation. To illustrate the model, consider a spill of 100 mL toluene (density = 870 mg mL^{-1}), such that M_0 = 87,000 mg. According to a published algorithm (Keil and Nicas, 2003), the evaporation rate constant α for a small spill of a hydrocarbon solvent (which might also contain oxygen) is

$$\alpha, \text{min}^{-1} = .000524\ Pv + .0108\ (SA \div VOL) \quad \text{[Equation 15-18]}$$

where P_V is the solvent's vapor pressure in mmHg at 20°C, and $SA \div VOL$ is the initial surface-area-to-volume ratio of the spill in cm^{-1}. For toluene, P_V = 21 mmHg. Assume the depth of the spill is 0.25 cm, such that SA = 400 cm^2 and $SA \div VOL$ = 4 cm^{-1}. Based on Equation 15-18, α = .054 min^{-1}. Assume that the spill occurs in a WMR with V = 100 m^3 and Q = 10 m^3 min^{-1}. The parameters C_{IN}, k_L, and C_0 are all zero. By inserting these input values into Equation 15-16, we obtain

$$C(t), \text{mg m}^{-3} = 1021 \times [\exp(-.054 \times t) - \exp(-0.1 \times t)]$$

Based on Equation 15-17, t_{peak} = 13.4 min. By inserting t = 13.4 min into the above expression for $C(t)$, we find that C_{peak} = 228 mg m^{-3} (equal to 61 ppm at NTP conditions). The solid line curve in Figure 15-2 labeled "Exponentially Decreasing Emission Rate" shows the predicted concentration profile from t = 0 to t = 90 min.

WMR with a Linearly Decreasing Emission Rate

Consider a contaminant source with a linearly decreasing emission rate (Equation 15-8). The source is located in a WMR with volume V (m^3) and supply/exhaust air rate Q (m^3 min^{-1}). For simplicity, we do not consider the previous terms C_0, C_{IN}, and k_L, although they could be included. The equation which predicts $C(t)$, for $0 \leq t \leq t_{max}$ is as follows:

$$C(t), \text{ in mg m}^{-3} = \frac{G_0 \times (Qt_{max} + V)}{Q^2\, t_{max}}\left[1 - \exp\left(-\frac{Q}{V} \times t\right)\right] - \frac{G_0}{Q\, t_{max}} \times t \quad \text{[Equation 15-19]}$$

where G_0 is the emission rate at time zero, and t_{max} is the time at which all the contaminant has been emitted. Because the emission rate is always decreasing, the contaminant level does not

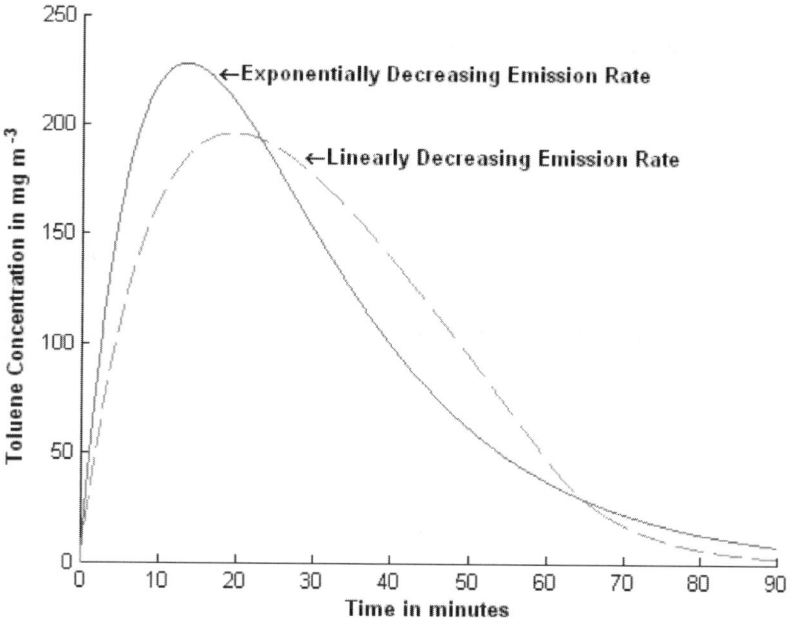

Figure 15-2. The predicted time series of toluene concentrations in a well mixed room with V = 100 m^3 and Q = 10 m^3 min^{-1}, for two alternative emission rate functions. One rate is exponentially decreasing with M_0 = 87,000 mg, and α = 0.054 min^{-1}, and corresponds to the solid line curve labeled "Exponentially Decreasing Emission Rate." The other rate linearly decreases to zero over the period zero to 60 minutes (the initial mass is M_0 = 87,000 mg), and corresponds to the dashed line curve labeled "Linearly Decreasing Emission Rate."

reach a steady-state value, but a peak concentration occurs at t_{peak} (min), or

$$t_{peak}, \text{ min} = -\frac{V}{Q} \times \ln\left(\frac{V}{Q\, t_{max} + V}\right) \qquad \text{[Equation 15-20]}$$

To illustrate the model, consider another scenario for a spill of 100 mL toluene. Assume that vapor emission from the spill linearly decreases, and becomes zero at t_{max} = 60 min. Given that 87,000 mg toluene was spilled, it can be shown that G_0 = 2,900 mg min^{-1}. The spill occurs in a WMR with V = 100 m^3 and Q = 10 m^3 min^{-1}. By inserting these input values into Equation 15-19, we obtain

$$C(t), \text{ in mg m}^{-3} = 338 \times [1 - \exp(-0.1 \times t)] - 4.8 \times t$$

This equation applies over the interval $0 \leq t \leq 60$ min. The dashed line curve in Figure 15-2 labeled "Linearly Decreasing Emission Rate" shows the predicted concentration profile from $t = 0$ to $t = 60$ min predicted by the above equation. Based on Equation 15-20, $t_{peak} = 19.5$ min. By inserting $t = 19.5$ min into the above expression for $C(t)$, we find that $C_{peak} = 196$ mg m^{-3} (equal to 52 ppm at NTP conditions). Once emission ceases at $t = 60$ min, there is exponential purging of toluene vapor from room air in the period $t = 60$ min to $t = 90$ min, referenced to time zero.

Note that in Figure 15-2, the predicted concentration profiles based on the exponentially decreasing versus linearly decreasing emission rates are similar. Part of the reason is that nearly all the spilled toluene mass evaporates in 60 minutes in both scenarios. For the linearly decreasing rate, 100% evaporates in 60 minutes, while for the exponentially decreasing rate, 96% evaporates in 60 minutes.

WMR with an Increasing Emission Rate Approaching Constant Emission

Consider a contaminant source with the emission rate function specified by Equation 15-9. The source is located in a WMR with volume V (m^3) and supply/exhaust air rate Q (m^3 min^{-1}). For simplicity, we do not consider the previous terms C_0, C_{IN}, and k_L, although they could be included. The equation which predicts $C(t)$, for $t \geq 0$, is as follows:

$$C(t),\ mg\ m^{-3} = \frac{I}{Q} \times \left[1 - \exp\left(-\frac{Q}{V}t\right)\right] + \frac{I}{Q - \alpha V} \times \left[\exp\left(-\frac{Q}{V}t\right) - \exp(-\alpha \times t)\right] \quad \text{[Equation 15-21]}$$

To illustrate the model, consider a process in which a toluene-containing coating is applied to a surface such that $I = 1000$ mg min^{-1} of toluene is applied. Assume that toluene evaporates from the surface with rate constant $\alpha = .054$ min^{-1}. The coating operation occurs from $t = 0$ to $t = 120$ min, in a WMR room with $V = 100$ m^3 and $Q = 10$ m^3 min^{-1}. By inserting these input values into Equation 15-21, we obtain

$$C(t),\ \text{mg m}^{-3} = 100 \times [1 - \exp(-0.1 \times t)] + 21.7 \times [\exp(-0.1 \times t) - \exp(-.054 \times t)]$$

The dashed line curve in Figure 15-1 labeled $G(t) = I \times [1 - exp(-\alpha \times t)]$ shows the concentration profile from $t = 0$ to $t = 120$ min predicted by the above $C(t)$ equation. Note that by t = 120 min, the toluene vapor concentration has reached the same steady-state level of 100 mg m^{-3} as in the scenario where $G = 1000$ mg min^{-1} from the outset. This circumstance results because the increasing emission rate function also reaches the constant value of 1000 mg min^{-1}.

The dashed line curve in Figure 15-1 shows that once the coating application ceases at $t = 120$ min, such that no further toluene is applied, the decrease in the toluene vapor concentration is more gradual than for pure exponential purging (shown by the solid line curve in Figure 15-1). The reason is that the toluene mass present on the coated surface at $t = 120$ min provides a continuing source of vapor emission into room air. It can be shown that 18,490 mg toluene is on the coated surface at $t = 120$ min. The toluene emission rate for $t > 120$ min can be modeled as exponentially decreasing with $M_0 = 18{,}490$ mg and $\alpha = .054$ min^{-1}. In the period 120 min $< t \leq 180$ min shown in Figure 15-1, the *C(t)* equation corresponding to the dashed line curve is

$$C(t), \text{mg m}^{-3} = 217 \times (\exp[-.054 \times (t - 120)] - \exp[-0.1 \times (t - 120)]) + 100 \times \exp[-0.1 \times (t - 120)]$$

In the above expression, the first term on the right-hand side accounts for new emission of toluene vapor from the coated surface, and the second term accounts for the exponential purging from room air of the vapor concentration that existed at $t = 120$ min. Again, the reason that 120 is subtracted from t in the power of the exponent terms is to maintain the original reference time scale.

WMR with a Decreasing Emission Rate due to Backpressure

Consider a contaminant source with the emission rate function specified by Equation 15-10. The source is located in a WMR with volume V (m^3) and supply/exhaust air rate Q (m^3 min^{-1}). For simplicity, we do not consider the previous terms C_0, C_{IN}, and k_L, although they could be included. The equation which predicts *C(t)*, for $t \geq 0$, is as follows:

$$C(t), \text{mg m}^{-3} = \frac{G_0}{Q + \dfrac{G_0}{C_{sat}}} \left[1 - \exp\left(-\frac{G_0 + Q \times C_{sat}}{V \times C_{sat}} \times t\right)\right] \quad \text{[Equation 15-22]}$$

The coefficient in front of the brackets contains a denominator term G_0/C_{sat}, which is unusual. However, this term has an *equivalent ventilation* unit of m^3 min^{-1}.

To illustrate the model, consider a process that emits mercury (Hg) vapor at an initial rate $G_0 = 10{,}000$ µg min^{-1} into the air of a WMR room with $V = 100$ m^3 and $Q = 10$ m^3 min^{-1}; this supply/exhaust air rate corresponds to 0.1 air change per minute. The saturation concentration of Hg vapor in air is approximately 13,000 µg m^{-3}. At time zero, the initial Hg concentration in room air is zero. By inserting these input values into Equation 15-22, we obtain

$$C(t), \mu\text{g m}^{-3} = 929 \times [1 - \exp(-0.108 \times t)]$$

According to Equation 15-11, the Hg emission rate will decrease to the constant value of 9,290 µg min^{-1}; the corresponding steady-state concentration in air will be $C_{SS} = 929$ µg m^{-3}. In contrast, if the backpressure effect were ignored, the predicted steady-state concentration would be the initial emission rate G_0 (10,000 µg min^{-1}) divided by the supply/exhaust rate Q (10 m^3 min^{-1}), or 1,000 µg m^{-3}. The latter value is 7.6% greater than the C_{SS} estimate which incorporates the backpressure effect.

Near Field/Far Field with a Constant Emission Rate

Assume that a room with volume V (m^3) and supply/exhaust air rate Q (m^3 min^{-1}) contains a point source emitting contaminant at constant rate G (mg min^{-1}). The NF zone surrounding the source has volume V_{NF} (m^3), in which case $V_{FF} = V - V_{NF}$. The airflow rate between the NF and FF zones is β (m^3 min^{-1}). For simplicity, we assume that $C_{IN} = 0$ and $k_L = 0$, and that the initial contaminant concentration in both zones is zero. Let $C_{NF}(t)$ in mg m^{-3} denote the near field contaminant concentration. The equation which predicts $C_{NF}(t)$, for t ≥ 0, is:

$$C_{NF}(t)\text{, mg m}^{-3} = \frac{G}{Q} + \frac{G}{\beta} + G \times \left(\frac{\beta Q + \lambda_2 V_{NF}(\beta + Q)}{\beta Q\, V_{NF}(\lambda_1 - \lambda_2)}\right)\exp(\lambda_1 \times t) - G \times \left(\frac{\beta Q + \lambda_1 V_{NF}(\beta + Q)}{\beta Q\, V_{NF}(\lambda_1 - \lambda_2)}\right)\exp(\lambda_2 \times t), \quad \text{[Equation 15-23]}$$

where

$$\lambda_1\text{, min}^{-1} = 0.5\left[-\left(\frac{\beta V_{FF} + V_{NF}(\beta + Q)}{V_{NF} V_{FF}}\right) + \sqrt{\left(\frac{\beta V_{FF} + V_{NF}(\beta + Q)}{V_{NF} V_{FF}}\right)^2 - 4\left(\frac{\beta Q}{V_{NF} V_{FF}}\right)}\right]$$

$$\lambda_2\text{, min}^{-1} = 0.5\left[-\left(\frac{\beta V_{FF} + V_{NF}(\beta + Q)}{V_{NF} V_{FF}}\right) - \sqrt{\left(\frac{\beta V_{FF} + V_{NF}(\beta + Q)}{V_{NF} V_{FF}}\right)^2 - 4\left(\frac{\beta Q}{V_{NF} V_{FF}}\right)}\right]$$

The terms λ_1 and λ_2 are ventilation removal rate constants. The absolute value of λ_1 reflects the rate of contaminant removal from the FF zone, and the absolute value of λ_2 reflects the rate of contaminant removal from the NF zone. Because the absolute value of λ_2 is usually greater than that of λ_1, removal from the NF zone proceeds more quickly than removal from the general room.

There is an analogous equation for the chemical concentration in the FF, denoted $C_{FF}(t)$. The latter expression is not given here because the values predicted by the $C_{FF}(t)$ equation are quite similar to those predicted by the WMR equation for the same values of G, Q and V. Those readers desiring the exact equation for $C_{FF}(t)$ can consult Nicas (1996).

If the emission process continues for a long time such that $t \rightarrow \infty$, the contaminant concentrations in the NF and FF zones reach steady-state values given by, respectively:

$$C_{NF,SS}\text{, mg m}^{-3} = \frac{G}{Q} + \frac{G}{\beta} \quad \text{[Equation 15-24]}$$

$$C_{FF,SS},\ \text{mg m}^{-3} = \frac{G}{Q} \qquad \text{[Equation 15-25]}$$

The FF steady-state concentration is the same value predicted using the WMR model with a constant emission rate. If $\beta \leq Q$, it follows that $C_{NF,SS} \geq 2\ C_{FF,SS}$; in other words, the NF concentration is at least twice the FF concentration (and twice the value that would be predicted by the WMR model). This condition exists in a room with a low air speed near the source, or in a large room with a large Q value. In general, as the room size increases, Q increases, but β is minimally affected such that $\beta << Q$; in turn, the NF/FF model predicts much higher exposure intensity near the emission source compared to the WMR model. On the other hand, if $\beta >> Q$ because there is high air speed near the source (for example, due to the use of cooling fans), then $C_{NF,SS} \cong C_{FF,SS}$. For $\beta >> Q$, the rapid dispersion of contaminant away from the source makes the room behave as if it were well mixed.

To illustrate the model, consider an operation on a bench top that emits toluene vapor at constant rate $G = 1000$ mg min^{-1}, in a room with $V = 100$ m^3 and $Q = 10$ m^3 min^{-1}. An individual's breathing is always located within 0.9 m (3 feet) of the toluene emission. The NF zone is conceived as a hemisphere of radius $r = 0.9$ m. The NF zone has $FSA = 2\ \pi\ (0.9)^2 = 5.1$ m^2. The random air speed is 3.7 m min^{-1} (12 fpm), which is the median air speed value found in a study of indoor workplaces (Baldwin and Maynard, 1998). Based on Equation 15-12, the interzonal airflow rate $\beta = 9.4$ m^3 min^{-1}. By inserting these input values into Equation 15-23, we obtain

$$C_{NF}(t),\ \text{mg m}^{-3} = 206.2 - 103.3\ \exp(-\ 0.10 \times t) - 102.9\ \exp(-\ 6.26 \times t)$$

The solid line curve in Figure 15-3 labeled "Near Field" shows the toluene concentration profile from $t = 0$ to $t = 60$ min predicted by the above $C_{NF}(t)$ equation. Note that by $t = 60$ min, the toluene vapor concentration has reached a steady-state level of 206 mg m^{-3}, which is the value predicted based on Equation 15-24. In contrast, the dashed line curve in Figure 15-3 labeled "Well Mixed Room" shows the concentration profile predicted by assuming that room air is well mixed; the corresponding steady-state concentration is 100 mg m^{-3}. In this example, the NF/FF model predicts a steady-state concentration twice the value predicted by the WMR model. If we assume that the NF/FF better captures reality, Figure 15-3 also shows that the WMR model underestimates exposure intensity for persons located near an emission source. In a larger room with a larger Q value, the difference between $C_{NF,SS}$ and $C_{WMR,SS}$ would be greater still.

The rate constant $|\lambda_1| = 0.10$ min^{-1} governs contaminant removal from the FF zone, and the rate constant $|\lambda_2| = 6.26$ min^{-1} governs contaminant removal from the NF zone into the FF zone. Note that $|\lambda_2|$ is approximately equal to β (9.4 m^3 min^{-1}) divided by the FF volume (1.5 m^3), which is to say that $|\lambda_2|$ represents the number of air changes per minute in the NF zone. In general, air in the NF zone turns over rapidly, so that once contaminant emission ceases, the NF zone contaminant concentration rapidly decreases to equal the FF zone contaminant con-

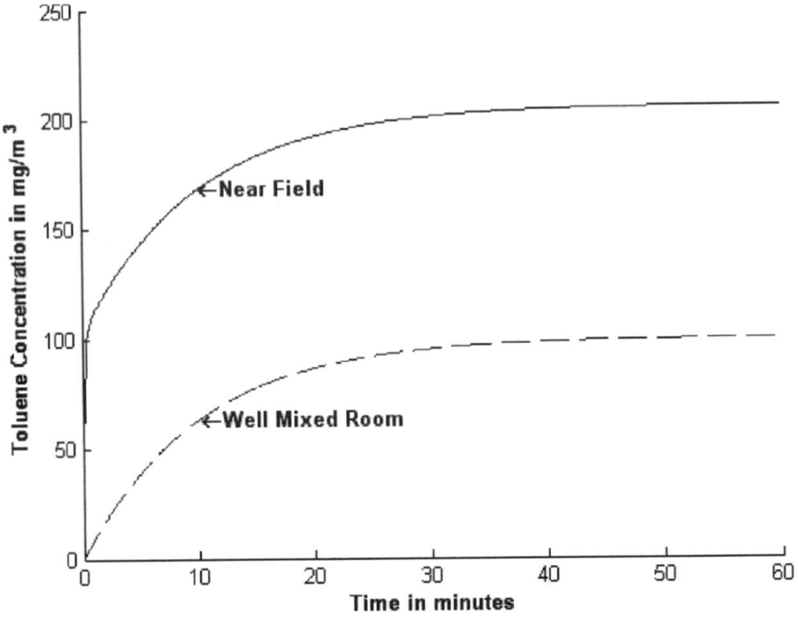

Figure 15-3. The predicted time series of toluene concentrations in a room with $V = 100$ m^3 and $Q = 10$ m^3 min^{-1}, and with a fixed point source emitting toluene vapor at the constant rate 1000 mg min^{-1}. The solid line curve labeled "Near Field" corresponds to the near field/far field dispersion model. The near field is a hemisphere of radius $r = 0.9$ m, the random air speed is 3.7 m min^{-1}, and the interzonal airflow rate $\beta = 9.4$ m^3 min^{-1}. The dashed line curve labeled "Well Mixed Room" corresponds to the well mixed room dispersion model.

centration (that is, the general room concentration). This rapid decrease (usually within one minute) is the basis for the previous statement that the WMR dispersion construct is adequate to estimate exposure for all individuals in a room about two minutes after emission from a point source ceases.

Near Field/Far Field with an Exponentially Decreasing Emission Rate

Assume that a room with volume V (m^3) and supply/exhaust air rate Q (m^3 min^{-1}) contains a point source emitting contaminant with an exponentially decreasing rate $G(t) = \alpha\, M_0 \exp(-\alpha \times t)$ in mg min^{-1}. There is a near field zone surrounding the source with volume V_{NF} (m^3), in which case $V_{FF} = V - V_{NF}$. The airflow rate between the near field and far field zones is β (m³/min). For simplicity, we assume that $C_{IN} = 0$ and $k_L = 0$, and that the initial chemical concentration in both zones is zero. The equation which predicts $C_{NF}(t)$, for $t \geq 0$, is:

$$C_{NF}(t),\ \text{mg m}^{-3} = \frac{\alpha M_0 \left[\left(\frac{\beta+Q}{V_{FF}}-\alpha\right)\left(\frac{\beta}{V_{NF}}+\lambda_2\right) - \frac{\beta^2}{V_{NF}\ V_{FF}}\right]}{(\lambda_1-\lambda_2)\,V_{NF}\left[\left(\frac{\beta+Q}{V_{FF}}-\alpha\right)\left(\frac{\beta}{V_{NF}}-\alpha\right) - \frac{\beta^2}{V_{NF}\ V_{FF}}\right]}\exp(\lambda_1 \times t)$$

$$-\frac{\alpha M_0 \left[\left(\frac{\beta+Q}{V_{FF}}-\alpha\right)\left(\frac{\beta}{V_N}+\lambda_2\right) - \frac{\beta^2}{V_{NF}\cdot V_{FF}} + (\lambda_1-\lambda_2)\left(\frac{\beta+Q}{V_{FF}}-\alpha\right)\right]}{(\lambda_1-\lambda_2)\,V_{NF}\left[\left(\frac{\beta+Q}{V_{FF}}-\alpha\right)\left(\frac{\beta}{V_{NF}}-\alpha\right) - \frac{\beta^2}{V_{NF}\ V_{FF}}\right]}\exp(\lambda_2 \times t)$$

$$+\frac{\alpha M_0 \left(\frac{\beta+Q}{V_{FF}}-\alpha\right)}{V_{NF}\left[\left(\frac{\beta+Q}{V_{FF}}-\alpha\right)\left(\frac{\beta}{V_{NF}}-\alpha\right) - \frac{\beta^2}{V_{NF}\ V_{FF}}\right]}\exp(-\alpha \times t)$$

[Equation 15-26]

where λ_1 and λ_2 are computed as shown in Equation 15-23.

There is an analogous equation for the contaminant concentration in the FF zone, $C_{FF}(t)$. The latter expression is not given here because the values predicted by the $C_{FF}(t)$ equation are quite similar to those predicted by the WMR model with an exponentially decreasing emission rate, given the same values for Q and V and the same $G(t)$ function. Readers desiring the exact equation for $C_{FF}(t)$ can consult Keil and Nicas (2003).

Because the emission rate is always decreasing, $C_{NF}(t)$ does not reach a steady state value (other than zero). A peak concentration develops, but the time at which it occurs cannot be analytically determined. Instead, the peak concentration is found by inspection of the time series of $C_{NF}(t)$ values.

To illustrate the model, consider an operation on a bench top in which 100 mL toluene is spilled onto a surface, such that $M_0 = 87{,}000$ mg and $\alpha = .054$ min⁻¹. The operation occurs in a room with $V = 100$ m³ and $Q = 10$ m³ min⁻¹. An individual's breathing is always located within 0.9 m (3 feet) of the liquid toluene. The NF zone is conceived as a hemisphere of radius r = 0.9 m and $FSA = 5.1$ m². The random air speed is 3.7 m min⁻¹ (12 fpm), and the interzonal airflow rate $\beta = 9.4$ m³ min⁻¹. By inserting these input values into Equation 15-26, we obtain

$$C_{NF}(t), \text{ mg m}^{-3} = -1055.3 \exp(-0.10 \times t) - 487.7 \exp(-6.26 \times t) + 1543.0 \exp(-.054 \times t)$$

The solid line curve in Figure 15-4 labeled "Near Field" shows the toluene concentration profile from $t = 0$ to $t = 60$ min predicted by the above $C_{NF}(t)$ equation. A peak concentration of 538 mg m^{-3} (equal to 143 ppm at NTP conditions) develops at 5.1 minutes. In contrast, the dashed line curve in Figure 15-4 labeled "Well Mixed Room" shows the concentration profile predicted assuming that room air is well mixed, and is quite similar to the concentration profile predicted by the $C_{FF}(t)$ equation. For the WMR concentration profile (based on Equation 15-16), the peak concentration is 228 mg m^{-3} (equal to 61 ppm at NTP conditions) and develops at 13.4 minutes. In this example, the NF/FF model predicts a peak exposure concentration about 2.3-fold greater than predicted by the WMR model. The NF/FF model also predicts an earlier peak concentration than does the WMR model. If we assume that the NF/FF better captures reality, Figure 15-4 shows that the WMR model underestimates exposure intensity for persons located near an emission source.

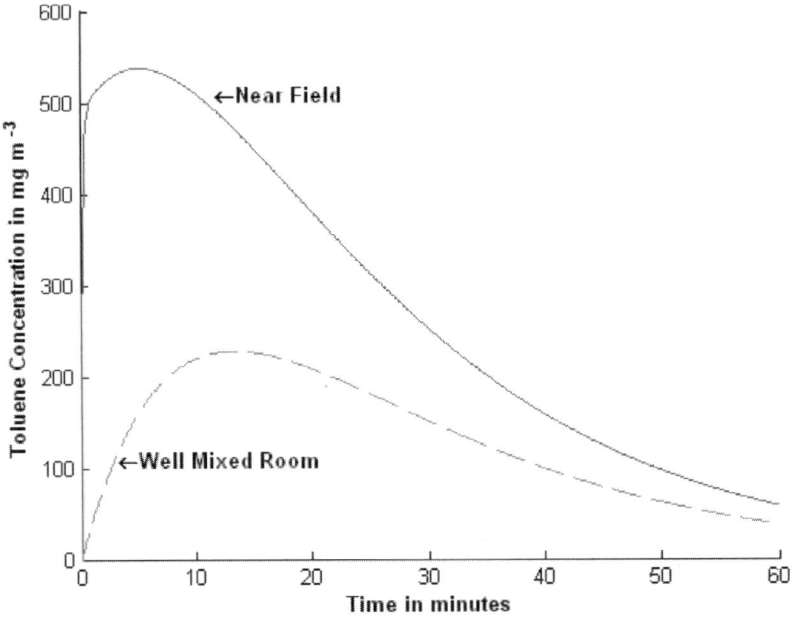

Figure 15-4. The predicted time series of toluene concentrations in a room with $V = 100$ m^3 and $Q = 10$ m^3 min^{-1}, and with a fixed point source emitting toluene vapor at an exponentially decreasing rate with $M_0 = 87{,}000$ mg and $\alpha = 0.054$ min^{-1}. The solid line curve labeled "Near Field" corresponds to the near field/far field dispersion model. The near field is a hemisphere of radius $r = 0.9$ m, the random air speed is 3.7 m min^{-1}, and the interzonal airflow rate $\beta = 9.4$ m^3 min^{-1}. The dashed line curve labeled "Well Mixed Room" corresponds to the well mixed room dispersion model.

Hemispherical Turbulent Diffusion without Advection and with a Constant Emission Rate

Assume that a room contains a source emitting a chemical at constant rate G (mg min^{-1}). The source sits on a flat surface such as a bench top. The release position is taken as the origin (0, 0, 0) in a three-dimensional coordinate system. Assume that mass disperses away from the origin with radial symmetry via turbulent diffusion with coefficient D_T (m^2 min^{-1}). Assume there is no advective airflow near the source that affects the pattern of dispersion, or U = 0. Let $C(x, y, z, t)$ in mg m^{-3} denote the chemical concentration at position (x, y, z) relative to the source at the origin. Given that the source is on a flat surface for which $z = 0$, the z coordinate for the worker's position must be greater than or equal to zero. The equation which predicts $C(x, y, z, t)$, for $t \geq 0$, is

$$C(x, y, z, t), \text{ mg m}^{-3} = \frac{G}{2\pi D_T\sqrt{x^2 + y^2 + z^2}} \times \left[1 - erf\left(\sqrt{\frac{x^2 + y^2 + z^2}{4 D_T t}}\right)\right] \quad \text{[Equation 15-27]}$$

The term *erf* in the brackets denotes the *error function*, which is an integral expression. Many mathematical software packages compute *erf* values. If software is not available, the following approximation provides values close to those computed by Equation 15-27:

$$C(x, y, z, t), \text{ mg m}^{-3} \cong \frac{G}{2\pi D_T\sqrt{x^2 + y^2 + z^2}} \times \left[1 - \sqrt{1 - \exp\left(-\frac{\sqrt{x^2 + y^2 + z^2}}{\pi D_T t}\right)}\right] \quad \text{[Equation 15-28]}$$

If the emission process continues for a long time such that $t \to \infty$, the chemical concentration reaches a steady state value, or

$$C(x, y, z)ss, \text{ mg m}^{-3} = \frac{G}{2\pi D_T\sqrt{x^2 + y^2 + z^2}} \quad \text{[Equation 15-29]}$$

Due to the assumed radial symmetry in dispersion, the radial distance r from the emission point can be used in Equations 15-27 through 15-29, where $r = \sqrt{x^2 + y^2 + z^2}$. Note that the Equations 15-27 through 15-29 are intended to estimate exposure intensity for an individual located close to the emission source, say, within one meter. Assuming the lack of contaminant reflection by the ceiling and walls, and failing to explicitly account for dilution ventilation, make these equations unrealistic at long distances from the emission source.

To illustrate the model, consider an operation on a bench top that emits toluene vapor at constant rate G = 1000 mg min^{-1}. An individual's breathing zone is always located at a position (x = 0.7 m, y = 0 m, z = 0.6 m) relative to the point of toluene emission (at the origin), which is a radial distance of 0.9 m (3 feet). Assume that the turbulent diffusion coefficient D_T

= 1 m^2 min^{-1}. Note that the room parameters V and Q play no role in the computation. By inserting these input values into Equation 15-27, we obtain

$$C(t), \text{mg m}^{-3} = 177 \times \left[1 - erf\left(\sqrt{\frac{0.81}{4t}}\right)\right]$$

The solid line curve in Figure 15-5 labeled "Hemispherical Diffusion" shows the toluene concentration profile from $t = 0$ to $t = 60$ min predicted by the above $C(t)$ equation. The concentration profile is approaching a steady-state value of 177 mg m^{-3} (equal to 47 ppm at NTP conditions), which is the value predicted based on Equation 15-29. For comparison, the dashed line curve in Figure 15-5 labeled "Near Field" shows the concentration profile predicted by the NF/FF model with $r = 0.9$ m, $\beta = 9.4$ m^3 min^{-1}, $V = 100$ m^3, $Q = 10$ m^3 min^{-1}, and $G = 1000$ mg min^{-1} (the same parameter set used for Figure 15-3). In this example, the hemispherical diffusion and NF/FF models predict a similar concentration profile, although a slightly higher steady-state exposure value (206 mg m^{-3}) is predicted by the NF/FF model.

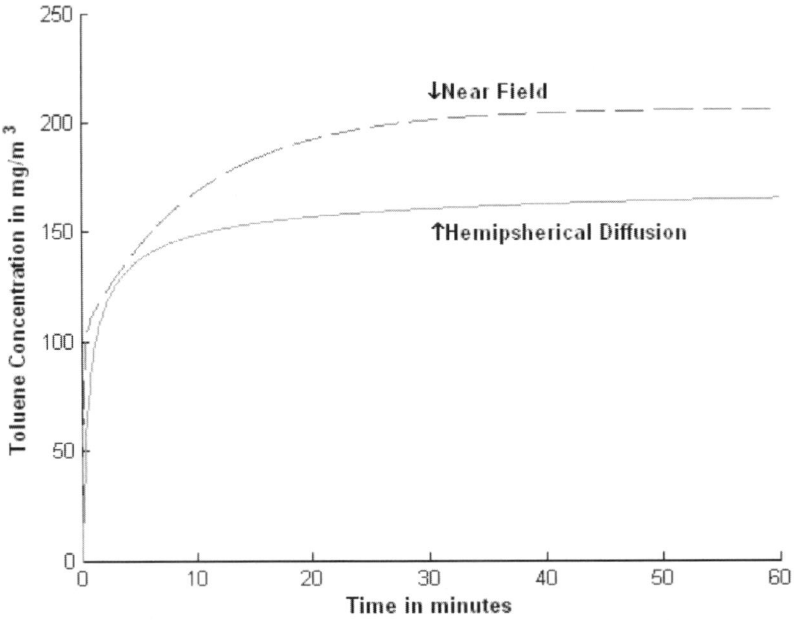

Figure 15-5. The predicted time series of toluene concentrations in a room with $V = 100$ m^3 and $Q = 10$ m^3 min^{-1}, and with a fixed point source emitting toluene vapor at the constant rate 1000 mg min^{-1}. The solid line curve labeled "Hemispherical Diffusion" corresponds to dispersion via hemispherical turbulent diffusion with $D_T = 1$ m^2 min^{-1}. The receptor position is at radial distance $r = 0.9$ m from the emission point. The dashed line curve labeled "Near Field" corresponds to the near field/far field dispersion model. The near field is a hemisphere of radius r = 0.9 m, the random air speed is 3.7 m min^{-1}, and the interzonal airflow rate $\beta = 9.4$ m^3 min^{-1}.

In fact, the Figure 15-5 concentration profiles could be made more similar by setting D_T = 0.86 m² min⁻¹, because the latter value corresponds to a steady-state level of 206 mg m⁻³ in the hemispherical diffusion model. However, this idea segues to the question of how to assign the D_T value in the first place. The author does not know of an algorithm that reliably predicts D_T, and turbulent diffusion coefficient values estimated in different indoor settings span a 100-fold range from 0.1 to 10 m² min⁻¹ (Keil, 2000). The net result is substantial uncertainty in specifying the D_T value appropriate for a given setting, and substantial uncertainty in the predicted exposure concentration. In the previous example, the author assigned D_T by using the geometric mean of the range limits, or $\sqrt{0.1 \times 10} = 1$ m² min⁻¹. Due to the present uncertainty in specifying D_T, the author prefers using the NF/FF model for estimating exposure intensity near an emission source.

It should be noted that the turbulent diffusion model presented above has several drawbacks. It does not maintain a mass balance of contaminant in room air, it does not account for removal of airborne chemical by exhaust ventilation and other mechanisms such as sorption or deposition onto surfaces, and it does not account for *reflection* of molecules from room surfaces. Furthermore, as the distance from the source goes to zero, the predicted concentration increases to infinity, as seen in the steady-state Equation 15-26.

These problems are addressed in numerical models devised by Drivas et al., (1996) and Nicas (2001). The latter is based on a probabilistic method termed a discrete-time Markov chain. In brief, a room is divided into small imaginary cells. The probabilities by which a molecule (or particle) moves to neighboring cells in a small time step are assigned based on first-order rates of dispersion due to turbulent diffusion, advection and gravitational settling (for particles). The method accounts for reflection from and/or deposition onto surfaces and the location of supply air inlets and exhaust air outlets. The method also bounds the airborne concentration near the emission point to a finite value. The details are beyond the scope of this chapter, but are provided in Nicas (2001).

A NUMERICAL METHOD

If a closed-form equation is not available for a model, a numerical method implemented by a computer code can be used instead (Nicas and Armstrong, 2003). The continuous emission process is divided into a series of pulse releases in one-second intervals. Each pulse contributes a component of the exposure concentration that develops at some future time of interest T (sec). The latter concentration $C(T)$ is the sum of these components. The method is explained using the WMR dispersion construct. The pulse release or purging equation for a well mixed room is:

$$C(t),\ \text{mg m}^{-3} = \frac{M_0}{V}\exp\left(-\frac{Q}{V}t\right) \qquad \text{[Equation 15-30]}$$

where M_0 is the contaminant mass (mg) in room air at time zero, V is the room volume (m³), Q is the room supply/exhaust air rate (m³ sec⁻), and $t \geq 0$. This equation can be considered a pulse release expression if mass M_0 is treated as instantaneously emitted at $t = 0$.

Next, consider the continuous emission rate function $G(t)$. In the interval from $t = i - 1$ sec to $t = i$ sec, the mass emitted M_i is the definite integral of $G(t)$, or

$$M_i = \int_{t=i-1}^{t=i} G(t)\, dt \qquad \text{[Equation 15-31]}$$

Assume emission begins at $t = 0$. Denote the mass emitted in the first interval running from 0 to 1 sec as M_1. We treat M_1 as being released as a pulse at 0.5 sec. By itself, this pulse creates a room concentration M_1/V. At future time T, the room concentration due to M_1 is

$$C_1(T) = \frac{M_1}{V} \exp\left(-\frac{Q}{V}(T - 0.5)\right)$$

That is, at time T, an interval of $T - 0.5$ sec has elapsed, and a simple exponential decay has operated on the concentration M_1/V. The subscript "1" on $C_1(T)$ is an index to track which pulse is involved. In general, the mass M_i emitted in the ith interval, which runs from $i - 1$ sec to i sec, is specified by Equation 15-31. M_i is treated as a pulse release at the interval midpoint, $i - 0.5$ sec. By itself, this pulse creates a room concentration equal to M_i/V. At future time T, the room concentration due to pulse M_i is

$$C_i(T) = \frac{M_i}{V} \exp\left(-\frac{Q}{V}(T - i + 0.5)\right)$$

If future time T is an integer number of seconds, the room concentration $C(T)$ is the sum of the individual $C_i(T)$ values (corresponding to the individual pulse releases), or

$$C(T),\ mg\ m^{-3} = \sum_{i=1}^{T} C_i(T) = \sum_{i=1}^{T} \frac{M_i}{V} \exp\left(-\frac{Q}{V}(T - i + 0.5)\right) \qquad \text{[Equation 15-32]}$$

NF/FF Dispersion

If a pulse release of mass M_0 occurs at time zero in the NF zone, the initial near NF concentration is M_0/V_{NF}, and the initial FF concentration is zero. The following equation describes the decay in the NF concentration for $t \geq 0$:

$$C_{NF}(t),\ \text{mg m}^{-3} = \frac{M_0}{V^2_{NF}(\lambda_1 - \lambda_2)}\left[-(\beta + \lambda_2 V_{NF})\exp(\lambda_1 \times t) + (\beta + \lambda_1 V_{NF})\exp(\lambda_2 \times t)\right]$$

[Equation 15-33]

where λ_1 and λ_2 are computed as shown in Equation 15-23.

Given initial concentrations of zero in both zones, a continuous emission source in the NF zone, and M_i defined by Equation 15-31, the NF zone concentration at future time T (an integer number of seconds) is computed via the numerical method, or

$$C_{NF}(T),\ \text{mg m}^{-3} = \qquad \text{[Equation 15-34]}$$

$$\sum_{i=1}^{t_R} \frac{M_i}{V_{NF}^2(\lambda_1 - \lambda_2)}\left[-(\beta + \lambda_2 V_{NF})\exp(\lambda_1 \times (T - i + 0.5)) + (\beta + \lambda_1 V_{NF})\exp(\lambda_2 \times (T - i + 0.5))\right]$$

Hemispherical Turbulent Diffusion with Advection

Consider that a pulse release of mass M_0 occurs at time zero on a flat surface in a room. The release position is taken as the origin (0, 0, 0). Assume there is an advective flow from left to right at speed U (m sec^{-1}) along the room's x-axis. The concentration evolves over time at position (x, y, z) relative to the release position as

$$C(x, y, z, t),\ \text{mg m}^{-3} = \frac{M_0}{4(\pi D_T t)^{1.5}} \times \exp\left(-\frac{(x - U \times t)^2 + y^2 + z^2}{4 D_T t}\right) \qquad \text{[Equation 15-35]}$$

Given an initial room concentration of zero, a continuous emission source at the origin and M_i defined by Equation 15-31, the concentration at position (x, y, z) at future time T (an integer number of seconds) is computed via the numerical method, or

$$C(x, y, z, T),\ \text{mg m}^{-3} =$$

$$\sum_{i=1}^{t_R} \frac{M_i}{4(\pi D_T (T - i + 0.5))^{1.5}} \times \exp\left(-\frac{[x - U \times (T - i + 0.5)]^2 + y^2 + z^2}{4 D_T (T - i + 0.5)}\right) \qquad \text{[Equation 15-36]}$$

If there is no advective airflow, we simply set $U = 0$ in Equations 15-35 and 15-36.

A Computer Code

Assume we want to look at exposure over a one-hour period. First, we create a 3600 × 1 matrix named PULSES in which we store the contaminant masses emitted in each of the 3600 one-second intervals in an hour. Next, we create a 3600 × 3600 matrix named DATA which tracks the concentration component due to each of the 3600 pulses over subsequent times. The 3600 rows correspond to the 3600 pulses, and the 3600 columns correspond to the 3600 values of T referenced to the start of the process at time zero. The first row tracks component C_1

due to pulse M_1 for $T = 1$ sec to $T = 3600$ sec. The second row tracks component C_2 due to pulse M_2 for $T = 2$ sec to $T = 3600$ sec. The third row tracks component C_3 due to pulse M_3 for $T = 3$ sec to $T = 3600$ sec, and so forth. The concentration at time T is found by summing down the row entries in the column corresponding to time T. $C(T)$ values are stored in a 3601×1 matrix named CONCENTRATIONS; the first row entry is assigned a value of zero corresponding to time 0.

Appendix 15-1 is a code written for MATLAB software which implements the hemispherical turbulent diffusion model with advection given an exponentially decreasing contaminant emission rate. Input values are: $G_0 = 100$ mg/sec, $\alpha = .0033$ sec^{-1}, $D_T = .017$ m^2 sec^{-1}, $U = .025$ m sec^{-1}, $x = 1$ m, $y = 0$, $z = 1$ m. The code assigns the input values, then creates the PULSES matrix and assigns the cell values. The code creates the DATA matrix and computes the cell values, and then computes the entries for the CONCENTRATIONS matrix.

A caveat for the numerical method is that it must be reasonable to treat the $G(t)$ function as independent of the airborne contaminant concentration. This assumption might be violated if a gas-phase chemical has a very low saturation vapor pressure, such that low concentrations in air decrease the net emission rate. In general, this concern does not apply for chemicals with vapor pressures > 1 mmHg.

QUESTIONS

15.1. A bottle of benzene is left uncapped in a laboratory. The bottle has a cylindrical neck with diameter 2 cm and length 2 cm. The laboratory temperature is 25°C, and the ambient pressure is one atmosphere. Benzene's vapor pressure at 25°C is 96 mmHg. What is your estimate of the constant benzene emission rate from the uncapped bottle? Use the Fuller equation to estimate benzene's molecular diffusion coefficient in air.

15.2. At time zero, 50 mL n-pentane is spilled in a well-mixed room that has volume 100 m^3 and a supply/exhaust airflow rate of 8 m^3 min^{-1}. The supply air contains no n-pentane vapor, there are no significant contaminant removal mechanisms other than exhaust air, and the initial concentration in room air is zero. For n-pentane at 20°C, $P_V = 420$ mmHg and liquid density = 0.63 g mL^{-1}. Assume the depth of the spill is 0.20 cm, and the surface area of the spill is 250 cm^2. Assume that the vapor emission rate is exponentially decreasing.

a. What is your estimate of the first-order evaporation rate constant α (min^{-1}) for the n-pentane spill?

b. At what time (min) do you predict the peak concentration of n-pentane in room air will occur?

c. What is your estimate of the peak concentration (mg m^{-3}) of n-pentane in room air?

15.3. An individual applies an adhesive to the surfaces of small boards, and then attaches items to the board surfaces. The work is done without local exhaust ventilation at a table that is 3 feet long, 3 feet wide, and 3.5 feet high. The individual's nose and mouth are never more than 2 feet higher than the table surface. The work is performed in a room with a volume of 3530 ft^3. The room has a supply/exhaust airflow of 6 nominal ACH. The random air speed near the table is 15 ft min^{-1}. Over the course of a two-hour work period, the individual uses two pints (946 mL) of adhesive that contains n-hexane at 20% volume-by-volume. The liquid density of n-hexane is 0.66 g/mL. The n-hexane quickly evaporates after the adhesive is applied.

a. If the emission of n-hexane vapor is assumed to be constant, what is your estimate of the n-hexane emission rate G, in mg min^{-1}?

b. If the near field/far field dispersion model is used, the near field zone might be considered a rectangular box of air on the table top. What are the dimensions of this near field box? What is the free surface area of this near field box?

c. Given the free surface area estimated in part (b), what is your estimate of the value of the interzonal airflow rate β, in m^3 min^{-1}?

d. What is your estimate of steady-state n-hexane exposure concentration (mg m^{-3}) in the near field zone?

e. What is your estimate of the steady-state n-hexane exposure concentration (mg m^{-3}) in the far field zone?

REFERENCES

Baldwin, P.E.J.; Maynard, A.D.: A survey of wind speeds in indoor workplaces. Ann. Occup. Hyg. 42:303-313 (1998).

Drivas, P.J.; Valberg, P.A.; Murphy, B.L.; Wilson, R.: Modeling indoor air exposure from short-term point source releases. Indoor Air 6:271-277 (1996).

Fuller, E.N.; Ensley, K.; Giddings, J.C.: Diffusion of halogenated hydrocarbons in helium: the effect of structure on collision cross sections. J. Phys. Chem. 73:3679-3685 (1969).

Haberlin, G.M.; Heinsohn, R.J.: Predicting solvent vapor concentrations from coating inside of bulk storage tanks. Am. Ind. Hyg. Assoc. J. 54:1-9 (1993).

Heinsohn, R.J.: Industrial Ventilation: Engineering Principles, pp. 251-252. John Wiley and Sons, Inc., New York (1991).

Hemeon, W.C.: Plant and Process Ventilation, Second Edition, pp. 235-245. Industrial Press, Inc., New York (1963).

Hummel, A.A.; Braun, K.O.; Fehrenbacher, M.C.: Evaporation of a liquid in a flowing airstream. Am. Ind. Hyg. Assoc. J. 57:519-525 (1996).
Jayjock, M.A.: Back pressure modeling of indoor air concentrations from volatilizing sources. Am. Ind. Hyg. Assoc. J. 55: 230–235 (1994).
Keil, C.B.: A tiered approach to deterministic models for indoor air exposures. Appl. Occup. Environ. Hyg. 15:145-151 (2000).
Keil, C.B.; Nicas, M.: Predicting room vapor concentrations due to spills of organic solvents. Am. Ind. Hyg. Assoc. J. 64:445-454 (2003).
Nicas, M.: Estimating exposure intensity in an imperfectly mixed room. Am. Ind. Hyg. Assoc. J. 57:542-550 (1996).
Nicas, M.: Modeling turbulent diffusion and advection of indoor air contaminants by Markov chains. Am. Ind. Hyg. Assoc. J. 62:149-158 (2001).
Nicas, M.: Estimating methyl bromide exposure due to off gassing from fumigated commodities. Appl. Occup. Environ. Hyg. 18:200-210 (2003).
Nicas, M.; Armstrong, T.W.: Computer implementation of mathematical exposure modeling. Appl. Occup. Environ. Hyg. 18:556-571 (2003).
Nicas, M.; Plisko, M.J.; Spencer, J.W.: Estimating Benzene Exposure at a Solvent Parts Washer. J. Occup. Environ. Hyg. 3:284-291 (2006).

Appendix 15-1. A MATLAB code to implement the numerical method for the combination of hemispherical turbulent diffusion with advection and an exponentially decreasing chemical emission rate function.

```
% Assign model parameters based on a 1-sec time unit.
Gzero = 100; alpha = .0033; D = .017; U = .025;

% The position of the worker relative to the source at the origin
x = 1; y = 0; z = 1;

% The contaminant masses released in successive 1-sec intervals are computed.

PULSES = zeros(3600,1);
for t = 1:3600
PULSES(t,1) = (Gzero/alpha)*(exp(– alpha*(t–1)) – exp(– alpha*t));
end

% A given row of DATA tracks the airborne concentration over time due to a given pulse.

DATA = zeros(3600,3600);
for T = 1:3600
```

```
  temp1 = 1/(4*((pi*D*(T – 0.5))^1.5));
  temp2 = exp(– ((x-U*(T – 0.5))^2 + y^2 + z^2) /(4*D*(T – 0.5));
  DATA(1,T) = temp1*temp2;
end

for row = 2:3600
  for T = row:3600
    DATA(row,T) = DATA(row–1,T–1);
  end
end

for t = 1:3600
  temppulse = PULSES (t,1);
  DATA(t,:)  =  temppulse*DATA(t,:);
end

% Summing down a column of DATA (corresponding to a given time T) equals the
  airborne concentration at time T due to all the pulses released prior to time T.

CONCENTRATIONS = zeros(3600,1);
for T =1:3600
  CONCENTRATIONS (T+1,1) = sum(DATA(:,T));
end
```

16

QUANTITATIVE SURVEYING – STATISTICAL MODELING OF OCCUPATIONAL EXPOSURES

DISTRIBUTIONS

Perhaps Chapter 14 left the reader with some unanswered questions. The worst-case sampling approach, developed by NIOSH and used by OSHA, is an approach[1] that may allow an immediate decision, but gives no information for future decisions or for calculating parameters of the exposure distribution. So, what is a sound approach? For acute toxicants with irreversible (or even reversible) effects, a continuously reading instrument is often available and probably the best way to monitor exposures. An alarm system can be used to avoid hazardous exposures. But what about chronic toxicants?

For chronic toxicants it has been argued that the long-term average of exposures is more important than the variability of exposures as the ultimate factor driving the risk of disease (Roach, 1966; Rock, 1982; Rappaport, 1985, 1987, 1991; Rappaport and Selvin, 1987). Rappaport shows clearly that for many chronic toxicants such as lead, mercury, asbestos, and other fibrogenic dusts, the body burden driving disease is far more dependent on average exposures than the day-to-day variation in exposures. Consequently, it is the long-term average of exposures that should be in compliance with an OEL and not individual exposures on a given day. In fact, OSHA PELs (in comprehensive standards) appear to relate the risk of disease to lifelong exposure at the PEL (see, e.g., benzene, OSHA, 1987). This would imply that daily exposures may be above or below the PEL as long as the lifelong average is at or below the PEL. So how should the population be sampled and the average calculated?

We can confidently categorize chemicals, such as asbestos, whose effects depend on the lifelong accumulation of body burden, as chronic toxicants. These toxicants are cumulative. Peak or short-term exposures appear not to be related to the propensity to cause health effects.

---

[1] We will see in Chapter 18 that it may be the best approach for determining compliance. The point here is that, for the practitioner, there are better approaches.

Although they may have short half-lives in the blood, they are often sequestered in bone, fat, the lungs, or another organ. Then there are the majority of chemicals that appear to have both chronic and acute effects at exposure levels that are likely to occur in the workplace. For these chemicals short-term exposure peaks as well as long-term averages may drive differing parts of the disease process or risk. There is not overwhelming evidence for or against this idea; however there may be situations where peak exposures may effect the risk of disease for a *chronic* toxicant (Checkoway and Rice, 1992). There is even some evidence that the most consistent trend for dose and response for benzene is for maximum peak exposure as the dose index rather than cumulative dose or average exposure × time (Wong, 1987). For these chemicals perhaps the values of the exposure distribution at the high end drive risk and should be controlled. Then, of course, there are the regulatory agencies who, regardless of the toxicology of the chemical, will enforce against exceedances of the OEL; again this argues for control of the high end of exposures. How does one determine what the high end of the exposures looks like?

One of the most useful tools available in statistics is the probability distribution. Certain populations have a tendency to be distributed in a characteristic manner. Continuous data that tend to be distributed randomly about an average value will fit a normal frequency distribution (as described in Chapter 7).

If a set or sample of data *fit* a particular distribution, that distribution can be used to model the population, or estimate its parameters, e.g., average, variability, extreme values, etc. For example, if a set of data is approximately normally distributed, then the normal distribution model can be used to estimate parameters of the population. When the wrong model assumptions are applied to a data set, the resulting analysis can be misleading. As an example of inappropriate use of a distribution, consider the interpretation of the arithmetic average or mean of a group of data. Remember from Chapter 7 that for a normal distribution, mean = median = mode. Consider a college classroom with these ages of students: 20, 20, 20, 20, 20, 20, 20, 20, 21, and 69. The arithmetic average (mean) of ages in this class is approximately 25. If one were to report that the *average* age in this class is 25, most people would assume that about half the class is younger and half older than 25. In effect, most people would be inferring a normal distribution (mean = median). However, in this case the distribution is not symmetrical, and the median or mode (20) would be a more descriptive statistic of the *typical* age of the students. Figure 7-8 shows that mean, median, and mode are not equivalent for skewed distributions; so the application of the correct distribution is important in how we interpret the average age. Making the best application of distribution type also has other implications, to be discussed later.

In order to determine the mean exposure, the exposure variability, and the number of exposures to measure (sample size), the appropriate distribution must be used. Exposure data taken over time usually are adequately modeled by a lognormal distribution, whereas random errors associated with repeated monitoring and analysis of a single air sample usually are adequately modeled by a normal distribution. Consequently, these two distributions are most important in occupational exposure assessment.

The Normal Distribution

For occupational exposure assessment, the normal distribution is important in describing random errors that occur in replicate measurements of the same exposure concentration. In measuring an exposure concentration it is very unlikely, because of sampling and analytical error, that a single measurement will give the true concentration. However, if multiple measurements of the exposure concentration are made, then the mean of the exposure concentration measurements should be equivalent to the true exposure concentration, provided that there is no measurement bias. The distribution of the measured exposure concentration values should be normal, as it is equally likely that an exposure concentration measurement will be above, as that it will be below, the true value. This variation of measurements about their mean is an indication of the random error or the precision of the monitoring method.

Leidel and Busch (1985) give the following examples in which the normal distribution is usually an adequate model.

1. Populations of replicate analyses – repeated analyses of a given air contaminant sample.

2. Populations of replicate measurements – repeated measurements (air sampling and analyses) of the same true value in a simulated or laboratory condition such as an exposure chamber.

3. Populations of replicate, full-period (say, 8-hour) measurements of exposure (see Figure 16-1) – repeated simultaneous measurements of the same, or an equivalent, exposure concentration.

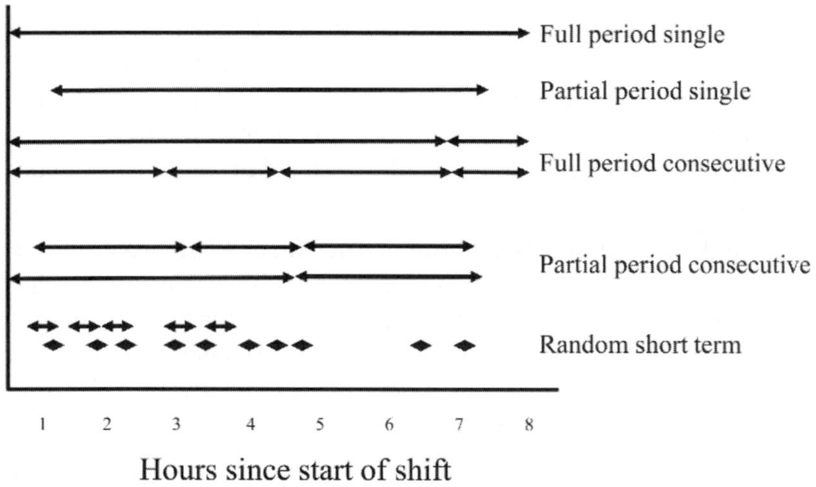

Figure 16-1. Types of exposure measurements that could be made to estimate an 8-hr TWA exposure (adapted from NIOSH, 1977).

A normal distribution can be completely characterized through the estimation of the two parameters, the standard deviation and mean. The following equations, which have some similarities and are often confused, utilize these parameters.

$$95\ \%tile = \bar{x} + 1.65\sigma \qquad \text{[Equation 16-1]}$$

$$95\%\ UCL = \bar{x} + 1.65\ \sigma/\ n^{0.5} \qquad \text{[Equation 16-2]}$$

$$95\%\ UCL = \bar{x} + t_{(\alpha,n-1)}s/n^{0.5} \qquad \text{[Equation 16-3]}$$

$$95\%\ UTL = \bar{x} + k_{(P,\alpha,n-1)}s \qquad \text{[Equation 16-4]}$$

$$95\%\ UCL = MC + 1.65\ (CV)\ (STND) \qquad \text{[Equation 16-5]}$$

Equation 16-1 calculates the 95th percentile of a normal distribution. The value of Z (+1.65) is specific for that percentile. A Z-score of zero would indicate that $\bar{x}$ is an estimator of the 50th percentile, which is true. For percentiles less than 50, a negative value of Z is used. Note that σ is used, not s for the standard deviation. If σ is not known or when $n < 20$ (often the case for hygiene work), then the estimate of the percentile may be in considerable error, particularly the farther the estimated percentile is from the 50th percentile.

Equation 16-2 estimates the upper confidence limit on the mean. Notice that it is sample-size-dependent; as *n* increases, *UCL* decreases. For ease of understanding, one may think of the calculation of a 95% *UCL* as indicating that there is a 5% chance that the true mean is greater than the 95% *UCL*. (This is not technically precise.) A confidence interval *(CI)*, having both upper and lower bounds, may be calculated also; for example, a 95% *CI* would indicate a 5% chance that the true mean would lie outside the *CI*, 2.5 % at either end.

In Equation 16-3 a *t-score* is used to calculate the *UCL* rather than the Z-score (1.65), as the standard deviation *(s)* is estimated. The sample size used in the estimate is *n*, and *n*-1 is the number of degrees of freedom. The *t-score* varies with sample size *(n)* and α (in this case α = 1-0.95 = 0.05), and may be found in any statistics text. For example, if a sample size of 10 were used in the calculation, the t-value would be 1.83. For any *UCL*, *t* approaches Z as *n* approaches infinity.

In Equation 16-4 an upper tolerance limit is calculated. A tolerance limit is a confidence limit placed on a percentile *(P)* other than the 50th percentile, or mean. The chosen value of *k* is dependent on three parameters: the percentile, the α-level for the confidence limit, and the sample size (*n*). Thus, if the sample size is 20, a 95% confidence limit can be calculated for the 95th percentile using a *k*-score of 2.396. Notice the *k*-score gives a *UTL* considerably larger than the 95th percentile, which is based on Z, or 1.65. *K*-scores can be found in many statistics books.

NIOSH (1977) has devised a technique to determine if a single measured concentration *(MC)* is less than the upper 95 %tile of the distribution of replicate measurements of the true concentration at the OEL (Equation 16-5). In effect, the question being asked is, what confidence do I have in this single estimate, given the associated measurement error? This procedure is most frequently used by federal and state compliance officers, as each monitoring result is usually compared to the PEL. Equation 16-5 is based upon Equation 16-2, but note that since $n = 1$, the mean is estimated by the single measurement (*MC*, measured concentration). Sample size *(n)* is one. The standard deviation cannot be calculated from the datum, so it is determined by the total coefficient of variation for the measurement method ($CV = s/\bar{x}$) multiplied by the OEL. The latter relationship ($CV{*}OEL = s$) is derived as follows. Most sampling and analytical methods have been laboratory-validated by NIOSH at three concentrations: 0.5, 1, and 2 times the OEL. (See Chapter 25 for details.) The mean of these three concentrations is about equal to the OEL. A variance can be calculated for each of these concentrations based on the six replicate measurements made at each. From each of the variances and corresponding mean concentrations, a CV is calculated, and these values are then pooled so that a single estimate of CV is derived. So for a population of replicate exposure measurements near the OEL, the standard deviation is approximately equal to the product of the pooled CV and the OEL, since $s = CV \times \bar{x}$. A Z-score (1.65) is used because it is assumed that σ has been adequately estimated by *s*.

Single measurement estimates of exposure should be used rarely by hygienists, as they give us information for that day and that worker only. If resources and time are available to take one measurement, then generally resources and time should be made available to take more than one measurement randomly over time and space, so that a distribution can be characterized as discussed in the rest of this chapter.

The compliance officer would use Equation 16-5 with a minus sign to obtain an *LCL*. In this way when $LCL > PEL$, one would be 95% confident of a true exposure value $> LCL$. A corporate hygienist would use Equation 16-5 as is, the so-called employer's test. Figure 16-2 demonstrates uses and misuses of these tests.

The Lognormal Distribution[2]

Consider a similar exposure group (SEG, as defined in Chapter 14) made up of 10 workers performing similar tasks on a continuous basis during an 8-hour shift. In defining the population of exposures for this group of 10 workers, one must consider how the distribution may change with time. A *stationary* distribution does not change with time. Assume for the moment that the worker exposures (i.e., the average and the variability of exposures) remain approximately constant throughout the year. This means that there are no major process changes, like a new raw material or a new ventilation system, that would shift the distribution. Then, using a random sampling strategy, one could sample the population of exposures throughout the year. Those exposure monitoring results might appear as in the frequency distribution plot in Figure

[2] The reader should review Chapter 7 on the lognormal distribution before continuing.

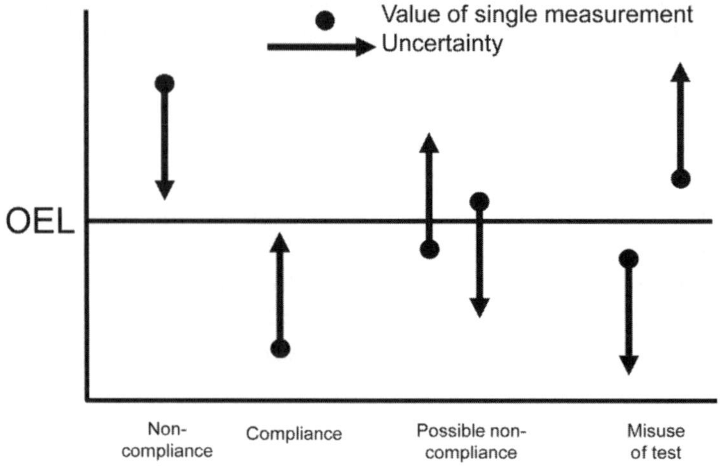

Figure 16-2. Possible statistical tests and outcomes based on a point estimate of exposure.

16-3. The results are likely skewed to the right and, when this occurs, the lognormal usually is a reasonable approximation for the true distribution. Rappaport et al. (1981) evaluated the use of the 2-parameter lognormal distribution and reviewed previous studies that indicated a lognormal distribution was often appropriate for the following types of data: workers' short-term exposures during a single day; workers' average daily exposures over weeks, months, or years; and air concentrations at fixed locations in the work area.[3]

A possible exception to the application of the lognormal distribution to exposure results is for noise exposures (Behar and Plener, 1984; Brunn et al., 1986; Hopson, 1990). Decibels (the units that noise exposure results are expressed in) are the logarithms of sound power. Hence, noise exposure results are actually logarithms of exposures to sound power, and consequently the normal distribution may adequately characterize them, as the logarithms of lognormal data are normally distributed. A more technically accurate approach based on using sound power rather than decibels and then applying the lognormal distribution has been derived (Grzebyk and Thiery, 2003).

Several important characteristics tend to cause exposure data to be lognormally distributed.[4]

[3] Others have confirmed this (see, e.g., Kumagai and Matsunga, 1994; and Osvoll and Woldbæk, 1999). Based on theoretical physical models (such as dilution ventilation) it can be shown that exposure distributions should be skewed. However, other more complex, skewed distributions such as the beta and 4-parameter lognormal are the solutions to these theoretical considerations (Flynn 2004a and b).

[4] See also Limpert et al., 2001 for an excellent review of the lognormal and its general application to the biological, chemical, and physical sciences.

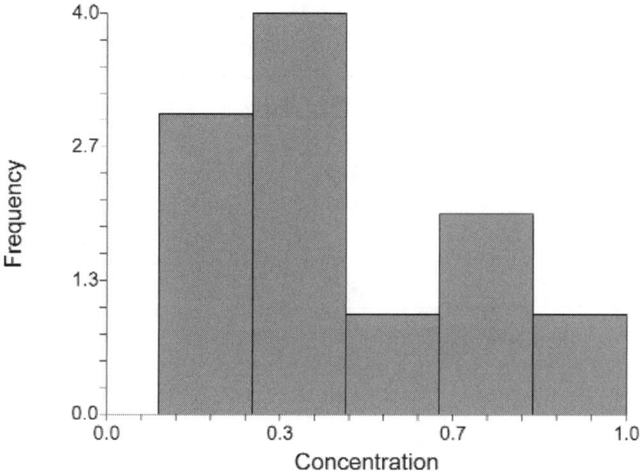

Figure 16-3. A histogram for a set of 11 exposure measurements showing typical skew or asymmetry.

1. An exposure concentration cannot be below zero; thus the distribution is bounded on the left-hand side.
2. The median of exposures is usually low (e.g., < 100 ppm), relative to the possible range of exposures (e.g., 0–1,000 ppm).
3. The range of exposures can be quite high, with the standard deviation up to 10 times the mean (see Table 16-1 for typical ranges).
4. Variables that affect exposure concentrations can occasionally cause very high concentrations. This would include process upsets, equipment malfunction, worker error, and other variables.
5. Variability in exposures is actually a function of several variables, which themselves may be normally or lognormally distributed. The product or nonlinear interaction of these variables will give skewed or lognormal exposure results.

The parameter[5] often used to relate the variability (spread and skewness) of a lognormal distribution is geometric standard deviation (GSD). A GSD of one represents no variability in

[5] The lognormal as referred to here is characterized by a location parameter, such as the mean, and a shape parameter, such as the standard deviation. A 4-parameter lognormal also is defined by minimum and maximum values. The latter have application to occupational exposures as for example, the concentration of a volatile organic compound cannot exceed its saturated vapor pressure. Thus, concentrations beyond this will be *assumed* by the long tail of the asymptotic 2-parameter distribution, but eliminated in the 4-parameter distribution.

Table 16-1
The extent of variability in lognormally distributed populations

GM	μ_C	GSD	Range[1]
1	1.1	1.5	0.4–2.3
	1.3	2.0	0.3–4
	1.5	2.5	0.2–6.3
	1.8	3.0	0.1–9
	2.2	3.5	0.1–12.3
10	11	1.5	4–23
	13	2.0	3–40
	15	2.5	2–63
	18	3.0	1–90
	22	3.5	0.8–123
50	54	1.5	22–113
	64	2.0	13–200
	76	2.5	80–313
	90	3.0	6–450
	110	3.5	4–613
100	109	1.5	44–225
	127	2.0	25–400
	152	2.5	160–625
	183	3.0	11–900
	219	3.5	8–1225

[1] The approximate 2.5th and 97.5th percentiles of the distribution (i.e., exp. ($\overline{x}_L \pm 25_L$).

the population. (In this case, the standard deviation of the log values is zero (see Chapter 7, Equation 7-9.) NIOSH (1975) reported that:

1. GSDs of two and above represent relatively high workplace environmental variability.

2. Exposure distributions having GSDs of 1.44 and below may be approximated by normal distributions.

3. Most occupational environments will possess distributions with geometric standard deviations ranging from 1.5 to 3.5.

However, it should be pointed out that geometric standard deviations may appear on either side of this range. Geometric standard deviations of 6 and above have been reported in the petroleum industry (Rappaport and Selvin, 1987). Buringh and Lanting (1991) found

GSDs ranging from 1.2 to 10, with a mean of 2.7 in Dutch factories. Limpert et al. (2001) found GSDs ranging from 1.1 to 33 for a side range of biological, chemical, and physical processes, but the majority in the range of 1.4 to 3. On the other hand, the way that the population is defined will determine the GSD. So the more heterogeneous a population of workers, with respect to their tasks, the control of their workplace, and other factors, the greater the GSD. More on this later.

Testing for Lognormality and Normality

Although we commonly assume lognormality for exposure data and normality for repeated measurements, the assumption may not always be correct. Lognormality can be examined using log-probability paper (NIOSH, 1977). Therefore, we should test the usual hypothesis that data are normally distributed. For data suspected of being lognormally distributed, the distribution of the logarithms is tested for normality.

Hypothesis tests can be performed using one or more of several goodness-of-fit tests that have been derived (Gilbert, 1987 or Waters et al., 1991). A common test is the non-parametric Kolmogorov-Smimov test. However, this test is not valid if the parameters used in the test are estimated from the data set (Gilbert, 1987, p. 158). The Lilliefors test was derived to avoid these problems. However, the W-test of Shapiro and Wilk for $n < 50$ or the D-Agostino test for $n \geq 50$ are quantitative tests as opposed to the graphical Lilliefors test (Gilbert, 1987). Also the I-test (Martinez and Iglewicz, 1981) is sensitive to long tails on one or both sides, a problem with environmental data sets, and performs similarly to the W- or D-tests. These tests are found in many statistical software packages.

Any of the latter three tests are recommended for testing the null hypothesis; however, it should be realized that departure from normality or lognormality commonly can occur by chance with small ($n < 20$) data sets. Furthermore, with large data sets, small, but real departures from lognormality may be detected, but they may not be important enough to warrant the fitting of another distribution such as the beta or 4 parameter lognormal (Gilbert, 1987). Therefore, it is wise to look at a frequency plot and/or a probability plot to get a *feel* for the distribution.

Also quick tests can be used to *sense* the skewness of the data. For example, NIOSH (1977) states that GSDs < 1.44 are approximately normal. Gilbert (1987) states that if the ratio of the highest to lowest datum exceeds 20, expect skewed or non-normal data. Finally, if the *CV* exceeds 50% one should expect non-normality, as calculations with these CV values will lead to negative estimated values for lower percentiles if the normal distribution is assumed.

Hypothesis tests, graphical tests, and qualitative tests such as those above should be used jointly to reject normality. If normality is rejected, but lognormality cannot be confirmed, one must consider the consequences of assuming one or the other. In a case where several data sets are lognormal, one is normal, and one is neither, it is reasonable to assume that all are lognormal if they are related (e.g., exposure distributions for workers from the same SEG).

Otherwise, nonparametric statistical approaches that do not assume a distribution, but are less powerful, are necessary.

SAMPLING ERROR VERSUS MEASUREMENT ERROR

Figure 16-4 shows a lognormal distribution with a normal distribution added at a point along the lognormal distribution. This is to indicate that for each estimate of a true exposure from the lognormal distribution, there will be a normally distributed random error associated with the measurement method (often called sampling and analytical error, but do not confuse it with statistical sampling error). For measurements taken in a workplace environment, both statistical sampling and measurement (monitoring) errors exist. Both those errors contribute to the difference between the estimated mean and the true mean. We can examine the relative importance of these two sources by defining several populations.

As examples of definitions of populations of exposures, consider the following problems.

1. Estimating an employee's 8-hr TWA exposure on a given day.
2. Estimating an employee's average 8-hr TWA exposure over an amount of time longer than a day, perhaps up to a working lifetime.
3. Estimating the average 8-hr TWA exposure for a group of workers in the same SEG on a given day.

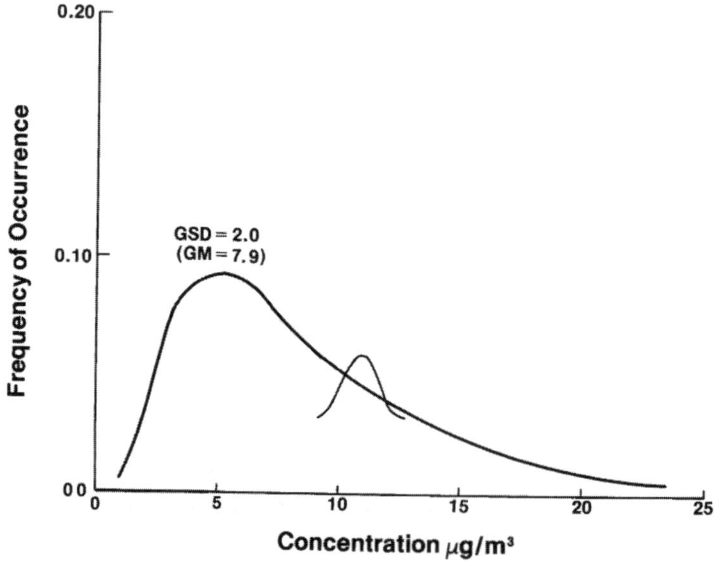

Figure 16-4. Lognormal distribution representing an exposure distribution with a superimposed normal distribution representing measurement error for any single measurement.

4. Estimating the average 8-hr TWA exposure for a SEG over a period of time longer than one day up to a working lifetime.

A sample size of 1 would estimate each mean, with increasing average error as one descended down the list. For population 1, the source of error in the estimate would be random and/or systematic error in the measurement technique. Other random and systematic errors owed to space and time environmental or exposure variability (i.e., statistical sampling error) would be included in estimates of means of populations 2 to 4.

Thus environmental variability and measurement error will affect the accuracy of exposure population parameter estimates. The bottom line is that as the magnitude of total variability increases, the error associated with estimating the parameter (e.g., mean) also increases. If the total variability cannot be reduced, then the confidence interval around the parameter (the estimation error or uncertainty) can be reduced only by increasing the sample size. So, to improve our estimate of the parameter, should we put effort into reducing variability (caused either by measurement error or environmental variability), or increasing sample size, or both?

Random and Systematic Errors Caused by Measurement Technique

Each of the monitoring procedures described in Parts V and VI of this text is subject to measurement error. The measurement errors are both systematic and random. Random errors are quantitated as the coefficient of variation typically given for validated sampling and analytical methods (see Chapter 25). Random error is inversely related to precision. Where sampling and analytical methods have not been validated, the coefficient of variation or random error is not known.

Systematic errors in validated sampling and analytical measurements are quantitated as bias. Bias is the difference between the expectation of the observed value and the true value, relative to the true value (see Chapter 7, Equation 7-2). Although both sources of error can occur with any monitoring technique, random error is most common. Systematic error can be corrected; and when systematic errors are significant, monitoring techniques typically are revised in order to reduce the systematic error.

Examples of measurement error include:

1. Pump flow rate fluctuations
2. Drops in pump battery voltage
3. Failure to correct for
    a. Temperature and pressure effects on calibration devices
    b. Collection efficiency of the sampling medium
    c. Desorption efficiency

4. Variation in
    a. Electrical current to the analytical instrument
    b. Supply of fuel to an analytical instrument such as a flame ionizing detector
5. Randomly varying climatic factors, including temperature, pressure, and humidity
6. Pump calibration errors
7. Air sample degradation during storage
8. Humidity effects on air sample collection
9. Errors within a laboratory due to built-in biases such as improper training, failure to follow standard techniques, improper calibration, and inadequate quality control
10. Differences between laboratories caused by the use of differing equipment, different training techniques, different climatic conditions, and different background levels of pollutants
11. Interferences from other analytes, which may mask or enhance the relevant analyte.

Random and Systematic Variation in Workplace Exposure Levels

In the simplest case, one may define an exposure population as the 8-hr TWA exposure concentration for a given worker on a given day. If we assume no measurement error in this simple case, the population ($N = 1$) has no variation. The measurement result is the exposure for that worker on that particular day.

Had the exposure population been defined as the average 15-min TWA exposure to a worker on a given day, then there would be environmental variability from one time period to the next during that day. Worker exposures will vary over any averaging period.

Random variation includes variation in exposure levels within a worker (i.e., within a day and from one day to the next), and variation from one worker to another (between workers) in a given job category or SEG. If a population is defined using the SEG concept[6], exposures will

[6] That is to say that the workers' mean exposures over time are equivalent. If this is not the case (and it probably is not), then there will be systematic differences between workers' mean exposures caused by determinant variables, like differences in the way work is performed. If these differences are large (more on this in the "Defining the SEG" section) then the SEG concept does not hold for the group of workers.

randomly fluctuate with day (time) and person (space). This random variation may be due to work practices, the way workers respond to domestic problems, weather changes, arguments with the boss, process variables, and worker health problems; the list is limited only by the imagination.

Systematic variables that affect occupational exposure levels are also termed determinant variables. Determinant variables should be considered in defining a population or SEG. For example, defining a population for a given season may not allow exposure estimates from that population to be extrapolated to other seasons. If determinant variables change after a population is defined, then the population will not be stable or stationary (i.e., the mean and variance are changing).

According to Leidel and Busch (1985), variation in worker exposure levels may be caused by variables that have been grouped into five categories: process, environmental and meteorological, temporal and seasonal, behavioral, and incidental (see Table 16-2).

The Contribution of Measurement Error to Total Variability

Thus far, we have five important pieces of information. First, the sources of error associated with a single measurement have been given. Second, factors that affect environmental variability and, consequently, affect statistical sampling error, have been stated. Third, we know that the two sources of error account for the total error or total variability in the estimation process. Fourth, the measurement error is available for many measurement methods (see Chapter 25), and is usually quantitated as a *CV*. The *CVs* (also, relative standard deviations) for typical measurement methods can be found in Table 25-1. Fifth, the total variability associated with estimating, say, an exposure mean is characterized by the GSD, which usually varies from 1.5 to 3.5 but may vary by as much as 1.2 to 10.

The environmental variability[7] is the difference between total variability and measurement variability. But the relationship is actually not quite so straightforward. Total variability is lognormally distributed, whereas measurement variability is normally distributed. The relationship between these variables has been worked out by Nicas et al. (1991). The total variability of the exposure concentrations is related by the variance[8] or Var[c] as:

$$Var[c] = \mu^2 * [\exp[\ln^2 GSD_e + \ln(1 + CV^2_m)] - 1] \qquad \text{[Equation 16-6]}$$

where GSD_e is the environmental GSD and CV_m is the measurement error. Thus, realistic values of GSD_e and CV_m can be input into Equation 16-6 to derive Var[*c*].

---

7 We can also think of this as the exposure variability across space and time, where the variability from worker to worker is the spatial variability. Environmental variability causes statistical sampling error and, therefore, the two terms are used almost interchangeably.

8 Remember that the variance is the square of the standard deviation.

Table 16-2
Determinant variables causing variation in individual worker exposure levels (modified from Leidel and Busch, 1985)

Process Variables

- Process type and operation
- Physical state and properties of material used (e.g., vapor pressure, size distributions of particulates and aerosols)
- Rate of operation (e.g., mass or volumetric rate, revolutions per minute, linear rate, items in a given time period)
- Energy conditions of operation (e.g., temperature, pressure)
- Degree of process automation
- Emissions from adjacent operations
- Airflow patterns around workers (e.g., from exhaust ventilation, from adjacent operations)
- Heating and ventilation airflows
- Exposure control methods (e.g., local exhaust ventilation, respirators)

Environmental Variables

- Meteorological conditions
- Age, size, and physical layout of plant
- Job category (e.g., responsibilities, work operations, work area, time spent at each)

Time Variables

- Contaminant build-up in the workplace air from morning to afternoon
- Exponential clearance due to air flushing and dilution during non-working hours
- Cyclical process operations
- Work shift
- Season of year
- Year or decade

Behavioral Variables

- Worker job practices, movements, habits
- Worker training
- Worker attitudes
- Management and supervisory attitudes
- Presence of exposure measurement equipment, industrial hygiene personnel, or supervisory personnel

Incidental Variables

- Spills due to falls, punctures, tears, corrosion, etc.
- Equipment maintenance or lack thereof
- Failure of process equipment prone to corrosion and leakage (e.g., pump packings, tank vents)
- Interruption of utilities to process equipment or exposure control systems
- Interaction due to accidental mixing or simultaneous release of two vessels' contents
- Interruption or increase in flow of one or more process streams
- Vessel failure
- Accidental over-pressurization, overheating, or over-cooling of process equipment
- Sudden plant flooding, violent storms, or earthquakes
- Operator errors or instrument failure

When $CV_m = 0$, then GSD_e is totally responsible for Var[c]. So the percent of the total variability caused by measurement error ($\%_m$) can be derived:

$$\%_m = (Var[c] - Var[c]_{CV=0}) / Var[c] * 100\% \qquad \text{[Equation 16-7]}$$

where $\text{Var}[c]_{CV=0}$ is Equation 16-6 evaluated at $CV_m = 0$, or the environmental variability.

Figure 16-5 was derived by using Equations 16-6 and 16-7. The figure shows that at a typical GSD of 2.7 and CV of 0.1, measurement error accounts for very little of the total error. This tells us that, in most cases, duplicate measurements (e.g., side-by-side quality control checks) are a waste of resources. The resources used to make the duplicate measurements would be put to better use in taking additional measurements on other workers. The latter has the effect of increasing sample size and thereby improving the accuracy of estimating the mean. That is not to say that duplicate measurements are never needed, particularly if the laboratory is suspected of being biased. However, routine duplicates are probably wasteful in most cases where GSD > 2.0. Figure 16-5 also indicates that when GSD exceeds roughly 2, then cheaper but less accurate (higher CV_m) methods will probably allow better estimates of the mean for the same overall costs because more measurements can be taken for the same amount of money (n is larger).

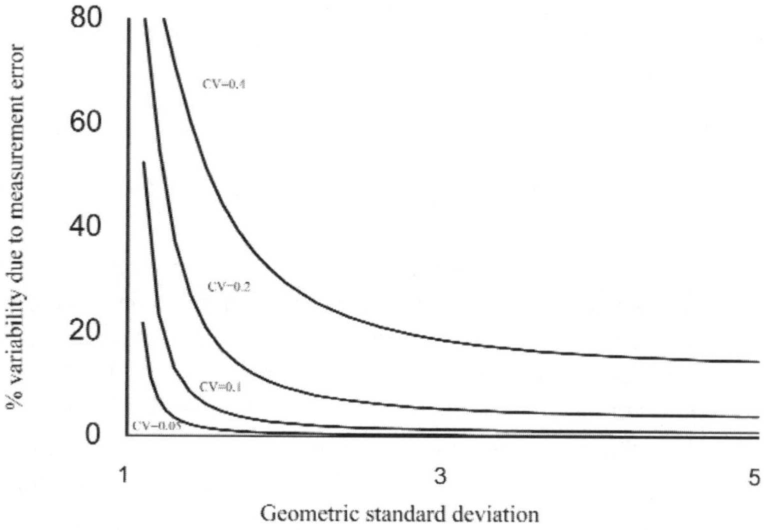

Figure 16-5. The percentage of overall exposure estimate variability caused by measurement error and its dependence on the geometric standard deviation of the exposure population and the coefficient of variation of the measurement technique.

ESTIMATING THE MEAN OF AN EXPOSURE DISTRIBUTION

On the surface it may seem that the problem of estimating the mean is an easy one to solve – take a few measurements and calculate the mean. The mean is usually taken to be the sample mean or $\bar{x}$,

$$\bar{x} = \frac{(\sum x_i)}{n} \quad \text{[Equation 16-8]}$$

that is, add up the values and divide by the number of values. However, there are other ways to estimate a mean. If readers perused Chapter 7, Table 7-1 carefully, they noticed that one of the equations was:

$$\mu_c = \exp\,(\mu_L + 0.5\sigma^2_L) \quad \text{[Equation 16-9]}$$

where the subscripts *c* and *L* refer to the concentrations and their logarithms, respectively. This equation is used to calculate μ_c from a lognormal distribution when μ_L and σ_L are known (i.e., not estimated). However, the so-called maximum likelihood estimator (*MLE*):

$$MLE = \exp\,(\bar{x}_L + 0.5s^2_L) \quad \text{[Equation 16-10]}$$

can be used as a *biased* estimator of the mean of concentrations (μ_c), if μ_L and σ_L are estimated from a sample. There has been much confusion over which equation to use to estimate the mean.[9]

There is yet a third estimator of the mean of a lognormal distribution, and although it is the best estimator, it is more difficult to use because it requires that a table be consulted. This minimum variance unbiased (*MVU*) estimator is:

$$MVU = [\exp\,(\bar{x}_L)]\,\psi_n \quad \text{[Equation 16-11]}$$

where ψ_n is the expansion of an infinite series and dependent on *n* and $0.5s_L{}^2$; the latter term is referred to in Table 16-3 as *t*.

When a choice of estimators is possible, it is important to ponder two factors – bias and precision. Consider an exposure population with μ_c = 10 ppm. Now imagine sampling that population many times (1,000,000 times), each time with *n* of 50. From each sample we could calculate a mean, by any technique we chose. If the mean of the 1,000,000 sample means was 10 ppm, then we would conclude that our technique is not biased. If, on the other hand, the mean of means was 11 ppm, we would conclude that our technique had a bias of 10%.

But what about precision? We could also calculate a variance of the 1,000,000 sample means. If we compared the variance of our technique and found that it was smaller than for all

[9] The geometric mean (GM) has also been used inappropriately to estimate μ_c (e.g., Breslin et al., 1967). The GM is equal to the median, and Figure 16-1 shows that the median and mean of a lognormal are not equal. Therefore, GM is always a negatively biased estimator of μ_c.

Table 16-3
Multiplying factor $\psi_n(t = 0.5\ s_y^2)$ for estimating the lognormal mean and variance (from Gilbert, 1987)

	Number of Samples (n)																
t	2	5	8	10	13	15	20	25	30	50	70	90	100	150	200	500	∞
0.05	1.025	1.041	1.045	1.046	1.047	1.048	1.048	1.049	1.049	1.050	1.050	1.051	1.051	1.051	1.051	1.051	1.051
0.10	1.050	1.082	1.091	1.093	1.096	1.097	1.099	1.100	1.101	1.103	1.103	1.104	1.104	1.104	1.105	1.105	1.105
0.15	1.076	1325	1.138	1.143	1.147	1.149	1.152	1.154	1.155	1.158	1.159	1.160	1.160	1.160	1.161	1.161	1362
0.20	1.102	1.169	1.187	1.194	1.200	1.203	1.202	1.210	1.212	1.216	1.217	1.218	1.218	1.219	1.220	1.221	1.221
0.25	1.128	1.214	1.238	1.247	1.155	1.259	1.265	1.268	1.271	1.276	1.278	1.180	1.280	1.281	1.282	1.283	1.284
0.30	1.154	1.260	1.291	1.302	1.312	1.317	1.325	1.330	1.333	1.340	1.342	1.344	1.345	1.346	1.347	1.349	1.350
0.35	1.180	1.307	1.345	1.359	1.372	1.378	1.387	1.393	1.398	1.406	1.410	1.412	1.412	1.415	1.416	1.418	1.419
0.40	1.207	1.356	1.401	1.418	1.433	1.441	1.453	1.460	1.465	1.476	1.480	1.483	1.484	1.486	1.488	1.490	1.492
0.45	1.234	1.406	1.459	1.479	1.498	1.506	1.521	1.530	1.536	1.548	1.554	1.557	1.558	1.562	1.563	1.566	1.568
0.50	1.261	1.457	1.519	1.542	1.564	1.514	1.592	1.602	1.610	1.625	1.631	1.635	1.637	1.641	1.643	1.646	1.649
0.55	1.288	1.509	1.581	1.608	1.633	1.645	1.666	1.678	1.687	1.705	1.713	1.717	1.719	1.724	1.726	1.730	1.733
0.60	1.315	1.563	1.645	1.675	1.705	1.719	1.743	1.757	1.768	1.789	1.798	1.803	1.805	1.811	1.814	1.819	1.822
0.65	1.343	1.618	1.711	1.746	1.780	1.796	1.823	1.840	1.852	1.876	1.887	1.893	1.896	1.907	1.905	1.912	1.916
0.70	1.371	1.675	1.779	1.818	1.857	1.876	1.907	1.926	1.940	1.968	1.981	1.988	1.990	1.998	2.002	2.009	2.014
0.25	1.399	1.733	1.849	1.894	1.938	1.958	1.994	2.016	2.032	2.064	2.079	2.087	2.090	2.099	2.103	2.111	2.117
0.80	1.427	1.792	1.922	1.971	2.021	2.045	2.085	2.110	2.128	2.165	2.182	7.191	2.194	2.205	2.210	2.219	2.226
0.85	1.456	1.853	1.996	2.052	2.108	2.134	2.179	2.208	2.228	2.270	2.289	2.300	2.304	2.316	2.322	2.332	2.340
0.90	1.485	1.915	2.074	2.135	2.197	2.227	2.278	2.310	2.333	2.381	2.402	2.414	2.419	2.432	2.439	2.451	2.460
0.95	1.514	1.979	2.153	2.221	2.291	2.323	2.380	2.417	2.442	2.496	2.521	2.534	2.540	2.554	2.562	2.708	2.586
1.00	1.543	2.044	2.235	2.310	2.387	2.424	2.487	2.528	2.556	2.617	2.644	2.660	2.666	2.683	2.692	2.576	2.718
1.05	1.573	2.111	2.320	2.403	2.487	2.528	2.598	23644	2.676	2.744	2.774	2.792	2.798	2.818	2.828	2.845	2.858
1.10	1.602	2.180	2.407	2.496	2.591	2.636	2.714	2.765	2.800	2.876	2.911	2.930	2.938	2.959	2.970	2.990	3.004
1.15	1.632	2.250	2.497	2.596	2.698	2.748	2.834	2.891	2.930	3.014	3.053	3.076	3.083	3.108	3.120	3.143	3.158
1.20	1.662	2.321	2.589	2.698	2.510	2.864	2.960	3.022	3.066	3.159	3.203	3.228	3.237	3.263	3.277	3.303	3.320
1.25	1.693	2.395	2.685	2.803	2.926	2.985	3.090	3.159	3.207	3.311	3.359	3.387	3.397	3.427	3.442	3.471	3.490
1.30	1.724	2.470	2.783	1.911	3.045	3.111	3.226	3.301	3.354	3.470	3.523	3.554	3.565	3.599	3.616	3.648	3.669
1.35	1.754	2.547	2.884	3.023	3.169	3.241	3.367	3.450	3.508	3.636	3.695	3.729	3.741	3.779	3.798	3.833	3.857
1.40	1.786	7.626	2.988	3.139	3.298	3.376	3.514	3.604	3.669	3.809	3.875	3.912	3.926	3.968	3.989	4.028	4.055
1.45	1.817	2.706	3.096	3.259	3.431	3.515	3.666	3.766	3.836	3.991	4.063	4.105	4.120	4.166	4.189	4.233	4.263
1.50	1.849	2.788	3.206	3.382	3.569	3.661	3.825	3.933	4.011	4.181	4.260	4.306	4.323	4.374	4.400	4.448	4.482
1.55	1.880	2.873	3.320	3.510	3.711	3.811	3.990	4.108	4.193	4.379	4.467	4.518	4.536	4.592	4.621	4.675	4.712
1.60	1.913	2.959	3.437	3.642	3.859	3.967	4.161	4.291	4.363	4.587	4.683	4739	4.759	4.821	4.853	4.912	4.953
1.65	1.945	3.047	3.558	3.777	4.012	4.129	4.339	4.480	4.581	4.804	4.910	4.971	.993	5.062	5.097	5.162	5.207
1 30	1.977	3.137	3.682	3.918	4.171	4.297	4.525	4.678	4.787	5.031	5.147	5.215	5.239	5.314	5.352	5.424	5.474
1.75	2.010	3.229	3.810	4.062	4.334	4.471	4.717	4.883	5.003	5.769	5.395	5.469	5.496	5.578	5.621	5.700	5.755
1.80	2.043	3.323	3.942	4.212	4.504	4.651	4.917	5.097	5.228	5.517	5.655	5.736	5.766	5.856	5.903	5.990	6.050
1.85	2.077	3.420	4.077	4.366	4.680	4.838	5.125	5.320	5.461	5.776	5.928	6.016	6.048	6.147	6.198	6.294	6.360
1.90	2.110	3.516	4.216	4525	4.861	5.031	5.341	5.552	5.705	6.048	6.212	6.309	6.344	6.453	6.509	6.613	6.686
1.95	2.144	3.619	4.359	4.685	5.049	5.732	5.566	5.794	5.959	6.331	6.511	6.616	6.655	6.773	6.834	6.949	7.029
2.00	2.178	3.721	4.506	4.857	5.243	5.439	5.799	6.045	6.224	6.628	6.823	6.938	6.980	7.109	7.176	7.302	7.389

other techniques, we would consider it to be the minimum variance estimator. Thus, a minimum variance unbiased estimator is the most accurate.

The following facts are important in sorting out any confusion about calculating the mean of a lognormal:

1. The *MVU* and $\bar{x}$ are unbiased estimators.
2. The *MLE* is a biased estimator.
3. The *MVU* has the smallest (minimum) variance of the three.

If you are not convinced, see Attfield and Hewett (1992). For example, they show that for $n = 10$ and GSD = 3, the variances are $\bar{x} = 3.13$, *MLE* = 3.29, and *MVU* = 2.73. The bias for *MLE* is 1.036. Are these values significant? As the GSD increases and n decreases, the answer to that question is *yes*. Therefore, because Table 16-3 will allow one to do so, the *MVU* should be calculated whenever possible if you are interested solely in estimating the mean.

Exact Confidence Limits for the Mean of a Lognormal

The above equations for estimating the mean of a lognormal tell us nothing about the accuracy of that estimate. To determine the accuracy we must calculate confidence limits. There is a method, but once again it requires that a value be found from a table, and previously the table was obscure. Land (1975; see also Gilbert, 1987) derived functions for the upper (UL_1-α) and lower ($LL\ \alpha$) confidence limits on the mean of a lognormal distribution as

$$UL_{(1-\alpha)} = \exp\left(\bar{x}_L + 0.5s_L^2 + \frac{s_L H_{1-\alpha}}{\sqrt{n-1}}\right) \quad \text{[Equation 16-12]}$$

and:

$$LL_{(\alpha)} = \exp\left(\bar{x}_L + 0.5s_L^2 + \frac{s_L H_{\alpha}}{\sqrt{n-1}}\right) \quad \text{[Equation 16-13]}$$

where H_α and $H_{1-\alpha}$ are values dependent on n and s_L as shown in Table 16-4. Note that these equations are very similar to Equation 16-3, for the *CL* for a normal distribution based on a *t*-score. The first two right-hand terms in the equations represent the *MLE* estimate of the mean (Equation 16-10), and the last term substitutes an *H*-score for a *t*-score. Hence, confidence limit equations for the mean require the use of the *MLE*. Land has created a small program that allows one to calculate these confidence limits for any values of n, s_L, and α. This program is free of charge and available on the Internet (Lyon and Land, 2005). In addition, Hewett and Ganser (1997) have produced graphs for finding H_α (their C) and Grzebyk and Thiery (2003) have produced complete tables for confidence limits with particular application to sound pressure levels.

Table 16-4a
Values of $H_{1-\alpha} = H_{0.90}$ for computing a one-sided upper 90% confidence limit on a lognormal mean

s_y \ n	3	5	7	10	12	15	21	31	51	101
0.10	1.686	1.438	1.381	1.349	1.338	1.328	1.317	1.308	1.301	1.295
0.20	1.885	1.522	1.442	1.396	1.380	1.365	1.348	1.335	1.324	1.314
0.30	2.156	1.627	1.517	1.453	1.432	1.411	1.388	1.370	1.354	1.339
0.40	2.521	1.755	1.607	1.523	1.494	1.467	1.437	1.412	1.390	1.371
0.50	2.990	1.907	1.712	1.604	0.567	1.532	1.494	1.462	1.434	1.409
0.60	3.542	2.084	1.834	1.69E	1.650	1.606	1.558	1.519	1.485	1.454
0.70	4.136	2.264	1.970	1.800	1.743	1.690	1.631	1.583	1.541	1.504
0.80	4.742	2.503	2.119	1.914	1.845	1.781	1.710	1.654	1.604	1.560
0.90	5.349	2.736	2.280	2.036	1.955	1.880	1.797	1.731	1.672	1.621
1.00	5.955	2.980	2.450	2.167	2.073	1.985	1.889	1.812	1.745	1.686
1.25	7.466	3.617	2.904	2.518	2.391	2.271	2.141	2.036	1.946	1.866
1.50	8.973	4.276	3.383	2.896	2.733	2.581	2.415	2.282	2.166	2.066
1.75	10.48	4.944	3.877	3.289	3.092	2.907	2.705	2.543	2.402	2.279
2.00	11.98	5.619	4.380	3.693	3.461	3.244	3.005	2.814	2.648	2.503
2.50	14.99	6.979	5.401	4.518	4.220	1938	3.629	3.380	3.163	2.974
3.00	18.00	8.346	6.434	5.359	4.994	4.650	4.270	3.964	3.697	3.463
3.50	21.00	9.717	7.473	6.208	5.778	5.370	4.921	4.559	4.242	3.965
4.00	24.00	11.09	8.516	7.062	6.566	6.097	5.580	5.161	4.796	4.474
4.50	27.01	12.47	9.562	7.919	7.360	6.829	6.243	5.769	5.354	4.989
5.00	30.01	13.84	10.61	8.779	8.155	7.563	6.909	6.379	5.916	5.508
6.00	36.02	16.60	12.71	10.50	9.751	9.037	8.248	7.607	7.048	6.555
7.00	42.02	19.35	14.81	12.23	11.35	10.52	9.592	8.842	8.186	7.607
8.00	48.03	22.11	16.91	13.96	12.96	12.00	10.94	10.08	9.329	8.665
9.00	54.03	24.87	19.02	15.70	14.56	13.48	12.29	11.32	10.48	9.725
10.00	60.04	27.63	21.12	17.43	16.17	14.97	13.64	12.56	11.62	10.79

Adapted from Gilbert (1987) based on work of Land (1975).

Table 16-4b
Values of $H_\alpha = H_{0.10}$ for computing a one-sided lower 10% confidence limit on a lognormal mean

s_y \ n	3	5	7	10	12	15	21	31	51	101
0.10	-1.131	-1.320	-1.296	-1.285	-1.281	-1.279	-1.277	-1.277	-1.278	-1.279
0.20	-1.350	-1.281	-1.268	-1.266	-1.266	-1.266	-1.268	-1.272	-1.275	-1.280
0.30	-1.289	-1.252	-1.250	-1.254	-1.257	-1.260	-1.266	-1.272	-1.280	-1.287
0.40	-1.245	-1.233	-1.239	-1.249	-1.254	-1.261	-1.270	-1.279	-1.289	-1.301
0.50	-1.213	-1.221	-1.234	-1.250	-1.257	-1.266	-1.279	-1.291	-1.304	-1.319
0.60	7.190	-1.215	-1.235	-1.256	-1.266	-1.277	-1.292	-1.307	-1.324	-1.342
0.70	-1.176	-1.215	-1.241	-1.266	-1.278	-1.292	-1.310	-1.329	-1.349	-1.370
0.80	-1.168	-1.219	-1.251	-1.280	-1.294	-1.311	-1.332	-1.354	-1.377	-1.403
0.90	-1.165	-1.227	-1.264	-1.298	-1.314	-1.333	-1.358	-1.383	-1.409	-1.439
1.00	-1.166	-1.239	-1.281	-1.320	-1.337	-1.358	-1.387	-1.414	-1.445	-1.478
1.25	-1.184	-1.280	-1.334	-1.384	-1.407	-1.434	-1.470	-1.507	-1.547	-1.589
1.50	-1.217	-1.334	-1.400	-1.462	-1.491	-1.523	-1.568	-1.613	-1.063	-1.716
1.75	-1.260	-1.398	-1.477	-1.551	-1.585	-1.624	-1.677	-1.732	-1.790	-1.855
2.00	-1.310	-1.470	-1.562	-1.647	-1.688	-1.733	-1.795	-1.859	-1.928	-2.003
2.50	-1.426	-1.634	-1.751	-1.862	-1.913	-1.971	-2.051	-2.133	-2.223	-2.321
3.00	-1.560	-1.817	-1.960	-2.095	-2.157	-2.229	-2.326	-2.427	-2.536	-2.657
3.50	-1.710	-2.014	-2.183	-2.341	-2.415	-2.499	-2.615	-2.733	-2.864	-3.D07
4.00	-1.871	-2.221	-2.415	-2.596	-2.681	-2.778	-2.913	-3.050	-3.200	-3.366
4.50	-2.041	-2.435	-2.653	-2.858	-2.955	-3.064	-3.217	-3.372	-3.542	-3.731
5.00	-2.217	-2.654	-2.897	-3.126	-3.233	-3.356	-3.525	-3.698	-3.889	-4.100
6.00	-2.581	-3.104	-3.396	-3.671	-3.800	-3.949	-4.153	-4.363	-4.594	-4.849
7.00	-2.955	-3.564	-3.904	-4.226	-4.377	-4.549	-4.790	-5.037	-5.307	-5.607
5.00	-3.336	-4.030	-4.418	-4.787	-4.960	-5.159	-5.433	-5.715	-6.026	-6.370
9.00	-3.721	-4.500	-4.937	-5.352	-5.547	-5.771	-6.080	-6.399	-6.748	-7.136
10.00	-4.109	-4.973	-5.459	-5.920	-6.137	-6.386	-6.730	-7.085	-7.474	-7.906

Adapted from Gilbert (1987) based on work of Land (1975).

Table 16-4c
Values of $H_{1-\alpha} = H_{0.95}$ for computing a one-sided upper 95% confidence limit on a lognormal mean

s_y \ n	3	5	7	10	12	15	21	31	51	101
0.10	2.750	2.035	1.886	1.802	1.775	1.749	1.722	1.701	1.684	1.670
0.20	3.295	2.198	1.992	1.881	1.843	1.809	1.771	1.742	1.718	1.697
0.30	4.109	2.402	2.125	1.977	1.927	1.882	1.833	1.793	1.761	1.733
0.40	5.220	2.651	2.282	2.089	2.026	1.968	1.905	1.856	1.813	1.777
0.50	6.495	2.947	2.465	2.220	2.141	2.068	1.989	1.928	1.876	1.830
0.60	7.807	3.287	2.673	2.368	2.271	2.181	2.085	2.010	1.946	1.891
0.70	9.120	3.662	2.904	2.532	2.414	2.306	2.191	2.102	2.025	1.960
0.80	10.43	4.062	3.155	2.710	2.570	2.443	2.307	2.202	2.112	2.035
0.90	11.74	4.478	3.420	2.902	2.738	2589	2.432	2.310	2.206	2.717
1.00	13.05	4.905	3.696	3.103	2.915	2.744	2.564	2.423	2.306	2.205
1.25	16.33	6.001	4.426	3.639	3.389	3.163	2.923	2.737	2.580	2.447
1.50	19.60	7.120	5.154	4.207	3.896	3.612	3.311	3.077	2.881	2.713
1.75	22.87	8.250	5.960	4.795	4.422	4.081	3.719	3.437	3.200	2.997
2.00	26.14	9.387	6.747	5.396	4.962	4.564	4.141	3.812	3.533	3.295
2.50	32.69	11.67	8.339	6.621	6.067	5.557	5.013	4.588	4.228	3.920
3.00	39.23	13.97	9.945	7.864	7.191	6.570	5.907	5.388	4.947	4.569
3.50	45.77	16.27	11.56	9.118	8.326	7.596	6.815	6.201	5.681	5.233
4.00	52.31	18.58	13.18	10.38	9.469	8.630	7.731	7.024	6.424	5.908
4.50	58.85	20.88	14.80	11.64	10.62	9.669	8.652	7.854	7.174	6.590
5.00	65.39	23.19	16.43	12.91	11.77	10.71	9.579	8.688	7.929	7.277
6.00	78.47	27.81	19.68	15.45	14.08	12.81	11.44	10.36	9.449	8.661
7.00	91.55	32.43	22.94	18.00	16.39	14.90	83.31	12.05	10.98	10.05
8.00	104.6	37.06	26.20	20.55	18.71	17.01	15.18	13.74	12.51	11.45
9.00	117.7	41.68	29.46	23.10	21.03	19.11	17.05	15.43	14.05	12.85
10.00	130.8	46.31	32.73	25.66	23.35	21.22	18.93	17.13	15.59	14.26

Adapted from Gilbert (1987) based on work of Land (1975).

Table 16-4d
Values of $H_\alpha = H_{0.05}$ for computing a one-sided lower 5% confidence limit on a lognormal mean

s_y \ n	3	5	7	10	12	15	21	31	51	101
0.10	-2.130	-1.806	-1.731	-1.690	-1.677	-1.666	-1.655	-1.648	-1.644	-1.642
0.20	-1.949	-1.729	-1.678	-1.653	-1.646	-1.640	-1.636	-1.636	-1.637	-1.641
0.30	-1.816	-1.669	-1.639	-1.627	-1.625	-1.625	-1.627	-1.632	-1.638	-1.648
0.40	-1.717	-1.625	-1.611	-1.611	-1.613	-1.617	-1.625	-1.635	-1.647	-1.662
0.50	-1.644	-1.594	-1.5%	-1.603	-1.609	-1.618	-1.631	-1.646	-1.663	-1.683
0.60	-1.589	-1.573	-1.584	-1.602	-1.612	-1.625	-1.643	-1.662	-1.685	-1.711
0.70	-1.549	-1.560	-1.582	-1.608	-1.622	-1.638	-1.661	-1.686	-1.713	-1.744
0.80	-1.521	-1.555	-1.586	-1.620	-1.636	-1.656	-1.685	-1.714	-1.747	-1.783
0.90	-1.502	-1.556	-7.595	-1.637	-1.656	-1.680	-1.713	-1.747	-1.785	-1.826
1.00	-1.490	-1.562	-1.610	-1.658	-1.681	-1.707	-1.745	-1.764	-1.827	-1.874
1.25	-1.486	-1.596	-1.662	-1.727	-1.758	-1.793	-1.842	-1.893	-1.949	-2.012
1.50	-1.508	-1.650	-1.733	-1.814	-1.853	-1.896	-1.958	-2.020	-2.091	-2.169
1.75	-1.547	-1.719	-1.819	-1.916	-1.962	-2.015	-2.088	-2.164	-2.247	-2.341
2.00	-1.598	-1.799	-1.917	-2.029	-2.083	-2.144	-2.230	-2.318	-2.416	-2.526
2.50	-1.727	-1.986	-2.138	-2.283	-2.351	-2.430	-2.540	-2.654	-2.780	-2.921
3.00	-1.880	-2.199	-2.384	-2.560	-2.644	-2.740	-2.874	-3.014	-3.169	-3.342
3.50	-2.051	-2.429	-2.647	-2.855	-2.953	-3.067	-3.226	-3.391	-3.574	-3.780
4.00	-2.237	-2.672	-2.922	-3.161	-3.275	-3.406	-3.589	-3.779	-3.990	-4.228
4.50	-2.434	-2.924	-3.206	-3.476	-3.605	-3.753	-3.960	-4.176	-4.416	-4.685
5.00	-2.638	-3.183	-3.497	-3.798	-3.941	-4.107	-4.338	-4.579	-4.847	-5.148
6.00	-3.062	-3.715	-4.092	-4.455	-4.627	-4.827	-5.106	-5.397	-5.721	-6.086
7.00	-3.499	-4.260	-4.699	-5.123	-5.325	-5.559	-5.886	-6.227	-6.608	-7.036
8.00	-3.945	-4.812	-5.315	-5.800	-6.031	-6.300	-6.674	-7.066	-7.502	-7.992
9.00	-4.397	-5.371	-5.936	-6.482	-6.742	-7.045	-7.468	-7.909	-8.401	-8.953
10.00	-4.852	-5.933	-6.560	-7.168	-7.458	-7.794	-8.264	-8.755	-9.302	-9.918

Adapted from Gilbert (1987) based on work of Land (1975).

Other Approaches to Estimating Confidence Limits

Prior to the use of Land's equation for *CL*, other approaches were tried. For example, the normal *CL* based on *t* has been modified as:

$$UCL = \exp\left(\frac{\bar{x}_L + t_{1-\alpha}\, s_L}{n^{0.5}}\right) \qquad \text{[Equation 16-14]}$$

However, this equation biases the *UCL* because it is based on the geometric mean ($\exp \bar{x}$), which is always smaller than the mean. Another approach is to center the *CL* around the *MLE*, or:

$$UCL = \exp\left(\frac{\bar{x}_L + 0.5 s_L^2 + t_{1-\alpha}\, s_L}{n^{0.5}}\right) \qquad \text{[Equation 16-15]}$$

However, Attfield and Hewett (1992) have shown that *CLs* calculated this way are still biased compared to Land's. The general order of *CLs* based on Equation 16-15 (and a similar equation for the *LCL*) and Land's is $LCL_t < LCL_L < UCL_t < UCL_L$, where the *L* and *t* subscripts refer to Land's and the *t*-score (Equation 16-15), respectively. Therefore, using even the modified approach in Equation 16-15 gives lower and upper limits that are too small. As the GSD increases and the sample size decreases, the differences can be quite large. Still other approaches have been suggested but Land's confidence intervals perform best (Hewett, 1995 and 1997). Hence, the use of Land's equations is warranted.

ESTIMATING SAMPLE SIZES

Sample Size for Estimating a Lognormal Mean

For a normal distribution, a sample size for estimating the mean can be derived from a probability statement as:

$$n = \left(\frac{Z_{\alpha/2}\, \sigma}{\delta}\right)^2 \qquad \text{[Equation 16-16]}$$

where σ is usually estimated,[10] and δ is the difference one is willing to accept or $\mu - \bar{x}$, the difference between the appropriate (upper or lower) confidence limit and the estimate of the mean, or how wide we are willing to allow the confidence limits to be.

A similar equation can be derived for the lognormal by using Land's confidence limits. Taking the natural logarithm of Equation 16-12, we have:

---

[10] Ideally one should use a t-score, which requires iteration since the t-scores cannot be found without using a value for n. (See Gilbert, 1987 for details.)

$$lnUL_{(1-\alpha)} = \bar{x}_L + 0.5s_L^2 + \frac{s_L H_{1-\alpha}}{\sqrt{n-1}}$$

Now the "*MLE*" term can be moved to the left:

$$lnUL_{(1-\alpha)} - (\bar{x}_L + 0.5s_L^2) = \frac{s_L H_{1-\alpha}}{\sqrt{n-1}}$$

Actually this "*MLE*" term is ln *MLE*; so:

$$lnUL_{(1-\alpha)} - \ln MLE = \frac{s_L H_{1-\alpha}}{\sqrt{n-1}}$$

The difference of two logs is the log of the quotient; and

$$\ln\frac{UL_{(1-\alpha)}}{MLE} = \frac{s_L H_{1-\alpha}}{\sqrt{n-1}} \quad \text{[Equation 16-17]}$$

We cannot solve directly for n, since $H_{1-\alpha}$ is dependent on n. Thus, we can work backward to determine the accuracy associated with a given n. Assume that a GSD of 3 (moderate-high variability) is expected, and we want to see how accurately the mean can be estimated from a sample size of 20. From Chapter 7, Equation 7-9, for a GSD of 3.0, $s_L = 1.1$. From Table 16-4, the *H*-score for n = 21, $s_L = 1.1$, and a one-tailed upper 95% CL is about 2.7.[11] Substitution reveals that the ratio of *UCL* to *MLE* is 1.94. This implies that if we sample the population ($n = 20$) and the GSD is in fact 3.0, then our estimate of the sample mean will be about 1/1.94, or about one-half or less of the *UCL*. In other words, the *UCL* will be about $\bar{x}$ + 94% of $\bar{x}$. If more accuracy is desired, then the sample size should be increased. For a slightly different approach to confidence limits, also based on Land's work, see Armstrong (1992).

TESTING THE NULL HYPOTHESIS: $\mu_c \geq$ OEL

The above use of confidence limits represents a reasonable approach to testing the null

[11] Cubic interpolation (4-point Lagrangian) is required for accurately using these tables. This means the use of still another table. However, for our applications, a rough estimate is appropriate since other parameters are also estimated with less accuracy than can be gained from eyeballing the tables.

hypothesis that the mean of exposures is ≥ an OEL.[12] If the OEL is not contained in the confidence bounds around the mean, then we can reject the null with 1-α confidence. Again for the sake of understanding, we can state approximately that at 95% confidence there is still a 1 in 20 chance that the mean and the OEL are equivalent, but generally we are willing to accept this 5%, Type I or *public health* risk. But what if we fail to reject the null? What errors are associated with this decision? As Figure 16-6 shows, there is a β, or Type II error, associated with this decision. This error is sometimes referred to as consumer's risk (Gilbert, 1987), but here it might be referred to as *employer's risk.* It is the chance that the workplace will be declared out of compliance, when in fact it should not be. The complement of this error, 1-β, is the probability of declaring the workplace safe or rejecting the null when we should do so. It is the *power* of the test.

Other tests, based upon approximations, have been derived for comparing the mean of a

Test Conclusion \ Truth	μ = OEL	μ < OEL
μ = OEL	1-α	β
μ < OEL	α	1-β = Power

α is the probability of concluding μ < OEL when it is not (public heath cost)
β is the probability of concluding μ = OEL when it is not (employer's cost)
Power is the probability of concluding compliance when you should (1-β)
1-α is the probability of concluding non-compliance when you should

Figure 16-6. Errors associated with testing the null hypothesis, H_0: $\mu \geq$ OEL.

---

[12] Null hypotheses are typically statements with an equality, while alternate hypotheses are statements of inequality. For example,

$$H_0: \mu_c = \text{OEL and } H_a: \mu_c \neq \text{OEL}.$$

However, in the case of the corporate hygienist (the employer's test), one is only interested in the alternative when μ_c< OEL. In other words one would like to reject the null and accept H_a. So in this *one-tailed* test the null defaults to

$$H_0: \mu_c \geq \text{OEL}.$$

For a compliance officer's test the opposite would be true.

lognormal exposure distribution with an exposure standard (Rappaport and Selvin,1987; Lyles and Kupper, 1996; Hewett, 1995 and 1997). These tests perform reasonably well but are approximations; only Land's test will give the β and α errors that are predicted for a given sample size when run in computer simulations of thousands of tests (Hewett, 1997).

Sample Size for a Null Hypothesis Test of the Mean

The power of the test and the chances for Type I and II errors are dependent on sample size. Previously sample sizes based on α only were calculated by working backwards. (The *H*-score was based only on α). Avoiding a Type I error should certainly be the hygienist's first priority. However, unnecessary employer expenditures also should be avoided, particularly those for expensive engineering controls. When testing a null hypothesis on the mean (as opposed to a simple estimate of a mean), both β and α need to be considered. For example, for hypothesis tests of the mean of a normal distribution, Equation 16-16 can be modified to account for β (Snedecor and Cochran, 1967),

$$n = \left(\frac{(Z_\alpha + Z_\beta)\ \sigma}{\delta} \right)^2 \qquad \text{[Equation 16-18]}$$

but this equation will obviously not work for a lognormally distributed mean.

Table 16-5 shows approximate sample sizes based on an α-level of 5% and a β-level of 10% (Hewett, 1997).[13] The ratio μ_c/OEL and the *GSD* must be estimated in order to enter the table. For example, if one wishes to differentiate a mean as high as 75% of the *OEL*, from the *OEL*, in an employer's test, then the ratio is 0.75. The smaller the difference between the mean and the *OEL* that the hygienist or inspector tries to discern, the closer the ratio is to unity, and the larger the sample size that is needed. For example, when the ratio is estimated to be 75%, the necessary sample size may be as high as 435 for a *GSD* of 4. However, if the *GSD* is 1.5, then only 23 air samples are necessary. For a *GSD* of 1.5 and ratio of 0.1, only four samples would be necessary in order to reject the null, if it should be rejected. These sample sizes are about 2-3 times those obtained with Equation 16-17, which is based on estimating a mean and not testing a null hypothesis.

Although many of the sample sizes in Table 16-5 are large, there are good reasons (such as uncertainty in OELs) for keeping average exposure levels at some low fraction of the OEL, such as 0.25. It can be seen from Table 16-5 that when $\mu_c/OEL = 0.25$, sample sizes are reasonable. Finally, it is important to point out that only a compliance officer, performing a lower, one-tailed test would enter the table with $\mu_c/OEL > 1$.

---

[13] It should be noted that Lyles and Kupper also derived sample sizes based on a new test that they derived and for Rappaport and Selvin's test. The values for these as well as Hewett's are close. Hewett's results are based on the exact Land's test and so should be considered most accurate for obtaining the α and β cited.

Table 16-5
Sample size calculations[1] for testing the mean exposure from a lognormal distribution of 8-hr TWAs when $\alpha \approx 0.05$ and $\beta \approx 0.10$[2] based on Land's confidence interval (from Hewett, 1997)

σ_g / μc/OEL	1.5	2.0	3.0	3.5
0.10	4	6	10	15
0.50	7	17	49	90
0.75	23	73	235	435
1.50	9	27	85	170
2.00	4	9	26	51
3.00	3	4	10	18

[1] Based upon the employer's test for μ_c/OEL < 1 [H_o: $\mu_c \geq$ OEL, H_A: $\mu_c <$ OEL (acceptable exposure)]; and upon inspector's test for μ_c/OEL > 1 [H_o: $\mu_c \leq$ PEL, H_A:$\mu_c >$ OEL (unacceptable exposure)].

[2] α is the significance level; 1-β is the power, σ_g is the geometric standard deviation of the distribution of 8-hr TWAs.

It can be argued, in most tests of H_0: $\mu_c = OEL$, that small, true differences between μ_c and *OEL* are not important. The reason is that the *OEL* usually has great uncertainty associated with it. For most OELs, we know little or nothing, quantitatively, about associated risks. Therefore, it is unreasonable to look for small differences; those differences may be real numerically but not real biologically. In other words, the *OEL* may be in error such that undesirable effects are occurring at levels below the *OEL*. This is another reason for controlling exposures to some fraction of the *OEL*, e.g., 0.25. The difficulty, of course, lies in explaining to management or others why a sample mean of, for instance, 0.43 *OEL*, might be statistically insignificantly different from the *OEL* but also biologically insignificantly different.

ESTIMATING AN EXCEEDANCE FRACTION FOR AN EXPOSURE DISTRIBUTION

As was pointed out in the introduction there are toxicants for which the toxicokinetics indicate that the risk of disease is determined by a lifetime accumulation of exposure. For such toxicants the day to day variability is irrelevant to risk as the product of the average daily dose and number of days of exposure drives the lifetime dose. Thus, the average dose is the parameter of interest. We cannot derive the actual dose to, for example, the lung or blood but use the exposure as a surrogate.

It was also pointed out that for many toxicants both short-term peaks and long-term averages may drive disease risk. Rappaport (1991) has pointed out that in order for this to be likely, two conditions must be met.

1. The toxicant must be eliminated rapidly from the body. Most toxicants, certainly those with molecular weights up to 200 Daltons, are eliminated with half lives of less than 24 hours and would meet this definition.

2. There must be nonlinear rates of biological damage during peak exposures. While there are fewer examples that meet this criteria, realistic examples can be generalized. For example, a peak exposure overwhelms an enzyme system allowing the compound to remain unmetabolized in the blood for longer periods, or causing an alternative enzyme system to be activated which produces a more toxic metabolite.

For these toxicants the mean exposure may very well drive disease risk most of the time. However, peak exposures or higher daily average exposures may also be indicators of risk. We do not know. But we do know that compliance, at least in the U.S., is not based upon whether the mean of exposures is > OEL, but whether the exposure for that worker on that day > OEL. Of course, to control exposures such that the likelihood is *zero* that any worker is exposed > OEL on any given day is not realistic.

How often should we allow potential exposures to exceed the OEL? There is no known biological basis for answering this question, and of course such considerations are not included in the setting of the OEL, including TLVs®. Ideally we would know, for example, that exceedance of an OEL at say 5% probability causes an excess disease risk of 0.001. Realizing that no exposure scenario is risk free, we (or the regulatory body) would then need to decide if that excess risk was acceptable. Whether peaks are even important has not been assessed frequently in epidemiological studies. What aspects of peaks might be important and unique, such as frequency, height, and duration has only recently been considered (Preller et al., 2004). Thus, we are distant from a solution to the question of how often exceedances should be allowed, even though it is likely that they will occur in even the best controlled workplace.

The term exceedance fraction (γ) has been introduced. This is the fraction of the exposure distribution that is predicted to exceed the OEL. Based on the parameters of the distribution, whether it is assumed to be normal or lognormal, we can calculate this fraction. We can also test this fraction against some limit (L), e.g., 5%. Other limits have been proposed (e.g., Rappaport et al., 1995 proposed 10%) , but again these are based largely on convenience rather than biology.

As shown in Figure 16-7, the exceedance fraction is the exposure probability that lies to the right of the OEL. Remembering that the logged exposure values are normally distributed, we can use a Z-table from any statistics book to calculate this probability by logging the parameters and then calculating the Z distance or number of standard deviations that the OEL is from the mean of the logs ($\bar{x}_L$).

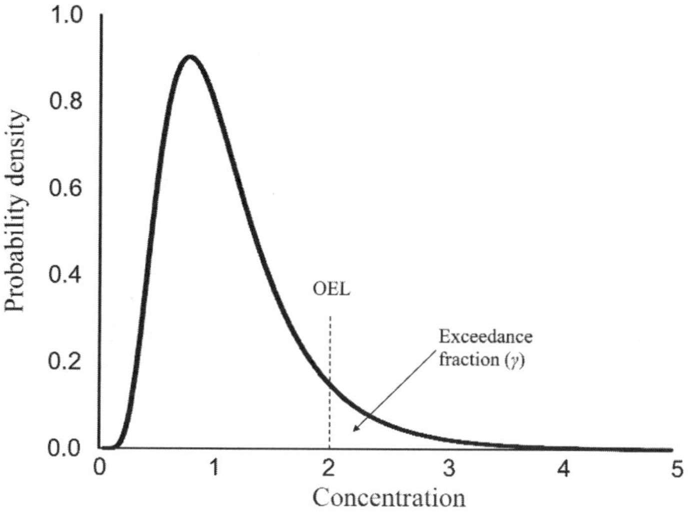

Figure 16-7. A lognormal distribution that models an 8-hr TWA exposure distribution showing the OEL and the portion of the distribution that is likely to exceed the OEL, the exceedance fraction (γ).

$$z = \frac{\ln OEL - \bar{x}_L}{s_L} \qquad \text{[Equation 16-19]}$$

In Chapter 18 we explore the relationship of the exceedance fraction to the likelihood that a worker's mean exposure is > OEL. In other words, if $\gamma = 5\%$ for the exposures in a similarly exposed group of workers, what is the likelihood that the mean exposure for any one of the workers is > *OEL*?

Testing the Null Hypothesis: $\gamma \geq L$

If we use a limit, L of 5%, and the exceedance fraction (γ) is say, 4%, then γ does not exceed the limit. However, just as is the case of estimating the mean, there is also uncertainty associated with estimating any other percentile of the distribution. In fact, in a typical sample, the uncertainty is larger in the tail since there is less known about the tail. The uncertainty is caused by measurement and statistical sampling errors. Therefore, we need to address this uncertainty and can test a specific hypothesis H_0: $\gamma \geq L$. As practitioners protecting a workforce (as opposed to compliance officers) we wish to reject this null in favor of an alternative H_0: $\gamma < L$, which would indicate that the probability for an exposure from our population of exposures to exceed the OEL is less than some limit, L.

Just as we can calculate confidence limits around a mean based on sample size and variance, we can do so for any other percentile. We do this by calculating tolerance limits as demonstrated in Equation 16-4. Equation 16-4 allows the calculation, for example, of the 95% upper confidence limit on the 95th %tile; this is termed the 95% upper tolerance limit on the 95th %tile. This is not what we want. We need the 95% UCL on the exceedance fraction; the latter may be any %tile value depending on the parameters of the distribution of exposures. Thus, the exceedance fraction is a %tile, but it is not usually a convenient number like 5% (which would coincide with the 95th %tile of the distribution). Tables for the *k*-factor that we need to calculate the tolerance limit (Equation 16-4) are usually based on percentiles like the 1, 5, and 10 %tiles.

A more general solution is needed. Based on sample size, Figure 16-8 allows the calculation of a 95% UCL on any exceedance fraction by entering the Z-score corresponding to the exceedance fraction (as calculated in Equation 16-19). First change the sign of the *z*, calculated with Equation 16-19, enter Figure 16-8 with -z, and find the 95% UCL based on the sample size. If this UCL ≥ the value set for *L*, say 5%, then we fail to reject the null hypothesis.

Multiple Exposure Limits

The exceedance fraction problem can be applied to another issue. Many substances have multiple exposure limits, (e.g., 15-min STEL and 8-hr PEL, or PEL and AL). It would be fortuitous if we could control average exposures to one level (e.g., the 8-hr PEL), and be assured that exposures would ever exceed another level (e.g., the STEL), at some reasonably acceptable rate, say, 5%. Spear et al. (1986) first considered this problem. They showed that the mean of lognormal distributions of exposures is independent of averaging time (e.g., the mean of 15-

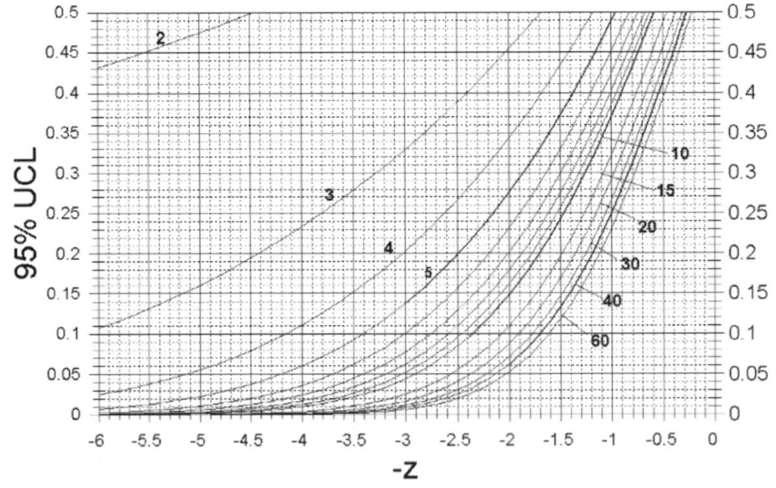

Figure 16-8. Upper confidence limit (95 percent) for the exceedance fraction (γ) versus the negative calculated z-value. Use the -z and the sample size to read the 95% UCL (from Hewett and Ganser, 1997, with permission).

min TWAs = the mean of 8-hr TWAs), but the variance increases as averaging time decreases. Thus the mean of a series of many 15-min TWAs should be the same as the mean for the integrated values whether they are 1-hr or 8-hr TWAs.

Rappaport et al. (1988) further demonstrated that although variance changes with averaging time, the arithmetic mean is correlated with the frequency of exceedances beyond some limit. If the mean (μ_c) is controlled to some value (e.g., action level = 0.5 ppm), the maximum fraction of exposures greater than some other value (e.g., the OEL = 1 ppm) can be calculated as:

$$\left(\frac{OEL}{\mu_c}\right)_{max} = \exp\left(0.5\ z^2_{(1-f)}\right) \qquad \text{[Equation 16-20]}$$

where z is from tables of the normal distribution, and f is the fraction of exposures exceeding the *OEL* (θ, as defined in the previous section). In the example above, $(OEL/\mu_c)_{max} = 2$; so $z_{(1-f)} = 1.17$, which is the 88.1st %tile of the normal distribution. Hence 1-*f* or a *maximum* of 11.9% exposures are expected to exceed the OEL if the mean is half the OEL. Table 16-6 shows values of OEL/μ_c and 1-*f*. Note that *OEL* may equal a *STEL* or any other value. So, for example, if the AL = μ_c = 0.5 ppm (i.e., 8-hour exposures are controlled at 0.5 ppm) and the *STEL* = 5 ppm (as for ethylene oxide), then a maximum of 1.5% (1-*f*) of 15-minute exposures are expected to exceed the STEL. Rappaport et al. showed that this relationship held for samples from several benzene and toluene diisocyanate exposure populations.

It is interesting to note that under the current OSHA scheme (AL = 0.5 PEL), as many as 12% of exposures may exceed the PEL if μ_c = AL. A hygienist may elect to set an internal AL according to the risk that he or she is willing to accept of being out of compliance for any single monitoring episode. For example, if μ_c is controlled at 0.25 PEL, then a maximum of only 5% of all exposures would be expected to exceed the PEL, and during a compliance inspection there would be a maximum 5% chance of OSHA detecting a noncompliance exposure. Of course, adequate monitoring data that document normal compliance should also be available. (More is said about this in Chapter 18.) Equally good reasons for controlling exposures at some fraction of the PEL are to also comply with a STEL or to add a safety factor to the PEL. In the history of OSHA (or recent history ACGIH®, for that matter), an OEL has never been revised upward, only downward.

It should be pointed out that this approach does have one drawback – lack of identifiability. In other words, although one may predict that only a maximum fraction of exposures will exceed some value, no information is gained about the actual magnitude of the overexposures.[14] This could be important for some toxicants. Of course, careful characterization of the exposure distribution should allow an estimate of the length of the tail of the distribution.

---

[14] The larger the GSD of the population having the exceedances, the larger the magnitude of the exceedances, i.e., the longer the distribution tail will be that extends beyond the OEL.

Table 16-6
Values of OEL/μ_c[1] that predict the maximum fraction (1-*f*)[2] of exposures that exceed the OEL

OEL/μ_c	1–*f*
1	50%
2	12%
3	7%
4	5%
5	4%
6	3%
10	1.5%

[1] OEL may be any occupational exposure limit for any averaging time, and μ_c is the arithmetic average of randomly sampled independent exposures from a lognormal distribution.

[2] This is the maximal value of 1-*f*. It varies with the GSD.

OTHER EXPOSURE ASSESSMENT ISSUES

Temporal and Spatial Correlation

The sample sizes calculated in the previous sections are based on independent measurements. Certainly the meaning of the word *independent* is understood in the general context; however, its meaning in statistics can be somewhat elusive. What does it mean for two measurements to be independent? It means there is no information redundancy; in no way does one predict the other.

For example, if two 15-minute measurements were taken every day in a plant at 8 AM and 8:15 AM, we might expect on average the second measurement to be partially predicted by the first. If we plotted all of the 8 AM measurements against the 8:15 AM measurements, we might expect to see a relationship rather than a random scatter of the points.

Similarly, if the daily exposures of two employees from the same well-defined SEG were monitored for a year, we might expect a relationship between the two employees' exposures. If one exposure were high on a given day, we would generally expect the other to be high, but not necessarily always.

In both of the above examples, the members of the population are not independent. Information contained in one measurement is at least partially contained in another; this is redundancy. On the surface, it may seem that redundancy is good, that it allows us to more tightly estimate the mean, for example. However, it is not.

In previous sections, sample sizes were determined that would provide a certain amount of information, say, $n = 20$, pieces of information. If some of the information is redundant, then perhaps, the effect of making 20 measurements is to get only 18.5 *pieces of information*. There are two potential penalties to be paid for this redundancy (Gilbert, 1987). Most important, the variance of the mean ($s_{\bar{X}}^2$) is underestimated; that is, confidence limits will be too tight.[15] Second, potentially the sample may be biased, causing the mean to be biased (low or high). The latter problem is particularly likely for small sample sizes, for example, five 15-minute samples taken in succession during a workday.

The correlation (partial or complete predictability) of measurements between time periods is termed temporal, serial, or auto (self) -correlation. The correlation of measurements between workers can be thought of as spatial correlation. Both can be considerable problems in ambient air and stream pollution studies. For the normal distribution, equations have been derived allowing the estimate of $s_{\bar{X}}^2$ and sample size based upon estimates of the autocorrelation coefficient (ρ_l) or the spatial correlation coefficient (ρ_s; see Gilbert, 1987).

The autocorrelation can occur at different *lags*. If consecutive 15-minute samples (X_i) are taken, the values may be paired at *one* time lag as follows:

X_1	X_2
X_2	X_3
X_3	X_4
X_4	X^5
X_5	X_6
.	.
.	.
X_{n-1}	X_n

The correlation coefficient for these data would be r_1, or the coefficient at a lag of one. Other lags can also be calculated, and other time periods can be used. For example, comparison of 8-hr TWAs at a lag of seven would compare results from the same day of the week, such as Monday. For environmental data a lag of one is most likely to be significant, but other lags may also be significant.

If five workers are monitored over consecutive time periods, and r_1, r_2, r_3, and r_4 calculated for each, this would yield 20 coefficients (4 for each worker). If the 20 estimates are truly zero, but subject to random error, we would expect one of them to be significant at the $\alpha = 0.05$ level by chance alone. We can define significance as follows. It can be shown that r_k is nor-

[15] The $s_{\bar{X}}^2$ is $s^2/(n-1)$. The square root of this quantity is the standard error and part of a confidence limit equation. One way of looking at the problem is that in the standard error term the n in the denominator is too large if measurements are not independent and, therefore, the standard error is underestimated.

mally distributed with $\mu = 0$ and $\sigma^2 = 1/N$, where N is the population of consecutive measurements (Chatfield, 1984, p. 25). Thus, 19 of 20 values of r would fall in the range $\pm$ 1.96/$n^{1/2}$ or 2/$n^{1/2}$. Then for a consecutive sample size of $N = 25$, r is significant when outside the range of $\pm$ 0.4. By chance alone we would expect 1 out of 20 r_k values to be outside this range, even if there were no autocorrelation.

Does either type of correlation exist in occupational exposure populations? There is some evidence for autocorrelation. Francis et al. (1989) discussed the overall problem and showed some evidence for autocorrelation of 8-hr TWA mercury exposures for workers in a chloralkali plant and daily 2-hr TWA exposures to an organic vapor. Carter (1990) found no evidence of autocorrelation for four styrene-exposed workers (8-hr TWA) in a cultured marble plant, and Hopson (1990) found no evidence for noise exposures (8-hr TWA) in the same plant.

Kumagai and co-workers (1993, 1994, 1995) found evidence for short-term (7.5 minutes, 15 minutes, 30 minutes, and 60 minutes) autocorrelation for workers exposed to solvents in various operations. For 16 worker-chemical combinations, the number of significant values of r_1, was approximately 13, 10, 8, and 7 for 7.5, 15, 30, and 60-min TWAs, respectively. For 8-hr TWAs on 10 of the workers, the number of consecutive measurements for cobalt and some solvents was generally < 30. Based on a significance level of $-0.4 < r < 0.4$, 2 of 10 estimates of r_1 and 1 of 10 estimates of r_2 were significant. The results for 8-hr TWAs are therefore mixed, but autocorrelation is likely for short-term samples.

Symanski and Rappaport (1994) found evidence of autocorrelation in 53 workers from five different data sets comparing 123 time series. The effect was isolated to 30% of the workers and 29% of the time series. They concluded that autocorrelation did not appear to be pervasive in shift-long samples. George et al. (1995) found some evidence for autocorrelation in shift long samples for solvent exposures in an auto assembly plant. However, they found that a lack of stationarity of the population of exposures (changes in the mean and variance over relatively short periods of time) was a bigger issue. This issue is further discussed in the next section.

Two observations are important. First, in order to find significant autocorrelation, large (at least $n = 30$) continuous data sets are needed; otherwise the significance level for r is too large. These large data sets are common in air pollution work but not in hygiene work. Second, autocorrelation is more likely to occur as the averaging time decreases. Because hygienists rarely take 30 consecutive short-term samples, the lack of evidence is not particularly meaningful.

But what of spatial correlation? This effect seems likely among workers who are purposely grouped because they are thought to have similar exposures, if they are monitored on the same day. McBride (1991) could not find spatial correlation among centrifugally spun pipe casting operators' dust exposures. Although the workers were doing very similar jobs, it was thought that worker remoteness from the equipment and the large size of the building made worker exposures more dependent on random and small variations (GSDs were ~ 1.5) in background concentrations rather than the piece of machinery that the workers operated. There appear to be no other investigations of this effect.

Whether or not there is a lack of independence in the workplace in question, what can be done to avoid it? The answer is simple if the hygienist has the luxury of time. When sampling a SEG, if only one air sample per day is taken from the SEG, and air samples are not taken on consecutive days,[16] then the hygienist should benefit from the full effect of the sample size. Plant hygienists can afford this luxury, but of course others have time constraints. Nevertheless, the hygienist should at least be aware of the potential for the effects of non-independence.

Population Stationarity

When we sample a population of exposures we are typically doing so to estimate its parameters, e.g., the mean and variance. But what if those change over time as we are collecting the sample? For example, say we collect a sample over a month and during that time there is a major equipment failure and an old piece of equipment is brought on-line. Three months later the equipment has been repaired. What happened to exposures during that time? They likely shifted. The mean may have become larger and fluctuations in the older piece of equipment or in the way workers used it could have caused the workers' exposures to be more erratic from day to day; this would cause the variance to increase. So what population were we measuring?

As was previously mentioned, we must define a population in terms of space (which workers) and time before we begin sampling. We must include in the definition the conditions under which the workplace operates. For example, when a process is shut down for maintenance and the operators also partake in the maintenance, as is common in the petrochemical industry, we should consider this distribution of exposures to be different from normal operating exposures, at least until we prove they are equivalent. In some workplaces it may be necessary to examine several different determinant variables (e.g., season) to see if they cause a shift in the mean and variance.

How do we know if a population is stationary over time? This is not an easy determination as it requires large data sets. Like correlation, it is better to design a sampling strategy or in this case define a population such that one is likely to avoid non-independence or non-stationarity, rather that test for it.

As for autocorrelation, stationarity is also a time series problem. Over time we expect measurements to randomly vary about the mean. If the variation is systematic, e.g., there is a drift up or down or the scatter decreases or increases around the mean, or there is a shift that

---

[16] Samples taken every other day will avoid autocorrelation effects at a lag of one day (period), which for environmental data is often the most significant correlation coefficient. However, other lags associated with process cycles can also be significant. So monitoring only on Mondays is not a good idea either.

may be followed days later by another shift, then this is systematic change, not random. Thus, there are visual methods for examining stationarity. George et al. (1995) found that ventilation and process changes as well as down time lead to time series that were not stationary over periods of about 3 months. This hindered their ability to examine the data for autocorrelation. Visual examination of the data hinted at non-stationarity and was as reliable as two statistical tests that were used.

Symanski et al. (1996) used an analysis of variance (ANOVA) approach (2-way and nested, mixed effects models) to discern non-stationarity (shifts in the geometric means) over periods generally from 1–2 years for 12 worker groups in five industries. This approach requires relatively large sample sizes and a specific design. It is discussed in detail as a method for both assessing and controlling workplace exposures in the next chapter.

Defining the SEG

As previously noted, placing workers into an exposure group implies that they have equivalent risk levels. That is, for chronic toxicants their mean exposures are equivalent, and for chronic toxicants for which occasional high exposures may be important, the likelihood of any worker in the group exceeding the OEL is similar. This grouping shortcut allows random sampling of any of the members of the SEG in order to characterize the overall population. For example, a SEG of 10 workers sampled over a period of one year (250 working days) would imply a population of 2,500 – 8-hr TWA exposures. A random sample of only 10 of these exposures may be enough to characterize the mean exposure and, therefore, the risk for the entire group.

But what if workers are misclassified? How can this occur? What does it mean to have *about the same risk*? Is a twofold difference in risk or mean exposure level or in 95th %tiles of exposure close enough? In order to assess these questions, we must further explore the nature of environmental variability. Leidel and Busch (1985) state that if the variability in the exposure sample distribution exceeds a geometric standard deviation of about 3, then the factors used to define the SEG should be reassessed. This is a reasonable idea, but one must be careful about redefining SEGs. High geometric standard deviations may be the result of many factors, such as inconsistent or unwritten work practice or procedural requirements, old and faulty machinery, and a large day-to-day random variability in the frequency of tasks that each worker in the exposure group is expected to perform.

Periodic maintenance (also discussed under stationarity), may occur for only two weeks out of the year, but may be quite a bit higher than *normal* exposures. In essence, there may be two overlapping exposure distributions. Reassigning workers to new SEGs across space or workers may not reduce GSDs. Redesigning SEGs across time for the maintenance issue would reduce GSDs, for the two new SEGs compared to the old, combined one.

High variability across a SEG may be caused by workers themselves (within-worker variability) rather than variability in ambient environmental concentrations from worker to worker (between worker variability). Therefore, reassigning workers to new SEGs is not an easy task, particularly if it is based only on qualitative knowledge.

Environmental variability in a SEG has two components: between-worker and within-worker. (Examine Figure 16-9.) The average exposure for the SEG is 50 ppm. However, the individual worker average exposures vary from about 30 to 70, about a twofold difference in risk. Although this may seem significant, it represents a very tight SEG. The between-worker variability might be thought of as the range of the seven means. The within-worker variability might be thought of roughly as the average length of the vertical bars or the average exposure range.[17]

Now examine the SEG in Figure 16-10. The ranges of exposures (GSD) for both the workers and the SEG are larger than in Figure 16-9. In addition, although the average exposure for the group is around 200 ppm, two workers have average exposures approaching 500 ppm while the majority (4) have average exposures of around 50 ppm, a tenfold difference. Somehow it does not seem reasonable to classify these workers as having similar risks.

So how can we define a SEG in a quantitative way? Rappaport (and many co-workers) have approached the problem using statistical analysis of variance models and other techniques. In addition, other authors have sought to look at the determinants of exposure. These approaches are discussed in the next chapter.

Figure 16-9. SEG of 7 workers, showing individual worker exposure ranges (2.5 %tile – 97.5 %tile) and having an overall mean exposure of 50, and a small GSD.

[17] Actually it is a bit more complicated than this as will be demonstrated in Chapter 17.

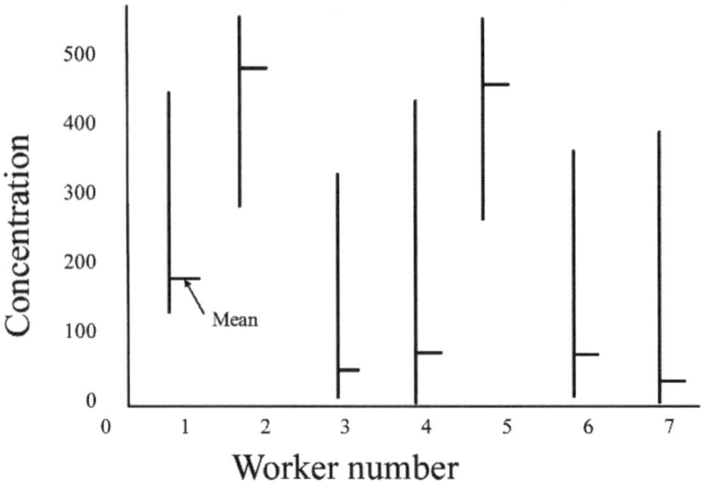

Figure 16-10. SEG of 7 workers, showing individual worker exposure ranges (2.5 %tile – 97.5 %tile) and having an overall mean exposure of 200, and a medium GSD.

Measurements < LOD

A common problem with summarizing sets of environmental data is that often some of the concentrations that are monitored are not detectable. This does not mean that they are zero, but only that our measuring devices cannot differentiate the levels that are present from the background noise of the instrument. These data are termed less than the limit of detection (< LOD). The overall distribution is said to be left-hand censored at the LOD; that is, this portion of the distribution is missing (see Figure 16-11). Then what is to be done with these data when a mean, for example, is calculated?

Several approaches have been tried. Hornung and Reed (1990) examined the bias associated with estimating the mean exposure when a single value was applied to all of the < LOD values. Two values, $0.5 \times$ LOD and $2^{0.5} \times$ LOD were compared at several levels of censoring (0–60%) and variability (GSD = 1.5, 2, 3).[18] They found that, in general, $2^{0.5} \times$ LOD performed best at GSD = 1.5 or 2 (biases for estimating the mean increased from -2% to 8% as censoring increased). At GSD = 3 the two values were essentially equivalent except at high levels of censoring (> 40%) where $0.5 \times$ LOD performed better.

Although the above approach is reasonable for a quick estimate of the mean, it is clear that an estimate of any variance term such as the GSD will be biased since the values that are being substituted are all the same. There are statistical approaches to solving this problem.

[18] Other approaches such as assigning all censored values = LOD or = zero are commonly used, but are obviously biased.

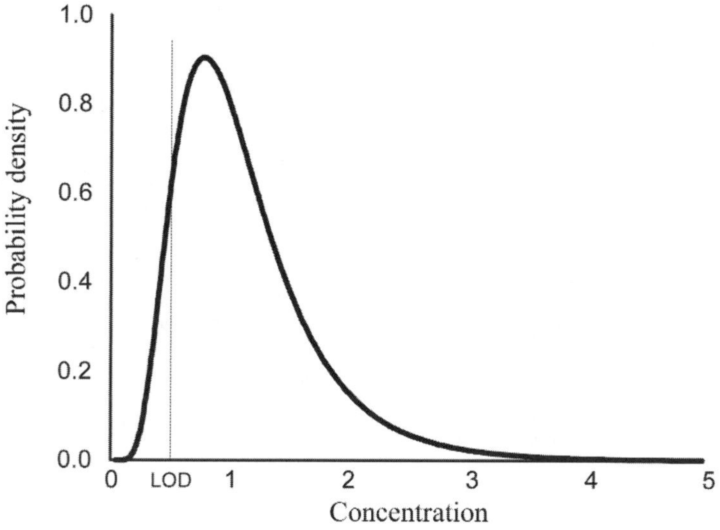

Figure 16-11. A lognormal exposure distribution showing the limit of detection (LOD). Values to the left of the LOD are said to be censored as their value cannot be accurately estimated by the analytical technique.

Cohen (1961, 1988) has devised a technique for estimating the mean and the standard deviation from a censored normal distribution. Although the technique appears to have general applicability to environmental data, the limitation is that many environmental data sets are not censored at a single point. The concentration of contaminant in air is a quotient, *detected mass/air volume sampled*. There is typically a single censoring point for the detected mass, the mass LOD (although this may vary as well if different labs are involved in the analysis), but the air volume sampled tends to vary from measurement to measurement. Therefore, each measurement has an individual concentration LOD. Attempts have been made to work around this issue (Perkins, 1990). However, there are better approaches.

For example, there is a technique for estimating the parameters of censored data sets based on regression analysis of expected order statistics (Newman et al., 1989 and 1992). This log probit regression (LPR) method is also described in AIHA (2006). However, a method of maximum likelihood (MLE) may be the best approach (Finkelstein and Verma, 2001). It requires iteration to find a result, but modern computers and spreadsheets make this task easy. In essence, the known values sampled from the distribution and the number of < LOD values are used to estimate a mean and standard deviation such that a distribution is constructed that most likely contains the < LOD values in the area of the tail below the LOD. The technique is based on a normal distribution and so the measured values and LOD are logged and the $\bar{x}_L$ and s_L are estimated.

Figure 16-12 shows a spreadsheet representation of how the calculations are performed.

	A	B	C
1	**Conc**	**Ln**	**ln liklihood**
2	53	=LN(A2)	=LN((1/((2*PI())^0.5*B$22))*EXP(-(1/2)*((LN(A2)-B$21)/B$22)^2))
3	66	=LN(A3)	=LN((1/((2*PI())^0.5*B$22))*EXP(-(1/2)*((LN(A3)-B$21)/B$22)^2))
4	76	=LN(A4)	=LN((1/((2*PI())^0.5*B$22))*EXP(-(1/2)*((LN(A4)-B$21)/B$22)^2))
5	101	=LN(A5)	=LN((1/((2*PI())^0.5*B$22))*EXP(-(1/2)*((LN(A5)-B$21)/B$22)^2))
6	110	=LN(A6)	=LN((1/((2*PI())^0.5*B$22))*EXP(-(1/2)*((LN(A6)-B$21)/B$22)^2))
7	133	=LN(A7)	=LN((1/((2*PI())^0.5*B$22))*EXP(-(1/2)*((LN(A7)-B$21)/B$22)^2))
8	161	=LN(A8)	=LN((1/((2*PI())^0.5*B$22))*EXP(-(1/2)*((LN(A8)-B$21)/B$22)^2))
9	188	=LN(A9)	=LN((1/((2*PI())^0.5*B$22))*EXP(-(1/2)*((LN(A9)-B$21)/B$22)^2))
10	208	=LN(A10)	=LN((1/((2*PI())^0.5*B$22))*EXP(-(1/2)*((LN(A10)-B$21)/B$22)^2))
11	227	=LN(A11)	=LN((1/((2*PI())^0.5*B$22))*EXP(-(1/2)*((LN(A11)-B$21)/B$22)^2))
12	265	=LN(A12)	=LN((1/((2*PI())^0.5*B$22))*EXP(-(1/2)*((LN(A12)-B$21)/B$22)^2))
13	295	=LN(A13)	=LN((1/((2*PI())^0.5*B$22))*EXP(-(1/2)*((LN(A13)-B$21)/B$22)^2))
14	321	=LN(A14)	=LN((1/((2*PI())^0.5*B$22))*EXP(-(1/2)*((LN(A14)-B$21)/B$22)^2))
15	766	=LN(A15)	=LN((1/((2*PI())^0.5*B$22))*EXP(-(1/2)*((LN(A15)-B$21)/B$22)^2))
16	1547	=LN(A16)	=LN((1/((2*PI())^0.5*B$22))*EXP(-(1/2)*((LN(A16)-B$21)/B$22)^2))
17	**35**	=LN(A17)	=LN(NORMDIST(LN(A17),B$21,B$22,TRUE))
18	**38**	=LN(A18)	=LN(NORMDIST(LN(A18),B$21,B$22,TRUE))
19	**40**	=LN(A19)	=LN(NORMDIST(LN(A19),B$21,B$22,TRUE))
20	**66**	=LN(A20)	=LN(NORMDIST(LN(A20),B$21,B$22,TRUE))
21	=AVERAGE(B2:B20)	4.9336	=SUM(C2:C20)
22	=STDEV(B2:B20)	1.01658	
23	MLE		
24	=EXP(B21+0.5*B22^2)		
25			
26			

Figure 16-12. A spreadsheet application of the maximum likelihood method for estimating the mean and standard deviation of logged measurements. Values in column A are raw data, with cells A17–A20 being the LOD values for those censored measurements. The sum in cell C21 is maximized, using a function like Excel Solver by allowing the computer to iterate by changing the estimates that are in cells B21 and B22. The initial estimates that are manually typed into B21 and B22 are obtained in cells A21 and A22. An estimate of the MLE for the mean is obtained in A24.

The difficulty is making sure that the formulas are entered correctly. Twenty measurements were taken. The four in bold near the bottom of column A are < LOD. Note that these bold values are the LODs for each sample. Since the sample volumes were not constant, the LODs varied. Note as well that one of the LODs (66) is actually larger than some of the data that were quantitated. The sample volume for this measurement was low.

In column B the concentration values and LODs are logged. In column C the likelihoods are calculated. For the quantitated data, Chapter 7, Equation 7-5 is used to calculate the probability density from a normal distribution given a value of x (in this case ln conc). For the LOD data, an Excel function is used that returns the cumulative density for the value x (ln LOD), given a value for the mean and standard deviation. The TRUE parameter in this function is simply a toggle that returns the cumulative rather than the probability function. The values for mean and standard deviation are in cells B21 and B22. Initially these cells are estimated based on the values calculated in cells A21 and A22 (the mean and standard deviation of the column of logs). The values in A21/A22 are typed into B21/B22. (Do not enter the formulas used to derive them).

The likelihoods in column C are summed in cell C21. It is this value that we want to maximize. The spreadsheet does so by changing the values of mean and standard deviation in B21 and B22. It changes these values very rapidly – hundreds or even thousands of times – until it finds a maximum in the sum of likelihoods. In Excel, this is accomplished using an *add in* tool (you must install it separately) called *Solver*. Once *Solver* is run and the likelihood sum is maximized based on *Solver* changing the estimates of the mean and standard deviation, the last values to appear in cells B21 and B22 are the maximum likelihood estimates of the mean and standard deviation of the logged values. In cell A24 a formula has been added that calculates the MLE of the mean of concentrations (Equation 16-10). The GSD is estimated by exponentiating (inverse ln) the final value in cell B22. An MVU for the mean (Equation 16-11 may be calculated as well as confidence limits on the mean or exceedance fraction.

This method can be used for all exposure data (gases and vapors, mold counts, aerosols, radiation, noise, etc.) It will work for multiple LODs. It can be used for any amount of censoring. However, it is probably not wise to rely on any method when censoring exceeds 50%, i.e., more than half the measurements are < LOD.

FINAL NOTE

This chapter has presented various tools for exposure assessment. It should be apparent that the problem is more complicated than is perhaps commonly thought. Moreover, the tools must somehow be tied together or applied in a logical fashion, which is the subject of chapter 18. However, the reader must be warned that there is no reliable, simplified method for doing this. In other words, an understanding of the basis, application, and limitations of these methods is a prerequisite for performing economically efficient and technically sound exposure assessment. Finally, it should be obvious to the reader that a single reading of this chapter will not allow for the formulation of sound exposure assessment strategies. Many have stated that professional judgment plays a significant role in exposure assessment. It should be apparent from Chapters 12 through 18 that this must be the case and that thorough professional knowledge and understanding of the tools outlined in these chapters must be coupled with experience to maximize worker protection.

QUESTIONS

16.1. For the following normally distributed data (7.7, 6.1, 10.3, 10, 13.1, 8.3, 13.8, 7.2, 10.6, 10.7, 9.4, 12.2, 10.3, 12.9, 7.4, 7.7, 10.8, 7.0, 10.3, 6.8) calculate:

a. The 95th percentile.

b. The 95% UCL on the mean, assuming σ is estimated.

c. The 95% UCL on the 95th percentile.

d. Why is the answer for (c) so different from that for (a)?

16.2. An OSHA officer measures a concentration of benzene of 1.1 ppm for a personal sample. The PEL is 1 ppm, and the *CV* for the benzene measurement method is 10%. Should the officer issue a citation based on this information?

16.3. For the following lognormally distributed data (10, 6, 8, 25, 3, 15, 21, 10, 8, 5):

a. What is the arithmetic mean?

b. What is the GSD?

c. What is the *MLE* of the mean?

d. What is the *MVU* estimate of the mean?

16.4. Based on Table 16-5, calculate an acceptable sample size if you suspect a geometric standard deviation for the population of concentrations of about 1.7 and a mean of concentration of about 25 ppm, and if the OEL is 100 ppm.

16.5. If you collect the calculated number of measurements for the scenario in question 16.4., what power will you have to reject the null hypothesis, if in fact it should be rejected and your assumptions (GSD) are correct?

16.6. If the 8-hr TWA OEL is 20 ppm, the mean of these exposures is controlled to 10 ppm, and the 15-min STEL is 50 ppm, then the STEL will be exceeded what maximum percentage of measurements?

16.7. Using the data in question 16.3., assume that the LOD is 6.5, so that some of the data given are actually < LOD. Recalculate the GSD and *MVU* using the maximum likelihood method (you will need spreadsheet software). Are the MLE, MVU and GSD reasonably close to the original values in answer 16.3? What causes this difference, regardless of its size?

16.8. A sample of 8-hr TWA measurements is taken for comparison with an OEL of 10. The MLE for the mean 8-hr TWA is 3, and the estimate of GSD (s_g) is 2. Estimate the s_L and $\bar{x}_L$ from equations in Chapter 7, Table 7-1. From these, use Equation 16-19 to find the z-score for the exceedance fraction (γ). Consulting a Z table find that the probability that an exposure will exceed the OEL of 10.

16.9. For the previous problem assume that the sample size was 10. Use Figure 16-8 to determine the 95% upper confidence limit on the γ calculated in question 16.8.

REFERENCES

AIHA: A strategy for assessing and managing occupational exposures, 3rd ed, B. Bullock and J.S. Ignacio (eds.). American Industrial Hygiene Association, Fairfax, VA (2006).

Armstrong, B.G.: Confidence intervals for arithmetic means of log normally distributed exposures. Am. Ind. Hyg. Assoc. J. 53:481-485 (1992).

Attfield, M.D.; Hewett, P.: Exact expressions for the bias and variance of estimators of the mean of a lognormal distribution. Am. Ind. Hyg. Assoc. J. 53:432-435 (1992).

Behar, A.; Eng, P.; Plener, R.: Noise exposure-sampling strategy and risk assessment. Am. Ind. Hyg. Assoc. J. 45:105-109 (1984).

Breslin, A.J.; Ong, L.; Glauberman, H.; George, A.C.; Leclare, P.: The accuracy of dust exposure estimates obtained from conventional air sampling. Am Ind. Hyg. Assoc. J. 28: 56-61 (1967).

Brunn, I.O.; Campbell, J.S.; Hutzel, R.L.: Evaluation of occupational exposures: A proposed sampling method. Am. Ind. Hyg. Assoc. J. 47:229 (1986).

Buringh, E.; Lanting, R.: Exposure Variability in the Workplace: Its Implications for the Assessment of Compliance. Am. Ind. Hyg. Assoc. J. 52(1):6-13 (1991).

Carter, S.R.: An estimation of autocorrelation and spatial correlation from 8-hour time-weighted average exposures. Research Project, The University of Alabama at Birmingham, School of Public Health (1990).

Chatfield, C.: The Analysis of Time Series: An Introduction, 3rd ed. Chapman and Hall, New York (1984).

Checkoway H.; Rice, H.C.: Time-weighted averages, peaks, and other indices of exposure in occupational epidemiology. Am. J. Ind. Med. 21:25-33 (1992).

Cohen, A.C., Jr.: Tables for maximum likelihood estimates: Singly truncated and singly censored samples. Technometrics 3:535-541 (1961).

Cohen, A.: Censored, truncated, and grouped estimation, In Lognormal Distributions, E.L. Crow, K. Shimizu (eds.). Marcel Dekker, New York (1988).

Flynn, M.R.: A stochastic differential equation for exposure yields a beta distribution. Ann. Occup. Hyg. 48:491-497 (2004a).

Flynn, M.R.: The 4-parameter lognormal (SB) model of human exposure. Ann. Occup. Hyg. 48:617-622 (2004b).

Finkelstein, M.M.; Verma, D.K.: Exposure estimation in the presence of nondetectable values: another look. AIHAJ 62:195-198 (2001).

Francis, M.; Selvin, S.; Spear, R.; Rappaport, S.: The effect of autocorrelation on the estimation of workers' daily exposures. Am. Ind. Hyg. Assoc. J. 50:37-43 (1989).

George, D.K.; Flynn, M.R.; Harris, R.L.: Autocorrelation of interday exposures at an automobile assembly plant. AIHAJ 56:1187-1194 (1995).

Gilbert, R.O.: Statistical Methods for Environmental Pollution Monitoring. Van Nostrand Reinhold, New York (1987).

Grzebyk, M.; Thiery, L.: Confidence intervals for the mean of sound exposure levels. AIHAJ 64:640-645 (2003).

Hewett, P.: Mean Testing: II. Comparison of Several Alternative Procedures. Appl. Occup. Environ. Hyg. 12:347-355 (1997).

Hewett, P.: Sample size formulae for estimating the true arithmetic of geometric mean of lognormal exposure distributions. AIHAJ 56:219-225 (1995).

Hewett, P.; Ganser, G.H.: Simple procedures for calculating confidence intervals around the sample mean and exceedance fraction derived from lognormally distributed data. Appl. Occup. Environ. Hyg. 12:132-142 (1997).

Hopson, N.L.: A study of autocorrelation in a population of occupational noise exposures. Research Project, The University of Alabama at Birmingham, School of Public Health (1990).

Hornung, R.W.; Reed, L.D.: Estimation of average concentration in the presence of nondetectable values. Appl. Occup. Env. Hyg. 5:46-51(1990).

Kumagai, S.; Matsunaga, I.; Kusaka, Y.: Autocorrelation of short-term and daily average exposure levels in workplaces. Am. Ind. Hyg. Assoc. J. 54:341-350 (1993).

Kumagai, S.; Matsunaga, I.: Approaches for estimating the distribution of short-term exposure concentrations for different averaging times. Ann. Occup. Hyg. 38:815-825 (1994).

Kumagai, S.; Matsunaga, I.: Changes in the distribution of short-term exposure concentration with different averaging times. AIHAJ 56:24-31 (1995).

Land, C.E.: Tables of confidence limits for linear functions of the normal mean and variance. Selected Tables in Mathematical Statistics 3:385-419 (1975).

Leidel, N.A.; Busch, K.A.: Statistical design and data analysis requirements, In Patty's Industrial Hygiene and Toxicology, Vol. 3A, 2nd ed. Theory and Rationale of Industrial Hygiene Practice: The Work Environment, L.J. Cralley, L.V. Cralley (eds.). Wiley Publishing, New York (1985).

Limpert, E.; Stahel, W.A.; Abbt, M.: Lognormal distribution across the sciences: keys and clues. BioScience 51:341-352 (2001).

Lyles, R.H.; Kupper, L.L.: On strategies for comparing occupational exposure data to limits. AIHAJ 57:6-15 (1996).

Lyon, B.F.; Land, C.E.: Confidence Limits for Linear Functions of the Normal Mean and Variance http://conflimit.nci.nih.gov/ (2005).

Martinez, J.; Iglewicz, B.: A test for departure from normality based on a biweight estimator of scale: Biometrika 68:331-333 (1981).

McBride, J.B.: A study of spatial correlation in a workplace exposure zone, Master's Research Project. School of Public Health, The University of Alabama at Birmingham (1991).

Newman, M.C.; Dixon, P.M.; Looney, B.B.; Pinder, J.E., III: Estimating mean and variance for environmental samples with below detection limit observations. Water Resources Bull. 25:905-915 (1989).

Newman, M.C.; Greene, D.; Dixon, P.M.; Looney, B.B.; Segal, C.: UNCENSOR@v.3.0. Savannah River Ecology Laboratory, Aiken, SC (Nov. 1992).

Nicas, M.; Simmons, B.P.; Spear, R.C.: Environmental versus analytical variability in exposure measurements. Am. Ind. Hyg. Assoc. J. 52:553-557 (1991).

NIOSH: Exposure Measurement Action Level and Occupational Environmental Variability, N.A. Leidel, K.A. Busch, W.E. Crouse (eds.). National Institute for Occupational Safety and Health, Cincinnati, OH (1975).

NIOSH: Occupational Exposure Sampling Strategy Manual, N.A. Leidel, K.A. Busch, J.R. Lynch (eds.). National Institute for Occupational Safety and Health, Cincinnati, OH (1977).

OSHA: Occupational exposure to benzene; final rule. 29 CFR Part 1910. Fed. Reg. 52:34540 (1987).

Osvoll, P.O.; Woldbæk, T.: Distribution of skewness of occupational exposure sets of measurements in the Norwegian industry. Ann. Occup. Hyg 43:421-428 (1999).

Perkins, J.L.; Cutter, G.N.; Cleveland, M.S.: Estimating the mean, variance, and confidence limits from censored (<Limit of Detection), lognormally-distributed exposure data. Am. Ind. Hyg. Assoc. J. 51:416-419 (1990).

Preller, L.; Burstyn, I.; De Pater, N.; Kromhout, H.: Characteristics of peaks of inhalation exposure to organic solvents. Ann. Occup. Hyg. 48:643-652 (2004).

Rappaport, S.M.: Smoothing of exposure variability at the receptor: Implications for health standards. Ann. Occup. Hyg. 29:201 (1985).

Rappaport, S.M.: Biological considerations for designing sampling strategies. Symposium on Advances in Air Sampling, Pacific Grove, California. American Conference of Governmental Industrial Hygienists, Cincinnati, OH (1987).

Rappaport, S.M.: Review: Assessment of long-term exposures to toxic substances in air. Ann. Occup. Hyg. 35:61-121 (1991).

Rappaport, S.M.; Lyles, R.H.; Kupper, L.L.: An exposure-assessment strategy accounting for within- and between-worker sources of variability. Ann. Occ. Hyg. 39:469-495 (1995).

Rappaport, S.M.; Selvin, S.: A method for evaluating the mean exposure from a log normal distribution. Am. Ind. Hyg. Assoc. J. 48:374 (1987).

Rappaport, S.M.; Selvin, S.; Spear, R.C.; Keil, C.: An evaluation of statistical schemes for air sampling, In Chemical Hazards in the Workplace. American Chemical Society, New York (1981).

Rappaport, S.M.; Selvin, S.; Roach, S.A.: A strategy for assessing exposures with reference to multiple limits. Appl. Ind. Hyg. J. 3:310 (1988).

Roach, S.A.: A more rational basis for air sampling programs. Am. Ind. Hyg. Assoc. J. 27:1 (1966).

Rock, J.C.: A comparison between OSHA-compliance criteria and action-level decision criteria. Am. Ind. Hyg. Assoc. J. 43:297 (1982).

Snedecor, G.W.; Cochran, W.G.: Statistical Methods, 6th ed., p.113, Iowa State University Press, Ames, IA, USA (1967).

Spear, R.C.; Selvin, S.; Francis, M.: The influence of averaging time on the distribution of exposures. Am. Ind. Hyg. Assoc. J. 47:365 (1986).

Symanski, E.; Rappaport, S.M.: An investigation of the dependence of exposure variability on the interval between measurements. Ann. Occup. Hyg. 38:361-372 (1994).

Symanski, E.; Kupper, L.L.; Kromhout, H.; Rappaport, S.M.: An investigation of systematic changes in occupational exposure. AIHAJ 57:724-735 (1996).

Waters, M.A.; Se1vin, S.; Rappaport, S.M.: A measure of goodness-of-fit for the lognormal model applied to occupational exposures. Am. Ind. Hyg. Assoc. J. 52:493-499 (1991).

Wong, O.: An industry wide mortality study of chemical workers occupationally exposed to benzene II. Dose-response analyses. Br. J. Ind. Med. 44:382-395 (1987).

17

QUANTITATIVE SURVEYING – APPLICATION OF VARIANCE COMPONENT MODELS TO EXPOSURE ASSESSMENT

Elaine Symanski, PhD

VARIATION IN EXPOSURE FOR A SINGLE WORKER

Let's begin by evaluating for a single worker the variation in shift-long exposure to an airborne contaminant over time. Figure 17-1 plots the logged exposures for a worker exposed to inorganic lead over a period of approximately 30 days (data from Cope et al., 1979). For the i-th worker, the following model could be constructed to represent the logged exposures over time:

$$Y_j = ln(X_j) = \mu_i + e_j \quad j = 1,2,..., n \text{ days} \qquad \text{[Equation 17-1]}$$

where,

X_j = The shift-long exposure concentration of the contaminant for an individual worker on the j-*th* day

Y_j = The natural logarithm of the shift-long exposure concentration of the contaminant for an individual worker on the j-*th* day

μ_i = Mean exposure level for the i-*th* worker on the logarithmic scale

e_j = Deviation of the observed value from the true mean, μ_i

In this model, the natural logarithm of the measured concentration of a contaminant for a worker is the sum of two components: a component related to the worker's true mean value on a logarithmic scale and a component related to random fluctuations over time. In Equation 17-1, we assume that e_j is normally distributed with a mean 0 and variance σ^2_W. Using statistical notation, $e_j \sim N(0, \sigma^2_W)$. It directly follows that $Y_j \sim N(\mu_i, \sigma^2_Y)$. Note that $\sigma^2_Y = \sigma^2_W$; σ^2_W represents the variation in a worker's shift-long exposures about his or her true mean.

It is worthwhile to note that a similar formulation of Equation 17-1 might be constructed

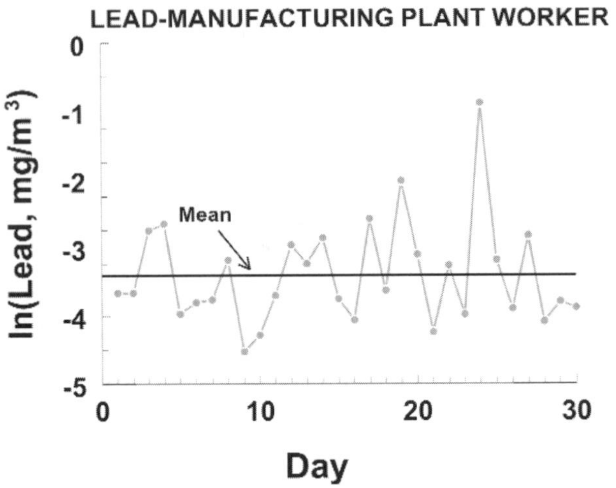

Figure 17-1. Time plot of shift-long exposures to inorganic lead (on a logarithmic scale) (data from Cope et al., 1979).

to represent a situation whereby a group of workers are all measured once. In this case, the error term reflects variation between workers, as well as variation within workers (i.e., variation in shift-long exposures from day-to-day). To evaluate the relative contributions of these two sources of variation, we would collect multiple measurements on a group of workers – this is the topic of the next section.

VARIATION IN EXPOSURE FOR A GROUP OF WORKERS

Let us assume that we are focused on making statements about exposure for a group of workers. In this context, an important question must be addressed. Are we interested in making inferences about exposures for the workers, and only those workers, on whom exposure measurements have been collected? Or, are we interested in making inferences about the occupational group of workers from which a random sample of individuals was selected? In the former case, inferences would be focused on evaluating whether there are differences in the mean exposure levels among workers represented in the data and in the latter instance, inferences would be focused on assessing the degree of variability among workers' mean levels within the population from which the sample of workers was drawn.

Analysis of Variance (ANOVA) Models

In either case, we might consider applying analysis of variance (ANOVA) models to examine variability in exposure and attribute sources of variability to different categorizations of the data. ANOVA models are well described in statistical textbooks (e.g., see Neter et al.,

1996). Briefly, in categorizing the data, the classification variables are referred to as factors and are specified by levels of the factor. For example, in our context, a factor under investigation might be 'worker' and levels of the factor would be particular workers, i.e., Worker 1, Worker 2, etc., where multiple exposure measurements would be collected on each worker.

Fixed-Effects ANOVA Model

Let's assume that we are only interested in the exposures received by a *specific* set of workers. To represent the exposures received by these workers, the following model could be constructed:

$$Y_{ij} = ln(X_{ij}) = \mu_Y + \alpha_i + e_{ij} \qquad \textit{for } i = 1,\ 2,...,k \textit{ workers and } j = 1,2,...,n_i \textit{ days} \qquad \text{[Equation 17-2]}$$

where,

X_{ij} = The shift-long exposure concentration for the i-*th* worker on the j-*th* day

Y_{ij} = The natural logarithm of the shift-long exposure concentration for an individual worker on the j-*th* day

μ_Y = Mean exposure level for the group of k workers; i.e.,

$$\mu_Y = \frac{\mu_1 + \mu_2 + ... + \mu_k}{k}$$

α_i = The 'fixed' effect due to the i-*th* worker; thus, a worker's mean value is represented by $\mu_i = \mu_Y + \alpha_i$

e_{ij} = The deviation of the j-*th* measurement from the i-*th* worker's true mean value, μ_i

An important property of the model is that the fixed effects, α_i, are population parameters where $\alpha_1, \alpha_2, ..., \alpha_k$ sum to zero; i.e., $\sum_{i=0}^{k} \alpha_i = 0$. Also, we assume the e_{ij}'s are independently and normally distributed with zero mean and variance, σ_W^2, i.e., $e_{ij} \sim N(0, \sigma_W^2)$. It follows that $E(Y_{ij}) = \mu_Y + \alpha_i$ and $\mathrm{Var}(Y_{ij}) = \mathrm{Var}(e_{ij}) = \sigma_W^2$. In other words, $Y_{ij} \sim N(\mu_Y + \alpha_i, \sigma_W^2)$. In this model, the observations on each of the workers come from normal probability distributions that are characterized by a common variance. It directly follows that the $Cov(Y_{ij}, Y_{ij'}) = \sigma_W^2 \ for\ j = j'$ and zero otherwise. In other words, the observations collected on the same worker are independent, as well as observations collected on different workers. Finally, note that Equation 17-2 accommodates unbalanced data (i.e., unequal numbers of observation for the k workers on whom measurements have been collected). In the case where the data are balanced, $n_1 = n_2 = ... = n$.

The Structure of the Data

We might view the structure of the multiple measurements collected on k workers as shown in Table 17-1. In an analysis of variance (ANOVA) table, we can partition the total variability in exposure into two sources of variability – variation between workers and variation within workers (see Table 17-2). The analysis of variance develops because we make different estimates of the variance which are functions of the sums of squares associated with each source of variation. Under the assumptions of the model, the Mean Square Within (MSW) and the Mean Square Between (MSB) are independent random variables because the MSW depends only on the variance of the samples whereas the MSB depends only on their means. Since the mean and variance of a sample from a normal population are independent, it follows that the MSW and MSB are also independent.

Focusing on our question regarding whether mean exposure levels differ between the k workers in our sample, we can apply an F-test. The F-test is a test of the null hypothesis that the population mean values of the k workers are all identical, i.e., H_0: $\alpha_1 = \alpha_2 = \ldots = \alpha_k$. We can construct an F-statistic as a ratio of the mean square between to the mean square within, i.e., $F = \frac{MSB}{MSW}$ with k-1 and N-1 degrees of freedom with k equal to the number of workers and N equal to the total number of measurements. The value of the F statistic should be around 1 when the null hypothesis holds, and should become large when the mean values differ substantially. Using hypothesis testing, this then allows for a formal evaluation of differences among mean exposure levels for workers on whom measurements were collected. Note, however, that rejection of the null hypothesis that the mean values are not the same does not provide us with information about which pair or pairs of mean values differ. To evaluate this latter question, multiple comparison procedures must be used – readers are referred to Neter et al. (1996) (Chapter 17) for details regarding the options that are available.

Table 17-1
Structure of the monitoring data for use in an ANOVA model

Worker	Sample size	Observations	Total (Sum)	Sample mean
1	n_1	$Y_{11}, Y_{12}, \ldots, Y_{1n_1}$	$T_1 = Y_1$	$\bar{Y}_1 = T_1/n_1$
2	n_2	$Y_{21}, Y_{22}, \ldots, Y_{2n_2}$	$T_2 = Y_2$	$\bar{Y}_2 = T_2/n_2$
3	n_3	$Y_{31}, Y_{32}, \ldots, Y_{3n_3}$	$T_3 = Y_3$	$\bar{Y}_3 = T_3/n_3$
…	…			
k	n_k	$Y_{k1}, Y_{k2}, \ldots, Y_{kn_k}$	$T_k = Y_k$	$\bar{Y}_k = T_k/n_k$
	$N = \sum_{i=1}^{k} n_i$		$G = T_1 + T_2 + \ldots T_k$	$\bar{Y} = G/N$

Table 17-2
The Analysis of Variance Table

Source	Degrees of freedom (d.f.)	Sum of Squares	Mean Square	F*
Between workers	k–1	$SSB = \sum_{i=1}^{k} n_i (\bar{Y}_i - \bar{Y})^2$	$MSB = \frac{SSB}{(k-1)}$	$\frac{MSB}{MSW}$
Within workers	N–k	$SSW = \sum_i \sum_j (Y_{ij} - \bar{Y}_i)^2$	$MSW = \frac{SSW}{(N-k)}$	
Total	N–1	$TSS = \sum_{i=1}^{k} \sum_{j=1}^{n_i} (Y_{ij} - \bar{Y})$		

* $F \sim F_{N-k}^{k-1}$. Under the null hypothesis of no differences in worker's mean levels, F is distributed according to Fisher's F-distribution with k-1 and N-k degrees of freedom for the numerator and denominator, respectively.

Clearly, applying the one-way fixed analysis of variance model has implications regarding our sampling strategy, namely that all members of a population are monitored. Thus, this approach becomes more inefficient as the population size increases and would require a considerable measurement effort if inferences are focused on a large number of individuals.

Random-Effects ANOVA Model

An alternative approach would be to identify an occupational group of workers (however classified) and collect measurements on a representative sample from the group. In this instance, inferences are to be made about the population from which the workers are sampled. To evaluate variation in exposure between and within workers, the following one-way random effects model could be applied:

$$Y_{ij} = ln(X_{ij}) = \mu_Y + b_i + e_{ij} \qquad for\ i = 1,\ 2,...,k\ workers\ and\ j = 1,2,...,n_i\ measurements$$ [Equation 17-3]

where,

X_{ij} = The shift-long exposure concentration for the i-*th* worker on the j-*th* day

Y_{ij} = The natural log-transformed value of the exposure concentration

μ_Y = Overall fixed job group mean (mean of Y_{ij})

b_i = The random effect measuring the deviation of the i-*th* worker's true mean exposure from μ_Y, and

e_{ij} = The random effect measuring the deviation of j-*th* measurement from the i-*th* worker's mean exposure ($\mu_Y + b_i$)

Note that the fixed-effects analysis of variance model (Equation 17-2) and the one-way random-effects model (Equation 17-3) appear the same. However, they are quite different because of the way in which we define the terms of the model. In other words, we define the 'worker' effect, b_i in the one-way random-effects model as a random variable where workers are viewed as a random sample from a larger population; in contrast, in Equation 17-2, the 'worker' effect, α_i is defined as a fixed effect.

Under the one-way random-effects model, the logged value of the shift-long exposure concentration for any individual in the group on a given day is represented as the sum of three components, a component related to the overall mean exposure for the group, which is a fixed but unknown constant, a component related to random variation between individuals and a component related to random variation from day-to-day. It is assumed under the model that both the random worker effect b_i and the random day-to-day effect e_{ij} are distributed normally with zero means and variances, σ_B^2 and σ_W^2. Note that σ_B^2 and σ_W^2 refer to the between- and within-worker variance components, respectively. It follows that $E(Y_{ij}) = \mu_Y$ *for all i,j* and that $Var(Y_{ij}) = \sigma_Y^2 = \sigma_B^2 + \sigma_W^2$. Thus, the one-way random-effects model is an example of a variance component model since it partitions the total variance into two components. It directly follows that $Cov(Y_{ij}, Y_{ij'}) = \sigma_B^2$ *for* $j \neq j'$. Thus, under the one-way random-effects model, measurements collected on the same person are correlated. Specifically, the correlation between pairs of measurements from the same worker is the same (and positive) regardless of the time interval separating them, i.e., the $Cov(Y_{ij}, Y_{ij'}) = [\sigma_B^2 / (\sigma_B^2 + \sigma_W^2)] > 0$ *for* $j \neq j'$. Note that $\sigma_B^2 / (\sigma_B^2 + \sigma_W^2)$ is also referred to as the intra-class correlation coefficient. Under this correlation structure, which is referred to as compound symmetry, we assume independence among measurements collected on different workers but dependence among measurements collected on the same worker. Other covariance structures may be more appropriate and are discussed later in the chapter. For more discussion of the random-effects model, readers are referred to Searle et al., 1992.

To apply the one-way random-effects model, it becomes necessary to collect repeated measurements on a random sample of workers in an occupational group, however that group is classified. Here, 'repeated' implies multiple measurements collected on the same individual – thus, a group of workers are monitored repeatedly over time. The group may be defined by job title, location or a combination of any factors that identify a particular set of workers. Thus, how we define the group has implications for the degree to which exposures within the group vary between workers. Once exposure measurements are collected on a random sample of workers, the data are used to obtain estimates of the mean exposure for the group (*i.e.*, μ_Y) as well as the within- and between-worker variance components (i.e., σ_W^2 and σ_B^2). Moreover, in focusing on our question regarding the degree of variation between workers within an occupational group, the same F statistic that was used to evaluate the hypothesis, H_0: $\alpha_1 = \alpha_2 = ... = \alpha_k$ in the fixed-effects model (Equation 17-2) is used in the random-effects model to evaluate H_0: $\sigma^2_B = 0$. A cautionary note is needed here, however, since the ratio of the mean square between to the mean square within (i.e., the F statistic) does *not* have an F distribution when using unbalanced data (unless $\sigma^2_B = 0$) (Searle et al., 1992). Thus, inferential testing regarding the hypothesis that the between-worker variance component is zero is restricted to balanced data.

Applying One-Way Random-Effects Models to Characterize Exposure Variability

In a comprehensive investigation of exposure variability, Kromhout et al. (1993) utilized nearly 14,000 personal exposure measurements and applied the one-way random-effects model to characterize the between- and within-worker sources of variation in exposure for 165 groups of workers classified on the basis of job title and location (factory). Across all groups of workers, the median value of the between-worker variance component was smaller, on average, as compared to the within-worker variance component suggesting that there was typically greater variability from one sampling shift to another than between workers for an occupational group. In a multivariate analysis, Kromhout et al. (1993) also reported that job groups performing work outdoors with an intermittent process generating the exposure had higher levels of within-worker variability than job groups in indoor workplaces, irrespective of whether the process was intermittent or continuous ($R^2 = 0.41$). In contrast, the only variable that contributed significantly to between-worker variability was the nature of the process, with lower levels for job groups characterized by continuous rather than intermittent processes ($R^2 = 0.13$).

Following the work carried out by Kromhout and colleagues, numerous other investigations have also partitioned exposure variability using similar modeling techniques (Burdorf and van Tongeren, 2003). A recent systematic review of the literature provides a cogent summary of studies that have applied the one-way random-effects model to quantify the sources of variation in exposure to chemical agents (Symanski et al., 2006). While the majority of studies have focused on variability in airborne exposures, investigations have also been extended to evaluate dermal exposures (Kromhout and Vermeulen, 2001), as well as biological monitoring data (Symanski and Greeson, 2002). Additionally, within- and between-worker sources of variability have been examined in the context of physical agents (Kromhout et al., 1995) and physical work load (Burdorf, 1992).

Should 'Worker' be Evaluated as a Fixed or Random Effect?

In determining whether 'worker' should be viewed as fixed or random, the purpose of the exposure assessment and the nature of the sampling strategy used to determine which workers were (or are to be) sampled must be considered. As noted above, if the workers that are (were) identified for sampling are (were) viewed as the population of interest, then they are (were) selected for specific reasons and we would refer to the worker effect as 'fixed' in the analysis of variance model. In this instance, we would be interested in making inferences regarding the mean exposure levels of the workers on whom measurements were collected; thus, a fixed-effects analysis of variance model would be appropriate. On the other hand, if the workers are viewed as coming from a larger (and possibly infinite) set of workers in an occupational group, of which only a random sample occur in the data, then we would refer to the worker effect as 'random'. In this case, interest would lie in estimating the variance of the population from which the workers were drawn and a one-way random-effects model would be applied. Thus, decisions regarding whether to apply a fixed- or random-effects model will depend on how we define the population and what inferences are to be made. Table 17-3 summarizes salient characteristics of both models, providing details regarding where the important differences lie.

Averaging Time in the Context of ANOVA Models

As discussed in Chapter 14, exposure measurements may be collected over different averaging times depending on the purpose of the exposure assessment. While Equations 17-1, 17-2, and 17-3 have been illustrated in the context of shift-long exposure measurements, these models could easily be applied to monitoring data collected over shorter intervals. Examples of studies evaluating between-worker and within-worker (i.e., within-shift) sources of variability using short-term measurements include Burdorf et al. (1994), Eduard and Bakke (1999),

Table 17-3
Characteristics of the one-way classification for the fixed-effects model and the random-effects model to evaluate exposures in an occupational group of workers

Feature	Fixed-Effects Model	Random-Effects Model
Model equation	$Y_{ij} = \mu_Y + \alpha_i + e_{ij}$	$Y_{ij} = \mu_Y + b_i + e_{ij}$
Mean of Y_{ij}	$E(Y_{ij}) = \mu_Y + \alpha_i$	$E(Y_{ij}) = \mu_Y$
'Worker' factor	Fixed effect, unknowable constant: α_i	Random effect: $b_i \sim N(0, \sigma_B^2)$
Error term	$e_{ij} = Y_{ij} - E(Y_{ij}) = Y_{ij} - (\mu_Y + \alpha_i)$ $e_{ij} \sim N(0, \sigma_W^2)$	$e_{ij} = Y_{ij} - E(Y_{ij}) = Y_{ij} - (\mu_Y)$ $e_{ij} \sim N(0, \sigma_W^2)$
Variance of Y_{ij}	$Var(Y_{ij}) = \sigma_W^2$	$Var(Y_{ij}) = \sigma_W^2 + \sigma_B^2$
Covariance of Y_{ij}, Y_{ij}	$Cov(Y_{ij}, Y_{ij'}) = \begin{cases} \sigma_W^2 & for\ j = j' \\ 0 & Otherwise \end{cases}$	$Cov(Y_{ij}, Y_{ij'}) = \begin{cases} \sigma_W^2 + \sigma_B^2 & for\ j = j' \\ \sigma_B^2 & for\ j \neq j' \\ 0 & Otherwise \end{cases}$

and van der Woord et al. (1999). Note, however, that applying models to measurements of mixed sampling times in a single analysis is not appropriate since the underlying population of exposures experienced over full work shifts differs from that for exposures with shorter averaging times (Symanski et al., 2006).

METHODS FOR ESTIMATING $\sigma_B^2 + \sigma_W^2$ UNDER THE ONE-WAY RANDOM-EFFECTS MODEL

In Chapter 18, different methods of estimation of the parameters of the lognormal distribution are discussed. Similarly, a variety of different methods of estimation are available for generating estimates of the between- and within-worker variance components, including method of moments estimation (the Analysis of Variance (ANOVA) method), Minimum Norm Estimation, Maximum Likelihood (ML) method of estimation, and Restricted Maximum Likelihood (REML) method of estimation. Based on a review of the literature in which variance component estimates were obtained using the one-way random-effects model (Symanski et al., 2006), investigators in the occupational arena have typically applied either REML or the ANOVA method of estimation. What method of estimation should be applied? The answer regarding which method is preferable may depend on which property or set of properties is viewed as more important, e.g., unbiasedness, consistency or sufficiency. In the balanced case, ANOVA and REML estimators are identical and share both unbiasedness and minimum variance properties. For unbalanced data, Searle et al. (1992) (Sec. 6.8) summarize the properties of estimators of variance components in which they favor maximum likelihood estimators over ANOVA estimators because they are consistent and asymptotically normal; however, they do not make a strong case in favor of REML over ML or vice versa.

Method of Moments Estimation

The ANOVA method of estimation has intuitive appeal since it is based on deriving the expected values of the Sum of Squares Between and the Sum of Squares Within, equating the observed and expected values and then solving for the estimators. In other words, we construct an ANOVA table (as before under the fixed-effects ANOVA model), derive the expected values of the expected mean squares, which are linear functions of the variance components, and then use those relationships to derive estimators for the variance components (see Table 17-4).

Let's first examine the situation in which we have balanced data (i.e., when there are equal numbers of measurements collected on each worker). Here, the expected values of the sum of squares between and sum of squares within are shown below (Searle et al., 1992):

$$E(SSB) = (k-1)(n\sigma_B^2 + \sigma_W^2) \qquad \text{[Equation 17-4]}$$
$$E(SSW) = k(n-1)\sigma_W^2$$

where, as before, k refers to the number of workers and n refers to the number of measure-

Table 17-4
The expected value of the Mean Squares under the one-way random effects model (balanced and unbalanced cases)

Source of Variation	d.f.*	Sum of Squares	Mean Square (MS)	Balanced case	Unbalanced case
Between workers	k–1	$\sum_{i=1}^{k}\sum_{j=1}^{n_i}(Y_i - \bar{Y})^2 = n_i\sum_{j=1}^{n_i}(\bar{Y}_i - \bar{Y})^2$	$MSB = \frac{SSB}{k-1}$	$\sigma_W^2 + n\sigma_B^2$	$\sigma_W^2 + \frac{N - \frac{\sum_{i=1}^{k} n_i^2}{N}}{k-1}\sigma_B^2$
Within workers	N–k	$\sum_{i=1}^{k}\sum_{j=1}^{n_i}(Y_{ij} - \bar{Y}_i)^2$	$MSW = \frac{SSW}{N-k}$	σ_W^2	σ_W^2
Total	N–1	$\sum_{i=1}^{k}\sum_{j=1}^{n_i}(Y_{ij} - \bar{Y})^2$			

$^{*}N = \sum_{i=1}^{k} n_i$

ments collected on each worker (note that k*n refers to the total number of measurements). Thus, by equating the observed and expected values of the mean squares, we obtain the following ANOVA estimators of σ_W^2 and σ_B^2:

$$\hat{\sigma}_W^2 = MSW$$
$$\hat{\sigma}_B^2 = \frac{1}{n}(MSB - MSW) \qquad \text{[Equation 17-5]}$$

In Equation 17-5, note that $\hat{\sigma}_B^2$ is negative when MSB < MSW. Since variances can never be negative, obtaining a negative value for an estimate of $\hat{\sigma}_B^2$ suggests either that the model is not appropriate or that the true value of the between-worker variance component is zero (Searle et al., 1992). In practice, $\hat{\sigma}_B^2$ is typically set to '0' or another method of estimation is applied.

In the unbalanced case, the ANOVA estimators of σ_W^2 and σ_B^2 become:

$$\hat{\sigma}_W^2 = MSW$$
$$\hat{\sigma}_B^2 = \frac{MSB - MSW}{\left(N - \sum_i n_i^2 / N\right) / (k-1)} \qquad \text{[Equation 17-6]}$$

where n_i equals the number of measurements collected on the i-*th* worker and N equals the total number of measurements collected on k workers $\left(N = \sum_{i=1}^{k} n_i\right)$.

For information on interval (rather than point) estimates, readers are referred to Burdick and Graybill (1992).

EVALUATING HOMOGENEITY IN EXPOSURE LEVELS

How Different are Workers' Mean Exposure Levels?

Early in the occupational literature, the concept of 'homogenous exposure groups', which has since morphed to 'similar exposure groups' (see, for example, Mulhuasen and Damiano, 1998) was introduced whereby workers' exposure profiles were assumed to be the same on the basis of workplace factors such as ventilation or process equipment, the specific agents with which they came in contact and the tasks they performed (Corn and Esmen, 1979). Yet, does classifying workers on the basis of what work is performed and in which location of the workplace *necessarily* produce groups with similar exposures? This is a question that deserves attention. First, what is meant by similarity? Further, how different can workers' exposures levels be within an occupational group and still be deemed 'homogenous'? Secondly, how can we evaluate whether a *priori* classification results in a group of workers with homogenous or heterogeneous exposures?

Building on the one-way random-effects model, Rappaport (1991) defined a metric, the so-called ${}_BR_{95}$ value, which provides a means by which uniformity of exposure within an occupational group of workers could be formally evaluated. This value is defined as the ratio of the 97.5^{th} to the 2.5^{th} percentiles of the log-normally distributed mean exposures of a group of workers (which are viewed as random variables in the one-way random-effects model) – this distribution has been referred to as the 'between-worker' distribution. Also, note that the chosen percentiles of 2.5 and 97.5 ${}_BR_{95}$ values capture 95% of the distribution.

Under the one-way random-effects model, the ${}_BR_{95}$ value is represented as follows:

$$
\begin{aligned}
{}_BR_{95} &= \frac{v_{0.975} \text{ of the between-worker distribution}}{v_{0.025} \text{ of the between-worker distribution}} \\
&= \frac{\exp[\mu_Y + 1.96\sigma_B]}{\exp[\mu_Y - 1.96\sigma_B]} \\
&= \exp[\mu_Y + 1.96\sigma_B - (\mu_Y - 1.96\sigma_B)] \qquad \text{[Equation 17-8]} \\
&= \exp[3.92\sigma_B]
\end{aligned}
$$

where,

$v_{0.975}$ = the 97.5th percentile of the underlying lognormal distribution of workers' mean exposure levels

$v_{0.025}$ = the 2.5th percentile of the underlying lognormal distribution of workers' mean exposure levels

μ_Y = The mean of the logged exposures

σ_B = The between-worker standard deviation

To identify a 'monomorphic group', Rappaport et al. (1991) suggested a cut-point of 2 for the ${}_BR_{95}$ value; i.e., when 95% of the individual mean exposures lie within a factor of 2 (in other words, σ_b is less than 1.2). Since a cut-point of 2 permits relatively little heterogeneity in exposure among workers' mean exposures in an occupational group, other cut-points could arguably be used to identify a group of workers whose exposure profiles are similar. In evaluating 183 job groups classified either by job title and location or by location alone (combining several job groups together), Rappaport et al. (1993) reported ${}_BR_{95}$ values ≤ 2 for 22% of the groups analyzed in their study and ${}_BR_{95}$ values ≥ 50 for approximately 10% of the groups. These findings suggest that observational approaches alone are unsuited for distinguishing groups of workers with similar exposure levels and that quantitative assessment using the techniques described herein provide powerful tools for examining this question in a more rigorous fashion.

Utilizing the ${}_BR_{95}$ value as a metric for assessing homogeneity in exposure within an occupational group has been applied in numerous other investigations; a couple of which we will discuss as examples. In a study of bakery workers (Houba et al., 1997), the authors reported a much greater degree of variability in exposure among workers for wheat allergens (${}_B\hat{R}_{95}$ = 1578) than for inhalable dust exposure (${}_B\hat{R}_{95}$ = 34.2), suggesting that the degree to which exposures vary between and within workers may depend upon the specific agent under investigation. In this same study, differences were detected depending on the nature of the bakery in which work was performed. For example, doughmakers in bread bakeries appeared to experience more heterogeneous inhalable dust exposures than in confectioneries (${}_B\hat{R}_{95}$ = 18.2 *vs.* 4.5, respectively), pointing possibly to differences in the process and/or the workplace that influence workers' exposures. It is worth noting, however, that similar differences were not observed for exposure to wheat allergens (${}_B\hat{R}_{95}$ = 52.2 *vs.* 59.6, respectively). While the aforementioned study evaluated homogeneity in exposure across bakeries, other investigations have been restricted to single workplaces in which differences were also detected in the degree to which exposure levels vary among workers. For example, Scheepers et al. (2003) reported an estimated 25-fold difference in respirable dust mean exposure levels among workers who were engaged in tasks above the ground at a coal mine as compared to a 2.6 fold difference among underground workers.

Indeed, in evaluating the effect of the classification scheme to identify groups of workers, we would expect differences based upon how refined the scheme is to classify workers. In their systematic review of the literature, Symanski et al. (2006) examined the degree of variability of exposures among groups classified according to different schemes and reported that the variation in exposure between workers, on average, tended to increase as workers were aggregated across jobs and location. Median ${}_B\hat{R}_{95}$ values ranged from 4.8 for groups classified by job and location to 65.0 for groups classified across jobs and locations. Nonetheless, even for workers classified by job and location (n = 306 groups), 25% of the groups were characterized by ${}_B\hat{R}_{95}$ values that were 10 or higher and 5% had ${}_B\hat{R}_{95}$ values that were 45 or higher, which confirmed earlier findings (Kromhout et al., 1993) that the same classification scheme can produce groups with both homogenous as well as heterogeneous exposures.

How Different are Exposure Levels from Shift to Shift?

An underlying assumption of the one-way random-effects model is that the degree to which exposures vary from shift to shift is the same for all workers within an occupational group; in other words, σ_W^2 is the same for all workers in the group. A R_{95} value can also be constructed to gauge the degree to which exposures vary from shift to shift: The ${}_W R_{95}$ value is represented as follows:

$$\begin{aligned} {}_W R_{95} &= \frac{v_{0.975} \text{ of the within-worker distribution}}{v_{0.025} \text{ of the within-worker distribution}} \\ &= \frac{\exp[\mu_Y + 1.96\sigma_W]}{\exp[\mu_Y - 1.96\sigma_W]} \\ &= \exp\left[\mu_Y + 1.96\sigma_W - (\mu_Y - 1.96\sigma_W)\right] \\ &= \exp[3.92\sigma_W] \end{aligned} \qquad \text{[Equation 17-9]}$$

where,

- $v_{0.975}$ = the 97.5th percentile of the underlying lognormal distribution of a worker's shift-long exposure levels
- $v_{0.025}$ = the 2.5th percentile of the underlying lognormal distribution of a worker's shift-long exposure levels
- μ_Y = The overall mean of the logged shift-long exposures
- σ_W = The within-worker standard deviation

In evaluating the ${}_W R_{95}$ value among the 306 groups classified by job and location in Symanski et al. (2006), the 25th, 50th, and 75th percentiles of the weighted empirical cumulative distribution function for the ${}_W R_{95}$ values were 9.0, 26.2, and 65.0, respectively. These findings suggest that exposures, on average, vary considerably from day-to-day. Such variability in exposure complicates efforts to estimate an individual worker's mean exposure level or evaluate exposure relative to occupational exposure limits.

EVALUATING EXPOSURE RELATIVE TO OCCUPATIONAL EXPOSURE LIMITS

Given that exposures vary both within and between workers in an occupational group of workers, two questions that pertain to exposures relative to occupational exposure limits are raised (Tornero-Velez et al., 1997):

a. What is the probability that a single measurement obtained from a randomly selected worker in an occupational group on a randomly selected day exceeds an occupational exposure limit (OEL)? (probability of exceedance)

b. What is the probability that the mean exposure for a randomly selected worker exceeds an OEL? (probability of overexposure)

The probability of exceedance can be expressed as a function of the mean and variance components of the underlying distribution of the natural log-transformed exposures for an occupational job group. Under the one-way random-effects model (Equation 17-3), we have already defined the expected value (mean) and variance of Y_{ij} (the log-transformed variate); namely, $E(Y_{ij}) = \mu_Y$ and $Var(Y_{ij}) = \sigma_Y^2 = \sigma_B^2 + \sigma_W^2$. Thus, the exceedance probability (γ) can be represented as follows:

$$\begin{aligned}\gamma &= P\{X_{ij} > OEL\} \\ &= P\left\{\frac{\ln X_{ij} - \mu_Y}{\sigma_Y} > \frac{\ln OEL - \mu_Y}{\sqrt{\sigma_B^2 + \sigma_W^2}}\right\} \\ &= P\left\{Z > \frac{\ln OEL - \mu_Y}{\sqrt{\sigma_B^2 + \sigma_W^2}}\right\} \\ &= 1 - \Phi\left\{\frac{\ln(OEL) - \mu_Y}{\sqrt{\sigma_B^2 + \sigma_W^2}}\right\}\end{aligned} \qquad \text{[Equation 17-10]}$$

where $\Phi\{z\}$ denotes the probability that a standard normal variate falls below the value of z.

To develop the expression for the probability of overexposure, we first note that the mean and variance of the untransformed exposures (i.e., $\{X_{ij}'s\}$) are $\mu_X = exp(\mu_Y + 0.5\sigma_Y^2)$ and $Var(X_{ij}) = \mu_X^2\,[exp(\sigma_Y^2) - 1]$, respectively. We also note that the conditional mean of Y_{ij} for a fixed b_i can be written as $\mu_{Y_i} = \mu_Y + b_i + 0.5\sigma_W^2$ and, analogously, on the original lognormal scale as $\mu_{X_i} = exp(\mu_Y + b_i + 0.5\sigma_W^2)$. In this application, the reader is reminded that μ_{X_i} is a random variable since b_i is a random variable. Thus, the probability of overexposure (θ) can be represented as follows (Rappaport et al., 1995; Lyles et al., 1997a):

$$\begin{aligned}\theta &= P\{\mu_{X_i} > OEL\}\\ &= P\{\exp(\mu_Y + b_i + 0.5\sigma_W^2) > OEL\}\\ &= P\left\{\frac{(\mu_Y + b_i + 0.5\sigma_W^2) - \mu_Y - 0.5\sigma_W^2}{\sigma_B} > \frac{\ln(OEL) - \mu_Y - 0.5\sigma_W^2}{\sigma_B}\right\}\\ &= P\left\{Z > \frac{\ln(OEL) - \mu_Y - 0.5\sigma_W^2}{\sigma_B}\right\}\\ &= 1 - \Phi\left\{\frac{\ln(OEL) - \mu_Y - 0.5\sigma_W^2}{\sigma_B}\right\}\end{aligned} \quad \text{[Equation 17-11]}$$

where $\Phi\{z\}$ denotes the probability that a standard normal variate falls below the value of z.

One question that arises is how similar is the probability of exceedance to the probability of overexposure? This question was raised in the literature almost 20 years ago (Spear and Selvin, 1989) and was later evaluated by Tornero-Velez et al. (1997). Utilizing data collected on 179 groups of workers, Tornero-Velez et al. (1997) reported that the point estimates of the probability of exceedance were larger than that for the probability of overexposure for 83% of the groups, which might suggest that $\hat{\gamma}$ is a conservative estimate of θ. In further analyses, however, they present a graphical display of the relationship between θ and γ for two different exposure scenarios ($\sigma_Y^2 = 0.164$ and $\sigma_Y^2 = 2.59$) and point out that γ serves as an *upper* bound for θ when the exceedance probabilities are at the 'low' end, but a *lower* bound when the exceedance probabilities are 'high'. Moreover, the point at which the exceedance probability makes the transition from a lower to an upper bound depends on the magnitude of the intraclass correlation coefficient (i.e., the ratio of the between-worker variance to the sum of the within- and between-worker variances). Thus, having information solely on the probability of exceedance is not sufficient to make a corresponding statement about the probability of overexposure (see Rappaport et al., 1998 for additional discussion of this issue), providing another example of how information about the between- and within-worker sources of variation can be utilized in exposure assessment. Readers are referred to a number of investigators that have examined probabilities of exceedance and overexposure for exposures to manganese, nickel and chromium in the construction industry (Rappaport et al., 1999), inhalable dust in the carbon black manufacturing industry (van Tongeren et al., 2000), mercury in the chloralkali industry (Symanski et al., 2001c) and dust in a cement factory (Mwaiselage et al., 2005) (probability of overexposure).

Formal Test for Overexposure

To evaluate 'overexposure' (θ) relative to an occupational exposure limit (OEL), Rappaport et al. (1995) proposed methods for balanced data that relied upon a Wald-type statistic and allowed for the evaluation of whether the probability that a randomly selected worker's mean exposure level exceeds some specified level (A). Given that the mean exposure for an occupational group of workers (on the untransformed scale) is represented as $\mu_X = exp[\mu_Y + 0.5(\sigma_B^2 + \sigma_W^2)]$, Rappaport et al. (1995) defined a quantity R as follows:

$$R = ln\left(\frac{\mu_X}{OEL}\right) - \frac{\sigma_B^2}{2} + \sigma_B z_{1-A} \qquad \text{[Equation 17-12]}$$

where z_{1-A} is a value from the standard-normal distribution such that $P(Z > z_{1-A}) = A$. Under the null hypothesis, H_0: R = 0, we note that

$$\frac{\mu_X}{OEL} = exp\left(\frac{\sigma_B^2}{2} - \sigma_B z_{1-A}\right) = exp\left[\frac{1}{2}\left((\sigma_B - z_{1-A})^2 - z^2_{1-A}\right)\right] \qquad \text{[Equation 17-13]}$$

In defining a new value, L as $exp\left[\frac{1}{2}\left((\sigma_B - z_{1-A})^2 - z^2_{1-A}\right)\right]$, the null and alternative hypotheses can be represented as follows:

$$H_0 : \frac{\mu_X}{OEL} \geq L \ \ (unacceptable\ exposure)$$

$$H_1 : \frac{\mu_X}{OEL} < L \ \ (acceptable\ exposure) \qquad \text{[Equation 17-14]}$$

This formal test for overexposure is illustrated in Rappaport et al. (1995a) for four groups of workers exposed to nickel aerosols in the nickel-producing industry. Additionally, interested readers are referred to Rappaport et al. (1995a) for more details on the hypothesis testing procedure, Lyles et al. (1997a) for an evaluation of the behavior of the Wald statistic in addition to other statistics that might be used to make the determination regarding overexposure (i.e., likelihood ratio and score-type test statistics) and Lyles et al. (1997b) for further development of the testing protocols when different numbers of exposure measurements are collected on each worker in an occupational group.

THE ERROR STRUCTURE/ERROR STRUCTURE/COMPOUND SYMMETRY (CS) VERSUS FIRST-ORDER AUTOREGRESSIVE [AR(1)]

Previously, we considered a compound symmetric structure in describing the covariance between repeated measurements (see Equation 17-3). Under compound symmetry (CS), it is assumed that the covariance between measurements collected on the same worker is the same irrespective of the time interval separating them. In other words, the covariance between measurements separated by one day is considered the same as measurements separated by many days or many weeks. However, this may not be a reasonable assumption. Instead, it is possible that the covariance between measurements collected on the same worker may depend on the interval that separates them and, in this instance, we may choose to apply a first-order autoregressive [AR(1)] model. Under an AR(1) model, the correlation function decays exponentially as the interval between measurements increases (i.e., $\rho^\tau = \alpha^\tau$), where τ is the number of time units between measurements. Under this structure, exposure measurements that are collected closer together in time are more highly correlated than those further apart. We represent the covariance and correlation structures as $Cov(Y_{ij}, Y_{ij'}) = \sigma_B^2 + \rho^\tau \sigma_W^2$ and $\mathrm{Corr}(Y_{ij}, Y_{ij'}) = (\sigma_B^2 + \rho^\tau \sigma_W^2)/\sigma_Y^2$ for $j \neq j'$.

As discussed in Chapter 16 using classical time series analysis, there is limited evidence of autocorrelation in shift-long airborne exposure measurements (Francis et al., 1989; Kumagai et al., 1993; Symanski and Rappaport, 1994; George et al., 1995). Serial correlation between measurements with shorter averaging times (e.g., 7.5, 15, 30 or 60 minutes) has also been reported for both chemical (organic solvents) (Kumagai et al., 1993) and physical (magnetic fields) agents (van der Woord et al., 1999). We are unaware of studies that have examined serial correlation in biological measures of exposure, which would require as they did for the personal sampling data relatively long series of measurements collected consecutively over time ($\geq$ 30 measurements). However, studies that have incorporated an AR(1) error structure in random-effects models reported moderate to substantial levels of serial correlation in some of the biological monitoring data that were analyzed (Symanski et al., 2001a; Symanski and Greeson, 2002; Lin et al., 2005). The extent to which biomonitoring data will be autocorrelated will depend on the residence time of the contaminant in the body relative to the interval between pairs of measurements collected on the same individual. In general, the degree of serial correlation will increase for biomarkers with longer half-lives and shorter intervals between repeated measurements.

Does the specification of the error structure matter? In evaluating the results of the model with compound symmetry as compared to the model with an AR(1) error structure (Symanski et al., 2001a; Symanski and Greeson, 2002), ignoring serial correlation (when it is present) produced estimates of the within-worker variance component that were smaller than they should be, which is an expected consequence of a positively autocorrelated process (Diggle, 1990). These findings underscore the importance of properly specifying the covariance structure of exposure data to minimize biases that may be introduced in applying models that are too simple. We will address the issue of adequate specification of the model as it relates to the mean structure in a later section.

A NOTE OF CAUTION IN USING INFORMATION CONTAINED IN ESTIMATES OF VARIANCE COMPONENTS TO MAKE MEANINGFUL STATEMENTS ABOUT EXPOSURE

It is clear from the above discussion about random-effects models that useful information can be gleaned from our evaluation of the relative magnitude of the variation between and within workers. As demonstrated in previous sections, we can use such information contained in the variance components to assess homogeneity in exposure among a group of workers and evaluate compliance relative to occupational exposure limits. Yet, a question should always arise whether sufficient data are available upon which our assessments are made. In other words, is the precision in our estimates of the variance components sufficient to allow for judgments about exposure whether it relates to homogeneity within a group of workers, compliance with exposure limits or some other question? Unfortunately, confidence intervals associated with point estimates of the variance components are rarely reported in the literature (and not easily computed with unbalanced data (Burdick and Graybill, 1992)). Nonetheless, greater attention should be paid to the adequacy of our sample both in terms of workers monitored and measurements collected (Symanski et al., 2006; Symanski et al., 2007).

MIXED-EFFECTS AND TWO-WAY RANDOM-EFFECTS MODELS – EXTENSIONS OF THE ONE-WAY RANDOM-EFFECTS MODEL

Thus far, we have restricted our discussion largely to the one-way random-effects model in which ‘worker’ is viewed as a random effect. However, this model is rather simplistic and we may wish to evaluate other factors (either as random or fixed) as well. In the situation where we add one or more fixed effects, we construct what is referred to as a mixed-effects model. Mixed-effects models are appropriate to assess the influence of specific determinants on exposure levels while accounting for the between- and within-worker sources of variation in exposure.

Non-Stationarity in the Mean Level – Shifts in Exposure Levels Over Time

In the first application of mixed-effects models in the occupational arena, Symanski et al. (1996) examined shifts in exposure levels between two survey periods while accounting for the between- and within-worker sources of variation in exposure. In this application, a fixed survey effect was added to the one-way random-effects model to evaluate differences in mean exposure levels between surveys. Using balanced data, two different modeling approaches were applied – the first applied a mixed model to data collected on workers in two survey periods and the second applied a nested-design mixed model where workers contributed measurements to only one survey period. To illustrate one of the models that was applied, let’s focus on the two-way mixed-effects model:

$$Y_{ijk} = ln(X_{ijk}) = \mu_Y + \alpha_i + b_j + e_{ijk} \quad \textit{for } i=1,2 \textit{ surveys, } j=1,2,...,m \textit{ workers, and } k=1,2,...,n \textit{ days/worker/survey} \qquad \text{[Equation 17-15]}$$

where,

X_{ijk} = The shift-long exposure concentration for the j-*th* worker on the k-*th* day during the i-*th* survey

Y_{ijk} = The natural log-transformed value of the exposure concentration

μ_Y = Overall mean (mean of Y_{ijk})

α_i = Fixed effect on exposure due to the measurement having been collected during the i-*th* survey

b_j = The random effect measuring the deviation of the j-*th* worker's true exposure from $\mu_Y + \alpha_i$, and

e_{ijk} = The random effect measuring the deviation of the j-*th* worker's exposure from $(\mu_Y + \alpha_i)$ on the k-*th* day during the i-*th* survey

In Equation 17-15, it is assumed that the α_i sum to zero with a population variance defined as $\sum_{i=1}^{a} \alpha_i^2/(a-1)$, where a=2 in the model applied in Symanski et al. (1996) (i.e., there were two survey periods). It is further assumed that b_j and e_{ijk} are mutually independent and normally distributed with zero means and variances σ_B^2 and σ_W^2, respectively. Thus, $E(Y_{ijk}) = \mu_Y + \alpha_i$ *for all i, j, k*; $Var(Y_{ijk}) + \sigma_Y^2 = \sigma_B^2 + \sigma_W^2$ *for all i, j, k*; and $Cov(Y_{ijk}, Y_{ijk'}) = \sigma_B^2$ *for* $k \neq k'$.

In summarizing the findings of Symanski et al. (1996), there were 40 occupational groups of workers who had sufficient data to examine whether exposures varied across the two survey periods using either the two-way mixed model shown in Equation 17-15 or a modification of the model in which different workers were sampled in each survey period. There was a 1.5 to 3-fold difference in geometric means between survey periods in nearly 70% of the occupational groups. Over 80% of the groups with significant changes in exposure occurred over intervals of one year or longer, which suggests that non-stationary behavior in exposure levels should be evaluated when monitoring data are collected over extended periods of time.

The authors were also interested in assessing the bias in estimates of the variance components in instances when the mean structure is improperly modeled. In other words, they asked a question regarding the magnitude of the error that would be expected when changes in exposure levels occur but are not taken into account in evaluating the within- and between-worker sources of variation in exposure. They derived equations for the bias in estimates of the variance components obtained when changes in exposure levels occur between two survey periods and the one-way random-effects model is applied instead of the two-way mixed model (which properly specifies the mean structure):

$$Bias\left(\hat{\sigma}_W^2\right) = \frac{n\sum_{i=1}^{2}\alpha_i^2}{(2n-1)}$$

$$Bias\left(\hat{\sigma}_B^2\right) = \frac{-\sum_{i=1}^{2}\alpha_i^2}{(4n-2)} \qquad \text{[Equation 17-16]}$$

As shown in Equation 17-16, the bias is a function of the number of repeated measurements per worker (n) and the magnitude of the change in exposure levels between surveys (α_i), but operates in opposite directions depending on which variance component is referenced. Specifically, applying the one-way random-effects model results in an underestimation of the between-worker variance component and an overestimation of the within-worker variance component. Thus, ignoring a fixed effect due to calendar time tends to produce biased estimates of the within-worker (between-worker) variance that are smaller (larger) than they should be. Given that the magnitude of the bias can be substantial when the mean structure is not properly specified, selecting statistical models that properly characterize the behavior of exposures over time becomes important to ensure that valid inferences are made.

Non-Stationarity in the Mean Level – Trends in Exposure Levels Over Time

There is ample evidence of trends in exposure levels over time (Symanski et al., 1998; Teschke et al., 1999, Burstyn et al., 2000; Symanski et al., 2000a; Hall et al., 2002; Raaschou-Nielsen et al., 2002). The mean structure in mixed effects models may be modified to account for such trends. In using an example from the nickel industry, Symanski et al. (2001b) evaluated monitoring data that had been collected during the period 1973-1995 from 10 different facilities using a mixed-effects model that included trend components (i.e., the intercept and linear slope terms) as fixed effects and specified workers, job groups, buildings and plants as random effects. This specific model assumed that the effect of time is the same for all job groups of workers; an alternative model might have specified the time effect differently. Given that the data were hierarchical in nature with measurements nested within workers, workers nested within job groups, job groups nested within building and buildings nested within facilities, the covariance structure between measurements was more complicated than the models presented in this chapter, i.e., there was dependence not only among measurements collected on the same worker, but also for measurements collected on workers in the same job title, measurements collected on job groups performing work in the same building and measurements collected across buildings at the same facility.

Evaluating Differences in Exposures Across Job Groups

While we can apply the one-way random-effects model on a group-by-group basis, we are often interested in making inferences regarding more than one occupational group of workers. While we can apply the one-way random-effects model to each group separately, pooling data offers advantages in obtaining more efficient estimates of the model parameters. In this context, the one-way random-effects model can be expanded to evaluate groups of workers together in a single analysis either by modeling 'job group' as a 'random' or 'fixed' effect, depending on the purpose of the exposure assessment and the manner in which the data are collected. Similar to the question that is asked in modeling worker as a 'random' or 'fixed' effect, if interest is focused on making statements about the underlying population of all occupational groups, then job group would be modeled as a random effect in a two-way random-effects model. Here, since occupational groups are assumed to have come from a population, 'job group' accounts for another source of random variation in the data. Thus, the total variability in exposure is partitioned among the between-group, between-worker and within-worker sources of variation. For example, in a study of total dust exposure in flour mills and bakeries, Nieuwenhuijsen et al. (1995) combined 346 measurements collected on workers in 13 groups and reported substantial variability in exposure between groups (53% of the total) as compared to variability between workers (18%) or between work shifts (29%). Other examples from the literature investigating random variation among job groups include investigations in myriad workplaces including the rubber industry (Kromhout and Heederik, 1995), electric power industry (Kromhout et al., 1995), the carbon black industry (van Tongeren et al., 1997) and fish processing (Jeebhay et al., 2005), to name a few. As was the purpose of many of these studies, partitioning the variability into its core components (between-group, between-worker and within-worker) is useful when evaluating optimal methods for classifying groups of workers.

On the other hand, if interest is focused only on those groups represented in the sample, then the one-way random effects model can be expanded to evaluate job group as a fixed effect in a mixed-effects model. In this scenario, we model differences among the mean exposure levels of the groups along with the between- and within-worker sources of variation. Readers are referred to Rappaport et al. (1999) and Symanski et al. (2001a) for applications of the mixed model to assess whether mean exposure levels vary among different groups of workers. Other studies have examined not only differences on the basis of job group, but on the basis of tasks and the use of specific products (Burstyn and Kromhout, 2000).

Evaluating Determinants of Exposure

Exposure arises as a consequence of factors related to the agent, the process, the work environment and the worker. Thus, evaluating factors that influence exposure is an important aspect of exposure assessment. Burstyn and Teschke (1999) provide a review of studies that have relied on fixed-effects models to identify determinants of exposure. To evaluate specific determinants of exposure, exposure measurements collected on a group of workers can be coupled to information gathered by questionnaire regarding the nature of work carried out while personal sampling is conducted. In instances when repeated measurements are collected on workers along with information about work practices, work tasks, use of personal protection equipment and other determinants of exposure, mixed-effects models can be applied. Burdorf (2005) provides an illustrative summary of exposure modeling, especially in the context when repeated measurements on workers are collected.

Let's look at a general representation of a mixed-effects linear regression model that evaluates determinants of exposure:

$$Y_{ij} = \alpha_0 + \sum_{m=1}^{p} \alpha_m (C_{mij}) + b_i + e_{ij} \qquad \textit{for } i = 1, 2, \ldots, k \textit{ workers,}$$

$$j = 1, 2, \ldots, n_i \textit{ measurements, and} \qquad \text{[Equation 17-17]}$$

$$m = 1, 2, \ldots, p \textit{ determinants of exposure}$$

In Equation 17-17, Y_{ij} is defined as before in this chapter; α_0 is the intercept of the linear model, α_m represents the regression coefficient representing the fixed effect of the m-*th* determinant, C_{mij} represents the value of the m-*th* determinant associated with the j-*th* measurement on the i-*th worker,* b_i represents the random worker effect and e_{ij} represents the random effect of the j-*th* measurement for the i-*th* worker (i.e., the error term). Under this model, it is assumed that b_i and e_{ij} are normally distributed with zero means and variances σ_B^2 and σ_W^2. Thus, Y_{ij} is assumed to be normally distributed with mean $\alpha_0 + \sum_{m=1}^{p} \alpha_m (C_{mij})$ and variance σ_Y^2 where $\sigma_Y^2 = \sigma_B^2 + \sigma_W^2$.

In situations where multiple measurements on workers are collected, mixed-effects linear regression models (such as the one depicted in Equation 17-17) offer clear advantages because they provide an opportunity to evaluate determinants of exposure in addition to the within- and between-worker sources of variation (Rappaport et al., 1999). Thus, Equation 17-17 is similar to traditional regression models in that it incorporates regressor effects, but different because it accounts for the correlation between measurements collected on the same person (or, more generally, on the same level of any factor declared as random in the model equation). Interested readers are referred to Rappaport et al. (1999), Burstyn et al. (2000), Peretz et al. (2000), Hines and Deddens (2001) and Vermeulen et al. (2004) for examples of studies that have applied mixed models to evaluate determinants of exposure, and to Symanski et al. (2001b) and Peretz et al. (2002) who provide evidence that erroneous inferences about exposure determinants may be made if correlated data are not properly modeled.

RANDOM- AND MIXED-EFFECTS MODELS – CHANGING THE VARIANCE-COVARIANCE STRUCTURE

In the mixed-effects models that we have discussed thus far in evaluating differences in exposure levels across job groups (or across the various levels of any determinant of exposure), an important assumption of the models that have been applied is that the between- and within-worker variance components are the same across job groups. However, we must ask ourselves whether this assumption is tenable. We would expect an association between the mean and variance of a lognormally distributed variate. Thus, in instances when there are clear differences in mean exposure levels across job groups, it becomes critical to evaluate the assumption of equal variances across groups.

Let's consider a situation where we wish to evaluate differences in the mean exposure levels; thus, we might construct a mixed-effects model that includes 'job group' as a fixed effect. Under this model, is the degree to which exposures vary from shift to shift (i.e., the within-worker source of variation) the same across job groups? Likewise, is the degree to which worker mean exposure levels vary within job groups (i.e., the between-worker source of variation) the same?

In this scenario, three different variance-covariance structures could be evaluated:

(1) The between- and within-worker variances are distinct across all job groups;

(2) The between-worker variance component is distinct across all job groups; the within-worker variance component is common across groups; and

(3) The between- and within-variance components are the same across job groups.

Clearly, option one is the most flexible and option three the least flexible as it relates to the variance-covariance structure of the measurements. To evaluate whether it is appropriate to pool variance components across groups, a series of likelihood ratio tests can be used to make comparisons between different models. To apply the likelihood ratio test, a test statistic is computed as the difference in the –2 log likelihood values between the two models being compared. This statistic is approximately distributed as a chi-squared variate with degrees of freedom specified by the difference in the number of parameters estimated between the models and is evaluated at a specified level of significance (typically $p = 0.05$). Thus, we may compare the model with different between-worker variances, but a common within-worker variance across groups to the model with common between- and within-worker variances across groups. For I groups, this comparison tests the hypothesis that H_0: $\sigma^2_{B,1} = \ldots = \sigma^2_{B,I}$. We may also compare the model with distinct between- and within-worker variances to the model with different between-worker variances, but a common within-worker variance. This comparison tests the hypothesis that H_0: $\sigma^2_{W,1} = \ldots = \sigma^2_{W,I}$

Rappaport et al. (1999) evaluated different variance structures when evaluating aerosol exposures among four groups of construction workers and reported that the within-worker variances, but not the between-worker variances, could be pooled across job groups. A later study (Symanski et al., 2001c) examined heterogeneity in exposure variability among four groups of mercury-exposed workers using both personal and biological monitoring data. While there was no evidence of heterogeneity in the variation over time or between workers for airborne measures of exposure to mercury (Hg), groups did not share common levels of variability when the assessment was carried out using blood or urinary Hg measurements. Other studies have examined heterogeneity in the variances in two-way random effects models (modeling job group as a random effect) (Symanski et al., 2001c) and in mixed models with fixed effects for survey period and factory (van Tongeren et al., 2006).

GROUP VERSUS INDIVIDUAL-BASED EXPOSURE ASSESSMENT

In exposure assessment for epidemiology, either an individual-based or group-based approach can be adopted. An individual-based approach is utilized when every worker in the study is monitored and the data used to estimate their exposure levels. Alternatively, in a group-based approach, representative measurements are collected on a random sample of workers in each group. Using monitoring data, an estimate of the mean exposure for the group is applied to each worker whether they have been sampled or not. In both cases, errors are introduced but the error structure is quite different (Armstrong, 1998). In assigning group-specific mean levels for all individuals in a job, a type of error referred to as 'Berkson error' may be present whereas a classical measurement error structure may be present in assigning individual-level estimates of exposure. How is this related to variance component models? Let's look at some examples.

Individual-Based Exposure Assessment

In the case where a simple linear regression describes the relationship between a continuous health outcome and exposure (on the log-scale) and individual-based estimates of exposure are used, the degree of bias in an estimated regression coefficient is a function of the relative magnitude of the within- to between-person variance components (Thomas et al., 1993). To see how we arrive at that determination, we assume that a classical measurement error model (Fuller, 1987) holds – in other words, a true predictor is unobservable and instead we must rely on an imperfect measure. In our context, true exposure is the mean exposure level and the imperfect measure is an estimate of exposure obtained from monitoring data (both on the logged scale). In this scenario, a simple model for the exposure-response relationship becomes:

$$Y_i = \alpha + \beta lnX_i + v_i \quad \text{for i=1,2...,k workers} \qquad \text{[Equation 17-19]}$$

where

Y_i	=	The continuous response for the i-*th* individual
X_i	=	The true mean exposure for the i-*th* individual
α	=	The intercept
β	=	The slope
v_i	=	The response error term

In Equation 17-19, we assume that lnX_i's $\sim N(\mu_X, \sigma_B^2)$, $v_i \sim N(0,\sigma_v^2)$ and given that $X_i = x_i$, Y_i is distributed normally with mean $\alpha + \beta \ln x_i$ and variance σ_Y^2.

Since the true mean exposure for the i-*th* person (X_i) is not observable, we must rely on Z_i, which represents an imperfect measure of X_i where:

$$lnZ_i = ln\, X_i + \varepsilon_i \qquad \text{[Equation 17-20]}$$

In Equation 17-20, ε_i represents measurement error. Note that the measurement error arises not only due to errors in sampling and analysis, but also due to factors that influence fluctuations in shift-long exposure levels about a worker's mean value. In assuming that $\varepsilon_i \sim N(0, \sigma_W^2)$, it directly follows that $lnZ_i \sim N(\mu_X, \sigma_Z^2)$ where $\sigma_Z^2 = \sigma_B^2 + \sigma_W^2$.

Under expectation, it has been shown (Thomas et al., 1993) that the ratio of the 'observed' coefficient (β') (obtained using the observed Z_i values) to the true coefficient (β) (obtained using the X_i, if they were observable) can be expressed as follows:

$$\frac{\beta'}{\beta} = \frac{\sigma_B^2}{\sigma_B^2 + \sigma_W^2}$$ [Equation 17-21]

Thus, in regressing Y_i on the imperfect measure lnZ_i, instead of lnX_i, we obtain a slope coefficient β' that is attenuated. The factor $\sigma_B^2/(\sigma_B^2+\sigma_W^2)$ represents the degree of attenuation of the true slope coefficient β.

Similarly, we may observe n randomly-collected measurements of the surrogate measure on j occasions for each X_i (*i.e.*, $n\ Z_{ij}$*'s*) and compare the true exposure-response regression with the regression of Y_i's on $\overline{\ln Z_i}$ $\left(where\ \overline{\ln Z_i} = \sum_{j=1}^{n} \ln Z_{ij} / n \right)$. Since $E(ln\ X_i | lnZ_{i1}, ..., lnZ_{in}) = E(ln\ X_i | lnZ_i)$, the ratio of the observed coefficient to the true coefficient becomes:

$$\frac{\beta'}{\beta} = \frac{\sigma_B^2}{\sigma_B^2 + \frac{\sigma_W^2}{n}}$$ [Equation 17-22]

As before, regressing Y_i on the logarithm of measured exposure lnZ_i, instead of the log-scale of the true exposure X_i, yields a slope coefficient that is attenuated. Under expectation, the factor represents the degree of attenuation of the slope coefficient β.

Several studies in the literature have relied upon estimates of the variance components to evaluate the degree of attenuation that would be likely in a regression coefficient including Heederik et al. (1991), Rappaport et al. (1995), and Symanski et al. (2000; 2001a). Using Equation 17-22 and substituting the estimates of the variance components for the corresponding parameters, Figure 17-2 depicts the degree to which the regression coefficient (from a simple linear regression) is underestimated for three different measures of exposure to mercury (figure from Symanski et al., 2000b). As can be seen from the figure, the effects of measurement error are greatest for airborne mercury and least for urinary mercury (corrected for creatinine). Secondly, it is worthwhile noting that increasing the number of repeated measurements collected on each individual will diminish the attenuating effects of imperfectly measured exposure, but cannot eliminate them altogether even when large sample sizes are collected.

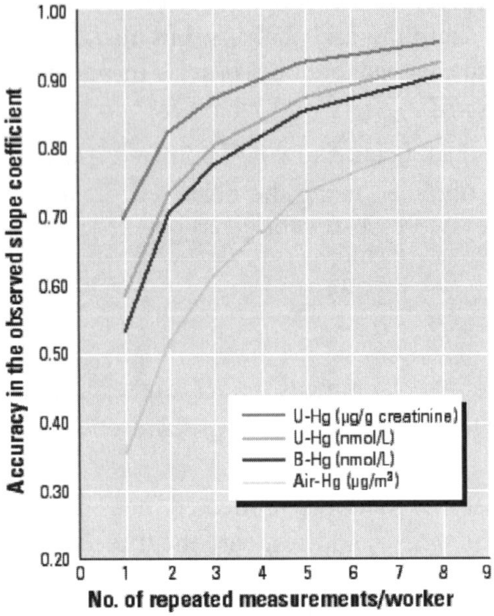

Figure 17-2. The estimated degree of attenuation in an estimated regression coefficient for different exposure measures collected on workers at a chloralkali plant plotted as a function of the number of repeated measurements collected on each worker (U-Hg = Urinary mercury; B-Hg = Blood mercury) (from Symanski et al., 2000b).

Group-Based Exposure Assessment

In models where job group is modeled as a random effect in a two-way random-effects analysis of variance model (see section on *Evaluating differences in exposures across job groups* above), Kromhout and Heederik (1995) proposed a measure of contrast among groups that relied upon the inter-group and intra-group variance components, which was referred to as 'elasticity' according to the following:

$$\frac{\hat{\sigma}_G^2}{\hat{\sigma}_G^2 + \hat{\sigma}_B^2} \qquad \text{[Equation 17-18]}$$

In Equation 17-18, $\hat{\sigma}_G^2$ is an estimate of the between-group variance and $\hat{\sigma}_B^2$ is an estimate of the within-group (i.e., between-worker) variance. Here, it is expected that an optimal classification scheme would maximize differences between groups and minimize differences within groups.

Individual- Versus Group-Based Assessment: How Do We Decide?

Clearly, decisions regarding whether to apply an individual-based or group-based exposure assessment will depend heavily on the question under investigation, as well as the availability of monitoring data. Nonetheless, it is of interest to better understand the error that is likely to be introduced in either case. Using Monte Carlo simulations of 100 persons divided into four exposure groups, Seixas and Sheppard (1996) explored the attenuating effects of the exposure-response relationship using individual exposure estimates and group mean exposure estimates, as well as estimates using the James-Stein shrinkage estimator. The latter is a weighted mean that utilizes information from the estimates for both the group and the individual with the weight determined by the relative magnitude of the variance components. Specifically, as the within-worker variance increases relative to the between-worker variance, the group mean is weighted more heavily whereas more weight is placed on the individual means when the reverse situation applies. Their simulation results confirmed the bias that is introduced when an individual-based approach is applied, as well as the loss of precision in applying a group-based approach. Moreover, the James-Stein estimator performed reasonably well in that the simulated regression coefficients contained less bias and were more precise compared to the estimates generated using the individual and group means, respectively.

Interested readers are referred to Kromhout et al. (1996), van Tongeren et al. (1999), Heederik and Attfield (2000), Jansen and Burdorf (2003), Schlünssen et al. (2004) and Teschke et al. (2004) for investigations that have examined the effects of different grouping schemes on measures of effect in epidemiologic investigations. Tielemans et al. (1998) also provides a nice summary of error effects for group-based and individual-based strategies in assessing exposure.

FINAL NOTE

This chapter has presented a framework for examining exposure variability in the context of both fixed and random effects in analysis of variance models. Such models provide us with powerful tools to accurately assess exposure and make meaningful inferences about worker exposure. Although the underlying principles and concepts may appear difficult upon first reading, readers are encouraged to re-visit this chapter taking into account lessons learned in biostatistics, epidemiology and industrial hygiene to help reinforce the material that is presented. While pertinent references have been provided for each topical area in the chapter, the list is by no means complete and other examples could have been selected as well. Thus, the reader is encouraged to become familiar with the work that has evolved in this area in the past 20 years, as well as to seek out new contributions that will continue to further enhance the field of exposure science.

QUESTIONS

17.1. Assume that you are interested in evaluating whether mean exposure levels for five workers exposed to benzene (mg/m^3) differ. You collect the following personal sampling data:

Worker	Shift-long exposure
1	0.032
1	0.288
2	0.224
2	0.320
3	6.49
3	0.511
4	0.895
4	1.34
5	0.160
5	5.11

a. Run a one-way fixed effects model to evaluate whether the individual worker's mean values are different. What conclusions do you make?

b. What hypothesis is being tested in using the F-statistic (the ratio of the mean square within to the mean square between) when applying the fixed effects model?

17.2. The table below compiles personal monitoring data collected on laminators exposed to styrene at a boat manufacturing plant (mg/m^3) [abstracted from Lyles et al. (1997)]:

Worker (i)	Y^*_{ij}	Worker (i)	Y^*_{ij}
1	3.071	8	4.477
1	3.871	8	4.807
1	2.965	8	5.345
2	4.319	9	5.060
2	4.396	9	5.271
2	5.045	9	5.454
3	5.221	10	5.188
3	4.876	10	4.499
3	5.058	10	5.340
4	4.572	11	5.970
4	5.116	11	5.660
4	5.578	11	5.175
5	5.351	12	5.619
5	3.925	12	1.843
5	4.217	12	5.545
6	5.889	13	4.200
6	4.893	13	5.294
6	4.775	13	4.945
7	5.192		
7	4.457		
7	5.097		

*Y_{ij} represents the natural log of the j-*th* shift-long personal exposure measurement on the i-*th* worker

a. Using a one-way random-effects model, estimate the mean exposure for the group of laminators (i.e., μ_Y) and the between- and within-worker variance components (i.e., σ^2_B *and* σ^2_W).

b. Using the estimated values of σ^2_B *and* σ^2_W, compute the ${}_BR_{95}$ and ${}_WR_{95}$ values.

c. Is this group of laminators homogenously exposed? Explain your answer.

17.3. A study was carried out by Nieuwenhuijsen et al. (1995) to evaluate exposures to inhalable dust among workers at four different types of bakeries (based upon which type of product was produced): 1) wheat bread bakeries; 2) confectioneries; 3) crisp bakes factories; and 4) rye bread bakeries. Job titles were common across the different types of bakeries and were used to classify groups of workers. The following table abstracts data for two workers.

Job title	**N**	**k**	**AM**	**GM**	**GSD_B**	**GSD_W**	**$_BR_{95}$**
All-round staff workers							
Wheat bread bakeries	37	16	1.4	0.9	2.0	2.3	13.8
Confectioneries	4	2	7.5	1.2	--	--	--
Crisp bakes factories	29	12	1.3	1.1	1.2	1.8	1.8
Dough makers							
Wheat bread bakeries	37	15	6.8	4.5	2.1	1.7	18.2
Confectioneries	40	18	2.8	2.2	1.5	1.8	4.5
Crisp bakes factories	40	16	3.2	2.8	1.3	1.5	3.1
Rye bread bakeries	7	3	1.0	0.9	--	--	--

*N = number of measurements in a group; k = number of workers in a group; AM = Arithmetic Mean; GM = Geometric Mean; GSD_B = Between-worker geometric standard deviation; GSD_W = Within-Worker geometric standard deviation; $_BR_{95}$ = Ratio of the 97.5th and 2.5th percentiles of the between-worker distribution.

a. You are an occupational hygienist at the plant and must write a brief report that summarizes the exposure data that were collected in the bakeries. In discussing the results indicated in the table with your boss, she notes that the mean exposure level of 'All-round staff workers' in wheat bread bakeries is nearly the same as for 'All-round staff workers' in crisp bakes factories and recommends that you add the following statement in your report: "The exposure profiles for All-round staff workers in Wheat bread bakeries and crisp bakes factories are similar." Do you follow the recommendation? Explain your response.

b. Your boss also notes that your table is incomplete since there are no estimates of the GSD_B, GSD_W, and $_BR_{95}$ for All-round staff workers at confectioneries and for Dough makers at rye bread bakeries. Provide an explanation for the missing data in the table.

c. After some consideration, you decide that it is necessary to evaluate homogeneity in day-to-day exposures experienced by workers in each occupational group. Calculate the $_WR_{95}$ value for the group of dough makers in wheat bread bakeries.

17.4. Assume that you have run a one-way random-effects model to evaluate the relative magnitude of the between- and within-worker sources of variation on personal monitoring data collected on a group of workers exposed to benzene. You obtain estimated values of 1.63 and 2.89 for σ_B^2 and σ_W^2, respectively, and -0.580 for μ_Y. Estimate the probability of exceedance and overexposure for this group of workers. Assume an OEL of 1.63 mg/m³.

REFERENCES

Armstrong, B.: Effect of measurement error on epidemiological studies of environmental and occupational exposures. Occup. Environ. Med. 55:651-656 (1998).

Burdick, R.K.; Graybill, F.A.: Confidence intervals on variance components. Marcel Dekker, Inc., New York, NY (1992).

Burdorf, A.: Sources of variance in exposure to postural load on the back in occupational groups. Scand. J. Work Environ. Health 18:361-367 (1992).

Burdorf, A.: Identification of determinants of exposure: consequences for measurement and control strategies. Occup. Environ. Med. 62:344-350 (2005).

Burdorf, A.; Lillienberg, L.; Brisman, J.: Characterization of exposure to inhalable flour dust in Swedish bakeries. Ann. Occup. Hyg. 38:67-78 (1994).

Burdorf, A.; van Tongeren, M.: Variability in workplace exposures and the design of efficient measurement and control strategies. Ann. Occup. Hyg. 47: 95-99 (2003).

Burstyn, I.; Kromhout, H.: Are the members of a paving crew uniformly exposed to bitumen fume, organic vapor, and benzo(a)pyrene? Risk Anal. 20:653-663 (2000).

Burstyn, I.; Kromhout, H.; Kauppinen, T.; Heikkila, P.; Boffetta, P.: Statistical modeling of the determinants of historical exposure to bitumen and polycyclic aromatic hydrocarbons among paving workers. Ann. Occup. Hyg. 44:43-56 (2000).

Burstyn, I.; Teschke, K.: Studying the determinants of exposure: A review of methods. Am. Ind. Hyg. Assoc. J. 60:57-72 (1999).

Cope, R.; Panacamo, B.; Rinehart, W.E.; Ter Haar, G.L.: Personal monitoring for tetraalkylead in an alkyl lead manufacturing plant. Am. Ind. Hyg. Assoc. J. 40:372-379 (1979).

Corn, M.; Esmen, N.A.: Workplace exposure zones for classification of employee exposures to physical and chemical agents. Am. Ind. Hyg. Assoc. J. 40:47-57 (1979).

Diggle, P.J.: Time series. A biostatistical introduction, pp. 90-91. Clarendon Press, Oxford, (1990).

Eduard, W.; Bakke, B.: Experiences with task-based exposure assessment in studies of farmers and tunnel workers. Norsk Epidemiologi 9:65-70 (1999).

Francis, M.; Selvin, S.; Spear, R.; Rappaport, S.M.: The effect of autocorrelation on the estimation of workers' daily exposures. Am. Indus. Hyg. Assoc. J. 57:599-606 (1989).

Fuller, W.A.: Measurement error models. John Wiley and Sons, New York, NY (1987).

George, D.K.; Flynn, M.R.; Harris, R.L.: Autocorrelation of interday exposures at an automobile assembly plant. Am. Ind. Hyg. Assoc. J. 56:1187-1194 (1995).

Hall, A.; Teschke, K.; Davies, H.; Demers, P.; Marion, S.: Exposure levels and determinants of softwood dust exposures in BC lumber mills, 1981-1997. Am. Ind. Hyg. Assoc. J. 63:709-714 (2002).

Heederik, D.; Boleij, J.S.M.; Kromhout, H.; Smid, T.: Use and analysis of exposure monitoring data in occupational epidemiology: An example of an epidemiological study in the Dutch animal food industry. Appl. Occup. Environ. Hyg. 6, 458-464 (1991).

Heederik, D.; Attfield, M.: Characterization of dust exposure for the study of chronic occupational lung diseases: A comparison of different exposure assessment strategies. Am. J. Epidemiol. 151:982-990 (2000).

Hines, C.J.; Deddens, J.A.: Determinants of chlorpyrifos exposures and urinary 3,5,6-trichloro-2-pyridinol levels among termiticide applicators. Ann. Occup. Hyg. 45:309-321 (2001).

Houba, R.; Heederik, D.; Kromhout, H.: Grouping strategies for exposure to inhalable dust, wheat allergens and α-amylase allergens in bakeries. Ann. Occup. Hyg. 41:287-296 (1997).

Jansen, J.P.; Burdorf, A.: Effects of measurement strategy and statistical analysis on dose-response relations between physical workload and low back pain. Occup. Environ. Med. 60:942-947 (2003).

Jeebhay, M.F.; Robins, T.G.; Seixas, N.; Baatjies, R.; George, D.A.; Rusford, E.; Lehrer, S.B.; Lopata, A.L.: Environmental exposure characterization of fish processing workers. Ann. Occup. Hyg. 49:423-437 (2005).

Kromhout, H.; Heederik, D.: Occupational epidemiology in the rubber industry: Implications of exposure variability. Am. J. Ind. Med. 27:171-185 (1995).

Kromhout, H.; Loomis, D.P.; Mihlan, G.J.; Peipins, L.A.; Kleckner, R.C.; Iriye, R.; Savitz, D.A.: Assessment and grouping of occupational magnetic field exposure in five electric utility companies. Scand. J. Work Environ. Health 21:43-50 (1995).

Kromhout, H.; Symanski, E.; Rappaport, S.M.: A comprehensive evaluation of within- and between-worker components of occupational exposure to chemical agents. Ann. Occup. Hyg. 37:253-270 (1993).

Kromhout, H.; Tielemans, E.; Preller, L.; Heederik, D.: Estimates of individual dose from current measurements of exposure. Occup. Hyg. 3:23-39 (1996).

Kromhout, H.; Vermeulen, R.: Temporal, personal and spatial variability in dermal exposure. Ann. Occup. Hyg. 45: 257-73 (2001).

Kumagai, S.; Matsunaga, I.; Kusaka, Y.: Autocorrelation of short-term and daily average exposure levels in workplaces. Am. Ind. Hyg. Assoc. J. 54:341-350 (1993).

Lin, Y.S.; Kupper, L.L.; Rappaport, S.M.: Air samples versus biomarkers for epidemiology. Occup. Environ. Med. 62:750-760 (2005).

Lyles, R.H.; Kupper, L.L; Rappaport, S.M.: Assessing regulatory compliance of occupational exposures via the balanced one-way random effects ANOVA model. J. Agricul. Bio. Environ. Stat. 2:64-86 (1997a).

Lyles, R.H.; Kupper, L.L.; Rappaport, S.M.: A lognormal distribution-based exposure assessment method for unbalanced data. Ann. Occup. Hyg. 41:63-76 (1997b).

Mulhausen, J.R.; Damiano J.: A strategy for assessing and managing occupational exposures, 2nd ed. Fairfax,VA: AIHA Press. ISBN 0-932627 (1998).

Mwaiselage, J.; Bratveit, M.; Moen, B.; Yost, M.: Variability in dust exposure in a cement factory in Tanzania. Ann. Occup. Hyg. 49:511-519 (2005).

Neter, J.; Kutner, M.H.; Nachtsheim, C.J.; Wasserman, W.: 1996. Applied Linear Statistical Models. McGraw-Hill, New York (1996).

Nieuwenhuijsen, M.J.; Lowson, D.; Venables, K.M.; Taylor, A.J.N.: Flour dust exposure variability in flour mills and bakeries. Ann. Occup. Hyg. 39:299-305 (1995).

Peretz, C.; Froom, P.; Pardo, A.; Goren, A.: Exposure to benzene in fuel distribution installations: Monitoring and prevention. Arch Environ Health 55:439-446 (2000).

Peretz, C.; Goren, A.; Smid, T.; Kromhout, H.: Application of mixed-effects models for exposure assessment. Ann. Occup. Hyg. 46:69-77 (2002).

Raaschou-Nielsen, O.; Hansen, J.; Thomsen, B.L.; Johansen, I.; Lipworth, L.; McLaughlin, J.K.; Olsen, J.H.: Exposure of Danish workers to trichloroethylene, 1947-1989. Appl. Occup. Environ. Hyg. 17:693-703 (2002).

Rappaport, S.M.: Assessment of long-term exposures to toxic substances in air. Ann. Occup. Hyg. 35: 61-121 (1991).

Rappaport, S.M.; Kromhout, H.; Symanski, E.: Variation of exposure between workers in homogeneous exposure groups. Am. Ind. Hyg. Assoc. J. 54:654-662 (1993).

Rappaport, S.M.; Lyles, R.H.; Kupper, L.L.: An exposure-assessment strategy accounting for within- and between-worker sources of variability. Ann. Occup. Hyg. 39:469-495 (1995a).

Rappaport, S.M.; Lyles, R.H.; Kupper, L.L.: Letter to the editor re: A lognormal distribution-based exposure assessment method for unbalanced data. Ann. Occup. Hyg. 42:417-419 (1998).

Rappaport, S.M.; Symanski, E.; Yager, J.W.; Kupper, L.L.: The relationship between environmental monitoring and biological markers in exposure assessment. Environ. Health Perspect. 103(suppl. 3):49-54 (1995b).

Rappaport, S.M.; Weaver, M.; Taylor, D.; et al.: Application of mixed models to assess exposures monitored by construction workers during hot processes. Ann. Occup. Hyg. 43:457-469 (1999).

Scheepers, P.T.J.; Micka, V.; Muzyka, V.; Anzion, R.; Dahmann, D.; Poole, J.; Bos, R.P.: Exposure to dust and particle-associated 1-nitropyene of drivers of diesel-powered equipment in underground mining. Ann. Occup. Hyg. 47:379-388 (2003).

Schlünssen, V.; Sigsgaard, T.; Schaumburg I.; Kromhout, H.: Cross-shift changes in FEV_1 in relation to wood dust exposure: the implications of different exposure assessment methods. Occup. Environ. Med. 61:824-830 (2004).

Searle, S.R.; Casella, G.; McCulloch, C.E.: Variance Components. John Wiley and Sons, New York, NY. ISBN 0 471 62162 5 (1992).

Seixas, N.S.; Sheppard, L.: Maximizing accuracy and precision using individual and grouped exposure assessments. Scand. J. Work Environ. Health 22:94-101 (1996).

Spear, R.C.; Selvin, S.: OSHA's Permissible Exposure Limits: Regulatory compliance versus health risk. Risk Anal. 9:579-586 (1989).

Symanski, E.; Bergamaschi, E.; Mutti, A.: Inter- and intra-individual sources of variation in levels of urinary styrene metabolites. Int. Arch. Environ. Occup. Health. 74:336-344 (2001a).

Symanski, E.; Chang, C.; Chan, W.: Long-term trends in exposures to nickel aerosols. Am. Ind. Hyg. Assoc. J. 61:324-333 (2000a).

Symanski, E.; Chan, W.; Chang, C.: Mixed-effects models for the evaluation of long-term trends in exposure levels with an example from the nickel industry. Ann. Occup. Hyg. 45:71-81 (2001b).

Symanski, E.; Greeson, N.M.H.: Assessment of variability in biomonitoring data using a large database of biological measures of exposure. Am. Ind. Hyg. Assoc. J. 63:390-401 (2002).

Symanski, E.; Greeson, N.M.H.; Chan, W.: Evaluating measurement error in estimates of worker exposure assessed in parallel by personal and biological monitoring. Am. J. Ind. Med. 50:112-121 (2007).

Symanski, E.; Kupper, L.L.; Kromhout, H.; Rappaport, S.M.: An investigation of systematic changes in occupational exposure. Am. Ind. Hyg. Assoc. J. 57:724-735 (1996).

Symanski, E.; Sällsten, G.; Barregård, L.: Variability in airborne and biological measures of exposure to mercury in the chloralkali industry: Implications for epidemiologic studies. Environ. Health Perspect. 108:569-573 (2000b).

Symanski, E.; Kupper, L.L.; Hertz-Picciotto, I.; Rappaport, S.M.: Comprehensive evaluation of long-term trends in occupational exposure: part 2. Predictive models for declining exposures. Occup. Environ. Med. 55:310-316 (1998).

Symanski, E.; Maberti, S.I.; Chan, W.: A meta-analytic approach for characterizing the within- and between-worker sources of variation in exposure to workplace contaminants. Ann. Occup. Hyg. 50:343-357 (2006).

Symanski, E.; Rappaport, S.M.: An investigation of the dependence of exposure variability on the interval between measurements. Ann. Occup. Hyg. 38:361-372 (1994).

Symanski, E.; Sällsten, G.; Chan, W.; Barregård, L.: Heterogeneity in sources of exposure variability among groups of workers exposed to inorganic mercury. Ann. Occup. Hyg. 45:677-687 (2001c).

Teschke, K.; Spierings, J.; Marion, S.A.; Demers, P.A.; Davies, H.W.; Kennedy, S.M.: Reducing attenuation in exposure-response relationships by exposure modeling and grouping: The relationship between wood dust exposure and lung function. Am. J. Ind. Med. 46:663-667 (2004).

Teschke, K.; Marion, S.A.; Vaughan, T.L.; Morgan, M.S.; Camp, J.: Exposures to wood dust in U.S. industries and occupations, 1979-1997. Am. J. Ind. Med. 35:581-589 (1999).

Thomas, D.; Stram, D.; Dwyer, J.: Exposure measurement error: Influence on exposure disease relationships and methods for correction. Annu. Rev. Public Health 14:69-93 (1993).

Tielemans, E.; Kupper, L.L.; Kromhout, H.; Heederik, D.; Houba, R.: Individual-based and group-based occupational exposure assessment strategies: Some equations to evaluate different strategies. Ann. Occup. Hyg. 42:115-119 (1998).

Tornero-Velez, R.; Symanski, E.; Kromhout, H.; Yu, R.C.; Rappaport, S.M.: Compliance versus risk in assessing occupational exposures. Risk Anal. 17:279-292 (1997) [erratum appears in Risk Anal. 17:657 (1997)].

van der Woord, M.P.; Kromhout, H.; Barregård, L.; Jonsson, P.: Within-day variability of magnetic fields among electric utility workers: Consequences for measurement strategies. Am. Ind. Hyg. Assoc. J. 60: 713-719 (1999).

van Tongeren, M.; Burstyn, I.; Kromhout, H.; Gardiner, K.: Are variance components of exposure heterogeneous between time periods and factories in the European carbon black industry? Ann. Occup. Hyg. 50:55-64 (2006).

van Tongeren, M.K.; Kromhout, H.; Gardiner, K.: Trends in levels of inhalable dust exposure, exceedance and overexposure in the European carbon black manufacturing industry. Ann. Occup. Hyg. 44:271-280 (2000).

van Tongeren, M.K.; Gardiner, K.; Calvert, I.; Kromhout, H.; Harrington, J.M.: Efficiency of different grouping schemes for dust exposure in the European carbon black respiratory morbidity study. Occup. Environ. Med. 54:714-719 (1997).

van Tongeren M.J.; Kromhout, H.; Gardiner, K.; Calvert, I.A.; Harrington, J.M.: Assessment of the sensitivity of the relation between current exposure to carbon black and lung function parameters when using different grouping schemes. Am. J. Ind. Med. 36:548-556 (1999).

Vermeulen, R.; Li, G.; Lan, Q.; Dosemeci, M.; Rappaport, S.M.; Bohong, X.; Smith, M.T.; Zhang, L.; Hayes, R.B.; Linet, M.; Mu, R.; Wang, L.; Xu, J.; Yin, S.; Rothman, N.: Detailed exposure assessment for a molecular epidemiology study of benzene in two shoe factories in China. Ann. Occup. Hyg. 48: 105-116 (2004).

18

QUANTITATIVE SURVEYING – FORMING THE STRATEGY

The purpose of this chapter is to discuss exposure monitoring strategies. Of course one possible strategy is to not monitor. As odd as that may seem, there may be legal reasons for it (at least in the U.S.), although these are certainly not grounded in ethics or morals. Based on the previous chapters, we should be convinced that compliance with an OEL 100% of the time is not likely; so how can we define compliance in a way that we can actually achieve it? Before a strategy can be formed, one must also interpret what the applicable OEL is trying to prevent. It is trying to prevent disease of some kind, but is it disease that occurs after a short exposure, long exposure, or both?

THREE TOXICANT TYPES

We can think of examples of toxicants with OELs that are set based upon clearly acute effects (H_2S, NH_3, Cl_2, phosgene) and chronic effects (asbestos, lead, cadmium). For perhaps 10–20 compounds in each class, the risk assessment, TLV® *Documentation*, or other basis for the OEL clearly indicates that either an exposure of very short duration or a long-term mean (chronic) exposure drives the disease that we are trying to prevent. Between these two extremes there are many toxicants for which the rationale for an OEL is not so well developed, but nonetheless, these substances are meant to be monitored with 8-hr TWAs. However, the biological half-life for a particular disease outcome for any of these chemicals may be short enough that exceeding the 8-hr TWA OEL might cause harm to the worker. Moreover, for most of these substances, we do not have solid evidence that only a long term mean drives disease, i.e., that they are clearly chronic hazards. We will call this in-between toxicant type *subchronic* for lack of a better term. Any strategy for exposure monitoring must consider the toxicodynamics of all three of these toxicant types.

TO MONITOR OR NOT?

Rappaport (1984) argued that, from a purely economic and short-sighted viewpoint, monitoring in the U.S. is not cost-effective. This conclusion is based on several factors. First, the OSHA comprehensive standards, as well as the OSH Act, state that any worker exposed beyond a PEL[1] on any given day constitutes a violation of the Act. Because occupational exposures are not constant in time or space and are usually skewed to the left, every employer is, at least in theory, out of compliance. The degree of likelihood of this *out of compliance* situation is dependent on the mean and the GSD of the exposure distribution. Therefore, the more an employer monitors, the greater the risk of self-incrimination. Since there are about 2,000 (state and federal OSHA) inspection officers and more than 6 million U.S. workplaces, the odds of being inspected are small. Finally, if an inspection does occur, the odds of OSHA finding an out-of-compliance situation may be small, as OSHA will not have time to monitor many workers, and the majority of exposures may be less than the PEL. Therefore, why should the employer bother?

Several reasons will be given as to why the employer should bother, but another question is, why doesn't OSHA make its compliance rules more realistic, more reflective of the inherent variability in exposure distributions, at least those for chronic toxicants like lead and asbestos for which it is thought that a single day beyond the PEL is not what drives lifetime risk of disease? At least one attempt has been made to do just that. In the preamble to the benzene standard (OSHA, 1987), four possible ways to change OSHA's compliance procedures were proposed.

1. Adopt a 40-hr, rather than 8-hr, TWA. This averaging time would be more reasonable based on Roach's (1977) body burden half-life arguments. It would *average-out* any peaks that might occur in the five 8-hr TWAs. Some members of industry, as well as OSHA, objected to the difficult logistics of requiring a single employee to wear a monitoring apparatus on five consecutive days.

2. Increase the 8-hr TWA-PEL (e.g., from 1 to 2 ppm) so that there is a lower probability of detecting outliers. OSHA and the AFL-CIO workers' union argued that in many plants where only minimal monitoring takes place, this would allow the average exposure level to drift upward, thus increasing risk.

3. Require OSHA to find two consecutive noncompliance values before issuing a citation. If the probability of noncompliance (portion of exposures exceeding the PEL) were, say, 20%, then under this scheme the probability of a citation would be 0.2 × 2 or 4%, a reasonable value if one assumes that the mean exposure drives disease. But OSHA gave three reasons why it did not like the approach. First, upon

---

[1] In this chapter, the generic term OEL is used except in a few cases where it is clear that a legal OSHA PEL is referred to.

coming back for a second confirming result the employer could temporarily reduce exposures. Second, if the probability of exceeding the PEL were, say, 50%, then the probability of a citation would only be 25%. Third, given constant resources, fewer inspections could be made.

4. Allow the use of employer data to rebut the presumption of noncompliance based on a single noncompliance value. OSHA pointed out that the employer could alter the exposure average in various ways. For example, the exposure group could be diluted with persons having low exposures, or the employer might purposely monitor only on days when exposures were expected to be low.

All four alternatives were dismissed, and it would appear that hygienists (at least in the United States) must control exposures for clearly chronic toxins in order to reduce the probability of a citation to some acceptable gamble (say, 5%). More recent work has shown that OSHA's approach, basing the grounds for a citation on one exceedance, is in fact reasonable, and these results further indicate that monitoring is indeed a worthwhile effort.

The *OSHA Technical Manual* (OSHA, 1995) also directs the compliance officer to consider employer data in deciding whether to issue a citation. This is somewhat similar to number 4 above. The employer data can be of any type – random or worst case – but random sampling does allow the employer to calculate any percentile as well as the mean. Thus OSHA's policy does not prohibit random sampling, nor does it absolutely require a citation for every exposure above the PEL. Most important, it does justify, on economic grounds, employer expense for monitoring. Of course, costs of monitoring are always justified for protecting worker health or future legal considerations. Certainly all details of the monitoring program should be clearly documented, such as how SEGs are defined, how random sampling is performed, and how quality control is maintained.

EXCEEDANCE, COMPLIANCE, AND OVEREXPOSURE

Should a PEL never be exceeded? Can a PEL be exceeded and regulatory compliance still be achieved? More importantly, can a PEL be exceeded and health be protected or overexposure be avoided? If OSHA cites any exceedance that it finds, is it overprotecting, i.e., is it citing employers who are controlling exposures well, but have occasional upsets that OSHA blindly finds? Remember that we have admitted that even in well-controlled situations there will be occasional exposures that exceed an OEL. The question we would like to answer is, under what conditions, when we exceed an OEL, is there a risk of harmful effect that is greater than a risk that has been deemed acceptable? Where a risk assessment has been performed, the excess risk at the OEL is stated and the discussion can proceed. Where the risk is not stated, as is mostly the case (see Chapters 10 and 11), the discussion is undefined. Nonetheless, we have to realize that most monitoring and control strategies have to allow for occasional exceedances of an OEL; it is difficult to determine when these exceedances are causing harm for the vast majority of *sub-chronic* toxicants. We can define three terms that are important to differentiate:

- the exceedance rate (γ) is the fraction of exposures that ever exceed a PEL
- the probability of an overexposure (θ) is the probability of an exposure that causes a disease risk that exceeds the risk implied by the PEL or any other *acceptable* risk level
- non-compliance is the probability that OSHA will detect an exceedance.

To answer the above questions we need to know at least something about the time dependent nature of the toxicant and something about how the PEL was set. The questions must be answered differently for an acute versus a chronic toxicant versus the many *sub-chronic* toxicants that are not clearly in either group. For a toxicant that acts very rapidly and has either a ceiling or a STEL, exceeding the PEL might damage the workers' health. For many toxicants in this category, the risk of harming the worker's health by exceeding the PEL is not worth taking, and consequently safeguards must be in place to assure that this does not occur. The best way of doing this is to install direct reading equipment with alarm systems that warn the workers when it is unsafe to stay in the environment. On the other hand, for toxicants that are chronic in nature, where the effect is caused by accumulation of the toxicant over time rather than any short-term peak in exposure, exceeding the 8-hr TWA will not likely harm the employee's health. It is the long-term average of 8-hr TWAs that drives the disease process. In other words, the probability of an overexposure (θ) is the probability that a *long term mean* exposure causes a disease risk that exceeds the risk implied by the PEL. For all other toxicants (the majority) the picture is not so clear.

In the U.S., OSHA's policy is to issue a citation when a single exceedance is found. Chapters 10 and 11 discussed that for chronic toxicants such as asbestos and lead, OSHA's PELs are based on epidemiological work that relates the average exposure of workers to a particular outcome, such as cancer. Thus, in essence, the chronic PEL is based upon the principle that the long-term average exposure drives the disease. This implies that exceeding the PEL is acceptable (will not cause overexposure), so long as the average exposure does not exceed the PEL.

However, one must admit that the epidemiological evidence often lacks consideration of the possibility that shorter-term exposures might also drive the disease process. In addition, there are two parts to exceeding the PEL, how often and by how much. It could be that an 8-hr TWA that exceeds the PEL by some large margin might actually drive the disease process in an unexpected manner. Thus we need to understand that when we are attempting to control average exposures, we should limit exceedances even though the more we limit exceedances (either shift the entire distribution to the left or limit factors that may contribute to exposures in the tail), the more we reduce the average exposure.

Remember also from previous discussions that exposure standards never increase. Thus by controlling to an average value that is < PEL, or by controlling some upper percentile to below the PEL, we err on the safe side relative to any future findings that a current long term PEL is inadequate. Also, even for chronic toxicants, as the mean exposure approaches the PEL, it becomes increasingly difficult to detect the difference between a mean exposure and the PEL; if we control average exposures for a chronic toxicant to some fraction of the PEL, it becomes statistically easier to confirm this difference.

Is OSHA's Approach Sound for Chronic Toxicants?

Is OSHA overzealous in citing workplaces for only a single exceedance when the toxicant has only *chronic* effects? Studies have shown that, fortuitously, OSHA's citation policy yields a high likelihood that it will detect a workplace where a single worker's average exposure exceeds the PEL, i.e., there is an overexposure. Spear and Selvin (1989) show that when the fraction of measurements that may exceed the PEL (γ) is low, the probability of no overexposures (θ) i.e., the fraction of individual worker-mean exposures that exceeds the PEL is also small or a bit smaller than γ. Conversely, when γ is large, θ is the same or larger. Thus, in the latter case, when OSHA detects an exceedance it is probably detecting an overexposure.

This relationship between γ and the θ is dependent upon the variability between and within workers in an exposure group, as discussed in previous chapters. Spear and Selvin considered the case where OSHA finds an exceedance and they ask the question, "Is OSHA justified to give a citation?" Tornero-Valez et al. (1997) ask the inverse: "If OSHA does not find an exceedance, is it reasonable to assume that overexposure has not occurred?" Based on data for 179 exposure sets they found that for 10% of them $\gamma > 0.25$. The probability of compliance is

$$(1 - \gamma)^n$$

where n is the number of measurements made by OSHA. Thus, the probability of OSHA concluding compliance when exceedance rate, $\gamma = 0.25$ ranges from 0.32 to 0.75 for typical n of 1–4, even though actual overexposure values (θ) were quite high (> 25%) in these 18 datasets. This, of course, means that it is likely that OSHA is not detecting workplaces that are non-compliant and where individual workers are overexposed.

Tornero-Valez, et al. (1997) also noted that controlling exceedance rates should not be a substitute for controlling θ, i.e., one should not assume that by holding the grand mean exposure (for the exposure group) to say 0.25 PEL and thus limiting exceedances to 5%, that overexposures (i.e., mean exposures for individual workers) are being avoided. Again this depends on the total GSD and on how between and within variance is distributed in the exposure group. For large GSDs (> 5), θ may well exceed 5%, when $\gamma = 5\%$. On the other hand, in more consistently controlled workplaces ($GSD_T < 3.5$) controlling the exceedance rate to 5% and thus the grand mean of exposures to 0.25 PEL should also control individual risk to less than that implied at the PEL.

Again, the above relates only to chronic toxicants. For the majority of toxicants that may have sub-chronic effects, the situation is less comfortable. For example, if a substance has a 1 in 1000 risk of causing at least debilitating effects when exposures exceed the PEL for one day, and a workplace exceeds the OEL 50% of the time, then a single OSHA measurement has only a 50% chance of detecting what amounts to a series of overexposures.[2]

THE AIHA STRATEGY

Convinced that monitoring should be performed, the next question is, how? The NIOSH (1977) strategy, discussed in Chapter 14, Figure 14-1, represents one answer to the question, but it is a short-term solution, as it tells us that under present conditions, based on today's results, no more monitoring is necessary, more monitoring is necessary, or controls and more monitoring are necessary. This approach may be acceptable when exposures are quite low, e.g., exposure group mean < 0.1 PEL for the day, or for more acute toxicants, when $\hat{\gamma} < 1\%$. (Again for the most acute toxicants like hydrogen sulfide, direct reading instruments with alarms are warranted.) However, the NIOSH approach is really established to give OSHA guidance on its sampling strategies for a single day or a few days in a plant. It is not a random sampling strategy across time.

A committee of AIHA has made a proposal for a random sampling strategy that has some characteristics of the NIOSH strategy (AIHA, 1991, 1998).[3] However, they propose a strategy based on random sampling of carefully defined homogeneous exposure groups (HEGs, now called similar exposure group or SEG).[4] The individual monitoring events should be staged on random days for single workers within each SEG in order to avoid correlation effects. A minimum sample size of six is recommended, but this is based on a normal distribution.[5] The authors (AIHA) state that "a reasonable approximation to an exposure distribution is usually

---

[2] Say there are 1) 10 workers, 2) 50% of 8-hr TWAs exceed the OEL, 3) each exceedance has on average a 1 in 1000 chance of an undesirable effect. Thus, on a given day, the odds of an undesirable effect are $0.5 \times 10 \times 0.001 = 0.005$. So on about 1–2 days per year (assuming only 250 shifts per year) the effect would occur, *on average*. If the employee developed respiratory disease as a result of this exposure and missed 3 days of work, would this be an undesirable effect?

[3] The AIHA document is a good introductory text for the overall subject of exposure and risk assessment. It is recommended reading for any hygienist charged with establishing or maintaining such a program.

[4] The term homogeneous, previously used by AIHA, has been dropped based on the difficulty in establishing groups with truly homogeneous risk.

[5] Lognormally based sample size calculations were presented in Chapter 16, showing that when the distribution approaches normality (GSDs < 2.0) and the true mean is about one-half the standard or less, six would be an adequate size to reject a null hypothesis involving the mean and an OEL. More on this later.

possible with 10 samples; however, for rigorous goodness of fit testing for a distribution, 30 samples or more are often needed." Therefore, the final decision about sample size is necessarily based upon professional judgment, the problem at hand...testing a hypothesis, estimating the mean, estimating a percentile, diagnosis of control issues, task monitoring, and other factors such as the toxicity of the contaminant, costs per measurement, etc.

The AIHA strategy is shown in Figure 18-1. Its similarities to the NIOSH strategy (yes, no, maybe) are apparent. The cyclical nature of the process is really the key point. Exposure assessment is not a start-end process. We gather information that allows us to make decisions. Even if the decisions are eventually *acceptable* or *unacceptable*, these outcomes lead to other events, like controlling the exposure, which then require further assessment, in this case to determine how well controls are working. In the case of an acceptable exposure outcome, reassessment is eventually required to determine that process, worker, and other changes have not changed the conclusion.

The basic characterization step is essentially covered in Chapters 12 and 13. The expo-

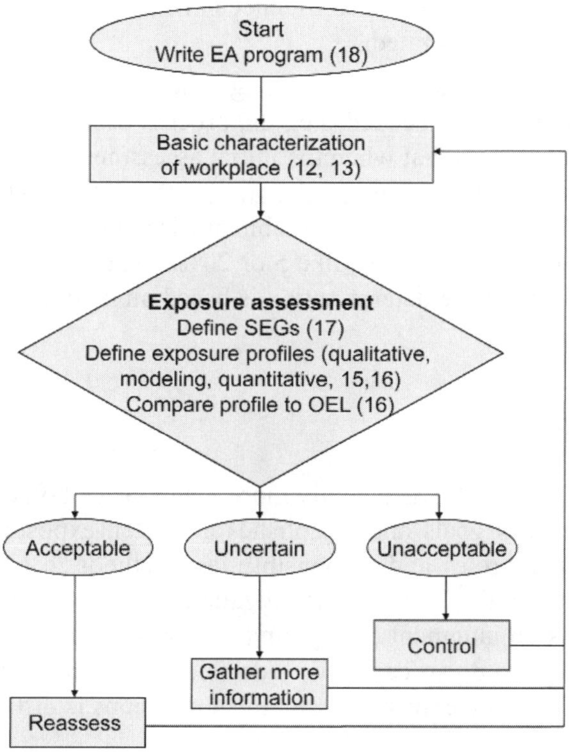

Figure 18-1. The AIHA exposure assessment strategy. Numbers in parentheses refer to the chapter in this volume where topic is discussed. Adapted from AIHA (2006) with permission.

sure assessment step is equivalent to the discussion in chapters 14, 15, 16, and 17, but the AIHA exposure assessment process may be qualitative, semi-quantitative, or quantitative. For example, in a workplace there may be, say 150 chemical agents. Perhaps three of these are pervasive...they are a part of most processes and they clearly enter the air. The other 147 are used sporadically, used in very small amounts (e.g., a liter a month), are not volatile or mechanically processed in such a way as to become airborne...etc. These 147 chemicals nonetheless require a written determination of exposure potential. This will likely be made with no measurements or modeling, modeling only, a few measurements that may include presumed or simulated worst case conditions, or some combination of modeling and a few measurements. The other three chemicals are really the subject of most of the discussion in Chapter 14 and all of the discussion in Chapters 16 and 17.

To perform exposure assessment in a non-quantitative fashion for these 147 chemicals requires use of professional judgment (PJ). AIHA expends a lot of effort in describing PJ. Unfortunately, it is difficult to explain how past experience may allow one hygienist to look at a situation in a walkthrough survey and determine that exposures are very unlikely to exceed the OEL, and be right, while another hygienist predicts that the mean of exposures will be about the OEL, and be really wrong. Nonetheless that is what PJ is all about. There is a balance that every hygienist strikes between confidence in PJ and monitoring. Over time and with experience, PJ becomes more calibrated.

Often cutoff values are used for determining whether quantitative exposure profiling (AIHA terms this a full profile) is needed, i.e., the process described in Chapters 16 and 17. For example, AIHA recommends that when the initial assessment,[6] which may be based on PJ only, indicates that either STELs or 8-hr TWAs are $\geq$ 10% of the OEL, then a full (quantitative) profile should be developed. How comfortable one is with a PJ driven determination may cause one to use a lower or higher cutoff, like 5 or 20%. Clearly other factors like the toxicity of the contaminant, whether the determination is based on modeling data, and other factors come into play as well.

BEFORE YOU START

In Figure 18-1 the start button is actually more than a place holder. AIHA states that this is where the hygienist states goals and establishes a written exposure assessment program which might include goals, roles and responsibilities, methods to be used, SEG definition processes, exposure acceptability criteria, prioritization processes and criteria, etc. The point is that anytime one relies on judgment of any kind, it is wise to have a rational and organized basis for doing so. Failure to do so can lead to decisions that are difficult to defend later. A step-by-step process for making exposure assessment decisions is difficult because individual

[6] The exposure assessment process is iterative as the cyclical arrows in Figure 18-1 indicate. Thus, there may be an initial assessment followed by successively more refined assessments.

situations often warrant different approaches. However, in all cases, certain things are necessary, and a checklist is valuable for assuring that all components of a strategy have been considered. The following checklist may not be complete for every situation, but it should provide an adequate starting place.

1. Where does one start? Unless documentation on current conditions is needed, or regulations require it, make simple, inexpensive and/or necessary control adjustments first. If there is an obvious lack of control procedures for a particular process, the sampling effort should not be wasted (unless of course a historical documentation of exposures is warranted); the control procedures should be corrected first. Often it is possible to correct work procedures with little or no cost to the company. A simple change in work procedures may reduce exposures considerably. Inadequate training of workers can lead to high exposures; this situation must be corrected.

2. What question must be answered? The question should be stated in the form of a hypothesis (even if you are not going to collect monitoring data) and include a definition of the statistical parameter of interest and the population. In defining the SEG, time and space are important. The scientific method should be used. Here are some examples of hypotheses or questions.

   a. What is the mean exposure over a one-year period among workers performing a defined task?

   b. For a chronic toxicant is the mean exposure over a one-year period, for a defined SEG, less than the OEL?

   c. Is the 95 %tile of 8-hr TWA exposures over a one-year period, for a defined SEG, less than the OEL?

   The first question asks for an estimate of a population parameter; the last two are hypothesis-testing questions. Often a parameter must be estimated in order for us to form and test a hypothesis. For example, the GSD must be estimated in order to calculate a sample size for testing hypothesis (b), above. This can be done by guessing or by taking a small number of random measurements in a pilot study. It may also be necessary to estimate both the mean and the variance (GSD) in order to define SEGs in a uniform manner. The point is that if the question is not first posed, the data that are collected may represent a wasted effort, in that they cannot be used to answer the question.

3. Has a rational time period been determined? For assessing long-term exposures to chronic toxicants, a lifetime average exposure is a biologically rational parameter. However, this implies that the mean is not calculated until the end of the period. Furthermore, we need to assume a stationary distribution, and it is unreasonable

to expect the mean of an exposure distribution to remain stable for years. Therefore, a more reasonable statistical sampling period might be one year for a sub-chronic or chronic toxicant although any decision in this regard often will be arbitrary.

For acute toxicants the issue is the reversibility of the effects. For toxicants with irreversible effects, monitoring needs to be continuous. For those with reversible effects, monitoring should take place on a frequency commensurate with the seriousness of the effects and should occur as frequently as changes in variables occur that might change concentrations. In all cases, choose a defensible time period. Data collected over time periods longer than one year can be analyzed using control charting procedures discussed later.

4. Does the toxicant have chronic, sub-chronic, and/or acute affects? The body burdens for chronic toxicants are driven by average exposures, whereas for acute toxicants body burdens are considerably affected by peaks or short-term spikes in the exposure profile. There may be some exceptions to this rule for chronic toxicants, such as certain carcinogens that have alternate metabolic pathways once exposures exceed a certain level. For carcinogens, the chronic effect is not necessarily owed to the chronic nature of exposure, but perhaps to the chronic nature of the biological process that eventually allows a tumor to develop. In some cases, peaks above a certain level as well as long-term averages could be important. For most toxicants the health outcome derived from the animal and/or human data upon which the OEL is based may be fairly clear, but its toxicokinetics (i.e., can the effect be caused by a few relatively high exposures, a long-term mean or some other combination) are usually not clear. While a PEL or other regulatory standard may state that the required averaging time is say 8 hours, the hygienist should understand, to the extent that data are available, the biological half-lives of the toxicant. Of course when there are 1000s or even 10s of toxicants in the workplace this is difficult at best, but toxicokinetics should be studied for the higher risk substances.

5. Does the acute toxicant have warning properties? Acute toxicants essentially fall into two categories: irritants and non-irritants. Irritants such as ammonia, tear gas, and stannous chloride are extremely irritating to individuals at levels less than those that might cause irreversible effects. Consequently, these levels are rarely encountered without an escape effort. On the other hand, compounds such as hydrogen sulfide and sulfur dioxide do not have adequate warning properties. These latter toxicants require direct-reading instrumentation with an alarm system.[7]

---

[7] This is a good rule of thumb; but see later discussion under Acute Toxicants.

6. Are exposures continuous or intermittent? This dichotomy is largely related to the nature of the tasks that the worker performs. A worker in a petroleum refinery may be constantly exposed to benzene and may encounter higher exposures to it when entering certain process units or when opening certain equipment. A laboratory worker in a properly controlled environment can be exposed to non-detectable levels of benzene, perhaps until such time as benzene is used for a task, in a hood. The task may involve the filling of a beaker or an extraction procedure. Consequently, exposure to benzene in this case is intermittent, and its duration is equal to that of the task. Control techniques and sampling strategies for these two types of exposures may be different. For more information on task-based sampling strategies, see Nicas and Spear (1993a, b). Note that the procedures outlined in these references allow one to do task monitoring and from that calculate parameters such as full shift means. The advantage is that task data can point to those tasks that require controls.

7. Have differences in worker habits (e.g., work practices) and slight differences in work procedures and settings that might cause significant between-worker variability been considered in defining the SEG? Even though workers in the SEG may be doing the same job and may be exposed to the same background environment, the exposures may be substantially different among workers in the SEG.

8. Is a worst-case approach or a negative documentation approach needed for monitoring employee exposures? If so, make more than one exposure estimate. Tuggle (1981) has recommended the use of two worst-case estimates, but three or four may be appropriate, depending on concentrations and toxicity. Where possible, worst-case exposure estimates should be simulated as absolutely worst-case. As an example, consider a SEG in which employees perform five different tasks. On any given day each worker may perform approximately two tasks. Therefore, there are several combinations of two tasks that a worker may perform. If the nature of the task allows the possibility that a single worker may perform more than two tasks in one day, then a simulated worst-case estimate will take this into consideration. An initial monitoring approach might be to estimate exposures for each task and construct a worst-case day. Approaches to simulating a worst case obviously will vary with the process. Simulations should be well documented. Keep in mind that worst-case simulations will not allow the construction of an exposure distribution, and this may be important at some time in the future. However, task data do allow easier determinations of where controls are needed.

9. When a full exposure profile is indicated, PJ takes on another role... e.g., the determination of sample size. For example, if a sample size of 30 is calculated to estimate a mean, based upon the methods presented in Chapter 16, but resources are only available for $n = 10$, then the effect of a sample size of 10 on the accuracy of the estimate of the mean can be calculated (see Chapter 16). If the accuracy is not

acceptable, then the hygienist should realize that it would be best to find more resources or develop a new goal, as an expenditure on the small sample size could represent a waste of resources. For instance, large sample size may be needed if the GSD is expected to be, say 3.5. In this case, regardless of the mean, the hygienist perceives that the workplace is quite variable. Thus, even if the mean of exposures is = 0.5 OEL, a large number of exposures will exceed the OEL, and some by a very large margin. We will likely conclude what we may have already suspected...the workplace is out of control. What tasks are causing the presumed variability? Can we control them?

Other Considerations

The following are presented as strategy considerations. Two major dichotomies are used: acute versus chronic and sub-chronic toxicants and continuous versus discontinuous (or task-oriented) exposure. Such a strategy should be flexible. For example, a random sampling strategy to assess 8-hour exposures to a chronic toxicant may seem frivolous when the worker is rarely exposed (e.g., once per week). Using a random approach, those days selected on which no exposure occurs could be *included* by not sampling them and counting exposure as zero. Alternatively, if it were not known whether the task would be performed and, in fact, it was not, then the monitoring result would be < LOD, a marginally useful result, unless something were known about what the worker did that day. The hygienist can also construct an 8-hr TWA by monitoring tasks only and assigning zero exposure for the remainder of the shift, where appropriate.

Acute Toxicants

A NOTE

It has been argued that for acute toxicants with irreversible effects (e.g., H_2S) that continuous direct reading instrumentation with alarms should be used. That recommendation is made repeatedly below. There are no doubt many workplace situations where acute toxicants (e.g., H_2S) are present and well controlled and no direct reading instrumentation is used. In these cases, the hygienist should carefully document the IDLH (concentration which is considered to be immediately dangerous to life and health, see e.g., the NIOSH website, http://www.cdc.gov/niosh/idlh/intridl4.html) and its biological basis, the distribution of concentrations present in the workplace, and, most importantly, the likelihood of process upsets, no matter how unlikely, that might lead to IDLH concentrations. One has to ask the question, is it impossible for the concentration of the toxicant to reach IDLH in this process? Or is the process that employs the acute irreversible toxicant reliable 100% of the time? If the answer is yes then continuous monitoring is not needed.

DISCONTINUOUS RARE TASK EXPOSURES

In many cases exposures to acute toxicants can occur very infrequently and be task-oriented. The best example of this is maintenance tasks. For example, on a daily basis a worker has to maintain a cooling system that operates with liquid ammonia. The maintenance operation lasts 15 minutes. Or a worker replaces chlorine, ethylene oxide (a mutagen with a short-term limit), or other gas cylinders. This latter task may be performed on a weekly basis and last less than an hour. Consequently, it would be difficult to construct a reasonably sized database for such rare exposures that would allow extrapolation to the right tail. (The sample size should be at least 20 to have a reasonable likelihood of a result in the highest 5%.) When tasks are rare and short, perhaps the best approach is to control the situation on the basis of assumed worst-case possibilities, on the knowledge that estimating a 99th or some other *worst-case* percentile of the distribution would be difficult and unreliable. The use of self-contained breathing apparatus respirators may apply in this situation. The fine line between use of respirators and installation of engineering controls will depend on the frequency of the task, technical, and economic feasibility of control, and toxicity of the agent. For highly toxic agents (e.g., chlorine gas) strict control measures (e.g., respirators) and continuous reading instruments are necessary because even the most remote chance of overexposure is not tolerable (see footnote 7).

FREQUENT DISCONTINUOUS TASK EXPOSURES

The strategy must be based on the toxicity of the acute toxicant. Highly acute toxicants without adequate warning properties will require continuous monitoring. For lesser toxicants with reversible effects, as tasks become more frequent within a day (e.g., three to four times daily), perhaps the best approach is to evaluate simulated worst-case exposures. Depending on the toxicity of the acute agent, two to four simulated worst-case exposures may be evaluated.

CONTINUOUS EXPOSURES

Highly toxic compounds require continuous monitoring, which is the most straightforward approach for all continuous exposures to acute toxicants. For less toxic compounds, as tasks occur more frequently and for longer periods during the workday, exposure becomes, or should be considered, continuous. Therefore, direct-reading instruments or a statistically based sampling strategy for 15-minute time periods is appropriate. If discrete tasks within the workday can be easily identified, then one should consider monitoring tasks for purposes of constructing a worst-case simulated exposure. If tasks cannot be easily identified and monitored during the workday, then for low-toxicity compounds with very adequate warning properties occasional exposures greater than the standard may be tolerated. Consider the calculation of a tolerance limit on some reasonable percentile, such as the 95th percentile, based on $n > 20$.

Obviously situations in between high and low toxicity require more thought than this for a sampling strategy. As the toxicity increases and warning properties become less evident, the use of continuous monitoring becomes more necessary. One should always read the documentation for the OEL to determine if adequate safety factors have been applied in setting the OEL. Training workers to recognize warning properties and take appropriate emergency action is also necessary.

Chronic and Subchronic Toxicants

DISCONTINUOUS EXPOSURES

Where tasks are well-defined and occur only a few times per day, the 8-hr TWA exposures are thought to be low, and exposure assessment is of low to medium priority, it may be best to measure task exposure and construct a simulated worst-case. The number of simulated worst-cases constructed should depend upon the toxicity of the agent and closeness of estimates to the OEL. If estimates are above some cut-off fraction of the OEL,[8] one should go to the next step and measure full-period 8-hr TWA exposures for purposes of estimating the mean or 95th %tile.[9]

There are several approaches to task monitoring for chronic toxicants. Random 8-hr TWA monitoring can be effective in delineating task contributions to exposure if performed in conjunction with observations of task length, task types are few and infrequently performed, the potential for exposure during the remainder of the day is observed, and rough estimates are acceptable. Otherwise, stratified (by task) random monitoring of individual tasks allows calculation of an 8-hr TWA and task averages, again if observations are made by the hygienist. Nicas and Spear (1993a, b) show that task-oriented stratified random sampling can give a tighter estimate of the mean (and by calculation any percentile) for a given budget, or a lower cost for a given estimation accuracy, compared to simple random sampling over the full 8 hours.

ESSENTIALLY CONTINUOUS EXPOSURES

Where exposures are essentially continuous and a professional judgment estimate of the mean (chronic toxicants) or 95th %tile (subchronic toxicants) exposure is greater than one-tenth of the OEL, the hygienist should estimate, or perform a screening study to estimate the GSD and mean. This information then is used to find an adequate sample size for an accurate characterization of the exposure distribution (estimate the GSD, mean, 95th %tile, etc.). For chronic toxicants the mean then is compared to the OEL (or preferably some fraction of the

---

[8] For example, 0.1 as described earlier. These values should be applied based on a consistent documented rationale.

[9] Remember from Chapter 16 that controlling a mean exposure to 0.25 OEL limits exceedances to no more than 5%. This is equivalent to controlling a 95th or higher %tile to not exceed the OEL.

OEL, such as 0.25 OEL) by using statistical procedures (see Chapter 16). For subchronic toxicants the mean may be compared to 0.25 OEL so as to control exceedances at the 95th %tile or the 95th % tile may be compared directly to the OEL.

It is clear that these two scenarios represent two extremes for chronic/subchronic exposures. Obviously as the number of tasks and their duration during a day increase, the exposure to the agent becomes more continuous.

BAYESIAN APPROACHES

It has recently been recognized that many exposure assessment problems often lead to answers that are less than palatable (Hewett et al., 2006). For example, three 8-hr TWA measurements give 0.05, 0.1, 0.2 to be compared to an OEL of 1 ppm. On the surface things look controlled. If we assume lognormality, the MVU estimate of the mean is 0.12 and the 95th %tile is 0.31. However, because the sample size is only three, the Land's UCL estimate on the mean is 13+ and the UTL on the 95th %tile is 20+. Now if one knew nothing about this exposure distribution (e.g., a rookie consultant walking into a fiber glass operation for the first time having not even read the NIOSH literature (e.g., Health Hazard Evaluations) on these types of operations) then she might do the statistics and assume the worst. But in reality hygienists often collect n = 3 samples for such operations. Based on the typical AIHA cutoff, discussed earlier, a PJ assessment would be accurate in guessing a mean of 0.1 and assuming low priority for any monitoring of this operation.

If we read the literature or have experience with these operations, then we might feel justified to say good enough, because we know that the mean and 95th %tile exposures in these operations are typically around those we have just calculated. Thus, we are in a predicament – the statistics say one thing because the low sample size yields little confidence, and experience says something else. There is a solution that is particularly applicable to these cases when n is small and the hygienist has some prior information – Bayesian Decision Analysis (BDA). For a good overview of this approach see Hewett et al. (2006).

Simplistically put, in BDA, one defines a prior distribution that may be based on previous exposure data from this operation, other similar operations, other operations that have similar tasks but different compounds, modeling data (Chapter 15), a panel of experts, or just a wild guess,[10] etc. One then calculates a likelihood distribution, which is similar to maximum likelihood techniques, discussed elsewhere for calculating the mean of a lognormal and also the mean and other parameters when some of our data are < LOD. The likelihood distribution is based on the data we collect now. The prior and likelihood distributions are combined to create a posterior distribution that represents the probability based on both the prior and current information. The process is similar to the hygienist saying "This operation typically has low

[10] This is being a bit facetious. In reality, if we wildly guess low or high then we really gain nothing by using BDA. It works as it should when prior information is based in reality.

exposures but let me just check." Upon checking suspicions are confirmed based on n = 3.

Typical outputs appear in Figure 18-2. Here the probabilities are in terms of an exposure rating as shown in Table 18-1. In other words, the Bayesian *posterior* distribution in Figure 18-2 shows that there is about an 85% probability that these exposures are well controlled and a 12% probability that they are category 3. There is only a small probability of about 1% that the exposures are category 4 and keep in mind that this would indicate an even smaller probability that the mean exceeds 0.5 OEL. This is close to what we thought based on the prior distribution, but the likelihood indicates that this guesstimate (the prior) based on past experience was a little low. Thus, we might conclude that this SEG is okay for now but needs more investigation later when other more pressing priorities are not demanding resources.

Unfortunately, calculation of the three distributions shown in Figure 18-2 is not easy. Hewett et al. refer to software that they developed and can now be downloaded from http://www.oesh.com. However, it should not be used without training, in particular on interpretation of results and the limitations of setting up priors.

There is also an Internet site (BUGS, 2004) that allows one to calculate the distributions but its use is not straightforward. Hopefully user friendly software will be developed soon. The Bayesian approach not only mirrors the usual decision logic used in hygiene work but the charts are much easier for a plant engineer or manager to interpret than a statement like, "The estimate of the mean is 0.4, however, it could be as high as 9.2." That one does not go over too well.

CONTROL CHARTING

Once an adequate exposure assessment program is established, and data begin to accumulate, there are some helpful analyses that can be performed beyond estimating the mean or another percentile or testing null hypotheses such as H_0: $\mu_c \geq 0.25$ OEL or 95th %tile ≥ OEL. As the program matures, it is often important to examine changes in exposure concentrations over time. As an example, consider a program to assess exposures to cadmium in a SEG. Let us say that the SEG is monitored randomly about 36 times a year or three times a month. By calculating monthly averages, the hygienist could, after a suitable time period (say, one year) begin to look for trends, cycles, or changes that might be caused by improved control efforts.[11] He or she might also be able to detect evidence that previously successful controls, such as work practices, were no longer being used. As was discussed in Chapter 16, this approach can also be used to look for shifting means and variances, i.e., non-stationarity.

The ability to do these things can be aided by a statistical method known as control charting or quality control, which has long been applied to laboratory assay work (see, e.g., Verma et al., 2006) and should be a part of any industrial hygiene laboratory quality control program. Its application to occupational exposure data has been suggested (Hawkins and Landenberger, 1991).

---

[11] One could also calculate other %tiles, but the important thing here is trends.

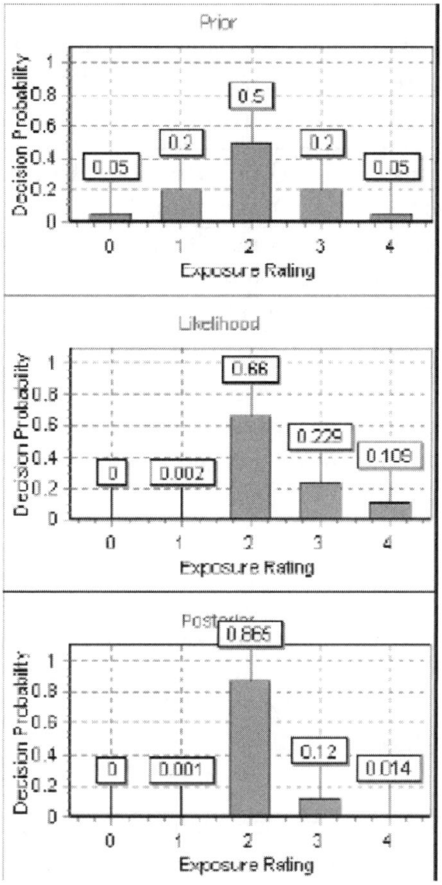

Figure 18-2. Bayesian decision charts for a) an informative prior scenario where the probability of each category is estimated based on past experience, b) the likelihood is based on measurements of 0.05, 0.1 and 0.2 ppm, and c) the posterior is calculated from these two distributions and an OEL = 1 ppm (from Hewett et al., 2006, with permission).

There are several types of statistical process controls. The most useful for exposure data is Shewhart charts[12] that show trends in the mean exposure or the variance of exposures (see Figures 18-3 and 18-4). Each point in these figures represents the mean or a variance parameter (the range or standard deviation) of a *rational subgroup*. The variance charts utilize range

[12] For a good overview of these charting techniques the reader should see Gilbert (1987). The technique requires the use of several constants that must be found in tables. Although these tables can be found in Gilbert, ASTM (2002) gives a complete description of them and of their use and derivation.

Table 18-1
AIHA exposure categorization scheme. An Exposure Category can be assigned to a SEG whenever the true 95th %tile exposure ($X_{0.95}$) or mean ($\bar{x}$) falls within the specified range (adapted from Hewett et al., 2006)

Exposure Category	Rule-of-thumb Description*	Qualitative Description	Recommended Statistical Interpretation**	Notes
0	Exposures are trivial to nonexistent – employees have little to no exposure, with little to no inhalation contact.	Exposures, if they occur, infrequently exceed 1% of the OEL.	$X_{0.95} < 0.01$ OEL $\bar{x} < 0.0025$ OEL	1
1	Exposures are highly-controlled – employees have minimal exposure, with little inhalation contact.	Exposures infrequently exceed 10% of the OEL.	0.01 OEL $< X_{0.95} <$ 0.1 OEL 0.0025 OEL $< \bar{x} <$ 0.025 OEL	2
2	Exposures are well-controlled – employees have frequent contact at low concentrations and rare contact at high concentrations.	Exposures infrequently exceed 50% of the OEL and rarely exceed the OEL.	0.1 OEL $< X_{0.95} <$ 0.5 OEL 0.025 OEL $< \bar{x}$ $<<$ 0.125 OEL	2, 3, 4
3	Exposures are controlled – employees have frequent contact at low concentrations and infrequent contact at high concentrations.	Exposures infrequently exceed the OEL.	0.5 OEL $< X_{0.95} <$ OEL 0.125 OEL $< \bar{x} <$ 0.25 OEL	2, 4
4	Exposures are poorly-controlled – employees often have contact at high or very high concentrations.	Exposures frequently exceed the OEL.	$X_{0.95} >$ OEL $\bar{x} >$ 0.25 OEL	4

* The *Rule-of-thumb* descriptions were based on similar descriptions (AIHA, 2006).

** $X_{0.95}$ = the true group 95th %tile exposure. $\bar{x}$ is the true arithmetic or MVU mean and inequality values are based on Table 16-6 (Chapter 16).

1 Category 0 was added to distinguish between highly-controlled exposures and situations where exposures are either nonexistent or trivially low. It was included in AIHA (1991).

2 *Infrequently* refers to an event that occurs no more than 5% of the time.

3 *Rarely* refers to an event that occurs no more than 1% of the time.

4 *High concentrations* are defined as concentrations that exceed the OEL.

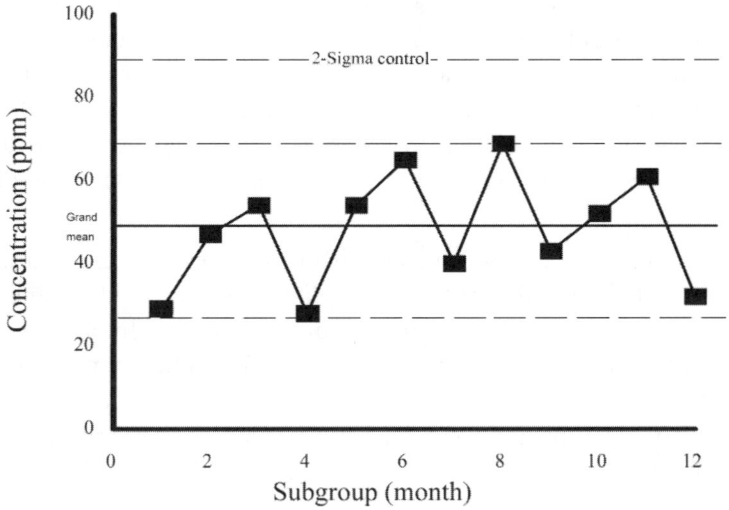

Figure 18-3. Control chart for means with 95% (± 2 standard deviation) control limits. Control limits and center line based on first 7 months.

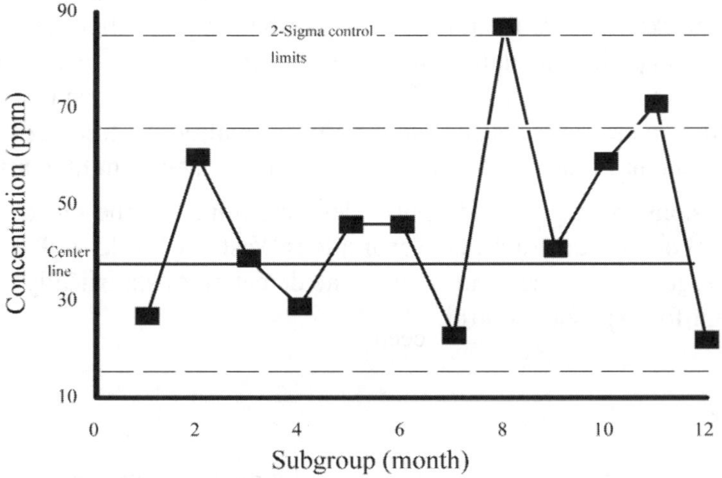

Figure 18-4. Control chart for standard deviation with 95% (± 2 standard deviation) control limits. Control limits and center line based on first 7 months.

data when sample sizes (n_s) of each rational subgroup are ≤ 10, or the standard deviation when the sample sizes are > 10. A rational subgroup is a group of data collected over some defined space or time period or both. Thus, measurements of an SEG over a one-month period could represent a rational subgroup. The analysis techniques are easiest when all n_s are equal, but unequal subgroups can be manipulated. In addition, many statistical software packages will perform the analysis, but some may not have the ability to deal with complications such as unequal n_s.

An important assumption of control charts is that of normality. In Figure 18-3 the outside horizontal lines are called control limits and are placed on either side of the grand mean (the mean of subgroup means), based on the variance of these means. In other words, if the means of the subgroups vary considerably, then the limits are placed far apart. Rules are then used to determine an *out of control* situation, for example, two means in succession that are outside of the control limits. If the means are not normally distributed, then the variance of the means may cause the control limits to be inflated. Then one would be less likely to find an out-of-control situation.

Exposure data are not likely to be normally distributed. Hawkins and Landenberger (1991) have argued that because of the central limit theory of statistics, this should not be a problem. The theory essentially states that the sampling distribution of the grand mean, no matter how the underlying data are distributed; tends toward normality. Or the means of the subgroups should be approximately normally distributed even if the individual measurements are drawn from a lognormal distribution. However, the reasonableness of this approximation is dependent on the skewness of the underlying distribution, the subgroup size, and the number of subgroups used in the calculation of the variance of the grand mean. The best approach is to test the population of means for normality. If it is not normal, then use a log transformation of the individual measurements and recheck for normality. If a transformation is used, the central horizontal line in Figure 18-4 will represent the geometric, not the arithmetic, mean.

Control decisions must be based upon rules determined at the outset. Many statistical software packages contain such rules. One example is Westgard's rules (1981) (cited in Hector, 1986), shown in Figure 18-5. These rules allow the detection of statistically significant trends, which is important for exposure control.

CONCLUSION

It should be obvious to any reader that the development of sampling strategies is not easy. There are no *right* answers, only best ones for the circumstances, and the differences in outcomes between two strategies may be marginal or large. However, if the health professional follows the principles outlined earlier and then classifies the agent according to toxicity, acute/chronic nature, and duration and frequency of exposure, the process should proceed relatively rationally. Ultimately, attainment of the goal of safeguarding employee health without wasting resources will be realized.

Westgard's Rules

Rule 1_{2s} The warning rule demands further examination of the data when a control sample value falls outside the range $\bar{x} \pm 2s$.

Rule 1_{3s} Calls for immediate rejection of the run when a control sample value falls outside the range $\bar{x} \pm 3s$.

Rule 2_{2s} Calls for rejection of the run when two consecutive control sample values fall outside the range $\bar{x} \pm 2s$. It applies to control samples from consecutive runs and to two samples within the same run.

Rule R_{4s} Calls for rejection of the run when the difference between two control sample values within a run exceeds 4s.

Rule 4_{1s} Calls for rejection of the run when four consecutive control sample values exceed $\bar{x} \pm 1s$ or $\bar{x} - 1s$ (lie on the same side of the mean at a distance from it that is greater than 1s).

Rule $10_{\bar{x}}$ Calls for rejection of the run when a control sample gives the tenth consecutive result that has fallen on the same side of the mean value.

Figure 18-5. Westgard's Rules for interpreting QC data in Figures 18-2 and 18-3.

QUESTIONS

18.1. State the positive and negative attributes (relative to worker protection and employer fairness) of OSHA's current compliance program for airborne hazards.

18.2. Assuming a lognormal distribution and an estimate of the mean exposure value equal to 25% of the OEL (or a willingness to accept all differences as not statistically different when $OEL/\mu_c < 4$), what is the smallest GSD for which AIHA's recommended sample size of six is satisfactory for testing the null hypothesis H_0: $\mu_c \geq OEL$? Set $\alpha = 0.05$ and disregard β by using an equation developed in Chapter 16.

18.3. The mean of the exposures is constant for a stable population regardless of averaging time (Chapter 16). It can also be shown that the average of means for shorter averaging times taken randomly from specific time periods (say two 4-hr TWAs or four 2-hr TWAs) is equal to the 8-hr TWA mean. How can this knowledge be used to design a task-based monitoring scheme that will yield both 8-hr TWA and task distributions?

REFERENCES

AIHA: A Strategy for Occupational Exposure Assessment, N.C. Hawkins, S.K. Norwood, J.C. Rock (eds.). American Industrial Hygiene Association, Fairfax, VA (1991).

AIHA: A strategy for assessing and managing occupational exposures, 2nd ed., J.R. Mulhausen, J. Damiano (eds). American Industrial Hygiene Association, Fairfax, VA (1998).

AIHA: A Strategy for Assessing and Managing Occupational Exposures, 3rd ed., J.S. Ignacio, W.H. Bullock (eds.). American Industrial Hygiene Association, Fairfax, VA (2006).

ASTM: Manual on Presentation of Data and Control Chart Analysis, Manual 7A, Seventh Edition, Prepared by Committee E-ll on Quality and Statistics. Philadelphia, PA (2002).

BUGS Project: Bayesian Inference Using Gibbs Sampling. Medical Research Council, Biostatistics Unit, Cambridge, England. http://www.mrc-bsu.cam.ac.uk/bugs/ (2004).

Gilbert, R.O.: Statistical Methods for Environmental Pollution Monitoring. Van Nostrand Reinhold, New York (1987).

Hawkins, N.C.; Landenberger, B.D.: Statistical control charts: A technique for analyzing industrial hygiene data. Appl. Occup. Env. Hyg. 6:689-695 (1991).

Hewett, P.; Logan, P.; Mulhuasen, J.; Ramachandran, G.; Banerjee, S.: Rating Exposure Control Using Bayesian Decision Analysis. J. Occ. Env. Hlth. 3:568-581 (2006).

Hector, T.: Assessment and Control of Biochemical Methods. Wiley Publishing, New York (1986).

Nicas, M.; Spear, R.C.: A task-based statistical model of a worker's exposure distribution: Part I – Description of the model. Am. Ind. Hyg. Assoc. J. 54:211-220 (1993a).

Nicas, M.; Spear, R.C.: A task-based statistical model of a worker's exposure distribution: Part II – Application to sampling strategy. Am. Ind. Hyg. Assoc. J. 54:221-227 (1993b).

NIOSH: Occupational Exposure Sampling Strategy Manual, N.A. Leidel, K.A. Busch, J.R. Lynch (eds.). National Institute for Occupational Safety and Health, Cincinnati, OH (1977).

OSHA: Occupational exposure to benzene; final rule. 29 CFR Part 1910. Fed. Reg. 52:34540 (1987).

OSHA: Technical Manual, TED 1-0.15A, Appendix II:1-6. Sampling and analytical errors (SAEs), Environmental Variables. Occupational Safety and Health Administration, Washington, DC, US (1995).

Rappaport, S.M.: The rules of the game: An analysis of OSHA's enforcement strategy. Am. J. Ind. Med. 6:291 (1984).

Roach, S.A.: A most rational basis for air sampling programs. Ann. Occup. Hyg. 20:65-84 (1977).

Spear, R.C.; Selvin, S.: OSHA's permissible exposure limits: regulatory compliance versus health risk. Risk Analyses, 9:579-586 (1989).

Tornero-Valez, R.; Symanski, E.; Kromhout, H.; Yu, R.C.; Rappaport, S.M.: Compliance versus risk in assessing occupational exposures. Risk Analysis 17:279-292 (1997).

Tuggle, R.M.: Assessment of occupational exposure using one-sided tolerance limits. Am. Ind. Hyg. Assoc. J. 42:493-498 (1981).

Verma, D.K.; Ferguson, K.; McCollin, S.A.; Shaw, L.; des Tombe, K.: Analytical Performance Criteria: A PC based QC program for an industrial hygiene laboratory. J. Occ. Env. Hlth. 3:D56-D66 (2006).

Westgard, J.O.; Barry, P.L.; Hunt, M.R.; Groth, T.: A multi-rule Shewhart chart for quality control in clinical chemistry. Clin. Chem. 27:493-501 (1981).

V

AIR SAMPLING AND LABORATORY ANALYSIS

19

LABORATORY ANALYTICAL TECHNIQUES[1]

INTRODUCTION

There are two distinct methodologies for the identification and quantification of airborne contaminants, regardless of either the type or source of the air contaminant(s): monitoring methods, and methods based on the collection of discrete samples followed by analysis (which can be in the field, the laboratory, or both). *Monitoring* refers specifically to the use of real-time or near-real-time instruments in the field. Collection of air samples by any means followed by the transportation to and analysis in a laboratory is not the same as *monitoring*, although the two methods are both frequently but erroneously called *monitoring* and both yield measurements. Air monitoring, which will be discussed in Chapters 27 through 29, will not be considered here. Instead the laboratory analytical techniques that are used to assay samples that are field collected will be examined here. However, there is overlap between some of the analytical devices that are used in the lab versus in the field as will be seen in Chapters 27 through 29.

After an air sample is collected, it must be analyzed. The analysis is likely to be both qualitative (i.e., what species exist) and quantitative (i.e., how much of each exists, an *assay*). In some cases the total amount of air contaminant assayed by a non-specific method is assumed to be one likely species, or all the sum total of all species are combined to calculate a single concentration of a surrogate analyte.

The analysis, in most cases, occurs in a laboratory setting, and so the air sample must be submitted to the laboratory from the field location where the sample was collected. Whether the air sample is analyzed in the field or in the laboratory, a handful of analytical techniques are used most commonly to identify and quantify the majority of airborne contaminants (see Verma and Tombe, 1999 for a typical history of the use of analytical methods and Schlecht and Cassinelli, 1997 for a survey of analytical methods used in hygiene work). These are: chromatography, spectroscopy, electrochemistry, titrimetry and gravimetry. Chromatographic

[1] The author is deeply indebted to Professor Steven Levine for substantial input to this chapter.

methods include gas chromatography (GC) and liquid chromatography (LC). GC is the workhorse for analysis of contaminants in environmental air. However, LC is important for the analysis of organic compounds that are nonvolatile or unstable at elevated temperatures, which precludes their analysis by GC. Nonvolatile and thermolytically unstable compounds, as a class, usually include pesticides, herbicides, pharmaceutical and biological species, so LC methods are of primary importance now and only likely to get more so in the future.

Spectroscopic methods include methods that rely either on absorption or emission of photonic energy at discrete wavelengths. Theoretically, this can include absorption or emission of light anywhere in the spectrum, but practically, the wavelengths of greatest interest lie in the infra-red (IR), visible and ultraviolet (UV) portions of the spectrum. Spectroscopic methods commonly used to analyze samples of gases, vapors, liquids and solids in an industrial hygiene laboratory include IR spectroscopy; colorimetry (in the visible light region of the spectrum); atomic absorption spectroscopic (AAS) analysis of individual metals; and inductively coupled plasma atomic emission spectroscopic (ICP-AES) analysis of mixtures of metals.

Electrochemical techniques, such as the ion-selective electrode, are also in widespread use in the hygiene laboratory. Finally, gravimetry, or weighing, the simplest of analytical techniques, still is widely used to assay several general classes of dusts.

In keeping with the objectives of this book, explanations of analytical methods and instruments will be focused on basic applications relevant to hygiene practice, rather than to the rubric of analytical chemistry and the interests of analytical chemists. Thus, the subject of chromatography constitutes about one-third of this chapter. This seemingly disproportionate share of the text devoted to this subject is because chromatographic methods are so widely used for the analysis of industrial hygiene samples.

This book chapter should not be considered to be a definitive description of analytical methodology or of analytical methods. The core reference for such information is the *NIOSH Pocket Guide to Chemical Hazards* (2005). This document references methodology from both NIOSH and OSHA that are also cited in the reference section.

- NIOSH Manual of Analytical Methods http://www.cdc.gov/niosh/nmam/
- OSHA Sampling and Analytical Methods http://www.cdc.gov/niosh/nmam/
- SKC Inc. Air Sampling Guides http://www.skcinc.com/guides.asp, and http://www.skcinc.com/reference.asp, and SKC Catalog (printed annually) wherein SKC, Inc. has been chosen here because of the quality of their printed and web-based analytical guides that include, NIOSH, OSHA, EPA, and other methods.

CHROMATOGRAPHY

Chromatography denotes a method in which components of a mixture are separated through the use of a column containing a stationary phase through which a gas or liquid is made to flow. The differential partition of the components of the mixture between the stationary and mobile phase is the mechanism whereby the components are separated. If all parameters are kept constant including

- the temperature, composition, length, diameter and relevant physical characteristics of the stationary phase
- the composition and flow rate of the mobile phase, and
- the geometry of the column,

then each component of any mixture will emerge from the end of the column at a time characteristic of that compound. That is called the *retention time* (see Figure 19-1) which can be used as a first step in identifying each compound, even without additional information other than calibration of the system with a mixture of known components.

Of course, there seems to be almost as many detectors and interpretation methods as there are column and phase combinations. These various detectors can be used as a powerful and sometimes definitive complement to simply using the retention time to identify each component of each mixture. Not only that, but these detectors allow the investigator to quantify each analyte.

Gas Chromatography

Gas chromatography (GC) is an abbreviation of the correct form gas-liquid chromatography (GLC) or gas-solid chromatography (GSC) wherein the liquid or solid designation refers

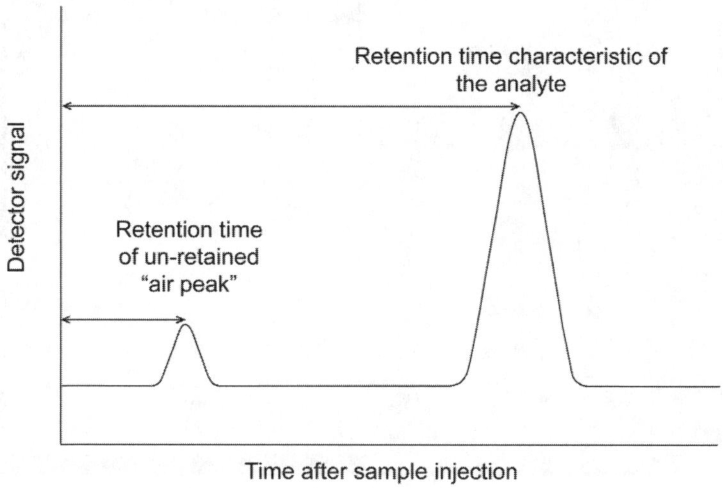

Figure 19-1. A typical graphical output from a GC for a single analyte in air, no solvent.

to the stationary phase. In many if not most cases, the stationary phase is a polymeric or other low volatility liquid or semi-solid that is coated onto a thermally-stable high-surface area solid support. GC is almost always the form of the acronym that is used. In GC, the analyte in a *matrix* can be injected into the system as a gas or a liquid (which is then volatilized in a heated *injection port*). This gaseous mixture is carried through the column by a carrier gas such as helium, the most commonly used, which constitutes the mobile gas phase. The coated stationary phase is placed in a tube, commonly called a column, the length of which affects the efficiency of separation of the components of the mixture. All other things being equal, the efficiency of the separation increases as the square root of the increase in column length. Column dimensions vary widely, but lengths most commonly used in GC range from two meters for a 1/8 inch (0.3 cm) diameter packed column to tens of meters for capillary columns (shown in Figure 19-2).

A schematic of a chromatograph is shown in Figure 19-3. A recorded output from a GC after injection of a single analyte in air was shown in Figure 19-1. Ideally, the analyte(s) of interest are baseline separated from the non-analytes and solvent matrix of the mixture and elute from the column at retention times characteristic of that analyte and with readily-quantified peaks in detector response. Calibration would have occurred prior to analysis of mixtures of unknowns, wherein the calibration would establish both retention times (for identification of the analyte) and the response factor (i.e., peak area units per microgram of injected analyte) for quantification. If the analyte cannot be resolved from interferences and/or from the solvent in which it was dissolved and injected into the GC, then software can be used to deconvolve the analyte peak from the interferents. More useful is the approach of using a variety of highly specific detectors, as dealt with in the section immediately below.

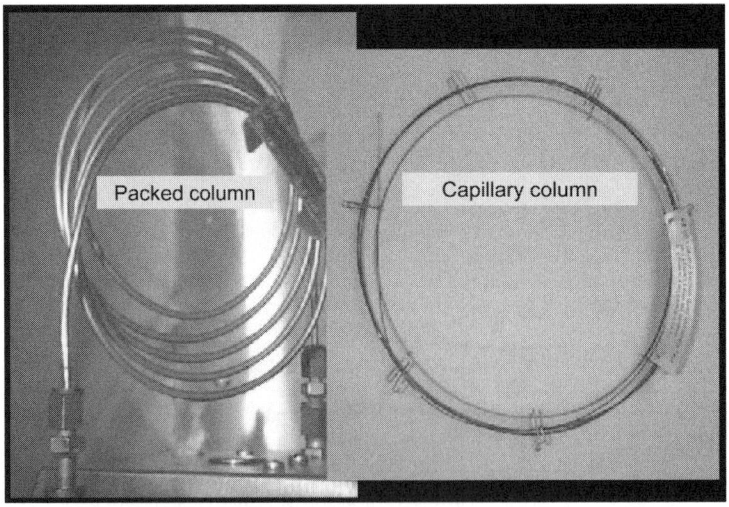

Figure 19-2. Packed (left) and capillary (right) GC columns.

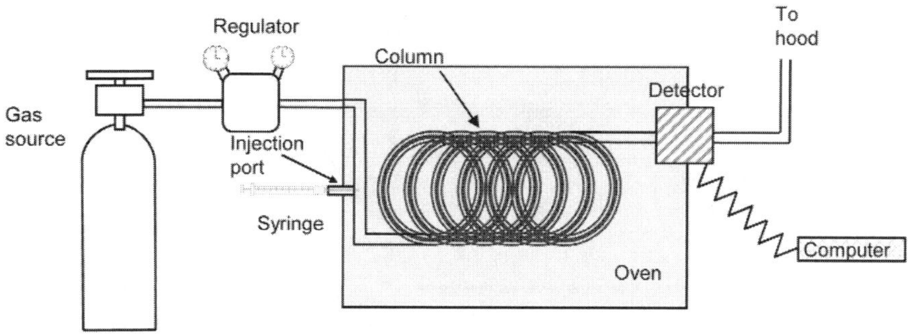

Figure 19-3. Schematic of a gas chromatograph.

GC DETECTORS

A wide range of detectors are used in chromatography work. In GC, the flame ionization detector (FID) enjoys the greatest number of applications in a hygiene laboratory. The response factor of each analyte must be predetermined and the instrument plus detector combination calibrated under the conditions of use.

Flame Ionization Detector (FID)

A hydrogen-oxygen flame ionizes the constituents of the chromatographed mixture as each emerges from the end of the column (Figure 19-4). The resulting plasma provides a small current that can be amplified to indicate the presence of an analyte. Hydrocarbons tend to produce a signal that is proportional to the number of carbon atoms, although S-, O-, or N-containing functional groups may reduce the signal. When an FID is operated and maintained properly, it has a very high signal-to-noise ratio for hydrocarbons. It is, therefore, the detector of choice for many compounds of interest to industrial hygienists, particularly when sulfur, nitrogen, and phosphorus groups are not present on the molecules.

It is convenient that the positive signal on an FID from a molecule containing one carbon is offset by the same negative signal on an FID from two sulfur atoms, resulting in zero signal. It is also convenient that the molecule CS_2 (carbon disulfide) is volatile and therefore, poorly retained by most GC columns and emerges at the very beginning of a chromatogram, well in advance of the elution time of most analytes. In addition, CS_2 is an excellent solvent for a wide variety of analytes, and an excellent eluent for many analytes that have been adsorbed by activated carbon, such as that carbon used in adsorption tubes that are used to evaluate exposures of personnel to airborne contaminants. Thus, CS_2 is used in many, if not most, methods for measuring exposure of workers to organic solvents in air.

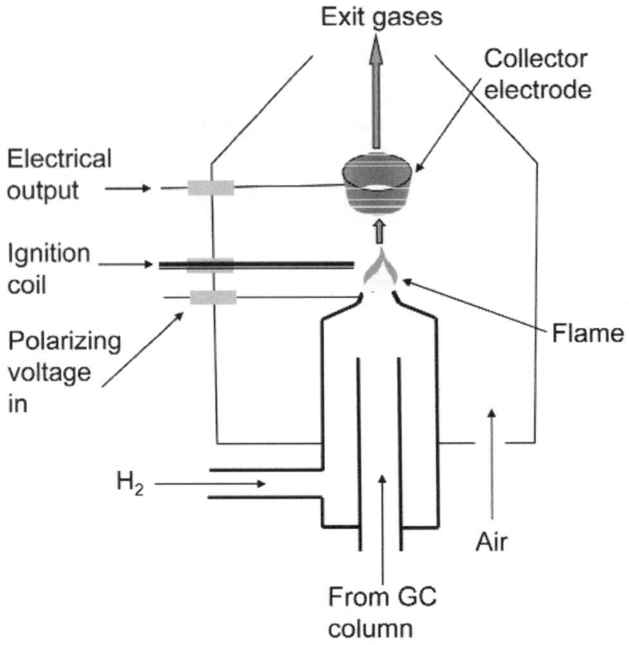

Figure 19-4. Schematic of a flame ionization detector.

Photoionization Detector (PID)

The PID is commonly considered as a possible substitute for the FID. In the PID, a lamp that emits ultraviolet light is used to ionize the gas stream (as contrasted to the flame that serves this purpose in an FID). The most commonly used PID lamp emits light with a photonic energy of 10.2 eV, which is ultraviolet (UV) light at a wavelength of 121.6 nm. The resulting positive ions are attracted to a negative electrode, and the current necessary to be supplied to negate the positive charge is monitored. This detector has a broad application to many compounds similar to that of the FID. However, for aromatic and some other substances it is more sensitive than the FID. It is also nondestructive, meaning that it can be placed in series with other detectors.

Thermal Conductivity Detector (TCD)

A heated wire or a thermistor provides a certain resistance to electrical current flow, and this resistance is dependent upon the thermal conductivity of the surrounding gas. Two detectors are used in parallel, with the second one acting as a control to provide a flat baseline. The low sensitivity of TCD is its chief disadvantage. The TCD is also nondestructive.

Electron Capture Detector (ECD)

When a nucleus emits high energy electrons, those electrons are termed beta (β) radiation. Nickel-63 (^{63}Ni) is such a nucleus, and is therefore used as a source of ionizing radiation in an ECD. When the mobile GC phase (carrier gas) contains an analyte, in which molecules include electronegative atoms, that analyte captures the beta-electrons (Figure 19-5). That electron capture can be detected with incredibly high sensitivity, often allowing assays to be performed down to the ppb or ppt concentration level in air, water, or other materials. Halogens, peroxides, quinones, and nitro groups capture the electrons effectively; hydrocarbons, alcohols, and amines do not. The ECD has its greatest application in the detection of highly chlorinated compounds such as the chlorinated pesticides. Chlori-, fluori, and brominated compounds are in this group. Frequently, compounds that cannot be detected by GC-ECD are derivatized with perfluoro functional groups, thus allowing the sensitive detection of those derivatives.

Thermionic Detector (TD or NPD)

These detectors are also referred to as nitrogen-phosphorus detectors (NPD). They are similar to the FID, but the FID is not very sensitive to nitrogen- and phosphorus-containing compounds. With the NPD, after the carrier gas stream leaves the flame, it flows around an electrically heated rubidium silicate bead. The heated bead forms a high temperature plasma

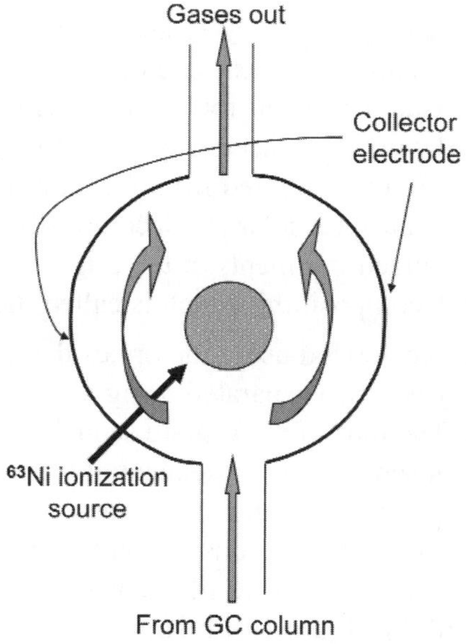

Figure 19-5. Schematic of an electron capture detector.

thereby ionizing the P- and N-containing molecules so they can be detected with high specificity and sensitivity. Examples of the uses of this detector are the analysis of solvent-based acrylic paints for residual acrylonitrile, and assessment of exposure of workers to elemental phosphorus captured in an impinger with a xylene impinger-reagent.

Flame Photometric Detector (FPD)

Like the TD, the FPD is based on flame ionization. In this case, the chemiluminescence of ions is detected (chemiluminescence is the emission of light as a result of a chemical reaction). Sulfur or phosphorus atoms have a high potential for chemiluminescence; so the FPD is made for them. In fact, a specific detector is made for each of the two atoms in chromatographable compounds.

Mass Spectroscopy Detector (GC-MS)

Mass spectrometers and mass spectroscopy are both abbreviated MS. MS can be considered as the single most powerful GC detector. No other detector competes when it comes to identifying unknown compounds. For the quantification of analytes, GC-MS may or may not be the best choice when compared to other detectors. In cases in which the MS detector for the GC and another detector are both applicable, it is best to choose the least complex and least expensive, which is almost never the GC-MS. That said, the argument can be turned around. That is, the GC can be looked at as a sample inlet device for the MS. It just depends on which side of the instrument and of the application you stand on.

In GC-MS, solutions containing the analytes are injected into a GC and are separated, usually using a capillary GC column. At the exit end of the GC column, the carrier gas delivers analytes in a more-or-less Gaussian peak form. This carrier gas with the analytes flows through a calibrated *pinhole inlet* into the vacuum system of the MS inlet to the MS's ionization source. Thus, the analytes are concentrated from the carrier gas, and in the ion source of the MS, are ionized and undergo uni-molecular ion decomposition that is characteristic of each molecule. This process yields ionized fragments of the original molecule plus, in many cases, an ion that has not undergone decomposition, which is called the *molecular ion*.

The ions are focused and accelerated using ion optics that are roughly analogous to lenses in optical spectroscopy. They are then separated using either a magnetic field (Figure 19-6) or a quadrupole (Figure 19-7). The ions of interest are sequentially focused on the detector. As an example of the mechanism of separation and sequential focusing, consider that the magnetic field strength (Figure 19-6) might focus an ion of just the *right mass* on the detector (e.g., mass 43 which might be the propyl group). Ions of lighter mass would be bent more in any given magnetic field and would be focused toward the inside of the curve (e.g., mass 29 which might be the ethyl group or "A" in Figure 19-6); similarly ions heavier than that focused on the detector would be bent less in the magnetic field (e.g., mass 57, which might be the butyl

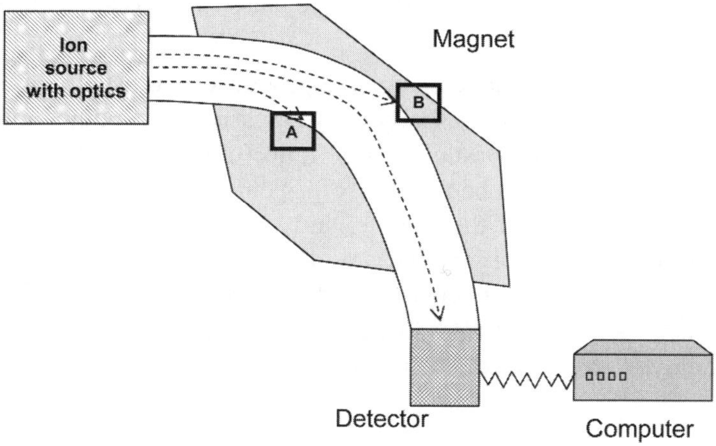

Figure 19-6. The magnet of a mass spectrometer showing how a low mass ion A) would be pulled downward before the detector and a high mass ion B) would not make the turn. The magnetic field is scanned to sequentially focus ions of different mass on the detector. Only an ion that corresponds perfectly to the magnet's setting will reach the detector.

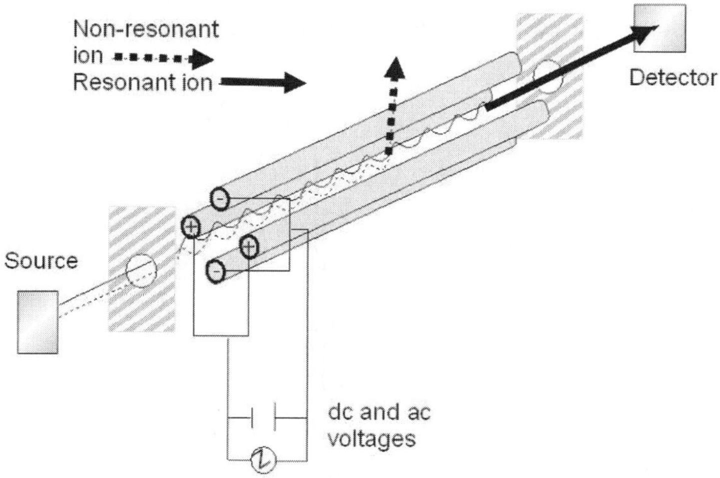

Figure 19-7. The quadrapole of a mass spectrometer showing how a nonresonant ion will not reach the detector. Voltages are scanned to allow ions of one mass through at a time.

group or "B" in Figure 19-6). Figure 19-8 shows the mass spectroscopic output one might get from n-pentane.

Ions of both positive and negative charge are produced, but usually only the unit-charge positive ions are analyzed. There are usually many ions for each compound with the ion of highest mass, that is highest mass-to-unit-charge ratio (m/e) representing the *molecular ion* (Figure 19-8; i.e., 72, which would be the MW of n-pentane). The molecular ion informs the analyst of the molecular weight of the analyte, and ions characteristic of various fragments inform the structure of the analyte. In this way, through spectral interpretation and library matching, the identity of the unknown analyte is discovered. The methodology just described grossly simplifies what actually can and does happen. Suffice it to say that an MS laboratory almost always employs a mass spectroscopist.

Liquid Chromatography

In order for GC to be effective, a compound must exhibit at least a few mmHg (several hundred Pa) of vapor pressure at a temperature just below its degradation (pyrolysis) temperature. For many organic compounds that are of interest, this is not the case. Compounds that have excessively high boiling points (low vapor pressure) and/or are highly polar are generally unsuited for GC analysis, unless they have been derivatized with a methylating or trimethylsilylating agent. Since these derivatizing reactions may be complex and/or incomplete, in many cases, a better choice may be the use of liquid chromatography (LC).

LC is an acronym that is commonly used to describe gravity fed mobile (liquid) phases in columns containing a stationary phase. The columns may be short and wide-bore to long and narrow bore. Long, narrow bore columns yield high efficiency (high plate number) separations but the instrumentation required is complex because of the very high pressures that may be

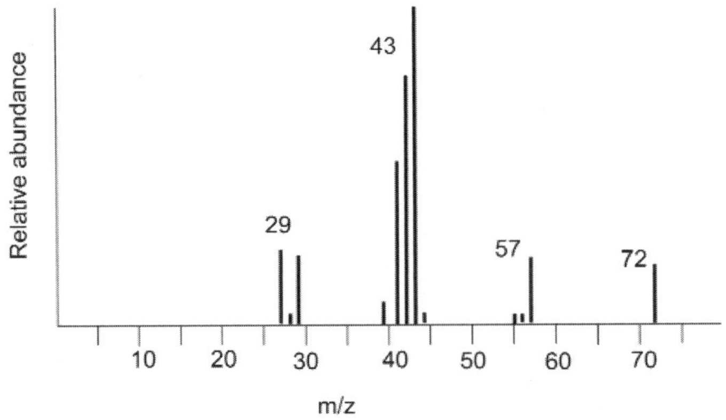

Figure 19-8. Simplified mass spectrum of pentane, or H_3-CH_2-CH_2-CH_2-CH_3.

required to move the mobile (liquid) phase through the column (up to six thousand psi or 40,000 kPa). For that reason, these LC methods are commonly known by the acronym HPLC, which has come to be known as both high *pressure* liquid chromatography and high *performance* liquid chromatography. Many, if not most, assessments of airborne contamination in workplace or environmental situations involve the use of HPLC rather than the simpler gravity-fed mobile phases of the lower pressure methods (see Table 19-1). A liquid solvent reservoir is substituted for the carrier gas used in GC, and the solvent (mobile phase) flow is regulated by a high pressure pump. Several detectors are used.

DETECTORS

Ultraviolet (UV)

The UV detector is the one most commonly used for HPLC. Many analytes that have been subject to LC and HPLC analysis are UV absorbers, and even more analytes can be made to become UV-absorbing through derivatization with an aromatic moiety. The simplest UV detector emits and detects either one wavelength or a region of wavelengths of UV light. However, simplicity does not always suffice, and for greatest specificity UV detectors may have a variable wavelength light source and detector. In that way, the spectrum and spectral purity of a peak can be determined.

Fluorescent (fl)

UV absorption may not always give a sufficiently low limit of detection for certain analyses. Therefore, analytes may be derivatized with functional groups that fluoresce at a known wavelength following absorption of a UV photon. This may seem counterintuitive, but it in fact results in a detector with higher sensitivity and specificity than is found for a UV detector alone. Frequently, you will find an HPLC equipped with both a UV detector and a fluorescence detector. Figure 19-9 shows the HPLC of isocyanate monomers using OSHA Method 42 with a fluorescence detector.

Mass Spectroscopy (MS)

A GC-MS has already been explained. The same principles and practices apply to an HPLC-MS. However, this instrument is more complex because instead of helium or hydrogen for a mobile phase, the HPLC has water or a non-aqueous solvent as the mobile phase. In addition, the HPLC is almost always relatively low resolution when compared to that obtainable by use of a GC. Therefore, analyte peaks obtained using an HPLC may not be pure. This puts a greater burden on the MS in that it must make several scans of each LC chromatographic peak and then separately identify then quantify the compounds in that peak. This is the point at which multiple detectors might be used, such as LC-UV-MS-MS. In Figure 19-10, the ini-

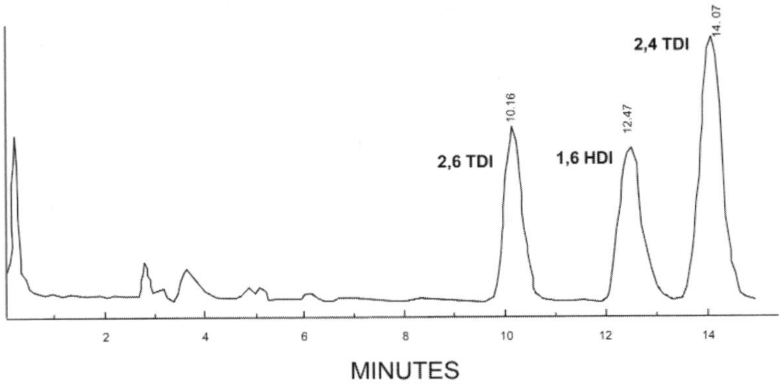

Figure 19-9. HPLC-UV chromatogram of diisocyanates using OSHA Method 42.

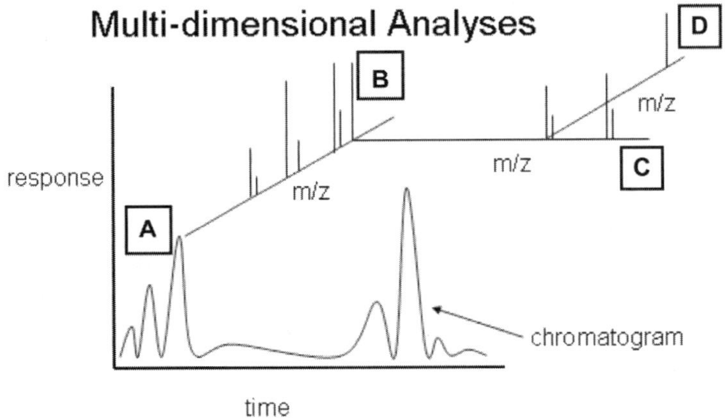

Figure 19-10. Multi-dimensional analysis of an HPLC-MS output showing the chromatogram in the x-axis and the sequential mass spectra of HPLC peak A (the third peak in the chromatogram) followed by sequential mass spectra (B, C, D) of individual mass peaks obtained with an HPLC-MS-MS-MS instrument (courtesy S. Markey, National Institutes of Health).

tial HPLC-UV tracing or chromatogram might be rendered for any analysis with such an instrument (or a GC for that matter). Suspecting that the third peak detected with the UV detector (Figure 19-10, peak "A") is not pure, one could obtain a mass spectrum using the HPLC-UV-MS (output "B" in Figure 19-10). The final peak (line) in that spectrum could also be checked for purity (output "C"); that is, is it from a single compound, and so on (output "D") using HPLC-UV-MS-MS.

ION CHROMATOGRAPHY AND ION EXCHANGE CHROMATOGRAPHY

Ion chromatography is a sub-type of liquid chromatography. It is commonly used to assay organic and inorganic cations (mainly metals) and anions (mainly acid and base species). For this reason it is treated separately from GC and HPLC, which are used to assay molecules. In ion exchange chromatography (IEC), an ion exchange resin is used as a solid phase. A common resin is XAD-2, which consists of polymer beads composed of polystyrene, crosslinked by divinyl benzene. The beads are modified by reaction with acidic or basic ions such as $SO_3^-H^+$, COO^-H^+, $NH_3{}^+OH^-$, $N(CH_3)_3{}^+OH^-$. Anion analytes are exchanged for OH^- groups and cation analytes for hydrogen ions. A separate column must be used for each group. The liquid phase is generally an acid or a base that also has an affinity for either the cation or the anion analyte, respectively. Therefore, an equilibrium is established that is shifted far toward the mobile liquid phase, but the stationary solid phase impedes the passage of the analytes, depending on the affinity of the solid phase sites for each analyte. Hence, anions are separated based on approximate affinities. The order of relative affinity of anions on strong base anion exchangers follow the general order of: citrate > salicylate > ClO_4 > I^- > $S_2O_3^{2-}$ > CrO_4^{2-} > SO_4^{2-} > SO_3^{2-} > HPO_4^{2-} > NO_3^- > Br^- > NO_2^- > CN^- > Cl^- > HCO_3^- > $H_2PO_4^-$ > CH_3COO^- > IO_3^- > $HCOO^-$ > BrO_3^- > ClO_3^- > F^- > OH^-. Similar sequences can be written for the mono- and di-valent metals.

ATOMIC SPECTROSCOPY

Atomic Absorption Spectroscopy (AAS)

With AAS, assays are limited to the quantification of one element at a time. Thus, this method, while highly sensitive and quantitative, cannot be considered to be a tool that is useful for identification of unknown species in mixtures. At temperatures of between approximately 1500°C to 3000°C (associated with flames of various fuel/oxidant combinations), only a very small fraction of the available atoms is promoted to an energy state above the ground state. Those that remain in the normal state are available to absorb energy at the same wavelength characteristic of the element and at that which the excited atoms emit.

The instrument has a light source, called a hollow-cathode (HC) lamp (shown in Figure 19-11) that emits the radiation characteristic of a single element. There are as many HC lamps as there are elements for which the AAS instrument will be used. For example, the light from an HC source designed for analysis of sodium emits a wavelength of light that is emitted by

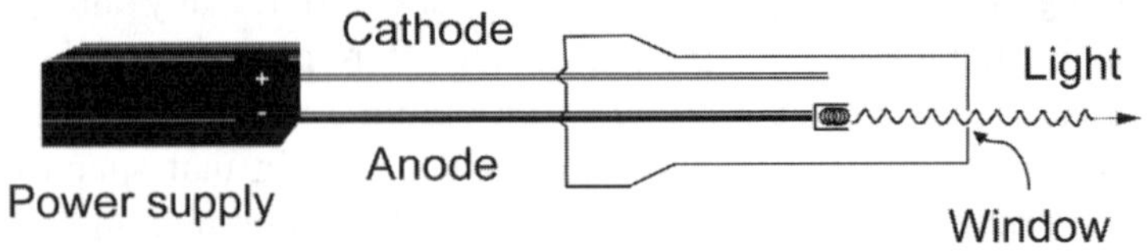

Figure 19-11. A hollow cathode lamp for use in an atomic absorption spectrometer (AAS).

excited sodium atoms. The source is typically a lamp containing a tungsten anode and a cathode composed of the element in question or coated with that element. The lamp contains neon or argon, which is ionized by a current between the electrodes, and the gaseous cations then bombard the cathode. This causes some cathode atoms to be dislodged, and a portion of these emit the characteristic light.

The sample solution is aspirated using a nebulizer into a flame (commonly hydrogen or acetylene mixed with air or oxygen, Figures 19-12 and 19-13) which serves to heat the sample. After passing the flame, the light of interest must be separated from extraneous light emitted from the flame. This is accomplished with a monochromator (Figure 19-14). The absorption of light from the HC lamp is proportional to the concentration of sodium in the sample.

The most common alternative to the flame is the electrothermal atomizer, and the most common of this type is the graphite furnace (Figure 19-15). With this device the sample is placed in a graphite tube attached to two electrodes, across which an inert gas flows to prevent oxidation of the graphite as well as the sample. A current is applied, and at first the volatiles such as water are driven from the sample; then the current is increased, and the sample is ashed. Finally the current is stepped up considerably, and within a few seconds the entire sample is atomized, a step that improves sensitivity and detectability. However, the technique is not as reproducible as flame AAS, yielding coefficients of variation of 5 to 10% versus 1 to 2% for flame AAS.

This is very useful for certain samples, such as blood, which can be placed directly in the furnace without being prepared. However, workplace airborne dust samples are collected on filters used for sampling air, and the filters must be digested by using acids. This decomposition process is the most time-consuming portion of AAS work. In addition, it may lead to the majority of errors associated with AAS. The acids, filter material, and extraneous material collected on the filter form a matrix that is placed in the flame or graphite furnace along with the elements of interest. Matrix effects can cause interferences (biases) and decrease the reproducibility. They tend to be greater for the graphite furnace, hence its poorer precision relative to the flame. Both can be caused by incomplete breakdown of the matrix to elemental form, particularly when halogens are present. In addition, large quantities of Ti, Zr, or W can interfere by forming oxides. In either case, the interfering compounds block the transmission of

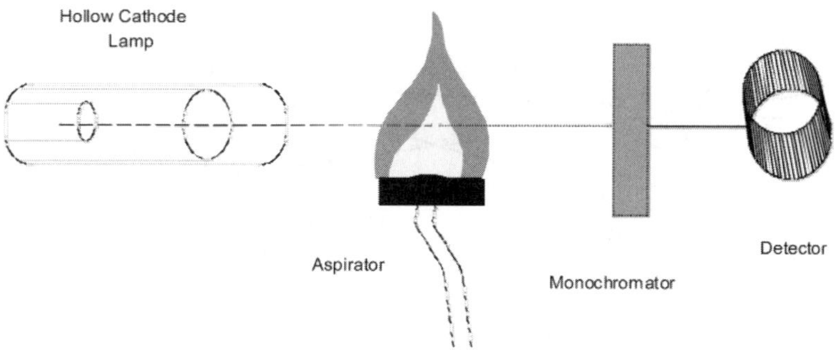

Figure 19-12. Schematic of an atomic absorption spectroscopy unit.

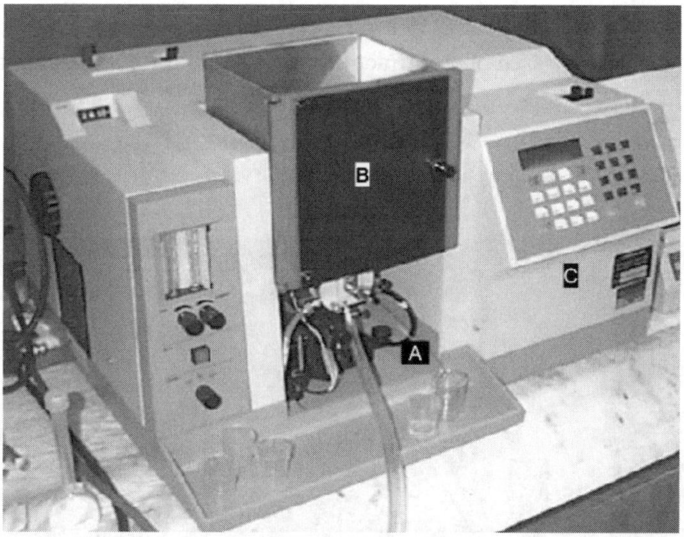

Figure 19-13. An atomic-absorption spectrometer with flame inlet showing A) sample solution inlet tube aspirating sample from a beaker, B) the window behind which the flame is located, and C) the position behind which the hollow cathode lamp is located.

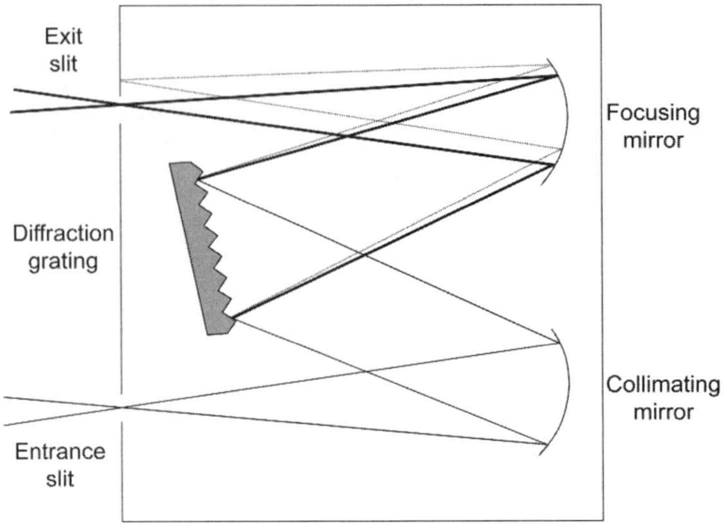

Figure 19-14. A monochromator based on a diffraction grating useful for wide variety of wavelengths.

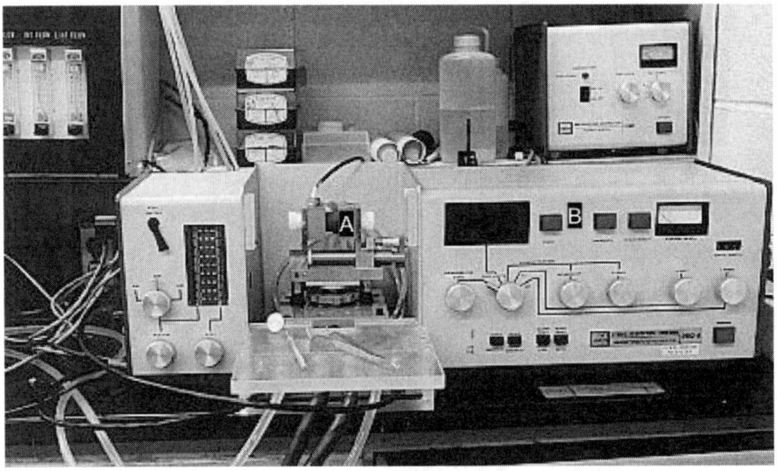

Figure 19-15. Atomic-absorption spectrometer with graphite furnace inlet, showing A) furnace for vaporizing sample and B) position behind which the hollow cathode lamp exists.

light, thus forming a positive interference.

Some compounds of certain metals are difficult to solubilize. Thus, some laboratories may require a separate filter sample for these metals so that alternate preparation techniques may be used. A worker exposed to Co, W, and Ni could be sampled with one pump, but the flow would need to be split to two filter heads so that one filter could be prepared for Ni and Co, the other for W. Of course, W would be contained on either filter and thus could be an interferent in the analysis of Ni or Co, particularly if the graphite furnace were used.

Atomic or Optical Emission Spectroscopy (OES) and Inductively Coupled Plasma (ICP)

When inorganic compounds are heated, they are changed to their ionic or elemental state, depending on the temperature and the chemical environment. There are numerous ways of providing the energy needed to cause the reduction, including combustion, electrical current, and radio frequency or other electromagnetic energy. When the elements are heated, electrons in the shells become excited, and some will advance to higher energy states. When the elements return to their normal energy states, electromagnetic energy, often visible light, is emitted. This light often is spectrally quite pure and is associated with a particular electron shift in a given atom. Therefore, the wavelength of the light can be used to identify the element, and its intensity can be used to quantitate the element. This process is termed atomic emission spectroscopy (AES) (Figure 19-16).

Only a small percentage of the total number of atoms of many elements will become excited and emit radiation in a flame (1500–3000°C). For these elements, flame optical emission spectroscopy (OES) lacks sensitivity, and thus for many years OES did not have general applicability to environmental work. However, in the mid-1970s the inductively coupled plasma (ICP) or inductively coupled argon plasma (ICAP) technique was developed, which heats the elemental solution to temperatures exceeding 7000°C (Figure 19-17). In addition, unlike AAS, AES can be used to develop spectra for many elements (up to 60) with a single sample injection. Thus a powerful tool was developed. Originally the technique was referred to as atomic emission spectroscopy (see for example NIOSH method 7300), but manufacturers now

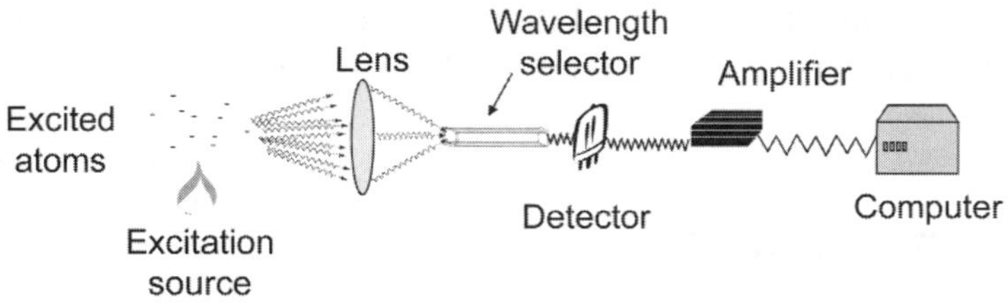

Figure 19-16. Schematic of an atomic emission spectrometer (AES).

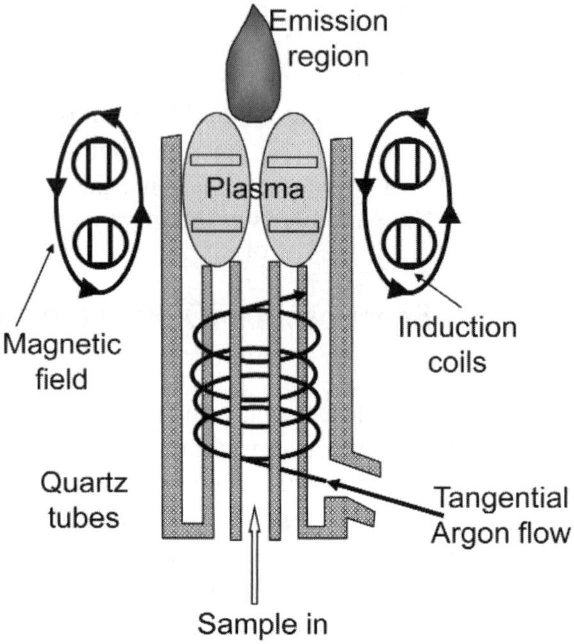

Figure 19-17. Schematic of an inductively-coupled plasma excitation source for AES.

refer to it as optical emission spectroscopy. In addition, ICAP instruments are now commonly supplied with mass spectrometers rather than optical emissions detectors. These ICAP-MS systems can differentiate isotopes and have other advantages.

A liquid sample is injected into the ICAP, thus sample preparation is still necessary. From there, an argon flow carries the sample to a water-cooled induction coil powered by a radio frequency generator that ionizes the argon as well as the sample. These ions and electrons are confined by an induction coil that forms a magnetic field. The environment is a hot, reducing environment in the argon. Therefore, many of the oxide-interferents found in AAS are not present in ICAP. Linear calibration curves of several orders of magnitude are possible. Compounds of Ca, La, Mg, Sb, Sn, W, and Zr may have low recoveries. Although the majority of metals analyses can be performed with ICAP, graphite furnace and flame AAS are still widely used.

MOLECULAR SPECTROSCOPY

Like atomic spectroscopy, molecular spectroscopy operates by examining the emission or the absorption of electromagnetic radiation, in this case from molecules. Absorption is used in most cases. The previously discussed ultraviolet (UV) detector, widely used in HPLC, is based on molecular absorption spectroscopy. Aromatic compounds absorb UV well. Absorption spectroscopy also is widely employed in applications in the visible and the infrared range.

Colorimetry

Absorption spectroscopy in the visible range is often referred to as colorimetry. Colorimetry (Figure 19-18) is the measurement of *color*, or specifically the measurement of the molecular absorption of specific wavelengths of light. Commonly, both long-wave UV (> 185 nm) and visible light are combined for these purposes. Thus both UV and visible spectroscopy are termed colorimetry. All organic compounds absorb electromagnetic radiation, causing valence electrons to be excited to higher levels. Colorimetry is most productive for certain functional groups that are termed chromophores. In some cases analytes that do not contain the necessary chromophores can be made to react (be derivatized) with compounds that do, thus expanding the application of colorimetry. It is, therefore, often necessary to perform *wet chemistry* separations, reactions, and so on. For these reasons, the application of colorimetry is probably on the decline for hygiene work.

Perhaps the easiest way to describe the overall process is to use an example (NIOSH, 1994b). Formaldehyde has long been analyzed by using colorimetry. It is extracted from air by passing the air through a solution of water and sodium bisulfite. Then chromotropic acid and concentrated sulfuric acid are added, and the mixture is heated at 95°C to cause a reaction of formaldehyde and chromotropic and sulfuric acids, to form a chromophore absorbing at 580 nm. The colorimeter usually is a simple and relatively inexpensive device that is easy to calibrate. The difficulty and the expense of the technique lie in the trouble and the time needed to prepare samples and standards.

Infrared Spectroscopy

A molecule can be pictured as a group of spheres (atoms) held together with springs. The spheres oscillate relative to each other. The frequency of these oscillations, and of some rotations, can correspond to the frequencies of infrared (IR) energy (wavelength of about 1-40 μm). Radiant heat is in the infrared region and is typically polychromatic. When heated, mol-

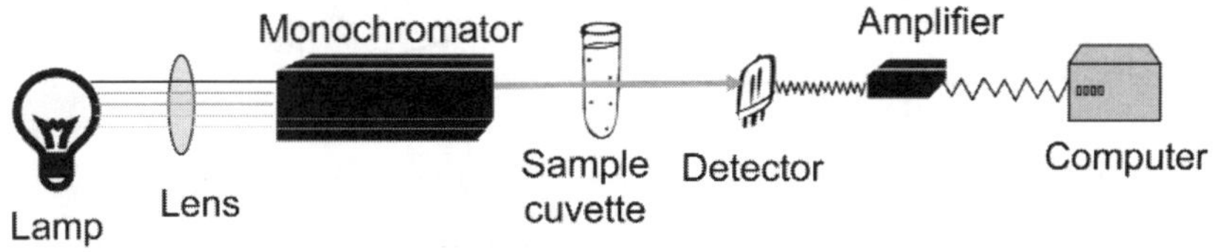

Figure 19-18. Schematic of a UV-visible spectrometer.

ecules oscillate or vibrate. When IR is monochromatic or confined to certain wavelengths, then only certain groups of spheres will vibrate – those that have vibration frequencies corresponding to the emitted IR. Thus, if IR is emitted at each wavelength in succession, some of the IR may be absorbed, according to the molecules present. The plot of IR absorption versus wavelength is called a spectrum (see Wilks, 1976 and numerous Internet sites). The relationships most often used involve Beer's Law, or

$$A = abc$$

where A is the absorbance and scales linearly with the concentration (c), a is the absorptivity and is constant for each molecule at each wavelength, and b is the pathlength. For gases, b is most often in meters. In order to maximize A at a given value of c (usually in the parts-per-million range), the pathlength (b) is increased (i.e., to 10 meters) using the folded-path cell (Figure 19-19).

IR spectroscopy can be used to qualitatively and quantitatively analyze organic and many inorganic compounds. The spectrum produced for a given analyte is often quite complex (see Figure 19-20). On the positive side, each organic compound has a unique spectrum although differences between compounds can be subtle. Computer programs are often necessary for sorting out the various spectra and making a qualitative identification of an unknown(s). However, for the analysis of the vapors of organic compounds in air, the spectrum of water vapor is by far the dominant spectral pattern. For the highest resolution and highest sensitivity, Fourier Transform IR (FTIR) is used for the most efficient identification of unknowns. A simple schematic of the Fourier transform optical system is shown in Figure 19-21.

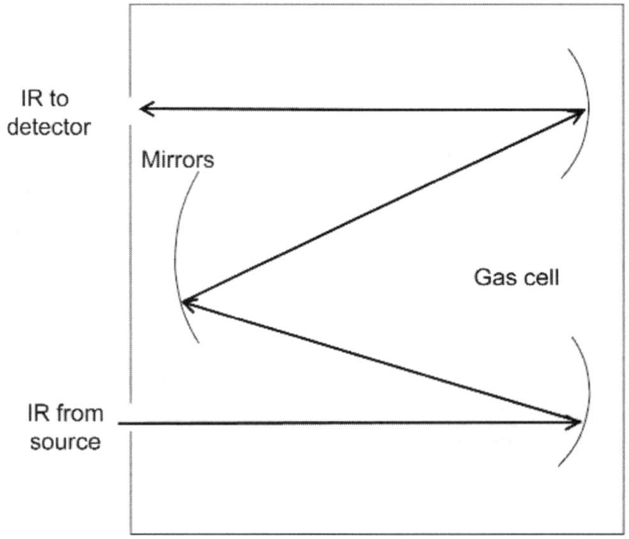

Figure 19-19. Use of mirrors to increase the pathlength and thus the sensitivity in an IR Spectrometer.

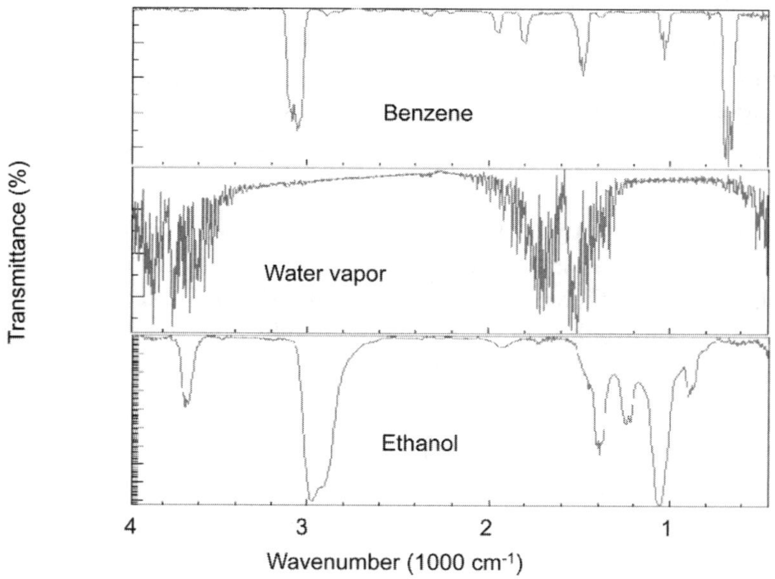

Figure 19-20. IR spectra for three compounds showing overlap of portions of the ethanol and benzene spectra with the water spectrum (adapted from NIST, 2006).

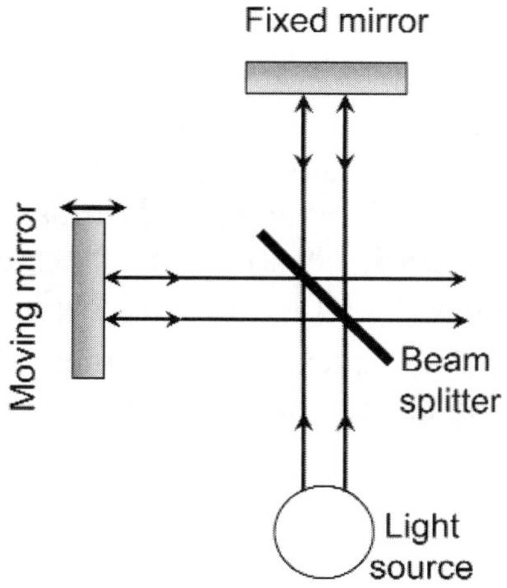

Figure 19-21. A simple schematic of the Fourier transform optical system.

Although infrared spectroscopy has been widely used to analyze compounds in solution and thus could be used to analyze air contaminants extracted from a collection media such as charcoal, the problem is finding a solvent that is transparent in the regions of the spectra where the analyte absorbs best. For complex mixtures this is essentially impossible, and even for single analytes the problem does not lend itself to a universal or near universal solvent such as the CS_2 used in GC work.

One application is very important because of the importance of the analyte (crystalline silica). After collection of coal dust containing crystalline silica, the sample is subjected to low temperature ashing, rinsing, then redeposition on a filter (NIOSH Method 7603, issue 3, 2003); or, after collection of construction dust, the sample is subjected to low temperature ashing followed by formation of a pellet (NIOSH Method 7602). These are assays of solid samples, as opposed to liquid or vapor.

Air is relatively transparent to IR (with the exception of water vapor, the most important interferent); so analysis of compounds from direct input of contaminated air is the most common application of IR in hygiene work. It is often cumbersome to collect bags or other vessels of air for laboratory analysis (see Chapter 22). Conversely, IR instruments have not been miniaturized to the point that they can be used easily to collect breathing zone samples. However, portable and field IR units have been widely used to monitor area concentrations at fixed locations. For this reason IR and FTIR instruments are discussed in more detail in Chapter 27. Generally, these portable instruments are best used to quantitate known gases and vapors, rather than to identify them. The big problem facing all investigators using IR for direct analysis of contaminants in air is that the spectrum of water vapor overwhelms the spectra of most other compounds, except when the instrument has adequate sensitivity, selectivity, and stability.

GRAVIMETRY

Not much can be said about weighing. It is applied to only a few analytes, most notably dusts. However, these very often are sampled by hygienists. The method obviously is quantitative only; so the identity of the dust usually is made on the basis of the industrial process and materials. Laboratory balances typically used for analysis have detectabilities on the order of 30 μg for sensitivities of 1 μg (NIOSH, 1994a). For a 1,000 L air sample (about 2 L/min for 8 hours) this means a detection limit of a bit less than 0.1 mg/m^3.

Of concern is the potential for interferences, particularly adsorbed water vapor. The potential for this problem to occur and techniques to control it vary with the filter material and so are discussed in Chapter 24.

MICROSCOPY

Microscopy once was the method of choice for analyzing all particulate matter. The particulates usually were collected in water with an impinger or a bubbler (Chapter 22). A portion of the suspension was placed under the microscope on a counting slide such as a Dunn cell; the slide allowed the liquid to be divided into small areas, in each of which the total number of particles could be counted. A count for the entire volume of water could be extrapolated from the count of the small sample, and this was divided by the air volume sampled to obtain a concentration, often expressed as millions of particles per cubic foot (mmpcf). Fortunately, gravimetry has been substituted, but microscopy still plays an important role, especially in the analysis of asbestos and other fibers such as man-made mineral fibers (MMMF), the most common of which is fibrous glass. The word *fortunately* applies to the issue of the efficiency of analysis.

Polarized Light Microscopy (PLM)[2]

PLM is commonly used to identify and quantify asbestos and other minerals in bulk samples. The samples are commonly taken from building materials such as spray-on insulation, ceiling tiles, pipe and vessel insulation, and steel beam insulation. The purpose of the analysis is to determine if asbestos exists, and therefore if removal may be necessary. NIOSH (1994c) reported that the limit of detection for this method is < 1% for any of the asbestos minerals.

Polarized light is light that has been filtered so that all of the waves are in parallel planes. This is accomplished with a device called a polarizer, which is built into the microscope. Many different minerals can be identified with PLM, and generally each is identified by eliminating others in a dichotomous fashion, much as one would use a key to identify a species of insect.

Fibers and particles, both manufactured and natural, can be classed as isotropic and anisotropic, terms based on refractive index. As light passes from one transparent material through another, say, from air to asbestos, its velocity (direction and/or speed) changes if the two materials have different density. The ratio of light velocities (analyte:air) is the refractive index. When the structure of a material is such that it varies systematically (i.e., the atomic components are not randomly spaced as in non-stressed glass or perfectly spaced in all directions as in a cubic crystal, e.g., all axes are the same, as in diamond), then the substance will have more than one refractive index and will be anisotropic. Such substances usually are crystalline with more than one axis (non-cubic), or the repeating units are partially aligned by manufacturers (polyester fibers) or by nature (cellulose fibers).

In order to find the anisotropic features of a particle or fiber, polarized light must be directed parallel to the particle's major and minor axes. White, polarized light directed in this way will yield colors that are characteristic of the substance for that axis. These principles are used to identify particles, but they are often augmented by a procedure known as dispersion staining.

---

[2] For a thorough discussion of this subject the reader should see McCrone et al., 1979, and McCrone, 1980, on which most of this material is based.

Dispersion staining is not a straightforward coloring process as one might expect, but it does convey color to the particles – hence the name. Asbestos and many particles are transparent; they absorb no light and therefore are colorless. Refraction of light through a transparent particle can be seen at the edges only, as a result of recombining or interference of diffracted light waves, which yields a change in contrast. Thus, some mechanism is needed to impart color to the particles, and, most important, to do so in a way that gives relatively unique color combinations. Slight differences in refractive index (η) of particles and media can cause this. Refractive index values vary with the color of the light, with most refractive indices based on the refraction of a specific wavelength of light, the so-called D-line in the spectrograph of sodium, or about 589 nm. Chrysotile asbestos has a refractive index of 1.55 for the sodium D-line, but in white light the refraction varies with wavelength. If the chrysotile is placed in a liquid medium (the dispersion stain) also with η of 1.55, and the total refractive spectrums across all wavelengths of the medium and particle are not exactly the same (see Figure 19-22), then these differences will impart color to the particle boundaries that is unique for the combination of refractive indices. The color can be seen only as a result of a *stop* or light-blocking diaphragm, which is inserted into the microscope.

Together optical crystallography and dispersion staining are used to identify asbestos and other particles. Quantitation results from a simple visual estimate of the percentage of the total particles that are of a specific type, such as chrysotile asbestos.

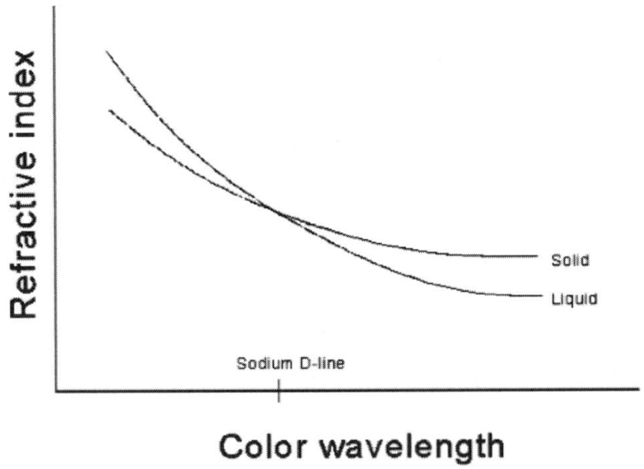

Figure 19-22. Plot of refractive index versus color for a solid particle and a dispersion staining liquid, having equivalent refractive indices at the sodium D-line.

Phase Contrast Microscopy (PCM)

PCM is commonly used to quantitate airborne fiber shapes that have been collected on filters. A fiber usually is defined as a particle having an aspect (length to width) ratio of 3:1 (NIOSH, 1994d). However, this value is arbitrary, and other techniques define the ratio as 5:1 (see for example the B rules in NIOSH method 7400). This is the only qualitative criterion used in the technique; so asbestos is not differentiated from fibrous glass, cellulose, hair, and many other substances. Nonetheless, the technique should give a reasonable estimate of asbestos exposure when the main source of dust is asbestos. When this is not the case, as in asbestos abatement operations where other building products also are major sources, it is not necessarily reasonable to expect PCM to provide a good indication of asbestos exposure. Nevertheless, it is the accepted OSHA method for asbestos analysis (OSHA, 1991), and is used in virtually all types of exposure situations.

If one collects asbestos or any other particle on a filter, the filter must be made transparent if one is to count the particles under a microscope. The mixed-cellulose ester filter (MCE, see Chapter 24) is used for such purposes. However, when the filter is made transparent, the particle becomes indifferentiable from the filter. In other words, both asbestos particles and filter are transparent and colorless; they do not differentially absorb the wavelengths of the color spectrum. In addition, they both absorb negligible amounts of the total spectrum; so there is no contrast between the two. However, some of the light passing the edges of a particle is diffracted, and a phase shift of a quarter of a wavelength takes place relative to those waves that are not diffracted (see McCrone et al., 1979, pp. 61-64).

This phase shift still cannot be seen because when the two wave types are focused in the eyepiece of the microscope, no wave interference takes place. Interference would cause a change in wave amplitude (either negative or positive), which could be seen as contrast.

The PC microscope causes the diffracted waves to be diffracted once more, so that now the shift is half a wavelength. When these waves are focused with the non-diffracted waves, there now is interference, which can be seen as contrast or higher wave amplitude; the particles are brighter than their surround. Thus, in the dissolved filter environment, asbestos particles are invisible unless phase contrast is employed.

Electron Microscopy

Light microscopy only can resolve particles down to about 0.3 μm, following the equation

$$X = \frac{0.61\lambda}{NA} \qquad \text{[Equation 19-1]}$$

where X is the resolution distance, λ is the wavelength of light, commonly assumed to be filtered blue light of about 0.45 μm, and NA is the numerical aperture of the microscope objective, which is maximally 0.95 for a dry (non-immersion) lens (McCrone et al., 1979). Many asbestos fibers have smaller diameters than this and thus are not seen in the light microscope.

If we are in occupational settings where the main exposure to dust is known to be asbestos, and if it is reasonable to assume that analysis based on PCM is an accurate index or predictor of exposure, and thus risk, in those settings, then the lack of ideal resolution should not matter. In other words, in these environments we know that PCM counts are biased low, but the bias is assumed to be consistent across concentration. However, in less traditional exposure situations, such as immediately after an asbestos abatement or removal operation, it might be expected that a larger proportion of the remaining suspended fibers would be small. In this case, the PCM negative bias might be even larger. In addition, in abatement and other situations, there is some concern that PCM counts might be inflated by the many non-asbestos sources of fibers.

Thus, a technique is needed that has better resolution than PCM and is qualitative as well as quantitative; two types of electron microscopy have been used for these purposes. The electron microscope uses electrons rather than light to resolve objects. The resolution of a space between two particles is a function of the light passing through that space in the case of light microscopy – so the wavelength of the light is important. Smaller wavelengths pass through smaller spaces (see Equation 19-1); still we are limited to blue light. Electron microscopes (EM) record the absorption and passage of electrons; here the path widths are considerably smaller than those of blue light waves. Of course, EM also is considerably more expensive than light microscopy, but the cost of analyzing samples with EM has become more reasonable relative to PCM.

Transmission electron microscopy (TEM) can be coupled with analytical techniques known as selected area electron diffraction (SAED) and energy dispersive X-ray detection (EDX), which allow identification of asbestos minerals (NIOSH, 1977). The initial indication that the asbestos mineral is in fact a fiber is still based upon arbitrary aspect ratio rules such as 3:1. SAED involves the scatter of electrons in regular patterns based upon the properties of the crystal. These patterns are visible on the viewing screen of the microscope and can be used to differentiate chrysotile asbestos from the amphiboles (amosite, crocidolite, anthophyllite, tremolite, and actinolite). However, individual amphiboles cannot be distinguished.

EDX involves the generation of a spectrograph based on the emission of X-rays from the crystal as a result of its being bombarded with electrons. Typically ratios of iron, magnesium, and silica, identified from X-rays, are used to identify the asbestos minerals. However, anthophyllite and fibrous talc can be mistakenly identified as chrysotile. Thus, SAED is also needed to avoid this interference (NIOSH Method 7402, 1994e).

Scanning electron microscopy (SEM) allows high resolution and thus aids in the counting of small fibers. However, it cannot be coupled with SAED, and therefore is not as qualitative as TEM.

X-RAY DIFFRACTION

X-ray diffraction is an X-ray spectroscopic method, as are X-ray emission, absorption, scattering, and fluorescence. Energy dispersive X-ray detection or EDX, described in the preceding section, is an X-ray fluorescence method. SAED, from the preceding section, is a diffraction method; however, electrons are diffracted, not X-rays. X-ray diffraction is used most often for analysis of quartz and other crystalline forms of silica.

When a monochromatic (one wavelength) beam of X-rays strikes a crystal, the X-rays are diffracted in a pattern that is specific for the crystal. Qualitative analysis is achieved by comparing the pattern to patterns of known crystals. The intensity of the pattern is indicative of concentration; so quantitation can be achieved with calibration standards of known quantity. X-ray diffraction is time-consuming and thus relatively costly. The samples and the standards require considerable preparation prior to analysis.

ELECTROCHEMICAL METHODS[3]

There are many electrochemical methods including amperometry, coulometry, voltammetry/polarography, potentiometry, and conductometry. These are often used in direct reading instruments and are discussed more fully in Chapter 27. The analyte usually becomes part of a solution or of the electrolyte of an electrochemical cell. Therefore, the analyte is usually ionic or converted to an ionic form in solution. It is reduced or oxidized. This allows for the differentiation of valence states such as Cr+3 versus Cr+6.

The two main laboratory-based techniques used to analyze air samples are polarography, which is a form of voltammetry, and potentiometry.[4] Only a few laboratory analysis methods important to hygiene work are based on these techniques, among them methods for such analytes as chlorides, fluorides, ammonia, cyanides, bromides, and nitric acid using potentiometry; and chromic acid, formaldehyde, hydrogen peroxide, and uranium compounds using polarography.

The ion selective membrane electrode is the most common electrochemical device used in hygiene laboratory work, and is quite common in any laboratory, as the pH meter falls into this category. It uses a potentiometric method; a voltmeter reads the voltage across two electrodes, one placed in the analyte solution and the second in a standard solution, as shown in Figure 19-23. A membrane composed of silicate glass, a crystalline substance (referred to as solid state), or a polymer separates the analyte from the standard solution or electrolyte. The

---

[3] This section is largely taken from Skoog et al., 2007.

[4] Other electrochemical methods are used in conjunction with chromatographic separation techniques; however, these are the most widely used for hygiene analyses. Again see Skoog et al., for a complete discussion.

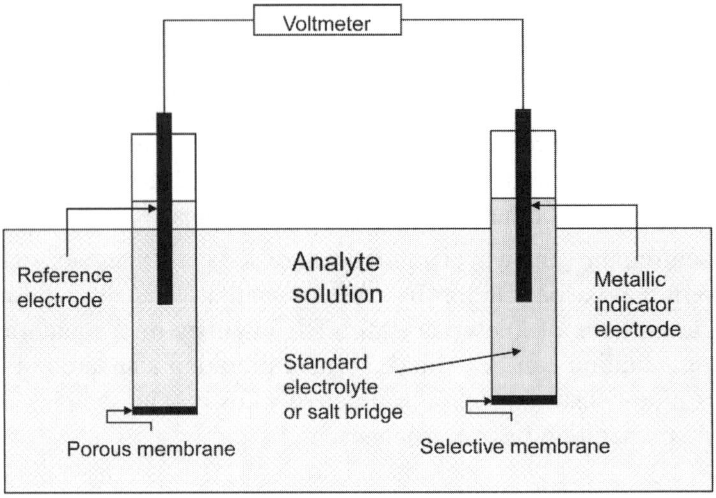

Figure 19-23. Schematic for a selective membrane electrode potentiometric device as it would be used to measure an ionic species in a liquid. For measuring pH (H+), the reference electrode might be the standard saturated calomel electrode with known potential and the indicator electrode composed of silver. The selective membrane would be thin glass. In most pH meters these electrodes are combined into one probe.

analyte ions are exchanged across the membrane, causing an equilibrium change between ions in the standard solution and the membrane. The movement of ions causes a change in voltage. As the principle is based on ion exchange, interferences are expected from other ions of like charge, and this is a concern with ion selective membranes, to the extent that the membrane is not completely selective.

Polarography is the earliest type of voltammetry and is distinguished by its use of the dropping mercury electrode. Like all voltammetric methods, it is the measurement of current through an electrochemical cell at some *applied voltage*. The latter typically varies during the experiment, thus both qualitative and quantitative information about the analyte can be obtained from the plot of current versus voltage. The purpose of the application of the voltage is to completely polarize or charge the solution. Although it is a widely used method in chemistry, polarography is used in methods for only a few substances of hygiene importance. Most of these substances (listed above) have alternate methods of analysis.

TITRATION

Titration once was a very commonly used technique for analyzing a wide range of compounds of hygiene interest, including most acids and bases and associated gases, hydrogen sul-

Table 19-1
Some compounds for which HPLC methods exist (from SKC, 2006) (Note that many are high molecular weight and/or polar.)

Acetaldehyde
Amine Compounds (various)
Benzidine Dyes
Bisphenol A
Dichloroaniline
Dinitrobenzene
Hydroquinone/Quinone
Pesticides (various)
Polynuclear aromatic hydrocarbons

fide, formaldehyde, and carbon dioxide. Today there are a few methods validated by OSHA or NIOSH that are based on titration, including those for sodium hydroxide, sulfur dioxide, sulfuric acid, and alkaline dusts.

In these techniques, commonly a pH-sensitive dye is added to an aqueous solution of the analyte. Acids are then added if the solution is basic and vice versa, in known quantities, so that the mole concentration of the analyte can be determined stoichiometrically. The end point is usually dependent on a color change of the dye.

QUESTIONS

19.1. Of the chromatography detectors, which are nondestructive?

19.2. What are the major differences in the application of liquid versus gas chromatography?

19.3. What are the advantages and disadvantages of emission versus absorption spectroscopy for the analysis of metal elements?

19.4. List the analytical techniques that you think would be most applicable to the following analytes: benzene, arsenic, a complex organophosphorus pesticide, the mass of dust on a filter, sulfuric acid captured on a filter, aniline, and total number of particles on a filter.

REFERENCES

McCrone, W.C.: The Asbestos Particle Atlas. Ann Arbor Science Publishers, Ann Arbor, MI (1980).

McCrone, W.C.; McCrone, L.B.; Delly, J.G.: Polarized Light Microscopy. Ann Arbor Science Publishers, Ann Arbor, MI (1979).

NIOSH: Review and Evaluation of Analytical Methods for Environmental Studies of Fibrous Particulate Exposures. National Institute for Occupational Safety and Health (1977).

NIOSH: NIOSH Manual of Analytical Methods, Particulates not otherwise regulated, total and respirable, P.C. Schlecht, P.F. O'Connor (eds.). National Institute for Occupational Safety and Health, Cincinnati, OH, Methods 500 (issue 2, 1994a) and 600 (issue 3, 1998) http://www.cdc.gov/niosh/nmam/method-2000.html.

NIOSH: NIOSH Manual of Analytical Methods. Formaldehyde by Vis, Method 3500, issue 2, P.C. Schlecht, P.F. O'Connor (eds.). National Institute for Occupational Safety and Health, Cincinnati, OH. http://www.cdc.gov/niosh/nmam/pdfs/3500.pdf (1994b).

NIOSH: NIOSH Manual of Analytical Methods, Asbestos (bulk) by PLM, Method 9002, issue 2, P.C. Schlecht, P.F. O'Connor (eds.). National Institute for Occupational Safety and Health, Cincinnati, OH. http://www.cdc.gov/niosh/nmam/pdfs/9002.pdf (1994c).

NIOSH: NIOSH Manual of Analytical Methods, Asbestos and other fibers by PCM, Method 7400, issue 2, P.C. Schlecht, P.F. O'Connor (eds.). National Institute for Occupational Safety and Health, Cincinnati, OH. http://www.cdc.gov/niosh/nmam/pdfs/7400.pdf (1994d).

NIOSH: NIOSH Manual of Analytical Methods, Asbestos by TEM, Method 7402, issue 2, P.C. Schlecht, P.F. O'Connor (eds.). National Institute for Occupational Safety and Health, Cincinnati, OH. http://www.cdc.gov/niosh/nmam/pdfs/7402.pdf (1994e).

NIOSH: NIOSH Manual of Analytical Methods, Elements by ICP (Nitric/Perchloric Acid Ashing), Method 7300, issue 3, P.C. Schlecht, P.F. O'Connor (eds.). National Institute for Occupational Safety and Health, Cincinnati, OH. http://www.cdc.gov/niosh/nmam/pdfs/7300.pdf (2003).

NIOSH: Pocket Guide to Chemical Hazards, DHHS (NIOSH) Publication No. 2005-151, National Institute for Occupational Safety and Health, Cincinnati, OH. http://www.cdc.gov/niosh/npg/ (2005).

NIST: NIST/EPA Gas-Phase Infrared Database, Standard Reference Database, 35 JCAMP Format, National Institute for Standards and Technology, Gaithersburg, MD. http://www.nist.gov/srd/nist35.htm (2006).

OSHA: Volume 29 Code of Federal Regulations 1910. 1001. Occupational Safety and Health Administration, Government Printing Office, Washington, DC (1991).

Schlecht, P.C.; Cassinelli, M.E.: Survey of analytical methods. Parts I and II. The Synergist (Oct/Nov 1997).

SKC: SKC guide to NIOSH/OSHA air sampling standards. Comprehensive Catalog and Air Sampling Guide. SKC, Eighty Four, PA (2006).

Skoog, D.A.; Holler, F.J.; Crouch, S.R.: Principles of Instrumental Analysis, 6th ed. Thomson Brooks/Cole (2007).

Verma, D.K.; Tombe, K.D.: Measurement of benzene in the workplace and its evolution process, Part I: Overview, history, and past methods. Am. Ind. Hyg. Assoc. J. 60:38-47 (1999).

Wilks, P.A.: Air analysis with quick, reliable results. Env. Sci. Technol. 10:1204–1208 (1976).

20

PUMPS AND THEIR CALIBRATION

Having set priorities for taking samples of workplace air for analysis, the hygienist has a wide array of available sampling methods. Some of these methods employ instruments that allow the industrial hygienist to monitor changes in concentration of airborne contaminants in the workplace as they occur. This instrumentation is termed *direct-reading* and gives real time analysis. However, the vast majority of samples of workplace air are collected and then submitted to laboratories for subsequent analysis. Although the actual parcel of air may be taken to the laboratory, the most common practice is to remove the contaminants from the air and submit the concentrated contaminants to the laboratory for analysis.

Whether the airborne contamination is measured in the field or in the laboratory, some method typically is needed to move the air into the sampling instrument or into the *sampling head* that extracts the contaminant from the air. The sampling head may be a filter, a charcoal tube, or one or a combination of several other extraction media (Chapters 21, 22, and 24). The method for moving the air may be passive, requiring no additional energy input other than the kinetic energy inherent in the air molecules (Chapter 23). However, the most common procedure is to use a dynamic method, which requires an external source of energy for pumping the air through the instrument or the sampling head – the subject of this chapter.

PUMPS

In the 1958 edition of Patty's classic *Industrial Hygiene and Toxicology* (Patty, 1958), no mention is made of pumps in the chapter devoted to sampling and analysis of contaminants. Certainly pumps were used prior to that time; however, air sampling and analysis were much more difficult then than they are now, and other aspects of the task overshadowed the importance of pumps (Charsha and Linch, 1957). More importantly, the types of pumps available to the industrial hygienist were extremely limited. The hygienist could choose between alternating current (AC) pumps, which require an external power source, and hand-operated pumps, which required that the hygienist or a technician hand-crank the pump in order to move air through the sampling head. If internal power sources were used, the batteries were quite heavy. In any case, the pumps were often heavy and bulky and required either that fixed air samples be taken or that the hygienist try to follow the worker and collect the best possible *breathing zone* sample. Personal samplers, attached to the worker, were virtually impossible.

Early efforts to overcome the bulkiness and lack of portability of pumps were creative (Charsha and Linch, 1957). It had long been recognized that compressed air moving through a small annular opening (a Venturi) would induce airflow through a second opening connected to the Venturi. An easily portable compressed gas (Freon®) was used to induce the Venturi effect and create what may have been the first portable, personal pump, weighing only about 1 pound or 454 g (see Figure 20-1). One can of Freon would aspirate 7 to 10 cubic feet (200-300 L) of air. Limiting orifices of various sizes were used to control the flow rate. However, the flow rate was sensitive to temperature, which influenced the Freon pressure and therefore the rate of flow of the Freon.

The first pump, from which current models of pumps have evolved, was battery-operated. The advent of lightweight, rechargeable batteries allowed this development. In 1960, Sherwood and Greenhalgh reported the development of this first personal air sampler (see also Ogden, 2001). Although other battery-powered pumps had been developed (Lippmann et al., 1959), they were not of adequate size or weight to be worn by a worker. Two pumps were developed by Sherwood and Greenhalgh. The first weighed only 1 lb (454 g), and the batteries gave it an operating life of 30 hours. The second device could operate for 100 hours without being recharged; however, it weighed 2 lb (910 g). Both obtained airflows of approximately 0.5 L/min and could be worn by a worker. This pump was later produced commercially, and other pumps of similar design were produced (Ingram, 1964). Further development and evaluation of portable air pumps continued at a rapid pace until the mid-1970s, when low flow pumps (e.g., Sipin, 1971) and variable-flow pumps were developed. These are further discussed later.

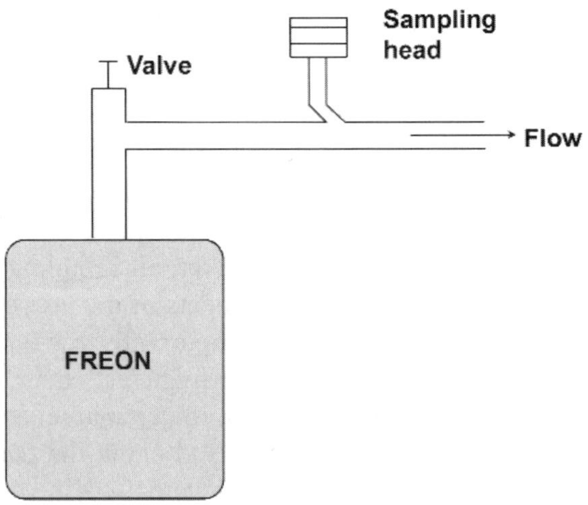

Figure 20-1. Diagram of the Charsha-Linch compressed Freon® driven pump.

The NiCad Battery

Although the standard carbon battery was available in 1960, and perhaps its capacity could have been used to run the early pumps, it had as a main disadvantage that it was not rechargeable. Several rechargeable battery systems became available, including the nickel-cadmium (NiCad), the silver-cadmium, and the mercury battery. The early pumps of Sherwood and Greenhalgh were operated on the mercury battery; however, most portable industrial hygiene equipment now operates on NiCad batteries, with newer nickel metal hydride (NiMH), lithium ion (Li-ion), and lithium ion polymer (Li-pol), and even throw away alkaline batteries seeing increasing use. For this reason, it is very important that every hygienist have an understanding of the specifications and limitations of NiCad batteries.

Typical NiCad battery systems are arranged from cells that are connected in series. The cells are typically sealed by the manufacturer to form a battery pack, which has positive and negative electrodes. Most equipment contains an external recharging jack so that it is not necessary that the batteries be handled unless they are no longer rechargeable.

The most important problem that develops for nickel-cadmium batteries is a process called memory effect (Myers, 1981). This term is widely applied to any NiCad battery problem. There are, however, several different problems that can arise. Although most of the nickel-cadmium battery systems used in portable industrial hygiene equipment will allow operation of the equipment for periods of up to 14 hours (longer in some more recent models), if the equipment is consistently used for periods less than this (for example, 4 hours) and then recharged, the battery *memorizes* the time period and will lose power after that period (about 4 hours in the example). Infrequent use of batteries will also cause immature capacity loss. There is a phenomenon called voltage depression which means the voltage drops while the battery is still capable of producing adequate current. Thus, the instrument (laptops, cell phone, etc.) thinks the battery needs to be charged. This is apparently not a problem for pumps.

A partial solution to these problems is to utilize the equipment for a period of at least 8 to 10 hours at least occasionally. It was once recommended that if the pump were only needed for 1 hour, one should later discharge the pump for 8 hours in the lab prior to recharge. However, the battery life is limited to so many hours of use and so if one does this consistently, then the battery pack life is shortened. It is also important that the battery not be *completely* discharged after each use, or its lifetime can be shortened considerably.

Nickel-cadmium batteries also tend to lose their capacity if stored for long periods of time. For example, a fully charged instrument placed on the shelf for a period of two months will likely contain only half of its charge when it is taken down for use (Kovein and Abell, 1986). This means that maintenance is required so that instruments that are only infrequently used are discharged and charged periodically. Again, if the discharge and charging cycle does not occur, the memorization process will occur, and the battery charge/discharge cycle will be shortened. Or more importantly the battery may completely discharge after 6 months of storage and then it may not be rechargeable.

These problems can be fully avoided if the following three steps are followed (Myers, 1981,[1] see also SKC, 2006a for other recommendations):

1. Periodic discharge to a voltage of $0.9N_c$ volts, where N_c is the number of cells in the battery path. Note batteries should not be fully discharged to 0 volts as this will likely ruin them. Batteries will fully discharge in about 3 to 5 months if not left on a trickle charger.

2. Immediate recharge (14 hours or whatever the manufacturer recommends) at a rate recommended by the manufacturer.

3. After the recharge period, continuous trickle charge at $0.02C$ to $0.05C$ or that recommended by the manufacturer where C = amp-hour capacity. In older systems using the trickle charger for long periods (months) was not recommended. Instead, the instrument needed to be pulled from the charger and stored. To keep it from completely discharging, it needed to be exercised (used for 8-10 hours and then recharged) about monthly. Some newer charging systems have the ability to *pulse trickle*, thus allowing the instrument to be left on the charger.

Most current or recently manufactured equipment has a charging apparatus that allows the above to be accomplished.[2] Regarding point (1) above, a cell that is repeatedly discharged to below $0.9N_c$ can be permanently damaged. This can happen accidentally if the batteries are discharged by turning on pumps in the laboratory and leaving them to discharge. The pumps may be forgotten, so that the batteries are allowed to fully discharge. The automatic cycler prevents this. An additional advantage of some cyclers is that they will sometimes allow at least partial restoration of a defective battery (one whose voltage has gone below 0.9) after several charge/discharge cycles.

Myers recommends that a battery pack should go through the charge/discharge cycle every week. However, in describing a new and improved battery cycler, Kovein and Abell recommend every two months because of the expected 50% charge available after this period. A period of four to six weeks would seem reasonable, as after such a period of non-use the battery still should have capacity for about 8 hours of operation (most instruments are developed to run up to 12 hours on a full charge).

Although NiCad batteries have a finite life span, on the order of 500 cycles, and a cycling device is not capable of restoring all NiCad batteries, the device nevertheless can be a valuable tool. Most nickel-cadmium battery packs cost \$100 to \$300; so a cycle device can soon pay for itself.

---

[1] There are hundreds of Internet sites that discuss this subject. After reading about 10 of them, one gets a pretty good idea of fact versus fiction. However, there is still much mythology about NiCad batteries. Hopefully this discussion has not contributed to it.

[2] A battery cycler/charger can be built for \$400 to \$600 (Kovein and Abell, 1986).

The use of NiMH, Li-ion, and Li-pol batteries is being seen on some equipment, particularly newer models. They have the advantage of no (Li) or minimal (NiMH) memory problems and long run time. Memory issues are not of much concern if new NiCad (post about 1998) battery-operated pumps are purchased with a recent (post 2000) cycling device, and the device is used.

Finally, there are new personal pump models and even direct reading equipment that have been designed to run on throw away alkaline batteries. These have the advantage of being maintenance free. There are direct costs. Three new AA alkaline batteries might cost $1.00 and last for one full-shift pump run. A NiCad battery pack for a pump might cost $100 – $150 and give 500 cycles. There are also indirect costs that include the effects of throwing away 1500 alkaline batteries versus one NiCad battery pack. One final note – it seems to be advantageous to take alkaline batteries out of equipment before storage. The batteries seem to last longer and there is no chance for leakage.

Pumping Mechanisms

Pumps used in industrial hygiene for purposes of sampling air can be classified into three main categories: diaphragm, piston, and vane. The vane pump mechanism is generally applied to line voltage (AC) as opposed to battery operation. The vast majority of battery-operated pumps are diaphragm pumps; a few battery-operated pumps are piston-driven. For industrial hygiene air sampling, the sampling head (where the contaminant is collected) is located upstream of the pumping device.

DIAPHRAGM PUMPS

Figure 20-2 shows a schematic drawing of the diaphragm mechanism, which is found in most battery-operated, personal sampling pumps. The figure shows that a rod connects a motor shaft to the diaphragm. Because the motor shaft is eccentric or not perfectly centered, as the motor shaft turns, the connecting rod moves up and down. This in turn causes the diaphragm to move up and down. When the diaphragm moves upward, a partial vacuum is created in the newly opening air chamber. This causes air to flow in through the inlet valve. When the connecting rod is moved downward, air is pushed out of the air chamber through the outlet valve. Both inlet and outlet valves can be quite simple, consisting of nothing more than flaps of plastic constructed so that when air moves in one direction, they collapse over the opening, but when air moves in the opposite direction, the valves are pushed forward and flutter with passage of the air.

It is important to point out that the valves and the diaphragm in the diaphragm pump are constructed of polymeric materials that can wear out and absorb atmospheric gases and vapors. Generally, because the sampler head removes contaminants from the airstream and is placed upstream of the diaphragm, certain sampler heads remove many contaminants. However, fil-

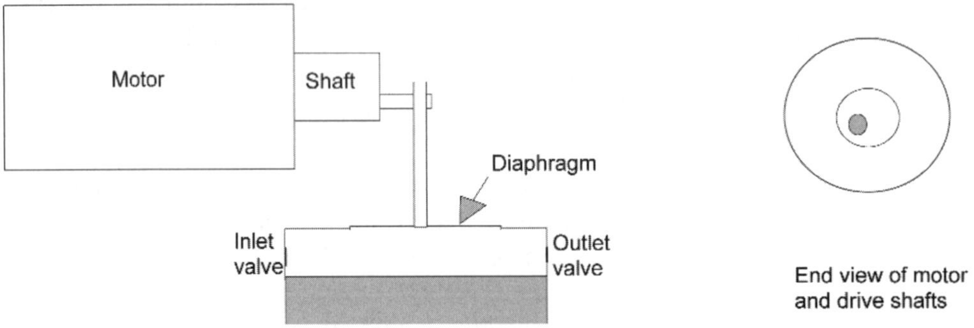

Figure 20-2. Side view of a diaphragm pumping mechanism.

tering heads allow gases and vapors to pass, and adsorption sampling heads such as charcoal allow some aerosols to pass. If a pump begins to fail to operate as efficiently as it did when new, it may be that the diaphragm has become cracked and is leaking, or that wear is occurring on the shaft and connecting rod. Neither of these possibilities is particularly important in terms of the actual airflow that the calibrated pump delivers; the pump should be calibrated against what actually moves through the sampling head. On the other hand, if the hygienist must move at least 2 L/min, and the pump will only perform at 1.9 L/min, then the pump needs to be repaired if it originally delivered > 2.0 L/min.

PISTON PUMPS

Figure 20-3 shows a diagram of the piston pump. It operates essentially the same as the diaphragm pump with the difference being that the air chamber is typically larger, and instead of a diaphragm moving up and down about its center, a rigid piston moves up and down in much the same way as a piston in a combustion engine. Typically, the top of the piston head is flat. The piston tends to displace a greater amount of air on each stroke than is moved in the diaphragm pump; so the magnitude of the pulsations produced by it are greater than for the diaphragm pump, and dampeners become especially important. Piston pumps are easier to start under load than diaphragm pumps, as the area of the diaphragm is larger than the head of a piston for equivalent volumes. Also because the piston is more rigid than a diaphragm, it is less sensitive to back pressure caused by a filter. Therefore, each stroke of a piston (provided that leakage is minimal) is more likely than the diaphragm to deliver a constant volume regardless of upstream pressure. Nevertheless, these pumps (piston and diaphragm) perform the same work for the same energy input. Only a few pumps sold commercially for industrial hygiene purposes operate on the piston principle.

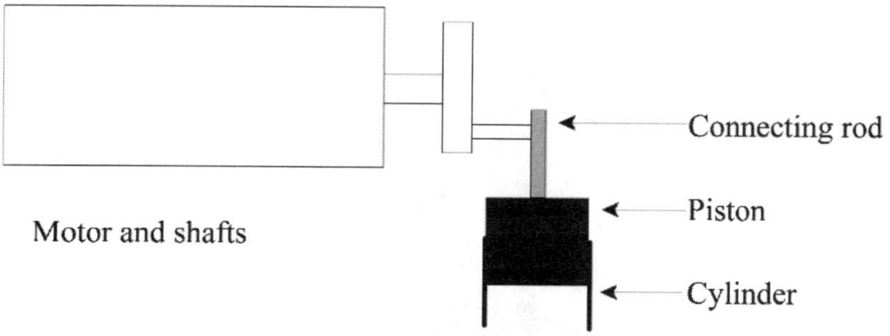

Figure 20-3. Side view of a piston pumping mechanism.

VANE PUMPS

Figure 20-4 shows a photo of a vane or rotary pumping mechanism. Alternating current usually is applied to such pumps; so they are not typically portable. However, they do have applications for collecting fixed-location air samples, as well as filling plastic bags, cylinders, and chambers, and other purposes in the laboratory. Vane pumps, which are often called vacuum pumps, typically can create vacuums of about one-half an atmosphere. Large-size, AC-operated diaphragm and piston pumps can be purchased with similar operating parameters to those of vane pumps (Gast, 2006). However, vane pumps tend to be more economical than those pumps. As illustrated in the figure, the vanes are attached to a plate that moves about in a circular fashion inside a chamber. Because the center of the plate is not the center of the chamber, an expanding air volume is created by a part of the chamber and the vanes as they turn. As a vane slides past the inlet valve, it opens from its previously closed position and air enters to fill the partial vacuum. As the vane continues to move, the space begins to close, and the air is forced out of the outlet valve. Obviously, vane pumps are subject to wear along the contact points of the vane and the outside of the cylinder. However, these parts typically can be easily replaced.

Flow Pulsation

Each of the three major types of pumps has a problem that may need correction for industrial hygiene work. Over a sufficiently small period of time the actual flow rate of air through the pump versus time would look somewhat as shown in Figure 20-5 (Hall, 1999). As can be seen, pulsations of air are coming through the pump, as opposed to a smooth, continuous flow. For most industrial hygiene sampling work, this is not important because only the total amount of air that comes through the pump matters. On the other hand, in rare cases it may be important that the air passes at a relatively constant rate so that each small time increment of the entire sampling period is sampled equally. The pulsations caused by diaphragm pumps are not

Figure 20-4. Front view of a vane or rotary pumping mechanism (photo, R.K. Oestenstad).

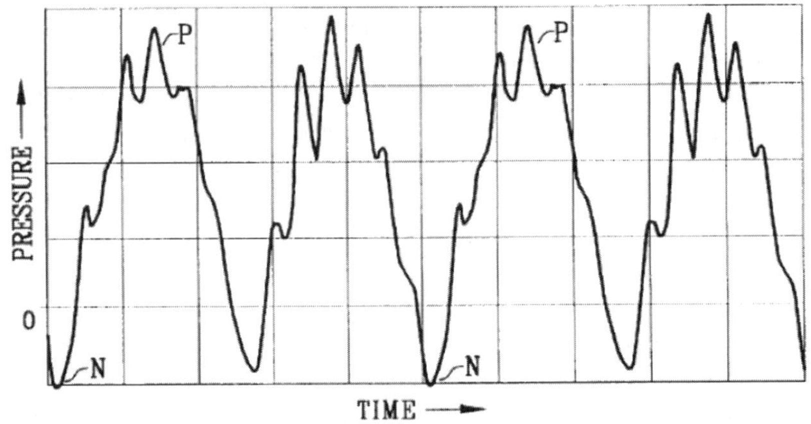

Figure 20-5. Airflow (in terms of pressure) through a diaphragm or piston driven pump without a pulsation dampener. Note the negative pressure for a portion of the cycle (adapted from Hall, 1999).

necessarily important in this regard, as they are extremely small relative to workplace fluctuation in concentration. It is more important that the pulsations can cause difficulties when aerosols are sampled – in particular, when only a certain fraction of the dust is sampled based upon its particle size. Separation of the particles according to size is accomplished by causing the air to be maintained at a particular velocity. Obviously, in the case where there are pulsations, the velocity will vary, and the separation will not occur appropriately. For this reason, manufacturers place dampening devices in the pumps to dampen out the pulsations. In older models, these dampeners may not be effective (see Chapter 24 generally and Figures 24-16 and 24-17 specifically). Pumps designed for other purposes and adapted to air sampling may not have dampening devices.

A more important issue relative to flow pulsation is constant flow control. As will be seen later, most modern pumps have the ability to maintain constant volume flow, even as the load (e.g., the build-up on a filter) increases or as the pressure and temperature change. In order to do this, most measure flow in some manner and use that signal to control the pump speed. Clearly the pulsation phenomenon can have an effect on the flow that is measured and then effect the feedback signal to the pump. Thus the functioning of flow dampeners is an important aspect of most pumps.

Adjustable Flow

The pump invented by Sherwood and Greenhalgh (1960) represented a turning point in industrial hygiene. The first personal battery-operated pump, it could be worn by the worker and thus be used to sample the environment to which the worker was actually exposed. The battery operation and its size and weight allowed this pump to be a purely personal device. However, it soon became apparent that this first pump had two problems related to flow rate that would be solved in future commercial pumps: the need for adjustable settings of flow and the need to maintain those settings as constant during the course of a sampling event.

The Sherwood and Greenhalgh pump could be operated only at one beginning flow rate. The motor operated at only one speed, and there was no provision to bypass a portion of the flow away from the sampling head to reduce flows. In later commercial pumps that became available in the late 1960s, adjustable flow was accomplished in one of at least two ways. The first technique was in use in 1964 (Filtronics pump; Ingram, 1964) and later utilized in a pump produced by the Sipin Company in a contract with the National Institute for Occupational Safety and Health (Sipin, 1971). It employed the use of a potentiometer to change the voltage supplied to the pump motor. Today, either voltage or current to the motor is usually used to control flow rate.

The second technique was to use an assembly of needle bypass valves (MSA). The principle of the bypass system is to allow a portion of the flow to come from the atmosphere, not the sample head (see Figure 20-6). An arrangement that is similar in principle to this is used in most high flow pumps (0.5-5 L/min). The system consists of two valves. A sample valve is closed to reduce flow through the sample head, but this increases back pressure, causing battery drain and motor strain. Therefore, a bypass valve is used that allows air to enter directly from the atmosphere, bypassing the sample head. The combined flows then enter the pump.

There is a lower limit of flow (about 0.5-1 L/min) for high flow pumps. Below this, flow is erratic, or the motor is subject to overheating. This situation has led to the evolution of two pump types, high flow and low flow. The low flow pumps (see Figure 20-7) operate in the range of about 0 to 0.5 L/min and utilize a motor potentiometer to control flow. High flow pumps (see Figure 20-8) typically operate in the range of 0.5 or 1 L/min up to 3, 4, or 5 L/min.

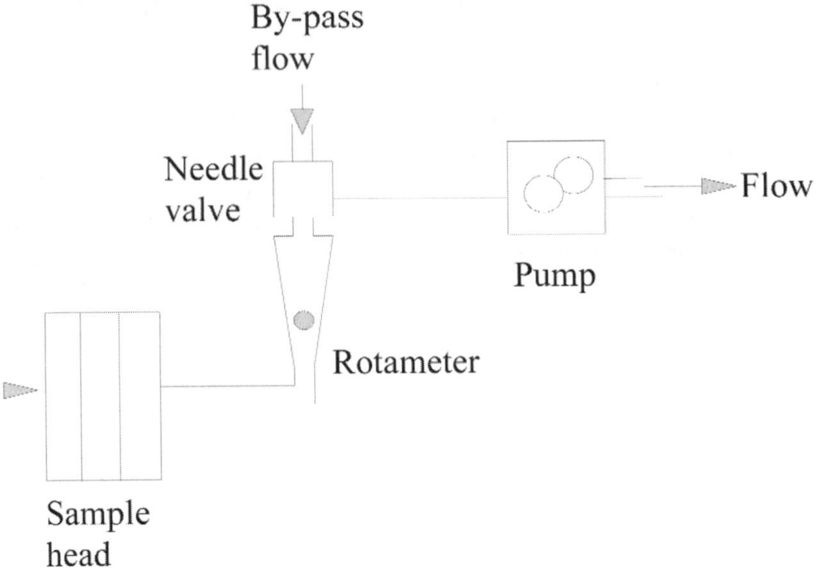

Figure 20-6. Sampling pump configuration in which a needle by-pass valve is used to reduce flow through the sampling head.

Figure 20-7. Some typical low flow pumps (photo, R.K. Oestenstad).

Figure 20-8. Some typical high flow pumps (photo, R.K. Oestenstad).

Constant Flow

Although the early commercial pumps did offer the ability to operate over a range of flows, there was no further control of flow rate. Consider a pump, calibrated to operate at 2 L/min, attached to a sampling head containing a filter. As the filter became more and more loaded, the resistance to airflow increased. Typically this meant that the actual flow of air through the filter decreased, as the amount of work (flow × resistance) the pump performed remained constant (see Baker et al., 1981).

Therefore, the industrial hygienist periodically needed to monitor pumps in order to record airflow changes or adjust airflow to its original value. Typically, this was done at 0.5, 1, and then every 2 hours (Weiss et al., 1982). The average air flow rate could be approximated as that air flow rate to which the adjustment was made at each time increment, or it could be taken as the average of the highest rate versus some number representing the lower rates, such as the average of all the lower rates. Typically the former was done, which tended to slightly overestimate flow rate. A second approach was to allow flow rate to change over the entire sampling shift. The average airflow was commonly taken to be the mean of the beginning and ending calibrated airflows.

In 1975, a second generation of pumps was introduced in the United States by the DuPont Company, which overcame the problem of flow rate variation caused by changing resistance. The new generation of pumps had a *constant-flow* feature that electronically controlled the flow rate of the pump to approximately ± 5% or less of the original calibrated flow rate (Linch, 1981). This was and is commonly done by changing the motor speed (voltage and/or current) via feedback from a flow sensor (Hall, 1999). The advantage of these new pumps was that the industrial hygienist no longer had to be concerned about periodically checking the flow rate of the pump and making adjustments.

Figure 20-9 shows a diagram for one type of constant-flow mechanism first used by DuPont. The pressure sensor (*A*) measures the difference in upstream and downstream pres-

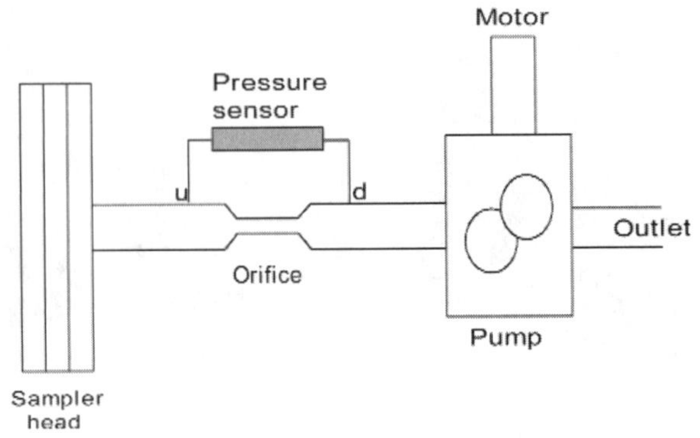

Figure 20-9. A diagram of a limiting orifice meter used to control *constant* flow.

sures (P_u - P_d) at pressure taps on either side of a narrow constriction. If the sensor detects that M is different from its original value, then the motor speed is increased or decreased through an electronic feedback loop. This device may be called a ΔP controller, but is based on a long-used laboratory device known as the limiting orifice meter.

Orifices

The flow rate of air can be controlled by forcing it through an orifice. There are two types of orifices used in industrial hygiene, which are differentiated by the upstream to downstream pressure ratio. They are the limiting orifice and the critical orifice. Figure 20-10 depicts a sampling configuration that contains an orifice, and shows the shapes of the two most common orifices and a diverging-converging nozzle. The latter is also known as a Venturi, and it is a convenient place to start the discussion of orifices.

As flow through a Venturi throat increases, the velocity increases until it becomes sonic or equal to the speed of sound for the medium, in this case air.[3] At this point it can move no

[3] The reason for this maximum velocity is interesting. Imagine a sudden decrease in pressure at a point upstream from the throat of the Venturi, such as might be caused mechanically by adjusting a pump to increase the flow. This pressure change would advance toward the throat as a wave, the speed of which is the acoustic velocity or speed of a sound wave (c). Now realize that the air is moving in the opposite direction. So as long as the velocity of the air is less than c, the pressure wave will advance, and the decreased pressure is transmitted forward, causing an increase in velocity. However, the velocity can only increase up to c at the throat; for in order for it to increase beyond c, a pressure wave would have to traverse the throat, and the pressure wave moving at speed c would not be able to progress against an air velocity greater than c. The controlling variable here is the small constant diameter of the throat, which effectively acts as a dam (Daugherty and Ingersoll, 1954).

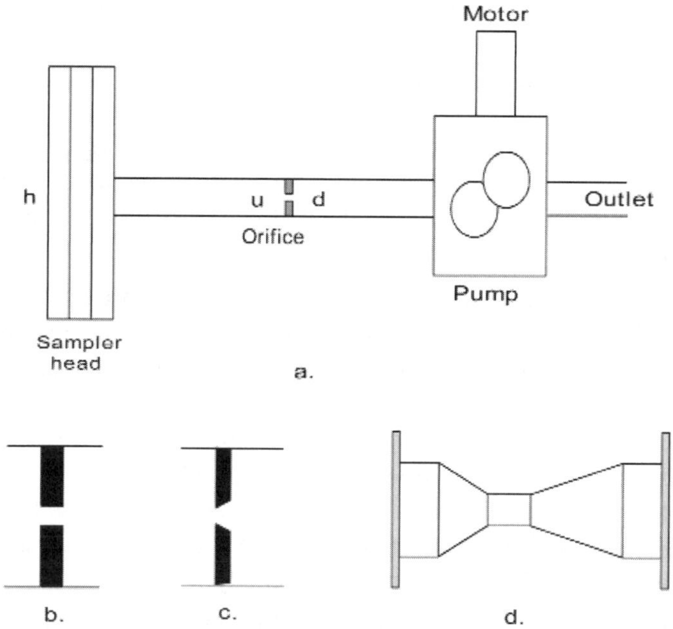

Figure 20-10. a) An in-line square-edged orifice showing upstream (u), downstream (d), and sampling head (h) positions; b) a square-edged orifice; c) sharp-edged orifice and d) Venturi or diverging-converging nozzle.

faster no matter how hard the pump works, that is, regardless of how small the downstream pressure, P_d, becomes. This condition is described as critical flow, and it will be maintained as long as the ratio P_d/P_u is less than some critical ratio (r_c). The r_c is commonly reported to be 0.53, but it varies with the shape of the orifice from about 0.4 to 0.9, the latter being for certain Venturi configurations (Zimmerman and Reist, 1984). For the typical critical orifice used in hygiene sampling trains the value for the square edge (Figure 20-10b) is 0.44 and that for the sharp edge (Figure 20-10c) is 0.46.

It is important to note that if the r in practice exceeds r_c then the orifice only *limits* flow, depending on the value of r_c. Any orifice or Venturi operating at $r > r_c$ is a limiting orifice (Daugherty and Ingersoll, 1954; Daugherty and Franzini, 1965).

LIMITING ORIFICES

Several pumps used for industrial hygiene air monitoring utilize limiting orifice meters (Figure 20-9) to maintain flow at a *constant* rate. However, neither the volumetric flow nor the mass flow is exactly maintained as constant if the density of the air changes. (See later discussion.) This can be seen from the following equation relating volumetric and mass flow rate to pressure and density for a limiting orifice.

$$W = Q_u \rho_u = CA[2\,(P_u - P_d)\,\rho_u]^{0.5} \qquad \text{[Equation 20-1]}$$

where Q is the volumetric flow rate (cc/min), ρ is the air density (g/cc), W is the mass flow rate (g/sec), P is the pressure (g/cm^2), C *is a* dimensionless constant for a given orifice shape and diameter, A is the cross-sectional area of the orifice (cm^2), and the subscripts u and d are for upstream and downstream and can be related to Figure 20-9.

The principle of using a limiting orifice in conjunction with a pressure sensor, to maintain *constant* flow is that P_u - P_d (ΔP) is maintained constant. Because C and A are constants, W and Q are constant if density (ρ) is constant and ΔP is held constant. In other words, all terms are either constant or, in the case of P_u - P_d are held constant. Thus a constant-flow pump with an orifice meter calibrated at 25°C (77°F) and 1 atm of pressure will operate at the calibrated flow rate if the upstream density does not change. A change in density can be caused by a change in atmospheric pressure or by an increase in resistance across the sampling head. If the resistance across the sampling head increases, then a greater vacuum (relative to atmospheric pressure) will occur at P_u and the upstream density will decrease.

If the density is not constant, then its effect on flow can be derived as follows. If all terms are constant in Equation 20-1 except Q_u and ρ_u, then from Equation 20-1, $Q_u\rho_u$ is proportional to $(\rho_u)^{0.5}$ and thus Q_u is proportional to $(1/\rho_u)^{0.5}$. Therefore:

$$Q_1 / Q_2 = (\rho_2/\rho_1)^{0.5} \qquad \text{[Equation 20-2]}$$

where 1 and 2 are subscripts relating two sets of conditions for the upstream location. From the ideal gas law, $\rho = MP/RT$ where M is molecular weight and R the ideal gas constant. Therefore, ρ is proportional to P/T (M and R are constants), and by substitution into Equation 20-2:

$$Q_1/Q_2 = \left(\frac{P_2T_1}{T_2P_1}\right)^{0.5} \qquad \text{[Equation 20-3]}$$

For most constant-flow pump operations, the increased resistance of the sampling head over time will cause only a small change in the upstream density and therefore Q. Hence, manufacturers claim *constant* (± 5%) flow. However, under changing ambient conditions, the error can be greater.

Example 20-1. Calculate the change in volumetric flow if a constant flow pump calibrated at NTP (25°C and 1 atm) is used in the field at 40°C and 740 mmHg. From Equation 20-3:

$$Q_1/Q_2 = \left(\frac{740 \times 298}{760 \times 313}\right)^{0.5} = 0.936$$

So the flow will be maintained constant at about 94% of its laboratory calibrated value. (Note: If density is not constant during the sampling period, then Q will not be exactly constant.)

The limiting orifice meter is used internally as a constant-flow control mechanism in some older pumps (see, e.g., some DuPont, SKC, and Gillian pumps). With pressure changes, the pumps are not truly constant-flow in terms of mass or volume, but they do generally meet the manufacturer's claim that they are accurate to ± 5%. Newer pumps that have recent advances in microprocessor control and suitable pressure and temperature sensing, can measure ambient pressure and temperature and correct for the above effect (see, e.g., the SKC Leland Legacy and MSA Escort ELF Pumps).

For illustration, Figure 20-11 shows the effect of altitude on flow rate for DuPont (no longer manufactured) P2500 and P4000 pumps (Treaftis et al., 1986). From these curves (also predicted by Equation 20-3) a P2500 pump calibrated at 200 m (650 ft) but used at 1,000 m (3,300 ft) would have an actual volumetric flow rate about 7% greater than calibrated. (Figure 20-12 shows pressure and elevation or altitude relationships.) To avoid the pressure error, calibrate these pumps at altitudes similar to those where they are used. Even so, a maximum typical error that might occur with temperature would be a 60°F (33°C) temperature change, from 70°F (21°C) in the calibration laboratory to 10°F (-12°C) outdoors. This would cause a 5.8% decrease in the actual volumetric flow rate. When pressure and temperature change in opposite directions, the error can be appreciable. Of course, to minimize the density error, calibrate at operating conditions. In most cases, these considerations are not important. However, as soon as they are forgotten, a Sydney or Los Angeles-based hygienist will calibrate pumps and then head for a tin mine located in Bolivia or Colorado at an altitude of 10,000 ft (3,000 m). Pump error can be considerable.

For small pressure changes, such as those caused by filter loading, the DuPont P2500 and P4000 gave flows of ± 5% of original calibrated values. This was true at pressure drops up to 30 in. H_2O (7.5 kPa) for the P2500 and 40 in. H_2O (9.95 kPa) for the P4000 (Treaftis et al., 1986).

These resistance-related pressure drops are not likely to be exceeded at flows of 4 L/min or less.

CRITICAL ORIFICES

In some cases, critical orifices are used to control flow, often being added to the sampling train by the hygienist (Figure 20-10a). Under critical flow conditions (i.e., pressure ratio < r_c), only the upstream conditions need to be known to determine flow rate. When air is considered to be a perfect gas, we have:

$$W_{\max,u} = Q_{\max,u}\,\rho_u = 0.0405\,C\,A\,\frac{P_u}{\sqrt{T_u}} \qquad \text{[Equation 20-4]}$$

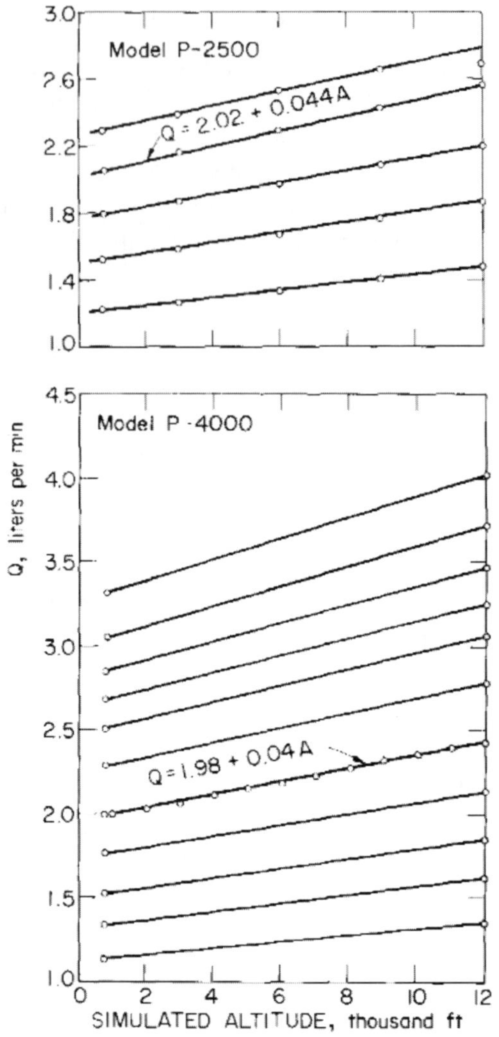

Figure 20-11. Effect of altitude on volumetric flow rates (originally set at 740 ft or 226 m) of DuPont pumps (from Treaftis et al., 1986, with permission).

where *W, A, Q, P,* and *T* are as defined for Equation 20-1, C is a dimensionless coefficient of discharge, and subscript *max* indicates critical conditions. Numerous authors (Corn and Bell, 1963; Zimmerman and Reist, 1984; Nelson, 1992) have used the constant 0.388 rather than 0.0405 in conjunction with SI units. However, 0.0405 should be used with SI units (Perry and Green, 1984).[4] Note that W_u and Q_u are the mass and volumetric flow rates through the upstream section of the sampling train, as shown in Figure 20-10a.

From Equation 20-4, it can be seen that W_u will vary as P_u varies. However, we are usually interested in the volumetric flow through the sampling head (Q_h, not Q_u). From the ideal gas law, density is proportional to *P/T*. So substituting in Equation 20-4 for ρ, we have:

$$Q_u K \frac{P_u}{T_u} = 0.0405\, C A \frac{P_u}{\sqrt{T_u}}$$

where *K* is a proportionality constant. This shows that Q_u is independent of P_u (terms cancel), but dependent on $\sqrt{T_u}$. So Q_u will be constant for constant *T*. Again, we are not interested in

---

[4] The complete equation for the critical orifice is (Daugherty and Franzini, 1965, p 267):

$$G_{max} = \frac{AP_1}{\sqrt{T_1}} \left(\frac{gK}{R} \left(\frac{2}{K+1} \right)^{K+1/K-1} \right)^{0.5}$$

where G is in lb/sec or kg/sec, A is m² of ft², *P* is n/m² or lb/ft², *T* is °R or °K, and g is the gravitational constant used to convert lb-force to lb-mass in the English system. It is superfluous in the SI system. *R* is the ideal gas constant, 8.314 n-m-°K^{-1}-mole^{-1}, and *K* is the critical pressure to critical volume ratio, 1.4 for air. For the SI units we have when $\sqrt{T_1}$ is moved into the main radical:

$$G = \frac{m^2 n}{m^2} \left(\frac{1.4°K\text{-}mol}{8.314 n\text{-}m\text{-}°K} \times \frac{kg\text{-}m}{n\text{-}sec^2} \times \frac{0.029\, kg}{mol} \times \left(\frac{2}{2.4} \right)^{2.4/0.4} \right)^{0.5}$$

$$G(Kg/sec) = \frac{0.0405\ AP_1}{\sqrt{T_1}}$$

Note that the second term in the main bracket is unity, i.e., from force = mass × acceleration, *ma/F* = 1 or in SI units: $\frac{kg\text{-}m}{n\text{-}sec^2} = 1$ Also for air, molecular weight is 29 g/mol. For English units, *g* is 32.2 ft²-lbm/lbf-sec² and R is $\frac{1545.12\ lbf\text{-}ft^2}{°R\text{-}lbmol}$. The constant is then 0.533.

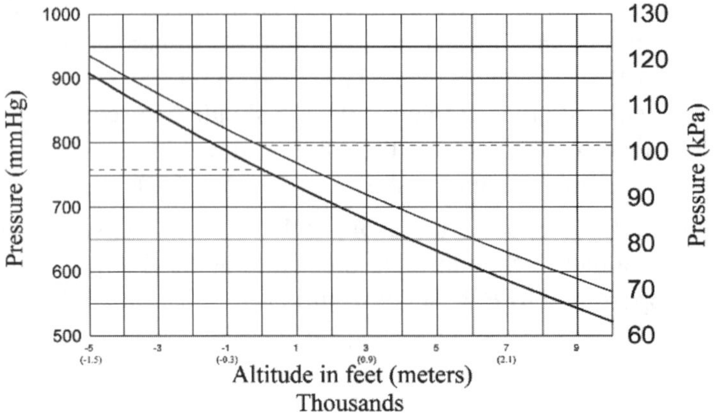

Figure 20-12. The relationship of atmospheric pressure to altitude or elevation showing standard or normal pressure at sea level (Data from NOAA (1976) using the relationship Elevation (m) = 44,331.514 – 11,880.516 $P^{0.1902632}$, P in kPa).

Q_u (the volumetric flow between the orifice and the head), but in Q_h, the flow through the sampling head. From the ideal gas law, $P_1 V_1 = P_2 V_2$, and because V is proportional to Q (volume flow rate), we have:

$$Q_h P_a = Q_u P_u \qquad \text{[Equation 20-5]}$$

where P_a is atmospheric pressure at the sampling head. In other words, because P_u is less than atmospheric pressure, the volumetric flow through the filter is less than Q_u. Consider that the mass flow of air at u or h is the same because of conservation of mass. Because at u the density must be less than atmospheric (partial vacuum), Q_u must be > Q_h in order to conserve mass.

The point is that as a filter is loaded with aerosol, the resistance to flow across it increases, so P_u decreases. As a consequence, over time the flow Q_h, is not constant, even when conditions are considered to be critical.

Example 20-2. A sampling train with critical orifice is calibrated at 10 L/min. The upstream pressure is -30 in. H_2O (-7.5 kPa) relative to atmospheric or 378 in. H_2O (94.6 kPa) absolute pressure. After 8 hours, P_u has dropped to -40 in. H_2O (-10 kPa) or 368 in. H_2O (91.4 kPa) absolute pressure. What is the flow at the end of the shift? From Equation 20-5 and assuming normal atmospheric pressure is constant at 407.7 in. H_2O (101.325 kPa):

$$10 \times 408 = Q_u \times 378$$

So at time 0, Q_u = 10.8 L/min. At 8 hours,

$$Q_h \times 408 = 10.8 \times 368$$

and thus, Q_h at 8 hours is 9.74 L/min. So if one assumes constant flow (10 L/min), a 1.3% error will be made relative to an average of pre- and post-calibration (9.87, the average of 10.0 and 9.74). Although this is a relatively small pressure drop, pressure changes associated with filtering dirty environments at 10 L/min (a typical flow rate for asbestos monitoring in abatement projects) may be considerably greater than 10 in. H_2O.

Critical orifices typically are not built into pumps but are used by the hygienist for *assumed* constant flow control. For example, they are used widely with ultra-high volume pumps ($Q > 5$ L/min) in asbestos abatement work where flows of 10 L/min are commonly used (see Figure 20-13). In many cases, these orifices may not be operating *critically* ($r < r_c$) because the motor of the pump cannot cause r_c to be achieved. Even if r_c is achieved, the above example shows that true constant flow is not achieved, but the error may be < 5%. Critical and limiting orifices can be made from common lab materials such as syringe needles and pipette tips (Zimmerman and Reist, 1984. Nelson, 1992).

Mass Flow Meters

A second device used to regulate flow in *constant-flow* pumps is the hot wire or thin film (see, e.g., TSI Sidepack[5]). A hot wire (e.g., platinum) is placed in the airstream, and the change in resistance across the wire caused by the calibrated flow rate is held constant. In other words, if the mass flow either increases or decreases, then an electronic circuit senses a change in current resistance, and the pump motor speed is decreased or increased, respectively.

Mass flow meters maintain constant *mass* flow (W) not volumetric flow (Q). If ambient

Figure 20-13. Some ultra-high flow air pumps (photo, R.K. Oestenstad).

5 Using mass flow control technology, sensors and microprocessors, this pump controls for pressure and temperature changes and constantly calibrates itself, keeping a record of the actual during-shift flow rates that can be downloaded.

P or *T* changes appreciably over the sampling period or from conditions of calibration, then adjustments will be made to keep *W* constant, causing a change in *Q*. In newer pumps pressure and temperature sensors may be used to avoid these effects; however, for older models, Figure 20-14 (Treaftis et al., 1986) shows the relationship between altitude (pressure) and *Q* when a MSA FIXT – FLO (no longer marketed) pump is calibrated at one altitude and used at another. For example, an MSA pump calibrated at 200 m (650 ft) above sea level and used at 1,000 m (3,300 ft) would operate at a 12% higher *Q* than calibrated.

Note that these errors may not be detected. For example, a pump calibrated in the laboratory at 21°C (70°F) and used at 40°C (104°F) would actually operate at a *Q* higher than that set in the lab. If it were post-calibrated in the lab, *Q* would return to its original calibration value, and no error would be seen. However, the field *Q* would be 313°K/293°K = 1.07, or approximately 7% higher than expected.

Constant Motor Speed and Current Technology

As a filter loads, resistance to flow increases, and the pump motor will slow unless given more energy in the form of electrical current or voltage. Therefore, one way to maintain con-

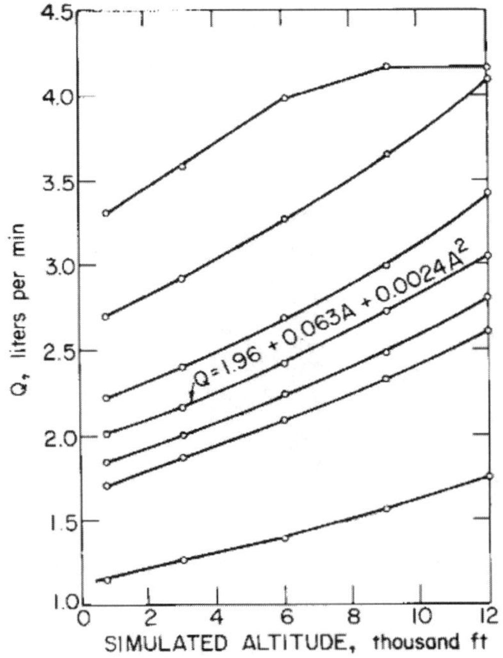

Figure 20-14. Effect of altitude on volumetric flow rate (originally set at 740 ft or 226 m) of the MSA FIXT-FLO pump (from Treaftis et al., 1986, with permission).

stant volumetric flow is to monitor the motor speed and keep it constant. Because each revolution of the motor causes the same number of strokes of the pump, the volume displaced and thus Q are constant. Of course, the mass flow rate (W) will vary with density. It should also be noted that the constancy of Q is dependent on the rigidity of the diaphragms in the pump and the airtightness of the cylinder. If the diaphragm sags appreciably at higher resistances, then the volume per stroke is not constant. Piston pumps do not present this problem if the piston to cylinder seal is good.

The discontinued Bendix BDX-60 pump used the constant motor speed mechanism. As shown in Figure 20-15 (Treaftis et al., 1986), its flow characteristics were essentially constant for Q at varying pressures (altitudes). The BDX-60 pump was able to maintain constant Q at filter pressure drops[6] up to 15 in. H_2O (3.8 kPa); this is satisfactory for most applications at 4 L/min or less.

As of 2006 pump technology has gone beyond those pumps studied by Treaftis et al. (1986). Many of those older models may still be in use but most models that were newly marketed after 1995 incorporate sensors for pressure and temperature as well as a volumetric or mass flow meter. The inputs are used to control motor speed via a microprocessor that comes programmed with equations for the relationships between all the monitored variables. These new features not only allow flow control, often to ± 2%, but reduce the need to carefully calibrate the pumps, although spot calibrations are recommended.

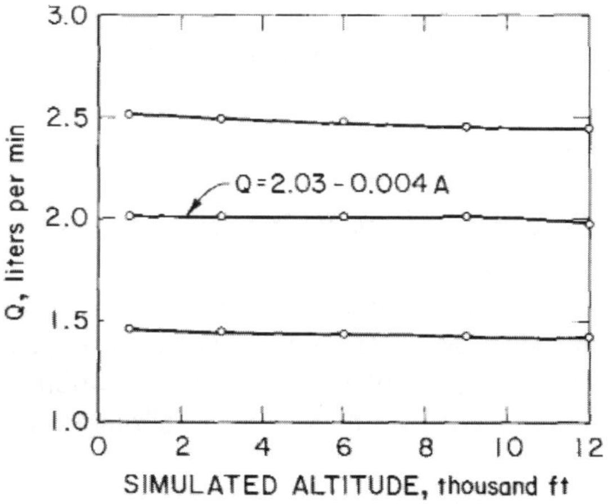

Figure 20-15. Effect of altitude on volumetric flow rate (originally set at 740 ft or 226 m) of the Bendix BDX-60 pump (from Treaftis et al., 1986, with permission).

[6] The dustiness or filter loading and the flow determine the resistance or pressure drop. A 15 in (3.8 kPa) pressure drop at 4 L/min rarely would be achieved, but at 10 L/min (only attainable on an ultra-high flow pump) this pressure drop would be achieved in a dusty environment.

Typical Flow Rates

Industrial hygiene air sampling flow rates usually are well demarcated into two areas. Aerosol sampling (usually filters or impingers) typically requires a higher flow rate than sampling for gases and vapors. There is very little overlap.

Typically aerosol sampling requires flow rates to be in the range of 1 to 3 L/min with the lower rates used for liquid aerosols and the higher rates for solids. Most commercial personal pumps that are advertised as high flow will have rates from 0.5 to 4 or as high as 5 L/min; however, under the conditions of a filter load, some pumps may be unable to consistently achieve this advertised upper flow rate.

In some cases, particularly fixed area sampling for asbestos, flow rates considerably higher than this are desirable. Flow rates as high as 15 L/min are used for asbestos sampling. In the past, these ultra-high flow rates could only be obtained with portable AC pumps that require an external power source (see Figure 20-13); however, newer battery technologies such as lithium ion allow personal sampling pumps to be built that can operate at >10 L/min for 8 hours (see, e.g., the SKC Leland Legend).

Typical sampling for organic vapors requires flow rates from as low as 1 cc/min up to approximately 200 cc/min; for some inorganic gases, rates up to 1 L/min may be required. This is the reason for low flow pumps.

The industrial hygienist must consider the classes of contaminants to be sampled in order to purchase the proper pumps. Typically, personal pumps are available in three categories, based upon the range of flow rates available: high flow for aerosols and some gases (0.5–5 L/min), low flow for gases and vapors (2–200 cc/min), and multiple flow (0.020–5 L/min).

Multiple Flow Pumps

With the second generation of pumps, the hygienist gained the advantage of constantly controlled flow rates; however, some of the simplicity of the first generation of pumps was lost. Troubleshooting of the first generation of pumps could usually be done by the hygienist or a technician. With the introduction of constant-flow mechanisms, pumps became more complicated, having gained the flow controller. The third generation of pumps introduced still more complications; however, once again they gave the hygienist a greater degree of flexibility.

The third generation of pumps was introduced by DuPont (Linch, 1981) in the late 1970s and offered a wider range of flow rates that allowed the hygienist to accomplish sampling for gases, vapors, and aerosols with one pump (see Figure 20-16). To accomplish this, the pump became more complicated both in number of parts and components and in number of settings and controls. The operation manual became more important.

To further add to these developments, subsequent models of third generation pumps made available sophisticated microprocessor-controlled timing and memory devices. These allow

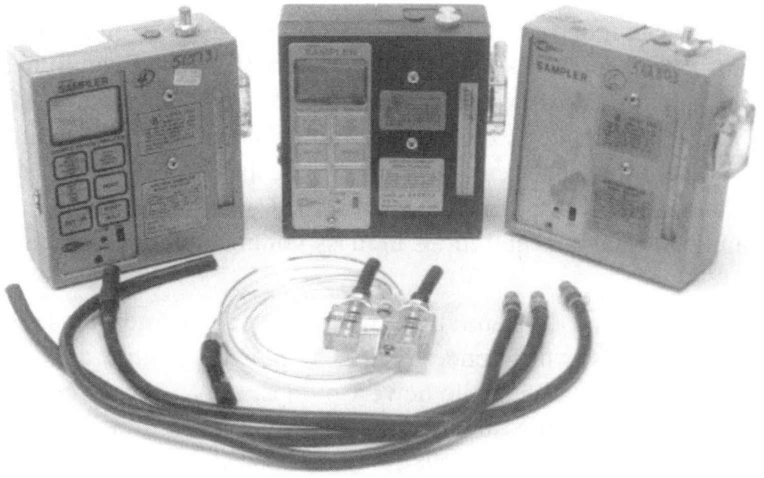

Figure 20-16. Some multi-flow pumps with external, attachable orifices for obtaining low flow (photo, R.K. Oestenstad).

the hygienist to program the pump according to the time characteristics of the sampling regime. For example, the hygienist can program a pump to go on and off at preset times. In addition, the on-off cycles can be set for varying lengths throughout the workday.

The main feature of the third generation of pumps, the ability to offer both high and low flow rate ranges, is accomplished by several differing mechanisms. The first DuPont multi-flow pumps (P4000) required that a motor drive belt be changed between two pulley sizes on the motor shaft to obtain the two flow modes (Linch, 1981).

Typically, current pump models have one operating mode for a flow of approximately 0.5 to 4 or 5 L/min and a second mode for flows of ~0.010 to 0.5 L/min. In order to operate in the low flow rate mode, in some models, the pump is set to the lowest flow rate that can be achieved in the high range, typically 0.5 to 1 L/min. A pressure switch then is activated, causing a constant resistance (usually 20 in. H_2O or 5 kPa) to be applied against the flow. Then a limiting orifice is used to obtain the desired flow (see, e.g., SKC, 2006b).

The limiting orifices may be internal or external, continuous range or discrete. The first ones used were small (0.5 cm diameter) metal plates with a discrete hole size. These were attached to the pump externally, usually behind the sampling head, in order to obtain a discrete flow rate (see Figure 20-16). More recent external versions use a single adjustable limiting orifice, allowing continuous rates from 0.001 to 0.2 L/min. Further developments have led to internal, adjustable limiting orifices (see, e.g., MSA; SKC, 2006b; Gillian; Buck, 2006; TSI; Casella; and others).

Pressure, Temperature, and Density Changes

Most pumps are calibrated and then used in the field under conditions similar to those where calibration took place. Occasionally a pump may be used at conditions quite different from where it was calibrated, for example, at a mine at 10,000 ft (~ 3000 m) in the Andes or in a meat locker at 0°C. In Chapter 10 (Unusual Ambient Conditions), the adjustment of TLVs® was discussed relative to unusual densities. In general, there is no need to correct TLVs®, or any OEL, for unusual densities based on presumed worker breathing rates. The TLV® Committee recommends that concentrations in mass/volume, rather than ppm, be compared to the TLV®.

Now in determining the concentration, we need the volume that the pump sampled. Should we correct it for density differences relative to how the pump was calibrated before we use it to calculate concentration? For the newest pumps, the answer is *no*. These pumps will sense ambient temperature and pressure and adjust motor speed to maintain constant volume flow. On older pumps, as Treaftis et al. (1986) pointed out, there may be errors in the calculated volume when the pump is used at conditions considerably different from calibration. However, even in older pumps differences between actual and calibrated flow rates are likely to be < 5% in most cases, and correcting for T and P is not warranted. This is particularly true when environmental conditions fluctuate and, therefore, determining what new T and P to use is also fraught with error or requires considerable collection of data. There are extremes like those cited above that would warrant correction. A later example related to rotameters further makes this point.

Desirable Features

When selecting a pump, the hygienist must consider all of the available features and then determine what he or she is willing to pay for personal sampling. Pumps can be purchased for prices ranging from approximately $300 to $1200. In addition to cost, one might consider ruggedness, and ease of operation. A discussion of other features that should be considered follows.

Weight. Low flow pumps typically have weights of approximately 4 to 10 oz. (110–280 g). On the other hand, high flow and multi-flow pumps have weights often twice that of low flow pumps and sometimes exceeding 35 oz. (1,000 g). Still, even the heaviest of the high flow pumps can be worn by a worker, although its size and weight at times can be inconvenient. Under certain circumstances, the hygienist may need for the worker to wear two pumps and, in the case of high flow pumps, this can become particularly cumbersome. If aerosol sampling must be performed, then there is no choice but to use high flow pumps. If both high and low flows are needed, consider purchasing two sets of pumps rather than one set of multi-flow pumps. The costs are about equal, and this will allow the use of low weight pumps at least some of the time. Of course, it also means twice as many pumps to maintain. Weight is an important consideration in the purchase of pumps because comfortable workers tend to be more agreeable study subjects.

Size. Low flow pumps are also approximately one-third the size or volume of high flow pumps. Even so, the bulkier high flow pumps have a profile (thickness) of less than 2 in. (5 cm) and consequently can be worn on the worker's belt, particularly on the hip, without putting any restriction on movement. Typically, those pumps that have the smallest width and lightest weight are most comfortable to wear. Some low flow pumps can be worn in the shirt pocket.

Flow Rates. Obviously, a hygienist who wishes to perform sampling for both aerosols and gases or vapors will have to make a decision between purchasing separate sets of low and high flow pumps and obtaining a single set of multi-flow pumps. Again, a distinct advantage of purchasing separate units is that the low flow pumps offer reduced weight and size. Multi-flow pumps should have flow rates ranging from ~ 0.010 to 4 or 5 L/min, whereas high flow pumps should have flow rates ranging from 0.5 to 4 or 5 L/min. The flow rate ranges for the low flow pumps is variable. However, the hygienist should seek a pump with flow rates of at least 50 to 200 cc/min. Some low flow pumps have ranges that are more restricted than this (say, 30 – 70 cc/min) and should be purchased only for repeated but specific sampling tasks.

Controls. Many aspects of the settings and controls on pumps need to be considered. The manner in which the controls are accessed is important. On the one hand, the access needs to be somewhat restricted so that pump settings cannot be tampered with during the sampling process. On the other hand, the mechanism by which the industrial hygienist accesses the controls need not be so cumbersome as to make calibration and pump adjustments difficult. The manner in which access to controls is accomplished on various pumps should be compared before purchase.

Likewise, the complexity of steps necessary to change controls, particularly the change from high to low flow in multi-flow pumps, should be examined. Manufacturers have differing ways of accomplishing this, and the industrial hygienist will find that changing 10 multi-flow pumps from high to low range can be quite tedious for certain pump models.

Inlet/Outlet. All pumps have the ability to provide a vacuum as well as to pressurize a vessel. For sampling purposes, usually the suction side of a pump is all that is needed. However, most hygienists will find that an outlet from the pump will be of great benefit in tasks such as filling air sampling bags. It may seem as though all pumps must have inlets and outlets, but one should realize that on some pumps the exhaust air is simply dumped into the internal workings of the pump. As a consequence, there is no way for the hygienist to access a port on the exhaust in order to pressurize a vessel. Furthermore, because needle valve and bleed valve mechanisms sometimes cause flow bypasses to occur, inlet and outlet flows may not be equal at a given setting. Pumps should always be calibrated in the way in which they are used.

Noise. Most of the newer makes and models of personal air sampling pumps operate at noise levels that are quite acceptable in most work environments. Some can even be used in offices and other indoor settings. Where noise is important, the hygienist should investigate the sound produced by various pump makes. Virtually all personal pumps use a rubber diaphragm rather than a rigid piston. As a rule, diaphragms are quieter than pistons.

If the hygienist needs to perform indoor sampling using ultra-high flow pumps (> 5 L/min), noise can be an extremely important factor. Mufflers are available for some of the larger ultra-high flow pumps to help dampen the sound. Noise levels for ultra-high flow pumps are generally on the order of piston > rotary vane > diaphragm (Gast, 2006).

Intrinsic Safety. In many industrial hygiene applications, a spark produced by an electric motor on a pump can be extremely dangerous if appropriate concentrations of explosive gases, vapors, or aerosols are in the atmosphere. This aspect is particularly important in mine operations where explosive levels of methane can occur. Most personal air sampling pumps are given the designation *intrinsically safe*. This means that the pump has been certified to operate safely in either explosive or flammable environments. In the United States, *UL Listed* means that the pump has passed an Underwriter's Laboratories test in a given type of explosive or flammable atmosphere. There are three classes and several groups of UL atmospheres. Often US-made pumps will include an OSHA approval for use in coal mines. This also indicates that the pump is intrinsically safe and has passed a series of tests.

Radio Frequency (RF) Interference. RF can interfere with the electronic controls of pumps (ANSI, 1991). The phenomenon has been demonstrated for noise dosimeters (Bohl and Clark, 1993). RF sources include *walkie-talkies* or other communications equipment, and of course cellular telephones. Electromagnetic fields from electric motors and high voltage equipment can also cause interference. A test of 10 instruments against RF interference using the ANSI (1991) standard showed that nine failed. These devices included several pumps (Feldman, 1993). Some manufacturers now offer RF-shielded cases.

Advanced Features. With each new generation of pumps came an approximate doubling of the price. Consequently, the industrial hygienist should think carefully about the actual necessity of each of the various features that have been introduced. The merits of multi-flow pumps have been discussed. Similarly, the electronic gadgets available on some of the newer pumps for time control, and so on, often can save time and, in many cases, allow one hygienist to do the work of two or more. Nevertheless, careful consideration is needed before they are purchased.

CALIBRATION

Airborne concentrations can be expressed as either mass or volume of contaminant per total volume of air. Either way, the denominator is calculated by using the volume of air that has been sampled. Hence, in order to calculate the concentration, a relatively accurate determination of the sampled volume of air must be made. How can this be done? What information is available to the hygienist? The previous sections have discussed flow rates of various pumps. This information can be determined and must be known for calculation of air sample volume. Flow rates are stated in terms of volume per unit time; so the product of the flow rate and time yields volume. Therefore, to calculate volume, the *average* flow rate over the total sampling time must be known. Hence, the concept of *constant flow* is important.

The actual flow rate compared to the recommended value of the monitoring method is not important in most applications, but the two should be close. However, some monitoring methods depend on a specific flow rate. In these cases, a *size-selective* sampler is to be used to sample aerosols, or a long-term colorimetric indicator tube is to be used to sample for gases or vapors. Impingers used for size selective sampling operate most efficiently in a relatively narrow flow rate range. Only in these cases is the exact flow rate setting critical, but the actual flow rate used must always be known.

Some pumps offer the hygienist the ability to read flow rate directly from a device built into the pump, called a rotameter. Rotameters also must be calibrated if they are to be used in the field.

Pumps should be calibrated before and after each use. Pumps wear out, batteries fail, and pump mechanical parts may fail because of extremes of the environment and for other reasons; so flow rates can change over a shift even with constant-flow pumps (see Baker et al., 1981). Also, shift-long changes in flow rate can be compared to recommended rules for pump accuracy (described later) in order to determine if an air sample should be declared invalid. Some newer pumps conveniently record flow rate at set intervals electronically; however this does not negate the need to at least occasionally check the electronic output against another calibration device.

Pumps should be calibrated with the actual sampling train that will be used in the field. This is particularly true for sampling trains with filter cassettes (see "Cassette Leakage," Chapter 22). Finally, the pre- and post-calibration of pumps serves as a permanent record of quality control (Shotwell, 1979). Long-term data for a given pump can show that batteries are beginning to fail, or that some other mechanical or electronic problem may be occurring.

Primary Calibration Instruments

Instruments used for measuring or calibrating airflow or volume are often termed *primary* or *secondary*. Primary instruments measure volume based on the measured dimensions of a physical space. These measured dimensions are traceable to a national or international standard such as those of the National Institute of Standards Technology of the United States. For example, a liter of water can be measured quite accurately if a line is accurately placed around the top of a container and the water filled to that line. Also, the volume of air moved by a pump in a specified period of time can be measured accurately with a similar volume-type device. For a primary device, the placement of the line or the definition of the dimensions of the container is traceable to a standard. Secondary devices are not traceable to a standard but can be calibrated (say, annually) with a primary device and have been shown to be reliable between calibrations.

The calibration of a pump with a primary standard requires not only the volumetric measuring device, but also a timing device, such as a stopwatch. In calculating the flow rate of the pump, a mathematical operation is performed, such as that done in calculating the total volume sampled by a pump. The volume measured by the volumetric calibration device is divided by the total time that it takes for the volume of air to be displaced, yielding the flow rate.

BUBBLE METER

Probably the most commonly used device for calibrating pumps in industrial hygiene is the bubble meter. It has also been called the frictionless piston meter; and this name, though somewhat in error, rather aptly describes how it works. It is a very simple device composed of a glass cylinder with openings at the top and the bottom (see Figure 20-17). The bottom opening is typically convergent so that the tube becomes larger moving upward and eventually forms a straight-sided cylinder like a burette. The pump is connected with tubing to the top opening, and a soap solution is placed at the bottom of the cylinder. When the pump is activated, the partial vacuum in the cylinder causes a soap bubble to form at the bottom of the cylinder, and it will rise in the cylinder as the pump evacuates the air that is above the soap bubble. The bubble, to the extent that it has no friction in its contact with the glass surface, forms the frictionless piston. The movement of the bubble from the bottom to the top of the cylinder can be timed, and the volume is known from the lines marked on the cylinder. Bubble meters of varying volumes are available for calibrating both low and high flow pumps.

Typically, three to six measurements are made of the time necessary for the bubble to traverse the cylinder. However, three readings are probably sufficient. If the three readings are in considerable agreement, then the pump has probably reached a stable flow, and differences in the three readings can be attributed to slight fluctuations in pump flow rates (but more likely are owed to fluctuations in the triggering of the stopwatch by the operator). Pumps should be turned on and allowed to operate for at least 5 minutes prior to calibration, as many pumps typically have somewhat erratic flow for the first few minutes of operation (Baker et al., 1981), and a representative calibration may not be attained otherwise. Weiss et al. (1982) recommend 15 minutes.

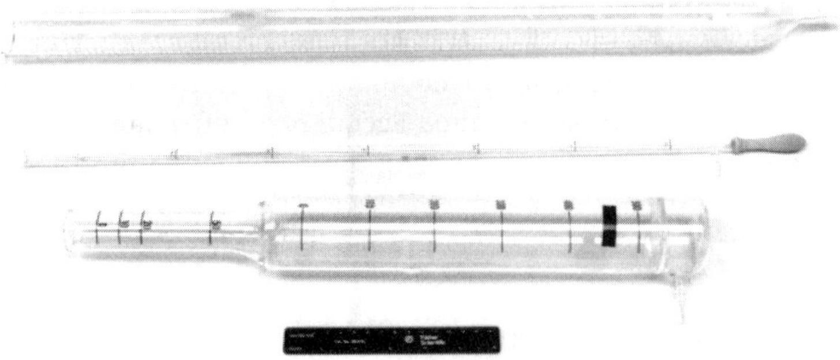

Figure 20-17. Several types of bubble meters (photo, R.K. Oestenstad).

The calibration of 10 pumps in three replicates using a bubble meter can be quite tedious, and depending on the flow rate, would probably take about one hour. However, this process has been speeded considerably with the introduction of electronically timed bubble meters produced by various manufacturers, as shown in Figure 20-18. Because an electronic timing device is employed, the bubble can be timed over a much shorter distance (i.e., a much smaller volume), and the calibration time can be considerably reduced. At least one manufacturer markets a frictionless piston meter that does not employ a bubble (see Figure 20-19). The

Figure 20-18. Several types of electronic bubble meters (photo, R.K. Oestenstad).

Figure 20-19. The 2007 Bios calibrator. It is about the size of a high flow pump (courtesy Bios International).

DryCal (Bios International Corporation) uses a near frictionless graphite carbon piston. Middendorf et al. (2006) showed that four models of electronic flow rate meters were less variable than the standard bubble meter and gave results that were within 2% of each other.

Bubble meters are considered primary standards because their volumes have been marked by using a device that is traceable to a standard. Errors associated with volumes determined in this way are quite small; so the major error in calibrating a pump with a bubble meter is in the operation of the timing device. Manual operation of stopwatches is certainly not error-free and can be subject to considerable bias based on the reflex time of the operator. On the other hand, electronically controlled timing devices are more accurate overall than manual stopwatches. It is advisable that electronic devices be compared to a manually controlled stopwatch/bubble meter from time to time just to make certain that the electronic device is still operating as accurately as a manual device. In addition, electronic devices should be calibrated by the manufacturer yearly.

SPIROMETERS

Another primary type of volumetric calibration device is the spirometer, which is typically seen in a physician's clinic to measure the ventilatory capacity of the human lung. Like bubble meters, spirometers provide an accurate method for determining volume and, from this, flow rate. The spirometer works by the movement of air into a volume within a drum. The bottom of the drum is sealed with water. As the air volume increases, the height of the drum increases, and this height is the dimension that is accurately measured. The drum must be counterbalanced so that there is essentially no force involved in moving it upward. Industrial hygiene use of spirometers is far less common than the use of bubble meters or the use of secondary devices.

MASS FLOW METERS

Mass flow meters, which employ thin film or wire technology, were discussed previously under constant flow control. These meters are widely used in the laboratory to measure airflows. They can also be used to calibrate pumps. Although they measure mass flow, a calibrator can also be designed to measure temperature and pressure and thus report volumetric flow for any set of environmental conditions. TSI makes a series of mass flow meter calibrators that have reported 2% accuracy, are traceable to a primary standard, and have the advantage of being very small and instantaneous in their output (see Figure 20-20).

Secondary Calibration Devices

Unlike primary devices, secondary devices do not have a fixed volume. Also, the volume (or some other characteristic) is usually subject to change as a result of wear on the instrument. The volume dimensions are not traceable to a standard. Hence, secondary calibration devices

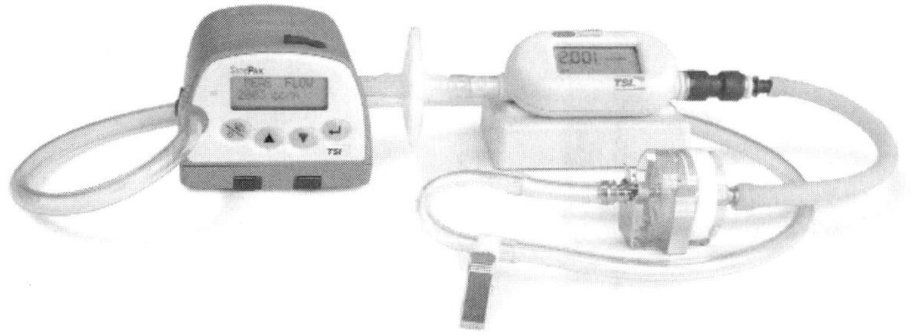

Figure 20-20. A mass flow meter shown in calibration setup with a TSI Sidepack® personal pump and 37 mm cassette (courtesy TSI and REPSS).

must be periodically calibrated with a primary device. However, they can offer considerable savings in time and money and also afford one the ability to measure very high volumetric flow rates (> 30 L/min) not possible on most bubble devices.

WET-TEST METER

Figure 20-21 shows a photograph and a schematic of a wet-test meter. The wet-test meter works similarly to the rotating vane pump. The chambers within the wet-test meter are filled and emptied of air as a result of the rotation of the vanes; this is caused by the movement of air through inlet and outlet valves. The level of water in the meter must be exact, and the meter is subject to corrosion and must be maintained properly. The wet-test meter counts the revolutions of the vanes, each revolution accounting for a given volume. It has the advantage of being able to measure large volumes over extended periods of time; thus, it can measure very high volumetric flow rates over a long time period. This gives an average flow rate for the period. Bubble meters also give average flow rates, but wet-test meters can be used to average flow rates over very long periods (say, 8 hours) if desired.

The wet-test meter is attached to the pump, and a beginning reading is made from the meter. The pump is then activated. The wet-test meter can then be read at any subsequent time that is convenient. Hencc, unlike the bubble meter, it does not have to be constantly attended.

DRY GAS METER

The dry gas meter (see Figure 20-22) is similar to gas meters utilized across the United States to meter natural gas use in homes. It is a relatively complicated instrument, which operates with a rubber bellows that empties and fills as a result of gas pressuring either the inside or the outside of the bellows. The bellows is attached to a number of levers that are in turn attached to a series of gears. The gears move dials that indicate the number of fillings of the

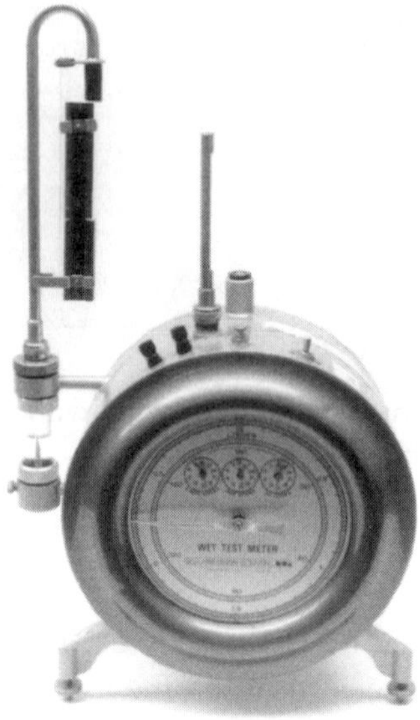

Figure 20-21. A photograph of a wet test meter (photo, R.K. Oestenstad).

Figure 20-22. Dry gas meters, one partially showing the inner workings (photo, R.K. Oestenstad).

bellows. There are a number of moving parts that are subject to wear – hence the need for periodic calibration of the dry gas meter with the primary device. However, the devices appear to have a long history of acceptable operation at homes.

The dry gas meter, like the wet-test meter, can afford several advantages over bubble meters. In addition, the dry gas meter has the advantage over the wet-test meter of not requiring the hygienist to be concerned about water levels. Unlike the wet-test meter, the dry gas meter can be stored without concern for corrosion problems. For more information on dry and wet-test meters and other less common calibration standards, see Lippmann (1989).

ROTAMETERS

The rotameter, like the mass flow meter, directly measures flow rate as opposed to volume. It is a simple device (see Figure 20-23) constructed of a tapered tube containing a floating ball or object of another shape. Lines on the glass tube indicate the height of the ball and the flow rate. When the outlet side of a pump is connected to the bottom of the rotameter and activated, it attempts to push the ball from the glass tube. However, an equilibrium point is reached, as the tapered tube eventually allows the flow of air around the ball and out the top. This equilibrium point is determined by the forces acting on the ball. The force acting downward on the ball is gravity. The most important force acting upward on the ball is the impaction of air molecules, which are being moved by the pump. This is the drag force. Consequently, as the flow rate of the pump increases, more air molecules will hit the bottom of the ball, causing it to rise higher.

If a pump is operating at a specific volumetric flow rate, for example, 2 L/min, and the temperature of the air changes considerably, will the rotameter still indicate a flow rate of 2 L/min? The ideal gas law and a rotameter equation can be employed to demonstrate that the height of the ball will change with temperature and with pressure of the airstream, given that the pump calibration or setting has not changed. This is owed to changes in density of the air. As it is the force of molecules or the drag force (i.e., velocity, mass, and number) that causes the ball to rise or fall, a change in density or mass flow rate will cause the ball to rise or fall. In other words, if the pump setting remains unchanged while the density changes, the rotameter will indicate a change in flow even though volumetric flow rate has not changed.

The hygienist is usually interested in volumetric flow rate. The relationship between the actual volumetric flow rates at two densities as indicated by a rotameter was approximately derived by Leidel et al. (1977). However, a more elegant approach was taken by Caplan (1985); (see also Perry and Green, 1984, for a good discussion). The complete equation for a rotameter is (Schoenborn and Colburn, 1939):

$$Q = C\left[\frac{D_t - D_f^2}{D_f}\right]\left[\frac{g\pi}{2\rho}\ V_f(\rho_f - \rho)\right]^{0.5} \qquad \text{[Equation 20-6]}$$

a)

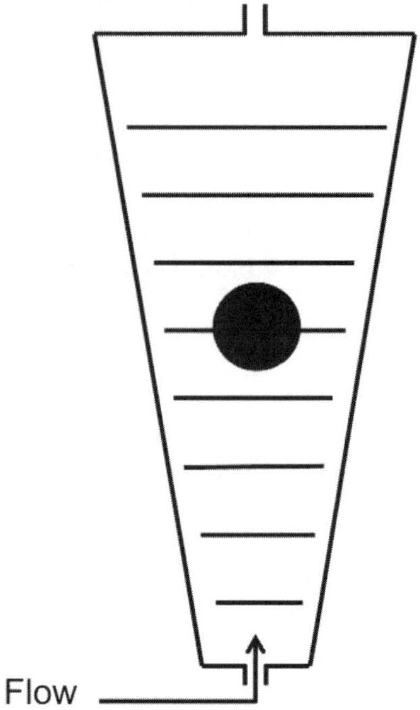

b)

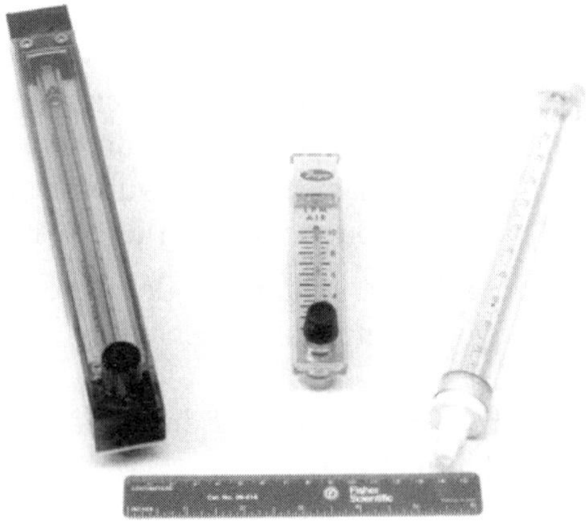

Figure 20-23. A rotameter schematic a) and several actual rotameters b) (photo, R.K. Oestenstad).

where Q is volumetric flow rate, f indicates float, t indicates tube, D is diameter, V is volume, g is the gravitational constant, and C is a discharge coefficient dependent on the Reynolds number and therefore viscosity. Caplan argues that C has a constant value for a given rotameter and for air because viscosity of air varies little over the intended range of use (Q). If all constants are combined into K, we have:

$$Q = K(D_t^2 - D_f^2)(1/\rho)^{0.5}$$

Note that D_t is not a constant but varies because the tube is tapered. Caplan argues that D_t varies with the scale reading, which is what we are interested in. Actually $(D_t^2 - D_f^2)$ is a function of annular area and, therefore, is proportional to the scale reading or R, and:

$$Q = K_l R(1/\rho)^{0.5}$$

or:

$$R = K_l Q(\rho)^{0.5}$$

Because K_l is a constant, we can set up a proportion for two conditions as follows:

$$\frac{R_1}{R_2} = \frac{Q_1\sqrt{\rho_1}}{Q_2\sqrt{\rho_2}} \qquad \text{[Equation 20-7]}$$

If we calibrate a pump with a rotameter and then use both at a different ρ, then we can set $Q_l = Q_2$ and solve for R_2 to see what the rotameter should read in the field if Q is to remain constant. Because ρ is proportional to P/T (ideal gas law), we have:

$$R_2 = R_1\left[\frac{P_2}{T_2}\frac{T_1}{P_1}\right]^{0.5} \qquad \text{[Equation 20-8]}$$

Equations 20-7 and 20-8 are approximations used widely by hygienists. They should be used only over narrow density changes, as the value of C in Equation 20-6 is not a true constant. Furthermore, Nelson (1992) has pointed out that Equation 20-7 should *not* be used for different gases. For example, rotameters calibrated for air but used in a system containing sulfur hexafluoride might be corrected by Equation 20-7. Nelson states that he calibrated a rotameter for both chlorine (ρ = 3.2 g/L) and hydrogen chloride (ρ = 1.6 g/L) and the two curves were almost identical, but Equation 20-7 would have predicted otherwise. This is no doubt owed to the implications of constant viscosity when C is taken as a constant.

Isabel (1994) found that the use of a rotameter operating at about 20 to 100 mL/min and calibrated for air gave results that were 60% to 200% of actual flow for helium, neon, and sulfur hexafluoride, even when density was corrected by Equation 20-7. The results for He and Ne could be corrected with the following relationship:

$$Q_c = \frac{Q_r}{\left(\frac{\eta_g}{\eta_a}\right)^{0.3}\left(\frac{\rho_g}{\rho_a}\right)^{0.2}}$$ [Equation 20-9]

where Q_c is the corrected flow rate, Q_r is the indicated flow rate, and η is the viscosity, with *a* representing air and *g* the gas in question. Equation 20-9 was derived based on flow in the transition region (see Chapter 9, Figure 9-1). Equation 20-9 did not correct for SF_6, for which flow was in the Newton region.

Another commonly encountered error associated with rotameters involves their relative position in the sampling train. The principle of operation for a rotameter is that the difference in pressure above and below the ball is constant, $P_a - P_b$ = constant. If the bottom of the rotameter is open to the atmosphere, then P_b is atmospheric. Therefore, as Q increases, the ball rises, and when it reaches equilibrium, $P_a - P_b$ goes back to its original value. If a rotameter is calibrated so that P_a or P_b is atmospheric but is used as shown in Figure 20-6, then neither side has P equal to atmospheric, and correction should be made based upon the true pressure. The correction can be based on Equation 20-8 over small pressure differences. A better approach is to calibrate the rotameter as it will be used, which is the case for pumps with built-in rotameters.

Some rotameters that are built into pumps operate as in Figure 20-6, and thus are subject to changes in back pressure caused by sampling media, particularly filters, as the media are loaded with contaminants. Used in such settings, they are only approximate indicators of flow and not a suitable calibration device.

Example 20-3. A pump and filter sample train is calibrated in the lab at NTP and 2 L/min. The pump is taken to the field where the rotameter on the pump (configured as in Figure 20-6) indicates a flow of 1.9 L/min, and conditions are 735 mmHg (30.4 in. H_2O) and 41°C (105°F). Is the rotameter reading correct if the volumetric flow is actually still 2.0?

$$R_2 = 2\ Lpm \left[\frac{735\ mmHg}{760\ mmHg}\ \frac{460°F + 70°F}{460°F + 105°F}\right]\sqrt{0.5} = 1.9\ Lpm$$

thus indicating that the pump volumetric flow rate had not changed. This calculation would be valid only prior to loading of the sampling media with contaminants. If the rotameter reading were made at the end of the day, then that reading would be affected by *T* and *P* as above, but, in addition, the pressure upstream from the rotameter would have changed owing to medialoading, which also would affect the reading.

Example 20-4. For the conditions in the above example, the initial *P* caused by a filter is -30 mmHg (relative to atmospheric), and at the end of the sampling period the *P* caused by the loaded filter is -60 mmHg. The rotameter is calibrated by the manufacturer without any sample head in place: that is, upstream pressure is atmospheric. When the rotameter reads 2 L/min at the end of the period, what is the actual flow rate at the end of the period?

$$R_2 = 2.0\ Lpm\left[\frac{735\ mmHg}{760\ mmHg}\ \frac{460°F + 70°F}{460°F + 105°F}\right]^{0.5}\left[\frac{700}{730}\right]^{0.5} = 1.86\ Lpm$$

The first term corrects the reading for the changed atmospheric conditions, and the second term corrects for the loaded filter pressure drop. Example 20-4 illustrates a point but is impractical. It would be more cumbersome to measure the pressure loss across the filter and use Equation 20-8 than to use a primary calibrator and measure the actual flow in the field. A constant-flow pump should be able to overcome the pressure drop caused by the loaded filter and maintain the preset flow (volumetric or mass flow). The rotameter should be used only as a field check to see that the flow is approximately unchanged.

The difficulty with memorizing equations is demonstrated in these examples. Which subscript in Equation 20-8 goes on top, 1 or 2? It is easier to construct this equation rationally than to memorize it. As the air density decreases, the rotameter ball will fall for constant Q and indicate a value of R_2 that is too low relative to R_1. Therefore, the fraction under the square root sign must be arranged to cause R_1 to be *adjusted* upward. When temperature increases, density decreases, and when pressure increases, density increases. For these reasons, subscripts of the fractions for T and P are always opposite. All one needs to remember is the form of the equation, that is, a square root times the indicated rotameter flow rate.

In many applications the density correction is insignificant. At sea level up to low elevations of 300 m or 1,000 ft (most major world cities are < 300 m in elevation), normal atmospheric pressure changes, on average, from 760 to 733 mmHg or -3.5%. There are unusual low pressure situations that can occur such as cyclones, hurricanes, tornadoes, and severe winter cyclonic storms. However, it is not probable that air sampling and rotameter use would occur under such conditions; and if sampling results were averaged over days, then normal low pressure weather would be averaged with high pressure weather. Temperature can be more variable but a range of 0°C to 45°C (32-113°F) would include most industrial hygiene applications. Hence, an approximate worst-case rotameter deviation would be for a -27 mmHg pressure deviation and a +25°C temperature change from NTP. This would cause a rotameter reading of:

$$\frac{R_2}{R_1} = \left[\frac{733\ mmHg}{760\ mmHg} \times \frac{293°K}{318°K}\right]^{0.5} = 0.94$$

or about a 6% error in estimated flow rate, if no adjustments were made and filter pressure drop were ignored – another reason to use rotameters as approximate flow indicators.

A more drastic condition can occur at higher elevations. For example, at an elevation of 1,524 m (5,000 ft), about the elevation of several western U.S. cities, standard atmospheric pressure is only 83% of that at sea level. Hence, pumps calibrated in Houston (sea level) with a rotameter and used in Denver will give a rotameter reading approximately 9% lower in

Denver on average.

Precision rotameters that are delivered with their own individually determined factory-calibration curves are available. These rotameters are slightly more expensive than non precision rotameters but can be quite useful for industrial hygiene work. They are particularly useful for field calibration checks on ultra-high, non-constant-flow pumps and for the accurate metering of gas or vapor streams for laboratory calibration work. Other rotameters used on small battery-operated pumps should be used as approximate flow checks only. A final note on the possible effect of pump pulsations on rotameters is worth considering. Although not reported in the literature, OSHA technicians have noticed that pump pulsations can cause rotameter errors, even in pumps with pulsation dampeners. Errors in the precision rotameter reading from -14% to +45% have been observed and can usually be corrected with a 5-micron sampling filter between the pump and the precision rotameter (personal communication, Ray Feldman, OSHA, 1993). Thus, hygienists should take separate reference readings in the lab for each pump to be used with that rotameter, since this error can vary from pump to pump, and also varies with temperature.

Calibration of Calibrating Devices

Rotameters and wet and dry test meters, being secondary calibration devices, must be periodically calibrated. The most important aspect of calibrating these devices is the calibration configuration. A good calibration configuration is shown in Figure 20-24a. As illustrated, the secondary device, particularly in the case of rotameters, should be at the upstream end of the calibration train.

If the rotameter were placed as shown in the calibration set up in Figure 20-24b, then a slightly negative pressure (relative to atmospheric) would be formed on both sides of the rotameter. This probably would not be the way the rotameter would be used. Dry gas and wet-test meters may be subject to the same pressure-induced errors, and consequently should be calibrated in the way they are used. One way to check the calibration of any of these secondary devices is to use a pump that has been previously calibrated with a primary device.

Pump Error

The error associated with air sampling or the pump and calibration error is commonly assumed to be 5% (Leidel et al., 1977). This value is derived by NIOSH, but the exact source is not clear. The original low flow pump was designed under contract to NIOSH (Sipin, 1971). Among the specifications was that the contractor should deliver a pump with "an indicating device, such as stroke or revolution counter, which gives a number that can be converted to total volume of air sampled within ± 5%." Many pump manufacturers now claim that their pumps have an accuracy of ± 5 %, and for whatever reason this number has become an accepted standard even though no definitive study of *sampling error* is known to have been made.

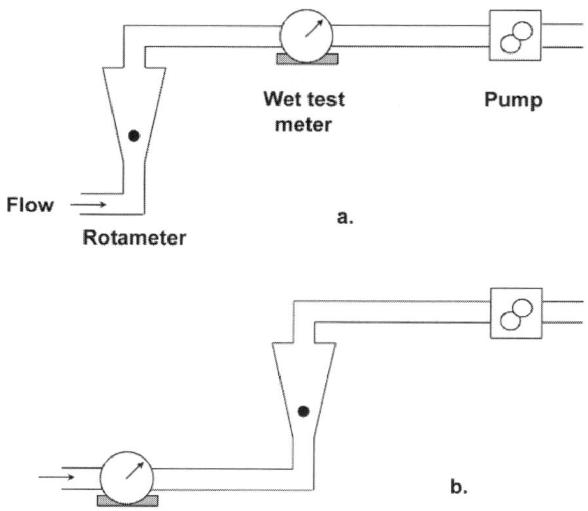

Figure 20-24. A good a) calibration setup for a rotameter which will be used in the field with the upstream end open to the atmosphere. A poor b) setup.

A number of errors can be associated with the determination of air sample volume, such as:

1. calibrator bias
2. electronic fluctuations in pump current
3. pressure and/or temperature effects on Q in constant-flow pumps
4. nonlinear change in Q over time when pre- and post-Q values are averaged
5. bias introduced when the medium used for calibration is not that used in the field
6. tubing that is crimped while in the sampling train.

One way that the 5% criterion is tested (even though this application does not consider all the above factors) is in the pre- and post-calibration comparison. Generally, if the post-calibration of a pump is within 5% of the pre-calibration, then some hygienists may disregard the post-calibration and simply use the pre-calibration value of Q in calculations; more often the average is used. Where the pre- and post-comparison differ by more than 5%, the sample may be declared invalid, but this decision is arbitrary because 5% bias has no basis relative to any empirical results. For a non-constant-flow pump, a difference of < 5% may be difficult to achieve when filters are used, as filter loading may slow the pump considerably. For this reason non-constant-flow pumps have historically been adjusted throughout the day, even though this practice may result in inaccuracy of estimated flows (Weiss et al., 1982).

One way to further assure that pumps are working properly throughout the shift is with a pump reliability tester (Abell et al., 1989). This device allows a laboratory check of the flow (± 1.5%) of a pump over an 8-hour period and provides a record of Q versus time. However, it costs several thousand dollars to build and is not known to be commercially available. TSI markets pumps that continuously record flow rates in memory throughout the run period.

Regardless of how confusing the ± 5% criterion and its application may be, the following important conclusions can be drawn.

1. A pre- and post-calibration performed in the lab may not detect variations in Q caused by the *field* environment. Calibration performed in the field will more accurately determine actual flows. Altitude effects alone can cause errors of > 5% unless calibration and use occur at the same conditions.

2. A pre- and post-calibration (particularly under field conditions) should allow reasonably accurate Q values to be determined.

3. The total errors caused by air sampling and analysis (Chapter 25) are small relative to environmental fluctuations that cause statistical sampling error (Chapter 17). Therefore, the former should not be given undue attention, but should be kept in perspective.

4. Proper maintenance of pumps should allow one to avoid pre- and post-calibration differences of > 5%, but, in any case, the sample should not be invalidated without reason. Useful information still may be obtained from the results.

QUESTIONS

20.1. As a hygienist, you have a technician who reports to you and is responsible for maintaining all hygiene equipment including pumps. Define at least four points of a good pump maintenance and recordkeeping program. What should the technician do to properly maintain the pumps, and what records are important?

20.2. So-called constant-flow pumps can be controlled by one of three mechanisms. State the mechanisms, what if anything they hold constant, and any advantages or disadvantages.

20.3. Why do manufacturers get by with claiming *constant flow*? What does the fine print say?

20.4. A sampling train with critical orifice and filter is used in the same configuration as that of Figure 20-10a and Example 20-2. The train is precalibrated and assumed to be constant flow. What is the maximum tolerable pressure drop to be caused by loading of the filter if the value of (Pre Q – Post Q)/2 is to be kept at 95% of Pre Q?

20.5. State one advantage and two disadvantages of multi-flow pumps.

20.6. A pump is pre-calibrated at 2.1 L/min and post-calibrated at 2.0 L/min. A 427-minute air sample is collected. What is the air sample volume?

20.7. A bubble meter is used to calibrate a pump. The bubble traverse time is 28 seconds, and the bubble meter volume is 500 cm^3. What is the flow rate in L/min?

20.8. A non-constant-flow pump and separate rotameter are calibrated at NTP. The pump flow at NTP is 3.5 L/min. The pump is taken to Denver (5,280 ft or 1550 m) and used in a meat packing plant at -5°C (23°F). Assuming that the pressure at Denver is average for that altitude, and the rotameter is used with one side open to the atmosphere, what pump flow rate will the rotameter indicate?

20.9. In a few sentences, describe an experiment that would test the contribution of one aspect of pump error. Describe the aspect or error source, and be as specific as possible about how you would test it. Your results should be as general as possible; that is, the more situations they can apply to, the better.

REFERENCES

Abell, M.T.; Agee, J.P.; Hentz, P.A.; Anastas, M.Y.; Boyd, J.E.: Personal pump reliability tester. Appl. Ind. Hyg. 4(6):143-146 (1989).

ANSI: Electromagnetic Fields, Safety Level with Respect to Human Exposure to Radiofrequency, ANSI C95.1-1991. American National Standards Institute, New York (1991).

Baker, W.B.; Clark, D.G.; Lautenberger, W.J.: Industrial hygiene air sampling with constant flow pumps, In Chemical Hazards in the Workplace, G. Choudhary (ed.). American Chemical Society, Washington, DC (1981).

Bohl, C.D.; Clark, W.W.: The effect of hand-held radio operation on audio dosimeters. Am. Ind. Hyg. Assoc. J. 54(3):121-126 (1993).

Buck, A.P.: VSS and Basic Pumps, Orlando FL http://www.apbuck.com/shop/category.asp?catid=54 (2006).

Bios International Corporation: DryCal Primary Gas Flow Calibrators. Butler, NJ, USA http://www.drycal.com/drycal/index.htm (2006).

Caplan, K.J.: Rotameter corrections for gas density. Am. Ind. Hyg. Assoc. J. 46:B10-B16 (1985).

Casella: Personal air sampling pumps. Casella CEL Ltd, Regent House, Wolseley Road, Kempston, Bedford MK42 7JY, UK http://www.casellacel.com/Products/menu.asp?id=45 (2007).

Charsha, R.C.; Linch, A.L.: "Freon" powered portable air sampling kit. Ind. Hyg. Q.:135-138 (1957).

Corn, M.; Bell, W.: A technique for construction of predictable low capacity critical orifices. Ind. Hyg. J.: 502-503 (1963).

Daugherty, R.L.; Franzini, J.B.: Fluid Mechanics with Engineering Applications, 6th ed., pp. 363-364. McGraw-Hill, New York (1965).

Daugherty, R.L.; Ingersoll, A.C.: Fluid Mechanics with Engineering Applications, 5th ed., p. 152. McGraw-Hill, New York (1954).

Feldman, R.F.: Degraded instrument performance due to radio interference: Criteria and standards. Appl. Occup. Environ. Hyg. 8(4):351-355 (1993).

Gast: Vacuum Pumps. Gast Manufacturing Corp., Benton Harbor, MI http://www.gastmfg.com/productinfo.html (2006).

Gillian: Air Sampling Systems. Division of Sensidyne, Clearwater, Fl. USA http://209.18.104.171/prodcat.php?ID=3 (2006).

Hall, P.M.: Isothermal flow controller for air sampler. U.S. Patent 5892160, U.S. Patent Office, Washington, DC, USA http://www.uspto.gov/patft/index.html (1999).

Ingram,W.T.: Personal air-pollution monitoring devices. Ind. Hyg. J.: 212-216 (1964).

Isabell, M.T.: The effect of viscosity on the operation of a precision rotameter. Master's thesis, University of Alabama at Birmingham, School of Public Health (1994).

Kovein, R.J.; Abell, M.T.: Benefits of a nickel-cadmium battery cycler. Appl. Ind. Hyg. 1:29-33 (1986).

Leidel, N.A.; Busch, K.A.; Lynch, J.R.: Occupational Exposure Sampling Strategy Manual. National Institute for Occupational Safety and Health, Cincinnati, OH (1977).

Linch, A.L.: Evaluation of Ambient Air Quality by Personnel Monitoring, Vol. II: Aerosols, Monitor Pumps, Calibration, and Quality Control, 2nd ed. CRC Press, Boca Raton, FL (1981).

Lippmann, M.: Calibration of air sampling instruments, In Air Sampling Instruments, 7th ed., S. V. Hering (ed.). American Conference of Governmental Industrial Hygienists, Cincinnati, OH (1989).

Lippmann, M.; Cristofano, E.E.; Graveson, R.T.: Lightweight air sampling pumps for field use, Ind. Hyg. J.: 212-216 (1959).

Middendorf, P.J.; McIntosh, D.L.; Tow, L.V.; Williams, P.L.: Performance of electronic flow rate meters used for calibration of air sampling pumps. AIHAJ 62:472-476 (2001).

MSA: Peresonal Sampling Pumps. Mine Safety Appliances Company, Pittsburgh, PA http://www.msanorthamerica.com/catalog/catalog536.html (2006).

Myers, G.E.: An automatic cycler/charger for nickel-cadmium batteries. Am. Ind. Hyg. Assoc. J. 42:542-545 (1981).

Nelson, G.O.: Gas Mixtures; Preparation and Control, Chap. 3. Lewis Publishers, Chelsea, MI (1992).

NOAA: U.S. Standard Atmosphere, 1976, National Oceanic and Atmospheric Administration, U.S. Government Printing Office, Washington, DC (1976).

Ogden, T.: R.J. Sherwood, Obituary. Ann. Occ. Hyg. 45:323-325 (2001).

Patty, F.A.: Industrial Hygiene and Toxicology, 2nd rev., Vol. I. Interscience Publishers, New York (1958).

Perry, R.H; Green, D.W. (eds.): Perry's Chemical Engineers' Handbook, 6th ed., pp. 5-12-5-18. McGraw-Hill, New York (1984).

Schoenborn, E.M., Jr.; Colburn, A.P.: The flow mechanism and performance of the rotameter. Am. Inst. Chem. Eng., pp. 359-389 (1939).

Sherwood, R.J.; Greenhalgh, D.M.S.: A personal air sampler. Ann. Occup. Hyg. 2:127-132 (1960).

Shotwell, H.P.: Quality control in air sampling pump calibration. Am. Ind. Hyg. Assoc. J. 40:249-253 (1979).

Sipin, A.J.: Personal Sampler Pump for Charcoal Tubes, Contract HSM 71. Available from NIOSH, Cincinnati, OH (1971).

SKC: Maintaining SKC pump NiCad battery packs, Eighty Four, PA http://www.skcinc.com/instructions/1363.pdf (2006a).

SKC: Personal Sample Pumps, Eighty Four, PA http://www.skcinc.com/pumps.asp (2006b).

Treaftis, H.N.; Tomb, T.F.; Gero, A.J.: Effect of altitude on the volumetric flow rate of constant flow pumps. Appl. Ind. Hyg. 1(1):54-58 (1986).

TSI: Personal Exposure Monitors. Shoreview, MN http://www.tsi.com/JoinCategories.aspx?Cid1= 153&Cid2=155 (2006).

Weiss, S.A.; Beaulieu, H.J.; Buchan, R.M.: Air volume measurement error in pumps with rotameters and high range constant flow pumps. Am. Ind. Hyg. Assoc. J. 43(10):754-758 (1982).

Zimmerman, N.J.; Reist, P.C.: The critical orifice revisited: A novel low pressure drop critical orifice. Am. Ind. Hyg Assoc. J. 45(5):340-344 (1984).

21

GASES AND VAPORS – SOLID MEDIA

Today, the utilization of solid media to remove gaseous airborne contaminants for analysis is by far the most important and widely used technique. However, this has not always been the case. As late as 1965, the measurement of gas and vapor concentrations in air was difficult, time-consuming, and not very accurate. Prior to that time, the use of solid media that depends on the phenomenon of adsorption was likely of the same order of popularity as the use of liquid absorption media (see Chapter 22). However, advances in gas chromatography, whose application to industrial hygiene began to be developed about 1965, caused an increase in the development and use of solid sorbent media methods of collection. By the mid-1970s solid media were the method of choice for most volatile organics. Today they are also used for many semivolatile organics as well as inorganic gases.

As laboratory techniques employing gas chromatography for the analysis of gases and vapors were being developed, one of the most important considerations was how to collect the gases and vapors in order to transport them to the laboratory. Early on, both absorptive and adsorptive techniques[1] were tried. Impinger techniques, which rely on the absorption of gases and vapors from an airstream that passes through a liquid medium, were examined. These techniques had been popular for a number of gases and vapors that, prior to 1965, were analyzed with wet chemical laboratory techniques such as titration and visible spectroscopy. Although impinger techniques still have their role in industrial hygiene and are necessary, particularly for certain types of reactive gases, their use has largely been replaced by the use of solid media. Impingers are bulky, cumbersome, time-consuming to prepare, and potentially hazardous. Similarly, bags and rigid containers still have a role in hygiene, but their use is secondary to that of solid media.

---

[1] Both depend on the interactive forces of attraction between molecules as described in Chapter 8. Absorption is the mixing of one molecule in another liquid or gaseous group of molecules to form a solution. Adsorption is a surface phenomenon; one molecule becomes attracted to the surface of a solid. Airborne toluene is absorbed into liquid heptane, but airborne toluene is adsorbed onto charcoal. Some phenomena are not so easy to describe. We often speak of skin absorption of chemicals, but because the process takes place in the stratum corneum (a solid matrix of dried cells composed of lipid and keratin, large solid molecules), adsorption would probably be a better term.

HISTORICAL USE OF SOLID MEDIA

Prior to 1936, filtration techniques were available for assaying aerosols from industrial atmospheres. Certainly silica, lead, asbestos, coal dust, and other aerosols had shown that they could damage health or kill thousands of workers; so monitoring their presence was important. However, organic vapors also were causing diseases among workers (Hamilton, 1943). In the United States the organic solvents industry was expanding, and it was well established in Germany, France, and the United Kingdom (Aftalion, 1991). However, volatile vapors could not be collected on filters; and although direct-reading methods (see Chapter 26) were available, they were nonspecific, difficult to integrate over long time periods (e.g., 8 hours), and, most important, costly.

The adsorptive properties of charcoal, silica gel, and alumina had been documented by the mid-1930s. Cook and Coleman (1936) proposed the use of charcoal to serve as an adsorber in a gravimetric technique for assaying volatile organic vapors. The technique was not specific for a given compound, but it was inexpensive. However, it suffered from other drawbacks.

Early attempts to use charcoal (see, e.g., Fieldner et al., 1919; Greenburg, 1926; Elkins et al., 1937; and Verma and Tombe, 1999) were affected by its propensity to adsorb water, which usually constitutes considerably more mass in the airstream than the contaminants. A gravimetric technique could not distinguish the two.

Cook and Coleman used water preadsorbers, but found that fingerprints, condensed water, and other contaminants on the outside of the glass tube holding the charcoal were gravimetric interferences. They tried rubbing the tubes with clean cloths using a carefully controlled regimen, but carbon, being a good conductor, helped establish an electrical charge on the glass that caused weights to fluctuate.

Moskowitz and Burke (1938) found that tubes filled with silica gel did not suffer from weight fluctuation. Silica gel adsorbs great quantities of water from the air and is inefficient in adsorbing non-polar organics, but these problems were solved by increasing the number of silica gel tubes and by using a water preadsorber. The results from numerous determinations were in excellent agreement with averaged readings from a direct-reading instrument (the interferometer). Similar techniques remained popular into the 1950s (Patty, 1958).

By the 1940s, (Cralley et al., 1943; Danner and Goldenson, 1947) charcoal and silica gel were being used to adsorb vapors for qualitative laboratory analysis. Halogenated organic vapors could be assayed by using relatively complex titration techniques, which were not specific but measured total halides.

In order to develop a method that could distinguish specific compounds, it was necessary to separate the compounds from one another for individual analyses. The separation technique, gas chromatography (GC), had been available since 1952, but was first given specific application to industrial hygiene in 1958 (Rushing). Rushing noted in his review that one important problem that must be overcome was that of how best to concentrate the contaminants. At that time, the contaminated air was injected directly into the GC apparatus; so concentrations only as low as 100 ppm commonly could be detected.[2] Rushing pointed out that charcoal or silica gel could be used to capture and concentrate the contaminants, but he then described a complex technique for desorbing the contaminants to introduce them to the GC. During the next few years the contaminants were commonly desorbed by using heat.

Silica gel became more popular than charcoal because the silica gel released the contaminants more easily with heat (Altshuller et al., 1962). Nevertheless, silica gel also had the disadvantage of a low capacity for polar compounds due to its high capacity for water vapor.

Furthermore, heat desorption allowed only one analysis of the sample, and the ability to perform replicated analyses was thought important at that time.

No doubt many desorption solvents were tried, but they interfered with the analysis. Only in 1964 was a report made of the use of carbon disulfide to remove vapors from charcoal (Otterson and Guy, 1964). A solution of carbon disulfide and the contaminants then was analyzed by using gas chromatography. The advantage of carbon disulfide (CS_2) was that it did not interfere with the analysis of the other compounds by being eluted from the GC column at the same time that they were, or by causing a large detector response because of its much greater concentration than that of the analytes. CS_2 is now the solvent of choice for the millions of analyses done each year and in most government *approved* methods. Interestingly, the Otterson and Guy paper was never published.

Once this discovery was made, many things began to fall in place making charcoal the solid sorbent of choice. During this period of time an industrial hygienist wishing to perform an analysis of a plant atmosphere for a particular compound was often a pioneer; there was usually very little information available to the hygienist about how to accomplish the task. The published literature was available, and at times it may have contained specific methods that the hygienist could follow (e.g., in 1968 Reid and Halpin published GC/charcoal methodologies for 10 compounds). However, more often than not, it contained no directly applicable or analogous information specific to the problem. So a great deal of research on the part of the hygienist or the laboratory was necessary.

While researchers continued to experiment with the use of silica gel, molecular sieves, and packing materials used in GC columns (Altshuller et al., 1962; Cropper and Kaminsky, 1963; Whitman and Johnston, 1964), others concentrated on developing specific methods employing charcoal as the collection medium. By 1970, when the National Institute for Occupational Safety and Health (NIOSH) was formed, enough basic research on charcoal had been performed to allow NIOSH to begin the development of numerous methods employing charcoal/GC for specific compounds (White et al., 1970). A standard NIOSH tube was described in 1971 (Kupel and White). By 1974, NIOSH had developed enough of these methods that, in many cases, the industrial hygienist needed only to consult a manual of methods in order to find a method that could be applied to the sampling and analysis problem (NIOSH, 1974).

[2] With advances, direct air injection will now allow detection of about 1 ppm.

CHARCOAL

What Kind of Charcoal

Although there are many types of solid sorbents available for the collection of vapors from airstreams, charcoal is the most widely used. However, there are many types of charcoal. Such diverse substances as sugar, coconut, petroleum, cotton, Douglas fir, and paraffin (Cookson, 1978) have been used to make commercial charcoal. One could add to the list olive pits (used in Spain) and almond stones. While the main ingredient is, of course, carbon, an equally important ingredient is the presence of a natural porosity (enhanced by activation). For *synthetic* petroleum charcoal the porosity is induced by mixing the raw material with solvent and a surfactant.[3] For industrial hygiene work, two types of charcoal have been used most frequently, and one of these, coconut-based charcoal, is utilized in the vast majority of NIOSH methods employing charcoal. The second charcoal, derived from petroleum, is used less frequently.

Preparation and Activation

The material used by hygienists and commonly called charcoal is an amorphous (i.e., noncrystalline) material composed almost entirely of carbon. Though impurities do exist, the process by which charcoal is made causes most of them to be volatilized from the carbon. In order to manufacture charcoal, a carbon source such as coconut shell is placed in a hot environment that is deprived of oxygen, in a process called pyrolysis or destructive distillation. The manufacture of charcoal has been conducted like this for thousands of years. If the starting material is something other than wood (e.g., petroleum or coal), then the name applied to the end product may be different. For example, in the petroleum refining industry, heavy hydrocarbons are heated to form a product called coke. Similarly, coal can be used to form coke. The end product of these two processes then is used as a source of carbon or fuel to smelt iron. However, for the development of solid air sampling media, the charcoal must be further processed.

The pyrolysis process does not yield a product having the high sorptive properties necessary for removing contaminants from airstreams. In order to obtain these properties, the charcoal must be *activated*. The generic term for the product is activated carbon. However, hygienists often refer to the activated sorbent, regardless of its source, as charcoal.

The activation process is accomplished by heating the carbon to 800°C to 900°C, usually in the presence of steam, or sometimes carbon dioxide. This results in a honeycomb-like structure containing many sorption sites. A gram of activated carbon contains about 1,000 m^2 (10,000 ft^2) of surface area (Lewis, 1996).

[3] Personal communication, M. Harper, NIOSH (2006).

Structure

Cookson (1978) provides a good review of the structure of charcoal. A reasonable starting point for understanding the structure of activated carbon is the structure of graphite. However, graphite is a crystalline structure, composed of layers of six-carbon rings fused together, as shown in Figure 21-1. Activated carbon is amorphous overall but contains microcrystalline regions. In addition, the activation process causes voids or pores to develop in the carbon.

Activated carbon has often been referred to as a nonpolar material (see, e.g., Rushing, 1958; Soule, 1978; Olishifski and Kerwin, 1988), and relative to silica gel this is true. The activation process does cause considerable oxidation to occur (water or carbon dioxide being the oxidant), and this yields carboxyl groups, phenolic hydroxyl groups, carbonyl groups, lactones, carboxylic acid anhydrides, and cyclic peroxides (Cookson, 1978). These groupings are polar and can be reactive.

As a consequence, activated carbon can hardly be said to be nonpolar. However, unlike silica gel it is heterogeneous, having polar and predominantly nonpolar sites. At relative humidities less than 50% it is essentially hydrophobic. Its dual hydrophilic nature is revealed at higher humidities. However, the process of coming into equilibrium with high humidity is slow; thus, charcoal is a relatively poor adsorbent for water. In simplest terms, this means that activated carbon has a wide range of applications to the adsorption of air pollutants.

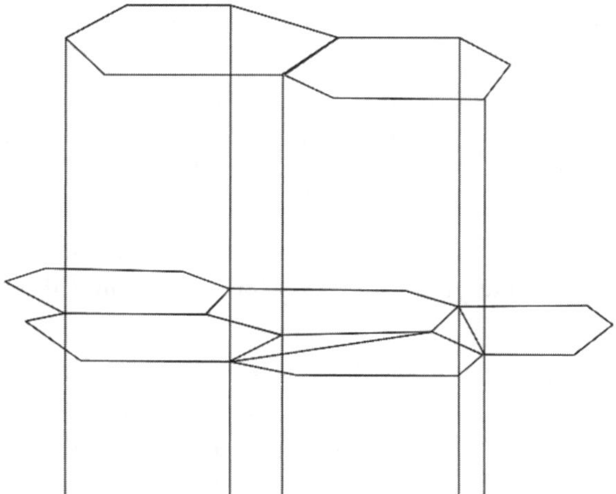

Figure 21-1. Carbon atom arrangement in a graphite crystal (adapted from Cookson, 1978).

Specifically, the applicability of activated carbon must be characterized for each atmosphere based on the polarity of the contaminants. Although the level of airborne water vapor is not so predominant as with silica gel, it has an influence on the efficiency of activated carbon as an adsorbent (see later discussion). Water is very polar, and the interactions that it forms are quite strong (hydrogen bonds). Water interacts with polar sites, where the abundance of water can cause the exclusion of other polar compounds.

Other Charcoal Variables

There are variables in the manufacture of activated charcoal; charcoal sold by one supplier is often different from that sold by another. Also, even though a charcoal manufacturer may use the same process and basic raw material for each lot of charcoal that is processed, the source material can vary.

The source for most industrial hygiene analytical charcoal is coconut shell. All coconut shells are not created equal. Even coconut shells from small areas of the world such as Malaysia cannot be expected to be consistent across time. The manufacturers and suppliers of charcoal tubes are very much aware of this, and therefore they process and sell charcoal tubes by the lot.[4] A lot is a well-mixed batch of charcoal that is processed under a particular set of conditions. Charcoal tubes are always given a lot number, and it is important that certain laboratory tests be performed on each lot in order to *characterize* the lot. The most important of these characteristics is the ability of the charcoal to release the adsorbed compound (desorption efficiency or DE). The results of DE tests can be applied to the results of individual chemical analyses in order to make any necessary corrections. ASTM (2001a) lists DE values for a number of analytes.

Early on, NIOSH planned a certification program that would evaluate the considerable variation in charcoal (McCammon, 1976). However, apparently because the methods for analysis of specific compounds developed by both NIOSH and OSHA have been followed very closely by laboratories throughout the United States, and because those methods require specific types of charcoal (coconut or petroleum), the certification program was never developed. In effect, standardized analytical methods have promoted the standardization of charcoal.

Part of this standardization includes the size and the shape of the tube in which the charcoal is contained and the mesh size of the charcoal. Charcoal comes in various granule sizes, which can be screened in order to obtain sizes of a particular size range. The charcoal tube developed by NIOSH (White et al., 1970; Kupel and White, 1971) and now in common usage contains charcoal granules that will pass through a number 40 U.S. standard sieve (about 0.84

---

[4] For example, SKC's Coconut Shell Charcoal sorbent Lot 120 was available for nearly 15 years before it was replaced by Anasorb CSC Lot 2000 in 1996. Lot 2000 is expected to have the same longevity. Lab tests run on Lot 2000 show that its sorbent quality is equivalent to Lot 120.

mm) but will not pass through a number 20 sieve (about 0.42 mm; Dean, 1979). Consequently, the term 20/40 mesh has been given to the charcoal contained in most charcoal tubes. Other mesh sizes are available and are used in special circumstances; but the *standard* or NIOSH tube contains the 20/40 mesh charcoal. Fraust and Hermann (1969) showed that this mesh size was more efficient than smaller mesh sizes. Particle size does have an effect on the adsorption process (Rehrmann and Jonas, 1978).

The size and the shape of the tube also were specified by NIOSH. Figure 21-2 shows that the tube holds more than charcoal. It contains a section of glass wool and two polyurethane (foam) spacers, which serve to keep the charcoal in place and separate the two sections of charcoal. In addition, small screens and clips are included, which hold the entire assembly in place. In some cases the foam may adsorb the analyte;[5] so tubes are available that contain only glass wool separators.

The standard charcoal tube consists of 150 mg of activated charcoal, 100 mg in a front section and 50 mg in a back section. The two sections are separated in order to detect *breakthrough*, which is discussed later. The direction of airflow through the tube is important; the smaller section should always be downstream of the larger section. Other charcoal tube sizes can be obtained, including the jumbo size, second in popularity. It consists of 400 mg and 200 mg sections of charcoal. The standard NIOSH, tube used widely in the United States, has been adopted on an international scale and is used in Europe as well as other parts of the world. The SKC Corporation (2006) is one of the world's leading suppliers of charcoal tubes. See Table 21-1 for details on SKC carbons. Supelco (2006) offers a relatively narrower range of charcoal and other sorbent tubes and Matheson-Kitagawa (2004) and MSA (1998) apparently also make a similar charcoal tube, although details on the charcoal source are not readily available. Drager (2006) makes a smaller cylindrical tube. One should make certain that desorption effi-

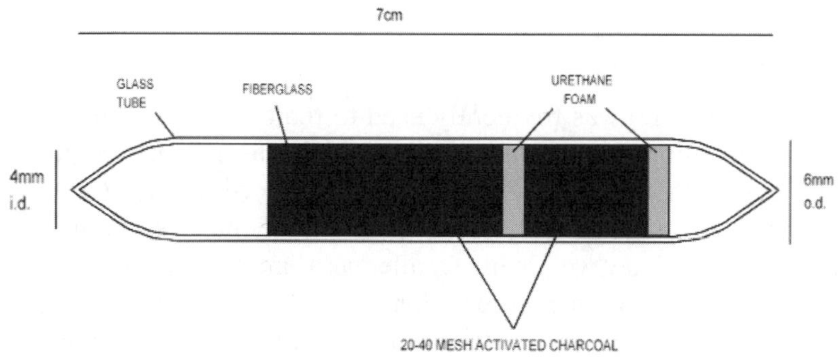

Figure 21-2. The NIOSH standard charcoal tube (drawing not to scale).

[5] See OSHA methods for methylene chloride and ethylene dichloride.

Table 21-1
Properties of SKC carbons

Name	Source	Advantages	Disadvantages
Coconut	Coconut	NIOSH Standard Most researched Large surface area	Most hydrophilic Reactive
Petroleum	Petroleum	Less hydrophilic Less reactive Good for chlorinated hydrocarbons	Little micropore space, lower breakthrough volume Less surface area
JXC, Carbide	Petroleum Jelly Sludge	Most hydrophilic Least ash – reactivity High surface area	Cost

ciency, blank and other laboratory studies are performed by the lab with the type of charcoal that is used in the field.

Recovery of Analytes

Because of its structure, charcoal has proved to be a very effective medium for adsorbing many vapors and some gases from the airstream. However, as important as adsorbing the gases and vapors from the air is the removal of the gases and vapors from the charcoal so that they may be analyzed.

THERMAL DESORPTION

Thermal desorption (TD) was frequently used to remove analytes from charcoal prior to 1970. However, as mentioned previously, silica gel became more popular than charcoal because the silica gel released the contaminants more easily with heat (Altshuller et al., 1962). As solvents were discovered that would desorb charcoal, thermal desorption practices were de-emphasized in the 1970s. However, with smaller and smaller concentrations becoming of interest and the development of other solid sorbents (to be discussed later), thermal desorption has regained popularity. A typical apparatus has a mechanism for holding the tubes while they are heated and purged simultaneously with an inert carrier gas. The temperature and time of desorption and carrier gas flow rate are adjustable. There are accommodations to purge air from the sample tube and analytical system before heat is applied to prevent oxidation of sorbent and analyte. There should also be a cold trap in the transfer line that is used to concentrate the desorbed sample. The desorbed sample, contained in the carrier gas, goes to the gas chromatograph and capillary column by way of a heated transfer line (ASTM, 2003).

The chief advantage of TD versus conventional solvent extraction is a 100 fold improvement in detection limits. The sample tubes, unaffected by solvent are also reusable, there is no chromatographic interference or artifacts from the desorption solvent, and economics are improved. Then why is TD not used more? Actually in some parts of the world it is used more than solvent desorption. For ambient air it is used widely in the U.S. as there are methods (EPA) that require it. There are no quantitative NIOSH methods based on TD. Beyond these reasons, the chief reason of the past was that only one analysis could be performed. In addition, sorbent tube dimensions vary, interfaces with the GC or GC-MS vary, and there are other issues. Most of these, however, have been solved in modern TD apparatus.[6] Thus, there is now a wide array of methods from around the world, both ambient and workplace air, that rely on TD for recovery of analytes (see e.g., ASTM, 2003 and Markes, 2003). It should also be noted that TD is applicable to desorption of other sampling devices, such as passive dosimeters (Chapter 23) and rigid containers (Chapter 22).

SOLVENT DESORPTION

The removal of most workplace air analytes from charcoal is still accomplished by using carbon disulfide (CS_2). Thus, there is a large amount of literature on this technique. Typically, 1 ml is added to a vial containing either the front or the back section of the charcoal tube, and this solution is periodically agitated for 30 minutes. However, the removal of the analytes may not be 100% efficient.

The desorption solvent must do a good job of desorbing the analytes, but must not interfere in the analysis by being eluted (emerging from the GC column) at the same time as the analyte(s). It must be small enough to be able to penetrate the narrow necked pores of charcoal, have a higher heat of adsorption than the adsorbed molecule, and be a good solvent for the desorbed molecules. Carbon disulfide fits all these criteria (except as a solvent for polar compounds), and the most outstanding feature is probably the poor FID response.

The efficiency of recovery of the analytes from the charcoal is ascertained in a desorption study. There has been some controversy about the technique used to perform a desorption efficiency (DE) study. During sampling, the contaminant comes in contact with the charcoal in the vapor state and in a dynamic or moving atmosphere. Therefore, it would be reasonable to conduct a desorption efficiency study by loading the charcoal in the laboratory with a known amount of contaminant under the same conditions in which loading would occur in the field. However, historically this has not been done.[7]

---

[6] For further information the reader is referred to the excellent publications of Markes International, Limited at http://www.markes.com/downloads/publications.aspx.

[7] It should be noted that charcoal itself adsorbs some of the desorbing solvent, leading to some desorption efficiencies greater than 100%.

Instead, a small quantity (similar in mass to that which field samples are suspected to hold) is *spiked* onto the charcoal in the liquid state (Fracchia et al., 1977) and then desorbed at a later time. A microsyringe is used for this. This technique is obviously easier than setting up a dynamic airborne exposure apparatus. But does it adequately mimic exposure? Some researchers have suspected that it does not, and so other loading techniques have been tried. The method is recommended by ASTM (2001b).

A second method is to expose a charcoal blank to a 1 ml solution of CS_2 and the analyte – the phase-equilibrium approach (Dommer and Melcher, 1978). The temperature of the solution has been reduced. The charcoal and the solution then approach equilibrium as they warm to room temperature. The method makes it possible for one set of standard solutions to be prepared for the desorption study and for calibration of the GC. It was thought that the cooling of the solution was necessary to gain an equilibrium similar to that obtained with a gas phase/solid phase equilibrium, but apparently that idea has been abandoned, as seen below.

Commercial and government laboratories apparently have adopted a combination of the direct spike and a modified phase equilibrium or a solution spike approach (contacts regarding this were made with OSHA, NATLSCO, and Clayton laboratories). There are several reasons for this. The direct spike method is convenient and easy. However, when small ($< 1\ \mu L$) aliquots are spiked onto the charcoal, errors in measuring the volume with the syringe cause irreproducible results. Therefore, at low loadings (the criterion for low varies with the lab), a solution of CS_2 and the analyte is spiked. This allows a much larger volume of solution to be spiked than would be spiked for the pure analyte. The amount of CS_2 in the solution is usually considerably less than the 1 ml used in the phase equilibrium approach, and no chilling takes place.

The underlying question is, how well do any of these three methods mimic vapor loading? One study (Krajewski et al., 1980) showed that for 10 compounds, the desorption efficiencies were significantly different from dynamic vapor loading in seven cases for direct spiking, in three cases for solution spiking, and in eight cases for phase equilibrium. They concluded that solution spiking was best.

Thomas and Cohen (1995) devised a simple but elegant way to achieve dynamic vapor loading without the need for a test chamber. They assembled a sampling train with a filter cassette (see Chapter 24) followed by a charcoal tube and a pump. The liquid analyte was placed on the filter with the pump running, causing the analyte to evaporate from the filter and be absorbed by the charcoal. For seven analytes of varying polarity near 100% of the analyte evaporated (2-ethoxyethanol was the exception), DE values were near 100% (again 2-ethoxyethanol was the exception) and no different from DE values derived from liquid spiking.

A technique has been proposed for passive dosimeters (see Chapter 23) that is similar to that of Thomas and Cohen. Rodriguez et al. (1982) attempted to mimic vapor loading. A 3M organic vapor monitor contains a flat round disc of charcoal about one inch (2.5 cm) in diameter. They placed a one inch diameter Whatman #5 filter paper onto a spacer plate that resides above the charcoal. The compound was spiked to the paper and then the dosimeter sealed. The liquid should then evaporate from the paper and be adsorbed to the charcoal; so they theorized that *more reliable* DE values should be obtained. Comparisons were made only with the phase equilibrium approach. Generally, the two methods agreed well. The phase method gave the higher result. 3M recommends this vapor spike technique for its organic vapor monitors (3M, no date).

For polar compounds, DE values using CS_2 can be quite low. They can be improved by adding a small percentage (typically less than 10%) of a polar compound to the CS_2. Water (Langvardt and Melcher, 1979), amyl or hexyl alcohol (Kenny and Stratton, 1989), 2-(2-butoxyethoxy) ethanol (Beck et al., 1990), and methanol (Fracchia et al., 1977) as well as methanol, propanol, butanol, and acetone (NIOSH, 1977) have been used with CS_2. In addition, single solvents such as dimethylformamide (Rudling and Bjorkholm, 1986) and methylene chloride (Rodriguez et al., 1982) have been used.

Effect of Loading on Desorption Efficiency

Regardless of the solvent and method, early studies indicated that the amount of analyte adsorbed on the charcoal or the loading level did not have an effect on desorption efficiency (White et al., 1970; Fracchia et al., 1977; Dommer and Melcher, 1978; Krajewski et al., 1980; Posner, 1980). In other words, the higher the loading was, the more analyte desorbed, so that DE was a constant. However, it became obvious that for some compounds, such as acetone and styrene, the relationship was not linear. Furthermore, as interest in lower loadings increased (owing to declining environmental levels caused by lower standards and better controls), DEs for other compounds began to show non-linearity. One solution to this was a calibration curve, as shown in Figure 21-3.

Posner (1981) made some important points about DE loading relationships:

- If they could be linearized, then random errors associated with DE determination could be treated more straightforwardly.
- Existing adsorption theory explains the curvilinear relationship.

Posner demonstrated how to use Freundlich or Langmuir theory (see Appendix 21-1 for details) to form a straight-line relationship between DE and loading. First,

$$DE = \frac{C_L W_L}{C_L W_L + C_s W_s} \qquad \text{[Equation 21-1]}$$

where C = concentration in solid (s) or liquid (L) and W = weight of the solid (carbon) or liquid (carbon disulfide) phase. Because CW = mass of the analyte, DE = mass of the analyte in the liquid divided by the total mass of analyte in both phases. The equation has three unknowns (DE, C_L, C_s). GC analysis of the liquid phase determines C_L, leaving DE and C_s. Langmuir theory states:

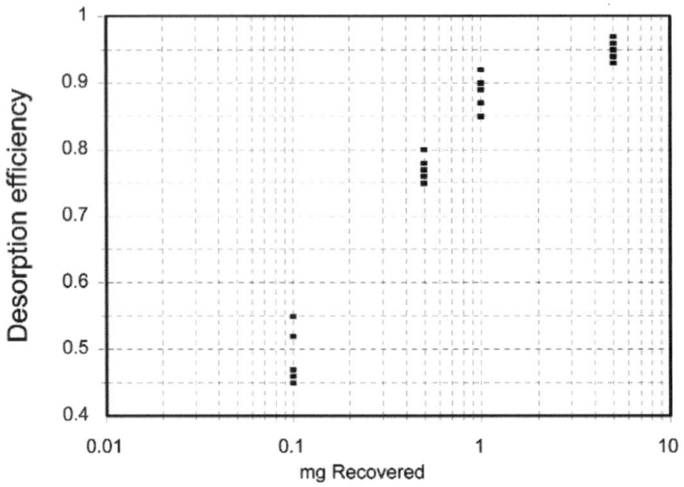

Figure 21-3. Non-linearized calibration curve for desorption efficiency showing N = 5 replicates at each of four loadings.

$$\frac{1}{C_s} = \frac{1}{QbC_L} + \frac{1}{Q} \qquad \text{[Equation 21-2]}$$

where *b* and *Q* are constants for the analyte and a given type of charcoal. By rearranging Equation 21-1 and letting D = DE, we have, $\frac{D}{1-D} = \frac{C_L W_L}{C_s W_s}$ and substituting from Equation 21-2:

$$\frac{D}{1-D} = \frac{W_L}{W_s Qb} + \frac{C_L W_L}{W_s Q}$$

$$\frac{1}{C_s}$$

Now if we use 1 ml of CS_2 (times density yields W_L) and 100 mg of charcoal (W_s), then W_L, W_s, Q, and b are constants and $D/(1-D)$ is linearly related to C_L, as shown in Figure 21-4. Neither the intercept ($W_L/W_s Qb$) nor the slope ($W_L/W_s Q$) is significant to the problem. From Figure 21-4, a DE can be determined for any loading.

A similar derivation can be made from Freundlich theory (see Posner, 1981, and Appendix 21-1), which gives a straight line when ln[*(1-D)/D)*] is plotted against C_L. Notice that according to either theory, C_L is proportional to the mass loading, so that loading mass (mg) or volume (μL) may be plotted. The relationship (Freundlich or Langmuir) giving the

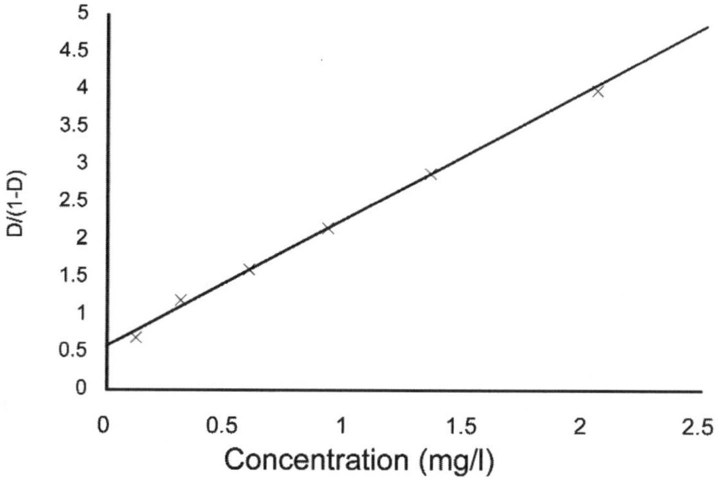

Figure 21-4. Langmuir linearization of desorption efficiency data showing best fit line.

highest correlation coefficient should be used. For single compounds, the DE experiment is relatively simple, and Figure 21-4 can be constructed if DE values vary across a wide range of loadings. (At higher loadings only, a single DE may suffice.)

Posner also showed that for some compounds there was no effect of loading on desorption efficiency at loadings down to 1 mg. However, the volume or weight of the liquid (CS_2) did have an effect on desorption efficiency. He then developed the following relationship,

$$\frac{1}{D_n} = \frac{1}{n}\left(\frac{1}{D_1} - 1\right) + 1$$

where *N* is the number of multiples of the original volume used, D_l is the desorption efficiency obtained using the unit volume, and D_n is the DE obtained using *n* times the unit volume. For example, if one uses 2 ml of CS_2 and obtains a DE of 0.4 (40%), then 13.5 times as much CS_2 or 27 ml will be needed in order to obtain a DE of around 90%. There are tradeoffs involved, in that if 27 ml were used to desorb the analyte, then the contaminants would be in such a diluted state that they might not be detectable by the GC.

Solvent Desorption of Mixtures

DE can be maximized by finding the best solvent or solvents (not necessarily CS_2). This has been determined for many single compounds as discussed earlier (see, e.g., NIOSH, 1977). However, airborne mixtures are often found in industry, and some are quite complex. Thus, 5% 2-(2-butoxyethoxy) ethanol in CS_2 was used successfully in sampling paint solvents containing up to 42 compounds of varying polarity (Beck et al., 1990). Lopes (1997) showed that 4% dimethyl sulfoxide in CS_2 would successfully, simultaneously desorb a wide range of paint compounds that are used in the automotive industry and often appear together in workplace air. Similarly, amyl or hexyl alcohol in CS_2 was used to desorb 12 compounds from charcoal (Kenny and Stratton, 1989). In each case, the supplemental solvent is eluted very late from the GC column. Of course, if one is interested in only a few compounds that are eluted in, say, 5 minutes, then before another analysis can be performed the supplemental solvent must be eluted. This might take an additional 30 minutes. Therefore, there is no single, universal combination of desorption solvents.

If one is interested in sampling a complex mixture, then it should be quite clear that, in many cases, a single charcoal tube will do the job, provided that the correct desorption solvents are chosen. However, in cases where a complex mixture (more than two components with widely varying polarity) is to be sampled, DE determination may be difficult.

The determination of DE values for components in a mixture is not so simple as determining a DE value separately for each component, unless the polarities of the components are similar. Several studies have shown that DE values are affected when polarities vary in a mixture. Originally it was thought that polarity had no effect (White et al., 1970), but later studies found different DE values, depending on whether the component was single or in a mixture (Fracchia et al., 1977; Mueller and Miller, 1979; Krajewski et al., 1980). Typical effects are shown in Table 21-2 for data from Fracchia et al. For the nonpolar component (toluene) the DE values are essentially unchanged. However, for the polar components the values increase in the mixture. Therefore, if individually determined DEs are used to adjust contaminant assay values, rather than mixture-determined DEs, concentrations are overestimated.

Water vapor also affects DE values (Foerst, 1979). If the compounds in Table 21-2 are sampled under conditions of 80% relative humidity, the DE values for toluene should remain unchanged, but those for the other compounds should decrease. Rudling and Bjorkholm (1986) showed that adsorbed water (up to 20 μL, equivalent to 80% RH at typical sampling conditions) caused DE values for SKC charcoal to decrease as shown in Figure 21-5a. Figure 21-5b shows that the effect of water vapor can vary with the charcoal manufacturer. Twenty μL of adsorbed water had no effect on DE for toluene, 1,1,1-trichloroethylene, or propyl acetate. Figure 21-5a indicates that if ethyl cellosolve were sampled at 80% RH and a DE determined at 0% RH, the corrected mass would be underestimated. It is clear that water vapor can lower desorption efficiencies. Since the problem is one of partitioning, the solution can be to use a polar co-solvent or a drying agent, or both. Drying agents may block syringes, however (Harper, 1993).

Example 21-1. Assume that 10 μL (~10 mg) of ethyl Cellosolve® is found on a charcoal tube. If the effect of the adsorbed water is not considered, what would the corrected mass be?

$$\frac{mass\ found}{DE} = \frac{10}{0.75} = 13.3\ mg$$

Table 21-2
DE values determined individually or as part of a mixture of the four compounds (Fracchia et al., 1977).

Component	μL Added to Tube	DE from Mixture	DE as Individual Compound
Toluene	8.7	0.94	1.02
M-Butanol	5.5	0.49	0.34
Cellosolve® Acetate	3.7	0.73	0.47
N-Butyl Cellosolve®	2.7	0.24	0.07

However, since the air sampling took place at 80% RH, the calculation should be:

$$\frac{10}{0.4}=25\ mg$$

The actual mass then is underestimated by a factor of two when water vapor is not considered. Under these extreme conditions, only the hygienist knows if polar compounds are sampled under high RH values, and the hygienist should consult with the laboratory and make appropriate adjustments.

Rudling (1988) developed a computer model for estimating DEs for mixtures. It requires the estimation of Freundlich or Langmuir constants for the single compounds. For routine analysis of mixtures, construction of such a model would seem appropriate; but there are far too many possible mixtures for this to be practicable for a commercial laboratory.

What happens when a charcoal tube is submitted to a commercial laboratory and the hygienist asks for toluene, butanol, dioxane, and methyl isobutyl ketone analyses? With any luck, the laboratory will contact the hygienist to explain its procedure and manifestations thereof. If not, the hygienist will be unaware of the possible errors.

Most hygienists will never perform a DE experiment, but the majority will sample a complex mixture. Hence, the above discussion is important. Laboratory techniques usually follow recipes or set procedures. Laboratory supervisors and directors are usually so involved in administration that they cannot provide special instructions for each sample, unless asked to do so! For complex mixtures ask the following questions prior to sampling.

- What desorption solvents will be used?
- Are the DE versus loading curves linear for my compounds at the concentrations I expect?

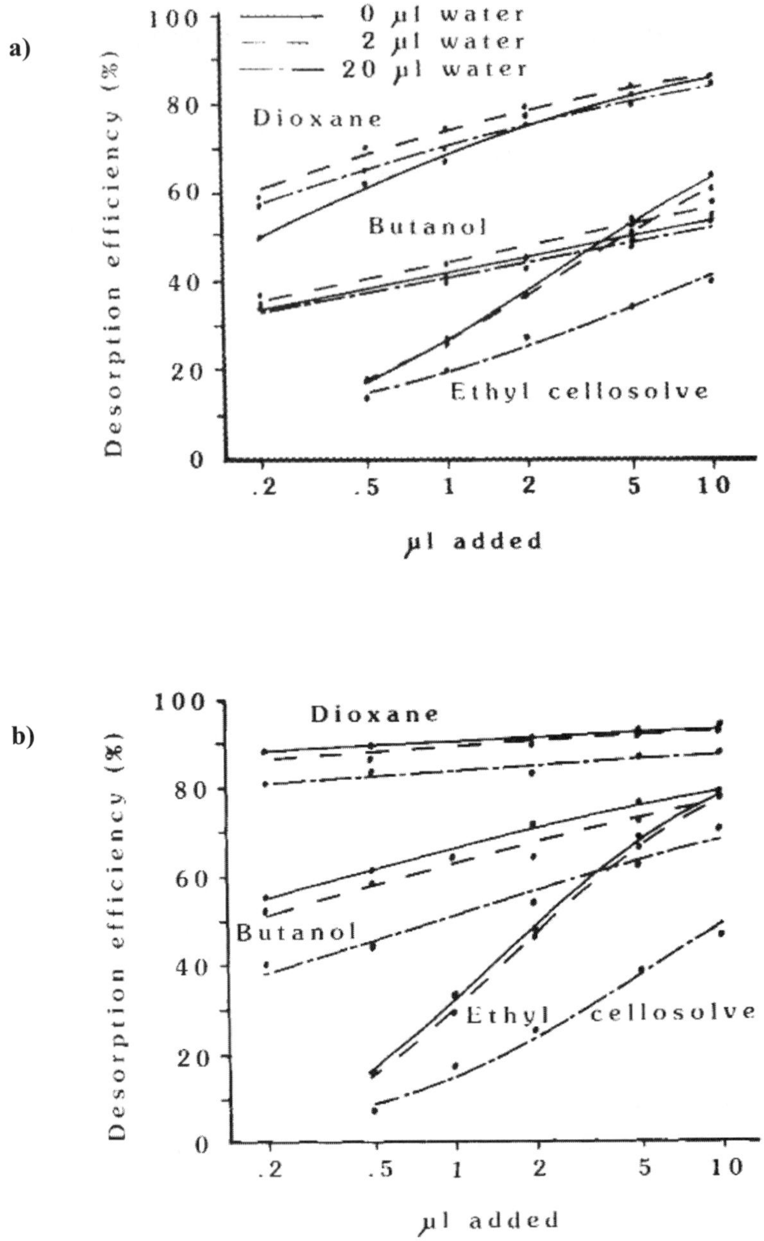

Figure 21-5. Desorption efficiency as a function of added amounts of dioxane, butanol, and ethyl cellosolve® on Merck a) or SKC b) charcoal with pre-absorbed water (see legend) and using CS_2 as desorber (from Rudling and Bjorkholm, 1986, with permission).

- Are DE experiments for mixtures performed?
- What are the potential errors associated with my mixture?
- Can I do something different to aid the process (e.g., use two parallel sorbent tubes)?
- Can the laboratory do something different to improve DE values (e.g., use another solvent, or desorb using two solvents in series)?
- How much will it cost?

Reactions with Charcoal[8]

Some polar compounds can bind irreversibly to charcoal or oxidize on it (Johansen and Wendelboe, 1981; Elskamp and Schultz, 1983; Folke et al., 1984). For example, although the ketones for the most part do not react on charcoal, three ketones in particular have been noted for their ability to react on charcoal over time and thus cause analyte loss. These compounds are 2-butanone, cyclohexanone, and mesityl oxide. When DE studies are performed with 2-butanone, they are commonly quite good, greater than 90%. However, if charcoal is loaded with 2-butanone and then placed in storage, the amount of 2-butanone extracted from the charcoal decreases with storage time (Saalwaechter et al., 1977; Elskamp and Schultz, 1983).

Most commercial laboratories should be aware of these problems and consequently counsel a hygienist not to use coconut charcoal for 2-butanone; or, if this advice is too late, the laboratory should do the analysis as rapidly as possible. However, a laboratory whose technicians are not particularly accustomed to performing air sampling analysis may not be aware of these and other problems associated with the analysis of contaminants on charcoal; so it is important that the industrial hygienist be aware of these problems and properly educate the laboratory whenever necessary. For a good approach to sampling ketones, see Harper et al. (1993).

Charcoal Sampling Variables

FLOW RATE

The most important question that a hygienist must ask when using a charcoal tube is "What volume of air should be sampled?" The charcoal tube has a finite capacity to hold the analyte. However, as Figure 21-6 shows, there are different paths to reaching that capacity, depending upon variables such as concentration and flow rate through the tube.

However, saturation capacity (W_s) is not really the variable of interest. As can be seen in

[8] There are occasional reactions of the analytes with the desorbing solvent as well. For example, amines react with CS_2 (personal communication, M. Harper, NIOSH, 2006).

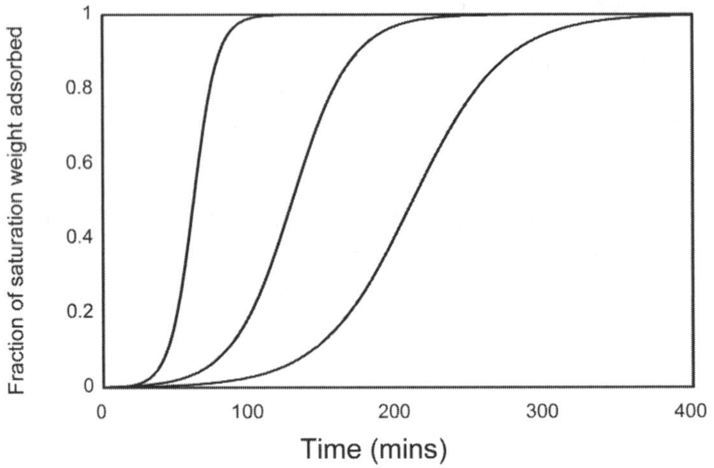

Figure 21-6. Weight change of a charcoal tube as a contaminated gas stream is flowed through it for differing conditions of concentration, flow rate, and/or relative humidity showing changes in shape of curve.

Figure 21-7, when the outlet concentration (C_o) equals the inlet concentration (C_i), saturation has occurred. However, if the hygienist samples until time of saturation (t_s), then significant loss of analyte will have occurred earlier. Ideally, one should stop sampling before any loss occurs, that is, before t_s or W_s. However, the lower end of any curve such as Figure 21-7 is asymptotic; in theory as soon as flow through the tube begins, some analyte passes from the outlet of the tube. At the early stages, this amount is, in most cases, far smaller than we can detect (see Figure 21-8). But an important question is "When does C_o become big enough to be of concern, that is, to cause an appreciable error or loss?"

In laboratory studies of charcoal, a C_o concentration of 5% C_i often is considered significant for defining *breakthrough*. (Breakthrough is a relative term because, in theory at least, one molecule breaks through instantaneously.) But the 5% breakthrough concept is not of much use in the field. We cannot monitor downstream of the tube and determine when 5% C_i is reached (nor is C_i constant).[9]

For this reason, the sorbent tube has two sections. When significant analyte is found in the second section, detectable breakthrough from the first has occurred. We need not assume that the tube is invalidated if *any* analyte is found in the second section, as we can add the mass from the two sections together. However, when enough mass is found in the second section, we must assume that a significant amount was lost from that section to the atmosphere. How then do we define *enough*?

[9] Hori et al. (1999) have rigged a sensor downstream of the charcoal tube to detect breakthrough while in use. In addition to being too sensitive, it is quite cumbersome.

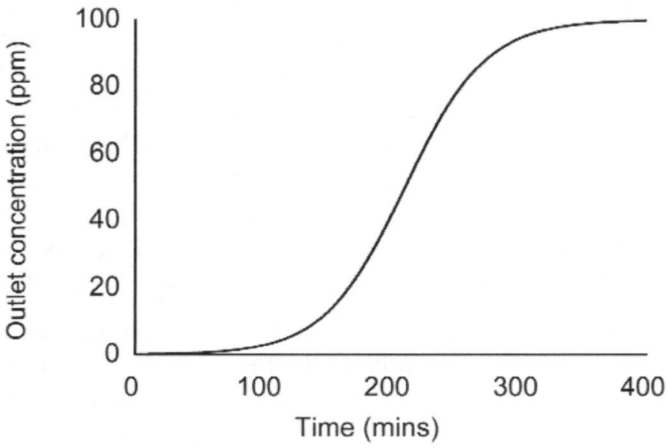

Figure 21-7. Concentration on the outlet side of a charcoal tube as contaminated air at inlet concentration (C_i = 100 ppm) is flowed through it.

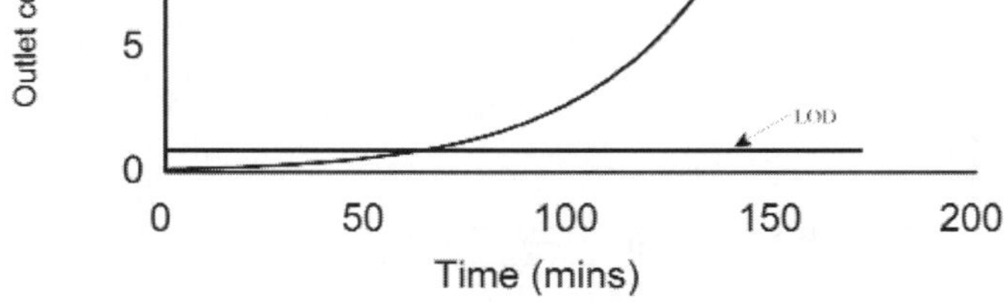

Figure 21-8. Data from Figure 21-7 showing only concentration up to 20% of C_i and showing that the limit of detection (LOD) restricts observation of some early portion of the experiment.

When the second section holds 50% of the first ($W_B/W_F = 0.5$), saturation has occurred, as the amount of charcoal is 100 mg in the front and 50 mg in the back section. So 50% loading on the back section relative to the front is a maximum. Figure 21-9 shows that the slope of the loading curve versus distance dictates whether a certain back-to-front mass ratio may indicate significant loss from the tube. For both curves in Figure 21-9, the mass of analyte that has escaped or broken through is 5% (the area under the tail). However, the front-to-back mass ratios are different for the two curves. For steep fronts (curve *a*) a larger breakthrough to the back section may be allowed before significant loss occurs from the tube. For flatter fronts (*b*), a small mass on the back section relative to the front may indicate significant breakthrough.

Figure 21-10 shows actual breakthrough data for a charcoal respirator bed for polar and non-polar adsorbates from paint spray. Think of the four curves as advancing from left to right with time. Xylene has broken through last, while the two polar compounds broke through first. Note also that the charcoal has far less capacity for the polar compounds as the masses on the y-axes differ by two orders of magnitude. Although the toluene curve might indicate less affinity for toluene versus xylene, there may be much more toluene in the paint spray.

Several recommendations have been made for acceptable back-to-front section loading ratios. NIOSH (1984) recommends reporting possible loss when $W_B/W_F = 10\%$. Melcher et al. (1978) stated that if $W_B/W_F < 10\%$, no significant loss has occurred, or if $W_B/W_F > 25\%$, loss probably has occurred, and results should be reported as minimum amount present. Saalwaechter et al. (1977) recommended that W_B/W_F should be < 20% for a reliable sample, as did Brown and Woebkenberg (1989).

A laboratory should report the actual W_B, and this should be added to W_F to calculate the

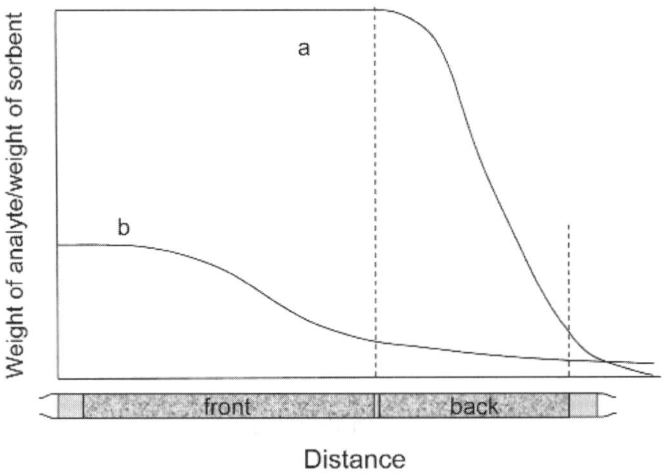

Figure 21-9. Mass of two adsorbed analytes, a and b, versus distance from charcoal tube inlet at the end of the sampling period.

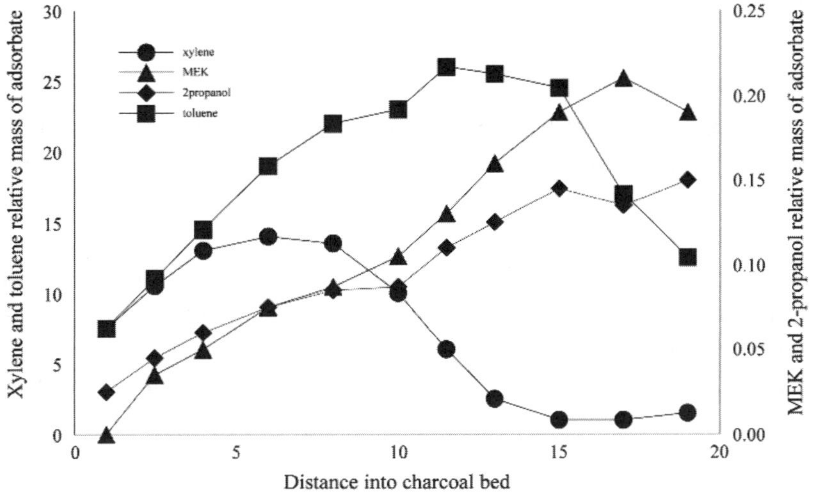

Figure 21-10. Observed mass of four adsorbed analytes from paint spray versus distance in a respirator charcoal bed: a) toluene, b) xylene, c) MEK, d) 2-propanol (adapted from Brown, 2001, with permission).

concentration. Occasionally this is not done. Instead, if $W_B/W_F < 10\%$ or some other criterion, then the laboratory may only report that no perceptible breakthrough occurred. This precludes the ability to sum W_F and W_B.

The above approach is strictly retrospective. How can a hygienist estimate how long to sample and at what rate? The factors affecting the slope of the curves in Figure 21-6 and their inflection points are related in a relatively complex way. Models have been developed (see Appendix 21-1) that give the relationships of the variables and allow prediction of one variable when all but one of the starting values of inlet concentration (C_i), flow rate (Q), and time (t_b) to reach an outlet concentration of C_o (such as 5% C_i) are defined.

Although it is possible (with difficulty) to apply the models to the question of how long one can sample at a given rate, NIOSH and others have taken another approach (Hull, 1984; Harper, 1993). NIOSH defines a breakthrough volume in its analytical methods for relatively worst case conditions. Breakthrough volume (V_b) is the maximum volume that can be sampled, when RH = 0% and $C_i = 2 \times$ the occupational exposure limit (OEL), so that C_o does not exceed 5%. In addition, a safety factor of 0.67 is applied to this laboratory-derived V_b (NIOSH, 1994).

Example 21-2. The breakthrough volume for 1,2-dichloropropane on petroleum charcoal is 6 L. At what flow rate (Q, not to be confused with the constant Q in Equation 21-2) should one sample in order to obtain an 8-hour (t) sample without breakthrough, at a concentration around the OEL?

$Q = V_b/t$

$Q = 6 \text{ L}/480 \text{ min}$

$Q = 12.5 \text{ cm}^3/\text{ min}$

NIOSH determines the V_b at a relatively high flow rate. In the above example, flow rates for determining V_b were 70 to 170 cm³/min. Because a Q of 12.5 cm³/min was calculated, smaller than those values used to determine V_b, this should be low enough not to cause breakthrough.

In the above example, it is implied that V_b is constant with Q. In other words, if we want to sample at a higher flow rate, can we just decrease the sample time, and V_b remains constant? Table 21-3 shows that the assumption that V_b is constant with Q is not necessarily correct. (This is also shown in Appendix 21-1.) The results in Table 21-3 show that any attempt to explain

Table 21-3
Effect of increasing Q on activated charcoal V_b

Author(s)	Compound(s)	Q (Lpm)	V_b
Gideon, 1984	n,t-butylchloride t-butylbenzene	0.2 – 1	Increase[a]
Gideon, 1984	ethylbenzene	0.2 – 1	Decrease[a]
Gideon, 1984	s,n,i-butylbenzene s,i-butylchloride m-xylene n-propylbenzene toluene		Constant[a]
Easley and Suggs, 1985	hexane	0.3 – 0.7	Constant
Fraust and Herman, 1969	acetates	0.1 – 1	Constant
Nelson and Harden, 1972	toluene	0.75 – 3.8[b]	Constant
Melcher et al., 1978	vinyl chloride	0.02 – 0.2	Decrease
Hill et al., 1976	vinyl chloride	0.05 – 1	Decrease
Severs and Skory, 1975	methylene chloride	0.1 – 0.5	Decrease
Foerst, 1979	vinylidene chloride	0.2 – 1	Increase[c]
Levine and Schneider, 1982	benzene	0.25 – 5	Increase

[a] Effect of Q was measured against adsorption capacity and then determined from the Langmuir isotherm.

[b] Extrapolated to charcoal tubes, actually used 75 mm diameter respirator cartridge at 14-71 Lpm.

[c] Experiments done at 85-95% relative humidity.

why V_b increases (with Q) for some compounds must be compatible with V_b decreasing or remaining constant for other compounds. NIOSH (1994) typically recommends 0.01 to 0.2 Lpm as the flow rate for methods employing charcoal tubes. However, Table 21-3 does not seem to justify either end of this range. If V_b is insensitive to Q or increases as Q increases, then there should be no upper limit on Q. On the other hand, if V_b decreases with increasing Q, there may be an important lower limit on Q.

A problem related to Q is diffusion bias at low flow rates. Heitbrink (1983) showed that at very low flow rates (0.001 Lpm) charcoal tubes tended to oversample compared to the same tubes at Q of 0.02 L/min. He showed that this could be caused by diffusion of the analyte into the tube. Very low flow rates may be attempted in trying to extend the sample time, particularly with solid sorbents having low V_b (e.g., some synthetic polymers, which are discussed later). For these reasons, it probably is wise to stay within the NIOSH recommended limits for Q of 0.01 to 0.2 L/min, unless reference to data similar to those in Table 21-3 can be made.

CONCENTRATION

V_b is also a complex function of C_i (as shown in Appendix 21-1). We would expect V_b generally to decrease as C_i increases for the following reasons. At a constant flow rate, as C_i increases, the t_b should decrease. $Q \times t_b$ equals V_b; so V_b should also decrease. The consequence is relatively clear; at higher ambient concentrations we must sample for shorter periods of time to avoid breakthrough, if Q is constant.

However, the adsorption capacity will increase with C_i. Therefore, the relationship between C_i and t_b is favorable to the hygienist. In other words, an increase in C_i, say, twofold, does not decrease t_b by 2, but by something less than 2. The consequence of this is that a small error in estimating the concentration of a contaminant for purposes of calculating Q and t (flow rate and how long to sample) should not invalidate that estimate.

TEMPERATURE

Temperature is a thermodynamic property. As molecules are heated, they move faster and are, therefore, more likely to escape an adsorptive attraction. This depends on the polarity or the strength of the attraction. Nelson et al. (1976) calculated that a 10°C increase in temperature would cause a 1% to 10% decrease in t_b, not a particularly large amount. Most charcoal sampling methods are validated at laboratory temperatures of 20°C to 25°C. Temperature has the greatest effect on the more weakly absorbed molecules.

MANUFACTURER OF CHARCOAL

Several studies have shown that the variability in properties (i.e., V_b, DE) with different manufacturers of coconut charcoal is considerable (Nelson and Harder, 1972; Saalwaechter et al., 1977; Rudling and Bjorkholm, 1986). In the validation of OSHA and NIOSH methods from which V_b is calculated, the manufacturer of the charcoal is not consistently reported. Hence, the effective V_b for the user could be different from that reported by NIOSH, depending on the charcoal used.

COMPETITION FOR SORPTION SITES

It was mentioned previously that charcoal has both polar and nonpolar sites. Superficially then it would seem that polar compounds and non-polar compounds would have their *places*, and no competition would occur. Although there certainly are non-polar compounds such as hexane and very polar compounds such as water, many compounds are somewhere in between. Furthermore, in the case of water, the ambient concentration relative to that of organic contaminants is very high. For example, at 80% RH and 25°C the water concentration is around 30,000 mg/m³. Typical organic vapor concentrations to which workers are exposed are < 1,000 mg/m³.

Thus, water will simply displace many polar compounds at high humidities (Hill et al., 1976; Foerst, 1979) and will limit the adsorption of non-polar compounds at very high relative humidities (> 80%) (Nelson et al., 1976; Saalwaechter et al., 1977). Nelson et al. showed that when RH was 80 to 90%, t_b was reduced relative to t_b at 20% or 50% RH for benzene, 1-chlorobutane, ethyl acetate, carbon tetrachloride, and hexane. These are all relatively non-polar compounds. At low relative humidities, water vapor does not adsorb to the same sites as organic compounds. However, at higher humidities, water absorbs to itself by hydrogen bonding, forming *lakes* of water that eventually coalesce, blocking access for hydrophobic molecules such as those above.

Water is not the only polar compound in the work atmosphere; moreover, many of the others are important. Goller (1985) showed that when an atmosphere of acetone, isopropanol, ethanol, toluene, and xylenes was sampled, using two charcoal tubes in series at flow rates from 50 to 150 cc/min, both polar and non-polar (toluene and xylene) compounds were lost, as shown in Table 21-4. The total volume sampled was about 20 to 60 L; so sample times were around 4 hours. For an average molecular weight of 100, the average total concentration is in the neighborhood of 60 ppm, not particularly high. It can be seen that under these very reasonable sampling parameters, polar compounds and, in some cases, non-polar compounds escaped to the backup tube.

However, the more important results are shown in Table 21-5, also from Goller's (1985) work. (First, remember there were two tubes in series, each with a front and back section.) It shows that the NIOSH 10% rule (W_B/W_F) may need further examination. Because $N = 100$, the values in Table 21-5 can be read as percent. Therefore, 42% of samples having > 10% of the front section mass found in the back section of the first tube also showed the compound in the second tube. This would be expected and indicates breakthrough as defined by NIOSH (1984). However, in addition, when $W_B/W_F > 10\%$, some 20% of tubes showed no breakthrough to the second tube. These would be false positives. Invalidation of these tubes by the 10% rule would

Table 21-4
Percentage of total mass of polar or non-polar compounds collected on front tube under varying loading conditions when a mixture of five compounds (see text) was sampled (Goller, 1985)

Condition	% Collected	Sample Size
Total load < 8 mg	Polar, 88-100	6
Polar load > 6 mg	Polar, 62-82	10
Non-polar > 6 mg Polar < 4 mg	Non-polar, 80-100	4
Polar > 7 mg Non-polar > 7 mg	Polar, 58-72	5

Table 21-5
Validity of the NIOSH 10% Rule (derived from Goller, 1985)

		> 10% in back section of first tube? Yes	No
Compound(s) detected in second tube?	Yes	42	4
	No	20	34

represent an unnecessary loss of data. In 4% of cases, $W_B/W_F < 10\%$, and in these four tubes breakthrough occurred to the second tube. These would be false negatives. Overall, in 76% (42% + 34%) of cases the 10% rule did prove effective. However, in 80% (20/24) of the tubes indicating breakthrough, no breakthrough occurred. The NIOSH rule did tend to err on the safe side in that there were 20% false positives (tube thrown out when it should not have been) and only 4% false negatives. One solution, in sampling concentrations of mixed polar compounds (> 50 ppm), would be to use jumbo (400/200 mg) charcoal tubes.

STORAGE

Typically, after air samples are collected in the field, there must be some period of time during which the samples are stored. It is rare that the industrial hygienist is able to take the sample directly to a chemist who will immediately perform the analysis. Even if the samples were taken to the private lab of the hygienist's employer or placed directly into overnight mail and analyzed by the receiving laboratory immediately, at least one to two days would elapse before the samples would be analyzed. More likely, perhaps a week or more would pass prior to analysis. During a long research project covering several months samples may be accumulated so that the analyses can be performed in one batch. What happens to the analytes during storage? It has already been shown that storage of some ketones and other reactive compounds on charcoal will lead to reaction losses. Other questions may arise.

The ends of charcoal tubes are heat-sealed by the manufacturer. Once sampling is completed, the broken ends are covered with small polyethylene caps. Several authors have noted loss or gain of contaminants through end caps. Teass et al. (1976) showed that approximately 3% of adsorbed dichloromethane escaped through the end caps of charcoal tubes after a room temperature storage period of 60 days.

Also, it has been shown that charcoal tubes can take up substantial amounts of toluene through the plastic end caps during storage (Dharmarajan and Smith, 1981). Charcoal tube ends that were broken and then capped were placed in a toluene-containing atmosphere along with two other sets of tubes that were broken and uncapped, and unbroken, respectively. The rise in toluene concentration in the capped tubes was dramatic. After a period of 48 hours the concentration in the capped tubes was equivalent to that in the uncapped tubes; both had reached saturation capacity. Although the ambient toluene concentration was not stated, the experiment still makes a point: end caps are permeable.

When results from analysis of charcoal tubes indicate $W_B/W_F = 50\%$, this may indicate that significant breakthrough has occurred. However, it can also indicate that migration has taken place during storage, and that the analyte has simply reached equilibrium throughout the charcoal. If a sizable proportion of the tubes in an analysis batch have the 0.5 ratio, then it is likely that migration has occurred during storage. On the other hand, if only a small portion of the tubes have such a result, then it could be that breakthrough has occurred only in those tubes. Complete migration of dichloromethane upon room temperature storage of tubes for a period of 20 days has been shown (Teass et al., 1976; Harper et al., 1994). Migration effects also have been shown for tubes loaded with dichloromethane and refrigerated during storage (Saalwaechter et al., 1977). There is no doubt that migration increases with the vapor pressure of the compound; refrigeration reduces volatility, and less volatile compounds are less likely to migrate. Migration has also been observed with acetone and diethyl ether, but not too many other compounds.[10]

Several conclusions can be drawn from this small amount of research on the effects of storage:

a. Laboratories should store exposed charcoal tubes under refrigeration.

b. Highly volatile compounds should be analyzed within a few days of collection.

c. Tubes containing extremely volatile compounds should be shipped to the labora-

---

[10] Personal communication, M. Harper, NIOSH, 2006.

tory on dry ice. Although a more precise definition of *extremely volatile* needs to be found, perhaps the definition *compounds with vapor pressures > 200 mmHg at 25°C* would be a good starting place.

d. During the sampling period, field blanks (broken tubes with capped ends) should not be stored in contaminated areas. However, after sampling, blanks and samples should be stored together.

SILICA GEL

Until the application of GC to analysis of workplace atmospheres and the development of the NIOSH standard charcoal tube, silica gel and charcoal were perhaps evenly used for the bulk analysis (gravimetric) of organics and later the specific analysis of compounds using wet chemistry methods (see Moskowitz and Burke, 1938; Cralley et al., 1943; Pernell, 1944). Later silica gel was touted equally with charcoal as a collection medium for compounds analyzed by GC (Altshuller et al., 1962; Whitman and Johnston, 1964; Van Mourik, 1965; Campbell and Ide, 1966). Silica gel is less reactive and easier to desorb with heat than charcoal and so was recommended by these authors.

Silica gel is an amorphous (noncrystalline) silicate prepared by *gelling* or allowing to dry a solution of sodium silicate and sulfuric acid. Washing leaves > 95% silicon, oxygen, and hydrogen. Surfaces are thought to consist of silanols (Si-O-H), which explains the polar nature of silica gel (Willey, 1985; Lewis, 1996). Since it is amorphous, it does not have the detrimental toxicological effects on the lungs of crystalline silica.

In early uses, silica gel tubes were usually preceded by drying tubes containing magnesium perchlorate (Cralley et al., 1943), Ascarite[11] (Moskowitz and Burke, 1938), or the 5A molecule sieve (Whitman and Johnston, 1964). In later studies, the drying agents were replaced by multiple-section tubes of silica gel, which could be used to detect breakthrough caused by displacement of analytes by water vapor.

Current methods employing silica gel are for polar compounds easily displaced by water; methanol is the most common. In some cases, large silica gel beds are employed (e.g., 260/520 mg). Table 21-6 shows that most compounds collected with silica gel contain nitrogen (amines and nitro compounds).

Apparently no systematic (or otherwise) study has been undertaken with silica gel to determine its properties relative to sorption and desorption, such as desorption efficiency, storage, adsorption capacity, competition, and so on, no doubt because of its use as a specialty sorbent relative to charcoal. Because of its affinity for water, care should be used in extending the recommended sampling time in a given method, particularly in relative humidities > 50%.

---

[11] Today Ascarite is sodium hydroxide coated silica. It rapidly and quantitatively adsorbs CO_2 and acid gases.

Table 21-6
Some sampling methods employing silica gel

Chemical Name	Method
n,n-dimethylacetamide, 1,1,2,2-tetrabromomethane, o-toluidine, n,n-dimethylaniline, nitrobenzene, diethylamine	MDHS[1] 96
aromatic amines	NIOSH[2] 2002
1,1,2,2-tetrabromoethane, acetylene tetrabromide	NIOSH 2003
n,n-dimethylformamide (dmf), dimethyl acetamide	NIOSH 2004
nitroaromatic compounds	NIOSH 2005
aliphatic amines	NIOSH 2010
phenyl ether – biphenyl mix	NIOSH 2013
p-chlorophenol	NIOSH 2014
2-butanone (methyl ethyl ketone)	OSHA[3] 16
n-methyl dicyclohexylamine, 2-(2,4-dinitrophenoxy) ethanol, n-ethylmorpholine, mercaptoethanol, p-nitrochlorobenzene, o-chlorophenol, nitrotoluene (m-isomer), diethyl sulfate, dimethylaminobenzene, dimethyl acetamide, formamide, o-chloroaniline, 2-amino-2-methylpropanol, xylidine	OSHA CSI[4]
2,2-dichloropropionic acid, trichloroacetic acid	OSHA[3] PV2017
acetamide	OSHA[3] PV2084
p-chloroaniline	OSHA[3] PV2109

[1] http://www.hse.gov.uk/pubns/mdhs/index.htm
[2] http://www.cdc.gov/niosh/nmam/
[3] http://www.skcinc.com/nioshdbs/oshameth/oshameth.htm
[4] http://www.osha.gov/dts/chemicalsampling/toc/toc_chemsamp.html

OTHER SOLID SORBENTS

The remaining solid sorbents can be placed in two groups: inorganic ones, which have minor usage relative to silica gel, and organic polymers, which are receiving increased attention. All of the sorbents discussed below have been used as packing materials for the solid phase of gas chromatography columns, and in many cases their sorbent properties were acknowledged prior to their use in air sampling. Also, some have uses in industry as desiccants, molecular sieves, or other means of separating or purifying product streams. A good overview of their usefulness and applicability (particularly the organic ones) can be found in Stanetzek et al. (1996) and Vahdat et al. (1995).

Inorganic

Table 21-7 shows six inorganic solid sorbents and the limited number of methods based on them. Because they are used occasionally, a description of each is important. Alumina is an activated, porous, granular form of aluminum oxide used as a desiccant in industry. Firebrick is ground organic refractory used to line kilns, ovens, boilers, and so forth. It is an inorganic substance based on silica, dolomite, or alumina. Florosil is activated magnesium silicate. Molecular sieves belong to the group of minerals known as zeolites. Their crystalline structure makes for very uniform pore sizes (30-100 nanometers), giving them the ability to easily separate small molecules from large ones. They are widely used to remove water and other small contaminants from product streams, to soften water, and to separate hydrocarbons. Each of these four sorbents is widely used in industrial wastewater applications.

Hopcalite is a catalyst based on copper, manganese oxides, and perhaps cobalt and silver oxides; it is widely used to convert carbon monoxide to the dioxide in respirator cartridges and for cleaning bottled and compressed breathing air. Lime is calcium oxide. All six of the definitions are from Lewis (1996).

Organic

The organic synthetic solid sorbents are polymers in the form of spheres, coalesced spheres, or amorphous masses. There are several important series including the Amberlite® XAD series (Rohm and Haas, 1978, now licensed by Supelco), the Chromosorb® Century series (Johns-Manville), the Porapak series (Waters), Tenax (Enka, NV) and HayeSep series (Hayes Separations). HayeSep D is an important sorbent for chemical warfare agent monitoring. An important polymeric sorbent that is a single piece rather than spheres, coalesced spheres, or amorphous masses is polyurethane foam. Then there are carbon based sorbents that involve further processing than the more common charcoal sorbent. Anasorb 747 can be

Table 21-7
Chemicals with sampling methods based on miscellaneous inorganic sorbents

Inorganic sorbent	Chemical(s)
Alumina	Amines
Florisil	PCBs
Firebrick	Hydrazine
Hopcalite	Mercury
Lime	Hydrogen cyanide
Treated Molecular Sieve	Nitrogen dioxide

described as a petroleum-based charcoal. Anasorb CMS is a carbon molecular sieve. There are a number of these (Spherocarb, Carboxen 1000, Carboxen 1004, Carbopack V, Carbopack X, etc.) and one or another is found in almost all multiple sorbent-thermal desorption tubes. See Table 21-8 for a listing of organic solid sorbents. ASTM (2003) is an excellent application document and is recommended for anyone considering organic solid sorbents other than charcoal.

There are several advantages of using the polymeric sorbents over charcoal-based sorbents (McCammon, 1984). 1) They can be desorbed efficiently, even at small loadings; thus low environmental concentrations can be monitored (Dietrich et al., 1978; Dressler, 1979) (particularly when using thermal desorption) (ASTM, 2003). 2) They tend to be very consistent from lot to lot. Although surface area tends not to be so large as that of activated charcoal, many have moderate to large surface areas. 3) They are not generally susceptible to the effects of high relative humidity although the more polar ones can be affected. On the other hand, some (Porapak *Q*) can be affected by interactions with reactive gases, but this can be used to an advantage in developing methods for reactive gases such as NO_2 (Gruber et al., 1972). Nothing seems poised to replace the charcoal-based sorbents as inexpensive, all-around sorbents for general application to air sampling of organics, particularly mixtures of organics; but synthetic porous polymer use will increase as uses for more exotic chemicals in the workplace are discovered and implemented.

Although no single polymer-based sorbent can take the place of charcoal, Tables 21-9 and 21-10 make it clear that XAD-2 and other sorbents based on styrene-divinyl benzene copolymer and similar non-polar compounds have widespread application. In several cases, XAD-2 and other polymers have been coated and, therefore, serve as a substrate for a reaction. For example, XAD-2 coated with 2-hydroxymethyl piperidine is used to sample aldehydes. The aldehydes react with the coating, and the product is analyzed.

Table 21-11 shows the application of polar polymeric sorbents to air sampling methods. SKC Inc. now markets several additional solid sorbents. Their characteristics are shown in Table 21-12. These sorbents have been used to fill some very special needs, such as sampling reactive ketones.

Polyurethane foam (PUF) is receiving wide use for adsorbing a range of heavier organics. These include pesticides and PCBs. EPA (U.S. EPA Method TO-4) uses a foam packed tube to monitor for and differentiate between the vapor state of semi-volatile organics and the solid state.

Table 21-8
A description of organic solid sorbent types (adapted from ASTM, 2003)

Type	Composition	Example	Manufacturer
B	carbon molecular sieve	Carboxen 569	Supelco Inc., USA
I	carbon molecular sieve	Unicarb, formerly Spherocarb	supplied by Markes International, UK
L	carbon molecular sieve	Carbosieve S-III	Supelco Inc., USA
M	carbon molecular sieve	Carboxen 1000	Supelco Inc., USA
R	carbon molecular sieve	Anasorb CMS	SKC Inc., USA
E	ethylvinylbenzene/ divinyl benzene	Porapak Q	Waters Associates Inc., USA
C	graphitized poly(diphenyl-p-phenylene oxide)	Tenax GR	Enka Research Institute NV, NL
F	graphitized carbon	Carbopack B	Supelco Inc., USA
G	graphitized carbon	Carbotrap	Supelco Inc., USA
J	graphitized carbon	Carbotrap C	Supelco Inc., USA
K	graphitized carbon	Carbopack C	Supelco Inc., USA
P	graphitized carbon	Anasorb GCB1	SKC Inc., USA
Q	graphitized carbon	Anasorb GCB2	SKC Inc., USA
S	graphitized carbon	Carbograph 1	Carbochimica Romana, I
T	graphitized carbon	Carbograph 4	Carbochimica Romana, I
U	intermediate composition between graphitized carbon and carbon molecular sieve	Carbopack X	Supelco Inc., USA
D	poly(diphenyl-p-phenylene oxide)	Tenax TA	Enka Research Institute NV, NL
A	polystyrene	Chromosorb 106	Manville Corp., USA
O	styrene/divinylbenzene	Chromosorb 102	Manville Corp., USA
H	vinylpyrrolidone	Porapak N	Waters Associates Inc., USA

Table 21-9
Chemicals with NIOSH or OSHA sampling methods based on styrene-divinyl benzene co-polymer

XAD – 2	***Coated SAD – 2***
Adipic Acid	Acrolein
Anisaldehyde	Aldehydes, formaldehyde
Anisidine(s)	Amino alcohols
Azo-bis formamide	Amines, 1°, 2°, 3°, ethyl and methyl
Benzofuran	Cyanogen
Benzothiazole in asphalt fume	Cyanogen chloride
4,4'-Bipyridine	Cyanamide
2- Chloro-6-trichloromethyl pyridine	Diazomethane
Coumarins	Maleic anhydride
2,5-Dimethoxyaniline	Phosgene
Dinitrobenzene	Piperazine
Dodine	***Porapak P***
5-Fluorouracil	Diethyl carbamoyl chloride
Fungicides	Dimethyl sulfate
Herbicides	***XAD – 4***
Hexachlorobutadiene	Benzyl alcohol
2-Hydroxy-4-methoxyacetophenone	Butyl isocyanate
Insecticides	Butylphenol
Maleic anhydride	Carbon tetrabromide
Menadione	Catechol (pyrocatechol)
Methacrylate, ethyl and methyl	Chloroacetyl chloride
6-Methoxytetralone	Chloropicrin
3-Methyl-2-cyclopentene-2-ol-one	Cresol
Methyl cyclopentadienyl Mn tricarbonyl (as Mn)	-Diol acrylates
Methyl and ethyl silicate	1-Dimethylamino ethanol and propanol
Methyltetrahydrophthalic anhydride	4,6-Dinitro-o-sec-butyl phenol
4-Nitrodiphenyl	Diphenyl ether
Nicotine	3-Ethenylpyridine and nicotine
Nitroethane	Iodoform
Organotin compounds	Isophorone
Piperonyl butoxide	Methoxyphenol
Poly nuclear aromatic hydrocarbons	Methyl ethyl ketone peroxide
Polychloro diphenyls, benzenes, naphthalenes	Methyl isocyanate
Quinone	Nicotine
Tetraethyl lead	Nitropropane
Vinyl cyclohexene dioxide	Phenol, Nonylphenol
Chromosorb 102	Phenyl ether
Chlordane	Tetraethyl tin
Tetraethyl pyrophosphate	Trichloronitromethane
	Chromosorb 101
	Bis-chloromethyl ether

Table 21-10
NIOSH or OSHA sampling methods based on other non-polar solid polymeric sorbents

Porapak Q (Ethyl Styrene – Divinyl Benzene Copolymer)	***Chromosorb 106 (Polystyrene)***
Benzoic acid	Indoor VOC (EPA TO-17)
Dimethyl sulfate	Dipropyl disulfide
Furfuryl alcohol	Indene
Methyl cyclohexanone	Naphthalene
Methyl methacrylate	Chloroprene
Anasorb 727 (Polystyrene)	Nitromethane and propane
Methyl methacrylate	Allyl propyl disulfide
	Cyclohexanone
	Di(ethyleneglycol)Ethyl ether acrylate
	1,6-Hexanediol diacrylate
	Isobutyric acid
	Tetrahydrofurfuryl acrylate
	2,2,4-Trimethyl-1,3-pentanediol diisobutyrate

QUESTIONS

21.1. What were the problems associated with using charcoal for vapor sampling prior to 1964?

21.2. As silica gel is not reactive, why did it not assume the place of charcoal in modern hygiene work?

21.3. Why must desorption efficiency studies be performed at several loadings?

21.4. Typically, laboratories do not perform DE studies on mixtures. How may this omission affect reported airborne concentration results?

21.5. Name the four most important factors likely to affect breakthrough volume for a particular compound on a given manufacturer's charcoal. Which of these will typically be the easiest to control or predict?

21.6. The following compounds have been collected separately on charcoal and must be sent overnight to the laboratory. Which, if any, will you ship in a cooled atmosphere? Justify your reasoning.

a. ethylene oxide

Table 21-11
Sampling methods based on polar polymeric sorbents

Tenax (Poly (2,6-diphenyl-p-phenylene oxide)	***XAD – 7 (Acrylate ester polymer)***
Indoor VOC (EPA TO-17)	Amino alcohols
α-Acetophenone	Cresol(s)
Allyl glycidyl ether	Butylated hydroxytoluene
Amino pyridine	Caprolactam
Anhydrides	Catechol
Benzotrichloride	Chloroacetyl chloride
Biphenyl	Glycols
Chlorophene	Hexamethylenetetramine (HMTA)
2,3-Dimethyl-2,3-dinitrobutane	Isocyanate, butyl and methyl
Dinitrotoluene	Chlorophenols
Epoxyethylbenzene	Polyfunctional aziridine
Glycol di and trinitrates	Resorcinol
Ethylene dibromide	Tergitol NP-33
Octanethiol	***Coated XAD – 7***
Phenylphenol	Aliphatic amines
Phosphorous	n-Aminoethylpiperazine
Phthalates	Aniline(s)
Pindone	Chlorophenols
Sodium o-phenyl phenate	Cyanoacrylate, methyl and ethyl
Styrene oxide	Hydroquinone
o-Toluidine	Morpholine(s)
Chromosorb 104 (Acrylonitrile/divinyl benzene copolymer)	***Chromosorb 108 (Acrylate polymer)***
Butyl mercaptan	Chloronitropropane
Chlordane	***Porapak R (Vinyl pyrrolidone)***
Endrin	Pentachloroethane
Tetraethyl pyrophosphate	***Porapak T (Polyethylene glycol dimethyl-acrylate)***
	Hexachlorocyclopentadiene

Table 21-12
Proprietary SKC solid sorbents

Name	Description	Characteristics
Anasorb^R 727	Polystyrene-based	• Very non-reactive, ketones stable • Compounds with BP > 200C not adsorbed • H_2O will not displace polar compounds
Anasorb^R 747	Proprietary activated carbon beads	• Similar to charcoal • OSHA and NIOSH methods for wide range of compounds
Anasorb^R CMS carbon molecular sieve	Beaded activated carbon with micropores	• Adsorbs low molecular weight compounds – CH_3C1, CH_2C1; acetone, MEK stable • EPA method TO-2

b. toluene

c. methylene chloride

d. butyl aldehyde

21.7. Make a list of the variables that affect the adsorption of organics on charcoal as given by the Wheeler and Dubinin equations. Pair each variable with either or both equations. Are there important differences? Can you explain this? (Optional, based on Appendix 21-1.)

APPENDIX 21-1. ADSORPTION THEORY

Introduction

If one exposes charcoal (or any other solid sorbent) to varying concentrations of an airborne chemical, a plot of the amount adsorbed per unit weight of adsorbent (the concentration on the sorbent) versus the concentration in the solvent (in this case air) can be made (see Figure 21-A1). As can be seen, the relationship is not linear throughout the range of concentrations. At low airborne concentrations, the adsorption is proportionally lower. At higher airborne concentrations, it approaches a maximum value: the capacity of a charcoal for the chemical or adsorbate.

It would be convenient if we could predict the extent of adsorption at varying concentrations. One could fit a relationship to the data mathematically, or one could develop a physical-chemical model that tried to explain what is happening and allowed prediction of the dependent variable (amount adsorbed) when given an independent variable(s).

It is known that the airborne concentration of a chemical is proportional to the partial pressure it exerts, which is dependent on temperature. Also the amount adsorbed to carbon from the air is affected by temperature. Consequently, curves such as those shown in Figure 21-A1 are derived at one temperature and are known as isotherms. By developing experiments at constant temperature, temperature is controlled in the model.

In addition to adsorbing a chemical from air, we are interested in desorbing the chemical from the sorbent so that it can be analyzed. For charcoal, carbon disulfide is often used for this. The CS_2 is adsorbed by the charcoal in this process, displacing the analyte, and the analyte

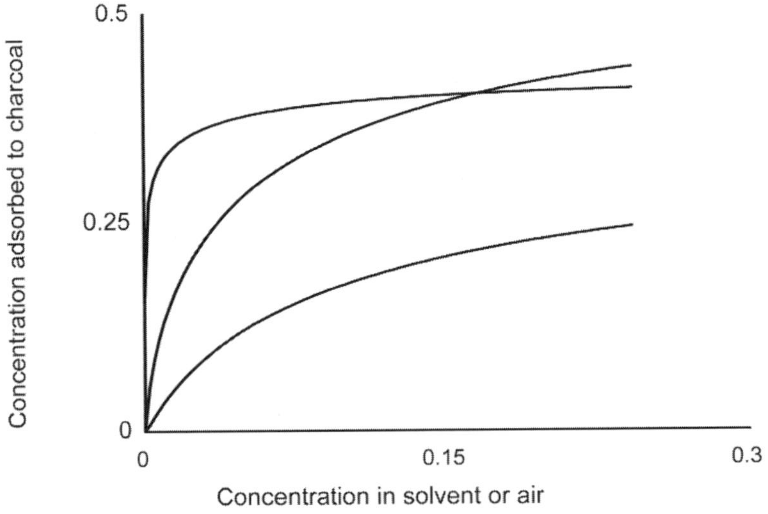

Figure 21-A1. Adsorption isotherms for three different chemicals.

establishes an equilibrium between charcoal and CS_2. The CS_2 is a good desorbing solvent because it is a small and relatively non-polar molecule. Therefore, it fits into the micropores of the charcoal and has a high affinity for it, leading to its displacement of other sorbed species. However, when the other sorbed species are in low concentration on the charcoal, a smaller proportion is displaced into the solvent than at high sorbed species concentrations. So there is also an effect of concentration in the case of liquid desorption, similar to gaseous adsorption. Because the amount of chemical that can be desorbed has an effect on the analysis result for an airborne concentration, it is important to be able to model liquid desorption as well.

It should be said that the two cases outlined above are static. They consider the amount adsorbed or desorbed at equilibrium conditions. This is fine in the case of desorption, but in the case of adsorption, air is moving through the sorbent tube at some velocity and thus conditions are not static. We should, therefore, consider what is happening *kinetically*.

Figures 21-7 and 21-8 show plots of the concentration measured downstream in a respirator charcoal cartridge versus time – kinetically. When significant quantities of chemical are detected downstream, breakthrough has occurred. One may define significant breakthrough as that for which the outlet concentration is 1, 5, or 10% of the inlet concentration. As with static adsorption it would be helpful if we could predict, for a set of conditions, when C_0 equals 1% or some other percentage of C_i.

It is known that the static capacity of the charcoal for the chemical is constant. However, in the case of kinetic adsorption, what is important is how that capacity is approached (the slope of the middle portion of the curve in Figure 21-7). The relative position along the x-axis for the curve is also important. The time to breakthrough (t_b) is dependent on initial concentration (C_i), speed through the charcoal bed (flow rate, Q is proportional to velocity for a fixed area), temperature (T), and competing species. We will also see that t_b is dependent on the diffusion rate of the adsorbing species and the grain size of the solid sorbent.

Static Adsorption-Isotherms

The simplest model of adsorption capacity is the Freundlich isotherm (Freundlich, 1926).

$$C_s = AC^{1/n} \quad \text{[Equation 21-A1]}$$

where C_s is the concentration in the sorbent, C is the concentration in the liquid or gas (solvent), and A and n are constants. This equation is basically an attempt at curve fitting. It remains a bit more convex to the concentration axis than those curves in Figure 21-A1 and does not show an upper limiting value. However, a logarithmic transform of the equation allows a slope ($1/n$) to be found that can be interpreted as the intensity of adsorption. In addition, an intercept (lnA) can be interpreted as a measure of the adsorbent capacity (Adamson, 1990, p. 425).

Unlike the Freundlich, the Langmuir isotherm (Langmuir, 1918) is based on a physical model. It assumes that the chemical is adsorbed as a monolayer on the sorbent surface area. At equilibrium, molecules are evaporating from filled spaces at the same rate as that at which they are condensing on free spaces. The greater the concentration in the gas or liquid phase, the greater the tendency toward condensation. So at equilibrium:

$$K_e S_o = K_c C(S_t - S_o) \qquad \text{[Equation 21-A2]}$$

where C is the concentration in the solvent, K_e and K_c are evaporation and condensation rate constants, and S_t and S_o are total and occupied surface areas. The ratio S_o/S_t can be thought of as the mass or number of moles adsorbed relative to the maximum number of moles that can be adsorbed. This is equivalent to C_s/C_{sm}, where C_s is the concentration on the sorbent and C_{sm} is the maximum concentration on the sorbent. Therefore, Equation 21-A2 can be rearranged as follows with the additional substitution of b for the ratio of constants K_e/K_c:

$$C_s = \frac{C_{sm} bC}{1 + bc} \qquad \text{[Equation 21-A3]}$$

The maximum concentration on the sorbent (C_{sm}) is also the monolayer capacity. Equation 21-A3 can be further rearranged in a linear form to:

$$\frac{1}{C_s} = \frac{1}{C_{sm} bC} + \frac{1}{C_{sm}} \qquad \text{[Equation 21-A4]}$$

So, if the concentration on the sorbent (C_s, or the amount that leaves the gaseous or liquid phase) is plotted as a function of the concentration in the gaseous or liquid phase (C), then the slope is $1/(C_{sm}\,b)$ and the intercept $1/C_{sm}$, and so both C_{sm} and b can be calculated.

Another theoretical approach is that of Dubinin and coworkers (Dubinin, 1960). Dubinin adopted earlier work of Polanyi (around 1914; see Dubinin, 1975), which stated that there is a *potential* field at the surface of a solid that attracts adsorbates to the surface. In micropores, which have diameters just larger than those of a vapor (6-7 angstroms) and account for greater than 95% of the adsorption on carbon (Dubinin, 1975), those force fields of attraction are significant; because the molecules come close to the surface. In larger macropores, molecules would not necessarily come close to the surface, and so the force fields would not be important. However, there are relatively few of the macropores; so they are disregarded.

Thus, the number of molecules adsorbed is dependent on the number of sorption sites or the volume of micropores and the adsorption potential. From thermodynamics, the adsorption potential for an ideal gas may be written as:

$$\Delta G^\circ = RT \ln (P_0/P) \qquad \text{[Equation 21-A5]}$$

where ΔG° is the amount of work per mole adsorbed (also known as the Gibbs free energy), R is the ideal gas constant, T is the absolute temperature, P_0 is the saturated vapor pressure or maximum concentration at that temperature, and P is the partial pressure exerted by the vapor or the concentration of the vapor.

Dubinin noted that the relationship between the amount adsorbed and the heat of adsorption ($\Delta G°$) took the shape of curves in Figure 21-6, somewhat similar to the negative half of the normal distribution. This function has the form:

$$Y = K \exp(-K'X^2) \qquad \text{[Equation 21-A6]}$$

In this case, the dependent variable (Y) is the amount of volume that becomes filled with the adsorbent (V, proportional to the concentration on the sorbent, C_s). The independent variable is $\Delta G°$. As the volume fills, it approaches a maximum adsorption volume (V_0) at any concentration; so at any one time the volume that has been filled is some portion of V_0, up to 100%. By substitution of Equation 21-A5, V, and V_0 into the general form in Equation 21-A6, we have:

$$V = V_0 \exp(-KR^2T^2\ln^2(P_0/P) \qquad \text{[Equation 21-A7]}$$

As can be seen, we have the exponent of a negative value or a fraction; so the occupied volume approaches the maximum volume as the temperature decreases or the concentration (partial pressure) increases.

One last step can be used to relate the adsorption potential for one substance to that of another. It would be useful to predict V for one compound from a known V for another compound based on the relative affinities (attraction potentials) of the two compounds for charcoal. If β is the affinity and a reference compound has an affinity of one, then:

$$V = V_0 \exp\left(\frac{-KR^2T^2}{\beta^2}\ln^2(P_0/P)\right) \qquad \text{[Equation 21-A8]}$$

Because adsorption space or volume is not determined but rather the amount (mass) adsorbed is determined, one must assume that the micropores are filled with condensed liquid. Indeed that is one of the bases of Dubinin theory. Thus, the volume or space filled can be converted to mass (M) if we know the density of the liquid:

$$M = V\rho$$

or

$$M_0 = V_0\rho \qquad \text{[Equation 21-A9]}$$

where ρ is the density of the liquid. Finally, then, a predictive equation based upon experimental data for the mass adsorbed (M), the partial pressure to which the sorbent is exposed, and the temperature can be written as follows:

$$\ln M = \ln M_0 - KR^2T^2\ln^2(P_0/P)/\beta^2 \qquad \text{[Equation 21-A10]}$$

So, for carbon exposed to various vapor concentrations at constant temperature, if $\ln M$ is plotted against $\ln^2 P_0/P$, then the slope is $-KR^2T^2/\beta^2$, and the intercept is $\ln M_o$.

The Kinetics of Adsorption

The kinetics of adsorption are shown in Figure 21-7, and this form has been constructed mathematically by using either empirical or theoretical means. The theoretical approaches are derived from Bohart and Adams (1920), Mecklenburg (1926, 1930), Klotz (1946), and Wheeler and Robell (1969). The work of Jonas and coworkers (Jonas and Rehrmann, 1972, 1973; Jonas et al., 1979, 1983, 1985; see also Reucroft et al., 1971, Sansone and Jonas, 1981, and Sansone et al., 1979) in modifying the Wheeler equation, explaining the dependence of its parameters on physical variables, and applying it to the adsorption process has been impressive. The modified Wheeler equation states that:

$$t_b = \frac{M_e W}{C_i Q} - \frac{\rho_\beta M_e}{C_i K_v} \ln\,(C_i/C_o) \qquad \text{[Equation 21-A11]}$$

where t_b is the time to breakthrough (min), C is concentration (g/cc) at the inlet (i) and outlet (o), M_e is the adsorption capacity (g sorbate/g sorbent), ρ_β is the bulk density of the packed bed of carbon (g/cm^3), Q is the volumetric flow rate through the bed (cm^3/min), W is the weight of the carbon, and K_v is a rate coefficient (min^{-1}). When t_b to reach a certain C_o is measured over varying weights (W) of charcoal, then t_b is plotted against W, the slope is M_e/C_iQ, and the intercept is $-(M_e\rho_\beta/K_vC_i)\ln\,(C_i/C_o)$ This allows the calculation of M_e and K_v.

Approaches other than the Wheeler approach have been tried. The shape of the curve in Figure 21-7 is that of a cumulative lognormal distribution function. Grubner and Burgess (1979, 1981) have used this feature in empirical curve fitting through the use of the theory of statistical moments. Also the effects of humidity have been considered in the Wheeler equation (Lodewyckx and Vansant 1999, 2000).

Yoon and Nelson (1990) and Yoon et al. (1996) have used a similar approach to derive an empirical equation that fits Figure 21-7. However, as Wood and Moyer (1989) have pointed out, it makes only a slight refinement in the Wheeler equation:

$$t_b = \frac{M_e W}{C_i Q} - \frac{\rho_\beta M_e}{C_i K_v} \ln\left(\frac{C_i - C_o}{C_o}\right) \qquad \text{[Equation 21-A12]}$$

The refinement can be seen in the last term, C_o (the outlet concentration) is subtracted from C_i, which allows the top portion of the curve in Figure 21-7 to be predicted. (In the Wheeler equation the top part of the S shape is not predicted.) However, we are generally only interested in what happens at low C_o, say, 1 % to 10% of C_i. When C_o equals 1% of C_i then the above correction is only a 0.2% difference. Thus, the Yoon and Nelson equation is a more nearly perfect, but perhaps unnecessary, refinement in most applications.

Applications

The application of Freundlich and Langmuir isotherm equations to desorption studies has been an important one. In the 1960s and early 1970s, concentrations in the workplace caused relatively high loadings to occur on charcoal tubes. Consequently, the amounts desorbed from the charcoal were relatively high; only the upper portion of the isotherm curves (Figure 21-A1) were seen, and the DE values appeared constant. Furthermore, the desorption efficiencies were usually near 100%. However, as lower concentrations were monitored in the workplace, it became obvious that, in some cases, desorption efficiency values were not constant over all concentrations.

Posner demonstrated a clever way to fit desorption data to a straight line using the Freundlich or Langmuir equations. As desorption efficiencies typically are calculated at only three or four concentrations, his approach allows easy prediction of desorption efficiencies at any concentration, as well as calculation of the associated experimental errors from a regression model. First, we can write desorption efficiency as:

$$D = \frac{C_L W_L}{C_L W_L + C_s W_s} \qquad \text{[Equation 21-A13]}$$

where C is concentration and W is weight for either the liquid or the solid phase. Each term in Equation 21-A13 has mass as units; so the desorption efficiency is the mass in the liquid divided by the total mass for both phases. Equation 21-A13 can be arranged as follows:

$$\frac{1-D}{D} = \frac{C_s W_s}{C_L W_L} \qquad \text{[Equation 21-A14]}$$

Equation 21-A14 can be substituted into the Langmuir equation (Equation 21-A4) to get:

$$\frac{D}{1-D} = \frac{W_L}{W_s C_{sm} b} + \frac{C_L W_L}{W_s C_{sm}} \qquad \text{[Equation 21-A15]}$$

So a plot of $D/1 - D$ versus C_L yields a linear relationship. By substitution into the Freundlich equation (Equation 21-Al), we can obtain:

$$\frac{1-D}{D} = AC_L^{1/n}\ (W_s/W_L)$$

or: [Equation 21-A16]

$$\ln\left(\frac{1-D}{D}\right) = \ln\left(\frac{AW_s}{W_L}\right)\ \frac{1}{n}\ln(C_L)$$

So, once again, a plot of $\ln((1 - D)/D)$ versus $\ln C_L$ gives a straight-line relationship.

Jonas and coworkers have been most instrumental in fitting both the Wheeler and the

Dubinin equations to a wide range of charcoals and various adsorbates. The basic approach is as follows. As can be seen from the Wheeler equation (Equation 21-A11), K_V, the rate constant, and M_e, the adsorption capacity of the charcoal for the vapor in question, can be calculated through experimentation. The object is to minimize (or eliminate) the amount of experimentation that must be performed; consequently, the goal of the work of Jonas et al. has been to estimate K_V through the use of equations. Consequently, a reference M_e is calculated and the chemical's β is set to one for the Dubinin equations to work. The process is as follows:

1. Using various bed weights of charcoal, calculate M_e and K_v for a reference compound.

2. Using the Lorenz-Lorentz equation, calculate the β, or affinity coefficient, for some other vapor. The Lorenz-Lorentz equation is:

$$P = \left(\frac{\eta^2 - 1}{\eta^2 + 2} \right) \frac{M_W}{\rho} \qquad \text{[Equation 21-A17]}$$

where η is the refractive index determined at the sodium D-line, ρ is the liquid density, M_w is the molecular weight of the liquid, and P is the electronic polarization. The ratio of the polarizations relates the affinity coefficient as follows:

$$\beta = P/P_{ref} \qquad \text{[Equation 21-A18]}$$

So the affinity of a compound is assumed to be related to its induced polarity, and one must know the refractive indices for the calculation. Other approaches such as parachors also have been tried, but were less successful than electronic polarization (Reucroft et al., 1971).

3. Determine the value of M_e for the non-reference compound using the Dubinin equation (Equation 21-A10) and the β value calculated above.

4. Calculate a value of K_v, the rate constant, for the new vapor. The rate constant expresses the migration of the adsorbate to the sorbent. It is thus dependent on Q (or, more specifically, on the superficial velocity across the carbon beads)[12], the diffusion rate of the adsorbate, the temperature, and the adsorbent grain size. This can be written as:

$$K_v = 10(V/M_w \, dP_T)^{0.5} \qquad \text{[Equation 21-A19]}$$

where V is the velocity, M_w is the molecular weight of the vapor (which is approximately proportional to the diffusion rate), d is the adsorbent particle diameter, and P_T is the total pressure. If V, d, and P_T are constant from one adsorbate to another, then:

[12] The flow rate through the charcoal bed divided by the bed area is the superficial velocity.

$$K_v M_w^{0.5} = (K_v M_w^{0.5})_{ref}$$ [Equation 21-A20]

Now, K_v, calculated above, and M_e, calculated from the Dubinin equation, can be substituted into the Wheeler equation for any vapor, and the breakthrough time calculated. Sansone et al. (1979) and Jonas et al. (1979) found small associated errors for K_v and M_e calculated in the above manner, as shown in Table 21-A1. A wide variety of vapors was included in the experiments.

When the Wheeler parameters (K_v and M_e) are experimentally determined, usually t_b is plotted against varying values of bed weight (W). Wood and Moyer (1989) derived two other forms of the equation for use with different experimental approaches. For example, if C_i is varied instead of bed weight, then:

$$\ln\left(\frac{C_i - C_o}{C_o}\right) = \frac{K_v C_o t_b}{M_e \rho_\beta} + \frac{K_v W}{Q \rho_\beta}$$ [Equation 21-A21]

can be used to find K_v and M_e. Likewise, a residence time approach (Ackley, 1985) can be used, in which Q is varied. The residence time T (sec) is:

$$T = V_s/Q$$ [Equation 21-A22]

where V_s is the volume of the sorbent. Because V_s is equal to sorbent weight divided by the bed density, we have:

Table 21-A1
Comparison of experimentally determined versus calculated adsorption parameters (from Sansone et al., 1979 and Jonas et al., 1979)

	Adsorption Capacity, M_e (g/g)			Adsorption Rate Constant, k_v (min^{-1})		
Vapor	expt	calc'd	% dev	expt	calc'd	% dev
CCL_4	0.741			735		
chloroform	0.728	0.693	+5.1	780	834	-6.5
benzene	0.404	0.409	-1.2	1029	1031	-0.2
ρ-dioxane	0.476	0.483	-1.4	1083	971	+11.5
sec-butylamine	0.331	0.337	-1.8	928	1066	-12.9
1,2-dichloroethane	0.616	0.583	+5.7	1048	916	+14.4
chlorobenzene	0.545	0.527	-3.4	799	859	-7.0
acrylonitrile	0.404	0.375	+7.7	1160	1251	-7.3

$$T = \frac{0.06\ W}{Q\ \rho_\beta}$$ [Equation 21-A23]

and:

$$t_b = \frac{M_e \rho_\beta}{C_i}\left(\frac{T}{0.06} - \frac{1}{K_v}\ln\left(\frac{C_i - C_o}{C_o}\right)\right)$$ [Equation 21-A24]

The 0.06 value is used to convert Q from L/min to cm³/sec. So, t_b is plotted against T. Note that Equations 21-A21 and 21-A24 are the Yoon and Nelson modification to the Wheeler equation.

The three approaches were used for charcoal challenged by acetone (Wood and Moyer, 1989). The results are shown in Table 21-A2. Not surprisingly, the adsorption capacities (M_e) are quite similar across the three methods and probably are within the experimental error. However, the K_v values are quite variable across the columns. This is so because, as shown in Equation 21-A19, K_v is dependent on velocity. This explains why the residence time approach (in which velocity is varied) gives differing values of K_v. If the values obtained by Wood and Moyer (Table 21-A2) were corrected for velocity according to Equation 21-A19, perhaps less variability in K_v would be observed. Finally, in the breakthrough curve-fitting approach, C_o is varied, which should also affect K_v; because K_v is dependent on diffusion rate, according to Fick's Law it would also be dependent on the concentration gradient from the gas stream to the charcoal surface. Equation 21-A19 does not account for the effect of concentration gradient on K_v. However, as C_i, the inlet concentration, increases, the gradient (C_i – concentration at surface) will be greater and, therefore, K_v will increase. This is shown in Table 21-A2.

In summary, much work has been done to theoretically and empirically explain the

Table 21-A2
Comparison of modified Wheeler parameters obtained by three approaches

	Adsorption Capacity, M_e (g/g)			Adsorption Rate Coefficient, k_v (min^{-1})		
Average Airflow Rate (L/min)	Bed Weight Variation	Residence Time Variation	Breakthrough Curve Fitting	Bed Weight Variation	Residence Time Variation	Breakthrough Curve Fitting
32.2	0.106	0.106	0.092	4730	5050	6580
40.4	0.110	0.108	0.107	4880	5920	6870
63.9	0.116	0.116	0.114	7940	7860	9160
83.0	0.115	0.113	0.109	7440	7860	10410
94.5	0.113	0.113	0.108	8670	8770	13090
112.5	0.109	0.110	0.107	9460	8730	12780

adsorption phenomenon, but much work remains. Obviously, K_v needs further elucidation, but even greater challenges remain for application of the equations to charcoal tube air sampling. Most of the application work cited above has been for respirator cartridge efficiency determination. In this case, a 10% breakthrough determination and the associated t_b are appropriate measures of efficiency. A similar criterion may be suitable for charcoal tubes, but some modification would be necessary. In the case of charcoal tubes, it might be acceptable that 1% of the combined mass contained on the front and back sections was lost. But how can this be related to a ratio of inlet and outlet concentrations? The relationship between breakthrough time at 1% and the ratio of weights in the two charcoal tube sections is dependent on the shape of the breakthrough curve. As we have seen, the latter is dependent on flow rate or superficial, velocity, temperature, molecular weight, inlet concentration, and competing species or other organics. Thus, if a relationship could be found to predict the time at which some unacceptable amount of the mass had been lost from the charcoal tube (say, 1% of the total mass), then the hygienist would know how long to sample to avoid breakthrough, and maximum breakthrough volumes V_b as predicted by NIOSH would not need to be used. In some cases, the V_b are oversimplifications that cause the hygienist to stop the sampling period when considerable capacity is still left in the charcoal tube. If a relationship between the front and back section weight ratios and breakthrough time could be found (or even if the limits could be expressed), the hygienist would know if an observed ratio meant significant loss from the tube. These ratios would vary for each analyte or group of analytes. Until then, we must rely on the NIOSH general 10% rule; though this discussion should indicate that the above rule is not necessarily applicable in every case.

REFERENCES

Ackley, M.W.: Residence time model for respirator sorbent beds. Am. Ind. Hyg. Assoc. J. 46:679-689 (1985).

Adamson, A.W.: Physical Chemistry of Surfaces, 5th ed. Wiley Publishing, New York (1990).

Aftalion, F.: A History of the International Chemical Industry. University of Pennsylvania Press, Philadelphia (1991).

Altshuller, A.P.; Bellar, T.A.; Clemons, C.A.: Concentration of hydrocarbons on silica gel prior to gas chromatographic analysis. Ind. Hyg. J. 46:164-166 (1962).

ASTM: Standard Practice for Sampling Atmospheres to Collect Organic Compound Vapors (Activated Charcoal Tube Adsorption Method), ASTM D3686-95, ASTM International, West Conshohocken, PA, USA (2001a).

ASTM: Standard Practice for Analysis of Organic Compound Vapors Collected by the Activated Charcoal Tube Adsorption Method, ASTM D3687-01, ASTM International, West Conshohocken, PA, USA (2001b).

ASTM: Standard Practice for Selection of Sorbents, Sampling, and Thermal Desorption Analysis Procedures for Volatile Organic Compounds in Air, D 6196 –03. ASTM International, West Conshohocken, PA, USA (2003).

Beck, S.W.; Stock, T.R.; Whitehead, L.W.: Improved efficiency and desorption of oxygenated solvents from activated charcoal using a new polar additive to carbon disulfide. App. Occ. Env. Hyg. 5:171-177 (1990).

Bohart, G.S.; Adams, E.Q.: Some aspects of the behavior of charcoal with respect to chlorine. J. Am. Chem. Soc. 42:523-544 (1920).

Brown, R.C.: Filtration in industrial hygiene. AIHAJ 62:633-643 (2001).

Brown, R.H.; Woebkenberg, M.L.: Gas and vapor sample collectors, In Air Sampling Instruments, 7th ed., S.V. Hering (tech. ed.). American Conference of Governmental Industrial Hygienists, Cincinnati, OH (1989).

Campbell, E.E.; Ide, H.M.: Air sampling and analysis with microcolumns of silica gel. Am. Ind. Hyg. Assoc. J. 47:323-331 (1966).

Cook, W.A.; Coleman, A.L.: Determination of injurious constituents in industrial atmospheres. II. Determination of solvent vapors in air by means of activated charcoal. J. Ind. Hyg. Tox. 18:194-210 (1936).

Cookson, J.T.: Adsorption mechanisms: The chemistry of organic adsorption on activated carbon, In Carbon Adsorption Handbook, P.N. Cheremisinoff, F. Ellerbusch (eds.). Ann Arbor Science, Ann Arbor, MI (1978).

Cralley, LV.; Shea, T.E.; Cralley, L.J.: A modification of the silica gel method for the determination of atmospheric organic solvent vapors. J. Ind. Hyg. Tox. 25:172-174 (1943).

Cropper, F.R.; Kaminsky, S.: Determination of toxic organic compounds in admixture in the atmosphere by gas chromatography. Anal. Chem. 35:735-743 (1963).

Danner, C.E.; Goldenson, J.: Determination of halogenated organic vapors in air by adsorption on specially prepared activated charcoal and preparation of charcoal. J. Ind. Hyg. Tox. 29:218-224 (1947).

Dean, J.A. (ed.): Lange's Handbook of Chemistry, 12th ed., Sieves and Screens. McGraw-Hill, New York (1979).

Dharmarajan, V.; Smith, R.N.: Permeation of toluene through plastic caps on charcoal tubes. Am. Ind. Hyg. Assoc. J. 42:691-693 (1981).

Dietrich, M.W.; Chapman, L.M.; Mierure, J.P.: Sampling for organic chemicals in workplace atmospheres with porous polymer beads. Am. Ind. Hyg. Assoc. J. 39:385-392 (1978).

Dommer, R.A.; Melcher, R.G.: Phase equilibrium method for determination of desorption techniques. Am. Ind. Hyg. Assoc. J. 39:240-246 (1978).

Drager: Activated Charcoal – Tubes. Pittsburgh, PA http://www.draeger.com/ST/internet/US/en/Products/Detection/Drager-Tubes/SamplingTubes/ActivatedCharcoal/pd_activated_charcoal.jsp (2006).

Dressler, M.: Extraction of trace amounts of organic compounds from water with porous organic polymers. J. Chrom. 165:167-206 (1979).

Dubinin, M.M.: The potential theory of adsorption of gases and vapors for adsorbents with energetically nonuniform surfaces. Chem. Rev. 60:235-241 (1960).

Dubinin, M.M.: Physical adsorption of gases and vapors in micropores. Prog. Surf. Memb. Sci. 9:1-70 (1975).

Easley, L.D.; Suggs, H.J.: The effect of low flow rates on the collection efficiency of hexane on charcoal. Am. Ind. Hyg. Assoc. J. 46:150-153 (1985).

Elkins, H.B.; Hobby, A.K.; Fuller, J.E.: The Determination of Atmospheric Contaminants I: Organic Halogen Compounds. J. Ind. Hyg. 19, 474–485 (1937).

Elskamp, C.J.; Schultz, G.R.: An alternate sampling and analytical method for 2-butanone. Am. Ind. Hyg. Assoc. J. 44:201-204 (1983).

Enka Research Institute, NV; Amhem, Netherlands.

Fieldner, A.C.; et al.: Methods of testing gas masks and absorbents. J. Ind. Eng. Chem. 11:519 (1919).

Foerst, D.: A sampling and analytical method for vinylidene chloride in air. Am. Ind. Hyg. Assoc. J. 40:888-893 (1979).

Folke, J.; Johannsen, I.; Cohr, K.H.: The recovery of ketones from gas-sampling charcoal tubes. Am. Ind. Hyg. Assoc. J. 45:231-235 (1984).

Fracchia, M.; Pierce, L.; Graul, R.; Stanley, R.: Desorption of organic solvents from charcoal collection tubes. Am. Ind. Hyg. Assoc. J. 38:144-146 (1977).

Fraust, C.L.; Hermann, E.R.: The adsorption of aliphatic acetate vapors onto activated carbon. Am. Ind. Hyg. Assoc. J. 30:494-499 (1969).

Freundlich, H.: Colloid and Capillary Chemistry. Methuen, London (1926).

Gideon, J.A.: The adsorption of selected structural organic isomers and homologs of charcoal under industrial hygiene sampling conditions, Dissertation. University of Cincinnati (1984).

Goller, J.W.: Displacement of polar by non-polar organic vapors in sampling systems. Am. Ind. Hyg. Assoc. J. 46:170-173 (1985).

Greenburg, L.: Benzol poisoning as an industrial hazard. U.S. Pub. Hlth. Rep. 41:1 (1926).

Grubner, O.; Burgess, W.A.: Simplified description of adsorption breakthrough curves in air cleaning and sampling devices. Am. Ind. Hyg. Assoc. J. 40:169-178(1979).

Grubner, O.; Burgess, W.A.: Calculation of adsorption breakthrough curves in air cleaning and sampling devices.

Env. Sci. Tech. 15:1346-1351 (1981).
Grubner, O.; Lynch, J.J.; Cares, J.W.; Burgess, W.A.: Collection of nitrogen dioxide by porous polymer beads. Am. Ind. Hyg. Assoc. J. 33:201-206 (1972).
Hamilton, A.: Exploring the Dangerous Trades. Northeastern University Press Boston, MA (1943).
Harper, M.: Evaluation of solid sorbent sampling methods by breakthrough volume studies. Ann. Occup. Hyg. 37:65-88 (1993).
Harper, M.: Personal communication. NIOSH, Cincinnati, Ohio (2006).
Harper, M.; Kimbeland, M.L.; Orr, R.J.; Guild, L.V.: An evaluation of sorbents for sampling ketones in workplace air. Appl. Occup. Env. Hyg. 8:293-304 (1993).
Harper, M., et al.:Validation of sorbent sampling methods for methylene chloride. Appl. Occup. Environ. Hyg. 9: 198-205 (1994).
Heitbrink, W.A.: Diffusion effects under low flow conditions. Am. Ind. Hyg. Assoc. J. 44:453-462 (1983).
Hill, R.H., Jr.; McCammon, C.S.; Saalwaechter, A.T.; Teass, A.W.; Woodfin, W.J.: Gas-chromatographic determination of vinyl chloride in air samples collected on charcoal. Anal. Chem. 48: 1395-1399 (1976).
Hori, H.; Noritake, Y.; Murobushi, H.; Higashi, T.; Tanaka, I.: A method for detecting breakthrough of organic solvent vapors in a charcoal tube using semiconductor gas sensors. Appl. Occup. Envron. Hyg.14(8):558-564 (1999).
Hull, R.D.: Development and evaluation of methods, In NIOSH Manual of Analytical Methods, 3rd ed., P.M. Eller, (ed.). National Institute for Occupational Safety and Health, Cincinnati, OH (1984).
Johansen, I.; Wendelboe, J.F.: Dimethylformamide and carbon disulphide desorption efficiencies for organic vapours on gas-sampling charcoal tube analyses with a gas chromatographic backflush technique. J. Chrom. 217:317-326 (1981).
Johns-Manville: Chromosorb. Celite Division, Greenwood Plaza, Denver, CO.
Jonas, L.A.; Rehrmann, J.A.: The kinetics of adsorption of organo-phosphorus vapors from air mixtures by activated carbons. Carbon 10:657-663 (1972).
Jonas, L.A.; Rehrmann, J.A.: Predictive equations in gas adsorption kinetics. Carbon 11:59-64 (1973).
Jonas, L.A.; Tewari, Y.B.; Sansone, E.B.: Prediction of adsorption rate constants of activated carbon for various vapors. Carbon 17:345-349 (1979).
Jonas, L.A.; Sansone, E.B.; Farris, T.S.: Prediction of activated carbon performance for binary vapor mixtures. Am. Ind. Hyg. Assoc. J. 44:716-719 (1983).
Jonas, L.A.; Sansone, E.B.; Farris, T.S.: The effect of moisture on the adsorption of chloroform by activated carbon. Am. Ind. Hyg. Assoc. J. 46:20-23 (1985).
Kenny, J.; Stratton, G.: An improved desorption solvent for organic solvent mixtures. Am. Ind. Hyg. Assoc. J. 50:A431-A434 (1989).
Klotz, I.: The adsorption wave. Chem. Rev. 39:241-267 (1946).
Krajewski, J.; Gromiec, J.; Dobecki, M.: Comparison of methods for determination of desorption efficiencies. Am. Ind. Hyg. Assoc. J. 41:531-534 (1980).
Kupel, R.E.; White, L.D.: Report on a modified charcoal tube. Am. Ind. Hyg. Assoc. J. 32:456 (1971).
Langmuir, I.: The adsorption of gases on plane surfaces of glass, mica and platinum. Journal of American Chemical Society 40:1361–1403 (1918).
Langvardt, P.W.; Melcher, R.G.: Simultaneous determination of polar and nonpolar solvents in air using a two-phase desorption from charcoal. Am. Ind. Hyg. Assoc. J. 40: 1006-1012 (1979).
Levine, M.S.; Schneider, M.: Flow rate associated variation in air sampling of low concentrations of benzene in charcoal tubes. Am. Ind. Hyg. Assoc. J. 43:423-416 (1982).
Lewis, R.J., Sr.: Hawley's Condensed Chemical Dictionary, 12th ed. Van Nostrand Reinhold, New York (1996).
Lodewyckx, P.; Vansant, E.F.: Influence of humidity on adsorption capacity from the Wheeler-Jonas Model for prediction of breakthrough times of water immiscible organic vapors on activated carbon beds. Am. Ind. Hyg. Assoc. J. 60:612-617 (1999).
Lodewyckx, P.; Vansant, E.F.: The influence of humidity on the overall mass transfer coefficient of the Wheeler-Jonas Equation. Am. Ind. Hyg. Assoc. J. 61:461-468 (2000).
Lopes, D.P.: DMSO/CS_2 Mixture: A solvent system for desorption of polar and nonpolar paint solvents from acti-

vated charcoal. Am. Ind. Hyg. Assoc. J. 58:603-607 (1997).

Markes: National and International Standard Methods Related to Speciated Monitoring of Vapour-phase Organic Chemicals in Air, Publication 03. Markes International Limited, Llantrisant, RCT, UK http://www.markes.com/downloads/publications.aspx (2003).

Matheson-Kitagawa: Detector Tube Listing for Matheson-Kitagawa Toxic Gas Detector System, Matheson Tri Gas Technical Bulletin 102-1 Issue 3 https://www.mathesontrigas.com/pdfs/litcenter/specequipmentdatasheets/model%208014%20kitagawa%20tube%20list.pdf (2004).

McCammon, C.S., Jr.: The NIOSH charcoal tube and other solid sorbent sampling tube certification program. Am. Ind. Hyg. Assoc. J. 33:489-491 (1976).

McCammon, C.S.: General considerations for sampling airborne contaminants, In NIOSH Manual of Analytical Methods, 3rd ed., Peter M. Eller (ed.). National Institute for Occupational Safety and Health, Cincinnati, OH (1984).

Mecklenburg, W.: Layer filtration, a contribution to the theory of the gas mask. Chem. Abstr. 20:531 (1926).

Mecklenburg, W.: Layer filtration, a contribution to the theory of gas masks II. Chem. Abstr. 24:4561 (1930).

Melcher, R.G.; Langner, R.R.; Kagel, R.O.: Criteria for the evaluation of methods for the collection of organic pollutants in air using solid sorbents. Am. Ind. Hyg. Assoc. J. 39:349-361 (1978).

MSA: Sampling Pump Accessories, Pittsburgh, PA http://media.msanet.com/NA/USA/PortableInstruments/PersonalSamplingPumps/PersonalSamplingPumpAccessories/08-09-01.pdf (1998).

Moskowitz, S.; Burke, W.J.: Determination of organic vapors using the method of adsorption. New York State Department of Labor Industrial Bull. 17:168-169 (1938).

Mueller, F.X.; Miller, J.A.: Determination of airborne organic vapor mixtures using charcoal tubes. Am. Ind. Hyg. Assoc. J. 40:380--387(1979).

Nelson, G.O.; Harder, C.A.: Respirator cartridge efficiency studies: IV. Effects of steady-state and pulsating flow. Am. Ind. Hyg. Assoc. J. 33:797-805 (1972).

Nelson, G.O.; Correia, A.N.; Harder, C.A.: Respirator cartridge efficiency studies: VII. Effect of relative humidity and temperature. Am. Ind. Hyg. Assoc. J. 37:280-288 (1976).

NIOSH: NIOSH Manual of Analytical Methods. National Institute for Occupational Safety and Health, U.S. Government Printing Office, Washington, DC (1974).

NIOSH: NIOSH Manual of Analytical Methods, 2nd ed. National Institute for Occupational Safety and Health, Cincinnati, OH (1977).

NIOSH: NIOSH Manual of Analytical Methods, 3rd ed., P.M. Eller (ed.). National Institute for Occupational Safety and Health, Cincinnati, OH (1984).

NIOSH: NIOSH Manual of Analytical Methods (NMAM®), 4th ed., P.C. Schlecht, P.F. O'Connor (eds.), DHHS (NIOSH) Publication 94-113. Cincinnati, OH (1994), Supplements: 1st, 96-135, 2nd, 98-119, 3rd 2003-154. http://www.cdc.gov/niosh/nmam/

Olishifski, J.B.; Kerwin, M.A.: Air-sampling instruments, In Fundamentals of Industrial Hygiene, 3rd ed., B.A. Plog (ed.). National Safety Council, Chicago, IL (1988).

Otterson, E.J.; Guy, C.U.: A method of atmospheric solvent vapor sampling on activated charcoal in connection with gas chromatography. Transactions of the Twenty-Sixth Annual Meeting of the American Conference of Governmental Industrial Hygienists, Philadelphia, PA, p. 37. American Conference of Governmental Industrial Hygienists, Cincinnati, OH (1964).

Patty, F.A.: Industrial Hygiene and Toxicology, 2nd rev. ed., Vol. 1, General Principles. Interscience, New York (1958).

Pernell, C.: Collection and analysis of halogenated hydrocarbon vapors employing silica gel as an adsorbing agent. J. Ind. Hyg. Tox. 26:331-333 (1944).

Posner, C.: Comments on "phase equilibrium method for determination of desorption efficiencies" and some extensions for use in methods development. Am. Ind. Hyg. Assoc. J. 41:63-66 (1980).

Posner, J.C.: Desorption of organic analytes from activated charcoal II: Dealing with the problems. Am. Ind. Hyg. Assoc. J. 42:647-652 (1981).

Rehrmann, J.A.; Jonas, L.A.: Dependence of gas adsorption rates on carbon granule size and linear flow velocity. Carbon 16:47-51 (1978).

Reid, F.H.; Halpin, W.R.: Determination of halogenated and aromatic hydrocarbons in air by charcoal tube and gas chromatography. Am. Ind. Hyg. Assoc. J. 29:390-396 (1968).

Reucroft, P.J.; Simpson, W.H.; Jonas, L.A.: Sorption properties of activated carbon. J. Phys. Chem. 75:3526-3531 (1971).

Rodriguez, S.T.; Gosselink, D.W.; Mullins, H.E.: Determination of desorption efficiencies in the 3M 3500 organic vapor monitor. Am. Ind. Hyg. Assoc. J. 43:569-574 (1982).

Rohm and Haas: Amber-hi-lites No. 159. Philadelphia, PA (1978).

Rudling, J.: Multicomponent adsorption isotherms for determination of recoveries in liquid desorption of mixtures of polar solvents adsorbed on activated carbon. Am. Ind. Hyg. Assoc. J. 49:95-100 (1988).

Rudling, J.; Bjorkholm, E.: Effect of adsorbed water on solvent desorption of organic vapors collected on activated carbon. Am. Ind. Hyg. Assoc. J. 47:615-620 (1986).

Rushing, D.E.: Gas chromatography in industrial hygiene and air pollution problems. Ind. Hyg. J. 19:238-245 (1958).

Saalwaechter, A.T.; McCammon, C.S.; Roper, C.P.; Carlberg, K.S.: Performance testing of the NIOSH charcoal tube technique for the determination of air concentrations of organic vapors. Am. Ind. Hyg. Assoc. J. 38:476-485 (1977).

Sansone, E.B.; Jonas, L.A.: Prediction of activated carbon performance for carcinogenic vapors. Am. Ind. Hyg. Assoc. J. 42:688-691 (1981).

Sansone, E.B.; Tewari, Y.B.; Jonas, L.A.: Prediction of removal of vapors from air by adsorption on activated carbon. Env. Sci. Technol. 13:1511-1513 (1979).

Severs, L.W.; Skory, L.K.: Monitoring personnel exposure to vinyl chloride, vinylidene chloride and methyl chloride in an industrial work environment. Am. Ind. Hyg. Assoc. J. 36:669-676 (1975).

SKC: SKC Comprehensive Catalog & Air Sampling Guide. SKC Inc., Eighty Four, PA http://www.skcinc.com/ (2006).

Soule, R.D.: Industrial hygiene sampling and analysis, In Patty's Industrial Hygiene and Toxicology, 3rd ed., Vol. I. General Principles, G.R. Clayton, F.E. Clayton (eds.). Wiley Publishing, New York (1978).

Stanetzek, I.; Giese, U.; Schuster, R.H.; Wunsch, G.: Chromatographic characterizations of adsorbents for selective sampling of organic air pollutants. Am. Ind. Hyg. Assoc. J. 57:128-133 (1996).

Supelco: Chromatography Supplies. Bellefonte, PA http://www.sigmaaldrich.com/Brands/Supelco_Home.html (2006).

Teass, A.W.; McCammon, C.S.; Roper, C.P.; Saalwaechter, A.T.: The sampling of organic vapors using activated charcoal. Second NIOSH Solid Sorbents Roundtable, E.V. Ballou (ed.), pp. 55-73. NIOSH, Cincinnati, OH (1976).

Thomas, M.L.; Cohen, B.S.: A simple method for vapor dosing of charcoal sorbent tubes. Am. Ind. Hyg. Assoc. J. 56(1):70-73 (1995).

3M: Technical Data Bulletin 1028, Organic Vapor Monitor Sampling and Analysis Guide, Occupational Health and Environmental Safety Division, St. Paul, MN http://multimedia.mmm.com/mws/mediawebserver.dyn?6666660Zjcf6lVs6EVs666CswCOrrrrQ- (no date).

Vahdat, N.; Swearengen, P.M.; Johnson, J.S.; Priante, S.; Mathews, K.; Neidhardt, A.: Absorption capacity and thermal desorption efficiency of selected absorbents. Am. Ind. Hyg. Assoc. J. 56(1):32-38 (1995).

Van Mourik, J.H.C.: Experiences with silica gel as adsorbent. Am. Ind. Hyg. Assoc. J. 26:498-509 (1965).

Verma, D.K.; Tombe, K.D.: Measurement of benzene in the workplace and its evolution process, Part I: Overview, history, and past methods. Am. Ind. Hyg. Assoc. J. 60:38-47 (1999).

Waters: Millipore Corporation, Waters Chromatography Division, Cidra, Puerto Rico (1993).

Wheeler, A.; Robell, A.J.: Performance of fixed-bed catalytic reactors with poison in the feed. J. Catalysis 13:299-305 (1969).

White, L.D.; Taylor, D.G.; Mauer, P.A.; Kupel, R.E.: A convenient optimized method for the analysis of selected solvent vapors in the industrial atmosphere. Am. Ind. Hyg. Assoc. J. 31:225 (1970).

Whitman, N.E.; Johnston, A.E.: Sampling and analysis of aromatic hydrocarbon vapors in air: A gas-liquid chromatographic method. Ind. Hyg. J.:464-469 (1964).

Willey, J.D.: Amorphous silica, in Kirk-Othmer's Concise Encyclopedia of Chemical Technology. Wiley Publishing, New York (1985).

Wood, G.O.; Moyer, E.S.: A review of the Wheeler equation and comparison of its applications to organic vapor respirator cartridge breakthrough data. Am. Ind. Hyg. Assoc. J. 50:400-407 (1989).

Yoon, Y.H.; Nelson, J.H.: Contaminant breakthrough: A theoretical study of charcoal sampling tubes. Am. Ind. Hyg. Assoc. J. 51:319-325 (1990).

Yoon, Y.H.; Nelson, J.H.; Lara, J.: Respirator cartridge service-life: Exposure to mixtures. Am. Ind .Hyg. Assoc. J. 57:809-819 (1996).

22

GASES AND VAPORS – BAGS, RIGID CONTAINERS, AND IMPINGERS

The three subjects of this chapter are loosely related. All are gas and vapor sampling devices; however, impingers are also used to sample aerosols such as acid mists and bioaerosols, but are seldom used to sample dusts. All three now see limited use in typical hygiene work; but rigid containers, in particular, are seeing increased use in ambient air pollution monitoring of mixed gases and vapors over relatively long periods. Thus, the three are certainly not so important as solid sorption media or filters, but still they have certain uses for which there is no current substitute.

GAS SAMPLING BAGS

Gas sampling bags are simply two separate layers of plastic that have been seamed at the edges to form a bag. Usually they also have a closable port or valve, through which they can be filled or purged, and a septum. The septum is usually a thick piece of plastic-coated rubber about the size of a small coin (0.5-1 cm diameter), which is used to inject gases or liquids into the bag with a syringe (see Figure 22-1). These bags have been used since the 1950s for two main purposes in hygiene and ambient air quality work: to collect air samples and to prepare known concentrations of gases and vapors to use in calibrating direct-reading and other analytical instruments. They are still used much for the latter, and some approved sampling and analytical methods utilize them (see SKC, 2006a for a list that includes NIOSH, OSHA and other methods).[1]

---

[1] There are methods for CO, CO_2, methyl acetylene, nitrous oxide, sulfur hexafluoride, and sulfuryl fluoride that utilize gas sampling bags. They are the only means for collecting the anaesthetic agent nitrous oxide other than passive monitoring and for reduced sulfur gases, such as H_2S.

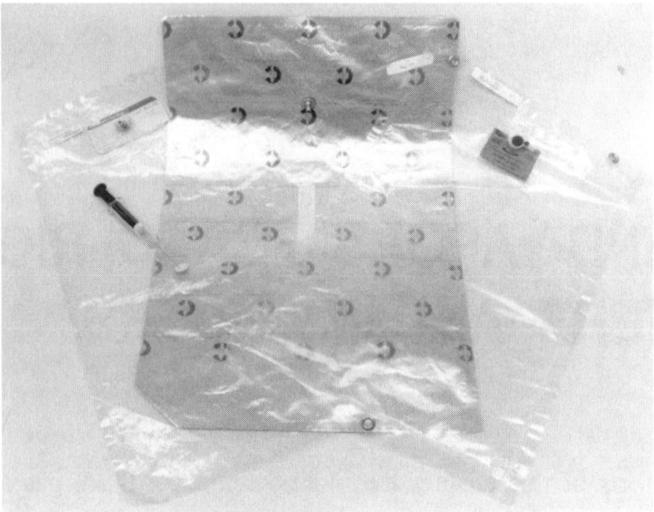

Figure 22-1. A variety of gas sampling bags with ports for filling and evacuating, septa for injecting analytes for standard preparations, and a syringe commonly used for such purposes. Note that the different polymers cannot be differentiated by sight (photo, R.K. Oestenstad).

Mechanisms for Losses of Gases and Vapors

The main concern in such uses is loss of the gas(es) or vapor(s). Plastic materials are subject to the permeation described in Chapter 8. First the substance is absorbed or goes into solution with the bag; then the substance diffuses across the bag thickness. The rate of diffusion is dependent on the concentration that goes into solution at the exposed surface, and the sorbed concentration is dependent on the concentration in the bag. Thus, the higher the concentration is in the bag, the more the vapor permeates the bag. If one tested for the percentage of the original bag content lost over time, one would expect the percentage to be constant with respect to concentration.

However, in some cases the diffusion coefficient may be concentration-dependent – it increases with concentration. In these cases the percentage loss is not constant with concentration, but generally increases with concentration. This increasing effect of concentration on percentage loss can generally be expected to be small and insignificant at low concentrations. In other words, for bag concentrations that are about 0.1 saturated vapor concentration or less, percentage losses should be constant. (The reader is referred to Chapters 2 and 3 of Comyn, 1986, for considerably greater detail on the process of permeation.)

The permeation process is very much dependent on the chemical nature of the plastic and the gas or the vapor (permeant) in the bag. Permeants with chemistries similar to that of the plastic will be easily absorbed by the plastic; like dissolves like. Table 22-1 lists the polymer materials that are used in gas sampling bags. Non-polar permeants (e.g., toluene) would be

Table 22-1
Gas sampling bags typically used for hygiene work

Bag Type	Chemical Composition
5 Layer	Inside – Polyethylene/Polyamide/Aluminum/Polyvinylidene Chloride/Polyester – Outside
[1]Al Mylar	Aluminum Coated Mylar
FlexFoil	SKC proprietary, 4 layer *foil*, inner layer *inert*
FlexFilm	SKC proprietary
Halar®	Ethylene, chloro-trifluoroethylene copolymer
Mylar®	Polyester[1] (copolymer of a dicarboxylic acid and a glycol)
Nalophan Melinex®	PET polyethylene terephthalate (used in measuring odors as a result of a European Standard CEN TC 264 WG 2 N 251, *Air quality, determination of odour concentration by dynamic olfactometry*)
[1]Nylon	
[1]Saran®	Polyvinylidene chloride
[2]Tedlar®	Polyvinyl fluoride
Teflon®	FEP – Fluorinated ethylene, propylene

[1] Cited in literature, apparently not available commercially.
[2] Apparently manufacturer of Tedlar (Dupont) will no longer do so as of 2007.

more likely to permeate non-polar plastics such as Saran or Mylar, and vice versa. The strategy behind using multiple layers of different polarities is that a molecule may easily permeate the first layer, but will not likely easily permeate subsequent layers. However, this only disallows the permeant from crossing the bag; it is nonetheless lost to the bag and, therefore, the concentration in the bag is affected.

The physical molecular structure of the plastic is also important. For example, a polar permeant may be adsorbed to a polar plastic such as Tedlar, but the polar forces between the Tedlar molecules are strong, holding these polymer molecules close together. Thus, diffusion is impeded by lack of space for the permeant molecules. So even though permeation of the permeant may be slow or almost nonexistent, adsorption to the plastic surface still may occur, and this would be seen as loss of the gas or the vapor.

In addition to surface absorption and permeation, other causes of loss include leaks and reactions. The latter can be particularly important if reactive gases or vapors are sampled. However, it is particularly hard to predict. When low concentrations of pollutants in ambient air are collected (e.g., NO_2 and hydrocarbons) reactions can be expected to occur just as they would in the atmosphere. Sunlight and plastic surfaces may aid this process, or the substances may react with the plastic. For this reason, stainless steel canisters that have been specially treated are used to sample pollutants over long time periods (Oliver and Pleil, 1986; Jayanty, 1989: Rossner and co-workers, 2004a, 2004b, 2005). In addition, some bags such as Tedlar® are available in black to block sunlight (SKC, 2006b).

In summary, losses from gas sampling bags may be caused by permeation, adsorption, reactions, and leaks.[2] The factors affecting these phenomena include the chemical nature of the plastic and the permeant, the temperature (permeation is temperature-dependent), the concentration of the gas(es) or the vapor(s), the time that the contaminants must be stored in the bag, the reactive nature of the contaminant(s) relative to the plastic, the surface-to-volume ratio of the bag, humidity, and the nature of the mixture in which the gas or the vapor of interest resides (Schultz, 1967). Regarding the last factor, a laboratory study may indicate that a particular contaminant can be efficiently collected and stored in a gas sampling bag for 8 hours, but the other contaminants that exist with the one of interest, in the ambient or workplace air, may induce reactions or affect permeation of the contaminant of interest in ways that are not seen in the laboratory. In light of all of the above, one might ask "How do I determine the appropriate sampling bag for my application from Table 22-1?" The answer is not easily given.

Laboratory Tests of Gas Sampling Bags

Table 22-2 shows gases and vapors that have been evaluated for compatibility with various plastics and the concentrations tested. However, it would be difficult to list all the other variables in these experiments, such as storage time and mixtures, and the associated losses. The table does show some references that have been published,[3] and it is obvious that much more work could be done. If gas sampling bags were as popular as charcoal tubes, no doubt much more literature would exist.

The study by Posner and Woodfin (1986) merits discussion. They examined five polymers for losses of six different organics (see Table 22-2). Four of the organics were relatively non-polar, whereas acetone and methanol were quite polar. They concluded that there was a loss of analyte with time for each combination. Thus, none of the bags is perfect. It was concluded that the Tedlar bag performed best over short storage times. The losses after 4 hours were 0 to 5% of original concentrations. Losses were greatest for the two polar compounds, as would be expected. (In the case of benzene, losses were 11%, but storage was for 6 hours rather than 4). However, after 24 hours the losses ranged from 5 to 32%. Thus, the Tedlar bag could be used for short-term sampling and calibration tasks but not tasks requiring long stor-

[2] The materials of the fittings are also important in losses over time. Stainless steel fittings were not as good for long-term storage of aromatic hydrocarbons as plastic fittings (Personal communication, M. Harper, NIOSH, 2006).

[3] There are no doubt other works published on this subject, but many more studies likely exist that are not published.

Table 22-2
Compounds and the concentrations for which bags have been laboratory tested

Bag	Compounds (Concentrations, ppm)	Reference
5-layer	CO (1-8)	Polasek and Bullin (1978)
	Acetone (100, 700), benzene (107), 1,3-butadiene (95), butane (1050), methanol (200), 1,1,1-trichloroethylene (100)	Posner and Woodfin (1986)
Alum. Mylar	NO_2, SO_2, butyraldehyde, ethylene, 2-methylpentane, benzene, acetone (all nominally 65 ppm)	Baker and Doerr (1959)
	see Polasek and Bullin	
FlexFoil®	CH_4 (50), NO (10), CO (100), CO_2 (800), H_2 (1–2%)	SKC (2006b)
FlexFilm®	1,2-dichloroethane, methyl ethyl ketone (MEK), heptane, isopropyl alcohol, benzene, toluene, ethyl acetate, trichloroethylene, acetone, methylene chloride, propylene oxide, allyl chloride, acrylonitrile, vinylidene chloride, bromoethane, acetonitrile (conc. not given)	SKC (2006b)
Halar®	see Posner and Woodfin	
Mylar®	SO_2 (1), NO_2 (1), ozone (1), mixed hydrocarbons (7–20)	Conner and Nader (1964)
	Various hydrocarbons (0.03–0.33)	Clemons and Altshuller (1964)
	Benzene (10, 20, 30, 50), methanol (100, 200, 300, 400), dichloromethane (250, 500, 650), methylisobutyl ketone (50, 100, 150, 200)	Smith and Pierce (1970)
	see Baker and Doerr	
Nylon	N_2O (900)	Austin et al. (1978)
Saran®	Acetone, ethanol, ethyl acetate, benzene, toluene, styrene, CCl_4, methyl chloroform, trichloroethylene, perchloroethylene (all concentrations were 200, 500, or 1000 ppm)	Desbaumes and Imhoff (1971)
	Acetone (2000, 500), benzene (200, 12.5), CS_2 (107, 11), CCl_4 (100, 5), chloroform (100, 25), dioxane (200, 50), ethanol (2000, 500), ethylether (800, 200), Freon 12® (1000, 500), methylchloride (50), methylethylketone (400, 100), perchloroethylene (200, 50), styrene (200, 50), methyl chloroform (700), trichloroethylene (200, 50), vinyl chloride (1000, 250), vinylidene chloride (200, 50)	Schaffer et al. (1967)
	see Baker and Doerr	
	see Smith and Pierce	
	see Posner and Woodfin	
	see Polasek and Bullin	
Tedlar®	Methanol (77–99), styrene (77–103), allyl alcohol (82–101), ethylbenezne (87–107), propylene oxide (78–95), MtBE (84–99)	McGarvey and Shorten (2000)
	see Posner and Woodfin	
	see Polasek and Bullin	
Teflon®	see Conner and Nader	
	see Posner and Woodfin	

age of any of the compounds, with the exception of 1-butene and butadiene.

The five-layer bags showed considerable losses, but the losses tended to level off after some time period. We would expect this, as the aluminum effectively stops permeation. The losses are no doubt owed to adsorption in or on the polyethylene and polyamide layers. Because the losses in the five-layer bag leveled off with time, Posner and Woodfin suggested that this bag could be used to make standards that could be carried to the field, after they were first equilibrated and analyzed in the laboratory.

It should be pointed out that the Posner and Woodfin results in no way mean that the other bag types are useless. Indeed, the Halar bag performed equivalently to the Tedlar bag on a short-term basis, and both Halar and Teflon outperformed Tedlar in the case of benzene. Thus, it is difficult to make generalizations; any of the bags may be appropriate for a given application.

In a follow-up study on the utility of Tedlar, McGarvey and Shorten (2000) attempted to account for the fate of six compounds after three weeks of storage. They found that MtBE (methyl tert-butyl ether) was not absorbed by the Tedlar. Losses for allyl alcohol, ethyl benzene, and propylene oxide were up to 20% while for styrene the three week loss was > 40% and for methanol it exceeded 80%. Interestingly, the authors also performed a cleaning flushing study and found that ethyl benzene and styrene persisted in the bags after flushes, the last preceded by heating at 27°C. For methanol, the authors' results coupled with those of Pau et al. (1991) indicate that methanol does not appear to permeate through the bag and does not appear to back permeate from the bag after flushes and time; it appears to react or irreversibly hydrogen bond to the polymer.

Finally, bags have been used to sample exhaled breath (see, for example, the NIOSH method for perchloroethylene in exhaled breath). Groves and Zellers (1996) showed that water vapor in the breath can condense in the bag and absorb polar contaminants such as methanol.

Polymer Contaminants

It is important to point out that each of the plastic materials contains unbound compounds that are added to the resin to impart various properties to the finished plastic, such as reduction of brittleness. These latter additives are called plasticizers. In addition, there is also residual, unreacted monomer. Normally these compounds are of low quantity and not of any consequence when the bags are used for calibration or to sample single compounds in relatively high concentrations. However, if they are used to sample ambient air for a wide range of compounds, biases can occur. The biases can be caused when the same compound is sampled from the air that is offgassing from the bag, or when the offgassing compound has the same gas chromatography retention time as a compound of interest. Lonneman et al. (1981) found a wide range of contaminants in both new and used Teflon and Tedlar bags. Heat treatment at 190°C for 52 hours reduced the contamination but did not eliminate it. The total hydrocarbon contamination was as high as 4.5 ppm (as equivalent carbon). Surprisingly, hydrocarbons such as acetylene, isobutane, and propylene were found in the mix. Perhaps these compounds are absorbed from ambient air during manufacturing and storage of the bags. Other contaminants such as hexafluoropropene are added in the manufacturing process. Other researchers have found similar results (see Kopczynski et al., 1975; Cox et al., 1982).

In summary, plastic gas sampling bags can be convenient to use. However, the sources of both positive and negative biases should be understood, and where they cannot be avoided, they should at least be accounted for. It is a good idea to flush plastic bags with clean air and, whenever possible, to analyze their contents prior to use. Finally, one should avoid entraining particulates in the bags as the particulates can act as both sink and source for gaseous contaminants.

RIGID CONTAINERS

Glass bottles (van Houten and Lee, 1969), evacuated glass bulbs, evacuated aerosol cans (Miller et al., 1976), steel, stainless steel, and aluminum canisters, and a variety of other rigid containers have been used to collect air samples. A glass bulb that can be evacuated, taken to the field, and opened to collect a grab air sample is shown in Figure 22-2. It also can be used to make calibration standards. Many rigid containers are less likely to demonstrate the problems associated with gas sampling bags. Certainly permeation within the glass is not a problem. However, adsorption to the wall and reaction are still concerns.

Only in the case of light hydrocarbons (C_1-C_4), and perhaps a few other difficult-to-sample gases, would rigid containers be a method of choice for workplace air sampling (Tombe et al., 1991). However, rigid containers are perhaps the only sampling technique available for these compounds. For workplace air sampling, rigid containers are most likely to be used for calibration. In the case of calibration standards, it should be noted that, unlike the situation with bags, as samples are withdrawn from a rigid container, its sides do not collapse. If the rubber septum leaks because of the reduced inside pressure, then the concentration in the contain-

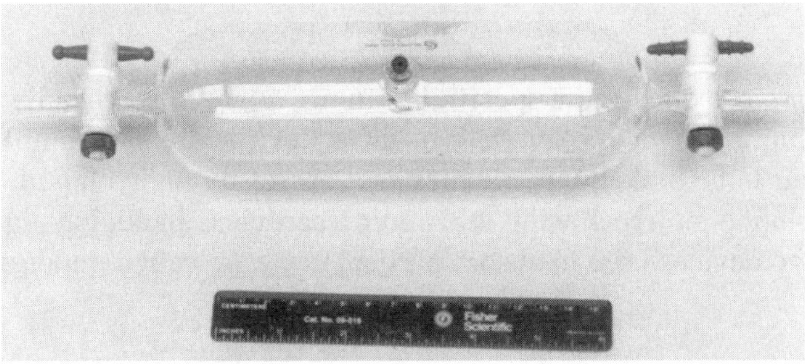

Figure 22-2. A rigid glass container with two stopcocks for filling and a septum for withdrawing contents (photo, R.K. Oestenstad).

er will be diluted. Otherwise, as samples are withdrawn, the pressure is reduced, and this will reduce the mass per volume that is delivered in subsequent samples taken from the rigid container.

Rigid containers see their greatest use in ambient air pollution work. Evacuated canisters can be fitted with regulators in order to slowly draw in air over long periods, or they can be used as grab samplers. Aluminum canisters (Lecky and Andrews, 1977; Gholson et al., 1990) have been used for such purposes; however, they do not provide a satisfactory reaction-free environment. A stainless steel canister with a specially treated surface, called a passivated SUMMA canister, has proved to be quite acceptable for collecting and storing low concentrations of many organic and inorganic ambient air pollutants for as long as two weeks (Oliver et al., 1986; Jayanty, 1989). These canisters are available from several suppliers; however, complete systems (including regulators) cost in the neighborhood of U.S. $2,000 each.[4]

A small personal cannister has been introduced and tested (Rossner and co-workers 2004a, 2004b, and 2005) that has advantages for sampling long periods (say, 40 hours). For flow control it relies on a capillary device that is not constant flow, but declines from ~ 0.225 ml/min to ~ 0.16 ml/min over a 12-hour period. Field tests show that it performs equivalently to passive samplers. However, peak concentrations could lead to biases as a result of the non uniform sampling rate. In 2006, it was not clear that this device was marketed.

A fused silica coating with the polished canister technology in a small MiniCan of 390-mL capacity has been designed for use as a personal sampler in industrial hygiene (Entech Inc., Simi Valley, CA). It is designed to be carried in a holster and fitted with an air sampling orifice. OSHA (2000) conducted a brief study of the canisters in which they concluded that the devices can all be effectively cleaned and carryover of contaminants eliminated, the system can detect down to 1 ppb of volatile organic vapors, contaminants in air samples are stable for at least two weeks (for the compounds tested), and the system's precision is satisfactory. OSHA further concluded that for these small canisters there is no sampling error analogous to active sampling and no sampling pump to calibrate and transport. OSHA now has a tentative method based on the device (OSHA, 2004).

IMPINGERS

The word *impinger* once had a specific meaning; today it is generally applied to all liquid absorption sampling devices. A variety of these devices are shown in Figure 22-3. They work by causing air to be broken into bubbles as it passes through a liquid. Gases and vapors either go into solution or react with the liquid; particles, including liquid particles, are absorbed. Years ago devices used to absorb particulates were called impingers, whereas those

---

[4] Note that canisters, valves, and other parts are now available that are lined with Silico-steel® or Silonite® and claimed to be even less reactive.

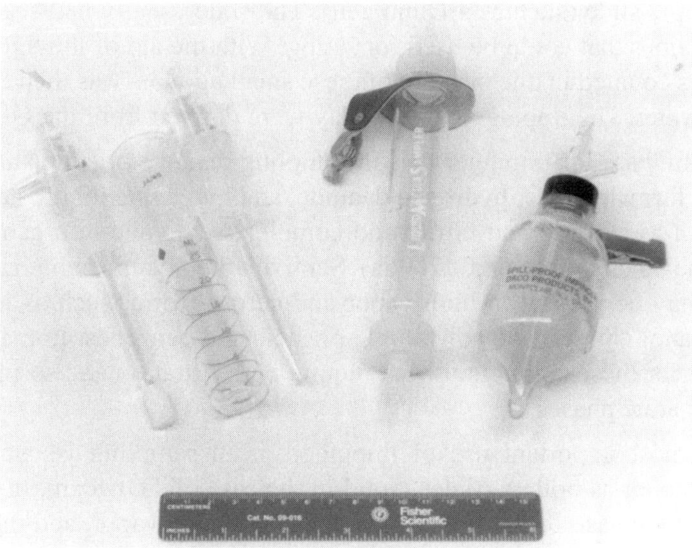

Figure 22-3. A variety of midget impingers (photo, R.K. Oestenstad).

used to absorb gases were called gas washing bottles or bubblers. However, this terminology was confusing because some of the devices were suitable for either purpose. In addition, the sampling of most solid particulates with impingers became obsolete as filtration methods were perfected. The exception to this is their use to sample bioaerosols[5] and a few other mixed media contaminants such as isocyanates mentioned later (see e.g., Lin et al., 1997; Lundholm, 1982; Silas et al., 1986; Jensen et al., 1991; Fangmark et al., 1991, and Chapter 5, Volume 2).

In the 1960s many occupational exposure limits for total and respirable dusts were written in units of millions of particles per cubic foot (mppcf),[6] and required the counting of particles with a microscope. The particles were usually collected in a Greenburg-Smith impinger at a flow rate of 1 ft^3/min or 28 Lpm. Such a high flow rate was used because lower flow, battery-operated pumps were not available. This flow rate and the size of the impinger meant that the technique was a fixed, area sampling technique. Often the sampling times were short (5 minutes) so that the liquid (in this case water) in the impinger did not become so dirty as to prohibit counting. A small portion of the liquid was removed from the impinger and placed on

---

5 Bioaerosols are bacteria and other living and biologically derived material suspended in air.

6 One problem in performing retrospective epidemiological studies is that old exposure measurements may be in different units (mppcf) than newer ones (mg/m^3). Conversion constants can be applied, but the constant is dependent on the mass and size distribution of the particles (see Chapters 9 and 22).

a special microscopic slide (such as a Dunn cell). The slide usually had segregated areas and etched circles and lines that could be used for sizing. With the aid of the light microscope, the number of particles counted in one or more of these small aliquots was then extrapolated to the entire volume of water. The impingers used today were derived from the Greenburg-Smith.

Today the main uses for impingers are in sampling reactive organic and inorganic gases, such as ammonia, formaldehyde, hydrogen cyanide, and hydrazine; liquid aerosols, especially those of acids such as sulfuric and nitric; and complex, semivolatile organic compounds, of which there may be 50 or so (see SKC, 2006a). Semivolatile organic compounds include many compounds that may be present in both vapor and aerosol form, such as isocyanates, pesticides, amines, ethanol amines, and substituted phenols and benzenes. In many cases, alternative methods exist for these compounds that require both filtration and sorption tubes for collecting the two separate phases.

Perhaps the most important use of impingers is in sampling isocyanates. Many isocyanates exist in the air as both particles (solid in the case of TDI-foam, liquid in the case of HDI-paints, dust in the case of dry-powder coatings, etc.) and vapor, and the isocyanate moieties on the particles are very reactive (with other isocyanates, curing agents, or just plain humidity), the reagent-containing impinger is the best sampler in which to ensure near complete capture or all isocyanate forms. Fine particles and vapors not collected in the impinger are collected on a reagent-coated filter after the impinger (see NIOSH methods 5521, 5522, 5525).

Hygienists generally do not like impingers. Even though the devices have been made spill-proof over the years, apparently people have a phobia about handling the liquid; impingers are thought to be less convenient than other sampling methods. Evaporation from impingers is a very important problem and can limit sample time. In that regard, there should always be a moisture trap between an impinger and a pump. However, in many cases there is no current substitute.

Liquid Absorption Theory

By making several assumptions, Elkins et al. (1937) derived an equation for predicting the percentage recovery from an impinger. They answered this question: "If a volume of air (V) is pulled through the impinger, and it has some concentration (C) *of* contaminant, what mass (m) of contaminant will be expected in the liquid?" If the process is *100%* efficient, then

$$m = CV \qquad \text{[Equation 22-1]}$$

First, it must be noted that if the contaminant reacts with the liquid and the contact time is reasonable (i.e., the flow rate is not extremely high), then efficiency should be near *100%*, and Equation 22-1 will apply. However, if, as is true in most cases, the process is not reactive but absorptive, then some mass will be lost, depending on the equilibrium between the contaminant and the liquid.

Elkins et al. assumed that the contaminant behaved as an ideal gas, that the solution formed was ideal (i.e., it obeyed Henry's law), and that the concentrations of the contaminant in the air and in solution were in equilibrium (see, also, Neale and Perry, 1959). Gage (1960) modified the Elkins relationship to:

$$E = \frac{vK}{V}\left(1 - e^{\frac{-VK}{v}}\right) \qquad \text{[Equation 22-2]}$$

where E is the efficiency, V is air volume, v is liquid volume, and K is a measure of the volatility of the contaminant from solution (i.e., related to Henry's constant). Equation 22-2 shows that one can increase the efficiency by increasing the liquid volume, decreasing the air sample volume, or choosing a liquid with greater affinity (less volatility) for the contaminant.

Gage proved the above in experiments with ethylene oxide and chlorobenzene. However, he also found that flow rate (Q) affected E beyond a certain value of Q. The flow rate does not appear in Equation 22-2 because Equation 22-2 is a thermodynamic rather than a kinetic approach. In other words, it examines the energy relationship of the phases and the contaminant but does not examine time. Time is important in considering how long it will take a contaminant molecule to diffuse from the center of the bubble to the liquid. If the bubble vertically traverses (bottom to top) the liquid in a smaller time than it takes for diffusion, then some of the contaminant will not be absorbed. Thus, at short diffusion times efficiency is nominally 100%, but at some critical diffusion time it begins to decrease. Flow rate affects the time factor in two ways. As flow rate increases, the bubble size increases (Gage, 1960); so the diffusion distance is longer. Second, larger bubbles are more buoyant and have a shorter residence time than smaller bubbles.

Generally, a Q of 1 Lpm is recommended in many impinger methods. This Q will yield the efficiencies reported in the standard method. Higher values of Q also may yield efficient results, but the user must find the critical value if higher flows are needed. Published methods often report acceptable values of V and v. By increasing the sampling period, V will increase and therefore E will decrease. However, it should be noted that Equation 22-2 is exponential. In other words, if the affinity of the contaminant and the liquid is strong,[7] it may be possible to increase V with little effect on E. One may also increase both V and v to maintain E, but increasing v will dilute the contaminant in the liquid, perhaps affecting one's ability to detect it. Finally, one may use two impingers in series, which is equivalent to increasing v but allows the calculation of E.

For calculating impinger collection efficiency of two or more impingers, the following approximation is often used:

---

[7] As an example of the effect of affinity on performance, Petreas et al. (1986) compared two absorbents for formaldehyde under similar conditions. For para-rosaniline, the efficiency was 92%, but increased to 99% for chromotropic acid. As V and v were held constant, this demonstrates the effect of the affinity coefficient (K).

$$E = \frac{X}{T} \qquad \text{[Equation 22-3]}$$

where X is the mass found in the first impinger, and T is the total mass in all impingers combined. The equation is an approximation because T cannot represent the true total unless efficiency is 100%. A more appropriate approach equates the efficiencies in the first two impingers:

$$\frac{X}{T} = \frac{Y}{T-X} \qquad \text{[Equation 22-4]}$$

where Y is the total mass in the second impinger. By using this equation, the true value of T can be found, and then the efficiency of the impingers can be determined.

Types of Impingers

Some older hygiene texts show several types of impingers not shown in Figure 22-3. For example, some have glass beads in the bottom of the impinger. This technique seems logical, but the efficiency can be reduced because of channeling of the bubbles (Gage, 1960). In addition, many attempts have been made to add imperfections to the sides of the impinger to increase the surface area (see, e.g., Neale and Perry, 1959). These efforts rarely lead to any improvement on the basic design. Both OSHA and NIOSH now rely on two impingers for all impinger methods: the midget impinger (see Figure 22-3), which is simply a midget version of the Greenburg-Smith, and the fritted glass bubbler.[8] The only difference between the two is in the opening at the end of the tube that brings air to the liquid. In the midget impinger it is a simple opening, as shown in Figure 22-3. In the fritted glass bubbler it is composed of fritted glass, which is a fused glass structure having small, porous openings.

One might think that the fritted structure would create smaller bubbles and thus always give higher efficiency. However, such is not the case. In the case of formaldehyde, the fritted bubbler (98%) outperformed the midget (88%; Septon and Ku, 1982); but in the case of ammonia the opposite was true. In NIOSH and OSHA methods the two types are not always compared; so if a change is to be made by the user, a laboratory comparison should be made first.

---

[8] Teflon impingers are also available and required in certain EPA methods for reactive gases.

QUESTIONS

22. 1. Explain the procedure by which a gas sampling bag might be used to provide stable concentrations of a contaminant that could be used for field calibration of a direct-reading instrument.

22. 2. If the bag in question 22.1 were used to collect field samples after it had been used to calibrate field instruments, what special precautions would need to be taken? Hint: Think about the process that causes the calibration standard concentration to become stable.

22. 3. What would be the advantages and disadvantages of a glass bulb over a plastic bag for field sampling?

22. 4. Two impingers are used in series to collect ammonia. The first collects 9 mg and the second collects 0.9 mg. Use both Equations 22-3 and 22-4 to find the efficiency of the impinger and compare the results.

22. 5. An impinger and absorbing liquid have a reproducible 85% efficiency for collecting phosgene. Should this technique be used? If so what adjustments need to be made to the results?

REFERENCES

Austin, J.C.; Shaw, R.; Crichton, R.; Cleaton-Jones, P.E.; Moyes, D.: Comparison of sampling techniques for studies of nitrous oxide pollution. Brit. J. Anaesth. 50:1109-1112 (1978).

Baker, R.A.; Doerr, R.C.: Methods of sampling and storage air containing vapors and gases. Int. J. Air Poll. 2: 142-158 (1959).

Clemons, C.A.; Altshuller, A.P.: Plastic containers for sampling and storage of atmospheric hydrocarbons prior to gas chromatographic analysis. J. Air Poll. Control Assoc. 14:407-408 (1964).

Comyn, J.: Polymer Permeability. Elsevier Applied Science Publishers, New York (1986).

Conner, W.D.; Nader, J.S.: Air sampling with plastic bags. Am. Ind. Hyg. Assoc. J. 25:291-297 (1964).

Cox, R.D.; McDevitt, M.A.; Lee, K.W.; Tannahill, G.K.: Determination of low levels of total nonmethane hydrocarbon content in ambient air. Env. Sci. Technol. 16:57-61 (1982).

Desbaumes, E.; Imhoff, C.: Use of Saran bags for the determination of solvent concentration in the air of workshops. Staub-Reinhalt Luft 31 :36-41 (1971).

Elkins, H.B., Hobby, A.; Fuller, J.E.: The determination of atmospheric contaminants. J. Ind. Hyg. 19:474 (1937).

Fangmark, I.; Wikstrom, L.E.; Henningson, E.W.: Collection efficiency of a personal sampler for microbiological aerosols. Am. Ind. Hyg. Assoc. J. 52:516-520 (1991).

Gage, J.C.: The efficiency of absorbers in industrial hygiene air analysis. Analyst 85:96-103 (1960).

Gholson, A.R.; Jayanty, R.K.M.; Storm, J.F.: Evaluation of aluminum canisters for the collection and storage of air toxics. Anal. Chem. 62:1899-1902 (1990).

Groves, W.A.; Zellers, E.T.: Investigation of organic vapor losses to condensed water vapor in Tedlar bags used for exhaled breath sampling. Am. Ind. Hyg. Assoc. J. 57:257-263 (1996).

Harper, M.: Personal communication. NIOSH, Cincinnati, Ohio (2006).

Jayanty, R.K.M.: Evaluation of sampling and analytical methods for monitoring toxic organics in air. Atmos. Env.

J 23:777-782 (1989).
Jensen, P.A.; Davis, G.N.; Todd, W.F.: Evaluation of Sampling Alternatives for Bioaerosols. Phase II: Viable Microorganism Sampling Efficiency. NIOSH, Cincinnati, OH (1991).
Kopczynski, S.L.; Lonneman, W.A.; Winfield, T.; Seila, R.: Gaseous Pollutants in St. Louis and Other Cities. J. Air Poll. Control Assoc. 25:251 (1975).
Lecky, J.H.; Andrews, R.: A trace gas sampling cartridge. Int. Lab., pp. 63-65 (1977).
Lin, X.; Willeke, K.; Ulevicius, V.; Grinshpun, S.A.: Effect of sampling time on the collection efficiency of all-glass impingers. Am. Ind. Hyg. Assoc. J. 58:480-488 (1997).
Lonneman, W.A.; Bufalini, J.J.; Kuntz, R.L.; Meeks, S.A.: Contamination from fluorocarbon films. Env. Sci. Tech. 15:99-,103 (1981).
Lundholm, I.M.: Comparison of methods for quantitative determination of airborne bacteria and evaluation of total viable counts. Appl. Env. Microb. 44:179-183 (1982).
McGarvey, L.J.; Shorten, C.V.: The effects of adsorption on the reusability of Tedlar air sampling bags. Am. Ind. Hyg. Assoc. J. 61:375-380 (2000).
Miller, R.W.; Stroz, D.; Principe, A.H.: Vacu-samplers, a simple method for taking accurate quantitative gas and vapor samples using partially evacuated steel containers. Am. Ind. Hyg. Assoc. J. 37:315-319 (1976).
Neale, E.; Perry, B.J.: The sampling and analysis of toxic vapours in the field. Analyst 84:226-232 (1959).
Oliver, K.D.; Pleil, J.D.: Sample integrity of trace level volatile organic compounds in ambient air stored in SUMMA polished canisters. Atmos. Env. 20: 1403-1411 (1986).
OSHA: Evaluation of Entech Canister System, Y. Chan, P. Hearty http://www.entechinst.com/pdf/Entech_final_report.pdf (March 2000).
OSHA: Volatile Organic Compounds in Air, Method No. PV2120, U.S. Occupational Safety and Health Administration, Salt Lake City, UT http://www.osha.gov/dts/sltc/methods/partial/pv2120/ pv2120.html#five (2004).
Pau, J.C.; Knoll, J.E.; Midgett, M.R.: A Tedlar bag sampling system for toxic organic compounds in sources emission sampling and analysis. J. Air Waste Mgmt. Assn. 41:1095-1097 (1991).
Petreas, M.; Twiss, S.; Pon, D.; Imada, M.: A laboratory evaluation of two methods for measuring low levels of formaldehyde in indoor air. Am. Ind. Hyg. Assoc. J. 47:276-280 (1986).
Polasek, J.C.; Bullin, J.A.: Evaluation of bag sequential sampling technique for ambient air analysis. Env. Sci. Technol. 12:708-712 (1978).
Posner, J.C.; Woodfin, W.J.: Sampling with gas bags I: Losses of analyte with time. Appl. Ind. Hyg. 1:163-168 (1986).
Roberts, L.R.; McKee, H.C.: Evaluation of absorption sampling devices. J. Air Poll. Control Assoc. 9:51-53 (1959).
Rossner, A.; Farant, J.P.: A novel personal air sampling device for collecting volatile organic compounds: A comparison to charcoal tubes and diffusive badges. J. Occup. Environ. Hyg. 1: 69-81 (2004a).
Rossner, A.; Warner, S.D.; Vyskocil, A.; Tardif, R.; Farant, J.P.: Performance of small evacuated canisters equipped with a novel flow controller for the collection of personal air samples. J. Occup. Environ. Hyg. 1: 173-181 (2004b).
Rossner, A.; Wick, D.P.: A field study to assess the long-term sampling feasibility of evacuated canisters and the development of a mathematical model to analyze potential sampling bias. J. Occup. Environ. Hyg. 2: 474-480 (2005).
Schaffer, A.W.; Stolpe, L.J.; Hoyle, H.R.: Industrial hygiene applications for plastic bags made from Saran: A summary of practical experience, paper presented at the 8th Conference on Air Pollution and Industrial Hygiene Studies, Oakland, CA (Feb. 1967).
Schultz, F.J.: Plastic bags for collection of gas samples. Atmos. Env. 1:515-519 (1967).
Septon, J.C.; Ku, J.C.: Workplace air sampling and polarographic determination of formaldehyde. Am. Ind. Hyg. Assoc. J. 43:845-852 (1982).
Silas, J.C.; Harrison, M.A.; Carpenter, J.A.: Evaluation of particulate air samplers for airborne aflatoxin B_1. J. Tox. Env. Hlth. 18:215-220 (1986).
SKC: The SKC sampling guides, In SKC Comprehensive Catalog & Sampling Guide. SKC, Inc., Eighty Four, PA (2006a).

SKC: FlexFoil Grab Bags for Sampling Low Molecular Weight Compounds, SKC, Inc., Eighty Four, PA http://www.skcinc.com/prod/245-01.asp (2006b).

Smith, B.S.; Pierce, J.O.: The use of plastic bags for industrial air sampling. Am. Ind. Hyg. Assoc. J. 31 :343-348 (1970).

Tombe, K.D.; Verma, D.K.; Stewart, L.; Reczek, E.B.: Sampling and analysis of light hydrocarbons (C_1-C_4)-a review. Am. Ind. Hyg. Assoc. J. 52:136-144 (1991).

van Houten, R.; Lee, G.: A method for the collection of air samples for analysis by gas chromatography. Am. Ind. Hyg. Assoc. J. 30:465-469 (1969).

23

GASES AND VAPORS – PASSIVE MONITORING

INTRODUCTION

Passive monitoring, passive dosimetry, diffusional monitoring, diffusional dosimetry, diffusional sampling, and personal dosimetry are all terms that have been applied to the process of collecting air samples without the aid of a pump through the physical process of diffusion. Dosimetry also is a term commonly applied to radiation monitoring, and in that case it actually is an indication of dose because ionizing radiation striking any part of the body is potentially striking a target organ and thus is a part of the overall dose that may cause disease. On the other hand, the monitoring of airborne concentrations provides only an index of dose, as even in the case of inhalation of the contaminant only a portion is absorbed, and only a fraction of that reaches the target organ(s). Thus, perhaps passive monitoring is the best term.

Passive monitoring monitors have been on the market since the early 1980s. The concept was introduced in the hygiene literature by the work of Palmes et al. (1976) for the monitoring of nitrogen dioxide and Tompkins and Goldsmith (1977) for H_2S, SO_2, NO_2, and organic compounds. Soon DuPont, 3M, Perkin-Elmer (in the United Kingdom), and other major companies were marketing passive devices. There are now far more devices and companies offering devices than can be enumerated (see Figure 23-1 for a few examples). In addition, devices and companies enter and leave the market with surprising turnover. Nonetheless, the ease of use of passive monitoring and a multitude of studies showing that they work well has meant even greater use.

Some potential and proven problems associated with the use of passive monitors have been discovered and reported. However, solutions are also being found with regularity. As a result, the number of monitor models available for a given substance and the reliability of each usually are not fixed. A user of passive monitors must stay current on the literature and be prepared to undertake expensive validation tests if a new unproven monitor is to be tried. Although passive devices do offer ease of use, they also require some up-front work. The fre-

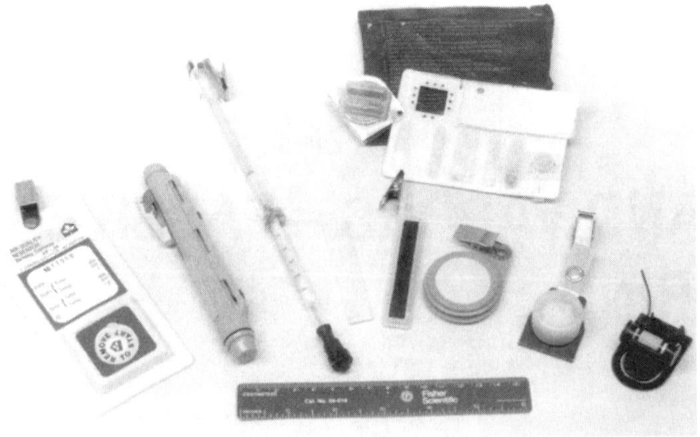

Figure 23-1. Various passive monitoring devices, some of which are no longer manufactured, from left to right, formaldehyde, N_2O, NH_3 (length of stain), hydrocarbon monitor with cover removed showing charcoal strip, last three are also hydrocarbon monitors with charcoal. At the top is U.S. Army passive detector unit M-256 for warfare agents (photo, R.K. Oestenstad).

quent user can, and should, conduct a thorough validation study unless the manufacturer or others have done so. A thorough validation study is yet to be defined in any legislative or regulatory way, but several protocols have been presented and will be discussed in this chapter.

THEORY OF OPERATION

Diffusion

All passive monitors work on the principle of diffusion.[1] In many cases this is a straightforward application of Fick's laws as discussed in Chapter 8 (see Equation 8-12). In these

[1] There is a group of devices that do not consistently rely on diffusion but rather convective mass transfer. These devices are sometimes called indicator badges. They are generally non-quantitative, colorimetric direct-reading devices that can be fairly reliable qualitative indicators. In diffusion devices, a pathlength for diffusion is always defined, whether it be defined by the geometry of the device, the thickness of a plastic membrane, or by variable but predictable length (length of stain devices). In convective mass transfer devices, the pathlength is not defined; the sorbent often sets on a card-like surface with no housing. When the ambient air is very stagnant, a diffusion path will develop in front of the sorbent. As the wind speed increases across the face of the sorbent, this diffusion path disappears, and transfer of the contaminant to the sorbent occurs by convective air currents, or wind. Thus, only in a marginal sense can these devices be termed diffusion monitors, and so they warrant no further discussion.

devices the diffusion process occurs across a space of defined dimensions as in the Palmes tube, one of the first passive monitors (see Figure 23-2). The diffusion path ends at the sorbent, in many cases a solid sorbent, whose efficiency usually has been demonstrated earlier in an active method. For example, passive monitors for organics commonly employ charcoal or, to a lesser extent, one of the synthetic sorbents described in Chapter 21. In the case of inorganics, reagents may be impregnated onto solid sorbents to promote a reaction.

Some monitors are said to work on the principle of diffusion even though there may be a liquid sorbent and a membrane separating the sorbent from air. Many samplers use membranes to act as a wind shield even thought the sorbent is solid, e.g., charcoal. The membrane is a very thin layer of plastic with the double role of eliminating air distortions (eddies) and preventing particle matter and water droplets from depositing on the collection media. It may interact with the contaminant in one of three ways.

- There is little or no interaction with the contaminant. The contaminant molecules and the mean free path of the molecules is small compared to the pore size. Fick's law describes the diffusion process; for all practical purposes the contaminant is diffusing through air and the membrane is not significant. The wind screen on the 3M OVM (Organic Vapor Monitor) is an example, even though the 3M OVM contains a solid sorbent, charcoal. For liquid sorbents, the pores in the membrane cannot be so large as to allow the passage of the sorbent. In the case of a water-based sorbent, a hydrophobic membrane is used.

- The mean free path of the molecules is big compared to the pore size. Thus, there is frequent contact of the contaminant molecules with the pores in the membrane. This is termed Knudsen diffusion and is modeled differently from Fickian diffusion.

- The pore size is not much larger than the contaminant molecule. The contaminant may have little electrostatic attraction for the membrane wall (similar to O_2 diffusing through polyethylene) or there may be significant electrostatic attraction.

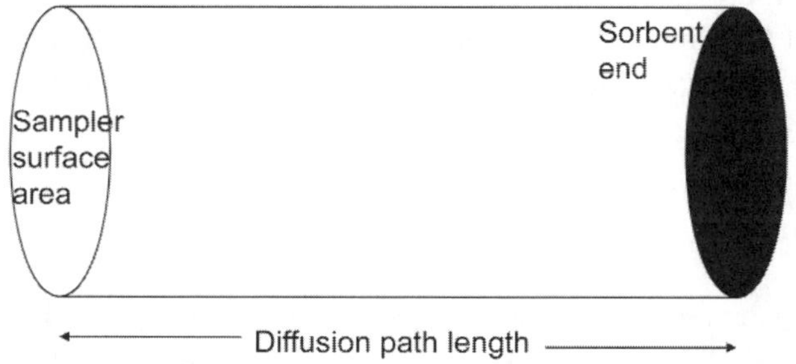

Figure 23-2. A schematic of a Palme's Tube, showing the dimensions (area and length) that affect mass flux rate.

Either case is permeation and is modeled by the equations in Chapter 8. In the first case above, the contaminant diffuses through air. In this case, the contaminant must diffuse through the membrane.

FICK'S LAW

Fick's first law was given in Equation 8-12. It can be integrated to give

$$F = \frac{-D(C_a - C_o)}{L} \qquad \text{[Equation 23-1]}$$

where L is the diffusion pathlength as shown in Figure 23-2, C_a is the ambient concentration at the front of the pathlength, and C_o is the concentration at the end of the pathlength, or at the sorbent. If the sorbent is 100% efficient (i.e., every contaminant molecule that contacts the sorbent is irreversibly captured) then $C_o = 0$. This assumption is commonly made although in many cases it is not strictly obeyed, especially for charcoal. The flux rate across the ambient end of the monitor is in units of mass/area/time. Thus, when $C_o = 0$, Equation 23-1 can be written as

$$\frac{M}{At} = -D\frac{C_a}{L} \qquad \text{[Equation 23-2]}$$

and by rearranging we can solve for C_a, the primary variable of interest:

$$C_a = \frac{ML}{tDA} \qquad \text{[Equation 23-3]}$$

DA/L has units of volume/time and is equivalent to the sampling rate of a pump. By finding the mass of contaminant on the monitor, knowing the time of monitoring, and knowing the sampling rate, the concentration can be found if all assumptions are met.

The sampling rate is dependent on the geometry (*A, L*) of the monitor and the diffusion coefficient. In the past, diffusion coefficients often were estimated by using various estimation equations (Fuller and Giddings, 1965; Reid and Sherwood, 1966; Lugg, 1968). However, for various reasons to be discussed later, it is now more common, and the best practice, to expose the monitors to a known C_a and then calculate D.

Temperature

Diffusion coefficients are temperature-dependent. Generally, in the industrial hygiene literature all vapor and gas diffusion coefficients are said to vary with $T(°K)^{1.5}$. This relationship is derived from the Hirschfelder equation for non-polar gases (Hirschfelder et al., 1964, p. 539) introduced to the hygiene literature by Palmes et al. (1976). However, other experimenters have used exponents including 1, 1.75, 1.78, 1.81, and 1.823, and in one case D varied with temperature as the viscosity of air at that temperature to the 2.74 power (Fuller and Giddings, 1965; Reid and Sherwood, 1966, pp. 530-531). Fuller and Giddings found that their prediction equation employing 1.75 as an exponent of T gave the best results. Thus, the exact relationship of D to T is not certain.

It also should be noted that D is said to be inversely proportional to the absolute pressure (Hirschfelder et al., 1964, p. 539). However, C_a in Equation 23-3 is in units of mass/volume and so is proportional to pressure. In other words, if pressure increases, volume shrinks, and the ratio mass/volume increases. In summary, C_a is proportional to P/T, and D is approximately proportional to $T^{1.5}/P$; therefore, the mass taken up by the monitor is approximately proportional to $T^{0.5}/P^0$ and is usually corrected on the basis of this relationship.

The temperature dependence of diffusion, as well as permeation and solubility coefficients also can be very accurately related over a narrow temperature range of 0°C to 40°C, by an Arrhenius relationship (Perkins and You, 1992), or

$$D = D_o \exp(-ER/T) \qquad \text{[Equation 23-4]}$$

where D_0 is the *pre-exponential factor*, E is the activation energy, T is absolute temperature, and R is the ideal gas constant. By logarithmically transforming the equation and determining D at a minimum of two temperatures, a straight-line relationship of ln D and $1/T$ can be determined, where ln D_0 is the intercept, and $-ER$ is the slope. A comparison of the accuracies of the two approaches (Arrhenius versus $T^{1.5}$) would seem instructive, but one does not appear to have been performed. Certainly the $D \propto T(°K)^{1.5}$ relationship is simple and easy to use.

Response Time

When a diffusion monitor is first opened to the atmosphere, a period of time is necessary before the equilibrium concentration gradient develops inside the monitor. One might think of a concentration front progressing along the length of the monitor, as depicted in Figure 23-3. Based on this *response time phenomenon*, one might expect a negative bias to be associated with sampling for short periods of time.

However, consider a monitor that is opened to ambient air for only one second and then capped. During the one second, a small amount of the contaminant eventually will enter the monitor. If the sorbent is 100% efficient, all of the contaminant eventually will be captured. Now consider the same scenario except that one second after the monitor is opened, it is not closed but instead is removed to a *clean* environment. Now approximately one-half of the one second contaminant pulse will diffuse back out to the clean environment. Which scenario is likely to give the better indication of the time weighted average concentration for the one second period? Intuitively, it might seem that because the concentration gradient does not have time to get established, we would need to trap all of the pulse to have the best chance of estimating the 1 s TWA concentration.

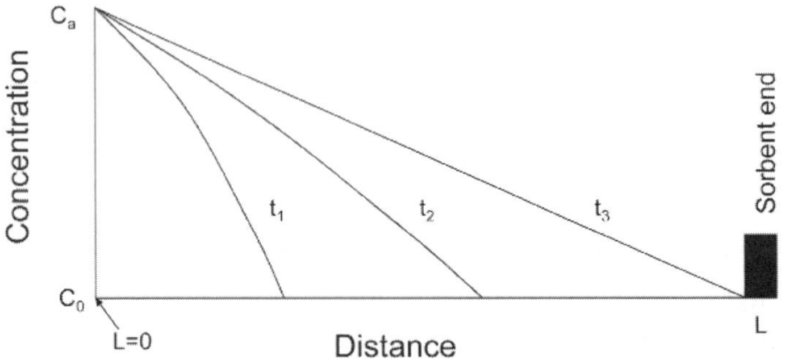

Figure 23-3. Plot of concentration vs. distance along the pathlength of a passive monitor showing the concentration profile for several short time periods leading to time t3, at which an equilibrium concentration gradient is established.

Now, however, imagine a series of pulses occurring throughout a sampling period. Given the normal nature of environmental variability, we would expect some type of fluctuating concentration profile. Each time the ambient concentration decreases, the concentration gradient in the monitor is in both directions – toward the sorbent and toward the environment. Hence, it is a normal part of the diffusion process for the gradient fluctuations to occur, and pulses occurring at the beginning of a period should be accounted for in the same way as those occurring in the middle.

Originally it was argued that the use of Equation 23-3 assumed that the time needed to reach the equilibrium concentration gradient had to be small (see, e.g., Tompkins and Goldsmith, 1977; Hearl and Manning, 1980; Einfield, 1983; Persoff and Hodgson, 1985); and if it were too long, then negative biases could be associated with short sampling periods. However, it is now recognized that a short pulse followed by capping of the monitor does not end the monitoring process. Instead, the contaminant caught in the diffusion zone eventually is trapped by the sorbent. In fact, Underhill (1983) and Bartley et al. (1983) independently showed that capping the monitor caused oversampling, as under normal conditions a portion of the pulse would diffuse back to the environment. They recommended using a cap containing a sorbent for closing the monitors at the end of the sampling period. This would essentially eliminate the positive bias by absorbing about half of the contaminant. The idea of a sorbent cap has not caught on among manufacturers. However, the hygienist can correct the situation, as a sorbent cap is equivalent to removing the monitor to a clean environment and then capping it after a period of a few minutes (see, e.g., Hori and Tanaka, 1993).

This positive bias for short sample periods may be ignored by some because it errs on the safe side. However, it can be substantial. The magnitude of the error is dependent on the geometry of the monitor, the diffusion coefficient, and the length of the sampling period. The Palmes tube (Figure 23-2) is a worst-case situation because of the long pathlength (several cm).

Bartley et al. (1983) predicted errors as high as 40% for the Palmes tube. Similar errors were later confirmed (Bartley et al., 1987). Cassinelli et al. (1985) found errors that exceed about 12% for sampling times as long as one hour for SO_2 monitors having considerably shorter pathlengths than the Palmes tube. Hori and Tanaka (1993) confirmed that the error could be negated. Instead of capping organic vapor monitors with a sorbent cap, they removed the monitors to a clean environment for one minute after sampling and then capped them.

Transient Conditions

Related to the response time phenomenon and errors associated with small sampling times are errors associated with transient conditions (widely fluctuating concentrations) over longer sampling times. For example, a constant exposure of a monitor for a one minute sample period will give the same error as a one minute peak of the same average concentration occurring at the end of an 8-hour sampling period. Although this type of transient behavior tends to give positive errors as previously discussed, a peak at the beginning of a long-term sampling period tends to give a slight negative error. These two extremes are the only transient behaviors that require testing in order to characterize transient errors (Bartley et al., 1987). Hearl and Manning (1980) provided correction factors for sinusoidally varying transient conditions, but they would be difficult to apply, as a knowledge of the frequency of the peaks and their phase relative to the start time is necessary.

Bartley et al. (1987) showed mathematically that transient concentration sampling errors varied according to when the peaks occurred. The errors associated with transient exposure conditions reach their absolute value maximums when single pulses of short duration relative to the sampling period occur at either the end or the beginning of the sampling period, and when these pulses represent the only exposures during the sampling period (see Figure 23-4). The error is theoretically infinite when the peak occurs at the end of the period and gives a maximum negative bias (about -8%) when the peak is at the beginning of the period. However, under more typically variable environments, they showed with real data that the average error was zero. In addition, the maximal error was about 40% as previously stated. It should be noted that the calculations based on real data were for the Palmes tube, and that the potential errors are most severe at this long pathlength. Finally, the calculations were based on the fact that the dosimeter was capped immediately after sampling. Using a sorbent cap would decrease the positive errors associated with latc-occurring peaks. Thus, hygienists should be aware that the errors can be large in extreme conditions (i.e., long pathlength, exposure extremes at the beginning or the end of the sampling period, and no sorbent cap).

Weak Sorbents – $C_0 \neq 0$

All of the discussion of passive monitors, to this point, has been based on the assumption that the sorbent is 100% efficient and the sorption is irreversible; that is, that C_0 is zero. Indeed, the mathematical treatment of Fick's law in Equation 23-3 makes that assumption. However,

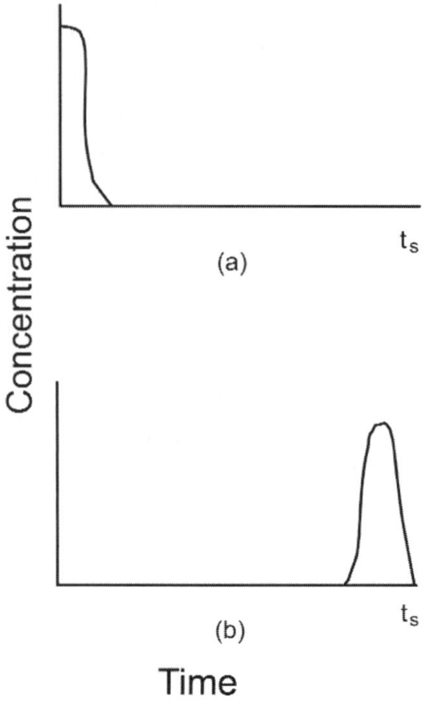

Figure 23-4. Two worst-case exposure profiles for transient conditions for passive monitors, where t_s = sample time.

in the case of charcoal, this is not always a good assumption as an equilibrium concentration (C_e) will exist above the charcoal. Therefore, $C_0 = C_e$. Charcoal is used in all diffusion monitors that are sold as general-purpose organic vapor monitors (OVM).[2] Binding to charcoal is reversible, and the strength of the attraction is based on the forces of molecular interactions (see Chapter 8).

In Chapter 21 we learned that charcoal has a retention volume, or saturation capacity, for each chemical that is dependent on the polarity and the volatility of the chemical and competition from other contaminants such as water vapor. The more volatile a contaminant is, the smaller its retention volume on charcoal. Put another way, a volatile contaminant is more weakly held by charcoal than a nonvolatile one. A polar compound also will have a smaller

[2] Several monitors used for specific organic compounds do not employ charcoal, including monitors for formaldehyde, ethylene oxide, and vinyl chloride. These are reactive or volatile organics that are difficult to monitor with charcoal. In addition, several manufacturers are marketing semi-general-purpose OVM. These contain synthetic polymeric sorbents and are used to take advantage of heat desorption for GC analysis.

retention volume, as the relative availability of polar sites is less than that for non-polar sites on charcoal. Relative humidity plays a large role in this. However, in either an active or a passive system, charcoal does not retain every molecule that it contacts up to the retention volume and then adsorb no more of the chemical. Instead, some loss occurs along the way, usually long before the retention volume or saturation is reached, and this loss increases as the retention volume is approached. Thus, if one plots the mass absorbed by a passive monitor versus time, the uptake is not linear. The deviation from linearity depends on the retention volume, the mass of charcoal used, the sampling rate of the monitor (*DA/L*), and the ambient concentration (Posner and Moore, 1985).

As the equilibrium concentration (C_e) above the charcoal increases, due to decreasing sorbent weight or retention volume, the departure from linearity increases. The concentration gradient at constant C_a (in this case C_a to C_e) decreases with time and, therefore, the actual sampling rate of the monitor is decreasing because C_e increases. Equation 23-3 assumes that $C_0 = 0$. It is important to point out that this change in sampling rate with time is independent of transient conditions.

Each manufacturer has a recommended capacity[3] for a given chemical and monitor. These recommendations should not be exceeded, or the departure from linear uptake becomes too great. The presence of competitors, such as water vapor, affects the capacity, but it is difficult to predict these potential effects in order to precisely adjust sampling times downward.

It is important to note that the loss of sorbed contaminant from charcoal also happens in active charcoal tube systems, but in those systems the released analyte is moved along with the airstream to cleaner charcoal where it is trapped again. If this happens too often in an active system, then breakthrough is significant. This event generally can be detected by analyzing the backup section of the charcoal tube. This is not the case for passive monitors employing charcoal.

BACKUP SORBENT PADS

An approach to account for the biases due to the non-linear sampling rates caused by violation of the assumption that $C_0 = 0$, and to some extent those caused by transient conditions, is the addition of a backup sorbent pad. This is another layer of charcoal, separated from the first by a short diffusion path. Thus, at least a part of the adsorbed contaminant released from the first layer diffuses to the second layer. The backup system was first proposed (Lautenberger et al., 1981) as a means of extending the sampling time and detecting breakthrough, both problems related to violation of the assumption that $C_0 = 0$. However, the theoretical basis was not

[3] It is important to note that the sampling rate is assumed to be approximately linear. Therefore, manufacturers have recommended capacities (the product of concentration sampled and time). These capacities should be based on the approximated linear portion of the uptake curves shown in Figure 23-5 and should not be exceeded, or undersampling will occur. Capacity is expressed as ppm/hr or mass (mg).

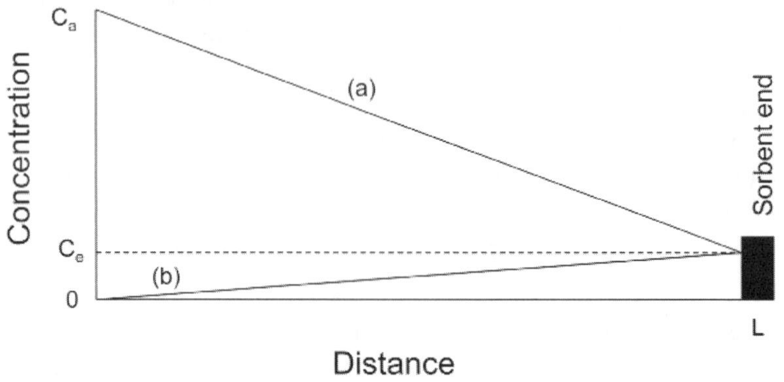

Figure 23-5. A plot of concentration vs. the pathlength or distance showing a) a concentration gradient between some ambient concentration and the equilibrium concentration, and b) the concentration gradient between an outside concentration of 0 and the equilibrium concentration above the weak sorbent.

given until later (Bartley, 1983). Bartley also argued that backup layers could afford a degree of accountability for biases introduced by transient conditions, if the masses found on the two layers were used appropriately in calculations of the TWA concentration.

Bartley showed that the masses found on the two layers of charcoal could be used to find the average airborne concentration, but that the relationship was not so easy as adding the two masses together and then applying the sum to Equation 23-3. Instead, the mass found on the second layer is dependent on the concentration gradient found in the second diffusion zone. This concentration is a function of the isotherm for the charcoal and the contaminant. A typical isotherm is shown in Figure 23-6 and can be modeled by the Langmuir form (see also Chapter 21, Appendix 21-1):

$$C = \frac{(m/a)}{(1 - m/m_s)} \qquad \text{[Equation 23-5]}$$

where C is the airborne concentration in the air above the first sorbent layer, m is the mass contaminant in that sorbent layer, a is the equivalent of the Henry's law constant, and m_s is the theoretical saturated mass that the sorbent can hold at infinite airborne concentration. Figure 23-6 and Equation 23-5 are based on static exposures of the sorbent to the corresponding airborne concentration. In essence, Equation 23-5 is an isotherm equation.

Bartley used Equation 23-5 to find the constant a, which was used to find another constant, the relaxation time (t_l):

$$t_l = aL_1/(AD) \qquad \text{[Equation 23-6]}$$

where L_l is the pathlength in front of the first layer, and A and D were defined in Equation 23-

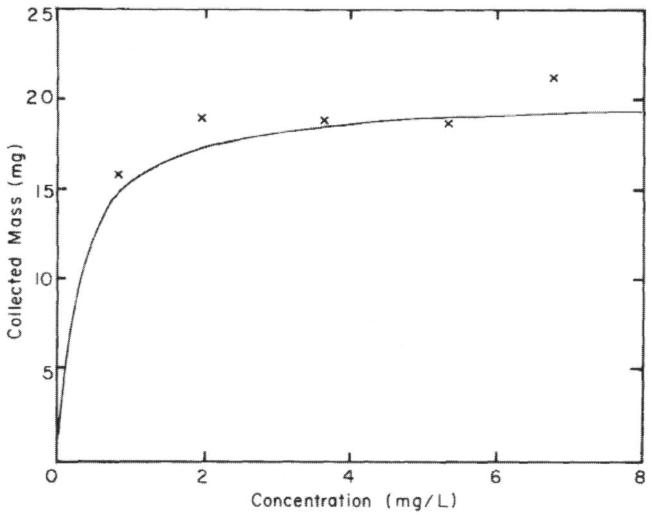

Figure 23-6. Experimental adsorption isotherm for methochloroform on 300 mg DuPont Pro-Tek activated carbon at 298°K, 760 mmHg, and 70% relative humidity. The solid line indicates empirical Langmuir isotherm (with permission from Bartley, 1983).

3. Next, a calibration constant, *k*, was determined as:

$$k = a\,[1 - \exp(-t_s/t_1)] \qquad \text{[Equation 23-7]}$$

where t_s is the sampling period. This constant then was used to calculate the average airborne concentration between the two layers (C_2):

$$C_2 = k m_1 t_s \qquad \text{[Equation 23-8]}$$

and the TWA concentration *(C)* could be estimated as:

$$C = C_2 + \frac{L_2}{ADt_s} m_1 t_s + m_2 t_s \qquad \text{[Equation 23-9]}$$

where m_1 and m_2 are the determined masses in the two layers.

Bartley found that this model estimated the true TWA concentration quite well for experimental data involving methyl chloroform, even though the isotherm data in Figure 23-6 were somewhat incomplete. However, it is not known that any manufacturers recommend, or commercial laboratories perform, the isotherm experiments needed to calculate the coefficients. Instead, the two masses are simply added together and used in Equation 23-3. This may cause considerable estimation errors, as the offgassing of the charcoal layers is not completely accounted for. The error increases for more polar and volatile contaminants that are sampled for time and concentration combinations that exceed their capacities. Although oversampling may seem like an easy problem to avoid, this is not the case where complex mixtures of competing organics (as well as water vapor) occur at unknown concentration.

Transient Conditions or Concentration Peaks and Valleys

So long as the ambient concentration (C_a) is greater than C_e, the gradient is inward, and the dosimeter will sample according to a curve such as that in Chapter 21, Figure 21-5a. If C_e becomes greater than C_a (i.e., C_a approaches zero, as in a transient condition), then the gradient will be toward the outside, and the calculated concentration from Equation 23-3 will be negatively biased (see Chapter 21, Figure 21-5b).

Unfortunately C_e is not a constant,[4] but if, on average, it is small relative to the average C_a, then the sampling bias is minimal under most transient conditions. Such a small value for C_e (at a given value of C_a, say, the PEL) can be expected for many compounds, including most hydrocarbons, aromatic and chlorinated hydrocarbons, and esters. Alcohols, ketones, epoxides, aldehydes, other polar compounds, and volatile compounds such as methylene chloride and pentane will have higher values of C_e. In these cases, sampling error may be significant, particularly if relative humidity is greater than 50%. Therefore, if C_a widely fluctuates and approaches zero for periods of several minutes, sampling error is likely to occur, particularly for the latter compounds.

EXPERIMENTAL DATA FOR TRANSIENT EFFECTS

Numerous studies have been reported in an attempt to confirm the effects of violation of the $C_0 = 0$ assumption and transient effects (including those at short sample times) for charcoal monitors. Unfortunately the two sources of error often are not differentiated, and so it is sometimes difficult to assign the observations to one source or the other. In addition, one would expect the greatest errors to be associated with sampling of mixtures containing polar compounds. However, polar compounds, let alone mixtures, have rarely been addressed.

Einfield (1983) examined transient behavior as exemplified by constant-height trichloroethylene peaks occurring approximately at 2-minute or 5-minute intervals and also a random transient behavior. In none of the three cases did concentrations fall below about 40 ppm; so it is unlikely that the concentration gradient was reversed at any point in the experimentation. The results indicated that the monitors performed equivalently under these *transient* versus steady state conditions, as might be expected. Interestingly, one monitor had an overall sampling bias of 8%, and the second had a bias of -12%. Charcoal tubes had a bias of 5%.

Feigley and Chastain (1982) found extremely mixed results when three makes of moni-

[4] As the amount absorbed increases, the C_e also increases.

tors were exposed to ethyl benzene, methyl ethyl ketone, *n*-hexane, and 1,2,3,3-tetrachloroethane, separately. Constant exposures, as well as beginning and ending peak exposure regimens, were examined. Thus, both transient and $C_0 \neq 0$ errors should have existed. One would expect ending peaks to cause positive biases and beginning peaks to cause negative biases, relative to the constant exposure regimen. Overall, the data showed no general trend, and biases were extremely variable, ranging from 86% to -50%. The *n*-hexane data were the most consistent and did show the expected trend when all monitor makes were grouped, but inconsistent results were seen for individual monitor makes.

Compton et al. (1984) examined the effect of square wave concentration pulsations of ethyl benzene varying from zero to 150 ppm and having periods of 2, 6, and 10 minutes over 3-hour runs. For a non-polar compound such as ethyl benzene, one would expect little bias to occur for this exposure regime, as C_e would be near zero, and each peak represents a small portion of the overall loading. In addition, the overall loadings (approximately 200 ppm-hr) were well within the manufacturer's recommendations. As a result, no significant sampling biases were found.

Hori and Tanaka (1993) examined the effects of very short and transient exposures on sampling bias for a single monitor make exposed to benzene or ethyl acetate for alternating periods of 10 s interspersed with periods of no exposure for 10 s. At these loadings one would not expect C_e to be significant, and thus only the positive errors associated with short period sampling would be expected. However, even these errors were not observed because the authors removed the monitors to a clean environment for 1 minute prior to capping.

Perkins (1981) exposed several monitor makes to variable toluene concentrations and 80% relative humidity. Even when the monitors were exposed to 100 ppm for 4 hours followed by zero ppm for 4 hours, there was no apparent effect on sampling rate relative to the sampling rate at a constant loading. This points out that for toluene there is a strong affinity for charcoal, and therefore C_e is essentially zero under these loading conditions.

In the model that Bartley (1983) developed for calculating concentrations from the two masses found on a monitor with a backup pad, he predicted that such a backup sorbent would substantially reduce errors associated with sampling peaks at the beginning or the end of a sampling period. For methyl chloroform the model predicted errors of -30% or 10% for a DuPont monitor, depending on whether the peak occurred at the beginning or the end, respectively, of the sampling period. The peak times were 0.1 of the sampling period. These errors were reduced to -5% and 3% with a backup sorbent, but would be expected to increase for more polar compounds having less linear isotherms. It should be noted that the above calculations were made at 80% relative humidity, and thus speak well for the added efficiency of backup sorbents when Bartley's calculations are used to find the TWA concentration.

SUMMARY OF RESPONSE TIME AND SORBENT FACTORS

In summary, several types of errors can be examined with respect to response time. All monitors are subject to errors when sampling times are short. The errors increase as sampling times decrease and diffusion pathlengths increase. For a Palmes tube sampling NO_2 for 15 minutes, they can be important. However, they are generally positive and can be offset by capping at the end of the period with a sorbent cap or removing the monitor to clean air for a minute or two prior to regular capping. In addition, all monitors are subject to errors associated with transient concentrations, and these errors are greatest when isolated peaks that account for most of the loading occur at the beginning or the end of the sampling period. They are similar in magnitude to short sampling time errors, and the positive ones (peak occurs at the end of period) can be corrected in the same manner (sorbent capping).

The potentially most serious errors probably have received the least attention and are associated with weak (mainly charcoal) sorbents. For these sorbents, the assumption that the concentration at the sorbent is kept at zero can be violated. The severity of the violation is dependent on the sorbent capacity for the contaminant and the strength of the intermolecular forces holding the contaminant to the charcoal. When these forces are weak or when the capacity, as defined as the near linear portion of the uptake curve is exceeded, then errors associated with transient conditions and sampler overloading may be large. Backup sorbents aid in correcting these errors, and the calculations of Bartley (1983) give the best approach for calculating the TWA concentration from the two masses. This requires that an isotherm be developed for each contaminant. This does not seem to be common practice. Finally, the errors are most likely and yet most unpredictable in sampling mixtures of polar and non-polar compounds. In this case, Bartley's method does not hold unless mixture isotherms are generated; an infinite number of combinations exist. For sampling complex mixtures (a good example would be paint solvents), monitors should have backup pads, and calculated capacities should be reduced considerably. This will mean shorter sampling times and, therefore, the use of more monitors. Because water vapor will compete with polar contaminants, relative humidity levels probably should not exceed 50% (see later discussion of this subject).

External and Internal Mass Transfer, Face Velocity or Boundary Effects

The Palmes tube shown in Figure 23-2 and, indeed, all diffusion-driven monitors work on the principle that the airborne concentration at the face or the beginning of the diffusion path is the ambient concentration. We also assume that transfer of contaminant to the inside of the monitor is by diffusion alone, not by wind or convective currents. If the Palmes tube were situated so that the opening faced into a stiff breeze, then mass would be transferred into the monitor by means other than diffusion. This condition is called *convective internal mass transfer*, and it causes the diffusion pathlength of the monitor to vary and be less than normal; the monitor will oversample.

If, on the other hand, there are absolutely no convective currents from any direction on the exterior of the monitor, then as diffusion causes molecules to enter the monitor, the only means of resupplying those molecules is diffusion on the exterior of the monitor. Therefore, the diffusion pathlength is increased beyond the face of the monitor. It is said that the monitor is starving, and it will undersample. This condition is called resistance to external mass transfer. Both of these situations have been experimentally verified, but they usually can be avoided.

Williams et al. (1995) showed that the pathlength in the stagnant air zone of the monitor was the most important variable affecting variability in sampling rate. For a chlorine device with a windscreen used in their experiments, there was little effect of wind speed at any orientation when the pathlength was about 1 cm. Tompkins and Goldsmith (1977) showed that for the Abcor monitor, a 15 ft/min (7.5 cm/s) face velocity across the monitor opening was sufficient to overcome the external mass transfer problem. Hori and Tanaka (1996) showed that for the 3M OVM the minimum perpendicular face velocity necessary to avoid eternal mass transfer was 0.25 m/s (50 ft/min).

Oversampling of monitors caused by internal mass transfer was demonstrated by Samimi and Falbo (1983, 1985), who directed velocities of 22 and 44 cm/s perpendicular to the face of the Abcor monitor. Sampling rates increased at the higher velocity by 18, 22, and 35% for butyl acrylate, ethyl acrylate and styrene, respectively. Hori and Tanaka (1996) showed that the DuPont sampler would oversample when velocities were perpendicular to the face and as low as 0.5 m/s (50 ft/min). The 3M device did not oversample even when face velocities were up to 2 m/s.

Feigley et al. (1996) used a mannequin to empirically study the effects of wind speed and direction on oversampling and undersampling in the 3M OVM. They determined that wind speeds of at least 70 cm/s (140 ft/min) were necessary to keep the sampling rate of OVMs on the mannequin at no more that ± 10% of the sampling rate for OVMs in unobstructed airflow. However, they noted that the mannequin was fixed and that airflow from behind the mannequin was not considered, two unlikely conditions. They concluded that when the sampler is on a body it will always undersample compared with the sampler in an unobstructed flow situation.

It is important to mention that passive monitors are increasingly used as environmental monitors (outdoors and even indoors). In these applications, starvation of the sampler is more likely to occur. However, maintaining a certain airflow around the sampler and using corrections could remedy most problems. It would seem, however, that more work on face velocity effects is warranted. It can be assumed that in fixed locations or when workers are likely to be oriented consistently perpendicular to the wind, oversampling can occur. For more likely conditions sampling rates may be affected at low wind speeds (low worker movement), but these conditions have not been accurately defined for all monitors. Obviously the greater the movement of the worker the less likely the monitor will be starved.

Permeation

Permeation (see Chapter 8) and Knudsen diffusion are special cases of diffusion. The rate-controlling step for mass transport is diffusion through a polymer membrane rather than diffusion through the air. For permeation, unlike the case of Knudson diffusion, the contaminant molecules have significant molecular interactions with the membrane molecules. This interaction causes some degree of solubility of the contaminant in the membrane, which can be dependent on the airborne concentration.[5] As Figure 23-7 shows, C_a drives C_i, and from Chapter 8, Equation 8-13, $C_i = k_H P$. If the sorbent is 100% efficient, then Equation 8-14 holds, and

$$Flux = \frac{M}{At} = P\frac{C_a}{L} \qquad \text{[Equation 23-10]}$$

It can be seen that if *M, A, t,* and *L* are known and C_a is controlled in a laboratory calibration, then *P* can be determined.

In early publications on permeation devices, a constant *K* was used (Reiszner and West, 1973; West and Reiszner, 1978). The constant *K* was determined by exposing the monitor to

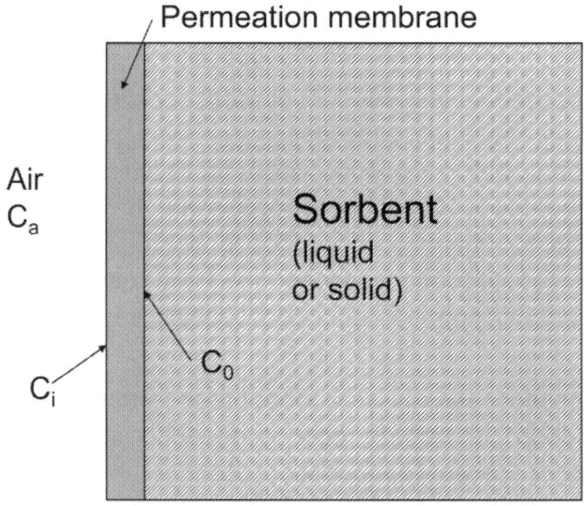

Figure 23-7. A schematic of a permeation device showing the ambient concentration air (C_a), the diffusion driving concentration at the surface of the exposed membrane (C_i), and the concentration at the membrane surface in contact with the sorbent (C_0). $C_0 = 0$ when the sorbent is 100% effective.

[5] Some gaseous molecules have minimal molecular interactions with polymer membranes; for example, oxygen movement through a polyethylene membrane is largely a diffusional process. Even so, the rate of diffusion in air would be different from that in polyethylene. When molecular interaction is minimal, the permeability coefficient (see Chapter 8) is constant for that polymer and contaminant regardless of the ambient concentration of the contaminant and the thickness of the membrane; but when interaction is significant, the permeability coefficient is not constant with concentration. (Yasuda and Stannett, 1975).

known concentrations,

$$K = \frac{C_a\, t}{M}$$ [Equation 23-11]

Thus, from Equation 23-10 we see that

$$K = \frac{L}{PA}$$ [Equation 23-12]

K is similar to the inverse of the sampling rate (DA/L) for the diffusion monitor. In the calibration of permeation monitors against different concentrations and exposure times, the monitors are assayed for the absorbed mass, and mass is plotted against the product of concentration and time. If Equation 23-10 is rearranged to

$$C_a\, t = \frac{L}{PA}\, M$$ [Equation 23-13]

then $L/P\,A$ or K is the slope of this calibration curve.

There are several advantages of permeation monitors. The membranes in some cases may have advantageous selectivity properties, although this can be a difficult thing to accomplish because interfering contaminants often have similar chemical properties to the contaminant of interest and also would likely permeate the membrane. A liquid sorbent can be placed behind the membrane without sorbent loss. The monitor's housing and membrane can be recycled by replacing the sorbent.

Finally, it should be noted that several assumptions are made in deriving Equations 23-12 and 23-13. It has been noted that we assume that $C_a = k_H P$. First, we are assuming that Henry's law holds. Second, the assumption is made that the diffusion coefficient for the contaminant through the membrane is not concentration (C_a) -dependent. These are probably reasonable assumptions at low concentrations. As C_a approaches saturation, these assumptions are likely to be violated.

Length of Stain Diffusion Devices

In Chapter 29, length of stain or detector tubes is described. These tubes are about the size of charcoal tubes and contain a solid, reagent-coated sorbent. The contaminant reacts with the sorbent to cause a color change, the length of which indicates concentration. Normally air moves into the tube via a small hand-held pump. However, some of these tubes have been redesigned or calibrated for passive use. In the Palmes tube and most other diffusion devices, the diffusion pathlength is set. This is important because, as will be seen later, the sampling rate of these devices is proportional to the pathlength. However, in the length of stain monitor the pathlength changes as the reaction of the contaminant consumes the solid sorbent. Thus, calibration of these devices is somewhat more complicated than that for set pathlength devices. Other passive devices employing non-colorimetric sorbents such as charcoal, in long tubes, have been manufactured. The pathlength for these devices also changes as sorption takes place. The rate of change of the diffusion pathlength is a function of the surface-area-to-depth ratio of the sorbent.[6] Perhaps a better name to characterize these devices would be *varying pathlength devices*.

In length of stain detector tubes and passive monitors, the coating is derived such that a specific colorimetric, chemical reaction will occur for a particular chemical or group of chemicals. Interferences are common, but the tubes are relatively inexpensive and have the advantage of giving immediate quantitative results. As air enters the tube, the contaminant reacts with the sorbent, depleting the coating through reaction. As a result, the contaminant reacts farther down the tube, and the length of the color change or stain is proportional to the concentration if the sampling rate is maintained at some prespecified flow rate.

Apparently, Hill and Fraser (1980) first published an approach for using detector tubes passively. They attempted to use Fick's law to predict the performance of the tubes, but it was not predictive. Pannwitz (1984; see also May, 1989) pointed out that Fick's law must be adjusted in the case of detector tubes because, as the reaction proceeds down the tube, the diffusion pathlength increases. For detector tubes, a small change in the mass collected is proportional to the increment in the length of stain, or:

$$dm = dL\,kA \qquad \text{[Equation 23-14]}$$

where k is a constant known as the absorption capacity of a layer element (mg/cm^3), and the other terms are the same as those for passive monitors. Fick's law can also be written in differential form as

$$dm = \frac{DAC_a\,dt}{L} \qquad \text{[Equation 23-15]}$$

assuming that the concentration at the site of reaction is zero. We can assume that the diffusion pathlength is the length of stain, and that the two *dm* terms are equivalent, that is, that any mass entering the tube (as a result of Equation 23-14) is reacted or adds to the length of stain. Therefore, the two equations can be equated as

$$LdL = \frac{DCdt}{k}$$

Integration yields

[6] Even in devices employing very thin layers of charcoal, there is a theoretical change of pathlength over time. However, in practice this change can be ignored.

$$\frac{L^2}{2} = DC_a \frac{t}{k}$$

Thus, the concentration is proportional to the square of the stain length.

Pannwitz (1984) showed that by plotting the product *Ct* versus L^2 from calibration experiments, a straight-line relationship is obtained. The slope should be *2D/k*. Such a calibration chart should accompany the detector tube sold as a passive monitor, or the concentration readings printed on the tube should be non-linear, becoming farther apart as one proceeds from the open end of the tube.

Sefton et al. (1982) devised a combination length of stain-permeation device that is claimed to have the following advantages: The membrane provides some selectivity, and a narrower tube makes for a longer and easier-to-read stain than in other length of stain devices. The addition of the permeation membrane makes the mathematical modeling of the device a bit more difficult. However, a suitable model was derived, calibration curves were generated, and reasonably low calibration errors (< ± 20% at the 95% confidence level) were seen. It is not known if the technique is being used commercially.

A list of chemicals for which these dosimeters is available along with measurement ranges is given by ASTM (2003).

OTHER FACTORS IN OPERATION

Relative Humidity

It has already been mentioned that charcoal-based passive monitors are affected by water vapor in ways similar to the effects on charcoal tubes. In addition, it was recommended that these monitors not be used when relative humidity (RH) exceeds 50%. This is a reasonable recommendation, based on the following research. Gregory and Elia (1983) exposed two types of monitors to methyl chloroform (MC) concentrations of roughly 250, 500, and 700 ppm or about 0.5, 1, and 2 times the then-current OSHA PEL of 350 ppm for one hour. The monitors were then exposed to either 10, 55, or 70% RH uncontaminated air for 1, 3, or 6 hours.

Figure 23-8 summarizes some of the results. Sample losses were very high at 70% RH, and significant but smaller at 55% RH. The near-zero losses at 10% RH indicate that, in relatively dry environments, charcoal's affinity for MC is quite high. Therefore, sample losses were caused by water displacing MC and not by a change in C_e caused by the zero ambient concentration of MC. Similar experiments were run for methylene chloride, toluene, isopropanol, and methyl isobutyl ketone. Losses were insignificant for toluene and MIBK. The first result is expected, but the second a bit curious in light of the polarity of ketones. Isopropanol losses were 35% after 3 hours of exposure to 70% RH. Methylene chloride losses were 20% after one hour and 70% after three hours; however, similar losses were reported for charcoal tubes after similar exposures, indicating, the extremely low retention volume of

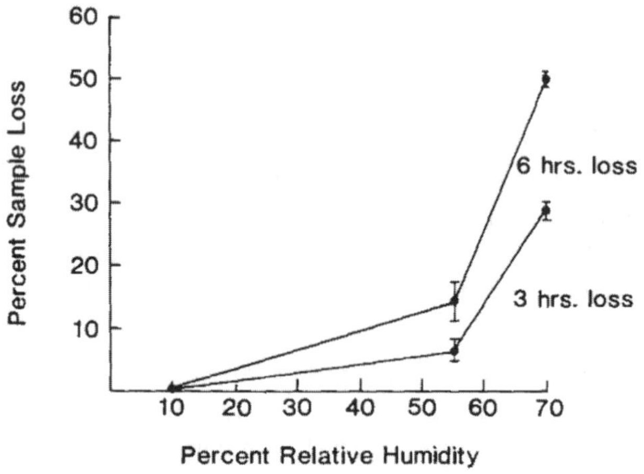

Figure 23-8. The effect of relative humidity on sample loss of the Pro-Tek G-AA sampler at 5 mg loading of methylchloroform (with permission from Gregory and Elia, 1983).

charcoal for this volatile compound.

Perkins (1981) conducted slightly different experiments involving toluene and RH. He exposed three types of OVMs to 100 ppm toluene in conjunction with 10, 50, or 80% RH for seven hours. All of the OVMs showed slight losses (< 10%) at 50% RH. However, losses at 80% RH were about 20% for at least one OVM model, indicating that even for a non-polar compound such as toluene, the type of charcoal may lead to significant RH interference.

One would not expect that water vapor presence during sampling would be an important factor in monitors employing strong (reactant) sorbents, and it probably is not. However, it may affect the sorbent during storage and prior to use of the monitor. More specifically, Kennedy and Hull (1986) found that 3M formaldehyde monitors gave consistent and unbiased results when they were first conditioned at 80% RH prior to exposure. Without the conditioning, the results were erratic and had considerable negative bias. Their interpretation was that the devices were losing water during storage (prior to being exposed). The manufacturer has since taken care of the problem, and one hopes that this event caused all manufacturers to look at the effects of water, or the loss of it, on performance. Another related aspect is the fact that generating formaldehyde in the laboratory for calibration purposes is not straightforward. Para-formaldehyde appears to yield entirely formaldehyde monomer, whereas evaporated formalin yields a mix of formaldehyde and methoxymethanol (Pengelly et al., 1996).

Lindenboom and Palmes (1983) reported that the Palmes tube had an apparent loss of sampling rate at low pressures (equivalent to 18,000 ft or about 5,000 m). There should not be an effect of pressure on sampling rate (see earlier discussion under *Temperature*). Later Palmes and Johnson (1987) reported that the anomaly could be explained by the dehydration effects of reduced pressures on the sorbent used in the NO_2-absorbing Palmes tube.

Other Interferents

Once again the performance of charcoal in sampling one contaminant is most likely to be affected by the presence of others, but only a few experiments have been made to confirm this suspicion. Gregory and Elia (1983), in their experiments of the effects of RH on methyl chloroform retention, also examined how the presence of toluene at a concentration of 780 ppm affected methyl chloroform retention. For one device the losses of methyl chloroform were 10% and 30% after exposures to toluene for 1 hour and 3 hours, respectively. The other device showed no significant losses.

Perkins (1981) exposed three OVM models to a complex mixture of five compounds at quite low concentrations for a period of 7 hours. The results shown in Table 23-1 indicate that none of the monitors performed acceptably. When a t-test was performed comparing monitor mean versus active charcoal tube mean, one monitor failed all five tests. A more lenient test that compared the monitor mean to a range (charcoal tube mean ± 25%) gave essentially the same results. It should be noted that the loadings (ppm-hr) for the five compounds were well within manufacturer's guidelines. The combined loading only exceeded the recommended loading for methylene chloride. Two of the monitors are no longer manufactured and it is likely that, in the intervening years, manufacturers have improved their product based on these and other early results.

Table 23-1
Monitor performance summary for precision and accuracy of a complex mixture (adapted from Perkins, 1981)

Compounds	Proposed Concentration (ppm)	ACT[1] Mean ±25%(ppm)	CT Mean vs. Monitor Mean ($\alpha = 0.05$)			Monitor $\bar{x}$ Within ACT $\bar{x}$ ± 25% (95% Confidence)		
			Monitor Type					
			3M[2]	AB	RZ	3M	AB	RZ
Toluene	50	53.2 (39.9 – 66.5)	A[3]	N	A	A	A	A
Trichloroethylene	25	26.0 (19.5 – 32.5)	A	N	A	A	N	A
n-Hexane	20	20.9 (15.7 – 26.1)	A	N	N	A	N	A
Acetone	10	7.6 (5.7 – 9.5)	N	N	N	N	N	N
Methylene chloride	10	10.0 (8.2 – 13.6)	N	N	A	N	N	A

[1] ACT = Activated Charcoal Tube
[2] RZ = Reiszner MiniMonitor, 3M = 3M OVM, AB = Abcor Monitor
[3] A = Acceptable N = Not acceptable

Dust Interference

In a study performed by Stewart et al. (1987), it was found that passive monitors for formaldehyde often gave concentrations that could not be accounted for on the basis of human irritation. Concentrations of 5 ppm were reported, but irritation occurs in most people at levels of 0.12 to 1.6 ppm. They suspected that dust from resin operations, in which a formaldehyde resin was used, was the cause. The dust was present on the diffusion surfaces of the monitors, and CH_2O was perhaps offgassing from the dust prior to analysis. During the winter sampling period the dust was wiped off, and lower concentrations were found. (As an aside, they wondered if the CH_2O was not also offgassing from the dust in the lungs.) Although the presumed cause of the high concentrations has not been confirmed, it is plausible, and hygienists should be on the watch for similar anomalies that can also occur during active sampling. In addition, the dose received by the worker, as it is partitioned to different physical states, needs further attention in many cases.

VALIDATION OF MONITORS

Given the potential sources of error cited above, it is surprising that a standardized validation protocol for passive monitors was not devised early on and adhered to by every manufacturer. Such a protocol for active sampling methods was offered by NIOSH in the 1970s and is discussed in Chapter 23. However, the potential errors associated with passive monitoring took about 10 years to be discovered (provided that we found them all). In the meantime, manufacturers and distributors came and went in rapid succession. As a consequence only in recent years have government agencies begun to offer protocols that are thorough enough to examine all of the known sources of error. Early proposals were made (see, e.g., Coulson, 1981; Brown et al., 1984); but even though they gave the protocols for experiments, they did not give a basis for interpretation of results. More recent protocol designs have attempted to do that.

The NIOSH Protocol

Cassinelli et al. (1987) of NIOSH proposed one of the earliest protocols. The protocol is published in a somewhat obscure conference proceeding; so although manufacturers are no doubt aware of it, most hygienists probably are not. By becoming familiar with the sources of error previously described and the protocol experiments, the hygienist can get a better idea of the kind of experimental results that manufacturers should be offering.

The essentials of experimental design and interpretation of results are given in Table 23-2. It can be seen that the effort to complete the protocol is no small undertaking, and yet each of the experiments seems reasonable in light of the potential errors. A total of 238 monitors are required to complete the experiments, not to mention 40 to 50 active sampler results in the field comparison tests.

Table 23-2
NIOSH protocol for the validation of passive monitors (adapted from Cassinelli et al., 1987)

Characteristics	Experimental Design	Interpretation of Results
1. Analytical Recovery	Spike 16 monitors, 4 at each of 4 concentration levels (0.1, 0.5, 1.0 and 2.0 × STD). Equilibriate about 12 h, and analyze.	For the 3 higher levels require ≥ 75% recoveries with $\bar{s}_r \leq 0.1$.
2. Sampling Rate and Capacity	Expose monitors (4 per time period) for ⅛, ¼, ½, 1, 2, 4, 6, 8, 10 & 12 h to 2 × STD. 80% RH and 20 cm/s face velocity. Plot concentration vs. time exposed. Determine MRST and SRST.	Verify sampling rate. State useful range at 80% RH & 2 × STD. Capacity = sample loading corresponding to the downward break in uptake vs. time curve from constant concentration (Fig. 21-5). SRST = time linear uptake rate achieved MRST = 0.67 × capacity (1 analyte) MRST = 0.33 × capacity (multi-analyte)
3. Reverse Diffusion	Expose 20 monitors to 2 × STD and 0.5 × MRST. Analyze 1st set within 1 day, 2nd set after 2 weeks storage at about 25°C, 3rd set after 2 weeks storage at about 5°C.	Require ≤ 10% difference between means of the two monitor sets at the 95% CL.
4. Storage Stability *Laboratory Analyzed:*	Expose 3 sets of monitors (10 per set) at 80% RH, 1 × STD and 0.5 × MRST. Analyze first set within 1 day, second set after 2 weeks storage at about 25°C, third set after 2 weeks storage at about 5°C.	Require ≤10% difference at the 95% CL between means of stored monitor sets and set analyzed within 1 day.
Direct Reading:	Expose 10 monitors to storage test conditions. Take readings within 5 minutes, then at 15, 30, 60 & 120 minutes.	State maximum time after exposure when readings are ≥ 90% of initial readings.
5. Factor Effects	Test the following factors at the levels shown. Use a 16-run fractional factorial design (4 monitors per exposure) to determine significant factors.	Indicate any factor that causes a statistically significant difference in recovery at the 95% CL. Investigate further to characterize its effect.

Factor	Test Levels
analyte concentration	0.1 & 2 × STD
exposure time	SRST & MRST
face velocity	10 & 150 cm/s
relative humidity	10 & 80% RH
interferent	0 & 1 × STD
monitor orientation	parallel & perpendicular (to airflow)

Table 23-2 (Cont.)
NIOSH protocol for the validation of passive monitors (adapted from Cassinelli et al., 1987)

Characteristics	Experimental Design	Interpretation of Results
6. Temperature Effects	Expose monitors (10 per temp.) to 0.6 × STD at 10, 25 & 40°C for 0.5 × MRST.	Define temperature effect and verify correction factor, if provided.
7. Accuracy and Precision	Calculate precision and bias for monitors (≥10 per conc. level) exposed to 0.1, 0.5, 1 & 2 × STD at 80% RH for ≥ 0.5 × MRST. Use data from previous experiments.	Require bias within ± 25% of true value at 95% CL with precision $\bar{s}_r$ ≤ 10.5% for 0.5, 1 & 2 × STD levels.
8. Shelf Life	Observe monitors throughout evaluation for changes in blank values, physical appearance, etc. Test monitors from more than one lot, if possible.	Note shelf storage time at which changes begin to occur. Indicate whether correctable or not.
9. Behavior in the Field	Consider problems not predictable from laboratory experiments.	Record temperature, humidity, air velocity, other contaminants, etc.
Area Sampling	Expose passive monitors and independent method samplers (13 each) to the same environment.	Calculate precision and bias. Compare with laboratory results.
Personal Sampling	Conduct personal sampling with ≥ 25 sampler pairs. Place pairs of passive monitors and independent samplers on the same lapel of each worker.	Calculate bias. Compare with laboratory results.

STD – The appropriate exposure standard (OSHA PEL, ACGIH® TLV® or NIOSH REL).
MRST – Maximum recommended sampling time.
$\bar{s}_r$ – Pooled relative standard deviation.
RH – Relative humidity.
SRST – Shortest recommended sampling time.
CL – Confidence level.

The protocol was used by NIOSH in tests of monitors for SO_2, NH_3, and CH_2O (Kennedy and Hull, 1986). The tests turned up such interesting outcomes as the effect of water loss during storage of the 3M CH_2O monitor and the unexpected oversampling of sulfur dioxide monitors for *short* periods that were as long as 1 hour. Those two results alone were worth the costs of the tests.

A modification has been proposed to the protocol, which seems reasonable for charcoal-based OVMs (Guild et al., 1992; Harper and Guild, 1996; Brown et al., 1989). The proposal is to perform a full validation on so-called key compounds (the lowest molecular weight members) in a homologous series. Other compounds would receive less testing. The idea is reasonable because it is likely that reverse diffusion, storage stability and capacity, and relative humidity effects would be most remarkable for the most volatile compound in the series. On the other hand, the effects of starvation and temperature changes are less predictable, but also not so important as other effects. At least one manufacturer (SKC) is using this proposed terminology, calling it the NIOSH protocol level 5, and the shortened protocol for other homologous series' members is called level 5B.

The Health and Safety Executive (HSE) of Great Britain also proposed an evaluation protocol (Firth, 1987; Brown, 1987). There are some experimental design differences relative to the NIOSH method, however, the protocols do have a number of similarities. In addition, validation standards have been published by ISO (1999), CEN (1995) and ASTM (2002). There are similarities between all and it has been proposed (Harper and Guild, 1996) that a manufacturer who passes the NIOSH standard, for example, should receive reciprocity to the European standard (CEN).

Field Studies and PAT Program

An important and controversial part of the NIOSH and HSE protocols is the requirement for side-by-side field comparisons of the passive and active methods. Many field studies have been published for passive monitors, and they should be referred to before further field studies or actual field use takes place (see Table 23-3). In addition, an excellent free publication presents up-to-date passive monitor results from around the world (HSL). Field studies of

Table 23-3
Chemicals and classes of chemicals for which field studies of passive monitors have been published

Chemicals	Classes
ammonia	acetates
ethylene oxide	aromatic hydrocarbons
formaldehyde	alcohols
hydrogen fluoride	alkanes
methylene chloride	anesthetic gases
nitrogen dioxide	chlorinated solvents
nitrous oxide	ketones
sulfur dioxide	mixed solvents
styrene	nitriles
toluene diisocyanate	oil mists
vinyl chloride	paint solvents
	polynuclear aromatic hydrocarbons
	particles

monitor performance under conditions of probable use are conceptually the ultimate test of monitor performance, as all confounding variables of importance in that operation should be occurring. However, as Firth (1987) noted, interpretation of such results is not without difficulty. There are also many possible field combinations of important variables. Which is best? Finally, manufacturers have trouble finding a proprietor who is willing to help generate published data that may be incriminating (Harper and Guild, 1996).

A typical analysis approach used in many studies of personal, paired side-by-side results is regression analysis. Such an analysis usually assumes that the independent variable (usually assigned as the active method) is error-free. If it is not, then estimates of the strength of the relationship (both the slope and the correlation coefficient) will be low (Fuller, 1987). Another way of looking at the problem is that a regression of active on passive and vice versa should give equivalent results (Tuggle and Hawkinson, 1981). If the results are not equivalent, then each method has a different relationship with the true concentrations.

Simple *t*-tests compare the central tendency only, not differences at the extremes. As Tuggle and Hawkinson have pointed out, equivalence of means is a necessary but not sole condition for equivalence of methods. However, *F*-ratio tests, for comparison of variances, often are overshadowed by the large component of the variance caused by the environment relative to each method (see Chapters 16 and 30). Finally, it is important to point out that a *t*- or *F*-test that points to equivalence (i.e., the null hypothesis of equality is not rejected) when performed at an α level of 5% does not imply 95% confidence about the result. Instead β error, not α error, is important in a fail-to-reject outcome. The β error and the complementary power of the test should be back-calculated for this outcome. Tuggle and Hawkinson suggest that a power of 80% should be necessary as a condition for excepting the test result. A method for simultaneously testing both bias and precision of two methods relative to one another that is more powerful than *F*- and *t*-tests separately is available (Blackwood and Bradley, 1991; Russell et al., 1994) (see Chapter 30 for details). Other approaches for comparing paired method results have been proposed but require extremely large sample sizes (Rockette and Wadsworth, 1985).

There is some obvious duplication in the NIOSH and HSE requirements to take both fixed and personal samples in the field. Other than perhaps the effects of face velocity and a wider range of sampled concentrations in the personal samples, other factors such as temperature, RH, and interferences should be similar. The comparison of fixed sampling results suffers from the same statistical problems as those for personal sampling results. However, in the case of fixed sampling, regression analysis may not be reasonable because concentration will be near constant from sampler to sampler at a given location. Relative bias and precision can be calculated and compared to laboratory-derived results. However, the matter of acceptable differences in the field and laboratory precision and bias results needs careful consideration; no guidelines were given by NIOSH.

In summary, passive monitors offer an overriding advantage in ease of use. Where hygiene field strength is in short supply, workers can be trained to conduct their own monitoring. In addition, when consideration of the cost and the maintenance (time and materials) of pumps is considered, passive systems may be less expensive than active systems (see

Nothstein et al., 2000, for an excellent presentation of a cost analysis). The potential confounding variables described here should not be a roadblock to the use of passive monitors. Rather, this discussion should heighten the hygienist's awareness of the need for adequate validation studies from manufacturers and field studies on the part of the user.

AIHA conducts a Proficiency in Analytical Testing (PAT) program that involves AIHA accredited laboratories (see Chapter 26 for details). Included in the program is an option to analyze dosimeters from 3M, SKC, and Assay Technology that have been exposed under controlled conditions to benzene, toluene, and o-xylene. Until 1999 NIOSH analyzed results of the PAT program on an *as requested* basis. The last report from NIOSH indicated that passive dosimeter results agreed better with PAT charcoal tube results than with calculated chamber concentrations (Feng and Schlecht, 1999). Further PAT details are in Chapter 26.

QUESTIONS

23.1. What is the difference between permeation and diffusion? Is one a special case of the other?

23.2. What happens to the rate of diffusion or the flux rate in a passive device if the sorbent is less than 100% efficient? Explain your answer in terms of Fick's law (Equation 23-1).

23.3. Name at least five factors that could cause a bias in the measured concentration of a polar organic compound when using a passive diffusion device containing charcoal. State whether each bias is likely to be positive, negative, or both.

23.4. Assuming that the diffusion coefficient is proportional to absolute temperature to the 1.5 power and that the ambient concentration (mass per volume) is indirectly proportional to absolute temperature, use Equation 23-3 to show that the mass taken up by a diffusion device is proportional to $T^{0.5}$. Show unit cancellation.

23.5. Explain why the backup pad in a charcoal passive device is not analogous to the backup section of a charcoal tube.

23.6. In terms of Fick's law (Equation 23-1), what variable is affected when the air at the face of a passive device is stagnant?

REFERENCES

ASTM: Standard Practice for Evaluating the Performance of Diffusive Samplers, D 6246. ASTM International, West Conshohocken, PA, USA (2002).Order at www.astm.org.

ASTM: Standard Practice for Measuring the Concentration of Toxic Gases or Vapors Using Length-of-Stain Dosimeters, ASTM D4599-03. ASTM International, West Conshohocken, PA, USA (2003).

Bartley, D.L.: Passive monitoring of fluctuating concentrations using weak sorbents. Am. Ind. Hyg. Assoc. J. 44:879-885 (1983).

Bartley, D.L.; Doemeny, L.J.; Taylor, D.G.: Diffusive monitoring of fluctuating concentrations. Am. Ind. Hyg. Assoc. J. 44:241-247 (1983).

Bartley, D.L.; Deye, G.J.; Woebkenberg, M.L.: Diffusive monitor test: Performance under transient conditions. Appl. Ind. Hyg. 2:119-122 (1987).

Blackwood, L.G.; Bradley, E.L.: An omnibus test for comparing two measuring devices. J. Quality Technol. 23: 12-11: (1991).

Brown, R.H.: Applications of the HSE diffusive sampler evaluation protocol, in Diffusive Sampling: An Alternative Approach to Workplace Air Monitoring, A. Berlin, R.H. Brown, K.J. Saunders (eds.). Royal Society of Chemistry, London (1987).

Brown, R.H.; Harvey, R.P.; Purnell, C.J.; Saunders, K.J.: A diffusive sampler evaluation protocol. Am. Ind. Hyg. Assoc. J. 45:67-75 (1984).

Brown, R.H.; Saunders, K.J.; Fields, B.; et al.: A proposal for a harmonized approach to the evaluation of diffusive, samplers, paper presented at the American Industrial Hygiene Conference, St. Louis, MO (1989).

Cassinelli, M.E.; Hull, R.D.; Cuendet, P.A.: Performance of sulfur dioxide passive monitors. Am. Ind. Hyg. Assoc. J. 46:599-608 (1985).

Cassinelli, M.E.; Hull, R.D.; Crable, J.V.; Teass, A.W.: Protocol for the evaluation of passive monitors, In Diffusive Sampling: An Alternative Approach to Workplace Air Monitoring, A. Berlin, R.H. Brown, K.J. Saunders (eds.). Royal Society of Chemistry, London (1987).

CEN: Workplace atmospheres – Diffusive samplers for the determination of gases and vapours – Requirements and test methods EN 838:1995. Comite Europeen de Mormalisation, Brussels (1995).

Compton, J.R.; Dwiggins, G.A.; Feigley, C.E.; Ludwig, D.A.: The effect of square wave exposure profiles upon the performance of passive organic vapor monitoring badges. Am. Ind. Hyg. Assoc. J. 45:446-450 (1984).

Coulson, D.M.: Method validation, In Annals of the American Conference of Governmental Industrial Hygienists Dosimetry for Chemical and Physical Agents. Vol. 1, pp. 83-90, W. D. Kelley (ed.). ACGIH®, Cincinnati, OH (1981).

Einfield, W.: Diffusional sampler performance under transient exposure conditions. Am. Ind. Hyg. Assoc. J. 44:19-35 (1983).

Feigley, C.E.; Chastain, J.B.: An experimental comparison of three diffusion samplers exposed to concentration: profiles of organic vapors. Am. Ind. Hyg. Assoc. J. 43:227-234 (1982).

Feigley, C.E.; Underhill, D.W.; Lungu, C.T.; Riley, T.D.: The sampling rates of diffusive samplers measured in the laboratory and in simulated use as personal samplers. Am. Ind. Hyg. Assoc. J. 57:551-556 (1996).

Feng, H.A.; Schlecht, P.: Proficiency analytical testing (PAT) program (November 23, 1998). Am. Ind. Hyg. Assoc. J. 60:281-284 (1999).

Firth, J.G.: HSE protocol for assessing performance, In Diffusive Sampling: An Alternative Approach to Workplace Air Monitoring, A. Berlin, R.H. Brown, and K.J. Saunders (eds.). Royal Society of Chemistry, London (1987).

Fuller, W.A.: Measurement Error Models, pp. 1-99. Wiley Publishing, New York (1987).

Fuller, W.A.; Giddings, J.C.: A comparison of methods for predicting gaseous diffusion coefficients. J. Gas Chrom. 3:222-227 (1965).

Gregory, E.D.; Elia, V.J.: Sample retentivity properties of passive organic vapor samplers and charcoal tubes under various conditions of sample loading, relative humidity, zero exposure level periods, and a competitive solvent. Am. Ind. Hyg. Assoc. J. 44:88-96 (1983).

Guild, L.V.; Myrmel, K.H.; Myers, G.; Dietrich, D.F.: Bi-level passive monitor validation: A reliable way of assuring sampling accuracy for a larger number of related chemical hazards. Appl. Occup. Env. Hyg. 7:310-317 (1992).

Harper, M.; Guild, L.V.: Experience in the use of the NIOSH diffusive sampler evaluation protocol. Am. Ind. Hyg. Assoc. J. 57:1115-1123 (1996).

Hearl, F.J.; Mauning, M.P.: Transient response of diffusion dosimeters. Am. Ind. Hyg. Assoc. J. 41:778-783 (1980).

Hill, R.H.; Fraser, D.A.: Passive dosimetry using detector tubes. Am. Ind. Hyg. Assoc. J. 41:721-729 (1980).

Hirschfelder, J.O.; Curtiss, C.F.; Bird, R.B.: Molecular Theory of Gases and Liquids, pp. 578-596. Wiley Publishing, New York (1964).

Hori, H.; Tanaka, I.: Response characteristics of the diffusive sampler at fluctuating vapor concentrations. Am. Ind. Hyg. Assoc. J. 54:95-101 (1993).

Hori, H.; Tanaka, I.: Effect of face velocity on performance of diffusive samplers. Ann. Occup. Hyg. 40(4):467-476 (1996).

HSE: Protocol for assessing the performance of a diffusive sampler Health and Safety Executive Occupational Medicine and Hygiene Laboratory Methods for the Determination of Hazardous Substances MDHS 27, revised. Health and Safety Executive, London, UK (February 1994).

HSL: The Diffusive Monitor. HSE Health and Safety Laboratory, London, UK (since 1988).

ISO: Workplace Atmospheres – Protocol for Evaluating the Performance of Diffusive Samplers. International Organization for Standardization (ISO), ISO/FDIS 16107, www.iso.ch (1999).

Kennedy, E.R.; Hull, R.D.: Evaluation of the DuPont Pro-Tek formaldehyde badge and the 3M formaldehyde monitor. Am. Ind. Hyg. Assoc. J. 47:94-105 (1986).

Lautenberger, W.J.; Kring, E.V.; Morello, J.A.: A New Passive Organic Vapor Badge with Backup Capability. ACS: Symp. Series No. 149 (1981).

Lindenboom, R.H.; Palmes, E.D.: Effect of reduced atmospheric pressure on a diffusional sampler. Am. Ind. Hyg. Assoc. J. 44: 105-108 (1983).

Lugg, G.A.: Diffusion coefficients of some organic and other vapors in air. Anal. Chem. 40:1072-1077 (1968).

May, W.J.: Gas detection by diffusive samplers – theory and application to the Draeger range of samplers. Ann. Occup. Hyg. 33:69-78 (1989).

Nothstein, G.L.; Hahne, R.M.A.; Spence, M.W.: Evaluation of the cost-effectiveness of various passive monitors and active monitors for industrial hygiene sampling. Am. Ind. Hyg. Assoc. J. 61:64-68 (2000).

Palmes, E.D.; Johnson, E.R.: Explanation of pressure effects on a nitrogen dioxide (NO_2) sampler. Am. Ind. Hyg. Assoc. J. 48:73-75 (1987).

Palmes, E.D.; Gunnison, A.F.; DiMattio, J.; Tomczyk, C.: Personal sampler for nitrogen dioxide. Am. Ind. Hyg. Assoc. J. 37:570-577 (1976).

Pannwitz, K.H.: Direct reading diffusion tubes. Praeger Rev. 53:11-17 (June 1984).

Pengelly, I.; Groves, J.A.; Levin, J.O.; Lindahl, R.: An investigation into the differences in composition of formaldehyde atmospheres generated from different source materials and the consequences for diffusive sampling. Ann. Occup .Hyg.40(5):555-567 (1996).

Perkins, J.: Laboratory evaluations of passive organic vapor monitors, in Dosimetry for Chemical and Physical Agents, W. D. Kelley (ed.). American Conference of Governmental Industrial Hygienists, Cincinnati, OH (1981).

Perkins, J.L.; You, M.J.: Predicting temperature effects on chemical protective clothing permeation. Am. Ind. Hyg. Assoc. J. 53:77-83 (1992).

Persoff, P.; Hodgson, A.T.: Correction for external mass-transfer resistance in diffusive sampling. Am. Ind. Hyg. Assoc. J. 46:648-652 (1985).

Posner, J.C.; Moore, G.: A thermodynamic treatment of passive monitors. Am. Ind. Hyg. Assoc. J. 46:277-285 (1985).

Reid, R.C.; Sherwood, T.K.: The Properties of Gases and Liquids: Their Estimation and Correlation, 2nd ed., pp. 520-570. McGraw-Hill, New York (1966).

Reiszner, K.D.; West, P.W.: Collection and determination of sulfur dioxide incorporating permeation and West-Gaeke procedure. Env. Sci. Technol. 7:526-532 (1973).

Rockette, H.E.; Wadsworth, H.M.: Equivalency of alternative cotton dust samplers to the vertical elutriator. Am. Ind. Hyg. Assoc. J. 46:716-719 (1985).

Russell, C.M.; Williamson, D.F.; Bartko, J.J.; Bradley, E.L.: A simulation study of reliability indicators applied to paired measurements. Am. J. Hum. Bioi. 6:311-320 (1994).

Samimi, B.; Falbo, L.: Validation of Abcor (NMS) organic vapor dosimeter under various concentrations and air velocity conditions. Am. Ind. Hyg. Assoc. J. 44:402-408 (1983).

Samimi, B.; Falbo, L.: Comparison of standard charcoal tubes with Abcor (NMS) gasbadges within controlled atmosphere. Am. Ind. Hyg. Assoc. J. 46:49-52 (1985).

Sefton, M.V.; Kostas, A.V.; Lombardi, C.: Stain length passive dosimeters. Am. Ind. Hyg. Assoc. J. 43:820-824 (1982).

Stewart, P.A.; Cubit, D.A.; Blair, A.; Spirtas, R.: Performance of two formaldehyde passive dosimeters. Appl. Ind. Hyg. 2:61-65 (1987).

Tompkins, F.C.; Goldsmith, R.L.: A new personal dosimeter for the monitoring of industrial pollutants. Am. Ind. Hyg. Assoc. J. 38:371-377 (1977).

Tuggle, R.M.; Hawkinson, R.W.: Incorrect use of t-tests. Am. Ind. Hyg. Assoc. J. 42:325-326 (1981).

Underhill, D.W.: Unbiased passive sampling. Am. Ind. Hyg. Assoc. J. 44:237-239 (1983).

West, P.W.; Reiszner, K.D.: Field tests of a permeation type personal monitor for vinyl chloride. Am. Ind. Hyg. Assoc. J. 39:645-650 (1978).

Williams, C.E.; Pintauro, P.N.; Rando, R.J.: A transient model of mass transfer and kinetics in a passive vapor sampler. Am. Ind. Hyg. Assoc. J. 56:1074-1082 (1995).

Yasuda, H.; Stannett, V.: Permeability coefficients, In Polymer Handbook, 2nd ed., J. Brandrup, E.H. Immergut (eds.). Wiley Publishing, New York (1975).

Youngquist, G.R.: Diffusion and flow of gases in porous solids. Industrial and Engineering Chemistry 62:52-63 (1970).

24

AEROSOLS

INTRODUCTION

Aerosols[1] can be defined as particles, liquid or solid, suspended in air. This obviously is a broad definition that can include particles of varying chemical and physical properties. Certainly particle size, shape, optical properties, polarity, volatility, and other variables should come to mind. All of these are important in the sampling of aerosols. Aerosols can be characterized further according to their physical state (solid or liquid) and the way they are generated (dusts, mists, fumes, etc.). The latter characteristic also is qualitatively indicative of particle size.

Historically, aerosols were sampled with impingers, such as the Greenburg-Smith or the midget, based on a method developed in 1925 (see Ayer, 1969). Later, it became well known that impingers were not efficient (typically less than 50%; Patty, 1958) at smaller particle sizes (< 0.7 μm). Employing impingers in series will not substantially increase the collection of the smallest particles. Efficiencies for two impingers in series, sampling lead fumes, ranged from about 50% to 90% (Patty, 1958). Efficiencies varied with the particle size distribution of each particular aerosol.

Certain solid aerosols were analyzed microscopically by counting particles, yielding results in million particles per cubic foot (mppcf).[2] Silica, in particular, and coal dust were important in this group of aerosols. TLVs® set prior to 1976 (ACGIH®, 1976) and exposure limits of many countries were based on studies of disease in which monitoring had been per-

---

[1] There are several excellent volumes on aerosol applications to hygiene. These include Hinds (1999) and Baron and Willeke (2001).

[2] These were the so-called mineral dusts (see any *TLVs® and BEIs®* book prior to 1987) and included many minerals such as silica and all of its crystalline and amorphous forms, coal dust, diatomaceous earth, graphite, mica, perlite, Portland cement, soapstone, talc, and many *nuisance* particles (see Chapter 10).

formed using the impinger method.[3] However, studies in the late 1940s of the contents of diseased lungs led to the observation that only dusts of small particle sizes appeared to cause certain pneumoconioses, particularly silicosis. In addition, it appeared that the total mass of this particle size fraction was of toxicological importance, not the total number of particles. This portion of the total dust concentration that entered the deepest parts of the lungs was termed the *respirable fraction* and contained very few particles larger than 10 μm. Impingers collected larger particles very efficiently, but the smaller, more harmful particles were collected less efficiently. For these and other reasons (see Ayer, 1969), it was realized that a new, particle size-selective method of sampling was needed for some particles, with results reported in mass/volume or mg/m^3.

A shift to such a gravimetric technique is not straightforward, since conversion of the mppcf exposure limits to mg/m^3 depends on the density and the size distribution of the particles. Nonetheless, in 1976, ACGIH® recommended[4] conversion of all mineral dust standards using the factor 6 mppcf = 1 mg/m^3. This equality was based on a single study (Jacobson and Tomb, 1967) and some calculations. Subsequent studies have shown a wide range of conversion factors, from 3 to 22, for a number of minerals (Rice et al., 1984; Sheehy and McJilton, 1987; Tomb and Haney, 1988; Montgomery et al., 1991).

By the 1960s U.S. hygienists were realizing the utility of membrane filters for sampling *mineral dusts* as well as other aerosols (see, e.g., Renshaw et al., 1969). Thus, even before exposure limits changed from mppcf to mg/m^3, filtration became a common method for dusts. To compare results with the mppcf limits, the filters typically were dissolved so particles could be counted under the microscope. However, counting represented a considerable effort. Today only fiber counting (mostly asbestos and fiber glass) is performed on particles, as the exposure limits for these substances are expressed in fibers/cm^3 of air; but even for these substances, the impinger is a relic of the past – the filter has taken its place.

FILTRATION

Filtration Principles

Some of the filters used to collect aerosols are composed of matted fibers such as fiber glass. Others are membranes that have relatively uniform pores, in both spacing and size. Still there are membranes with *pores* that follow a more tortuous path. Regardless of the filter type,

[3] In the United Kingdom, the method of monitoring was the thermal precipitator, and in South Africa the konimeter (Patty, 1958; Elkins, 1959). Yet in each case, the results were reported in mppcf, mp/m^3, or particles/cm^3.

[4] The history on this subject is somewhat confusing. Even though ACGIH® made notice of the recommended change to the use of the conversion factor 6 mppcf = 1 mg/m^3 in the 1967 *TLVs® and BEIs®* book, it did not provide the rationale for doing so until the 1976 *TLVs® and BEIs®* book. Indeed, when OSHA adopted the 1968 TLVs® as PELs, it adopted this change in units also.

the mechanisms by which it collects particles can be envisioned by imagining a particle coming near the surface of the filter, such as one of the fibers in a fiber glass filter.

There are five possible mechanisms by which a particle can be *caught* by a filter (see Figure 24-1). These same mechanisms were used to explain the deposition of particles in the lungs in Chapter 6. The first three – interception, sedimentation, and impaction – are effective in capturing large particles, whereas diffusion and electrostatic charge are effective for smaller particles. In interpreting Figure 24-1, one should realize that a small particle will travel with the streamlines of laminar airflow, even as the streamline changes direction. However, as the particle size becomes larger, its inertia becomes greater and, therefore, the particle tends to escape the streamline as the latter makes a change in direction. If the inertia is large enough, the particle makes no or little change in direction and crashes into the filter – impaction. If the

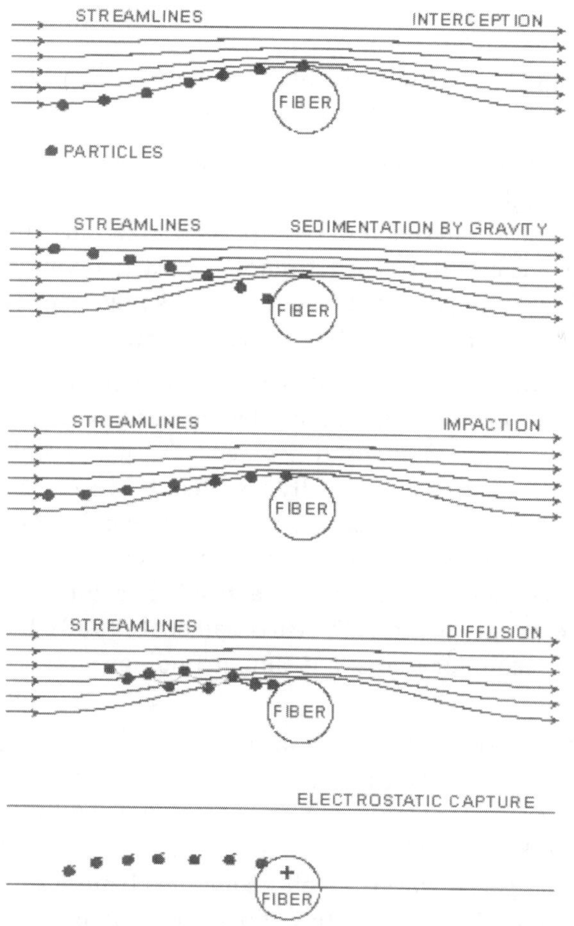

Figure 24-1. The five mechanisms of aerosol filtration (adapted from Japuntich, 1984).

inertia is such that the particle just deviates from the streamline, then the particle and filter surfaces may just interact – interception. If the particle enters an area of the filter where the air velocity becomes slower, it will fall to the filter if the force of gravity on its mass exceeds its forward inertia – sedimentation.

Smaller particles will tend to stay with the streamlines unless those streamlines bring them close to the filter. When this happens, Brownian motion or diffusion may cause the particle to move toward the filter. Brownian motion is the movement of particles as a result of gas molecules impacting on the particles. For this force to be significant, the particle's inertia must be quite small. Similarly, if the particles have a charge opposite to that of the filter, they will be attracted to the filter.

The laws of physics dictate that the first three mechanisms work best on large particles; and as the particle size decreases, the efficiency of these mechanisms decreases. On the other hand, the last two mechanisms work best on small particles; and as particles get larger, these mechanisms are less effective. The intersection of these two efficiency curves occurs, in theory, at an aerodynamic diameter of about 0.3 μm. In other words, particles of 0.3 μm diameter are a bit too small for impaction, sedimentation, and interception to be optimally effective and a bit too big for electrostatic charge and diffusion to be optimally effective.

Thus, filters are often graded on how they perform when challenged with a 0.3 μm diameter aerosol. The HEPA (high efficiency particulate arrester or air) filter refers to any filter that is at least 99.97% efficient in filtering a 0.3 μm particle size. The filter can be made of anything as long as it meets this criterion. The terminology is actually most important in the application of filtration to air cleaning in ventilation and for respirators. However, it is an important reference point even for air sampling; no filter is 100% efficient.

In reality, it is important to point out that the count median aerodynamic diameter (see Chapter 9) of lowest efficiency may be smaller than 0.3 μm. Recent studies have shown that although the efficiency of HEPA filters is at least 99.97%, the particle size of minimum efficiency is less than 0.25 μm, and more typically less than 0.15 μm. In addition, for other filter types, the particle size of minimum efficiency is smaller than theory predicts, as are the efficiencies. These two parameters (particle size and efficiency) vary with flow rate, type of filter material, manufacturer, aerosol type, humidity, and other factors (Brown, 2001; Moyer and Stevens, 1989a, b; Stevens and Moyer, 1989; Brosseau et al., 1989). It should be noted that this work was performed on respirator filters, but similar findings would be expected for air sampling filters.

As filters become loaded, their efficiencies improve for reasons that should be apparent from the above discussion. However, the resistance to flow across the filter also increases. Therefore, more energy (battery reserve in a personal pump) is needed to maintain the same airflow through the filter. When the battery cannot supply the energy, the flow decreases, even in some *constant* flow pumps. However, a well-maintained battery should be able to overcome this resistance and maintain the *constant* flow for most particle loadings encountered in the workplace.

Filter Types and Uses

There are basically two types of filters used for hygiene work: screen or membrane filters and depth or fiber filters. At one time, paper or cellulose fiber filters were used, but now the only fiber filters used significantly are quartz and glass fiber filters. All other filters are composed of films, usually of polymer and in one case silver, and are referred to as membrane filters. Table 24-1 shows the five major types used for hygiene work and some descriptive data for each.

Polyvinyl chloride (PVC), mixed cellulose ester (MCE), and Teflon (PTFE or polytetrafluoroethylene) are relatively uniform films or screen filters having interconnected structures (see Figure 24-2). MCE is a blend of nitrocellulose and cellulose acetate. Glass fiber filters look much different from membrane filters (see Figure 24-2); the structure of glass fiber and other depth filters makes it clear why they have a high capacity and relatively low pressure drop.

Glass fiber filters were the first filters to be developed to meet the specific needs of air pollution assessment (Gelman and Marshall, 1975), and they are also recommended for use in gravimetric analyses. However, this recommendation is most applicable to high volume air sampling used in ambient air quality work. In hygiene work, it is possible to break the fibers

Table 24-1
Filter descriptions

Type	Pore Size (µm)[1]	Number of Methods/ Analytes[2]	Analytes	Notes
Polyvinyl chloride (PVC)	0.8, 5	25	Inorganic, respirable and total dust, few organics	Resistant to acids/bases Low moisture pickup
Mixed cellulose ester (MCE)	0.45, 0.8	25	Fibers, elements, inorganics, a few organics	Easily dissolved or digested
Teflon (PTFE)	0.5, 0.45, 1, 2, 5	40	Organics	Hydrophobic Solvent resistant Inert
Silver	0.45, 0.8	3	Polynuclear aromatic hydrocarbons, Br, Cl	Uniform pore size No organic background
Glass fiber (AE)	1	150	Organics, PNAs, dyes	High capacity

[1] This is a nominal pore size.

[2] An approximate number; methods with multiple analytes or analytes with multiple methods are counted once (SKC, 2007).

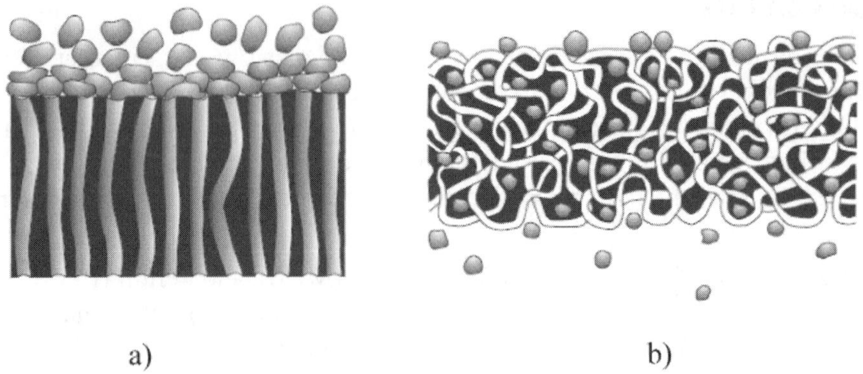

Figure 24-2. Schematic cross sections showing a) depth-type filter and random entrapment of particles and b) screen-type or membrane filter and particles retained on or near the surface (from Millipore, 1998, with permission).

with the cassette (filter holder) edges and cause a weight loss. In high volume air sampling, a rubber gasket holds the filter in place and minimizes fiber breakage.

As Table 24-1 shows, there is some overlap in the way that the filters are used. Glass fiber filters are widely used to collect organics such as pesticides, polynuclear aromatic hydrocarbons (PNA or PAH), dyes, and other high molecular weight compounds. However, Teflon is used to monitor similar organics, and the silver membrane is used in one method for PNAs. Teflon may be used when it is known or suspected that glass may cause a reaction of the analytes. Silver can be heat-treated to make it essentially organic-free, thus providing a low background.

Similarly, MCE and PVC are used for inorganic compounds and elements. Generally, PVC is used in gravimetric methods because of its limited interaction with airborne water vapor. MCE, on the other hand, is easily acid-digestible for elemental analysis, and it can be made to dissolve or become invisible so that fibers can be microscopically counted.

All of the filters are quite efficient. For example, typical aerosol retention efficiencies (dioctyl phthalate 0.3 μm @ 32 L/min) for PVC, Teflon, and glass fiber are 99.94, 99.79, and 99.98%, respectively (Pall, 2006).

Effects of Water Vapor

The effect of absorbed water vapor on gravimetric techniques has long been a source of error. There are at least three approaches currently in use: (1) ignore the problem; (2) dessicate the filter each time that it is weighed, that is, before and after the sample is collected; and (3) allow the filter to equilibrate to *room* relative humidity each time before it is weighed.

The first alternative is not a solution, for reasons to be shown. The second is problematic because when the filter is removed from the desiccator and placed on the balance, it immediately begins to gain weight and cause erratic readings. Currently, NIOSH recommends the third alternative in the Methods 0500 and 0600 and defines room relative humidity as 50 ± 5% (NIOSH, 1994, 1998a). These methods utilize PVC filters, which are less affected by water vapor than MCE or glass fiber. Nevertheless, this precaution is warranted.

The effect of water absorption can be significant even for PVC (Charell and Hawley, 1981). When 47 mm PVC filters from two unnamed manufacturers were exposed to varying concentrations of water vapor, the weight gains were 30 and 200 μg/% absolute water concentration. As a reference, 50% RH at 21°C (70°F) constitutes a water concentration of 1.23%, and 70% RH at 32°C (90°F) constitutes 3.32% water. The effects were even greater on MCE filters (1,200-1,500 μg/% water concentration), as expected. Of course, the effect on 37 mm PVC filters, as opposed to the 47 mm filters used in the study, would be less than this because of the smaller surface area.

Note that the collected particles would also gain weight from the water. Tierney and Conner (1967) concluded from a study of glass fiber filters, used to collect urban particles, that RH values up to 100% (at 72°F or 22°C) had virtually no effect on the tare (blank) weights of the filters. Similarly, no effect was found on the weights of the particulate matter at RH values up to 50%. However, above 50%, the weight of the particles began to increase significantly. At 70% RH the increases in particulate weights were less than 5%. At 100% RH the weight of the particles was increased by 40% to 90%, and the relationship was approximately linear in between. Thus, the results of these two studies warrant the use of PVC filters and controlled humidity weighing procedures in gravimetric methods.

PARTICLE SIZE SELECTION

The same phenomena driven by particle mass and size (relating to inertia and gravity) that affect the efficiency of filtration also affect the efficiency of a sampling *head* in collecting particles from the atmosphere. The human nose and mouth are sampling heads, and Chapter 10 (see discussion of inhalable fraction) showed how they do not collect all particles from the atmosphere. Smaller particles are collected more efficiently than larger particles.[5] The efficiency depends on the breathing rate, the shape of the head and the orifice (mouth or nose) among other factors. Similarly, a filter used to collect aerosols must be placed in some type of holder, and this holder will exhibit certain collection efficiencies, depending on its shape and the velocity through it.

Perhaps a good starting point in this discussion is to introduce the concept of isokinetic

---

[5] It is important that the reader not be confused by the concepts of particle collection and deposition. What is *collected* by the mouth and the nose is not necessarily deposited in the alveoli. The portion that is deposited will be deposited in different sections, depending on particle size (aerodynamic diameter), inertia, and gravitational force (mass × gravitational acceleration). Some portion will not be deposited even though it is collected.

sampling.[6] Isokinetic means same speed. If you want to know the concentration of aerosols being emitted from an industrial exhaust or stack, say, the stack of a coal-fired power plant, you must place some type of probe into the stack. The probe should point into the flow. In this case, you do not want any particle size selection; you want to collect particle sizes in the same distribution in which they exist in the gas stream. In addition, you want to estimate the total mass concentration accurately. If the velocity in the probe (V_p) is greater than the velocity in the stack (V_s), then the distribution of particle sizes collected is skewed toward smaller sizes. The concentration is underestimated, as the volume collected is inflated relative to the total, mass collected; the larger particles are less able to change course and enter the probe. If $V_p < V_s$, the small particles will follow contours around the probe, but larger particles will continue into the probe. The distribution is then skewed to the large particle sizes, and the mass concentration is overestimated. Only when the probe velocity is the same as the stack velocity will each particle size be sampled representatively.

This isokinetic concept is important for stack sampling, but hygienists must be concerned with what the worker breathes, and must recognize that the human head does not sample isokinetically. Actually, the hygienist must be concerned not only with what particle sizes the worker breathes but also where the target site of the toxicant is, and what particle sizes are available to that target site. As Chapter 6 discussed, if a toxicant only has an effect at the deepest portions of the lung, the alveoli, then what particle sizes are breathed by the worker is not so important as what particle sizes are deposited in that region. Silica is a good example of such a toxicant, as it has its only effect in the gas exchange region. On the other hand, if the toxicant is absorbed across the respiratory tract and has an effect at, say, the blood-forming organs, then it does not matter where the particles are deposited; any particle deposited in the respiratory tract may be absorbed and enter the bloodstream. Lead-containing particles are an example of such a toxicant. For silica we need a sampler that will collect only those particles that are likely to reach the gas exchange region; for lead we need a sampler that will function like the human head.

Particle Size Selection Conventions

There has been considerable work in the area of size-selective conventions. ACGIH® (ACGIH®, 1993), the International Standards Organization (ISO, 1991), the U.S. ASTM (2001a), and the European Standardization Committee (CEN, 1993) have adopted identical standards for the performance of aerosol samplers for personal workplace air sampling (see Figure 24-3). The adoption of uniform standards was first recommended by Soderholm (1989). These developments have been reviewed extensively in Vincent (1999) and Kenny (2000). However, the proposal to consider three separate deposition regions and appropriate particle size-selective criteria for each was made by the International Standards Organization (ISO, 1981). These uniform particle size-selection conventions are based on the sampling of air in a

[6] For a good review of particle size-selective sampling, see Lippmann (1989).

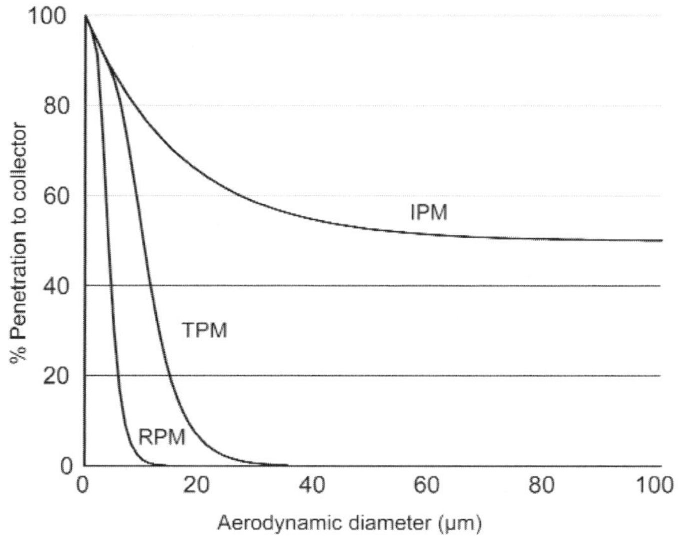

Figure 24-3. The Universal size-selective sampling convention for inhalable (IPM), thoracic (TPM), and respirable (RPM) particulate masses.

manner similar to the way that particles are sampled by the human head and the ability of the various particle sizes to penetrate to the three subregions of the respiratory tract. These subregions are the total respiratory tract, represented by the Inhalable Matter Fraction (IMF); the sub-larynx region, represented by the Thoracic Matter Fraction (TMF); and the gas exchange region, or alveoli, and non-ciliated regions, represented by the Respirable Matter Fraction (RMF).

Prior to this recent adoption of uniform conventions , the RMF was size-selected by using two different criteria. Many parts of the world used British conventions recommended by the British Medical Research Council (Davies, 1952). Other parts of the world used conventions adopted by the U.S. Atomic Energy Commission (ICRP, 1960) and ACGIH® (Aerosol Technology Committee, 1970). These standards and recommendations as well as more recent recommendations, including the universally adopted standard of Soderholm, are summarized in Table 24-2.

It should be realized that historically (and in many cases to the present) the IMF usually was approximated by the *total mass* and was sampled by using a variety of devices, most of which had no known relationship to the way that the human head samples. The new conventions considerably improve the scientific basis for aerosol sampling; however, at least in the U.S., we are still very much in a transition period. In the United States, OSHA shows no sign of adopting the new conventions, and many hygienists are confused about the need to change to new devices or sampling practices. In Europe, (at least as of 2000) there is some confusion. The Chemical Agents Directive (European Union, 1998) seems to have strengthened European

Table 24-2
Cut-points[1] for size-selective sampling for various aerosol fractions

Source	IMF	TMF	RMF
1952 BMRC[2]	none	none	5.0
1968 ACGIH®[3]	none	none	3.5
1981 ISO[4]	~ 30	8.5	5.0
1985 ACGIH®[5]	Up to 100	10.0	3.5
1989 Universal[6]	Up to 100	10.0	4.0

[1] Mass median aerodynamic diameter (μm) of the selected fraction; 50% of particles are smaller or larger
[2] As cited in Davies (1952)
[3] Equivalent to the Los Alamos, U.S. Atomic Energy Commission, or International Committee on Radiation Protection Criteria of 1960 (see Aerosol Technology Committee, 1970)
[4] ISO (1981), see also Vincent and Mark (1987)
[5] Phalen et al. (1986)
[6] Soderholm (1989)

exposure limits while at the same time weakened the status of an RPM or IPM convention apparently leaving the use of those conventions up to individual member states (Kenny and Ogden, 2000).

It is important for the reader to realize that the conversion to the new conventions is a multi-step process. The first step is to recommend the conventions. This has been done. Second, devices must be manufactured and tested (or old devices or techniques modified) that will perform according to the criteria. A plethora of new devices has come onto the market since 1990. Third, practitioners must accept the new devices and techniques. This has been successful given the widespread marketing of the new devices. Finally, governments must adopt the changes as legally binding. This is occurring at a faster pace in Europe than in the United States (Kenny, 2000). Therefore, it is important to examine both the old and the new practices.

INHALABLE PARTICLE MASS[7] OR TOTAL MASS

IPM samplers

Basically, a particle can enter or be aspirated by a sampler by sedimentation, capture (being sucked in), or its own inertia (as when the sampler faces into a strong wind, or a parti-

[7] Note that while this and the other fractions often are referred to as mass fractions, they can just as easily be assayed as number or surface area fractions. The term *mass* refers to the way the fraction is assayed, usually weighted.

cle is ejected: toward the sampler at close range). Early studies and theoretical considerations show that the orientation of a sampler (including the human head) to the wind is important relative to sampler efficiency. Of course, the wind speed, velocity through the sampler orifice, sampler shape, and particle size also are quite important (see, e.g., Raynor, 1970; Ogden and Wood, 1975; Breslin and Stein, 1975; Wood and Birkett, 1979). These early studies compared samplers to the ideal, that is, isokinetic efficiency. But the problem is to design a sampler that mimics the human head; if a strong wind causes the human head to inhale larger particles that it normally would not, then the sampler should do likewise. Thus samplers should be evaluated relative to inhalable particle size-selective criteria that are based on the human. Work on the sampling performance of the human head was summarized by Vincent and Armbruster (1981), whose data were adopted in the recommendations of Soderholm (1989) and are summarized as the IPM data in Figure 24-3.

CASSETTES

Historically, and in many cases today, aerosol samples taken in the United States were collected with a clear polyethylene two- or three-piece, 37 mm outside diameter cassette, as shown in Figure 24-4. However, apparently only in the case of the comprehensive OSHA standard for asbestos is the cassette actually required.[8] For most aerosols, the cassette usually is

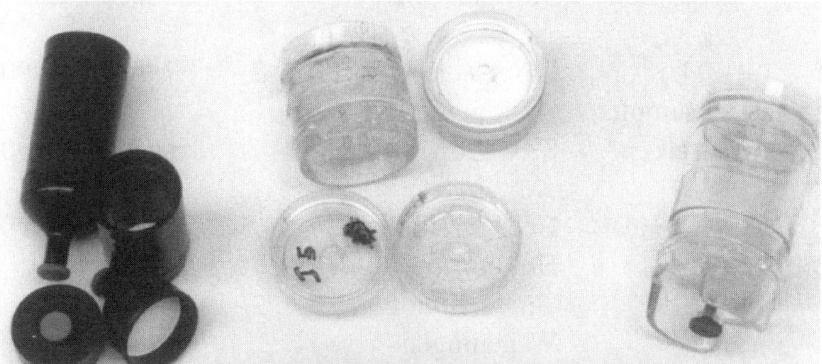

Figure 24-4. U.S. filter cassettes: from left, conducting 25 mm with extension cowl, non-conducting 25 mm without cowl, 37 mm with cowl (photo, R.K. Oestenstad).

[8] Actually the similar but smaller 25 mm cassette is required. In addition, OSHA health inspectors follow the OSHA Technical Manual (OSHA, 2006a) and the OSHA Chemical Sampling Information (OSHA, 2006b) which require the use of the 37 mm cassette.

used in the three-piece mode; that is, the top is left in place, leaving a small (4 mm) opening through which the air flows. This configuration is termed *closed face*. For some dusts, such as asbestos and fiberglass, the top piece of the cassette is removed before air sampling to obtain a more even distribution of these fibers on the filter.[9] In these cases, a sample of the filtered fibers is counted; in other cases, the filter is weighed, or the filter is digested and the total analyte assayed. In the latter two cases; it does not matter where the particles fall on the filter.

THE IOM AND OTHER IPM DEVICES

Today a wider variety of devices is used, some of which are required or recommended by other governments (see Table 24-3). These are shown in Figure 24-5. Some of these, like the plastic cassette, were not designed on the basis of an inhalable dust particle size-selective sam-

Table 24-3
Personal inhalable size selective samplers (references below[1])

Sampler	Other names	Source	Q L/min	Orifice (mm)	Orientation[2]	Filter (mm)
IOM sampler[3]		SKC	2	15	90°	25
Cassette, open or closed[4]	Total sampler	Millipore, SKC	1–4	4 or 35	135–180°	37 or 25
7-hole	SHS, UK or HSE sampler	Casella, JS Holdings	2	7 @ 4	90°	25
Button	windscreen	SKC	4	many @ 381	90°	25
CIS[3]	GSP, conical	BGI, Strohlein, JS Holdings	3.5	8	90°	37
PAS-6	conical	University of Wageningen	2	6	135–180°	25

[1] Kenny et al., 1997; Davies et al., 1999; Li et al., 2000
[2] Relative to body where 0° is up and 180° is down
[3] Has internal cassette
[4] Clear plastic polystyrene models are non-conducting

[9] In the case of a 37 mm cassette, there is a 33 mm opening or an exposed filter surface area of 855 mm^2. For the 25 mm cassette there is a 22.2 mm opening or an exposed filter surface area of 385 mm^2. However, these areas are not constant from one cassette to the next. See later discussion under depositional patterns.

Figure 24-5. Inhalable particle mass fraction samplers: from far left, clockwise CIS or conical, 7-hole sampler, IOM sampler; button sampler (photos courtesy SKC, BGI, and CasellaCel).

pling criterion. The IOM sampler was the first to be specifically designed for this purpose.

Mark and Vincent (1986) performed seminal work on assessing the performance of three samplers commonly used in the UK (7-hole, single hole, and open face) to meet the inhalable mass convention by placing them in a wind tunnel on a mannequin. The mannequin was rotated during the sampling period to average the effects of wind direction, and a sampler in the mouth provided a reference value. Previous work had shown differences in aspiration caused by mouth versus nose breathing to be insignificant, at least for the wind conditions tested (Ogden and Birkett, 1977; Armbruster and Breuer, 1982). Later results have shown that mouth versus nose breathing is important[10] (Kenny, 2003). All three devices tended to undersample; higher wind speeds (wind speeds were 0.5, 1, and 2.6 m/s) caused the greatest biases; and the open face cassette performed best of the three.

They then went about the task of designing and testing a device that would not be biased relative to the mouth-breathing mannequin. This device, now called the IOM (for the Institute of Occupational Medicine in Edinburgh, UK) sampler, was found to be relatively unbiased. It consisted of a 15 mm opening (see Figure 24-6) and sampled at 2 L/min. The opening feeds to

[10] More recent work has shown that the mannequin per se is not necessary. A so-called bluff body, essentially a small rectangular box, can be used in the place of the mannequin. This also allows the use of smaller wind tunnels (Witschger et al., 1998; Paik and Vincent, 2004). In addition, considerable mathematical modeling work has been performed, both semi-empirical (see, e.g., Tsai et al., 1996a and Sreenath et al., 1999) and based on more theoretical models such as computational fluid dynamics (Savio et al., 2003).

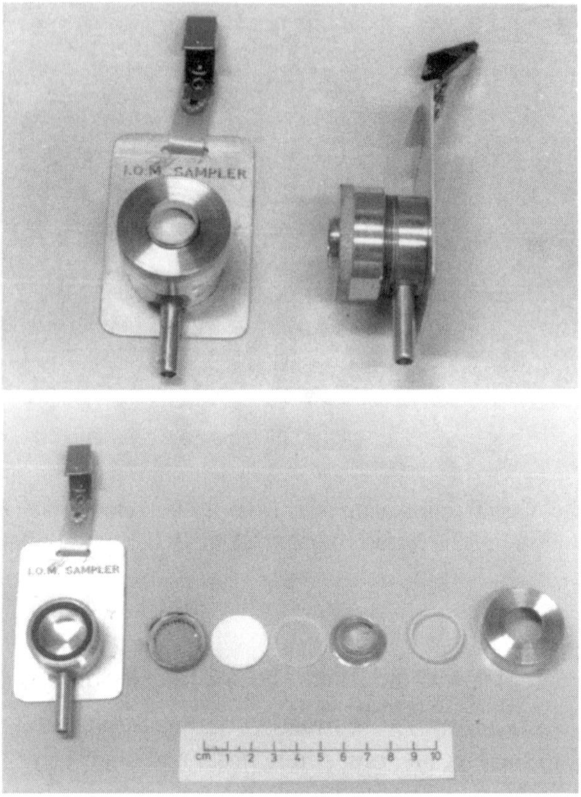

Figure 24-6. An exploded view of the IOM personal inhalable dust sampler (courtesy, SKC, Inc, with permission).

an internal aluminum cassette that houses a 25 mm filter. The cassette is pre- and post-weighed with the filter in place. This feature was added as Mark and Vincent found significant particle mass (up to 40% of the total mass sampled) on the inside surfaces of other devices. They also noted that the open face cassette was subject to dust loss due to handling.

Laboratory Tests

Since this seminal work, the IOM device and others have received considerable laboratory (Aitken et al., 1999; Aizenberg et al., 2000; Kenny et al., 1997; Kenny et al., 1999; Li et al., 2000; Teikari et al., 2003; and Witschger et al., 2004) and field testing. The summary laboratory results from several studies are shown in Figure 24-7a, b, c. Note that these figures are estimated curves based on replicate experimental data that in some cases is actually quite variable. None of the samplers completely mimics the IPM convention at high wind speeds and

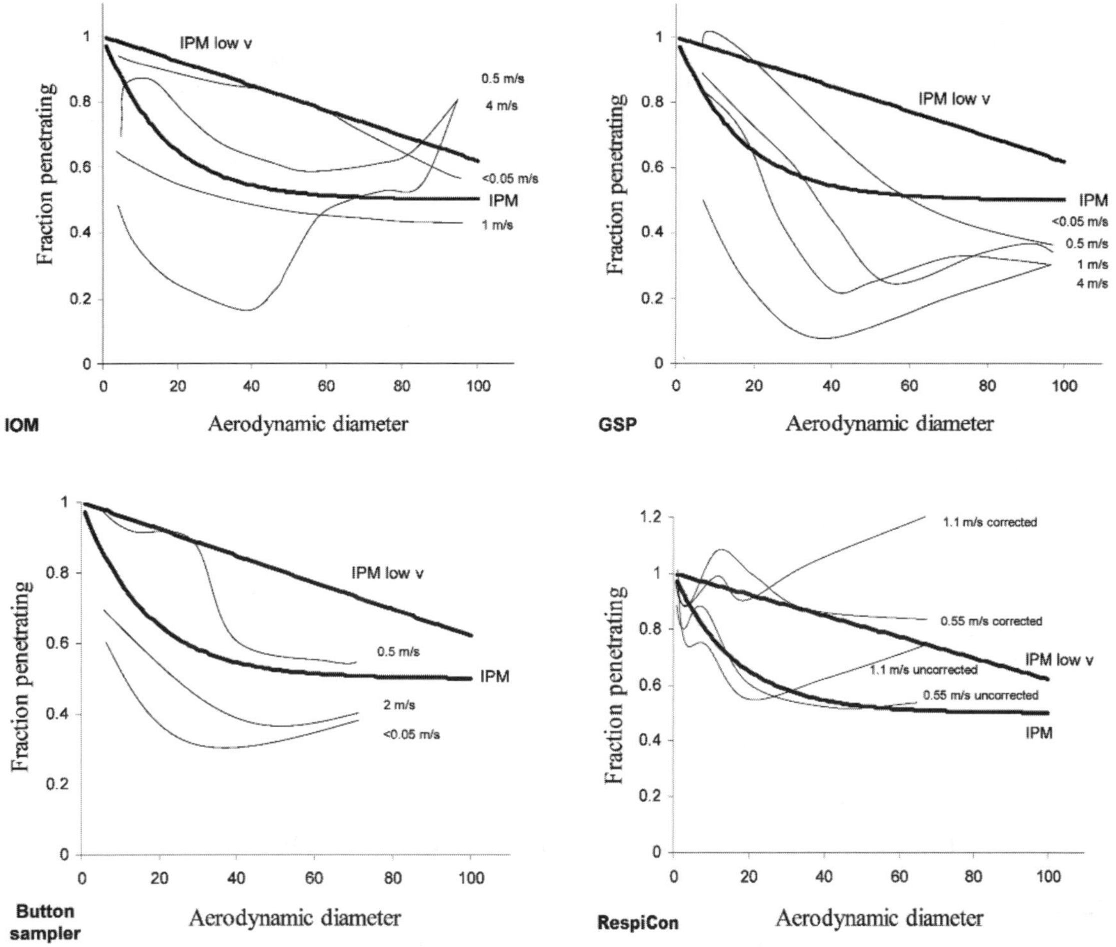

Figure 24-7a. Approximate average performance of 4 inhalable mass samplers compared to the IPM and proposed IPM low wind velocity convention. Data for Respicon are from freely suspended samplers (Li et al., 2000). Button sampler suspended on mannequin at 2 high wind speeds (Aizenberg et al., 2000) and at lowest speed, samplers freely suspended or suspended from simplified torso and facing dust source (Witschger et al., 2004). For IOM and GSP data, samplers were suspended from mannequin at 3 high speeds (Kenny et al., 1997) and from mannequin or freely at lowest speed (Kenny et al., 1999).

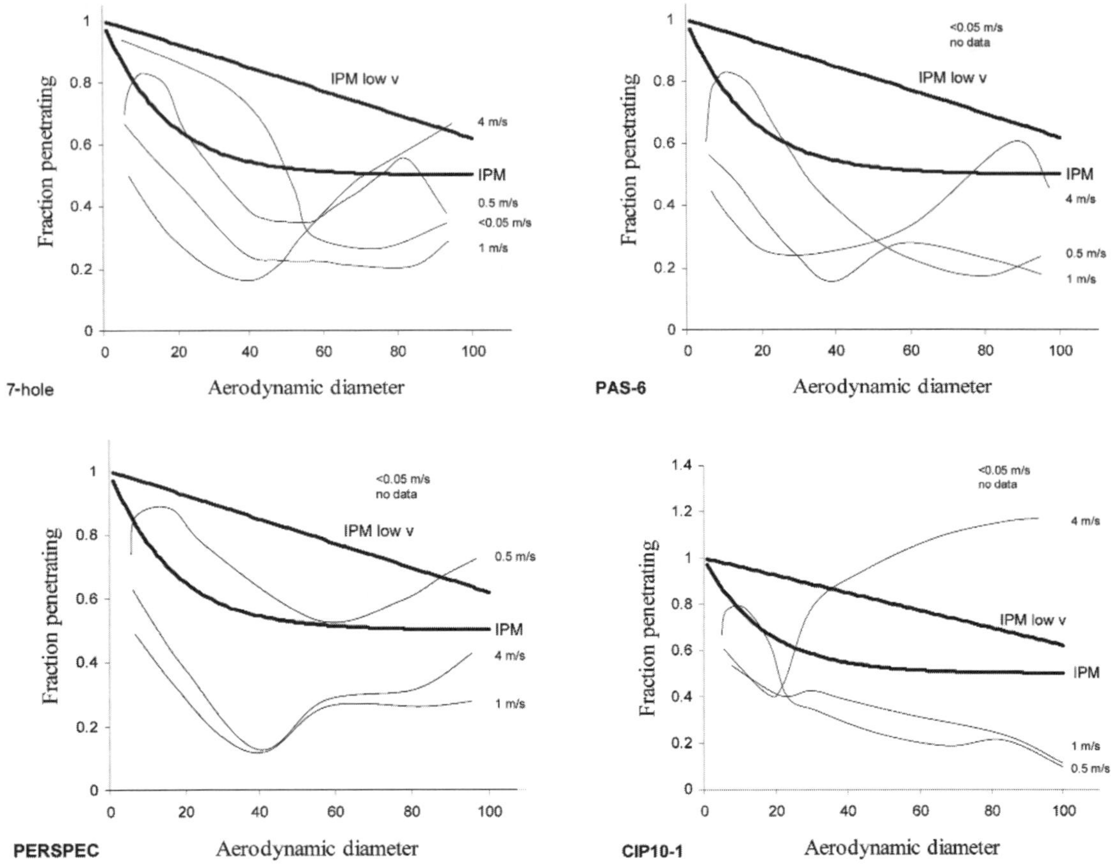

Figure 24-7b. Approximate average performance of 4 inhalable mass samplers compared to the IPM and proposed IPM low wind velocity convention. Data are from samplers suspended from mannequin at 3 high wind speeds (Kenny et al., 1997) and from mannequin or freely at lowest speed for 7-hole only (Kenny et al., 1999).

none completely mimics the proposed low wind speed convention for the lowest wind speed (0.1 m/s). The samplers in Figure 24-7a appear to give the best performance. As will be seen, the results in field tests are better than the lab simulations would indicate (Kenny, 2003). The IOM has become the standard (Bartley, 1998). The RespiCon® is a virtual impactor and is described later.

While the data in Figure 24-7 indicate the aspiration efficiencies in the samplers for various wind speeds, there is also the issue of precision. Table 24-4 shows the precision for seven of the devices at 3 wind speeds (Kenny et al., 1997). The button sampler (not shown in Table 24-4 had precision equal to or better than the GSP, IOM, and 37 mm cassette in separate test-

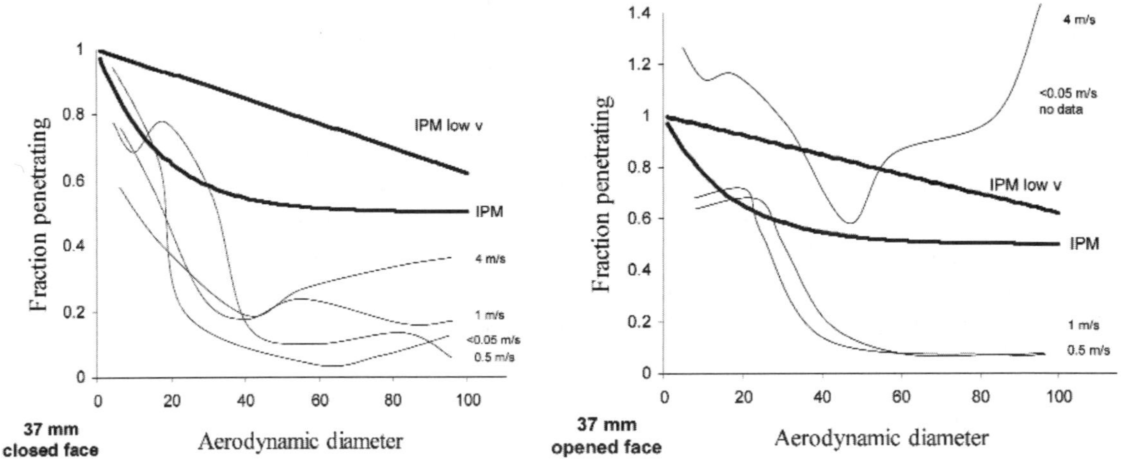

Figure 24-7c. Approximate average performance of the 37 mm cassette, inhalable mass sampler, compared to the IPM and proposed IPM low wind velocity convention. Data are for samplers suspended from a mannequin at 3 high speeds (Kenny et al., 1997) and from mannequin, or freely at lowest speed for closed face only (Kenny et al., 1999).

Table 24-4
Sampler types ranked in order of precision by wind speed (residual variance: most precise first, adapted from Kenny et al., 1997)

0.5 ms^{-1}	1.0 ms^{-1}	4.0 ms^{-1}
GSP	GSP	PERSPEC
PAS-6	PAS-6	GSP
PERSPEC	CIP 10-I	37-mm closed
37-mm open	37-mm open	PAS-6
CIP 10-I	37-mm closed	IOM
IOM	PERSPEC	7-hole
37-mm closed	IOM	CIP 10-I
7-hole	7-hole	37-mm open

ing (Aizenberg et al., 2000). Of course accuracy is the combination of bias and precision and accuracy maps such as that in Figure 24-8 can be created. However, these must be made for each wind speed (or averaged) and are somewhat tedious to compile. Thus, they rarely appear in the literature.

How does one sort through all this data to determine the best sampler? The first issue is wind speed. In most workplaces, the wind speed is low; it rarely exceeds 0.3 and is typically < 0.1 m/s or about 20 f/min (Baldwin and Maynard, 1998); this is near the lower range of most hand-held anemometers. However, outside workers experience higher wind speeds. For example a fairly common 18 km/hr wind (about 10 mph) would be 5 m/s. So for most situations, the lower wind speeds in Figure 24-7 and Table 24-4 apply.

Field Tests

Another way to assess the performance of the samplers is to compare them in the field in actual workplace atmospheres. While this approach takes into consideration such factors as actual wind speeds, orientation to the source, and worker movement, it is of course difficult to obtain a representative sample of workplaces with respect to these important variables.

One of the first field studies utilized a remote-controlled mannequin (Botham et al., 1991) in 23 areas of nine industrial sites to closely follow and mimic the movements of workers (Vaughan et al., 1990). Two each of the IOM and 7-hole (7H) devices were attached to the mannequin at midchest, and always in the horizontal orientation of IOM-7H-impactor-7H-

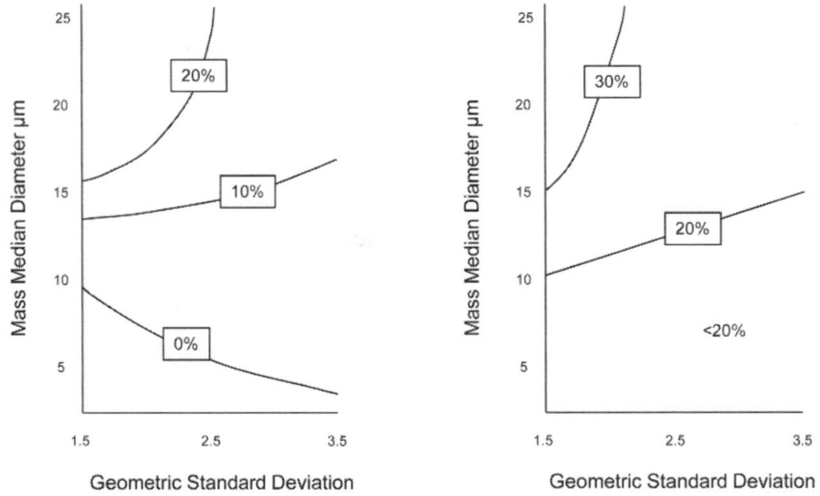

Figure 24-8. Estimated bias (left) and accuracy maps for the IOM sampler at 0.5 m/s (from Kenny et al., 1997, with permission).

IOM. The impactor was a port leading to a particle size fractionating sampling device. The results indicated that the IOM device was positively biased (1.14), and the 7H was only slightly negatively biased (0.98), compared to the mouth sample from the mannequin. It was suspected that, because of the orientation of the IOM sampler (slightly pointed upward), it may have oversampled particles that fell into it. In addition, its larger hole size would have collected more projected particles than either the 7H or the mannequin, as the mannequin head was always farther from the source. These observations led the United Kingdom to adopt either device for sampling inhalable dust (HSE, 1989).

Tatum et al. (2001) and Davies et al. (1999) found that the CIS and the 7-hole samplers consistently sampled more wood dust than the closed face 37 mm cassette (CFC). Hauck et al. (1997) found the button sampler to be equivalent to the CFC; however, particle sizes were not given (see later discussion for importance of particle size).

Many studies have compared the 37 mm cassette to the IOM. Teikari et al. (2003) compared the 37 mm cassette and the RespiCon® to the IOM, a particle counter and a 2 stage impactor (these are described later), in field and lab studies. The IOM had very good internal consistency within its three versions[11] and compared well with the particle counter and impactor in both lab and field. The IOM tended to sample more than two times the 37 mm cassette. Many other field studies have shown that the 37 mm cassette undersamples compared to the IOM (Calzavara et al., 2003; Demange et al., 2002; Kerr et al., 2002; Liden et al., 2000; Lillienberg and Brisman, 1994; Martin and Zalk, 1998; Noto et al., 1996; Predicala and Maghirang, 2003; Spear et al., 1997; Tatum et al., 2001; Tsai et al., 1995; 1996b, c; Vaughan et al., 1990; Wilsey et al., 1996).

However, one issue with the 37 mm cassette is wall losses (Baron, 1998 and 2003a). In fact, the IOM comes with a built in cassette that is weighed with the filter to account for wall losses. In one field study the IOM and 37 mm cassette were equivalent if the dust on the inside of the 37 mm cassette was included in the analysis (Demange et al., 2002). The particle sizes in this study were relatively small (15 μm mass median aerodynamic diameter); it should be expected from Figure 24-7 that when sampling larger particle size distributions, the IOM to 37 mm cassette ratio will increase. MSA sells a cassette having an aluminum cartridge crimped onto the filter[12] (Baron, 1998 and 2003a).

Typically the cassette is used in the closed face mode in the U.S. Research by Cohen et al. (1984) showed that the open and closed face modes gave extremely similar results for sam-

---

[11] The IOM can be used in conjunction with porous plastic foam inserts to measure the other particle size fractions as will be described later.

[12] This is apparently used in the U.S. mining industry. There is also a device called the Accu-Cap that is made for any 37 mm cassette (Omega, 2006a). It is marketed for fibrous contaminants, i.e., those like wood dust that are electrostatically charged, not those like fiber glass that must be counted from a uniform deposit on the filter.

pling beryllium particles with mass median aerodynamic diameters (MMAD) of 2 to 9 µm. However, work by Beaulieu et al. (1980) showed that the sampling results of open face cassettes were higher than those for closed face when chromic acid and paint mists and perlite, Portland cement, grain, wood, and cellulose dusts and welding fumes were sampled with pairs of cassettes. The open-to-closed-face paired ratios ranged from 0.5 to 7.5, and the averaged paired ratios within each aerosol type ranged from 1.0 to 3.8.

Thus, these studies on the 37 mm cassette make it clear that for smaller particle sizes (say < 15 µm MMAD), the cassette samples reasonably well if the particles lost to the walls are included in the assay. However, beyond these particle sizes, the cassette is grossly biased low. A correction factor (1.5?) might be used but its value would likely change with particle size.

Other Cassette Issues

CASSETTE LEAKAGE

One potential problem with the 37 mm cassette is bypass leakage (Baron 2003b). Figure 24-9 shows that the leakage may occur in two ways. Along either path, aerosol goes around the filter edge and is therefore not filtered. Frazee and Tironi (1987) studied the problem by preparing cassettes in one of two ways: manual assembly or using a specially built mechanical press. Vacuums of 13 in. of water (3.2 kPa) were applied to the cassettes and the top opening was closed; essentially no external leaks (path B) were detected even though the PVC filters were visibly loose in many of the hand-assembled cassettes. Shrink bands, often used

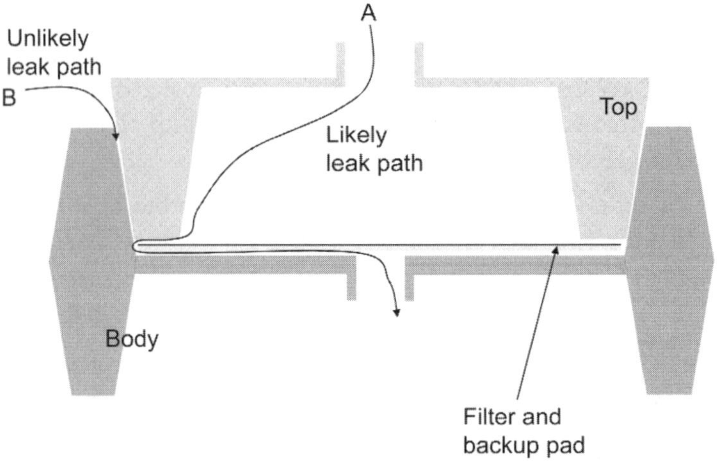

Figure 24-9. A 37 mm filter cassette showing potential leakage paths (adapted from Frazee and Tironi, 1987, with permission).

around cassettes to prevent this presumed problem, was not used.[13] The vacuum used was thought to be extreme.

Internal (path A) leakage was tested by placing a thin flow barrier between the filter and the filter support pad, and then applying vacuum. At vacuums of only < 1 in. H_2O (< 0.24 kPa), 16 of 20 cassettes loaded with PVC filters and manually pressed leaked at flow rates of 100 to 135 ml/min. The cassettes were manually re-pressed and showed leakage at 26 in. H_2O (6.3 kPa), whereas mechanically pressed cassettes showed no leaks at this latter pressure. Thus, mechanical pressing appears to be important.

More recently, Baron (2003b) tested several manufacturers' pre-assembled cassettes and found them well-sealed (see, e.g., SKC). However, hand pressed and even press assembled cassettes containing unusual filter combinations gave high rates of leakage. A press and leak tester are available commercially (Omega, 2006b). Baron recommends a leak test program based on pressure drop (van den Heever and Tiernan, 1999) where the pressure drop across well-sealed cassettes is measured and that variability is used as a quality control parameter.

Often sampling trains consisting of a pump and filter cassette are calibrated by using the same cassette for each pump to be calibrated under the assumption that all cassettes/filters are alike. If leakage only occurs via path A in Figure 24-9, then the flow to the pump through the cassette is the same as that entering the cassette, as opposed to path B. There could be some differences in flow from one cassette to the next, since pressure drops across cassettes with leaks at path A would be different than those without leaks. However, there also are likely to be differences in the filters themselves that lead to different pressure drops and, therefore, differing flows. Ten 25 mm cassettes, with MCE filters and an extension cowl (see section on electrostatic effects) were purchased from each of three manufacturers. One at a time the cassettes were attached to a sampling pump as they would be used in the field, and the train was calibrated with an electronic bubble meter. All 10 cassettes were applied in succession, and then the process was repeated for a total of five calibration runs. Table 24-5 shows the results. The average within coefficient of variation for the five replicate calibrations of each cassette indicates that the calibration and pump fluctuation error is quite small compared to the *CV* between the cassettes. Differences in filter resistances and perhaps differences in internal leak rates likely account for these errors. The result is particularly important to hygienists who use a *calibration cassette* rather than calibrating each pump with the cassette that will be used in the field. The calibration cassette could lead to unnecessary calibration errors.

ELECTROSTATIC CHARGE EFFECTS, WALL LOSSES, AND HEAVY FILTER LOADINGS

Airborne particles (Johnston et al., 1985), filters, in particular PVC, polycarbonate, and Teflon® (Mark, 1974; Liu et al., 1985), and samplers, including plastic and metal cassettes (Turner and Cohen, 1984; Baron and Deye, 1990a, b) can carry significant static charges.

[13] It is recommended that these bands be used to prevent tampering and for labeling (Baron, 2003a).

Table 24-5
Variability in flow rates[a] through filter cassettes presumably caused by leakage

Manufacturers	$\bar{Q}$ (L/min)	$\overline{CV}_B$ (%)[b]	$\overline{CV}_W$ (%)[c]
A	1.54	6.0	0.2
B	1.61	3.1	0.1
C	1.58	5.3	0.1

[a] Determined with an SKC electronic bubble meter; unpublished data from D. Hicks and author
[b] The coefficient of variation for flow through 10 different cassettes with the pump operating at constant conditions
[c] The pooled CV, for five replicate calibrations of each cassette, across all 10 cassettes

There have been numerous field observations of particle deposition on the inside (Speight and Marsh, 1984; Peck et al., 1985; Mark and Vincent, 1986; Seixas et al., 1987; Puskar et al., 1991) and outside (Baron and Deye, 1990a) of plastic cassettes. Often electrostatic charges have been blamed for these depositions.[14] In addition, in both the United States and Europe, a conductive cowl (see Figure 24-4) is required on all cassettes used to monitor asbestos and other fibers that require microscopic counting. The cowl is not required for other aerosols. United States OSHA uses the polystyrene cassette for all aerosols except fibers. For fibers, a carbon-filled, conductive polystyrene cassette is used. Finally, the IOM and CIS samplers contain an internal cassette that is pre- and post-weighed with the filter so as to include in the weighing any particles not on the filter. What is the rationale for all of this?

In order to explore the possible explanations, some important facts must be kept in mind. First, relative humidity does have a significant effect on static charge (see, e.g., Baron and Deye, 1990a, b). Second, as particles tend to pile onto a filter, those not interacting directly with the filter will be more likely to be dislodged from it for a variety of reasons, including rough handling during shipment, removal of the filter during weighing, and possibly electrostatic attraction to other surfaces after sampling has ceased. Third, for most types of aerosol sampling, filter depositional patterns are irrelevant; for microscopically counted aerosols, depositional patterns are paramount and can be affected by charge (Johnston et al., 1987). Fourth, particles that are aspirated through a small opening such as that on a closed face cassette lose inertia as a result of the loss of velocity upon entering the sampler; devices with orifices similar in size to the depositional surface do not cause this effect.

From these facts and the research to date we can make some observations and conjec-

[14] In fact, an electrostatically charged polymer sheet has been developed as a passive dust sampler (Brown et al., 1994).

tures. However, we must discuss microscopically counted aerosols separately from the others. Field studies have shown high particle losses to clear polystyrene cassettes during asbestos sampling (Peck et al., 1985). A second U.S. study (Seixas et al., 1987) involved the cowl and the conductive cassette. However, the filters were heavily loaded. Similar results were not seen for man-made mineral fibers (Kauffer and Eypert-Blaison, 2004). Baron (1987) has argued that this loading effect causes the counts from the filters to be low. Perhaps as few as 10% of the fibers are actually counted, as the others are obscured. Also the heavy loading causes some of the fibers to be dislodged. Therefore, it is argued that these two factors together account for what appears to be a large proportion of fibers (7-50%) on the cowl. These results seem to appeal for use of the conductive cassette and the cowl, and for keeping filter loadings low.[15]

Baron and Deye (1990a, b) conducted extensive laboratory studies and from these and theoretical calculations they argued that the electrostatic charge on a cowl should be concentrated along the edge of the opening. Therefore, like-charged particles should be repelled from the opening and opposite-charged particles should be attracted to this edge and to the outside, not the inside, surface. Baron and Deye also showed that conductive cassettes gave more even deposition patterns than non-conductive ones. Hygienists are warned to be aware of the effects of low RH (certainly < 30% RH should be of concern), not to overload filters, and to take care in shipping cassettes and removing filters from them. To avoid these and other effects, Chen and Baron (1996) have argued for the use of high flow rates, avoidance of areas with high wind velocities, and minimization of the angle between the inlet and external air motion.

For non-fibrous aerosols, generally much higher loadings are achieved in practice, particularly for *nuisance dusts*. In these cases, one would expect losses from the filter for all of the reasons cited above. However, such losses can occur even at low loadings, particularly if RH is low. In one study (Puskar et al., 1991), loadings were < 20 μg, and RH ranged from 15% to 25%. Losses to the cassette walls and to the backup pads (thought to be related to removing the filter for weighing) were 78% of the total mass on average.

In a more recent study, the loss of carbon black from shipped filters that had been loaded in the laboratory via a SIMPEDS cyclone (see next section) was examined. The filters were mailed to various locations and then returned unopened. Of 120 loaded filters, the dust in 31 was grossly dislodged, and they were not further analyzed. For nine filters with loading preweights of ≥2 mg, the losses were quite variable, up to 25%. For the remaining 80 filters, the post-weight was about 93% of the pre-weight. The correlation coefficient for the latter group was 0.93 (Van Tongeren et al., 1994).

The IOM and CIS samplers correct these losses by using the internal cassette (see Figure 24-6); however, for other devices the only choices are to extract the lost aerosol from the surfaces or to build or use a commercial internal cassette. Puskar et al. (1992) have developed a tool for stamping thin aluminum cartridges that can be placed in the 37 mm cassette and loaded

[15] The NIOSH Method (7400) for fibers calls for loadings of 100-1,300 fibers/mm^2 of filter area, and requires that no more than 50% of the surface should be covered with particles.

with a filter.[16] In addition, it would seem logical to use a conductive cassette rather than the clear polystyrene cassette. Conductive cassettes are available in several sizes.

DEPOSITIONAL PATTERNS AND FIBER COUNTING

A number of studies of depositional patterns of asbestos fibers on filters have been performed, and the overwhelming conclusion is that patterns often are not uniform. The methods for analyzing asbestos (see, e.g., NIOSH 7400) require a maximum of 100 microscopic fields to be examined and the fibers counted. The standard field has an area of 0.00785 mm^2, and the filter surface area is nominally 385 mm^2. Therefore, only 0.2% of the filter surface is examined. For this reason, sampling bias in the selection of the fields can be important.

Baron and Deye (1990a) showed that electrostatic charge can have a significant effect on deposition. They recommend that if few particles are found along the edges of the filters, then the fields should be selected from the middle of the filter. This is sometimes difficult when the filter is cut radially.

Baron et al. (1994) have examined deposition patterns at various flow rates (2-10 L/min) with the cassette oriented at various angles relative to a moderate flow field of 0.14 m/sec (28 ft/min). The patterns indicate clearly that at higher sampling rates, non-uniform distributions prevail. Higher rates are, however, warranted so as to avoid complete loss of fibers to wall deposition.

A number of studies by Feigley and coworkers (Hook et al., 1983; Jenkins et al., 1992; Feigley et al., 1992) confirm higher densities near the middle and a number of other irregular patterns. In the most recent study (Feigley et al., 1992), nine cassette types were examined. Interestingly, three gave effective filter surface areas of 420, 404, and 401 mm^2 rather than the assumed value of 385 mm^2. Failure to account for this would lead to underestimates of the concentrations. At pump flow rates of 9.5 L/min,[17] they found a clear trend of high to low fiber densities from the center to the edge of the filter. The effects were minimized for a cassette with a bell-shaped entry.[18] They concluded that if the effects are caused by electrostatic losses to the walls, then counting fields in the middle of the filter would be reasonable. However, if the patterns are owed to air streamlines caused by the shape of the cassette entry, then counting only fields in the middle would give positively biased results. If the analyst is not sure what is occurring, an approach that randomly selects fields from across the entire filter (or wedge) is probably most pragmatic. One such approach is shown in Figure 24-10.

---

[16] See earlier comment; this is apparently available from MSA.

[17] Flow rates of this magnitude are commonly used in the United States for stationary measurements performed to release for occupation a previously abated area, usually an office area where asbestos has been removed.

[18] Available from Envirometrics, Inc., Charleston, SC, USA, no website.

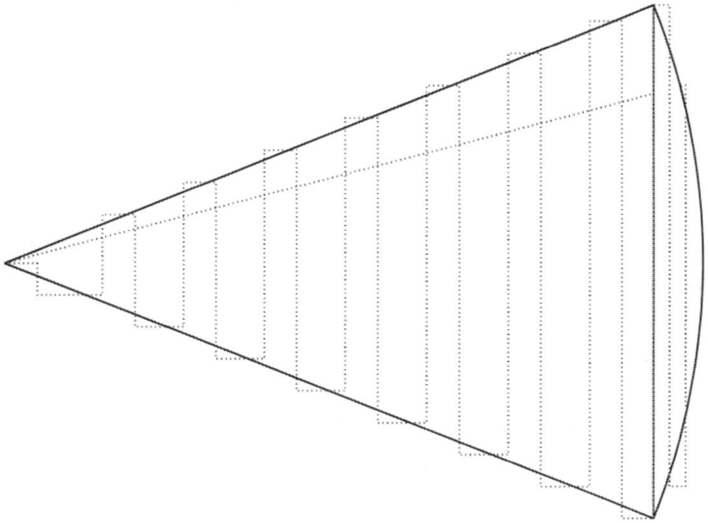

Figure 24-10. A diagram of a wedge cut from a filter and a counting pattern that can be used to cover the entire filter. The analyst has to become accustomed to flicking the wrist so that about 100 movements cover the entire wedge. Fields should be randomly selected.

WEIGHT INSTABILITY OF THE IOM AND CIS INTERNAL CASSETTE

The first cassettes in the IOM samplers were stainless steel and later were also made of plastic to cut costs. This lead to the discovery that the plastic cassettes were extremely subject to weight fluctuation caused by humidity (Smith et al., 1998). Liden and Bergman (2001) observed that the carbon black mixed into the plastic from both the SKC IOM and the CIS (conical inhalable sampler) to make them electrostatically conductive is the reason for the water absorption; however, they also observed that not all carbons have the same absorptive properties. A substitute conducting plastic was found to have 30 times less absorption of water.

The bottom line is that the plastic cassettes will absorb water and so careful pre- and post-handling of the cassettes is necessary before weighing, otherwise the cassette-absorbed water may contribute 1-2 mg of weight. This also means that, even with careful procedures, the differential water gain from one cassette to another will contribute to the variability of the method, which contributes to the determination of the smallest mass that can be detected (the LOD, see ASTM, 2006 for methodologies related to filter/cassette weighing) and the smallest mass that can be quantified (the LOQ, see next chapter). Paik and Vincent (2002) found these values to be 0.19 and 0.65, respectively. This is considerably larger than the LOD NIOSH reports for the weighing of filters only from 37 mm cassettes, 0.03 mg (NIOSH, 1994). In 2004, SKC changed the plastic material to address static and water vapor issues and the weight stability concerns raised by Paik and Vincent (2002). The cassette cannot be desiccated, only equilibrated at room humidity (personal communication, D. Dietrich, SKC).

RESPIRABLE PARTICLE MASS[19]

Cyclones

Cyclones are devices that separate particles according to their angular momentum. The cyclone imparts an angular acceleration[20] to the particle due to its cyclonic shape (see Figure 24-11). The momentum of the particle is the product of its mass and instantaneous velocity. Particles with higher masses will be intercepted by the sides of the cyclone, fall to the bottom, and not be collected on the filter. If the flow rate through the cyclone increases, the momentum of all of the particles increases; thus a larger portion is not collected on the filter. This is depicted in Figure 24-12. Note also that the slopes of the curves in Figure 24-12 are not steep. In other words, the cyclone is not 100% efficient up to some particle size and 0% efficient beyond that particle size. However, some cyclones give steeper curves than others (Saltzman and Hochstrasser, 1983; Saltzman, 1984) and all cyclones have steeper efficiency curves than the RPM curve (Trakumas and Hall, 2003 and see Figure 24-13).

Historically, two cyclones were used to collect *respirable dust*. In the United States, the 10 mm nylon cyclone (see Figure 24-14), also known as the Dorr-Oliver (name of the manufacturer, in Stamford, CT), was used, and is still used by OSHA, at a flow rate of 1.7 L/min to achieve a 3.5 μm cut-point and the ACGIH® criteria (see Table 24-6). It is used at a flow rate of 2.0 L/min to achieve the BMRC cut-point of 5 μm. The U.S. Mine Safety and Health Administration (MSHA) still uses the 2.0 L/min flow rate and the correction factor of 1.38, thereby attaining the BMRC criteria, rather than the old ACGIH® criteria.

In the United Kingdom, the BMRC criteria were used. Originally, a cyclone was designed to achieve the BMRC criteria (Higgins and Dewell, 1967), which was known as the British Cast Iron Association (BCIRA) or Higgins and Dewell cyclone. Later, the Safety in Mines Personal Equipment for Dust Sampling (SIMPEDS) cyclone was developed as a modified alternative, but over time it has been further modified. Because of the weight of the SIMPEDS (it was brass), a Scandinavian cyclone was developed out of aluminum (Liden, 1993). It was modified several times and is now manufactured by SKC (see Figure 24-15). A more original version of the SIMPEDS is manufactured by BGI (BGI-4).

---

[19] Again, while this and the other fractions are often referred to as mass fractions, they can just as easily be assayed as number or surface area fractions. The term *mass* refers to the way it is assayed, usually weighed.

[20] Velocity is a vector denoting both speed and direction. Acceleration is a change in speed or direction. Thus, when the particle speed is constant, but the particle has a circular motion, it is accelerating.

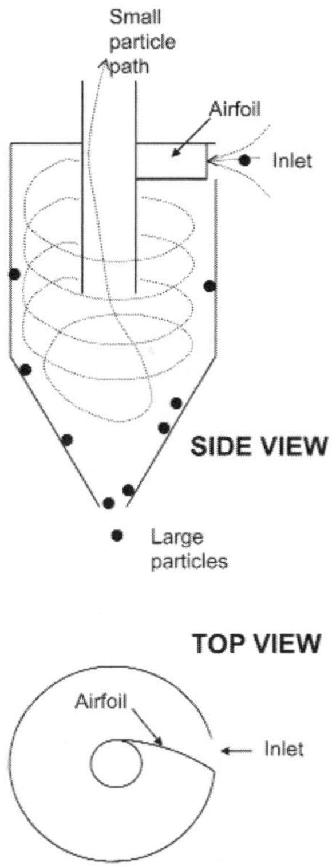

Figure 24-11. Two views of a cyclone. Entering air is given a cyclonic movement so that large particles are thrown against the cyclone side and fall out. The filter would be downstream from the top exit.

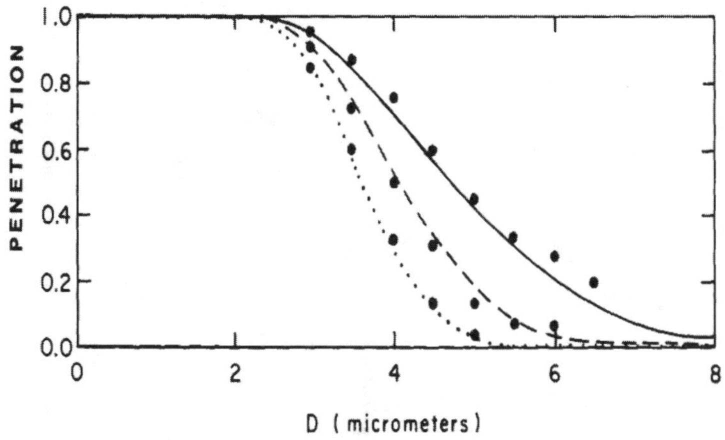

Figure 24-12. Modeled cyclone penetration curves and experimental data: solid = 1.4 L/min; dashed = 1.7 L/min; dotted = 2.0 L/min (from Bartley and Breuer, 1982, with permission).

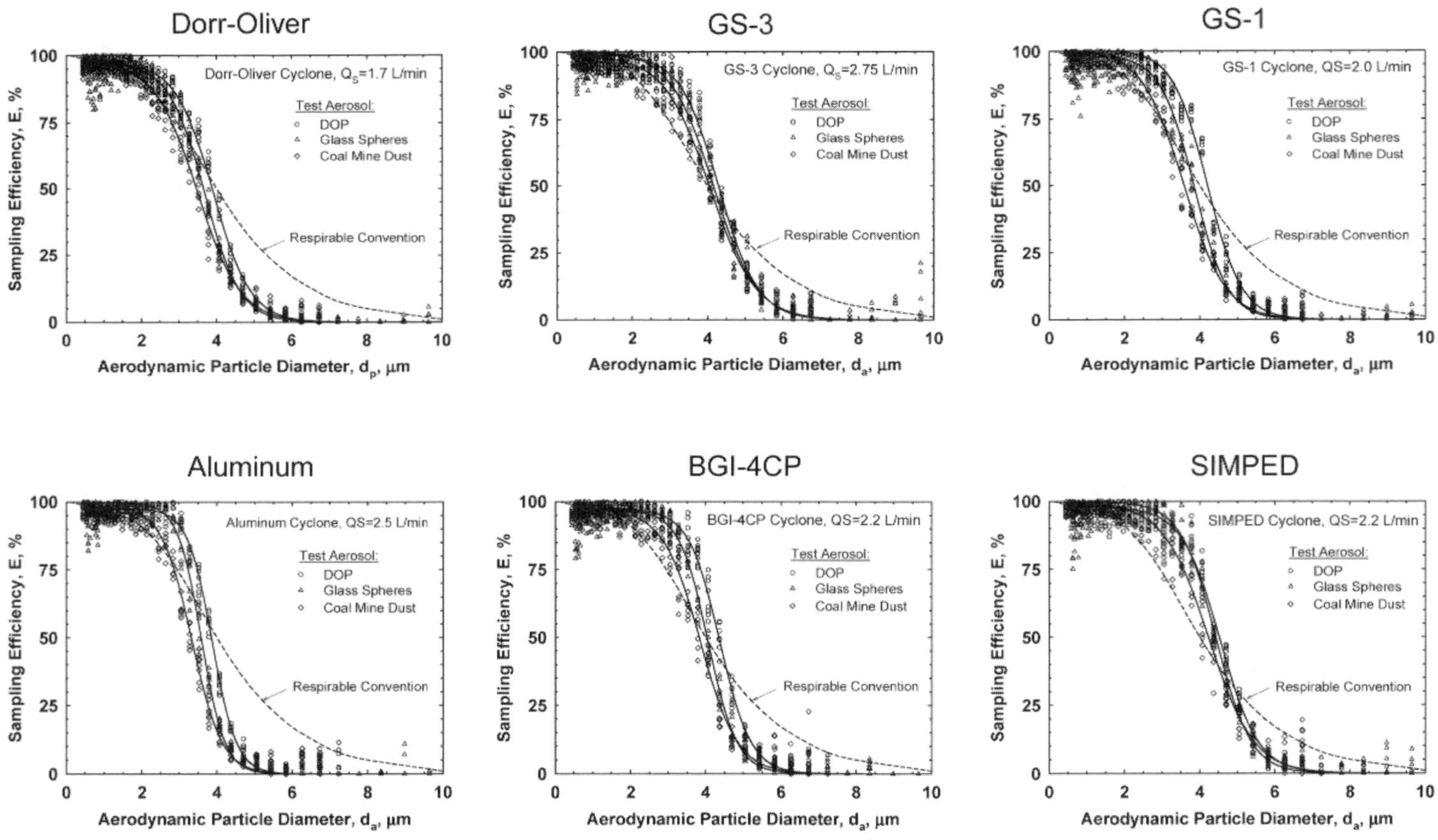

Figure 24-13. Penetration curves for 6 cyclones for three test aerosols (from Trakumas and Hall, 2003, with permission).

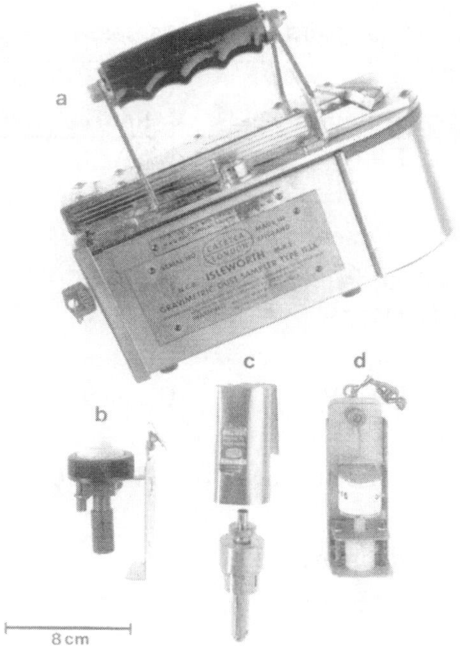

Figure 24-14. Historic size-selective aerosol samplers: a) MRE; b) SIMPEDS cyclone; c) BCIRA cyclone; and d) 10 mm nylon cyclone with cassette and holder (from Verma et al., 1992, with permission).

Figure 24-15. The SKC aluminum, the 1-hole inlet (simulates the Dorr-Oliver), and the 3-hole inlet cyclones (courtesy SKC, Inc.).

Table 24-6
Personal respirable size selective samplers (references below[1])

Sampler	Other names	Source	Q_4 L/min	$Q_{3.5}$ L/min	Filter (mm)
nylon cyclone[2]	Dorr Oliver (D-O)	MSA, Sensidyne	1.7[3]	1.4[4]	37
GS-3 cyclone	D-O like, but 3 inlets	SKC	2.75	3.7	25/37
GS-1 cyclone	D-O equivalent, 1 inlet	SKC	2.0	3.0	25/37
SKC aluminum	Modified SIMPEDS	SKC	2.5	2.8	25/37
BGI-4 or CAS-4 cyclone[3]	BCIRA or Higgins Dewell-like, SIMPEDS Casella plastic cyclone	BGI, CasellaCel	2.2		25/37
PEM impactor	Marple-Miller, 1 stage inertial impactor	MSP, SKC	2, 4, or 10[5]		37
Respirable PPI	Parallel particle impactor	SKC	2		37
Polyurethane foam	Modified IOM, dual fraction	SKC	2		25

[1] MSP, 2003; Gorner et al., 2001; SKC, OSHA (2003), MSHA (2003), BGI (2006), MSA (1998)
[2] non-conducting and required for OSHA/MSHA inspection officers
[3] MSHA uses 2.0 and a correction factor of 1.38 for an equivalent cut-point of 5 μm
[4] OSHA calibrates at 1.7, but allows 1.5–2 for a valid sample
[5] available at 2.5 μm cut only

Considerable research on these cyclones has been prompted by the Universal particle size-selective convention. The research shows that the existing cyclones can be operated to meet the new criteria and, in some cases, that the operating parameters of the past (and present for OSHA) may not have been achieving the old criteria (see Table 24-6 for the nylon cyclone). Bartley and Breuer (1982) calculated that the 10 mm nylon cyclone optimally met the old ACGIH® criteria when operated at 1.4 to 1.5 L/min, rather than 1.7 L/min (see also Gorner et al., 2001). In addition, it met the BMRC criteria better if it was operated at 1.2 L/min with a correction factor (K) of 0.91, rather than at 2.0 L/min with $K = 1.38$ as used by MSHA. Liden and Kenny (1993) found that the nylon cyclone was optimized for the BMRC criteria at a flow of 1.5 L/min, no correction factor. Ironically, it has been shown that the nylon cyclone meets the new, universal criteria when operated at a flow rate of 1.6 to 1.8 L/min (Liden and Kenny, 1993). While most U.S. hygienists (including OSHA, but with the exception of MSHA) have been operating at 1.7 L/min, they have been unwittingly using the new particle size-selective criteria all along. Optimization of the performance of the SIMPEDS and the

SKC[21] cyclones has also been calculated (Liden, 1993; Liden and Kenny, 1993).

OPTIMIZATION CRITERIA

Optimizing the performance of a cyclone is not a straightforward task. In past studies, many flow rates have been examined, because the shape of the penetration curve will change a bit with each flow rate. Generally, however, we cannot adjust the flow rate so that the universal RPM criteria and the cyclone penetration curves exactly coincide, and of course even if we could, Figure 24-13 shows that the curves would change for different aerosols. The work summarized in Figure 24-13 shows the variability in efficiency for three different aerosols. This is simply the nature of cyclones.

How then do we adjust Q so that the performance of the cyclone is most like the criteria? We could calculate the root mean square[22] difference for various points along the two curves; but as we ultimately are interested in the total weight of particles that is selected by the cyclone, differences at the right-hand tail (heavier particles) are more important than those at the left. Therefore, aerosol size distributions that are *large* will be sampled with a different bias from that of an aerosol that is distributed at the *small* end of the spectrum. Bartley and Breuer (1982) proposed that the cyclone be optimized, by finding the flow rate that gave the smallest average bias in the weight of particles collected when compared to the appropriate particle size-selective criteria. They proposed that the average bias could be calculated across a range of aerosol distributions representative of all industry, or that preferential weighing could be given to a *region* of dust distributions, such as those found in the coal industry.

To this end, Liden and Kenny (1993) recommended three testing methods that comprised up to 153 different aerosol distributions or combinations of mass median aerodynamic diameters (MMAD) and geometric standard deviations (GSD) including one approach consisting of 34 distributions based on the actual workplace measurements of Hinds and Bellin (1988). As these proposed evaluations were across a large number of distributions, a laboratory study is not feasible. Instead, Bartley and Breuer (1982) proposed the use of mathematical models that yield penetration curves for any flow rate that are based on empirical penetration data derived at one flow rate for that cyclone. These modeled penetration curves are studied to determine the optimal flow rate for the range of aerosols likely to be encountered.

From the optimal flow rate, one can then find the bias associated with each aerosol size distribution, a bias map. Figure 24-16 shows such an example for the nylon cyclone. Generally, the bias is less than 10%, but in some cases it can be quite high. An important conclusion can

---

[21] Since 1986 there have been at least 4 generations of what is now the SKC aluminum cyclone.

[22] The root mean square is similar in concept to the standard deviation. It is an attempt to derive an average difference by neutralizing the effect of the algebraic sign of the individual differences. Thus, each difference is squared, the square differences are averaged, and the square root of the mean is taken.

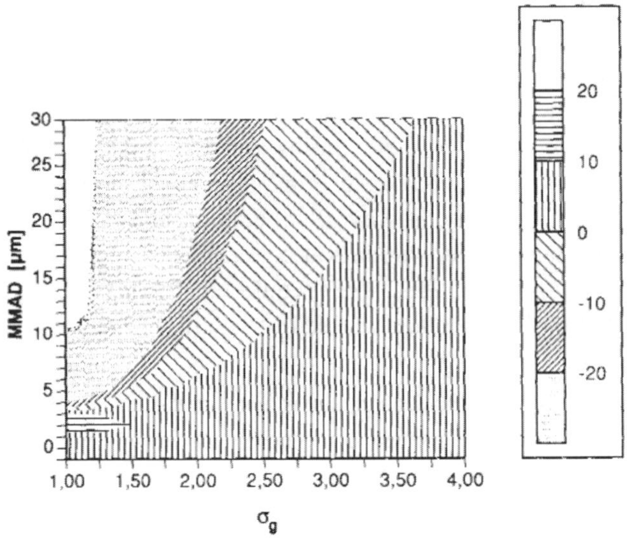

Figure 24-16. Modeled optimization graph for various lognormal aerosol size distributions (based on the original data of Blachman and Lippmann, 1974) showing contours of equal bias (%) for the 10 mm nylon cyclone (Q = 1.7 L/min in calm air and exposed to ferric oxide monodisperse aerosols) relative to the universal RPM convention (from Liden and Kenny, 1993, with permission).

be drawn: the error caused by the use of a non-optimal flow rate may be small compared to the error caused by the aerosol distribution. Thus, a relatively inexpensive way of determining the actual aerosol distribution would be important so that correction factors can be made (from a bias map) for cyclonic sampling. Finally, there have been a number of studies that determined optimum flow rates for cyclones to meet the Universal RPM (see, e.g., Kar and Gautum, 1995; Liden and Gudmundsson, 1996; Tsia and Shih; 1995 as well as other papers cited above); however, in the past decades manufacturers have changed designs, recommended and then retracted the use of correction factors, and changed recommended flow rates. It is, therefore, best to use the flow rate recommended by the manufacturer at the time the device was purchased or inquire directly with the manufacturer where there is a discrepancy between the original and a current recommended flow rate.

PRECISION AND BIAS

It should be apparent now that particle size-selective samplers are more difficult to evaluate for precision and bias than are other sampling methods such as charcoal tubes. A different bias and precision is possible for each aerosol size distribution. It is difficult to devise an overall statement of the accuracy (precision and bias) of particle size-selective samplers for comparison with the often cited NIOSH requirement that a method have an accuracy of ± 25% at the 95% confidence level for a single measurement (or, roughly stated, the absolute value of the bias plus two times the *CV* should not exceed 25%). For a review of thinking on the subject, the reader should see Bartley and Doemeny (1986), Kenny and Liden (1989), Liden and Kenny (1992), and Bartley and Fischbach (1993). Approaches for calculating the bias have been proposed, which can be accomplished with mathematical modeling. Precision assessments, however, require replicate sampling and analyses in differing aerosol environments (including varying and representative wind speeds, particle charge, and relative humidity effects). The CEN (2001) has devised an accuracy criterion that takes into account the cyclone bias plus the pump and weighing or analytical error. Apparently, it does not include variability from one sampler to another. A parallel ASTM method does (ASTM, 2001b).

The most comprehensive study of RPM samplers to date examined the bias and accuracy of 15 RPM samplers used throughout the world (Gorner et al., 2001). The samplers were assessed against generated and modeled aerosols of 1 to 25 μm (1 μm steps) MMAD and 2.0 to 3.5 (0.25 steps) GSD for a total of 175 aerosols. A bias criterion of < 10% (absolute value) was considered acceptable for each aerosol. The percent of aerosols sampled with less than 10% bias were: Dorr-Oliver (84%), SKC plastic (58%), Casella plastic (68%), and SKC aluminum (86%). As Figure 24-13 shows, often cyclones undersample the larger particles. Thus, for each of these cyclones, biases were greatest for the larger (> 10 μm MMAD) aerosols. This will cause an underestimation of the concentration for aerosol distributions dominated by larger particle sizes.

Chen et al. (1999) studied 25 replicates of SKC aluminum and 10 mm nylon cyclones for both liquid (dioctyl phthalate) and solid (potassium sodium tartrate) aerosols. Results showed that both cyclones were least precise at aerosol sizes around 5 mm (CVs in the 5 to 10% range) while at smaller particle sizes CVs were typically around 5%. Note that most of the variability was not within variation, i.e., variability caused by repeated measurements on the same cyclone, but rather due to between variation or lack of consistency between specimens of the same cyclone type. As a result, manufacturing quality control is questioned. In addition, cleaning cyclones after each use is extremely important. In terms of bias, as compared to the Universal RPM convention, both cyclones typically gave biases less than ± 5% for aerosols from MMAD 1 to 25 mm and GSD from 1 to 3.5. However, for solid aerosol sizes around 5 mm, biases peaked at – (10–15)% for the SKC and –10% for the nylon sampler. In both cases, the greatest biases occurred for very uniform particles (GSD < 1.5), indicating that cyclones should be rotated between SEGs; as particle size distributions become monodisperse and approach 5 mm, sample sizes should increase; and the hygienist should be aware that, for these same conditions, concentrations are more likely to be underestimated.

The precision of any cyclone will be affected by variations in pump flow rate, highly charged particles (depending on the charge distribution on the cyclone), and relative humidity.

Pump Fluctuations

Piston and diaphragm sampling pumps do not give smooth flow profiles over short time periods, even with pulsation dampeners. When the chamber empties, the flow is highest; and when the chamber fills, the flow is lowest (see Chapter 20, Figure 20-5). This periodicity is caused by the frequency of the motor as it turns the drive shaft. Because cyclone performance is dependent on flow rate, one might expect that the nature of the periodicity is important. Bartley et al. (1984) showed that the effect on cyclones was generally less than 5%, but could be as high as 22% (Berry, 1992) and when the flow variation was controlled to ±20% of mean flow, the error was < 2%. It is likely that newer *constant-flow pumps* adequately control these fluctuations.

Electrostatic and Humidity Effects

Cyclones can be non-conductive (10 mm nylon) or conductive (metal or carbon filled plastic). As with cassettes, the charge should be concentrated on the outside surface. One might expect that the most variable results would be obtained when charged nylon cyclones were used to sample charged aerosols, and that conducting cyclones would behave similarly, regardless of the aerosol charge.

Almich and Carson (1974) examined the effect of charge on the performance of nylon cyclones under worst-case (RH < 10%) conditions. They found that nylon cyclones retained more particles (i.e., the filter would undersample) when the particles (coal dust, 4-5 μm MMAD) were charged (67%) than when not (52%). A specially constructed stainless steel cyclone (presumably like the 10 mm nylon) also retained more charged (69%) than uncharged particles (56%). Liden (1993) found that the variation in collecting Carborundum particles, which were thought to be highly charged, was much greater than for foundry dusts. Briant and Moss (1984) also showed that the nylon cyclone allowed less penetration of charged polystyrene latex particles (0.5 1, 2, and 2.8 μm MMAD), but only if the cyclone was charged. A specially constructed, graphite-filled nylon cyclone was not affected by particle charge.

At low humidity levels, the nylon cyclone undersampled silica relative to the horizontal elutriator (see later discussion) even more than when humidity was high. The effect was not seen in sampling of wood dust (Sass-Kortsak et al., 1993). Chen and Huang (1999) found that the 50% cut-point for the 10 mm nylon cyclone became a 30% cut-point (the entire curve was shifted left) when sampling small particles (CMD, 3.5 μm; GSD, 1.3) under 10% humidity conditions and that the cyclones became dirty from deposits under these conditions, which was assumed to change their sampling dynamics. The nylon cyclone has sampled less than the SKC aluminum in field studies, possibly due to these effects (Verma et al., 1992; Sass-Kortsak et al., 1993; Groves et al., 1994).

The results indicate that charged particles (1) lead to less precision regardless of the conductive nature of the cyclone and (2) may lead to undersampling, certainly for the nylon cyclone (in the case of conducting cyclones, the cyclone would appear to need to carry a charge). Finally, the relevance of these results to workplace aerosols also is not clearly defined, but one should be aware that at low relative humidities, the nylon cyclone will probably have reduced penetration and, therefore, will undersample.

Impactors

When a particle cannot remain in an airstream as airflow lines move around an object, it is impacted on the object's surface; and this principle may be used in designing particle size-selective sampling devices. Figure 24-17 shows a two-stage inertial impactor. The first stage has a lower velocity than the second stage. The smallest particles do not have sufficient momentum to be impacted on the first stage. However, the increased velocity through the smaller, second jet imparts enough momentum to the smaller particles that they are impacted on the second stage. Many stages can be arranged in this manner, and thus many *cut-points* can be affected, one for each stage. Such multi-stage impactors are called cascade impactors and are discussed in a later section.

Table 24-6 lists two impactors that have been designed to sample the RPM fraction. The PEM device is a one stage inertial impactor; larger particles are collected on a greased impactor plate, while smaller particles proceed to be collected on a filter. The device does not

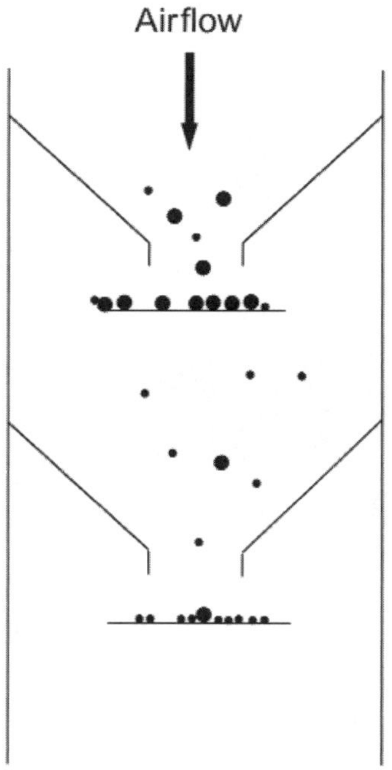

Figure 24-17. A two stage inertial impactor schematic.

follow the RPM but rather the U.S. EPA's definition for $PM_{2.5}$ and thus the 50% MMAD cut-point for this device is 2.5 rather than 4 μm, and the curve is steeper than for the RPM.

The second device, the PPI, is a 4-stage inertial impactor and is shown in Figure 24-18. All four stages deposit onto a single filter and thus the particle deposition is not uniform across the filter with respect to particle size. However, this is not important for weighing or chemical assay, only for techniques such as fiber counting. By carefully designing the cut-points for the four stages, the PPI actually gives a sampling efficiency curve that matches the RPM convention more closely than any cyclone is capable of (SKC, 2002).

Although not used to collect a RPM fraction, a single stage impactor has been designed to collect diesel particles in the presence of other, larger organic carbon, such as coal mine dust. Since the coal dust is generated mechanically, and the diesel particulate is a fume, the two can be separated fairly efficiently (Noll et al., 2005), even though chemically there are considerable similarities (at least from an analytical perspective).

Foams

Whether one thinks of the device as an impactor or a filter, it has been shown that polyurethane foams with certain pore sizes have the ability to act efficiently and accurately as particle size selective samplers. The initial work with foams as particle collectors goes back to the 1960s; however, the application to size selection originates with Mark et al. (1988). Gibson and Vincent (1981) showed that the foams collect particles via gravitational and inertial forces,

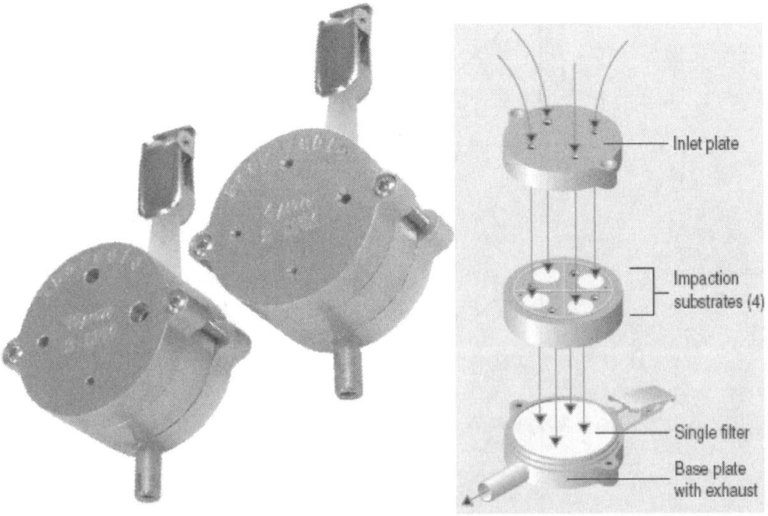

Figure 24-18. The SKC PPI (Parallel Particle Impactor), for RPM and TPM size fractions and a schematic showing how the four *stages* empty onto the same filter (courtesy SKC, Inc.).

while electrostatic forces are not important. However, there are no mathematical models for the physical properties that allow the foams to work. The foams are quite porous; only 3% of the structure is solid (Aitken et al., 1993).

To date, commercially available porous plastic foams have been fitted into the IOM cassette in front of the filter and in the CIS (GSP or conical sampler). A single foam fitted into the cassette of the IOM allows one to weigh the entire cassette (including filter and foam) to obtain the IPM and the filter alone to determine RPM. Two separate foam inserts of different pore sizes allow measurement of all three fractions.

Aitken et al. (1993) tested foams in the laboratory for particle size selective sampling. Others have successfully applied foams to welding fume (Chung et al., 1997), mineral dusts (Teikari et al., 2003), and primarily to bioaerosols (Kenny et al., 1998). Chen et al. (1999) found that foam outperformed aluminum and nylon cyclones in the laboratory.

Kenny et al., 2001 performed a wide ranging study of foam RPM samplers. They confirmed that sampling efficiency will decrease (the cut-point falls) with load; but this is only significant at high loadings of 20 mg or so (an 8-hr sample at 2 Lpm and 5 mg/m^3). However, in other studies of fine particulate (welding fume) this phenomena is more significant (Aitken et al., 1993; Chung et al., 1997). Across a very wide range of industries, the foam samplers gave equivalent results to cyclones. As expected, the foam samplers outsampled the cyclones at low concentrations (< 0.2 mg/m^3).

THORACIC PARTICLE MASS

There are only a few chemical agents for which the TPM convention appears to be applicable, i.e., the major site of exposure-related disease is the thoracic region. Asbestos has received the most attention, but cotton dust has a long history of being sampled with the vertical elutriator (see later discussion), a device that samples the thoracic fraction. In addition, sulfuric acid has effects in the thoracic region and particle size selective sampling of it has been recommended (Lippmann et al., 1987). Finally, metal working fluids have a NIOSH method that is based on the TPM fraction (NIOSH, 2003).

In the case of asbestos, historically only fibers with physical diameters of < 3 μm were counted. This is equivalent to sampling the fibers using the thoracic convention (Baron, 1996). In other words, rather than the microscopist needing to size particles to determine whether to count them, if a TPM size selective sampler is used, the microscopist should be able to count all fibers sampled.

Currently available TPM samplers are shown in Table 24-7. All of these are variations of a similar device that was described for RPM sampling. The BGI cyclone was first described in the UK and was designed so that it could be used for both the RPM and TPM using different flow rates (Kenny and Gussman, 1997). Maynard (2002) has shown that the sampler efficiently aspirates fibers with lengths of 10-60 μm and diameters of 0.4 to 3.5 μm.

Table 24-7
Personal thoracic size selective samplers (references[1])

Sampler	Description	Source	Q L/min	Filter (mm)
GK 2.69	Cyclone	BGI	1.6	37
IOM foam	Modified IOM	SKC	2	25
PEM impactor[2]	Marple-Miller, 1 stage inertial impactor	MSP, SKC	2, 4, or 10	37
Thoracic PPI	Parallel particle impactor	SKC	2	37

1 SKC, 2002; MSP, 2003; BGI, 2006
2 Cut-point is 10 µm, however, device samples according to EPA PM_{10} definition, which is a steeper curve than TPM convention.

The IOM TPM foam sampler is identical to the modified IOM used for the RPM fraction, with the exception that the foam pore size is different. Maynard found that the IOM TPM foam sampler undersamples longer fibers.

The PEM impactor performance is based on the U.S. EPA PM_{10} definition (EPA, 1987). That curve is similar to the TPM convention except that it continues on the same downward slope at large particles rather than having the reverse S shape of the TPM curve; thus, it undersamples large particles relative to the TPM. Baron and John (1999) has shown that the negative bias of the PM_{10} curve relative to the TPM is greatest (20%) for particles > 15 µm in aerodynamic diameter. Interestingly, in a study (mainly wood dust samples) comparing the PEM to 37 mm cassettes, the PEM sampled about 1.6 times more than the cassette. This likely indicated two things: the PEM oversampled, perhaps due to particle bounce; and the cassette undersampled, likely due to losses of dust to the cassette due to charge effects.

Jones et al. (2005) have performed the most inclusive study of TPM sampling relative to fibers. They examined the uniformity of sample deposit (critical for counting random fields) for the GK2.69, modified SIMPEDS, CATHIA, and the IOM foam samplers which had been exposed to 1 and 10 µm spherical particles and fiber glass that spanned the TPM size ranges. The modified SIMPEDS gave hotspots of deposition attributed to the inertial flow pattern through the sampler port. The other three gave patterns similar to the cowled sampler. The modified SIMPEDS, CATHIA, and GK2.69 sampled in good agreement to the TPM convention, with a slight oversampling for fibers < 7 µm aerodynamic diameter, similar to the cowled sampler. The IOM foam sampler slightly undersampled all fiber sizes relative to the TPM convention.

SIMULTANEOUS MEASUREMENT OF MULTIPLE FRACTIONS

There are devices that allow the simultaneous assaying of all three fractions (see Table 24-8). One of these, the CIP 10 requires a different head for each fraction, thus unlike the others it will not sample the three fractions simultaneously. Several of the devices in Table 24-8 work on the principle of impaction: conventional inertial impaction as in the Marple and virtual impaction and as in the RespiCon®. Impaction also occurs on the polyurethane foam surfaces, along with the other filtration processes described earlier, such as interception.

Impactors

Many impactor designs are available (see Hering, 2001). However, few of these are made in a size that can be used for personal sampling. The device shown in Figure 24-19 is also known as the Marple cascade impactor after its designer (Rubow et al., 1987). The device shown is an eight-stage device. It is also marketed as a six-stage device (stages 3 through 8) or a four-stage device (stages 1,2, 3, and 5).

One of the most important uses of impactors is to sample microbiological agents from air (see Volume 2). Finally, it should be noted that no U.S. standard is known to employ impactors

Table 24-8
Multiple size fraction samplers (inhalable, thoracic, respirable fractions from same device, references below[1])

Sampler	Source	Q L/min	Principle	Filter (mm)
CIP 10[2]	Arelco	7-10	impaction?	foam
IOM foam[3]	SKC	2	impaction	25
Marple[4]	Thermo	2	cascade impactor	34
PERSPEC	Lavoro y Ambiente	2	virtual impactor?	47 or 50
RESPICON	TSI	3.11	virtual impactor	37
Sioutas[5]	SKC	9	cascade impactor	25 & 37

[1] TSI, 1999; Arelco, 2006; Thermo, 2006; SKC, 2002; Baron and John, 1999
[2] No simultaneous sampling of all three fractions; requires attachments for each fraction, poor collection of particles < 1 μm
[3] A dual foam sampler is available from SKC that will allow RPM and IPM dual sampling.
[4] Also known as the Anderson 290; up to 8 stages including 3.5 and 9.8 μm cuts, special slits in filters
[5] 5 cuts at > 2.5, 2.5 to 1, 1 to 0.5, 0.5 to 0.25, and < 0.25 μm

Figure 24-19. The assembled and disassembled Sierra (Anderson) personal cascade impactor (photo, M. Strawn).

as a particle size-selection tool; however, they are invaluable for microbiological agents, for particles without OELs, such as may occur in the pharmaceuticals industry, and for characterizing particle size distributions.

There are two key issues for inertial impactors: particle bounce and reentrainment or blow-off (John, 1999). When a particle strikes a stage it may bounce depending on the properties of the particle and stage. If it does not land again on the stage, but escapes, then it goes to the next stage, even though in theory it is too large to pass. When a particle is trapped on the stage it may be reentrained due to constant bombardment of the particle by the incoming jet. Thus, all impactors (except virtual) require that the impaction surface be coated. Many manufacturers offer filters to serve as impaction surfaces, these degrade impactor performance and do not entirely prevent bounce (Hering, 2001). The Marple impactor (see Table 24-8) uses greased and/or porous surfaces to avoid these issues. The Sioutas impactor uses Teflon filters and is claimed to eliminate bounce issues. Hinds et al. (1985) showed that the substrate used has a strong influence on the resulting MMAD. They recommended a field evaluation of different substrates to determine the best one for the particles being sampled. The substrates (greased or ungreased as in use) are weighed before and after sampling in order to assay the mass in each particle size range. If substrates are not greased carefully, the grease may run during sampling, causing a weight loss.

Another solution to these issues is the virtual impactor. Figure 24-20 shows one design for the virtual impactor principle; there are others. The virtual impactor has reverse collection. The largest particles are caught at the bottom stage, not at the top as in the inertial impactor.

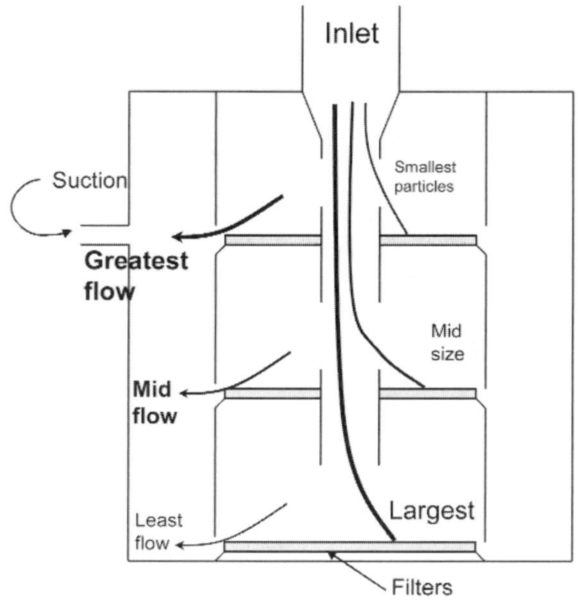

Figure 24-20. A schematic of one type of virtual impactor.

The larger particles at each stage have greater inertia than smaller ones and thus do not change course but continue straight down in the minor airflow. At the next stage the flow has decreased and some of the same particles that passed through the first stage now have inertia that has been reduced enough so that they follow the streamlines and fall out of the major flow due to decreasing velocity. Figure 24-20 is similar to the design of the RespiCon® sampler (see Figure 24-21).

Multi stage cascade impactors (e.g., the Sioutas or Marple) allow particle size distribution characterization. A frequency plot or histogram may be constructed by using the assayed masses from each stage of impactor.[23] The collected mass is plotted against the midpoint particle size for that stage (as opposed to frequency versus particle size, as would typically be used in a statistical distribution). Figure 24-22 shows such a distribution. However, the more frequent use of such data is to construct a probability plot (see Figure 24-23). Such a plot must be made on log-probability paper (alternatively the log-particle sizes may be plotted on normal probability paper). The mass median aerodynamic diameter (MMAD) and the geometric standard deviation (GSD) of the particle size distribution can be determined from this plot (see

[23] See Hinds (1986, 1999), Knutson and Lioy (1989), Baron and Willeke (2001), and Hewitt and McCawley (1991), Xu et al., 2002, and Chapter 9 of this text for further discussions of this subject.

Figure 24-21. The RespiCon® virtual impactor for IPM, TPM, and RPM size selective sampling (courtesy TSI, Inc).

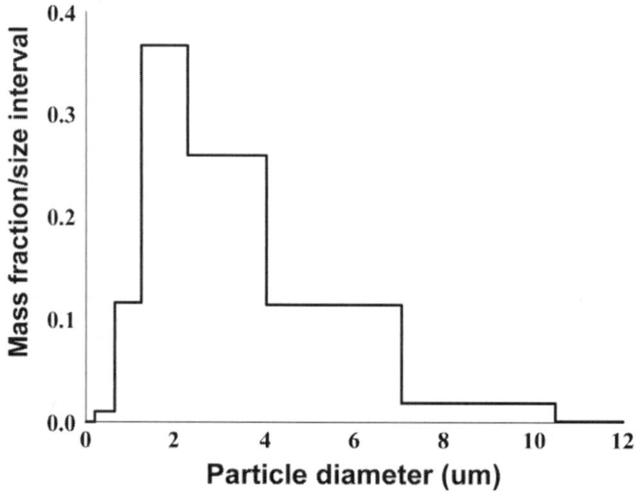

Figure 24-22. A typical histogram of particle size distribution using cascade impactor data.

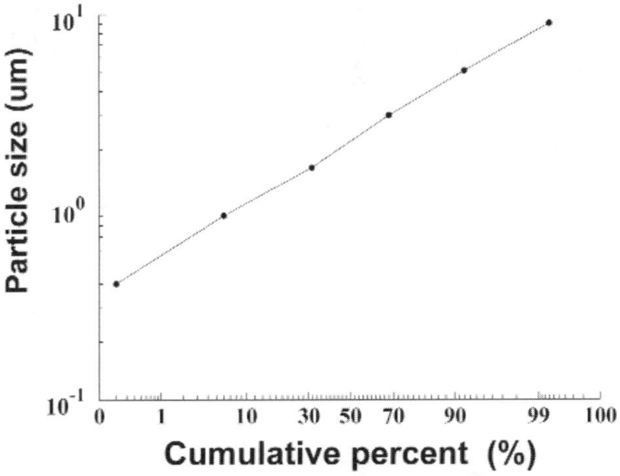

Figure 24-23. A cumulative frequency plot of the cascade impactor data shown in Figure 24-22. The 50th (MMAD), 16th and 84th percentiles can be estimated.

Chapters 7 and 9 for a review of these terms). If the points plotted on the log-probability paper are nearly in a straight line, then the distribution of particle sizes may be considered lognormal and the MMAD is the median or the 50th percentile of the probability plot.[24] The GSD can be determined as follows:

$$\frac{\dfrac{84th\ \%tile}{50th\ \%tile} + \dfrac{50th\ \%tile}{16th\ \%tile}}{2} = GSD$$

The cut-points for each stage are particle sizes at which the collection efficiency is 50%. As shown in Figure 24-24, there is some overlap of particle sizes collected on the various stages, particularly at the larger particle sizes (2 to 25 μm). Another purpose of cascade impactors is to determine the three major fractions of the Universal convention. One simple way to do this is to assume that the mass of everything collected on and below the stage corresponding to the convention, say for example 10 μm, is the TPM. In other words, if one of the stages has a cut-point of 10 μm, then the mass on that stage and all stages below it can be compared to the TPM. This technique disregards the particle crossover from one stage to the next. It is not an important error for small particle sizes, but errors increase at larger particle sizes, since the impactor stages do not have sharp cuts. There will be some particles 10 μm and smaller that will be on the previous stage and not assayed; there will be some particles larger

[24] Fifty percent of particle sizes have smaller masses and 50% have more mass than this median size, also called the cut-point.

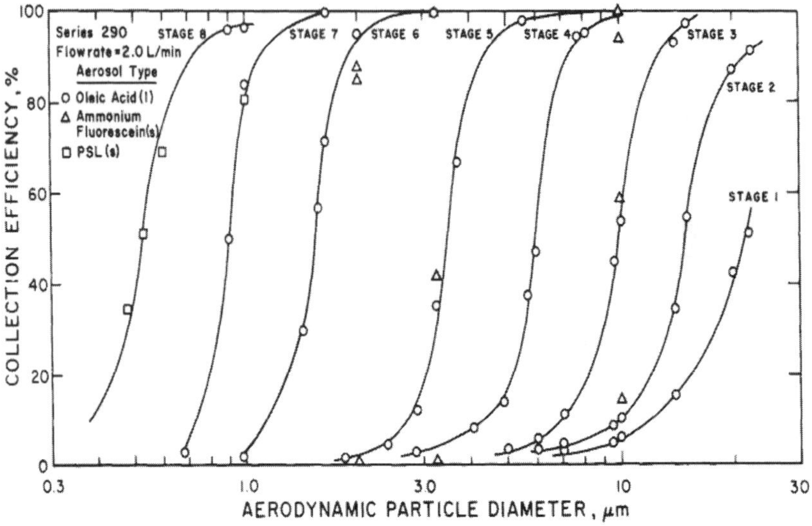

Figure 24-24. Efficiencies of the stages of the Marple or Sierra personal cascade impactor for three aerosols (from Rubow et al., 1987, with permission).

than 10 μm that will pass to the 10 μm stage and should not be assayed. There are ways (mathematically complex) to invert the particle size distribution and impute a better estimate of the TPM or RPM concentration (Ramachandran and Vincent, 1997).

For the IPM fraction, the issue is whether the inlet on the impactor samples accordingly (see, e.g., Li et al., 2000 for the effects of various inlets on the Marple impactor). The Marple fitted with a standard visor inlet consistently undersampled particles with aerodynamic diameters > 20 μm. There have been designs of cascade impactors that have IPM compliant sampling heads such as the Personal Inhalable Dust Spectrometer, or PIDS (Gibson et al., 1987). Thus, the sum of all the stages in this device is compared directly to the IPM. The PIDS has been used in various studies (see, e.g., Ramachandran et al., 1996; Spear et al., 1997; Thomassen et al., 1999; Lundgren et al., 2006); however, it does not appear to be marketed.

RespiCon®

The RespiCon® (TSI, 1991a) seems to be a very useful device (see Figure 24-21). There have been several studies of the RespiCon®. Li et al. (2000) found a reasonable match in laboratory studies for the IPM fraction when the manufacturer's correction factor was not used (see Figure 24-7a). Even better matches were found for the measured TPM and RPM fractions compared to the convention. Teikari et al. (2003) confirmed Li et al.'s results in laboratory and field testing for the TPM and RPM fractions. They found that the RespiCon® sampled about 50% of what the IOM sampled for the IPM fraction; however, there was no mention of the use of a correction factor. Tatum et al. (2002)found that the RespiCon® IPM fraction compared well with the CIS sampler (the 1.5 correction factor was used) when sampling wood dust, and the RPM fraction compared well with the BGI-4 cyclone. In addition, a microscopic examination of particles in each stage indicated that particles were being selected appropriately. Rando et al. (2005) found good agreement between the RespiCon® and the IOM, again with application of a correction factor. For the RPM, the RespiCon® comparison to an SKC cyclone was equivocal due to low dust loadings. The manufacturer does not recommend the use of a correction factor based on the work of Li et al. (2000) (TSI, 1996b).

CIP10 and PERSPEC

These devices are marketed in Europe and have seen considerable research. The CIP10 works by rotating a cup that contains polyurethane foam. The rotation creates a vacuum that causes flow, thus no pump is needed. Collection and particle size selection occurs on foam pieces. The CIP10 (Courbon et al., 1988) was invented in France; it will not simultaneously sample all three fractions; a different head is required for each. The CATHIA sampler (an area sampler and the original version of the personal CIP10) has seen the most testing (Gorner et al., 2001; Maynard, 2002; Kauffer et al., 1996). The thoracic version of the CIP10 yields low biases across a wide range of aerosol sizes and has been recommended for cotton dust sampling (Fabries et al., 1998). The CIP10 RPM version also performed well in laboratory testing (Gorner et al., 2001).

The PERSPEC is an Italian device that cleverly separates the particles on one filter, largest to smallest from inside to outer edge (Prodi et al., 1988, 1992). The design resembles a venturi with a wider outlet than inlet. A clean air steam flows down the middle of the venturi entrance and the particle-laden flow on the outside. Perhaps this has some characteristics of virtual impaction. A cutting device is then used to cut the filter in to a center circle and two rings so as to obtain the three fractions by addition of masses. The sampled IPM fraction was compared to the convention (see Figure 24-7b and Kenny et al., 1997). It is not clear if the device is still marketed.

Passive Devices

While none are apparently available commercially, at least three passive sampling devices have been devised. These work on the basis of electrical charge attraction, or electret (Brown et al., 1995), sticky surfaces (Vinzents, 1996) and simple gravitational, inertial and diffusional deposition (Wagner and Leith, 2001). The first two are assayed by weighting, while the latter requires microscopic sizing. All appear to undersample (Schnieder et al., 2002) and they can be calibrated to provide results that correlate reasonably well with IPM, RPM, or TPM samplers (Wagner and Macher, 2003 and Schneider et al., 2002). Their chief attraction is that many can be deployed for less cost than active samplers. Thus, even if precision is poorer than for active samplers, the overall error can be reduced by increasing sample size.

NON PERSONAL PARTICLE SIZE SELECTORS – ELUTRIATORS

There are two types of elutriators used for size selection of aerosols: the horizontal and the vertical. The horizontal type separates particle sizes based upon Stokes's law (see Chapter 9) or horizontal stopping distance. Aspirated particles have a horizontal component to their movement due to the flow of air through the elutriator. They also have a downward component caused by the force of gravity (see Figure 24-25). If the downward component causes them to settle onto the elutriator plates before the horizontal component causes them to get past the plates, then they are not selected for filtration. The vertical type (see Figure 24-26) aspirates a wide range of particle diameters; and then as the cross-sectional area of the device increases, the velocity decreases so that heavier particles lose their momentum and are overcome by gravity. Therefore, particle size selection is flow rate or velocity dependent for both devices.

Unfortunately, they will not work properly unless oriented correctly with respect to gravity; so they are not adaptable to personal sampling, but are used as area devices. One horizontal device has been widely used in coal mines, particularly in the UK. The MRE 113A (Mining Research Establishment of the United Kingdom; see Dunmore et al., 1964) horizontal elutriator (see Figure 24-13) is designed to meet the BMRC size criteria for RPM fraction. In fact, the particle size-selection curve of the BMRC was based on the theoretical performance of an elutriator in calm air (Liden and Kenny, 1991 and 1993), and consequently its performance in calm air closely mimics those particle size-selection criteria. It also performs well against the Universal RPM convention (Gorner et al., 2001). Mark et al. (1993) have shown that the MRE performance is variable when it faces a wind of 0.25, 2, or 5 m/sec (50-1,000 ft/min), velocities that may be commonly encountered in coal mines.

Vertical Elutriator

In the United States, a vertical elutriator has been used in textile mills for monitoring cotton dust since the early 1970s (Lynch, 1971). The Lumsden-Lynch device is shown in Figure 24-26. The purpose of the elutriator is to provide particle size selectivity with a 100% cut-off of 15 μm (Neefus et al., 1977). The median cut-off for sampling in calm air is 10 μm (Grimm, 1981; Rubow et al., 1983). Thus, this fraction is equivalent to the thoracic particulate matter (TPM) fraction, and apparently is the only current size-selective standard based on TPM in an occupational setting. The vertical elutriator suffers from poor precision; the *CV* for side-by-side elutriator measurements being on the order of 30% to 40% (Suh et al., 1980; McFarland et al., 1987).

SEMIVOLATILES AND SORBED VAPORS

Several questions that are sometimes worth asking are often overlooked in workplace aerosol sampling. First, "Is the compound in aerosol or vapor form, or both?" Second, "If it is

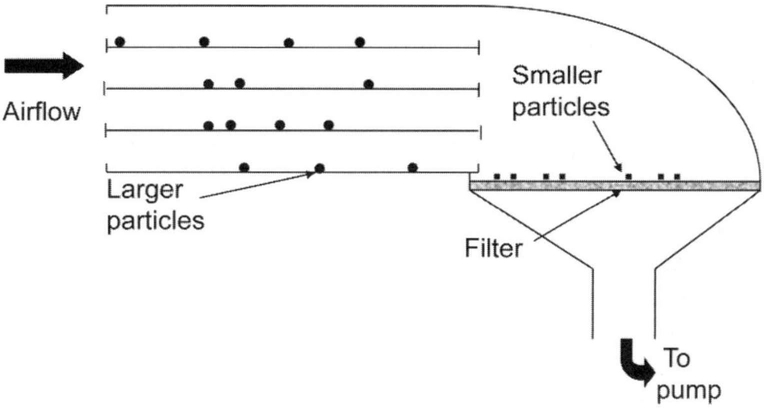

Figure 24-25. Schematic of a horizontal elutriator.

a)

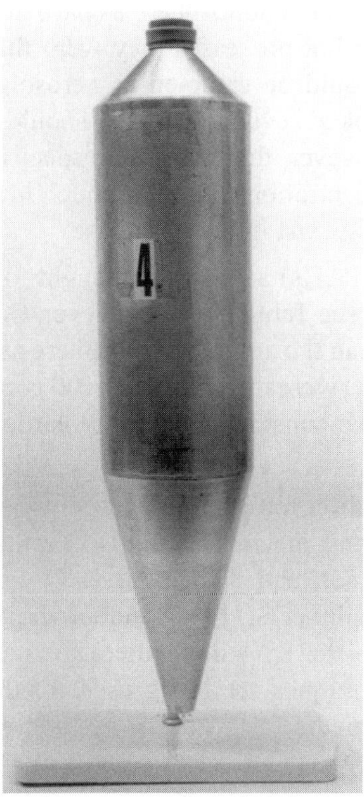

b)

Figure 24-26. a) Vertical elutriator and b) Diagram of sampling train (photo, R.K. Oestenstad).

in particulate form, will it evaporate from the sample collector?" Third, "Can significant amounts of vapors or gases be sorbed to particles and therefore not sampled by using typical gas/vapor sorption techniques?" A corollary to the last question is, "Are these compounds biologically available to the lung in a similar manner to that of the free vapor/gas?" This question will not be considered here; but, to the extent that these particle-adsorbed compounds are of biological significance, they should be assayed along with the free gases and vapors.

Many organic compounds are thought to be airborne as aerosols because they exist as solids at room temperature or as viscous liquids of low volatility. These compounds often are sampled in the workplace as aerosols, by use of common aerosol sampling techniques such as filters. However, as air passes over the filter collecting the particles, are the compounds of interest stable, or do they evaporate? It should be noted that the adsorption of volatile compounds onto particles can occur. For example, volatile organics such as aldehydes may be adsorbed to carbonaceous particles in diesel emissions, or unreacted gaseous monomers may be sorbed to polymer particles.

One way of considering the propensity of a compound to volatilize is to examine its vapor pressure. Compounds having vapor pressures less than 1 mmHg were historically thought to be nonvolatile (Lewis, 1986). They were thought to exist in the atmosphere as particles generated in a mechanical process, or after evaporation from hot processes they were thought to quickly condense onto airborne particles where they could be sampled as aerosols. In the workplace, where small concentrations are often overlooked (even when they should not be), this model of volatility is reasonable in many cases. However, there are some specific examples of when it is not. Ambient air sampling, where concentrations are often much lower than in the workplace, has revealed that these compounds may exist in the gas phase.

For example, polynuclear aromatic hydrocarbons (PAHs) are widely thought to exist in the aerosol phase, as their vapor pressures are quite low (see Table 24-9). However, Coutant et al. (1986) concluded that three- and four-ring PAHs exist in the ambient atmosphere as vapors. Similar observations (Cautreels and Cauwenberghe, 1978) were made for over 100 compounds that were found in the ambient environment and could be considered as semivolatile according to Table 24-9.

In the occupational environment, the phase distribution for a number of compounds and mixtures has been questioned. These include cutting and machining fluids (Krantz, 1983; Nilsson and Nygren, 1987), aromatic amines (Gunderson and Anderson, 1987), fluorides (Einfeld and Horstman, 1979), diesel emissions (Westerholm et al., 1991), indoor air pollutants (Wilson et al., 1989), and pesticides (Lewis and Jackson, 1982). No doubt there are many other examples, but it is not likely that special sampling techniques are being used for these and other compounds because NIOSH and other *standard* methods do not always address more subtle issues involved in sampling semivolatiles.

Evaporative losses from filters have been shown for sulfates, nitrates, and many organic compounds (see Raynor and Leith, 1999 for a review). Models have been developed to show that metalworking fluids evaporate readily from filters (see Raynor and Leith, 1999; Raynor et

Table 24-9
Saturation vapor pressures of some organic compounds at 25°C

Vapor Pressure mmHg	Chemical Class			
	Chlorinated	Oxygenated or Nitrogenous	Normal Alkanes*	Polynuclear Aromatic
1	Monochlorotoluenes	Benzaldehyde	Decane	
10^{-1}	p-Dichlorobenzene Hexachlorobutadiene	Toluene-2,4-diisocyanate Aniline Toxaphene	Dodecane	Quinoline
10^{-2}	Tetrachlorobenzenes 2,4,5-Trichlorophenol 2,4-D isopropyl ester	Nicotine p-Chloroaniline	Tetradecane	Naphthalene
10^{-3}	Pentachlorobenzene	p-Nitroaniline	Hexadecane (16)	Anthracene
10^{-4}	Aroclor 1242 Pentachlorophenol Heptachlor	Diazinon Butyl phthalate	Octadecane (18)	Phenanthrene
10^{-5}	Hexachlorobenzene Chlordane Aroclor 1260	Parathion Carbaryl	Eicosane	
10^{-6}	Dieldrin 2,3,7,8-Tetrachloro-dibenzodioxin		Hexacosane (26)	Fluoranthene
10^{-7}	DDT	Di(2-ethylhexyl)-phthalate		Benz(a)anthracene
10^{-8}			Nonacosane (29) (Cyrstal)	Chrysene
10^{-9}				Benzo(a)pyrene
10^{-10}				Benzo(ghi)perylene
10^{-12}				Coronene

*Number of carbons in ( ).
Adapted from Lewis, 1986.

al., 2000). They concluded that "the use of filters for quantifying occupational exposures to metalworking fluid mists is inappropriate" (Volckens et al., 2000). However, the NIOSH method is still based on filtration (NIOSH, 2003).

For example, asphalt fumes and coke oven emissions are monitored by collecting them on a filter, from which the organics are extracted with cyclohexane or benzene (NIOSH, 1998b). The emitted particulate matter, as expected, is loaded with organics, many of which would be classified as semi volatile. It might be expected that some would volatilize from the particles and the filter and be lost. Others, in vapor form, would be trapped poorly. Yet, Schwartz et al. (1981) found that the cyclohexane-extracted fraction increased disproportionately with sampling time. This indicates that vapors were sorbing to the filtered particles at an increasing rate as particle loading increased. On the other hand, the dichloromethane-extracted fraction decreased disproportionately with time, indicating a dipole charge differentiated adsorption-desorption process.

The decision regarding whether to consider multiple phases when sampling apparently is not an easy one, but Perez and Soderholm (1991) attempted to treat the problem in a systematic fashion. They reasoned that the problem was dependent not only on the vapor pressure of the chemical but also on the concentration in the workplace. Thus, a compound with a very low vapor pressure, say 10^{-5} mmHg, would be expected to exist in low vapor concentrations in the workplace. However, if its OEL were also quite low, lower than its saturated vapor pressure ($10^{-5}/760$ = 13 ppb), then significant amounts of vapor would be expected to be present. If the OEL was much higher, say, 10 ppm (expressed as mg/m^3), then the vapor fraction probably would not be significant enough to sample. Based on this concept, they devised two tables of compounds. The first table (see Table 24-10) contains compounds that the TLV® Committee lists only in units of mg/m^3. This implies that the airborne compound should exist most significantly in the aerosol state. The rationale for this is based on the evaporation rate from various particle sizes, given a vapor pressure (see TLV® Committee, 1983). However, the compounds may become airborne by means other than mechanical, and certainly the vapor pressures in Table 24-10 are well within those shown in Table 24-9. All of the compounds in Table 24-10 have saturated vapor concentrations that are larger than their respective TLVs®. As this ratio becomes larger, the likelihood increases that the vapor state will constitute an important part of the total exposure.

Perez and Soderholm (1991) also published a list of compounds having a TLV® in units of ppm and mg/m^3 and a saturated-vapor-to-TLV® concentration ratio ranging from 1 to 100 (Table 24-11). These compounds would be expected to exist in the vapor phase and would likely be sampled accordingly. However, Perez and Soderholm caution that when the vapors coexist with aerosols in which they may be soluble, then the compounds may be found in large proportion on the particles. Calculations for ethylene glycol (60%), dipropylene glycol methyl ether (80%), and 2-butoxyethanol (99%) showed impressive proportions on the particle phase when relative humidities were in the percentages in parentheses or higher. Another example would be the propensity of hydrocarbon vapors to be absorbed into oil mists.

Table 24-10
Chemicals with TLVs® listed in mg/m³ only that have saturated vapor concentrations greater than their TLVs®

Chemical Name	TLV® (mg/m³)	VP (mmHg)	SVC (mg/m³)	SVC/ TLV®	TLV® *Documentation* Notes[a]
Acrylamide	0.03	7.0×10^{-3}	$2.7\times10^{+1}$	9×10^{2}	WS
Ammonium perfluorooctanoate	0.1	7.0×10^{-5}	$1.6\times10^{+0}$	2×10^{1}	WS powder
Benzoyl peroxide	5	1.0×10^{-1}	$1.3\times10^{+3}$	3×10^{2}	SWS VP > 0.1, faint odor of benzaldehyde
Cadmium	0.05	2.0×10^{-5}	1.2×10^{-1}	2×10^{0}	NWS as Cd, significant VP
Chlorinated camphene	0.5	3.0×10^{-1}	$6.7\times10^{+3}$	1×10^{4}	NWS mild Cl/camphor odor, VP averaged
Chlorodiphenyl 42%	1	1.0×10^{-4}	$1.4\times10^{+0}$	1×10^{0}	NWS mild hydrocarbon odor
Chlorpyrifos	0.2	1.9×10^{-5}	3.5×10^{-1}	2×10^{0}	b
Diazinon	0.1	1.4×10^{-4}	$2.3\times10^{+0}$	2×10^{1}	SWS faint odor
Dibutylphthalate	5	1.0×10^{-1}	$1.5\times10^{+3}$	3×10^{2}	NWS odorless liquid[b]
Dimethylphthalate	5	1.0×10^{-1}	$1.1\times10^{+3}$	2×10^{2}	SWS oily liquid, VP < 0.1[b]
Dinitro-o-cresol	0.2	5.0×10^{-5}	5.3×10^{-1}	3×10^{0}	SWS odorless solid
Dinitrotoluene	1.5	$1.0\times10^{+0}$	$9.8\times10^{+3}$	7×10^{3}	VSWS oily liquid (isomer mixture)
Disulfoton	0.1	1.8×10^{-4}	$2.7\times10^{+0}$	3×10^{1}	Liquid
EPN	0.5	3.0×10^{-2}	$5.2\times10^{+2}$	1×10^{3}	SWS solid or crystals
Fenthion	0.2	3.0×10^{-5}	4.5×10^{-1}	2×10^{0}	NWS liquid, garlic odor
Heptachlor	0.5	3.0×10^{-4}	$6.1\times10^{+0}$	1×10^{1}	NWS waxy solid, camphoraceous odor
Hexachloronaphthalene	0.2	$1.0\times10^{+0}$	$1.8\times10^{+4}$	9×10^{4}	NWS solid, aromatic odor
Hydroquinone	2	1.0×10^{-3}	$5.9\times10^{+0}$	3×10^{0}	NWS crystalline solid
Mercury vapor	0.05	1.8×10^{-3}	$2.0\times10^{+1}$	4×10^{2}	NWS
Methyl parathion	0.2	5.0×10^{-1}	$7.1\times10^{+3}$	4×10^{4}	NWS liquid, garlic-like odor
Naled	3	2.0×10^{-3}	$4.1\times10^{+1}$	1×10^{1}	NWS slightly pungent odor
Nicotine	0.5	4.3×10^{-2}	$3.7\times10^{+2}$	7×10^{2}	fishy odor

Table 24-10 (Cont.)
Chemicals with TLVs® listed in mg/m³ only that have saturated vapor concentrations greater than their TLVs®

Chemical Name	TLV® (mg/m³)	VP (mmHg)	SVC (mg/m³)	SVC/ TLV®	TLV® *Documentation* Notes[a]
p-Nitroaniline	3	$1.0\times10^{+0}$	$7.5\times10^{+3}$	2×10^{3}	NWS VP < 1.0
Octachloronaphthalene	0.1	$1.0\times10^{+0}$	$2.2\times10^{+4}$	2×10^{5}	NWS waxy solid
Oxalic acid	1	1.0×10^{-3}	$4.9\times10^{+0}$	5×10^{0}	WS (dihydrate) odorless crystals, VP at 18.5°C
Pentachloronaphthalene	0.5	$1.0\times10^{+0}$	$1.6\times10^{+4}$	3×10^{4}	NWS VP < 1.0
Pentachlorophenol	0.5	1.7×10^{-4}	$2.4\times10^{+0}$	5×10^{0}	WS solid, phenolic odor
Phorate	0.05	2.0×10^{-3}	$2.8\times10^{+1}$	6×10^{2}	Liquid, VP at 21°C
Phosphoric acid	1	3.0×10^{-2}	$1.6\times10^{+2}$	2×10^{2}	VWS
Phosphorus	0.1	2.6×10^{-2}	$1.7\times10^{+2}$	2×10^{3}	SWS waxy solid
Propoxur	0.5	1.0×10^{-2}	$1.1\times10^{+2}$	2×10^{2}	SWS odorless crystals
Ronnel	10	8.0×10^{-4}	$1.4\times10^{+1}$	1×10^{0}	NWS powder
Sesone	10	1.0×10^{-1}	$1.7\times10^{+3}$	2×10^{2}	WS odorless solid
Sulfotep	0.2	1.7×10^{-4}	$3.0\times10^{+0}$	1×10^{1}	NWS liquid, garlic odor
Sulfuric acid	1	1.0×10^{-3}	$5.3\times10^{+0}$	5×10^{0}	WS[b]
Sulprofos	1	$7.9\times10^{+0}$	$1.4\times10^{+5}$	1×10^{5}	SWS liquid, sulfide odor
Tetrachloronaphthalene	2	$1.0\times10^{+0}$	$1.4\times10^{+4}$	7×10^{3}	NWS solid, aromatic odor, VP < 1.0
Tetraethyl lead	0.1	2.0×10^{-1}	$3.5\times10^{+3}$	4×10^{4}	NWS liquid, musty odor
Tetramethyl lead	0.15	$2.2\times10^{+1}$	$3.2\times10^{+6}$	2×10^{7}	NWS liquid, musty odor
Trichloronaphthalene	5	$1.0\times10^{+0}$	$1.3\times10^{+4}$	2×10^{3}	MWS solid, aromatic odor, VP < 1.0
m-Xylene α,α'-diamine	0.1	3.0×10^{-2}	$2.2\times10^{+2}$	2×10^{3}	WS solid

[a] Key: SVC, saturated vapor concentration VSWS, very slightly water soluble; SWS, slightly water soluble; MWS, moderately water soluble; WS, water soluble; NWS, not water soluble.
[b] No vapor hazard due to its VP.
Adapted from Perez and Soderholm, 1991.

Table 24-11
Chemicals with dual TLVs® that have SVC/TLV® ratios from 1 to 100

Chemical Name	TLV® ppm	TLV® mg/m³	VP mmHg	Calc' TLV® mg/m³	SVC mg/m³	SVC/ TLV®	TLV® *Documentation* Notes[a]
n-amyl acetate	100	530	$4.0\times10^{+0}$	$5.3\times10^{+2}$	$2.8\times10^{+4}$	$5\times10^{+1}$	VSWS liquid
sec-amyl acetate	125	665	$7.0\times10^{+0}$	$6.7\times10^{+2}$	$4.9\times10^{+4}$	$7\times10^{+1}$	Liquid
Benzyl chloride	1	5	1.0×10^{-0}	$1.3\times10^{+2}$	$6.8\times10^{+3}$	$5\times10^{+1}$	NWS liquid
Bromacil	1	10	8.0×10^{-4}	$1.1\times10^{+1}$	$1.1\times10^{+1}$	$1\times10^{+0}$	WS
2-Butoxyethanol	25	120	7.6×10^{-1}	$1.2\times10^{+2}$	$4.8\times10^{+3}$	$4\times10^{+1}$	WS liquid, ether odor
n-Butyl acetate	150	710	$1.0\times10^{+0}$	$7.1\times10^{+2}$	$6.3\times10^{+3}$	$9\times10^{+0}$	SWS liquid, fruity odor
sec-Butyl acetate	200	950	$1.0\times10^{+1}$	$9.5\times10^{+2}$	$6.3\times10^{+4}$	$7\times10^{+1}$	PNWS liquid, fruity odor
o-cresol	5	22	2.5×10^{-1}	$2.2\times10^{+1}$	$1.5\times10^{+3}$	$7\times10^{+1}$	Phenolic odor
m-cresol	5	22	1.5×10^{-1}	$2.2\times10^{+1}$	$8.8\times10^{+2}$	$4\times10^{+1}$	Phenolic odor
p-cresol	5	22	1.1×10^{-1}	$2.2\times10^{+1}$	$6.4\times10^{+2}$	$3\times10^{+1}$	Phenolic odor
Cyclohexanol	50	200	$3.5\times10^{+0}$	$2.1\times10^{+2}$	$1.9\times10^{+4}$	$9\times10^{+1}$	WS (4.3 g/ 100 g) liquid, camphor odor
2-n-dibutyl-aminoethanol	2	14	1.0×10^{-1}	$1.4\times10^{+1}$	$9.4\times10^{+2}$	$7\times10^{+1}$	WS (0.4%) liquid, amine odor
o-Dichlorobenzene	50	300	$1.2\times10^{+0}$	$3.0\times10^{+2}$	$9.5\times10^{+3}$	$3\times10^{+1}$	Liquid
p-Dichlorobenzene	75	450	6.4×10^{-1}	$4.5\times10^{+2}$	$5.1\times10^{+3}$	$1\times10^{+1}$	SWS (0.0079 g/ 100 g), VP at 20°C
Diethanolamine	3	15	1.0×10^{-2}	$1.3\times10^{+1}$	$5.7\times10^{+1}$	$4\times10^{+0}$	VWS
Diisobutyl ketone	25	150	$1.7\times10^{+0}$	$1.5\times10^{+2}$	$1.3\times10^{+4}$	$9\times10^{+1}$	NWS liquid
Dipropylene glycol methyl ether	100	600	3.0×10^{-1}	$6.1\times10^{+2}$	$2.4\times10^{+3}$	$4\times10^{+0}$	WS liquid, ethereal odor
Ethyl alcohol	1000	1900	$4.3\times10^{+1}$	$1.9\times10^{+3}$	$1.1\times10^{+5}$	$6\times10^{+1}$	WS
Ethyl benzene	100	435	$7.1\times10^{+0}$	$4.3\times10^{+2}$	$4.1\times10^{+4}$	$9\times10^{+1}$	SWS
Ethyl butyl ketone	50	230	$1.4\times10^{+0}$	$2.3\times10^{+2}$	$8.6\times10^{+3}$	$4\times10^{+1}$	NWS
Ethylene glycol	50	125	6.0×10^{-2}	$1.3\times10^{+2}$	$2.0\times10^{+2}$	$2\times10^{+0}$	WS (vapor?)
Furfuryl alchohol	10	40	5.3×10^{-1}	$4.0\times10^{+1}$	$2.8\times10^{+3}$	$7\times10^{+1}$	WS, unstable

Table 24-11 (Cont.)
Chemicals with dual TLVs® that have SVC/TLV® ratios from 1 to 100

Chemical Name	TLV® ppm	TLV® mg/m³	VP mmHg	Calc' TLV® mg/m³	SVC mg/m³	SVC/ TLV®	TLV® *Documentation* Notes[a]
Glutaraldehyde	0.2	0.7	1.5×10^{-2}	8.2×10^{-1}	$8.2\times10^{+1}$	$1\times10^{+2}$	50% H_2O solution
Glycidol	25	75	9.0×10^{-1}	$7.7\times10^{+1}$	$3.7\times10^{+3}$	$5\times10^{+1}$	WS, saturated vapor 1180 ppm
sec-Hexyl acetate	50	300	$3.0\times10^{+0}$	$3.0\times10^{+2}$	$2.3\times10^{+4}$	$8\times10^{+1}$	NWS
Isoamyl acetate	100	525	$4.0\times10^{+0}$	$5.3\times10^{+2}$	$2.8\times10^{+4}$	$5\times10^{+1}$	SWS
Isoamyl alcohol	100	360	$2.0\times10^{+0}$	$3.6\times10^{+2}$	$9.5\times10^{+3}$	$3\times10^{+1}$	SWS
Isooctyl alcohol	50	270	3.6×10^{-1}	$2.7\times10^{+2}$	$2.5\times10^{+3}$	$9\times10^{+0}$	NWS, SVC @ room T = 180 ppm
Isophorone	5	25	2.0×10^{-1}	$2.8\times10^{+1}$	$1.5\times10^{+3}$	$5\times10^{+1}$	VSWS
Isophorone diiso-cyanate	0.005	0.046	3.0×10^{-4}	4.6×10^{-2}	$3.6\times10^{+0}$	$8\times10^{+1}$	
Methacrylic acid	20	70	1.0×10^{-1}	$7.0\times10^{+1}$	$4.6\times10^{+2}$	$7\times10^{+0}$	WS
Methyl n-amyl ketone	50	235	$2.6\times10^{+0}$	$2.3\times10^{+2}$	$1.6\times10^{+4}$	$7\times10^{+1}$	VSWS, VP at 25
α-Methyl styrene	50	240	$1.9\times10^{+0}$	$2.4\times10^{+2}$	$1.2\times10^{+4}$	$5\times10^{+1}$	NWS
Naphthalene	10	50	8.7×10^{-2}	$5.2\times10^{+1}$	$6.0\times10^{+2}$	$1\times10^{+1}$	NWS
Nitroglycerin	0	0.5	2.6×10^{-4}	4.6×10^{-1}	$3.2\times10^{+0}$	$7\times10^{+0}$	SWS
Nitrotoluene	2	11	1.5×10^{-1}	$1.1\times10^{+1}$	$1.1\times10^{+3}$	$1\times10^{+2}$	Almost NWS, solid or liquid
Phenol	5	19	3.5×10^{-1}	$1.9\times10^{+1}$	$1.8\times10^{+3}$	$9\times10^{+1}$	WS
Phenyl ether (101-84-8)	1	7	2.1×10^{-2}	$7.0\times10^{+0}$	$2.0\times10^{+2}$	$3\times10^{+1}$	NWS
Phenyl glycidyl ether	1	6	1.0×10^{-2}	$6.1\times10^{+0}$	$8.1\times10^{+1}$	$1\times10^{+1}$	
Phenylhydrazine	5	20	1.0×10^{-1}	$2.2\times10^{+1}$	$5.8\times10^{+2}$	$3\times10^{+1}$	SWS
Phthalic anhydride	1	6	5.0×10^{-2}	$6.1\times10^{+0}$	$4.0\times10^{+2}$	$7\times10^{+1}$	SWS
Propyl alcohol	200	500	$1.4\times10^{+1}$	$4.9\times10^{+2}$	$4.5\times10^{+4}$	$9\times10^{+1}$	WS

Table 24-11 (Cont.)
Chemicals with dual TLVs® that have SVC/TLV® ratios from 1 to 100

Chemical Name	TLV® ppm	TLV® mg/m³	VP mmHg	Calc' TLV® mg/m³	SVC mg/m³	SVC/ TLV®	TLV® *Documentation* Notes[a]
Trimellitic anhydride	0	0.04	4.0×10^{-6}	3.9×10^{-2}	4.2×10^{-2}	$1\times10^{+0}$	
Turpentine	100	560	$5.0\times10^{+0}$	$5.6\times10^{+2}$	$3.7\times10^{+4}$	$7\times10^{+1}$	NWS
Vinyl cyclohexene dioxide	10	60	1.0×10^{-1}	$5.7\times10^{+1}$	$7.6\times10^{+2}$	$1\times10^{+1}$	VSWS
Vinyl toluene	50	240	$1.1\times10^{+0}$	$2.4\times10^{+2}$	$7.0\times10^{+3}$	$3\times10^{+1}$	VSWS

[a] Key: VSWS, very slightly water soluble; SWS, slightly water soluble; WS, water soluble; PNWS, practically not water soluble; NWS, not water soluble.
Adapted from Perez and Soderholm, 1991.

Sampling

A number of NIOSH sampling and analytical methods employ media in series for collecting gas and particle phases. For example, pentachlorophenol is sampled with a filter followed by a bubbler. Polynuclear aromatic hydrocarbons are sampled with a filter followed by a sorbent tube (XAD-2). Thus, in both cases the particles are captured first, and vapors, either originally airborne or desorbing from the particles, are captured second. The two together give the total airborne compound. It is important to repeat that the second medium does not necessarily give the airborne vapor, as some compounds will desorb from the particle phase (see, e.g., Coutant et al., 1986).

In some cases, it may be important to differentiate the two phases, as the toxicological impacts may be different. At this time, no OELs are known to require this differentiation (However, the need has been discussed; see TLV® Committee, 1983). A particle saturated with a compound could have a greater local impact on the lung than an equivalent amount of vapor, or the particle might have a greater propensity to be absorbed in the blood than the equivalent amount of vapor. Little is known about either phenomenon.

However, there are some methods for monitoring ambient air that do differentiate the vapor or the gas from the aerosol phase. For example, gaseous and particulate inorganic sulfur species have been distinguished (Lodge, 1989). In order to do so, the usual process must be reversed; the gas phase must be sampled first. Xiong et al. (1998) successfully tested a device composed of a filter cassette preceded and succeeded by charcoal tubes so that vapor phase was collected, followed by particle phase, followed by vapors evaporated from the filtered particle phase. However, typical sorption tubes will sample some particles (see, e.g., Fairchild and Tillery, 1977). Thus, a device is needed that will separate the particles cleanly from the vapor. Such a device is the denuder tube.

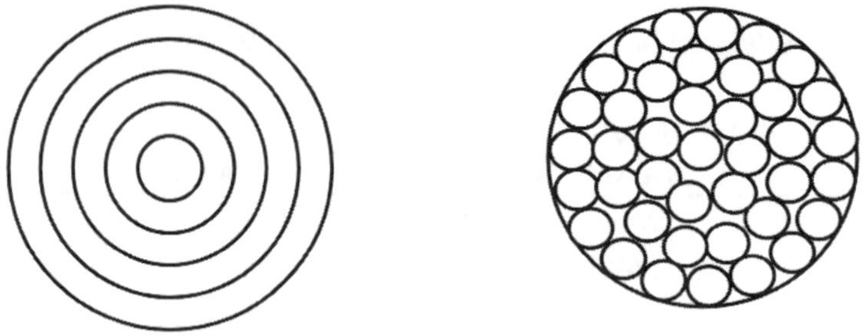

Figure 24-27. Cross-sectional schematics of concentric and parallel denuder tubes.

Denuders

The denuder tube is actually a series of parallel tubes into which the flow enters. The tubes may be parallel or concentric (see Figure 24-27). Splitting the flow makes it become laminar, and thus gas molecules will diffuse by molecular diffusion rather than turbulent diffusion. The former can be predicted. Additionally, the diffusion rate of particles caused by molecular collisions can also be calculated. Molecular diffusion is relatively fast. Thus if the tube walls have an affinity for the gas phase, it will be removed. The trick is to construct the tube length such that particles do not have time to diffuse to the tube wall before they exit to a filter or other collection device, but the gas molecules do. Downstream of the filter an absorber is sometimes necessary if the compound of interest readily desorbs from the particles captured on the filter. Denuders have been used successfully in ambient air sampling work for quite some time (Lane et al., 1988; Cheng, 1989; Krieger and Hites, 1992). The need to separate methylene diisocyanate and hexamethylene diisocyanate physical forms has been recognized, and this has been successfully accomplished with a denuder in the industrial environment (Rando and Poovey, 1994 and 1999).

QUESTIONS

24.1. Why is there not one filter that can be universally applied to the collection of aerosols?

24.2. Should one be concerned about the effects of water vapor in gravimetric filter procedures? If so, what would be a way to avoid or minimize the effects?

24.3. Why is particle size selective sampling in a somewhat confused or at least transient state at the time of writing (2007)?

24.4. State the particle size fraction names according to the Universal criteria and the key devices that are available for measuring each.

24.5. In a screening survey, exposures to a mixed aerosol are measured as IPM with an IOM sampler. However, it is known that some of the aerosol species have their primary effect deep within the respiratory tract. The results indicate that none of the measurements exceeds 0.25 of the smallest TLV® for the various species. What can be concluded?

24.6. ACGIH®, OSHA, and other organizations around the world have set OELs for agents such as lead and cadmium based on *total dust* measurements that were taken with 37 mm cassettes at the time workers were exposed. These measurements have been compared with disease experiences in those workers (epidemiological studies). Comment on these OELs relative to the current IPM convention and the current use of IPM samplers such as the IOM.

24.7. What sampling devices are available to sample the respirable fraction? What are the advantages and disadvantages of each?

24.8. Airborne polynuclear aromatic hydrocarbons are sampled with the following train: denuder, filter, charcoal tube. What does the assay of each component of the train tell us?

REFERENCES

ACGIH®: Annual reports of the committees on threshold limit values and biological exposure indices. Report to the ACGIH® Membership, Salt Lake City, Utah, May 21, 1991. Appl. Occup. Env. Hyg. 6:794-797 (1991).

ACGIH®: Threshold Limit Values for Chemical Substances and Physical Agents and Biological Exposure Indices. American Conference of Governmental Industrial Hygienists, Cincinnati, OH (1993).

ACGIH®: TLVs® Threshold Limit Value for Chemical Substances in Workroom Air. American Conference of Governmental Industrial Hygienists, Cincinnati, OH (1976).

Aerosol Technology Committee, American Industrial Hygiene Association: Guide for respirable mass sampling. Am. Ind. Hyg. Assoc. J. 32:133-137 (1970).

Aitken, R.J.; Vincent, J.H.; Mark, D.: Application of porous foam as size selectors for biologically relevant samplers. Proceeding of the international symposium on air sampling instrument performance. Appl. Occup. Env. Hyg. 8(4):363-369 (1993).

Aitken, R.J.; Baldwin, P.E.J.; Beaumont, G.C.; Kenny, L.C.; Maynard, A.D.: Aerosol inhalability in low air movement environments. J. Aerosol Sci. 30(5):613-625 (1999).

Aizenberg, V.; Grinshpun, S.A.; Willeke, K.; Smith, J.; Baron, P.A.: Performance characteristics of the button personal inhalable aerosol sampler. Am. Ind. Hyg. Assoc. J. 61:398-404 (2000).

Almich, B.P.; Carson, G.A.: Some effects of charging on 120-mm nylon cyclone performance. Am. Ind. Hyg. Assoc. J. 35:603-612 (1974).

Arelco, CIP10 Personal dust sampler, Arelco, Fontenay sous Bois Cedex, France http://www.arelco.fr/PDF/CIP_10_GB.pdf (2006).

Armbruster, L.; Breuer, H.: Investigations into defining inhalable dust, In Inhaled particles V, W.H. Walton (ed.). Ann. Occup. Hyg. 26:21-32 (1982).

ASTM: Standard Guide for Personal Samplers of Health-Related Aerosol Fractions [Metric], ASTM D6062M-96. ASTM International, West Conshohocken, PA, USA (2001a).

ASTM: Standard Practice for Evaluating the Performance of Respirable Aerosol Samplers, ASTM D6061-01.

ASTM International, West Conshohocken, PA, USA (2001b).

ASTM: Standard Practice for Controlling and Characterizing Errors in Weighing Collected Aerosols ASTM D6552. ASTM International, West Conshohocken, PA, USA (2006).

Ayer, H.E.: The proposed ACGIH® mass limits for quartz: Review and evaluation. Am. Ind. Hyg. Assoc. J. 30:117-125 (1969).

Baldwin, P.E.J.; Maynard, D.: A survey of wind speeds in indoor workplaces. Ann. Occup. Hyg. 42(5):303-313 (1998).

Baron, P.A.: Overloading and electrostatic effects in asbestos fiber sampling I. Am. Ind. Hyg. Assoc. J. 48:A709-A710 (1987).

Baron, P.A.: Application of the thoracic sampling definition to fiber measurement. Am. Ind. Hyg. Assoc. J. 57:820-824 (1996).

Baron, P.A.: Personal aerosol sampler design: A review. Appl. Occup. Evn. 13(5):313-320 (1998).

Baron, P.A.: Factors affecting aerosol sampling, Chapter O, NIOSH Manual of Analytical Methods. National Institute for Occupational Safety and Health Division of Physical Sciences and Engineering, Cincinnati, OH http://www.cdc.gov/niosh/nmam/pdfs/chapter-o.pdf (2003a).

Baron, P.A.: Aerosol sampling: minimizing particle losses from cassette by-pass leakage, Chapter N, NIOSH Manual of Analytical Methods. National Institute for Occupational Safety and Health Division of Physical Sciences and Engineering, Cincinnati, OH http://www.cdc.gov/niosh/nmam/pdfs/chapter-n.pdf (2003b).

Baron, P.A.; Deye, G.J.: Electrostatic effects in asbestos sampling I: Experimental measurements. Am. Ind. Hy Assoc. J. 51:51-62 (1990a).

Baron, P.A.; Deye, G.J.: Electrostatic effects in asbestos sampling II: Comparison of theory and experiment. Am. Ind. Hyg. Assoc. J. 51:63-69 (1990b).

Baron, P.; John, W.: Sampling for thoracic aerosol, In Particle Size-Selective Sampling for Particulate Air Contaminants, J. H. Vincent (ed.). ACGIH®, Cincinnati, OH (1999).

Baron, P.A.; Willeke, K.: Aerosol Measurement: Principles, Techniques, and Applications, 2nd ed. Wiley Publishing, New York (2001).

Baron, P.A.; Chen, C.C.; Hemenway, D.R.; O'Shaughnessy, P.: Nonuniform air flow in inlets: The effect on filter deposits in the fiber sampling cassette. Am. Ind. Hyg. Assoc. J. 55:722-732 (1994).

Bartley, D.L.: Inhalable aerosol samplers. Appl. Occ. Env. Hyg. 13:274-278 (1998).

Bartley, D.L.; Breuer, G.M.: Analysis and optimization of the performance of the 10 mm cyclone. Am. Ind. Hyg. Assoc. J. 43:520-528 (1982).

Bartley, D.L.; Doemeny, L.J.: Critique of 1985 ACGIH® report on particle size-selective sampling in the workplace. Am. Ind. Hyg. Assoc. J. 47:443-447 (1986).

Bartley, D.L.; Fischbach, T.J.: Alternative approaches for analyzing sampling and analytical methods. Appl. Occup. Env. Hyg. 8:381-385 (1993).

Bartley, D.L.; Chen, C.C.; Song, R.; Fischbach, T.J.: Respirable aerosol sampler performance testing. Am. Ind. Hyg. Assoc. J. 11:1036-1046 (1994).

Bartley, D.L.; Breuer, G.M.; Baron, P.A.; Bowman, J.D.: Pump fluctuations and their effect on cyclone performance. Am. Ind. Hyg. Assoc. J. 45:10-18 (1984).

Beaulieu, H.J.; Fidino, A.V.; Arlington, K.L.B.; Buchan, R.M.: A comparison of aerosol sampling techniques: *Open* versus *closed-face* filter cassettes. Am. Ind. Hyg. Assoc. J. 41:758-765 (1980).

Berry, R.D.: The effect of flow pulsations on the performance of cyclone personal respirable dust samplers. J. Aerosol Sci. 22(7):887-899 (1992).

BGI: Respirable Dust Cyclones, BGI, Inc., Waltham, MA, USA http://www.bgiusa.com/ihi/bgi4l.htm (2006).

Blachman, M.W.; Lippmann, M.: Performance characteristics of the multicyclone aerosol sampler. Am. Ind. Hyg. Assoc. J. 35(6):311-326 (1974).

Botham, R.A.; Hughson, G.W.; Vincent, J.H.; Mark, D.: The development of a test system for investigating the performances of personal aerosol samplers under actual workplace conditions. Am. Ind. Hyg. Assoc. J. 52:423-427 (1991).

Breslin, J.A.; Stein, R.L.: Efficiency of dust sampling inlets in calm air. Am. Ind. Hyg. Assoc. J. 36:576-583 (1975).

Briant, J.K.; Moss, O.R.: The influence of electrostatic charge on the performance of 10-mm nylon cyclones. Am.

Ind. Hyg. Assoc. J. 45:440-445 (1984).
Brosseau, L.M.; Evans, J.S.; Ellenbecker, M.J.; Feldstein, M.L.: Collection efficiency of respirator filters challenged with monodisperse latex aerosols. Am. Ind. Hyg. Assoc. J. 50:544-549 (1989).
Brown, R.C.; Hemingway, A.; Wake, D.; Thompson, J.: Field trials of an electret-based passive dust sampler in metal processing industries. Ann. Occup. Hyg. 39(5):603-622 (1995).
Brown, R.C.; Wake, D.; Thorpe, A.; Hemingway, M.A.; Roff, M.W.: Preliminary assessment of a device for passive sampling of airborne particulate. Ann. Occup. Hyg. 38:303-318 (1994).
Brown, R.C.: Filtration in industrial hygiene. Am. Ind. Hyg. Assoc. J. 62:633-643 (2001).
Calzavara T.S.; Carter, C.M.; Axten, C.: Air Sampling Methodology for Asphalt Fume in Asphalt Production and Asphalt Roofing Manufacturing Facilities: Total Particulate Sampler versus Inhalable Particulate Sampler. App. Occ. Env. Hyg. 18: 358-367 (2003).
Cautreels, W.; Cauwenberghe, K.: Experiments on the distribution of organic pollutants between airborne particulate matter and the corresponding gas phase. Atmospheric Environment 12:1133-1141(1978).
CEN, European Committee for Standardization: Size Fraction Definitions for Measurement of Airborne Particles in the Workplace. EN 481:1993, Brussels (1993).
CEN, European Committee for Standardization: Workplace atmospheres. Assessment of performance of instruments for measurement of airborne particle concentrations. EN 13205:2002, Brussels (2001).
Charell, P.R.; Hawley, R.E.: Characteristics of water adsorption on air sampling filters. Am. Ind. Hyg. Assoc. J. 42:353-360 (1981).
Chen, C.C.; Baron, P.A.: Aspiration efficiency and inlet wall deposition in the fiber sampling cassette. Am. Ind. Hyg. Assoc. J. 57:142-152 (1996).
Chen, C.C.; Huang, S.H.: Shift of aerosol penetration in respirable cyclone samplers. Am. Ind. Assoc. J. 60:720-729 (1999).
Chen, C.C.; Lai, C.Y.; Shih, T.S.; Hwang, J.S.: Laboratory performance comparison of respirable sampler. Am. Ind. Hyg. Assoc. J. 60:601-611 (1999).
Cheng, Y.: Diffusion batteries and denuders, In Air Sampling Instruments, S.V. Hering (ed.). ACGIH®, Cincinnati, OH (1989).
Chung, K.Y.K.; Aitken, R.J.; Bradley, D.R.: A development and testing of a new sampler for welding fume. Ann. Occup. Hyg. 41(3):355-372 (1997).
Cohen, B.S.; Harley, N.H.; Lippmann, M.: Bias in air sampling techniques used to measure inhalation exposure. Am. Ind. Hyg. Assoc. J. 45:187-192 (1984).
Courbon, P.; Wrobel, R.; Fabries, J.F.: A new individual respirable dust sampler; the CIP 10. Ann. Occup. Hyg. 32:129-143 (1988).
Coutant, R.W.; Brown, L.; Chuang, J.; Lewis, R.G.: Field evaluation of phase distribution of PAH, In Proceedings of the 1986 EPA/APCA Symposium on Measurement of Toxic Air Pollutants, Raleigh, NC, Apr. 27-30, 1986, EPA Report No. 600/9-86-013, pp. 146-155. Environmental Monitoring Systems Laboratory, U.S. Environmental Protection Agency (1986).
Davies, C.N.: Dust sampling and lung disease. Brit. J. Ind. Med. 9:120-126 (1952).
Davies, H.W.; Teschke, K.; Demers, P.A.: A field comparison of inhalable and thoracic size selective sampling techniques. Ann. Occup. Hyg. 42(6):381-392 (1999).
Demange, M.; Gorner, P.; Elcabache, J.M.; Wrobel, R.: Field comparison of 37-mm closed-face cassettes and IOM samplers. Appl. Occup. Env. Hyg. 17(3):200-208 (2002).
Dunmore, J.H.; Hamilton, R.J.; Smith, D.S.G.: An instrument for the sampling of respirable dust for subsequent gravimetric assessment. J. Sci. Instrum. 41:669-672 (1964).
Einfeld, W.; Horstman, S.: Investigation of a dual filter sampling method for gaseous and particulate fluoride. Am. Ind. Hyg. Assoc. J. 40:626-632 (1979).
Elkins, H.B.: The Chemistry of Industrial Toxicology, 2nd ed. Wiley Publishing, New York (1959).
EPA: 40 CFR 53 Subpart D: Procedures for testing performance characteristics of methods for PM_{10}. Environmental Protection Agency. Fed. Reg. 52:24,729-24,735 (1987).
European Union: Council Directive of 7 April 1998 on the protection of the health and safety of workers from the risks related to chemical agents and work (14th individual directive within the meaning of Article 16(1) of

Directive 89/391/EEC). Official journal of the European communities May 1998, No. L131, pp. 11-23.
Fabries, J.F.; Gorner, P.; Kauffer, E.; Wrobel, R.; Vigneron, J.C.: Personal thoracic CIP10-T sampler and its static version CATHIA-T. Ann. Occup. Hyg. 42(7):453-465 (1998).
Fairchild, C.I.; Tillery, M.I.: The filtration efficiency of organic vapor sampling tubes against particulates. Am. Ind. Hyg. Assoc. J. 38:277-283 (1977).
Feigley, C.E.; Maguire, K.E.; Hussey, J.R.: Asbestos fiber deposition patterns on various 25-mm cassette filters at a high flow rate. Appl. Occup. Env. Hyg. 7:749-757 (1992).
Frazee, P.R.; Tironi, G.: A filter cassette assembly method for preventing bypass leakage. Am. Ind. Hyg. Assoc. J. 48:176-180 (1987).
Gautam, M.; Sreenath, A.: Performance of a multi-inlet cyclone sampler. J. Aerosol Sci. 28(7):1265-1282 (1997).
Gelman, C.; Marshall, J.C.: High purity fibrous air sampling media. Am. Ind. Hyg. Assoc. J. 36:512-517 (1975).
Gibson, H.; Vincent, J.H.: The penetration of dusts through porous foam filter media. Ann. Occ. Hyg. 24:205-215 (1981).
Gibson, H.; Vincent, J.H.; Mark, D.: A personal inspirable dust spectrometer for applications in occupational hygiene research. Ann. Occup Hyg. 31:463-479 (1987).
Gorner, P.; Wrobel, R.; Micka, V.; Skoda, V.; Denis, J.; Fabries, J.F.: Study of fifteen respirable aerosol samplers used in occupational hygiene. Ann. Occup. Hyg. 45(1):43-54 (2001).
Grimm, C.B.: Characterization of the Lumsden-Lynch vertical elutriator. Master's Thesis. Texas A&M University (1981).
Groves, W.A.; Hahne, R.M.A.; Levine, S.P.; Schork, M.A.: A field comparison of respirable dust samplers. Am. Ind. Hyg. Assoc. J. 55:748-755 (1994).
Gunderson, E.C.; Anderson, C.C.: Collection device for separating airborne vapor and particulates. Am. Ind. Hyg. Assoc. J. 48:634-638 (1987).
Hauck, B.C.; Grinshpun, S.A.; Reponen, A.; Reponen, T.; Willeke, K.; Bornschein, R.L.: Field testing of new aerosol sampling method with a porous curved surface as inlet. Am. Ind. Hyg. Assoc. J. 58:713-719 (1997).
Heever, D.J.V.; Tiernan, G.: Quantification of bypass leakage in filter holders with cyclone elutriators. Am. Ind. Hyg. Assoc. J. 60:545-547 (1999).
Hering, S.V.: Impactors, cyclones and other particle collectors, In Air Sampling Instruments, 9th ed., B.S. Cohen and C.S. McCammon (eds.). ACGIH®, Cincinnati, OH (2001).
Hewitt, P.; McCawley, M.A.: A microcomputer spreadsheet technique for analyzing multimodal particle size distributions. App. Occup. Env. Hyg. 6:865-873 (1991).
Higgins, R.I.; Dewell, P.: A gravimetric size-selecting personal dust sampler, In Inhaled Particles and Vapour II, pp. 575-585, C. N. Davies (ed.). Pergamon Press, London (1967).
Hinds, W.C.: Aerosol Technology: Properties, Behavior, and Measurement of Airborne Particles. Wiley Publishing, New York (1999).
Hinds, W.C.; Bellin, P.: Effect of facial-seal leaks on protection provided by half-mask respirators. Appl. Ind. Hyg. 3:158-164 (1988).
Hinds, W.C.; Liu, W.V.; Froines, J.R.: Particle bounce in a personal cascade impactor: A field evaluation. Am. Ind. Hyg. Assoc. J. 46:517-523 (1985).
Hinds, W.C.: Data analysis, In Cascade Impactor Sampling and Data Analysis, J.P. Lodge, T.L. Chan (eds.). Am. Ind. Hyg. Assoc., Akron, OH (1986).
Hook, M.B.; Feigley, C.E.; Ludwig, D.A.: Interwedge variation in the membrane filter method for airborne asbestos fibers. Am. Ind. Hyg. Assoc. J. 44:542-546 (1983).
HSE: General Methods for the Gravimetric Determination of Respirable and Total Inhalable Dust. Health and Safety Executive, MDHS 14, HMSO, London (1989).
ICRP (International Commission on Radiation Protection): Health Physics, Vol. 3. (1960).
ISO, Committee TC 146: Size definitions for particle sampling. Am. Ind. Hyg. Assoc. J. 42:64-69 (1981).
ISO: Air Quality-Particle Size Fraction Definitions for Health-Related Sampling, Approved for publication as CD 7708. International Organization for Standardization, Geneva (1991).
Jacobson, J.; Tomb, T.F.: Relationship between gravimetric respirable dust concentration and midget impinger number concentration. Am. Ind. Hyg. Assoc. J. 28:554-556 (1967).

Japuntich, D.A.: Respiratory particulate filtration. J. Int. Soc. Resp. Proto 2:137-169 (1984).

Jenkins, S.L.; Feigley, C.E.; Jackson, K.L.: Fluorometric analysis of the uniformity of deposition on cassette membrane filters. Appl. Occup. Env. Hyg. 7:665-671 (1992).

John, W.: Sampling for respirable and fine aerosol, In Particle Size-Selective Sampling for Particulate Air Contaminants, J.H. Vincent (ed.). ACGIH®, Cincinnati, OH (1999).

Johnston, A.M.; Jones, A.D.; Vincent, J.H.: The effect of electrostatic charge on the aspiration efficiencies of airborne dust samplers: With special reference to asbestos. Am. Ind. Hyg. Assoc. J. 48:613-621 (1987).

Johnston, A.M.; Vincent, J.H.; Jones, A.D.: Measurements of electric charge for workplace aerosols. Ann. Occup. Hyg. 29:271-284 (1985).

Jones, A.D.; Aitken, R.J.; Fabries, J.F.; Kauffer, E.; Liden, G.; Maynard, A.; Riediger, G.; Sahle, W.: Thoracic size-selective sampling of fibers: performance of four types of thoracic sampler in laboratory tests. Ann. Occ. Hyg. 49:481-492 (2005).

Kar, K.; Gautam, M.: Orientation bias of the isolated 10-mm nylon cyclone at low stream velocity. Am. Ind. Hyg. Assoc. J. 56:1090-1098 (1995).

Kauffer, E.; Eypert-Blaison, C.: Effect of using a cowl when measuring the fiber number concentration by the membrane filter method. J. Occup. Env. Hyg. 1:289-295 (2004).

Kauffer, E.; Vigneron, J.C.; Fabries, J. F.; Billon-Galland, M.A.; Brochard, P.: The use of a new static device based on the collection of the thoracic fraction for the assessment of the airborne concentration of asbestos fibres by transmission electron microscopy. Ann. Occup. Hyg. 40(3): 311-319 (1996).

Kenny, L.C.: Scientific principles and pragmatic solutions for the measurement of exposure to inhalable dust. Ann. Occup. Hyg. 47(6):437-440 (2003).

Kenny, L.C.: The international conventions for health-related sampling of aerosols – a review of current status and future evolution. Appl. Occup. Env. Hyg. 15(1): 68-71 (2000).

Kenny, L.C.; Gussman, R.A.: Characterization and modeling of a family of cyclone aerosol preseparators. J. Aerosol Sci. 28:677-688 (1997).

Kenny, L.; Liden, G.: The application of performance standards to personal airborne dust samplers. Am. Occup. Hyg. 33:289-300 (1989).

Kenny, L.C.; Ogden, T.L.: Twenty-five years of inhalable dust. Ann. Occup. Hyg. 44(8):561-563 (2000).

Kenny, L.C.; Aitken, R.; Chalmers, C.; Fabries, J.F.; Gonzalez-Fernandez, E.; Kromhout, H.; Liden, G.; Mark, D.; Riediger, G.; Prodi, V.: A collaborative European study of personal inhalable aerosol sampler performance. Ann. Occup. Hyg. 41(2):135-153 (1997).

Kenny, L.C.; Stanecliffe, J.D.; Crook, B.; Stagg, S.; Griffiths, W.D.; Stewart, I.W.; Futter, S.J.: The adaptation of existing personal inhalable aerosol samplers for bioaerosol sampling. Am. Ind. Hyg. Assoc. J. 59:831-841 (1998).

Kenny, L.C.; Aitken, R.J.; Baldwin, P.E.J.; Beaumont, G.C.; Maynard, A.D.: The sampling efficiency of personal inhalable aerosol samplers in low air movement environments. J. Aerosol Sci. 30(5):627-638 (1999).

Kenny, L.; Chung, K.; Dilworth, M.; Hammond, C.; Wynn Jones, J.; Shreeve, Z.; Winton, J.: Applications of low-cost, dual-fraction dust samplers. Ann. Occup. Hyg. 45(1):35-42 (2001).

Kerr, S.M.; Muranko, H.J.; Vincent, J.H.: Personal Sampling for Inhalable Aerosol Exposures of Carbon Black Manufacturing Industry Workers. App. Occ. Env. Hyg. 17:681-692 (2002).

Knutson, E.O.; Lioy, P.J.: Measurement and presentation of aerosol size distributions, In Air Sampling Instruments, 7th ed., S.V. Hering (ed.). ACGIH®, Cincinnati, OH (1989).

Krantz, S.: Methods of measuring air pollutants from metalworking fluids, paper presented at the Swedish-German Seminar on Metalworking Fluids in Work Environment, Swedish Work Environment Funds, Stockholm, June 15-16, pp. 103-107 (1983).

Krieger, M.S.; Hites, R.A.: Diffusion denuder for the collection of semivolatile organic compounds. Am. Chem. Soc. 26:1551-1555 (1992).

Lane, D.A.; Johnson, N.D.; Barton, S.C.; Thomas, G.H.S.; Schroeder, W.H.: Development and evaluating novel gas and particle sampler for semivolatile chlorinated organic compounds in ambient air. Am. Chem. Soc. 22:941-947 (1988).

Lewis, R.G; Jackson, M.D.: Modification and evaluation of a high-volume air sampler for pesticides and semivolatiles industrial organic chemicals. Anal. Chem. 54:592-594 (1982).

Lewis, R.G.: Problems associated with sampling for semivolatile organic chemicals in air, In Proceedings of the 1986 EPA/APCA Symposium on Measurement of Toxic Air Pollutants, Raleigh, NC, Apr. 27-30,1986, EPARep No. 600/9-86-013, pp. 134-145. Environmental Monitoring Systems Laboratory, U.S. Environmental Protection Agency (1986).

Li, S.N.; Lundgren, D.A.; Rovell-Rixx, D.: Evaluation of six inhalable aerosol samplers. Am. Ind. Hyg. Assoc. J. 61:506-516 (2000).

Li, S.N.; Lundgren, D.A.; Rovell-Rixx, D.; Ray, A.E.: Effect of impactor inlet efficiency on the measurement of wood dust size distribution. Am. Ind. Hyg. Assoc. J. 62:19-27 (2001).

Liden, G.: Evaluation of the SKC personal respirable dust sampling cyclone. Appl. Occup. Env. Hyg. 8:178-189 (1993).

Liden, G.; Bergman, G.: Weighting imprecision and handleability of the sampling cassettes of the IOM Sampler for inhalable dust. Ann. Occup. Hyg. 43(3):241-252 (2001).

Liden, G.; Gudmundsson, A.: Optimization of a cyclone to the 1993 international sampling convention for respirable dust. Appl. Occ. Env. Hyg. 11(12):1398-1408 (1996).

Liden, G.; Kenny, L.: Comparison of measured respirable dust sampler penetration curves with sampling conventions. Am. Occup. Hyg. 35:485-504 (1991).

Liden, G.; Kenny, L.: The performance of respirable dust samplers: Sampler bias, precision and inaccuracy. Ann. Occup. Hyg. 36:1-22 (1992).

Liden, G.; Kenny, L.: Optimization of the performance of existing respirable dust samplers. Appl. Occup. Env. Hyg. 8(4):386-391 (1993).

Liden, G.; Melin, B.; Lidbolm, A.; Lindberg, D.; Noren, J.O.: Personal sampling in parallel with open-face filter cassettes and IOM samplers for inhalable dust-implications for occupational exposure limits. Appl. Occup. Env. Hyg. 15(3):263-276 (2000).

Lillienberg L.; Brisman, J.: Flour dust in bakeries—a comparison between methods. Ann. Occ. Hyg. 38:571–75 (1994).

Lippmann, M.: Size-selective health hazard sampling, In Air Sampling Instruments, 7th ed., S. V. Hering (ed.). ACQ, Cincinnati, OH (1989).

Lippmann, M.; Gearhart, J.M.; Schlesinger, R.B.: Basis for a particle size-selective TLV for sulfuric acid aerosols. Appl. Ind. Hyg. 2(5):188-199 (1987).

Liu, B.Y.H.; Pui, D.Y.H.; Rubow, K.L.; Syzmanski, W.W.: Electrostatic effects in aerosol sampling and filtration Ann. Occup. Hyg. 29:251-269 (1985).

Lodge, J.P. Jr.: 713. Semi-continuous determination of atmospheric particulate sulfur, sulfuric acid and ammonium sulfates, In Methods of Air Sampling and Analysis, 3d ed. Lewis Publishers, Chelsea, MI (1989).

Lundgren, L.; Skare, L.; Liden, C.: Measuring dust on skin with a small vacuuming sampler – a comparison with other sampling techniques. Ann. Occup. Hyg. 50(1):95-100 (2006).

Lynch, J.R.: Air sampling for cotton dust, In Transactions of the National Conference on Cotton Dust and Health, pp. 33-43. University of North Carolina, School of Public Health, Chapel Hill, NC (1971).

Mark, D.: Problems associated with the use of membrane filters for dust sampling when compositional analysis required. Ann. Occup. Hyg. 17:35-40 (1974).

Mark, D.; Vincent, J.H.: A new personal sampler for airborne total dust in workplaces. Ann. Occup. Hyg. 30:89-101 (1986).

Mark, D.; Borzucki, G.; Lynch, G.; Vincent, J.H.: The development of a personal sampler for inspirable, thoracic, and respirable aerosol, In Aerosols, Their Generation, Behavior, and Applications, 2nd Conference, the Aerosol Society, Bournemouth, UK (1988).

Mark, D.; Lyons, C.P.; Upton, S.L.: Performance testing of the respirable dust sampler used in a British coal mill. Appl. Occup. Env. Hyg. 8(4):370-380 (1993).

Martin, J.R.; Zalk, D.M.: Comparison of total dust/inhalable dust sampling methods for the evaluation of airborne wood dust. App. Occ. Env. Hyg. 13(3):177-182 (1998).

Maynard, A.D.: Thoracic size-selection of fibres: Dependence of penetration on fibre length for five thoracic sampler types. Ann. Occup. Hyg. 46(6):511-522 (2002).

McFarland, A.R.; Hickman, P.D.; Parnell, C.B. Jr.: A new cotton dust sampler for PM_{10} aerosol. Am. Ind. Hyg Assoc. J. 48:293-297 (1987).

Millipore: Millipore Particle Monitoring Guide, Millipore Corporation, Bedford, MA http://www.millipore.com/publications.nsf/dda0cb48c91c0fb6852567430063b5d6/b76a969e6d73cbd5852568c5006434c9/$FILE/ATTKZK5J/ADO30.pdf (1998).

Montgomery, J.A.; Horstman, S.W.; Breslow, N.E.; Heyer, N.; Stebbins, A.; Checkoway, H.: A comparison of sampling methods for airborne silica in the diatomaceous earth industry. Appl. Occup. Env. Hyg. 6:696-702 (1991).

Moyer, E.S.; Stevens, G.A.: "Worst case" aerosol testing parameters: II. Efficiency dependence of commercial respirator filters on humidity pretreatment. Am. Ind. Hyg. Assoc. J. 50:265-270 (1989a).

Moyer, E.S.; Stevens, G.A.: "Worst case" aerosol testing parameters: III. Initial penetration of charged and neutralized lead fume and silica dust aerosols through clean, unloaded respirator filters. Am. Ind. Hyg. Assoc. J. 50:271-274 (1989b).

MSA: MSA Sampling pump accessories (08-09-01), MSA International, Pittsburgh, PA, USA http://media.msanet.com/NA/USA/PortableInstruments/PersonalSamplingPumps/PersonalSamplingPumpAccessories/08-09-01.pdf (1998).

MSHA: Verification of Underground Coal Mine Operators' Dust Control Plans and Compliance Sampling for Respirable Dust; Proposed Rule, Federal Register 68(44):10784 http://www.msha.gov/REGS/FEDREG/PROPOSED/2003PROP/03-3941.PDF (March 6, 2003).

MSP: Personal environmental monitor (PEM™), MSP Corp., Minneapolis, MN, USA http://www.mspcorp.com/products.htm (2003).

Neefus, J.D.; Lumsden, J.C.; Jones, M.T.: Cotton dust sampling, II – vertical elutriation. Am. Ind. Hyg. Assoc. 38:394-400 (1977).

Nilsson, C.; Nygren, O.: Determination of oil aerosol levels in air using zinc in additives as a tracer. Am. Occup. Hyg. J. 31:241-247 (1987).

NIOSH: Benzene-soluble fraction and total particulate (asphalt fume), method 5042, Issue I, In NIOSH Manual of Analytical Methods. NIOSH Division of Physical Sciences and Engineering, Cincinnati, OH (1998b).

NIOSH: Metalworking fluids (MWF) all categories, method 5524, Issue 1, In NIOSH Manual of Analytical Methods, National Institute for Occupational Safety and Health Division of Physical Sciences and Engineering, Cincinnati, OH (2003).

NIOSH: Particulates not otherwise regulated, respirable, method 0600, Issue 3, In NIOSH Manual of Analytical Methods, National Institute for Occupational Safety and Health Division of Physical Sciences and Engineering, Cincinnati, OH (1998a).

NIOSH: Particulates not otherwise regulated, total, method 0500, Issue 2, In NIOSH Manual of Analytical Methods, National Institute for Occupational Safety and Health Division of Physical Sciences and Engineering, Cincinnati, OH (1994).

Noll, J.D.; Timko, R.J.; McWilliams, L.; Hall, P.; Haney, R.: Sampling results of the improved SKC diesel particulate matter cassette. J. Occup. Env. Hyg. 2:29-37 (2005).

Noto, H.; Halgard, J.; Daae, H.; Bentsen, R.K.; Eduard, W.: Comparative study of an inhalable and a total dust sampler for personal sampling of dust and polycyclic aromatic hydrocarbons in the gas and particulate phase. Analyst 121(9):1191-1196 (1996).

Ogden, J.L.; Wood, J.D.: Effects of wind on the dust and benzene-soluble matter captured by a small sampler. Ann. Occup. Hyg. 17:187-195 (1975).

Ogden, J.L.; Birkett, J.L.: The human head as a dust sampler, In Inhaled Particles IV, W.H. Walton (ed.), pp. 93-104. Pergamon Press, Oxford (1977).

Omega: ACCU-CAP™ Cassette Insert for Gravimetric Analysis of Fibrous Contaminants. SKC Omega Specialty Division, Houston, TX, USA http://www.omegaspec.com/products/FeaturedProducts.asp?CatNo=PC043712-CP (2006a).

Omega: Accu-Press™ Pneumatically Activated Cassette Closure and Omega Filter Cassette Leak Testers. SKC Omega Specialty Division, Houston, TX, USA http://www.omegaspec.com/products/lab_equip.asp (2006b).

OSHA: Gravimetric Determination, Occupational Health and Safety Administration, Salt Lake Technical Center,

Salt Lake City, UT, USA http://www.osha.gov/dts/sltc/methods/partial/pv2121/pv2121.html (2003).

OSHA: Chemical Sampling Information, Occupational Health and Safety Administration, Salt Lake Technical Center, Salt Lake City, UT, USA http://www.osha.gov/dts/chemicalsampling/toc/toc_chemsamp.html (2006b).

OSHA: OSHA Technical Manual. Occupational Health and Safety Administration, Washington, DC http://www.osha.gov/dts/osta/otm/otm_toc.html (2006a).

Paik, S.; Vincent, J.H.: Filter and cassette mass instability in ascertaining the limit of detection of inhalable airborne particulates. Am. Ind. Hyg. Assoc. J. 63:698-702 (2002).

Pall: Filtration products for air monitoring and sampling. Pall Life Sciences, Ann Arbor, MI http://www.pall.com/pdf/01.0253_Air_Monitoring_BRO.pdf (2006).

Patty, F.A.: Industrial Hygiene and Toxicology, 2d rev. ed., Vol. I. Wiley Publishing, New York (1958).

Peck, A.S.; Serocki, J.J.; Decker, L.C.: Airborne asbestos measurement: Preliminary findings identify a new source of variability in the membrane filter method. Industrial Hygiene Forum. Am. Ind. Hyg. Assoc. J. 46:14-16 (1985).

Perdicala, B.Z.; Maghirang, R.G.: Field comparison of inhalable and total dust samplers for assessing airborne dust in swine confinement barns. Appl. Occup. Env. Hyg. 18:694-701 (2003).

Perez, C.; Soderholm, S.C.: Some chemicals requiring special consideration when deciding whether to sample the particle, vapor, or both phases of an atmosphere. App. Occup. Env. Hyg. 6:859-864 (1991).

Prodi, V.; Sala, C.; Belosi, F.: PERSPEC, Personal size separation sampler: Operational experience and comparison with other field devices. Appl. Occup. Environ. Hyg. 7(6):368-373 (1992).

Prodi, V.; Belosi, F.; Mularoni, A.; Lucialli, P.: PERSPEC: A personal sampler with size characterization capabilities. Am. Ind. Hyg. Assoc. J. 49(2):75-80 (1988).

Puskar, M.A.; Harkins, J.M.; Moomey, J.D.; Hecker, L.H.: Internal wall losses of pharmaceutical dusts during closed-face, 37mm polystyrene cassette sampling. Am. Ind. Hyg. Assoc. J. 52:280-286 (1991).

Puskar, M.A.; Fergon, S.M.; Harkins, J.M.; Hecker, L.H.: Gravimetric determination of airborne dust by using a filter cartridge inside a closed-face, 37-mm polystyrene cassette. Am. Ind. Hyg. Assoc. J. 53:692-698 (1992).

Ramachandran, G.; Vincent, J.H.: Evaluation of two inversion techniques for retrieving health related aerosol fractions from personal cascade impactor measurements. Am. Ind. Hyg. Assoc. J. 58:15-22 (1997).

Ramachandran, G.; Werner, M.A.; Vincent, J.H.: Assessment of particle size distributions in workers' aerosol exposures. Analyst 121:1225-12235 (1996).

Rando, R.J.; Poovey, H.G.: Development and application of a dichotomous vapor/aerosol sampler for HDI-derived total reactive isocyanate group. Am. Ind. Hyg. Assoc. J. 60:737-746 (1999).

Rando, R.J.; Poovey, H.G.: Dichotomous sampling of vapor and aerosol of methylene-bis-(phenylisocyanate) [MDI] with an annular diffusional denuder. Am. Ind. Hyg. Assoc. J. 55:716-721 (1994).

Rando, R.; Poovey, H.; Mokadam, D.; Brisolara, J.; Glindmeyer, H.: Field performance of the RespiCon for size-selective sampling of industrial wood processing dust. J. Occup. Env. Hyg. 2:219-226 (2005).

Raynor, G.S.: Variation in entrance efficiency of a filter sampler with air speed, flow rate, angle and particle size. Am. Ind. Hyg. Assoc, J. 31:294-304 (1970).

Raynor, P.C.; Leith, D.: Evaporation of accumulated multicomponent liquids from fibrous filters. Ann. Occup. Hyg. 43(3):181-192 (1999).

Raynor, P.C.; Volckens, J.; Leith, D.: Modeling evaporative loss of oil mist collected by sampling filters. Appl. Occup. Env. Hyg. 15(1):90-96 (2000).

Renshaw, F.M.; Bachman, J.M.; Pierce, J.O.: The use of midget impingers and membrane filters for determining particle counts. Am. Ind. Hyg. Assoc. J. 30:113-116 (1969).

Rice, C.; Harris, R.L.; Lumsden, J.C.: Reconstruction of silica exposure in the North Carolina dusty trades. Am. Ind. Hyg. Assoc. J. 45:689-696 (1984).

Rubow, K.L.; Marple, V.A.; Robert, K.Q. Jr.: Sampling characteristics of the SRRC cotton dust sampler in calm air, In Proceedings of the Seventh Cotton Dust Research Conference sponsored by the National Cotton Council and The Cotton Foundation, Memphis, TN (1983).

Rubow, K.L.; Marple, V.A.; Olin, J.; McCawley, M.A.: A personal cascade impactor: Design, evaluation and calibration. Am. Ind. Hyg. Assoc. J. 48:532-538 (1987).

Saltzman, B.E.; Hochstrasser, J.M.: Design and performance of miniature cyclones for respirable aerosol sampling.

Env. Sci. Technol. 17:418-424 (1983).

Saltzman, B.E.: Generalized performance characteristics of miniature cyclones for atmospheric particulate sampling. Am. Ind. Hyg. Assoc. J. 45:671-680 (1984).

Sass-Kortsak, A.M.; O'Brien, C.R.; Bozek, P.R.; Purdham, J.T.: Comparison of the 10-mm nylon cyclone, horizontal elutriator, and aluminum cyclone for silica and wood dust measurements. Appl. Occup. Env. Hyg. 8:31-37 (1993).

Schneider, T.; Schilunessen, V.; Vinzents, P.S.; Kildeso, J.: Passive sampler used for simultaneous measurement of breathing zone size distribution, inhalable dust concentration and other size fractions involving large particles. Ann. Occup. Hyg. 46(2):187-195 (2002).

Schwartz, G.P.; Daisey, J.M.; Lioy, P.J.: Effect of sampling duration on the concentration of particulate organics collected on glass fiber filters. Am. Ind. Hyg. Assoc. J. 42:258-263 (1981).

Seixas, N.; Siwinski, G.; Ghosh, A.: Bias introduced by NIOSH method 7400 quantitative fiber exposure assessment, industrial hygiene forum. Am. Ind. Hyg. Assoc. J. 48: A242-A243 (1987).

Sheehy, J.W.; McJilton, C.E.: Development of a model to aid in reconstruction of historical silica dust exposures in the taconite industry. Am. Ind. Hyg. Assoc. J. 48:914-918 (1987).

SKC: Parallel particle impactors. SKC, Inc., Eighty Four, PA http://www.skcinc.com/instructions/1683.pdf (2002).

SKC: SKC Comprehensive Catalog and Sampling Guide. SKC, Inc., Eighty Four, PA (2007).

SKC: SKC Cyclones. SKC, Inc., Eighty Four, PA.

Smith, J.P.; Bartley, D.L.; Kennedy, E.R.: Laboratory investigation of the mass stability of sampling cassettes for inhalable aerosol samplers. Am. Ind. Hyg. Assoc. J. 59:582-585 (1998).

Soderholm, S.C.: Proposed international conventions for particle size-selective sampling. J. Occup. Hyg. 33:301-320 (1989).

Spear, T.M.; Werner, M.A.; Bootland, J.; Harbour, A.; Murray, E.P.; Rossi, R.; Vincent, J.H.: Comparison of methods for personal sampling of inhalable and total lead and cadmium containing aerosols in a primary lead smelter. Am. Ind. Hyg. Assoc. J. 58:893-899 (1997).

Speight, R.G.; Marsh, A.H.: Use of a cowl in asbestos air sampling, letters to the editor. Ann. Occup. Hyg. 28:353359 (1984).

Stevens, G.A.; Moyer, E.S.: "Worst case" aerosol testing parameters: I. Sodium chloride and dioctyl phthalate aerosol filter efficiency as a function of particle size and flow rate. Am. Ind. Hyg. Assoc. J. 50: 257-264 (1989).

Suh, M.W.; Neefus, J.D.; Jones, M.G.; Shofner, F.M.; Miller, A.C. Jr.; Kreikebaum, G.: Statistical analysis CAM/LVE data in the major cotton manufacturing process areas, In Proceedings of the Fourth Special Session Cotton Dust Research, sponsored by the National Cotton Council and The Cotton Foundation, Memphis, TN (1980).

Tatum, V.L.; Ray, A.E.; Rovell-Rixx, D.C.: The performance of personal inhalable dust samplers in wood-products industry facilities. Appl. Occup. Env. Hyg. 16(7):763-769 (2001).

Tatum, V.; Ray, A.E.; Rovell-Rixx, D.: Performance of the respicon personal aerosol sampler in forest products industry workplaces. Am. Ind. Hyg. Assoc. J. 63:311-316 (2002).

Teikari, M.; Linnainmaa, M.; Laitinen, J.; Kalliokoski, P.; Vincent, J.; Tiitta, P.; Raunemaa, T.: Laboratory and field testing of particle size-selective sampling methods for mineral dusts. Am. Ind. Hyg. Assoc. J 64:312-318 (2003).

Thermo: Marple 290 Personal Cascade Impactors. Thermo Electron Corporation, Boston, MA, USA http://www.thermo.com/com/cda/product/detail/1,1055,22497,00.html (2006).

Thomassen, Y.; Nieboer, E.; Ellingsen, D.; Hetland, S.; Norseth, T.; Odland, J.O.; Romanova, N.; Chernova, S.; Tchachtchine, V.P.: Characterization of workers' exposure in a Russian nickel refinery. J. Environ. Monit. 1:15–22 (1999).

Tierney, G.P.; Conner, W.D.: Hygroscopic effects on weight determinations of particulates collected on glass-filters. Am. Ind. Hyg. Assoc. J. 28: 363-365 (1967).

TLV® Committee: Chemical substances TLV committee study papers. Ann. Am. Conf. Ind. Hyg. 4:153-157 (1983).

Tomb, T. F.; Haney, R.A.: Comparison of number and respirable mass concentration determinations, In Advances in Air Sampling. American Conference of Governmental Industrial Hygienists. Lewis Publishers, Chelsea, MI (1988).

Trakumas, S.; Hall, P.: Performance Assessment of Personal Respirable Cyclone Samplers, paper presented at American Industrial Hygiene Conference and Exposition, Dallas, TX https://www.aiha.org/membersonly/MeetingsEducation/html/abs03/po127.htm or http://www.aiha.org/aihce03/handouts/po127trakumas_files/frame.htm (Internet Explorer browser only) (2003).

Tsai, C.J.; Shih, T.S.: Particle collection efficiency of two personal respirable dust samplers. Am. Ind. Hyg. Assoc. J. 56: 911-918 (1995).

Tsai, P.J.; Vincent, J.H.; Wahl, G.; Maldonado, G.: Occupational exposure to inhalable and total aerosol in the primary nickel production industry . Occ. Env. Med. 52: 793-799 (1995).

Tsai, P.J.; Vincent, J.H.; Mark, D.: Semi-Empirical model for the aspiration efficiencies of personal aerosol samplers of the type widely used in occupational hygiene. Ann. Occup. Hyg. 40(1):93-113 (1996a).

Tsai, P.J.; Werner, M.A.; Vincent, J.H.; Maldonado, G.: Exposure to nickel-containing aerosol in two electroplating shops: comparison between inhalable and 'total' aerosol. App. Env. Occ. Hyg.11: 484-492 (1996b).

Tsai, P.J.; Vincent, J.H.; Wahl, G.A.; Maldonado, G.: Worker exposures to inhalable and total aerosol during nickel alloy production. Ann. Occ. Hyg. 40: 651-669 (1996c).

TSI: RespiCon® Particle Sampler. TSI, Inc., St. Paul, MN, USA http://www.tsi.com/documents/resp_bro.pdf (1999a).

TSI: Revised Correction Factor for the RespiCon™ Particle Sampler, Application Note ITI-078. TSI, Inc St. Paul, MN, USA http://www.tsi.com/documents/iti_078.pdf (1999b).

Turner, S.; Cohen, B.S.: Effects of electrostatic charge on aerosol collection with polystyrene filter cassettes. Am. Ind. Hyg. Assoc. J. 45: 745-748 (1984).

van den Heever, D.J.; Tiernan, G.: Quantification of bypass leakage in filter holders with cyclone elutriators. Am. Ind. Hyg. Assoc. J. 60:545-547 (1999).

Van Tongeren, M.J.A.; Gardiner, K.; Calvert, I.A.: An assessment of the weight-loss in transit of filters loaded with carbon black. Ann. Occup. Hyg. 38(3): 319-323 (1994).

Vaughan, N.P.; Chalmers, C.P.; Botham, R.A.: Field comparison of personal samplers for inhalable dust. Ann. Occup. Hyg. 34: 553-573 (1990).

Verma, D.K.; Sebestyen, A.; Julian, J.A.; Muir, D.C.F.: Field comparison of respirable dust samplers. Ann. Occup. Hyg. 36:23-34 (1992).

Vincent, J.H. ed.: Particle Size-Selective Sampling for Particulate Air Contaminants, ACGIH®, Cincinnati, OH, USA (1999).

Vincent, J.H.; Mark, D.: The basis of dust sampling in occupational hygiene: A critical review. Ann. Occup. Hyg. 24: 375-390 (1981).

Vincent, J.H.; Armbruster, L.: Short communication on the quantitative definition of the inhalability of airborne dust. Ann. Occup. Hyg. 24:245-248 (1981).

Vincent, J.H.; Mark, D.: Comparison of criteria for defining inspirable aerosol and the development of appropriate samplers. Am. Ind. Hyg. Assoc. J. 48: 454–457 (1987).

Vinzents, P.S.: A passive personal dust monitor. Ann. Occup. Hyg. 40(3):261-280 (1996).

Volckens, J.; Boundy, M.; Leith, D.: Mist concentration measurements II: Laboratory and field evaluations. Appl. Occup. Env. Hyg. 15(4):370-379 (2000).

Wagner, J.; Leith, D.: Passive aerosol sampler. Part II: Wind tunnel experiments. Aer. Sci. Tech. 34:193-201 (2001).

Wagner, J.; Macher, J.M.: Comparison of a passive aerosol sampler to size-selective pump sampler in indoor environments. Am. Ind. Hyg. Assoc. J. 64:630-639 (2003).

Westerholm, R.N.; Almen, J.; Li, H.; Rannug, J.U.; Egeback, K.E.; Gragg, K.: Chemical and biological characterization of particulate, semivolatile, and gas-phase-associated compounds in diluted heavy-duty diesel exhausts: comparison of three different semivolatile-phase samplers. Env. Sci. Technol. 25: 332-338 (1991).

Wilsey, P.W.; Vincent, J.H.; Bishop, M.J.; Brosseau, L.M.; Greaves, I.A.: Exposures to inhalable and "total" oil mist aerosols by metal machining shop workers, Am. Ind. Hyg. Assoc. J. 57:1149-1153 (1996).

Wilson, N.K.; Kuhlman, M.R.; Chuang, J.C.; Mack, C.A.; Howes, J.E., Jr.: A quiet sampler for the collection of semivolatile organic pollutants in indoor air. Env. Sci. Technol. 23: 1112-1116 (1989).

Witschger, O.; Grinshpun, S.A.; Fauvel, S.; Basso, G.: Performance of personal inhalable aerosol samplers in very slowly moving air when facing aerosol source. Ann. Occup. Hyg. 48(4):351-368 (2004).

Witschger, O.; Willeke, K.; Grinshpun, S.A.; Aizenberg, V.; Smith, J.; Baron, P.A.: Simplified method for testing personal inhalable aerosol samplers. J. Aerosol Sci. 5:855-874 (1998).
Wood, J.D.; Birkett, J.L.: External airflow effects on personal sampling. Ann. Occup. Hyg. 22: 299-310 (1979).
Xiong, J.Q.; Chingping, F.; Cohen, B.S.: A portable vapor/particle sampler. AIHAJ 59:614-621 (1998).
Xu, Z.; Gautam, M.; Mehta, S.: Cumulative frequency fit for particle size distribution. Appl. Occup. Env. Hyg. 17(8):538-542 (2002).

25

INTERPRETING ANALYTICAL METHODS AND SELECTING A LABORATORY

INTRODUCTION

When preparing for an industrial hygiene sampling campaign, the hygienist must make two important decisions – what sampling and analytical method will be used to collect the samples and what laboratory will be used to analyze the samples. There are many ways to accomplish an exposure monitoring goal. In the U.S., the actual method that is used is left to the discretion of the hygienist and the laboratory. Even for substances having an OSHA comprehensive standard, OSHA allows the use of equivalent methods. Since it is not uncommon for laboratories to develop their own nuances to *established* methods, it is imperative that the hygienist understands the methods used and remains in close communication with the analysis laboratory to avoid being surprised by method variances that may affect sampling results.

OSHA (2006) and many other governmental and professional agencies around the world (see in references ASTM[1], 2006; ISO, 2006; HSE, 2006; and EPA, 2003[2]) have published standardized sampling and analytical methods for ambient and workplace air, for surfaces, and for biological fluids. The NIOSH methods (NIOSH, 1977, 1984, 1994) are offered in a consistent format, have been validated by using one set of criteria (Bush and Taylor, 1981; Kennedy et al., 1995), and have been available to the public for the longest period of time. Each method's text covers most of the important aspects of sampling and analysis methodology and thus serves as a good model for methods. This chapter will discuss the type of information that can be found in a sampling and analytical method, using NIOSH methods as a model, and the data that should be gathered to help the hygienist select the right laboratory for analyzing the collected samples.

---

[1] See also review article by Ashley and Harper (2005).

[2] In addition to EPA methods, EPA (2006) also has a technology verification program that evaluates the performance of devices such as size selective samplers, portable GC and IR devices, and other direct reading equipment used by hygienists.

INTERPRETING THE NIOSH ANALYTICAL METHOD

Figure 25-1 is page one of the NIOSH sampling and analytical method for dioxane. There is nothing unique about the dioxane procedure; it was picked at random as a representative method. The best way to understand how to use methods is to discuss each section in turn. It is important to point out that virtually all of the parameters given in a method are guidelines that can be modified if the hygienist understands the basis for the parameter. Thus, the purpose of this section is to explain the basis for each parameter.

Introductory Information

The descriptive information at the beginning of each method should not be overlooked as it identifies the analyte and the method. Synonyms may be helpful; however, the CAS (Chemical Abstract Services) number may be the only way to be sure of the chemical's identity. Method revision (issue) number should be noted to make sure the latest revision is used. The date of the revision is important as it can help determine if OELs provided are current or out of date. Level of evaluation (full, partial, etc.) is also important as it indicates the amount of validation effort spent by the organization authoring the method.

Sampling

An exact description of the sampler (medium and equipment, such as impinger) is given. Modification of the method by use of another medium will require a new validation.

FLOW RATE

The flow rate is usually given as a range. For charcoal-based methods it generally is 0.01 to 0.2 L/min. The lower limit is based on limiting the effects of diffusion of the analyte, as opposed to active sampling of the analyte, into the tube. The upper limit is a rough guide to decrease the likelihood of analyte breakthrough from the charcoal. However, it can be exceeded in some circumstances, if necessary, without decreasing collection efficiency. These limits were discussed in detail in Chapter 21. The flow rates for filtration of metals usually are set at 1 to 4 L/min. These too are rough guidelines that can be modified as discussed in Chapter 24. For semivolatile organics (e.g., benzidine), flow rates are usually much lower (0.2 L/min and lower) to reduce volatilization or to avoid breakthrough when a sorbent tube is used in combination with the filter. One should be careful when modifying these flow rates.

SAMPLE VOLUMES

The volumes are the minimum and maximum volumes of air that should be passed

DIOXANE 1602

$C_4H_8O_2$ MW: 88.11 CAS: 123-91-1 RTECS: JG8225000

METHOD: 1602, Issue 2 **EVALUATION: FULL** **Issue 1: 15 May 1985**
Issue 2: 15 August 1994

OSHA : 100 ppm (skin)
NIOSH: 1 ppm/30 min; carcinogen
ACGIH: 25 ppm (skin); carcinogen
(1 ppm = 3.60 mg/m^3 @ NTP)

PROPERTIES: liquid; d 1.0337 g/mL @ 20 °C;
BP 101 °C; MP 12 °C
VP 3.9 kPa (29 mm Hg; 3.8% v/v) @ 20 °C;
explosive range 2 to 22% v/v in air

SYNONYMS: dioxan; p-dioxane; 1,4-diethylene dioxide; diethylene ether

SAMPLING

SAMPLER: SOLID SORBENT TUBE (coconut shell charcoal, 100 mg/50 mg)

FLOW RATE: 0.01 to 0.2 L/min

VOL-MIN: 0.5 L @ 100 ppm
-MAX: 15 L

SAMPLE STABILITY: >1 week @ 25 °C
>6 weeks @ -10 °C

BLANKS: 2 to 10 field blanks per set

ACCURACY

RANGE STUDIED: 157 to 660 mg/m^3 [1] (10-L samples)

BIAS: + 9%

OVERALL PRECISION ($\hat{S}_{rT}$): 0.054 [1]

ACCURACY: ± 18.7

MEASUREMENT

TECHNIQUE: GAS CHROMATOGRAPHY, FID

ANALYTE: dioxane

DESORPTION: 1 mL CS_2, 30 min

INJECTION VOLUME: 5 µL

TEMPERATURE-INJECTION: 225 °C
-DETECTOR: 250 °C
-COLUMN: 75 °C

CARRIER GAS: He, 30 mL/min (60 psig)

COLUMN: 6 m x 3-mm stainless steel packed with 10% FFAP on 80/100 mesh Chromosorb W-HP Chromosorb W-HP

CALIBRATION: solutions of dioxane in CS_2 with internal standard

RANGE: 0.2 to 7 mg per sample [1,2]

ESTIMATED LOD: 0.01 mg per sample [3]

PRECISION ($\bar{S}_r$): 0.017 [1]

APPLICABILITY: The working range is 5.5 to 190 ppm (20 to 700 mg/m^3) for a 10-L air sample.

INTERFERENCES: None known. An alternate GC column is 30 m x 0.32-mm ID fused silica capillary coated with 1 µm DB-5 [3].

OTHER METHODS: This method combines and replaces S360 [4] and P&CAM 127 [5] for dioxane. A similar method appears in the dioxane criteria document [6].

NIOSH Manual of Analytical Methods (NMAM), Fourth Edition, 8/15/94

Figure 25-1. The NIOSH sampling and analytical method (1602) for dioxane, page 1 (NIOSH, 1994).

through the medium. The lower value is based on the limit of detection (the smallest amount that can be reliably measured by the method) and the PEL (see later discussion of the LOD under "Measurement"). For dioxane, the PEL is 100 ppm or 360 mg/m³ and the method LOD is 0.01 mg. At the minimum sample volume of 0.5 L we have

$$\frac{360\ mg}{m^3} \times 0.5\ L \times \frac{1\ m^3}{1{,}000\ L} = 0.18\ mg$$

Thus, when the minimum volume is sampled at 100 ppm, the mass collected will be 18 times the LOD and reliably measured. Lower concentrations, when sampled at the minimum volume, will be detectable; however, measurement uncertainty will increase until at a concentration of approximately 5 ppm, the analyte will no longer be detectable (< LOD).

The maximum volume is based on the avoidance of significant breakthrough or analyte loss. In this case, it is 15 L. At the end of the method (see "Evaluation" discussion below) it is stated that breakthrough was detected in the validation study after a 40 L sample was taken. Therefore, it appears that there is plenty of room to modify the method by extending the sample time or increasing the flow rate. However, the validation study was done in dry air, and this is an important point. Dry air was used in many of the validation studies, but a relative humidity of 50% at NTP might be more representative. Dry air does provide a baseline, but it is difficult to know how to adjust the baseline (see discussion of breakthrough volume in Chapter 21). In addition, the breakthrough volume is not linearly related to concentration or flow rate (equivalent to velocity through the tube). In the validation study, a 651 mg/m³ concentration and a 0.19 L/min flow rate were used. Finally, other organics in the atmosphere will reduce the breakthrough volume. Thus, for differing RH, *Q*, interferents, or *C*, we have no way of knowing how to adjust the maximum volume. In fact, we cannot even be sure that obeying the maximum volume will always avoid breakthrough. Thus, adhering to the maximum volume guideline is generally a good idea. One way to avoid breakthrough when uncertainties exist is to use a jumbo (600 mg) charcoal tube. However, the LOD will be increased by a factor of about 3, as additional solvent will be needed to extract the analyte, thereby diluting it. Other types of sampling media (filters, impingers, etc.) will be affected by similar sampling considerations. However, each method will identify an optimal sampling range, defined by sample time and flow rate.

SHIPMENT AND STABILITY

This section provides general guidance on how the sample must be handled after collection and how quickly it must be analyzed. As noted for dioxane, shipping and storage temperatures are important. For this compound, freezing the samples extends the period of stability significantly. Use of preservatives, protection from light, use of static-free shipping materials, or other special shipping and handling conditions may be important for other compounds or sampler types. In some instances, special handling and hold times are just as important for the unused media as for the samples once they have been collected. To ensure sample integrity, the hygienist must be familiar with and prepared to follow all shipping and handling requirements. In the case of filters, certain particles (e.g., wood dusts) are more likely to dislodge from cassette-held filters and be lost to weighing if the cassette contents are not included. The IOM cassette avoids this issue as the entire cassette is weighed.

BLANKS

There are at least two types of blanks, termed here the laboratory (also known as the media or QC, quality control, or analysis method blank) and the field blank. The first is based on the laboratory analysis of several media blanks. The results account for background signal caused by the media or any other aspect of the method, such as a contaminated extraction solvent. NIOSH (1998) recommends that the laboratory analyze at least six media or lab[3] blanks (more if the media is coated). The lab blank results typically should be less than the limit of detection. Consistent and significant laboratory blank values may be subtracted from all analytical field results[4] (see section on "Calculations").

Field blanks are media taken to the field but not used. The media are handled just like the used media, but they are not exposed to the air. For charcoal tubes, the ends are broken and then the plastic end caps are placed in the tube ends. During the sampling, these are left in an area similar to where the samples are taken, but where it is unlikely that they will be tampered with. If the used media are placed in the trunk of a car for a week after sampling, so are the field blanks. If the media are exposed to 90% relative humidity or interferents that may penetrate the charcoal tube end caps, so are the field blanks. Thus, any significant mishandling or unknown exposure of the samples should be reflected in the field blanks. Filters will adjust their equilibrium with water vapor in air. The humidity will penetrate small openings in cassettes, even if they are not opened. Since water vapor adsorption is an important bias for filtration and lab blanks will not account for this bias, field blanks are particularly important for determining this and similar biases.

NIOSH recommends that the number of field blanks be at least two because of the way the blanks are interpreted. Generally, if the field blanks are contaminated, the entire sample set is assumed to be contaminated. If laboratory blanks are consistently less than the limit of detection, and a single field blank were 2 × LOD, would this be grounds for throwing the sam-

---

[3] If one is using a commercial lab, this will be done for routine analyses for each media batch opened by the lab, at no cost to the hygienist, provided that the hygienist uses media supplied by the lab. If the hygienist uses her own media, then she will need to supply these blanks and may have to pay to have them analyzed. The hygienist should ask for enough of the media to perform the sampling and provide the necessary field blanks. She will pay for the field blank analysis.

[4] This may be done by the lab. Ask the lab how it handles lab blank corrections. AIHA accredited labs are required to indicate whether corrections are made and by what amount.

ples out? If you had only one field blank, then maybe so. If, on the other hand, the second blank were less than the limit of detection, then there would be grounds to conclude that the first blank was aberrant. More importantly, if the second blank were 3 × LOD, then contamination certainly would be suspected.

In routine sampling programs, using as many as 10 field blanks probably is a waste of resources (especially because the analysis costs are the same as for exposed sample media). If well-thought-out procedures are used to guard the media against contamination, the exposed samples are shipped immediately, the historical information on field blanks indicates that the system works (i.e., contamination is insignificant), then why is a field blank needed? Some laboratories insist on them. In many cases where field contamination is unlikely, again based on a history, we are better off spending money on a result rather than repeating something we already know. See Chapter 16 for a discussion on variability in methods versus environmental variability for further justification. It also seems quite illogical to *correct* field results with results from a few (and in some cases only one) field blanks. However, some laboratories do so.

For non-routine analysis, two field blanks are usually sufficient to indicate sampling, handling, and shipping conditions of the samples. Valid field blank data will ensure defensibility of results, where this is important (e.g., legal work).

If the hygienist is providing the media, additional blanks not exposed to field conditions may be required for the laboratory to evaluate media background and to perform routine QC functions. The laboratory and the hygienist will determine how many blanks will be needed and whether there will be a charge for analysis of these additional QC blanks.

Use of an AIHA accredited laboratory will help ensure that media and field blank data are provided, as reporting this information is a requirement for accreditation. It is also important for the hygienist to look at the blank data and determine if or when blank corrections are important and should be applied, as only the hygienist knows what the actual sampling conditions were. When in doubt as to the validity of a blank value, it is probably a good policy to ignore it in the calculation of results, as this errs on the safe side. On the other hand, field blanks that clearly indicate contamination, require an investigation of procedures and resampling.

Accuracy

To discuss the accuracy of the method, we must first discuss how methods are validated by NIOSH.[5] The original scheme (Bush and Taylor, 1981) was part of what was called the Standards Complection Program. It was replaced by an updated version (Kennedy et al., 1995). Let us take the case of a charcoal tube method validated under the former version (this applies to dioxane). In simple terms, a chamber would be filled three times at three different concentrations – 0.5, 1, and 2 times the PEL with air mixed with the contaminant. Six charcoal tube

[5] OSHA, ASTM and other organizations have their own protocols.

samples would be taken from the chamber at each concentration. For dioxane, the PEL is 360 mg/m³; so the target concentrations would be 180, 360, and 720 mg/m³. The range tested was actually 157 to 660 (see "Evaluation of Method" section).

From these tests, three standard deviations and their corresponding coefficients of variation (*CV*) can be calculated. These *CV* results can be pooled (essentially averaged) if they are equivalent. This is determined with a statistical test.[6] This pooled *CV* then represents the analytical precision (this value is actually at the bottom of the second column in that part of the method and is 0.017 or 1.7 % for dioxane). The *overall precision* is estimated by adding 5% sampling (pump) error, or

$$CV_a^2 + CV_s^2 = CV_T^2 \qquad \text{[Equation 25-1]}$$

where *s* and *a* indicate the sampling or pump and analytical precision, respectively. In the case of dioxane we have

$$\sqrt{0.05^2 + 0.017^2} = 0.054$$

In addition, the bias can be calculated by comparing the results with what is assumed to be the true concentration (see Equation 7-2). This is the SAE value, or OSHA Sampling and Analytical Error, that is equal to the *CV* in Equation 16-5 and used to adjust OSHA monitoring results before determining that a citation is warranted. Some typical method precision values with added sampling (pump) error are shown in Table 25-1.

The bias for dioxane was 9%. Such a high value actually is unusual for organics. Unfortunately, the method provides no speculation about its cause. Normally known and consistent biases are used to correct the analytical results. Based on the discussion in Chapter 7 (Equation 7-6) and the original NIOSH validation procedure, the overall accuracy would be $2 \times 5.4 + 9 = 19.8\%$ and would meet the minimum criterion set for accuracy by NIOSH to validate a method. This is 25% at the 95% confidence level.

Statistically this is a bit of over simplification. In NIOSH's new validation procedure a much more rigorous approach is taken that considers the variability in the estimates of the bias and the precision. In other words, the bias and precision are not known, they are estimates. Thus, we need to calculate a double confidence limit, one set of limits around the bias and precision and one set for the actual measurement. The above calculation only considers the latter and is, therefore, an underestimation of the accuracy. For more details see Fischbach et al. (1996) and Bartley (2004). Bartley ties the NIOSH validation protocol to a new approach that is taken by ISO which seeks to identify the components of precision and bias and the uncertainties for each. These are then combined statistically into an accuracy statement. There is a

[6] It is generally observed that, for airborne concentrations, the variability about a mean observation varies with the mean; that is, the higher the concentration, the larger the standard deviation. But the $CV = s/\bar{x}$; so it tends to be constant across concentration.

Table 25-1
Approximate method[1] precision values as reported in NIOSH Manual of Analytical Methods

Analyte(s)	Precision CV (as %)
Aliphatic alkanes (C_6-C_{12})	< 5
Aromatic hydrocarbons (benzene)	1–2
Formaldehyde (VIS spectroscopy)	3
Inhalable particulates[2]	2.6
Lead (flame atomic absorption)	3
Metals (ICAP, PVC filter)	2–10, some much higher, e.g., Aluminum 21%
Metals (ICAP, MCE filter)	5 (Sb, Sn, W, Zr, 15–58%)
Silica[3] (XRD)	8
Asbestos (PCM)	12

[1] Does not include sampling error.
[2] Multi-lab study 4–15% for glass fiber filters and 1–12% for PVC, with much higher values for both at low loadings (Stacey et al., 2002).
[3] PAT results for inter-laboratory CV, 25–30% and typically 10–20% for one lab (Eller et al., 1999).

nice piece of software on the Internet that guides laboratories through the process of developing a validation for a new method (NIOSH, 1996; Abell and Kennedy, 1997). NIOSH plans to rely more on volunteer labs for validation work (Schlecht and Cassanelli, 1997). The website is also instructive for the hygienist who would like to better understand the process. Finally, for those interested in how the European agencies are going about the process of methods development, Brown (2000) provides an informative article.

Measurement Applicability, Interferences, Other Methods

The information in this section is a summary of the *nuts and bolts* of the method. Most of the information is straightforward.

RANGE

The range here is not the same as the range under the section entitled "Accuracy" above. Here the values represent the range of sample masses present in the solutions that were used to calibrate the GC in the validation study. The analytical chemist determines this range based on the desorption volume and the optimal response range of the instrument being used. The range can then be adjusted as necessary to ensure adequate calibration of the GC or other instrument being used. Instrument linearity and LOD studies help define the upper and lower limits of the full instrument calibration range (see LOD discussion below).

APPLICABILITY OR WORKING RANGE

This range is the same as the range under the section entitled "Measurement Applicability, Interferences, Other Methods." Now, however, the total mass per sample (0.2 to 7 mg) has been converted to an air concentration represented by a 10 L air sample, or

$$\frac{0.2\ mg}{10\ L} \times \frac{1000\ L}{m^3} = 20\ \frac{mg}{m^3}$$

This format helps the hygienist to determine what sample volume to collect for the concentration range of analyte expected to be present. There is nothing magical about a 10 L sample or a mass range of 0.2 to 7 mg. If the GC were calibrated outside this range, say, 0.1 to 10 mg, and the air sample volume, using a jumbo tube, were 96 L (0.2 L/min for 480 min), then the working range would be only about 1 to 100 mg/m^3 because of the increased sample volume.

LOD

The estimated limit of detection deserves special discussion. The definition of the LOD[7] is the smallest concentration "that can be detected with reasonable certainty for a given analytical procedure" (Anonymous, 1980). This general definition and others like it have been used by the International Union of Pure and Applied Chemists (IUPAC) to formulate a working definition. The problem becomes a statistical one based on the term *reasonable certainty*.

If we injected a blank sample into the GC repeatedly, we would not get the same result but would have a range of results. Usually we would expect these results to fit a normal distribution. Because the distribution is asymptotic, and is affected by many variables, there is a small chance that even a fairly large signal is in fact indistinguishable from a blank or the *background*. We need to limit this chance to something reasonable. In other words, if we obtain a small signal from an environmental sample, we want to be fairly certain that we do not call it a true signal when in fact it is a background signal. On the other hand, we do not want our definition of a background signal to be so broad that we fail to detect true signals too frequently. The IUPAC defines the largest background signal (x_L) as the 99.86th %tile of the distribution of blank values, or

$$x_L = \bar{x}_b + 3s_b \qquad \text{[Equation 25-2]}$$

or the mean plus three standard deviations of the background signal. It may seem that this is a large number, or that two standard deviations (the 97.7th %tile) would suffice. However, the

7 Also referred to as the Detection Limit or DL.

calculation assumes normality. If the distribution is not normal, then the percentile represented by 3*s* may be as low as 94.5%.[8] Thus, there could be a 5.5% probability that a signal we call true is in fact background.

Regression analysis also has been used to determine LOD values. For a thorough review of the subject see Winefordner and Long (1983). It is the method used by NIOSH (Kennedy et al., 1995). One first spikes the media[9] at several low levels (say 5), not to exceed 10 × the suspected LOD. The response is plotted against the mass and from a regression analysis the standard error of the regression is attained. The standard error is used in Equation 25-2. For other nuances to the technique see Kennedy et al.

The results from an LOD derivation and how the LOD is used is shown in Figure 25-2. In this example, the determined LOD is an instrument response of 140 units. For a blank with zero mass the instrument response is 80 units. The right hand curve is for a true environmental signal of 140 units. Upon analyzing that sample one time, we have a 50% probability of not detecting the true environmental signal (i.e., calling it background). We also have a 0.1% probability (see the small black tail of the left curve) that this signal is in fact 0. If the true signal is 2 × LOD, we will report a detected value 99% of the time, again assuming a normal distribution. This 2 × LOD value has been called the reliable LOD (Abell and Kennedy, 1997).

Note that the precision of a measurement near the LOD is typically poorer than at higher levels. In addition, for GC methods, the recovery of analyte from the media is poorer at lower concentrations. Thus, another possible question is, what is the smallest signal that we can *precisely* say is environmental, not *background*? Contrast this with the LOD definition, the largest signal that we are willing to call background. Thus, the lowest mass that can be reported with acceptable precision has been called the limit of quantitation (LOQ) and defined as

$$LOQ = \bar{x}_b + 10s_b \qquad \text{[Equation 25-3]}$$

There are other definitions such as the mass above which recovery is 75%, for GC methods (Kennedy et al., 1995). In some methods, NIOSH reports an LOQ, and some laboratories report an LOQ. It is particularly important when the majority of values for a group of monitoring results are in the region of the LOD and the LOQ. Results should not be extrapolated beyond the instrument calibration curve (Botnick and White, 1998).

---

[8] This is based on Tchebyscheff's theorem that the lower bound for the probability that a random variable *Y* falls in the interval of $\mu \pm k\sigma$ is 100 $(1 - 1/k^2)$% for an unknown distribution. Therefore, for $k = 2$, the probability is 75% and there is no more than a 12.5% chance that *Y* is larger than the interval, or the upper percentile is the 87.5th. For $k = 3$ the interval covers at least 88.9% and so the upper tail %tile is 100 – 5.5 = 94.5.

[9] One could inject small aliquots directly into the instrument but that would not take into consideration recovery from the media. It is important to point out that there are several types of LODs. There is one for the method that takes recovery, sample preparation, media variability, etc. into consideration. By injecting mass directly to the instrument, we can determine an instrument LOD, in which, by itself, the hygienist is generally not interested.

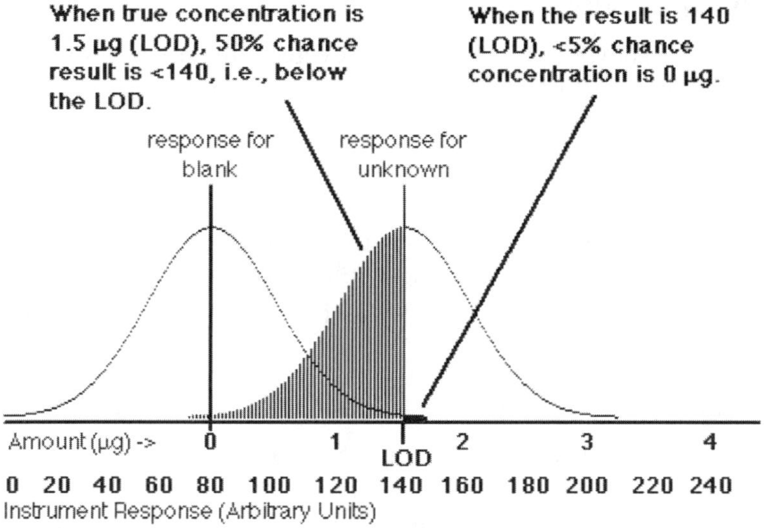

Figure 25-2. Distributions for a background signal based on 0 mass (left) and a true environmental signal of 140 units (right) coming from a mass of 1.5 µg. The LOD or 99.9th %tile of the blank distribution is shown, also 140 units.

It has been recommended that the manner in which the LOD is determined be given, such as $LOD_{k=3}$. This is necessary because the history of the development of LOD nomenclature is rather long and complicated; hence, several techniques have been used.[10] AIHA accredited laboratories may report LOQ as defined above or report a *reporting limit*, or RL that may be the lowest standard that was used in the calibration or a value like LOD, but the coefficient in front of the s_b in Equation 25-2 might range from 2 through 10. Apparently there are no consistent rules required by accreditation. Minimally, hygienists should expect that laboratories report the method mass associated with the RL, when values are less than the RL. In addition, it can be useful for the laboratory to report the actual result; for example, 0.005 mg <RL, RL = 0.01 mg.

Finally, it should be noted that the method (LOD, LOQ, RL, or whatever the laboratory uses) is reported as a mass. For every possible sample volume there is a related concentration (as opposed to mass) value (mass/volume) as was described in Chapter 16. For example, in Figure 25-2, the mass LOD is 1.5 µg (same as 140 units), but the concentration LOD will vary with the sample volume. Thus, for a volume of 10 L, the concentration LOD is 1.5 µg/l0 L = 15 µg/m^3. So, given that sample volume and mass LOD, 15 µg/m^3 is the smallest concentration that can be detected.

[10] In fact, other techniques still are commonly used with direct-reading instruments, as discussed in Chapter 28.

Reagents and Equipment

These sections are shown in Figure 25-3. They generally are clearly interpreted and give the details necessary for selecting the correct sampling devices, including the media.

Sampling

Generally these instructions deal with the pump and are routine. However, if the sample requires special treatment such as cooling, or other field preparation for shipping, such as might be necessary for impinger samples, these requirements would be detailed here.

Analysis – Calibration, QC, Sample Preparation, and Measurement

The next three sections (Figures 25-3 and 25-4) are basically for the analytical chemist, but it is important for the hygienist to understand the ramifications of some of the recommendations. In particular, the quality control section is important. In the U.S. many labs are accredited by AIHA, which has a set of QC requirements. (See later discussion.)

In addition, for organic analytes, the desorption efficiency studies can have ramifications for the hygienist. As discussed in Chapter 21, desorption efficiencies depend on several factors, and mixtures may be more important than we realize. DE curves usually are constructed only for single compounds. If the hygienist suspects that DE might be causing a bias, then special DE studies should be discussed with the lab.

Calculations

Generally the calculations (Figure 25-4) are simple. In general, for the charcoal tube, the mass in both sections is added together. Thus, the general form is

$$\frac{\textit{total mass} - \textit{background mass}}{\textit{sample volume}}$$

In some cases, the background, justifiably, may be ignored, for reasons cited earlier.

For asbestos or fiber results, the concentration is reported in fibers per cubic centimeters. To determine the total number of fibers on the filter, we use the total number of fibers counted divided by the number of microscope fields examined, for example, 45 f/100 fe. This is multiplied by the area of a field to give the number of fibers per unit area. A typical field area is 0.00785 mm^2. Then the total area of the filter must be known. For a 25 mm filter, the exposed area[11] is 385 mm^2. The total number of fibers is then divided by the volume in cubic centimeters. A typical calculation is

---

[11] However, this can vary; see Chapter 24.

DIOXANE: METHOD 1602, Issue 2, dated 15 August 1994 - Page 2 of 4

REAGENTS:

1. Eluent: carbon disulfide,* chromatographic quality, containing 0.1% (v/v) octane, decane, or other suitable internal standard.
2. Dioxane, reagent grade.*
3. Desorption efficiency (DE) stock solution, 0.1 mg/µL. Dilute 1 g dioxane to 10 mL with pentane.
4. Helium, purified.
5. Hydrogen, prepurified.
6. Air, compressed, filtered.

* See SPECIAL PRECAUTIONS.

EQUIPMENT:

1. Sampler: glass tube, 7 cm long, 4-mm ID, flame-sealed ends with plastic caps, containing two sections of activated (600 °C) coconut shell charcoal (front = 100 mg; back = 50 mg) separated by a 2-mm urethane foam plug. A silylated glass wool plug precedes the front section and a 3-mm urethane foam plug follows the back section. Pressure drop across the tube at 1 L/min airflow must be less than 3.4 kPa. Tubes are commercially available.
2. Personal sampling pump, 0.01 to 0.2 L/min, with flexible connecting tubing.
3. Gas chromatograph, FID, integrator, and column (page 1602-1).
4. Vials, 2-mL, PTFE-lined caps.
5. Syringes, 10-µL and other convenient sizes for preparing standards, readable to 0.1-µL.
6. Volumetric flasks, 10-mL.
7. Pipet, 1-mL.

SPECIAL PRECAUTIONS: Carbon disulfide is toxic and a serious fire and explosion hazard (flash point = -30 °C).

Dioxane is toxic, causing central nervous system depression and necrosis of liver and kidneys, as well as a skin irritant [6], and a suspect carcinogen [7]. Use personal protective equipment and work only in a hood.

SAMPLING:

1. Calibrate each personal sampling pump with a representative sampler in line.
2. Break the ends of the sampler immediately before sampling. Attach sampler to personal sampling pump with flexible tubing.
3. Sample at an accurately known flow rate between 0.01 and 0.2 L/min for a total sample size of 0.5 to 15 L.
4. Cap the samplers. Pack securely for shipment.

SAMPLE PREPARATION:

5. Place the front and back sorbent sections of the sampler tube in separate vials. Discard the glass wool and foam plugs.
6. Add 1.0 mL eluent to each vial. Attach cap to each vial.
7. Allow to stand 30 min with occasional agitation.

CALIBRATION AND QUALITY CONTROL:

8. Calibrate daily with at least five working standards over the range 0.01 to 7 mg dioxane per sample.

NIOSH Manual of Analytical Methods (NMAM), Fourth Edition, 8/15/94

Figure 25-3. The NIOSH sampling and analytical method for dioxane (1602), page 2 (NIOSH, 1994).

DIOXANE: METHOD 1602, Issue 2, dated 15 August 1994 - Page 3 of 4

a. Add known amounts of dioxane to eluent in 10-mL volumetric flasks and dilute to the mark. By serial dilution, prepare solutions containing 0.01 to 7 mg dioxane/mL.
b. Analyze together with samples and blanks (steps 11 and 12).
c. Prepare calibration graph (ratio of peak area of analyte to peak area of internal standard vs. mg dioxane).

9. Determine desorption efficiency (DE) at least once for each lot of charcoal used for sampling in the calibration range (step 8). Prepare three tubes at each of five levels plus three media blanks.
   a. Remove and discard back sorbent section of a media blank sampler.
   b. Inject a known amount (1 to 20 µL) of dioxane or DE stock solution directly onto front sorbent section with a microliter syringe.
   c. Cap the tube. Allow to stand overnight.
   d. Desorb (steps 5 through 7) and analyze together with working standards (steps 11 and 12).
   e. Prepare a graph of DE vs. mg dioxane recovered.
10. Analyze three quality control blind spikes and three analyst spikes to ensure that the calibration graph and DE graph are in control.

MEASUREMENT:

11. Set gas chromatograph according to manufacturer's recommendations and to conditions given on page 1602-1. Inject sample aliquot manually using solvent flush technique or with autosampler.
    NOTE: If peak area is above the linear range of the working standards, dilute with eluent, reanalyze, and apply the appropriate dilution factor in calculations.
12. Measure peak area. Divide the peak area of analyte by the peak area of internal standard on the same chromatogram.

CALCULATIONS:

13. Determine the mass, mg (corrected for DE) of dioxane found in the sample front (W_f) and back (W_b) sorbent sections, and in the average media blank front (B_f) and back (B_b) sorbent sections.
    NOTE: If $W_b > W_f/10$, report breakthrough and possible sample loss.
14. Calculate concentration, C, of dioxane in the air volume sampled, V (L):

$$C = \frac{(W_f + W_b - B_f - B_b) \cdot 10^3}{V}, \ \text{mg/m}^3.$$

EVALUATION OF METHOD:

Method S360 [4] was issued on March 14, 1975, and validated over the range 155 to 651 mg/m^3 using atmospheres generated by calibrated syringe pump [1]. Desorption efficiency averaged 0.88 in the range 1.8 to 7.3 mg dioxane per sample. Breakthrough was observed when a test atmosphere containing 651 mg/m^3 of dioxane in dry air was sampled at 0.19 L/min for 210 min (40-L air sample volume). The front section held 26 mg of dioxane. Precision and accuracy are listed on page 1602-1. A similar method, without internal standard, was collaboratively tested in the range 6 to 190 ppm dioxane [2,8].

REFERENCES:

[1] Documentation of the NIOSH Validation Tests, S360, U.S. Department of Health, Education, and Welfare, Publ. (NIOSH) 77-185 (1977), available as GPO Stock #017-033-00231-2 from

NIOSH Manual of Analytical Methods (NMAM), Fourth Edition, 8/15/94

Figure 25-4. The NIOSH sampling and analytical method for dioxane (1602), page 3 (NIOSH, 1994).

$$\frac{\dfrac{45\ fibers}{100\ fields} \times \dfrac{1\ field}{0.00785\ mm^2} \times 385\ mm^2}{\dfrac{2\ L}{\min} \times \dfrac{1{,}000\ cm^3}{L} \times 480\ \min} = 0.02\ f/cm^3$$

Notice how the units cancel each other. When the laboratory average blank value is reliably determined, it can be subtracted from the fiber/fields term. High field blank values are reasons for concern that the entire sample set is contaminated.

Evaluation of Method

This is perhaps the most important but most under-read section of any method. It gives limitations indicating what method recommendations can be modified. It turns out that many charcoal methods are quite limited. As discussed earlier, the breakthrough volume is a very important parameter, yet typically we know little from the validation study about the effects of concentration, flow rate, or other adsorbents, including water. Previous work has shown that the performance of charcoal is severely affected by 80% RH (Kring et al., 1984). However, the effects at more normal conditions need further inquiry. The evaluation section is just as important in other methods involving filters and impingers.

CHOOSING A LABORATORY

Introduction

It is important to contact the laboratory while making plans for sampling, not after the samples have been collected. Laboratory personnel should be able to provide guidance on issues such as whether multiple solvents or metals can be collected on a single sample, or are multiple samples required because different desorption or dissolution techniques are required for some of the desired compounds; are there interferences that need to be considered; is the time available for sampling compatible with the method limits of detection; and economics. The more closely the hygienist and laboratory personnel work on developing the sampling strategy, the better the chance for generation of good measurement data. Finally, many laboratories will supply the sampling media free of charge; so the hygienist can get exactly what is needed and feel confident that it is fresh.

We can think of a number of questions that are important to consider in choosing a laboratory:

1. Does the laboratory typically analyze:

    a. Industrial hygiene samples?

    b. The analytes of interest?

2. Does it have an available QC plan?
3. Does the laboratory use recognized standard methods?
4. Does it have qualified personnel knowledgeable in hygiene methods?
5. Does it have adequate facilities?
6. Is the laboratory AIHA-accredited?
7. Will the laboratory supply the necessary media?
8. What is the usual turn-around time?
9. Will the laboratory supply results via telephone, fax, or email?
10. Are prices competitive?
11. Can the laboratory be flexible in analysis procedures, and at what cost?

The first six questions can be answered to some degree by checking the AIHA list of accredited laboratories (AIHA, 2006a). The accreditation requirements (AIHA, 2006b), discussed below, have a bearing on these questions. In areas outside North America, the questions may require individual answers. For example, there may be many environmental or forensic laboratories, but all will not analyze air samples. Many industrial plant QC laboratories analyze bulk products containing the analytes of interest, for example, benzene. But this does not mean they are experienced at the desorption of charcoal and all other necessary preparations for analyzing benzene from air.

It is necessary for the laboratory to have QC plans that minimally involve routines for analyzing standards, blind analysis of knowns, duplicate analysis of unknowns, and statistical analyses of these results on an inter- and intralaboratory basis. The QC plans often involve other features that are discussed later.

Questions 7 through 11 are independent of AIHA accreditation. Usually the laboratory will ship the necessary media. Check to see if it pays the cost. If any shipped media are not used, ask if they can be returned. Check the cost of the media; it may be cheaper to buy your own, particularly for routine monitoring, but remember you need to supply the lab with samples from each batch to use in lab blank or QC analyses. The turnaround time typically should not exceed five to 10 working days.

Prices vary considerably. Most laboratories will charge one price for the first analyte from a charcoal tube, say, \$30. The next analyte(s) from the same tube might cost \$20 each. This is an important consideration if multiple analytes are present. In some cases, the necessary analyses do not normally appear on the pricing sheet. These analyses can be extremely variable in cost, and in some cases the laboratory may refuse to do them. Many laboratories are not set up for research projects; if you are told this by a laboratory representative, consider it an honest answer that will probably save you a lot of headaches. Laboratory technicians usually are trained to do routine analyses, and many are not experienced in performing non-routine work.

AIHA Accreditation

Since 1972 AIHA has had a laboratory accreditation program. By1974, 32 laboratories were accredited. Today, more than 200 laboratories are accredited;[12] about 20% do not accept samples commercially. To be accredited, a laboratory must implement and maintain a quality management system compliant with ISO/IEC 17025:2005 (ISO, 2005) and AIHA modules (AIHA, 2006b) applicable to the laboratory's scope of analysis. The system typically will be described in a quality manual and associated policies and procedures. These documents must address 15 management system requirements such as organizational structure, management personnel criteria, document control, records management, procurement, preventive and corrective actions, and internal assessments. A Technical Manager must meet experience and education requirements and must be available to staff and customers 50% of each day. There must be a QC coordinator. In smaller laboratories these two roles may be filled by the same person, provided that person meets all of the criteria.

In addition, there are a number of technical requirements addressing method selection and validation, equipment calibration, routine quality control (QC), determination of method uncertainty, traceability of measurements, reporting of results, etc. The QC program requires records demonstrating regular analysis of QC samples and statistical treatment of QC data (see, e.g., the QC procedures in Chapter 18; AIHA, 1988; and Burkart et al., 1984). Laboratory compliance with these requirements is verified during biannual on-site assessments of each laboratory. Participation in and maintenance of proficiency in the appropriate proficiency testing programs is also required.

PROFICIENCY TESTING PROGRAMS

A key component of the AIHA accreditation program is the requirement for laboratories to participate in appropriate proficiency testing (PT) programs to demonstrate their ability to provide acceptable results. The Proficiency Analytical Testing (PAT) program, which provides PT samples for fiber counting, analysis of metals from filters, analysis of silica from filters, and analysis of solvents from sorbent tubes and/or passive monitors,[13] was begun in 1972 by

---

[12] This number is for the Industrial Hygiene Laboratory Accreditation Program (IHLAP). There are also programs for environmental lead (AIHA, 2006a, 131 labs), microbiology (96), and food (2). A related program is the Asbestos Analyst Registry (AAR) (Kelley, 1987; AIHA, 2006b). The AAR was devised to register accredited analysts. There were about 1,200 analysts participating in the Asbestos PAT program in 1993 (Groff and Schlecht, 1993).

[13] In addition, a laboratory proficiency program for blood lead is monitored by the Centers for Disease and Prevention (see Maher et al., 1979; Donadio, 1982; Parsons, 1992).

NIOSH to support government laboratories. It was expanded to include other laboratories through the AIHA accreditation program. In the mid 1980s, AIHA took over the program from NIOSH. PAT samples are prepared by a private contractor and are submitted to laboratories four times each year. Laboratories accredited in the four analytical areas or fields of testing (FoTs) must participate in the PAT program. Non-accredited laboratories may also pay to participate in this, and other, AIHA proficiency programs.

As the AIHA accreditation program has expanded to other analytical areas, the types of PT samples also increased. An Environmental Lead Proficiency Analytical Testing Program (ELPAT) provides quarterly soil, paint, wipe, and air samples for analysis of lead. The Bulk Asbestos Proficiency Testing Program (BAPT) provides quarterly samples of building materials for asbestos analysis using polarized light microscopy. The Beryllium Proficiency Testing Program (BePAT) provides filters for beryllium analysis three times a year. Environmental Microbiological Proficiency Analytical Testing (EMPAT) samples are offered for fungal and bacterial analysis of air, wipe, and bulk samples. For non-culturable (direct) analysis of air samples, AIHA offers the EMPAT Fungal Direct Exam PT program that uses digital images of fungal spores. All EMPAT PT samples are submitted to laboratories three times a year.

AIHA accreditation policies now require laboratories to participate in some type of PT program for every (FoT) for which accreditation is sought. For areas where AIHA does not offer a PT program, laboratories have several options. In some cases, they may purchase PT samples from another provider, e.g., the British Workplace Analysis Scheme for Proficiency Testing (WASP) program offers a variety of PT samples, including formaldehyde by liquid chromatography analysis and solvents by thermal desorption and GC/MS analysis. In other cases, a group of laboratories performing similar analyses may develop a focused round robin program. AIHA guidelines have been established for this type of program for pharmaceutical laboratories and laboratories analyzing breathing air samples. The third option for laboratories is establishment of their own, internal, blind PT program for FoTs for which no outside samples are available.

Following the completion of each round of an AIHA or other provider PT program, participating laboratories receive a summary of their performance as it relates to that of other participating laboratories (see NIOSH, 1990 for details). Each PT program has its own set of criteria for determining the acceptable *performance limits* for each analyte, or the range of values for a particular PT sample that is considered to be acceptable or passing. Target concentrations for each PT sample and the performance of a group of reference labs for that sample are two of the pieces of information used to establish performance limits. The reference laboratories have had no outliers in the previous round for any analyte. The reference value is the mean for these laboratories; for metals and solvents, performance limits are usually ± 3 standard deviations of the mean. An outlier falls outside the performance limits. Typical results are shown in Table 25-2.[14]

---

[14] For reviews of the performance of laboratories from a historical perspective, see Abell and Doemeny (1991) and Schlecht and Shulman (1986), and for the last report from NIOSH see Feng and Schlecht (2000).

Table 25-2
IH PAT round 166 statistics from 2006 (generated by the American Industrial Hygiene Association)

Contaminant	#	Reference value[1]	RSD (%)[2]	Lower Limit[3]	Upper Limit	# of Participating Labs	Low outlier[4]	High outlier
Cadmium	1	0.0117	5.7	0.0097	0.0137	197	6	4
(mg)	2	0.0060	5.4	0.0050	0.0070	197	3	6
	3	0.0090	4.9	0.0077	0.0103	197	3	3
	4	0.0029	6.7	0.0023	0.0035	197	2	3
Chromium	1	0.1202	5.6	0.1000	0.1405	196	8	8
(mg)	2	0.0906	4.8	0.0774	0.1038	196	9	8
	3	0.0623	5.3	0.0525	0.0722	196	8	5
	4	0.1513	5.1	0.1282	0.1745	196	7	7
Lead (mg)	1	0.0699	4.5	0.0604	0.0794	199	4	3
	2	0.0397	5.5	0.0331	0.0462	199	5	7
	3	0.1360	5.7	0.1129	0.1591	199	6	3
	4	0.1083	5.3	0.0910	0.1256	199	6	3
Silica (talc)	1	0.1819	15.9	0.0953	0.2686	61	2	0
(mg/m^3)	2	0.0611	17.1	0.0298	0.0923	61	0	2
	3	0.1371	15.8	0.0721	0.2021	61	0	1
	4	0.1113	14.4	0.0633	0.1593	61	3	0
Asbestos	1	191	20.0	93	322	743	13	31
(f/mm^2)	2	238	20.0	116	402	743	10	26
(amosite)	3	342	20.0	167	577	743	17	8
	4	83	20.0	41	140	743	36	23
Benzene	1	0.1509	5.1	0.1280	0.1738	162	5	7
(mg)	2	0.0762	6.8	0.0606	0.0918	162	8	5
	3	0.3912	5.0	0.3327	0.4496	162	6	5
	4	0.2926	4.9	0.2498	0.3355	162	7	5
o-xylene	1	0.6118	6.6	0.4906	0.7329	162	2	10
(mg)	2	1.2002	6.6	0.9641	1.4364	162	4	7
	3	1.4772	9.0	1.0784	1.8761	162	11	2
	4	0.9129	6.5	0.7356	1.0903	162	5	8
Toluene	1	1.1818	4.9	1.0097	1.3539	162	6	10
(mg)	2	1.5659	5.3	1.3158	1.8160	162	10	6
	3	0.6356	6.1	0.5185	0.7526	162	5	7
	4	2.0521	6.2	1.6694	2.4348	162	10	6

[1] Average value of results from a subset of labs who meet certain proficiency standards over time
[2] Relative standard deviation or coefficient of variation over the reference labs
[3] Limits are reference value ± 3 RSD × reference value
[4] Number of results that were below or above the limits

For IHLAP and Environmental Microbiological accredited laboratories to maintain their accreditation status for a given FoT, they must be rated proficient in two of the three most recent PT rounds. For example, a laboratory participating in metals will report 12 results for each round. To be proficient for that round, the laboratory must have results inside the performance limits for at least 9 of the 12 results. For Environmental Lead accredited laboratories there must be no more than 25% outliers in the four most recent rounds or no outliers in the two most recent rounds.

It is important when selecting a laboratory to check its proficiency status for the FoTs of interest. Table 25-3 shows PAT results for the latter half of 2005 and the first half of 2006. While all laboratories participating are not accredited, it is still obvious that even accredited labs will have outliers. Once a hygienist has selected a laboratory, it is important to continue to monitor the lab's performance. Some laboratories post this information on their websites. Others will need to be contacted on a regular basis. As an additional test of a lab's performance, the hygienist can submit spiked samples that are blind to the laboratory. When doing this, it is important to submit multiple samples and to devise the criteria that will be used to determine acceptable performance before the samples are sent.

EQUIVALENT METHODS

Earlier, it was mentioned that OSHA allows the use of equivalent methods in some of its comprehensive standards. That raises the question, what is equivalency? This question actually does not have an easy answer. If a method, or a new device as a part of a method, meets some accuracy criterion should it not be acceptable? Or are we to admit that methods (including the devices that are part of them) have flaws or biases and regardless of how well the new method works, it should perform like the old method. Interestingly, in some cases this may mean *no better than the old method*. For example, OSHA typically states that, for a sample size of no less that 100 side by side measurements, 90% of measurements made with a new method or device (such as a cyclone) should be within ± 25% of the standard method or device.

Perhaps OSHA does in fact mean that a hygienist can take 100 side by side measurements and if 90 or more of the results for the new method are within 25% of the old paired result, the new method is acceptable. However, what if the hygienist did this experiment one week and found 92 satisfactory new method results, but the following week repeated the experiment and got only 82 satisfactory new results, and the following week, etc. In other words, this is actually a statistical sampling problem if it is treated correctly. Krishnamoorthy and Mathew (2002) show that the actual hypothesis to be tested is that the mean of differences in the paired samples ± a factor $k \times$ the standard deviation of those differences $\leq$12.5 %, for the specific OSHA problem. They have derived exact values for k that approach Z scores as n approaches infinity.

However, there is another issue. The more biased and imprecise the old method is and the more unbiased and precise a new method is, the more likely the new method will fail this test. Thus, a performance standard, such as the ± 25% accuracy criterion that OSHA and NIOSH

Table 25-3
Non-proficient (NP, see text for definition) results for IH PAT Rounds 163 – 166 (2005/2006, generated by the American Industrial Hygiene Association)

Analyte/round	Total number of labs participating[1]	Number of NP labs[2]	% NP
Asbestos			
Round 163	764	109	14.3
Round 164	752	81	10.8
Round 165	747	90	12.0
Round 166	743	77	10.4
Metals			
Round 163	205	7	3.4
Round 164	200	7	3.5
Round 165	199	7	3.5
Round 166	198	6	3.0
Organic solvents			
Round 163	172	11	6.4
Round 164	169	9	5.3
Round 165	167	9	5.4
Round 166	162	9	5.6
Silica			
Round 163	61	2	3.3
Round 164	61	2	3.3
Round 165	62	2	3.2
Round 166	61	3	4.9

[1] For metals and solvents there are usually 2-3 analytes; see Table 25-2. A participating lab is not necessarily accredited as many non-accredited labs participate in PAT.
[2] Indicates that analysis result was outside of the ± 3 RSD control limits; see Table 25-2.

use for their own methods development, would be a fairer standard. There is also an ISO standard on equivalence of measurement procedures (ISO, 2002).

EUROPEAN EFFORTS

As cited throughout this text, the European Union has developed a number of standards relative to hygiene and occupational health (see CEN, 2006). Relative to analytical methods, many, if not all, European countries have developed methods. Thus, the CEN chose to prepare a standard (EN 482:1994) for the performance of analytical methods rather than to prepare methods in the fashion that NIOSH has. Further, it undertook a project more recently to determine how methods used by EU members were performing relative to the standard (Breuer et al., 2006). The study was focused on methods for the 126 substances that are the most frequently analyzed in Europe, with emphasis on German and French databases, and have European indicative occupational exposure limit values (IOELV, see Chapter 11). Methods were categorized into A1, A2, B, and C categories or ratings. Most (95%) were given a suitable rating (A – meets requirements, or B – has potential to do so). A database containing the methods evaluation results is now available (BGIA, 2006).

Earlier a proficiency testing program was mentioned that is being used by some U.S. laboratories where they are doing tests that are not covered in AIHA PT programs, e.g., thermal desorption of organics. This program, Workplace Analysis Scheme for Proficiency (WASP) is organized by the Health and Safety Executive of the UK, however, more than half the participating laboratories are outside the UK. An analysis was presented of PT data from this program for cadmium, lead, and chromium (Stacey, 2006). The analysis showed the classic increasing relationship of standard deviation between labs versus average result, that inductively coupled plasma-atomic emission spectroscopy gave better between laboratory standard deviations than flame atomic absorption spectroscopy, especially for chromium, that performance has improved in the last 11 years, and that most labs are producing results that fall within ± 8.3% at the 95% confidence level.

QUESTIONS

25.1. Discuss the rationale for setting upper and lower limits on flow rates through charcoal tubes. What factors control these? What limits would you recommend?

25.2. What limits on flow rates through filters would you recommend?

25.3. The LOQ for a method is 0.005 mg cadmium. If exposures are thought to occur at the action level (0.005 mg/m^3), what minimum air volume is necessary in order to quantitate these exposures? What minimum flow rate should be used if sampling will occur over 7 hours?

25.4. If exposures are thought to occur at 0.1 f/cm^3, what is the minimum volume necessary to just detect these exposures? (Hint: see NIOSH fiber method.)

25.5. The OEL for toluene is 50 ppm. If a 24 L volume (0.05 L/min × 480 min) is collected at the OEL, how much contamination would the mean laboratory blank need to have in order to represent 1% of the sample mass?

25.6. For a fiber glass sample, 10 fibers/100 fields are counted. The microscope field area is 0.008 mm^2? Does the fiber count exceed the LOD? (See NIOSH 7400 method.)

25.7. Assuming that the performance of reference laboratories adequately represents that of all laboratories, what is the probability that a laboratory will have an outlier by chance alone on a given result? For simplicity, assume the results are independent (Labs are likely to produce dependent results, i.e., if a lab produces an outlier in its first analysis, it is more likely to produce another outlier than a lab that does not produce an outlier on the first analysis). What is the probability that a laboratory will be non-proficient and risk losing accreditation for a given analyte? This requires that the laboratory have at least 1/4 outlier results over 2 of 3 consecutive rounds. For example, for silica say the outlier results are 0/4, 2/4, 2/4. For metals the results might be 5/12, 0/12, 5/12. This is a binomial problem. The equation for the binomial is

$$p(y) = \frac{n!}{y!\ (n-y)!} p^y q^{n-y}$$

where:

n = the number of trials

y = the number of successes

$n-y$ = the number of failures

p = the probability of success

q = the probability of failure.

REFERENCES

Abell, M.T.; Doemeny, L.J.: Monitoring the performance of occupational health laboratories. Am. Ind. Hyg. Assoc. J. 52:336-339 (1991).

Abell, M.T.; Kennedy, E.R.: A computer program to promote understanding of the monitoring method evaluation guidelines used at NIOSH. Am. Ind. Hyg. Assoc. J. 58:236-241 (1997).

AIHA: Quality Assurance Manual for Industrial Hygiene Chemistry. American Industrial Hygiene Association, Fairfax, VA (1988).

AIHA: Asbestos Analysis Registry. American Industrial Hygiene Association, Fairfax, VA http://www.aiha.org/Content/LQAP/AAR/aar.htm (2005).

AIHA: AIHA Accredited Laboratory Listings. American Industrial Hygiene Association, Fairfax, VA http://www.aiha.org/Content/LQAP/accred/AccreditedLabs.htm (2006a).

AIHA: Laboratory Accreditation Policy Modules. American Industrial Hygiene Association, Fairfax, VA http://www.aiha.org/Content/LQAP/documents/accredpolicymods.htm (2006b).

Anonymous: Guidelines for data acquisition and data quality evaluation in environmental chemistry. Anal. Chem. 52:2242 (1980).

Ashley, K.; Harper, M.: ASTM international standards for monitoring chemical hazards in workplaces. J. Occup. Env. Hyg. 2: D44-D47 (2005).

ASTM: Committee D22.04 on Workplace Atmospheres. ASTM International, West Conshohocken, PA http://www.astm.org/cgi-bin/SoftCart.exe/COMMIT/SUBCOMMIT/D2204.htm?L+mystore+xoqd6962+1156434490 (2006).

Bartley, D.L.: Reconciling traditional accuracy assessment with the ISO guide to the expression of uncertainty in measurement (IOS/GUM). J. Occup. Env. Hyg. 1:D37-D41 (2004).

BGIA: GESTIS Analytical methods. Berufsgenossenschaftliches Institut für Arbeitsschutz, Sankt Augustin, Germany http://www.hvbg.de/e/bia/gestis/analytical_methods/index.html (2006).

Botnick, E.; White, K.T.: Discussion of terms used by laboratories when speaking about detection, quantification and reporting limits. The Synergist, Oct:15-16 (1998).

Breuer, D.; Quintana, M.J.; Howe, A.: Results of the EU project entitled "Analytical methods for chemical agents" for the evaluation of methods for analysis of hazardous substances in workplace air. J. Occ. Env. Hlth. 3:D126-D136 (2006).

Brown, R.H.: New European legislation and its relation to methods performance criteria and standardization. Appl. Occup. Env. Hyg. 15(1): 100-105 (2000).

Burkart, J.A.; Eggenberger, M.; Nelson, J.H.; Nicholson, P.R.: A practical statistical quality control scheme for the industrial hygiene chemistry laboratory. Am. Ind. Hyg. Assoc. J. 45:386-392 (1984).

Bush, K.A.; Taylor, D.G.: Statistical protocol for the NIOSH validation tests, In Chemical Hazards in the Workplace, G. Choudhary (ed.), ACS Symposium Series 149. American Chemical Society, Washington, DC (1981).

CEN: Occupational Health and Safety Standards. European Committee for Standardization, http://www.cenorm.be/cenorm/businessdomains/businessdomains/healthandsafety/occupational+health+and+safety/workprogrammeohs.asp (2006).

Donadio, R.E.: OSHA criteria for laboratory proficiency in blood lead analysis. Pub. Hlth. Rep. 97:175-176 (1982).

Eller, P.M.; Feng, H.A.; Song, R.S.; Key-Schwartz, R.J.; Esche, C.A.; Groff, J.H.: Proficiency analytical testing (PAT) silica variablility, 1990-1998. Am. Ind. Hyg. Assoc. J. 60.533-539 (1999).

EPA: Environmental Test Methods and Guidelines. U.S. Environmental Protection Agency, Washington DC http://www.epa.gov/Standards.html (2003).

EPA: Environmental Technology Verification Program. U.S. Environmental Protection Agency, Washington DC http://www.epa.gov/etv/index.html (2006).

Feng, H.A.; Schlecht, P.C.: Proficiency Analytical Testing (PAT) Program (August 1999). Am. Ind. Hyg. Assoc. J. 61: 144 (2000).

Fischbach, T.; Song, R.; Shulman, S.: Some statistical procedures for analytical method accuracy test and estimation. Am. Ind. Hyg. Assoc. J. 57: 440-451 (1996).

Groff, J.H.; Schlecht, P.C.: Proficiency Analytical Testing (PAT) program. Appl. Occup. Env. Hyg. 8:837-839 (1993).

HSE: Methods for the Determination of Hazardous Substances (MDHS) guidance, Health and Safety Executive, London, UK, SE1 9HS http://www.hse.gov.uk/pubns/mdhs/index.htm (2006).

ISO: Air quality – Evaluation of the suitability of a measurement procedure by comparison with a required measurement uncertainty ISO 14956:2002. International Organization for Standardization, Geneva, Switzerland (2002).

ISO: General requirements for competence of testing and calibration standards, ISO 17025:2005, 2nd ed. International Organization for Standardization, Geneva, Switzerland (2005).

ISO: Standards: 13 Environment. Health protection. Safety. International Organization for Standardization, Geneva, Switzerland http://www.iso.ch/iso/en/CatalogueListPage.CatalogueList?ICS1=13&ICS2=&ICS3=&scopelist= (2006).

Kelley, W.D.: Asbestos Analysis Registry. Appl. Ind. Hyg. 2:R8-R9 (1987).

Kennedy, E.R.; Fischbach, T.J.; Song, R.; Eller, P.M.; Shulman, S.A.: Guidelines for Air Sampling and Analytical Method Development and Evaluation, DHHS (NIOSH) Publication No. 95-117, National Institute for Occupational Safety and Health, Cincinnati, Ohio (1995).

Kring, E.V.; Ansul, G.R.; Henry, T.J.; Morello, J.A.; Dixon, S.W.; Vasta, J.F.; Hemingway, R.E.: Evaluation of the standard NIOSH type charcoal tube sampling method for organic vapors in air. Am. Ind. Hyg. Assoc. J. 45:250-259, (1984).

Krishnamoorthy, K.; Mathew, T.: Statistical methods for establishing equivalency of a sampling device to the OSHA standard. Am. Ind. Hyg. Assoc. J. 63: 567-571 (2002).

Maher, C.C.; Roettgers, D.M.; Conlon, H.J.: Interlaboratory comparison of blood lead determinations. Am. Ind. Hyg. Assoc. J. 40:230-237 (1979).

NIOSH: NIOSH Manual of Analytical Methods, 2nd ed. National Institute for Occupational Safety and Health, Cincinnati, OH (1977).

NIOSH: NIOSH Manual of Analytical Methods, 3rd ed. National Institute for Occupational Safety and Health, Cincinnati, OH (1984).

NIOSH: Laboratory Reports and Rating Criteria for the Proficiency Analytical Testing (PAT) Program. National Institute for Occupational Safety and Health, Cincinnati, OH (1990).

NIOSH: NIOSH Manual of Analytical Methods (NMAM®), 4th ed., P.C. Schlecht, P.F. O'Connor, eds., DHHS (NIOSH) Publication 94-113, 1st Supplement Publication 96-135, 2nd Supplement Publication 98-119, 3rd Supplement 2003-154. Cincinnati, OH http://www.cdc.gov/niosh/nmam/ (1994).

NIOSH: Analytical Method Evaluation Software. National Institute for Occupational Safety and Health, Cincinnati, OH http://www.cdc.gov/niosh/nmam/sampanme.html (1996).

NIOSH: McCammon, C.S.; Woebkenberg, M.L.: General considerations for sampling airborne contaminants, In NIOSH Manual of Analytical Methods (NMAM). National Institute for Occupational Safety and Health, Cincinnati, OH http://www.cdc.gov/niosh/nmam/pdfs/chapter-d.pdf (1998).

OSHA: OSHA Sampling and Analytical Methods. Occupational Safety & Health Administration, Washington, DC http://www.osha.gov/dts/sltc/methods/index.html (2006).

Parsons, P.J.: Monitoring human exposure to lead: An assessment of current laboratory performance for the detection of blood lead. Env. Res. 57:149-162 (1992).

Schlecht, P.C.; Cassinelli, M.E.: Survey of analytical methods, parts I and II. Synergist 26-27, 28-30 (1997).

Schlecht, P.C.; Shulman, S.A.: Performance of asbestos fiber counting laboratories in the NIOSH Proficiency Analytical Testing (PAT) program. Am. Ind. Hyg. Assoc. J. 47:259-269 (1986).

Stacey, P.R.: The performance of laboratories analyzing heavy metals in the workplace analysis scheme for proficiency (WASP). Ann. Occ. Hyg. 50:417-425 (2006).

Stacey, P.; Revell, G.; Tylee, B.: Accuracy and repeatability of weighting for occupational hygiene measurements: Results from an inter-laboratory comparison. Ann. Occup. Hyg. 46(8):693-699 (2002).

Winefordner, J.P.; Long, G.L.: Limit of deduction: A closer look at the IUPAC definition. Anal. Chem. 55:712A-724A (1983).

VI

REAL TIME ANALYSIS

26

AN OVERVIEW OF DIRECT-READING INSTRUMENTATION[1]

HISTORY

Perhaps the first direct-reading instrument was the canary, as caged canaries were taken into mines to serve as indicators of methane or toxic gas exposure. Later miner's lamps (the Davy lamp) were used; these devices were approved by the U.S. Bureau of Mines in 1924 (Patty, 1958). Changes in the color of the flame or its intensity was an indication that the methane level and/or the lack of oxygen was hazardous. Hot-wire devices were the first electronic devices to be available (AIHA, 1980), and these devices are still used in direct-reading instruments. Ness (1991) reports, without reference, that refinements to this instrument (the addition of a catalyst to the wire) were made by 1923, but that practical application of this feature was not available until 1965.

Patty (1958) describes only five direct-reading devices. Two of these, the catalytic hot-wire device mentioned above (used then and now to detect flammable vapors and gases) and the ultraviolet light detector (used then and now for Hg, trichloroethylene and other vapors/gases, O_3), are still used. The others, the interferometer (changes in refraction due to gases and vapors), the Orsat (CO, CO_2, O_2, N_2, CH_4), and the hopcalite (CO) monitor, are not. The interferometer measured the change in refractivity of light of a particular wavelength passing through clean air versus contaminated air. It was widely used. However, in later years as exposure levels began to decrease, it became obsolete because of its lack of sensitivity. For example, Patty reported that a change in refractivity of 3×10^{-6} was equivalent to about 1,800 ppm benzene. Elkins (1959) reported that vapors with a molecular weight of about 100 could be detected at 100 ppm.

---

[1] This chapter focuses on gases and vapors. While many of the principles and definitions are also appropriate for aerosols, the reader should see Chapter 28 for specific information on DRI for aerosols.

The Orsat instrument also was of low sensitivity. It was used to measure CO, CO_2, oxygen, methane, and some other combustibles. For example, detectable concentrations of CO were at levels immediately dangerous to life, and accuracy was ± 0.2%. The hopcalite CO monitor was better in these regards but was subject to several interferences. The principle is that hopcalite (a mixture of copper and manganese oxides; Elkins, 1959) catalyzes the oxidation of CO to CO_2. This releases heat which is detected and converted to an electronic signal. The presence of hydrogen, methane, or NO_2 caused a positive interference, as they too can be oxidized. Measurements as low as 20 to 30 ppm could be made, but were subject to effects of temperature and humidity unless special precautions were taken.

Elkins (1959) describes two other direct-reading devices, which are still used. The halide meter (apparently still produced; ACGIH®, 2001) measures the presence of halogenated compounds. The air passes an arc between copper electrodes, and any halogens burn green. The intensity of the green light is measured. The device will not differentiate among various halogen compounds. The second device is the colorimetric indicator tube, still widely used. Elkins refers to these colorimetric devices as *test paper methods*; many were homemade devices consisting of a chemical-impregnated paper that changed color on exposure to a chemical or a group of chemicals. They are more sophisticated and quantitative now, and are described in Chapter 27 as detector tubes.

Cook and Coleman (1936) described the thermal conductivity meter. The detector still is used in a number of direct-reading instruments, but also is often used as a detector for GC. Its principle is described in Chapter 19. Finally, Verma and Tombe (1999) give an excellent summary of the history of use of sampling and analytical instruments for benzene.

CURRENT TRENDS

Operating Principles

Today hundreds of direct-reading instruments are on the market, utilizing a wide and ever expanding array of technologies. Table 26-1 shows a compilation from a UK database (Gentry, 1993) of 1,178 instruments that no doubt has expanded. The instrument operating principles are detailed in the next three chapters and calibration techniques are explained in the last chapter. The listing in Table 26-1 appears a bit more overwhelming than it actually is. In reality, many of the classes in that table are subclasses. In addition, many of the operating principles are similar if not identical to those of laboratory-scale instruments described in Chapter 19. Thus, a more useful categorization might appear as in Table 26-2.

Table 26-2 also describes the general applications of the instruments. All of the direct-reading instruments are quantitative to some degree; they enumerate something. The question is, what? Most instruments are designed or used to be selective for a single substance; so it is assumed that the quantitation is for that substance. However, there can be interferences, or substances that cause an unwanted response. Thus, when the interferences are numerous, the term

Table 26-1
Operating principle of gas detection instruments (adapted from Gentry, 1993)

Total Number of Instruments = 1178 Total Number of Manufacturers = 240	
Electrochemical	480
Catalytic	246
Infrared	144
Semiconductor	127
Gas Chromatography & Ionization	80
Thermal Conductivity	69
Spectroscopy (Visible)	53
Paper Tape	36
Paramagnetic	31
Chemiluminescence	29
Ultraviolet	24
Adsorption for Later Analysis	17
Detector Tube	12
Colorimetry	11
Fluorescence	11
Mass Spectrometry	5
Atomic Absorption	3
Total	1378[1]

[1] Discrepancy occurs because some instruments use two or more principles.

semi-selective is applied to the instrument. For other single substance instruments, the number of interferences may be small, but the probability of an interference appearing in the same environment with the analyte is high. These instruments are also termed semi-selective.

Fewer instruments are designed to sense multiple substances simultaneously. Most of these do not differentiate; they quantitate all the substances together often with variable sensitivity. They are termed general survey instruments. These instruments respond to a wide range of compounds. Many of them respond to all types of hydrocarbons. If the hygienist knows that only one compound exists in the atmosphere in significant quantities, the instrument is specific for that compound by default. Mostly these instruments allow us to find *hot spots*, or to look for high, explosive concentrations of hydrocarbons.

Finally, there are two instrument types that sense multiple substances simultaneously and individually quantitate those substances from the same air sample. These are the gas chromatograph (especially GC with mass spectrometer detector) and the Fourier transform infrared

Table 26-2
Operating principles of direct-reading instruments for gases and vapors

Method	Applications[1]	Substances
ELECTROCHEMICAL		
Conductivity	SS, S-sel to Sel	CO, CO_2, $COCl_2$, COS, CS_2, CH_4, NH_3, HCl, SO_2, and others
MOS	SS, MS, S-sel to Sel	Hydrocarbons, CO, H_2S, refrigerants, NH_3
Gold film	SS, Sel	H_2S, Hg
Coulometry	SS, Sel	O_2
Potentiometry and voltammetry	SS, S-sel to Sel	Aldehydes, Amines, AsH_3, B_2H_6 Br_2, CH_2O, CH_4, Cl_2, ClO_2, CO, CO_2, $COCl_2$, COS, CS_2, ETO, Fl_2, GeH_4, H_2, H_2O_2, H_2S, HBr, HCl, HCN, HF, Hg, HNNH, NH_3, NOx, O_2, O_3, PH_3, Propylene oxide, SiH_4, SO_2
Ionization: FID/PID	GS	Organic vapors
SPECTROCHEMICAL		
IR	SS, S-sel to Sel	Organic compounds, CO, CO_2, many others
FTIR	MS, Sel	Organic compounds, CO, CO_2, many others
Ultraviolet/Visible	SS, S-sel to Sel	C_2H_2, Acrylonitrile, Amines, NH_3, Ar, AsH_3, Br_2, CO_2, Cl_2, ClO_2, CGM, Diborane, Fl_2, CH_2O, Freon, Glutaraldehyde, He, HNNH, H_2, HBr, HCl, HCN, HF, H_2S, I_2, Mercaptans, Ne, Nicotine, NOx, N_2O, O_3, $COCl_2$, PH_3, SiO_4, SO_2, TDI
Chemiluminescence	SS, Sel	NOx, O_3
Colorimetry/photometry		
Flame photometry	SS, S-sel to Sel	SO_2, H_2S, total S and P
UV Fluorescence	SS, Sel	Isocyanates, hydrazines COCl, SO_2, acid gases
Paper tape	SS, Sel	See Chapter 27, Table 27-5
Detector tubes	SS or GS, S-sel	Many compounds
THERMOCHEMICAL		
Thermal conductivity	GS	leak detection for CO_2, Freons, H_2, He, SF_6
Hot-wire/catalytic/MOS	GS	Combustible compounds
GC (many detectors)	MS, Sel	Most, depends on detector

[1] Single (SS) or multiple (MS) substances near simultaneously or in one instrument
Sel = selective, interferences are relatively minor in most applications or are known and controlled
S-Sel = Semi-selective – many potential interferences or some interferences likely to be present in same atmosphere
GS = General Survey – non-selective

spectrometer. These were originally large instruments, confined to the laboratory. Now they are smaller and portable. It is likely that GCs soon will be available that can be worn by the worker (see Chapter 27).

Selection Strategies

With so many instruments on the market, the hygienist needs all the help available and more for selecting instruments. The list from Table 26-1 was prepared by the British government. Such a database is not available in the United States. A publication by ACGIH® (2001) has compiled some information on many instruments. A good listing also was compiled by Stetter (1984), but it is now out of date. Grafstein and Levine (1993) described an expert system that could be used to select instrumentation.

With so many instruments and manufacturers, performance standards are needed. If nothing else, the manufacturers should report basic information in a standard format so that the hygienist can make easier comparisons. These standards are taking shape in Europe (see, e.g., CEN, 2006, and the discussion in Gentry, 1993), but not so in the United States.[2] The AIHA Committee on Gas and Vapor Detection Systems (AIHA, 1992) published a document that defines terms and gives specification requirements, but it is now out of print. The ASTM has also published a standard practice for use of DRI with definitions. Many of the definitions from these two documents are used in this chapter. Finally, the U.S. EPA (EPA, 2006; Waits et al., 2005) has sponsored the Environmental Technology Verification program since 1995, which tests a variety of technologies including direct-reading instruments. Although many of the instruments are more useful for ambient and stack gas emission monitoring, several have application to hygiene work. These include field portable gas chromatographs, gas chromatographs/mass spectrometers, x-ray fluorescence analyzers, and infrared monitors (optical open-path); ammonia sensors, personal impactor samplers, ion mobility spectrometers, ion selective electrodes, mercury continuous emission monitors, NO/NO_x portable analyzers, surface acoustic wave detectors, and portable multigas emissions analyzers.

DEFINITIONS

Direct-reading instruments (DRI) come in an array of sizes and shapes. Some are personal-size; they can be worn by a worker, although a standard definition for the size and the weight that constitutes *personal* has not been developed. Other DRI are portable; they can be carried around in the field by the hygienist but not worn by the worker. Portable instruments

---

[2] In the United States there was a NIOSH-funded program for the evaluation of many types of direct-reading instruments (see McCammon, 1976). From this program a number of reports were issued, but many are now out of date. There are also a few reports issued by AIHA (see, e.g., AIHA, 1993) covering a select group of instruments only. As a result, there is little to guide the hygienist in making informed decisions about instrument selection and purchase.

must also be rugged. Transportable instruments are those that are designed for a fixed location but can be moved from one location to another fairly easily. They also typically must be more rugged than a laboratory instrument. Finally, a fixed instrument may be any of the three. It is just an instrument that has a permanent location. Often acute toxicants are monitored with fixed instruments with alarms. The alarm will sound if a preset, potentially hazardous concentration is reached. Also, personal and portable instruments are battery-operated as defined here, but transportable instruments are not.

DRI operate on many different principles. Most are electronic; detector tubes are not. The devices all have one thing in common: the result is given in the field. The instrument does the sampling and the analysis. This is the meaning of *direct-reading*.

Another term that is often used is *real time*. Most DRI are real time; that is, they give a continuous readout of the concentration or some other signal that is reflective of change in concentration. In other DRI, the output is not considered real time. For example, a portable GC can give results no faster than the elution time for the analyte (see Chapter 19, Figure 19-2); the result obtained is a grab sample. Other DRI are designed to give only one reading, that being an integrated result. For example, detector tubes provide only one result at the end of a sampling period that may be as short as a minute or as long as 8 hours. Even real time instruments do not perfectly reflect concentration changes over time, as all sensors have a response time.

Response Time

The easiest way to explain response time is by examining Figure 26-1. This fictitious tracing of the output of a DRI shows how the DRI is challenged. During the first portion of the trace, the challenge to the DRI is constant, about 10 mV; and then a new challenge (higher concentration) begins (start). For some period of time, the DRI does not respond, although this is not readily discernable from the graph; this is the *lag time*. Then the DRI begins to respond, but the response is not complete until the end of the *rise time*. The reverse process occurs when the challenge is reduced, only now the rise time becomes the *fall time*; the lag time is still the lag time. The *response time* begins when the challenge changes and lasts until the full response occurs (or it is sometimes defined as 90% of the full response). Thus, it is lag plus rise time or lag plus fall time, about one minute from the graph. When the instrument is returning to a baseline or no challenge, this lag plus fall is also referred to as the *recovery time*.

Noise, Drift, and LOD

There are other important features shown in Figure 26-1 that apply to electronic DRI. First, notice that the signal, even when the challenge does not change, fluctuates over short time periods. This fluctuation is *noise*. In addition, when the instrument is allowed to run with a constant challenge for long periods of time (say, hours), the signal in Figure 26-1 shows a gradual rise. Sometimes the signal shows a gradual drop, either one is known as *drift*. If the

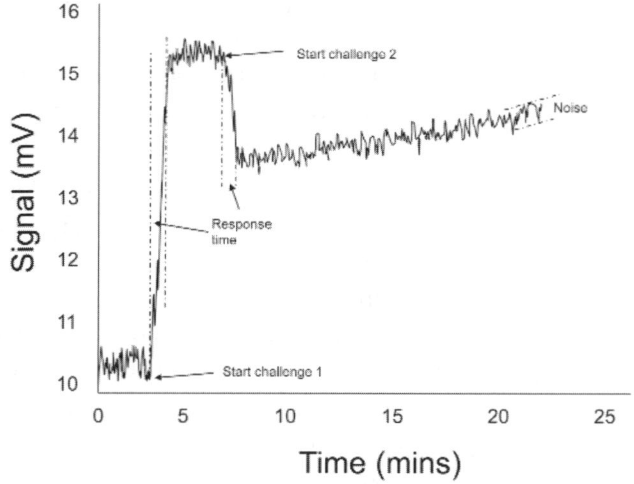

Figure 26-1. A fictitious tracing of a direct-reading instrument signal in response to different challenges versus time.

challenge is clean air, then it is called *zero drift*. If the challenge is a concentration that is at the maximum end of the range of the instrument, it is called *span drift*. Drift is reported as a rate, such as 1% full scale in 24 hours or 1 ppm per 24 hours.

The drift can be very important. If an instrument with an alarm is used to monitor an acute toxicant, then the alarm should be set at an appropriate value. If the instrument drifts upward 5% of full scale each day, and the drift is not corrected, then false alarms will eventually occur. Though this may seem like erring on the safe side, it could cause workers to begin to ignore alarms. Thus, the extent of drift and the means to correct it are important.

Noise is particularly important when a direct-reading instrument is used to make a single reading in time. For example, the meter is examined at a particular time, and the reading is 10 ppm, but because of the noise we do not know the true concentration. It could be higher or lower than the reading. Noise is indirectly proportional to precision. If the concentration that we are trying to determine is quite low, then it may be difficult to determine if the signal is from the environment or from noise. We need to know the *limit of detection*, or the LOD.

In the previous chapter, we defined the LOD for laboratory analytical techniques as three times the standard deviation of blank values (s_b), for a mean of zero. The s_b reflects the noise in the instrumentation as well as other variables, such as the GC injection technique of the analyst. However, DRI often give continuous results, as opposed to discrete results like those of a GC or an AA spectrophotometer. How then do we get discrete measurements from a continuous DRI in order to calculate s_b? We could sample the noise signal at discrete times; that is, just pick values from the noise signal at random.

An alternative technique has developed that is not well defined. In this approach, the noise signal at zero challenge is recorded for some period of time, say, 30 minutes. Parallel lines are drawn as a Figure 26-1, that just touch the maximum and minimum noise values (termed peak to valley or peak). The distance from one line to the other is the noise, and two or three times the noise is defined as the LOD. Whether two or three is used is somewhat arbitrary; but the longer the noise signal is monitored, the smaller the multiplier may reasonably be.

In Figure 26-2, noise outputs for the two instruments are given. For one instrument the peak to valley noise is about 10 mV, for the other, about 2 mV. The figure also shows a signal that we are trying to detect, 10 mV. Although this signal is clearly above both noise tracings, we must realize that the tracings are only for a period of 1 minute. Therefore, we multiply the peak to valley noise by a factor of 2 and obtain LODs of 4 and 20 mV. Thus, only one of the two instruments is analytically sensitive enough to detect the 0.1 ppm challenge concentration.

A rationale for a larger multiplier for the s_b can be given as follows. First, we assume that the noise signal is normally distributed. Now, if we assume that in monitoring the noise signal we have observed at least 99.7% of the noise values, then the peak to valley noise should be equivalent to $6s_b$, because $\pm 3s_b$ accounts for 99.7% of a normal distribution. Doubling or tripling this value (two or three times the noise) seems overcautious, as it would be equivalent to $12s_b$ or $18s_b$. On the other hand, if, in fact, we had observed only 95% of the noise distribution, then the peak to valley noise would account for $4s_b$, and observing 66% of the noise distribution would account for only $2s_b$. Finally, if the noise distribution is not normal, then these percentages could be much lower (see discussion of LOD in Chapter 23). The multiplier that is used must reflect the subjective nature of the user's judgment that the noise distribution has been adequately observed, and that it is normally distributed.

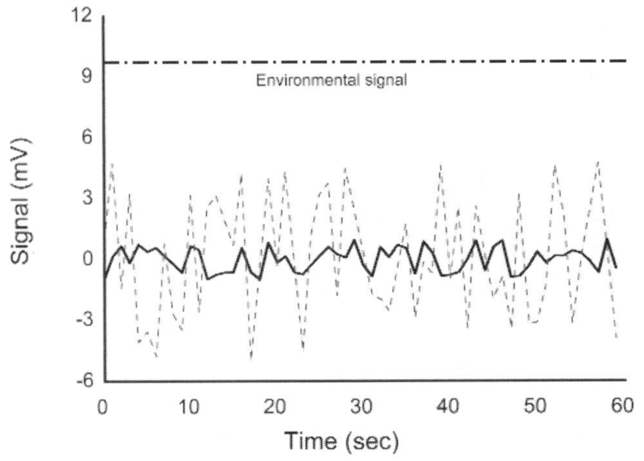

Figure 26-2. The noise or response of two direct-reading instruments to a zero concentration relative to a non-zero environmental signal. The average peak to valley distance for the two signals differs by 5-fold.

Sensitivity

Often one hears that a method or an instrument is sensitive. The implication of this is not always obvious. Does it mean that the instrument can detect small quantities? That it can discriminate between two close quantities? That it gives a large signal when challenged with a small quantity? That its LOD is low? The answer is a bit of all of these. Skoog et al. (2007) states that "Most chemists agree that the sensitivity of an instrument or method measures its ability to discriminate between small differences in analyte concentration." There are actually two aspects to this: noise or reproducibility (which determines LOD) and calibration slope. Because the calibration slope is sometimes called the calibration sensitivity, the overall sensitivity (consideration of slope plus noise) should be differentiated as the *analytical sensitivity*.

The calibration slope is the signal divided by the concentration. So if we calibrate an instrument by challenging it with about five different concentrations, then we can plot each concentration versus the instrument response (say, millivolts) and fit a line through the points. The slope of the line is the calibration sensitivity, in mV/ppm, for example.

Consider two instruments. Both have the same calibration slope, say, 100 mV/ppm, but the LODs differ by a factor of 5, say 5 mV and 25 mV (remember the LOD is a function of noise). For a challenge concentration of 0.1 ppm, the instruments should signal 10 mV (0.1 × 100), but the instrument with the 25 mV LOD does not have the analytical sensitivity to measure the signal.

Conversely, now consider two instruments having the same LODs, say, 10 mV (see Figure 26-3), but different calibration slopes, say, 1 mV/ppm (instrument 2) and 5 mV/ppm (instrument 1). The instruments are placed in an environmental challenge of 2.5 ppm. Figure 26-3 shows that for instrument 2 the signal produced (about 4 mV) is below the LOD. Instrument 1 yields a signal (~13 mV), which is greater than its LOD. Thus, even though the instruments have the same LOD, instrument one has greater calibration sensitivity and, therefore, greater analytical sensitivity.

Sensitivity involves the ability of the instrument to sense small concentrations by yielding a large signal and to have low noise output so that small signals can be differentiated from background.

The analytical sensitivity (y) has been defined, based on this concept, as:

$$\gamma = m/s_s \qquad \text{[Equation 26-1]}$$

where m is the calibration slope, and s_s is the standard deviation of the signal (Skoog et al., 2007). Since s_s varies with the level of the signal, and we are generally most interested in analytical sensitivity in the region of small signals, the term s_b, the standard deviation of the blank, is appropriate. Unfortunately, it is difficult to use this equation when the peak to valley noise is measured rather than S_b.

It is important to note that the LOD is 3 × the standard deviation of the signal blank (i.e., mV). To convert this to concentration, we have

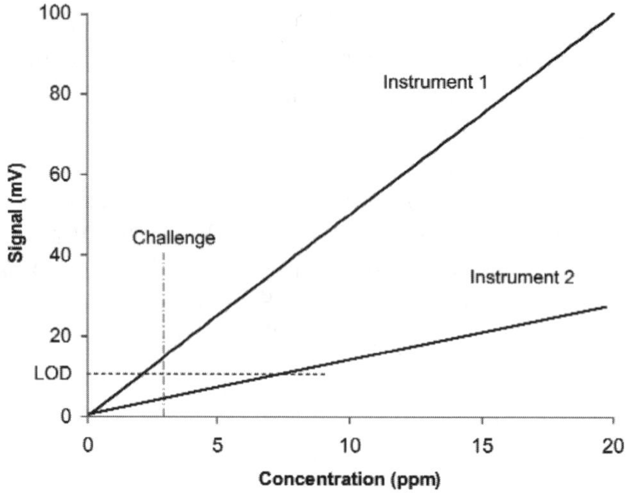

Figure 26-3. Calibration curves for two instruments having the same LOD (10 mV) and challenged by a concentration that instrument 1 can measure as the signal produced is above the LOD, while for instrument 2 the concentration cannot be measured as it produces a signal below the LOD. Note that instrument 1 has a calibration sensitivity 5 × that of instrument 2.

$$LOD_c = \frac{1}{m} \times LOD_s \qquad \text{[Equation 26-2]}$$

Because LOD_s is a function of the standard deviation of the blank (e.g., 2 × or 3 ×), LOD_c is proportional to:

$$\frac{1}{m} \times s_b \qquad \text{[Equation 26-3]}$$

Thus, LOD_c and analytical sensitivity are both functions of the same two variables at low concentrations – the calibration slope or calibration sensitivity and s_b. This means that at low concentrations, LOD_c and analytical sensitivity are proportional.

Range

Direct-reading instruments ideally should respond over a wide range of concentrations. The limitations on this *dynamic range* are the LOD on one end and the sensor saturation on the other. When the sensor is saturated, the instrument is said to *peg*, usually indicating that the needle on the gauge has gone all the way to the right. On instruments that give a numeric (digital) readout, there would appear to be no upper limit on the readout; nonetheless, the sensor can become saturated. At saturation, the instrument will respond no further. If given an even higher challenge concentration, it will continue to read at about the same level. The cause of this depends on the chemistry and the physics of the sensor.

It is important to know the shape of the calibration curve for the dynamic range of the instrument. In Figure 26-3, the two calibration curves are linear, at least for that portion of the dynamic range that is plotted. In some cases, the dynamic range is not linear; or there is not one calibration slope, but an infinite number. In Figure 26-4, a logarithmic calibration curve is shown, as well as a straight line approximating the lower range of the instrument. As can be seen, the straight-line assumption for the lower range does not appear to be in great error. Other straight lines could be drawn for the other portions of the curve. Of course, the more straight lines that are drawn, the better each approximates the exact curve.

Unfortunately, there are no standards that limit the amount of error that can occur with straight-line approximations. Nonetheless, manufacturers report linear ranges. For many types of sensors, the response is essentially linear throughout the dynamic range; but for others, this is not the case, and so the definition of linear range becomes important.

For some sensors it is known that the response is logarithmic rather than linear. This can be corrected in one of at least two ways. First, a non-linear meter can be used. This technique has largely disappeared with digital outputs and electronics. It is like plotting a logarithmic function on logarithm paper to yield a straight line. As can be seen in Figure 26-5, the meter can be difficult to read at lower values. Secondly, the signal can be modified by circuitry in the instrument. This is the best and most common way.

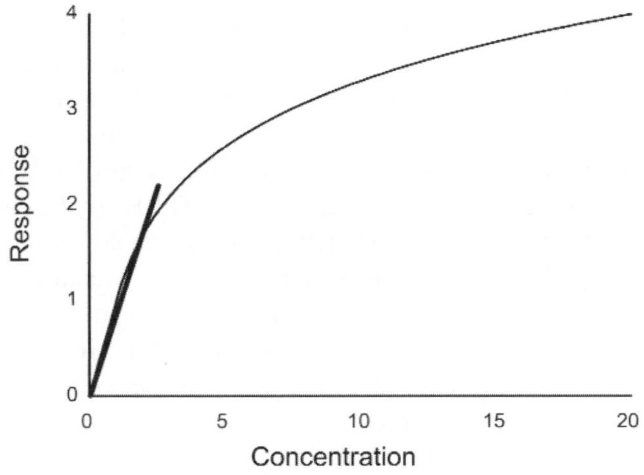

Figure 26-4. A logarithmic calibration curve showing how its lower region can be approximated by a straight line.

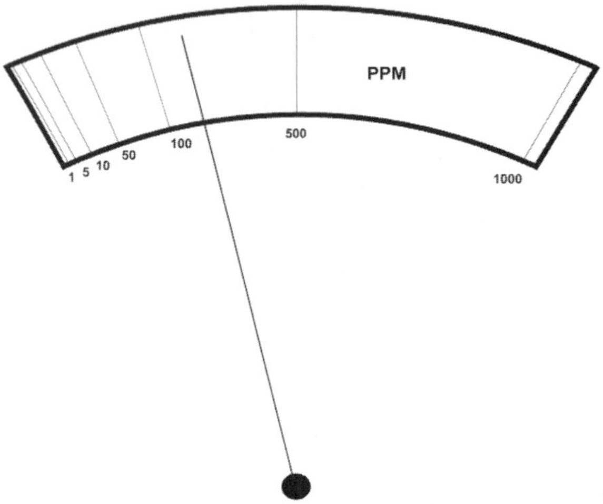

Figure 26-5. The face of a meter with a logarithmic scale.

Selectivity and Specificity

It is doubtful that any sensor responds to a single analyte and only that analyte. If a DRI is designed to measure SO_2, it is likely that it will respond to other gases or vapors. The ability of a DRI to respond to a select number of analytes is its selectivity. Analytes that cause an unwanted response are called interferents, and they can be positive or negative.

There are three important aspects to selectivity. First, are the interferents present in your application? If not, they can be ignored, but hygienists who use DRI in many different settings do not have this luxury. Second, what is the relative response to the interferents? Table 26-3 shows a very well-constructed chart of interferences for various DRI of a well-known manufacturer. The analyzer for H_2S shows that it will respond to SO_3; however, it takes more than 10^4 ppm of SO_3 to cause a 1 ppm response on the meter. Thus, this interferent is, in virtually all cases, a non-interferent, as such concentrations of SO_3 are not likely to exist when H_2S is the gas of interest. On the other hand, ethyl mercaptan (C_2H_2SH) will cause a 1:1 interference; 1 ppm ethyl mercaptan causes a 1 ppm meter H_2S response. Table 26-3 or similar information is a must when purchasing DRI. The third aspect is whether the manufacturer has tested a particular gas as an interferent. There are thousands of gases and vapors, and manufacturers cannot test them all. Often a manufacturer will list a few interferents. To many readers this implies that the list is exhaustive; there are no more. This is the wrong conclusion. It simply means that these are the only interferents tested by the manufacturer. It is always a good idea to ask what interferents have been tested.

In the instrument in the preceding paragraph, the response is generally assumed to be caused by H_2S, but it may not be. The output can be subject to different kinds of error: (1) the

Table 26-3
The approximate concentration (ppm) of interfering gas required to cause a 1 ppm deflection in the chosen analyzer (adapted from Interscan Corporation with permission)[aa].

Analyzer	Interfering Gas																
↓	C_2H_5SH	Cl_2	CO	H_2	HCl	HCN	HNNH[a]	H_2S	NH^3	NO	N_2O	NO^2	O_3	SHC_b	SO_2	SO_3	UHC[c]
CO	$>10^3$	$>10^{3d}$	—	8^e	$>10^3$	$>10^3$	$>10^3$	$>10^3$	$>10^3$	$>10^3$	$>10^{4d}$	$>10^{3d}$	$>10^3$	$>10^4$	$>10^3$	$>10^4$	17
Cl_2	3^d	—	400^d	6000^d	4^d	1.5^d	0.5^d	1^d	14^d	150	$>10^4$	1	1	$>10^4$	1^d	$>10^4$	>500
ClO_2	9^d	3	1200^d	$>10^{4d}$	$12^{d, e}$	4.5^d	1.5^d	3^d	42^d	450	$>10^4$	3	3	$>10^4$	3^d	$>10^4$	>1500
H_2	$>10^3$	$>10^{3d}$	0.1	—	$>10^3$	$>10^3$	$>10^3$	$>10^3$	$>10^3$	$>10^3$	$>10^4$	$>10^{3d}$	$>10^{3d}$	$>10^4$	$>10^3$	$>10^4$	3
H_2S[f]	3	11^d	40	10	11	10^e	4	—	220	4	$>10^4$	65^d	15^d	$>10^4$	4	$>10^4$	15
H_2S[g]	1	50^d	8000	5000	15	15	6	—	300	15	$>10^4$	60^d	20^d	$>10^4$	6	$>10^4$	>500
NO	$>10^3$	$>10^3$	$>10^3$	$>10^4$	$>10^3$	$>10^3$	$>10^3$	$>10^3$	$>10^3$	—	$>10^4$	$>10^3$	$>10^3$	$>10^4$	$>10^3$	$>10^4$	>500
NO_2	3^d	0.7	350^d	6000^d	$4^{d, e}$	2^d	0.7^d	0.7^d	15^d	150	$>10^4$	—	1	$>10^4$	1^d	$>10^4$	>500
O_3	3^d	1	400^d	6000^d	4^d	2^d	0.5^d	1^d	15^d	150	$>10^4$	1	—	$>10^4$	1^d	$>10^4$	>500
SO_2	1	3^d	700	5000^d	2	2	1	0.3	68	6^e	$>10^4$	24^d	2^d	$>10^4$	—	$>10^4$	>500

[aa] In many cases, specificity can be improved; response values are not absolute, and may vary depending on sensor formulation.
[a] Hydrazine
[b] Saturated hydrocarbons
[c] Unsaturated hydrocarbons
[d] Negative interference
[e] Rejection ratio can be improved electronically
[f] Data shown for H_2S models with ranges higher than 0-1999 ppb
[g] Data shown for H_2S models with range of 0-1999 ppb

identified analyte is not present – false positive; (2) the analyte present is not identified – false negative. Both errors occur at once when an analyte is identified incorrectly. These errors are related to specificity. An instrument has high specificity if, of the unknowns that are identified, the percentage of false positives and false negatives is low.

Accuracy, Precision and Bias

Just as laboratory-based sampling and analysis methods can be assessed for accuracy, bias, and precision, so can DRI. The same principles outlined in Chapters 7 and 25 apply. Many manufacturers report accuracy for a DRI, e.g., ±2%. However, if the components of accuracy are not defined then we may be comparing apples to oranges. An individual instrument may be calibrated and readings of known concentrations made with the instrument on the day of calibration. From those readings, the bias and precision at one or more concentrations can be determined. At multiple concentrations, the bias and precision can be pooled as is done for laboratory-based analytical methods evaluation.

However, DRI are not always calibrated every day, and so a DRI might be calibrated once monthly or weekly and used every day. Hence, we could also measure the response of the DRI over the calibration period and over multiple calibration periods to known concentrations. Finally, we could measure the accuracy of several DRI. Our accuracy measurement might then include components due to concentration, day of calibration period, calibration period, and instrument replicate. There would be uncertainty with each of these variance components and a statistical model can be developed to estimate the overall accuracy based on measurements taken with respect to each of the above factors.

Shulman and Smith (2002) performed such an analysis for a CO monitor and determined the upper 95% confidence interval for accuracy at 3 concentrations. The values were ±12, 20, and 27% for 10, 30, and 50 ppm. Thus, at 100 ppm, the data indicate that we are 95% confident that 95% of readings of 100 ppm will be within ± 12.5 ppm of 10 ppm. This implies that we cannot reliably detect 10 ppm from background. These values are considerably larger than the values claimed by the manufacturer (± 3% or 2 ppm, whichever is greater). The latter are probably based on an accuracy determination for a single instrument (or perhaps an average of several) on the same day as calibration. The authors provide a spreadsheet with formulas for making these calculations. They conclude that calibration should take place often as opposed to once per month as in the study, and that accuracy estimates will be tighter as more calibration periods are added to the analysis. Three were used in the study.

READING AND RECORDING OUTPUT

Analog Versus Digital

The best way to distinguish these two output modes is as continuous and discrete. An analog device gives a voltage signal that is continuous; the output is typically directed to a meter with a needle or to a recording device such as a chart recorder that gives a continuous tracing. With the advent of computers, analog output has become less common although it is still used. This change has occurred because digital computers process discrete bits of information; most instruments now contain digital microprocessors, so there has been a shift from *meters* to digital displays.

The microprocessor (computer) converts the analog information to discrete pieces of information in a process called analog to digital conversion. Generally, an instrument that has a digital display will have an output port for a digital signal. Analog output (at the back of some newer and most older instruments) may be converted to digital with an external A to D converter. These devices are often used on laboratory instruments to record the output of an analog device with a digital computer.

Chart Recorders

As late as 1995 chart recorders were the primary means of recording the output from a DRI. A chart recorder (see Figure 26-6) takes the analog output of an instrument and converts it to movement of a pen or other transcribing device. Chart recorder output is shown in Figure 26-7. The width of the chart paper corresponds to a maximum voltage that is controlled on the chart recorder. For example, if the recorder voltage range is set to 0 to 500 mV, then the recorder pen should be at the bottom of the paper when a signal of 0 mV is sent to the recorder and move to the top of the paper at 500 mV.

Recorders have advantages and disadvantages. The user is able to see and record the results as they happen; but unless the settings such as chart speed and range are recorded, tracings are difficult to interpret later. Moreover, integrating areas and compiling statistics can be tedious. Computers can eliminate many of these problems.

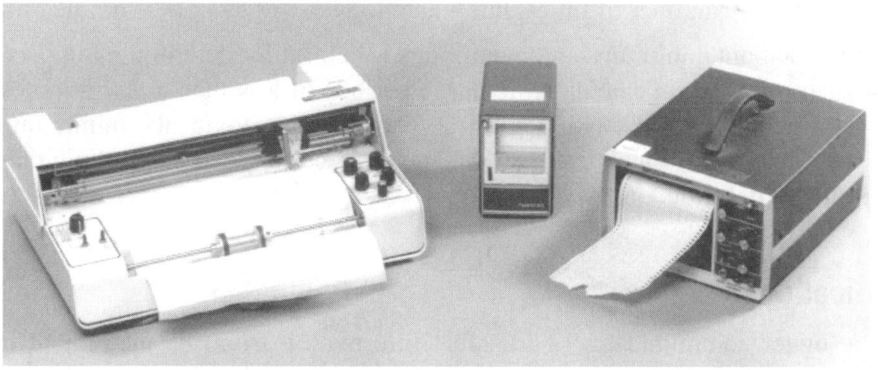

Figure 26-6. A variety of chart recorders (photo, M. Strawn).

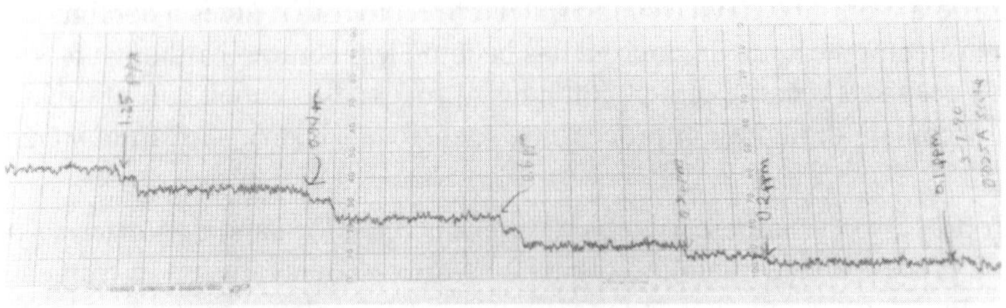

Figure 26-7. An example of chart recorder output, in this case for a calibration, from a MIRAN 1A infrared analyzer. Notes indicate where a new injection was made. Note the noise.

Data Loggers

By 1995, data loggers were available and many DRI had output ports that would allow connection to an external data logger. Today, most DRI have some capacity for internal data logging that can usually be downloaded to a computer. Data loggers have ports for input and output, a microprocessor that manipulates data and instructions and may include an A to D converter, and memory storage chips. On older instruments with continuous, analog signals, the data logger must sample the signal. Within limits, the user controls the frequency with which samples are taken.

Data loggers are small devices that can monitor the output of a variety of instruments. The manufacturer of the data logger does not have to be the manufacturer of the instruments. The output is stored in the memory of the data logger, and at the end of the sampling period, the data are sent to a printer or a computer. If exported to a microcomputer, the data can be further manipulated, and reports can be produced.

The amount of data that a data logger (internal or external) can store depends on the size of its memory and the number of statistics that the data logger computes and stores. Some typical statistics include minimum, maximum, and average values. A continuous analog signal can be sampled an infinite number of times, but a data logger will sample at defined intervals.

Information about minimums and maximums is helpful for determining impact noise. For chemicals, such information calculated for time periods of less than a few minutes would be useful only for acute toxicants with ceiling OELs. For other toxicants, minimum and maximum data would not be of much use, unless synchronized to video to determine tasks associated with peaks as described later.

Communications

A data logger communicates: it transfers information from the instrument to the user. However, the advent of remote sensing and multiple fixed monitoring stations has made it necessary for communications to be more sophisticated. Remote sensing is the sensing or detection of a contaminant at a considerable distance from where the contaminant is emitted. This implies that the contaminant occurs at a sensor, remote from the rest of the instrument or remote from where the data output occurs.

Also a sensor may be located remotely from the site of emission for at least two reasons. First, the instrument containing the sensor, or the person holding the instrument, may not be able to withstand certain environmental extremes. Therefore, the contaminated air is brought to the sensor, usually with a pump and a sampling line. Second, a single sensor may be used sequentially to detect concentrations at multiple sites, usually at fixed locations, via sampling lines.

Even though the term may be slightly misused, remote sensing has come to mean more than this. For example, multiple sensors may be located at various sites, each of them communicating back to a main instrument. The communications may be via wiring, even telephone, coaxial cable or fiber optic lines, for long-distance telecommunications (for a good overview see Smith, 1990). However, more sophisticated systems utilizing transmitters and receivers are also used. For example, sensors and transmitters on individual workers send information, via radio-frequency, to a central receiving point where the information is processed and monitored. The transmission could even occur via satellite. At hazardous waste sites, workers may swallow capsule-size sensor/transmitters that send temperature data to a central receiver where heat stress on the workers is monitored. Methylene chloride exposures have been monitored in this manner (Kovein and Hentz, 1991). Such systems are not inexpensive, but they can often solve a number of monitoring problems. Also the systems are likely to improve in both reliability and cost.

Remote sensing instruments based on open path IR or UV absorption are now common and are discussed further in Chapter 27. These instruments emit the electromagnetic radiation over many meters. The EMR received by a detector is related to the concentration of chemicals in the open atmosphere.

Remote sensing allows the real time display of data to alert public health officials, emergency responders and hygienists of unhealthful or dangerous atmospheres. Past episodes that have resulted in tragic deaths of both workers and the public due to the release of acutely toxic compounds might have been prevented had remote sensing been employed.

VIDEO

A somewhat related communications subject is the use of video equipment in monitoring exposures. There are at least two purposes. First, by coupling DRIs, usually with remote sensing, to video equipment, the hygienist can determine what tasks are causing the highest exposures (Gressel et al., 1993). This may seem like a result that could be easily obtained by simply observing the worker without even using monitoring equipment. No doubt it has been done this way for years with some success. However, in the short history of the use of video equipment, several enlightening results have been obtained. Rosen (1993) showed that the installation of a ventilation system for welders was not working because workers were not using it correctly for the variety of welding tasks. Kovein and Hentz (1991) found increased methylene chloride exposures related to job tasks, work practices, and emission sources in a furniture stripping operation.

A second use of video equipment has been to monitor the worker's skin exposure. By using a fluorescent aerosol tracer as part of the emission source, an ultraviolet light source, and a video camera to record the distribution of the tracer, those sites most susceptible to skin exposure can be ascertained. Also an estimate of total exposure has been correlated with the intensity of the video image (Fenske et al., 1986a, b). In another study, the video intensity was correlated with Malathion (a pesticide) metabolites in the urine (Fenske, 1988). These techniques are improving with respect to the response time of the instrument and the degree of autocorrelation (see Chapter 15) in the exposure data (Heitbrink et al., 1993).

DESIRABLE FEATURES

There are a few important considerations that one should ponder in purchasing or using any DRI. These are listed in Table 26-4, and most have been discussed in previous sections. However, a few require further elucidation.

Output Ports. Not all DRI come with output ports. If you want to send data to a chart recorder, then you will need an analog output port, usually two terminals. If you wish to send signals directly to a computer, then the digital DRI must have a suitable port such as an RS-232 or USB port. On more modern instruments, the output port on the instrument may be proprietary; however, a cable is provided that has a USB (Universal Serial Bus) fitting on one end that allows connection to a computer.

Size and Weight. Instruments to be worn by workers should have size and weight not exceeding those of the portable pump. Roughly this means a weight of less than 1.5 lb (0.7 kg),

Table 26-4
Factors to Consider in Purchasing DRI

Response time	Analog/Digital
Noise/precision	Output ports
LOD	Size and weight
Zero drift	Calibration effort
Sensitivity	Security
Range	RF shielding
Selectivity/Interference	Intrinsic safety
Specificity	

and a profile (dimension away from the body) that does not exceed about 2 in. (5 cm). Instruments that will be carried by the hygienist may be a bit heavier than this because they typically are not carried more than a few hours. However, any instrument that weighs more than 10 lb (4 kg) will likely require a backpack or other suitable transport mechanism if it is to be carried more than 30 minutes. As a rough guide, think of carrying milk home from the grocery store. A gallon of milk weighs about 8 pounds. Four liters weighs about 4 kg. Imagine carrying a gallon of milk for 30 minutes.

Calibration Effort. As will be seen in Chapter 30, many calibration procedures can be used. Some require a few minutes; others may take half a day. Some instruments require lengthy calibration efforts, whereas others can be adequately calibrated with one gas at one concentration. For most instruments, routine accuracy checks (with independent standards) should be used to verify that the instrument remains in calibration. Some manufacturers sell calibration kits that simplify the job. Be leery of any manufacturer that claims that its factory calibration is good for one year. There are simply too many things that can go wrong.

Security. In some cases, it is important for instruments to be designed so that they (and their settings) are tamper-proof. This has to be assessed on a case-by-case basis. On the other hand, avoid portable or personal DRIs that have protruding controls that, when bumped, change settings.

Radiofrequency (RF) Shielding. This subject was addressed in Chapter 20 for pumps. There have been at least two studies of the effect of RF on DRI performance. Cook and Huggins (1984) studied noise, combustible gas, and CO instruments. They concluded that combustible gas meters, which typically do not have signal amplification circuitry, should not be affected by normally occurring workplace RF; but the other instruments with more complicated circuitry would. Feldman (1993) studied combustible gas/oxygen meters, pumps, air velocity meters, and CO detectors. The instruments were tested in RF fields designed according to an ANSI standard for human exposure to RF (ANSI, 1986). Those that had integrated RF shielding passed the test. Of those that failed, modifications such as painting the surface with a nickel-based paint improved performance. Manufacturers of DRI now are generally aware of this problem and are responding. However, older DRIs could be susceptible to RF effects.

Intrinsic Safety. As was described for pumps in Chapter 20, DRI should not be taken into potentially flammable environments unless they have been tested for intrinsic safety. This means that the DRI has a built-in feature that will prohibit the spread of a flame from within the instrument. Several organizations in the United States do such testing, and the approvals are usually displayed on those models that pass. A description of the different atmospheres tested is given in Chapter 20.

QUESTIONS

26.1. Why would it be important to know that an instrument's response was integrated over a 5-minute period for a toxicant with a 2-minute biological half-life?

26.2. Tschebyscheff's theorem states that the lower bound for the probability that a random variable Y falls in the interval of $\mu \pm k\sigma$ is 100 $(1 - 1/k^2)$% for an unknown distribution, where k is the number of standard deviations that are added to the mean. If k is 2, then $\bar{x} + 2s$ = the 84.5th %tile, as opposed to the 95th %tile if the distribution were normal. What does this rule mean for the calculation of LODs using the 2 × peak to valley noise rule?

26.3. For the two instruments in Figure 26-3, assume that the peak to valley noise for each instrument is proportional to the standard deviation of a blank signal. Also assume that for each instrument the LOD was calculated as 2 × the peak to valley noise. Now show that the analytical sensitivities (Equation 26-1) for the two instruments are equal.

REFERENCES

ACGIH®: Air Sampling Instruments, 9th ed. American Conference of Governmental Industrial Hygienists, Cincinnati, OH (2001).

AIHA: Manual of Recommended Practice for Combustible Gas Indicators and Portable Direct-Reading Hydrocarbon Detectors, 1st ed. American Industrial Hygiene Association, Fairfax, VA (1980).

AIHA: Manual of Recommended Practice for Combustible Gas Indicators and Portable Direct-Reading Hydrocarbon Detectors, 2nd ed. American Industrial Hygiene Association, Fairfax, VA (1993).

AIHA: Metrological recommendations for air-monitoring instruments – a guidance document. Am. Ind. Hyg. Assoc. J 53:673-676 (1992).

ANSI: ANSI/ISA S12.13 Part 1 – 1986, American National Standard Performance Requirements, Combustible Gas Detectors, p. 19. ISA, Research Triangle Park, NC (1986).

ASTM: Standard Practice for General Ambient Air Analyzer Procedures, D3249-95. ASTM International, West Conshohocken, PA, USA (2005).

CEN: European Committee for Standardization: General Requirements for the Performance of Procedures for Workplace Measurements, Document EN 482. CEN, Brussels (2006).

Cook, C.F.; Huggins, P.A.: Effect of radio frequency interference on common industrial hygiene monitoring instruments. Am. Ind. Hyg. Assoc. J. 45:740-744 (1984).

Cook, W.A.; Coleman, A.L.: Determination of injurious constituents in industrial atmospheres. II. Determination of solvent vapors in air by means of activated charcoal. J. Ind. Hyg. Tox. 18:194-210 (1936).

Elkins, H.B.: The Chemistry of Industrial Toxicology, 2nd ed. Wiley Publishing, New York (1959).

EPA: Environmental Technology Verification Program. U.S. Environmental Protection Agency, Washington, DC http://www.epa.gov/etv/ (2006).

Feldman, R.F.: Degraded instrument performance due to radio interference: Criteria and standards. Appl. Occup. Env. Hyg. 8:351-355 (1993).

Fenske, R.A.: Correlation of fluorescent tracer measurements of dermal exposure and urinary metabolite excretion during occupational exposure to Malathion. Am. Ind. Hyg. Assoc. J. 49:438-444 (1988).

Fenske, R.A.; Leffingwell, J.T.; Spear, R.C.: A video imaging technique for assessing dermal exposure I. Instrument design and testing. Am. Ind. Hyg. Assoc. J. 47:764-770 (1986a).

Fenske, R.A.; Wong, S.M.; Leffingwell, J.T.; Spear, R.C.: A video imaging technique for assessing dermal exposure II. Fluorescent tracer testing. Am. Ind. Hyg. Assoc. J. 47:771-775 (1986b).

Gentry, S.J.: Instrument performance and standards. Appl. Occup. Env. Hyg. 8:260-266 (1993).

Grafstein, O.; Levine, S.P.: An expert system for guiding selection of direct-reading air monitoring instrument. Appl. Occup. Env. Hyg. 8:157–167 (1993).

Gressel, M.G.; Heitbrink, W.A.; Jensen, P.A.: Video exposure monitoring – a means of studying occupational air contaminant exposure, Part I – Video exposure monitoring techniques. Appl. Occup. Env. Hyg. 8:334-338 (1993).

Heitbrink, W.A.; Gressel, M.G.; Cooper, T.C.; Fischbach, T.; O'Brien, D.M.; Jensen, P.A.: Video exposure monitoring – a means of studying sources of occupational air contaminant exposure, Part 2 – Data interpretation. Appl. Occup. Env. Hyg. 8:339-343 (1993).

Kovein, R.J.; Hentz, P.A.: Real-time personal monitoring in the workplace using radio telemetry, In Field Screening Methods for Hazardous Wastes and Toxic Chemicals, pp. 473-481. Second International Symposium, NIOSH, Cincinnati, OH, Feb. 12-14 (1991).

McCammon, C.S. Jr.: A summary of the NIOSH evaluation program of portable direct-reading meters. Am. Ind. Hyg. Assoc. J. :289-295 (May 1976).

Ness, S.A.: Air Monitoring for Toxic Exposures: An Integrated Approach. Van Nostrand Reinhold, New York (1991).

Patty, F.A. (ed.): Industrial Hygiene and Toxicology, 2nd ed. Wiley Publishing, New York (1958).

Rosen, G.: PIMEX, Combined use of air sampling instruments and video filming: Experience and results during six years of use. Appl. Occup. Env. Hyg. 8:344-347 (1993).

Shulman, S.A.; Smith, J.P.: A method for estimation of bias and variability of continuous gas monitor data: application to carbon monoxide monitor accuracy. AIHAJ 63:559-566 (2002).

Skoog, D.A.; Holler, F.J.; Crouch, S.R.: Principles of Instrumental Analysis, 6th ed. Thomson Brooks/Cole (2007).

Smith, J.P.: Use of data communications with air monitoring data. Appl. Occup. Env. Hyg. 5:213-221 (1990).

Stetter, J.R.: Instrumentation to monitor chemical exposure in the synfuel industry. Ann. Am. Conf. Ind. Hyg. 11:225-245 (1984).

Verma, D.K.; Tombe, K.D.: Measurement of benzene in the workplace and its evolution process, Part I: Overview, history, and past methods. Am. Ind. Hyg. Assoc. J. 60:38-47 (1999).

Waits, A.; Hartzell, E.; Harten, T.: The U.S. EPA Environmental Technology Verification program – an overview. J. Occ. Env. Hlth. 2:D87-D90 (2005).

27

INSTRUMENTS FOR GASES AND VAPORS

INSTRUMENTS FOR SINGLE COMPOUNDS

In the preceding chapter it was explained that single compound instruments respond to the gas or the vapor that they have been designed for, but that the detector is also responsive to other gases and vapors, called interferences or interferents. Thus, all single substance DRI are selective to a degree. In chemistry terms, "A specific reaction or test is one that occurs only with the substance of interest, while a selective reaction or test is one that can occur with other substances but exhibits a degree of preference for the substance of interest. Few reactions are specific, but many exhibit selectivity" (Christian, 1994). IUPAC (Vessman et al., 2001) has stated that specificity is the ultimate of selectivity and from that perspective there are likely only two *specific* techniques that have the power to sort out one environmental signal from many others: GC (especially GC/MS) and FTIR. Both of these use a separation and/or a preconcentration step; these steps are discussed at the end of this chapter. Thus, DRI for single compounds are selective to varying degrees.

These single substance instruments often rely on electrochemical sensors or detectors. There are no consistent definitions for these sensor types, as many manufacturers apply the general term *electrochemical* to all their products, although there are at least three general types of electrochemical sensors and many subtypes. Therefore, it is sometimes difficult to determine what sensor is being used in a single substance instrument. For many hygienists this may seem unimportant, but there are advantages and disadvantages for each type. The literature contains only a few reviews of such information, mostly prepared by manufacturers[1] (e.g., Wang, 2000). A good first step in making it easier for hygienists to make inter-instrumental comparisons would be to devise a set of universally used standard definitions for the modes of

[1] However, the journal *Analytical Chemistry* publishes research reviews of categories such as gas chromatography and electrochemical sensors every two years or so, and these can have useful application information in addition to the more usual technical information dealing with technology advances.

operation of the sensors.[2]

The hygienist must contact individual manufacturers (literature or the Internet may be helpful) to find out the specifics for each instrument, a daunting task when there are hundreds of manufacturers. Another helpful approach would be to consult an exhaustive listing of instruments and their features that followed a well-defined format. The advantages of such approaches were mentioned in Chapter 26. In this chapter, an attempt is made to link specific sensor methodologies with specific chemicals or classes of chemicals.

Electrochemical

In Chapter 26, Table 26-2 lists most of the various sensors used in direct-reading instruments (DRI). The vast majority of single substance instruments use electrochemical sensors. At the end of this chapter is a compilation of the numerous companies that market sensors, and most of the chemicals that can be monitored with electrochemical and other sensors covered in this chapter. These devices come in all three sizes (personal, portable, and transportable), and there are literally hundreds of makes and models. Sensors that are marketed as *fixed* are not included in this chapter even though many of them are small and operate on the principles discussed here.

The discussion of laboratory based electrochemical cells or sensors in Chapter 19 is broadened here in an attempt to shed some light on this group, which has confusing nomenclature (see, e.g., Ness, 1991; Skoog et al., 2007; Stetter, 1984). The principles of electrochemistry have been used in the laboratory for many years to analyze air samples collected in the field and brought to the laboratory (see, e.g., Duffy and Pelton, 1965). Application to instruments that could be carried to the field and used in the direct measurement of pollutants in air was dependent on several developments, including the application of polymer membranes that could be used to keep the wet chemicals in the sensor and allow the controlled diffusion of analytes to the sensor (Kuempel and Schults, 1971) and also various electronics developments. Several of the earlier devices and perhaps a few in use today mixed the airstream with a flowing, liquid electrolyte stream (see, e.g., the discussion of conductometric devices); used a batch system of impingers, rather than continuous collection followed by field analysis (Sherwood, 1969); or allowed the electrolyte stream to flow in a thin layer down the electrodes as the air moved past (Diamond, 1963). However, by far the most common solution to the problem of introducing the air to the electrolyte is the selective membrane, usually a polymer.

The use of selective membranes is common in electrochemical devices used in hygiene *laboratory* work, e.g., the pH meter falls into this category as described in Chapter 19 (refer to Figure 19-23). In the pH meter, the voltmeter reads the voltage across two electrodes that are

---

[2] The Instrument Society of America (ISA, http://www.isa.org) has some standards related to DRI, e.g., H_2S, CO, combustible gases, O_2, and Cl_2, but not a comprehensive standard with sensor definitions and requirements for their identification in instruments by manufacturers.

separated by a salt bridge and a membrane (usually porous silicate glass). These allow ions to flow from the analyte solution to the electrolyte and then the indicator or working electrode without spilling the electrolyte into the analytical solution or vice-versa.

We typically employ the pH meter probe by placing it into a liquid solution, like water from a stream. For hygiene work, the analyte solution is the air, and we need to get the analyte of interest to move from the air into the electrolyte or internal solution (see Figure 27-1). Preferably we would like to minimize the migration of other constituents of the air. This is done by using selective membranes. These come in two types (Skoog et al., 2007). Hydrophobic or microporous membranes are composed of non-polar polymers like polypropylene or Teflon (Ross et al., 1973), that repel water, thus keeping the aqueous or gel internal solution in the sensor. However, they are porous (holes < 1 μm) thus allowing gases to cross the membrane. These are not particularly selective as they allow relatively free diffusion of smaller molecules. On the other hand, homogeneous membranes, like silicon rubber, work on

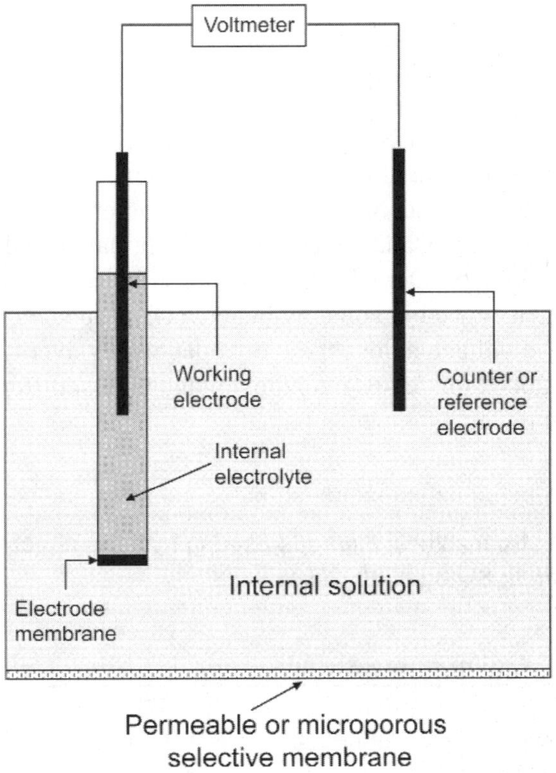

Figure 27-1. A potentiometric electrochemical cell (similar to the pH meter in Chapter 19, Figure 19-7) that has a polymer membrane allowing the analyte to enter the electrolyte from the air. The analyte is ionized in the internal solution. When it crosses the working electrode membrane it causes production of electrons creating a potential across the voltmeter.

the principle of permeation; the gaseous molecules must be soluble in the polymer in order to diffuse across it. Another solution is to use an electrolyte that is solid rather than liquid or gel; thus, the air may interact directly with the electrolyte.

Many instruments sample the air by allowing the diffusion of the analytes to this external membrane or the solid electrolyte; others contain a pump for moving the air to the polymer membrane. If the diffusion technique is used, then long sampling lines or tubes are impractical unless the sensor can be located at the far end of the sampling line away from the instrument. Some instruments allow for this setup.

Once the analyte dissolves into the electrolyte and then interacts with the working electrode, it must be measured. All electrochemical instrument measurements are based on some aspect of the flow of electrical current. Current may flow through electrodes made of metals or other conducting materials due to reduction or oxidation of the analyte, thus producing electrons.

The major electrochemical methods used in semi-selective to selective instruments are in bold in Table 27-1. The Table also shows the quantity that is measured. For some of the methods, Ohm's law is instructional, or

$$E = IR$$

where E is the potential (voltage), I is the current (amps), and R the resistance (ohms). Thus, in potentiometry we measure potential. In conductometry, resistance is measured, that is inversely proportional to the conductance of current. In coulometry, current is consumed over time to reduce or oxidize a sample, and the charge is accumulated so that coulombs (total electrons) are measured. Coulometry can take place under either a constant potential or constant current, but for the O_2 meters used in some hygiene work, it appears that voltage is held constant. In voltammetry, the voltage across the electrodes varies over time in one of four basic ways (see Skoog et al. (2007) for details). As the voltage varies, different ions in solution will produce varying current.

Table 27-1
Summary of common electroanalytical methods used in hygiene direct-reading instruments (adapted from information in Skoog et al., 2007)

Method	Quantity measured	Interaction site	Current	Potential
Potentiometry	*Potential*	Interfacial	Zero	Measured
Conductometry	*Resistance* = 1/conductance	Bulk	Varies with resistance	Constant
Coulometry	*Charge quantity* = f (time × current)	Interfacial	Accumulated	Constant
Voltammetry/ amperometry	*Current* as a f (potential)	Interfacial	Measured	Varies, but controlled

In electrochemistry, the electrode potential has been defined relative to a standardized reference electrode, the standard hydrogen electrode (SHE). In essence, this electrode potential is actually the potential of a electrochemical cell with a defined reference electrode (in this case the SHE). Thus, the electrode potential is a way of numerically quantifying the potential in the cell for current to flow, or its electron energy, or more specifically the relative strength of an indicator electrode to be an electron acceptor. The value can be + or - depending on whether the cell generates electrons (+, a battery or galvanic cell, or potentiometric sensor) or consumes them (negative potential, electrolytic, other sensors). While the details are not important for our considerations, these cell potentials have been determined for many electrodes and help electrochemists determine what electrodes are best for each analyte. The Nernst equation[3] relates cell potential to the chemical activities of the cell components.

Most of the methods measure ion activity that takes place in a thin layer at the surface of the electrodes. Skoog et al. (2007) refer to this layer as the *electrical double layer* since the ions tend to sort themselves, positive ions in the layer closest to a negative electrode and vice-versa. This is called interfacial interaction. On the other hand, in conductance measurements, resistance across the entire solution is measured.

POTENTIOMETRY

In order for us to understand potentiometry, an analogy might be useful. Potentiometry measures potential energy. Thus, we can think of changes in analyte concentration as causing more electrons to pile up at one electrode than the other. Think of those electrons as water behind a dam. That water is *wanting* to flow downhill. The electrons are wanting to flow (as current) also. However, in potentiometry there is no current flow; it is the *height* of electrons behind the dam that is being measured. Potentiometry is the measurement of potential or voltage change at an electrode after the ions from the analyte have come into equilibrium with an internal solution, in which the electrodes are placed. Essentially no current flows because the resistance across the reference electrode is high. The ion selective membrane electrode (ISE) is the most common electrochemical device used in laboratory work, and is quite common in

[3] One form of the Nernst equation is

$$E = E^0 - \frac{RT}{nF}\ln\left(\frac{A_{Red^b}}{A_{Ox^a}}\right)$$

where R = gas constant, T = temperature in Kelvin, n = number of electrons in half-reaction, i.e., at the indicator electrode relative to a standardized reference electrode, F = Faraday constant (96485 Coulombs/mol), A = activity (1 for a pure solid, liquid or solvent and expressed in mol/L for solutes and in atmospheres for gases), and E is potential where E^0 is the standard electrode potential or the potential when the activities are one and logarithm is thus 0.

any laboratory because the pH meter falls into this category.

For sensors used in hygiene work, the indicating or working electrode (see Figure 27-1) is placed in a standard solution (internal electrolyte) of known concentration, generally containing ions like those of interest. A membrane, like the silicate glass of the pH meter or a polymer, separates the analyte and internal electrolyte. The analyte ions are exchanged across the electrode membrane, causing an ion equilibrium change. The accumulation of ions at the indicator electrode causes a change in potential, read as voltage.

The degree of selectivity for an analyte of interest is dependent on the electrolytes and the selective membrane and electrode membrane compositions. There are apparently many combinations of these factors as potentiometry is used for a very wide array of gases and vapors (see list at end of chapter, Chapter 26, Table 26-2, and ACGIH®, 2000).

VOLTAMMETRY

Voltammetry employs a special means to improve selectivity. If a compound is adsorbed in a solution, say, an electrolyte, it may or may not be oxidized (give up electrons). However, if one of the electrodes is polarized beyond a certain threshold (i.e., has a different potential from that of the other electrode), then the analyte will migrate to the appropriate electrode and be oxidized. This causes electrons to flow. In one series of sensors, this current is converted to voltage and measured (Interscan, 2006). If the voltage applied to the electrode is constant and polarization is constant, the process is called *amperometry* (Skoog et al., 2007). Most detectors used for DRI appear to be amperometric, even though they are often referred to as voltammetric. If the voltage is varied in some way over time, we have voltammetry. The selected voltage makes the method specific, i.e., the analyte can be qualitated based on the voltage at which a current is produced. Quantification is based on the current produced being proportional to concentration (Interscan, 2006).

The polarized electrode aids the selectivity process. A given analyte will not be oxidized unless the potential is above the voltage that causes diffusion of the analyte through the electrolyte to the electrode – the diffusion current. Below this voltage, nothing happens. Above this voltage, oxidation occurs. Through careful choice of the electrode system, the electrolyte, the polymer membrane through which the analyte diffuses into the electrolyte, and an appropriate voltage, considerable selectivity is gained. The polymer membranes or other *filters* can be used to screen out interferences. The term *filter* may refer to carbon-coated minerals, including silica gel and alumina, and other solid substances (Bergman, 1986). Voltammetry is used by several companies in detectors for a wide range of vapors and gases.

COULOMETRY

Coulometric techniques have been in use in the laboratory for many years. Field-collected samples were brought to the laboratory and analyzed for ozone with coulometry (Duffy and

Pelton, 1965). This basic chemistry has been used in DRI for ozone (NIOSH, 1975a). Oxygen sensors are perhaps the most important application in air monitoring. Coulometric approaches to measuring O_2 have the advantage of measuring very low concentrations and a sensor that does not require replacement. A few other DRI operate on coulometry, but coulometry is not so important as potentiometry and voltammetry.

Current is measured in amps, essentially electrons per time. Thus, the integration of current over the time needed to reduce or oxidize the analyte is measured as coulombs. In many of the coulometric laboratory techniques, a batch or discontinuous process is used. A sample is added to a cell, and the consumption of electrons needed to convert the analyte to another species is measured.

CONDUCTOMETRY

In conductometric instruments the resistance to current or electron flow across a solid or liquid is measured. For example, a current passes with relative difficulty between two electrodes placed in deionized water; but when the deionized water is replaced with seawater, the flow is less restricted so the conductance is greater. Most instruments for hygiene work are based on conduction across a solid and are sometimes referred to as solid state[4] detectors or sensors. The most common of these is the metal oxide semiconductor (MOS). MOSs are also referred to as solid state as are polymeric semiconductors and piezoelectric sensors. However, there are portable sensors (see e.g., Calibrated Instruments at end of chapter) available for several gases that work on the conductometric principle using a liquid electrolyte. These are batch instruments in that they sample the atmosphere, the analyte is ionized and mixed with the electrolyte, and the change in conduction is measured. They are not specific since any ion will change conduction, but the use of polymers and other *filters* can make them more specific.

In addition, there are electrolytic conductometric detectors that are used in batch analysis of outputs from chromatographs that are useful for sulfur, nitrogen, and halogen-containing compounds. The compounds are ionized in a solution that is kept at 850-1000°C (Skoog et al., 2007). Similarly performing solid electrolyte, conductometric detectors are available that detect compounds in the gas phase. It is not clear that the latter have hygiene applications.

Metal Oxide Semiconductors

In MOS sensors, metal oxide microcrystals (oxides of tin, zinc, lanthanum, tungsten, and nickel are the most commonly used) are coated on a silica or alumina substrate. When the metal oxide is heated, oxygen is adsorbed on the surface and becomes negatively charged. Because electrons are transferred to the adsorbed oxygen, a positive charge is left on the semi-

[4] All of the devices discussed here under conductometric are also referred to as solid state. For an overview of the many uses of these devices see Capone et al., 2003.

conductor. Inside the sensor, electrical current flows, and resistance to this electrical current are caused by negatively charged oxygen. In the presence of a reducing gas, say trichloroethylene, a surface catalyzed combustion occurs and the equilibrium surface density of negatively charged oxygen decreases, thereby decreasing the resistance of the sensor (Stetter, 1984; Ho et al., 2001).

The most common device is based on tin oxide, and is sometimes referred to as the Taguchi (AIHA, 1993b) or TGS sensor. It is sometimes used in general survey combustible gas meters (see later discussion), and has been adapted to use for CO (Firth et al., 1975). An increasing number of MOS devices are entering the market with claims of good selectivity and freedom from poisoning. They include devices for chlorinated hydrocarbons (Cao and Stetter, 1990) and hydrogen sulfide (Smith and Shulman, 1988), and can be personal or portable devices. In an evaluation of five H_2S devices, temperature and humidity had varying degrees of effect, whereas CO and methane were not significant interferents. The response times, recovery times, and linearity of response also were quite variable (Smith and Shulman, 1988).

MOS detectors will respond to any reducible analyte. This lack of selectivity cannot be eliminated completely but it can be improved by the use of filters or chromatographic columns, catalysts, promoters, or more specific surface additives, the selection and preparation of the material for the sensing layer, by analyzing the transient sensor response, changing the temperature to maximize sensitivity to a particular analyte, or using a variable temperature operating mode (Capone et al., 2003).

There is one other important *solid state* conductometric technique that does not employ MOS. Instead, a gold film is used to detect changes in resistance/conductance caused by mercury vapor. The device (see Figure 27-2) is reported to have few interferences. Another such device, used for measuring H_2S and based on the gold film, is sold by the same company. The H_2S instrument has an operating range of 0.003–50 ppm and is, therefore, good for tracking down septic odors in buildings, as well as for use in refineries and other workplaces with large sources of H_2S. Both devices are portable, but sensors can be purchased for the Hg device that are worn by the worker and read at the end of the sampling period by using the main instrument.

Piezoelectric Chemical Sensors, SAW and QCM

The piezoelectric effect is the creation of molecular dipole moments in the lattice of a crystal dielectric material when it is deformed by applied pressure. The pressure can be caused by the mass of molecules that condense or are absorbed or adsorbed on the surface, or particles falling on the surface from the atmosphere. If electrodes are applied to the sides of a thin piezoelectric crystal, current will flow through an external circuit when the pressure is applied. If alternating voltage is applied to the electrodes, then mechanical oscillations occur and are stable at the natural resonant frequency of the crystal. These oscillations and the current they induce are modified as the pressure is applied. A chemical sensor can be made of a piezoelec-

Figure 27-2. The Jerome model J405 mercury vapor analyzer (courtesy, Arizona Instrument, LLC).

tric oscillator that responds to chemical concentration changes (mass) by changing either its resonant frequency or wave speed as mass accumulates. These sensors are usually coated with a polymeric film. The surface film adds some gas/vapor selectivity by selective adsorption or absorption. A quartz crystal used in such a configuration is referred to as a quartz crystal microbalance or QCM. The oscillations produced are a function of the physical properties of the crystal sheet or rectangular bar. In the QCM, shear vibrations are produced relative to the upper and lower layers of the crystal. In another type of piezoelectric crystal, surface elastic or acoustic waves are formed and measured. These are referred to as SAW devices. (Buck et al., 2004).

SAW sensors are used in DRI. They are generally non-selective, however, surface coatings can improve selectivity. More importantly, multiple, heterogeneously-coated SAW sensors used in arrays and coupled with microprocessor based output with comparison to calibrated patterns can considerably improve selectivity and quantitation (Groves and Zellers, 1996; Park et al., 2000). Still their best use is in niche applications such as determining the pattern produced by certain fragrances, bacteria, exhaled breath, and in industrial process control, e.g., quality control for food and cosmetics (see for example, http://www.alpha-mos.com/). Such arrays of chemical detectors have been termed *electronic noses*.[5] Advantages of SAW sensor arrays include small size, low temperature of operation, wide sensitivity, and a relatively high degree of selectivity, which can be adjusted by varying the set of sensor coatings.

[5] Electronic noses have employed many different types of detector arrays including SAW, QCM, MOS, and chemiresistors. The arrays may be of one type or mixes of different sensor types. The original intent of the devices was to mimic the human olfactory sense.

Groves and Achutan (2004) performed field and laboratory evaluations of an array of 6 SAW sensors, configured into a prototype, hand-held instrument and used for the detection of perchloroethylene in breath. The field results agreed well with the NIOSH 3704 GC method, and the LOD was 50 ppb.

Hsieh and Zellers (2004) evaluated a commercially available, but modified, hand-held device, the VaporLab that consisted of a small pre-concentrator tube (optional packing materials are possible), an array of four *working* SAW sensors, and an uncoated reference SAW sensor. Tests were performed on laboratory mixtures, and also Monte Carlo analysis simulations of mixtures were performed. The device, as originally sold, was not meant for quantitative analysis. Thus, the authors used a multivariate statistical method, extended disjoint principal components regression, for vapor recognition and quantification. The software, coupled with a stored library of calibrated responses and response patterns, can be used to recognize and discriminate individual vapors from other vapors in a library. Responses arising from a vapor mixture can also be recognized, and the components and their respective concentrations can be determined. The authors concluded that the VaporLab is capable of recognizing and quantifying a wide range of vapors and simple vapor mixture components at concentrations near respective TLV® values. Mixtures of more than two components could not be quantified. Modifications were recommended. Finally, it was concluded that, for analysis of complex mixtures, chromatographic separation is necessary. (Note that the VaporLab now includes a *miniature chemical separation system*, or GC separation.)

Conducting-Polymer Chemiresistors

Chemiresistors are solid-state conductometric devices that have been reported since at least the 1960s. Typically there is a thin film of a semiconducting material deposited onto a pair of electrodes, and it is this thin film that has evolved. Most recently this transducer film has incorporated nanometer-size metal particles each coated by an organic monomolecular layer, e.g., alkanethiol (the organic part) that is used to stabilize a gold (the metal) nanocluster material. As an analyte is sorbed by the organic monolayer, the swelling changes the resistance of the device. The coatings have been referred to as colloidal metal insulator metal ensembles (MIME) and monolayer protected nanoclusters (MPN). For the alkanethiol coatings, varying the alkane length, its thickness, and the operational temperature will affect strength of response to various organic vapors (Wohltjen and Snow, 1998).

While these sensors have been used in electronic noses, without preconcentration of the analyte and without species separation (see for example Rivera et al., 2003), their most promising use has been recently reported as detectors in micro GCs and will be discussed in the GC section.

CALIBRATION

Generally the manufacturers of each device sell a calibration kit or require factory calibration. In the case of O_2 monitors, fresh air is used to calibrate the device at a concentration of about 20.9%. This is a typical one-point calibration that is used for many monitors having a linear response. Those with a nonlinear response typically require two-point calibrations, which can be more difficult. Calibration kits typically consist of a small bottle of gas having a known concentration. They typically have a finite shelf life (see Chapter 30). Ozone instruments must be calibrated with an ozone generator, as ozone is very unstable and cannot be stored in a cylinder.

Some manufacturers state that their instruments need calibration only yearly; others give no information about calibration in their literature. One should be leery of either situation. For details on generating known concentrations for calibrating many of the instruments in this chapter see Chapter 30. Calibration of devices that allows the analyte to diffuse to the sensor can be difficult, and it usually requires a calibration adapter of some type to make a tight fitting from the calibration device to the opening of the instrument to the sensor.

A representative from at least one company (Shaw, 2005) states that proper calibration is 90 percent of successful toxic gas detection and cites the following guidelines in electrochemical DIR use

- far too much emphasis has been placed on experiences garnered in temperature, oxygen and combustible gas measurement and calibration; other gases are much more difficult
- while most gases are unstable and many advances have been made in improving their stability in bottles, care must be taken with bottled gases
- calibration with permeation devices is an excellent choice but requires care and investment
- pay careful attention to potential interferents, temperature, and pressure in the working environment.

APPLICATION AND EVALUATION

The early demise of the NIOSH program for evaluation of DRI (McCammon, 1976) left a vacuum in this area in the United States. Several NIOSH evaluations were reported in the mid-seventies, including SO_2 meters (NIOSH, 1974), NO_x (NIOSH, 1975b), and ozone (NIOSH, 1975a). However, they are of little practical use for today's instruments. Since that time, few evaluations have been performed.

Two evaluations of CO DRI have been done. Roberge and Goyer (1993) examined the data processing characteristics, effects of temperature, relative humidity, and interferences, drift, response time, and precision and accuracy for six instruments. The results point to the need for similar evaluations of other instrument types or, minimally, a standard requiring what manufacturers should test and report. Tables 27-2 through 27-5 indicate the discrepancies that may exist in what instruments are touted to do versus actual performance.

Smith and Shulman (1994) evaluated one brand of datalogging CO DRI. They too found results that point to the need for caution in the use of DRI. For example, they found that temperature affected response, and more importantly, the effect varied across multiple instruments from the same manufacturer and with concentration. They found that calibration results were

Table 27-2
Response time for six CO DRI (from Roberge and Goyer, 1993)

Instrument	Manufacturer's Specification(s)	90% Response Time(s)	10% Return Time(s)
Interscan 2000	15	60	65
Neotronics Multiwarn	30	45	50
Neotronics Pac II	60	15	15
Neotronics Datalogger 190	< 60	15	15
Metrosonics PM 7700	35	60	60
Dräger Exotox 75	< 80	50	50

Table 27-3
Response bias at six concentrations for six CO DRI (from Roberge and Goyer, 1993)

Instrument	Manufacturer's Specification(s)	Mean concentration read at known concentration (ppm), % of reading error in ( )[1]					
		12	30	69	91	125	400
Interscan	2% of reading	12 (0%)	34 (13%)	73 (6%)	99 (9%)	---	428 (7%)
Multiwarn	± 2 ppm or 3%	12 (0%)	28 (6%)	68 (1%)	93 (2%)	126 (1%)	---
Pac II	± 3%	13 (8%)	2 (6%)	75 (9%)	92 (1%)	125 (0%)	412 (3%)
Datalogger 190	3%	9 (25%)	25 (16%)	61 (12%)	90 (1%)	112 (10%)	385 (4%)
PM 7700	--	10 (16%)	28 (6%)	73 (6%)	91 (0%)	118 (6%)	412 (3%)
Exotox 75	± 1 ppm	10 (16%)	32 (6%)	73 (6%)	91 (0%)	131 (5%)	410 (2%)

Table 27-4
Measured concentration ($\bar{x}$, ppm) and reproducibility (coefficient of variation) of six CO DRI (from Roberge and Goyer, 1993)

	Challenge Concentration			
	92 ppm		0 ppm	
Instrument	$\bar{x}$	CV,%	$\bar{x}$	CV,%
Interscan	87.4	< 1	6.2	10
Multiwarn	92.0	0	4.0	17
Pac II	92.6	< 1	7.0	10
Datalogger 190	87.4	< 1	0.0	0
PM 7700	93.0	0	3.8	11
Exotox 75	88.2	2	1.6	33

Table 27-5
Interference responses (ppm) for six CO DRI (from Roberge and Goyer, 1993)

	Gas and challenge concentration (ppm)[1]					
Instrument	NO (40)	NO_2 (18)	H_2S (9)	SO_2 (21)	C_2H_4 (1%)	H_2 (499)
Interscan 2000	-22	-21	-13	-6	> 500	> 350
Multiwarn	89	5	45	21	> 200	88
Pac II	80	6	43	18	> 500	54
Datalogger 190	77	< 1	< 1	< 1	> 200	33
PM 7700	12	< 1	< 1	< 1	> 999	> 999
Exotox 75	13	< 1	< 1	< 1	> 999	7

[1] For 2.5% methane and 0.6% pentane response was < 1 ppm for all instruments

not stable for one month periods and recommended more frequent calibrations. Similarly, Park et al. (2003) found that bias of a hydrogen peroxide device increased with time from calibration; they recommended calibration every third day.

Infrared (IR) Spectroscopy

Most laboratory-scale IR instruments are dispersive; they have the ability to examine the absorption of IR radiation at discrete wavelengths by scanning across the entire IR spectrum (about 2.5–14.5 μm). Examination of this spectrum (see Chapter 19, Figure 19-20 for examples) is helpful in identifying unknowns. Technically, in dispersive instruments, reflection gratings or prisms are used to disperse or separate the wavelengths (Skoog et al., 2007). Non-dispersive instruments are commonly used for quantitation of specific compounds. They utilize the entire spectrum, but often filter out the unneeded portion. These instruments can be of three types: (1) those that use the entire spectrum with no wavelength filtering device, which are mainly used to monitor higher levels of CO and especially CO_2; (2) those that use a filter to give a narrow band of IR for quantitating an individual compound, e.g., the Foxboro model 203 and older versions; and (3) those that have a variable filter wheel that can be set to quantitate a specific contaminant or automatically rotated to give a complete spectrum that can be used to identify unknowns, the Foxboro IA and IB series (Egert et al.,1993).

CO AND CO_2 INSTRUMENTS

A schematic of one type of NDIR instrument is shown in Figure 27-3. There are many configurations. The IR may pass a chopper that is used to periodically block the IR (say, five times per second), allowing the instrument to detect other internal sources of IR[6] and preventing the light from passing both chambers simultaneously and hitting the detector. Alternatively, the IR radiation passes a reference cell containing non-contaminated air or the sample cell through which contaminated air is passed.

Other instruments use rotating filter wheels (gas filter correlation method), instead of choppers. These wheels contain gas-filled cells; these are different from the reference and sample cells in Figure 27-3. A typical arrangement of filter wheel cells includes a cell with the gas of interest, a cell with a gas like N_2 that does not absorb IR, and 2 open cells. In this arrangement only a sample cell (no reference cell as in Figure 27-3) is required, as when the filter wheel rotates the detector and sees three kinds of information on each cycle thereby allowing correction for interferences.

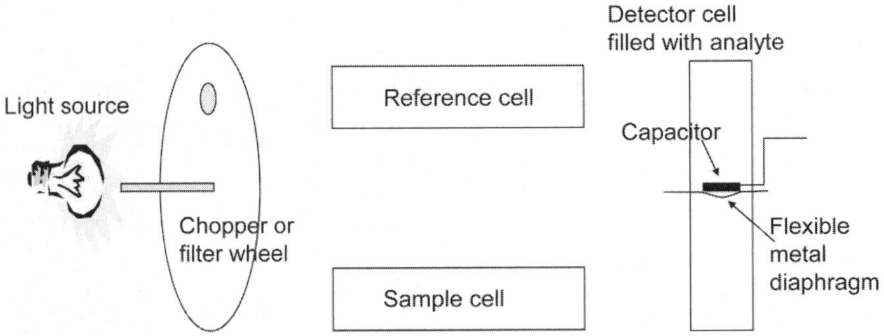

Figure 27-3. A schematic of a double beam NDIR instrument.

[6] Remember that any body, hotter than its surround, will emit IR.

After passing the sample and or reference cells, the IR goes to a detector. The detector depicted in Figure 27-3, contains only the gas to be analyzed; it has two halves. If the contaminant enters the sample cell, then it absorbs IR at its characteristic wavelength(s). Consequently, some of the IR radiation is blocked, and the gas on that side of the detector does not get heated as much as the reference side. Note that it does not matter that the entire IR spectrum enters the detector because the contaminant is only heated by those wavelengths that it absorbs. However, many instruments have filters to remove all but the wavelength of interest so as to increase sensitivity. Finally, the divider in the detector shown in Figure 27-3 is a flexible thin metal foil. As the pressure-temperature relationship on either side of it changes, it flexes. The foil also serves as one side of a capacitor so that, as it flexes, it moves with respect to the other side of the capacitor, and the capacitance changes. Detectors may also be solid state conductors that change conductance as the amount of IR falling on them changes (K2BW, 2006). In the MIRAN instruments, a pyroelectric lithium tantalate detector is used; it senses the heat caused by the IR radiation impinging on it.

There are NDIR devices on the market for CO, and recently a large number of devices have become available for CO_2 for indoor air quality studies. If a tight airspace is occupied by respiring humans, the CO_2 concentration will rise unless outdoor air is supplied; hence, the CO_2 level is indicative of ventilation adequacy. Typically, concentrations of 700 to 1,000 ppm CO_2 are thought to be indicative of inadequate fresh air ventilation (ASHRAE, 1989; Persily and Dols, 1990; AIHA, 1993a). Therefore, for indoor air quality work, devices reading up to 1,500 ppm are adequate; but in other applications, for example, the poultry industry, concentrations of 1% are possible.

Most manufacturers of CO_2 devices sell calibration kits that consist of one cylinder of nitrogen for setting the zero and one cylinder containing some concentration of CO_2. Thus, a two-point calibration is made. Fresh air cannot be used because the ambient concentration of CO_2 is at least 400 ppm and variable depending on nearby combustion sources. As the devices do not follow Beer's Law (see later discussion), they are not linear. Therefore, either the meter is not linear (see Chapter 26, Figure 26-4), or for digital devices, the electronics must be corrected for the non-linearity. Either way this is a potential source of error.

For these devices, accuracy is related to cost. Readings may be ± 50 ppm to as much as ± 500 ppm (± 10%), depending on the concentration being measured for the cheaper hand-held instruments, whereas precision of ± 1–3% of the reading at the temperature of calibration can be achieved in more expensive instruments. Careful analysis of this feature is necessary before a purchase is made. Though the CO_2 instruments were historically regarded as portable, some are approaching the size and the weight of personal monitors.

THE FOXBORO MODEL 203

The Foxboro Company made various instruments that it no longer manufactures. In fact, the sole Foxboro instrument is now marketed by Thermo Fisher Scientific Corporation.

However, the Internet is full of used models of these instruments and they are still widely used. In addition, the Wilks Company (see list at end of chapter) now manufactures an IR spectrophotometer for air analysis. At one time the Foxboro Company was called Foxboro-Wilks and so it is assumed that this company is related to the inventor of the MIRAN. The MIRAN model 203, a filtering NDIR (and previous models 101, 102, and 103) do not have any wavelength scanning ability. Instead, a filter is used to allow a narrow band of IR to enter the sample cell. The filter selection is based on the absorption bands for the compound of interest. In 1995, Foxboro offered 28 filters for standard compounds. The filters allow these instruments to be fairly selective. MIRAN instruments are in the transportable class, weighing about 27 lb or 12 kg. Foxboro claimed that the instrument requires no calibration beyond that from the factory. The new Wilks device appears similar to the old Foxboro 203.

THE FOXBORO MODELS 1A AND 1B

Technically, though they are not dispersive, these instruments do offer the ability to scan the entire spectrum and compare it with known spectra with use of a filter wheel. However, until the more recent models (lB2 and lBX) were designed to export these spectra to a microcomputer, and software was written to make the comparison, the analysis required a wealth of experience or guesswork. As a result, these instruments were largely used as quantitative devices for known compounds.

The original MIRAN (miniature infrared analyzer, see Figure 27-4) 1A is still used in hygiene work (try an Internet search for used equipment and equipment rentals). It is highly versatile because it allows one to dial in any wavelength and because of the previous work done by Foxboro to determine operating parameters for several hundred gases and vapors (a used instrument is somewhat less useful without a manual). In addition, other researchers have published information on important applications.

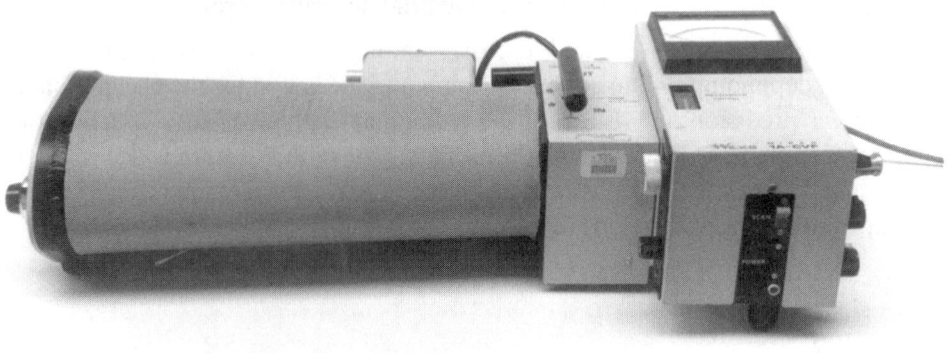

Figure 27-4. A Foxboro MIRAN 1A. Total length of the instrument is about 2 feet or 0.6 m (photo, R.K. Oestenstad).

The analog output lA requires that several parameters be set. The optimum wavelength must be found, and it is not always the wavelength at which absorption is at a maximum, as absorbance is measured indirectly by difference. If the contaminant concentration is small, the amount absorbed is small, and the difference in the light from the source and the detected sample is small. This causes a small signal-to-noise ratio, and the sample is not likely to be differentiated from background noise. Now, if the IR is passed through the cell multiple times with a system of mirrors, more and more of the source light will be absorbed. Thus, an instrument only about 0.4 m long can have a cell path length of 20.25 m by using a system of mirrors. The longer paths decrease the limit of detection and improve the calibration sensitivity for a given gas; so we might ask, why not leave the path length at 20.25?

The answer lies in the deviation of the MIRAN response from linearity, or its deviation from Beer's law:

$$A = \xi \times b \times c \qquad \text{[Equation 27-1]}$$

where A is the absorbance, ξ is a proportionality constant, sometimes referred to as the absorptivity constant, b is the path length, and c is the concentration. The law states that the amount of light absorbed is proportional to the concentration. However, deviations from Beer's law are the norm for a variety of reasons, including concentration-dependent interactions between molecules that alter their ability to absorb IR and instrumental effects such as stray radiation (Skoog et al., 2007). In short, spectroscopy becomes a compromise between achieving good calibration sensitivity and obtaining responses that are in the linear range of the instrument (i.e., low to medium absorption). As Beer's law states that absorption is related to path length, maximum path lengths may extend sensitivity at the expense of linearity.

The shape of a typical calibration curve for a MIRAN 1A is shown in Chapter 26, Figure 26-5. The response is measured and read (from a meter) in absorption units. Note that the relationship is fairly linear, usually up to about 0.5 absorption unit. There are three approaches to using the calibration curve: (1) use straight-line approximations as shown in Chapter 26, Figure 26-4; (2) plot the calibration curve with ample calibration points and read each unknown concentration off of the plot; and (3) use a curve-fitting computer program to fit a quadratic equation to the relationship, and use the equation to predict concentration from field absorbances. The first two approaches are now less common. The last one is used in the more advanced MIRANs (IB2 and IBX). In addition, the data from, say, a 1A can be input to a spreadsheet and a curve fitting function can be used to determine the relationship between concentration and absorption. Calibration procedures for the MIRAN are described in Chapter 30.

Unlike the model lA, the later 1B models were microprocessor-based and have digital (and analog) output. The various parameters, wavelength, path length, and calibration constants can be entered into the computer and stored. Parameters for about 100 compounds are stored during manufacture. The IR spectrum for a sample can be exported to a computer and analyzed with Foxboro software (Spectra/Match). The software gives the best possible matches based on a library of 400 compounds and is needed if the 1B is used.

MIRAN SapphIRe SERIES

There are currently three MIRAN models (DL, SL, XL) marketed by Thermo Fisher Scientific Corporation that can be upgraded from one to the other (see Figure 27-5) . The main difference in the models is the number of single gas applications that are pre-programmed into the device allowing for quantitative analysis of up to 120 gases. This is useful for situations where the instrument is likely to be used in one location (plant) and the compounds of interest are known or where one or a few analytes (e.g., ethylene oxide and nitrous oxide in hospitals) will be measured.

However, since all three instruments have the ability to draw contaminated air into the sample cell and perform an infrared scan of the air sample, they might be classified as multiple substance DRI. On the other hand, the number of *unknowns* that can be quantitated in this way is limited. The personal computer-based ThermoMatch® software can be used to calculate a correlation function against a library of known absorption spectra for more than 150 compounds. Spectral similarities between the unknown scan and the library of spectra are compared and the most appropriate match is reported. The reliability of these matches for complex atmospheres would be uncertain.

IR INTERFERENCES

There are potential interferences in using any of the described IR instruments, both dispersive and non-dispersive. When IR is absorbed by a molecule, the wavelength absorbed is

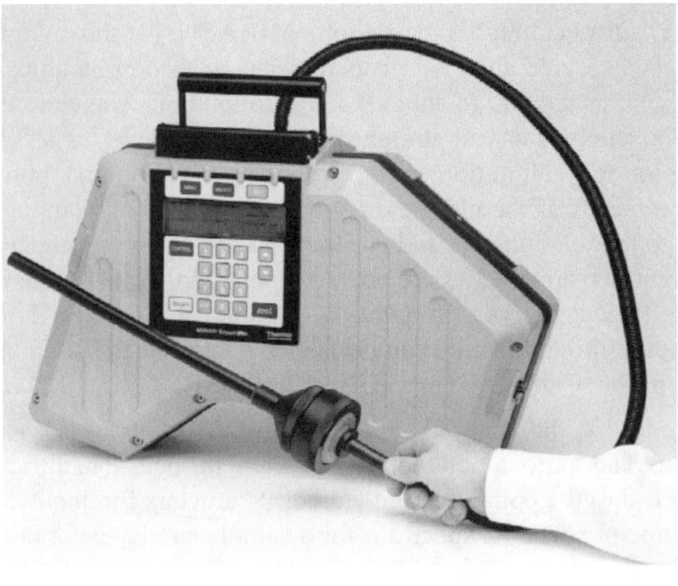

Figure 27-5. The MIRAN SapphIRe Portable Infrared Ambient Analyzer (photo, Thermo Fisher Scientific Corp.).

related to the functional groups contained in the molecule. The molecular vibration frequency has to match the frequency of the light. The absorption causes changes in rotational and vibrational energy; the amplitude of the vibrations changes, not the frequency. So generally only a relatively narrow band of IR will fit a particular functional group. For example, if an IR instrument were set to monitor at a wavenumber of 1050 cm^{-1} (9.52 µm), many alcohols would cause a response. Even compounds with more differing structures can be interferences for one another if the monitored wavelength corresponds to a functional group that the two have in common. For this and reasons discussed later (see FTIR), filtering IR analysis is semi-quantitative unless the sources of contaminants are well characterized. To reiterate, if one uses a MIRAN SapphIRe or 1A to scan an air sample that contains many substances, it is possible that the monitored wavelengths that are absorbed by the analyte of interest are unique, i.e., they are not absorbed by other compounds in the atmosphere. However, as the atmosphere increases in complexity this probability decreases; thus, IR is best applied for repeated monitoring in settings where the atmosphere, and potential interferences, have been well characterized.

APPLICATIONS

The MIRAN has been applied in many settings and for many different analytes. Literature exists for a variety of compounds, including ammonia (Manninen, 1988), styrene (Guillemin et al., 1978), ethylene oxide (Mortimer et al., 1986; Yoshida et al., 1989), acrylonitrile (Jacobs and Syrjala, 1978), and no doubt others. It has been used to monitor tracer gases in the study of reentrainment of exhausts (Barnett et al., 1983; Khoury et al., 1988). It is a particularly effective device for monitoring sulfur hexafluoride, a widely used tracer gas. In addition, the MIRAN has been used to monitor the release of solvents from laboratory fume hood carbon adsorption beds (Kelmig et al., 1991) and as an analytical device in the study of permeation of chemicals through industrial gloves (Nelson et al., 1981; Perkins and Ridge, 1986).

Perhaps one of the most interesting applications is in the monitoring of the anesthetic nitrous oxide (N_2O) in dental, veterinarian, and surgery suites (Hannifen et al., 1978; Bishop and Hossain, 1984; Burroughs and Woebkenberg, 1988; Potts and Craft, 1988; Burkhart and Stobbe, 1990). N_2O cannot be adsorbed or absorbed conveniently using active methods; there are passive devices on the market, and several of the studies utilizing the MIRAN have involved the validation of these passive devices. The sample is taken by using a pump and a gas sampling bag. The bag may be attached to the person (see Figure 27-6), or if the person, usually an anesthetist, is fairly immobile, the bag is fixed to a stand, and a sampling tube is connected to the person's lapel. Once the sample is collected, the MIRAN is used to analyze it. However, the sample has to be introduced to the MIRAN with some thought.

The sampling cell volume of the MIRAN is 2.3 L to 6 L, depending on the model. An attached pump is used to move air into the sample cell. Therefore, the response time of the instrument is related to the mixing properties of the cell. If a concentration of 100 ppm is contained in the cell, and then a new sample is pumped into the cell at a concentration of 50 ppm, some time will pass before the concentration in the cell becomes 50 ppm. If a bag sample were

Figure 27-6. Hospital workers wearing gas sampling bags.

to be introduced into the cell via the sample pump, the original air in the cell would gradually mix with the air from the bag. If there were enough air in the bag to complete the process, eventually the concentration in the cell would approximate that from the bag. How much air would need to be collected in the bag? The following equation can be used to estimate this:

$$\frac{C}{C_{\max}} = 1 - e^{\frac{-Qt}{V}} \qquad \text{[Equation 27-2]}$$

where C_{max} is the maximum concentration or the concentration in the bag, C is the concentration in the cell at any time, t ($C = 0$ at $t = 0$ for a cell that is uncontaminated), Q is the flow rate, and V is the volume. The MIRAN pump flow rate is 30 Lpm, and we can use a cell volume of 5.6 L (the model 1A). In theory, the concentration in the cell will never reach that in the bag, but will only approach it. We can calculate the time to reach some fraction of the bag concentration, say, 0.95. In this case, the time is about 0.5 minutes. This does not seem long, but at 30 Lpm we would need to collect a 15 L sample. If we wanted the concentration in the cell to reach 99% of that in the bag, we would need a 26 L air sample. Finally, these are minimum sample sizes because the equation assumes perfect mixing in the chamber, and the deviation from perfection is unknown.

The unknown effects of imperfect mixing and the bulky sample bags that would result from large samples make an alternative method necessary. The MIRAN 1A and 1B models have valves on the sample cell that are virtually leak-free down to 10^{-5} torr. Therefore, with a vacuum pump, we can evacuate the cell, close the valves, attach the bag with tubing (making sure we have at least 6 L), open the valve, and take the reading. This is a very simple approach and many samples can be analyzed in a short time. If other anesthetic gases need to be measured, 6 L of sample is needed for each. The instrument is purged, zeroed at the appropriate wavelength, and evacuated, and then a new sample is introduced. The MIRAN may also be used as a real time instrument in these situations; however, in order to obtain a TWA, the strip chart recording must be integrated or a datalogger must be used (the 1A has only analog output). It should be noted that the MIRAN instruments (except the 203 model) do require an AC power source, and that the 203 model cannot be evacuated.

Another application is in the monitoring of CO_2 in indoor air quality studies. There is a problem in calibrating a MIRAN 1A or 1B, however. Normally, clean air spiked with known amounts of contaminant is used for calibration, and to provide a zero background, organics can be filtered from room air with a charcoal canister. Because CO_2 ambient concentrations are at least 400 ppm, the trick is to find CO_2 free air. Bottled nitrogen could be used.[7] However, it would need to be humidified to roughly room RH to mimic the atmosphere, as water vapor is an absorber in virtually all regions of the spectrum. Burroughs and Woodfin (1988) used ascarite to scrub the CO_2 from room air and then used pure CO_2 to create known concentrations.

One last point about the MIRAN is important and concerns the potential effects of condensation within the sample cell. Water vapor or compounds with low vapor pressure may condense. Samimi (1983) concluded that calibration of the MIRAN for compounds with vapor pressures of < 15 mmHg could lead to up to 30% overestimation of concentration owing to the effects of condensation. Guillemin et al. (1978) found that in monitoring styrene, airborne polymer would polymerize on the optics. A filter is now available for avoiding this latter problem. In using the device to measure glove permeation rates, low vapor pressure compounds will condense on the optics at high concentrations (Perkins and Ridge, 1986). A heater jacket is available that allows one to evaporate the condensed compounds. In addition, the jacket can be used in the field to monitor condensible compounds. The silver bromide lenses can be damaged in some environments, and the Teflon lining of the cell can gain a memory because of adsorption of compounds that are then slowly released.

---

7 IR does not interact with molecules that will not form a dipole. Therefore, nitrogen, oxygen, and hydrogen are not part of the background absorption of the atmosphere.

Photoacoustic IR

Because molecules that interact with IR have increased, short-lived energy, they temporarily release heat. If the molecules are contained in a sealed chamber, the increased temperature corresponds to increased pressure; and if the IR source is intermittent, pressure changes will occur as a series of pulses. Pressure pulses can be sensed by the human ear as sound. Alexander Graham Bell discovered in 1880 that various substances placed in a test tube could be heard when they were pulsed intermittently with light. His cigar gave a large pop (Bruel and Kjaer, 1991). The microphone had not been invented yet, but it is now known to be an extremely sensitive, stable, linear detector of sound.

The Bruel and Kjaer Company of Denmark, long known for its noise monitoring equipment, introduced a photoacoustic IR device (apparently the instrument is now made and sold by Innova as shown in Figure 27-7 and there are now numerous other companies marketing a Photoacoustic IR). The IR source is focused at the sample cell, and is chopped; the chopper provides the pulsing that leads to the pressure waves; then the light is filtered. The filter wheel can contain up to five filters, including one that is used for water vapor. The sampling cell is fitted with two valves that are closed at the end of the sampling cycle. In order to purge the sampling line and cell, an air sample of about 140 cm³ is needed. The coupling of purge time and analysis time gives an overall response time of 0.5 to 2.5 minutes for analyzing one to five gases.

Because the microphone has a good signal-to-noise ratio, it is quite sensitive (LODs from 0.01 to 1 ppm). The response is linear over a very wide range (five orders of magnitude); so calibration is performed only at zero and at a known concentration that is at least 100 times the LOD. The microphone response is very stable so that calibrations remain effective for up to 12 months. The zero drift is reported to be ± LOD over 3 months.

Like the MIRAN B series, the instrument is selective to the extent that interferences are

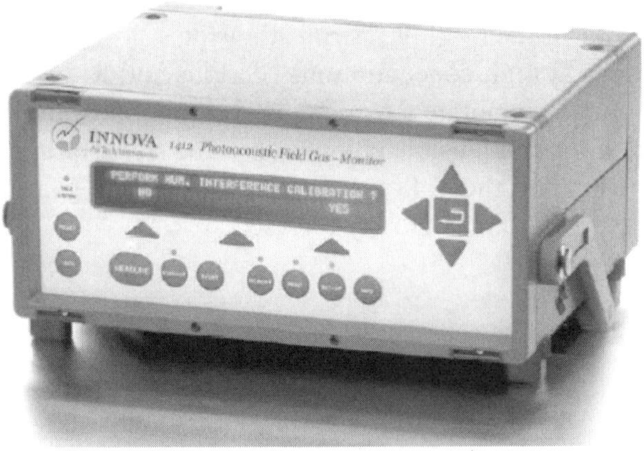

Figure 27-7. A photoacoustic IR instrument (courtesy, Innova).

known to exist in the environment in question. For example, in one application of the instrument (the measurement of N_2O) it was known that water and CO_2 were interferents (Ghittori et al., 1993). Consequently, filters for water and CO_2 were employed. The instrument stores the necessary calibration information and corrects for the known calibrated interferents. Assuming that these were the only interferents, the remaining signal was due to N_2O.

Colorimetric Devices

THE PAPER TAPE

It almost seems too simple to work well – a paper tape is impregnated with chemicals. The tape moves past an exposure chamber, where air is sucked through the tape and a reaction between the tape chemicals and the analyte(s) causes the tape to change color. The tape then passes a fiber optics light source. The reflected light is detected and compared to reflected light from an unexposed surface of the tape (see Figure 27-8). The paper tape is a method that has been around since 1972 (Honeywell, 2006; Denenberg and Kriesel, 1976). The device can be portable or transportable and remote sensing devices are available.

The list of chemicals for which tapes have been made is quite long (see Table 27-6) and is generally limited to reactive organic and inorganic substances. The process is applied to single substance analysis only. There are interferences. For example, triethylene diamine will interfere with the toluene diisocyanate (TDI) tape (Walker and Pinches, 1981). In another study, SO_2, NO, NO_2, and dichloride gases gave interferences for the TDI device (Dharmarajan and Rando, 1980). The repeatability of the device depends on the constancy of the tape movement as well as the pump flow rate. The manufacturer claims repeatabilities of better than 5% at the TLV®. In addition, the sensitivities are quite good (usually much less than 1 ppm), and response times are 10 to 240 seconds at the TLV® and 2 to 65 seconds at 10 × TLV®. However, at least in an older model, the response times for TDI were much longer (5–9 minutes; Dharmarajan and Rando, 1980), and repeatability was poor among nine instruments (Dharmarajan, 1996; Wanek, 1996). In addition, there were issues regarding a negative bias associated with high humidities (Dharmarajan, 1996a and b) and for a second manufacturer, summer temperatures experienced during shipping affected the tapes (Dharmarajan, 1996b). These issues are very important in light of the severely acute toxicities of some of the compounds in Table 27-6 and should be monitored carefully.

WET COLORIMETRIC DEVICES

The colorimetric technique was described in Chapter 19. The gas interacts with liquid chemicals, one of which has a color-producing region, a chromophore. The color change is detected by the absorbance of light at a specific wavelength. The method certainly seems like a laboratory method; so it is somewhat surprising that it has been adapted to a direct-reading, transportable (30 lb or 13 kg) device (see Figure 27-9). While TDI has been the primary ana-

a)

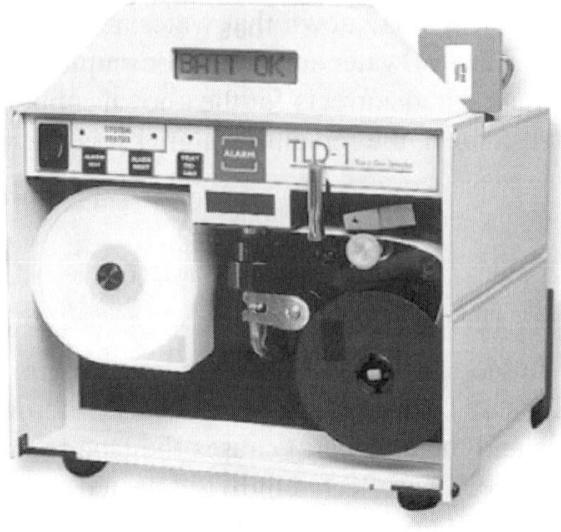

b)

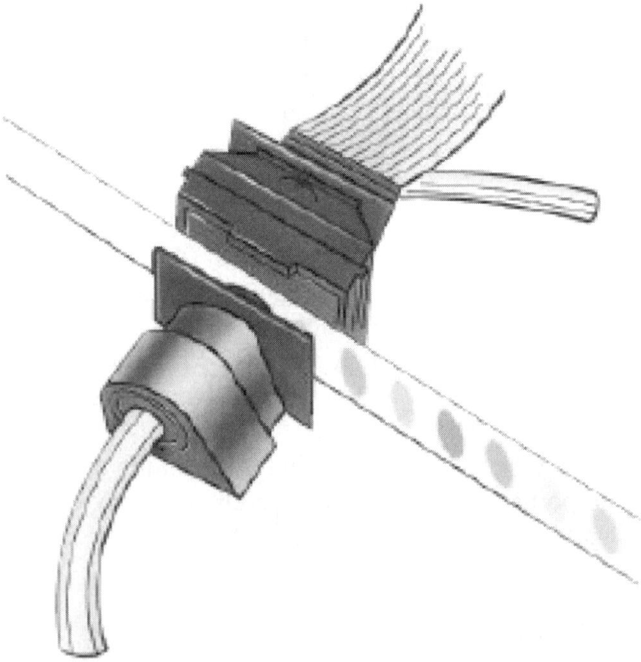

Figure 27-8. a) A portable paper tape device and b) a diagram showing the principle of operation (courtesy, Honeywell Analytics).

Table 27-6
Substances for which instruments using paper tapes have been commercialized

Amines	**Hydrides (cont.)**
Ammonia	Dislane
Methylene Dianiline	Germane
n-Butylamine	Hydrogen Selenide
p-Phenylene Diamine	Phosphine
Toluene Diamine	Silane
Trimethylamine	Stibine
Diisocyanates	t-Butylarsine
CHDI	t-Butylphosphine
HDI	Hydrogen Cyanide
HMDI	Hydrogen Sulfide
IPDI	**Mineral acids**
MDI	Hydrogen Bromide
NDI	Hydrogen Chloride
PPDI	Hydrogen Fluoride
TDI	Nitric Acid
TMDI	Sulfuric Acid
TMXDI	**Oxidizers**
XDI	Bromine
Hydrazines	Chlorine
N_2H_4	Chlorine Dioxide
MMH	Hydrogen Peroxide
UDMH	Nitrogen Dioxide
Hydrides	Ozone
Arsine	Phosgene
Diborane	Sulfur Dioxide

lyte for paper tape devices, the colorimetric device has evolved for formaldehyde monitoring. However, Table 27-7 shows that a wide range of applications are now available.

ELECTRONIC ANALYZERS BASED ON COLORIMETRIC DETECTOR TUBES

Recently, the Draeger Company has introduced a product called the Chip Measurement System (CMS®). The CMS® consists of substance specific chips that are composed of 10 capillaries containing granules based on Draeger's reagent system contained in hand-held colorimetric indicating tubes.[8] An analyzer contains optics, mass flow controller, pump system, data

[8] Colorimetric indicating tubes are discussed fully in Chapter 29.

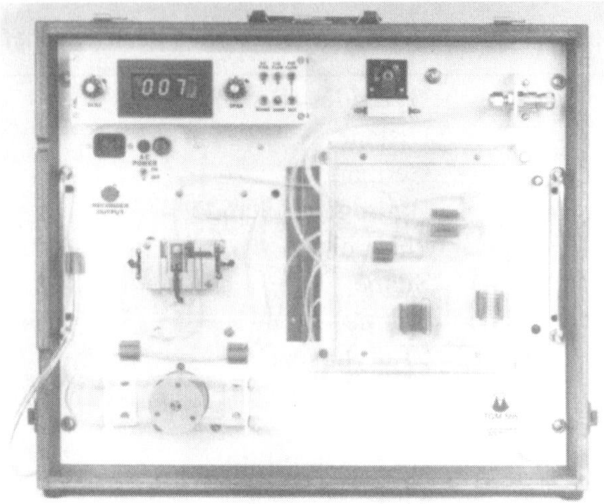

Figure 27-9. A transportable colorimetric device (courtesy, CEA Instruments).

logger, and electronics. The chips are calibrated at the factory and said to have a shelf life of two years. Like detector tubes, the measurements are TWAs for short time periods, typically 1–2 minutes at concentrations near OELs and up to 10 minutes for low concentrations. However, the device stores up to 50 measurements. Unlike the detector tubes, the various reagents in the capillaries control some interferents. The CMS® chips are said to have accuracies of ± 4–7%; however, there have been few independent studies. Verma et al. (2001) determined that, for the benzene system, precision was 20–44% in nine trials (n = 2–9) at concentrations from 0.4 to 9 ppm. Systems are available for about 35 different gases and vapors.

Chemiluminescence

Chemiluminescence is the generation of photons (light) as a result of a chemical reaction. This process produces light in the firefly and some other species. The phenomenon does not occur in every reaction and, in DRI, it is common only for measuring ozone and oxides of nitrogen. The devices are heavy and might be considered transportable rather than portable. In the NO_2 devices, ozone is generated to react with NO and yield luminescent NO_2. The total NO_2 is then thermally decomposed to NO, and again reacted with ozone. Thus, both NO and NO_2 can be determined by difference. For ozone analysis, the ambient air is mixed with a dye (rhodamine-b) on silica gel, which yields luminescence. Alternatively, ozone will react with ethylene to give off light. Both processes are reported to be quite selective for ozone. Hence, the ozone devices might be classified as quantitative rather than semi-quantitative. Chemiluminescence is a very sensitive method; the LOD for ozone is typically 1 to 10 ppb.

Table 27-7
Analytes and colorimetric methods used in DRI from CEA Instruments (see CEA for further details)

Parameter	Methodology	Maximum Full Scale Range (ppm)
Ammonia NH_3	Modified Bertholot	0–50
Chlorine CL_2	DPD Procedure	0–5
Ethylene Glycol MEG	Na Paraperiodate	0–50
Formaldehyde HCHO	Pararosaniline	0–10
Hydrazine N_2H_4	p-DMAB	0–20
Hydrogen Chloride HCl	Thiocyanate Method	0–50
Hydrogen Cyanide HCN	Chloramine T/Pyr-Barb	0–1
Hydrogen Fluoride HF	Zirconium-SPADNS	0–10
Hydrogen Peroxide H_2O_2	KI/Mo/DPD	0–5
Mercaptans	PAMA/$FeCl_3$	0–20
Nicotine NIC	CNBr-Aniline	0–10
Nitrogen Dioxide NO_2	Mod. Griess-Saltzman	0–10
Oxides of Nitrogen NOx	CrO_3 Oxidation	0–10
Phenol C_6H_5OH	4-Aminoantipyrine	0–10
Phosgene $COCl_2$	NBP/PBA	0–5
Sulfure Dioxide SO_2	West-Gaeke	0–10
Total Aldehydes TA	MBTH	0–10
Total Oxidants TOx	KI/DPD	0–5

GENERAL SURVEY INSTRUMENTS

These devices are typically used to measure total organics or total hydrocarbons. We will see that neither term is really concise, as the detectors usually respond to substituted hydrocarbons (e.g., ketones) as well as many inorganic compounds, and the response to each varies; thus, a mole of hexane, for example, does not cause the same response as a mole of methyl butyl ketone although both are six carbon compounds.

There are two main classes of these semi-quantitative instruments. Generally, the detectors used in the two types are different, but the main difference is the sensitivity of each. Combustible gas meters or indicators (CGM) are used to detect explosive or combustible atmospheres. They need not be very sensitive, and so they are not. General survey meters (GSM), also known as total hydrocarbon analyzers, usually employ different sensors and have LODs in the ppm or less range. They are used to find leaks and hot spots, to test the before and after effects of controls, and for many other uses. Even though neither is quantitative or selective, if the atmosphere is well characterized, the instruments can be *de facto* quantitative. Moreover, these detectors can be coupled with portable gas chromatographs and thus be highly selective and quantitative instruments.

Combustible Gas Meters (CGM)

A gas or a vapor is combustible if it will combine with oxygen in a continuous fashion when ignited or exposed to a source of sufficient activation energy such as a spark or a flame. The result is a fire; or if the gases are somewhat confined, the pressure will build initially, and an explosion will occur. Thus, some CGMs are referred to as explosivity meters. The propensity for combustion to occur depends on the concentration of the fuel[9] and the oxidant.[10] The fuel concentration must be within the range of the lower and upper flammable limits (LFL and UFL), sometimes referred to as explosive limits (LEL and UEL). These limits are given in Appendix A for several hundred compounds. They are typically determined at 20°C. When the temperature is higher, the limits widen (de Ris, 1981).

Any time a source of ignition will be generated, such as occurs in welding, the concentration of explosive gases should be known. Typically, the concentration is determined with the CGM (see Figure 27-10). There are three types of detectors used in CGMs: (1) The old hot wire may still be used, but the platinum wire must be operated at 800°C in order to burn methane. This causes the platinum to evaporate and the zero to drift greatly. (2) The catalytic device, referred to as the pellistor, is overwhelmingly the favorite. (3) Some MOS detectors are used.

THE PELLISTOR

This detector is sometimes referred to as a catalytic bead or solid state, but while it is solid it is not based on a semiconductor like the MOS. A platinum filament (wire) is covered with a catalyst-coated bead such as alumina. The catalyst may be palladium. Its use allows the combustion of methane at lower temperatures, say, 500°C, than does the hot-wire device. The device is referred to as the pellistor in Europe (Gentry and Walsh, 1986). Typically two detectors are used; one has been poisoned or has no catalytic coating, and thus the combustible gases are not oxidized. It may or may not be in the gas stream. At the other sensor, the gases are oxidized, and the heat of this reaction changes the resistance through the sensor. The two

---

[9] In our case, usually an organic gas/vapor is the fuel, but many gases and vapors will burn, including CO and H_2S. For that matter, some metals (e.g., sodium) will burn vigorously when exposed to oxygen.

[10] Again, in our case, we are speaking of oxygen, but many oxidizers, such as chlorine, can cause a reaction as violent as a fire.

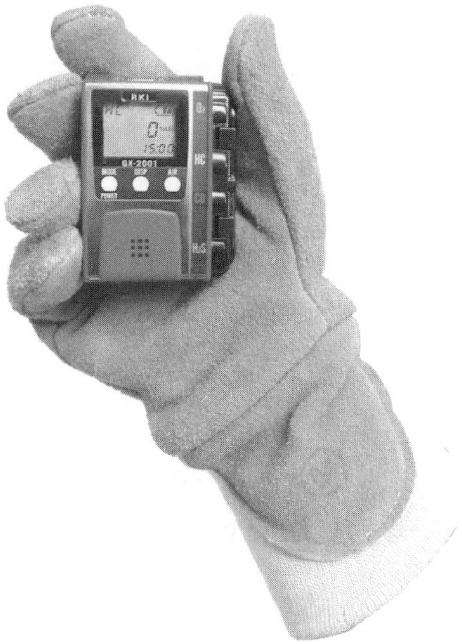

Figure 27-10. A very small four-gas personal CGM (courtesy, RKI Instruments and REPSS).

sensors are typically positioned as elements in a Wheatstone bridge (see Figure 27-11). When no combustible gases are in the atmosphere, the Wheatstone bridge is balanced, and the meter needle, coupled to the null detector, reads zero. As the concentration increases, the bridge becomes unbalanced, and the power source adjusts the null detector to rebalance the bridge; thus, the needle moves.

Readout

The CGM meters generally read percent of the LFL. As shown in Appendix A, the LFLs of compounds are in the low percent range. For example, the LFL for hexane is 1.2%. If a CGM calibrated for hexane reads 50% LFL in a hexane-contaminated environment, then the true concentration is $50\% \times 0.012 = 0.6\%$. A percent is a part per hundred, and there are $10^6/10^2$ or 10,000 ppm in 1%; so we can convert percentage to a decimal and multiply by 10^6 ppm, or $0.006 \times 10^6 = 6{,}000$ ppm, thus determining that 50% of the LFL of hexane is 6,000 ppm. This is not quite an explosive atmosphere, but it is definitely a toxic one. Even 1% of the LFL (120 ppm) is a chronic hazard, as the 8-hr TWA-TLV® for hexane is 50 ppm. Thus, CGMs are used to measure high concentrations of gases and vapors for prevention of fire and explosion hazards, but the readout can be converted to a concentration that can be related to toxic effects.

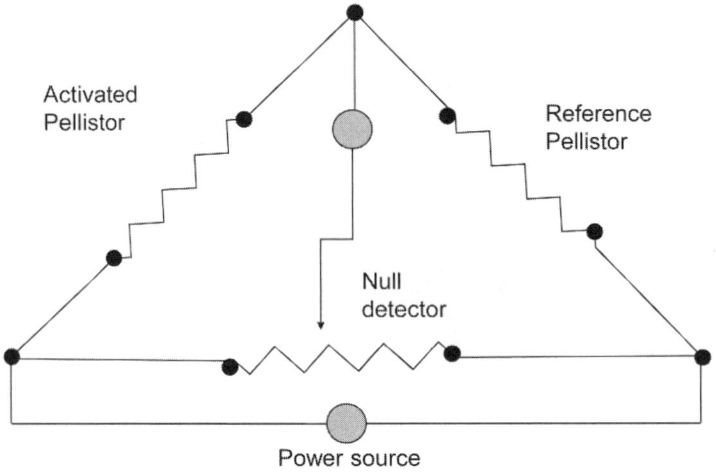

Figure 27-11. A Wheatstone bridge used in a CGM Pellistor device.

Relative Response

The pellistor-based CGM is essentially linear in response when the calibration gas is the gas being measured. Furthermore, it has been stated that the heat of combustion of a gas is proportional to its LFL and also proportional to the change in resistance in the sensor (AIHA, 1993b). Thus, when the meter reads 50% LFL, it should not matter what compound (hexane or methane) was used for calibration, even though the possible compounds have different LFLs. However, this is not the case in practice. Figure 27-12 shows the response of an instrument when calibrated with pentane. The response is linear, but the pentane-calibrated instrument does not respond linearly to other compounds. There are at least two reasons for this. First, the heat of combustion is mass-based, such as kJ/g, whereas the readout on the CGM is volume-based (%LFL is volume per volume). If the readout were in terms of mass, the linearity would be improved, but not perfect. The second reason is that a compound's heat of combustion is not directly proportional to mass. Substituted hydrocarbons such as dioxane and ethyl chloride will produce less energy from the Cl and O atoms than they would if a carbon were there. Therefore, if an instrument is calibrated with hexane or methane, which is usually the case, and used in an atmosphere containing methyl ethyl ketone or a mixture such as gasoline, the reading will be an approximation. However, the response should be within a factor of 5 of the true concentration. Thus, in poorly characterized atmospheres, taking action at 5% of the LFL, rather than the customary 10–20% LFL, provides a margin of safety.

Interferences

For CGMs the term *interferences* is applied here to mean any compound that causes a

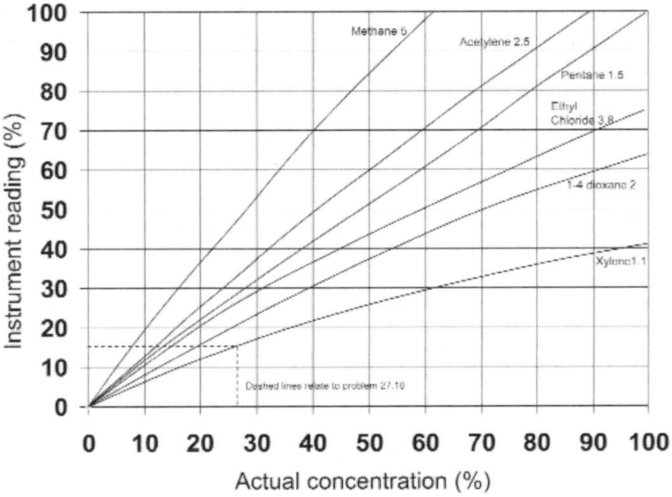

Figure 27-12. CGM response curves, when instrument calibrated with pentane, for several other compounds (numbers at end of names are LFLs). Y-axis = response as % LFL, X-axis = concentration as % LFL.

response (±) that will not equivalently affect the combustibility of an atmosphere. Water and oxygen are the two most important interferences for pellistors. Of course, oxygen is needed for combustion to occur, and so the sensor needs it. In theory, the amount of O_2 necessary is one molecule for each carbon atom. So a mole of hexane should require 6 moles of O_2; or because % by volume can be shown to be equivalent to % by mole, then hexane at 1.2% should require 7.2% O_2. Brief and Confer (1969) showed this was approximately true in laboratory tests. They recommended a minimum level of 10% O_2 as a general guideline. If the O_2 level is not sufficient, then the CGM may read too low. In practice, the meter may peg (go all the way to the right) and then return to O (Ness, 1991), or peg and then return to the low reading (Brief and Confer, 1969) when O_2 levels are below 10%. However, the health consequences of either result may be unpredictable. If the O_2 level is below 10%, then combustion may be more difficult to start (as when starting a campfire at high altitude), and would be incomplete. In summary, if the O_2 level exceeds 10%, then the meter should not be affected; but if the O_2 level is less than 10%, another detector that can be used in lower O_2 environments, such as the solid state (see later discussion), probably should be used.

When measuring large concentrations of organics, the pellistor should have little interference from water vapor. In one instrument so equipped the effect was small. The TLV® Sniffer (Scott Instruments) is a pellistor device that can be used to measure ppm concentrations of organics, presumably by amplifying the signal. The instrument is normally calibrated with hexane. When the instrument was calibrated in dry air at 2,500 ppm (about 20% LFL), and then similar concentration challenges were presented at relative humidities up to 85%, the effect was negligible, ± 4%. Similarly, at 750 ppm, the effect was only +8%. But at 20 ppm the effect was +100% (Chelton et al., 1983). This may be one reason why most instruments employing the pellistor have only the high (%LFL) scale.

Other Confounders

Temperature and flow rate also may confound pellistor-based CGM measurements. Although the sensor is heated to a high temperature and thus is relatively independent of ambient temperature fluctuations, the sample line length may be a factor. Many instruments use tubing and pumps to bring the sample to the detector. Thus, a hot sample of a vapor with a relatively low vapor pressure could partially condense in the cooler line before reaching the detector. Consequently, an explosive atmosphere of such a vapor could go undetected. Some devices allow the detector to be placed in the atmosphere to be tested (say, a tank) and tethered to the instrument. These sensors usually operate on diffusion rather than pumping of the atmosphere to the detector.

Many instruments use a small porous disk that limits the rate of flow to the detector according to the diffusion of the gas(es) through the disk. The disk is used to prevent flash fires from leaving the combustion chamber (Stetter, 1984). High and variable flow rates in these instruments should have no effect on the instrument, by either cooling the sensor or bringing more mass to it. On the other hand, very low flow rates could *starve* the sensor and cause a negative bias if the diffusion mass transfer across the disk were to exceed the mass flow rate into the instrument. It is doubtful that a pump-based instrument would suffer from such effects.

Sensor Poisoning

Several things can happen to the sensor to shorten its life. The catalyst may become chemically inactive (poisoned) as a result of reaction with contaminants, including arsenic and lead compounds (e.g., leaded gasoline) and other metals. Poison-resistant pellistors have been developed (Gentry and Walsh, 1986), but their resistance to poisoning may not be comprehensive. The catalyst may become coated by compounds that are not burned at the operating temperature of the device. These compounds may include silicones and silanes. Lastly, combustion of some compounds may result in products that can corrode the sensor, including halogenated hydrocarbons (HCI), H_2S and other sulfur compounds, nitro compounds, and acids (Ness, 1991).

Alarms

If a CGM is calibrated for hexane, and the concentration in a tank where welding is being performed increases until the CGM reads 100% LFL, then the actual concentration is 1.2% and a real problem exists. This situation needs to be detected before the concentration reaches

1.2%. Therefore, CGMs are equipped with alarms set to sound at a concentration < 100% LFL. Usually 20 or 25% LFL is the alarm level. However, there are reasons to set the alarm level as low as 10% LFL. First, the calibration may be inaccurate. Second, interferences such as water vapor and oxygen could give faulty readings. Third, and perhaps most important, the compound(s) detected may not be the ones used in the calibration. Therefore, using a level of 10% LFL as an alarm, emergency, or other action level seems reasonable, particularly when *hot work* is involved.

MOS SENSORS IN CGMs

The pellistor will not detect every combustible gas. Also its response may be poor for some gases when it is calibrated with hexane or methane. The MOS detector has a more universal response than does the pellistor. In addition, the MOS can be operated for short times (seconds) without O_2 (AIHA, 1993b). Eventually, though, the semiconductor surface must be repopulated with O_2. Then why are MOS devices not used exclusively? Because they too have disadvantages. The response at high concentrations is flat, and they are affected by relative humidities exceeding 70% (AIHA, 1993b). Also, their response is not linear, and thus they can be more difficult to calibrate. They require at least 10% relative humidity, and can become saturated. When this happens, the semiconductor must be *burned* or heated to reactivate it. This removes the contaminants and oxidizes the MOS. Finally, they are apparently more affected by elevated temperatures than pellistors (Ness, 1991).

Some instruments now come with a pellistor and MOS, or two MOS sensors; or a single MOS is operated at varying temperatures. The purpose of all of this is to vary selectivity. One manufacturer that combines the MOS with electrochemical cells (GfG) claims that the MOS in its instrument can detect numerous compounds and "ensures superior protection from hundreds of unknown toxic gases." The user should be aware that although the instrument may detect these compounds at ppm levels, it is not selective for them. The user must do that by characterizing the atmosphere with alternative methods.

The manner in which the variable temperature device (Enmet) is calibrated will change its response. Some knowledge of the environment that it will be used in will allow the user to carefully select the calibration gases and operating temperatures in order to improve selectivity. However, when used in these modes, MOS detectors are not being used as CGMs but rather as general survey meters, the subject of the next section.

Total Hydrocarbon Analyzers

It is important to point out again that the major difference in the CGM and the GSM is the range in which they operate. Typically, CGMs detect percentage concentrations, whereas GSMs detect ppm concentrations. The instruments usually use different detectors: pellistors or MOS in the CGM and MOS, flame ionization, or photo ionization detectors in the GSM; but we have already discovered that there is some overlap. For example, the TLV® Sniffer employs a pellistor but will detect concentrations in the ppm range although without much precision. In addition, the MOS can be used as a CGM or a GSM, but is not the primary detector used in GSMs. The most common GSM detectors are the photoionization detector and the flame ionization detector. In addition, a few instruments utilize thermal conductivity detectors to detect leaks or high concentrations.

Regardless of the detector, GSMs are non-selective and, as with the CGM, quantitation by the GSM is highly dependent on the way that it is calibrated. However, while they are not selective, they can be used in atmospheres where single compounds (leak detection, dry cleaning, for example) or simple mixtures or even complex mixtures are likely to occur. For the latter Coy et al. (2000) used a PID GSM to monitor paint vapors (ketones, esters, alcohols, and hydrocarbons) during spraying operations. The TWA PID results were highly correlated with the summed (across all compounds) charcoal tube/GC TWA results, and since the vapors had similar toxicities, the PID results were also highly correlated with the charcoal tube/GC results when applied to a mixture TLV® (see Chapter 10, Equation 10-1).

FLAME IONIZATION DETECTORS (FID)

These devices are the same as those used in laboratory GCs, described in Chapter 21 under gas chromatography. The sample stream comes in contact with a flame fueled by hydrogen. Carbon atoms become positively ionized when burned. They are attracted to a negatively charged platinum electrode, and a small current is induced.

Compounds that are not combustible will not ionize; nor will compounds without carbon. CO and CO_2 give insignificant responses. Water vapor response is also insignificant. The FID is a carbon counter. The FIDs used on laboratory GCs will give responses to various compounds roughly in proportion to their carbon content.

The detectors have good sensitivity; a few ppm generally can be detected precisely. However, the sensitivity decreases as the hydrocarbon molecule is substituted. Fluorine, oxygen, and nitrogen cause the greatest loss of sensitivity, but chlorine causes less. Most chlorinated hydrocarbons (depending on the degree of chlorination) cause a good response. However, in every case, as the molecule becomes smaller and the substitution more complete (e.g., carbon tetrachloride or ethylene glycol), the response decreases significantly. In some cases, no ions are yielded at all (Skoog et al., 2007).

The FID response is linear. Most laboratory GC FIDs have a linear response over seven orders of magnitude (Skoog et al., 2007); but portable units generally are not so linear. The degree of linearity of the response depends on the same molecular factors that affect the magnitude of the response, such as size and degree of substitution.

Apparently both FIDs and PIDs (see later discussion) are sensitive to O_2 in their response to hydrocarbons. In a study of the Thermo Fisher Scientific PID and an FID, both showed an increase in response with decreasing O_2. At an O_2 level of 15% the response of the FID to methane was about 150% of the response when O_2 was at 21%. This means that an FID calibrated at 21% O_2 would overestimate concentrations. Conversely, an instrument calibrated at 15%, for example, because of an error in the makeup of the calibration gas, would read much too low when the ambient O_2 level was 21% (Mouradian and Flannery, 1994).

There are several battery-operated, portable GSMs that operate on the FID principle;[11] some are approved for use in flammable environments, others are not (Maslansky and Maslansky, 1993). The ideal GSM should respond to all contaminants equally on a volume basis and only one calibration would be necessary. In addition, the relative concentrations of several vapors in a mixture might be estimated, based on their vapor pressures above the source. However, no GSM functions this way. Furthermore, as has been mentioned, the FID responds to carbon; so octane should give a response about eight times that of methane.

Table 27-8 shows the response factors for various groups of compounds relative to methane for the OVA[12] (see the TVA 1000, from which the OVA was derived, in Figure 27-13). The response for all other alkanes is ≥ that for methane. For example, a 100 ppm concentration of hexane will cause a response of 234 ppm on a TVA 1000 calibrated with methane. The relative responses are quite variable within a group of compounds. The response factor for methanol, for example, is 0.28 at 100 ppm, whereas that for 100 ppm n-butanol is 1.18. There are also steric effects; the response for tert-butanol is 1.05. Generally the aliphatic hydrocarbons have relative responses of < 1, while the relative response for aromatic compounds is > 1. As the response for the majority of compounds is < 1, there is a tendency to underestimate true concentrations.

The main differences in the TVA and the OVA are the updated electronics and the addition of a photoionization detector to the FID in the TVA. Although the TVA has replaced the OVA, several evaluations of the OVA have been performed, while no evaluations of the TVA have been reported. NIOSH (1976) found no effect of water vapor and a linear range for methane of 5 to 700 ppm. Que Hee and co-workers (Barsky et al., 1985; Lee et al., 1987) found the response in humid conditions was within -10% of the dry response.

A few notes about the use of TVAs are in order.[13] In field use, methane is sometimes a primary contaminant.[14] A way to subtract it from the rest of the contaminants is needed. This can be done with a charcoal respirator cartridge. The charcoal will not effectively capture methane (or ethane, ethylene, and other gaseous hydrocarbons). Thus, two readings can be made, one

---

[11] Numerous transportable, AC current, FID instruments are available (see ACGIH®, 2001).

[12] The OVA has evolved into the TVA (Toxic Gas Monitor) and is now owned by Thermo Fisher Scientific, who bought it from Foxboro, who bought it from Century Systems.

[13] For a good overview of practical considerations in using DRI, the reader should see Maslansky and Maslansky (1993). Most of the practical use considerations presented here are adapted from that source.

[14] This is particularly true at hazardous waste sites, landfills, petroleum refineries, or any place where bacterial breakdown of hydrocarbons is occurring in soil or water.

Table 27-8
Response factors for the TVA 1000 FID[1] for various compounds relative to methane

Organic Compound Class (number of compounds)	Response Factor[2]
Acetaldehyde	0.25
Acids (aliphatic) (2)	0.3 – 0.8
Alcohols (including phenol) (11)	0.12 – 1.2
Alkanes (13)	0.63 – 1.0
Alkenes (7)	0.29 – 0.85
Amines (aromatic and aliphatic) (7)	0.34 – 1.25
Aromatic hydrocarbons (8)	0.7 – 1.5
Chlorobenzenes (4)	0.5 – 2.0
Esters (20)	0.4 – 0.8
Ethers (5)	0.18 – 0.65
Ethyl Cellosolve®	0.22
Ethyl mercaptan	0.3
Ethylene oxide	0.7
Halogenated (Cl Br) aliphatics (40)	0.1 – 1.1
Hydrazines (2)	0.16 – 0.2
Ketones (10)	0.6 – 1.1
Nitriles (2)	0.7
Nitro compounds (2)	0.34 – 0.98

[1] Thermo offers a Response Factor Handbook containing 68 compounds that should be used rather than these summary data that apply to only one concentration and are for demonstration only.
[2] Multiply the methane-calibrated TVA 1000 reading by these values to get the concentration estimate.

with and one without the cartridge.

The device requires oxygen (air) to work. The manufacturer has found that the flame is extinguished at 14% O_2. However, its intensity is probably affected before that concentration is reached, and this may affect the calibration. The hydrogen tank will last for about one day of use. It can be refilled, but the user should be careful to use only ultra-high-purity fuel (99.999%) to avoid hydrocarbon contamination, which will cause a high background (Maslansky and Maslansky, 1993).

PHOTOIONIZATION DETECTORS (PID)

Many compounds can be ionized by electromagnetic (EM) energy, ultraviolet light being the most universal energy source in this regard. Once ionized these compounds can be detected by a means similar to FID. Appendix A gives the ionization potentials for many compounds.

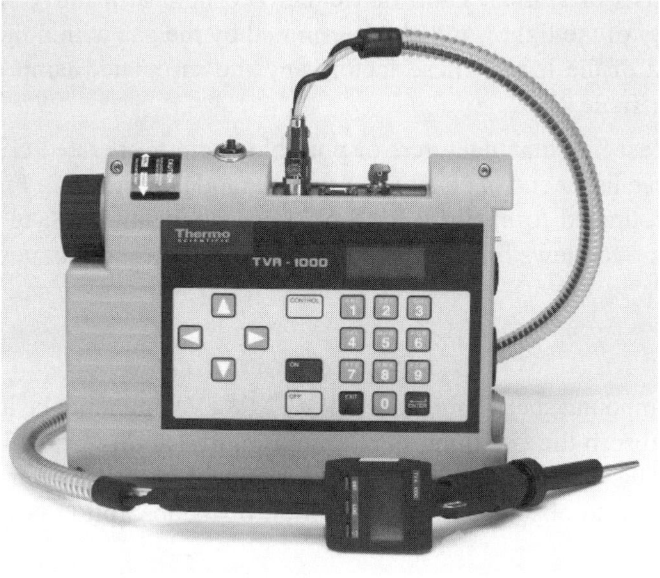

Figure 27-13. A Thermo Fisher TVA 1000 (courtesy, Thermo Fisher Scientific).

These values are the energies of the EM source (in electron volts or eV) necessary to cause ionization of the molecule. The energy of a particular wavelength of EM can be calculated from

$$E = h\,v$$

where E is the energy in eV, v is the frequency of the wave, and h is Planck's constant, 4.1356 $\times 10^{-15}$ eV sec. The wavelength can be determined from

$$c = \lambda\,v$$

where c is the speed of light, 2.9979×10^{10} cm/s, and λ is the wavelength. The shortest wavelengths have the highest energies. The UV spectrum covers a wide range of energies (3.1–124 eV; Shleien et al., 1984), but there are some practical limitations. For example, N_2 ionizes at 15.58 eV. Thus, a source in this range would cause a very high background. In addition, O_2 ionizes at 12.075 eV and water vapor at 12.6 eV. Any sources operating at wavelengths having energies exceeding 12 eV would be of little use as detectors. The wavelength associated with 12 eV is about 100 nm. It should be noted that while some constituents of air are not ionizable by these lamp energies, they can absorb UV when they are present in high concentrations. This absorption depletes the UV available to ionize the analytes of interest and can cause a low reading for those analytes as is described below for methane and water vapor.

The PID consists of a small lamp that emits UV light at a fairly specific wavelength, where the specificity of the light is mostly determined by the gas within the lamp and the window at the exit end of the lamp. These factors and the associated lamp energies for typical lamps are shown in Table 27-9.

There are at least five manufacturers of portable, battery-operated GSMs that operate on the PID principle (see list at end of chapter). A typical model is shown in Figure 27-14. Anyone using a PID is encouraged to obtain a RAE Systems publication (Haag and Wrenn, 2002), which has excellent overviews of PID technology, applications, and issues.

Methane Effects

Although a compound such as methane (IP = 12.98 eV) might not be ionizable with a 10.2 eV lamp, it might absorb the UV light, thus decreasing the ability of the instrument to ionize other compounds. This can lead to false negative results. Remember from Beer's Law that the ability of a compound to absorb light is a function of its absorption coefficient. Absorption

Table 27-9
Typical PID lamp gas and crystal characteristics (adapted from Haag and Wrenn, 2002)

Nominal Lamp Photon Energies (eV)	Fill gas	Major Emission Lines (eV)	Major Emission Lines (nm)	Relative Intensity[1]	Window Crystal	Crystal Transmittance λ Range (nm)[2]
11.7–11.8	Ar	11.83	104.8	10	LiF	105 – 5000
		11.62	106.7			
10.6	Kr	10.64	116.5	200	MgF_2	115 – 7000
		10.03	123.6	650		
10.2	H_2	10.20	121.6	1000	MgF_2	
		10.20	121.6	500		
9.8–10.0	Kr	10.03	123.6	650	CaF_2	125 – 8000
9.5–9.6	Xe	(9.92)	(125.0)	(250)	BaF_2	135 – 9900
		9.57	129.6	1000		
		8.44	147.0	600		
9.5	O_2	9.52	130.2	900	CaF_2	
		9.50	130.5	600		
		9.49	130.6	300		
8.4	Xe	8.44	147.0	600	Al_2O_3	145 – 4500
					SiO_2	145 – 2300

1 Relative emission intensity, from CRC Handbook of Chemistry and Physics.
2 Wavelength limits for 80% transmittance through a 1 mm thick crystal. Wavelength calculated as λ(nm) = 1239.84/E(eV).

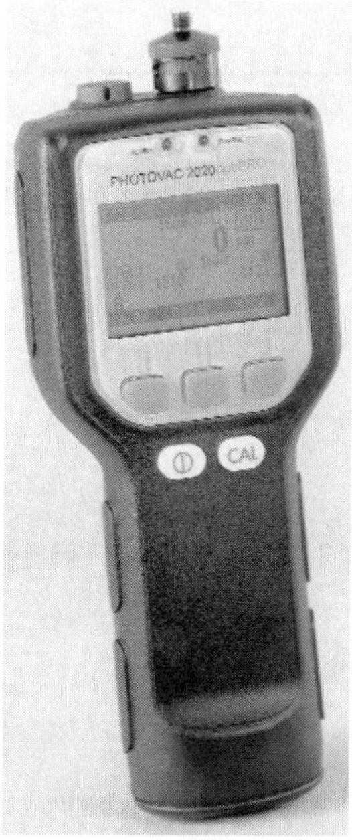

Figure 27-14. The Photovac 2020 ppb PRO handheld PID (courtesy, Photovac and REPSS).

coefficients are generally insignificant except for those of N_2O and methane. Three makes of PID were tested in a 100 ppm toluene environment in which the methane concentration was gradually increased from 0 to 5%. Detector sensitivity for the toluene decreased exponentially to almost 0 for two Photovac models, and for an HNU model sensitivity went to 0 at 3% methane. At 0.5% methane the response was reduced by roughly 30% (Nyquist et al., 1990). Thus, relatively large methane concentrations (the LEL is 5%) will affect PID output.

Water Vapor Effects

Water vapor concentration variations are one of the largest sources of error for PIDs. The 11.7 or 11.8 eV lamp should not, in theory, ionize water (IP = 12.6 eV). In one study, however, (Barsky et al., 1985), the AID PID (apparently no longer manufactured by AID), equipped with an 11.8 eV lamp, responded erratically in atmospheres of various contaminants when the RH was 90%. The instrument had been calibrated at 0% RH, as is usually the case since calibration gases are dry. Figure 27-15 shows the effects of water vapor on the response for one instrument. Because the RH level has an obvious effect on results, manufacturers provide a humidifier for calibration gases. It consists of a permeable polymer tubing that is used between the calibration gas and the instrument. The tube is permeable to water vapor, and allows the RH level in the calibration gas to reach that of the surrounding air. This implies that the instru-

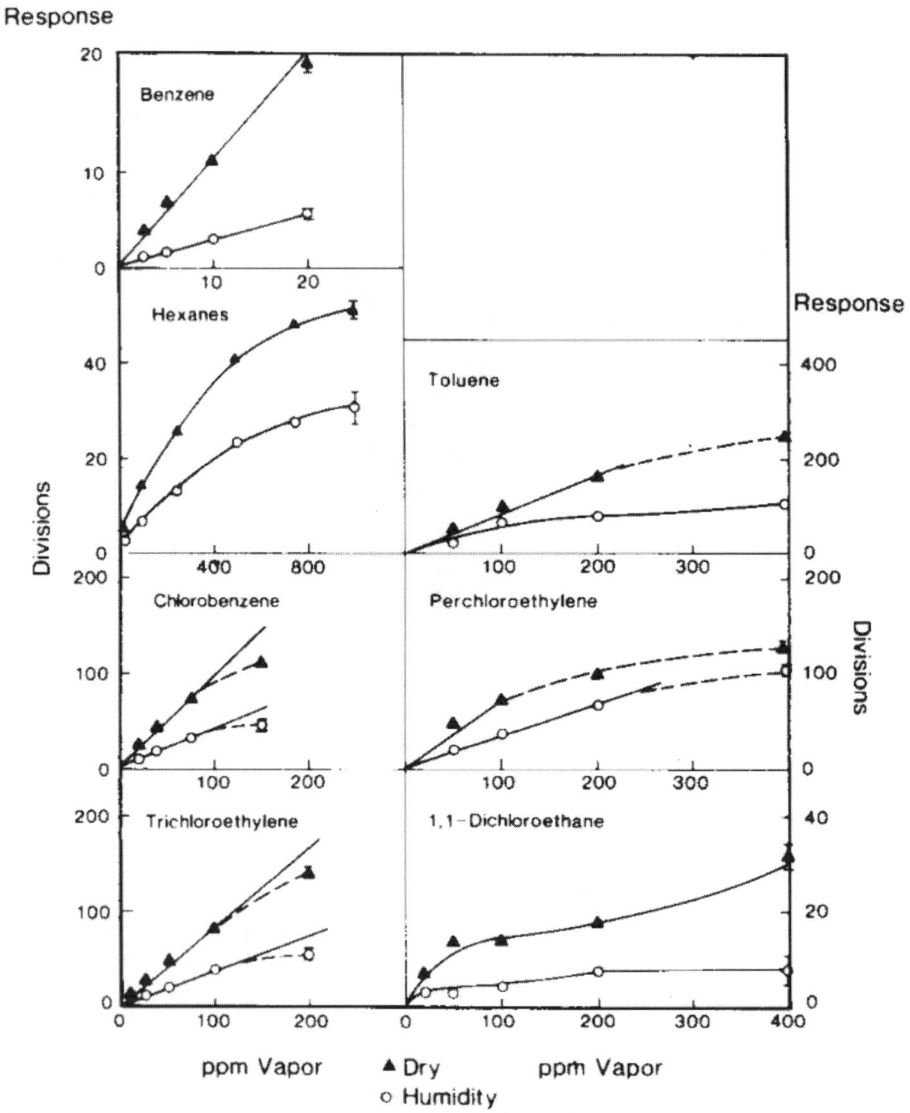

Figure 27-15. The response of the HNU 10.2 eV PID, calibrated at 0% RH, to selected organic compounds in the presence and absence of 90% relative humidity (from Barsky et al.,1985, with permission).

ment should be calibrated in the field for each condition of RH, and that the tubing is not permeable to any other compounds in the field atmosphere. Finally, it implies that the response of the instrument to the calibrant is affected by water in the same way that water affects the response to other compounds, e.g., benzene, that might be found in the field atmosphere. As illustrated in Figure 27-15, this is not the case.

Figure 27-16 shows that the humidity effects are likely at typical environmental conditions. For example, 90°F and 40% RH would cause a 70% reading. These conditions are common in the U.S. during summer.

There are a few other potential concerns about PID lamps. The lamps need to be kept clean. If the PID has a filter, it should be used. If not, do not add one,[15] but do try to keep the lamp clean. Manufacturers have lamp cleaning fluids, but use only the fluid designed for a given window. Fingerprints and grease decrease the output of the lamps; if the lamps are exchanged frequently, this could be a problem. Finally, some corrosive gases can affect the window. All of these problems can be overcome with careful attention to the manufacturer's instructions.

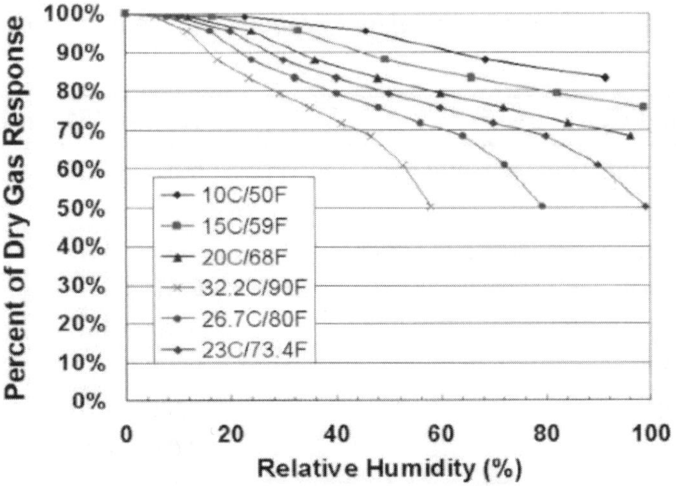

Figure 27-16. PID humidity effect curves for MiniRAE 2000 relative to calibration with any dry gas. Note that the effect is due to quenching or blockage of UV light by the absolute water vapor concentration (from Haag and Wrenn, 2002, and RAE Systems, with permission).

[15] A filter will change the flow to the detector and since the readout may be ppm, volume per volume, the reported concentration may be affected unless flow is directly measured in the instrument.

Response

The response of the FID is generally said to be linear over 6–7 orders of magnitude (Woebkenberg and McCammon, 2001). Figure 27-15 shows that the response of the PID is somewhat less linear than for the FID. Chelton et al. (1983) showed that the HNU instrument was linear for benzene from 1 to 100 ppm with a 15% error, even though the manufacturer claimed, at that time, that the linear range was 0.1 to 600 ppm. The assumption of linearity is not one that errs on the safe side (see, e.g., Chapter 26, Figure 26-4).

If a calibrated instrument (say, for benzene) is placed in an environment containing some other compound, the instrument may read low for one or all of three reasons: (1) the RH has an effect (see any of the compounds in Figure 27-15); (2) the concentration is in a nonlinear range for the compound in question (see, e.g., hexane in Figure 27-15); (3) the instrument is less sensitive for the compound than for benzene (see, e.g., hexane in Figure 27-15). The relative sensitivities of the 10.2 eV lamp are shown in Table 27-10. Interestingly, essentially all of the aliphatic compounds cause a response that is lower than for benzene. The message is

Table 27-10
Molar response of the photoionization detector to various classes of organic compounds relative to benzene[1]

Organic Compound Class (n)[2]	Relative Molar Sensitivity
C_6-C_{10} Alkanes (straight, branched, and cyclic) (18)	0.011–0.028
Chlorinated unsaturated hydrocarbons[3]	0.5–0.9
C_{10}-C_{22} Alkanes (straight, branched, and cyclic) (13)	0.37–1.13
C_7-C_{19} Alkenes (16)	0.51–1.17
C_4-C_9 Aldehydes (6)	0.30–0.53
C_3-C_9 Ketones (28)	0.35–0.82
C_4-C_{12} Alcohols (10)	0.023–0.36
C_4-C_{14} Esters (20)	0.01–0.82
C_7-C_{16} Aromatic hydrocarbons (13)	1.09–1.69
Chlorobenzenes (Cl_2-Cl_6) (10)	1.20–1.44
Chlorophenols (Cl-Cl_5) (19)	1.14–1.47
Polychlorinated biphenyls (10)	2.18–2.96
Phthalates (5)	0.56–1.78
Polycyclic aromatic hydrocarbons (2–4 rings) (12)	1.88–3.08

1 From Langhorst (1981) and Daisey (1988), using HNU PID (10.2 eV lamp)
2 Number of compounds tested
3 Data from HNU Systems, Inc.

this: the readout on the instrument is not ppm.[16] It should be thought of as *PID units* or equivalents of the calibration gas. It is clear that, in many cases (all those involving substituted or non-substituted aliphatic hydrocarbons), the PID will give a lower reading.

Oxygen

The previous discussion regarding the effects of O_2 levels on FID performance also applies to PIDs (Mouradian and Flannery, 1994). A Thermo Fisher Scientific PID calibrated at 21% O_2 with isobutylene gave a 20 to 25% higher response when the O_2 level was 15%. The importance of this relative to calibration is discussed in Chapter 30 under the heading "Compressed Gases." Of course this is only important when PIDs are used in confined spaces or other unusual atmospheres.

MULTIPLE SUBSTANCE INSTRUMENTS

It might be argued that the filtering or non-dispersive infrared MIRAN, when used in conjunction with software and a library of spectra, will detect multiple chemicals simultaneously under certain conditions. However, two types of instruments described in this section clearly can sample an atmosphere, sort out signals from multiple components, and provide results for multiple substances from the one sample. First, we examine the portable GC, including GC/MS. These units are portable because of the battery operation provided, but they are generally quite heavy. Smaller, personal-size GCs are being developed and the transportable Fourier transform infrared (FTIR) is described. Like its sister, the non-dispersive IR, it relies on spectra-matching software for identifying compounds. However, its increased resolution of those spectra make it more powerful than the non-FTIR instruments.

Portable Gas Chromatography

BASIC OPERATIONAL ASPECTS

Most of the GCs described here operate similarly to those used in laboratories and described in Chapter 19. They have packed or capillary columns through which a carrier gas flows and acts as the mobile phase. They use a variety of detectors, such as FID, PID, argon ionization, electron capture, far UV, thermal conductivity and mass spectroscopy or MS. However, there is a definite trend toward further downsizing of DIR GC, and in these newer (and in some cases prototype) devices, conductometric solid state sensor arrays, e.g., SAW, MOS, and chemiresistor, are being incorporated.

In GC, direct analysis of the air is performed with a pump and a sample loop – the latter

[16] Unfortunately, on many instruments, it is. The same can be said for any GSM detector, even FID.

is usually a small chamber that is filled by the pump for each sample cycle. The sample loop accumulates a sample, rather than continuously charging an air sample to the column. A more recent device is the pre-concentrator, which is charged with the sample and then thermally desorbed so as to increase the size of the sample. GCs require time to separate the components of the gas phase; so they cannot be truly continuous instruments. The response time, including the sampling time, can be a few seconds to many minutes, depending on the analyte, the column, and the column temperature.

Even if the analyte of interest elutes rapidly, there will be other components of the air that may not. These, like the analyte(s) of interest, will migrate from the column to the detector, possibly causing an increase in background noise, unless each analysis cycle permits them to fully elute. Most GCs have features that limit this problem. The entire column may be backflushed, or a pre-column is backflushed by reversing carrier gas flow. The pre-column is a short-length column with a liquid phase that will capture many higher molecular weight compounds. This is the preferred method, as it limits contamination of the main column and avoids lengthy backflushing of it.

Like laboratory-based GCs, the columns on DIR GCs are maintained at constant temperature, and for some models, the temperature can be *ramped*. This means that it is increased in stages so that analytes with long elution times at low temperatures elute faster than usual and in narrower peaks. Old portable GCs were not temperature-controlled.

The carrier gas is a problem for the manufacturer to overcome.[17] On FID units, the carrier gas may be H_2; thus, the carrier and the flame fuel come from the same tank. On other detectors, the carrier may be any number of gases, e.g., helium or air that is scrubbed with charcoal. The latter solves the problem of needing a compressed gas, provided the FID is not used. The carrier gas supply should last at least 8 hours. Also, portable GCs require batteries, which should last up to 8 hours. The battery charge life will depend on whether the GC is temperature-controlled.

While there are perhaps as many as 10 commercially available portable GCs (see manufacturer list at end of chapter), the technology is changing rapidly. This may be why recent reviews (Burroughs and Tabor, 1999) are scarce and comprehensive, published evaluations have not been performed.

CHROMATOGRAMS

It is important to reiterate from Chapter 19 that no GC column is universal; that is, none has the ability to separate all mixtures of compounds. The elution time is what determines identity or makes the instrument selective and specific. From this perspective, GC is not com-

[17] Compressed gases (for carrier gas or instrument calibration) cannot be taken aboard passenger aircraft either as checked or carry-on luggage, and the same is true for instruments with radioactive sources such as ECDs.

pletely selective, as two compounds can appear as the same peak, i.e., have the same elution time. More likely, however, is the shouldering phenomena shown in Figure 27-17; in this case, the two peaks are near synchronous. It may be difficult to differentiate them. Microprocessor-based GCs do have routines that detect shouldering and approximate the area under the two curves, but these techniques are not error-free. On some newer models, two columns are used in parallel. This gives two different elution times for each analyte and increases the selectivity.

The most frequent problems related to asymmetrical peak shapes are fronting and tailing. In fronting, the front side is lengthened and the tail is steep, likely caused by too large of a sample. Tailing is more frequent (see Figure 27-17) where the tail is lengthened and likely caused when the distribution of the analyte on the column is concentration dependent. Either leads to poor separation or resolution. Capillary columns limit these problems.

System contamination is usually responsible for *ghost* peaks or carryover problems. One source of contamination is when the materials from previous injections do not completely elute and are then passed to the detector on subsequent injections. Also the extra compounds may be from contamination in the injector. Impurities in solvents, vials, caps and syringes are other possible sources. Increasing the final temperature or time in the temperature program, a short bake-out after each run, or series of runs can minimize or eliminate ghost peaks.

There are numerous other issues that can arise such as 1) excessive baseline noise, which is often caused by contamination, 2) change in peak size over multiple runs, which can be caused by many things from contamination to changes in detector response, and 3) retention time shift, which can be caused by changes in temperature and carrier gas velocity.

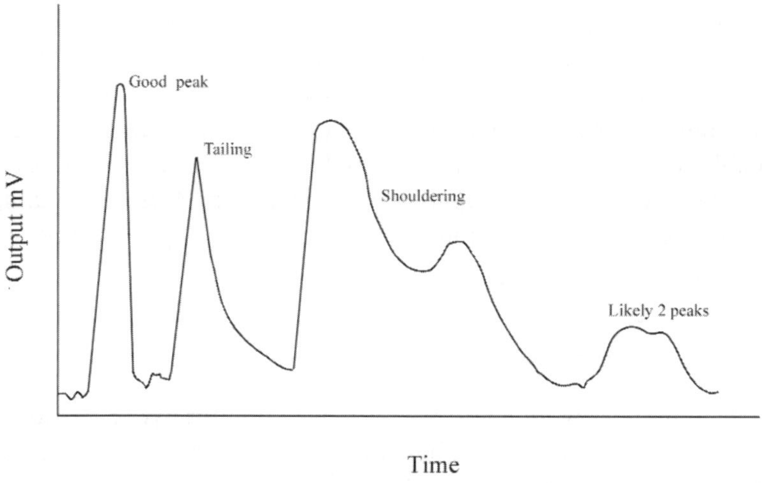

Figure 27-17. A chromatogram showing different types of peaks.

RECENT ADVANCES – TOWARD THE PERSONAL OR MICRO GC

Current portable GCs are significantly smaller and more compact than laboratory instruments. Advances in microprocessor-based electronics, and *micro* injectors, pre-concentrators, columns, and detectors are leading to smaller portable instruments. However, minimum weights still often exceed 10 lbs (4.5 kg) and can be 30–40 lbs; sizes can be *bigger than a shoe box*.

There are, however, at least two portable GCs that are based on SAW sensors, can be described as hand-held, and weigh less than 10 lbs. The hand-held VaporLab mentioned earlier under SAW detectors has now been equipped with GC separation capabilities. The introductory application of VaporLab is for benzene analysis. Subsequent applications will include chemicals such as vinyl chloride, ethylene oxide, 1,3-butadiene and others yet to be identified. Another device called the zNose® (to be differentiated from electronic noses, which typically do not employ separation technology) is based upon ultra-fast gas chromatography and SAW sensors (see ESTCal in manufacturer list for applications).

Newest developments offer the promise of achieving the performance of conventional instruments at truly personal sizes (Eiceman et al., 2006). Devices are already commercially available that can fit into the hand. Several of these are modular for controlling process gases (see for example http://www.c2v.nl/ and http://www.sls-micro-technology.de/); however, one device that has been developed by Sandia Laboratories is hand-held and commercially available (see Defiant Technologies in manufacturers list at end of the chapter). It is based on SAW sensors and uses scrubbed air as carrier gas (Manginell, no date).

Other μGCs have been developed to the prototype stage. Researchers from the University of Michigan have developed a μGC based on a SAW sensor (Whiting et al., 2001) and improved the device with a SAW array (Lu et al., 2006). Other important advances have included the pre-concentrator focuser (Lu and Zellers, 2002), the use of scrubbed air as carrier gas, and the ramping of temperature or tunable retention (Lu et al., 2003). The latest development is the incorporation of true micro-electro-mechanical system (MEMS)-based columns, injectors, pre-concentrators, and detectors in a prototype instrument (see Figure 27-18). This instrument incorporates an array of four chemiresistors and occupies a volume of < 10 cm^3. It carries its own calibrant to reduce drift and GC problems (see Figure 27-18, part a) and uniquely features micro-pumping. It is the first μGC demonstration of high-speed, high-resolution, multi-vapor recognition and quantification (15 widely varying vapors) based on retention time and sensor-array response patterns.

Fourier Transform IR (FTIR)

Transportable, remote sensing FTIR has been available for over a decade. Now there is a wide array of transportable devices. FTIR has several advantages and some disadvantages over NDIR or filtering IR devices (Levine et al., 1989). The advantages center around the interferometer used in FTIR that makes it a more quantitatively powerful instrument. The physics of

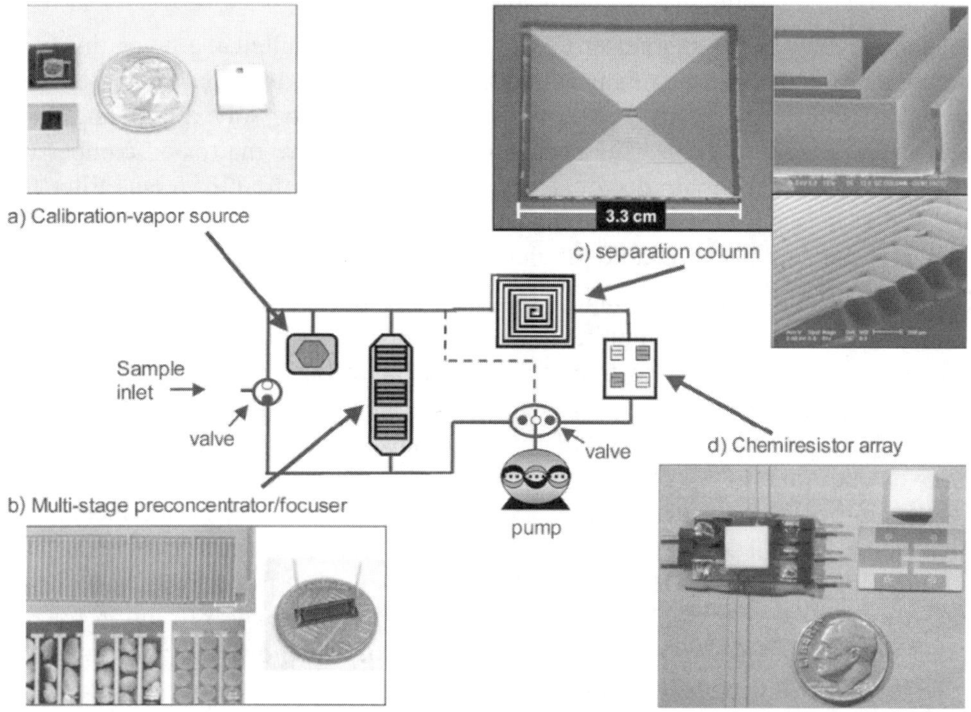

Figure 27-18. Block diagram of the MEMS μGC prototype analytical system: a) calibration-vapor source before (left) and after (right) assembly b) 3-stage adsorbent pre-concentrator focuser prior to loading and sealing (top left), with close-up scanning electron microscope images of each section loaded with adsorbents (lower left) and assembled structure with capillary interconnects on a U.S. penny; c) 3 m separation-column chip (left) with close up views of the channel cross-sections prior to (top right) and after (lower right) sealing; d) detector assembly with 4-chemiresistor array chip (right), Macor lid (white square structure), and sealed detector with connecting capillaries mounted on a custom mounting fixture (left). The dashed line is a flow-splitter (from Lu et al., with permission).

FTIR are discussed in Chapter 19. FTIR employs an interferometer (see Chapter 19, Figure 19-21), which is a system involving a moving mirror and a fixed mirror that change the amplitude of the reflected light. Constructive and destructive interference occur, so that when the moving mirror is positioned correctly, the power (amplitude) for a given frequency will go to zero. The frequency of the light that is affected is dependent on the mirror position, and the mirror moves quite rapidly, say, 1.5 cm/s, so that every wavelength is affected in a short period of time, i.e., the entire spectrum is scanned. The complete scan may be repeated many times for one analysis so as to increase the signal to noise ratio. As with NDIR, water vapor must be accounted for (since it absorbs IR well and is in high concentrations) and changes in water vapor concentrations during analysis can lead to greater noise.

The instrument records an interferogram, the power transmitted at a given time versus the distance traveled by the mirror; this is called the time domain. In order to convert this to the frequency domain (power versus frequency) an integral must be solved by using the Fourier transform from integral calculus. The necessary calculations are many and cumbersome, but easy with a computer. The main disadvantage then is the cost of FTIR versus filter IR; but if the filter IR is to be used to identify unknowns, it too will require a computer, as described earlier. The FTIR can do a more reliable job than the filter IR in identifying unknowns.

There are several other advantages (see Skoog et al., 2007; Levine et al., 1989). First, FTIR is fast; all wavelengths are monitored simultaneously. A complete scan can be performed in less than one second. This leads to a second advantage: Because the time scan is so short, multiple scans can be performed. A noise signal is reduced by a factor of $n^{0.5}$ (n = number of scans). So if 64 scans are performed, the noise is reduced by a factor of 8, and there is a similar improvement in the LOD. Third, the FTIR wavelength resolution is high. The resolution, or the width of the wave spectrum that is examined, is partly a function of speed; for higher resolution, some speed is lost. For a resolution of 2 cm^{-1}, the scan speed may be as fast as 1 s. The resolution of the filtering IR is about 1 % of the measuring frequency. Thus, for frequencies of 750 to 1300 cm^{-1} it would be 7.5 to 13 cm^{-1}. This improved resolution for the FTIR means that repeated scans can be used reliably to identify unknowns, even compounds with low IR absorption at low concentrations. This feature allows background scans to be subtracted accurately for better identification of unknowns. Fourth, the linear range is typically about four times that of filtering units (up to 2 absorbance units compared to about 0.5).

OPEN PATH – FOURIER TRANSFORM IR (OP-FTIR)[18]

In all the IR instruments described previously, the path for the IR is internal to the instrument; the sample must be brought to a sample cell. However, this does not need to be the case. The path of the IR could be the open air (i.e., an open path). In this case, the source of IR would be contained in one instrument and the detector in another (see Figure 27-19). Alternatively, a reflector could be used to reflect the beam back to an instrument containing a source and a detector. The beam must be focused at the detector; so a telescope is necessary. As the path between the detector and the source increases, the IR beam size increases, and thus some of the source does not hit the detector. But as the path length increases, the amount of absorption due to interaction with the contaminant also increases. As the path length increases, the signal-to-noise ratio decreases as a result of spreading of the IR beam, but the signal-to-noise ratio increases as a result of the increased absorption caused by the contaminant.

[18] Prior to the introduction of these devices, considerable work was done by Levine and co-workers on adapting FTIR remote sensing to the workplace environment (see Levine et al., 1989; Ying et al., 1989; Xiao et al., 1990, 1991; Franzblau et al., 1992; Yost et al., 1992).

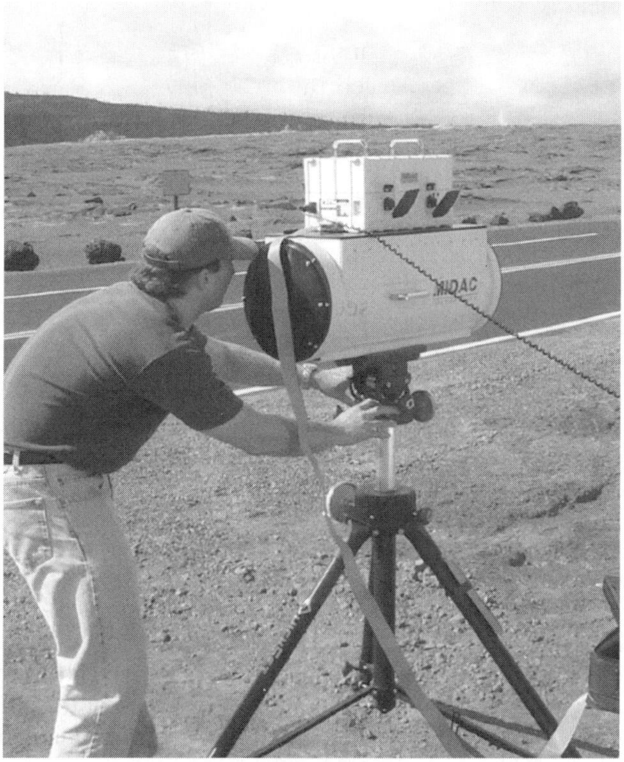

Figure 27-19. A remote sensing, open path FTIR (courtesy, MIDAC Corporation).

Applications of OP-FTIR have been made to detecting fugitive emissions, either at industrial plant fence lines, near plant sources, or at hazardous waste sites. In addition, a system of reflectors or multiple detectors could be used to make readings across several different paths at a plant site, at a waste site, or in a room. Each reading or scan would yield the average result for that path. The information from each path is exported to a computer, where computer mapping techniques (computed tomography, CT) yield a space-concentration profile as seen in Figure 27-20. Equipment and other sources of contaminants can be overlain on such a map to determine source-concentration relationships (Todd and Leith, 1990; ter Kuile et al., 1993; Samanta and Todd, 1996; Todd, 1996, 2000; Bhattacharyya and Todd, 1997; Farhat and Todd, 2000; Todd et al., 2001; Ross and Todd, 2002). More recent work has demonstrated the applicability of OP-FTIR to estimate actual worker personal exposures through the use of CT and bi-beam instruments (Wu et al., 2003; 2005a, b). Such an instrument detects gases across a space using a short and long path length; thus being able to detect concentration profiles in a segmented region. These instruments are considered to be transportable.

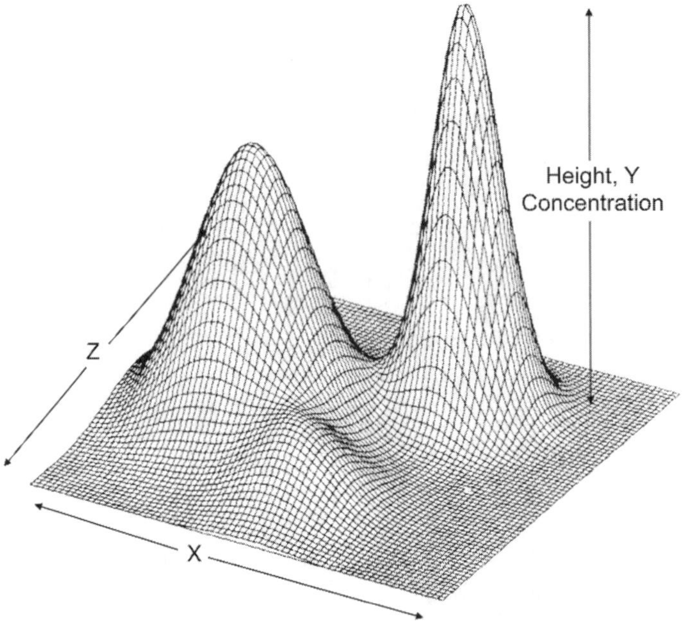

Figure 27-20. Computed tomography (computer mapping) of the signals obtained along several intersecting optical paths through a workspace, showing the concentration of the contaminant (height or y-axis) over space, the x- and z-axes (from Todd and Leith, 1990, with permission).

MANUFACTURERS OF INSTRUMENTS FOR GAS AND VAPOR DETECTION

In the list that follows, sensor methodologies are stated where manufacturers (or ACGIH®, 2001) have readily identified the basis for the sensor, i.e., CGM or combustible gas meter. Where individual gases or vapors are listed and no methodology is given, the sensor is likely to be potentiometric or perhaps voltammetric. Where a GSM or general survey meter is stated, this indicates that the manufacturer claims quantitative, selective vapor detection ability with one or more MOS, SAW, PID, FID or other detector without a separation device such as GC column.

Agilent Technologies http://www.chem.agilent.com/Scripts/PCol.asp?lPage=180 Portable GC, thermal conductivity detector

Analytical Specialists, Inc. http://www.analyticalspecialists.com/ Portable GC

Arizona Instrument LLC http://www.azic.com/ Hg, H_2S, pharmaceuticals

Bacharach, Inc. http://www.bacharach-inc.com/ Freon®, HHC, CSM, potentiometric: CH_4, NO_x, SO_2, N_2O

Baseline-Mocon http://www.baseline-mocon.com/products.php Portable GC, PID

BW Technologies Ltd. http://www.gasmonitors.com/main.cfm?deploc=1&depid=1&page=item1 PID GSM, CGM, potentiometric: H_2S,SO_2, PH_3, NH_3, NO_2, HCN, Cl_2, ClO_2

Casella London Limited http://www.casellameasurement.com/ PID

CEA Instruments, Inc. http://www.ceainstr.com/main.html colorimetric, portable instruments: C_2H_2, Acrylonitrile, Amines, NH_3, Ar, AsH_3, Br_2, CO_2, Cl_2, ClO_2, CGM, Diborane, Fl, CH_2O, Freon, Glutaraldehyde, He, HNNH, H_2, HBr, HCl, HCN, HF, H_2S, I_2, Ne, NOx, N_2O, O_3, $COCl_2$, PH_3, SiO_4, SO_2, SF_6

Calibrated Instruments, Inc. http://www.calibrated.com/ conductometric, liquid solutions, portable instruments: CO, CO_2, CH_4, NH_3, H_2S, SO_2, HCl, $COCl_2$, COS, CS_2, HCN, other hydrocarbons

Compur Monitors http://www.compur.com/ potentiometric?: GSM, CSM, Ar, $COCl_2$, HCN, Cl_2, NO_2, ClO_2, H_2S, Cl_2, HCl, As_2H_3, PH_3, N_2H_4, SO_2, He, Freon®

Constellation Technology http://www.contech.com/Chemical_Detection_Products.htm portable GC/MS

Crowcon Detection Instruments Ltd. http://www.crowcon.com/index.htm IR, CGM, potentiometric: H_2S, SO_2, Cl_2, NOx, NH_3, O_3, H_2, HCl, HCN, HF, Fl_2, PH_3, $COCl_2$, ETO

Defiant Technologies (Sandia MicroChemLab) http://www.defiant-tech.com/ hand-held GC

Draeger, Inc. http://www.draeger.com/index.html CGM, potentiometric: C_2H_2, Aldehydes, Alkenes, Acrylonitrile, Alcohols, Amines, NH_3, AsH_3, Br_2, CO_2, Cl_2, ClO_2, $COCl_2$, Diborane, Diethyl Ether, Dimethyl Sulfide, ETO, F_2, CH_2O, Freon, GeH_4, He, H_2, HNNH, HBr, HCl, HCN, HF, H_2S, H_2Se, I_2, Ne, NOx, N_2O, O_3, PH_3, Propylene Oxide, SiO_4, Styrene, SO_2, SF_6, Vinyl Acetate, Vinyl Chloride

Enmet Corporation http://www.enmet.com/index.php CGM, MOS GSM, MOS: H_2S, SO_2, NOx, HCl, HCN, NH_3, H_2, SiH_4, CO_2

ESTCal or Electronic Sensor Technology http://www.znose.com/index.htm 2 portable GCs with SAW detectors

Gasmet http://www.airqa.com/onsite.htm portable, closed cell, FTIR

GfG, Inc. http://www.gfg-inc.com/englisch/start.html CGM, MOS GSM, Cl_2, ClO_2, NH_3, ETO, HCl, HCN, H_2S, NOx, PH_3, SO_2, SiO_4

Genesis Lab Systems Inc. http://www.genlabsystems.com/ Portable Hg by AA

Gray Wolf Sensing Solutions http://www.wolfsense.com/ PID GSM, IAQ, CGM, SO_2, NOx NH_3, H_2S, HCN, HCl, H_2, O_3, EtO, Cl_2, Tetrahydrothiophen, SiH_4, AsH_3, PH_3, B_2H_6, ClO_2, CCl_2O, HF, F_2

Griffin Analytical http://www.griffinanalytical.com/griffin_family.html portable GC/MS

HNU Systems, Inc. http://www.hnu.com/ PID GSM, Portable GC, CGM

Industrial Scientific Corporation (ISO) http://www.indsci.com/ CGM, PID GSM , NH_3, CO_2, CO, Cl_2, ClO_2, H_2,HCN, H_2S, NOx, HCN, NH3, O_2, PH_3, SO_2

Inficon, Inc. http://www.inficonchemicalmonitoringsystems.com/en/index.html Portable GC and GC/MS, CGM, CO, Freon

Infitron http://www.infitron.com/ CGM, H_2S, SO_2, NO_2, Cl_2, H_2, HCN

Innova http://www.innova.dk/ PA-IR

Interscan Corp. http://www.gasdetection.com/ voltammetric: Br_2, CO, Cl_2, ClO_2, C_2H_4, ETO, HCHO, H_2, HNNH, HBr, HCl, HCN, H_2S, NO, NO_2, O_3, propylene oxide, and SO_2
IRGAS http://www.irgas.com/ portable, closed cell, FTIR
K2BW Environmental Equipment Co. http://www.k2bw.com/index.htm Portable GC
MDA Scientific http://www.honeywellanalytics.com/Home.aspx (see Table 27-6)
Metrosonics, Inc. http://www.metrosonics.com/ IAQ, CO, H_2S, NOx, HCN, NH_3, O_2
Microsensor Systems, Inc. http://www.microsensorsystems.com/ SAW, SAW GC
MIDAC http://www.midac.com/am.htm remote sensing, open path FTIR
Mine Safety Appliance (MSA) http://www.msanet.com/ CGM, PID GSM, Portable GC, PA-IR, SO_2, H_2S, NOx, Cl_2, ClO_2
MST, Inc. http://www.modsafe.com/ CO
Neotronics, Inc. http://www.nutech-australia.com.au/Neotronics/index.html CGM, CO_2, Cl_2, H_2S, NOx, NH_3, SO_2
Ocean Optics Inc. www.oceanoptics.com portable FTIR, remote sensing, open path FTIR
O.I. Analytical http://www.oico.com/default.aspx?id=productline&PLID=21 Portable GC
Photovac International, Inc. http://www.photovac.com/ Portable GC, PID and FID GSM
Quest Technologies www.quest-technologies.com IAQ, CGM, H_2S, SO_2, Cl_2, NOx, NH_3, HCN, ETO
RAE Systems, Inc. http://www.raesystems.com/ CGM, GSM PID, CO, CO_2, H_2S, SO_2, NO, NO_2, Cl_2, HCN, NH_3, PH_3
RKI Instruments www.rkiinstruments.com CGM, Electrochemical, IR and Paper tape: AsH_3, SO_2, Cl_2, Fl_2, HF, HCl NH_3, HCN, O_3, PH_3, SiO_4, ETO, O_2, CO, H_2S, CH_2O
SensIR Technologies http://www.sensir.com/s_Products/travelir.asp#Specs Portable FTIR
Thermo Fisher Scientific Environmental Instruments http://www.thermo.com/ OVA, PID, and FID GSM, IR (MIRAN), CGM, Cl_2, CO, HCN, H_2S, NH_3, NOx, PH_3, SO_2, AsH_3, SiH_4
TSI Inc. www.tsi.com GCM, IAQ
Varian, Portable GC, thermal conductivity detector http://www.varianinc.com/cgi-bin/nav?products/chrom/gc/microgc/mgc4900&cid=JNIJIQQJFP
Wilks Enterprise Inc. http://www.wilksir.com/index.html NDIR for single or multiple substances

Abbreviations:

CSM – confined space usually includes O_2, combustible gas, and CO detectors
GSM – general survey instruments, usually PID, FID,
HHC – halogenated hydrocarbons
IAQ – usually includes CO, CO_2, O_2
IR – infrared spectroscopy
NOx – NO and NO_2
PA – IR – photoacoustic IR

QUESTIONS

27.1. List 10 inorganic compounds that are monitored with electrochemical DRI.

27.2. What are the advantages and disadvantages of electrochemical DRI?

27.3. Among electrochemical methods, what factor is unique to voltammetry and adds to its selectivity?

27.4. Why is it a bit more difficult to calibrate for CO_2 than for other compounds?

27.5. A MIRAN 1A is calibrated at 300 ppm methylene chloride. How long will the pump need to run in order to purge the instrument down to < 1 ppm, using clean air?

27.6. What is the advantage of the photoacoustical detector over the temperature detector for IR? The advantage of FTIR over filtering NDIR?

27.7. What is the main difference between CGMs and GSMs?

27.8. What is the major detector in the CGM, and how does it work?

27.9. What are the two major detectors in the GSM? How are they different?

27.10. A combustible gas meter reads 15% LFL. It is calibrated for pentane, but it is known that the atmosphere contains xylene. From the calibration curve provided by the manufacturer (see Figure 27-12), find the correction factor. See Appendix A or Figure 27-12 for the xylene LFL.

27.11. The alarm for a CGM is set at 20% LFL. When the alarm sounds, what should the workers do? What procedures should be followed before re-entry and hot work are allowed?

27.12. What are the relative advantages and disadvantages of PID and FID GSMs?

27.13. What affects the response time of a portable GC?

REFERENCES

ACGIH®: Air Sampling Instruments, 9th ed. American Conference of Governmental Industrial Hygienists, Cincinnati, OH (2000).

AIHA: The Industrial Hygienist's Guide to Indoor Air Quality Investigations. American Industrial Hygiene Association, Fairfax, VA (1993a).

AIHA: Manual of Recommended Practice for Combustible Gas Indicators and Portable Direct-Reading Hydrocarbon Detectors, 2nd ed. American Industrial Hygiene Association, Fairfax, VA (1993b).

ASHRAE Standard: Ventilation for Acceptable Indoor Air Quality, 62-1989. American Society of Heating, Refrigerating and Air-Conditioning Engineers, Inc., Atlanta, GA (1989).

Barnett, J.L.; Richard, M.L.; Rose, V.E.: A tracer method for quantifying contamination of building supply air: Reentrainment of laboratory hood exhausts. Am. Ind. Hyg. Assoc. J. 44:580-582 (1983).

Barsky, J.B.; Que Hee, S.S.; Clark, C.S.: An evaluation of the response of some portable, direct-reading 10.2 eV and 11.8 eV photo ionization detectors, and a flame ionization gas chromatograph for organic vapors in high humidity atmospheres. Am. Ind. Hyg. Assoc. J. 46:9-14 (1985).

Bhattacharyya, R.A.; Todd, L.A.: Spatial and temporal visualization of gases and vapours in air using computed tomography numerical studies. Ann. Occ. Hyg. 41:105-122 (1997).

Bergman, I.: Portable gas monitors: Some problems of zero drift, response time and selectivity solved with "gas-modifier pumps." Ann. Occup. Hyg. 30:469-479 (1986).

Bishop, E.C.; Hossain, M.A.: Field comparison between two nitrous oxide (N_2O) passive monitors and conventional sampling methods. Am. Ind. Hyg. Assoc. J. 45:812-816 (1984).

Brief, R.S.; Confer, R.G.: Combustible gas indicator response in low oxygen atmospheres. Am. Ind. Hyg. Assoc. J. 30:576-581 (1969).

Bruel and Kjaer: Photoacoustics in Gas Detection. DK-2850. Naerum, Denmark (1991).

Buck, R.P.; Lindner, E.; Kutner, W.; Inzelt, G.: Piezoelectric chemical sensors, IUPAC Technical Report. Pure Appl. Chem. 76: 1139–1160 (2004).

Burkhart, J.E.; Stobbe, T.J.: Real-time measurement and control of waste anesthetic gases during veterinary surgeries. Am. Ind. Hyg. Assoc. J. 51:640-645 (1990).

Burroughs, G.E.; Woebkenberg, M.L.: Effectiveness of real-time monitoring, In Advances in Air Sampling. American Conference of Governmental Industrial Hygienists, Cincinnati, OH (1988).

Burroughs, G.E.; Woodfin, W.J.: Report on a microprocessor-based infrared analyzer. Appl. Ind. Hyg. 3:R2-R4 (1988).

Burroughs, G.E.; Tabor, M.W.: Instrument Performance Criteria, Portable Gas Chromatographs. App. Occ. Env. Hyg. 14: 159-162 (1999).

Cao, Z.; Stetter, J.P.: A selective solid state gas sensor for halogenated hydrocarbons, paper presented at Third International Meeting on Chemical Sensors, sponsored by the Edison Sensor Technology Center, Case Western Reserve University, Cleveland, OH (1990).

Capone, S.; Forleo, A.; Francioso, L.; Rella, R.; Siciliano, P.; Spadavecchia, J.; Presicce, D.S.; Taurino, A.M.: Solid state gas sensors: state of the art and future activities. J. Optoelectronics & Advanced Materials 5:1335 –1348 (2003).

Christian, G. (Ed.) Analytical Chemistry, 5th ed., p. 2. Wiley Publishing, New York (1994).

Chelton, C.F.; Zakraysek, N.; Lauiner, G.M.; Confer, R.G.: Evaluation of the performance and response of the Bacharach TLV Sniffer and Hnu Photoionization Gas Analyzer to common hydrocarbon solvents. Am. Ind. Hyg. Assoc. J. 44:710-715 (1983).

Coy, J.D.; Bigelow, P.L.; Buchan, R.M.; Tessari, J.D.; Parnell, J.O.: Field evaluation of a portable photoionization detector for assessing exposure to solvent mixtures. Am. Ind. Hyg. Assoc. J. 61:268-274 (2000).

Daisey, J.M.: Real-time portable organic vapor sampling systems: Status and needs, In Advances in Air Sampling. American Conference of Governmental Industrial Hygienists, Lewis Publishers, Chelsea, MI (1988).

Denenberg, B.A.; Kriesel, R.: Chemically impregnated paper tapes as a detection method in continuous personal and area monitors. Am. Ind. Hyg. Assoc. J. 37:246-251 (1976).

de Ris, J.: Chemistry and physics of fire, In Fire Protection Handbook, 15th ed. National Fire Protection Association, Quincy, MA (1981).

Dharmarajan, V.; Rando, R.J.: Critical evaluation of continuous monitors for toluene diisocyanate. Am. Ind. Hyg. Assoc. J. 41:869-878 (1980).

Dharmarajan, V.: Evaluation of personal continuous paper-tape monitors for toluene diisocyanate. Am. Ind. Hyg. Assoc. J. 57:68-71 (1996a).

Dharmarajan, V.: Humidity and isocyanate readings, Letter to the editor. Am. Ind. Hyg. Assoc. J. 57:665-668 (1996b).

Diamond, P.: New concepts in detection instrumentation. Am. Ind. Hyg. Assoc. J. 24:399-401 (1963).

Driscoll, J.N.; Spaziani, F.F.: Trace gas analysis by photoionization. Anal. Instrum. 13:111 (1975).

Duffy, T.L.; Pelton, P.L.: Some applications of coulometry to industrial hygiene analysis. Am. Ind. Hyg. Assoc. 26:544-548 (1965).

Egert, S.; Eyal, E.; Sivan, J.; Reichman, R.: Limitation of using multi-type sensors in combination for verifying atmospheric phenomena. Boundary-Layer Meteorology 62: 385-391 (1993).

Eiceman, G.A.; Gardea-Torresdey, J.; Dorman, F.; Overton, E.; Bhushan, A.; Dharmasena, H.P.: Gas Chromatography. Anal. Chem.78(12):3985-3996 (2006).

Farhat, S.K.; Todd, L.A.: Evaluation of open-path FTIR spectrometers for monitoring multiple chemicals in air. Appl. Occup. Environ. Hyg. 15(12):911-923 (2000).

Findlay, M.W.; Stetter, J.R.; Pritchett, T.: Sensor array based monitor for hazardous waste site screening, In Proceedings of HAZMAT 90 Central Conference, March 13-15, 1990. Environmental Hazards Management Institute, Durham, NH (1990).

Firth, J.G.; Jones, A.; Jones, T.A.: Solid state detectors for carbon monoxide. Ann. Occup. Hyg. 18:63-68 (1975).

Franzblau, A.; Levine, S.P.; Burgess, L.A.; Qu, Q.S.; Schreck, R.M.; D' Arcy, J.B.: The use of a transportable Fourier transform infrared (FTIR) spectrometer for the direct measurement of solvents in breath and ambient air – I: Methanol. Am. Ind. Hyg. Assoc. J. 53:221-227 (1992).

Gentry, S.J.; Walsh, P.T.: Calorimetric methods of gas detection. Analytical Proceedings 23:59-60 (1986).

Ghittori, S.; Fiorentino, M.L.; Imbrian, M.: An evaluation of a new portable multi-gas monitor. Appl. Occup. Env. Hyg. 8:283-287 (1993).

Groves, W.A.; Zeller, E.T.: Prototype instrument employing a microsensor array for the analysis of organic vapors in exhaled breath. Am. Ind. Hyg. Assoc. J. 57:1103-1108 (1996).

Groves, W.A.; Achutan, C.: Laboratory and field evaluation of a SAW microsensor array for measuring perchloroethylene in breath. J. Occup. Env. Hyg. 1:779-778 (2004).

Guillemin, M.P.; Bauer, D.; Hotz, P.A.; Lob, M.; Greuter, W.F.: Monitoring of styrene exposure in the polyester industry. Scand. J. Work Environ. Hlth. 4:14-21 (1978).

Haag, W.R.; Wrenn, C.: Theory and applications of direct reading photoionization detectors (PIDS), Doc. Number 000-4001-000. RAE Systems, Sunnyvale, CA, USA (2002).

Hannifen, M.A.; Reist, P.C.; Campbell, R.L.: Anesthetic waste gas exposure in dental survey. Am. Ind. Hyg. Assoc. J. 39:69-73 (1978).

Ho, C.K.; Itamura, M.T.; Kelley, M.; Hughes, R.C.: Review of Chemical Sensors for In-Situ Monitoring of Volatile Contaminants, Sandia report: sand2001-0643. Sandia National Laboratories, Albuquerque, New Mexico (2001).

Honeywell: MDA Scientific Chemcassette. Honeywell Analytics Distribution, Inc. Sunrise, FL USA http://www.honeywellanalytics.com/docGallery/564.PDF (2006).

Hsieh, M.D.; Zellers, E.T.: Adaptation and evaluation of a personal electronic nose for selective multivapor analysis. J. Occup. Env. Hyg. 1:149-160 (2004).

Industrial Scientific Corporation: Selecting gas sensors for personal safety instruments. ISHN (January 2000).

Interscan: Sensor Principle of Operation. Interscan Corporation, Chatsworth, CA, USA http://www.gasdetection.com/TECH/sensorprinciple2.htm (2006).

Jacobs, H.W.; Syrjala, R.H.: The use of infrared analyzers for monitoring acrylonitrile. Am. Ind. Hyg. Assoc. J. 39:161-165 (1978).

K2BW: Non-Dispersive Infrared and Gas Filter Correlation. K2BW Environmental Equipment Co., Willow Grove, PA, USA http://www.k2bw.com/5_c_18.htm (2006).

Kelmig, S.D.; Esmen, N.A.; Erdal, S.; Sansone, E.B.: Solvent desorption from carbon beds in ducted and nonducted laboratory fume hoods. Appl. Occup. Env. Hyg. 6:592-597 (1991).

Khoury, G.A.; Chang, S.N.; Lessley, D.A.; Abdelghani, A.A.; Anderson, A.C.: An investigation of reentrainment of chemical fume hood exhaust air in a heat recovery unit. Am. Ind. Hyg. Assoc. J. 49:61-65 (1988).

Kuempel, J.R.; Schults, W.D.: An electrochemical monitor for air pollutants. Analytical Letters 4:107-116 (1971).

Langhorst, M.L.: Photoionization detector sensitivity of organic compounds. J. Chromat. Sci. 19:98-103 (1981).

Lee, I.N.; Que Hee, S.S.; Clark, C.S.: Additivity of detector responses of a portable direct-reading 10.2 eV photoionization detector and a flame ionization gas chromatograph for atmospheres of multicomponent organics:

Use of PID/FID ratios. Am. Ind. Hyg. Assoc. J. 48:437-441 (1987).
Levine, S.P.; Li-Shi, Y.; Strang, C.R.I.; Hong-Kui, X.: Advantages and disadvantages in the use of Fourier transform infrared (FTIR) and filter infrared (FIR) spectrometers for monitoring airborne gases and vapors of industrial hygiene concern. Appl. Ind. Hyg. 4:180-187 (1989).
Lu, C.J.; Zellers, E.T.: Multi-adsorbent preconcentration/focusing module for portable-GC/microsensor-array analysis of complex vapor mixtures. Analyst 127:1061-1068 (2002).
Lu, C.J.; Whiting, J.; Sacks, R.D.; Zellers, E.T.: Portable GC with tunable retention and microsensor array detection for the determination of complex vapor mixtures. Anal. Chem. 75(6), 1400-1409 (2003).
Lu, C.J.; Jin, C.; Zellers, E.T.: Chamber Evaluation of a Portable GC with Tunable Retention and Microsensor-Array Detection for Indoor Air Quality Monitoring. J. Environmental Monitoring 8:270-278, 2006.
Lu, C.J.; Steinecker, W.H.; Tian, W.C.; Oborny, C.; Nichols, J.M.; Agah, M.; Potkay, J.A.; Chan, H.K.L.; Driscoll, J.; Sacks, R.D.; Wise, K.D.; Pang, S.W.; Zellers, E.T.: First generation hybrid MEMS gas chromatograph. Lab on a Chip 5:1123-1131 (2005).
Manginell, R.P.; Lewis, P.R.; Adkins, D.R.; Kottenstette, R.J.; Wheeler, D.; Sokolowski, S.; Trudell, D.; Byrnes, J.; Okandan, M.; Bauer, J.M.; Manley, R.G.: Recent Advancements in the Gas-Phase MicroChemLab™, http://www.sandia.gov/mstc/technologies/microanalytical/BAA/Recent_Advancements_in_the_Gas-Phase_MicroChemLab.pdf (no date).
Manninen, A.: Analysis of airborne ammonia: Comparison of field methods. Ann. Occup. Hyg. 32:399-404 (1988).
Maslansky, C.J.; Maslansky, S.P.: Air Monitoring Instrumentation. Van Nostrand Reinhold, New York (1993).
McCammon, C.S. Jr.: A summary of the NIOSH evaluation program of portable direct-reading meters. Am. Ind. Hyg. Assoc. J. 37:289-295 (1976).
Mortimer, V.D.; Kercher, S.L.; O'Brien, D.M.: Effective controls for ethylene oxide – a case study. Appl. Ind. Hyg. 1:15-20 (1986).
Mouradian, R.F.; Flannery, J.P.: Effect of oxygen content on calibration of portable VOC instruments. Am. Ind. Hyg. Assoc. J. 55:1084-1086 (1994).
Nelson, G.O.; Lum, B.Y.; Carlson, G.J.; Wong, C.M.; Johnson, J.S.: Glove permeation by organic solvents. Am. Ind. Hyg. Assoc. J. 42:217-225 (1981).
Ness, S.A.: Air Monitoring for Toxic Exposures. Van Nostrand Reinhold, New York (1991).
NIOSH: Evaluation of Portable Sulfur Dioxide Meters. National Institute for Occupational Safety and Health, Cincinnati, OH (1974).
NIOSH: Evaluation of Portable, Direct-Reading Ozone Meters. National Institute for Occupational Safety and Health, Cincinnati, OH (1975a).
NIOSH: Evaluation of Portable, Direct-Reading No-NOx Meters. National Institute for Occupational Safety and Health, Cincinnati, OH (1975b).
NIOSH: Evaluation of Portable, Direct-Reading Hydrocarbon Meters. National Institute for Occupational Safety and Health, Cincinnati, OH (1976).
Nyquist, J.E.; Wilson, D.L.; Norman, L.A.; Gammage, R.B.: Decreased sensitivity of photoionization detector total organic vapor detectors in the presence of methane. Am. Ind. Hyg. Assoc. J. 6:326-330 (1990).
Park, J.; Zhang, G.Z.; Zeller, E.T.: Personal monitoring instrument for the selective measurement of multiple organic vapors. Am. Ind. Hyg. Assoc. J. 61:192-204 (2000).
Park, J.; Plese, M.R.; Puskar, M.A.: Evaluation of a personal monitor employing an electrochemical sensor for assessing exposure to hydrogen peroxide in the workplace. Am. Ind. Hyg. Assoc. J. 64:360-367 (2003).
Perkins, J.L.; Ridge, M.C.: Use of infrared spectroscopy in permeation tests, In Performance of Protective Clothing, R.L. Barker, G.C. Coletta (eds.). American Society for Testing and Materials, Philadelphia, PA (1986).
Persily, A.; Dols, W.S.: The relation of CO_2 concentration to office building ventilation, In Air Change Rate and Airtightness in Buildings, ASTM STP-1067, M.H. Sherman (ed.), pp. 77-92. American Society for Testing and Materials, Philadelphia, PA (1990).
Potts, D.L.; Craft, B.F.: Occupational exposure of veterinarians to waste anesthetic gases. Appl. Ind. Hyg. 3:132-138 (1988).
Rivera, D.; Alam, M.K.; Davis, C.E.; Ho, C.K.: Characterization of the ability of polymeric chemiresistor arrays to quantitate trichloroethylene using partial least squares (PLS): effects of experimental design, humidity, and

temperature. Sensors and Actuators B 92:110–120 (2003).

Roberge, B.; Goyer, N.: Evaluation of personal carbon monoxide monitors. Appl. Occup. Env. Hyg. 8:275-278 (1993).

Ross, J.W.; Riseman, J.H.; Krueger, J.A.: Potentiometric gas sensing electrodes. Pure Appl. Chem. 36:473-487 (1973).

Ross, K.R.; Todd, L.A.: Field Evaluation of a transportable open-path FTIR spectrometer for real-time air monitoring. Appl. Occup. Environ. Hyg. 17(2):131-143 (2002).

Samanta, A.; Todd, L.A.: Mapping air contaminants indoors using a prototype computed tomography system. Ann. Occup. Hyg. 40(6):675-691 (1996).

Samimi, B.S.: Calibration of MIRAN gas analyzers: Extent of vapor loss within a closed loop calibration system. Am. Ind. Hyg. Assoc. J. 44:40-45 (1983).

Shaw, M.D.: Toxic Gas Detection in the Real World: Practicalities and Pitfalls. The Synergist 16(10):41-42 (October 2005).

Shleien, B.; Terpilak, M.S.: The Health Physics and Radiological Health Handbook. Nucleon Lectern Associates, Olney, MD (1984).

Sherwood, R.J.: Miniature air samplers for sulfur dioxide. Am. Ind. Hyg. Assoc. J. 30:614-619 (1969).

Skoog, D.A.; Holler, F.J.; Crouch, S.R.: Principles of Instrumental Analysis, 6th ed. Thomson Brooks/Cole (2007).

Smith, J.P.; Shulman, S.A.: An evaluation of H_2S continuous monitors using metal oxide semiconductor sensors. Appl. Ind. Hyg. 3:214-221 (1988).

Smith, J.P.; Shulman, S.A.: Evaluation of a personal data logging monitor for carbon monoxide. Appl. Occup. Env. Hyg. 9:418-427 (1994).

Stetter, J.R.: Instrumentation to monitor chemical exposure in the synfuel industry. Ann. Am. Conf. Ind. Hyg. 11:225-267 (1984).

ter Kuile, W.M.; Knole, B.; Hesselink, P.G.M.: Measurement and imaging of gases in industrial environments with the infrared gas cloud scanner. Appl. Occup. Env. Hyg. 8:47-54 (1993).

Todd, L.A.: Evaluation of an open-path Fourier transform infrared spectrophotometer using an exposure chamber. Appl. Occup. Environ. Hyg. 11(11):1327-1334 (1996).

Todd, L.A.: Mapping the air in real-time to visualize the flow of gases and vapors: Occupational and environmental applications. Appl. Occup. Environ. Hyg. 15(1):106-113 (2000).

Todd, L.; Leith, D.: Remote sensing and computed tomography in industrial hygiene. Am. Ind. Hyg. Assoc. J. 51:224-233 (1990).

Todd, L.A.; Farhat, S.K.; Mottus, K.M.; Mihlan, G.J.: Experimental evaluation of an environmental CAT scanning system for mapping chemicals in air in real-time. Appl. Occup. Environ. Hyg. 16(1):45-55 (2001).

Verma, D.K.; Saunders, G.A.; Cheng, W.K.: A laboratory evaluation of the accuracy and precision of the Photovac Snapshot portable gas chromatograph and the Draeger Chip Measurement System Monitor for benzene in air measurements. App. Occ. Env. Hyg. 16:832-840 (2001).

Vessman, J.; Stefan, R.I.; Van Staden, J.F.; Lindner, K.D.; Burns, D.T.; Fajgelj, A.; Müller, H.: Selectivity in analytical chemistry (IUPAC Recommendations 2001). Pure Appl. Chem. 73(8):1381–1386 (2001).

Walker, R.F.; Pinches, M.A.: Chemical interference effects in the measurement of atmospheric toluene diisocyanate concentrations when sampling with an impregnated paper tape. Am. Ind. Hyg. Assoc. J. 42:392-397 (1981).

Wanek, R.: Humidity and isocyanate readings, Letter to the editor. Am. Ind. Hyg. Assoc. J. 57:665-668 (1996).

Wang, A.Q.: Selecting gas sensors. ISHN January:34-38 (2000).

Webster's Tenth New Collegiate Dictionary. Merriam-Webster, Inc., Springfield, MA (1991).

Whiting, J.J.; Lu, C.J.; Zellers, E.T.; Sacks, R.D.: Toward a portable, high-speed, vacuum-outlet GC vapor analyzer employing air as carrier gas and surface acoustic wave detection. Anal. Chem. 73, 4668-4775 (2001).

Woebkenberg, M.L.; McCammon, C.S.: Direct-reading gas and vapor instruments, In Air Sampling Instruments, B.S. Cohen, C.S. McCammon, eds. American Conference of Governmental Industrial Hygienists, Cincinnati, OH, USA (2001).

Wohltjen H.; Snow, A.W.: Colloidal metal-insulator-metal ensemble chemiresistor sensor. Anal. Chem. 70(14):2856-9 (1998).

Wu, C. F.; Yost, M.G.; Hashmonay, R.A.; Larson, T.V.; Guffey, S.E.: Applying open-path FTIR with computed tomography to evaluate personal exposures. Part 1: Experimental studies. Ann. Occup. Hyg. 49(1):61-71 (2005b).

Wu, C.F.; Yost, M.G.; Varr, J.; Hashmonay, R.A.: Applying open-path FTIR with a bi-beam strategy to evaluate personal exposure in indoor environments: Experimental results of a validation study. Am. Ind. Hyg. Assoc. J. 64:181-188 (2003).

Wu, C.F.; Yost, M.G.; Hashmonay, R.A.; Larson, T.V.; Guffey, S.E.: Applying open-path FTIR with computed tomography to evaluate personal exposures. Part 1: Simulation studies. Ann. Occup. Hyg. 49(1):61-71 (2005).

Xiao, H.K.; Levine, S.P.; D' Arcy, J.B.; Kinnes, G.; Almaguer, D.: Comparison of the Fourier transform infrared (FTIR) spectrophotometer and the miniature infrared analyzer (MIRAN) for the determination of trichloroethylene (TCE) in the presence of freon-133 in workplace air. Am. Ind. Hyg. Assoc. J. 51:395-401 (1990).

Xiao, H.K.; Levine, S.P.; Herget, W.F.; D' Arcy, J.B.; Spear, R.; Pritchett, T.: A transportable, remote sensing, infrared air-monitoring system. Am. Ind. Hyg. Assoc. J. 52:449-457 (1991).

Ying, L.S.; Levine, S.P.; Strang, C.R.; Herget, W.F.: Fourier transform infrared (FTIR) spectroscopy for monitoring airborne gases and vapors of industrial hygiene concern. Am. Ind. Hyg. Assoc. J. 50:354--359 (1989).

Yoshida, K.; Bryant, H.E.; Visser, N.D.; Atiemo, M.A.: Determination of worker exposure to instantaneously emitted ethylene oxide in sterilization facilities. Appl. Ind. Hyg. 4:166-170 (1989).

Yost, M.G.; Xiao, H.K.; Spear, R.C.; Levine, S.P.: Comparative testing of an FTIR remote optical sensor with area samplers in a controlled ventilation chamber. Am. Ind. Hyg. Assoc. J. 53:611-616 (1992).

28

INSTRUMENTS FOR AEROSOLS

INTRODUCTION

With a few exceptions the direct-reading instruments (DRI) used to monitor aerosols employ the principles of optical physics, which were worked out mathematically by Gustav Mie in 1908. Mie's work was made possible by the work of Tyndall and Rayleigh and Maxwell's theory of electromagnetic radiation. Mie's equations give the exact relationship for the intensity of light scattering at a given angle from spherical, homogeneous particles, relative to the particle size and refractive index and the light source wavelength and intensity.[1] These instruments have been available since the 1940s (Hosey et al., 1960), but there have been many improvements, including their increased portability. The instruments generally report mass concentration, but mass is not read directly. They and their principle of operation are discussed in detail in this chapter.

In addition to light scattering, a few instruments operate on the optical principle of light extinction, which includes scattering and absorption as based on the Beer-Lambert law. All particles will scatter light, but only those composed of absorbing material will absorb it. The extinction of light is a combination of absorption and scattering. The latter causes light to be reflected or refracted at some angle away from the detector. Thus, even though it is not absorbed, the scattered light will not be detected by a detector located directly downstream from the parallel (polarized) light rays. Instruments operating on the extinction principle are not very sensitive for the same reason that infrared spectrophotometers are not. They sense the amount of light that passes through the sample and base the signal on the difference in the amount emitted and that sensed. This difference must be at least 5% to overcome background noise. This means that, as for IR spectrophotometers, the path length must be long or the concentration high. Light extinction instruments are sometimes called transmissometers or opacity meters. They are often used to measure exhaust stack emissions, but are not used as portable

[1] The theoretical discussions in this chapter are largely, except where noted, adapted from Hinds (1999), Chapter 16. Other very useful general references are Baron and Willeke (2001) and Baron (1998).

DRI in hygiene applications (Swift, 1989).

In addition to the optical instruments, there are two types of DRI for aerosols that operate on other principles, both of which directly measure the mass of collected particles. The first of these instruments to be developed was the GCA Technology[2] RDM (Respirable Dust Monitor) 101. The aerosol was deposited by impaction on a greased disk that had a beam of beta radiation focused across it. When particles were lodged on the disk, the beam was exponentially attenuated and a reduction in signal from the beta source was read by a Geiger counter (GCA, 1977; Baron, 1988a, b). The attenuation was directly related to mass. While this instrument and its sister, the RDM 201, are no longer manufactured, a few of these devices are still in use, and other models are produced by other manufacturers (Swift, 1989). For example, Andersen (now owned by Thermo Electron) produced a beta attenuation instrument and Thermo Electron now markets an instrument that incorporates both beta attenuation and photometry and is used for measuring ambient particulate matter (EPA PM_{10} and $PM_{2.5}$). Two other manufacturers listed at the end of the chapter market beta attenuation devices for ambient air quality work. Generally, these instruments collect particles on a filter tape that is advanced and read in increments such as one hour. Current beta attenuation instruments are mostly used for environmental work, where long sampling times and low concentrations are typical. These devices are not discussed further here.

A second non-optical technology has evolved around the piezoelectric crystal or quartz balance, mentioned in Chapter 27. It is discussed a the end of this chapter.

All of the instruments in this chapter are General Survey Meters (GSMs) and subject to the same limitations as the GSMs described in Chapter 27. Most importantly, they are not qualitative, only quantitative. Like the GSMs for gases and vapors, they can be calibrated for specific contaminants, and to the extent that only those contaminants exist in the environment to be measured, they can be made qualitative. If not appropriately calibrated, their main utility is in comparative situations. Specifically, they are used to find leaks or hot spots, to test the before and after effects of controls, to examine trends in exposure over time that might be correlated to specific tasks or operations, and for many other uses. The reader should refer to Table 28-1 throughout the chapter for a comparison of the various aerosol DRI.

OPTICAL MEASUREMENT OF AEROSOLS

In this section, we are interested in DRI that operate on the principle of light scattering. Light extinction instruments will not be discussed further. Chapter 9, Figure 9-9 shows that light interacting with a particle can be reflected, diffracted, refracted, and absorbed. Absorbed light may be re-emitted at a different electromagnetic frequency. Light scattering involves reflection, refraction, and diffraction. Figure 28-1 shows that the amount of light scattered in

[2] GCA Technology became MIE, Inc., which was sold to Thermo Environmental Corporation, which became Thermo Electron Corporation, and is now Thermo Fisher Scientific.

Table 28-1
Typical parameters for some direct-reading aerosol instruments (modified from Baron, 1988; Pui and Chen, 1999; and manufacturers)

Instrument Type[a]	Time Response[a] of Readout	Measured[b] Parameter	Important Aerosol[c] Properties	Particle Size Range (μm)	Main Advantages	Main Disadvantages
OPTICAL						
Photometers	Continuous	Integrated light scattering	Density, size distribution, refractive index	0.1–100	Continuous readout of mass concentration	Calibration changes with dust type
Optical particle counters	Continuous	Light scattering per particle	Density, size, refractive index	0.3–15 0.1–5 (laser)	Size distribution, high sensitivity	Relatively low resolution and accuracy in sizing
Condensation nuclei counters	Continuous	Light scattering per particle	Size	0.01–2	Good for surveying low concentrations	No sizing
MASS SENSING						
Piezo-balance	Discrete	Mass	Particle mass, water, shape	0.01–10	Direct measure of mass, measures fumes	Sensor cleaning[d] Internal losses
TEOM	Discrete	Mass	Adsorbed water	0–5,000	Direct measure of mass	Frequent filter replacement[d]
β-attenuation monitor	Discrete	Absorption of β-radiation	Atomic number	0–10	Direct measure of mass	Low sensitivity[d]

[a] Some instruments display a continuous indication of detected signal, but actually integrate over some relatively small time.
[b] Physical property that produces instrument signal.
[c] Parameters that significantly affect accuracy of measurement.
[d] May exhibit transient response to liquid aerosols or to high humidity.

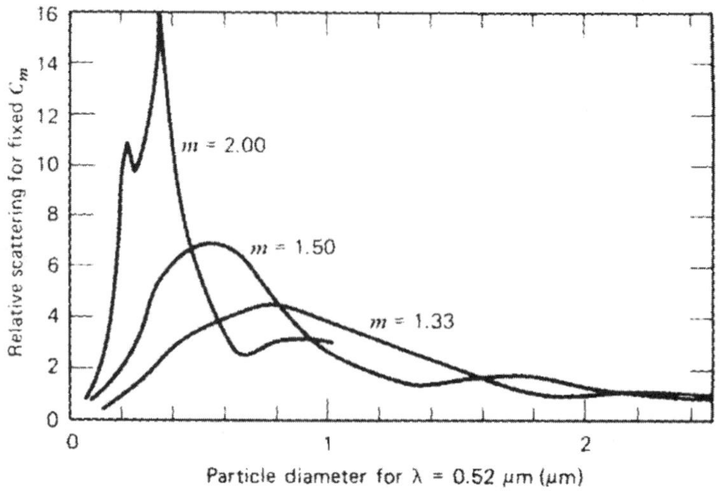

Figure 28-1. The relative scattering of light (0.52 μm) caused by one unit of aerosol mass divided into particles of various sizes and differing refractive indices (*v*), (from Hinds, 1999, with permission).

a volume containing many particles will be dependent on three properties: the wavelength of the light, the size of *r* the particle, and the refractive index of the particle. (Later when scattering is related to the mass of particles, particle density will be added to this list.) Notice that the particle size for maximum extinction (scattering plus absorption) is generally between 0.1 μm and 2 μm for the light wavelength used in Figure 28-1. This explains why particles in this size range are largely responsible for atmospheric haze, as the wavelength 0.52 μm is midway in the visible range (about 0.3–0.7 μm).

For a given aerosol generating process, if we assume that the refractive index and the size of the emitted particles are constant, and if we use an instrument with a given, constant light source, then as more particles enter the sensing chamber of the instrument, more light will be diverted or scattered. Thus, the intensity of the scattered light is indicative of the *number* of particles. In addition, we can relate the scattering to the total mass of the monodisperse particulate as:

$$\frac{I}{I_o} = \frac{3\ C_m\ Q_e\ L}{2\rho_p d}$$

where I/I_o is the ratio of the scattered to emitted light intensities, C_m is the total mass of the monodisperse particulate, Q_e is the extinction efficiency (dependent on the light wavelength and particle refractive index and size), L is the path length, ρ_p is the particle density, and d is the particle diameter. For the conditions stated, all are constants except C_m, ρ_p, and I, so in order for the equation to hold, the particle density, refractive index, and diameter must be known or somehow considered or approximated in the instrument calibration. In addition, the light source must be constant.

Obviously these restrictions are harsh because most industrial particulate is polydisperse. In addition, whereas particulate from one source may have a relatively constant density and refractive index, that from another source may have a different density and refractive index. Thus, the relationship of I/I_o to mass will change; that is, a new calibration will be needed for each particulate source.

We can lessen the dependence of the instrument on these variables by measuring the scattered light only at certain angles. For example, if we measure the scattered light at angles < 30° (the near forward scattering region), the effect of refractive index is lessened, as the light is mainly diffracted. However, we also lose sensitivity. Instrument design, therefore, is a compromise between detector sensing angle and sensitivity, among other variables. Figure 28-1 shows that when a constant mass in the sensing chamber is divided into larger and larger particles (e.g., beyond 2 µm), the amount of scattering changes little (i.e., refractive index has little effect). Figure 28-1 shows that, for a constant mass, the larger particle distribution will scatter less light than the one composed of smaller particles. For two sampled aerosols having the same concentration, measured with the same device, calibrated only once, the instrument will not give similar readings for the two aerosol concentrations unless the size distributions are similar. If the two MMADs are 0.5 and 7 µm, the concentration readings will be quite different. Field calibrations for each aerosol are required, unless one is only interested in relative readings. For example, if a change in a process reduced the concentration, one could quantitate the change provided that the size distribution was not changed.

Figure 28-2 shows that the sensed scattering at larger diameters is reduced relative to smaller diameters for one instrument (RAM-1[3]). It used infrared light (A = 880 nm) and sensing angles of 45° to 95° to attain the greatest sensitivity for particles in the 0.5 µm to 1 µm range. Figure 28-2 shows how the relative sensing compared to the RPM size convention.

Definitions Applied to Photometry

Figure 28-3 shows the various technologies that are applied to aerosol DRI. Not all are photometric and will be discussed later. The first important dichotomy is between measuring size distribution versus concentration. In hygiene work we are usually interested in the latter. Concentrations may be measured in terms of mass (mg/m^3) or count, i.e., number (e.g., fibers or particles per cm^3). As can be seen in the middle of the chart, there are instruments that will measure both count concentration and size distribution. There are also a few instruments that use software and will convert counts and other particle data to mass concentrations. This fairly recent development is not shown in the chart. Conversion software usually uses some sim-

[3] The instrument is no longer manufactured but its successors are sold by Thermo Fisher Scientific.

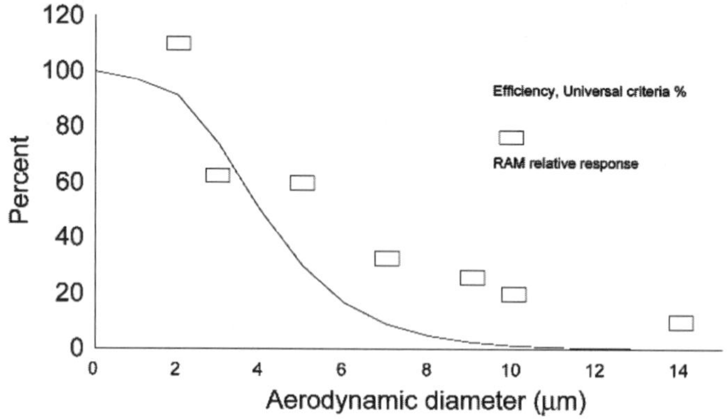

Figure 28-2. The relative response of a RAM photometer (symbols) to volcanic ash aerosol of various sizes compared to the Universal RMF sampling criteria (solid line) (adapted from Baron, 1988).

plifying assumptions that can make conversion to mass or surface area questionable. The primary assumptions are: spherical particles, known density and, sometimes, lognormal distribution. Thus, these outputs are best used in the same environment to indicate relative changes rather than absolute concentrations.

NEPHELOMETERS AND PHOTOMETERS

There is some confusion about the photometry nomenclature. Nephelometers are instruments that employ wide sensing angles, e.g., 8–170°. Instruments that sense many particles at once at a relatively narrow range of angles are photometers. The most popular convention seems to be that forward light scattering instruments measure scattering at < 30°. Instruments that operate at around 180° are termed back scattering light photometers. The majority of DRI used in hygiene work sense intermediate angles that include 90°, with a few near forward light scattering instruments in use.

Many aerosol DRI are photometers that read mass from light scatter and convert it to mass concentration based on the sensing chamber volume and flow rate. As noted above, this *calibration* works best if refractive index and size distribution varies little, as is usually the case in one workplace.

SINGLE, LARGE PARTICLE COUNTERS

Photometers measure scattered light from many particles at once, as opposed to single particle counters that measure light from individual particles. They rely on internal dilution of an air sample or low sample rates, so as to generally limit particle counts to < 100/cm^3 (Swift,

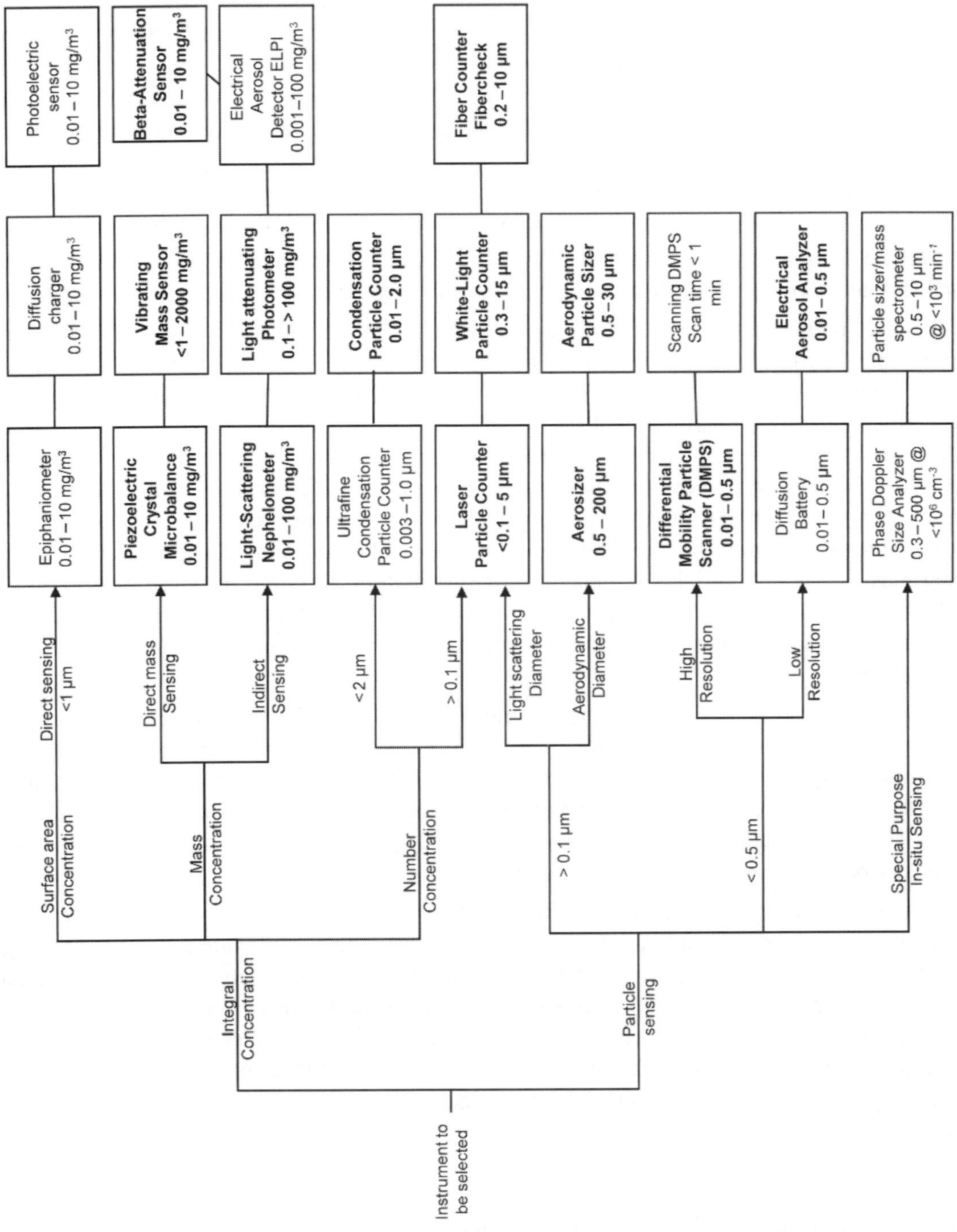

Figure 28-3. A flowchart for direct-reading aerosol instruments. Instruments in bold are mentioned in text (from Pui and Chen, 1999 and Baron and Willeke, 2001).

1989). This avoids the phenomenon known as coincidence error, which occurs when two or more particles are in the sensing chamber at once, and leads to an underestimation of the number concentration and an overestimation of size.[4]

Many portable DRI single particle counters that yield particle size distributions are available. They tend to see greatest use in the laboratory and in applications such as clean rooms where standards are written in particles per volume, but they have increasing use in the field to characterize particle size distributions and locate sources, leaks, hot spots, etc., particularly in indoor air quality work (Brauer et al., 1999). For example, the size distribution dependence of cyclone sampling efficiency was discussed in Chapter 24. Single particle counters can be used to characterize the size distribution and correct a cyclone bias.

Single particle counters generally work on the assumption that the intensity of the light from a single particle is size-dependent. The signal is then sent to a size channel. The channels are generally in increasing size ranges, such as 0.5 to 1, 1 to 2, and so on, with total size range generally 0.5 to 10 μm. Particle light scattering is size dependent; however, that dependence is not monotonic with size. The more spherical the particle, the worse the error; thus, spherical particles have resonance peaks in the scattering-size relationship that are very sharp. For all particles, light scattering cannot be related to particle size with high accuracy. This is why most *spectrometers* have only six size channels — more would be meaningless. Because the instrument must *see* many particles to generate the distribution, the time response can be on the order of minutes. Software may be applied to the counts to convert them to mass concentrations; thus, mass and count concentration are displayed, with the same caveats as before about this software.

Spectrometers is a term used by several manufacturers for photometers that size and count particles into many narrow size fractions. Generally, greater numbers of channels cost more money. There are many optical particle counters that perform sizing, but the term spectrometer is usually reserved for counters that have many channels or size ranges, up to 32 or more. They have special features that depend on principles other than light scattering (such as particle separation based on mass) that allow higher resolution of sizes.

Another technique for counting particles > 0.5 μm is *differential mobility particle sizing* using time of flight response (e.g., the Aerosizer or APS). The particles are accelerated through a nozzle one at a time; their speed will depend on their size. The velocity is measured by determining the time of flight through a known distance between two lasers. The advantage of this technique is that it gives sizes according to aerodynamic size in 52 channels; thus, it is called a spectrometer. Calibration is with unit density particles so the instrument bias is governed by actual departures from unit density. The Aerosizer has been used widely in work at NIOSH (O'Brien et al., 1986; Smith et al., 1987; Baron, 1988a, b). The Aerosizer had some problems

---

[4] Coincidence can also occur in photometers. When the concentration is quite high, the scattered light from one particle may be scattered by another one, thus changing the angle; or the second particle may absorb the light. However, the upper limit of the sensing range is often 100 mg/m^3 or higher.

relative to sizing and is now out of production. Improvements have been made to the newer APS. Generally, these spectrometers are quite expensive and out of the price range of most hygienists.

SINGLE, SUBMICRON PARTICLE COUNTERS

Photometry can detect particles larger than 0.3 µm if the light source is incandescent and down to 0.05 µm for laser sources. For most industrial aerosols measured as mass concentration either is quite adequate. However, in some instruments, particularly counters, other techniques are employed or combined to make counts over a wider range.

There are several techniques for counting submicron particles. A technique known as condensation nuclei (or particle) counting (CNC, CPC) allows detection limits to be lowered to 0.005 µm (Pui and Chen, 1999). In this technique, a particle passes through a saturated vapor such as butyl alcohol that is cooled causing the vapor to condense on the particle. Particles tend to grow to a similar size that is above the detection levels mentioned above. Of course, in its simplest form, the technique disallows particle sizing since the sizes have been modified. This technique is also used in a respirator fit tester (TSI, 2006).

Two other techniques for counting and sizing submicron particles listed in Figure 28-3 should be mentioned. The electrical mobility analyzer utilized the high mobility of submicron, charged particles (charged by the instrument) in a electrical field to separate them into fractions. Apparently it is no longer sold. Varying the separation voltage and the use of software allows size distributions to be obtained. Differential mobility analysis is an improvement on the electrical sizing technique (see Pui and Chen, 1999 for details).

Flow Rate Considerations

There are two reasons why the output of a photometer will not reflect the total aerosol in the atmosphere. First, the sensor angle and other instrument optics will be such that not all scattered light will be captured by the sensor, and the light at the edge of the laser beam will give a different signal than that at the center for the same particle, etc. This is an analysis problem. Second, the design of the sampling chamber and its entrance shape and flow rate into the chamber will affect the sampling efficiency. This is a sampling problem. The instrument may be purposely designed to cause undersampling of certain particle sizes, that is, to make the sampling efficiency mimic the respirable mass fraction (RMF) or the thoracic mass fraction (TMF). In other cases, the chamber is simply an open cavity with a wide entrance. Thus, particle losses may be relatively small.

The flow rate obviously will affect the sampling efficiency. On the other hand, the sampling rate does not affect the analytical response (Willeke and Degarmo, 1988). No matter how fast the particles enter the sensing chamber, the mass readout should not change. (However, the faster the particles enter the chamber, the faster the instrument can respond to changing concentrations.) The distinction between sampling and analysis dependence on flow rate is important because some manufacturers claim that their instruments are insensitive to flow rate, likely implying that the analytical response is not affected. As a result, some manufacturers use passive sampling, and others use relatively non-constant-flow pumps or fans (Kuusisto, 1983). However, every instrument's sampling efficiency is affected by flow rate. One example is shown in Figure 28-4 for the MINIRAM (discontinued, replaced by the personal DataRAM series), which normally operates in the passive mode. The figure also shows that although the effect of flow rate may be as much as 20%, the analytical variability associated with two different runs of the same instrument may be equally great.

Photometer and Particle Counter Instrument Models

There are numerous mass concentration measuring photometers on the market (see Mass Sensing Instruments section of this chapter and Figures 28-5, 28-6, 28-7 and 28-8). A number of them have been the subject of experimental investigation, and are discussed in the next section; however, many of the studied models have been discontinued, in some cases replaced by newer versions. For hygiene work, light attenuating photometers that measure mass concentration and laser particle counters that report count concentrations (see Figure 28-9 and Figure 28-10) are most often used. Either of these instruments may fractionate according to size. Photometers use sampling inlets to give size separation, such as an impactor or cyclone that will yield the RPM. Portable counters typically give 5–6 size ranges, which may or may not correspond to the size selection conventions. In some models, these concentrations are converted from count to mass with the use of software and an algorithm. Condensation nuclei counters are used for indoor air quality work where aerosols are composed of particles much

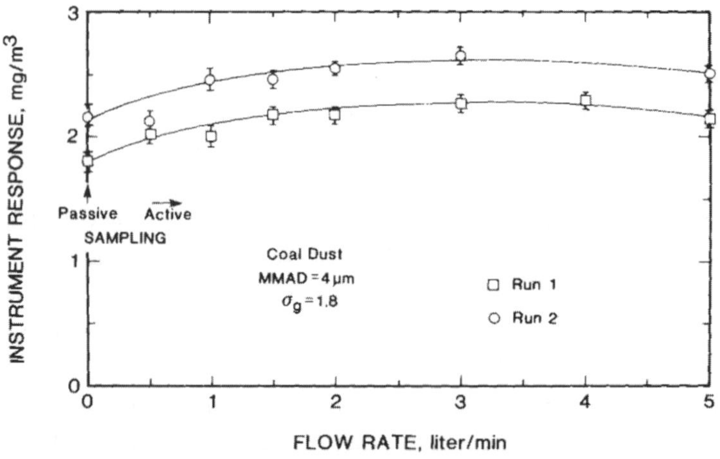

Figure 28-4. MINIRAM response to the same aerosol concentration at various flow rates including passive (from Willeke and Degarmo, 1988, with permission).

a)

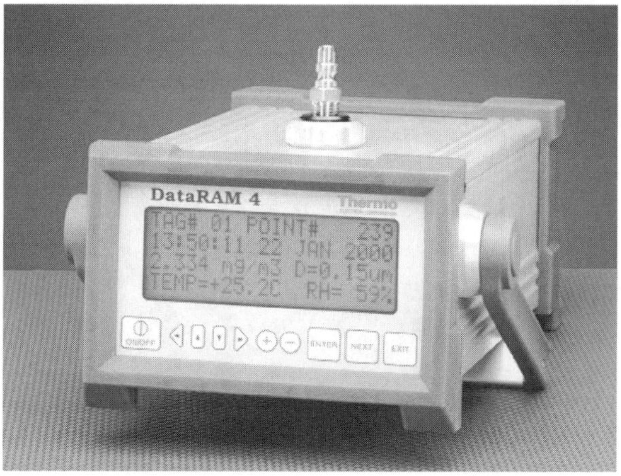

b)

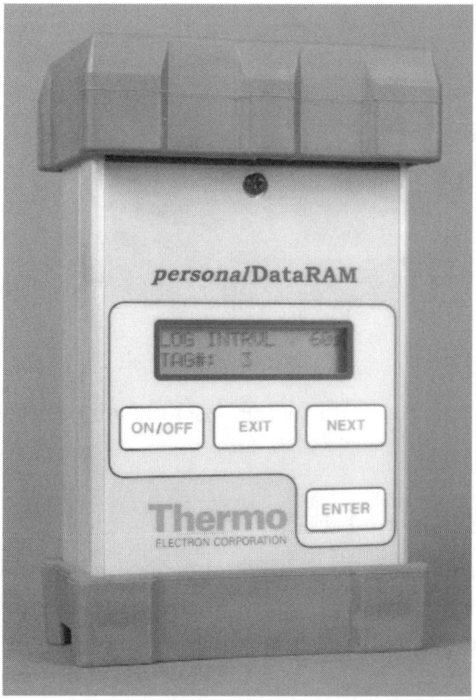

Figure 28-5. The a) DataRAM, b) personal DataRAM photometers (courtesy, Thermo Fisher Scientific).

Figure 28-6. DUSTTRAK® photometer capable of aerosol concentration measurements at PM_{10}, $PM_{2.5}$ and PM_1 μm cut-points (courtesy, TSI and REPSS).

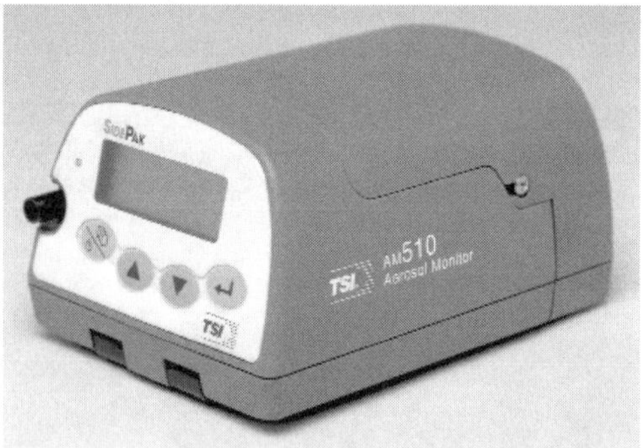

Figure 28-7. TSI AM-510 Personal Aerosol Monitor, laser photometer (courtesy, TSI and REPSS).

smaller than seen in typical industrial work.

Portable instruments are generally lightweight (one kg or so); however, some that employ advanced sizing techniques are up to 5 kg (11 lbs.) or more and might be considered transportable. There are a few manufacturers that offer personal photometers that may be worn by the worker (see Figures 28-5, 25-7, and 28-8). These may incorporate passive or active sampling.

Calibration is typically performed by the manufacturer with Arizona Road Dust (ARD). It is sold under a variety of names including Arizona Test Dust, Arizona Silica, AC Fine and AC Coarse Test Dusts, SAE Fine and Coarse Test Dusts, J726 Test Dusts, and most recently

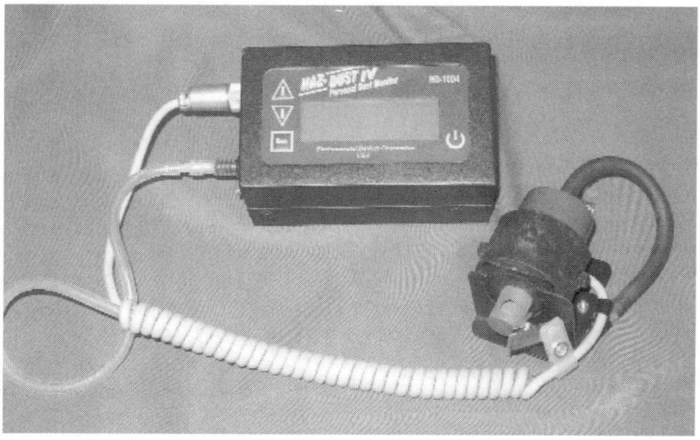

Figure 28-8. The HAZDUST IV personal photometer (courtesy, Environmental Devices Corporation).

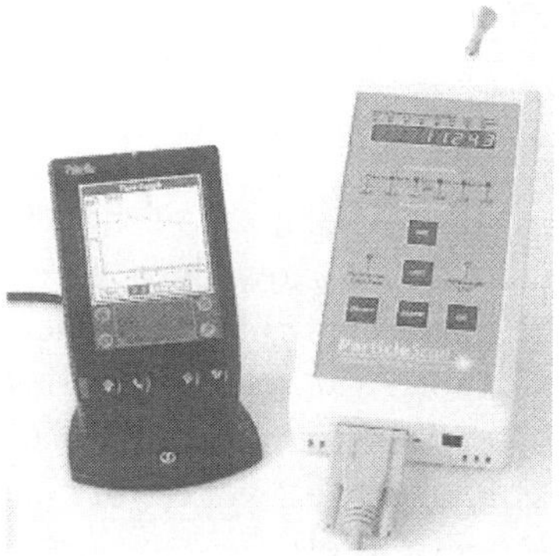

Figure 28-9. The ParticleScan particle counter connected to a PDA (courtesy, IQAir and REPSS).

ISO Ultrafine, ISO Fine, ISO Medium and ISO Coarse Test Dusts (Powder Technology, 2006). However, in most applications it is advisable to recalibrate photometers in side-by-side field comparisons using another monitoring system such as pump and filter (see later discussion).

Photometer Studies

RELATIVE RESPONSE

As previously discussed, all photometer responses are dependent on particle size, refractive index, and density. Therefore, even though the instruments are factory calibrated, they must be recalibrated for specific applications where absolute aerosol concentrations are need-

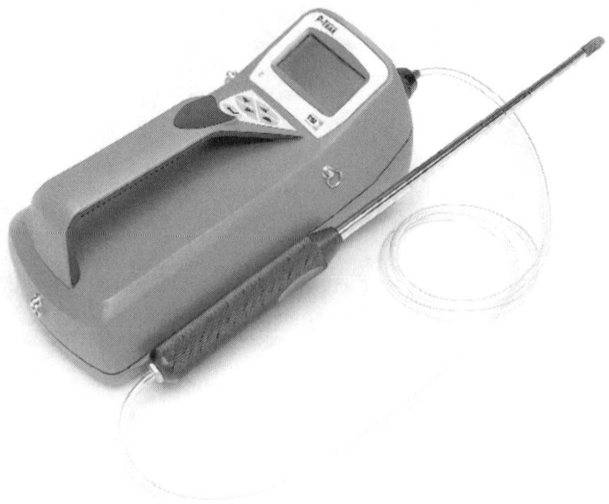

Figure 28-10. The TSI PTRAK® hand-held Ultrafine Particle Counter based on condensation nuclei technology (courtesy, TSI and REPSS).

ed. On the other hand, if only the relative response of the instrument is of interest, the factory calibration may be sufficient. Figure 28-11 shows how three instruments responded to fluctuating aerosol concentrations. The instruments had been factory-calibrated with Arizona road dust but were exposed to silicon carbide. The relative responses of each instrument were quite different, but the trends shown by each instrument are remarkably similar, indicating the utility of these instruments for assessing the effectiveness of controls, for monitoring a process to determine the effect of work practices on relative exposure, in walkthrough surveys to detect hot spots, and in monitoring the output of dust generators in laboratory studies, for example, toxicity studies (Chung and Vaughan, 1989). In walkthrough surveys, photometers can be utilized to gain a rough indication of concentration that can be used to set sample volumes for later pump and filter monitoring (Baron, 1988a). They may also be coupled with video cameras, as was discussed in Chapter 27, to monitor task and work practice effects on concentration (O'Brien et al., 1989). Finally, Peters et al. (2006) have compared two models of Grimm optical particle spectrometers (15 and 30 channel) with the TSI APS (considered the standard) for mono and polydisperse aerosols and found good agreement, generally. The Grimm products tended to underestimate number concentrations in the 0.7–2.0 μm size range for polydisperse aerosol compared to the APS.

CALIBRATION FOR SPECIFIC AEROSOLS

Most studies of photometers have involved calibration of the instruments for specific aerosols and atmospheres where absolute aerosol mass concentration measurements are need-

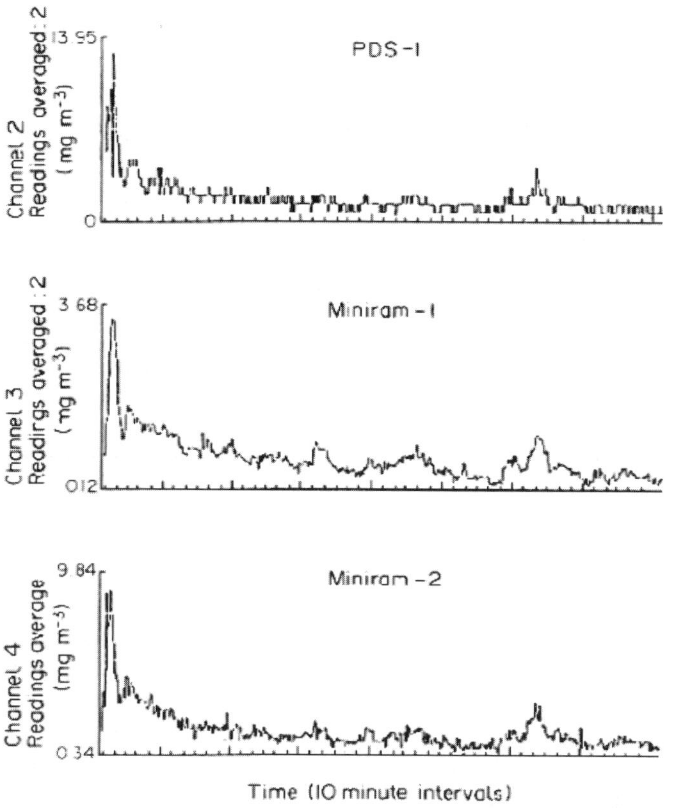

Figure 28-11. Continuous laboratory chamber dust concentration responses over an 8-hour period for generated silicon carbide dust (from Chung and Vaughan, 1989, with permission).

ed. These studies typically employ regression analysis of the instrument response against the measured concentration of a pump and filter or a cyclone.[5] Some studies have used more reliable reference devices such as cascade impactors. Some have been performed in the lab and others in the field. Tables 28-2 and 28-3 show the results of some of these studies. The actual instrument names were not given because the vast majority of the instruments tested in these studies are no longer on the market. However, in many cases, newer models have replaced

---

[5] Some photometers allow for a cyclone and filter cassette to be integrated into a train upstream and downstream, respectively, of the photometer. This allows the filter to collect what is sampled by the photometer and the cyclone if the RPM is of interest. If the photometer is biased with respect to sampling a particular size and does not allow for the addition of a size selector before the photometer, then a downstream filter will tell you what the photometer sees but not necessarily the correct concentration according to whatever fraction that you are interested in.

Table 28-2
Lab[1] comparisons of photometer responses to reference methods

Aerosols	Conc.[2] mg/m³	Size[3] μm	Reference method	Regression slope[4]	Authors
coal dust				1.0	Schnakenberg and Chilton, 1981
Arizona road dust	2–10			0.82	
limestone dust	2–6			0.81	
ARD			cyclone	1.0-1.06	Vinson and Williams, 1987
coal dust					
ARD		3.34	cyclone	1.07	Gero and Tomb, 1988
quartz		4.02		1.29	
limestone		2.91		1.69	
quartz			37 mm open face cassette or cyclone	0.095–1.96	Kuusisto, 1983
coal					
concrete grinding					
wood grinding					
plastic grinding					
stone	1–35	4–6.4	cyclone	0.9–1.2 (2.8)[5]	Thorpe and Walsh, 2002
white pine	1–12	16		1.9–2.6(6.77)	
fiberboard	1–19	10		1.5(4.8)	
white flour	0.5–4.5	1.7		2.7–3(6.37)	
ARD (SAE)	1.1–14	Fine		0.9–1.1(2.1)	
ARD SAE fine	1.8–17	2–2.36	MOUDI cascade impactor	0.75–1 (0.075)[6]	Tsai et al, 1996
coal	2.3–14	1.95–2.2		0.94–1.15(~10)	
quartz	1.9–16	1.7–2		0.67–0.92(5.9)	

[1] All aerosols were produced by aerosol generator except where noted.
[2] As measured by reference
[3] Average MMAD
[4] Photometer over reference
[5] Values in () are outliers for this study; all involved the Casella Microdust and were thought to be related to the factory calibration against total suspended particulate.
[6] Values in () are outliers for this study; all involved the Sibata PCD-1.

Table 28-3
Field[1] comparisons of photometer responses to reference methods

Aerosols	Conc.[1] mg/m³	Reference method	Regression slope[2]	Authors
welding, sintering, coal mines, diesel exhaust, copper smelting		37 open face cassette or cyclone	0.095–1.96	Kuusisto, 1983
welding fumes, 22 operations	1–112	cyclone	0.9–28	Glinsmann and Rosenthal, 1985
coal mine shear wall faces		cyclone	0.41–1.63	Page and Jankowski, 1984
borates MMAD 6.7–35		cassette	~ 0.25	Woskie et al., 1993
welding, grinding, cutting, boiler ash	0.02–4.8	PEM 2.5	0.7	Kim et al., 2004
RPM coal dust coal dust	0.06–10.8 0.2–76	cyclone cf cassette	0.99 0.84–0.92	Middendorf et al., 1999
swine production	~ 1–20	37 mm cf IOM	0.9 0.95	Taylor and Reynolds, 2001
machining mists	~ 0.5–5	ESP/37 mm cf cassette	~ 10[3]	Volckens et al., 2000

[1] As measured by reference
[2] Photometer over reference; where one instrument gave a highly aberrant value, it is in parenthesis.
[3] Slopes were ~0.7 when only the mists were measured with the reference, but when vapors were included obviously the photometers missed these.

these older models. In the lab studies most aerosols are generated with a machine such as the Wright Dust Feed, which is further discussed in Chapter 30. Large tunnels into which dusts are dispersed are also used. In a few cases, the dusts are generated using grinding or other processes that are similar to what might occur in industry.

Table 28-2 shows that the photometers perform quite well. Regression coefficients are generally around 1.0. Two points need to be made. Deviations from 1.0 are not entirely due to the photometers. The reference methods have errors as well. In addition, in many of the studies the sizes studied exceed 10 μm; photometers do not respond well to sizes in this range (see Figures 28-1 and 28-2). If the reference method is a cyclone, then the comparison is *apples to apples* since cyclones also do not respond well to particles > 10μm, MMAD, but inhalable or total samplers would sample these larger particles. In two studies in Table 28-2 (Thorpe and

Walsh, 2002; Tsai et al., 1996), two instruments gave responses that were vastly different from the other instruments in the study. In both cases, these instruments are no longer marketed.

It is possible to make corrections to the photometer response for an aerosol with different characteristics than the manufacturer's calibration aerosol to get a very accurate estimation of concentration. Since refractive index varies little for many industrial dusts and most photometers are forward light scattering so as to be least sensitive to refractive index, O'Shaughnessy and Slagley (2002) argued that, in most cases, refractive index could be ignored. They demonstrated that knowledge of the MMAD, density of the aerosol being measured and the calibration aerosol, and the response of the instrument to a given concentration, coupled with the response of the instrument to the calibration dust at the same MMAD as the aerosol being measured, could be used to define a very accurate correction factor.

Table 28-3 shows results for field comparisons and the importance of calibration. In particular, the studies of welding fume, mists, and borates stand out. In each case, there were likely large differences in the sizes of the assayed aerosol relative to the calibrant. The regression slopes were generally good for welding fume, but in some of the 22 operations, refractive indices were high (up to 2.4 compared to 1.5 for the calibrant) and particle sizes were $< 1\ \mu m$ (the size that the photometer is most sensitive to) and quite variable. In the case of borates, comparisons were made to cassette samplers and MMADs were quite high. Thus, the cassettes would be expected to sample large particles for which the photometer response would be poor, and that is what happened.

INTRA-MODEL PHOTOMETER VARIABILITY

Another source of error for any DRI is instrument-to-instrument or intra-model variability. In other words, how do side-by-side instruments of the same make respond compared to one another? We would expect that the instruments might be biased relative to the true concentration, but we would expect the bias to be fairly consistent from one instrument to the next. Kuusisto (1983) made a preliminary examination of this error using quartz at concentrations of 1 to 10 mg/m^3. He found that the regression slopes in comparisons of four RAM-1 instruments (one instrument was compared to the other three) were 0.78, 1.22, and 1.75, while two Sibata P5 instruments gave a slope of 0.75. A value of 1.0 would indicate perfect agreement between instruments. These relatively minor differences might be partially explained by differences in the true concentrations that the instruments were exposed to. However, the inlets were within 5 cm of each other.

Results from another study of intra-model variability also reflected a bias between instruments (Gero and Tomb, 1988). About 50 MINIRAMs were used in experiments that involved Arizona road dust and to a lesser degree quartz and limestone. One MINIRAM served as a reference for the others. To minimize the effect of variation in the dust generator concentrations, comparisons of the reference to the test MINIRAMs were not made until two consecutive 1-hour measurements of the test instrument were within 10% of each other. Even so, the ratio of test to reference instruments ranged from 0.7 to 1.6. About half of the instruments did not perform satisfactorily compared to the reference. The 25th and 75th %tiles were approximately 0.9 to 1.1. The range across all 50 instruments may indicate a quality control problem at the manufacturer.

PROCESS CONTROL MONITORING

There are a number of photometers on the market that are sold as process monitors. These instruments can be used for a variety of problems if they are designed correctly. For example, the monitoring of process dust generation may indicate product quality or the continued proper functioning of controls such as ventilation. In many industries, the ventilation air exhausted from the building has been heated or cooled, and its removal represents an energy loss. In some cases, this air can be recirculated to the building after it has been cleaned, in a process called exhaust gas recirculation. The performance of the air cleaner is paramount, and the use of photometers to monitor cleaning efficiency is appropriate. However, the photometer must be reliable in indicating out-of-control situations.

A number of laboratory and field studies (in a foundry recirculation system) were conducted to examine the reliability of three photometers for process monitoring (Smith, 1987; Smith et al., 1987). These instruments were the RAM-S, the PCAM, and the ATI 722. All were sold as process monitors and are no longer marketed. A number of real and potential problems arose. The user should be aware of the following points:

- A number of parts may fail, such as light sources and pumps. There is apparently no maintenance schedule for these items. In addition, some devices do not have indicators. The PCAM pump failed, and it had no low flow indicator. The ATI lamp failed, but it did have an indicator.

- Some instruments may have significant zero drift (the ATI) and require zero adjustment daily. Others (PCAM) may have automatic zeroing features.

- Some may suffer span drift due to such problems as dirty optics (ATI). A method of span checking such as a reference scattering device is needed and is contained with the PCAM.

- If the process effluent is subject to changes in size distribution, density, and refractive index, then the number of necessary calibrations may be impractical. For example, the above-cited work on welding fumes (Glinsmann and Rosenthal, 1985) shows the impracticality of applying photometers to such a variable aerosol. On the other hand, with many aerosols, only the size distribution is likely to change, and calibration curves for various sizes can be determined (see Smith et al., 1987).

- When components such as lamps are replaced, the calibration factors or responses of the instruments can change.

Single Particle Counters as Fiber Counters

Asbestos and other fiber concentrations are commonly reported in fibers per volume of air, f/cm³; so the single particle counter (SPC) is suited to the task of monitoring fiber concentrations. However, fibers usually are mixed with non-fibrous aerosols, and it is necessary to distinguish between the two, which the SPC does not normally do. However, the pattern of scattering produced by a fiber is different from that of an amorphous particle. The near forward scattered light from a fiber is oriented relative to the alignment of the fiber, and the shape of the signal from the pattern is related to the length of the fiber (Baron, 1979). Thus, the fibers must be aligned, only certain patterns must be sensed so as not to count short fibers (the OSHA rules require that only fiber lengths > 5 μm be counted), and signals from amorphous particles must be screened out by selecting only signals of a certain height.

The GCA Company (later MIE and now owned by Thermo Fisher Scientific) marketed a series of portable fiber monitors that are no longer sold (FAM5 and FM-7400). However, these instruments are still widely available for rental or as reconditioned monitors. In addition, a British instrument, the Fibercheck FC-2 is being marketed as a new monitor; at least there is an Internet site (see list at end of chapter). The FM-7400 discriminates fiber lengths by the signal pulse of the scattered light from individual, aligned fibers. The signal is a peak somewhat like that from a gas chromatograph output. It has a height or an amplitude and an area under the curve. The amplitude/area ratio or sharpness, proportional to fiber length, is controlled on the instrument by the user. Another discriminator controls the pulse height alone, which disallows noise from being counted.

Initial field studies of the prototype FAM showed that it tended to underestimate filter counts by about 35 to 40% (Baron, 1979). Subsequent tests showed good agreement between the FAM and filter samples for standard chrysotile and amosite and a *factory* chrysotile, all of which were generated in the laboratory (lles and Shenton-Taylor, 1986). The possible effects of non-fibrous dusts were investigated. Silicon carbide particles of about 5 μm physical diameter caused roughly a 10 f/ml response per 170 mg/m³ of silicon carbide. The response to talc, a fiber-shaped, non-crystalline particle, was essentially one to one, as might be expected (Iles and Shenton-Taylor, 1986). In another experiment (Phanprasit et al., 1988), a dust was generated that contained 60% chrysotile, 20% cellulose, 10% serpentine, and 10% *carbonate, etc.* A total of 33 tests gave a regression slope of 1.03 (95% CI = 0.87 to 1.19) for the FAM versus microscopically counted filters. Kauffer et al. (2003) found that the FM-7400 responded reasonably well to four types of fibers (chrysotile, glass and rock wool, and ceramic) yielding response ratios of 0.5–1 compared to phase contrast microscopy. On the other hand, the FC-2 overestimated concentrations of man-made fibers and failed to detect an atmosphere of fine chrysotile.

MASS SENSING INSTRUMENTS

The principle of direct measurement of particle mass using a piezoelectric crystal or quartz balance was first introduced in 1970 (Chuan). A quartz crystal will oscillate at a predictable frequency when an acoustic wave is induced through the crystal. If particles are deposited on the crystal's surface, the frequency will change in direct proportion to the mass deposited. The TSI model 3500 Piezobalance was the first DRI to employ this detector (Sem and Tsurubayashi, 1975). Instruments based on the principle have been developed that measure various dust fractions. The Piezobalance uses a single stage impactor. Instruments developed later employed cascade impactors with a quartz crystal microbalance at each stage (Swift, 1989). The most recent instrument, the tapered element oscillating microbalance (TEOM) employs a filter on a glass resonator that serves as the microbalance (Baron 1988a, b; Swift, 1989). It is widely used for ambient air pollution work, and is the only EPA-approved device that allows for continuous read out of concentration. The filter can be used in other analyses (e.g., metals).

NIOSH has been developing a TEOM RPM instrument for use in coal mines. It may be machine-mounted (NIOSH, 1997) or personal (Peters and Volkwein, 2003). The TEOM instrument is being developed specifically for coal mines and is expected to be available by 2008. It has passed numerous tests and appears to be at least as accurate as the standard coal mine dust personal sampler unit and more reliable. It provides real time feedback to the worker. It is, perhaps, one of the most significant aerosol instruments for hygiene work in the last 20 years.

The Piezobalance

The Piezobalance (originally sold by TSI and now marketed by Kanomax, Figure 28-12) uses a single stage impactor with a 3.5 μm cut-point. The particles then enter an electrostatic precipitator. The negative electrode is aligned with the flow and imparts a negative charge to the particles. They are then attracted to the piezoelectric crystal, which serves as the other electrode. As the particle loading increases on the crystal, it causes the crystal to resonate at a different frequency; and the frequency change is proportional to mass (Sem and Tsurubayashi, 1975; Daley and Lundgren, 1975).

The sensor will not be effective if it is heavily loaded, or if particles are chainlike and therefore do not lie flat on the crystal. The frequency-mass relationship is dependent on intimate contact of the particles with the crystal or each other. For this reason, 10 mg/m^3 concentrations and higher are not sensed well. This also means that the sensor must be periodically (~ about every 10–20 measurements) cleaned, which is accomplished with a built-in cleaning mechanism. Sampling times may range from 10–3600 seconds; however, at a concentration of 1 to 2 mg/m^3, the sensor becomes loaded at the end of a 2-minute sampling cycle and at the preset flow rate or 1 Lpm (original TSI data). Thus, repeated samples must be taken in order to approximate long-term TWAs. Up to 500 may be logged in memory of the device.

The effects of relative humidity and temperature are significant and must be corrected. A reference sensor is used to control background drift and noise caused by these factors (Daley and Lundgren, 1975). Changes in particle weight caused by particle water absorption will not be compensated. However, these short-term phenomena would not be common problems. On the other hand, it is not sensitive to particles larger than about 2–3 μm (Lundgren, 1976) and also the sampling inlet is sensitive to wind. It should work reasonably well indoors for small sticky particles, e.g., cigarette smoke.

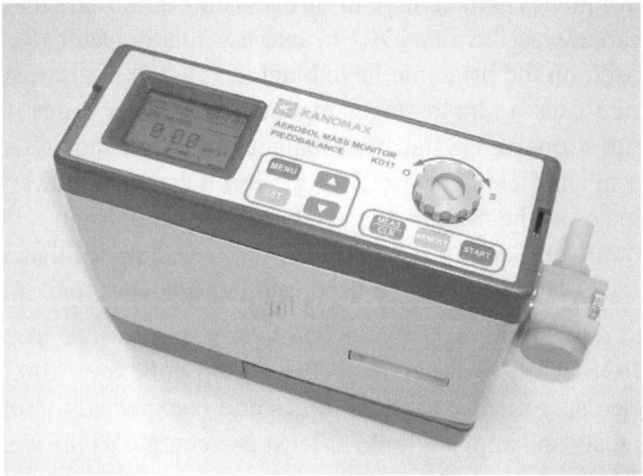

Figure 28-12. The Kanomax Piezobalance (courtesy, Kanomax).

Of the few published studies on the Piezobalance, one was conducted by the previous manufacturer (Sem et al., 1977). The investigators found excellent agreement between the Piezobalance and filters for the measurement of welding fumes in the 0.3 µm to 3.0 µm range after the instrument had been calibrated with the same welding fume. However, performance was equally good for 10 other aerosols after calibration with welding fumes, showing the wide range of applications for the instrument.

In a field study of cotton dust exposures, the Piezobalance agreed well with cyclone-filter results or vertical elutriator results. A preseparating device of 10 µm or 15 µm cutoff was used on the Piezobalance. In each case, the correlation coefficient exceeded 0.84 (Hersh et al., 1979). In a study of cigarette smoke, the Piezobalance also performed well (Ingebrethsen et al., 1988). It essentially gave one-to-one correspondence with a pump and 37 mm filter cassette. Finally, Kuusisto (1983) found that pairwise comparisons of three Piezobalances showed considerable intra-model variability. Regression slopes of 1.28, 1.27, and 0.53 were observed.

Quartz Crystal Microbalance (QCM) with Cascade Impactors

The plate of each impactor stage is a greased quartz crystal. A reference crystal is used for temperature and RH compensation. The devices measure size distribution and mass concentration. The sample periods are short (< 1 min) in order not to overload the sensors.

The PC-2 (California Instruments) has been the subject of two published studies. Fairchild and Wheat (1984) calculated the aerodynamic cut-points (D_{50}) from theory and calibration with various aerosols of relatively limited size distribution. These were compared with the manufacturer's D_{50} values, whose method of derivation had not been given. They found that, for all but the first two stages, the theoretical values fell within the 90% confidence limits of the calibration values. They attributed the discrepancy for the final two stages to impactor wall losses, which they characterized. The manufacturer's D_{50} values fell within the confidence interval for only three of the stages. O'Brien et al. (1986) found that the PC-2 underestimated the concentration of a complex grinding dust by a factor of about 2 throughout its range.

DRI FOR ELEMENTAL ANALYSIS

While the instruments discussed thus far are largely not qualitative, more recent instrumentation now allows for the direct reading or at least field analysis of the elemental content of particles, whether they are collected on a filter, or they exist on a surface. Note that these methods applied to aerosols collected on filters are not continuous, but allow the measurement of the elemental content of the filter in the field at the end of the sample period. More often these instruments are used to measure the elemental content of surfaces or contamination on surfaces. There are generally two types: field portable x-ray fluorescence (XRF) devices that may be used to measure elemental content on filters or surfaces and field portable visible and UV spectrometers that may be used in the field to assay elements such as Cr^{+6} that have been collected on filters.

Spectrometers are largely transportable due to their small size but not many are battery operated. NIOSH has published several methods for determination of elements in the field (e.g., NIOSH method 7703). Anodic stripping voltammetry has also been performed in the field for the determination of lead (Ashley, 1998; Ott and Williams, 2003).

A very convenient method for analyzing a wide range of elements is the portable XRF. It is applicable to elements of atomic number 22 and higher (this does leave out important elements such as beryllium) (Lawryk et al., 2005a and b) and most instruments will respond to some set number of pre-programmed elements with detection limits in the 10–150 ppm range. Application of the technology to filters used to sample lead-containing aerosols gave similar results to lab analyzed filters (slope = 0.96). The detection limit was 6.2 μg of lead per filter. Precision for multiple analyses was ±16% (Morley et al., 1999). Thus, the method is not as accurate as the traditional AA or ICAP method, but if many analyses need to be done, the instrument can prove cost effective. A recent review of portable elemental measurement systems is instructive (Lawryk et al., 2005a and b).

SOURCES FOR AEROSOL DRI

Manufacturers of instruments for aerosol monitoring are listed below.

Air Techniques, Inc. http://www.atitest.com/ 2 portable photometers

California Measurements, Inc. http://www.californiameasurements.com/index.html automated cascade impactor based on Quartz Crystal Microbalance

CasellaCel http://www.casellacel.com/home.asp one hand held photometer

Climet Instruments Company http://www.climet.com/ wide range (including hand held) of particle counters

Environmental Devices Corporation http://www.hazdust.com/ 6 photometers (hand held, personal, portable, and a wireless video based system)

GRIMM Aerosol Technik GmbH & Co. KG http://www.grimm-aerosol.com/ portable particle photometers, condensation particle counters, and spectrometers that measure counts and calculate concentrations

Harley Scientific http://www.btinternet.com/~harley.science/index.html Fibrecheck

Innov-X Systems, Inc. http://www.innov-xsys.com/ XRF

IQAir http://www.iqair.us/ 2 hand held particle counters, air cleaning equipment

KANOMAX USA, INC. http://www.kanomax-usa.com/ portable photometers and portable and handheld particle counters and condensation particle counters; Piezobalance Dust Monitor

Lighthouse Worldwide Solutions http://www.golighthouse.com/default.asp hand held and portable particle counters

Met One Instruments http://www.metone.com/particulate.htm portable, beta attenuation and photometric based dust monitors; portable particle counters, one that measures counts and calculates concentrations

MSP Corporation http://www.mspcorp.com/index.htm portable combination laser particle counter and differential mobility/condensation particle counter (Differential Mobility Spectrometer, DMS™ and Scanning Mobility Spectrometer, SMS™).

Particle Measuring Systems http://www.pmeasuring.com/ portable and hand held particle counters; broad size range particle counting spectrometer

QCM Research http://www.qcmresearch.com/cqcm.htm quartz crystal microbalances

SIBATA SCIENTIFIC TECHNOLOGY LTD. http://www.sibata.co.jp/~overseas/index.html portable, beta ray and photometric based, dust monitors; personal photometer; portable and hand held particle counters; spectrometer that measures counts and calculates concentrations

Thermo Fisher Scientific Corporation http://www.thermo.com/com/cda/home hybrid monitor that combines light scattering photometry and beta radiation attenuation; TEOM; personal, portable, and hand held photometers; hand held x-ray fluorescence; former MIE, Rupprecht & Patashnick, Niton, and Anderson products

TSI, Inc. http://www.tsi.com/Default.aspx integrating nephelometers, surface area concentration electrometer; personal, portable, and hand held photometers, particle counters, and condensation particle counters; Scanning Mobility Particle Sizer-based and electrometer- based spectrometers

QUESTIONS

28.1. Describe the limitations of aerosol DRI. For what tasks are they useful?

28.2. In optical light scattering instruments, what factors can be controlled by the manufacturer to affect the instrument response? What characteristics of the environment affect the instrument's response?

28.3. What makes the calibration of aerosol DRI different from gas and vapor DRI?

REFERENCES

Ashley, K.; Mapp, D.J.; Millson, M.: Ultrasonic extraction and field-portable anodic stripping voltammetry for the determination of lead in workplace air samples. Am. Ind. Hyg. Assoc. J. 59:671-679 (1998).

Baron, P.A.: Recent developments in fibrous aerosol measurements. J. Env. Path. Tox. 2:335-343 (1979).

Baron, P.A.: Modern real-time aerosol samplers. Appl. Ind. Hyg. 3:97-103 (1988a).

Baron, P.A.: Modern real-time aerosol samplers, In Advances in Air Sampling. Lewis Publishers, Chelsea, MI (1988b).

Baron, P.A.: Aerosol photometers for respirable dust measurements, In NIOSH Manual of Analytical Methods, National Institute for Occupational Safety and Health, Cincinnati, OH, USA http://www.cdc.gov/niosh/nmam/ pdfs/chapter-g.pdf (1998).

Baron, P.; Willeke, K.: Aerosol Measurement. Wiley and Sons, NY, NY (2001).

Brauer, M.; Hirtle, R.D.; Hall, A.C.; Yip, T.R.: Monitoring personal fine particle exposure with a particle counter. Journal of Exposure Analysis and Environmental Epidemiology 9:228-236 (1999).

Chuan, R.L.: An instrument for the direct measurement of particle mass. Aerosol Sci. 1:111 (1970).

Chung, K.Y.K.; Vaughan, N.P.: Comparative laboratory trials of two portable direct-reading dust monitors. Ann. Occup. Hyg. 33:591-606 (1989).

Daley, P.S.; Lundgren, D.A.: The performance of piezoelectric crystal sensors used to determine aerosol mass concentrations. Am. Ind. Hyg. Assoc. J. 36:518-532 (1975).

Fairchild, C.I.; Wheat, L.D.: Calibration and evaluation of a real-time cascade impactor. Am. Ind. Hyg. Assoc. J. 45:205-211 (1984).

GCA/Technology Division: Instruction Manual Respirable Dust Monitor Model ROM 201. Bedford, MA (1977).

Gero, A.J.; Tomb, T.F.: MINIRAM calibration differences. App. Ind. Hyg. 3: 110-114 (1988).

Glinsmann, P.W.; Rosenthal, F.S.: Evaluation of an aerosol photometer for monitoring welding fume levels in a shipyard. Am. Ind. Hyg. Assoc. J. 46:391-395 (1985).

Hersh, S.P.; Fornes, R.E.; Anand, M.: Short-term cotton dust sampling utilizing three non-gravimetric methods. Am. Ind. Hyg. Assoc. J. 40: 578-587 (1979).

Hinds, W.C.: Aerosol Technology. Wiley Interscience, New York (1999).

Hosey, A.D.; Jones, H.H.; Ayer, H.E.: Evaluation of an aerosol photometer for dust counting and sizing. Ind. Hyg. J. 21 :491-501 (1960).

Iles, P.J.; Shenton-Taylor, T.: Comparison of a fibrous aerosol monitor (FAM) with the membrane filter method for measuring airborne asbestos concentrations. Ann. Occup. Hyg. 30:77-87 (1986).

Ingebrethsen, B.J.; Heavner, D.L.; Angel, A.L.; Conner, J.M.; Steichen, T.J.; Green, C.R.: A comparative study of environmental tobacco smoke particulate mass measurements in an environmental chamber. JAPCA 38:413-417 (1988).

Kauffer, E.; Martin, P.; Grzebyk, M.; Villa, M.; Vigneron, J.C.: Comparison of two direct-reading instruments (FM-7400 and fibrecheck FC-2) with phase contrast optical microscopy to measure the airborne fibre number concentration. Ann. Occup. Hyg. 47(5):413-426 (2003).

Kim, J.Y.; Magari, S.R.; Herrick, R.F.; Smith, T.J.; Christiani, D.C.: Comparison of fine particle measurements from a direct-reading instrument and a gravimetric sampling method. Appl. Occup. Environ. Hyg. 1:707-715 (2004).

Kuusisto, P.: Evaluation of the direct reading instruments for the measurement of aerosols. Am. Ind. Hyg. Assoc. J. 44:86-874 (1983).

Lawryk, N.J.; Creek, K.; Hoover, M.D.: Field portable measurement of airborne beryllium, chromium, lead and other metals: Instruments, methods and future directions. Part I: Beryllium. The Synergist (2005).

Lawryk, N.J.; Ashley, K.E.; Drake, P.L.: Field portable measurement of airborne beryllium, chromium, lead and other metals: Instruments, methods and future directions. Part II: Chromium, lead and other metals. The Synergist (2005).

Lundgren, D.A.: Fine Particles: Aerosol Generation, Measurement, Sampling, and Analysis, B.Y.H. Liu (ed.). Academic Press (1976).

Middendorf, P.J.; Lehocky, A.H.; Williams, P.L.: Evaluation and field calibration of the miniram PDM-3 aerosol monitor for measuring respirable and total coal dust. Am. Ind. Hyg. Assoc. J. 60:502-511 (1999).

Morley, J.C.; Clark, C.S.; Deddens, J.A.; Ashley, K.; Roda, S.: Evaluation of a portable x-ray fluorescence instrument for the determination of lead in workplace air samples. Appl. Occup. Environ. Hyg. 14:306-316 (1999).

NIOSH: NIOSH Manual of Analytical Methods, 3rd ed., Vol. 1, Fibers, Method: 7400, Revision #3: 5/15/89. National

Institute for Occupational Safety and Health, Cincinnati, OH (1989).

NIOSH: Machine-Mounted Continuous Respirable Dust Monitor, Tech News No. 463. National Institute for Occupational Safety and Health, Cincinnati, OH http://www.cdc.gov/niosh/mining/pubs/pdfs/463.pdf (1997).

O'Brien, D.; Baron, P.; Willeke, K.: Size and concentration measurement of an industrial aerosol. Am. Ind. Hyg. Assoc. J. 47:386-392 (1986).

O'Brien, D.M.; Fischbach, T.J.; Cooper, T.C.; Todd, W.F.; Gressel, M.G.; Martinez, K.F.: Acquisition and spreadsheet analysis of real time dust exposure data: A case study. Appl. Ind. Hyg. 4:238-243 (1989).

O'Shaughnessy, P.T.; Slagey, J.M.: Photometer response determination based on aerosol physical characteristics. Am. Ind. Hyg. Assoc. J. 63:578-585 (2002).

Ott, S.D.; Williams, T.A.: NIOSH field portable analytical methods for airborne lead: Identifying and overcoming barriers to their use. The Synergist (2003).

Page, S.J.; Jankowski, R.A.: Correlations between measurements with RAM-1 and gravimetric samplers on longwall shearer faces. Am. Ind. Hyg. Assoc. J.. 45:610-616 (1984).

Peters, T.M.; Ott, D.; O'Shaughnessy, P.T.: Comparison of the Grimm 1.108 and 1.109 portable aerosol spectrometers to the TSI 3321 Aerodynamic Particle Sizer for dry particles. Ann. Occ. Hyg. 50:843-850 (2006).

Peters, T.M.; Volkwein, J.C.: Analysis of sampling line bias on respirable mass measurement. Appl. Occup. Environ. Hyg. 18(6):458-465 (2003).

Phanprasit, W.; Rose, V.E.; Oestenstad, R.K.: Comparison of the fibrous aerosol monitor and the optical fiber count technique for asbestos measurement. Appl. Ind. Hyg. J. 3:28-33 (1988).

Powder Technology: Arizona Test Dust Fractions. Powder Technology Incorporated, Burnsville, MN, USA http://www.powdertechnologyinc.com/docs/pages/product_list_4.html (2006).

Pui, D.Y.H.; Chen, D.: Direct-reading instruments for analyzing airborne particles, In Air Sampling Instruments, 9th ed. American Conference of Governmental Industrial Hygienists, Cincinnati, OH (1999).

Schnakenberg, G.H.; Chilton, J.E.: Direct reading systems. Ann. Am. Conf. Gov. Ind. Hyg. 1:273-293 (1981).

Sem, G.J.; Tsurubayashi, K.: A new mass sensor for respirable dust measurement. Am. Ind. Hyg. Assoc. J. 36:791-800 (1975).

Sem, G.J.; Kazuo, T.; Homma, K.: Performance of the piezoelectric microbalance respirable aerosol sensor. Am. Ind. Hyg. Assoc. J. 38:580-588 (1977).

Smith, J.P.: Use of scattered light particulate monitors with a foundry air recirculation system. Appl. Ind. Hyg. 2:74-78 (1987).

Smith, J.P.; Baron, P.A.; Murdock, D.J.: Response characteristics of scattered light aerosol sensors used for control monitoring. Am. Ind. Hyg. Assoc. J. 48:219-229 (1987).

Swift, D.L.: Direct-reading instruments for analyzing airborne particles, In Air Sampling Instruments, 7th ed. American Conference of Governmental Industrial Hygienists, Cincinnati, OH (1989).

Taylor, C.D.; Reynolds, S.J.: Comparison of a direct-reading device to gravimetric methods for evaluating dust aerosols in an enclosed swine production environment. Appl. Occup. Environ. Hyg. 16(1):78-83-2001).

Thorpe, A.; Walsh, P.T.: Performance testing of three portable, direct-reading dust monitors. App. Occup. Hyg. 46(2):197-207 (2002).

Tsai, C.J.; Shih, T.S.; Lin, J.D.: Laboratory testing of three direct reading dust monitors. Am. Ind. Hyg. Assoc. J. 57:557-563 (1996).

TSI: PORTACOUNT® Plus Respirator Fit Tester. TSI, Inc., St. Paul, MN http://www.tsi.com/Product.aspx?Pid=33 (2006).

Vinson, R.P.; Williams, K.L.: Performance Evaluation of Two Light-Scattering Dust Monitors. Information Circular 9162, United States Department of the Interior, Bureau of Mines, Pittsburgh, PA (1987).

Volckens, J.; Boundy, M.; Leith, D.: Mist concentration measurements II: Laboratory and field evaluations. Appl. Occup. Environ. Hyg. 15(4):370-379 (2000).

Willeke, K.; Degarmo, S.: Passive versus active aerosol monitoring. Appl. Ind. Hyg. 3:263-266 (1988).

Woskie, S.R.; Shen, P.; Finkel, M.; Eisen, E.A.; Smith, T.J.; Smith, R.; Wegman, D.H.: Calibration of a continuous-reading aerosol monitor (MINIRAM) to measure borate dust exposures. Appl. Occup. Environ. Hyg. 8:38-45 (1993).

29

DETECTOR TUBES

INTRODUCTION

Detector tubes or colorimetric indicator tubes are among the oldest direct-reading devices. The technology apparently was developed in 1917 for detecting carbon monoxide (AIHA, 1993). The principle is simple if one understands the chemistry. A contaminant of interest reacts with a chemical on a substrate and, in some cases, several reactions are involved. The result is a color change. In some systems the intensity of the change is indicative of the concentration, but in most colorimetric devices used today, the reaction takes place down the length of a glass tube, and the length of the reaction, or the length of stain, is indicative of the concentration. Air must be drawn into the tube, usually with a hand-operated pump; or, in passive devices (see Chapter 21), it diffuses into the tube. The tubes are generally a little longer and thinner than a jumbo charcoal tube, 12–14 cm long. There are differences in size for the two types of pumps that are used, and there can be small differences between manufacturers.

Tubes used with hand-operated pumps take grab samples, as the time needed to fill the pump is generally less than a minute. Thus, using grab samples to characterize exposures over a 15-minute or an 8-hour period would require many samples taken randomly and averaged. The needed sample size is subject to the same considerations discussed in Chapter 16, such as the required accuracy of the estimate and the variability of the concentration over the averaging time. Also, the results are indicative of area concentrations, for example, if the hygienist holds the tube near the worker's mouth, then at best, breathing zone concentrations are measured. For these reasons, a number of tube systems have been adapted to long-term monitoring using battery operated pumps that can be worn by workers.

Tube Fillers

There are now hundreds of detector tubes (see, e.g., Figure 29-1) sold by a number of manufacturers and their distributors (see "Manufacturers and Distributors" listing at end of chapter). Tubes for aerosols have even been developed (Leichnitz and Walton, 1981). The materials in the tubes do have some commonalities, however. For example, in many cases, the

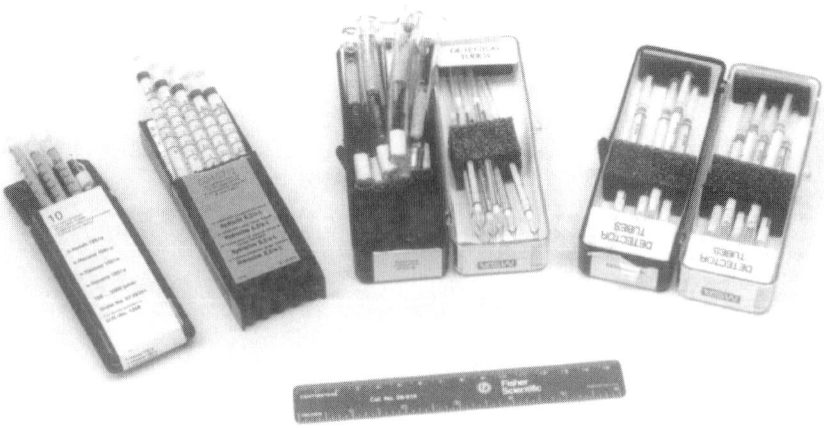

Figure 29-1. Some of the many detector tubes available.

substrate is silica gel; alumina, ground glass, pumice and resin are also used (Harper, 2001). Some tubes contain only an indicating layer. In these tubes, the silica gel or the substrate that is coated with the reactant chemical comes into direct contact with the air carrying the contaminant. In other tubes, pre-layers are employed. Pre-layers are generally used for two purposes: to convert the contaminant to another chemical species that will react with the indicator layer, or to absorb and/or to react with interferences. Interferences will react with the indicator layer and cause a color change, even though it may be less intense than, or of a different color from, that of the contaminant of interest.

There are several variations beyond these two basic types. Relatively recently, tubes with more than one indicating layer have been developed. These tubes indicate that only members of certain classes of compounds are present, and not in quantitative amounts. They were developed for hazardous materials response applications.[1] Also there are a few tube systems that keep the reagents separate until it is time to use the tube, as these reagents are not stable when combined. Sometimes one of the reagents is liquid. These systems may employ small, reagent-filled ampules that fit onto the front of the tube, or two tubes in series. In the latter case, the first tube may contain the pre-layer and the second the indicating layer. In use, the two tubes are opened and then connected in series.

[1] See also King et al. (1983) for a description of a prototype device that uses many tubes at once to sample for a wide range of contaminants. Current sampling manifolds offered by manufacturers have resulted from this application of detector tubes to emergency response operations.

Reaction Mechanisms

Although there are tubes available for many different chemicals, the reaction mechanisms are often the same for a large number of chemicals. This is shown in Table 29-1. For example, the reduction of chromate or dichromate can be used to detect a wide range of oxygen- and nitrogen-containing hydrocarbons, and reaction with tetra phenyl benzidine can be used to detect a number of chlorinated hydrocarbons. This simplicity of application of reactions works to the advantage of the tube designer and means there is a wide array of available tubes, but it also means a wide array of interferences exists for each tube. Basically, a single tube containing the chromate system is calibrated by the manufacturer for one of the various chemicals shown in Table 29-1; but in fact, the tube contents will react with any of the chemicals on the list, and perhaps others, regardless of the name of the chemical printed on the tube box. Manufacturers do try to limit interferences with pre-layers, but the detector tubes can only be used qualitatively if the hygienist eliminates interferences as a result of previous knowledge of the exact composition of the environment.

Pump Types

Although some detector tubes have been used passively, as discussed in Chapter 23, the majority of applications use a pump and, in most cases, a hand-held manual pump is used (see Figure 29-2). Generally, the pumps require that the operator pull a piston or squeeze a bulb in order to evacuate a chamber. Piston pumps require that the plunger be pulled out, typically one-half stroke, although this can vary. For bulb pumps, the bulb is squeezed and released, typically 10 times. The concentration determines the pump volume necessary to give an indication on the tube. In either case, the chamber refills with air that has passed through the tube, which is connected to an opening on the pump. Some manufacturers offer battery-operated pumps that use a mechanical action to operate the pump for the user. In other words, the pumping mechanism is the same, but the user does not have to do the work.

One might wonder, why not just use any battery-operated pump, instead of the hand-held or powered hand-held devices just described? The tubes are calibrated for a particular type of pump and a specific number of pump strokes. Each pump type gives a particular flow regimen; that is, the flow is not constant with time. As discussed later, deviations from these calibration conditions often void the calibration scale printed on the tube. In the U.S., certification of tubes and systems requires that the tube be used with the manufacturer's pump. This is not the case in Europe. Some manufacturers encourage this, while others do not (Harper, 2001).

For long-term tubes, low flow pumps (10–20 cm^3/min) are used. The manufacturer may make a special pump that will supply the correct flow rate for long-term tubes. Otherwise, low flow pumps may be used, but the calibration of the detector tube should be checked, for reasons that will be described later.

Table 29-1
Common colorimetric reactions in gas detector tubes (from Saltzman, 1989)

1. Reduction of chromate or dichromate to chromous ion:

Draeger: Acetaldehyde 100/a; acetone 500/a-L; alcohol 100/a; aniline 0.5a; cyclohexane 100/a; diethyl ether 100/a; ethanol 500/a-L, 1000/a-D; ethyl acetate 200/a, 500/a-D; ethyl glycol acetate 50/a; ethylene oxide 25/a; n-hexane 100/a; hydrocarbons 100/a-L; methanol 50/a; n-pentane 100/a; o-toluidine 1/a.

Gastec: Acetone 151; aniline 181; amyl acetate 147; butane 104; butyl acetate 142, 142L; n-butanol 115; cyclohexanol 118; ethanol 112, 112L; ethyl acetate 141, 141L; ethyl ether 161, 161L; ethylene oxide 163; gasoline 101, 101L; hexane 102H, 102L; hydrocarbons 103; isoamyl acetate 148; isoamyl alcohol 117; isobutyl acetate 144; isobutanol 116; isopropanol 113, 113L; isopropyl acetate 146; LP gas 100A; methanol 111, 111L, 111LL; methyl cyclohexanol 119; methyl ethyl ketone 152; methyl isobutyl ketone 153; methyl methacrylate 149; petroleum naphtha 106; propane 100B; propyl acetate 145; sulfur dioxide 5H; tetrahydrofuran 159; vinyl chloride 131.

Kitagawa: Acetone 102SA, 102SD; acrylonitrile 128SA, 128B; allyl alcohol 184S; butadiene 168SA, 168SC; butane 221SA; butanol 207U; butyl acetate 138SA, 138U; butyl acrylate 211U; butyl cellosolve 190U; cyclohexane 115S; cyclohexanol 206U; cyclohexanone 197U; diacetone alcohol 195U; dichloroethane 235S, 230S; dimethyl ether 123; dioxane 154; ether 107SA, 107U; ethyl acetate 111SA, 111U; ethyl alcohol 104A; ethylene oxide 122; furan 161; gasoline 110S; general hydrocarbons 187S; n-hexane 113; isobutyl acetate 153; isoprene 241U; isopropanol 150; isopropyl acetate 149, 111U; methyl acetate 148; methyl acrylate 211U; methyl alcohol 119; methyl ethyl ketone 139B; methyl isobutyl ketone 155; naphthalene 226U; organic gases 186; propyl acetate 151; propylene oxide 163; sulfur dioxide 103; tetrahydrofuran 162; vinyl chloride 132.

MSA: Ethanol 804136; ethylene 804428; hexane 497664.

2. Reduction of iodine pentoxide plus fuming sulfuric acid to iodine:

Draeger: Benzene 2/a, 5/b, 20/a-L; carbon disulfide 5/a, 10/a-L; carbon monoxide 2/a, 5/c, 8/a, 10/a-L, 10/b, 50/a-L, 0.001%/a, 0.3%/b; ethyl benzene 30/a; hydrocarbon 0.1%/a; methylene chloride 100/a*, 50/a-I*, natural gas*; petroleum hydrocarbons 100/a; polytest; toluene 5/a, 200/a-L.

Gastec: Acetylene 171; benzene 121, 121L, 121S, 121SL; carbon

Table 29-1 (Cont.)
Common colorimetric reactions in gas detector tubes (from Saltzman, 1989)

	monoxide 1HH, 1H, 1M, 1LK; chlorobenzene 126; o-dichlorobenzene 127; hydrocarbons 105; stoddard solvent 128; toluene 122, 122L; trichloroethylene 132HH; xylene 123.
Kitagawa:	Benzene 118SB, SC; carbon monoxide 106, chlorobenzene 178S; chloropicrin 172; dichlorobenzene 214S; ethylbenzene 179S; toluene 124SA, SB; xylene 134S.
MSA:	Carbon disulfide 492514; carbon monoxide 803943, 487334, 804423, 487335, 488906(HP); dichloromethane 804416, gasoline 492870, perchloroethylene 804429, 487337; qualitest 497665; toluene 803947; trichloroethane 487343.
3. Reduction of chemical with ammonium molybdate plus palladium sulfate to molybdenum blue:	
Draeger:	Ethylene 0.1/a, 50/a; methyl acrylate 5/a; methyl methacrylate 50/a.
Gastec:	Butadiene 174, 174L; ethylene 172, 172L.
Kitagawa:	Acetylene 101S; Butadiene 168SB; carbon monoxide 106A, 106B, 106C*; ethylene 108B, 108SA; hydrogen sulfide and sulfur dioxide 120C.
MSA:	Carbon monoxide 47134, ethyl mercaptan 804589
4. Reaction with potassium palladosulfite:	
Draeger:	Carbon monoxide 50/a-D.
Gastec:	Carbon monoxide 1L, 1La, 1LL; hydrogen 30; hydrogen cyanide 12H
Kitagawa:	Carbon monoxide 100, 106.
5. Color change of pH indicators (e.g., bromphenol blue, phenol red, thymol blue, methyl orange):	
Draeger:	Acetic acid 5/a, 5/a-L, 10/a-D; acetone 1000/a-D*; acid test; acrylonitrile 1/a*, 5/a*; amine; ammonia 2/a, 5/a, 10/a-L, 20/a-D; 0.5%/a; carbon dioxide 1000/a-L, 500/a-D, 1%/a-D; chlorobenzene 5/a*; cyanide 2/a*; cyclohexylamine 2/a; dimethylformamide 10/b*; formic acid 1/a; halogenated hydrocarbons 100/a**; hydrazine 0.25/a; hydrochloric acid 2/a, 10/a-D, 10/a-L, 50/a; hydrocyanic acid 2/a*, 20/a-D*; hydrogen fluoride 2/a-L; nitric acid 1/a; perchloroethylene 50/a-L; phosphine 0.01/a*; phosphoric acid esters 0.05/a*; sulfur dioxide 0.1/a*, 2/a-L; triethylamine 5/a; vinyl chloride 0.5/a*.
Gastec:	Acetaldehyde 92, 92M; acetic acid 81, 81L; acrolein 93; acrylonitrile 191, 191L; amines 180, 180L; ammonia 3H, 3M, 3L, 3La; arsine 19LA; tert-butyl mercaptan 75, 75L; carbon dioxide 2HH, 2H, 2L, 2LL; carbon disulfide 13, 13M; carbonyl sulfide 21; chlorine 8HH; diborane 22; 1,2-dichloroethylene 139; dimethylacetamide 184; dimethylformamide 183; ethyl mercaptan 72L, formaldehyde 91M, 91L; hydrazine 185; hydrogen chloride 14M, 14L;

Table 29-1 (Cont.)
Common colorimetric reactions in gas detector tubes (from Saltzman, 1989)

	hydrogen cyanide 12M, 12; hydrogen fluoride 17; mercaptans 70L; methacrylonitrile 192; nitric acid 15L; 2-pentenenitrile 193; perchloroethylene 133HA, 133M, 133L, 133LL; phosphine 7, 7La; pyridine 182; sulfur dioxide 5M, 5L, 5La, 5Lb; trichloroethylene 132HA, 132HM, 132L, 132LL; vinyl chloride 131La, 131L; vinylidine chloride 130L.
Kitagawa:	Acetaldehyde 133A, 133SB; acetic acid 216S; acetone 102SC; acrolein 136; acrylonitrile 128SC, 128SD; ammonia 105, 105B; arsine 121U; mercaptans 130U; carbon dioxide 126SA, 126SB, 126SC, 126SD, 126SH, 126UH; carbon disulfide 141SA, 141SB; carbonyl sulfide 239S; chloroprene; diborane 242S; dichloroether 223S; diethylamine 222S; formaldehyde 171SA; hydrazine 219S; hydrogen chloride 173SA; hydrogen cyanide 112B; nitric acid 233S; silane 240S.
MSA:	Acetic acid 804138; ammonia 804405, 800300, 804406, carbon dioxide 497606, 487333, 804419, 488907(HP); formaldehyde 497649; hydrogen chloride 803948; hydrogen cyanide 803945, hydrogen fluoride 804142; sulfur dioxide 497661; triethylamine 804134.
6. Reaction with o-tolidine:	
Draeger:	Chlorine 0.2/a, 0.3b, 1/a-L, 50/a; chloroform 2/a*; epichlorohydrin 5/b*; fluorine 0.1/a*; nitrogen dioxide 10/a-D; perchloroethylene 10/b, 200/a-D; trichloroethane 50/d*; trichloroethylene 2/a*, 10/a, 10/a-L*, 200/a-D; vinyl chloride 1/a*, 10/a-L*.
Gastec:	Chlorine 8H, 8Ha; chloroform 137; methyl bromide 136H, 136L, 136La; methylene chloride 138; nitrogen dioxide 9L; nitrogen oxides 11HA, 11S, 11L; nitric oxide 10; 1,1,1-trichloroethane 135, 135L.
Kitagawa:	Bromine 114; chlorine 109SA, 109SB; chlorine dioxide 116; nitrogen dioxide 117.
7. Reaction with tetraphenylbenzidine:	
MSA:	Part 82399 for bromine, chlorine, chlorine dioxide. Part 83099 for nitrogen dioxide. Part 85833** for chlorobromomethane; 1,1-dichloroethane; dichloroethylene (cis-1,2 and trans-1,2); ethyl bromide; ethyl chloride; perchloroethylene (tetrachloroethylene); trichloroethylene; 1,2,3-trichloropropane; vinyl chloride (chloroethylene). Part 85834* for chlorobenzene (mono); 1,2-dibromoethane (ethylene dibromide); dichlorobenzene (ortho); 1,2-dichloroethane (ethylene dichloride); dichloroethyl ether; 1,1-dichloroethylene (vinylidine chloride); methyl bromide; methylene chloride (dichloromethane); propylene dichloride (1,2-dichloro-

Table 29-1 (Cont.)
Common colorimetric reactions in gas detector tubes (from Saltzman, 1989)

propane); 1,1,2,2-tetrabromoethane; 1,1,2,2-tetrachloroethane; 1,1,3,3-tetrachloropropane; trichloroethane (beta 1,1,2); vinyl chloride (chloroethylene). Part 87042 for bromine; chlorine. Part 88536* for carbon tetrachloride; chlorobromomethane; 1-chloro-1,1-difluoroethane (Genetron 14213); chlorodifluoromethane (Freon 22); chloroform (trichloromethane); chloropentafluoroethane (Freon 115); chlorotrifluoromethane (Freon 13); 1,2-dibromoethane (ethylene dibromide); dichlorodifluoromethane (Freon 12); 1,1-dichloroethylene (vinylidine chloride); dichloroethylene (cis-1,2); dichlorotetrafluoroethane (Freon 114); fluorotrichloromethane (Freon 11); Freon 113; Freon 502; methyl bromide; methyl chloroform (1,1,1-trichloroethane); methylene chloride (dichloromethane); perchloroethylene (tetrachloroethylene); trichloroethane (beta 1,1,2); trichloroethylene; 1,1,2-trichloro-1,2,2-trifluoroethane (Freon 113); trifluoromonobromomethane (Freon 13131). Part 91624** for acetonitrile; acrylonitrile; 1-chloro-l-nitropropane; cyanogen; 1,1-dichloro-l-nitroethane; dimethylacetamide; dimethylformamide; fumigants (Acritet, Insect-O-Fume, Fume-I-Gate, termi-Gas, Termi-Nate); methacrylonitrile; nitroethane; nitromethane, 1-nitropropane; 2-nitropropane; n-propyl nitrate; pyridine; vinyl chloride. Part 460225 for chlorine. Part 460424* for nitric oxide.

8. Reaction of 2,4-dinitrophenylhydrazine forming a hydrazone:

MSA: Acetone 804141; methyl ethyl ketone 813334

9. Oxidation by an iodate/sulfuric acid reagent:

MSA: Aromatic hydrocarbons 804132; benzene 807024, 804411.

10. Oxidation of an aromatic amine:

MSA: Chlorine 803944; chlorine dioxide 804133, nitrogen dioxide 487341, 804435; nitrous fumes 487336, 804425, 803946, 804426; trichloroethane 487342; vinyl chloride 803950.

11. Reaction with a lead or silver compound forming the sulfide or metal.

MSA: Hydrogen sulfide 487399, 487340, phosphine 497101, 485680, 489119

12. Reaction of chemical with copper iodide.

MSA: Mercury 497663.

13. Oxidation of indigotine to isatine.

MSA: Ozone 804140

14. Reaction of chemical with sulfuric acid:

MSA: Phenol 813778, styrene 804135.

15. Reaction with an aromatic aminoaldehyde and an aromatic amine.

MSA: Phosgene 803949.

Table 29-1 (Cont.)
Common colorimetric reactions in gas detector tubes (from Saltzman, 1989)

16. Reaction of chemical with iodine, forming an iodide.	**17. Precipitation of selenium from selenium/sulfuric acid.**
MSA: Sulfur dioxide 487338, 497662, SF6 decomposition products 804433.	MSA: Water vapor 488908.
	* = Multiple reactions or multiple layer tube for improved specificity or preliminary reaction
	** = Pyrolyzer required
	HP = high pressure tube

Applications

Even though most detector tubes operate as grab samplers, they have a wide array of uses. The user must never forget that concentrations change over space and time, and these changes are hard to detect with detector tubes because they are not continuous-reading devices.

Walkthrough surveys. Detector tubes can be used to indicate whether certain contaminants may be present in an operation, if equipment is leaking, if controls are working, and a number of other parameters. They serve only as a rough guide, but that is the nature of information gathering on a walkthrough survey.

Sampling parameter estimation. Detector tubes can be used to estimate concentrations so that parameters such as flow rate, sample volume, media size, and sample size can be estimated for a later in-depth survey.

Confined space, hot work, explosive atmospheres, etc. In each of these situations, detector tubes may be used to gain information about the qualitative nature of an atmosphere. Continuous-read instruments have the advantage of sensing changes over time and space (if attached to a worker), but in some cases, detector tube systems can more easily help to identify the components of an atmosphere.

Emergency response and fire fighting. Detector tube kits are available that can be used systematically to determine the contaminants at a fire, spill, or accidental release. Positive results are not qualitative per se because of interferences, but negative results are a good indication that certain classes of compounds are not present.

Indoor air quality. Although used in such situations as responding to complaints, detector tubes are of dubious utility in many of these situations. Generally, the concentrations encountered are below detection limits, yet even these low concentrations are often of consequence to health in indoor environments, even if the contaminant is only a mild irritant. Detection limits for detector tubes are often higher than for other methods. Detector tubes may be used to monitor CO_2 levels, but the user should be aware that the application of indoor air quality standards that are based on CO_2 levels assumes steady state conditions over time and

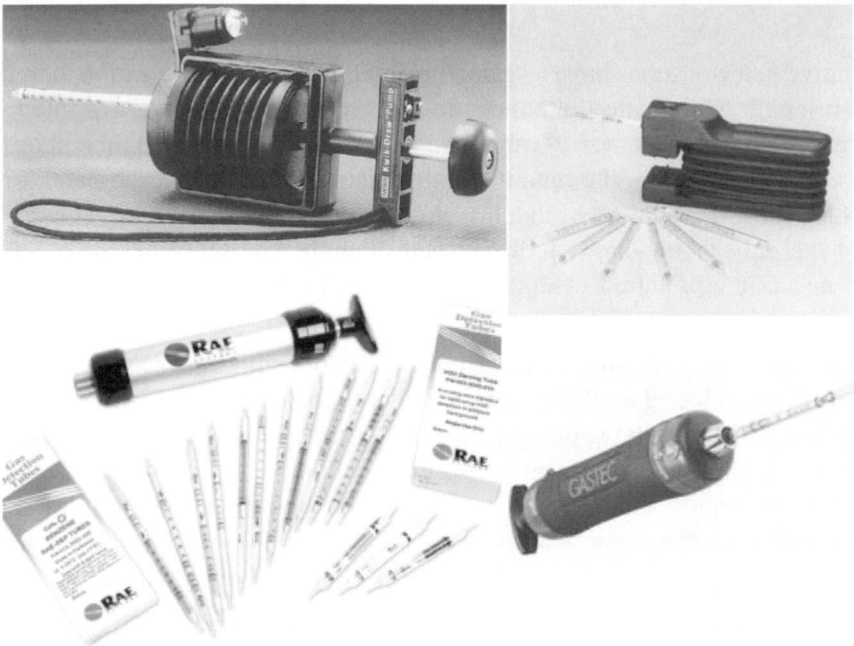

Figure 29-2. From top left corner, clockwise, the MSA Kwik-Draw®, the Drager Accuro®, the Gastec, and the RAE Systems detector tube pumps.

space. Detector tubes make it difficult to determine this. In short, a single reading or even several detector tube CO_2 readings may have little to do with whether a building is or is not adequately ventilated.

Breathing air. Test kits are available for monitoring breathing air as used for airline-supplied respirators. OSHA standard (29 CFR 1910.134) requires that the air meet requirements for CO, CO_2, O_2, oil mists, hydrocarbons, and water. The kits contain tubes for each.

TWA monitoring. There are long-term detector tubes (Gastec lists more that 30) that can be used in this application. However, a second alternative is to use a gas sampling bag to collect a sample from the employee's breathing zone and a detector tube to analyze the contents. Several tubes may be used to improve precision. The technique is applicable to personal monitoring when the worker is fairly immobile. The clear advantage is cost over typical pump and lab-analyzed techniques.

Breath analysis. Detector tubes may be used to confirm such gross overexposures as CO poisoning or alcohol intoxication or for routine analysis of exposure to chlorinated hydrocarbons (Droz et al., 1988) by monitoring exhaled air.

Process gas streams. Detector tubes are ideal for routine monitoring of process gas streams that are of known composition.

THEORY

Quantitative detector tubes have a scale imprinted on the side of the tube that is compared with the stain length to indicate the concentration. Some models have a printed calibration scale that is provided with each box of tubes and held next to the exposed tube in order to make the comparison. In either case, the manufacturer manually calibrates each batch of tubes and prints the scale. Thus, it would seem that anyone could use a detector tube. However, concentration is not the only factor that determines stain length. The other variables are important confounders in the interpretation of detector tube results. Knowledge of what controls stain length is important in the use of detector tubes.

Saltzman (1962) was the first to work out the relationship between stain length (L) and concentration (C, ppm) based on theoretical considerations from the consumption of catalysts. In effect, the reagent in the tube is being consumed as the stain proceeds. However, an important question is whether the reagent in one region is completely consumed before the stain advances to the next region, or does the gas traverse the tube faster than migration of the analyte to the reagent and subsequent reaction can take place?

If the reaction speed (kinetics) keeps up with the velocity of the analyte through the tube, then the analyte is in an equilibrium between the airstream and the reagent. A few detector tubes operate in this manner. In these cases, the reaction rate or the adsorption rate of the silica gel is relatively fast. In this simple case, L is directly proportional to C and the volume of air that flows through the tube (V):

$$L = kCV \qquad \text{[Equation 29-1]}$$

where k is a proportionality constant. In this case, the calibration scale is linear, and the flow rate (Q) of the air through the tube is irrelevant up to some unknown Q, at which point equilibrium will no longer exist. Where silica gel adsorption is important, in addition to reaction, the relationship is slightly different.

In most cases, equilibrium does not exist, and the relationship is more complex:

$$\frac{L}{H} = \ln\,(CV) + \ln\left(\frac{k}{H}\right) \qquad \text{[Equation 29-2]}$$

where H is a mass transfer proportionality constant that is proportional to Q raised to a power of 0.5 to 1.0. The reader should realize that because the tube diameter is, in theory, fixed, Q is proportional to the velocity of the air through the tube (see Chapter 8, Equation 8-30). Consequently, if Q changes, the proportion of mass transferred to the reagent at a given time also changes and, therefore, the stain length changes even though the concentration does not. This explains the importance of calibrating and using the tubes with a pump of the same flow rate characteristics.

FACTORS INFLUENCING PERFORMANCE

Temperature and Pressure

Because volume appears in the above equations, the stain length will be affected by changes in volume caused by deviations in pressure and temperature from the calibration conditions. For example, if the tubes are used at an elevation of 3,000 m, then the reduced pressure will cause each unit of volume that passes through the tube to contain fewer analyte molecules than found for the same volume and concentration at sea level. Thus, the stain length will be too short and will indicate a concentration that is biased low. It can be corrected by applying the ideal gas law. If a process stream at 400°C were monitored, the same phenomenon would occur. This temperature effect on volume can be corrected.

However, the temperature effect on the reaction rate cannot be corrected so easily. It is known that reaction rates are temperature-dependent. The relationship takes an Arrhenius form (see Chapter 23, Equation 23-4). This unknown effect on reaction rate may occur in sampling of process gas, confined spaces, or other situations where the temperature is outside the range of the tube. Therefore, both the tube and the gas temperature must be controlled to within a range specified by the manufacturer, typically 0° to 40°C.

The tube's temperature is controlled by keeping it in ambient air, and the gas temperature is controlled by using a special probe with a heat exchanger (see Figure 29-3). This is a stainless steel tube that is at ambient temperature before it is inserted into the gas stream. The heat exchanger will cool 650°C gas to temperatures below 40°C for up to 3 minutes when the ambient temperature is 20°C. In cases where the ambient temperature approaches 40°C, the heat exchanger would have to be cooled with ice or in some other way. One must be on guard for condensation in the heat exchanger tube. If condensible vapors are being sampled, the likelihood of condensation increases with temperature and concentration. In cases where the gas temperature is less than 0°, a hand wrapped around the tube will warm it sufficiently to keep

Figure 29-3. A heat exchanger probe used to cool hot gases before they enter the detector tube (courtesy, SKC, Inc., Part 801-03).

the gas and the tube in the acceptable range (Leichnitz, 1977).

These techniques can be used to keep the temperature within the range of 0° to 40°C. This inexact range will cause an error from calibration conditions that is probably less than 5%. However, in some cases, this range may in fact be too large, as large deviations may occur from the true concentration when temperatures are at the extremes of the range (see, e.g., Huebner, 1980; McCammon et al., 1982).

Humidity

Two examples of detector tube chemistry from Leichnitz (1977) should suffice to demonstrate the importance of airborne water on detector tube performance. The H_2S tube contains silica gel impregnated with lead acetate. The pores of the silica gel are filled with water in which lead acetate has been dissolved. Here the water in the pores is so extensive that even dry air flowing over the tube will not eliminate it. On the other hand, the CO detector tube contains a reagent system that must be dry. It is preceded with a drying layer to remove the water from air. These tubes are constructed in such a way as to avoid water interference. However, commercial tubes may be sold that do not have these mechanisms of defense from the effect of water. Unfortunately, the effect has not been examined very frequently. McCammon et al. (1982) found that NO_2, SO_2, and formaldehyde tubes gave significantly different responses when exposed to the same concentration at 50% or 90% RH. These humidities were within the ranges recommended by the manufacturer. CO, NO, CO_2, and H_2S tubes did not show significant effects. Tests of long-term CO detector tubes were generally favorable so long as the manufacturer's guidelines for humidity limits were observed (Keane, 1981).

Some manufacturers give humidity limitations in % RH, whereas others give them in

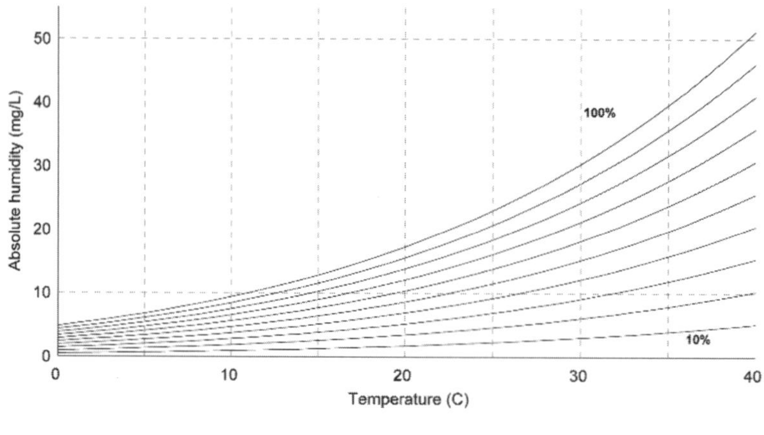

Figure 29-4. Absolute water concentration versus temperature (°C) for relative humidities from 10 to 100%.

absolute humidity (mg/l), the latter being more indicative of the potential for interference than % RH. However, RH data are much more commonly reported than absolute humidity data, and can be estimated. Figures 29-4 and 29-5 provide a means to convert the two values. These figures are derived from the relationship:

$$P_v = 6.112\ e^{\left(\frac{17.67\ T}{243.5\ +\ T}\right)} \qquad \text{[Equation 29-3]}$$

where P_v is the vapor pressure exerted by water (millibars), and T is in °C (Stock, 1986).

Jentzsch and Fraser (1981) examined the humidity dependence of long-term benzene, toluene, and trichloroethylene tubes. The manufacturer (National Draeger, Inc.) recommends that the tubes be used in humidities of 5 mg/L to 12 mg/L. Figure 29-4 shows that this corresponds to an RH of about 25% to 50% at 25°C, a very restricted range. Even so, when tubes were used within this range, humidity had an effect on the toluene tubes. Figure 29-5 shows the temperature-RH relationship in °F.

Flow Rate

As previously mentioned, pumps are an important part of the concentration-stain length relationship, as the velocity of the gas through the tube is related to stain length for many tube systems. It would seem that one could determine the flow rate of, say, a Draeger or a Gastec pump, and then use any other pump as long as it had the same flow rate. Unfortunately, this will not work because the hand-operated pumps do not have a constant flow rate. It varies with the resistance to flow and the pressure in the evacuated pump chamber. Resistance to flow is determined by the orifice sizes at either end of the tube when the tube ends are broken and by

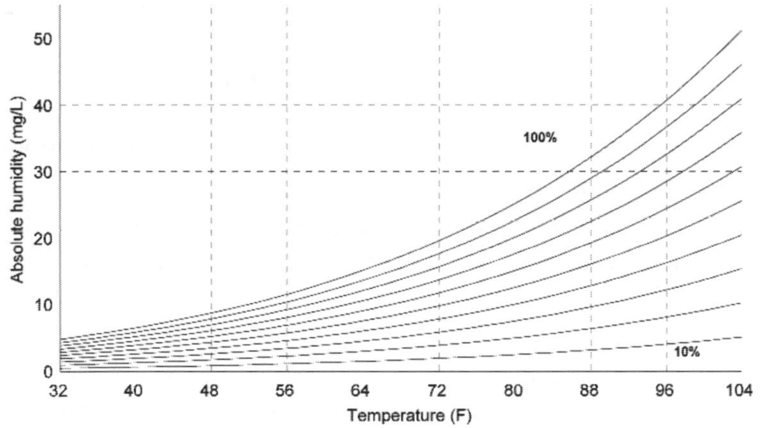

Figure 29-5. Absolute water concentration versus temperature (°F) for relative humidities from 10 to 100%.

the amount and the grain size of the tube packing. Thus, flow can vary from one type of tube to the next and certainly from one manufacturer to the next.

The pressure (negative, relative to atmospheric) established in the pump chamber at the beginning of a stroke is dependent on time. At the beginning of a stroke, the pressure is lowest and the flow highest. Near the end, pressure approaches atmospheric, and flow goes to zero. The relationship takes several forms, depending on the pump and the tube, but is not linear. Colen (1974) clearly showed that pressure and flow varied from pump to pump for six different manufacturers. Using CO tubes[2] and pumps of the six manufacturers, he showed that when tubes and pumps were interchanged, the resulting error could exceed 250%.

A problem also related to flow rate is channeling. The path of the airflow through the tube will be that of least resistance. The silica gel is composed of various grain sizes, which control the flow rate and, therefore, the reaction/absorption rate. During shipping or manufacture, the sizes may become separated, perhaps sorting from side to side of the tube. As a result, the airflow will be along the side with the larger particle sizes, a problem known as channeling. Channeling leads to an uneven stain front, thus making it difficult to read the length of stain. Certification programs for tubes (see later discussion) require that the difference in the two extreme stain lengths not exceed some percentage of the median stain length.

Observer Variability

Detector tubes have a reputation for being inaccurate, and the reading of the tube by a human is perhaps the biggest source of error (Harper, 2001). Certification programs usually require that they be ± 25% accurate at the OEL, but only 35% accurate at lower concentrations, e.g., 0.5 OEL (Gleason, 2006). Detector tubes are more difficult to read at lower concentrations than at higher ones, but even at higher concentrations the major source of error is interpretation of the stain front. A single observer tends to have good precision, but observer-to-observer variability can be great. In an analysis of two brands of detector tubes for seven different chemicals, McCammon et al. (1982) found that three observers were unlikely to be in agreement. In an analysis of variance, observer was a statistically significant factor for 11 of 14 tubes. For some of the tubes, the level of significance was quite high ($p \ll 0.001$), indicating large observer disagreement. The concentrations were TLV® and 2 × TLV®; larger differences would be expected at lower concentrations.

Kusnetz et al. (1960) also found that for three observers reading CO tubes, inter-observer differences were significant, but intra-observer differences were not. The tubes were 100

---

[2] It is interesting to note that CO was one of the first substances for which colorimetric devices were developed, and it is still the subject of most detector tube studies. Some of this notoriety is owed to the fact that there are no impinger, sorbent tube, or other active sampling methods for CO (NIOSH and OSHA publish direct read methods and OSHA also has a sampling bag/GC method). The hygienist's, and indeed OSHA's, choices are detector tubes (including long-term), direct-read instruments (electrochemical and infrared), and most recently passive monitors.

ppm tubes, and the readings were either 50, 60, 75, or 100 ppm. Their data also demonstrate the probable cause of large inter-observer differences. The calibration scale on detector tubes is discrete. Consider a tube that has lines in increments of 10 ppm, up to 50 ppm. The tubes are exposed to 12.5 ppm. Observers are likely to read either 10 ppm or 15 ppm, but probably not 12.5 ppm. If three observers read 10, 10, and 15 ppm, the *CV* would be 25%. If the stain length were poorly defined, the lack of precision would be greater. Because the *CV* has the mean in the denominator, *CVs* tend to be larger at lower concentrations for detector tube.

A few miscellaneous points are in order regarding the use of detector tubes. It should be noted that detector tubes should be read immediately. In many cases, the color begins to fade, and the front is less definable with time. In a few cases, detector tubes may be reused if the initial concentration is low or not detectable. However, if pre-layers exist that remove interferences such as water, the pre-layer may be exhausted and the indicator layer will give a false reading. Thus, caution is necessary.

Storage Time

Detector tubes are often used on a non-routine basis; so they are purchased and then stored for later use. It is not unusual to need the tubes only to find that they have expired in storage. Manufacturers print expiration dates on the tubes or the boxes, but there is always the temptation to use a slightly (or even grossly) expired tube. Also, many users store the tubes in refrigerators in order to extend the shelf life. However, there do not appear to be any studies of specific tubes to determine how long the shelf life may be extended.

Few data are available to demonstrate the effect of temperature on shelf life (see Table 29-2). These data fit an Arrhenius equation (see Chapter 23, Equation 23-4), as shown in Figure 29-6. The actual shelf lives used in the regression equation were 2 years, 0.5 years, and 2 weeks for the final three values in Table 29-2. The regression equation can be used to predict the effect of refrigeration on shelf life. At a temperature of 0°C, a shelf life of 16 years would be predicted. More research in this area seems indicated.

Table 29-2
Shelf life of Draeger carbon monoxide tubes (from Saltzman, 1989)

T°C	Shelf life
25	> 2 yr
50	> 1/2 yr
80	weeks
100	1 week
125	3 days
150	1 day

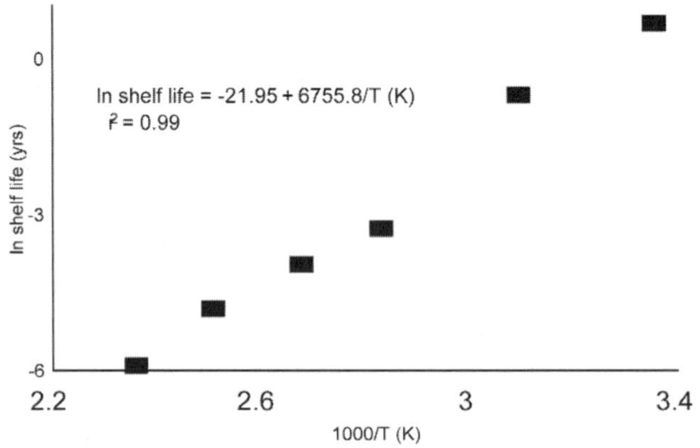

Figure 29-6. The natural logarithm of shelf life versus the inverse of absolute storage temperature for CO detector tubes and the regression equation in Arrhenius form (data from Saltzman, 1989).

Remote Sampling

As with other direct-reading instruments, detector tubes can be used to sample remote locations such as the bottom of tanks. Extension tubes as long as 60 ft (18 m) are available, which are placed between the tube and the pump. However, there are a few drawbacks. The use of the heat exchange tube is disallowed because the detector tube is placed directly in the environment. The tube cannot be observed during the sampling. The pump stroke will take longer to complete because the air must travel against greater resistance. Finally, the tubing must be airtight on the tube, and there must be no leaks in the tubing, as the air will otherwise bypass the tube, and the concentration will be underestimated.

ROUTINE MONITORING

In some cases, detector tubes are used on an occasional basis, largely to collect preliminary data that later will be verified by more accurate techniques. However, in many other cases they are used on a routine basis for everything from confined space entry to long-term monitoring of TWA concentrations. This is not a misuse of detector tubes. In fact, as Chapter 16 points out (see the discussion on measurement error and Chapter 16, Figure 16-5), the cost of detector tubes used in long-term sampling strategies can more than offset the poorer precision compared to other techniques. In other words, the reduced costs allow one to take more samples, which more than makes up for the lost accuracy. However, the use of detector tubes is not particularly wise if increased sample sizes are not used.

In the event that detector tubes are used on a routine basis, it is good idea to verify the calibration of each lot and to make inter-observer comparisons. The tubes should be purchased in large quantities so that each lot purchased can be evaluated one time. A random sample of tubes from the lot should be exposed to known concentrations and then be read by all observers. Statistical quality control techniques (see Chapter 18) and/or analysis of variance (Chapters 17 and 30) should be used to find non-conforming lots and aberrant observers. The results can be used to train observers and yield more precise inter-observer results. Various techniques for calibrating equipment including detector tubes are contained in Chapter 30.

In addition to the calibration of tubes, the pumps should be calibrated and visually checked (Kusnetz, 1960). A bubble meter can be used for such purposes. The meter should be 100 cm^3 in volume, as this is the single stroke volume of most pumps. The only trick is to get the bubble up to the zero mark. This can be done with a three-way valve. The leads from the valve are connected to the detector tube pump, the bubble meter, and a low flow pump. Once the low flow pump moves the bubble to the collect position, the valve is switched. Remember that a detector tube pump that is operating at a low flow rate due to a leak will directly cause a low reading. In addition, the changed velocity in the tube will affect the reaction and absorption rates, causing a change in the length of stain. Thus, the total effect on the indicated concentration can be larger than the size of the flow error.

CERTIFICATION PROGRAMS

A certification program for detector tubes was administered by NIOSH from 1973 to 1983 (Roper, 1974), but like many other programs of the 1980s it became a budget orphan and was abandoned. A U.K. program also was instituted in the early 1970s (BOHS, 1973). Fortunately, a private concern, the Safety Equipment Institute (SEI) in the United States (see Gleason, 2006) began certifying tubes in 1985. SEI uses fees paid by the manufacturers for the certification process as operating revenue. Private laboratories perform the tests, and a third party audits the laboratories. The program is voluntary. The criteria used in the original NIOSH program (*Federal Register*, 1973) have, for the most part, been incorporated into an ANSI standard (ANSI, 1990), which is used by SEI.

The certification criteria include the test concentrations and accuracy requirements, which are $\pm$ 25% at 1 $\times$ OEL to 5 $\times$ OEL and $\pm$ 35% at 0.5 OEL. Other criteria include stain length variability, limitations on channeling such that the difference in the maximum and minimum stain lengths within a tube must be no more than 20% of the median stain length, labeling and complete instruction requirements including known interferents, quality control, pump leakage, and unevenness of the zero point (i.e., the interface of the pre-layer and the indicating layer). Twenty tubes of each type are tested from batches that vary in size from 1,300 to 3,200 tubes (AIHA, 1993; Harper, 2001). A catalog can be obtained from SEI that contains a list of certified tubes (http://www.seinet.org/CPL/contents.htm). Draeger Safety, Inc., Gastec, Matheson/Kitigawa, and Mine Safety Appliances Company participate in the program and all of their pumps are certified. As of 2006, SEI certified tubes for 21 different chemical agents.

In Europe, certification is based on EN 1231 (CEN, 1996), which has important differences to ANSI 102 (Harper, 2001). The concentrations tested under ANSI are related to the OEL, whereas for EN they are what the manufacturer claims the tube is useful for. EN tests at several humidity and temperature regimens, whereas ANSI only 70°F (21°C) and 50% RH. For the EN standard, accuracy is calculated based on similar criteria as described in Chapter 25, i.e., $2CV \pm$ absolute value of bias. The ANSI standard is based on a U.S. Military Standard (414) that allows approximately 13% of the lot (lot varies in size from 1,300 to 3,200 tubes) to fail in order to meet the earlier cited criteria.

MANUFACTURERS AND DISTRIBUTORS (all web sites accessed December, 2006)

Gastec http://www.gastec.co.jp/
Matheson Gas Products (Kitigawa tubes) https://www.mathesontrigas.com/gashandlingequipment/index.aspx
Mine Safety Appliances Company http://www.msanorthamerica.com/catalog/product679.html
National Draeger, Inc. http://www.draeger.com/
(also sold by SKC, Inc.) http://www.skcinc.com/detector.asp
NEXTTEQ (Gastec) http://www.nextteq.com/products.aspx?category=1
RAESystems http://www.raesystems.com/products/tubes Note that RAE offers an informative handbook that is useful regardless of the detector tube system being used. RAE: *Gas detection tubes and sampling handbook*, RAESystems, San Jose, CA, USA (2005).
Sensidyne, Inc. (Kitigawa) http://www.sensidyne.com/prodcat.php?ID=1

QUESTIONS

29.1. Explain why detector tubes are subject to interferences.

29.2. Detector tubes have higher associated sampling and analytical error than most other measurement methods. In your opinion should they be used? Explain your answer. If you think they should only be used in special circumstances, explain why and state the circumstances.

29.3. According to Equation 29-1, the length of stain for most detector tubes is dependent on the concentration and what other variable?

29.4. Explain why the detector tube heat exchanger will cool 650°C air to below 40°C for only up to 3 minutes. Why not longer?

29.5. A detector tube has a stated humidity range of 5 mg/L to 24 mg/L. What relative humidity range is this at 40°C?

REFERENCES

AIHA: Direct-Reading Colorimetric Indicator Tubes Manual, 2nd ed., J.B. Perper, B.J. Dawson (eds.). American Industrial Hygiene Association, Fairfax, VA (1993).

ANSI: Gas Detector Tube Units – Short-Term Type for Toxic Gases and Vapors and Working Environments (ANSI/ISEA 102-1990). American National Standards Institute, New York (1990).

BOHS (British Occupational Hygiene Society): Chemical indicator tubes for measurement of the concentration of toxic substances in air. Ann. Occup. Hyg. 16:51-62 (1973).

CEN: EN 1231 Workplace atmospheres – Short term detector tube measurement systems – requirements and test methods. CEN, Brussels Belgium http://www.cenorm.be/CENORM/BusinessDomains/TechnicalCommittees Workshops/CENTechnicalCommittees/Standards.asp?param=6119&title=CEN%2FTC+137 (1996).

Colen, F.H.: A study of the interchangeability of gas detector tubes and pumps. Am. Ind. Hyg. Assoc. J. 35:686-694 (1974).

Droz, P.O.; Krebs, Y.; Nicole, C.; Guillemin, M.: A direct reading method for chlorinated hydrocarbons in breath. Am. Ind. Hyg. Assoc. J. 49:319-324 (1988).

Federal Register: Title 42, Code of Federal Regulations, Part 84: Certification of gas detector tube units, 38: 11458 (May 8, 1973).

Gleason, P.A: Gas detector tube certification. ISHN http://www.ishn.com/CDA/Articles/Feature_Article/bd2aa52a690ac010VgnVCM100000f932a8c0 (July, 2006).

Harper, M.: Performance evaluation of on-site colorimetric air sampling techniques. App. Occ. Env. Hyg. 16:1092-1096 (2001).

Huebner, D.J.: Evaluation of a carbon monoxide dosimeter. Am. Ind. Hyg. Assoc. J. 41:590-591 (1980).

Jentzsch, D.; Fraser, D.A.: A laboratory evaluation of long term detector tubes: Benzene, toluene, trichloroethylene. Am. Ind. Hyg. Assoc. J. 42:810-923 (1981).

Keane, M.J.: A performance evaluation of selected long-term gas detector tube units. Ann. Am. Conf. Gov. Ind. Hyg. 1:241-246 (1981).

King, M.V.; Eller, P.M.; Costello, R.J.: A qualitative sampling device for use at hazardous waste sites. Am. Ind. Hyg. Assoc. J. 44:615-618 (1983).

Kusnetz, H.L.: Air flow calibration of direct reading colorimetric gas detecting devices. Am. Ind. Hyg. Assoc. J. 21:340-341 (1960).

Kusnetz, H.L.; Saltzman, B.E.; Lanier, M.E.: Calibration and evaluation of gas detecting tubes. Am. Ind. Hyg. Assoc. J. 21:361-373 (1960).

Leichnitz, K.: Use of detector tubes under extreme conditions (humidity, pressure, temperature). Am. Ind. Hyg. Assoc. J. 38:707-711 (1977).

Leichnitz, K.; Walton, J.: Determination of aerosols by means of detector tubes. Ann. Occup. Hyg. 24:43-53 (1981).

McCammon, C.S.; Crouse, W.E.; Carroll, H.B. Jr.: The effect of extreme humidity and temperature on gas detector tube performance. Am. Ind. Hyg. Assoc. J. 43:18-25 (1982).

Palassis, J.; Bryant, C.; Zey, J.N.: Detector tubes, direct-reading passive badges, and dosimeter tubes, In Air Sampling Instruments, (9th ed.). American Conference of Governmental Industrial Hygienists, Cincinnati, OH (2001).

Roper, P.: The NIOSH detector tube certification program. Am. Ind. Hyg. Assoc. J. 35:438-442 (1974).

Saltzman, B.E.: Basic theory of gas indicator tube calibration. Am. Ind. Hyg. Assoc. J. 23:112-126 (1962).

Saltzman, B.E.: Detector tubes, direct-reading passive badges and dosimeter tubes, In Air Sampling Instruments, 7th ed., V. Hering (ed.). American Conference of Government Industrial Hygienists, Cincinnati, OH (1989).

SEI: Certified Product List. Safety Equipment Institute, Arlington, VA (1991).

Stock, T.H.: The use of detector tube humidity limits. Am. Ind. Hyg. Assoc. J. 47:241-244 (1986).

30

CALIBRATION OF DIRECT-READING INSTRUMENTATION

INTRODUCTION

Direct-reading instruments (DRI) require calibration, regardless of the occasional claim of manufacturers. Most manufacturers realize this and provide written guidance and equipment designed for easy calibration. However, a few claim that factory calibrations are good for a year or more and that no other calibration is required, or they design instrumentation that is virtually impossible to calibrate without special equipment.

The reasons for calibration are simple. No instrument reads concentration directly. Even though many may have analog meters or digital readouts with the terms ppm or mg/m^3 printed on them, they are actually detecting some attribute of an electronic signal, such as voltage, and converting it to ppm or mg/m^3. So regardless of the steps performed in a calibration, the end result is defining the relationship of concentration and an electrical signal (e.g., voltage). For detector tubes, the relationship is length of stain versus concentration. Beyond this preliminary need for calibration, all instruments suffer from drift and other phenomena that affect response. This can include aging of the electronics, contamination of GC columns and optics in photometric instruments, detector contamination, changes in temperature control and carrier gas flow rates, and many other reasons specific to each instrument type. This means that yesterday's response to a given environmental response may not be the same today. Thus, instruments must be periodically recalibrated.

Instrument calibration generally can be classified in one of two ways: multi-point calibrations and two-point calibrations for which one point is zero. For multi-point calibrations, the concentration-response relationship is defined with a calibration curve or a plot of concentration versus instrument output. If the instrument output is in concentration units, then the plot is known versus indicated concentration. To construct these plots, the instrument usually is challenged by several concentrations spanning those that will be of interest in the field.

For two-point calibrations, the assumption is made that the response of the instrument is linear with concentration,[1] and that when the instrument sees a zero concentration, the response, which will not likely be zero output, is equated to zero. In microprocessor-based instruments, the instrument is exposed to zero air, and the instrument records the response. The instrument is then exposed to a known concentration, and again the response is recorded. Essentially the instrument draws a straight line through these points, and the slope of the line is used by the instrument to indicate field concentrations.

For older, usually analog, instruments a span control is used to set the instrument response where it should be for a single concentration. First, the zero control is adjusted in non-contaminated air. This gives the instrument a correct starting point. Then the instrument is exposed to a known concentration, and the span control is adjusted so that the instrument reads that concentration. The span concentration used in calibration may have to be that which has been established by the manufacturer and, therefore, the manufacturer may supply the contaminant at this given concentration.

After a two-point calibration, one assumes that the instrument responds appropriately at all other concentrations, but over time this may cease to be the case; so periodic manufacturer calibration and adjustment is required. The hygienist could perform the detailed calibration and create a calibration curve, but actual adjustment of the electronics of the instrument to make it read appropriately throughout its range usually is not possible without contacting the manufacturer.

Two general classes of calibration technique have evolved. Static techniques involve creating a single known concentration, often in a vessel. The contents are then introduced into the instrument, usually by means of the pump that is part of the instrument. The air must move from the vessel to the instrument; so at some point, the gas must flow but one concentration is maintained. Static systems can also be continuous, closed-loop if the instrument is nondestructive when it analyzes the contaminant. In this case, the system volume includes the instrument sampling chamber and the gas stream continues to circulate from the calibration vessel to the instrument and back, but at one concentration.

In dynamic systems, a contaminated airstream is produced, but the contaminant and the diluent are constantly being generated and mixed, and thus the concentration can be caused to change constantly over time or in a step-wise fashion as greater or lesser amounts of analyte are introduced continuously or in step-wise fashion. The instrument samples this airstream. The main difference in the two systems is that, in the static system, the contaminated gas is made by the batch; so the concentration is thought to be constant but subject to decay. Decay can be caused by interaction (either reaction or sorption) of the analyte with parts of the calibration system or leakage. In many cases, the rate of decay is insignificant and so batch systems are viable alternatives for calibration. In the dynamic system the gas stream is continuously being mixed fresh; thus, leakage and interaction are less important, but there are other

---

[1] In some cases, the response is assumed to be log-linear, and the instrument makes the appropriate conversions in the process of calibration.

considerations related to a more complicated system that are discussed later.[2] Perhaps better terms would be *continuous* and *batch systems* (Cotabish et al., 1961).

STATIC OR BATCH CALIBRATION SYSTEMS

The main difference between batch systems is in the type of container or vessel that is used. Basically, there are plastic bags, rigid (usually glass or steel) containers, and compressed gas cylinders, which are usually purchased in a filled state. The first two systems require that the contaminant be introduced into a volume of air. Liquid or gas[3] contaminants can be introduced with syringes, or if large amounts of liquid are needed, pipettes can be used. The complete evaporation of a liquid contaminant is sometimes difficult, depending on vapor pressure and temperature. One must also remember that there is no turbulence in a container; as a result, layering of the components in the container according to molecular weight will be a problem that can be solved by mechanical mixing.

The calculations for determining the resulting concentration (C) in the container are dependent on the volume of clean or diluent gas (V_d) in the container and the gaseous volume (V_c) or mass (M_c) of added contaminant. The ratios obtained express concentration as a proportion or as mass per volume, or

$$C = \frac{V_c}{V_c + V_d} \qquad \text{[Equation 30-1]}$$

or

$$C = \frac{M_c}{V_c + V_d} \qquad \text{[Equation 30-2]}$$

The proportion obtained in the first equation can be changed to ppm, %, or ppb by using a unit multiplier of appropriate dimensions, such as 10^6 µL/L. If the contaminant is placed in the system as a gas, the use of Equation 30-1 is straightforward because the volume of the contaminant gas should be known. If a liquid is placed into the system and then allowed to evaporate, only the liquid volume of the contaminant is known. Thus, to find the mass and apply Equation

---

[2] The information in this chapter may be adequate for the calibration of DRI using batch systems. However, for dynamic systems, it may only provide an introduction, depending on the complexity of the vapor or aerosol generator and other factors. Therefore, anyone engaging in frequent use of dynamic calibration systems, especially those involving reactive compounds, should acquire the book by Nelson (1992) on this subject or another suitable reference.

[3] Although the chapter generally deals with gaseous systems and vapor generation, a short section on aerosol generation is included near the end.

30-2, one must multiply the liquid's density by the volume, observing units. If Equation 30-1 is used, the volume of vapor generated from the liquid must be found.

Example 30-1. 10 µl of liquid acetone is placed in 10 L of clean air. What is the resulting concentration in ppm? Assume NTP (as throughout the book, 1 atm and 25°C). The molar volume for an ideal gas at NTP and the molecular weight and density of acetone are used as follows:

$$10\ \mu L \times \frac{1\ ml}{10^3\ \mu L} \times \frac{0.7844\ g}{ml} \times \frac{mole}{58\ g} \times \frac{24.45\ L}{mole} = 0.0033\ L$$

This volume of acetone and the 10 L volume of the diluent can be substituted into Equation 30-1 and the result multiplied by 10^6 µL/L to obtain a concentration of 331 ppm. In most cases, the volume term for the contaminant that is in the denominator of Equation 30-1 or Equation 30-2 can be ignored, and the entire calculation can be performed in one step:

$$\frac{10\ \mu L_{lc}}{10\ L_d} \times \frac{1\ ml}{10^3\ \mu L_{lc}} \times \frac{0.7844\ g}{mL_{lc}} \times \frac{mole}{58\ g_{lc}} \times \frac{24.45\ L_{gc}}{mole} \times \frac{10^6\ \mu L_{gc}}{L} = 330.7\ \frac{\mu l}{L}\ or\ ppm$$

Thus, in this case, the error is insignificant in ignoring the contaminant volume term in the denominator. However, if % concentrations are being made, then this term becomes more significant. To avoid confusion, the subscript *lc* denotes liquid contaminant, and *d* denotes diluent. The *lc* becomes *gc* or gaseous contaminant when the molar volume is applied. Note that the units cancel appropriately.

Plastic Bags

Plastic bags have several advantages relative to rigid containers. They are light and easily portable. They do not break. It is easy to introduce the contaminant through a septum (see Chapter 22, Figure 22-1). It is easy to mix the contents by massaging the bag. Most important, as the contaminated gas is withdrawn from the bag, the pressure inside remains constant. This is not true for rigid containers. On the other hand, an important aspect of batch systems is the propensity of the container to adsorb, interact with, or promote an interaction with the contaminant. Bags are not immune from these concerns. The reader should see the discussion in Chapter 22 for more details on the interactions of plastic bags with their contents.

To make a bag *standard*, as it is often called, the required amount of dilution air is metered to the bag, and then the appropriate amount of contaminant is added. Because the vast majority of the bag contents will be dilution air, the amount of dilution air that is used will be determined by the total volume needed to calibrate the instrument. Generally, 10 L is enough, but it is easy to calculate. For example, if the instrument has a sampling rate of 0.5 L/min and the response time is less than one minute, then 0.5 L will suffice. However, one may want to get more than one reading, or to allow the instrument response to equilibrate for more than the stated response time, so that full response can be assured or checked.

The air is metered into the bag with a calibrated pump and stopwatch, or a flow meter can be conveniently used. Dry gas and wet test meters are ideal for this application. Wet test meters may add enough water vapor to the air to be of concern for some instruments. In order to fill a plastic bag, the bag must be at the most downstream point in the train; so the wet test meter cannot be moved downstream of the bag.

Bags should be purged before they are stored, or if they are immediately reused. One fills the bag with clean air, and then evacuates it by opening the valve and applying pressure to the bag or by using a pump in suction mode. This should be done more than once. Some authors recommend several times (Ness, 1991), but the exact number depends on the original concentration, the desired cleanliness in the bag, and the degree to which the previous contaminant adsorbed to or permeated into the bag. The latter factor is dependent on the residence time of the contaminant in the bag and other physical-chemical factors described in Chapter 22. If the residence time is short so that the bag has little *memory*, then three to five purges would seem reasonable based on simple considerations of serial dilution. For example, if a flattened bag contains as much as 5% of the original volume, then the concentration upon refilling it with clean air will be 5% of the original concentration. The second purge will yield 5% of 5%, or 0.25%. Algebraically,

$$C_n = (V_{fr})^n \, C_o \qquad \text{[Equation 30-3]}$$

where C_o is the pre-purge concentration, V_{fr} is the residual volume fraction of the original volume after purging, n is the number of purges, and C_n is the concentration after all purges. So for a 5% residual volume and five purges, 3×10^{-7} of the original concentration remains, provided there is insignificant memory.

Syringes

Liquid contaminants are commonly added to bags with microliter syringes; gaseous contaminants can be added with gas-tight syringes (see Figure 30-1). The latter have Teflon-tipped plungers that fit snugly into the glass barrel. Syringes are used in many different ways but each method is important if the amount delivered is to be unbiased. They should be kept very clean. Dirt will scratch the surfaces and cause leaks. A syringe contaminated with one solvent and then used with another may cause calibration problems. Syringes should be dried overnight, and can be dried in an oven at 50°C. This is preferable for viscous or low volatility compounds. Higher temperatures can be used if the plunger is pulled from the barrel during drying in order to avoid breaking the barrel. Reassembly should be conducted only after the two pieces have reached room temperature. Protect disassembled syringes from dirt and dust. Do not touch the plunger, as fingerprints may cause leaks or the freezing of a Teflon® plunger in the barrel (Hamilton, 2006).

To use a liquid syringe for adding analyte to a container with a septum,[4] place a small

[4] See Hamilton (2006) for more details on the use of syringes.

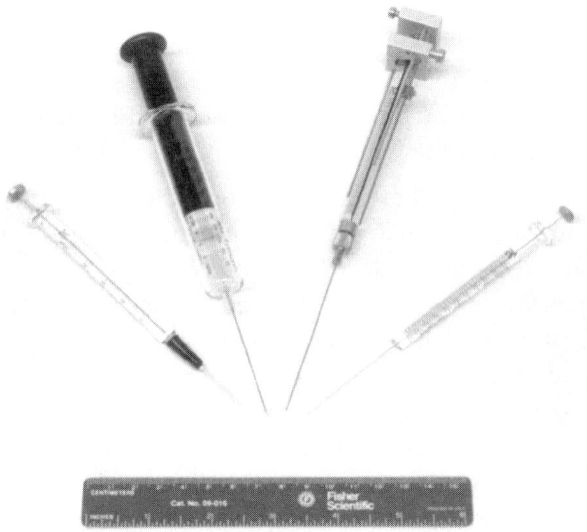

Figure 30-1. Gas tight and liquid microliter syringes (photo courtesy, R.K. Oestenstad).

amount of the liquid in a small vial and seal it with a septum top. Do not contaminate the main vessel of liquid by inserting the syringe into it, unless you are sure the syringe is relatively clean and small amounts of contamination are of no concern. Holding the syringe by the flange at the top of the barrel, pull back the plunger to about the volume desired. Invert the vial, and insert the needle into the septum and the liquid. Repeatedly press and withdraw the plunger button until no air bubbles are emitted from the needle. Hold the assembly vertically, upside-down, and withdraw the plunger until its tip is aligned with the top of that line matching the desired volume. Remove your fingers from the plunger, and withdraw the syringe. Blot any excess liquid from the needle, being careful not to draw liquid from the needle tip by capillary action. Insert the needle into the septum of the bag, and push the plunger in with one firm, but smooth, continuous stroke. Do not pump the plunger.

To fill a syringe with gas, several techniques can be used, one of which is shown in Figure 30-2. The gas flow just causes bubbling in the liquid. The syringe is inserted into the tubing and flushed several times, and is then filled. Ethylene oxide and ammonia will cause water (if it is the liquid) to travel up the tube and into the cylinder (Nelson, 1992). Other liquid-gas combinations may react; so the system is not perfect. Another technique is to have the gas flow through a rigid container such as that shown in Chapter 22, Figure 22-2; then the syringe is filled through the septum on the container. The container should be flushed with the gas for at least five volume changes, and the flow through the container should be high enough to cause turbulent mixing. Laminar flow will cause too much residual air to remain in the container, diluting the contaminant.

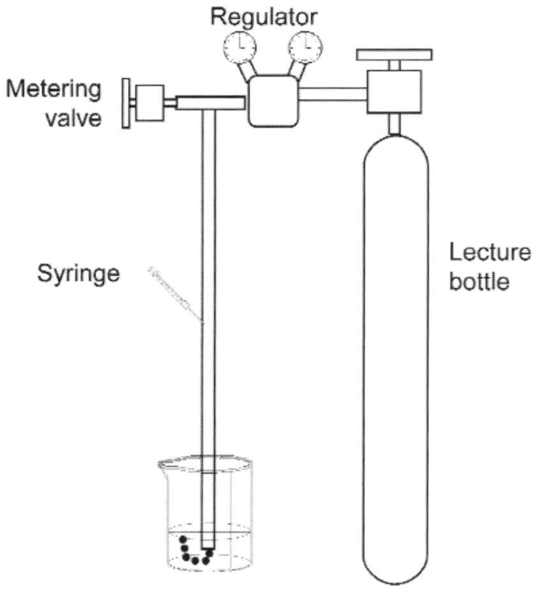

Figure 30-2. One technique for filling a syringe with gas (adapted from Nelson, 1992, with permission).

When the amount of contaminant needed is less than can be accurately delivered with a liquid syringe (e.g., 0.5 μL), then the vapor head space technique can be used. A small amount of liquid (e.g., 10 to 20 ml), is placed in a bottle of 250 to 500 mL. The bottle is sealed with a cap containing a built-in septum. (These bottles are commercially available.) The bottle is shaken gently to assure that the vapor produced by the liquid is saturated. A gas-tight syringe is then used to withdraw the appropriate volume of saturated vapor. The volume of the vapor in the bottle should be much greater than that of the syringe, or as the vapor is withdrawn, a significantly different pressure will be established.

Example 30-2. We wish to make a 1 ppm concentration of 1-propanol in a 1 L volume. Thus, we need 1 μL of 1-propanol vapor (μL_v), or

$$1\ \mu L_v = \mu L_v \times \frac{g\text{-}mole}{24.45\ L_v} \times \frac{60\ g}{g\text{-}mole} \times \frac{1\ mL_L}{0.7997\ g}$$

which equals 3×10^{-6} mL or 0.003 μL of propanol liquid. This amount is too small to be accurately delivered. The saturated vapor pressure of 1-propanol is 20.27 torr (20.27 mmHg or 2.7 kPa). This value is calculated from Appendix B for 25°C. Therefore, the saturated vapor concentration is 0.026671 or 26,671 ppm. Using Equation 30-1, we can determine the volume of saturated vapor needed; however, the contaminant volume must be adjusted by the saturated vapor concentration because this volume is not pure propanol. In addition, we will ignore the

V_c term in the denominator of Equation 30-1 because, for a 1 ppm concentration, it is insignificant. So,

$$C = \frac{V_c C_{c,f}}{V_d} ,$$

$$1 = \frac{V_c 0.026671}{1} , \textit{ and}$$

$$V_c = 37.5 \ \mu L ,$$

where C is in ppm and V_d is in L, so V_c is in μL. $C_{c,f}$ is the fractional concentration in the head space so that units cancel.

From the above discussion, it can be seen that each syringe technique has potential pitfalls. Whatever system is used, the user should avoid differential pressure effects, which can cause the syringe to become filled at a pressure different from that in the room. For example, the container, shown in Chapter 22, Figure 22-2, should be refilled after no more than 5% of its volume is removed. Otherwise negative pressure in the vessel occurs and is relieved by leakage through the septum, or future syringe filling will occur at the low pressure, and a significant error is made in the mass of contaminant thought to be delivered to the calibration system.

Closed-Loop Systems

If the instrument is nondestructive and has an identifiable outlet for the sampled gas, then a closed-loop system can be used (see Figure 30-3). The process is still a batch process, but in this case, the gas stream can be used over and over to make successively higher concentrations. The first concentration is made, the system is activated, and a reading is obtained. More contaminant is added, and the process is repeated. Although a plastic bag is not used, the old MIRAN 1A instrument is calibrated in this way by acting as a rigid container (see later discussion).

Rigid Containers

Glass bottles are generally non-reactive (with a few exceptions such as hydrofluoric acid), which makes them good rigid containers for calibration vessels. Unfortunately, the bottle must have an opening, which is usually closed with a rubber stopper or other polymer so that the advantage of using glass is diminished. A small surface area of rubber exposed to an organic calibrant may not lead to a substantial absorption error, but one should do everything possible to minimize this potential problem.

Just about any rigid vessel can be used as a calibration vessel, provided that one is cognizant of the potential for adsorption or reaction to occur. Ideally, the vessel should have one

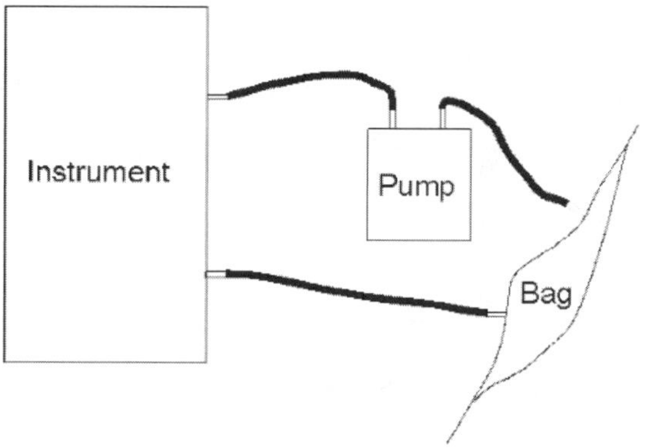

Figure 30-3. A schematic of a closed-loop calibration system using a plastic bag.

or more ports for filling, purging, and challenging the instrument, and a septum for the use of a syringe (see Chapter 22, Figure 22-2). If liquid contaminants are used, a septum is not necessary, since the syringe can be discharged through a relatively large opening and the opening quickly closed without appreciable error. Large carboy bottles used to hold bottled water can be adapted for use.

As mentioned earlier, rigid containers are subject to pressure effects. One way to alleviate the problem is to allow dilution air to enter as a sample is withdrawn, and then to make a calculation of the new concentration in the bottle. Such an approach requires the assumption of perfect mixing, and so a mechanism to accomplish mixing is desired. Aluminum foil strips can be placed in the container to act as stirrers as the container is shaken. Alternatively, a variety of manual or electronically driven fans and impellers are available commercially.

Another approach is the use of multiple vessels of the same volume, v_w, as shown in Figure 30-4. When the dilution air is allowed to enter the bottle on the left, mixing of the dilution air is greatest in that bottle, and the effect of the dilution air is less in each successive bottle. If we assume perfect mixing, which is approached as the bottle size is decreased, then an equation can be written (Setterlind, 1953; Nelson, 1992) for calculating the decay in concentration for the *nth* bottle in a system of *n* bottles of equal volume *V* and a total withdrawal volume of v_w:

$$C_n = C_o \exp\left(-\frac{v_w}{V}\right)\left[1 + \frac{1}{1!}\left(\frac{v_w}{V}\right) + \frac{1}{2!}\left(\frac{v_w}{V}\right)^2 + \ldots + \frac{1}{n-1!}\left(\frac{v_w}{V}\right)^{n-1}\right]$$

where C_n is the concentration in the final bottle, and C_o is the initial concentration in all bottles. The equation shows that for a system of five bottles, half of the total volume can be withdrawn (2.5 × the volume of one of the bottles), and the concentration in the last bottle will still be 89% of C_o. Time is also important. If the withdrawal were so slow that it approached the

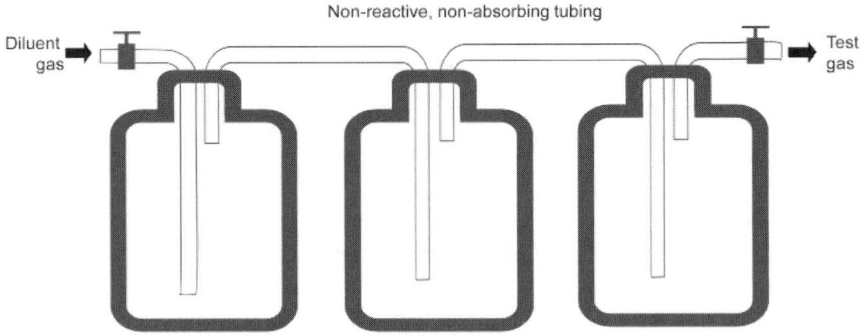

Figure 30-4. Three rigid containers connected in series (adapted from Nelson, 1992, with permission).

rate of diffusion of air, or if multiple withdrawals, separated by time intervals, were made, then the equation would not hold. A single withdrawal has to occur at a reasonable flow rate relative to the total volume.

Another rigid system for avoiding the effect of dilution depends on a plastic bag, as shown in Figure 30-5. The plastic bag should have low permeability for the contaminant, and the rubber band should be tight around the bag with few wrinkles. If the sample gas is made and used rapidly, this system should work well.

When containers are reused, they should be purged with clean air. The purge time will depend on the residual concentration in the container, the desired cleanliness, the flow rate of the purge air, and the volume of the container. These variables can be related by using a mass balance equation. Equation 30-1 was derived from the mass balance equation. Other questions regarding direct reading instruments, or DRI can be answered by using the mass balance equation. For example, what is the minimum response time of an instrument based on its sample volume and sampling rate? The equations for answering these questions are derived in Appendix 30-1. For example, see the *purge* equation.

Compressed Gases

A compressed gas cylinder can be used directly as a batch calibration system if the concentration in the cylinder is acceptable. In many cases, compressed gases can be ordered at specified concentrations. The size of the cylinder can be as large as typical laboratory cylinders with capacities of hundreds of liters or as small as the *lecture bottle* with a capacity of 0.44 L. The latter is also called a 1 lb (0.454 kg) bottle, based on its capacity to hold water.

The lecture bottle is very popular for both batch and continuous systems. The term *lecture bottle* is often applied to cylinders made from a wide range of materials with various types of construction, even though individual manufacturers may use their own nomenclature. Consequently, the amount of compressed gas contained in the cylinder is quite variable and

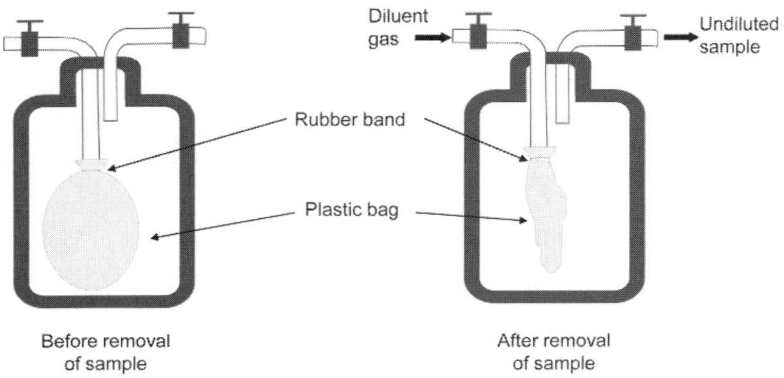

Figure 30-5. A technique for using a rigid container without causing sample dilution or partial vacuum (adapted from Nelson, 1992, with permission).

dependent on the above factors. One popular size holds approximately 17 L of compressed gas, and the cylinder is disposable. It is sometimes referred to as a cylinder of span gas. Several DRI manufacturers purchase specific span gases for their instruments from compressed gas manufacturers. These gases are then sold to DRI users for calibrating the span of the instrument in the previously described two-point calibration scheme. An adapter to get the gas into the instrument usually must be purchased.

Even aerosol cans can be purchased with gas mixtures (for example, Scott Specialty Gases). Many compressed gas companies will make gas mixtures based on the buyer's specifications. For a listing of specialty gas manufacturers, see the listing at the end this chapter. It should be noted that the National Institute of Standards and Technology offers a range of concentrations and mixtures of SO_2, NO_2, CO_2, CO, NO, ambient non-methane organics, propane, and methane in air and nitrogen (see the listing at the end this chapter).

PURITY

Different manufacturers use different terms to designate the purity of a single component or mixture of compressed gas(es). These include terms such as primary standard, certified, unanalyzed, ultra-pure, pure, zero-grade, zero-air, and so on. The user has to consult the manufacturers' catalogs to ascertain the actual concentrations and the related uncertainties. Table 30-1 shows several fictitious grades and the corresponding analytical data. The accuracy of the stated concentration of the components depends on concentration, and the tolerance of the manufacturing process varies with concentration. For example, Grade 1 in Table 30-1 can be blended or prepared with a tolerance of ± 2% when the concentration is in the range of 500 ppb to 999 ppb. This value is obtained by measuring the concentration of the blended gas stream repeatedly and comparing the results to what the stated or expected concentration is.

Table 30-1
Three fictitious grades of compressed gas mixtures and the typical parameters that are reported by the manufacturers

Product Grades	Concentration Range	Blend Tolerance	Analysis
Grade 1	10 to 99 ppb	± 10%	± 10%
	100 to 499 ppb	± 5%	± 5%
	500 to 999 ppb	± 2%	± 2%
Grade 2	1 ppm to 99 ppm	*zero*	± 5%
	100 ppm to 16%	*zero*	± 2%
Grade 3	1 ppm to 15%	± 10%	± 5%
	16% to 49%	± 5%	± 2%

In addition, each Grade 1 cylinder containing a component in the 500 to 999 ppb range is analyzed by the manufacturer before it is shipped. This analysis is usually printed on some kind of certificate. For the above variables, the analyses should be within ± 2% of what is stated on the certificate. When gases are accompanied by certificates, they are said to certified. The instruments used in the analyses may be calibrated with National Institute of Standards and Technology (NIST) Standard Reference Materials (SRM) in the United States (or similar agency in other countries), or with standards prepared by the manufacturer. The former are termed NIST traceable.

An example for the Grade 3 gases in Table 30-1 may be helpful. Say you ordered 95 ppm ammonia in air, Grade 3. Table 30-1 tells you that the manufacturer will make a cylinder of blended gas that will be 95 ppm ± 10% based on the manufacturer's quality control. Thus, you can expect a cylinder in the 85 to 105 ppm range. That cylinder is made and analyzed, and let us say that the certificate states 91 ppm. The 91 ppm is within the manufacturer's tolerance of 85 to 105, and the true concentration in the cylinder is 91 ± 5%. It can also be seen from Table 30-1 that if you need a blend that is exactly what you order, you should order Grade 2. Nevertheless, there will still be ± 2–5% uncertainty associated with the final analysis of a Grade 2 gas.

OXYGEN CONTENT

As was noted in Chapter 27, the O_2 level can affect the response of both PIDs and FIDs. Therefore, if either is calibrated at an O_2 level differing from ambient, the measurements will be in error. The bias will usually be negative because the calibration O_2 level is likely to be < 21%, and the instrument response increases with lower O_2 levels. Mouradian and Flannery

(1994) examined the O_2 content in 39 calibration-quality cylinders purchased from a *nationally recognized supplier* for use in calibrating for numerous volatile organic carbon compounds. The O_2 levels ranged from 14% to 23% and fell outside the range of 20% to 22% in 10 of the 39 cylinders. Thus, the O_2 content of span gases is important, and hopefully the manufacturers are addressing this issue.

STABILITY

Separation of mixtures, once they are thoroughly mixed and placed in the cylinders, apparently is not a problem (Matheson, 1978); however, manufacturers must thoroughly mix the gases prior to filling the cylinders. It is known that certain gases are unstable in that they react with container walls or with oxygen or water. They must be compressed in special containers or without these components present. A reputable manufacturer will know about the stability problems and try to prevent them. Poor stability causes a greater percentage error for gases in low concentration than for those at high concentration. Actually the amount of reaction that occurs is usually about the same in either case (i.e., the number of reaction sites on the container wall is constant), but at low concentration there is less gas to react, and so a larger proportion is lost.

Often a compressed gas is not used within a short time, and there is some question later about its stability. This problem can be solved if the gas is periodically analyzed by an independent instrument, one calibrated by another means. Such an analysis will indicate if the cylinder concentration is changing. However, such an analysis usually is not made, and when it is time to recalibrate an instrument, one wonders if the one-year-old span gas is stable.

The U.S. EPA has been interested in the stability of compressed gas mixtures used to calibrate a wide range of field and laboratory instruments used in air pollution work, particularly in analyses of incinerator and other hazardous waste operations effluents. They have found that a wide range of organic and inorganic mixtures, in the ppb and ppm range, are stable for periods of up to 10 years (Jayanty et al., 1992). As a result, some manufacturers are offering EPA gas mixtures (for example, Scott Specialty Gases). These gases must meet EPA-defined stability requirements, including analytical accuracy, NIST traceability, and tolerances on the maximum deviation in concentration over a defined time period. The maximum time period is 36 months, and for some low concentrations and certain mixtures, the period may be as short as 6 months. However, individual cylinders can be re-certified by the manufacturer.

It should also be noted that Jayanty et al. found that some gas mixtures were not stable. (Stability was defined as concentration varying by less than 10% over the time period.) These mixtures included aniline, cyclohexanone, p-dichlorobenzene, ethyl amine, 1,2-dibromoethylene, and formaldehyde at ppm levels in nitrogen. Ethylene oxide, propylene oxide, and 1,4-dichloro-2-butene were unstable in the sub-10 ppb range. Many manufacturers are now providing shelf life or stability data.

DYNAMIC, CONTINUOUS SYSTEMS

Dynamic, continuous flow calibration systems have the ability to change connotation with time. In order to calculate concentration, the flow, temperature, and humidity of both the diluent and analyte streams must be known and controlled. Flow control is quite important, as it is knowledge of flow rates that allows calculation and control of the final concentration. Figure 30-6 shows an example of a continuous system. Some of the components, such as strict humidity control, may not be necessary, depending on the application. The flow controllers can be replaced by meters, such as rotameters. However, flow metering of the dilution and contaminant (analyte) streams or inputs is required. Mixing of the two streams is also important. At high flow rates, some mixing will occur, but it is always a good idea to add a mixing chamber or other source of turbulence to ensure complete mixing.

The materials that are used in the system are important. As with batch systems, glass is generally a good choice. Teflon® is often used for valves and in short stretches of tubing that need to be somewhat flexible. However, Teflon is not very flexible. In fact, its rigidity is due to the strong intermolecular forces between the polymer molecules, which make it relatively impermeable. On the other hand, higher molecular wight compounds, e.g., naphthalene may condense or absorb on Teflon; thus, it is not a panacea. More flexible tubing, such as Tygon®, is made from plasticized polyvinyl chloride and is quite permeable. It should be used only in very small quantities. Stainless steel is a good choice for most organics, but may provide a reaction surface for many inorganics and some reactive organics and, of course, is not flexible.

The method for generating the diluent or clean gas must be considered. It will depend on

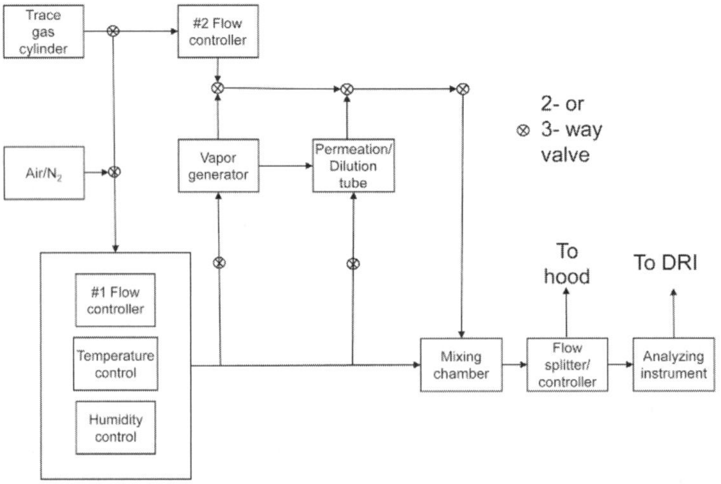

Figure 30-6. A continuous calibration system with means for using a vapor generator for high concentrations, permeation tubes for low vapor concentrations, and gas cylinders. Dilution flow can be controlled by a temperature, humidity, and flow controller.

the application. In some cases, air in the laboratory may be supplied with a pump. If the contamination of this air can be assumed acceptable, and its temperature and humidity control are not important, then laboratory air can be quite satisfactory. Otherwise, bottled air or nitrogen may be used, or laboratory air may be modified by cleaning and conditioning it, as discussed later. There are several choices for generating both gas/vapor and aerosol streams that will be discussed in detail. One of these choices is the use of compressed gases. The gas may be pure or diluted, and although the concentration needs to be known, it only needs to fall within a relatively wide range because the final concentration can be set by controlling the flow rates. Otherwise, the major consideration in using compressed gases is the expense, since, depending on the application, large quantities may be needed. Thus, no further discussion of the use of bottled gases will be given.

Calculations

One can calculate a *known*[5] concentration or use an independent measurement method to verify the final concentration, or both. The analyte stream may be vapor or gas mixed with air, particles mixed with air, or the analyte or contaminant may be directly introduced into the diluent. For example syringes or permeation or diffusion tubes, discussed later, may be used to introduce liquids to the diluent stream, which then evaporate. When the contaminant is *generated* within the diluent stream, the concentration is a function of the generation rate (G) of the contaminant and the flow rate (Q) of the clean air, as in the *steady state* equation of Appendix 30-1.

When the one is mixing a clean diluent stream with a contaminated analyte stream (e.g., bottled gas at 100 ppm CO_2), the calculation needed to determine the mixed concentration is straightforward if one pays close attention to units. If one gas stream containing the contaminant *(c)* is being diluted by a diluent *(d)* gas stream, the resulting or final concentration (C_f) is the concentration in the contaminant stream (C_c) adjusted by the ratio of the contaminant flow (Q_c) to the total flow:

$$C_f = C_c \frac{Q_c}{Q_c + Q_d} \qquad \text{[Equation 30-4]}$$

where Q_d is the flow rate in the dilution or clean gas stream.

[5] Each of the vapor/gas generation methods described below allows one to generate a known concentration from accurate knowledge of the variables in Equation 30-4 and Appendix 30-1 based on theoretical considerations or, in the case of compressed gases, based on the manufacturer's certification. However, even in the case of compressed gases, it usually is a good idea to check the calculated concentration with an independently calibrated instrument (e.g., by injection of the gas stream into a GC). Depending on the application, this may be a frequent requirement. In other cases, one check, just for peace of mind, may be sufficient.

Temperature, Humidity, and Flow Control

In any dynamic system, flow control is important, and temperature and RH control might be necessary. Control of flow can be accomplished in a number of ways; but it is necessary to point out that although there are many instruments for measuring flow, few will actually control it over time. Precision rotameters can be used to measure flow but not control it. This is fine for the short experiment where flow of a gas stream from a pump, compressed gas cylinder, or other means can be set once and will change little over the experiment. One should pay particular attention to the effects of gas density and viscosity on rotameter performance, as discussed in Chapter 20.

For controlling flow, critical or limiting orifices may be used with the stipulations given in Chapter 20. These devices must be coupled with a flow-measuring device, as they only control flow. The orifice meter can be used to measure pressure differential across an orifice and that is related to flow. In addition, it is possible through electronics to change the orifice size to control flow, as a result of electronic feedback used to maintain a constant pressure differential. Mass flow meters are also available, which can be used to control volumetric flow based on the ideal gas law (see Chapter 8). If we control, temperature and humidity (or at least they are considered to be relatively constant in a lab setting) then changes in mass flow will linearly affect volumetric flow. Typically, we want to know the volumetric flow since we calculate concentrations in ppm by volume or we use volume in the concentration density, e.g., mg/m^3.

Temperature control can be accomplished in various ways. A chamber can be built around the entire calibration system and control accomplished with a heat source such as a lamp, a thermostatic controller, and a fan to ensure even distribution of heat. If a lamp is used, precautions should be taken to avoid the localized effects on the system of radiant heat. Aiming the lamp at reflective surfaces such as aluminum foil can help to evenly distribute the radiant heat load. If only portions of the system require temperature control, then a water bath may be used, or a heat tape or heat pads. These devices also require thermostatic control.

Control of relative humidity is somewhat difficult. Nelson (1992) lists a number of saturated salt solutions that can be used to provide constant humidity levels, but they require continual maintenance. In addition, the corrosive environment produced by some of the salts can be harmful to instruments. There are instruments that will provide flow, temperature, and RH control simultaneously to a flowing airstream. These instruments are not inexpensive, but they can be well worth the cost, particularly if RH control is important (see, e.g., Miller-Nelson Research in the listing at the end of this chapter).

MASS FLOW METERS

Another way to measure flow that is widely used and commercially available depends on heat transfer. This principle is also used in velocity meters or hot wire anemometers. Two heated coils are placed around a tube through which a gas flows. The coils serve as heat sources, and the change in resistance to electrical current through the coils is indicative of temperature change. If no gas flows through the tube, then heat is not transferred from the upstream coil to the downstream coil. Thus, no temperature change or resistance change occurs, and no flow is indicated. However, if flow occurs, then heat from the upstream coil is transferred to the downstream coil, and the change in temperature causes a resistance change that is proportional to flow.

The more molecules that pass the coil, the more heat that can be exchanged. Thus, it is the mass flow that is measured and controlled, not the volume flow. Temperature and/or pressure changes that effect volumetric flow will not be detected if the mass flow is held constant.

Mass flow meters and controllers are extremely useful for flow measurement and control. They are compact (see Figure 30-7). They are quite precise, typically ± 0.2%. A word of caution is warranted, however, regarding their use with gases other than air. Because the response is dependent on the heat capacity of the gas, it will change for different gases. The manufacturer or the manual should be consulted in these cases, as correction factors for many gases are provided.

Generation of Saturated Vapor

One technique for generating a known concentration of a vapor is to saturate a gas stream with it. The vapor pressure of the chemical must be known, and the system temperature must be controlled because the vapor pressure is temperature-dependent. A gas stream such as air or nitrogen can be routed through an impinger or a bubbler, but the output from such an arrangement may not be saturated (Cotabish et al., 1961; Nelson, 1992). Several alternatives exist. First, an independently calibrated instrument can be used to determine the concentration. If temperature conditions in the generator are constant, then such a system can deliver repeatable vapor concentrations. A second alternative is to route the gas stream through more than one bubbler, maintaining the same temperature in each. Although this increases the likelihood of

Figure 30-7. Photograph of two mass flow sensors and their electronic controller (courtesy, Miller-Nelson Research, now Assay Technology).

saturation, it still leaves some doubt.

A third alternative is perhaps the best. In this case, the bubbler is followed by a condenser (Nelson, 1992) that can simply be a cooled glass tube or a second bubbler at a lower temperature (Cotabish et al., 1961). The evidence of liquid condensate indicates that the output is saturated. The second bubbler or the condenser should be maintained at a temperature for which vapor pressure data exist. Thus, if data only exist for 20°C, the first bubbler should be heated to a higher temperature. The higher this temperature, the more likely it is that condensate will form in the second apparatus.

A couple of precautions are worth noting. Temperature control is very important. If the room temperature is lower than the output from the generator, condensation will occur in the lines leading from the generator, and the concentration will be underestimated. It is a good idea to keep all tubing, from the generator to the point where dilution takes place, as short as possible. If necessary, the tubing is wrapped with heat tape and heated to avoid condensation.

Be aware of reactions in the generator. For example, ethanolamine will hydrolyze to form ammonia when humid air is passed through it (Saltzman, 1961). The concentration may be unpredictable, and important by-products may be formed. It is usually a good idea to use dry air in the generator and then add water vapor in the dilution air to simulate real conditions. In some cases, the presence of CO or CO_2 or even O_2 may cause undesirable reactions to occur (see, e.g., Saltzman, 1961). In these cases, the use of another vapor generation system is probably desirable.

For many volatile liquids, the saturated vapor concentrations are quite high. Leaks can prove to be explosion hazards, and even if they are controlled, high concentrations must be diluted. This may require double dilution or flow splitting in order to avoid either very high final flow rates or very low (and less controllable) flows through the generator. In a double dilution process, the contaminated air is diluted, and then, by using valves or flow splitters, part of it is discarded. The resulting new flow is first metered and then diluted one or more times until the desired concentration is achieved. For example, in Figure 30-6 the flow through the #2 controller dilutes the contaminant from the vapor generator or the permeation tube. This flow is then diluted by that from the #1 controller. If the resulting flow is excessive, it is split at the flow splitter, which can simply be a valve and a flow meter.

Generation of Contaminant Using Syringe Pumps

Syringes may be used to deliver a continuous feed of liquid or gas contaminant to a clean, continuously flowing gas stream. The filled syringe is inserted into a motor-driven unit (see listing at the end of this chapter for suppliers), which pushes the plunger forward at a known speed. In theory, it is possible to calculate the volume of gas or liquid delivered to the system from the speed of the syringe pump. It is also possible to weigh the pump before and after to determine the total mass that has been dispensed. This will give a second measured estimate of the *average* concentration during the dispensing time.

However, a number of potential problems arise if one wants a *constant* concentration. In some units, the motor drives a screw-like device, and the resulting output can be cyclical. The calculated concentration may be accurate on average, but at any given time may be in error by more than 5%. When gases are delivered, back diffusion of clean gas into the syringe can be significant if the delivery rate is low (Saltzman, 1961). In addition, incomplete vaporization of low volatility substances and surface tension effects (large drops may form before vaporization takes place) can cause errors.

When relatively low-boiling point liquids are delivered, lack of evaporation can be a factor, in which case liquid droplets fall off the end of the needle and concentrations are lower than anticipated. This problem can be solved by inserting the needle into a septum that is contained in a metal housing. The metal housing is then heated with a heating tape or other suitable device. Care should be taken not to heat the liquid to the point of decomposition and not to attempt concentrations that are beyond the room temperature saturation point. When liquid compounds of high boiling point are used, and low concentrations are attempted by using low syringe drive rates, then evaporation up into the needle may occur, thus causing concentrations to be too high. Thin-gauge needles will help avoid this problem.

Finally, some problems are overcome by using mixing chambers before the DRI. The chambers should be large enough to hold the output of the diluent stream and the vapor generated by several cycles or turns of the syringe. For higher concentrations (less diluent) this may be possible, but for low concentrations, the required size of the chamber may be too large.

Syringe pumps are not error-free, and it is a good idea to independently check the concentration delivery. They are not recommended for calibrating DRI unless one is interested in time weighted average concentrations, and even here the times necessary to estimate reasonably accurate average concentrations may be many minutes depending on the delivery rate of the syringe. On the other hand, they are easy to use, and required changes in concentration can be made quite rapidly.

Generation Using Diffusion Tubes

In Chapter 23 it was shown that if the concentration of a vapor at one end of a path is held constant and at the other end is held at zero, then the rate of diffusion along that path is also constant. The three variables are related (see Chapter 23, Equation 23-1) by a constant of proportionality, the diffusion coefficient (*D*) for the diffusant relative to the medium through which it is diffusing. *D* is affected by temperature and pressure. If the constant concentration is produced by the evaporation of a liquid at one end of the path, then this concentration is dependent on the liquid's vapor pressure, and it too is temperature-dependent. If temperature and pressure are held constant, or corrections are made when they change, the mass delivered at the end of a diffusion path, shown diagrammatically in Figure 30-8, can be quite constant. Its value can be known, provided that the diffusion coefficient of the vapor in question is known.

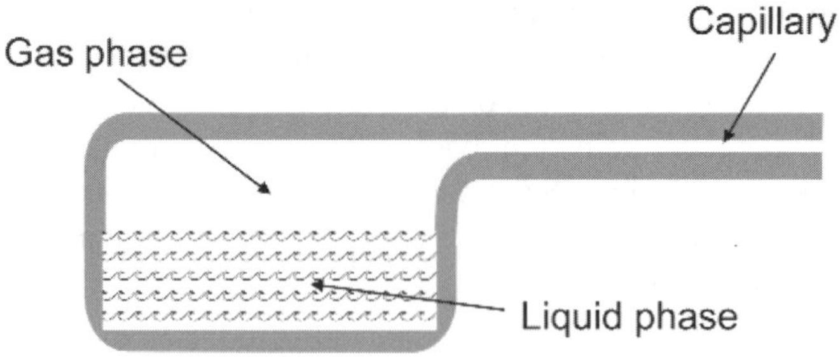

Figure 30-8. Schematic of one type of diffusion tube.

The process can be used to continuously supply vapor to a dynamic calibration apparatus. The idea apparently was first proposed in the 1950s (McKelvey and Hoelscher, 1957) and diffusion tubes or vials are now available commercially (see listing at the end of this chapter). The vials can be constructed in the laboratory, but they must then be calibrated with an independently calibrated instrument such as a GC, or the user must trust predictions of the diffusion rate from theoretical considerations (see later discussion). Some manufacturers supply certifications for the vials (Meyer, 1992) that will allow the hygienist to use them without calibration.

Whether the user calibrates a certified vial should also depend on a number of other factors, the most important of which is temperature control. The temperature dependence of diffusion of a vapor through air is important (the diffusion coefficient is dependent on temperature to the 3/2 power; see temperature effects on diffusion in Chapter 23). In addition, the vapor pressure that controls the evaporation of vapor from the liquid is also temperature dependent (see the vapor pressure equations in Appendix B and discussion in Chapter 8). Thus, temperature is typically controlled at ± 0.1°C. Other factors to consider are the nature of the diluent gas and the dimensions of the vial (Altshuller and Cohen, 1960). The diluent gas is usually air, the vials are usually calibrated with air, and diffusion coefficients are often given for vapors in air. Therefore, the use of another diluent gas, particularly one with a substantially different molecular weight from air, will cause error; but diffusion into air, nitrogen, or oxygen should be fairly constant. The dimensions of the vial need to be known for calculation of the diffusion rate; so they should be measured accurately.

CALCULATING DIFFUSION RATES

From Chapter 23, Equation 23-1 can be used to calculate the total mass fluxing from a diffusion vial. Multiplying both sides of the equation by the cross-sectional area of the diffusion tube will give the flux in units of generation rate (G) or mass/time:

$$F\,A = G = \frac{DA(C_a - C_o)}{L}$$

The negative sign has been removed as it simply indicates the direction of diffusion from high to low concentration. It is assumed that C_o is 0 if the outlet of the tube is continuously swept by the clean diluent. The concentration above the liquid (C_a) is dependent on the vapor pressure (P_a). However, concentration expressed as Pa/P is unitless or volume/volume, not mass/volume. To convert, the ideal gas law is used, and we have

$$G = \frac{DA}{L}\left(\frac{P_a}{P}\right)\frac{MP}{TR}$$

where *MP/TR* is molecular weight × total pressure over absolute temperature × the ideal gas constant, or the molecular weight over the molar volume. The term converts the concentration units to mass/volume. Thus, it can be seen that pressure units cancel and that temperature, which also affects both P_a and D, has a complex relationship with G. Therefore, the need to carefully control T is evident.[6] The mere evaporation of the liquid has a cooling effect on the system, and so temperature needs to be measured near the liquid. It should also be noted that D in the above equation remains dependent on P. In other words, the values of D from the literature are derived at given values of P and T, usually one atmosphere and 25°C.

In summary, one should know the diffusion coefficient (see Lugg, 1968; Nelson, 1992), the cross-sectional area and the length of the tube, and the vapor pressure in order to calculate the generation rate from the end of the diffusion tube. Alternatively, one may buy a certified tube from a manufacturer.

Generation Using Permeation Tubes

The movement of a vapor or a gas through a polymer can be closely controlled in order to supply a constant mass of a contaminant to a calibration system. The permeation process is driven by the concentration of the permeant at the up gradient or exposed side of the polymer;

---

[6] The relationship of P_a and G has been written in a slightly different way since at least 1928 (Lewis and Chang, 1928; Sherwood and Pigford, 1952; McKelvey and Hoelscher, 1957; Altshuller and Cohen, 1960; Worthington and Rey, 1991), or

$$G = \frac{DA}{L}\ln\left(\frac{P}{P - P_a}\right)\frac{MP}{TR}$$

The difference is in the pressure, that is, concentration fraction. The above equation is likely most technically correct, for reasons not clear to the author. The P_d/P term seems more logical relative to other applications of Fick's law. In any event, it can be shown that for vapor pressures < 100 mmHg (13.3 kPa) the relative difference in the two terms is less than 5%.

that is, the concentration of the permeant in the outermost layer of the polymer at that exposed side. That concentration is in turn controlled by the solubility of the permeant in the polymer, which is a function of the chemical similarity of the polymer and the permeant. Consequently, permeation is a solubility process that establishes a concentration gradient, which is followed by diffusion across the polymer thickness.

The idea of putting liquids into polymer tubes and then sealing both ends to create a permeation device was first proposed by O'Keeffe and Ortman (1966). The earliest efforts involved calibration standards made with liquefiable gases that were important ambient air pollutants, such as NO_2 and SO_2.

A short piece of plastic tubing (usually Teflon) was sealed at both ends, either by inserting a plastic plug and then crimping it in place with a wire or metal band or by forcing an oversized steel ball into the tubing. The tubing was then almost filled with the liquid by gently using pliers to cause a small opening at one end. Through this opening the liquid flowed from a compressed gas cylinder that had been connected to the Teflon tube (O'Keeffe and Ortman, 1966; Scaringelli et al., 1967, 1970; Saltzman et al., 1971). Today permeation tubes are also filled with room temperature liquids, liquefied gases, and even gases (see VICI Metronics). The latter are immersed in liquid N_2 during filling to prevent the loss of the gas.

TEMPERATURE EFFECTS

As with diffusion devices, permeation tubes deliver mass at a rate that is dependent on temperature. An Arrhenius relationship can be used to predict this effect (Saltzman et al., 1971; Perkins and You, 1992). Equation 23-4 from Chapter 23 can be used; the terms D and D_o for diffusion constants are replaced with P and P_o for permeation constants. For Teflon, the effect of temperature is severe enough to warrant temperature control of 0.1°C. A change of about 10% in the permeation rate is found per degree of temperature change (Scaringelli et a1., 1967).

Ho and Schlecht (1981) found that permeation tubes could be constructed of silicone rubber tubing, which partially overcame the temperature problem. They found that permeation of trichloroethylene, methyl ethyl ketone, and benzene increased only 2.5%, 1.7%, and 2.1 %, respectively, for a 2°C temperature change. They also found that the permeation rates of these chemicals were much higher in silicone rubber than in Teflon.

These results can be explained by the activation energies (E_p) of the polymer-permeant systems. The E_p values for permeation (see Chapter 23 for discussion of Equation 23-4) for the silicon system were around 2 kcal/g-mole, whereas Saltzman et al. (1971) found E_p values for Teflon and NO_2, butane, butene, propane, and propene in the range of 12.4 to 16.7 kcal/g-mole. This difference is mostly attributed to the polymer and not the permeant, as the activation energy can be thought of as the energy necessary to open a *hole* in the polymer large enough to allow a jump of the permeant molecule. In Teflon, the intermolecular forces are stronger than those for silicone and, therefore, more energy is needed.

Figure 30-9 shows how the activation energy accounts for temperature effects. E_p is the

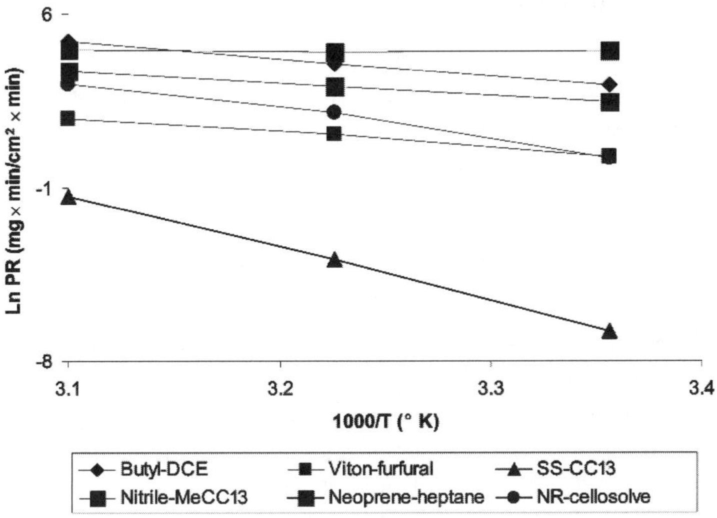

Figure 30-9. Arrhenius plots for thickness corrected, steady-state permeation rate or the permeability coefficient, where DCE is 1,2-dichloroethylene, MeCCl3 is methylchloroform, SS is Silver Shield®, and NR is natural rubber (from Perkins and You, 1992, with permission).

slope of the temperature-permeation relationship, and a shallow slope means that for a given temperature change the effect on permeation will be less than that for a steep slope (higher E_p). In Figure 30-9 the five polymers at the top are rubber-like and would, therefore, have weaker intermolecular forces than the Silver Shield® polymer at the bottom. Silver Shield, like Teflon, is a plastic, or a more inflexible polymer than the other five.

REACTANTS

As the chemical in the permeation tube is in pure form, and the permeation process is in both directions across the tube, reactions can occur between the tube, chemical, and normal or added atmospheric constituents that permeate into the tube. For example, Saltzman et al. (1971) found that water vapor affected the permeation rate of NO_2 by up to 30%. Scaringelli et al. (1967) found that O_2 reacted with H_2S, and so these tubes had to be used in nitrogen atmospheres. No atmospheric effects on the permeation rates of hydrocarbons were seen for the limited number of hydrocarbons examined. Manufacturers are aware of many of these problems and should be consulted as to the proper way to use their products.

CALIBRATING PERMEATION TUBES

As with diffusion tubes, the mass delivered from a permeation tube can be calculated from theory. The steady state permeation rate (*PR*, mass per time per length of tube) of a vapor through cylindrical tube walls can be given as

$$PR = \frac{2\pi DS\ (p_i - p_0)}{\ln \frac{r_o}{r_i}}$$

where *D* and *S* are the diffusion and solubility (Henry's Law) coefficients, *p* is the partial pressure of the permeant at the inner (*i*) and outer (*0*) surfaces of the polymer, and *r* is the inside (*i*) and outside (*o*) diameter of the tube (Ho and Schlecht, 1981). If the outside surface of the tube is continuously swept by a clean diluent gas, then p_o approaches 0. In addition, p_i is the saturated vapor pressure (p^*) of the permeant for both the liquid exposure and the saturated vapor exposure within the tube (i.e., the tube is not full of liquid). Therefore,

$$PR = \frac{2\pi DS(p^*)}{\ln \frac{r_o}{r_i}}$$

Although solubilities and diffusion coefficients for various chemicals in Teflon (and other polymers) do exist in the literature (see, e.g., Brandrup and Immergut, 1975), they vary with the exact composition of the polymer. Thus, polymers that have been prepared with additives, such as accelerants, may have different values from those without these additives. Teflon is probably less variable in this respect, but there are two commercial types of Teflon – Teflon FEP (fluorinated ethylene-propylene) and Teflon PTFE (polytetrafluoroethylene).

The most common way to determine the delivery rate of a permeation tube is to weigh it under temperature-controlled conditions at various time intervals. This can be done quite accurately with a microbalance. The only potential problem occurs when the tube is removed from its controlled environment to be weighed. At greater cost, even this can be overcome by using a temperature-controlled electrobalance. The latter also has the advantage of giving a continuous trace of weight versus time (Soda, 1980). Typically, the devices are calibrated at several temperatures, so the user can determine the permeation rate for the temperature of use. The data are best presented as an Arrhenius plot similar to those in Figure 30-9.

CERTIFICATION

Most manufacturers sell certified and uncertified state permeation tubes. Certified tubes have been calibrated by the manufacturer at one or more temperatures. The certification gives the calibration constants for a given temperature (mass delivered per time) and should state accuracy limits. For multiple temperature calibrations, a temperature versus permeation rate plot is usually provided. It should be noted that certifications are traceable to a government standard (National Institute of Standards and Technology) only to the extent that balances used to measure weight loss and temperature-sensing devices are traceable. The NIST offers certified tubes for SO_2. In addition, a standard practice for the calibration of tubes exists (ASTM, 2005), and an inquiry about whether this procedure or some other is followed by the manufacturer may be important.

Generation of Aerosols

The measurement of the weight (mass) or the volume occupied by a liquid or a gas is one of the more accurate measurements that can be made by hygienists. Standards for calibrating or manufacturing balances or volumetric devices are commonly available and traceable to government or international standards. Thus, a measured weight or volume of liquid can be evaporated into a measured volume of air, and the concentration so produced can be *known* quite accurately. When an instrument is calibrated against this known concentration, it usually is assumed that all of the error lies with the instrument, not with the process of preparing the known concentration. Although we shall see later that there are situations where this may not be a good assumption, for most applications the assumption works suitably.

There remains the problem of generating known aerosol concentrations. For gases and vapors, it is usually safe to assume that they exist as single molecules (they do not agglomerate), and that they uniformly distribute themselves. This makes the calculation of concentration easy, and *particle size* is essentially of no consequence. This is not true for aerosols. It might first appear that it would be relatively easy to weigh out a quantity of dust, disperse it continuously into air flowing at a known rate, and create a known concentration. Unfortunately, however, the dust will not likely form a distribution of particles that is easily defined or predicted, and the feed rate of the dust to the air is difficult to control. If the particles are too large, they will impact onto surfaces or fallout of the airstream by sedimentation.

All methods require a second or perhaps third independently calibrated instrument, as it is not possible to predict, with any certainty, the concentration or size distribution[7] produced by a dust generator using theoretical considerations. In essence, concentrations generated for aerosol measurements are not *known* in the same sense as those for gases and vapors.

As a result, DRI for aerosols are often calibrated by hygienists in the field by comparing the response of the DRI to a pump/filter result, as was discussed in Chapter 28. The filter result is considered to be the known, but as Chapter 24 points out, there are errors associated with filter sampling and analysis. If the calibration sample size is fairly large (say, > 20) then the random errors associated with the filter are effectively averaged out. However, any bias on the

[7] Unless one generates, for example, monodisperse aerosols from a suspension of sized, near homogeneous, polystyrene latex spheres.

part of the filter method relative to the DRI or the true concentration will not be decreased by increasing sample size, but can be estimated.

Manufacturers also calibrate aerosol DRI in field settings, but most calibrations are performed in a laboratory. These tend to be of two kinds. In many cases, each instrument is challenged with some aerosol[8] by using an aerosol generator (described below). The response of each instrument is compared to all others, and this check serves as a quality control measure. For more sophisticated instruments, particularly those that allow the estimation of particle size distributions, more complicated calibrations are performed. If the instrument employs cascade impactors, then an aerosol of differing size for each stage is used. These are usually monodisperse aerosols such as pollen grains or polystyrene spheres, or they are polydisperse aerosols whose size distribution is well characterized and consistent over time.

AEROSOL GENERATION TECHNIQUES

These techniques (which are also used for toxicology studies, filter and other exposure control method testing, respirator fit testing, and general aerosol research) may be divided into three categories: nebulizers, powder dispersers, and condensers. The nebulizers may be used to generate a purely liquid aerosol, or an aerosol composed of a solid central nucleus surrounded by a liquid. This latter type of aerosol usually goes through a drying stage so that only the solid remains. (For greater detail on these instruments, the reader should see Liu and Lee, 1975; Hinds, 1999; Leong et al., 1982; Lippmann, 1989.)

Perhaps the best-known nebulizer is the DeVilbiss, depicted in Figure 30-10. A liquid is drawn into a jet of compressed air, which causes the liquid to be fragmented. The jet creates a partial vacuum at the liquid feed tube, drawing the liquid into the jet. Larger droplets impact on the top of the device and flow back to the reservoir. A liquid of low volatility is typically used so that evaporation does not contribute significantly to a particle size change. Alternatively, a solution or a liquid suspension may be used. A solution of sodium chloride and water will yield a solid salt aerosol if the water is made to evaporate by passing the aerosol through a drying stage. A suspension containing a solid such as polystyrene spheres and water with an anionic surfactant to prevent coagulation can also be used. Once the aerosol is generated, the water is evaporated, leaving solid spheres of nearly unit density and known size. There are a few problems, such as the probability of particles that contain more than one sphere, but this is a convenient way to generate a monodisperse aerosol. Output concentrations of nebulizers are typically in the range of 5 g/m^3 to 50 g/m^3, with mass median aerodynamic diameters (MMADs, see Chapter 9) in the range of 1 μm to 6 μm and geometric standard deviations (GSDs) of 1.8 to 2.1 (Hinds, 1999).

Dry powders may be aerosolized with a powder dispersing device, but the challenge is to

[8] Arizona road dust is a favorite.

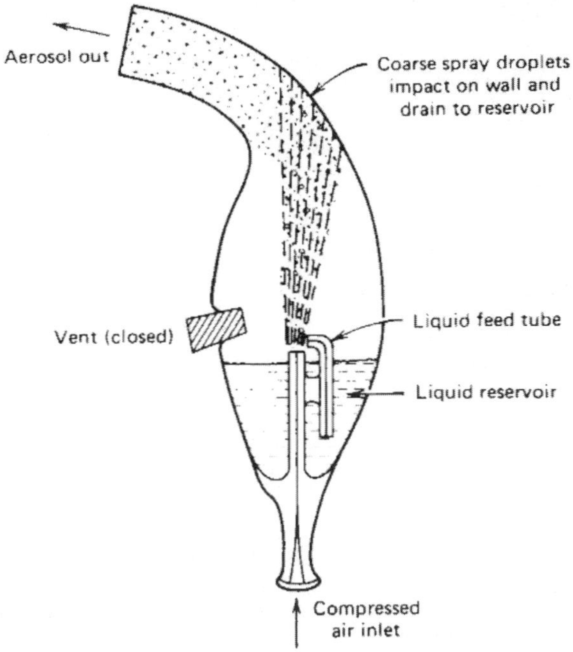

Figure 30-10. Diagram of a DeVilbiss model 40 nebulizer (from Hinds, 1999, with permission).

obtain a near-uniform feed rate. It has been determined that the best way to do this is to press the dust or the powder into a cake using consistent pressure, and then to scrape or grind this cake at a uniform rate. The Wright dust feed does this. Factors that can affect the output include the hydrophilicity of the dust and the humidity of the airstream. Dusts that absorb water tend to agglomerate. However, a dry airstream causes static charges that also can affect concentration. One solution is to use a radioactive source (beta, or for highly charged aerosols, alpha) that neutralizes excess charges (for more details see Hinds, 1999). Outputs of dispersing instruments are more controllable than those of nebulizers, ranging from 0.0003 g/m^3 to > 200 g/m^3. Achievable particle sizes also cover a larger range, from 0.5 μm to 100 μm.

Condensation generators can be used to form monodisperse or polydisperse submicron aerosols. They work by evaporating a viscous liquid with heat, and the subsequent cooling of the airstream causes condensation of the vapor. The particle size can be restricted by use of size-selection principles. The first two types of devices are more likely than condensers to be used for DRI calibration work.

DATA ANALYSIS USING LEAST SQUARES REGRESSION

Generally when an instrument is calibrated, a plot of signal or response versus concentration is made, similar to that shown in Figure 30-11.[9] It is now common to use the statistical method of regression analysis, specifically the method of least squares, for analyzing the calibration data and providing a line of best fit. This method gives both a line that fits the observed values and a measure of error around that line. The least squares method optimizes the line of best fit by minimizing the perpendicular distance of the actual data points from the predicted line ($\hat{Y}$; see Figure 30-11). If these distances are squared and then summed, then the line having the minimal sum is the least squares line.

The equation for such a line takes the form

$$\mu_{Y|X} = \beta_0 + \beta_1 X_i \qquad \text{[Equation 30-5]}$$

where $\mu_{Y|X}$ is the population mean value of a Y given the value of X, β_0 the Y-intercept, and β_1 is the slope of the line. The above equation represents the true model; from the n samples of the calibration the line, the estimated equation, where b estimates β is

$$\hat{Y}_i = b_0 + b_1 X_i$$

The true value of β_0 will be 0 if the instrument has no bias. β_1 may take on any value if the signal from the instrument is reported in units different from concentration. If the signal is in units

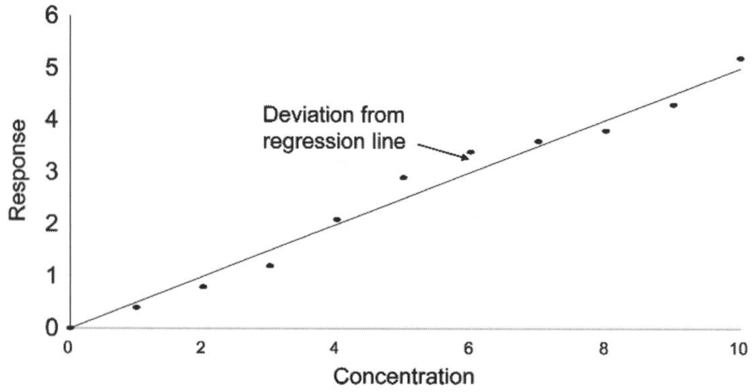

Figure 30-11. Typical calibration plot of concentration versus DRI response.

---

[9] Before computers or calculators with statistical functions were widely available, it was common to simply hand-draw a line of best fit through the data points. Such a line was an approximation, and such an analysis did not afford the user any information concerning the reliability or the precision of a measured concentration provided by a direct-reading instrument.

of concentration, and the relationship of the signal to the concentration is linear, then β_1 should have a value of 1. Regardless of the units of the signal, the agreement between signal and concentration should be good; that is, one should increase in proportion to the other. Therefore, the correlation coefficient should be close to 1 for an instrument that has a linear or a proportional response to concentration. Some instruments are not expected to have linear responses (e.g., some applications of infrared absorption devices), but the relationship can be made linear by transforming (e.g., taking the logarithm) of the concentration. In summary, we desire values of β_0 close to 0, β_1 close to 1 (when the signal is in units of concentration), and a correlation coefficient near 1. All of these values can be tested using null hypotheses.

Assumptions

Regression analysis is based on the following assumptions (Woolson, 1987, pp. 283-284):

a. The variance of *Y* at each value of *X* is constant.

b. The *X*s are measured without error.

c. If we are to test hypotheses regarding the slope and the intercept, the error in the expected value of the *Y*s is normally distributed.

d. Equation 30-5 is the correct model.

Assumptions *a* and *b* are commonly violated when one is performing calibrations. Assumption *d* may be violated in field calibrations of aerosol DRI. For details on the theoretical considerations involving violations of these assumptions, see Draper and Smith (1998, pp. 122, 291) or Fuller (1987, pp. 13, 74). In the next sections, approaches to performing the regression analysis, when assumptions *a, b,* and *d* are violated, are considered.

a. THE VARIANCE OF Y IS NOT CONSTANT WITH X

We would expect that the repeated analysis of a particular standard concentration, say, 10 ppm, would not always yield 10 ppm. Instead a range of values would be expected; they would vary because of random measurement error. If we have five standard concentrations and analyze each five times, then we might get a plot like that in Figure 30-12. Notice that the spread of the data at each value of concentration is about the same; that is, the value of s_y is the same at each *X*. However, it is likely that the value of s_y will in fact increase with *X*, as shown in Figure 30-13. When this occurs, we violate assumption a.

However, we can often correct this by transforming the *Y* observations so that s_y is constant, a process called weighted least squares. This is usually accomplished by transforming the *Y* values according to some function of the *X* values. For example, we might find that s_y is proportional to *X* or $X^{0.5}$. Most software programs will perform weighted least squares, but it is up to the user to define the weights that are applied to each value of *X*. This requires

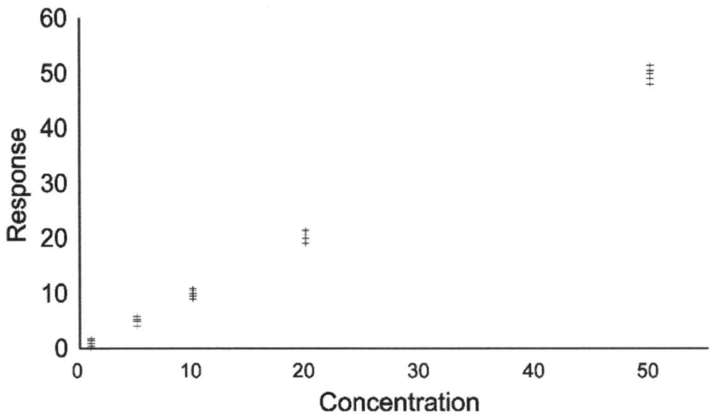

Figure 30-12. A calibration plot showing replicate values of response for fixed values of concentration that indicate constant variance in the *Y*s with *X*.

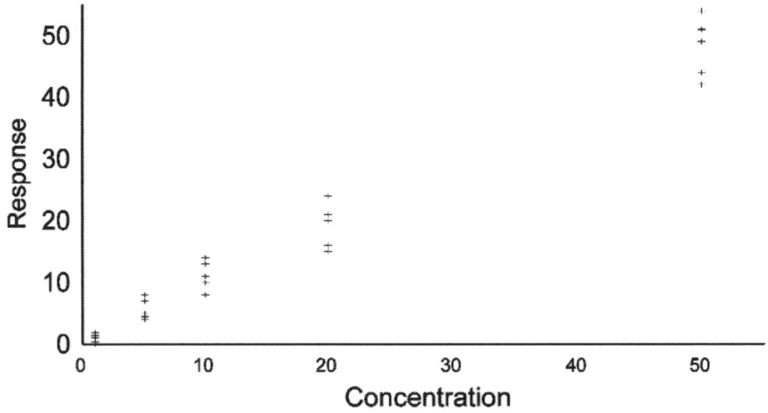

Figure 30-13. A calibration plot showing replicate values of response for fixed values of concentration that indicate non-constant variance in the *Y*s with *X*.

- observing more than one Y for each X,
- calculating s_y at each X, and
- applying a weight at each X equal to $1/s_y$.

Alternatively, one may avoid these steps and assume that the weight is proportional to

- $1/X$, equivalent to the variance of Y at each X is proportional to X,

- $1/X^2$, equivalent to the standard deviation of *Y* at each *X* is proportional to *X*, (i.e., the CV_y is constant), or
- a more complicated relationship (see, e.g., ASTM, 2001).

The problem of non-constant variability in the *Y*s can be overcome in an approximate fashion by making an educated guess at the weighting factors or by calculating them more exactly based on the actual replicate values of *Y*. It should be noted that when the sample size at each *Y* is small, it may appear that the s_y values are constant. In fact, a statistical test, such as Bartlett's test for homogeneity of variance, may show constant variance, but this may be due to a low test power caused by small sample size. Generally, we should expect that the variance is not constant with *Y* (see, e.g., Garden et al., 1980; Christian and Tucker, 1984; ASTM, 2001).

b. THERE IS MEASUREMENT ERROR IN X

The *known* concentrations used in a calibration are not exactly known and thus are subject to error. The amount of error involved in *knowing* these concentrations is variable, and so we violate assumption *b* to varying degrees. When we calibrate an aerosol DRI by comparing its response to filter samples that have been weighed, the error in values of *X* may be quite large, say, CV ~ 5% to 10%. In making known gas or vapor concentrations, it may be much smaller, say, < 2%.

There are actually two types of *X* variables or concentrations that may be measured or created. When *X* is fixed, it is created or set at a given value. For example, when calibration standards are prepared, *X* is essentially fixed at, say, 10 ppm, even though, in the process of preparing or fixing *X*, error is likely made such that the true concentration is, say, 9.9 ppm. This is different from the case where *X* is random. When an aerosol DRI is calibrated in the field with filter samples, the value of each filter sample is not fixed but random. We have no way to set it beforehand. This random value of *X* is estimated, again with error.

In the case of fixed *X*, the normal least squares analysis can be used if the error in measuring *X* is independent of the value of *X* (Fuller, 1987, pp. 80-81; Draper and Smith, 1998, p. 124). Unfortunately, this probably is not the case, as we might expect that the absolute error made in preparing the fixed environmental standard would likely increase with concentration, just as the error made by the DRI increases with concentration. It may be that the relative error (the *CV*, or s_x/X) is constant, but then the absolute error, s_x would not be. Thus, even in the case of fixed *X*, assumption b appears to be violated. When the value of *X* is random and measured with error, the problem is more difficult.

Solutions for Measurement Error in *X*

For the complex problem presented when the variance of *Y* at *X* is not constant and the values of *X* are not error-free, there do not appear to be any solutions derived in statistical textbooks. There are no simple methods. The normal approach to solving these problems is first to derive a model such as Equation 30-5. However, for this case, the model would have complex error terms for the *X*s and *Y*s. Then from this model an equation for the sum of squares may be derived. Next the parameters of the model, generally error terms, are estimated. Then the values for b_o and b_i are iterated, until the sum of squares is minimized. Such techniques appear in the literature (see, e.g., Christian and Tucker, 1982a, b, 1984, 1986), but the techniques do not appear to consider together all of the assumption violations discussed thus far.

c. NON-NORMALLY DISTRIBUTED ERROR IN ESTIMATING *Y* FROM *X*

In the special case where the values of *X* are derived from the environment (e.g., when filter samples are used to calibrate an aerosol DRI), the measured values of *X* (filter results) and *Y* (DRI response) are likely to be lognormal or skewed. In such cases, this is not a violation of assumption *d*. Assumption *d* simply requires that repeated DRI responses to the same environment be normally distributed. This is likely to be the case and, therefore, we generally need not worry about this assumption. Instead, violation of assumptions *a* and *b* will likely occur.

Applications of Regression Analysis

The following guidance can be offered for calibration work.

1. If one is interested only in the correlation of concentration and response, then none of the above assumptions applies. Simply calculate the value of *r*.

2. However, if the limit of detection (*LOD*) is to be calculated from the regression,[10] the slope and the intercept are to be tested against values of one and zero, or if the uncertainty of a predicted value of concentration based on the response of the instrument (e.g., 10 ± 3 ppm) is of interest, then the assumptions should be considered.

3. Minimally the weighted least squares techniques should be used in an attempt to satisfy assumption *a*. Weights should be determined from the data, or at least the assumption that the s_y at each *X* is proportional to *X* should be used, i.e., the *CVs* of the *Y* values at each *X* are constant.

4. In order to partially satisfy assumption *b*, the standards should be made as accurately as possible. If the error in *Y* is much greater than the error in *X*, then the matter becomes negligible (Draper and Smith, 1998, p. 124). Also the main problem

[10] This is equivalent to estimating the value of *X* that gives a signal that is on average not reliably differentiable from a Y-value of zero. It is a bit difficult to calculate from the regression data (see ASTM, 2001; Winefordner and Long, 1983), and the four assumptions should be met to do so.

with violating assumption *b* is a resulting bias in estimating the slope. The estimated slope (b_1) is the true slope (β) modified by an error term, or

$$b_1 = \frac{\beta}{1 + \frac{\sigma_e^2}{\sigma_x^2}}$$

where σ_e^2 is the variance at a value of *X*, that is, the measurement error for an *X*, and σ_x^2 is the variance across the true values of *X*. Thus, if σ_e^2 is small compared to σ_x^2, then the error in estimating the slope is small (Snedecor and Cochran, 1973, p. 165). As an example, if we calibrate the aerosol DRI across 25 random *X*s of 0.1 mg/m^3 to 10 mg/m^3, then the variance of the *X*s is about 6 (the range divided by 4, quantity squared; Snedecor and Cochran, 1973, p. 46). If the relative error (σ_x) in measuring or creating the *X* concentrations is, say, 20% of the mean concentration, and we assume that the average of the *X*s is about 2, then the value of σ_x^2 is 0.4^2, or 0.16. This value is probably underestimated, as the value of σ_e^2 is not constant with *X*. Even so, σ_e^2/σ_x^2 is about 0.16/6, or quite small; so it is apparent that the violation of assumption *b* can be minimized if the range of the *X*s is maximized, and *X* is created or measured relatively accurately.

Finally, much could be done in the way of providing solutions to the problem of simultaneous violation of assumptions *a* and *b*. In addition, computer simulations designed to show under what conditions these errors are negligible would be helpful.

SPECIFIC CALIBRATION EXAMPLES

The MIRAN 1A Closed-Loop Calibration

Infrared instruments are nondestructive, and this property can be used advantageously in calibrating the MIRAN lA. A closed-loop system can be used whereby the gas is continuously recirculated. One typically starts with a low standard (say 2 × *LOD)* and then adds analyte to the cell gas to make higher and higher standards. The pump that is normally used to draw air into the MIRAN is not airtight; so an alternative pump is supplied for calibration purposes. This pump has a metal bellows, and thus the use of polymer in the pump is minimized; it is subject to minimal memory caused by interaction of analytes with polymeric parts.

The calibration setup is shown in Figure 30-3, with the exception that the gas sampling bag is not needed because the MIRAN sampling cell serves as the calibration gas reservoir. The instrument has an internal sample cell of 5.6 L; so the total volume of the calibration system (pump, cell, and hoses) is 5.65 L. The analyte (gas or liquid) is injected, with a syringe, through a septum on the metal bellows pump. A typical calculation may be as follows. Say that a 100 ppm concentration of benzene is to be established in the system; and

$$100\ ppm = \frac{100\ \mu L_{b,v}}{1\ L_a} =$$

$$\frac{X\ \mu_{b,L}}{5.65\ L_a} \times \frac{mL_{b,L}}{10^3\ \mu L_{b,L}} \times \frac{0.87 g_{b,L}}{ml_{b,L}} \times \frac{g\text{-}mole}{78\ g\text{-}mole_{b,L}} \times \frac{24.45\ L_{b,V}}{g\text{-}mole} \times \frac{10^6\ \mu L_{b,V}}{L_{b,V}}$$

Notice that all units cancel, except ppm. The subscripts are used to keep air (*a*) differentiated from benzene liquid (*b, L*) and benzene vapor (*b, V*). Conditions of 25°C and 1 atmosphere are assumed. As the units cancel, all that is left is the arithmetic to solve for *X*, which is in units of µL, liquid benzene. The answer is 2.07 µL. Practically, one would inject 2.1 µL with a syringe and calculate the actual concentration proportionally as

$$\frac{2.1\ \mu L}{2.07\ \mu L} = \frac{X\ ppm}{100\, ppm}$$

The actual concentration is 101.4 ppm. Given the accuracy of the syringe, it is doubtful that the four digits are significant, and so in this case one could report 100 ppm as the concentration. However, when further injections are to be made and the concentration is calculated proportionally, the digits should be conserved. For example, if 5 µL are injected after the first 2.1 µL, to create a new concentration, then the new concentration is

$$\frac{7.1\ \mu L}{2.07\ \mu L} = \frac{X\ ppm}{100\ ppm}$$

or 343 ppm. For lower concentrations one must make up vapor standards in bags or bottles and then inject them into the MIRAN system, or use the vapor head space technique explained in the discussion of syringes in this chapter.

When the highest concentration used to calibrate the instrument is quite high (say, 1,000 ppm), then the purge equation in Appendix 30-1 should be used to calculate the time needed to purge the MIRAN. For example, if 1,000 ppm existed in the system and a new calibration were to be started, the user might want to purge with clean air down to 0.1 ppm. From the purge equation, we have for the 5.65 L volume and the approximate flow rate of the metal bellows pump of 15 L/min

$$\ln \frac{0.1}{1000} = -\frac{15\ t}{5.65}$$

and purge time, *t*, is about 3.5 minutes. This assumes perfect mixing, and the purged gas should be routed so as not to be reentrained.

Once the pairs of the concentration and the corresponding response of the instrument in absorption units have been found, these values may be plotted as in Figure 30-14. In this case, the concentration range extends so that the non-linearity of response is clear, as one would

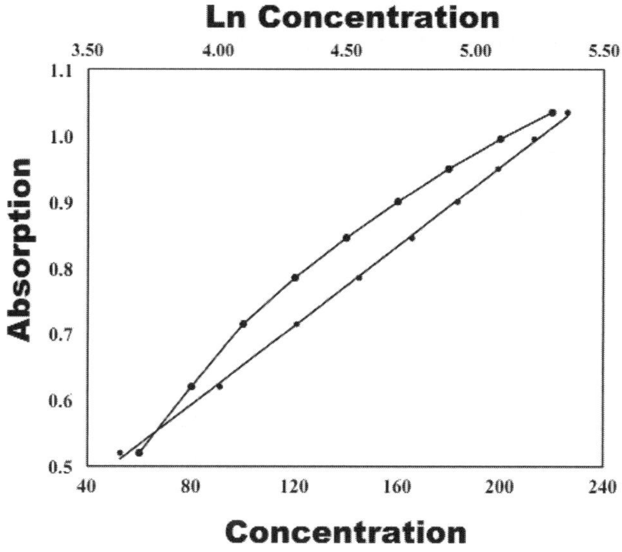

Figure 30-14. A MIRAN calibration curve for nitrous oxide showing non-linearity of response and a transformation of the absorption values to gain a linear relationship. Note logarithmic scale at top of graph.

expect from Beer's law. Here we have three choices. For low accuracy work, we may simply hand-draw a smooth curve through the points and read future concentrations from this curve. For greater accuracy, we can divide the curve into near-linear portions that give high correlation coefficients. If each portion contains only three points, then these correlation coefficients should be at least 0.99.[11] The regression equation for each portion can then be obtained. A spreadsheet package with regression tools is useful for this. For the most accurate work, the concentrations are plotted as logarithms to obtain a straight-line relationship. Such a plot is also shown in Figure 30-14. The correlation coefficient should not be significantly different from one, and the intercept should not be different from zero.[12] The straight-line relationship in Figure 30-14 ignores the non-constant variability in the *Y*s for each *X* and the measurement error in the *X*s. Regarding the first problem, weighted least squares regression analysis was performed, assuming that the variance in the *Y*s was proportional to X and to X^2. The results are shown in Table 30-2. As can be seen from Table 30-2, in this case the weighting factors

[11] A regression coefficient of ≤ 0.997 is not significantly different from 0 (1 degree of freedom, n = 3, $\alpha = 0.05$). For n = 4 the critical value for r is 0.95 (Snedecor and Cochran, 1973, p. 557).

[12] In the case of absorption devices such as the MIRAN, the intercept of the regression line for the logarithmic transformation of the concentrations may be zero, depending on whether near-zero concentrations are included in the regression. For example, in Figure 30-14, they are not, and so the Y-intercept is actually a negative value.

Table 30-2
Weighted least squares regression analysis of data presented in Figure 30-14 for nitrous oxide calibration of a MIRAN 1A

Conc	Ln Conc	Absorption Units	Conc^2
60	4.09	0.520	3600
80	4.38	0.620	6400
100	4.61	0.715	10000
120	4.79	0.785	14400
140	4.94	0.845	19600
160	5.08	0.900	25600
180	5.19	0.950	32400
200	5.30	0.995	40000
220	5.39	1.035	48400

Weight Factor	Slope	Intercept	Standard Error Slope	r^2
None	0.399	–1.122	0.00417	0.9992
1/Conc	0.403	–1.140	0.00405	0.9993
1/Conc^2	0.406	–1.158	0.00396	0.9993

made little difference in the outcome parameters. The value of r^2 was high (0.999) from the outset; thus, little measurement error in the *Y*s occurred. In cases where the unweighted r^2 value is lower (say, 0.95), weighting would be more likely to produce different results.

For calibrating the MIRAN, one generally has to assume one of the above relationships for weighted least squares, as it is difficult to actually measure the variability in the *Y*s. In order to do this, one would need to repetitively measure each calibration standard. One way to do this would be to set up a standard, read the MIRAN response, evacuate the standard to another chamber, fill the MIRAN with clean air and then evacuate it, and then refill the MIRAN with the same standard. This is obviously impractical. An alternate approach is to take random readings from the MIRAN over a long period (say, 10–20 minutes) for each calibration standard. However, this may not be as adequate as the former technique because it utilizes the same calibration injection repeatedly. Regarding the measurement error in the *X*s, one can only make the calibration standards as accurately as possible and remember that keeping the range of *X* large reduces this error.

Combustible Gas Meter – Batch System

Most CGMs are calibrated at two points by the user, at zero and at one other value termed the *span*. There is no reason why the devices cannot be calibrated at multiple points and a calibration curve constructed. A dynamic system could even be used. However, when the two-point calibration is performed, the instrument span is adjusted for the non-zero value, and it is assumed that all other concentrations within the range give a response that is proportional to the calibration concentration. The procedure is usually the following:

1. Turn on and warm up the instrument.

2. Set to zero in hydrocarbon-free air. For all practical purposes a few ppm of hydrocarbon is inconsequential, given that the LOD of the instrument is about 0.5% LFL or a few hundred ppm.

3. Prepare or obtain a calibration standard. Generally, this is methane at 10% to 50% of its LFL. The easiest way is to purchase a span gas from the manufacturer of the instrument. Alternatively, a bag or bottle standard can be prepared.

4. Expose the sensor to the span gas. If the gas is compressed, then it should flow over the sensor. In the case of a standard prepared in a bottle, it may be possible to place the sensor in the bottle. If not, then the gas must be pumped to the sensor. The same is true for a bag standard although it may be possible to press on the bag to cause gas to flow to the sensor. The important consideration here is whether the CGM operates passively or actively, and whether the instrument is flow rate-sensitive. The instruction manual should be consulted regarding desirable flow rates.

5. Adjust the span adjustment on the instrument so that it reads the same as the standard.

6. Repeat the process; no span adjustment should be needed.

Detector Tubes – Batch Calibration or Verification

Generally speaking, the purpose is to verify that, on average, the persons using detector tubes report a reading that is equivalent to that expected from the prepared standards. If biases are present, they can be corrected for individual readers. Precision for individual readers should not exceed those values given in Chapter 29. Alternatively, a more complicated scheme that examines other factors of importance in a particular operation or plant, such as temperature, RH, and interferences, may be termed a *calibration*. Because the particular set of conditions of interest may yield a bias across all readers, an adjustment of the calibration scale on the tubes will be needed.

To perform a verification, bag standards can be made at three different concentrations spanning the range of interest. Then a multifactorial experiment can be designed. For example, if there are three readers, then each reader might read five tubes taken from each bag standard – a total of 45 tubes. Some fictitious data are shown in Table 30-3. There is no easy way to analyze the data. The questions are: 1) Are the readers biased relative to the known concentration? 2) Are the reader's precision values acceptable relative to some standard, for example, ± 35% of the OEL or those given in Chapter 29? To answer the first question, there are several approaches. The mean of each group of five replicates could be compared to the expected (known) values, using a t-test. However, such t-tests would not be very powerful given n = 5. This is shown in Table 30-4. Alternatively, each reader's biases could be compared to a value of zero (the desired bias) in a hypothesis test. If the test is insignificant, then there is no bias; otherwise, the bias is significant. Table 30-4 shows that this approach is more powerful than the previous approach. The second reader is not significantly biased (all $p \geq 0.05$) except when his three biases are pooled ($p = 0.007$).

For the second question, again the precision (*CV*) for each set of five replicates can be calculated and compared to a target such as ± 35%. In this case, a *CV* of < 0.35/1.96 would be acceptable, as the target is usually stated at the 95% confidence level. The target variance would be $(CV \times \text{concentration})^2$. This target can be compared to the estimated variance by using a variance ratio or F-test. Table 30-5 shows the variances for the three readers. As can be seen, the variance for reader 3 is constantly high although not significant. Nonetheless, the results indicate that reader 3's performance needs improvement.

Table 30-3
Five replicate detector tube readings from three readers and known concentration

Reader or Known Concentration (ppm)			
One	**Two**	**Three**	**Known**
25	25	15	25
20	25	20	25
30	30	25	25
25	30	30	25
25	35	35	25
50	55	35	50
50	55	40	50
50	65	50	50
45	50	65	50
55	50	55	50
180	200	170	200
200	200	190	200
220	220	170	200
210	230	210	200
180	220	225	200

Table 30-4
Two-tailed significance levels (*p*) for H_0: $\bar{x}_c$ = expected concentration[1] and H_0: average bias across all concentrations = 0

Reader	Expected Concentration 25	50	200	Average Bias[2]
1	1.00	1.00	0.81	1.000
2	0.09	0.14	0.08	0.007
3	1.00	0.86	0.56	0.380

[1] Expected concentration is the known or prepared concentration and $\bar{x}_C$ is the average reader result (n = 5 for each reader) at that concentration.

[2] Average bias is the average of 15 biases (3 concentrations × 5 replicates) for each reader. It is tested against a value of 0.

Table 30-5
Variance values for combinations of reader and concentration

Reader	Concentration (target variance[a]) 25 (20)	50 (40)	200 (650)
1	12.5	12.5	320
2	25[b] (80)	37.5	180
3	62.5[b] (80)	142.5[b] (160)	595

[a] Target variance is defined based on an acceptable precision of 35% at 95% confidence level. Therefore, CV = 0.35/1.96 and target variance = (CV × concentration)^2.

[b] Variance is larger than target variance but not significant at p = 0.05 level, significant variance at p = 0.05 (one-tailed) shown in parentheses.

The bag standards are prepared by filling the bags with clean air to a known volume and then adding liquid or gas through the septum. Press on the bag to mix the gases well. The concentration is calculated by using Equation 30-1. After a bag has been used, it should be purged according to previous discussions and Equation 30-3. Tubes can be exposed by connecting the open end of the tube to the port on the bag with a very short piece of tubing, preferably Teflon or polyethylene. The connection must be airtight.

The persons who conduct the reading preferably should also operate the detector tube pump. However, the experiment should be blinded to the extent possible. The readers should not know the concentrations, and if possible and in a timely fashion, the tubes exposed by each reader should be collected, tagged, and then randomized by an independent person prior to being read. Exposure of the tubes also should be done randomly across readers, or the readers should take turns to avoid any time-related effects caused by concentration decay in the bags.

A Continuous System

Most DRI can be calibrated with batch systems. These are generally easier to set up than the continuous system. Although a few DRI require continuous systems, the more important reasons for using them include the ability to vary concentration over time to simulate real environments (see Woodfin, 1984, for an advanced design) and control temperature and RH (see Dixon et al., 1984, and the discussion in Nelson, 1992). In addition, when several instruments are to be calibrated, a continuous system can be established and left in place for future recalibrations. Generally, the contaminant can be quite easily changed by replacing a bottle of gas, the contents of a syringe pump or a vapor generator, or a diffusion or permeation tube. The following example involves the use of a permeation tube to establish a toluene concentration of 10 ppm at 40°C, 1 atmosphere, and 70% RH. The overall diagram for the system is shown in Figure 30-6. The system complexity shown is not necessary, but it does allow the use of permeation tubes to establish concentrations < 10 ppm, a syringe pump for higher vapor concentrations, and a mass flow meter for metering contaminant gases into the system. A Miller-Nelson flow-temperature-humidity control system controls all aspects of the diluent gas. In this example, we shall assume that the permeation tube delivers 5.4 µg/min, and the dilution flow controller has a range of 1–10 L/min. In order to use the steady state equation of Appendix 30-1, we will first convert the desired 10 ppm concentration based on conditions of 40°C, or

$$10\ ppm = \frac{10\ \mu L_{t,V}}{1\ L_a} \times \frac{10^3\ L_a}{1\ m_a^3} \times \frac{1\ L_{t,V}}{10^6\ \mu L_{t,V}} \times \frac{298\ °K}{313\ °K} \times \frac{92\ g}{g\text{-}mole} \times \frac{10^3\ mg}{1\ g} = \frac{3.5\ mg}{m^3}$$

Thus the dilution flow rate needed is GC^{-1}, or

$$\frac{5.4\ \mu g}{\min} \times \frac{1\ mg}{10^3\ \mu g} \times \frac{1\ m^3}{35.8\ mg} \times \frac{10^3\ L}{1\ m^3} = 0.1508\ \frac{L}{\min}$$

Since our flow controller will not deliver clean air outside the range of 1–10 L/min, we have two choices. We can split the diluent flow, before the clean air enters the permeation device, by using a second mass flow controller. This second device does not require temperature and humidity control, and the excess air is simply exhausted to the room using a *T* arrangement in the tubing. Alternatively, we can use another generator, such as a vapor or syringe pump generator, that has a higher generation rate.

In cases where syringe pumps or vapor generators are used, often the opposite problem to that for the permeation tube occurs; that is, the generator produces too much contaminant for the available dilution flow, and so the contaminant has to be diluted twice (see Figure 30-6). The first dilution establishes a concentration based on the steady state equation from Appendix 30-1. The second dilution produces a concentration based on Equation 30-4.

SOURCES OF EQUIPMENT MENTIONED IN THIS CHAPTER

Aerosol Generators

Air Techniques International http://www.atitest.com/html/products/aerosolgenerators.html
BGI http://www.bgiusa.com/agc/
Grimm http://www.grimm-aerosol.com/
TSI http://www.tsi.com/Category.aspx?Cid=115

Compressed Gases

Air Products and Chemicals, Inc. http://www.airproducts.com/index.asp
AirGas http://www.airgas.com/home.aspx?cat=32
Alphagaz, Air Liquide http://www.ca.airliquide.com/en/index.asp#
Linde http://www.us.linde-gas.com/international/web/lg/us/likelgus30.nsf/docbyalias/Homepage
Matheson Trigas https://www.mathesontrigas.com/index.aspx
MESA International Technologies http://www.cal-gas.com/
Norlab http://www.norlab-gas.com/
Portagas http://www.portagas.com/
Praxair http://www.praxair.com/
Scott Specialty Gases, Inc. http://www.scottgas.com/

Permeation and Diffusion Tubes

Advanced Calibration Designs, Inc. http://www.goacd.com/acd/Product/ProductInfo.html
Gastec http://www.gastec.co.jp/english/products/
VICI Metronics http://www.vici.com/calib/calib.php

Syringe Pumps

Braintree Scientific http://www.instechlabs.com/Pumps/syringe/
Harvard Apparatus http://www.instechlabs.com/Pumps/syringe/
KD Scientific http://www.kdscientific.com/
There are many more.

Temperature, RH, and Flow Controllers

Miller-Nelson Research, Inc. http://www.millernelson.com/
Gastec http://www.gastec.co.jp/english/products/
VICI Metronics http://www.vici.com/calib/calib.php

QUESTIONS

In all questions assume NTP unless otherwise stated.

30.1. A 100 ppm concentration of heptane is to be created in a plastic bag using a 10 L volume of air. How much liquid heptane should be injected into the air? Hint: The specific gravity of heptane is 0.63 g/ml, and 100 ppm is equivalent to 100 µL/L.

30.2. A 1 ppm concentration of methyl ethyl ketone is to be created in the MIRAN lA calibration system (volume = 5.65 L). Determine how much liquid MEK (density is 0.81 g/ml) is needed. If the amount is less than 0.1 µL, then calculate the volume of saturated head space vapor that is needed. Use the data in Appendices A and B.

30.3. A permeation tube generates 2 µg/min o-xylene. What dilution flow rate is necessary to achieve a 50 ppm concentration in a continuous system?

30.4. A vapor generator produces a 10% saturated vapor of nitrobenzene at 1 L/min. What additional dilution flow rate would be necessary to achieve a 10 ppm concentration in a continuous system? (See Appendices A and B).

30.5. A methane atmosphere is created in a plastic bag by injection of 0.1 L of the gas into a bag containing 10 L of air. What concentration is created?

30.6. It is estimated that when a plastic bag is emptied by rolling it, 2% (or 0.2 L) of the 10 L volume remains. In a particular application, a 1,000 ppm concentration of ethane is created in the bag. After the bag has been emptied three times and refilled with 10 L of air, what is the concentration of ethane in the bag?

30.7. When a syringe is emptied of a liquid, why should it not be pumped?

30.8. The stopcocks at either end of a 500 mL glass bulb are opened, and a gas stream of 100 ppm CO flows through at a rate of 10 L/min. Assuming perfect mixing, what is the concentration of CO in the bulb after 1 minute?

30.9. The pressure on a compressed gas cylinder reads 1,500 lb/in^2 (10,345 kPa). The cylinder has a volume of 6 ft^3 (170 L). What is the volume of the gas at NTP? Assume that the temperature is constant.

30.10. An instrument is being calibrated in a dynamic, continuous flow system at conditions of 10 ppm and 1 L/min. The sampling volume of the instrument is 1 L. The concentration is changed by the operator to 30 ppm. All other conditions remain the same. Assuming perfect mixing and that the instrument's electronic response time is instantaneous, the only variables affecting the instrument's overall response time are the sample volume, concentration change, and the flow rate. How long should it take for the instrument to read 29.9 ppm? If the sensor senses the average concentration across the entire sampling cell, will the response time be longer or shorter if the mixing is not perfect?

REFERENCES

Altshuller, A.P.; Cohen, T.R.: Application of diffusion cells to the production of known concentrations of gaseous hydrocarbons. Anal. Chem. 32:802–810 (1960).

ASTM: Standard Practice for Calibration Techniques Using Permeation Tubes, D 3609-00. Available form ANSI, American National Standards Institute, Washington, DC http://webstore.ansi.org/ansidocstore/ (2005).

ASTM: Standard Practice for Evaluation of Performance Characteristics of Air Quality Measurement Methods with Linear Calibration Functions, D 5280-96. ASTM International. West Conshohocken, PA (2001).

Brandrup, J.; Immergut, E.H. (eds.): Polymer Handbook. Wiley Publishing, New York (1975).

Christian, S.D.; Tucker, E.E.: Least squares analysis with the microcomputer, linear relationships, Part One. Am. Lab, pp. 36-41 (Sept. 1982a).

Christian, S.D.; Tucker, E.E.: Least squares analysis with the microcomputer, nonlinear relationships, Part Two. Am. Lab, pp. 31-36, (Sept. 1982b).

Christian, S.D.; Tucker, E.E.: Least squares analysis with the microcomputer, nonlinear relationships, Part Five. Am. Lab (Feb. 1984).

Christian, S.D.; Tucker, E.E.: Linear least squares analysis with dependent and independent variables subject to error. Am. Lab, pp. 33-36 (Aug. 1986).

Cotabish, H.N.; McConnaughey, P.W.; Messer, H.C.: Making known concentrations for instrument calibration. Am. Ind. Hyg. Assoc. J. 21:392-402 (1961).

Dixon, S.W.; Vasta, J.F.; Freeland, L.T.; Calvo, D.J.; Hemingway, R.E.: A multiconcentration controlled test atmosphere system for calibration studies. Am. Ind. Hyg. Assoc. J. 45:99-104 (1984).

Draper, N.R.; Smith, H.: Applied Regression Analysis, 2nd ed. Wiley Publishing, New York (1998).

Fuller, W.A.: Measurement Error Models. Wiley Publishing, New York (1987).

Garden, J.S.; Mitchell, D.G.; Mills, W.N.: Nonconstant variance regression techniques for calibration-curve-based analysis. Anal Chern. 52:2310-2315 (1980).

Hamilton: The Care and Maintenance of Hamilton Microliter/Gastight Syringes. Hamilton Company, Reno, NV http://www.hamiltoncompany.com/newdev/syringes/pdfs/L20058_care_use_guide.pdf (2006).

Hinds, W.C.: Aerosol Technology. Wiley Publishing, New York (1999).

Ho, J.S.Y.; Schlecht, P.C.: Permeation characteristics of organic vapors in silicone rubber tubing. Am. Ind. Hyg. Assoc. J. 42:70-76 (1981).

Jayanty, R.K.M.; Albritton, J.R.; Straley, Y.H.; von Lehmden, D.J.: Stability of PPB and PPM organic cylinder gases used for calibration and audits. J. Air Waste Mgmt. Assoc. 42:1198-1203 (1992).

Leong, K.H.; Wang, H.C.; Stukel, J.J.; Hople, P.K.: An improved constant output atomizer. Am. Ind. Hyg. Assoc. J. 43:135-135 (1982).

Lewis, W.K.; Chang, K.C.: Distillation. III. The mechanism of rectification. Trans. AIChE, 21:127-138 (1928).

Lippmann, M.: Calibration of air sampling instruments, In Air Sampling Instruments, 7th ed., S.V. Hering (ed.). American Conference of Governmental Industrial Hygienists, Cincinnati, OH (1989).

Liu, B.Y.H.; Lee, K.W.: An aerosol generator of high stability. Am. Ind. Hyg. Assoc. J. 36:861-864 (1975).

Lugg, G.A.: Diffusion coefficients of some organic and other vapors in air. Anal Chem. 7:1072-1077 (1968).

Matheson: Gas Mixtures – Facts and Fables. Matheson, Lyndhurst, NJ (1978).

McKelvey, J.M.; Hoelscher, H.E.: Appartus for preparation of very dilute gas mixtures. Anal. Chem. 29:123 (1957).

Meyer, P.C.: Letters to the editor. Am. Ind. Hyg. Assoc. J. 53:A298 (1992).

Mouradian, R.F.; Flannery, J.P.: Effect of oxygen content on calibration of portable VOC instruments. Am. Ind.

Hyg. Assoc. J. 55:1084-1086 (1994).
Nelson, G.O.: Gas Mixtures: Preparation and Control. Lewis Publishers, Ann Arbor, MI (1992).
Ness, S.A.: Air Monitoring for Toxic Exposures. Van Nostrand Reinhold, New York (1991).
O'Keeffe, A.E.; Ortman, G.C.: Primary standards for trace gas analysis. Anal. Chern. 38:760-763 (1966).
Perkins, J.L.; You, M.J.: Predicting temperature effects on chemical protective clothing permeation. Am. Ind. Hyg. Assoc. J. 53:77-83 (1992).
Saltzman, B.E.: Preparation and analysis of calibrated low concentrations of sixteen toxic gases. Anal. Chem. 33:1100 -1112 (1961).
Saltzman, B.E.; Burg, W.R.; Ramaswamy, G.: Performance of permeation tubes as standard gas sources. Env. Sci. Technol. 5:1121-1128 (1971).
Scaringelli, F.P.; Frey, S.A.; Saltzman, B.E.: Evaluation of Teflon permeation tubes for use with sulfur dioxide. Am. Ind. Hyg. Assoc. J. 27:260-267 (1967).
Scaringelli, F.P.; O'Keeffe, A.E.; Rosenberg, E.; Bell, J.P.: Preparation of known concentrations of gases and vapors with permeation devices calibrated gravimetrically. Anal. Chem. 42:871-876 (1970).
Setterlind, A.N.: Preparation of known concentrations of gases and vaoprs in air. Am. Ind. Hyg. Assoc. Q. 14:113 (June 1953).
Sherwood, T.K.; Pigford, R.L.: Absorption and Extraction. McGraw-Hill, New York (1952).
Snedecor, G.W.; Cochran, W.G.: Statistical Methods. Iowa State University Press, Ames, IA, USA (1973).
Soda, R.: Calibration of a permeation tube and a diffusion tube with an electrobalance. Ind. Hlth. 18:41-54 (1980).
Winefordner, J.D.; Long, G.L.: Limit of deduction: A closer look at the IUPAC definition. Anal. Chem. 55:712A-724A (1983).
Woodfin, W.J.: A gas and vapor generation system with programmed automatic cyclic output. Am. Ind. Hyg. Assoc. J. 45:138-142 (1984).
Woolson, R.P.: Statistical Methods for the Analysis of Biomedical Data. Wiley Publishing, New York (1987).
Worthington, B.; Rey, A.R.: Generation of dynamic standard test atmospheres for aromatic compounds by using the diffusion vial method. Am. Ind. Hyg. Assoc. J. 52:464-468 (1991).

APPENDIX 30-1 – EQUATIONS DERIVED FROM MASS BALANCE CONSIDERATIONS

The mass balance equation states that the rate at which mass changes with time in a vessel is equal to the generation rate minus the removal rate, or the rates can be written as:

$$Accumulation = Generation - Removal$$

or

$$\frac{VdC}{dt} = G - QC \qquad \text{[Equation A30-1]}$$

where symbols in the differential equation are defined as volume (V), concentration (C) at any time (t), and generation rate (G). Note that when concentration is in units of mass per volume and G is in mass per time, the units in the three terms are mass per time. In order for the equation to be useful it must be integrated; then boundary conditions can be set for specific applications. In addition, the equation can be simplified based on the application. For example, in the case of purging, the generation has ceased, some concentration, C_{max} has been reached, and we want to know how long it will take to reach some new concentration, C. It is important to note that C reaches zero only at infinite time. Thus, we typically pose a question such as, "How long will it take to purge 99% of the original mass, or purge down to 1% of the original concentration?" Finally, it should be noted that the equation assumes perfect mixing. The more that mixing departs from ideality, the greater the error.

For a purge problem when $G = 0$, we have:

$$\frac{VdC}{dt} = -QC$$

and rearranging,

$$\frac{dC}{C} = -\frac{Qdt}{V}$$

Integrating both sides, we have

$$\ln C = -\frac{Qt}{V} + k \qquad \text{[Equation A30-2]}$$

where k is an integration constant. Here k can be defined by setting boundary conditions. In our problem, when $t = 0$, $C = C_{max}$. These conditions must hold true, so they can be substituted into Equation A30-2 to determine the value of k, or $k = \ln C_{max}$. Thus, substituting for k, the purge equation is

$$\ln C = -\frac{Qt}{V} + \ln C_{max}$$

Algebraically this can be rearranged to yield

$$\ln \frac{C}{C_{max}} = -\frac{Qt}{V} \qquad \textit{PURGE}$$

If one wishes to purge 99% of the original mass, then the fraction C/C_{max} is set to 0.01.

In the case of buildup caused by the flow of a contaminated airstream into a vessel, the generation rate is not zero, but it can be defined. The concentration in the vessel starts at 0, and then contaminated gas enters the vessel having a concentration of C_{max}. Thus, the generation rate (mass per time) is QC_{max}. Substituting into Equation A30-1 we have

$$\frac{VdC}{dt} = QC_{max} - QC$$

Rearranging for integration and defining $C' = C_{max} - C$, we have

$$\frac{dC'}{C} = -\frac{Qdt}{V}$$

Integrating, we have

$$\ln C' = -\frac{Qdt}{V} + k' \qquad \text{[Equation A30-3]}$$

Defining the boundary conditions, we know that when $t = 0$, $C = 0$, and because $C' = C_{max} - C$, at the boundary, $C' = C_{max}$. Upon substituting into Equation A30-3, we can derive the value of k' as $\ln C_{max}$ and, substituting once again, Equation A30-3 becomes

$$\ln C' = -\frac{Qt}{V} + \ln C_{max}$$

Substituting for C', we have

$$\ln (C_{max} - C) = -\frac{Qt}{V} + \ln C_{max}$$

Rearranging,

$$\ln \frac{C_{max} - C}{C_{max}} = -\frac{Qt}{V}$$

Algebraically it can be shown that $(x - y)/x = (x/x) - (y/x)$. By making this substitution in the argument of the natural logarithm, taking the antilog of both sides, and rearranging we have

$$\frac{C}{C_{max}} = 1 - e^{\frac{-Qt}{V}} \qquad \textit{BUILDUP}$$

This equation can be used to solve for C, or the previous equation can be used to solve for t when C is given.

The purge equation is applied in situations when the dilution air is noncontaminated. For both the purge and buildup equations the value of C can be anything less than C_{max}. However, what of the case where the *dilution* air is at nonzero concentration? For example, an instrument is sensing 100 ppm, and then the concentration changes to 50 ppm. How long will it take for the instrument to respond? In this case, the dilution air is not at 0 ppm, but at 50 ppm. The problem could be turned around so that the concentration changes from 50 ppm to 100 ppm. In this case, the term dilution does not really apply, but for the sake of consistency we shall continue to call the air coming into the vessel or the instrument the dilution air.

We need to derive a separate equation for this set of conditions. Going back to our mass balance equation (A30-1), the generation rate can be defined as the flow rate of the dilution air times its concentration. We shall call the concentration in the dilution air C_d. It may be different from the concentration that we match with time and call C. Substituting into Equation A30-l, we have

$$\frac{VdC}{dt} = QC_d - QC$$

Let us define $C - C_d$ as C'' and for purging $C_d \leq C$. Thus,

$$\frac{VdC}{dt} = QC''$$

and

$$\frac{dC}{C''} = \frac{Qdt}{V}$$

Integrating both sides, we have

$$\ln C'' = -\frac{Qt}{V} + k'' \qquad \text{[Equation A30-4]}$$

Now, regarding the boundaries, when $t = 0$, $C = C_o$ and $C'' = C_o - C_d$, where C_o is the initial concentration. Substituting these conditions into Equation A30-4, we can derive the value of k'' as $\ln (C_o - C_d)$, Now substituting for k'' in Equation A30-4, we have

$$\ln C'' = -\frac{Qt}{V} + \ln (C_o - C_d)$$

Substituting for C'' we have

$$\ln (C - C_d) = -\frac{Qt}{V} + \ln (C_0 - C_d)$$

Rearranging yields

$$\ln\left(\frac{C - C_d}{C_o - C_d}\right) = \frac{Qt}{V} \qquad \textit{PURGING, NONZERO AIR}$$

When a concentration is being reduced with air of nonzero concentration, the concentration at any time C can be determined by knowing the starting concentration, C_o and the dilution concentration, C_d.

For the case where the concentration is increasing with air of nonzero concentration, we can define $C''' = C_d - C$, where for buildup $C_d \geq C$. From Equation A30-1,

$$\frac{VdC}{dt} = QC_d - QC$$

and as defined

$$\frac{VdC}{dt} = QC'''$$

Noting that $dC''' = -dC$ and integrating,

$$\ln C'' = -\frac{Qt}{V} + k'''$$

At $t = 0$, $C = C_o$, so $k''' = \ln (C_d - C_o)$, and

$$\ln C'' = -\frac{Qt}{V} + \ln (C_d - C_o)$$

$$\ln\left(\frac{C_d - C}{C_d - C_o}\right) = -\frac{Qt}{V} \qquad \textit{BUILDUP, NONZERO AIR}$$

In some applications we have steady state conditions. For example, if clean air is moving across a source of contamination that is being generated at a constant rate, then at a point downstream where mixing is assumed to have occurred, the concentration should be constant. There is no accumulation of contaminant; so the left-hand side of the original mass balance equation (A30-1) becomes zero, and the mass generated equals the mass removed. Thus we have

$$G = QC \qquad \textit{STEADY STATE}$$

If C is in units of ppm, then Q and G must be in corresponding units (e.g., L/min and µL/min, respectively). If C is in units of mg/m^3, then G must be in units of mg/min and Q in m^3/min.

APPENDICES

EQUATIONS DERIVED FROM MASS BALANCE CONSIDERATIONS

Appendix A

PHYSICAL–CHEMICAL CONSTANTS

Chemical (CAS number)	MW	Sp Gr	P_v	Flash	LEL	UEL	BP	Freeze	IP	Notes
Acetaldehyde (75-07-0)	44.1	0.79	740	-36	4	60	69	-190	10.22	
Acetic acid (64-19-7)	60.1	1.05	11	103	4	19.9	244	62	10.66	
Acetic anhydride (108-24-7)	102.10	1.08	4	120	2.7	10.3	282	-99	10	
Acetone (67-64-1)	58.1	0.79	180		2.5	13	133	-140	9.69	
Acetone cyanohydrin (75-86-5)	85.1	0.93	0.8	165	2.2	12	203	-4	?	
Acetonitrile (75-05-8)	41.1	0.78	73	42	3	16	179	-49	12.2	
2-Acetylaminofluorene (53-96-3)	223.30									Melt pt. 381
Acetylene (74-86-2)	26.00		44.20		2.5	100	s	-119s	11.4	Dair 0.91, Sol 2%
Acetylene tetrabromide (79-27-6)	245.70	2.97	0.02				474d	32	?	Sol 0.07
Acetylsalicylic acid (50-78-2)	180.20	1.35	~0				284d			Melt pt. 275, Sol 0.3%, MEC 40
Acrolein (107-02-8)	56.1	0.84	210	-15	2.8	31	127	-126	10.13	Sol 40%
Acrylamide (79-06-1)	71.1	1.12	0.007	280	?	?	347-572d	183	9.50	Melt pt. 184, Sol @ 86°F 216%
Acrylic acid (79-10-7)	72.1	1.05	3	121	2.4	8	286	55	?	Miscible
Acrylonitrile (107-13-1)	53.1	0.81	83	30	3.00	17	171	-116	10.91	Sol 7%
Adiponitrile (111-69-3)	108.2	0.97	0.002	199	1.7	5.0	563	34	?	Sol 4.5%
Aldrin (309-00-2)	364.9	1.60	8e-05				d	219	?	Melt pt. 219, Sol 0.003%
Allyl alcohol (107-18-6)	58.1	0.85	17	70	2.5	18.0	205	-200	9.63	Miscible
Allyl chloride (107-05-1)	76.5	0.94	295	-25	2.9	11.1	113	-209	10.05	Melt pt. 210, Sol 0.4%
Allyl glycidyl ether (106-92-3)	114.2	0.97	2	135	?	?	309	-148	?	Sol 14%
Allyl propyl disulfide (2179-59-1)	148.3	@ 59° 0.93	?	?	?	?	?	5	?	Insoluble
Alpha - Alumina (1344-28-1)	101.9	4.0	0				5396			Melt pt. 3632, insoluble
Aluminum (7429-90-5)	27.0	2.70	0				4221			Melt pt. 1220, insoluble
4-Aminodiphenyl (92-67-1)	169.2	1.16	@ 227°1	?	?	?	576	127	?	Melt pt. 127, Sol slight
2-Aminopyridine (504-29-0)	94.1	?	@ 77° 0.8	154	?	?	411	136	8.00	Melt pt. 137, Sol > 100%
Amitrole (61-82-5)	84.1	1.14	< 0.000008				?		?	Melt pt. 318, Sol @ 77°, 28%
Ammonia (7664-41-7)	17.0		8.5 atm	gas	15	28	-28	-108	10.18	Dair 0.60, Sol 34%

All temperatures in °F MW – Molecular weight BP – Boiling point at 1 atmosphere Sol – Solubility in water, weight %, 68° IP – Ionization potential, eV Dair – Gas density referenced to air	Flash – Flash point closed cup, unless noted oc for open cup P_v – Vapor pressure at 68°, mm Hg UEL – Upper explosive limit in air, % by volume at room T LEL – Lower explosive limit in air, % by volume at room T MEC – Minimum explosive concentration, g/m^3 Sp Gr – Specific gravity, 68° index to water at 39.2° (4°C)	d – decomposes e – explodes s – sublimes

Chemical (CAS number)	MW	Sp Gr	P_v	Flash	LEL	UEL	BP	Freeze	IP	Notes
Ammonium sulfamate (7773-06-0)	114.1	1.77	0				320d	268	?	Melt pt. 260, Sol 200
n-Amyl acetate (628-63-7)	130.2	0.88	4	77	1.1	7.5	301	-95	?	Sol 0.2%
sec-Amyl acetate (626-38-0)	130.2	0.87	7	89	1	7.5	249	-109	?	Sol slight
Aniline and homologs (62-53-3)	93.1	1.02	0.6	158	1.3	11	363	21	7.70	Sol 4%
o-Anisidine (90-04-0)	123.2	1.10	< 0.1	244	?	?	437	41	7.44	Sol @ 77°, 1%
p-Anisidine (104-94-9)	123.2	1.07	0.006	?	?	?	475		7.44	Melt pt. 135, Sol moderate
Antimony (7440-36-0)	121.8	6.69	0				2975	1166		Insoluble
ANTU (86-88-4)	202.3	?	low				d		?	Melt pt. 388, Sol 0.06%
Arsenic (7440-38-2)	74.9	5.73	0				s	1135		Insoluble
Arsine (7784-42-1)	78.0		14.9		5.1	78	-81	-179	9.89	Dair 2.69, Sol 20%
Asbestos (1332-21-4)	varies	?	0	0			d	1112		
Azinphos-methyl (86-50-0)	317.3	1.44	8 E-9				d	163	?	Sol 0.003%
Barium dinitrate (10022-31-8)	261.4	3.24	low				d	1094	?	Sol 9%
Barium dichloride (10361-37-2)	208.2	3.86	low				2840	1765	?	Sol 38%
Benzene (71-43-2)	78.1	0.88	75	12	1.2	7.8	176	42	9.24	Sol 0.07%
Benzidine (92-87-5)	184.3	1.25	low	?	?	?	752	239	?	Sol 0.04%
Benzoyl peroxide (94-36-0)	242.2	1.33	< 1	176	?	?	d, e	217	?	Sol < 1%
Benzyl chloride (100-44-7)	126.6	1.10	1	153	1.1	?	354	-38	?	Sol 0.05%
Beryllium (7440-41-7)	9	1.85	0				4532	2349		Insoluble
Boron oxide (1303-86-2)	69.6	2.46	0				3380	842	13.50	Sol 3%
Boron trifluoride (7637-07-2)	67.8		> 50 atm				-148	-196	15.50	Dair 2.38, Sol 106%
Bromine (7726-95-6)	159.8	3.12	172				139	19	10.55	Sol 4%
Bromoform (75-25-2)	252.8	2.89	5				301	47	10.48	Sol 0.1%
1,3-Butadiene (106-99-0)	54.1	0.65	2.4 atm		2.0	12.0	24	-164	9.07	Dair 1.88, Insoluble
2-Butanone (78-93-3)	72.1	0.81	78	16	1.4	11.4	175	-123	9.54	Sol 28%
2-Butoxyethanol (111-76-2)	118.2	0.90	0.8	143	@ 200° 1.1	@ 275° 12.7	339	-107	10.00	Miscible

All temperatures in °F MW – Molecular weight BP – Boiling point at 1 atmosphere Sol – Solubility in water, weight %, 68° IP – Ionization potential, eV Dair – Gas density referenced to air	Flash – Flash point closed cup, unless noted oc for open cup P_v – Vapor pressure at 68°, mm Hg UEL – Upper explosive limit in air, % by volume at room T LEL – Lower explosive limit in air, % by volume at room T MEC – Minimum explosive concentration, g/m^3 Sp Gr – Specific gravity, 68° index to water at 39.2° (4°C)	d – decomposes e – explodes s – sublimes

Chemical (CAS number)	MW	Sp Gr	P_v	Flash	LEL	UEL	BP	Freeze	IP	Notes
n-Butyl acetate (123-86-4)	116.2	0.88	10	72	1.7	7.6	258	-107	10.00	Sol 1%
sec-Butyl acetate (105-46-4)	116.2	0.86	10	62	1.7	9.8	234	-100	9.91	Sol 0.8
tert-Butyl acetate (540-88-5)	116.2	0.87	?	72	1.5	?	208	?	?	Insoluble
n-Butyl alcohol (71-36-3)	74.1	0.81	6	84	1.4	11.2	243	-129	10.04	Sol 9%
sec-Butyl alcohol (78-92-2)	74.1	0.81	12	75	@ 212° 1.7	@ 212° 9.8	211	-175	10.10	Sol 16%
tert-Butyl alcohol (75-65-0)	74.1	0.79	@ 77° 42	52	2.4	8.0	180	78	9.70	Miscible
Butylamine (109-73-9)	73.2	0.74	82	10	1.7	9.8	172	-58	8.71	Miscible
tert-Butyl chromate (1189-85-1)	230.3	?	?	?	?	?	?	32-23	?	Sol?
n-Butyl glycidyl ether (2426-08-6)	130.2	0.91	@ 77° 3	130	?	?	327	?	?	Sol 2%
n-Butyl mercaptan (109-79-5)	90.2	0.83	35	35	?	?	209	-176	9.15	Sol 0.06%
p-tert-Butyltoluene (98-51-1)	148.3	0.86	77° 0.7	155	?	?	379	-62	8.28	Insoluble
Cadmium (7440-43-9)	112.4	8.65	0				1409	610		Insoluble
Cadmium fume (1306-19-0)	128.4		0				d	2599		Melt pt. 25.99, SG 8.15 crystalline 6.95 amorphous
Calcium arsenate (7778-44-1)	398.1	3.62	0				d			Melt pt.?
Calcium oxide (1305-78-8)	56.1	3.34	0				5162	4662		Sol reacts
Camphor (76-22-2)	152.3	0.99	0.2	150	0.6	3.5	399	345	8.76	Insoluble
Carbaryl (63-25-2)	201.2	1.23	@ 77F° 0.00004				d	293	?	Melt pt. 203, Sol 0.01%
Carbon black (1333-86-4)	12.0	1.8-2.1	0				s			Insoluble
Carbon dioxide (124-38-9)	44.0		56.5 atm				s	-108	13.77	Melt pt. (-)109s, Sol @ 77°, 0.02%, Dair 1.53
Carbon disulfide (75-15-0)	76.1	1.26	297	-22	1.3	50.0	116	-169	10.08	Sol 0.3%
Carbon monoxide (630-08-0)	28.0		> 35 atm	gas	12.5	74	-313	-337	14.01	Sol 2%, Dair 0.97
Carbon tetrachloride (56-23-5)	153.8	1.59	91				170	-9	11.47	Sol 0.05%
Chlordane (57-74-9)	409.8	@ 77°1.6	1.0 E-5				d	217-228	?	Sol 0.0001%
Chlorinated camphene (8001-35-2)	413.8	1.65	@ 77° 0.4				d	149-194	?	Sol 0.0003

All temperatures in °F MW – Molecular weight BP – Boiling point at 1 atmosphere Sol – Solubility in water, weight %, 68° IP – Ionization potential, eV Dair – Gas density referenced to air	Flash – Flash point closed cup, unless noted oc for open cup P_v – Vapor pressure at 68°, mm Hg UEL – Upper explosive limit in air, % by volume at room T LEL – Lower explosive limit in air, % by volume at room T MEC – Minimum explosive concentration, g/m^3 Sp Gr – Specific gravity, 68° index to water at 39.2° (4°C)	d – decomposes e – explodes s – sublimes

Chemical (CAS number)	MW	Sp Gr	P_v	Flash	LEL	UEL	BP	Freeze	IP	Notes
Chlorinated diphenyl oxide (55720-99-5)								Properties vary depending on the specific compound		
Chlorine (7782-50-5)	70.9		6.8 atm				-29	-150	11.48	Sol 0.7%, Dair 2.47
Chlorine dioxide (10049-04-4)	67.5	@ 32° 1.6	> 1 atm	?	?	?	52	-74	10.36	Sol @ 77°, 0.3%, Dair 2.33
Chlorine trifluoride (7790-91-2)	92.5	@ 53° 1.77	1.4 atm				53	-105	13.00	Sol reacts, Dair 3.21
Chloroacetaldehyde (107-20-0)	78.5		100			?	186		10.61	Miscible; % sol in H_2O
40%		1.19		190				-3		
alpha-Chloroacetophenone (532-27-4)	154.6	1.32	0.0050	244		?	472	134	9.44	Insoluble
Chlorobenzene (108-90-7)	112.6	1.11	9	82	1.3	9.6	270	-50	9.07	Sol 0.05%
o-Chlorobenzylidene malononitrile (2698-41-1)	188.6	?	3.0 E-5	?	?	?	590-599	203-205	?	Insoluble, MEC 25
Chlorobromomethane (74-97-5)	129.4	1.93	115				155	-124	10.77	Insoluble
Chlorodiphenyl (42%Cl) (53469-21-9)	~258	@ 77° 1.39	0.001				617-691	-2	?	Insoluble
Chlorodiphenyl (54%Cl) (11097-69-1)	~326	@ 77° 1.38	6.0 E-5				689-734	50	?	Insoluble
Chloroform (67-66-3)	119.4	1.48	160				143	-82	11.42	Sol @ 77°, 0.5%
bis-Chloromethyl ether (542-88-1)	115.0	1.32	@ 72° 30	> 66	?	?	223	-43	?	Sol reacts
Chloromethyl methyl ether (107-30-2)	80.5	1.06	@ 70° 192	32oc	?	?	138	-154	10.25	Sol reacts
1-Chloro-1-nitropropane (600-25-9)	123.6	1.21	@ 77° 6	144oc	?	?	289	?	9.9	Sol 0.5%
Chloropicrin (76-06-2)	164.4	1.66	18				234	-93	?	Sol 0.2%
beta-Chloroprene (126-99-8)	88.5	0.96	188	-4	1.9	11.3	139	-153	8.79	Sol slight
Chromium metal (7440-47-3)	52.0	7.14	~0				4788			Melt pt. 3452, Insoluble
Cobalt metal (7440-48-4)	58.9	8.92	~0				5612			Melt pt. 2719, Insoluble
Copper (7440-50-8)	63.5	8.94	0				4703			Melt pt. 198,1 Insoluble
Copper fume (7440-50-8)	79.5	6.4 as CuO	0				d			Melt pt. 1879d, Insoluble
Crag herbicide (136-78-7)	309.1	1.70	0.1				d		?	Melt pt. 473d, Sol @ 77F°, 26%

All temperatures in °F
MW – Molecular weight
BP – Boiling point at 1 atmosphere
Sol – Solubility in water, weight %, 68°
IP – Ionization potential, eV
Dair – Gas density referenced to air

Flash – Flash point closed cup, unless noted oc for open cup
P_v – Vapor pressure at 68°, mm Hg
UEL – Upper explosive limit in air, % by volume at room T
LEL – Lower explosive limit in air, % by volume at room T
MEC – Minimum explosive concentration, g/m^3
Sp Gr – Specific gravity, 68° index to water at 39.2° (4°C)

d – decomposes
e – explodes
s – sublimes

Chemical (CAS number)	MW	Sp Gr	P_v	Flash	LEL	UEL	BP	Freeze	IP	Notes
Cresol-o (1319-77-3)	108.2	1.05	@ 77° 0.29	178	@ 300° 1.4	?	376	52	8.93	Melt pt. 88, Sol 2%
Crotonaldehyde (4170-30-3)	70.1	0.87	19	45	2.1	15.5	219	-101	9.73	Sol 18%
Cumene (98-82-8)	120.2	0.86	8	96	0.9	6.5	306	-141	8.75	Insoluble
Potassium cyanide (151-50-8)	65.1	1.55	~0				2957	1173		Sol @ 77°, 72%
Sodium cyanide (143-33-9)	49.0	1.60	~0				2725	1047		Sol @ 77°, 58%
Cyclohexane (110-82-7)	84.2	0.78	78	0	1.3	8	177	44	9.88	Insoluble
Cyclohexanol (108-93-0)	100.2	0.96	1	154	?	?	322	77	10.00	Sol 4%
Cyclohexanone (108-94-1)	98.2	0.95	5	146	@ 212° 1.1	9.4	312	-49	9.14	Sol 15%
Cyclohexene (110-83-8)	82.2	0.81	67	11	?	?	181	-154	8.95	Insoluble
Cyclopentadiene (542-92-7)	66.1	0.80	400	77oc	?	?	107	-121	8.56	Insoluble
2,4-D (94-75-7)	221.0	1.57	@ 320° 0.4				d		?	Melt pt. 280, Sol 0.05
DDT (50-29-3)	354.5	0.99	2.0 E-7	162-171	?	?	230d	227	?	Insoluble
Decaborane (17702-41-9)	122.2	0.94	0.2	176	?	?	415		9.88	Melt pt. 211, Sol slight
Demeton (8065-48-3)	258.3	1.12	3e-04	113	?	?	d	< -13	?	Sol 0.01%
Diacetone alcohol (123-42-2)	116.2	0.94	1	125	1.8	6.9	334	-47	?	Miscible
Diazomethane (334-88-3)	42.1		> 1 atm	gas	?	?	-9	-229	9.00	Sol reacts, Dair 1.45
Diborane (19287-45-7)	27.7		@ 62° 39.5 atm	gas	0.8	88	-135	-265	11.38	Sol reacts, Dair 0.97
1,2-Dibromo-3-chloropropane (96-12-8)	236.4	2.05	0.8	170oc	?	?	384	43	?	Sol 0.1%
Dibutyl phosphate (107-66-4)	210.2	1.06	1	?	?	?	212	?	?	Insoluble
Dibutylphthalate (84-74-2)	278.3	1.05	0	315	@ 456° 0.5	?	644	-31	?	Sol @ 77°, 0.001
o-Dichlorobenzene (95-50-1)	147.0	1.3	1	151	2.2	9.2	357	1	9.06	Sol 0.01%
p-Dichlorobenzene (106-46-7)	147.0	1.25	1.3	150	2.5	?	345	128	8.98	Sol 0.08%
3,3'-Dichlorobenzidine (91-94-1)	253.1	?	?	?	?	?	788	271	?	Sol 59°, 0.07%
Dichlorodifluoromethane (75-71-8)	120.9		5.7 atm				-22	-252	11.75	Sol @ 77°, 0.03%, Dair 4.2

All temperatures in °F MW – Molecular weight BP – Boiling point at 1 atmosphere Sol – Solubility in water, weight %, 68° IP – Ionization potential, eV Dair – Gas density referenced to air	Flash – Flash point closed cup, unless noted oc for open cup P_v – Vapor pressure at 68°, mm Hg UEL – Upper explosive limit in air, % by volume at room T LEL – Lower explosive limit in air, % by volume at room T MEC – Minimum explosive concentration, g/m^3 Sp Gr – Specific gravity, 68° index to water at 39.2° (4°C)	d – decomposes e – explodes s – sublimes

Chemical (CAS number)	MW	Sp Gr	P_v	Flash	LEL	UEL	BP	Freeze	IP	Notes
1,3-Dichloro-5,5-dimethylhydantoin (118-52-5)	197.0	1.5	?	346	?	?	?	270	?	Sol 0.2%
1,1-Dichloroethane (75-34-3)	99.0	1.18	182	2	5.4	11.4	135	-143	11.06	Sol 0.6%
1,2-Dichloroethylene (540-59-0)	97.0	@ 77° 1.27	180-265	36-39	5.6	12.8	118-140	-57 to -115	9.65	Sol 0.4%
Dichloroethyl ether (111-44-4)	143	1.22	0.7	131	2.7	?	352	-58	?	Sol 1%
Dichloromonofluoromethane (75-43-4)	102.9		@ 70° 1.6 atm				48	-211	12.39	Sol @ 86°, 0.7%, Dair 3.57
1,1-Dichloro-1-nitroethane (594-72-9)	143.9	1.43	15	136	?	?	255	?	?	Sol 0.3%
Dichlorotetrafluoroethane (76-14-2)	170.9		@ 70° 1.9 atm				38	-137	12.20	Sol 0.01%, Dair 5.93
Dichlorvos (62-73-7)	221.0	@ 77° 1.42	0.01	>175	?	?	d	?	?	Sol 0.5%
Dieldrin (60-57-1)	380.9	1.75	@ 77° 8E-7				d		?	Melt pt. 349, Sol 0.02
Diethylamine (109-89-7)	73.1	0.71	192	-15	1.8	10.1	132	-58	8.01	Miscible
2-Diethylaminoethanol (100-37-8)	117.2	0.89	1	126	?	?	325	-94	?	Miscible
Difluorodibromomethane (75-61-6)	209.8	@ 59° 2.29	620				76	-231	11.07	Insoluble
Diglycidyl ether (2238-07-5)	130.2	1.12	@ 77° 0.09	147	?	?	500	?	?	Sol?
Diisobutyl ketone (108-83-8)	142.3	0.81	2	120	@ 200° 0.8	@ 200° 7.1	334	-43	9.04	Sol 0.05%
Diisopropylamine (108-18-9)	101.2	0.72	70	20	1.1	7.1	183	-141	7.73	Miscible
Dimethyl acetamide (127-19-5)	87.1	0.94	2	158oc	@ 212° 1.8	@ 320° 11.5	329	-4	8.81	Miscible
Dimethylamine (124-40-3)	45.1	@ 44° 0.67	1.7 atm	@ 20° 20	2.8	14.4	44	-134	8.24	Sol @ 140°, 24%, Dair 1.56
4-Dimethylaminoazobenzene (60-11-1)	225.3	?	3 E-7	?	?	?	s		?	Melt pt. 237, Sol 0.001%
Dimethylaniline (121-69-7)	121.2	0.96	1	142	?	?	378	36	7.14	Sol 2%
Dimethyl-1,2-dibromo-2,2 dichlorethyl phosphate (300-76-5)	380.8	@ 77° 1.96	2 E-4				d		?	Melt pt. 80, Insoluble

All temperatures in °F MW – Molecular weight BP – Boiling point at 1 atmosphere Sol – Solubility in water, weight %, 68° IP – Ionization potential, eV Dair – Gas density referenced to air	Flash – Flash point closed cup, unless noted oc for open cup P_v – Vapor pressure at 68°, mm Hg UEL – Upper explosive limit in air, % by volume at room T LEL – Lower explosive limit in air, % by volume at room T MEC – Minimum explosive concentration, g/m³ Sp Gr – Specific gravity, 68° index to water at 39.2° (4°C)	d – decomposes e – explodes s – sublimes

Chemical (CAS number)	MW	Sp Gr	P_v	Flash	LEL	UEL	BP	Freeze	IP	Notes
Dimethylformamide (68-12-2)	73.1	0.95	3	136	@ 212° 2.2	15.2	307	-78	9.12	Miscible
1,1-Dimethylhydrazine (57-14-7)	60.1	0.79	103	5	2	95	147	-72	8.05	Miscible
Dimethylphthalate (131-11-3)	194.2	1.19	0.01	295	@358° 0.9	?	543	42	9.64	Sol 0.4
Dimethyl sulfate (77-78-1)	126.1	1.33	0.1	182	?	?	370d	-25	?	Sol @ 64°, 3%
ortho-Dinitrobenzene (528-29-0)	168.1	1.57	?	302	?	?	606		10.71	Sol 0.05%, Melt pt. 244
meta-Dinitrobenzene (99-65-0)	168.1	1.58	?	302	?	?	572		10.43	Sol 0.02%, Melt pt. 192
para-Dinitrobenzene (100-25-4)	168.1	1.63	?	?	?	?	570		10.50	Sol 0.01%, Melt pt. 343
Dinitro-o-cresol (534-52-1)	198.1	~1.1	5 E-5				594		?	Sol 0.01%, Melt pt. 190, MEC 30
Dinitrotoluene (25321-14-6)	182.2	1.32	1	404	?	?	572		?	Insoluble, Melt pt. 158
Di-sec-octyl phthalate (117-81-7)	390.5	0.99	< 0.01	420oc	@ 474° 0.3	?	727	-58	?	Sol @ 75°, 0.00003
1,4-Dioxane (123-91-1)	88.1	1.03	29	55	2.0	22	214	53	9.13	Miscible, Class IB flam liq, F1.P < 73° and BP ≥ 100°
Diphenyl (92-52-4)	154.2	1.04	0.005	235	@ 232° 0.6	@311° 5.8	489		7.95	Melt pt. 156, Insoluble
Dipropylene glycol methyl ether (34590-94-8)	148.2	0.95	0.5	180	@ 392° 1.1	3.0	408	-112	?	Miscible
Endrin (72-20-8)	380.9	1.70	Low				d		?	Insoluble, Melt pt. 392d
Epichlorohydrin (106-89-8)	92.5	1.18	13	93	3.8	21	242	-54	10.60	Sol 1.18
EPN (2104-64-5)	323.3	@ 77° 1.27	@ 212° 0.0003				?		?	Insoluble, Melt pt. 97°
Ethanolamine (141-43-5)	61.1	1.02	0.4	186	@ 284° 3.0	23.5	339	51	8.96	Miscible
2-Ethoxyethanol (110-80-5)	90.1	0.93	4	110	@ 200° 1.7	@ 200° 15.6	275	-130	?	Miscible
2-Ethoxyethyl acetate (111-15-9)	132.2	0.98	2	124	1.7	?	313	-79	?	Sol 23%
Ethyl acetate (141-78-6)	88.1	0.9	73	24	2.0	11.5	171	-117	10.01	Sol @ 77°, 10%
Ethyl acrylate (140-88-5)	100.1	0.92	29	48	1.4	14	211	-96	10.30	Sol 2%

All temperatures in °F MW – Molecular weight BP – Boiling point at 1 atmosphere Sol – Solubility in water, weight %, 68° IP – Ionization potential, eV Dair – Gas density referenced to air	Flash – Flash point closed cup, unless noted oc for open cup P_v – Vapor pressure at 68°, mm Hg UEL – Upper explosive limit in air, % by volume at room T LEL – Lower explosive limit in air, % by volume at room T MEC – Minimum explosive concentration, g/m^3 Sp Gr – Specific gravity, 68° index to water at 39.2° (4°C)	d – decomposes e – explodes s – sublimes

Chemical (CAS number)	MW	Sp Gr	P_v	Flash	LEL	UEL	BP	Freeze	IP	Notes
Ethylamine (75-04-7)	45.1	0.69	874	1	3.5	14.0	62	-114	8.86	Miscible, Dair 1.61
Ethyl benzene (100-41-4)	106.2	0.87	7	55	0.8	6.7	277	-139	8.76	Sol 0.01%
Ethyl bromide (74-96-4)	109.0	1.46	375	< 4	6.8	8.0	101	-182	10.29	Sol 0.9%
Ethyl butyl ketone (106-35-4)	114.2	0.82	4	115oc	?	?	298	-38	9.02	Sol 1%
Ethyl chloride (75-00-3)	64.5	0.92	1000	58	3.8	15.4	54	-218	10.97	Sol 0.6%, Dair 2.23
Ethylene chlorohydrin (107-07-3)	80.5	1.2	5	140	4.9	15.9	262	-90	10.90	Miscible
Ethylenediamine (107-15-3)	60.1	0.91	11	93	@212° 2.5	@212° 12	241	47	8.60	Miscible
Ethylene dibromide (106-93-4)	187.9	2.17	12				268	50	9.45	Sol 0.4%
Ethylene dichloride (107-06-2)	99.0	1.24	64	56	6.2	16	182	-32	11.05	Sol 0.9%
Ethylene glycol dinitrate (628-96-6)	152.1	1.49	0.05	419	?	?	387	-8	?	Insoluble
Ethylenimine (151-56-4)	43.1	0.83	160	12	3.3	54.8	133	-97	9.20	Miscible
Ethylene oxide (75-21-8)	44.1	@ 50° 0.82	1.46 atm	-201	3.0	100	51	-171	10.56	Miscible, Dair 1.49
Ethyl ether (60-29-7)	74.1	0.71	440	-49	1.9	36.0	94	-177	9.53	Sol 8%
Ethyl formate (109-94-4)	74.1	0.92	200	-4	2.8	16.0	130	-113	10.61	Sol @ 64°, 9%
Ethyl mercaptan (75-08-1)	62.1	0.84	442	-55	2.8	18	95	-228	9.29	Sol 0.7
N-Ethylmorpholine (100-74-3)	115.2	0.9	6	90oc	?	?	281	-81	?	Miscible
Ethyl silicate (78-10-4)	208.3	0.93	1	99	?	?	336	-117	9.77	Sol reacts
Ferbam (14484-64-1)	416.5	?	0	?	?	?	d		7.72	Sol 0.01%, Melt pt. > 356d, MEC 55
Ferrovanadium dust (12604-58-9)	106.8	?	0				?			Insoluble, Melt pt. 2696-2768°, MEC 1.3
Sodium fluoride (7681-49-4)	42.0	2.78	0				3099			Sol 4%, Melt pt. 1819°
Sodium aluminum fluoride (15096-52-3)	209.9	2.90	0				d			Sol 0.04%, Melt pt. 1832°
Fluorine (7782-41-4)	38.0		> 1 atm				-307	-363	15.70	Sol reacts, Dair 1.31
Fluorotrichloromethane (75-69-4)	137.4	@ 75° 1.47	690				75	-168	11.77	Sol @ 75°, 0.1%, Dair 4.74
Formaldehyde (50-00-0), gas	30		> 1 atm	gas	7.0	73	-6	-134	10.88	Miscible, Dair 1.04

All temperatures in °F MW – Molecular weight BP – Boiling point at 1 atmosphere Sol – Solubility in water, weight %, 68° IP – Ionization potential, eV Dair – Gas density referenced to air	Flash – Flash point closed cup, unless noted oc for open cup P_v – Vapor pressure at 68°, mm Hg UEL – Upper explosive limit in air, % by volume at room T LEL – Lower explosive limit in air, % by volume at room T MEC – Minimum explosive concentration, g/m^3 Sp Gr – Specific gravity, 68° index to water at 39.2° (4°C)	d – decomposes e – explodes s – sublimes

Chemical (CAS number)	MW	Sp Gr	P_v	Flash	LEL	UEL	BP	Freeze	IP	Notes
Formaldehyde, liquid (Formalin)	varies	@ 77° 1.08	1	185	7	73	214	?	?	Miscible
Formic acid (64-18-6) 90% solution	46.0	1.22	35	122oc	18	57	224	20	11.1	Miscible
Furfural (98-01-1)	96.1	1.16	2	140	2.1	19.3	323	-34	9.21	Sol 8%
Furfuryl alcohol (98-00-0)	98.1	1.13	@ 77° 0.06	149	1.8	16.3	338	6	?	Miscible
Glycidol (556-52-5)	74.1	1.12	@ 77° 0.9	162	?	?	320d	-49	?	Miscible
Graphite (7782-42-5)	12.0	2-2.25	0				s			Insoluble, Melt pt. 6602s
Hafnium metal (7440-58-6)	178.5	13.31	~0				8316			Insoluble, Melt pt. 4041
Heptachlor (76-44-8)	373.4	1.66	@ 77° 0.0003				293d		?	Sol 0.0006%, Melt pt. 203
n-Heptane (142-82-5)	100.2	0.68	@ 72° 40	25	1.05	6.7	209	-131	9.90	Sol 0.0003
Hexachloroethane (67-72-1)	236.7	2.09	0.2				s		11.22	Sol @ 72°, 0.005, Melt pt. 368s
Hexachloronaphthalene (1335-87-1)	334.9	1.78	< 1				650- 730		?	Insoluble Melt pt. 279
n-Hexane (110-54-3)	86.2	0.66	124	-7	1.1	7.5	156	-219	10.18	Sol 0.002
2-Hexanone (591-78-1)	100.2	0.81	11	77	?	8	262	-71	9.34	Sol 2%
Hexone (108-10-1)	100.2	0.80	16	64	@ 200° 1.2	@ 230° 8.0	242	-120	9.30	Sol 2%
sec-Hexyl acetate (108-84-9)	144.2	0.86	3	113	?	?	297	-83	?	Sol 0.08%
Hydrazine (302-01-2)	32.1	1.01	10	99	2.9	98	236	36	8.93	Miscible
Hydrogen bromide (10035-10-6)	80.9		20 atm				-88	-124	11.62	Sol 49%, Dair 2.81
Hydrogen chloride (7647-01-0)	36.5		40.5				-121	-174	12.74	Sol @ 86°, 67%, Dair 1.27
Hydrogen cyanide (74-90-8) 96%	27.0	0.69	630	0	5.6	40.0	78	7	13.60	Miscible
Hydrogen fluoride (7664-39-3)	20.0	@ 67° 1.00	783				67	-118	15.98	Miscible, Dair 1.86
Hydrogen peroxide (7722-84-1)	34.0	1.39	@ 86° 5				286	12	10.54	Miscible

All temperatures in °F MW – Molecular weight BP – Boiling point at 1 atmosphere Sol – Solubility in water, weight %, 68° IP – Ionization potential, eV Dair – Gas density referenced to air	Flash – Flash point closed cup, unless noted oc for open cup P_v – Vapor pressure at 68°, mm Hg UEL – Upper explosive limit in air, % by volume at room T LEL – Lower explosive limit in air, % by volume at room T MEC – Minimum explosive concentration, g/m^3 Sp Gr – Specific gravity, 68° index to water at 39.2° (4°C)	d – decomposes e – explodes s – sublimes

Chemical (CAS number)	MW	Sp Gr	P_v	Flash	LEL	UEL	BP	Freeze	IP	Notes
Hydrogen selenide (7783-07-5)	81.0		@ 70° 9.5 atm	gas	?	?	-42	-87	9.88	Sol @ 77°, 0.9%, Dair 2.80
Hydrogen sulfide (7783-06-4)	34.1		17.6 atm	gas	4.0	44.0	-77	-122	10.46	Sol 44%, Dair 1.19
Hydroquinone (123-31-9)	110.1	1.33	1 E-5	3291	?	?	545		7.95	Sol 7%, Melt pt. 339
Iodine (7553-56-2)	253.8	4.93	@ 77° 0.3				365		9.31	Sol 0.01%, Melt pt. 236
Iron oxide dust/fume (1309-37-1)	159.7	5.24	~0				?			Insoluble, Melt pt. 2664
Isoamyl acetate (123-92-2)	130.2	0.87	4	77	@212° 1.0	7.5	288	-109	?	Sol 0.3%
Isoamyl alcohol, 1° (123-51-3)	88.2	@ 57° 0.81	28	109	1.2	@212° 9.0	270	-179	?	Sol @ 57°, 2%
Isoamyl alcohol, 2° (528-75-4)	88.2	0.82	1	95oc	?	?	234	?	?	Sol?
Isobutyl acetate (110-19-0)	116.2	0.87	13	64	1.3	10.5	243	-145	9.97	Sol @ 77°, 0.6%
Isobutyl alcohol (78-83-1)	74.1	0.8	9	82	@123° 1.7	@202° 10.6	227	-162	10.12	Sol 10%
Isophorone (78-59-1)	138.2	0.92	0.3	184	0.8	3.8	419	17	9.07	Sol 1%
Isopropyl acetate (108-21-4)	102.2	0.87	42	36	@100° 1.8	8	194	-92	9.95	Sol 3%
Isopropyl alcohol (67-63-0)	60.1	0.79	33	53	2.0	@200° 2.5	181	-127	10.10	Miscible
Isopropylamine (75-31-0)	59.1	0.69	460	-35oc	?	?	91	-150	8.72	Miscible
Isopropyl ether (108-20-3)	102.2	0.73	119	-18	1.4	7.9	154	-76	9.20	Sol 0.2%
Isopropyl glycidylether (4016-14-2)	116.2	0.92	@ 77° 9	92	?	?	279	?	?	Sol 19%
Ketene (463-51-4)	42.0		> 1 atm	gas	?	?	-69	-238	9.61	Sol reacts, Dair 1.45
Lead (7439-92-1)	207.2	11.34	~0				3164			Insoluble, Melt pt. 621
Lindane (58-89-9)	290.8	1.85	1 E-5				614		?	Sol 0.001%, Melt pt. 235
Lithium hydride (7580-67-8)	7.95	0.78	~0				d			Sol reacts, Melt pt. 1256
LPG (68476-85-7)	52-58		> 1 atm	gas			> -44	?	10.95	Insoluble, Dair 1.45-2.00
Propane					2.1	9.5				
Butane					1.9	8.5				

All temperatures in °F MW – Molecular weight BP – Boiling point at 1 atmosphere Sol – Solubility in water, weight %, 68° IP – Ionization potential, eV Dair – Gas density referenced to air	Flash – Flash point closed cup, unless noted oc for open cup P_v – Vapor pressure at 68°, mm Hg UEL – Upper explosive limit in air, % by volume at room T LEL – Lower explosive limit in air, % by volume at room T MEC – Minimum explosive concentration, g/m^3 Sp Gr – Specific gravity, 68° index to water at 39.2° (4°C)	d – decomposes e – explodes s – sublimes

Chemical (CAS number)	MW	Sp Gr	P_v	Flash	LEL	UEL	BP	Freeze	IP	Notes
Magnesium oxide fume (1309-48-4)	40.3	3.58	~0				6512			Sol @ 86°, 0.009%, Melt pt. 5072
Malathion (121-75-5)	330.4	1.21	4 E-5	> 325oc	?	?	140d	37	?	Sol 0.02%
Maleic anhydride (108-31-6)	98.1	1.48	0.2	218	1.4	7.1	396		9.90	Sol reacts, Melt pt. 127
Manganese and compounds (7439-96-5)	54.9	7.2	~0				3564			Insoluble, Melt pt. 2271
Mercury vapor (7439-97-6)	200.6	13.6	1.2 E-3				674	-38	?	Insoluble
Mesityl oxide (141-79-7)	98.2	@ 59° 0.86	9	87	1.4	7.2	266	-57	9.08	Sol 3%
Methoxychlor (72-43-5)	345.7	1.41	Very low	?	?	?	d		?	Sol 0.00001, Melt pt. 171
Methyl acetate (79-20-9)	74.1	0.93	173	14	3.1	16	135	-145	10.27	Sol 25%
Methyl acetylene (74-99-7)	40.1		5.2 atm	gas	1.7	?	-10	-153	10.36	Insoluble, Dair 1.41
Methyl acetylene-propadiene mix.	40.1		> 1 atm	gas	3.4	10.8	-36 to -4	-213	?	Insoluble, Dair 1.48
Methyl acrylate (96-33-3)	86.1	0.96	65	27	2.8	25	176	-106	9.90	Sol 6%
Methylal (109-87-5)	76.1	0.86	330	-26oc	2.2	13.8	111	-157	10.00	Sol 33%
Methyl alcohol (67-56-1)	32.1	0.79	96	52	6.0	36	147	-144	10.84	Miscible
Methylamine (74-89-5)	31.1	@ 13° 0.70	3.0	14	4.9	20.7	21	-136	8.97	Sol soluble, Dair 1.08
Methyl(n-amyl) ketone (110-43-0)	114.2	0.81	3	102	@151° 1.1	@250° 7.9	305	-32	9.33	Sol 0.4%
Methyl bromide (74-83-9)	95.0	@ 32° 1.73	1.9 atm	gas	10	16.0	38	-137	10.54	Sol 2%, Dair 3.36
Methyl Cellosolve	76.1	0.96	6	102	1.8	14	256	-121	9.60	Miscible
Methyl Cellosolve acetate (110-49-6)	118.1	1.01	2	120	1.7	8.2	293	-85	?	Miscible
Methyl chloride (74-87-3)	50.5		5.0 atm	gas	8.1	17.4	-12	-144	11.28	Sol 0.5%, Dair 1.78
Methyl chloroform (71-55-6)	133.4	1.34	100	?	7.5	12.5	165	-23	11.00	Sol 0.4%
Methylcyclohexane (108-87-2)	98.2	0.77	37	25	1.2	6.7	214	-196	9.85	Insoluble
Methylcyclohexanol (25639-42-3)	114.2	0.92	@ 86°2	149-158	?	?	311-356	-58	9.80	Sol 4%
o-Methylcyclohexanone (583-60-8)	112.2	0.93	1	118	?	?	325	7	?	Insoluble

All temperatures in °F MW – Molecular weight BP – Boiling point at 1 atmosphere Sol – Solubility in water, weight %, 68° IP – Ionization potential, eV Dair – Gas density referenced to air	Flash – Flash point closed cup, unless noted oc for open cup P_v – Vapor pressure at 68°, mm Hg UEL – Upper explosive limit in air, % by volume at room T LEL – Lower explosive limit in air, % by volume at room T MEC – Minimum explosive concentration, g/m^3 Sp Gr – Specific gravity, 68° index to water at 39.2° (4°C)	d – decomposes e – explodes s – sublimes

Chemical (CAS number)	MW	Sp Gr	P_v	Flash	LEL	UEL	BP	Freeze	IP	Notes
Methylene bisphenyl isocyanate (101-68-8)	250.3		@ 77° 5 E-6	390	?	?	597		?	Sol 0.2%, Melt pt. 99
solid at 77F		1.23								
liquid at 122F		1.19								
Methylene chloride (75-09-2)	84.9	1.33	350	?	13	23	104	-139	11.32	Sol 2%
Methyl formate (107-31-3)	60.1	0.98	476	-2	4.5	23	89	-148	10.82	Sol 30%
5-Methyl-3-heptanone (541-85-5)	128.2	0.82	2	138	?	?	315	-70	?	Insoluble
Methyl hydrazine (60-34-4)	46.1	@ 77° 0.87	38	17	2.5	92	190	-62	8.00	Miscible
Methyl iodide (74-88-4)	141.9	2.28	400				109	-88	9.54	Sol 1%
Methyl isobutyl carbinol (108-11-2)	102.2	0.81	3	106	1.0	5.5	271	-130	?	Sol 2%
Methyl isocyanate (624-83-9)	57.1	0.96	348	19	5.3	26	139	-49	10.67	Sol @ 59°, 10%
Methyl mercaptan (74-93-1)	48.1	@ 32° 0.9	1.7 atm	0°oc liq	3.9	21.8	43	-186	9.44	Sol 2%, Dair 1.66
Methyl methacrylate	100.1	0.94	29	50oc	1.7	8.2	214	-54	9.70	Sol 1.5%
alpha-Methyl styrene (98-83-9)	118.2	0.91	2	129	1.9	6.1	330	-10	8.35	Insoluble
Mica, < 1% quartz (12001-26-2)	~797	2.6-3.2	~0							Insoluble, Melt pt.?
Molybdenum, insoluble (7439-98-7)	95.9	10.28	~0				8717			Insoluble, Melt pt. 4752
Monomethyl aniline (100-61-8)	107.2	0.99	0.3	175	?	?	384	-71	7.32	Insoluble
Morpholine (110-91-8)	87.1	1.007	6	98oc	1.4	11.2	264	23	8.88	Miscible
Naphtha, coal tar (8030-30-6)	~110	0.89 - 0.97	< 5	100 - 109	?	?	320-428	?	?	Insoluble
Naphthalene (91-20-3)	128.2	1.15	0.08	174	0.9	5.9	424		8.12	Sol 0.003%, Melt pt. 176
alpha-Naphthylamine (134-32-7)	143.2	1.12	@ 220° 1	315	?	?	573		7.30	Sol 0.002%, Melt pt. 122
beta-Naphthylamine (91-59-8)	143.2	@ 208° 1.06	@ 226° 1	315	?	?	583		9.71	Miscible hot H_2O Melt pt. 232
Nickel carbonyl (13463-39-3)	170.7	@ 63° 1.32	315	< -4	2	?	110	-13	8.28	Sol 0.05%
Nickel metal (7440-02-0)	58.7	8.9	~0				5139			Insoluble, Melt pt. 2831
Nicotine (54-11-5)	162.2	1.01	0.08	203	0.7	4.0	482d	-110	8.01	Miscible

All temperatures in °F MW – Molecular weight BP – Boiling point at 1 atmosphere Sol – Solubility in water, weight %, 68° IP – Ionization potential, eV Dair – Gas density referenced to air	Flash – Flash point closed cup, unless noted oc for open cup P_v – Vapor pressure at 68°, mm Hg UEL – Upper explosive limit in air, % by volume at room T LEL – Lower explosive limit in air, % by volume at room T MEC – Minimum explosive concentration, g/m³ Sp Gr – Specific gravity, 68° index to water at 39.2° (4°C)	d – decomposes e – explodes s – sublimes

Chemical (CAS number)	MW	Sp Gr	P_v	Flash	LEL	UEL	BP	Freeze	IP	Notes
Nitric acid (7697-37-2)	63.0	@ 77° 1.5	48				181	-44	11.95	Miscible
Nitric oxide (10102-43-9)	30.0		34.2				-241	-263	9.27	Sol 5%, Dair 1.04
p-Nitroaniline (100-01-6)	138.1	1.42	2 E-5	390	?	?	630		8.85	Sol 0.08%, Melt pt. 295
Nitrobenzene (98-95-3)	123.1	1.20	@ 77° 0.3	190	@200° 1.8	?	411	42	9.92	Sol 0.2%
4-Nitrobiphenyl (92-93-3)	199.2	?	?	290	?	?	644		?	Insoluble, Melt pt. 237
p-Nitrochlorobenzene (100-00-5)	157.6	1.52	@ 86° 0.2	261	?	?	468	182	9.96	Sol slight
Nitroethane (79-24-3)	75.1	1.05	@ 77° 21	82	3.4		237	-130	10.88	Sol 5%
Nitrogen dioxide (10102-44-0)	46.0	@ 68° 1.44	720				70	15	9.75	Sol reacts, Dair 2.62
Nitrogen trifluoride (7783-54-2)	71.0		> 1 atm				-200	-340	12.97	Sol slight, Dair 2.46
Nitoglycerine (55-63-0)	227.1	1.60	3 E-4	e			122 -140d	56		Sol 0.1%
Nitromethane (75-52-5)	61.0	1.14	28	95	7.3		214	-20	11.08	Sol 10%
1-Nitropropane (108-03-2)	89.1	1.00	8	96	2.2	?	269	-162	10.81	Sol 1%
2-Nitropropane (79-46-9)	89.1	0.99	13	75	2.6	11.0	249	-135	10.71	Sol 2%
N-Nitrosodimethylamine (62-75-9)	74.1	1.005	3	?	?	?	306	?	8.69	Sol soluble
o-Nitrotoluene (88-72-2)	137.1	1.16	0.1	223	2.2	?	432	25	9.43	Sol 0.07%
m-Nitrotoluene (99-08-1)	137.1	1.16	0.1	223	1.6	?	450	59	9.48	Sol 0.05%
p-Nitrotoluene (99-99-0)	137.1	1.12	0.1	223	1.6	?	460		9.50	Sol 0.04, Melt pt. 126
Octachloronaphthalene (2234-13-1)	403.7	2.00	< 1				770		?	Insoluble, Melt pt. 365
Octane (111-65-9)	114.2	0.70	10	56	1.0	6.5	258	-70	9.82	Sol @ 77°, 0.00007%
Oil mist, mineral (8012-95-1)	varies	0.90	< 0.5	380oc	?	?	680	0	?	Insoluble
Osmium tetroxide (20816-12-0)	254.2	5.10	7				266		12.60	Sol @ 77°, 6%, Melt pt. 105
Oxalic acid (144-62-7)	126.1	1.90	< 0.001	?	?	?	s		?	Sol 14%, Melt pt. 215s
Oxygen difluoride (7783-41-7)	54.0		> 1 atm				-230	-371	13.11	Sol 0.02%, Dair 1.88
Ozone (10028-15-6)	48.0		> 1				-169	-315	12.52	Sol @ 32°, 0.001%, Dair 1.66
Paraquat (1910-42-5)	257.2	1.24	1 E-7				d		?	Miscible, Melt pt. 572d

All temperatures in °F MW – Molecular weight BP – Boiling point at 1 atmosphere Sol – Solubility in water, weight %, 68° IP – Ionization potential, eV Dair – Gas density referenced to air	Flash – Flash point closed cup, unless noted oc for open cup P_v – Vapor pressure at 68°, mm Hg UEL – Upper explosive limit in air, % by volume at room T LEL – Lower explosive limit in air, % by volume at room T MEC – Minimum explosive concentration, g/m³ Sp Gr – Specific gravity, 68° index to water at 39.2° (4°C)	d – decomposes e – explodes s – sublimes

Chemical (CAS number)	MW	Sp Gr	P_v	Flash	LEL	UEL	BP	Freeze	IP	Notes
Parathion (56-38-2)	291.3	1.27	4 E-5	392oc	?	?	707	43	?	Sol 0.001%
Pentaborane (19624-22-7)	63.1	0.62	171	86	0.42		140	-52	9.90	Sol reacts
Pentachloronaphthalene (1321-64-8)	300.4	1.67	< 1				636		?	Insoluble, Melt pt. 248
Pentachlorophenol (87-86-5)	266.4	1.98	1 E-4				588d			Sol 0.001%, Melt pt. 374
n-Pentane (109-66-0)	72.2	0.63	420	-57	1.5	7.8	97	-202	10.34	Sol 0.04%
2-Pentanone (107-87-9)	86.1	0.81	27	45	1.5	8.2	215	-108	9.39	Sol 6%
Perchloromethyl mercaptan (594-42-3)	185.9	1.69	3				297d			Insoluble
Perchloryl fluoride (7616-94-6)	102.5		10.5 atm				-52	-234	13.60	Sol 0.06, Dair 3.64
Petroleum distillates, naphtha (8002-05-9)	~99	0.63 -0.66	~40	-40 to -86	1.1	5.9	86 to 460	-99	?	Insoluble
Phenol (108-95-2)	94.1	1.06	0.4	175	1.8	8.6	359		8.50	Sol 77°, 9%, Melt pt. 109
p-Phenylene diamine (106-50-3)	108.2	?	< 1	312	?	?	513		6.89	Sol @ 75°, 4%, Melt pt. 295
Phenyl ether, vapor (101-84-8)	170.2	1.08	@ 77° 0.02	239	0.7	6.0	498		8.09	Insoluble, Melt pt. 82
Phenyl ether-biphenyl mix., vapor (8004-13-5)	~166	@ 77° 1.06	@ 77° 0.08	239	?	?	495	54	?	Insoluble
Phenyl glycidyl ether (122-60-1)	150.1	1.11	0.01	248	?	?	473	38	?	Sol 0.2%
Phenylhydrazine (100-63-0)	108.1	1.10	@ 77° 0.04	190	?	?	470d	67	7.64	Sol slight
Phosdrin (7786-34-7)	224.2	1.25	0.003	347oc	?	?	d		?	Miscible
trans								44		
cis								70		
Phosgene (75-44-5)	98.9	@ 32° 1.43	1.6 atm				47	-198	11.55	Sol slight, Dair 3.48
Phosphine (7803-51-2)	34.0		41.3 atm		1.79	?	-126	-209	9.96	Sol slight, Dair 1.18
Phosphoric acid (7664-38-2)	98.0	@ 77°	0.03				415		?	Miscible, Melt pt. 108
pure		1.87								
50% solution		1.33								
Phosphorous, yellow (7723-14-0)	124.0	1.82	0.03	?		?	536		?	Sol 0.0003%, Melt pt. 111
Phosphorous pentachloride (10026-13-8)	208.3	3.60	@ 132° 1				s		?	Sol reacts, Melt pt. 324s

All temperatures in °F MW – Molecular weight BP – Boiling point at 1 atmosphere Sol – Solubility in water, weight %, 68° IP – Ionization potential, eV Dair – Gas density referenced to air	Flash – Flash point closed cup, unless noted oc for open cup P_v – Vapor pressure at 68°, mm Hg UEL – Upper explosive limit in air, % by volume at room T LEL – Lower explosive limit in air, % by volume at room T MEC – Minimum explosive concentration, g/m^3 Sp Gr – Specific gravity, 68° index to water at 39.2° (4°C)	d – decomposes e – explodes s – sublimes

Chemical (CAS number)	MW	Sp Gr	P_v	Flash	LEL	UEL	BP	Freeze	IP	Notes
Phosphorous pentasulfide (1314-80-3)	222.3/ 444.6	2.09	@ 572° 1	?	?	?	957		?	Sol reacts, Melt pt. 550
Phosphorous trichloride (7719-12-2)	137.4	1.58	100				169	-170	9.91	Sol reacts
Phthalic anhydride (85-44-9)	148.1		1.5 E-3	305	1.7	10.5	563		10.00	Sol 0.6%, Melt pt. 267
flake		1.53								
molten		1.20								
Picric acid (88-89-1)	229.1	1.76	@ 383° 1	302	?	?	> 572° e		?	Sol 1%, Melt pt. 152
Pindone (83-26-1)	230.3	1.06	Very low	?	?	?	d		?	Sol @ 77°, 0.002%, Melt pt. 230
Propane (74-98-6)	44.1		@ 70° 8.4 atm		2.1	9.5	-44	-306	11.07	Sol 0.01, Dair 1.55
beta-Propiolactone (57-57-8)	72.1	1.15	@ 77° 3	165	2.9	?	323d	-28	?	Sol 37%
n-Propyl acetate (109-60-4)	102.2	0.84	25	55	@ 100° 1.7	8	215	-134	10.04	Sol 2%
n-Propyl alcohol (71-23-8)	60.1	0.81	15	72	2.2	13.7	207	-196	10.15	Miscible
Propylene dichloride (78-87-5)	113.0	1.16	40	60	3.4	14.5	206	-149	10.87	Sol 0.3%
Propylene imine (75-55-8)	57.1	0.80	112	25	?	?	152	-85	9.00	Miscible
Propylene oxide (75-56-9)	58.1	0.83	445	-35	2.3	36	94	-170	9.81	Sol 41%
n-Propyl nitrate (627-13-4)	105.1	1.07	18	68	2	100	231	-148	11.07	Sol slight
Pyrethrum (8003-34-7)	316-374	~1	low	180-190	?	?	?	?	?	Insoluble
Pyridine (110-86-1)	79.1	0.98	16	68	1.8	12.4	240	-44	9.27	Miscible
Quinone (106-51-4)	108.1	1.32	@ 77° 0.1	100-200	?	?	s		9.68	Sol slight Melt pt. 240
Rhodium, insoluble metal (7440-16-6)	102.9	12.41	~0				6741			Insoluble, Melt pt. 3571
Ronnel (299-84-3)	321.6	@ 77° 1.49	@ 77° 8 E-4				d		?	Sol @ 77°, 0.004%, Melt pt. 106
Rotenone (83-79-4)	394.4	1.27	4 E-5	?	?	?	d		?	Insoluble, Melt pt. 330
Selenium (7782-49-2)	79.0	4.28	~0				1265			Insoluble, Melt pt. 392
Selenium hexafluoride (7783-79-1)	193.0		> 1 atm				-30	-59	?	Insoluble, Dair 6.66

All temperatures in °F MW – Molecular weight BP – Boiling point at 1 atmosphere Sol – Solubility in water, weight %, 68° IP – Ionization potential, eV Dair – Gas density referenced to air	Flash – Flash point closed cup, unless noted oc for open cup P_v – Vapor pressure at 68°, mm Hg UEL – Upper explosive limit in air, % by volume at room T LEL – Lower explosive limit in air, % by volume at room T MEC – Minimum explosive concentration, g/m^3 Sp Gr – Specific gravity, 68° index to water at 39.2° (4°C)	d – decomposes e – explodes s – sublimes

Chemical (CAS number)	MW	Sp Gr	P_v	Flash	LEL	UEL	BP	Freeze	IP	Notes
Silica, amorphous (7631-86-9)	60.1	2.20	~0				4046			Insoluble, Melt pt. 3110
Silica, crystalline (14808-60-7)	60.1	2.66	~0				4046			Insoluble, Melt pt. 3110
Silver (7440-22-4)	107.9	10.49	~0				3632			Insoluble, Melt pt. 1761
Soapstone, < 1% quartz	379.3	2.7-2.8	~0				?	?		Insoluble
Sodium fluoroacetate (62-74-8)	100.0	?	low				d		?	Miscible, Melt pt. 392
Sodium hydroxide (1310-73-2)	40.0	2.13	0 (approx)				2534			Sol 111%, Melt pt. 605
Stibine (7803-52-3)	124.8		> 1 atm	gas		?	-1	-126	9.51	Sol slight, Dair 4.31
Stoddard solvent (8052-41-3)	varies	0.78	?	102-110	?	?	309-396	?	?	Insoluble
Strychnine (57-24-9)	334.4	1.36	low	?	?	?	d		?	Sol 0.02%, Melt pt. 514
Styrene (100-42-5)	104.2	0.91	5	88	0.9	6.8	293	-23	8.40	Sol 0.03%
Sulfur dioxide (7446-09-5)	64.1		3.2				14	-104	12.30	Sol 10%, Dair 2.26
Sulfuric acid (7664-93-9) 96 -98% acid	98.1	 1.84	1 E-3				554	51		Miscible
Sulfur monochloride (10025-67-9)	135.0	1.68	7	245			280	-107	9.40	Sol decomposes
Sulfur pentafluoride (5714-22-7)	254.1	@ 32° 2.08	561				84	-134	?	Insoluble, Dair 8.77
Sulfuryl fluoride (2699-79-8)	102.1		@ 70° 15.18				-68	-212	13.04	Sol @ 32°, 0.2%, Dair 3.72
2,4,5-T (93-76-5)	255.5	1.80	1 E-7	?	?	?	d		?	Sol @ 77°, 0.03, Melt pt. 307
Tantalum (7440-25-7) metal powder	180.9	 16.65 14.40	~0				9797			Insoluble, MEC < 200
TEDP (3689-24-5)	322.3	@ 77° 1.20	2 E-4	?	?	?	d	?	?	Sol 0.0007%
Tellurium (7783-80-9)	127.6	6.24	~0				1814			Insoluble, Melt pt. 842
Tellurium hexafluoride (7783-80-4)	241.6		> 1 atm				s	-36s	?	Sol decomposes, Dair 8.34
TEPP (107-49-3)	290.2	1.19	2 E-4				d	32	?	Miscible
m-Terphenyls (92-06-8)	230.3	1.23	@ 200° 0.01	375oc	?	?	689		8.01	Insoluble, Melt pt. 192
o-Terphenyls (84-15-1)	233	1.1	@ 200° 0.09	325oc	?	?	630		7.99	Insoluble, Melt pt. 136

All temperatures in °F MW – Molecular weight BP – Boiling point at 1 atmosphere Sol – Solubility in water, weight %, 68° IP – Ionization potential, eV Dair – Gas density referenced to air	Flash – Flash point closed cup, unless noted oc for open cup P_v – Vapor pressure at 68°, mm Hg UEL – Upper explosive limit in air, % by volume at room T LEL – Lower explosive limit in air, % by volume at room T MEC – Minimum explosive concentration, g/m^3 Sp Gr – Specific gravity, 68° index to water at 39.2° (4°C)	d – decomposes e – explodes s – sublimes

Chemical (CAS number)	MW	Sp Gr	P_v	Flash	LEL	UEL	BP	Freeze	IP	Notes
p-Terphenyl (92-94-4)	230.3	1.23	very low	405	?	?	761		7.78	Insoluble, Melt pt. 415
1,1,2,2-Tetrachloro-1,2-difluoroethane (76-12-0)	203.80	1.65	40				199		11.30	Sol @ 77°, 0.01%, Melt pt. 77
1,1,1,2-Tetrachloro-2,2-difluoroethane (76-11-9)	203.8	1.65	40				197		?	Sol 0.01%, Melt pt. 105F
1,1,2,2-Tetrachloroethane (79-34-5)	167.9	@ 77° 1.59	5				296	-33	11.10	Sol 0.3%
Tetrachloroethylene (127-18-4)	165.8	1.62	14				250	-2	9.32	Sol 0.02%
Tetrachloronaphthalene (1335-88-2)	265.9	1.59-1.65	< 1	410oc	?	?	599-680		?	Insoluble, Melt pt. 360
Tetraethyl lead (78-00-2)	323.5	1.65	0.2	200	1.8		228d	-202	11.10	Sol 0.00002%
Tetrahydrofuran (109-99-9)	72.1	0.89	132	6	2	11.8	151	-163	9.45	Miscible
Tetramethyl lead (75-74-1)	267.3	2.00	23	100			212d	-15	8.50	Sol 0.002%
Tetramethyl succinonitrile (3333-52-6)	136.2	1.07	?	?	?	?	s		?	Insoluble, Melt pt. 338s
Tetranitromethane (509-14-8)	196.0	1.62	8				259	57		Insoluble
Tetryl (479-45-8)	287.2	1.57	< 1	e	?	?	356-374e		?	Sol 0.02%, Melt pt. 268
Thiram (137-26-8)	240.4	1.29	8 E-6	?	?	?	d		?	Sol 0.003% Melt pt. 312
Tin, inorganic not oxides (7440-31-5)	118.7	7.28	~0				4545			Insoluble, Melt pt. 449
Titanium dioxide (13463-67-7)	79.9	4.26	0				4532-5432			Insoluble, Melt pt. 3326 - 3362
Toluene (108-88-3)	92.1	0.87	21	40	1.1	7.1	232	-139	8.82	Sol @ 74°, 0.07%
Toluene-2,4-diisocyanate (584-84-9)	174.2	1.22	@ 77° 0.01	260	0.9	9.5	484		?	Insoluble, Melt pt. 71
o-Toluidine (95-53-4)	107.2	1.01	0.3	185	?	?	392	6	7.44	Sol 2%
Tributyl phosphate (126-73-8)	266.3	0.98	@ 77° 0.004	295oc	?	?	552d	-112	?	Sol 0.6%
1,1,2-Trichloroethane (79-00-5)	133.4	1.44	19	?	6	16	237	-34	11.00	Sol 0.4%
Trichloroethylene (79-01-6)	131.4	1.46	58	?	@ 77° 8	@ 77° 10.5	189	-99	9.45	Sol @ 77°, 0.1%
Trichloronaphthalene (1321-65-9)	231.5	1.58	<1	392oc	?	?	579-669		?	Insoluble, Melt pt. 199
1,2,3-Trichloropropane (96-18-4)	147.4	1.39	3	160	@ 248° 3.2	@ 302° 12.6	314	6	?	Sol 0.1%

All temperatures in °F MW – Molecular weight BP – Boiling point at 1 atmosphere Sol – Solubility in water, weight %, 68° IP – Ionization potential, eV Dair – Gas density referenced to air	Flash – Flash point closed cup, unless noted oc for open cup P_v – Vapor pressure at 68°, mm Hg UEL – Upper explosive limit in air, % by volume at room T LEL – Lower explosive limit in air, % by volume at room T MEC – Minimum explosive concentration, g/m³ Sp Gr – Specific gravity, 68° index to water at 39.2° (4°C)	d – decomposes e – explodes s – sublimes

Chemical (CAS number)	MW	Sp Gr	P_v	Flash	LEL	UEL	BP	Freeze	IP	Notes
1,1,2-Trichloro-1,2,2-trifluoroethane (76-13-1)	187.4	@ 77° 1.56	285	?	?	?	118	-31	11.99	Sol @ 77°, 0.02%
Triethylamine (121-44-8)	101.2	0.73	54	20	1.2	8.0	193	-175	7.50	Sol 2%
Trifluorobromomethane (75-63-8)	148.9		> 1 atm				-72	-267	11.78	Sol 0.03%, Dair 5.14
2,4,6-Trinitrotoluene (118-96-7)	227.1	1.65	2 E-4	?e	?	?	464 (e)		10.59	Sol @ 77°, 0.01%, Melt pt. 176
Triorthocresyl phosphate (78-30-8)	368.4	1.20	@ 77° 2 E-5	437	?	?	770d	52	?	Sol slight
Triphenyl phosphate (115-86-6)	326.3	1.29	@ 380° 1	428	?	?	776		?	Sol @ 129°, 0.002%, Melt pt. 120
Turpentine (8006-64-2)	~136	0.86	4	95	0.8	?	309-338	-58 to -76	?	Insoluble
Uranium, insoluble compounds (7440-61-1)	238.0	19.05	~0				6895			Insoluble, Melt pt. 2097, MEC 60
Vanadium pentoxide (1314-62-1)	181.9	3.36	~0				3182d			Sol 0.8%, Melt pt. 1274
Vinyl chloride (75-01-4)	62.5		3.3 atm	gas	3.6	33.0	7	-256	9.99	Sol @ 77°, 0.1%, Dair 2.21
Vinyl toluene (25013-15-4)	118.2	0.89	1	127	0.8	11.0	339	-106	8.20	Sol 0.009%
Warfarin (81-81-2)	308.3	?	@ 71° 0.09	?	?	?	d		?	Sol 0.002, Melt pt. 322
o-Xylene (1330-73-8)	106.2	0.88	7	90	0.9	6.7	292	-13	8.56	Sol 0.02%
m-Xylene (1330-73-8)	106.2	0.86	9	82	1.1	7.0	282	-54	8.56	Sol slight
p-Xylene (1330-73-8)	106.2	0.86	9	81	1.1	7.0	281	56	8.44	Sol 0.02%
Xylidine (1300-73-8)	121.2	0.98	< 1	206 (2,3-)	1.0 ortho	?	415-439	-33	7.65 (2,4-) 7.30 (2,6-)	Sol slight
Yttrium (7440-65-5)	88.9	4.47	~0				5301			Sol in hot water Melt pt. 2732
Zinc chloride fume (7646-85-7)	136.3	@ 77° 2.91	0				1350			Sol @ 77°, 435%, Melt pt. 554
Zinc oxide fume (1314-13-2)	81.4	5.61	0				?			Sol @ 64°, 0.0004%, Melt pt. 3587
Zirconium (7440-67-7)	91.2	6.51	~0				6471			Insoluble, Melt pt. 3375

All temperatures in °F MW – Molecular weight BP – Boiling point at 1 atmosphere Sol – Solubility in water, weight %, 68° IP – Ionization potential, eV Dair – Gas density referenced to air	Flash – Flash point closed cup, unless noted oc for open cup P_v – Vapor pressure at 68°, mm Hg UEL – Upper explosive limit in air, % by volume at room T LEL – Lower explosive limit in air, % by volume at room T MEC – Minimum explosive concentration, g/m^3 Sp Gr – Specific gravity, 68° index to water at 39.2° (4°C)	d – decomposes e – explodes s – sublimes

Appendix B

VAPOR PRESSURE CONSTANTS AND TEMPERATURE RANGES[1]

---

[1] Adapted from reference 1. Instructions for use follow table.

SUBSTANCE	EQ	t MAX	t MIN	A	B	C	REF
Acetaldehyde	1	-75	-45	7.3839	1216.8	250	2
	1	-45	70	6.81089	992	230	2
Acetic acid	2	-35	10		41.689	8.502	2
	1	0	36	7.80307	1651.2	225	2
	1	36	170	7.18807	1416.7	211	2
Acetic anhydride	4	35	164	16.3982	3287.56	-75.11	4
Acetone	4	-32	77	16.6513	2940.46	-35.93	4
	1	NA	NA	7.02447	1161	224	2
Acetonitrile	4	-13	117	16.2874	2945.47	-49.15	4
	1	NA	NA	7.11988	1314.4	230	2
Acetylene	2	-140	-82		21.914	8.933	2
Acetylene tetrabromide	5	NA	NA	514.35	22.212		7
Acetylsalicylic acid	5	NA	NA	413	21.548		5
Acrolein	4	-38	87	15.9057	2606.53	-45.15	4
Acrylamide	5	NA	NA	486.15	23.568		7
Acrylic acid	1	20	70	5.65204	648.629	154.683	2
	4	42	177	16.5617	3319.18	-80.15	4
Acrylonitrile	1	-20	140	7.03855	1232.53	222.47	2
Aldrin	0.000006 mmHg @ 20 C						5
Allyl alcohol	4	-13	127	16.9066	2928.2	-85.15	4
Allyl chloride	4	-43	73	15.9772	2531.92	-47.15	4
Allyl glycidyl ether (AGE)	4	NA	NA	427.05	22.032		7
Allyl propyl disulfide							
(alpha)-Alumina	2	1840	2200		540	14.22	2
Aluminum (as oxide)	2	1840	2200		540	14.22	2
4-Aminodiphenyl	5	NA	NA	575.15	24.15		7
2-Aminoethanol	4	71	204	17.8174	3988.33	-86.93	4
2-Aminopyridine	5	NA	NA	523.15	22.46		7
3-Amino 1,2,4,-triazole (Amitrole)							
Ammonia (solid)	1	NA	NA	9.96382	1617.907	272.55	2
(liquid)	1	NA	NA	7.3605	926.132	240.17	2
	2	-127	-78		31.211	9.9974	2
Ammonium chloride	1	NA	NA	9.3557	3703.7	232.04	2
	2	100	400		83.486	10.0164	2
Ammonium sulfamate							
n-Amyl acetate	5	NA	NA	669.15	22.76		7
sec-Amyl acetate	5	NA	NA	669.15	22.76		7
Aniline	4	67	227	16.6748	3857.52	73.15	4
	1	NA	NA	7.24179	1675.5	200	2

SUBSTANCE	EQ	t MAX	t MIN	A	B	C	REF
Anisidine (ortho-liquid)	3	61	218	8.9849	2997		3
(para-solid)	5	NA	NA	519.15	23.079		9
Antimony	2	1070	1325		189	9.051	2
Antimony trioxide	3	574	1425	3.9419	2318		3
ANTU							
Argon (liquid)	1	NA	NA	6.61651	304.227	267.32	2
Arsenic (solid)	3	440	815	10.8	6947		3
(liquid)	3	800	860	6.692	2460		3
trioxide (solid)	2	100	310		111.35	12.127	2
trioxide (liquid)	2	315	490		52.12	6.513	2
Arsine (liquid)	3	-143	-62	7.575	981.5		3
	2	-55	20	-2660.8	6.1579		2
Asbestos							
Asphalt (petroleum) fumes							
Atrazine							
Azinphos-methyl							
Barium	2	930	1130		350	25.765	2
Barium sulfate							
Benomyl							
Benzene	1	-12	3	9.1064	1885.9	244.2	2
	1	8	103	6.90565	1211.033	220.79	2
Benzidine	5	NA	NA	673.15	22.991		7
p-Benzoquinone							
Benzoyl peroxide							
Benzo(a)pyrene							
Benzyl chloride	3	22	179	1.689	0.0508		3
Beryllium							
Biphenyl	1	69	271	7.24541	1998.725	202.733	2
Bismuth telluride							
Borate, tetra, sodium (anhydrous, decahydrate, pentahydrate)							
Boron oxide							
Boron tribromide	2	-40	90		33.32	7.655	2
Boron trichloride	1	NA	NA	6.18811	756.89	214	2
Boron trifluoride	3	-154	-111	9.893	1174.9		3
	3	-101	-50	813.167	7.6148		3
Bromacil							
Bromine (solid)	1	NA	NA	9.7209	2041.3	260.1	2
(liquid)	1	NA	NA	6.8778	1119.68	221.38	2
Bromine pentafluoride							
Bromoform	5	NA	NA	422.65	21.8		8
1,3-Butadiene	1	-80	-62	7.0355	998.106	245.233	2
	1	-58	15	6.84999	830.546	238.854	2
Butane(n) liquid	2	-100	12		23.45	7.395	2
	1	NA	NA	6.83029	945.9	240	2
2-Butoxyethanol	5	NA	NA	444.35	22.116		7

SUBSTANCE	EQ	t MAX	t MIN	A	B	C	REF
n-Butyl acetate	1	60	126	7.12712	1430.418	210.745	2
	4	22	162	16.1836	3151.09	-69.15	4
sec-Butyl acetate	4	16	154	16.1714	3092.83	-66.15	4
tert-Butyl acetate	3	-21	118	1.697	0.02877		3
Butyl-acrylate	5	NA	NA	418.85	21.99		7
n-Butyl alcohol	3	-1	117	9.508	2582.8		3
	4	15	131	17.216	3137.02	-94.43	4
sec-Butyl alcohol	3	-12	99	9.636	2507.9		3
	4	25	120	17.2102	3026.03	-86.65	4
tert-Butyl alcohol	3	-20	82.9	9.956	2502.7		3
	4	20	103	16.8548	2658.29	-95.50	4
Butylamine	4	-14	100	16.6085	3018.7	-48.96	4
tert-Butyl chromate							
n-Butylglycidal ether (BGE)	5	NA	NA	437.15	22.081		7
n-Butyl lactate	5	NA	NA	461.15	22.194		7
Butyl mercaptan (tert)							
1-Butanethiol	1	-2	123	6.92754	1281.018	218.1	2
2-Butanethiol	1	-13	110	6.88698	1229.904	222.02	33
o-sec-Butylphenol	3	58	228	8.477	2797		3
p-tert-Butyltoluene	5	NA	NA	465.95	20.958		7
Cadmium	2	150	320.9		109	8.5564	2
	2	500	840		99.9	7.897	2
Cadmium oxide	3	1000	1559	9.451	12027		3
Calcium carbonate, cyanamide, hydroxide, oxide, silicate, sulfate							
Camphor (solid)	2	0	180		53.559	8.799	2
(liquid)	3	182	209	7.545	2249.9		3
Caprolactam (vapor)	3	89	137	9.032	3268.33		11
(dust)							
Captafol							
Captan							
Carbaryl	0.005 mmHg @ 20 C						7
Carbofuran							
Carbon black	1	NA	NA	9.3813	27240	264	2
Carbon dioxide	1	NA	NA	9.64177	1284.07	268.432	2
Carbon disulfide	1	-10	160	6.85145	1122.5	236.46	2
Carbon monoxide	1	-210	-165	6.2402	230.274	260	2
Carbon tetrabromide	3	97	189	7.971	2349		3
Carbon tetrachloride (sol.)	2	-70	-50		34.608	8.05	2
(liquid)	3	-50	20	7.7316	1709.53		3
	1	NA	NA	6.9339	1242.43	230	2
Carbonyl chloride (phosgene)	1	-68	68	6.84297	37.64	230	2
Carbonyl fluoride	5	NA	NA	190.15	20.136		7
Catechol (pyrocatechol)	3	119	245	8.656	2993		3
Cellulose							
Cesium hydroxide							

SUBSTANCE	EQ	t MAX	t MIN	A	B	C	REF
Chlordane							
Chlorinated camphene							
Chlorinated diphenyl oxide							
Chlorine (solid)	1	NA	NA	9.70512	1444.19	267.13	2
(liquid)	1	NA	NA	6.9379	861.34	246.33	2
Chlorine dioxide	2	-59	11		27.26	7.893	2
(liquid)	1	NA	NA	6.03611	590.09	176.15	2
Chlorine trifluoride (liquid)	1	NA	NA	7.36685	1096.28	232.63	2
Chloroacetaldehyde	5	NA	NA	358.15	22.275		7
Chloroacetophenone	3	-5	55	13.779	4740		11
Chloroacetyl chloride	5	NA	NA	378.15	21.571		7
Chlorobenzene	2	-35	-15		42.25	8.5	2
	1	0	42	7.1069	1500	224	2
	1	42	230	6.94504	1413.12	216	2
o-Chlorobenzylidene-malononitrile	5	NA	NA	613.15	22.364		7
Chlorobromomethane	1	16	68	6.49606	942.267	192.587	2
2-Chloro-1,3-Butadiene	1	20	60	6.1615	783.45	179.7	2
Chlorodifluoromethane	4	-48	-33	15.5602	1704.8	-41.3	4
Chlorodiphenyl (42% Cl)	0.00008 mmHg @ 20 C						7
Chlorodiphenyl (54% Cl)	0.000003 mmHg @ 20 C						7
1-Chloro-2,3-epoxy propane (epichlorhydrin)	5	NA	NA	388.35	21.83		7
2-Chloroethanol (ethylene chlorohydrin)	3	-4	128	8.732	2348		3
Chloroethylene	1	-65	-13	6.89117	905.01	239.48	2
Chloroform	1	-30	150	6.90328	1163.03	227.4	2
bis-Chloromethyl ether							
Chloromethyl methyl ether	3	16	66	7.871	1669.85		11
1-Chloro-1-nitropropane	5	NA	NA	443.75	22.322		7
Chloropentafluoroethane	4	-98	43	15.7343	1848.9	-30.88	4
Chloropicrin	3	-25.5	111.9	8.079	1993.25		11
(beta)-Chloroprene as chloroprene	1	20	60	6.1615	783.45	179.7	2
o-Chlorostyrene	1	NA	NA	6.86644	1541.1	198	2
o-Chlorotoluene	1	0	65	7.36797	1735.8	230	2
	1	65	220	6.94763	1497.2	209	2
2-Chloro-6-(trichloromethyl) pyridine							
Chlorpyrifos							
Chromium metal	3	1610	2840	7.695	13955		3
Chromyl chloride	3	-18	117	8.3508	2115		3
Chrysene	5	NA	NA	721.15	23.136		7
Clopidol							
Coal dust							
Coal tar pitch volatiles							
Cobalt (as Co)	2	NA	NA		309	7.571	2
Cobalt carbonyl (as Co)							

SUBSTANCE	EQ	t MAX	t MIN	A	B	C	REF
Cobalt hydrocarbonyl							
Copper (as Cu)	2	2100	2310		468	12.344	2
Cotton dust (raw)							
Cresol (o)	1	120	191	6.9117	1435.5	165.16	2
Cresol (m)	1	150	201	7.508	1856.36	199.07	2
Cresol (p)	1	128	202	7.03508	1511.08	161.85	2
Crotonaldehyde	5	NA	NA	375.15	22.17		7
Crufomate	5	NA	NA	390.15	21.841		7
Cumene (O)	1	10	90	7.14087	1687	205.81	2
(O)	1	90	270	6.99207	1599.61	198	2
(P)	1	NA	NA	6.98784	1623.51	197	2
Cyanamide	5	NA	NA	533.15	22.5		8
Cyanogen (solid)	2	-72	-28		32.437	9.6539	2
(liquid)	2	-32	-6		23.75	7.808	2
Cyanogen chloride	2	-5	40		27.1	7.84	2
Cyclohexane	1	-50	200	6.84498	1203.526	322.863	2
Cyclohexanol	1	94	161	6.2553	912.87	109.13	2
Cyclohexanone	1	84	203	6.88673	476.7	209.83	2
Cyclohexene	1	NA	NA	6.88617	1229.973	224	2
Cyclohexylamine	1	61	128	6.68954	1229.42	188.8	2
Cyclonite	3	45	105	12.32	6160		11
1,3-Cyclopentadiene	5	NA	NA	315.65	21.4		7
Cyclopentane	3	-68	49	7.942	1621.6		3
	1	-40	72	6.88676	1124.162	231.361	2
Cyhexatin							
2,4-D							
DDT	3	40	99	15.517	6572.46		11
Decaborane							
Demeton	0.001 mmHg at 33 C						7
Diacetone alcohol	1	28	115	8.50242	2400.56	263.79	2
1,2-Diaminoethane	4	19	152	16.482	3108.49	-72.15	4
Diatomaceous earth							
Diazinon							
Diazomethane	5	NA	NA	250.15	20.91		7
Diborane	3	-159	-87	7.408	829		10
1,2-Dibromoethane	1	52	131	6.72148	1280.82	201.75	2
2-N-Dibutylaminoethanol	5	NA	NA	497.15	21.298		7
Dibutyl phosphate							
Dibutyl phthalate	1	126	202	6.6398	1744.2	113.69	2
Dichloroacetylene							
Dichlorobenzene (O)	1	NA	NA	6.924	1538.3	200	2
(M)	1	NA	NA	6.88045	1496.2	201	2
(P)	1	NA	NA	6.89797	1507.3	201	2
3,3'-Dichlorobenzidine							
Dichlorodifluoromethane	5	NA	NA	243.35	19.864		7

SUBSTANCE	EQ	t MAX	t MIN	A	B	C	REF
1,3-Dichloro-5,5-dimethyl hydantoin							
1,1-Dichloroethane	1	-39	18	6.977	1174.02	229.06	2
(liquid)	2	0	30		31.706	7.909	2
1,2-Dichloroethane	1	-31	99	7.0253	1271.3	222.9	2
1,1-Dichloroethylene	1	-28	32	6.9722	1099.4	237.2	2
1,2-Dichloroethylene (cis)	1	0	84	7.0223	1205.4	230.6	2
(trans)	1	-38	85	6.9651	1141.9	231.9	2
Dichloroethyl ether	5	NA	NA	451.65	21.31		7
	3	14	183	8.454	2505.89		11
Dichlorofluoromethane	3	-91	28	8.042	1453.3		3
Dichloromethane	1	-40	40	7.4092	1325.9	252.6	2
1,1-Dichloro-1-nitroethane	4	NA	NA	397.15	22.086		7
1,2-Dichloropropane	1	45	96	6.987	1308.1	222.8	2
Dichloropropene	5	NA	NA	377.15	21.565		7
2,2-Dichloropropionic acid	4	NA	NA	458.15	22.18		7
Dichlorotetrafluoro-ethane	5	NA	NA	276.7	20.122		7
Dichlorvos	3	10	50	10.203	3552.56		11
Dicrotophos	5	NA	NA	673.15	22.99		7
Dicyclopentadiene	3	48	166	7.9365	2224.68		3
Dicyclopentadienyl iron	3	183	253	7.615	2470		11
Dieldrin							
Diethanolamine	1	194	241	8.1388	2327.9	174.4	2
Diethylamine	1	-30	100	6.83188	1057.2	212	2
Diethylaminoethanol	5	NA	NA	434.15	22.067		7
Diethylene triamine	5	NA	NA	479.85	22.278		7
Diethyl ether	1	-61	20	6.92032	1064.07	28.5	2
Diethyl ketone	1	NA	NA	6.85791	1216.3	204	2
Diethyl phthlate	5	NA	NA	571	22.64		7
Difluorodibromomethane	5	NA	NA	297.65	20.268		7
Diglycidal ether	5	NA	NA	558.15	22.596		7
Dihydroxybenzene	5	NA	NA	559.15	26.22		8
Diisobutyl ketone	5	NA	NA	441.25	21.684		7
Diisopropylamine	5	NA	NA	357.25	21.861		7
Dimethoxymethane	5	NA	NA	315.45	21.394		7
Dimethyl acetamide	5	NA	NA	439.15	22.09		7
Dimethylamine	1	-80	-30	7.42061	1085.7	233	2
	1	-30	65	7.18553	1008.4	227.353	2
Dimethylaminobenzene	1	NA	NA	7.1295	1675.84	201	2
Dimethylaniline	1	NA	NA	7.12954	1675.84	201	2
Dimethylbenzene (o)	3	-3.8	144.4	8.1426	2185.1		3
(Xylene) (m)	3	-6.9	139.1	7.9553	2123.89		3
(p)	3	-8.1	138.3	8.1224	2144.83		3
Dimethyl carbamoyl chloride	5	NA	NA	438.15	22.086		7
Dimethyl-1,2-dibromo-2-dichloroethyl phosphate							
Dimethylformamide	5	NA	NA	425.95	22.026		8

SUBSTANCE	EQ	t MAX	t MIN	A	B	C	REF
2,6-Dimethyl-4-heptanone	5	NA	NA	441.25	22.101		7
1,1-Dimethylhydrazine	5	NA	NA	336.15	21.528		7
Dimethylphthalate	3	101	283	8.745	3259.9		10
Dimethyl sulfate	5	NA	NA	461.15	20.938		7
Dinitolmide							
Dinitrobenzene (o)	5	NA	NA	592.15	22.935		7
(m)	5	NA	NA	575.95	22.876		7
(p)	5	NA	NA	572.15	22.862		7
Dinitro-o-cresol	3	7	60	13.596	5183.76		11
3,5-Dinitro-o-toluamide							
Dinitrotoluene (2,4)	5	NA	NA	573.15	22.866		10
Dioxane	3	-35	101	7.866	1867.49		3
Dioxathion							
Diphenyl	1	69	271	7.24541	1998.725	202.783	2
Diphenylamine	2	278	284		57.35	8.088	2
Diphenylmethane diisocyanate							
Dipropylene glycol methyl ether	5	NA	NA	462.15	22.199		7
Dipropyl ketone	5	NA	NA	416.85	21.98		7
Diquat							
Di-sec, octyl phthalate							
Disulfiram							
Disulfoton							
2,6-Di-tert-butyl-p-cresol							
Diuron	2.0 e-7 torr @ 30 C						7
Divinyl benzene	5	NA	NA	472.65	22.245		7
Emery							
Endosulfan							
Endrin							
Enflurane	5	NA	NA	329.65	21.486		9
Epichlorhydrin	3	-16	117	8.396	2148.19		3
EPN	0.0003 mmHg @ 100 C						7
1,2-Epoxypropane	5	NA	NA	307.05	21.539		8
2,3-Epoxy-1-propanol	5	NA	NA	435.15	22.072		7
Ethane	1	NA	NA	6.80266	656.4	256	2
Ethanolamine	4	71	204	17.8174	3988.33	-86.93	4
Ethion							
2-Ethoxyethanol	5	NA	NA	408.75	21.94		7
2-Ethoxyethyl acetate	5	NA	NA	429.45	22.044		7
Ethyl acetate	4	-13	112	16.1516	2790.5	-57.15	4
Ethyl acrylate	4	1	136	16.089	2794.94	-58.15	4
Ethyl alcohol (ethanol)	1	NA	NA	8.04494	1554.3	222.65	2
Ethylamine	1	-70	-20	7.09137	1019.7	225	2
	1	-20	90	7.05413	987.31	220	2
Ethyl amyl ketone	5	NA	NA	430.15	22.047		7
Ethyl benzene	1	NA	NA	6.95719	1424.255	213.206	2

SUBSTANCE	EQ	t MAX	t MIN	A	B	C	REF
Ethyl bromide	1	-50	130	6.89285	1083.8	231.7	2
Ethyl butyl ketone	3	19	150	9.3119	2727.47		3
Ethyl chloride	1	-65	70	6.8027	649.62	230	2
Ethylene	1	NA	NA	6.74756	585	255	2
	3	-4	128	8.732	2348		3
Ethylene chlorohydrin	3	-4	128	8.732	2348		3
Ethylenediamine	4	19	152	16.4082	3108.49	-72.15	4
Ethylene dibromide	2	10	150		38.082	7.792	2
Ethylene dichloride	3	-58	-59	8.174	1749.56		3
Ethylene glycol	1	25	112	8.2621	2197	212	2
	1	112	340	7.8808	1957	193.8	2
Ethylene glycol dinitrate	3	15	45	11.662	3820.76		11
Ethylene glycol methyl ether acetate	5	NA	NA	418.15	21.988		7
Ethylene oxide	1	-70	100	7.40783	1181.31	250.6	2
Ethylenimine	5	NA	NA	330.15	21.49		7
Ethyl ether	1	NA	NA	6.78574	994.195	220	2
	1	-61	20	6.92032	1064.07	228.8	2
Ethyl formate	1	-30	235	7.117	1176.6	223.4	2
Ethylidene chloride	4	-33	100	16.1764	2927.17	-50.22	4
Ethylidene norbornene							
Ethyl mercaptan	1	-40	100	6.95206	1084.531	231.385	2
Ethylmorpholine	5	NA	NA	411.15	21.952		7
Ethyl silicate	5	NA	NA	441.25	22.101		7
Fenamiphos							
Fensulfothion	5	NA	NA	411.15	20.71		7
Fenthion	5	NA	NA	378.15	20.54		7
Ferbam							
Ferrovanadium							
Fluorine	4	-214	-182	15.67	714.1	-6.0	4
Fluorotrichloromethane	4	-33	27	15.8516	2401.61	-36.3	4
Fonofos							
Formaldehyde	4	-88	-2	16.4775	2204.13	-30.15	4
Formamide	3	71	210	9.9238	3412.97		3
Formic acid	1	NA	NA	69.4459	1295.26	218	2
Furfural	3	19	151	8.9697	2616.95		3
Furfuryl alcohol	3	32	170	9.2155	2807.65		3
Gasoline							
Germanium tetrahydride	3	-163	-89	7.1623	788.31		3
Glutaraldehyde	3	28	187	8.35	2511.08		3
Glycerine (glycerol)	1	167	327	17.2392	4487.04	-140.2	2
Glycidol	5	NA	NA	435.5	22.072		7
Glycol monoethyl ether	5	NA	NA	408.75	21.94		7
Grain dust							
Graphite							
Gypsum							

SUBSTANCE	EQ	t MAX	t MIN	A	B	C	REF
Hafnium							
Halothane							
Helium	1	-269.3	-268.7	12.2514	33.7329	1.79	2
Heptachlor	1	NA	NA	16.1313	282.126	290	2
n-Heptane	1	NA	NA	6.9024	1268.115	216.9	2
2-Heptanone	5	NA	NA	423.75	21.808		7
3-Heptanone	3	19	150	9.3119	2727.47		3
Hexachlorobutadiene	5	NA	NA	483.15	22.082		7
Hexachlorocyclopentadiene	5	NA	NA	512.15	22.415		7
Hexachloroethane (solid)	3	33	185	7.9684	2300.32		3
Hexachloronaphthalene							
Hexafluoroacetone	5	NA	NA	246.15	20.872		7
Hexamethylene diisocyanate	3	120	180	10.229	3321.88		11
Hexamethyl phosphoramide	5	NA	NA	506.15	22.79		8
n-Hexane	2	-10	90		31.679	7.724	2
	1	NA	NA	6.87776	1171.53	224.366	2
2-Hexanone	3	8	127	9.6392	2700.43		3
Hexone	4	12	152	15.7165	2893.66	-70.75	4
Hexylene glycol	5	NA	NA	471.45	22.24		7
Hydrazine	1	-10	39	8.2623	1881.6	238	2
	1	39	250	7.77306	1620	218	2
Hydrogen							
Hydrogenated Terphenyls							
Hydrogen bromide (solid)	1	-120	-87	8.4622	1112.4	270	2
(liquid)	1	-120	-60	6.88059	732.68	250	2
Hydrogen chloride (solid)	2	-158	-110		19.588	8.443	2
(solid)	1	-127	-60	7.06145	710.584	255	2
Hydrogen cyanide	1	-85	-40	7.80196	1425	265	2
	1	-40	70	7.29761	1206.79	247.532	2
Hydrogen fluoride	1	-55	105	8.38036	1952.55	335.52	2
Hydrogen peroxide	2	10	90		48.53	8.853	33
Hydrogen selenide	2	-66	-26		20.21	7.4311	2
Hydrogen sulfide	2	-110	-83		20.69	7.88	2
	4	-83	-43	16.104	1768.69	-26.06	4
Hydroquinone	3	133	286	10.375	4139.4		3
4-Hydroxy-4-methyl-2-pentanone	1	28	115	8.50242	2400.56	263.79	2
2-Hydroxypropyl acrylate							
Indene	3	16.4	181.6	7.927	2297.2		3
Indium							
Iodine (solid)	1	NA	NA	7.26304	1697.87	204	2
	3	40	216	8.9226	2754.16		3
Iodoform							
Iron Oxide fume							
Iron pentacarbonyl	3	4.6	105	8.076	1965.3		3
Iron salts (soluble as Fe)							

SUBSTANCE	EQ	t MAX	t MIN	A	B	C	REF
Iron(+3) chloride	3	194	319	14.0361	6580.13		3
Iron(+2) chloride	3	700	1026	8.49	7265.35		3
Isoamyl acetate	3	0	142	8.4282	2295.57		3
Isoamyl alcohol	3	10	130.6	9.667	2726.37		3
Isobutyl acetate	3	-21.2	118	8.1369	2042.39		3
Isobutyl alcohol	3	-9	127.3	9.3646	2468.45	-100.3	3
Isooctyl alcohol	3	32.8	178.5	8.9436	2730.68		3
Isophorone	3	20	215.2	7.9515	2467.83		3
Isophorone diisocyanate							
Isopropoxyethanol							
Isopropyl acetate	3	-38.3	89	8.2133	1923.26		3
Isopropyl alcohol	3	-26.1	82.5	9.4339	2327.34		3
	4	0	101	18.6929	3640.2	-53.54	4
Isopropylamine	4	-34	64	16.3637	2582.35	-40.15	4
N-Isopropylaniline	5	NA	NA	479.15	23.535		7
Isopropyl ether	3	-57	67.5	7.9087	1702.26		3
Isopropyl glycidal ether	5	NA	NA	400.15	21.895		7
Kaolin							
Ketene	5	NA	NA	217.15	22.163		7
Lead (inorganic)	2	525	1325		188.5	7.827	2
Lead arsenate, chromate							
Lindane	3	30	100	13.163	5180.29		11
Lithium Hydride							
LPG							
Magnesite							
Magnesium oxide fume							
Malathion	0.00004 torr @ NTP						7
Maleic anhydride	3	44	202	8.5958	2702.57		3
Manganese	2	1510	1900		267	9.3	2
Manganese cyclopentadienyl tricarbonyl							
Marble/calcium carbonate							
Mercury	3	2	40	8.116	3235.5		2
Mesityl oxide	3	-8.7	130	8.3921	2213.63		3
Methacrylic acid	3	25.5	161	9.22	2745.89		3
Methane (solid)	1	-194	-184	7.6954	532.2	275	2
(liquid)	1	-174	-163	6.61184	389.93	266	2
Methanethiol	4	-73	27	16.1909	2338.38	-34.44	4
Methomyl							
Methoxychlor							
2-Methoxyethanol	3	-13.5	124.4	7.1001	1716.4		3
2-Methoxyethyl acetate	5	NA	NA	418.15	21.988		7
4-Methoxyphenol	5	NA	NA	516.15	22.855		7
Methyl acetate	1	NA	NA	7.20211	1232.83	228	2
	4	-28	87	16.1295	2601.92	-56.15	4
Methyl acetylene	4	-90	-6	15.6227	1850.66	-44.07	4

SUBSTANCE	EQ	t MAX	t MIN	A	B	C	REF
MAPP							
Methyl acrylate	4	-13	117	16.1088	2788.43	-59.15	4
Methylacrylonitrile	3	-44.5	90.3	7.7909	1775.34		3
Methylal	4	-3	42	15.8237	2415.92	-52.58	4
Methyl alcohol (methanol)	1	-20	140	7.87863	1473.11	230	2
Methylamine	1	-93	-45	6.91831	883.054	223.112	2
	1	-45	50	6.91205	838.116	214.237	2
Methyl amyl alcohol	5	NA	NA	404.95	21.92		7
Methyl n-amyl ketone	3	19	150	9.3119	2727.47		3
n-Methyl aniline	1	NA	NA	7.22584	1728.17	202	2
	4	47	207	16.3066	3756.28	-80.71	4
Methyl bromide	4	-58	53	16.0252	2271.71	-34.83	4
Methyl n-butyl ketone	3	7.7	127.5	9.639	2700.4		3
Methyl chloride	2	-49	-10		21.988	7.481	2
Methyl chloride	4	-93	-7	16.1052	2077.97	-29.55	4
Methyl chloroform	3	-52	74.1	7.9575	1752.95		3
Methyl 2-cyanoacrylate	5	NA	NA	321.15	21.432		7
Methylcyclohexane	1	0	210	6.82689	1272.864	221.63	2
Methylcyclohexanol	5	NA	NA	428.15	22.038		7
o-Methylcyclohexanone	5	NA	NA	433.15	274.21		7
2-Methylcyclopentadienyl manganese tricarbonyl							
Methyl demeton							
Methylene bisphenyl isocyanate (MDI)	5	NA	NA	467.15	25.22		10
Methylene chloride	4	-44	59	16.3029	2622.44	-41.7	4
4,4'-Methylene bis (2-chloroaniline)							
Methylene bis (4-cyclo-hexylisocyanate)							
4,4-Methylene dianiline	5	NA	NA	671.15	22.984		7
Methyl ethyl ketone (MEK)	1	NA	NA	6.97421	1209.6	216	2
	4	-16	103	16.5986	3150.42	-36.65	4
Methyl ethyl ketone peroxide							
Methyl formate	1	NA	NA	7.13623	1111	229.2	2
	4	-48	51	16.5104	2590.87	-4260	4
5-Methyl-3-heptanone	5	NA	NA	446.15	22.124		7
Methyl hydrazine	4	-3	127	15.1424	2319.84	-91.7	4
Methyl iodide	4	-13	52	16.0905	2639.55	-36.5	4
Methyl isoamyl ketone	5	NA	NA	417.15	21.983		7
Methyl isobutyl carbinol	3	-0.3	131.7	8.338	2404.32		3
Methyl isobutyl ketone	4	12	152	15.7165	2893.66	-70.75	4
Methyl isocyanate	4	-43	67	16.3258	2840.37	-56.31	4
Methyl isopropyl ketone	4	-2	133	14.1779	1993.12	-103.2	4
Methyl mercaptan	4	-73	27	16.1909	2338.38	-34.44	4
Methyl methacrylate	3	-30.5	101	8.1763	1975.4		3
Methyl parathion							
Methyl propyl ketone	3	-12	103.3	9.4323	2457.43		3

SUBSTANCE	EQ	t MAX	t MIN	A	B	C	REF
	4	2	137	16.0031	2934.87	-62.25	4
Methyl silicate	3	-130	-60	2903.2	17.3109	r = -.9	3
alpha-Methyl styrene	3	NA	NA	6.92366	1486.88		3
	3	7.4	165.4	7.9761	2237.16		3
Metribuzin							
Mevinphos							
MICA							
Mineral Wool Fiber							
Molybdenum (as Mo)	2	1800	2240		680	10.844	2
Monocrotophos	5	NA	NA	398.15	21.88		7
Morpholine	1	0	44	7.71813	1748.8	235	2
	1	44	170	7.1603	1447.7	210	2
Naled							
Naphthalene (solid)	2	0	80		71.401	11.45	2
(liquid)	2	120	200		47.362	7.927	2
beta-Naphthylamine	3	108	246.5	9.2238	3527.73		3
Neon	1	NA	NA	7.57352	183.34	285	2
Nickel carbonyl	3	-23	42.5	7.7449	1536.05		3
Nickel (metal)	3	1810	2732	9.452	19693.5		10
Nickel chloride	3	671	987	11.5212	10864.6		3
Nickel sulfide							
Nicotine	3	61.8	247.3	8.0939	2703.66		3
Nitrapyrin							
Nitric acid	3	0	90	8.4671	1992.37		3
Nitric oxide	2	-200	-161		16.423	10.048	2
	2	-163	-148		13.04	8.44	2
p-Nitroaniline	3	142.4	336	9.0455	3763.46		3
Nitrobenzene (liquid)	3	44.4	210.6	8.4085	2658.65		3
p-Nitrochlorobenzene	5	NA	NA	515.15	22.639		7
4-Nitrodiphenyl	5	NA	NA	603.15	22.974		7
Nitroethane	4	-58	43	17.0073	2618.73	-37.3	4
Nitrogen dioxide	4	-43	47	20.5324	4141.29	3.65	4
Nitrogen trifluoride	3	-170.7	-129	7.5163	667.51		3
Nitroglycerin	3	20	235	7.54	3016.8		3
Nitromethane	3	-29	101.2	8.294	2018.24	8.033	3
1-Nitropropane	3	-9.6	131.6	7.9806	2059.98		3
2-Nitropropane	3	-18.8	120.3	8.1666	2072.79		3
N-Nitrosodimethylamine	3	3.2	153	8.2005	2259.6		3
Nitrotoluene (o)	2	50	225		48.114	7.9728	2
(m)	2	55	235		50.128	8.0655	2
(p)	2	80	240		49.95	7.9815	2
Nitrotrichloromethane	3	-33.8	111.9	6.8693	1543.33		3
Nonane	3	1.4	150.8	8.1965	2243.79		3
Octachloronaphthalene	4	39	179	15.9671	3291.45	-71.33	4
Octane	3	-14	125	8.2859	2137.46		3

SUBSTANCE	EQ	t MAX	t MIN	A	B	C	REF
	4	19	152	15.9426	3120.29	-63.63	4
Oil, Mist, Mineral							
Osmium tetroxide (white)	3	-5.6	130	8.5785	2278.06		3
(yellow)	3	3.2	130	9.1887	2505.66		3
Oxalic acid (solid)	2	55	105		90.5026	12.2229	2
Oxygen difluoride							
Ozone	3	-180.4	-111.9	6.789	632.62		3
Paraffin, wax fume							
Paraquat							
Parathion	5	NA	NA	648.15	22.911		7
Pentaborane	3	-40	-158	8.029	1704.75		10
Pentachloronaphthalene							
Pentachlorophenol	3	211.2	309.3	9.2192	3694.6		3
Pentaerythritol	3	96	145	12.592	6059.67	r = -0.83	11
Pentane	2	-20	50		27.691	7.558	2
2-Pentanone	4	2	1137	16.0031	2934.87	-62.25	4
Perchloroethylene	3	-20.6	120.8	8.0331	2022.44		3
	4	34	187	16.1642	3259.29	-52.15	4
Perchloromethyl mercaptan							
Perchloryl fluoride	5	NA	NA	319.95	21.424		7
Perlite							
Phenol	3	40	181.9	9.2749	2884.8		3
	4	72	208	16.4279	3490.89	-98.59	4
Phenothiazine	5	NA	NA	644.15	22.898		7
N-Phenyl-beta-naphthalamine	5	NA	NA	668.15	22.975		7
m-Phenylene diamine	3	99.8	285.5	8.6758	3229.58		3
p-Phenylene diamine	5	NA	NA	540.15	24.227		10
Phenyl ether	3	120	301	9.1652	3605.59		3
	4	145	325	16.3459	4310.25	-87.31	4
Phenylethylene	3	-7	145	7.9304	2108.46		3
	4	32	187	16.0193	3328.57	-63.72	4
Phenyl glycidal ether (PGE)	5	NA	NA	518.15	22.439		7
Phenylhydrazine	5	NA	NA	516.65	22.433		7
Phenyl mercaptan	5	NA	NA	442.15	22.105		7
Phenylphosphine	5	NA	NA	433.15	22.062		7
Phorate							
Phosdrin							
Phosgene	1	-68	68	6.84297	941.25	230	2
Phosphene	3	-129.4	-87.5	7.2431	810.78		3
Phosphoric acid							
Phosphorus (yellow)	3	76.6	280	7.8254	2737.62		3
Phosphorus oxychloride	3	2	105	7.9073	1897.2		3
Phosphorus pentachloride	3	55.5	162	11.8294	3868.83		3
Phosphorus pentasulfide							
Phosphorus trichloride	3	-51.6	74.2	7.933	1751.07		3

SUBSTANCE	EQ	t MAX	t MIN	A	B	C	REF
Phthalic anhydride	4	136	342	15.9984	4467.01	-83.15	4
m-Phthalodinitrile							
Picloram							
Picric acid	3	185	334	12.194	5560		11
Pindone							
Piperazine dihydrochloride							
2-Pivalyl-1,3-indandione							
Platinum (metal)	3	2730	4407	8.0324	24084.8		3
Polytetrafluoroethylene							
Portland Cement							
Potassium hydroxide	3	719	1327	7.6003	7516.27		3
	2	1170	1327		136	7.33	2
Propane	4	-109	-24	15.726	1872.46	-25.16	4
Propane sultone							
Propargyl alcohol	5	NA	NA	387.15	21.826		7
beta-Propiolactone	5	NA	NA	428.15	22.038		7
Propionic acid	2	20	140		24.684	-8.715	2
Propoxur							
n-Propyl acetate	4	7	137	16.2291	2890.47	-64.15	4
n-Propyl alcohol	4	12	127	17.5439	3166.38	-80.15	4
n-Propyl nitrate	3	-9.6	131.6	8.2796	2176.18		3
Propylene	4	-113	-33	15.7027	1807.53	-26.15	4
Propylene dichloride	5	NA	NA	369.45	21.727		7
Propylene glycol dinitrate	3	15	50	10.195	3391.31		11
Propylene glycol mono-methyl ether	5	NA	NA	393.25	27.014		7
Propylenimine	5	NA	NA	339.15	21.547		7
Propylene oxide	3	-75	34.5	8.1067	1599.1		3
Propyne	5	NA	NA	250.05	19.721		7
Pyrethrum							
Pyridine	3	-18.9	115.4	8.346	2110.75		3
	4	12	152	16.091	3095.13	-61.15	4
Quartz							
Quinone	3	1	135	11.246	3630.89	r = -.89	11
RDX (see also cyclonite)	3	45	105	12.32	6160		11
Resorcinol	3	108.4	276.5	9.4201	3587.72		3
Rhodium (metal)							
Ronnel	8.0 e-4 torr @ 25 C						
Rosin core solder pyrolysis products							
Rotenone							
Rouge							
Rubber solvent							
Selenium compounds as Se							
Selenium hexafluoride	3	-118.6	-45.8	8.9763	1388.68		3
Sesone							
Silane	3	-179.3	-111.5	6.8681	645.53		3

SUBSTANCE	EQ	t MAX	t MIN	A	B	C	REF
	3	-112	-3.5	669.412	7.0155		3
Silica							
Silicon	2	1200	1320		170	5.95	2
	3	1724	2287	13.087	26128		3
Silicon carbide							
Silicon tetrahydride	3	-179.3	-111.5	6.868	645.53		3
Silver (metal)	2	1650	1950		250	8.762	2
Soapstone							
Sodium azide							
Sodium bisulfite							
Sodium-2,4-dichloro-phenoxyethyl sulfate							
Sodium fluoroacetate							
Sodium hydroxide	2	1010	1402		132	7.03	2
Sodium metabisulfite							
Starch							
Stearates							
Stibine (Sb)	3	886	1440	9.0032	10445.4		3
Stoddard solvent							
Strychnine							
Styrene monomer	3	-7	145.2	7.9304	2108.46		3
	4	52	187	16.0193	3328.57	-63.72	4
Subtilisins							
Sucrose							
Sulfotep							
Sulfur dioxide (liquid)	4	-78	7	16.768	2302.35	-35.97	4
	3	-10	414.9	7.7437	1274.96		3
Sulfur hexafluoride	4	-114	-53	19.3785	2524.78	11.16	4
Sulfuric acid	3	146	330	9.4359	3946.85		3
Sulfur monochloride	3	-7.4	138	8.1814	2164.37		3
Sulfur pentafluoride							
Sulfur tetrafluoride	5	NA	NA	233.15	20.757		7
Sulfuryl fluoride	5	NA	NA	217.75	20.614		7
Sulprofos							
Systox							
2,4,5-T							
Talc							
Tantalum							
TEDP (see also sulfotep)							
Tellurium	3	520	1087	6.9174	5472.87		3
Tellurium hexafluoride	3	-111	-38.6	9.2865	1499.79		3
Temephos							
TEPP	3	10	140	10.846	4295		11
1,1,1,2-Tetrachloro2,2-difluoroethane	3	-37.95	92	8.1323	191431		10
1,1,2,2-Tetrachloro-1,2-difluoroethane	3	-37.5	92	8.1323	191431		3

SUBSTANCE	EQ	t MAX	t MIN	A	B	C	REF
1,1,2,2-Tetrachloroethane	3	-3.8	145.9	8.0723	2168.54		3
Tetrachloroethylene	3	-20.6	120.8	8.0328	2022.35		3
Tetrachloromethane	3	-50	178.1	7.7316	1709.53		3
	1	NA	NA	6.9339	1242.43	230	2
Tetrachloronaphthalene							
Tetraethyl lead	3	38.4	183	9.1105	2829.06		3
Tetrahydrofuran	5	NA	NA	339.15	21.546		7
Tetramethyl lead	3	-29	110	6.7594	1505.93	r = -0.84	3
Tetramethyl succinonitrile							
Tetranitromethane	5	NA	NA	398.85	22.095		7
Tetrasodium pyrophosphate							
Tetryl (2,4,6-trinitro-phenylmethylnitramine)							
Thallium	3	825	1457	7.8934	8663.01		3
4,4'-Thiobis(6-t butyl-m-cresol)							
Thyoglycolic acid	3	-60	154	3.38	732.713	r=-0.93	3
Thionyl chloride							
Thiram							
Titanium dioxide	2	1450	2270		328	9.643	2
o-Tolidine							
Toluene	3	-26	262.5	8.5473	2218.34	r=-0.93	3
Toluene-2,4-diisocyanate	5	NA	NA	524.15	22.464		7
o-Toluidine	3	44	199.7	8.7746	2766.4		3
	4	102	227	16.67834	4072.58	-72.15	4
Toxaphene							
Tributyl phosphate	5	NA	NA	565.15	22.622		7
Trichloroacetic acid	3	51	195.6	9.3558	3025.18		3
1,2,4-Trichlorobenzene	3	38.4	213	8.027	2496.36		3
1,1,1-Trichloroethane	3	-52	74.1	7.9577	1753.02		3
1,1,2-Trichloroethane	3	-24	113.9	8.0851	2004.86		3
	4	29	155	16.0381	3110.79	-56.16	4
Trichloroethylene	4	-13	127	16.1827	3928.13	-43.15	4
Trichlorofluoromethane	4	-33	27	15.8516	2401.61	-36.3	4
Trichloromethane	1	-30	150	6.90328	1163.03	227.4	2
Trichloronaphthalene							
Trichloronitromethane	3	-33.8	111.9	6.8043	1526.73	r = -0.97	3
1,2,3-Trichloropropane	3	9	158	8.3351	2345.33		3
1,1,2-Trichloro-1,2,2-trifluoroethane	4	-23	87	15.8424	2532.61	-45.67	4
Tricyclohexyltin							
Triethylamine	4	-13	127	15.8853	2882.38	-51,15	4
Trifluorobromomethane	3	-116	-46	7.49	991.95		11
Trimellitic anhydride	3	275	330	8.256	3429.2		11
Trimethylamine	4	-58	32	16.0499	2230.51	-39.15	4
Trimethylbenzene (1,2,3)	3	16.8	176.1	6.0062	1472.42	r = -0.84	3

SUBSTANCE	EQ	t MAX	t MIN	A	B	C	REF
(1,2,4)	3	13.6	169.2	8.21	2343.27		3
(1,3,5)	3	9.6	164.7	8.1606	2299.7		3
Trimethyl phosphite	5	NA	NA	381.15	21.793		7
2,4,6-Trinitrophenol							
2,4,6-Trinitrophenyl-methylnitramine							
2,4,6-Trinitrotoluene	3	80	100	2659.86	6.1529		3
Triorthocresyl phosphate	5	NA	NA	693.15	23.052		7
Triphenyl amine	5	NA	NA	638.15	23.526		10
Triphenyl phosphate	3	193.5	413.5	9.0565	4215.54		3
Tungsten	3	3980	7670	9.0325	38145.5		3
Turpentine							
Uranium (natural)	3	2450	4420	8.5899	23351.4		3
Valeraldehyde	5	NA	NA	375.15	21.553		7
Vanadium (as V_2O_5)	3	2290	3800	9.6459	24661.5		3
Vegetable oil mist							
Vinyl acetate	4	-18	106	16.1003	2744.68	-56.15	4
Vinyl benzene	4	52	187	16.0193	3328.57	-63.72	4
Vinyl bromide	3	-95.4	15.8	7.4963	1329.51		3
Vinyl chloride	4	-88	17	14.9601	1803.84	-43.15	4
Vinyl cyanide	1	-20	140	7.03855	1232.53	222.47	2
Vinyl cyclohexene dioxide	5	NA	NA	500.15	22.365		7
Vinylidine chloride	5	NA	NA	310.15	20.55		7
Vinyl toluene (alpha)	3	7.4	165.4	7.9778	2237.81		3
(beta)	3	17.5	179	8.0782	2342.2		3
4-Vinyl toluene	3	16	175	8.1369	2347.21		3
VM&P Naphtha							
Warfarin							
Welding fume (NOC)							
Wood dust							
Xylene (o)	3	-3.8	144.4	8.1426	2185.1		3
(m)	3	-6.9	139.1	7.9553	2123.89		3
(p)	3	-8.1	138.3	8.1224	2144.83		3
m-Xylene alpha,alpha'-diamine							
Xylidine (2,4)	3	52.6	211.5	8.8029	2862.22		3
(2,6)	3	44	217.9	8.1197	2569.45		3
Yttrium							
Zinc chloride	3	428	732	9.5555	6688.08		3
Zinc chromates, oxide							

Instructions for Using the Vapor Pressure Table

There are one or more equations for each substance, each corresponding to a temperature range in the table (°C) for which the equation is applicable. For any temperature within the range(s) given, substitute the appropriate values from columns A, B, and C into the equation specified.

Definitions: T = temperature (°K)[2] or t = °C at which the estimation is *to* take place,

P = vapor pressure in torr.

Equations 1-5. The first four equations are based on empirical data and model fitting.

1. $\log_{10} P = A - [B/(t + C)]$
2. $\log_{10} P = [(-52.23B)/T] - C$
3. $\log_{10} P = A - (B/T)$

   If the correlation coefficient (r) of the constants generated by linear regression is less than $|0.99|$, then the value of r is given in column C in the table.

4. $\ln P = A - [B/(T + C)]$

The fifth equation is the Watson correlation equation, modified for this tabular format, where: *A* is normal boiling point, $B = K_f (8.75 + R \ln A)$, K_f is a constant from Chapter 14, Table 14-4 of Reference 5, and C is defined below under Equation 5. References in the table for Equation 5 refer to the source of the normal boiling point data.

5. $\ln P = (B/1.927)\, [1 - ((3-(2T/A)^c)/(T/A)) - 2C(3-(2T/A))^{C-1} \ln (T/A)]$,

   where for liquids $C = 0.19$,

   and for solids:

   if $T/A > 0.6$; C = 0.36

   if $0.6 > T/A > 0.5$; C = 0.8

   if $T/A < 0.5$; C = 1.19.

Example: Acetic acid at 25°C. From the table, the appropriate equation is Equation 1. Substituting the given values for A, B, and C:

$\log_{10} P = 7.80307 - [1651.2/(25 + 225)]$

$\log_{10} P = 1.19827$

(taking the antilog) Vapor pressure (*P*) = 15.7 torr

---

[2] Note °K = °C + 273

REFERENCES

1. Duncan, D.H.: A compilation of vapor pressure constants for substances having an ACGIH® threshold limit value, master's project. University of Alabama at Birmingham, Industrial Hygiene Program (1989).
2. Lange's Handbook of Chemistry, 13th ed., J.A. Dean (ed.). McGraw-Hill, New York (1985).
3. CRC Handbook of Chemistry and Physics, R.C. Weast (ed.). CRC Press, Boca Raton, FL, published annually.
4. Reid, R.C.; Prausnitz, J.M.; Sherwood, T.K.: The Properties of Gases and Liquids, 3rd ed., pp. 193-196. McGraw-Hill, New York (1977).
5. Lyman, W.J.; Reehl, W.F.; Rosenblatt, D.H.: Handbook of Chemical Property Estimation Methods, pp.14.2-14.7. McGraw-Hill, New York (1982).
6. Reid, R.C.; Prausnitz, J.M.; Sherwood, T.K.: The Properties of Gases and Liquids, 3rd ed., Appendix. McGraw-Hill, New York (1977).
7. Hawley's Condensed Chemical Dictionary, l2th ed., R.J. Lewis, Sr. (ed.). VanNostrand Reinhold, New York (1993).
8. Sax's Dangerous Properties of Industrial Matrials, 9th ed., R.J. Lewis, Sr. (ed.). Van Nostrand Reinhold, New York (1996).
9. The Merck Index, 10th ed., M. Windholz (ed.). Merck and Co., Rahway, NJ (1984).
10. Chemical Engineers' Handbook, 5th ed., R.H. Perry (ed.). McGraw-Hill, New York (1973).
11. Technical Database Services Vapor Pressure Database. Thermodynamics Research Center, Texas A&M University, College Station, TX (1988).
12. Verscheuren, K.: Handbook of Environmental Data on Organic Chemicals, 2nd ed. Van Nostrand Reinhold, New York (1983).
13. Handbook of Environmental Fate and Exposure Data for Organic Chemicals, Vol. Ill: Pesticides, P.H. Howard (ed.). Lewis Publishers, Chelsea, MI (1991).
14. NIOSH: NIOSH Pocket Guide to Chemical Hazards (DHEW/NIOSH Publication No. 90-117). National Institute for Occupational Safety and Health, Cincinnati, OH (1990).
15. CCOSH: CClNFODISC, MSDS [Computer Software]. Canadian Center for Occupational Health and Safety, Hamilton, Ontario, Canada (1995).
16. Handbook of Environmental Fate and Exposure Data for Organic Chemicals, Vol. I: Large Production and Priority Pollutants, P.H. Howard (ed.). Lewis Publishers, Chelsea, MI (1989).
17. NIOSH: Criteria for a Recommended Standard...Occupational Exposure to Waste Anesthetic Gases and Vapors (DHEW/NIOSH Publication No. 77-140). National Institute for Occupational Safety and Health, Cincinnati, OH (1977).

Appendix C

THREE DIMENSIONAL SOLUBILITY PARAMETERS $(J/cm^3)^{0.5}$

Adapted from E. Wallstrom and I. Svenningsen, Handbook of Solvent Properties, Scandinavian Paint and Printing Ink Research Institute, Horsholm, Denmark (1984), with permission.

Solvent	CAS-Number	D	P	H
Acetaldehyde	75-07-0	14.7	8.0	11.3
Acetic Acid	64-19-7	14.5	8.0	13.5
Acetic Anhydride	108-24-7	16.0	11.7	10.2
Acetone	67-64-1	15.5	10.4	7.0
Acetonitrile	75-05-8	15.3	18.0	6.1
Acetophenone	98-86-2	19.6	8.6	3.7
Acrylonitrile	107-13-1	16.4	17.4	6.8
Allyl Alcohol	107-18-6	16.2	10.8	16.8
Aniline	62-53-3	19.4	5.1	10.2
Anisole	100-66-3	17.8	4.1	6.7
Benzaldehyde	100-52-7	19.4	7.4	5.3
Benzene	71-43-2	18.4	0.0	2.0
1,3-benzenediol	108-46-3	18.0	8.4	21.0
Benzoic Acid	65-85-0	18.2	6.9	9.8
Benzonitrile	100-47-0	17.4	9.0	3.3
Benzyl Alcohol	100-51-6	18.4	6.3	13.7
Benzyl Butyl Phthalate	85-68-7	19.0	11.2	3.1
Benzyl Chloride	100-44-7	18.8	7.1	2.6
Biphenyl	92-52-4	21.4	1.0	2.0
Bromobenzene	108-86-1	20.5	5.5	4.1
Bromochloromethane	74-97-5	17.3	5.7	3.5
Bromoform	75-25-2	21.4	4.1	6.1
1-bromonaphthalene	90-11-9	20.3	3.1	4.1
Bromotrifluoromethane	75-63-8	9.6	2.4	0.0
Butane	106-97-8	14.1	0.0	0.0
1,3-butanediol	107-88-0	16.6	10.0	21.5
1-butanol	71-36-3	16.0	5.7	15.8
2-butanol	78-92-2	15.8	5.7	14.5
Butyl Acetate	123-86-4	15.8	3.7	6.3
Sec-butyl Acetate	105-46-4	15.1	3.7	6.3
Butyl Acrylate	141-32-2	15.6	6.2	4.9
Butylamine	109-73-9	16.2	4.5	8.0
Butyl Lactate	138-22-7	15.8	6.5	10.2
Butyraldehyde	123-72-8	14.7	5.3	7.0
Butyric Acid	107-92-6	14.9	4.1	10.6
Gamma Butyrolactone	96-48-0	19.0	16.6	7.4
Butyronitrile	109-74-0	15.3	12.4	5.1
Carbon Disulfide	75-15-0	20.5	0.0	0.6
Carbon Tetrachloride	56-23-5	17.8	0.0	0.6
Chlorobenzene	108-90-7	19.0	4.3	2.0

Solvent	CAS-Number	D	P	H
1-chlorobutane	109-69-3	16.2	5.5	2.0
Chlorodifluoromethane	75-45-6	12.3	6.3	5.7
Chloroform	67-66-3	17.8	3.1	5.7
3-chloro-1-propanol	627-30-5	17.5	5.7	14.7
M-cresol	108-39-4	18.0	5.1	12.9
Cyclohexane	110-82-7	16.8	0.0	0.2
Cyclohexanol	108-93-0	17.4	4.1	13.5
Cyclohexanone	108-94-1	17.8	6.3	5.1
Cyclohexylamine	108-91-8	17.2	3.1	6.5
Cyclohexylchloride	542-18-7	17.3	5.5	2.0
Cis-decahydronaphthalene	493-01-6	18.8	0.0	0.0
Trans-decahydronaphthalene	493-02-7	18.0	0.0	0.0
Decane	124-18-5	15.7	0.0	0.0
1-decanol	112-30-1	17.5	2.6	10.0
Diacetone Alcohol	123-42-2	15.8	8.2	10.8
Dibenzyl Ether	103-50-4	17.3	3.7	7.3
Dibutyl Phthalate	84-74-2	17.8	8.6	4.1
Dibutyl Sebacate	109-43-3	13.9	4.5	4.1
Dibutyl Stearate	123-95-5	14.5	3.7	3.5
o-dichlorobenzene	95-50-1	19.2	6.3	3.3
2,2-dichlorodiethyl Ether	111-44-4	18.8	9.0	5.7
Dichlorodifluoromethane	75-71-8	12.3	2.0	0.0
1,1-dichloroethane	75-34-3	16.5	8.2	0.4
1,1-dichloroethylene	75-35-4	17.0	6.8	4.5
Di (2-chloro Isopropyl) Ether	108-60-1	19.0	8.2	5.1
Dichloromonofluoromethane	75-43-4	15.8	3.1	5.7
1,2-dichlorotetrafluoroethane	76-14-2	12.6	1.8	0.0
Di-iso-butyl Carbinol	108-82-7	14.9	3.1	10.8
Diethylamine	109-89-7	14.9	2.3	6.1
2-(Diethylamino) Ethanol	100-37-8	14.9	5.8	12.0
Para-diethylbenzene	105-05-5	18.0	0.0	0.6
Diethyl Carbonate	105-58-8	16.6	3.1	6.1
Diethylene Glycol	111-46-6	16.2	14.7	20.5
Diethylene Glycol Butyl Ether Acetate	124-17-4	16.0	4.1	8.2
Diethylene Glycol Monobutyl Ether	112-34-5	16.0	7.0	10.6
Diethylene Glycol Mono Ethtyl Ether	111-90-0	16.1	9.2	12.2
Diethylene Glycol Monomethyl Ether	111-77-3	16.2	7.8	12.6
Diethylenetriamine	111-40-0	16.7	13.3	14.3
Diethyl Ether	60-29-7	14.5	2.9	5.1
Diethyl Ketone	96-22-0	15.8	7.6	4.7

Solvent	CAS-Number	D	P	H
Diethyl Phthalate	84-66-2	17.6	9.6	4.5
Diethyl Sulfate	64-67-5	15.7	14.7	7.1
Diethyl Sulfide	352-93-2	16.8	3.1	2.0
Di (Isobutyl) Ketone	108-83-8	16.0	3.7	4.1
Di (2-methoxyethyl) Ether	111-96-6	15.7	6.1	6.5
N,N-dimethylacetamide	127-19-5	16.8	11.5	10.2
Dimethylformamide	68-12-2	17.4	13.7	11.3
1,1-dimethylhydrazine	57-14-7	15.3	5.9	11.0
Dimethyl Phthalate	131-11-3	18.6	10.8	4.9
Dimethyl Sulfone	67-71-0	19.0	19.4	12.3
Dimethyl Sulfoxide	67-68-5	18.4	16.4	10.2
Dioctyl Phthalate	117-81-7	16.6	7.0	3.1
1,4-dioxane	123-91-1	19.0	1.8	7.4
Dipropylamine	142-84-7	15.3	1.4	4.1
Dipropylene Glycol	25265-71-8	15.9	20.2	18.4
Dipropylene Glycol Methyl Ether	34590-94-8	15.5	5.7	11.2
Dodecane	112-40-3	16.0	0.0	0.0
Eicosane	112-95-8	16.5	0.0	0.0
Epichlorohydrin	106-89-8	19.0	10.2	3.7
Ethanethiol	75-08-1	15.7	6.5	7.1
Ethanol	64-17-5	15.8	8.8	19.4
Ethanolamine	141-43-5	17.0	15.5	21.2
Ethyl Acetate	141-78-6	15.8	5.3	7.2
Ethyl Acrylate	140-88-5	15.5	7.1	5.5
Ethylbenzene	100-41-4	17.8	0.6	1.4
Ethyl Bromide	74-96-4	16.5	8.0	5.1
2-ethyl-1-butanol	97-95-0	15.8	4.3	13.5
Ethyl Chloride	75-00-3	15.7	6.1	2.9
Ethyl Chloroformate	541-41-3	15.5	10.0	6.7
Ethyl Cinnamate	103-36-6	18.4	8.2	4.1
Ethylene Carbonate	96-49-1	19.4	21.7	5.1
Ethylene Cyanohydrin	109-78-4	17.2	18.8	17.6
Ethylenediamine	107-15-3	16.6	8.8	17.0
Ethylene Dibromide	106-93-4	19.6	6.8	12.1
Ethylene Dichloride	107-06-2	19.0	7.4	4.1
Ethylene Glycol	107-21-1	17.0	11.0	26.0
Ethylene Glycol Monobutyl Ether	111-76-2	16.0	5.1	12.3
Ethylene Glycol Monoethyl Ether	110-80-5	16.2	9.2	14.3
Ethylene Glycol Monoethyl Ether Acetate	111-15-9	15.9	4.7	10.6
Ethylene Glycol Monomethyl Ether	109-86-4	16.2	9.2	16.4

Solvent	CAS-Number	D	P	H
Ethylene Glycol Monomethyl Ether Acetate	110-49-6	15.9	5.5	11.6
Ethyl Formate	109-94-4	15.5	8.4	8.4
2-ethyl Hexanol	104-76-7	15.9	3.3	11.8
Ethyl Lactate	97-64-3	16.0	7.6	12.5
Formamide	75-12-7	17.2	26.2	19.0
Formic Acid	64-18-6	14.3	11.9	16.6
Furan	110-00-9	17.8	1.8	5.3
Furfural	98-01-1	18.6	14.9	5.1
Furfuryl Alcohol	98-00-0	17.4	7.6	15.1
Glycerol	56-81-5	17.4	12.1	29.3
Heptane	142-82-5	15.3	0.0	0.0
Hexadecane	544-76-3	16.3	0.0	0.0
Hexamethylphosphoramide	680-31-9	18.5	8.6	11.3
Hexane	110-54-3	14.9	0.0	0.0
Hexylene Glycol	107-41-5	15.7	8.4	17.8
Isoamyl Acetate	123-92-2	15.3	3.1	7.0
Isobutyl Acetate	110-19-0	15.1	3.7	6.3
Isobutylalcohol	78-83-1	15.1	5.7	15.9
Isobutyl Isobutyrate	97-85-8	15.1	2.9	5.9
Isooctyl Alcohol	26952-21-6	14.4	7.3	12.9
Isopentane	78-78-4	13.7	0.0	0.0
Isophorone	78-59-1	16.6	8.2	7.4
Isopropyl Palmitate	142-91-6	14.3	3.9	3.7
Mesithylene	108-67-8	18.0	0.0	0.6
Mesityl Oxide	141-79-7	16.4	6.1	6.1
Methacrylonitrile	126-98-7	15.3	10.8	3.6
Methanol	67-56-1	15.1	12.3	22.3
o-methoxyphenol	90-05-1	18.0	8.2	13.3
Methyl Acetate	79-20-9	15.5	7.2	7.6
Methyl Acrylate	96-33-3	15.3	9.3	5.9
Methylal	109-87-5	15.0	1.8	8.6
Methyl Chloride	74-87-3	15.3	6.1	3.9
Methylcyclohexane	108-87-2	16.0	0.0	1.0
Methylene Dichloride	75-09-2	18.2	6.3	6.1
Methylene Diiodide	75-11-6	17.8	3.9	5.5
Methyl Ethyl Ketone	78-93-3	16.0	9.0	5.1
Methyl Isoamyl Ketone	110-12-3	16.0	5.7	4.1
Methyl Isobutyl Carbinol	108-11-2	15.4	3.3	12.3
Methyl Isobutyl Ketone	108-10-1	15.3	6.1	4.1
Methyl Methacrylate	80-62-6	17.5	5.5	4.3

Solvent	CAS-Number	D	P	H
1-methylnaphthalene	1321-94-4	20.6	0.8	4.7
Methyl Oleate	112-62-9	14.5	3.9	3.7
2-methyl-1-propanol	78-83-1	15.1	5.7	15.9
Methyl-2-pyrrolidone	872-50-4	18.0	12.3	7.2
Methyl Salicylate	119-36-8	16.0	8.0	12.3
Morpholine	110-91-8	18.8	4.9	9.2
Naphtha, High Flash	8030-30-6	17.9	0.7	1.8
Naphthalene	91-20-3	19.2	2.0	5.9
Nitrobenzene	98-95-3	20.0	8.6	4.1
Nitroethane	79-24-3	16.0	15.5	4.5
Nitromethane	75-52-5	15.8	18.8	5.1
2-nitropropane	79-46-9	16.2	12.1	4.1
Nonane	111-84-2	15.7	0.0	0.0
Nonyl Phenol	25154-52-3	16.5	4.1	9.2
Nonyl Phenoxy Ethanol	27986-36-3	16.7	10.2	8.4
Octane	111-65-9	15.5	0.0	0.0
Octanoic Acid	124-07-2	15.1	3.3	8.2
1-octanol	111-87-5	17.0	3.3	11.9
2-octanol	123-96-6	16.1	4.9	11.0
Oleic Acid	120-80-1	14.3	3.1	5.5
Oleyl Alcohol	143-28-2	14.3	2.6	8.0
Pentane	109-66-0	14.5	0.0	0.0
2,4-pentanedione	123-54-6	17.1	9.0	4.1
1-pentanol	71-41-0	15.9	4.5	13.9
Perfluoro (Dimethylcyclohexane)	355-27-3	12.4	0.0	0.0
Perfluoroheptane	335-57-9	12.0	0.0	0.0
Perfluoromethylcyclohexane	355-02-2	12.4	0.0	0.0
Phenol	108-95-2	18.0	5.9	14.9
Bis-(M-phenoxyphenyl) Ether	748-30-1	19.6	3.1	5.1
1-propanol	71-23-8	16.0	6.08	17.4
2-propanol	67-63-0	15.8	6.1	16.4
Propionitrile	107-12-0	15.3	14.3	5.5
Propylamine	107-10-8	16.9	4.9	8.6
Propyl Chloride	540-54-5	16.0	7.08	2.0
Propylene Carbonate	108-32-7	20.0	18.0	4.1
Propylene Glycol	57-55-6	16.8	9.4	23.3
Propylene Glycol Monomethyl Ether	107-98-2	15.6	6.3	11.6
Pyridine	110-86-1	19.0	8.8	5.9
2-pyrrolidone	616-45-5	19.4	17.4	11.3
Quinoline	91-22-5	19.4	7.0	7.6

Solvent	CAS-Number	D	P	H
Stearic Acid	57-11-4	16.3	3.3	5.5
Styrene	100-42-5	18.6	1.0	4.1
Succinic Anhydride	108-30-5	18.6	19.2	16.6
1,1,2,2-tetrabromoethane	79-27-6	22.6	5.1	8.2
1,1,2,2-tetrachloroethane	79-34-5	18.8	5.1	9.4
Tetrachloroethylene	127-18-4	19.0	6.5	2.9
Tetraethylorthosilicate	78-10-4	13.9	0.4	0.6
Tetrahydrofuran	109-99-9	16.8	5.7	8.0
Tetrahydronaphthalene	119-06-2	19.6	2.0	2.9
Tetramethylurea	632-22-4	16.7	8.2	11.0
Toluene	108-88-3	18.0	1.4	2.0
Tributyl Phosphate	126-73-8	16.3	6.3	4.3
Trichlorobiphenyl	25323-68-6	19.2	5.3	4.1
1,1,1-trichloroethane	71-55-6	16.8	4.3	2.0
Trichloroethylene	79-01-6	18.0	3.1	5.3
Trichlorofluoromethane	75-69-4	15.3	2.0	0.0
1,1,2-trichloro-1,2,2-trifluoroethane	76-13-1	14.7	1.6	0.0
Tricresyl Phosphate	1330-78-5	19.0	12.3	4.5
Tridecyl Alcohol	112-70-9	14.3	3.1	9.0
Triethanolamine	102-71-6	17.3	22.4	23.3
Triethylamine	121-44-8	17.8	0.4	1.0
Triethyleneglycol	112-27-6	16.0	12.5	18.6
Triethylene Glycol Monooleyl Ether	27323-44-0	13.3	3.1	8.4
Triethylphosphate	78-40-0	16.7	11.4	9.2
Trimethylbenzene	25551-13-7	17.8	0.4	1.0
2,2,4-trimethylpentane	540-84-1	14.1	0.0	0.0
Trimethylphosphate	512-56-1	16.7	15.9	10.2
Water	7732-18-5	15.5	16.0	42.3
Xylene	1330-20-7	17.6	1.0	3.1
o-xylene	75-47-6	17.8	1.0	3.1

Appendix D

ANSWERS TO STUDY QUESTIONS

Chapter 1. What Is an Industrial Hygienist?

1.1.

a. Biology, chemistry, engineering, epidemiology, law, management, mathematics, medicine, psychology, statistics, and toxicology.

b. Chemistry, biology.

1.2.

a. Environmental manager, environmental/health/safety manager/scientist, environmental health consultant, industrial health analyst, occupational hygienist, occupational technologist, occupational health scientist, industrial health specialist, health and safety investigator, workplace hazard inspector, hazard controller, industrial health engineer, risk manager.

b. It may cause confusion for management, workers, and the public at large regarding the expertise that an industrial hygienist can provide, as it may be confused with names used in other hygiene professions. For example, people often associate it with dental hygiene or sanitation. An undue emphasis on bacteria and infectious disease may be inferred.

1.3. An industrial hygiene professional may need to utilize many different disciplines in performing his or her job. Because of the variety of the subject matter, an undergraduate degree in either biology or chemistry, with a minor in the other, might better serve to fill a need for baseline knowledge in the most pertinent subjects, and a graduate degree in hygiene can build on this knowledge. In any case, a solid grounding in biology and chemistry is important.

Chapter 2. Industrial Hygiene – Historical Perspective

2.1. Occupational health nursing, occupational medicine, industrial hygiene, and occupational safety.

2.2.

a. Social Security Act of 1935, which made federal funds available for public health programs (including industrial hygiene).

b. The establishment of the American Conference of Governmental Industrial Hygienists and the American Industrial Hygiene Association around 1940.

c. The publication of *Industrial Hygiene and Toxicology* by Frank Patty in 1948.

This was the first large published work that dealt specifically with industrial hygiene principles and practices that defined the field.

2.3.

a. Some knowledge of engineering is helpful, especially in controlling harmful agents at the source, for example, local exhaust ventilation. Other areas include knowledge of noise and vibration as well as chemical or process engineering principles, which are often essential to an understanding of chemical emissions.

b. A good foundation in chemistry is important because the majority of hazards in the workplace are chemical in nature. A hygienist should have a basic understanding of the different classes of chemicals, and their structures and uses. Analytical chemistry provides the foundation for evaluation of chemical hazards.

c. A good foundation in biology is important, as the ultimate goal of an industrial hygienist is to preserve the health and well-being of the worker. So it follows that a good understanding of biological systems and the effects of exposure to hazardous agents on these systems is necessary in order for one to anticipate and recognize the occurrence of illness due to occupation.

2.4.

a. Occupational health prior to 1900: There were few practitioners of occupational health; certainly there was no formalized discipline. Early recordings of occupational diseases, (e.g., diseases associated with lead, mercury, and arsenic) were made, and a few physicians recognized the unhealthy conditions associated with many types of labor.

b. Industrial Revolution: A dramatic increase occurred in unhealthy and unsafe conditions in the workplace. Several pieces of legislation were passed in Great Britain primarily aimed at mitigating child labor abuses and reporting the occurrence of disease believed to be associated with occupation. Other notable events included the start of medical inspections of factories and initial studies on the association between disease and work.

c. 1910–1940: The most noteworthy events of this period include the work of Dr. Alice Hamilton, who was responsible for enlightening the American public on the unhealthy conditions of the workplace. She also was responsible for recognizing numerous associations between disease and occupational exposures to some organic compounds and metals. Other events included the passage of worker's compensation laws with the *no-fault* system of restitution and the

Social Security Act of 1935, which made funds available for public health programs, including occupational health.

d. 1940–environmental movement: During this period the Occupational Safety and Health Act was passed in the United States. It is considered to be the most important event in the history of U.S. occupational health.

2.5. Access to a safe and healthy workplace is a fundamental right owed to workers. However, this right must be balanced with what is reasonable, as risk is inherent in any workplace or occupation. At the very least, all reasonable precautions should be taken to safeguard the health and well-being of workers, and they should be made fully aware of the residual potential risks in their workplace.

2.6. The chief advantage to the employer of the worker's compensation system is that there is a cap on the amount of money that an employer will have to pay to an injured or diseased employee. The main advantage to the employee is that he or she will be compensated for a work-related injury regardless of the circumstances. In addition, the employee does not have to worry about losing his or her job because of lost workdays. The main disadvantage of the system for the employee is that he or she loses all rights to legal action against the employer (in virtually all of the states). Therefore, because the costs are well defined for the employer, there may be little incentive to correct truly negligent conditions. Finally, the chief disadvantage to the employer is the potential for abuse of the system by the employees.

2.7.

a. Fellow servant defense: Worker could not be compensated in full by the employer for an injury or illness if that injury or illness occurred because of the negligence of a fellow worker.

b. Assumption of risk: Employees assumed the risk inherent in the job; that is, the employee knew the hazards associated with the workplace or task when he or she applied for the job.

c. Contributory negligence: Employees would collect nothing if they were found negligent in any way, regardless of any negligence on the part of the employer.

2.8 AIHA: American Industrial Hygiene Association, a U.S.-based professional society for hygienists. ACGIH®: American Conference of Governmental Industrial Hygienists, a U.S.-based professional society for hygienists, and until recently, whose members had an affiliation with government. AIOH: Australian Institute of Occupational Hygiene, an Australian society for hygienists. BOHS: British Occupational Hygiene Society, a British professional society for hygienists.

OSHA: Occupational Safety and Health Administration, part of the U.S. Department of Labor, which administers regulations governing the health and safety of American workplaces. HSE: Health and Safety Executive, the rough equivalent to OSHA in the United Kingdom.

Chapter 3. Principles and Focus of Industrial/Occupational Hygiene

3.1. Ultimately, the goal of industrial hygiene is to protect worker health and well-being. Biology, the study of life, is the foundation for this goal and for our understanding of how and why disease agents affect human health. Without an understanding of the biological basis of health and disease, there is little rationale for evaluation of workplace hazards or their control. In such a situation, the impetus for evaluation and control is usually regulation-driven.

3.2.

a. Home/garden pesticide use: Use appropriate gloves and perhaps the appropriate respirator when applying pesticide. Follow manufacturer's instructions; carefully read warning label.

b. Electrical safety: Shut off power to make repairs to light fixtures, furnace, etc.

c. Household chemical use: Read warning labels; ventilate area when chemical is in use.

d. Lawn mower and other power equipment: Use goggles; keep equipment properly maintained.

e. Use of gasoline: Use safety storage container.

f. Sun toxicity: Use sunscreen.

g. Communicable diseases: Wash hands, avoid disease carriers, and sanitize surfaces.

h. Foodborne diseases: Cook and store food properly.

3.3. The hygienist has knowledge of both occupational diseases and injuries (engineers tend not to have this), as well as knowledge of the industrial process (physicians tend not to have this), thus effectively bridging the knowledge gap between professions.

3.4. The ecological perspective implies a knowledge of the interconnectedness of nature. Thus, solving an air pollution problem from an ecological perspective does not mean creating a water pollution problem. In the past, solving pollution problems in one medium commonly meant creating problems in another.

3.5. The primary responsibility of the industrial hygienist is to protect the health of the worker and the community.

Chapter 4. Disease Agents of the Workplace

4.1. Fungi, which often grow on damp organic material, are important biological agents affecting indoor air quality. Others include bacteria and the house dust mite.

4.2. Noise, a physical agent, is probably the most common single disease agent. OSHA estimated (1983) that 5.1 million American workers were exposed to noise levels that put them at risk for hearing loss.

4.3. Gravity. Although gravity does affect pressure propagation (noise and vibration) through media because of its effect on density, the overall effect is negligible because gravity is relatively constant on earth.

4.4.

a. Noble gases: He, Ne, Ar, Kr, Xe, Rn.

b. Halogens: I, Cl, Br, F, At.

c. Organic chemistry elements: C, N, O, S, P, H.

d. Metals and semiconductors: Si, B, As, Se, Te, Ge, Po, Sb, and the metals.

4.5. Monomers are simple compounds such as vinyl chloride or formaldehyde. Monomers are reacted to form polymers, that is, more than two repeating units of the same structure, usually of 100 to 1,000 repeating units. To complete the process, additional reactive monomer may be added, and a curing, vulcanizing, or crosslinking agent may be added to link the polymer into a network.

4.6. The formation of a metal fume is essentially an evaporation process whereby a sufficiently heated material (usually a metal) is vaporized and then subsequently oxidized upon entering the atmosphere. The aggregation of these molecules produces a fume. The particle size of a fume ranges from 0.001 μm to 1 μm.

Chapter 5. Common Industrial Processes and Their Associated Agents

5.1. Machining, forging

Welding, grinding

Inorganic metal surface treatment and/or painting–organic treatment

Electroplating

Buffing

5.2.

a. Vehicle: made up of binder and usually a solvent; the binder is usually an organic polymer that binds the pigment after curing.

b. Pigments: give paint color; examples include oxides of lead, cadmium, chromium, and iron.

c. Additives: a wide range of compounds; examples include biocides such as tributyl tin, copper and zinc naphthenate, and drying agents such as lead, barium, and zinc.

5.3. Because of the toxicities associated with commonly used degreasers such as trichloroethylene, perchloroethylene, and methylene chloride, industry changed to the less toxic chlorofluorocarbons such as Freon 113 and 1,1,1-trichloroethane. However, because of the volatility and the long-term atmospheric stability of these compounds, significant quantities are present in the stratosphere. It is here that the CFCs react with UV light to release halogen ions which react with ozone. The Montreal Protocol of 1987 requires a scheduled phase-out of many CFCs and other ozone depleters.

5.4.

a. Arc welding: The process relies on an electric arc or current as a heat source. The workpiece is one electrode, and the welding rod is the other. The heat causes the base metal and the rod to melt together. Types are:

Gas metal arc welding (MIG): A filler rod is the second electrode, and an inert gas, such as argon or CO_2 is used to keep the weld from being oxidized.

Gas tungsten arc welding (TIG): A non-consumable tungsten electrode is used, and a separate filler rod is fed to the weld. An inert gas such as helium is used to shield the weld.

Shielded metal arc welding (Stick): The filler rod (second electrode) is coated with a flux that prevents oxidation of the weld.

b. Submerged arc welding: The welding process is submerged in a granular flux that shields the weld and reduces the amount of fume, UV light, and other hazards generated.

c. Oxyfuel welding: A fuel and oxygen blend is used to join the metals; no filler rod is used.

d. Resistance welding: Resistance to current flow through a workpiece and added pressure produce a weld.

Chapter 6. Exposure Routes

6.1. *Assumptions*: The urban office worker lives in a large Southern city in an old apartment building close to work. The rural foundry worker lives in the South in a large agricultural region.

URBAN OFFICE WORKER		
Xenobiotic	**Route of entry**	**Source**
CO	Inhalation	Automobile emissions
NO_x	Inhalation	Automobile emissions
O_3	Inhalation	Photochemical smog
Bacteria	Inhalation/ingestion	Organic materials, humans
Fungi	Inhalation/ingestion	Organic materials, humans
Organics	Inhalation	Off-gassing of furniture paint, carpet, etc.; combustion and industrial sources
Pesticides	Inhalation/ingestion	Foods pest abatement
Metals	Inhalation/ingestion	Combustion, paint, industrial sources
RURAL FOUNDRY WORKER		
CO	Inhalation	Melting operations
Iron oxide	Inhalation/ingestion	Melting/pouring operations
Other metals	Inhalation/ingestion	Melting/pouring operations
Silica	Inhalation/ingestion	Casting operations
Ammonia	Inhalation	Coremaking
Epoxy resins	Dermal	Coremaking
Isocyanates	Inhalation	Coremaking
Pesticides	Ingestion/inhalation	Well water, foods, crop application

The relative magnitude of exposures to xenobiotics would probably be higher for the foundry worker; however, as the office worker may have similar exposures at home to those at work, his or her exposure may be more constant over time.

6.2.

a. $10\ mg/m^3 \times 10\ m^3 = 100\ mg$

b. probably not. The ability of a contaminant to be absorbed into the body and reach its target organ depends on several factors. For particulates, the determining factor is usually particle size. Most of the larger particles (10–50 μm) will be deposited in the head region; smaller ones that travel farther will be impacted or intercepted onto the surface of the larynx, pharynx, trachea, or bronchi. At this point, many particles will become trapped in mucus that is secreted by mucosal cells located in these regions and then swept upward by tiny hair cells to be either blown out or coughed and swallowed, where they may be absorbed by the gut.

   Smaller particles can travel farther down into the lung; some (usually < 3–4 μm) make it to the alveolar region. Mucosal and ciliary cells are not found here; so the alveolar tissue depends on macrophages to engulf foreign material. Once a macrophage engulfs a foreign material, it may migrate to the bronchiole region where it can be removed by the mucociliary system, or it can be absorbed through the alveolar tissue into the lymphatic system. Some particles are never engulfed by macrophages and will remain in the lung tissue indefinitely.

   For gases and vapors, the ability to reach the target organ is dependent on the absorptivity of the contaminant in the tissues of the respiratory tract. Those gases or vapors that are least water-soluble will more likely be inhaled and exhaled. As water solubility of the contaminant increases, so does absorption into the body.

   Metabolism and excretion also determine what happens to a contaminant upon entering the body. Once absorbed, the body responds by trying to metabolize the contaminant into a more water-soluble compound so that it may be excreted from the body. The efficiency with which the body accomplishes this will have a large impact on how much of the original compound ultimately reaches a target organ. Of course, some of the metabolites may reach other target organs or storage organs (sinks).

c. The *safe dose* usually is determined by taking only the inhalation route of exposure into consideration, with dermal routes more difficult to consider. If the skin is exposed to the same dose that is deemed *safe* for the lung, it is unlikely that the same amount of contaminant will reach the target organ owing to differences in sorption through the two membranes. Also, metabolism in the dermis may be different from that through the lung. Therefore, it may be inappropriate to apply the same safe dose to two routes of entry.

Chapter 7. Approximating the Truth

7.1. Acceptable values for precision and accuracy depend on what you are trying to measure or estimate. For sampling and analytical methods, uncorrectable bias is not acceptable and should be investigated. NIOSH sets a value for accuracy of ± 25% at the 95% confidence level. If bias is zero, then this sets a maximum for precision of about 12%.

7.2. $\bar{x} = 10$, $s = 0.41$, $CV = 0.04$

Accuracy $= 2CV + |\text{Bias}|$

$2 \times 4.1\% + 0 = 8.2\% < 25\%$

Yes, the method meets the NIOSH accuracy criterion.

7.3. $\bar{x} = 121.2$, $s = 4.56$, $CV = 0.03$

Although this method is precise ($CV = 3\%$), it is also biased. All of the repeated measurements are above the true concentration, suggesting a systematic error. The error should be investigated and corrected, or the method should be abandoned, unless it can be shown for other conditions (interferences, for example) that the bias is constant.

7.4. If the source of the bias in the first method can be determined and eliminated or corrected by subtraction, it would be the better choice because it is more precise than the second method (3% vs. 8%). If the bias cannot be corrected or eliminated, the second method is preferred.

7.5. Arithmetic mean ($\bar{x}_c$) = 16.8

Geometric mean ($\bar{x}_g$) = exp ($\bar{x}_L$) = 15.6, where $\bar{x}_L$ = mean of the logged data.

Geometric standard deviation (s_g) = exp (s_L) = 2.88, where s_L = the standard deviation of the logged data.

7.6. Precision is a function of random error about the mean, as seen in taking repeated measurements on the same attribute. Bias is a systematic error inherent in the system. Precision and bias are the two components of accuracy, which NIOSH relates by the following equation:

Accuracy = 2(Precision) + |Bias|, at 95% Confidence level

7.7. Measurement error is a random error associated with the collection of a parcel of air or extraction of contaminant from a parcel of air and the subsequent analysis or assay. This error usually constitutes only a small percentage of the total error associated with the estimation of average contaminant concentration. Statistical sampling error is a function of environmental variability in space and time and constitutes most of the error in estimates of the mean and other summary statistics from an exposure distribution. This latter error usually is not normally distributed but is skewed to the left.

Chapter 8. The Physical Nature of Gases and Vapors

8.1. $$760\ mmHg \times \frac{13.58\ g\ H_2O}{1\ g\ Hg} \times \frac{in}{25.4\ mm} = 407.7\ in,\ H_2O$$

8.2. $$\frac{25\ L}{8\ hr} \times \frac{1\ hr}{60\ \text{min}} \times 1000\ cc \times 1\ L = \frac{52\ cc}{\text{min}}$$

8.3. $$\frac{0.0013\ mmHg}{760\ mmHg} \times 10^6 = 1.71\ ppm$$

$$1.71\ ppm \times \frac{291\ g/mol}{24.45\ L/mol} = 14.05\ \frac{mg}{m^3}$$

This saturated concentration is less than the concentration expected if all the Hg evaporates:

$$\frac{28.3\ g}{90\ m^3} \sim \frac{300\ mg}{m^3}$$

Therefore, it will not all evaporate without dilution ventilation.

8.4. $$\frac{41\ mmHg}{760\ mmHg} \times 10^6 = 53{,}947\ ppm$$

8.5. $$n = \frac{PV}{RT}$$

$$n = \frac{1.2\ atm \times 1\ L}{0.0826\ atm\text{-}L/°K\text{-}mol \times 422\ °K} = 0.034\ \text{moles in one liter}$$

Then, determine the molecular weight of air. Air is composed of approximately 78% nitrogen and 21 % oxygen; the molecular weights are 28 and 32 grams per mole, respectively. Thus, the molecular weight of air is:

$$\left(0.78 \times \frac{28\ g}{mole}\right) + \left(0.21 \times \frac{32\ g}{mole}\right) = \frac{29\ g}{mole}$$

The density of air at 300°F and 1.2 atm is:

$$\frac{29\ g}{mole} \times \frac{0.034\ mole}{L} = \frac{0.986\ g}{L}$$

8.6. $0.05 \times 0.05 = 0.0025 \times 10^6 = 2{,}500\ ppm$

Realize that % is parts per hundred. Here both percentages are converted to fractions.

8.7. 50:50 by volume = ½ mole fraction of each component; this assumption is only correct if densities and molecular weights are the same, i.e.:

$$mL \times \frac{g}{mL} \times \frac{mole}{g} = moles$$

Use Raoult's law:

$P_i = P_{v,i} X_i$

$P_x = (5\ \text{mmHg})(0.5)$

$C_x = 2.5\ \text{mmHg}/760 \times 10^6 = 3{,}289\ \text{ppm}$

$P_y = (50\ \text{mmHg})\ (0.5)$

$C_y = 25/760 \times 10^6 = 32{,}895\ \text{ppm}$

Compound Y will have 10 times the airborne concentration of Compound X.

8.8. Use Graham's law:

$$\frac{R_{d_1}}{R_{d_2}} = \left(\frac{MW_2}{MW_1}\right)^{0.5}$$

The molecular weights of H_2S and SF_6 are 34 and 146 grams per mole, respectively; so:

$$\frac{100\ ft/\min}{x} = \left(\frac{146}{34}\right)^{0.05}$$

$x = 47.6\ ft/\min$

Based on this diffusion rate, it would take ~ 11 minutes to detect SF_6 500 feet away.

8.9. Hexane vs. ethyl benzene:

$A = 3.27$

Hexane vs. aniline

$A = 12.26$

Ethyl benzene vs. aniline:

$A = 10.01$

Hexane and ethyl benzene are the most soluble in one another.

8.10. $F_1 = 1\ mg/cm^2/sec.$

Using Fick's law:

$$F = -D\frac{dC}{dX}$$

With dX and D constant, F and dC are proportional; so:

$$\frac{1}{F_2} = \frac{100}{40}$$

$F_2 = 0.40$

The second flux rate is 0.4 mg/cm²/sec.

8.11. Using Graham's law, see question 8.8. The molecular weights of ethane and octane are 30 and 114 grams per mole, respectively; thus:

$$\frac{R_E}{R_O} = \left(\frac{114}{30}\right)^{0.5}$$

$$\frac{R_E}{R_O} = 1.9$$

Ethane diffuses 1.9 times faster than octane.

8.12. $$\frac{24.45\ L/mole}{72\ g/mole} \times 10\ \frac{mg}{m^3} = 3.4\ ppm$$

or:

$$\frac{10\ mg}{m_a^3} \times \frac{1\ m_a^3}{10^3 L_a} \times \frac{1\ g}{10^3\ mg} \times \frac{mole}{72\ g} \times \frac{24.45\ L}{mole} \times \frac{10^6 \mu L}{L} = 3.4\ \frac{\mu L}{L_a} = 3.4\ ppm$$

8.13. $$\frac{1\ mg}{20\ L} \times \frac{1{,}000\ L}{m^3} \times \frac{22.4\ L/mole}{78\ g/mole} \times \frac{313°K}{273°K} \times \frac{760\ mmHg}{790\ mmHg} = 15.8\ ppm$$

or use:

$$\frac{0.08203\ atm\text{-}L}{mole\text{-}°K} \times \frac{313°K}{790/760\ atm} = 24.7\ L/mole$$

8.14. 1 gallon = 8 pints; 3.79 L = 1 gallon

1 gallon = 8 pints; 3.79 L = 1 gallon

$$\frac{1\ pt_t}{1000\ m_a^3} \times \frac{1\ gal}{8\ pts} \times \frac{3.79\ L}{gal} \times \frac{10^3}{L} \times \frac{0.866}{ml} \times \frac{10^3\ mg}{g} = 410\ mg/m^3$$

$$\frac{410\ mg}{m^3} \times \frac{24.45\ L/mol}{92\ g/mol} = 109\ ppm$$

8.15. $\ln P_v = a - b/t$

$\ln P_v = 28 - (8{,}000/303°K)$

$P_v = 5$ mmHg

The saturated vapor pressure of propionic acid at 30°C is 5 mmHg.

(5 mmHg/760 mmHg) × 106 = 6,579 ppm

The concentration of propionic acid at saturation = 6,579 ppm.

8.16. Concentration is proportional to vapor pressure:

$$RH = \frac{C_a}{C_s} \times 100\%$$

so: $$RH = \frac{P_a}{P_s} \times 100\%$$

The vapor pressure of water vapor at saturation (at 70°F) is 18 mmHg, from Figure 8-2.

$$75\% = \frac{C_a}{18\ mmHg}$$
$$C_a = 13.5\ mmHg$$

The absolute humidity converting to mg/m³ is:

$$\frac{13.5\ mmHg}{760\ mmHg} \times 10^6 = 17{,}763\ ppm$$
$$17{,}763\ ppm \times \frac{18\ g/mol}{24.45\ L/mol} = 13{,}077\ \frac{mg}{m^3}$$

8.17. $$1\ ppm = \frac{1\ \mu L}{1\ L} \times \frac{\mu mol/24.45\ \mu L}{mole/24.45\ L} = \frac{\mu mole}{mole}$$

Chapter 9. Basic Aerosol Physics

9.1. Look at Figure 9-1.

Reynolds number = 0.001 to 1 for Stokes region

Reynolds number = 6.6 Vd, and 50 μm = 0.005 cm

1 = (6.6) (V) (0.005 cm)

$V = 30$ cm/sec

9.2. $V_{TS} = 0.003 \text{ SG } d_a^2$

$V_{TS} = 0.05$ cm/sec

9.3. Use Equation 9-8:

$$\frac{C_t}{C_o} = \exp\left(-V_{TS} \cdot \frac{t}{H} \right)$$

With the stirred settling model, the concentration after 1 hour would be 1.84 mg/m³. The concentration has decreased by:

$C_o - C_r = 5.0 \text{ mg/m}^3 - 1.84 \text{ mg/m}^3 = 3.16 \text{ mg/m}^3$

Therefore, the rate of sawdust generation required to maintain a concentration of 5 mg/m³ in a 1,000 m³ size room would be:

$3.16 \text{ mg/m}^3 \times 1{,}000 \text{ m}^3 = 3{,}160 \text{ mg} = 3.16 \text{ g/hr}$

9.4. First, determine if the particles are in the Stokes region. Use Equation 9-9 to determine the velocity of the particles being ejected from the sawblade:

$V = RPM\ \pi\ d_w$

$$V = \frac{10{,}000\ rev}{\text{min}} \times \frac{12.5\ cm}{rev} \times \frac{\text{min}}{60\ \text{sec}} = \frac{6{,}545\ cm}{\text{sec}}$$

Then, determine the Reynolds number for the initial velocity.

$Re_o = 6.6\ Vd$

$Re_o = 6.6 \times 6{,}545 \text{ cm/sec} \times 0.001 \text{ cm}$

$Re_o = 43.2$

These particles are outside of the Stokes region (see Figure 9-1); so use Equation 9-11 to determine the horizontal stopping distance.

$$S = \frac{\rho_p d}{\rho_g}\left[Re_o^{1/3} - \sqrt{6} \arctan\ (Re_o^{1/3}/\sqrt{6}\right]$$

$$S = 0.13\ cm[3.51 - 2.45(0.96)] = 0.153\ cm$$

Therefore, these size particles will not reach the employee's eye.

9.5.

a. For count mean, use Equation 9-16:

$$\bar{d} = \frac{\sum n_i d_i}{N}$$

Count mean = [(1)(4) + (3)(7) + (5)(4) + (8)(2) + (10)(1)]/18 = 3.9

b. For count median diameter, use Equation 9-14:

$$CMD = \exp \frac{\sum n_i \ln d_i}{N}$$

CMD = exp [(4 × 0 + 7 × 1.10 + 4 × 1.61 + 2 × 2.08 + 1 × 2.30)/18] = 3.1

(c) For mass median diameter, use Equation 9-13 where b = 3:

$$MMD = CMD \times e^{b \ln^2 \sigma_g}$$

(d) For geometric standard deviation, use Equation 9-15:

$$GSD = \exp \left[\frac{\sum n_i (\ln d_i - \ln CMD)^2}{N-1} \right]^{1/2}$$

$GSD = \exp [(4 \times 1.21 + 7 \times 0.001 + 4 \times 0.23 + 2 \times 0.90 + 1 \times 1.37)/17]^{1/2} = 2.1$

Chapter 10. Occupational Exposure Limits and Threshold Limit Values

10.1. You are asked to determine the level of risk based on exposures at the TLV® or the NOEL and uncertainty factors. For most TLVs®, this information is not given. The effect upon which the TLV® is based is usually in the *Documentation* but may be hard to discern.

10.2. We need to find the product of the number of workers exposed and the duration of exposure. Where we have a range for the duration, the average duration is usually estimated as the midpoint of the range. For the last category, we assume that the exposure duration is 45 years. Neither is a particularly good assumption, but this is the best we can do based on the information provided. Multiply years exposed by number of workers; then sum. Total person-years of exposure is 13,713. The risk of disease relative to this chemical is one for 4,500 person-years, i.e., 45 years × 100 workers. Therefore, one would expect about 13,713/4,500 or three cases among this group of 838 workers.

Years Exposed	# Workers	Person–Years
1	28	28
2	13	26
3	14	42
4	16	64
7	162	1,134
12	111	1,332
17	268	4,556
24.5	138	3,381
34.5	74	2,553
42	11	462
45	3	135
	Sum of Person–Years	13,713

10.3. $m_t = m_0 e^{-T_{0.5}k}$

$$\frac{m_t}{m_o} = 0.5 = e^{-T_{0.5}k}$$

$$0.5 = e^{-T_{0.5}k}$$

$$\ln(0.5) = -T_{0.5}k3$$

$$T_{0.5} = \frac{0.693}{k}$$

10.4. For the STEL (t = 15 minutes): The biological half-life should be greater than **5 minutes (t ~ 3T$^{0.5}$)**. For the 8-hr TWA: The biological half-life should be greater than 160 minutes. For compounds with $T_{0.5} < 160$, a 15-min STEL is adequate, but the averaging time could be longer. For example, if $T_{0.5} = 100$, then t could be as high as 300 minutes, but an 8-hr TWA (480 minutes) may miss important peaks.

10.5. $TLV_{mixture} = 1/(f_1/TLV_1 + f_2/TLV_2 + f_3/TLV_3)$

$TLV_{mixture} = 1/(0.9/100 + 0.09/50 + 0.01/.1)$

$TLV_{mixture} = 10\ mg/m^3$

10.6. Because the target region is thoracic, between 80.5% and 89% should be sampled based on the TPM size-selective criteria (see Table 10-2).

10.7. $$RF = \frac{8}{h} - \frac{24-h}{16} = 0.7$$

This technique (where h = shift length, in hours) only adjusts the TLV® for an 8-hour averaging time; the biological half-life is not considered. In this case, because the half-life is three days and the averaging time could be as long as nine days, we would not detect that the workers were working unusual shifts with a 9-day averaging time. In fact, if the workers worked 40 hours straight, the effect on the body burden should be no different than five 8-hour shifts. Thus, the adjustment is not needed when $3 \times T_{0.5}$ is long.

10.8. $25/50 + 75/100 + 10/400 > 1$

Yes, the additive TLV® for the mixture has been exceeded. It does not matter whether ppm or mg/m³ is used if one is consistent and the terms in the fraction cancel each other.

Chapter 11. Federal Legislation and Regulation Relative to Industrial Hygiene

11.1. The PELs currently in use were adopted from the 1968 ACGIH® TLVs® and ANSI standards (Z-2). They are listed in the Z Tables of the U.S. *Code of Federal Regulations*, Volume 29, Part 1910.1000.

In addition, there are 26 comprehensive standards, some for more than one compound. Unlike the standards found in the Z tables, these comprehensive standards contain detailed information on how an employer should go about achieving compliance.

There are also generic standards, such as the Hazard Communication Standard and the Respiratory Protection Standard. They (with exceptions that include the nuclear and construction industries) contain detailed information on how to create and maintain these types of occupational health programs.

On the whole, the OSHA health standards are less than adequate. The performance type standards (Z tables) are not updated frequently enough to reflect current knowledge, and there is very little information or guidance given to the employer on how to determine compliance with a specific standard.

On the other hand, the comprehensive standards are long on detail, but they may unnecessarily allow for the cessation of exposure monitoring, training, and medical surveillance if initial monitoring results are less than the action level for that particular substance.

11.2. On the basis that the worker is assuming risk (voluntary risk, i.e., the worker is fully aware of any risk inherent in his or her job and has had adequate training in minimizing this risk), it may seem reasonable for the worker to have a higher level of risk than the general public.

Assigned risk (involuntary risk) is generally unacceptable as the worker is assigned a job in which he or she may not know all the associated risks, or in which the particular job has been made a condition of employment. Putting the general public unknowingly at risk from exposure to a particular contaminant(s) would be analogous to assigning (involuntary) risk for workers. Thus, the risk level should be different for different populations based on awareness level, or at least that is the way that regulations seem to work. The general public is usually not aware of the risk associated with environmental contaminants and is limited as to any actions it can take in preventing exposure.

11.3. It would seem appropriate that OSHA continue issuing all three types of health standards. The chemical (vertical) and the industry (horizontal) standards provide detailed requirements for particularly toxic substances or industries where certain disease risks are high and associated almost entirely with that industry. It is, however, impractical to have standards issued in this manner for every compound or industry. Generic standards, such as the Hazard Communication Standard and the Respiratory Protection Standard, are very useful in that they target the greatest number of employees with the least amount of regulatory effort.

Chapter 12. Gathering Information About the Workplace

12.1. The background survey is generally part of the recognition step of industrial hygiene, as it prepares a hygienist to ask intelligent questions regarding the process, raw materials and by-products, existing control measures, nature of the work, etc., at a plant. With this background information, the hygienist is better able to recognize potential or existing occupational hazards during a walk-through survey.

12.2. The main differences in the two surveys include the following:

a. Corporate hygienists will be able to call numerous people for information and will have access to internal documents that the compliance officer will not know about.

The compliance officer will have to rely on open literature for process descriptions.

b. The compliance officer will have access to inspection data from similar industries.

The corporate hygienist is unlikely to have this information, but can access it through the OSHA IMIS system (http://www.osha.gov/pls/imis/industry.html).

c. The corporate hygienist will be able to find out about process uniqueness and peculiarities relative to textbook descriptions. The compliance officer is unlikely to have access to this information.

12.3. Useful documents for determining safe or acceptable exposures for chemical contaminants in the workplace include the OSHA standards, ACGIH® TLVs® and TLV® *Documentation*, NIOSH pocket guide, AIHA WEELs, NIOSHTIC database, and W.A. Cook's *Occupational Exposure Limits – Worldwide*.

12.4. Toxicity information on hazardous chemicals can be found in the NIOSH Criteria for Recommended Standards, International Agency for Research on Cancer's monographs on the *Evaluation of Carcinogenic Risk of Chemicals to Humans*, the National Toxicology Program's *Annual Report on Carcinogens*, the NIOSH *Registry of Toxic Effects of Chemical Substances*, and the NIOSHTIC and the TOXNET database systems.

12.5. Because the process is not patented, published texts on industrial processes will not be very useful. (Patents are public information.) However, Kirk-Othmer and Cralley and Cralley may be consulted to determine similar patented or common processes and chemicals used. Other useful documentation might include the information provided to the plant's operators and supervisors on the process, raw materials, by-products, standard operating procedures, and relevant work tasks. Other sources include previous industrial hygiene reports, the process patent itself, and interviews with the plant chemist or chemical engineer.

Chapter 13. Qualitative Surveying – The Walkthrough

13.1. What types of welding are typically done? How do they work? What types of base metals are welded? Are any exotic metals such as inconel or monel encountered? Are there MSDS available for them? What types of filler rods and flux are used? Is welding done on painted or chemically treated surfaces? Is brazing or soldering ever done?

Is welding done on an intermittent or a more continuous basis? Which groups of workers do most of the welding? Have there been any complaints of nausea, headache, or flu-like symptoms among these groups? Is there a hearing conservation program for welders?

Does welding occur on each shift? Is there any shift in which more welding is done than on others?

Is welding ever done in a confined space or a poorly ventilated area? If so, what precautions are taken?

Are welding hoods available? Are they typically used? Is regular maintenance performed on ventilation hoods? Are respirators worn? Are workers trained on the hazards associated with welding?

13.2. When the plant manager signs a safety directive, the employees see it as being backed by management. It is then useful to see if management enforces these directives through the supervisors, i.e., to get a feel for the sincerity and the strength of management's commitment throughout the company. In the case where safety personnel sign the directive, it is important to get a feel for the strength of the safety management. Are the directives simply pieces of paper composed to meet a legal obligation? Are the directives enforced or put into practice? Where the safety manager issues such directives, they may more likely be ignored by the production managers.

13.3. Yes, the VHI or the AIHA priorities scheme would be necessary because there are so many chemicals. Emphasis will probably be placed on hexane, PNAs, and benzene because of the potential for chronic toxicity associated with exposure to these chemicals and on hydrogen sulfide and ammonia because of their potential for acute toxicity. Other hydrocarbons are of considerably less toxicity, and many of the additives, catalysts, etc., are used on only a few units and would therefore be of lower initial priority.

Chapter 14. Quantitative Surveying – Exposure Assessment

14.1.

a. An asbestos abatement worker for a working lifetime of 8-hr TWA exposures to asbestos.

b. A group of outdoor painters exposed to lead (8-hr TWA) on the day shift during the summer.

c. A group of maintenance workers exposed to noise (8-hr TWA) at a petrochemical plant on the midnight shift for a period of one year.

d. A cockpit crew exposed to oxygen deprivation on transcontinental flights (flight TWA) for a period of three years.

e. A group of welders assigned to a petrochemical plant turnaround for a 2-week period on 12-hour shifts during the summer.

14.2. Probably not, for the following reasons:

a. The exposure limit set for lead was intended to prevent workers from being exposed to concentrations that will cause chronic disease. The chronic disease effects are caused by exposures over a period of months or years, not a single shift. Acute toxic responses to lead exposures that might happen after one 8-hr TWA exposure would occur only at much higher concentrations. The OSHA standard is based on risk assessments/epidemiological studies in which average exposures are used.

b. Due to the variability seen in environmental data, some high exposures are expected, but the average should remain below the PEL; one overexposure out of 30 exposures is expected due to chance.

However, the variability of human response to toxicants, the setting of acceptable risk, the extrapolation from animal toxicity data to humans, the subjectivity of determining safety margins, and the inexactness of epidemiologic studies imply that safety factors should be applied to OELs by hygienists whenever possible. Most importantly, OELs tend to decrease over time. These concerns are less important for a compound such as lead that has a comprehensive standard based on a thorough risk assessment. Nonetheless, the risk associated with a lifetime of such exposure is greater than 1 health effect in 1,000 exposed workers.

14.3.

a. Legal requirements: The sampling period should be of adequate length to compare with the appropriate legal standard, usually an 8-hr TWA, 15-min STEL, or ceiling limit.

b. Biological considerations: The sampling period could be set to no more than three times the time it takes for one-half of a dose to be cleared from the blood or the target organ.

c. Multiple standards considerations: Some contaminants have both acute and chronic effects within a range of frequently encountered concentrations, thus requiring standards that integrate exposure over different time periods (e.g., 8-hr TWA and 15-min TWA).

d. Task monitoring: Monitor for the duration of a short-term specific task to estimate worst-case exposure or the contribution of the task to the overall 8-hour exposure; an 8-hour integrated exposure might not indicate this.

e. Confined space monitoring: This may be an instantaneous measurement taken prior to entering a confined space, or it may be done continuously for the length of the task, if there is potential for an acute hazard to develop.

14.4. The NIOSH Sampling Strategy is a relatively easy *cookbook* method for determining what contaminants should be monitored and how frequently monitoring should be done. The OSHA officer can inspect monitoring records of companies in a relatively easy manner to determine compliance with any sampling requirements. As OSHA officers have little time and resources to spend in anyone workplace, the strategy allows determination of noncompliance with one measurement. Maximum risk employees are identified for monitoring with the assumption that if these exposures are below the action level, then monitoring can be discontinued; i.e., for the OSHA officer this implies no need to look further. Therefore, the total amount of data collected will be smaller than for random sampling, but of course the total exposure picture is not seen.

The NIOSH Sampling Strategy is a systematic method for estimating maximum risk occupational exposures. Individuals to be monitored are selected on the basis of the likelihood that they are the most heavily exposed among a population of workers. Because the results are biased, they cannot be used to estimate the parameters of the exposure distribution. Therefore, every employee who has the potential to be overexposed must be monitored.

On the other hand, random sampling takes into account the variability of exposure concentrations. If done properly, random sampling allows for the determination of unbiased estimates of the average exposure and the worst-case exposure (e.g., the 95th or 99th percentile) and enables hygienists to make inferences about exposures for the entire population.

Chapter 15. Quantitative Surveying – Application of Mathematical Modeling to Estimate Air Contaminant Exposure

15.1. We can use Equation 15-3 which, in turn, requires that we know benzene's molecular diffusion coefficient in air. The Fuller equation (Equation 15-4) is

$$D_M,\ m^2\ min^{-1} = \frac{0.6 \times T^{1.75}}{P_{atm}} \times \frac{\sqrt{0.0345 + \frac{1}{MW}}}{\left[2.70 + \left(\sum V_{chem}\right)^{1/3}\right]^2}$$

We are given T = 25°C (298.15 K), and P_{atm} = 1 atm (101,325 Pa). The molecular weight of benzene is 78.1 g mole^{-1}.

$\sum V_{chem} =$

(6 carbon atoms × 15.9 per carbon atom) + (6 hydrogen atoms × 2.31 per hydrogen atom)

= 109.26.

Therefore,

$$D_M = \frac{0.6 \times (298.15)^{1.75}}{101{,}325} \times \frac{\sqrt{0.0345 + \frac{1}{78.1}}}{\left[2.70 + (109.26)^{1/3}\right]^2} = 4.92 \times 10^{-4}\ m^2\ min^{-1}$$

A chemical's saturation vapor concentration in air is

$$C_{sat}\ \text{in mg m}^{-3} = \frac{1.605 \times 10^4 \times P_V \times MW}{T}$$

where P_v is the chemical's vapor pressure (mmHg) at temperature T (Kelvin units), and MW is the chemical's molecular weight (g).

Benzene's saturation vapor concentration in air is

$$C_{sat}\ \text{in mg m}^{-3} = \frac{1.605 \times 10^4 \times 96 \times 78.1}{298.15} = 4.15 \times 10^5\ \text{mg m}^{-3}$$

We have now estimated D_M and C_{sat}. The remaining two inputs for Equation 15-3 are the circular area of the diffusion pathway (diameter 2 cm) and the length of the diffusion pathway (2 cm = 2 × 10^{-2} m). The area in meters is

$$A = \pi \times \frac{d^2}{4} = \pi \times \frac{(2\ cm)^2}{4} \times \frac{1\ m^2}{10{,}000\ cm^2} = 3.14 \times 10^{-4}\ m^2$$

According to Equation 15-3, the constant vapor emission from the open bottle is

$$G = \frac{A \times D_M \times C_{sat}}{L}$$

$$= \frac{(3.14 \times 10^{-4}\ m^2)(4.92 \times 10^{-4}\ m^2\ min^{-1})(4.15 \times 10^5\ mg\ m^{-3})}{2\times10^{-2}\ m} = 3.2\ mg\ min^{-1}$$

Note that a reported experimental measurement of the D_M value for benzene at 25°C is 5.8 × 10^{-4} m^2 min^{-1}. Therefore, the D_M estimate based on the Fuller equation (4.9 × 10^{-4} m^2 min^{-1}) is approximately 15% too low.

15.2.

a. According to Equation 15-18, the evaporation rate constant α for a small spill of n-pentane is

$$\alpha = 0.000524\ P_V + 0.0108\ (SA \div VOL)$$

$$= (0.000524 \times 420) + 0.0108 \times (250 \div 50) = 0.27\ min^{-1}.$$

b. According to Equation 15-17, the time at which the peak concentration occurs is

$$t_{peak} = \frac{V}{\alpha V - Q} \times \ln\left(\frac{\alpha V}{Q}\right)$$

$$= \frac{100}{0.27\times100 - 8} \times \ln\left(\frac{0.27\times100}{8}\right) = 6.4\ min$$

c. The initial n-pentane mass spilled is

(50 mL) × (630 mg mL^{-1}) = 31,500 mg.

According to Equation 15-16, the peak concentration is

$$C(t),\ mg\ m^{-3} = \frac{\alpha M_0}{Q - \alpha V}\left[\exp(-\alpha \times t) - \exp\left(-\frac{Q}{V} \times t\right)\right]$$

$$= \frac{0.27 \times 31{,}500}{8 - (0.27 \times 100)}\left[\exp(-0.27 \times 6.4) - \exp\left(-\frac{8}{100} \times 6.4\right)\right] = 188\ mg\ m^{-3}$$

15.3.

a. The n-hexane mass that evaporates in two hours is

$(946 \text{ mL}) \times (0.2) \times (660 \text{ mg mL}^{-1}) = 124{,}872 \text{ mg}$

The average mass evaporating per minute is

$$G = \frac{124{,}872 \text{ mg}}{120 \text{ min}} = 1{,}040 \text{ mg min}^{-1}$$

b. The near field box is 3 feet long by 3 feet wide. Because the individual's face is never more than 2 feet above the table top, the height of the box is taken as 2 feet. This box has five surfaces through which air can move. The four vertical surfaces are each 3 ft × 2 ft, and the top horizontal surface is 3 ft × 3 ft. The free surface area is

$$4 \times (3 \text{ ft} \times 2 \text{ ft}) + (3 \text{ ft} \times 3 \text{ ft}) = 33 \text{ ft}^2$$

c.

$$\beta = \tfrac{1}{2} \times \text{FSA} \times s$$

$$= \tfrac{1}{2} \times (33 \text{ ft}^2) \times (15 \text{ ft min}^{-1}) = 248 \text{ ft}^3 \text{ min}^{-1} = 7.0 \text{ m}^3 \text{ min}^{-1}$$

d. The room volume is 3530 ft^3 = 100 m^3. The room receives 6 nominal ACH, so $Q = 10 \text{ m}^3 \text{ min}^{-1}$.

$$C_{NF,SS} = \frac{G}{\beta} + \frac{G}{Q} = \frac{1{,}040 \text{ mg min}^{-1}}{7.0 \text{ m}^3 \text{ min}^{-1}} + \frac{1{,}040 \text{ mg min}^{-1}}{10 \text{ m}^3 \text{ min}^{-1}} = 253 \text{ mg m}^{-3}$$

This is equivalent to 72 ppm at normal temperature and pressure conditions. Eight hours of such work would produce an 8-hr TWA n-hexane exposure level exceeding the current ACGIH® TLV®-TWA of 50 ppm.

e. $$C_{FF,SS} = \frac{G}{Q} = \frac{1{,}040 \text{ mg min}^{-1}}{10 \text{ m}^3 \text{ min}^{-1}} = 104 \text{ mg m}^{-3}$$

This is equivalent to 29 ppm at normal temperature and pressure conditions, and is 60% less than the predicted exposure level at the work table (based on the near field zone estimate). Therefore, if we were to use the well mixed room dispersion construct, we would judge the n-hexane exposure level to be less than the ACGIH® TLV®-TWA of 50 ppm.

Chapter 16. Quantitative Surveying – Statistical Modeling of Occupational Exposures

16.1.

a. The 95th percentile:

$95\ \%\text{tile} = \bar{x} + 1.65\ \sigma$

$\times = 9.4$

$\hat{\sigma} = 2.2$

$95\ \%\text{tile} = 13.0$

b. The 95% UCL on the mean, assuming sigma is estimated:

$\bar{x} = 9.4 \quad s = 2.2 \quad t = 1.74$

$95\%\ \text{UCL} = \bar{x} + t_{(\alpha,\ n-1)} s/n^{0.5}$

$95\%\ \text{UCL} = 10.0$

c. The 95% UCL on the 95 %tile:

$95\%\ \text{UTL} = \bar{x} + k_{(1-\alpha,p,n)} s$

$\bar{x} = 9.4 \quad s = 2.2 \quad k = 2.453$

95% UCL on the 95 %tile = 14.8

Why is the answer for (c) so different from that for (a)? 95% of the population values will be less than or equal to the value found at the 95 %tile if our estimates of the mean and the standard deviation are exact. The 95% UTL is the upper bound for 95% of the estimates of the 95 %tile upon repeated sampling from the same population. It takes into consideration sampling 11 error in the estimation of the mean and standard deviation, i.e., our estimates are not exact.

16.2. From Equation 16-5:

$LCL = 1.1 - 1.645\ (0.10)\ (1) = 0.94$ ppm

No, because the 95% LCL is below the standard (0.94 ppm < 1.00 ppm).

16.3.

a. What is the arithmetic mean? $\bar{x} = 11.1$

b. What is the GSD?

$s_L = 0.65$

$GSD = \exp(0.65)$

$GSD = 1.91$

c. What is the MLE of the mean?

$\text{MLE} = \exp(\bar{x}_L + 0.5 s_L^2)$

$\bar{x}_L = 2.2 \; s_L = 0.65$

$\text{MLE} = 11.1$

d. What is the MVU of the mean?

$\text{MVU} = [\exp(\bar{x}_L)]\psi_n$

$\psi_n = 1.194$

$\text{MVU} = 11.0$

16.4. μ_c/PEL = 25 ppm/100 ppm = 0.25, using simple interpolation of Table 16-5 (not accurate for this nonlinear function, but close enough) a rounded $n \sim 8$ is obtained.

16.5. Power = $1 - \beta = 1 - 0.1 = 0.90 = 90\%$

16.6. From Equation 16-20 or Table 16-7. 4% is the maximum percentage of exposures that will exceed the STEL.

16.7. The problem was set up in Excel as follows:

Conc	**Ln**	**ln likelihood**
8	=LN(A2)	=LN((1/((2*PI())^0.5*B$13))*EXP(-(1/2)*((LN(A2)-B$12)/B$13)^2))
8	=LN(A3)	=LN((1/((2*PI())^0.5*B$13))*EXP(-(1/2)*((LN(A3)-B$12)/B$13)^2))
10	=LN(A4)	=LN((1/((2*PI())^0.5*B$13))*EXP(-(1/2)*((LN(A4)-B$12)/B$13)^2))
10	=LN(A5)	=LN((1/((2*PI())^0.5*B$13))*EXP(-(1/2)*((LN(A5)-B$12)/B$13)^2))
15	=LN(A6)	=LN((1/((2*PI())^0.5*B$13))*EXP(-(1/2)*((LN(A6)-B$12)/B$13)^2))
21	=LN(A7)	=LN((1/((2*PI())^0.5*B$13))*EXP(-(1/2)*((LN(A7)-B$12)/B$13)^2))
25	=LN(A8)	=LN((1/((2*PI())^0.5*B$13))*EXP(-(1/2)*((LN(A8)-B$12)/B$13)^2))
3	=LN(A9)	=LN(NORMDIST(LN(A9),B$12,B$13,TRUE))
5	=LN(A10)	=LN(NORMDIST(LN(A10),B$12,B$13,TRUE))
6	=LN(A11)	=LN(NORMDIST(LN(A11),B$12,B$13,TRUE))

$\bar{x}_L$= (B2:B11) **2.07** SUM(C2:C11)

s_L= (B2:B11) **0.83**

MLE 11.183617676

The cells for the mean and standard deviation of the logs (bottom left) and the sum of the likelihoods (bottom right were modified to fit the table. As can be seen, the MLE for the mean of concentrations is 11.1. The MVU would be exp (2.07) × ψ = exp(2.07) × 1.35 = 10.7, and the GSD is exp(0.83) = 2.3.

Yes, these values are close to the answers in question 16.3. A statistical sample representing a population rarely gives a distribution exactly like that of its parent population (see Figure 7-3). This is statistical sampling error. When part of the sample is missing (a portion of the data is < LOD), then sampling error associated with the uncensored data further affects this difference. This introduces even more error.

16.8. $s_L = 0.7$ and $\bar{x}_L = 0.86$. Thus, the z-score is 2.1. Consulting a z-table we find about 1.8% probability that an exposure would exceed the OEL of 10.

16.9. In the previous example, the calculated z for the exceedance fraction, γ was 2.1. For a sample size of 10, from Figure 16-8, the UCL on the exceedance fraction is 14%. This exceeds an L value of 5%, and so if L were set to 5% then we fail to reject the null hypothesis: $\gamma \geq L$.

Chapter 17. Quantitative Suryveying – Application of Variance Component Models to Exposure Assessment

17.1.

a. The value of the F-statistic is 1.21 (p = 0.4091). Thus, there is no evidence that individual workers' mean values differ.

b. The F-test tells us if the individual worker mean values are different from one another (note that it does not tell us which mean values are different).

17.2.

a. $\hat{\mu}_Y = 4.81$

$\hat{\sigma}_B^2 = 0.129$

$\hat{\sigma}_W^2 = 0.566$

b. ${}_BR_{95} = 4.1$

${}_WR_{95} = 19.1$

c. Based upon the suggested cut-point of 2 to evaluate homogeneity within a group of workers (Rappaport, 1991), we would conclude that this group is not homogenously exposed.

17.3.

a. The exposure profiles for All-round staff workers in Wheat bread bakeries and Crisp bake factories differ in the degree of homogeneity within each group. There is far more heterogeneity in workers' mean level in wheat bread bakeries as compared to crisp bakes factories.

b. Given the extremely small number of workers sampled in each group, it would not be appropriate to run a random-effects model on these data. Estimates from such a model would be highly imprecise and unreliable.

c. ${}_WR_{95} = 8.00$

17.4. Probability of exceedance = 0.308

Probability of overexposure = 0.616

Chapter 18. Quantitative Surveying – Forming the Strategy

18.1. Currently, OSHA requires that a citation be issued based on the LCL from Equation 16-5. Unfortunately, this type of compliance program does not take into account the variability of exposures typically found in the workplace. Because most employers are potentially *out of compliance*, monitoring of employee exposures is actually discouraged, as the more samples taken, the greater the probability that the employer will find an overexposure. The result of all of this is that very few meaningful data are collected upon which exposure distributions can be calculated.

18.2. Using Equation 16-17 and iterating, the smallest GSD acceptable for a sample size of 6 would be approximately 2.5 under the assumptions.

18.3. A TWA can be determined for each task a worker performs in his or her job. From these data an 8-hr TWA mean can then be constructed based on the fact that the mean of exposures is constant regardless of the averaging time. In other words, the mean of the task means is the 8-hr TWA mean if the tasks cover the entire shift. By determining short-term task-based exposures rather than integrating the exposures over the entire day, a better estimate of the magnitude and of source variability can be determined, as the variability is dampened with increasing averaging times.

Chapter 19. Laboratory Analytical Techniques

19.1. The photoionization, thermal conductivity, and ultraviolet detectors are nondestructive detectors.

19.2. The GC only effectively separates those organic compounds exhibiting volatility or semi-volatility at a temperature that is below their degradation temperature. High performance liquid chromatography is better suited for those organic compounds with lower vapor pressures and higher molecular weights such as organophosphorus pesticides and more polar or water soluble compounds.

19.3. Atomic absorption (AA) using the graphite furnace can have poor reproducibility. The coefficient of variation is often 5–10%. Moreover, only one metal can be analyzed per run. It can have a lower LOD than emission spectroscopy in some cases.

Inductively coupled plasma optical emission spectroscopy (ICP-OES) allows for the analysis of several different metals at one time. Additionally, interferences found in atomic absorption techniques are reduced, if not eliminated, in ICP because of the high temperatures achieved. The instrument is considerably more expensive than an AA.

19.4. Benzene: Gas chromatography with a flame ionization detector.

Arsenic: Atomic absorption or emission ICP-OES spectroscopy.

A complex organophosphorus pesticide: High performance liquid chromatography with an ultraviolet detector.

The mass of dust on a filter: Gravimetric technique.

Sulfuric acid captured on a filter: Ion chromatography.

Aniline: Gas chromatography with a thermionic detector.

Total number of particles on a filter: Phase contrast microscopy.

Chapter 20. Pumps and Their Calibration

20.1. A good air sampling maintenance and recordkeeping program might contain the following elements: 1) standard operating procedures for the use and maintenance of the pumps, 2) a schedule for discharging and recharging the pumps and records of these activities, 3) periodic factory maintenance checks, and 4) records indicating how often each pump is used, how long it is used, and any problems encountered while it is in use.

20.2.

20.2.1. Orifices: Air is forced through an orifice and the volumetric flow rate is held constant by maintaining the difference between the upstream and the downstream pressure. The two major types of orifices are the limiting orifice and the critical orifice:

20.2.1.1.

a. Critical: The critical orifice is usually added to the sampling train as opposed to being used inside the pump as a constant-flow mechanism. These orifices operate at a downstream-to-upstream pressure ratio that is less than the critical ratio (r c), which is defined to be that point where the velocity of air moving through the orifice has reached the speed of sound and will increase no more regardless of the downstream pressure. These orifices are affected by changes in the air density, downstream of a filter, as a result of increased resistance to flow over time.

b. Limiting: The limiting orifice meter along with a pressure sensor device is used internally as a constant-flow control mechanism in some pumps. These orifices operate at a downstream-to-upstream pressure ratio that is greater than the critical ratio and thus do not operate critically. The limiting orifice will maintain a constant flow as long as the air density does not change. A change in density may be caused by a change in atmospheric pressure or by an increase in resistance of the sampling head. The major advantage of using an orifice to hold airflow constant is that despite a typical increase in resistance across the sampling head over time, the sample flow rate should be accurate to within ± 5% of the calibrated flow rate. As the ambient temperature and pressure change, so does the air density. Therefore, a disadvantage to using an orifice to achieve constant flow is that the flow rate over the sample period will change as ambient conditions change. This error can be greater than 5% if pumps are calibrated in one location and used in another; thus, this mechanism is not truly constant.

20.2.2.

Mass flow meters: These meters maintain constant mass, not constant volume flow, by maintaining the resistance across a wire caused by the calibrated flow rate. Mass flow will be maintained whenever there is a change in the atmospheric pressure, air temperature, and filter load. This will result in a volumetric flow change. The major disadvantage of mass flow meters is that any change in pressure or temperature during sampling will cause a change in the volumetric flow rate that will not be detected if post-sampling calibration conditions are the same as the pre-sampling calibration conditions, thereby causing an error in determining the volume of air collected. In other words, the user needs to know that mass, not volume flow, is held constant.

20.2.3.

Constant motor speed: Pumps using this mechanism maintain a true constant volumetric flow rate by maintaining a constant motor speed. Each revolution of the motor causes the same number of strokes of the pump, and each stroke has a constant volume. This is not a disadvantage unless one wants constant mass flow. This is usually not the case.

20.3. Absolute *constant flow* is not likely to be achieved because of the potential for an increase in resistance across the sampling head over time or because of changes in the ambient temperature or pressure. By using the ideal gas law, it can be shown that any error in sampling airflow rate due to this increase in resistance will be within 5% of the calibrated airflow. As this error associated with airflow is so small in comparison with other associated sampling errors, manufacturers claim that their pumps have *constant flow*. Note that constant motor speed pumps are not subject to this problem.

20.4. 95% of the pre-sampling calibration flow rate is 9.5 Lpm. Thus, the post-sampling calibration flow rate would have to be at least 9 Lpm to get an average Q of 9.5 Lpm. Using Equation 20-5:

$$Q_h P_a = Q_u P_u$$

$$10 \text{ Lpm} \times 408 \text{ in. } H_2O = 9 \text{ Lpm} \times P_u$$

$$P_u = 340 \text{ in. } H_2O$$

The maximum tolerable pressure drop caused by loading of the filter would be 378 – 340 or an additional 38 in. H_2O. This would be caused only by a very heavily loaded filter.

20.5. The main advantage of multi-flow pumps is that they allow for the sampling of gases, vapors, and aerosols with only one type of pump. The two main disadvantages of multi-flow pumps are: 1) They have become too complicated for most industrial hygienists or technicians to repair on their own, as some pumps have become too complicated to be operated without detailed study of the operation manual; and 2) they are considerably larger and heavier than low flow pumps.

20.6. $$\frac{2.1\ LPM - 2.0\ LPM}{2} \times 420 \text{ min} = 861\ L$$

20.7. $$\frac{500\ cc}{29 \text{ sec}} \times \frac{1\ L}{1000\ cc} \times \frac{60 \text{ sec}}{1\text{min}} = 1.07 \frac{L}{\text{min}}$$

20.8. Use Equation 20-8

$$\frac{R_1}{R_2} = \left[\frac{P_2\,T_1}{T_2\,P_1}\right]^{0.5}$$

From Figure 18-12, the average pressure at 5,280 ft is 630 mmHg; so:

$$R_2 = 3.5\ Lpm\left[\frac{630\ mmHg}{483°F} \times \frac{530°F}{760\ mmHg}\right]^{0.5} \qquad R_2 = 3.3\ Lpm$$

20.9. In order to investigate pump error, one might set up an experiment to determine how well pumps from different manufacturers perform and how well a particular type of pump from one manufacturer performs. In other words, you could investigate error between manufacturers and error within each particular type of pump.

In order to set up the experiment, you might have 10 runs for each pump from 10 different manufacturers. Each run would be conducted under the same ambient conditions and would last approximately 8 hours. Each pump would be monitored continuously (continuous measurement of flow rate would be needed, or a wet test meter could be used to get one average flow rate) to determine the true average flow rate. The error rate for each pump could then be computed by comparing the true flow rate with the one determined from pre- and post-calibration. Statistical analysis (analysis of variance) of the data will determine if most of the error is associated with a particular manufacturer(s), or if it is associated with a particular type of pump.

Chapter 21. Gases and Vapors – Solid Media

21.1. The early analytical methods prior to 1964 using charcoal to assay organic compounds were time-consuming, susceptible to interferences (mainly water vapor), nonspecific, and only sensitive down to concentrations of ~100 ppm.

21.2. The most likely reason is its high capacity for water vapor.

21.3. The desorption efficiency is not constant (i.e., not linear) throughout the relevant concentration range of all compounds. This is especially true at lower loadings. Therefore, a calibration curve of desorption efficiency vs. loading is necessary to accurately determine the total amount of contaminant adsorbed to the charcoal from the amount desorbed.

21.4. Studies have shown that the desorption efficiencies tend to increase for polar compounds in mixtures. Therefore, failure to account for this effect could result in an overestimation of contaminant air concentration.

21.5. The four most important factors affecting the breakthrough volume include: 1) the flow rate of air through the charcoal tube, 2) the ambient contaminant concentration, 3) the relative humidity, and 4) the ambient temperature. Other competing species also can be quite important, depending on concentration.

Of the four major factors affecting breakthrough volume, the sample flow rate is the one factor over which the industrial hygienist has the most control.

21.6. Ethylene oxide and methylene chloride should be stored on dry ice during shipping because of their high vapor pressures (> 200 mmHg), which could lead to migration from front to back sections of the tube.

21.7. Wheeler:

t_b – Breakthrough time

M_e – Adsorption capacity

W – Weight of the carbon

C_i – Concentration at the inlet

C_o – Concentration at the outlet

K_v – Rate coefficient

Q – Volume flow rate

ρ_β – Bulk density of the packed bed

Dubinin:

V_0 – Maximum adsorption volume at any concentration

K – Rate constant

R – Ideal gas constant

T – Absolute temperature

P_o – Saturated vapor pressure

P – Partial pressure

B – Relative affinity of compound for charcoal

The basic difference is that the Dubinin equation is a static adsorption isotherm based on thermodynamics; it has no time component. The Wheeler equation is kinetic; it has a time component and predicts adsorption at a given time.

Chapter 22. Gases and Vapors – Bags, Rigid Containers, and Impingers

22.1. It has been shown that the contaminant concentration in some bags (e.g., five-layer) levels off over time. Apparently the inner layers of the bag become saturated with the contaminant. Repeated analysis could determine a time at which concentrations become near stable. Once this is known, the measured concentration may be used in the field.

22.2. Heat treatment might be necessary to volatilize contaminants that have permeated the bag material. Additionally, the bags should be purged with clean air several times and then the contents analyzed for residual contamination to avoid any positive bias.

22.3. Permeation is not a problem associated with the use of a glass bulb for the collection of grab samples; however, adsorption to or reaction of the gas or vapor with the walls of the container may still occur. Also, reduced pressure as a result of withdrawing samples from the container may cause air to leak in through the rubber septum, thereby reducing the contaminant concentration in the glass bulb. This is not usually a problem with plastic bags, as the walls collapse as samples are being removed, keeping the pressure inside at atmospheric.

22.4.

$$E = \frac{X}{T}$$ X = mass found in first impinger; T = total mass in all impingers combined, assumed to be total sampled.

$$E = \frac{9}{9.9}$$ The efficiency of each impinger is 91%.

$$\frac{X}{T} = \frac{Y}{T-X}$$ Y = total mass in second impinger.

$$\frac{9}{T} = \frac{0.9}{T-9}$$

$T = 10$ The efficiency of each impinger is (9/10) or 90%.

The efficiency obtained in the second equation is more appropriate, as the T in the first equation cannot represent the true total unless the efficiency is 100%. However, as the efficiency approaches 100%, the error from the first equation approaches zero.

22.5. If the efficiency for this particular impinger and adsorbing liquid is indeed reproducible, then it would be appropriate to use this technique and then adjust the results by dividing by 0.85.

Chapter 23. Gases and Vapors – Passive Monitoring

23.1. Permeation is a special case of diffusion in that diffusion occurs in the permeation process, through a solid, after a chemical has been absorbed onto a solid, such as a polymer membrane.

23.2. By rearranging the integral of the Fick's first law equation, the ambient pollutant concentration can be calculated once the mass collected in the passive device has been determined.

$$F = \frac{-D(C_a - C_o)}{L}$$

C_o is assumed to be zero. However, if retention is not 100% or if the sorbent does not absorb 100% of the molecules it contacts, then a non-zero equilibrium concentration will exist above the charcoal such that $C_o = C_e$. Because the concentration gradient between C_e (C_o) and C_a drives diffusion, the sampling rate, or the flux rate, of the monitor will decrease as the concentration gradient between C_a and C_e decreases or as C_e increases.

23.3. Response time of the device, transient conditions in concentrations related to long sampling times, weak adsorption causing $C_o \neq 0$, high relative humidity, and other competing species are five factors that could result in a biased measurement of a polar compound in air. The first two apply whether the analyte is polar or not.

If a passive sampling device has a long response time and the sampling period is short, a positive bias may result if the passive device is immediately sealed after the sampling period is over.

Both negative and positive biases are possible when there are widely fluctuating contaminant concentrations over a long sampling period. For example, a negative bias may result when a peak occurs at the beginning of a long sampling period, and a positive bias may result when a peak occurs at the end of a long sampling period.

The overwhelming concentration of H_2O at high humidities compared to pollutant concentrations tends to cause H_2O sorption to dominate. In addition, charcoal has a greater affinity for non-polar compounds than polar ones; thus high concentrations of nonpolar compounds will reduce the efficiency of the charcoal for the polar compounds. Finally, for some compounds charcoal has little affinity. All of these factors can cause $C_o \neq 0$.

23.4. $C_a = \dfrac{mL}{tDA}$

$$m = \frac{C_a tDA}{L}$$

So:

$$m \propto DC_a$$

first,

$$D \propto t^{1.5}$$

second,

$$C_a = \frac{mass}{volume} \propto \frac{1}{t} \; since \;\; v \propto t$$

and so,

$$\therefore m \propto (t^{1.5})(t^{-1}) = t^{0.5}$$

23.5. In the charcoal tube, the backup section absorbs only vapors released from the front section in the equilibrium process. These vapors can only travel downstream. In the passive device, released vapor molecules may travel upstream to the atmosphere or to the backup pad. However, the backup section in the passive device can account for biases introduced by transient conditions if both of the masses found on the front and the backup layer are used together in Equation 23-9 to calculate the ambient concentration. In order to use this equation, however, one must estimate the concentration gradient found in the second diffusion zone. By assuming that the adsorption process is following the Langmuir isotherm, one can obtain estimates of several constants that are needed to determine the concentration between the two charcoal pads. In practice, this is not typically done. The backup section in a charcoal tube is used in conjunction with the front section to estimate that breakthrough has occurred, and can be used to estimate total mass collected.

23.6. The diffusion pathlength is increased to beyond the face of the monitor.

Chapter 24. Aerosols

24.1. Generally, the method of analysis dictates the type of filter to be used. For instance, mixed cellulose ester filters are used for the collection and the quantification of asbestos fibers because they can be made invisible after chemical treatment, thus allowing the counting of fibers. Polyvinyl chloride filters are used in gravimetric methods of analysis because they are less likely to interact with water vapor than glass fiber or MCE filters. Teflon filters may be used in place of glass fiber filters in sampling for certain organics as there is the potential that the glass filter will interact with the analytes.

24.2. There is potential for measurement bias due to absorbed water vapor in using gravimetric methods. Not only may the filter absorb water vapor, but the collected particles may do so as well. A possible solution for avoiding the interference of water vapor is to desiccate the filter before each weighing (before and after the sample is collected). A better solution is to allow the filter to equilibrate to room humidity, before each weighing as defined in the NIOSH Nuisance Dust Method. Field blanks are most important for filter methods. One should follow NIOSH rules on field blanks and where monitoring is routine, a historic database of field blanks is useful to look for trends.

24.3. Universal standards for particle size selection have been adopted by ACGIH®, the International Standard Organization, and the European Standardization Committee. However, it appears that OSHA will not be adopting these standards anytime in the foreseeable future. Therefore, there is no legal incentive for industrial hygienists to switch to new devices or to change their aerosol sampling practices although ethical and liability incentives exist.

24.4. There are three size fractions that are included in the universal criteria for the performance of personal aerosol samplers in a workplace setting. They are: the Inhalable Particulate Mass Fraction (IPM), which has a cut-point of 100 μm; the Thoracic Particulate Mass Fraction (TPM), which has a cut-point of 10 μm; and the Respirable Particulate Mass Fraction (RPM) which has a cut-point of 4 μm.

Devices used to measure IMF include the IOM, Button, conical, and 7-hole samplers and the two- or three-piece plastic cassette. The vertical elutriator is used to measure the TMF, but is not personal. Personal devices include the BGI cyclone, the CIP10, the RespiCon, IOM foam and the PPI sampler. The PEM sampler gives a slightly different result as it is based on EPA rather than universal criteria. Cascade impactors may be used as well. Devices used to measure the respirable fraction include cyclones, cascade impactor, the CIP10, the RespiCon, IOM foam, the horizontal elutriator, and the PPI sampler. The PEM sampler gives a slightly different result as it is based on EPA (cut-point 2.5 μm) rather than universal criteria.

24.5. One might conclude that the respirable fraction in this mixed aerosol is of low enough concentration to be of no concern, as the largest *total* dust measurement is one quarter of the smallest relevant RMF = TLV.

24.6. The answer depends on particle sizes in the workplaces where the exposures were measured. Assuming small particles such as fume, the 37 mm cassette would have undersampled true inhalable concentrations due to wall and other cassette losses. Thus, for a particular exposure that was determined in a risk analysis (see Chapter 10) to be acceptable, that acceptable exposure level is actually higher. In other words, an OEL set at say 100 μg/m³ might be twice as high for true inhalable dust fractions (see for example Liden et al., 2000 in the Chapter 24 references for more thinking on this). If particle sizes are large, say MMAD > 20 μm, the errors may be an order of magnitude or more and thus an OEL set for inhalable dust should be correspondingly higher than one set for total dust. Now what about using the IOM sampler for comparison to OELs set on total dust? We err on the safe side. In other words, the IOM will almost always sample more aerosol than the 37 mm cassette for the same reasons as above, thus giving us values that are higher than what the 37 mm cassette would give.

24.7. Sampling devices that can be used to determine the respirable fraction of a mixed aerosol include the nylon cyclone, the Simpeds cyclone, the SKC cyclone, the horizontal elutriator, the CIP10, the RespiCon, IOM foam, the cascade impactor, and the PPI sampler. The PEM sampler gives a slightly different result as it is based on EPA (cut-point 2.5 μm) rather than universal criteria.

There are disadvantages of using any of the cyclones: the sampling efficiency for a certain particle size range can only be approximated, results may be affected by electrostatically charged particles, and different bias and precision values are possible for each particle size distribution. There are also advantages: they can be used for personal sampling; they can be operated to achieve the new universal criteria; and for the flow rate, one can determine the bias associated with a given aerosol size distribution if it is known.

A major disadvantage of the horizontal elutriator is the fact that the device cannot be used for personal monitoring, as it must be properly oriented in order to operate correctly. Other disadvantages include the variability in performance for different wind velocities and different particle size distributions.

The two major advantages of the cascade impactor include its ability to characterize the particle size distribution of an aerosol and the fact that it can be used as a personal aerosol monitor. One disadvantage is that cascade impactors are susceptible to particle bounce, which can cause a bias in the weight of the particles in each stage.

The RespiCon and the IOM foam appear to be reliable and have test data to indicate satisfactory performance. The PPI seems ingenious, is small and easy to use. However, there appear to be no peer reviewed studies of it.

24.8. The denuder will separate the vapor or gas phase of a contaminant from the particulate phase, capturing the former and allowing the latter to pass through. From the analysis of the material collected in the denuder, a determination can be made of the fraction of contaminant that is in the vapor or gas phase. The particles that pass through the denuder will be captured by the filter. The fraction of the contaminant in the particulate phase is determined from the analysis of the filter. Finally, the charcoal tube will collect the semi-volatile organics that volatilize from the particles that have been collected on the filter. Thus, the sum of the last two quantities is the fraction of the contaminant found on particles in the airstream.

Chapter 25. Interpreting Analytical Methods and Selecting a Laboratory

25.1. Although the rationale for the setting of upper and lower flow rate limits is not usually given, limiting the effects of diffusive as opposed to active sampling for the contaminant is most likely the basis for the lower limit, and the assurance of adequate contact time between the contaminant and the media is most likely the basis for the upper limit. If the air sampling flow rate is not fast enough, diffusion of the contaminant into the tube could contribute a significant portion of the mass adsorbed onto the medium, resulting in a biased estimate of the contaminant concentration. If the flow rate is too high, the breakthrough occurs more quickly, possibly resulting in a biased estimate of the contaminant concentration.

Factors to consider in deciding on flow rates include expected contaminant concentration, analytical sensitivity, breakthrough volume, competitive species, and sampling time.

25.2. As the flow rate decreases, the velocity through the filter also decreases and a particle is more likely to stay in the air streamlines and not be filtered from the airstream. Although the effect of flow rate has not been examined at flows below 0.5 L/min (see the asbestos method), lower flows may be unbiased. At high flows, the most likely problem is overloading the filter, which may increase filter resistance too much for the pump to compensate or cause particles to be dislodged during shipping and handling. Flow rates up to 16 Lpm did not bias in asbestos sampling (see the NIOSH method).

25.3. $$\frac{0.005\ mg}{m^3} \times \frac{1\ m^3}{1000\ L} \times V(L) = 0.005\ mg$$

Minimum air volume (V) = 1000 L

$$\frac{1000\ L}{420\ \text{min}} \sim 2.4\frac{L}{\text{min}}$$

The minimum sampling rate for 420 minutes is about 2.4 L/min.

25.4. Because the exposed area of a 25 mm filter is 385 mm^2, the LOD per filter would be 7 fibers/mm^2 or 2695 fibers. Therefore, the minimum air volume needed to detect exposures at 0.1 f/cm^3 is:

$$\frac{0.1\ fiber}{cm^3} \times \frac{1\ cm^3}{0.001\ L} \times V\ (L) = 2{,}695\ fibers$$

minimum air volume (V) = 270 L.

25.5. The molecular weight of toluene is 92 g/mole. Therefore, 50 ppm toluene = 188 mg/m^3.

$$\frac{188\ mg}{m^3} \times \frac{1\ cm^3}{0.001\ L} \times 24\ L = 4.52\ mg$$

1% of 4.52 is approximately 0.05 mg. Therefore, 0.05 mg would be needed on the laboratory blank in order to represent 1% of the sample mass.

25.6. $$\frac{10\ fibers}{field} \times 1\ field \times 0.008\ mm^2 = 12.5\ \frac{fibers}{mm^2}$$

Yes, this fiber count does exceed the LOD of 7 fibers/mm^2.

25.7. ± 3 standard deviations include 99.73% of the estimates of the mean value, therefore, the theoretical probability of having an outlier on a given result, assuming the performance of the reference laboratories represents all laboratories and assuming that the occurrence of failures is a random event, is 0.27%. Note that the reference labs are likely better than the other labs and so the likelihood of an outlier for the other labs is probably somewhat greater, but we do not know what it is. Also the likelihood of failure is probably not random. Poor labs are more likely to fail multiple times than good labs; again we do not know that relationship.

For the second question, for the asbestos or silica analyses where there are 4 analysis per round and 3 rounds in question, we can treat it as a binomial experiment consisting of 12 identical, independent trials (analyses). Each trial results in one of the two outcomes: success or failure. To become non-proficient and risk losing accreditation (a lab can pay for a second set of samples to retest), a lab must have at least 2 outliers in 2 of three rounds. This means that at least 4 of 12 trials must be failures. However, a lab could have 4, 5, or even 6 failures out of 12 {1/4, 1/4 , 2/4} or {3/4, 1/4, 1/4} or {4/4, 1/4, 1/4} and still be proficient. We will calculate the probability of non-accreditation defining > 3 of 12 failures as non-proficient, knowing that this probability is an overestimate since a portion of labs with 4, 5, or 6 failures over 3 rounds will not risk non accreditation.

A result that falls within ± 3 standard deviations of the mean will be considered a success, and a result that falls outside this range will be considered a failure. The probability of success on a single trial is equal to p and remains the same from trial to trial. The probability of a failure is $(1-p) = q$. To solve this problem the formula for the binomial distribution is used:

$$p(y) = \frac{n!}{y!\ (n-y)!} p^y q^{n-y}$$

where:

n = the number of trials

y = the number of successes

$n–y$ = the number of failures

p = the probability of success

q = the probability of failure,

The probability of 3 failures or 9 successes in 12 trials is:

$$\frac{12!}{9!\ 3!} \times (0.9973^9 \times 0.0027^3) = 4.225214 \times 10^{-6}$$

Similarly, the probability of 10, 11, and 12 successes could be calculated. The sum of these probabilities (0.999999974152460) is the probability of a favorable outcome or pass. Thus, the probability of failure leading to non-accreditation (0.000000025847540) is small. This value is likely underestimated since independence is assumed, but is overestimated based on assuming that all conditions of 4, 5, and 6 failures lead to non-proficiency, when in fact they do not.

Chapter 26. An Overview of Direct-Reading Instrumentation

26.1. If the integration time of an instrument is longer than the biological half-life of a toxicant one is trying to measure, then a peak exposure associated with an adverse health effect (acute) may be missed or underestimated. The integration time of a direct-reading instrument (DRI) is similar to the biological half-life in that the response of the body to a challenge concentration is dampened because of associated absorption and metabolic rates in much the same way that the integration time associated with the DRI will dampen its output or response.

26.2. If the peak to valley noise is observed long enough and if the noise is normally distributed, we might assume that the peak to valley is about 6 standard deviations, as would be for a normal distribution. This would be 99.97% of the full range. So 2 times the peak to valley would be 12*s* or ± 6*s*. However, if the distribution is not normal, then according to Tschebyscheff's theorem ± 6*s* would yield:

$$100\left(1-\frac{1}{k^2}\right) = \textit{lower bound on inclusive interval}$$

$$\textit{for } 12\ s,\ k = 6$$

$$100\ (1-1/36)\ \% = 97.2\%$$

This means that 100 – (100 – 97.2)/2 = 1.4% in each tail or the 98.6 %tile is the smallest value for + ks_b and represents an LOD of $\bar{x} \pm 2.15\ s_b$ for a normal distribution (i.e., z-score corresponds to 2.15 for the 98.6 %tile). Thus 2 times the peak to valley is not as conservative as the ± 3*s* rule.

26.3. The analytical sensitivity is m/s_s, and at concentrations near the LOD this is m/s_b. Because s_b is proportional to the LOD signal, the analytical sensitivity is proportional to m/LODs.

Instrument	Slope, m	LOD_c	LOD_s
1	1	2	2
2	5	2	10

Therefore, for each instrument, the analytical sensitivity is 0.5; or for #1, 1/2 and for #2, 5/10.

Chapter 27. Instruments for Gases and Vapors

27.1. Just about any acid gas (SO_2, NO_2, HCN), diatomic halogens (e.g., Cl_2), basic gases (e.g., NH_3), O_2, O_3, H_2S.

27.2. The major disadvantages of using electrochemical direct-reading instruments include their susceptibility to interferences and lack of evaluation information and criteria. Advantages include the wide variety of instrumentation available, costs, applicable analytes, and ease of use.

27.3. In voltammetry, the indicating electrode is charged to supply a specific potential needed for the analyte of interest to diffuse through the electrolyte to the electrode. Once at the electrode, the analyte is oxidized, and the electric current is measured. Selectivity is achieved by appropriately choosing the electrode system, the electrolyte, the voltage, and the polymer membrane for the analyte to diffuse into the electrolyte. The ability to add the control voltage to the list of factors that determine selectivity is unique to voltammetry or amperometry.

27.4. In order to calibrate DRIs, they usually are first zeroed with clean air and then exposed to clean air spiked with known concentrations of the contaminant of interest. However, it is difficult to calibrate direct-reading instruments for carbon dioxide, as it occurs naturally in the atmosphere at a background concentration of approximately 400 ppm and higher in urban areas where there are combustion sources. One solution is to remove carbon dioxide from room air by using ascarite and then to use pure carbon dioxide to create known concentrations.

27.5. The volume inside the MIRAN lA is 5.6 L and the pump flow rate is 30 Lpm.

$$\ln \frac{C}{C_{max}} = -\frac{Qt}{V}$$

$$t = 1.065 \text{ min}$$

It will take about one minute for the concentration in the MIRAN to reach 1 ppm.

27.6. The temperature detector requires a larger sample volume than does the photoacoustical detector. Therefore, it takes a longer time to purge these instruments, there is a longer response time, and the linear range is smaller. NDIR uses a filtering device so that it is difficult to accurately identify more than one contaminant at a time. FTIR can monitor all wavelengths in a very short period of time. This allows for many scans to be repeated, so that there is an improvement in the signal to noise ratio and a decrease in the LOD. Also, owing to its high resolution, the FTIR can identify unknowns with better reliability, even those with low IR absorption at low concentrations. The linear range of the FTIR is four times that of filtering devices. The FTIR is more expensive.

27.7. The biggest difference between combustible gas meters and general survey meters for total hydrocarbons is the range in which they operate. CGMs detect percentage concentrations, whereas GSMs detect ppm (and in some cases ppb) concentrations through the use of different detectors.

27.8. The detector used most often in CGMs is the pellistor or catalytic bead, which consists of a wire that has been coated with a catalyst that allows a combustible gas to burn at a much lower temperature than it normally would. Typically, in a CGM, two detectors are used: one that has not been coated with a catalyst or for which the catalyst has been poisoned, and a second one that has been coated with a catalyst. The first wire will not cause a gas to combust at a given temperature; the coated wire will. As the concentration of a combustible gas increases, the second sensor will heat from the combustion reaction. Then a change of resistance in the sensor takes place, causing current to be sent to the first sensor in order to rebalance the Wheatstone bridge. As current flows from its source, a needle moves to indicate the percentage concentration of the combustible gas that is present.

27.9. The two major types of detectors used in general survey meters for total hydrocarbons are the flame ionization detector and the photoionization detector. Both rely on the ionization of the compound(s) of interest in order to measure the subsequent electric current produced. They are different, however, in the manner in which they cause the ionization of compounds. The FID burns compounds causing carbon atoms to become positively charged, whereas the PID causes compounds to become ionized by exposing them to ultraviolet light.

27.10. From the calibration curve, 15% of the lower flammable limit when the instrument is calibrated with pentane but senses xylene is equivalent to about 28% of the LFL for xylene. The LFL for xylene is 1.1%; therefore, the concentration of xylene is about 3000 ppm.

27.11. If a combustible gas concentration reaches 20% of the lower flammable limit, it is appropriate that the workers leave the area immediately. Before re-entry is allowed, the source of the hazard should be determined. Furthermore, any potential for an increase in the combustible gas concentration should be examined. Additional measurements should be taken in the area with a recalibrated CGM or another device such as a GSM or detector tube. All potential interferents such as low oxygen concentration and water vapor should be examined.

27.12. Disadvantages of general survey meters include the fact that they are non-selective. They are calibrated with one gas so that responses from other gases are all relative to the response from the calibration gas. PID GSMs are susceptible to bias due to water vapor, and their response is less linear than that of an FID GSM.

An advantage of PIDs is the fact that most compounds can be ionized by ultraviolet radiation and can therefore be detected by these instruments. Advantages of FIDs include less susceptibility to the effects of water vapor, and linear response over a wide range (and to methane, if this is important in the application).

27.13. The elution time of an analyte determines the response time of a portable GC. Factors that affect the elution time include size and polarity properties of the analyte, the type of column used, and the temperature of the column.

Chapter 28. Instruments for Aerosols

28.1. Aerosol direct-reading instruments generally are only used quantitatively. They can give the user no information as to the identity of the contaminant. However, they can be used *selectively* to the extent that only one contaminant or type of aerosol is in the environment. Aerosol DRI are useful in a variety of tasks such as detecting leaks or hot spots, determining the efficacy of control measures, or examining exposure trends associated with a specific process or task over time. Photometers can be calibrated to be reliable instruments for measuring mass concentration and some are personal-sized.

28.2. The response of optical light scattering instruments is dependent on light wavelength and particle size, density, and refractive index. The manufacturer can change the angle at which the scattered light is measured. The light wavelength also can be changed by the manufacturer. Therefore, these instruments must be calibrated for specific applications (size distribution, density and to a lesser extent refractive index) when precise aerosol concentration estimates are needed.

28.3. For gases and volatile vapors, known concentrations can be made with good precision and stability by using mass or volume measurements. For aerosols, this is much more difficult, as a given mass or volume of a liquid or a solid may be divided in many different ways. Using monodisperse aerosols, which unfortunately are unlike the aerosols we need to measure, is the only way to generate reasonably precise known concentrations. Thus, to calibrate an instrument in the lab (or the way the manufacturer does it) several monodisperse aerosols of varying sizes are generated separately to calibrate the instrument. Alternatively, one may take it to the field and measure aerosols side by side with non-direct read devices, keeping in mind that these devices have error as well.

Chapter 29. Detector Tubes

29.1. Detector tubes will respond to the presence of other chemicals as well as the chemical of interest if the reaction mechanism is the same for both. Detector tubes use reaction mechanisms that will detect a large number of chemicals regardless of the name of the chemical printed on the label.

29.2. Detector tubes are appropriate to use in estimating whether a chemical is present in the workplace or determining whether control measures are working. Also they are appropriate to use in special circumstances such as emergency response and firefighting, sampling of breathing air and confined spaces, and sampling parameter estimation. However, they are less appropriate to use in conducting a quantitative industrial hygiene survey due to their *grab sample* nature. On the other hand, because detector tubes are inexpensive compared to the more precise analytical techniques, one can afford to take more samples with the tubes, thereby increasing precision.

29.3. The length of stain in detector tubes is dependent on both the contaminant concentration and the volume of air that flows through the tube.

29.4. The detector tube heat exchanger will lose its ability to cool as it absorbs heat from the incoming air. Eventually, all capacity to cool will be lost when the temperature of the heat exchanger is equal to the temperature of the incoming air.

29.5. From Figure 27-5 the absolute humidity at 40°C is about 53 mg/L at saturation (100%).

Thus:

$$\frac{5\frac{mg}{L}}{53\frac{mg}{L}} \times 100 = 9\%\ relative\ humidity$$

$$\frac{24\frac{mg}{L}}{53\frac{mg}{L}} \times 100 = 45\%\ relative\ humidity$$

Thus, a water vapor reading of 5 mg/L at 40°C is equivalent to a relative humidity of 9%, and a water vapor reading of 24 mg/L is equivalent to a relative humidity of 45%.

Chapter 30. Calibration of Direct-Reading Instrumentation

30.1. $$\frac{100\ \mu L_{H,V}}{L_A} = \frac{X\ \mu L_{H,L}}{10\ L_A} \times \frac{1\ mL_{H,L}}{10^3 \mu L} \times \frac{0.63\ g}{mL} \times \frac{mole}{100\ g} \times \frac{24.45\ L}{mole} \times \frac{10^6 \mu L}{L_{H,V}}$$

Note that all units cancel to give:

$$\frac{\mu L_{H,V}}{L_A} = \frac{\mu L_{H,V}}{L_A},\ i.e.,\ ppm$$

Solving for X, we get 6.5 µl liquid heptane.

30.2. $$\frac{1\ \mu L_{M,V}}{L_A} = \frac{X\ \mu L_{M,L}}{5.65\ L_A} \times \frac{0.81\ g_M}{mL} \times \frac{mole}{72\ g_M} \times \frac{24.45\ L_{M,V}}{mole} \times \frac{10^6\ \mu}{L_M}$$

X = 0.02 µl MEK liquid.

As the volume of liquid MEK that is required is less than 0.1µl, the saturated head space vapor volume must be calculated by using a variation of Equation 30-1. V_c in the denominator is ignored. The saturated vapor pressure of MEK @ 25°C is 71 mmHg, and the saturated vapor concentration is 93,421 ppm.

$$C = \frac{V_c C_{c,s}}{V_d}$$

$$1\ ppm = \frac{V_c\ 93{,}421\ ppm}{5.65}$$

$$V_c = 0.0000605\ L\ or\ \ 60.5\ \mu L$$

The saturated head space vapor volume needed to achieve a concentration of 1 ppm of MEK in the MIRAN 1A is 60.5 μl.

30.3. For steady state conditions, use the steady state equation in Appendix 30-1: G = *QC*. You must convert 50 ppm to 217 mg/m³ or 217 μg/l so that the units cancel. Thus:

$$\frac{\frac{2\ \mu g}{\min}}{\frac{217\ \mu g}{L}} = Q = \frac{0.0092\ L}{\min}\ or\ \frac{9.2\ ml}{\min}$$

30.4. Use Equation 28-4.

$$C_f = C_c \frac{Q_c}{Q_c + Q_d}$$

The saturated vapor pressure of nitrobenzene at 25°C is calculated from Appendix B, Equation 3 as:

$\log_{10} P = A - B/T$

$\log_{10} P = 8.4085 - 2658.65/298.3$

P = 0.313 mmHg (note that 25°C is outside the tested temperature range so that value may be in slight error). Thus:

$$\frac{0.313\ mmHg}{760\ mmHg} \times 10^6 = 412\ ppm\ or\ \ 10\% = 41.2\ ppm$$

The dilution flow rate needed is:

$$10 = 41.2\left(\frac{1}{1+Q_d}\right)$$

$$\frac{41.2}{10} - 1 = Q_d$$

$$Q_d = 3.1\ Lpm$$

30.5. $\frac{0.1\ L\ methane}{10.1\ L\ air} \times 10^6 = 9{,}900\ ppm$ (Equation 30-1)

30.6. 0.02×1000 ppm = 20 ppm

0.02×20 ppm = 0.4 ppm

0.2×0.4 ppm = 0.008 ppm

The concentration of ethane in the bag after its third refill with 10 L of air is 0.008 ppm. A proportionality has been assumed, which can be proved as follows. After the bag is emptied once, the amount of ethane remaining is

$$0.2\ L \times \frac{1{,}000\ \mu L}{L}(ppm) = 200\ \mu L$$

Upon the first refill the concentration is

$$\frac{100\ \mu L}{L} \times \frac{10\ L}{\min} = \frac{1{,}000\ \mu L}{\min}$$

As step one is correct, the others are also, by analogy. Equation 30-3 can also be used.

30.7. Generally, syringes are made *to deliver* a certain volume. Therefore, some volume remains in the needle. If it is pumped, the volume in the needle is displaced, and too much is delivered.

30.8. Because a 100 ppm concentration is entering the bulb at 10 Lpm, there is a constant generation rate of

$$\frac{100\ \mu L}{L} \times \frac{10\ L}{\min} = \frac{1{,}000\ \mu L}{\min}$$

Thus, from the buildup equation

$$1-\exp\left(-\frac{QT}{V}\right) = \frac{C}{C_{max}}$$

and

$$1-\exp\left(-10\times\frac{1}{0.5}\right) = \frac{C}{100}$$

$$C = 99.99999979\ ppm$$

30.9. Use the following equation.

$$P_1V_1 = P_2V_2$$

Use 14.7 lb/in² or 101.379 kPa in one atmosphere at normal pressure. Then the volume of gas at NTP is calculated as,

$$1{,}500\frac{lb}{in^2}\times 6\ f^3 = 14.7\ \frac{lb}{in^2}\times V_2$$

$$V_2 = 612.24\ ft^3\ or\ 17{,}346.9\ L$$

30.10. Use the buildup, nonzero air equation from Appendix 30-1.

$$\ln\frac{C_d-C}{C_d-C_0} = -\frac{Qt}{V}$$

$$\ln\frac{30\ ppm-29.9\ ppm}{30\ ppm-10\ ppm} = -\frac{1\ L/\min\times t}{1\ L}$$

$$t = 5.3\ \min$$

If perfect mixing could not be assumed, then the response time would be longer, depending on the degree of imperfection. Also, the average concentration across the instrument's sensing path could be higher or lower than 29.9 ppm, depending on how concentrations were distributed in the rest of the sample volume.

INDEX

A

B

C

D

G

H

I

J

K

L

M

N

Q

R

W